ESSCIRC 2022 - IEEE 48th European Solid State Circuits Conference (ESSCIRC 2022)

Milan, Italy
19 – 22 September 2022

IEEE Catalog Number: CFP22542-POD
ISBN: 978-1-6654-8495-4

**Copyright © 2022 by the Institute of Electrical and Electronics Engineers, Inc.
All Rights Reserved**

Copyright and Reprint Permissions: Abstracting is permitted with credit to the source. Libraries are permitted to photocopy beyond the limit of U.S. copyright law for private use of patrons those articles in this volume that carry a code at the bottom of the first page, provided the per-copy fee indicated in the code is paid through Copyright Clearance Center, 222 Rosewood Drive, Danvers, MA 01923.

For other copying, reprint or republication permission, write to IEEE Copyrights Manager, IEEE Service Center, 445 Hoes Lane, Piscataway, NJ 08854. All rights reserved.

*** This is a print representation of what appears in the IEEE Digital Library. Some format issues inherent in the e-media version may also appear in this print version.

IEEE Catalog Number: CFP22542-POD
ISBN (Print-On-Demand):   978-1-6654-8495-4
ISBN (Online):   978-1-6654-8494-7

Additional Copies of This Publication Are Available From:

Curran Associates, Inc
57 Morehouse Lane
Red Hook, NY 12571 USA
Phone: (845) 758-0400
Fax: (845) 758-2633
E-mail: curran@proceedings.com
Web: www.proceedings.com

ESSCIRC 2022 TABLE OF CONTENTS

Joint Keynote 1

Date: Tuesday, September 20, 2022
Time: 9:20 - 10:00
Room: Aula Magna
Chair(s): Andrea Baschirotto, University of Milano-Bicocca

The Future of Short Reach Interconnect 1

Davide Tonietto

Huawei Technologies Co., Ltd., Canada

Joint Keynote 2

Date: Tuesday, September 20, 2022
Time: 10:00 - 10:40
Room: Aula Magna
Chair(s): Davide Giacomini, Infineon Technologies

**Semiconductors Take the Driver's Seat - Challenges and Opportunities for the Car
of the Future** 9

Tim Gutheit

Infineon Technologies AG, Germany

Joint Keynote 3

Date: Wednesday, September 21, 2022
Time: 8:30 - 9:10
Room: Aula Magna
Chair(s): Renato Lombardi, Huawei

Integrated Circuits as Key Enabler for Today's Smart MEMS Sensors 12

Dirk Droste[1], Horst Symanzik[1], Timo Gießelmann[1], Markus Ulm[1], Ivano Galdi[2], Riccardo Campagna[2]

[1]Bosch Sensortec GmbH, Germany; [2]Robert Bosch S.p.A., Italy

Joint Keynote 4

Date: Wednesday, September 21, 2022
Time: 11:10 - 11:50
Room: Aula Magna
Chair(s): Davide Pandini, STMicroelectronics

**The Next "Automation Age": How Semiconductor Technologies Are Changing Industrial
Systems and Applications** 17

Domenico Arrigo, Claudio Adragna, Vincenzo Marano, Rachela Pozzi, Fulvio Pulicelli, Francesco Pulvirenti

STMicroelectronics, Italy

ESSCIRC Keynote 1

Date: Tuesday, September 20, 2022
Time: 14:20 - 15:00
Room: Aula Magna
Chair(s): Andrea Baschirotto, University of Milano-Bicocca

Reminiscing Through 40 Years of CMOS Analog Circuit Design: from Audio to GHz 25

Rinaldo Castello

University of Pavia, Italy

ESSCIRC Keynote 2

Date: Wednesday, September 21, 2022
Time: 16:00 - 16:40
Room: Aula Magna
Chair(s): Andrea Mazzanti, University of Pavia

From Less Batteries to Battery-Less: Enabling a Greener World Through Ultra-Wide Power-Performance Adaptation Down to pWs 33

Massimo Alioto

National University of Singapore, Singapore

ESSCIRC Keynote 3

Date: Thursday, September 22, 2022
Time: 10:10 - 10:50
Room: Aula Magna
Chair(s): Joao Goes, FCT NOVA

Innovations for Intelligent Edge 41

Vida Ilderem, Stefano Pellerano, James Tschanz, Tanay Karnik, Vivek De

Intel Corporation, United States

Devices & Circuits for Quantum Computing

Date: Tuesday, September 20, 2022
Time: 11:10 - 12:50
Room: Aula 1B
Chair(s): Domenico Zito, Aarhus University
           Veeresh Deshpande, Helmholtz Zentrum Berlin

Coupling Control in the Few-Electron Regime of Quantum Dot Arrays Using 2-Metal Gate Levels in CMOS Technology 45

Bruna Cardoso Paz[1], Victor El-Homsy[1], David Niegemann[1], Bernhard Klemt[1], Emmanuel Chanrion[1], Vivien Thiney[2], Baptiste Jadot[2], Pierre-André Mortemousque[2], Benoit Bertrand[2], Thomas Bédécarrats[2], Heimanu Niebojewski[2], François Perruchot[2], Silvano De Franceschi[3], Maud Vinet[2], Matias Urdampilleta[1], Tristan Meunier[1]

[1]Institut Néel, France; [2]CEA-Leti, France; [3]CEA-IRIG, France

Multi-Gate FD-SOI Single Electron Transistor for Hybrid SET-MOSFET Quantum Computing 49

Fabio Bersano[1], Franco De Palma[1], Fabian Oppliger[1], Floris Braakman[2], Ionut Radu[3], Pasquale Scarlino[1], Martino Poggio[2], Adrian Mihai Ionescu[1]

[1]École Polytechnique Fédérale de Lausanne, Switzerland; [2]University of Basel, Switzerland; [3]Soitec, France

Cryogenic Comparator Characterization and Modeling for a Cryo-CMOS 7b 1-GSa/s SAR ADC 53

Gerd Kiene, Aishwarya Gunaputi Sreenivasulu, Ramon W.J. Overwater, Masoud Babaie, Fabio Sebastiano

Delft University of Technology, The Netherlands

A Cryogenic SRAM Based Arbitrary Waveform Generator in 14 nm for Spin Qubit Control 57

Mridula Prathapan[1], Peter Mueller[1], Christian Menolfi[1,2], Matthias Brändli[1], Marcel Kossel[1], Pier Andrea Francese[1], David Heim[1], Maria Vittoria Oropallo[1], Andrea Ruffino[1], Cezar Zota[1], Thomas Morf[1]

[1]IBM Research Zürich, Switzerland; [2]Cisco Systems Inc.

A 28nm 6.5-8.1GHz 1.16mW/qubit Cryo-CMOS System-on-Chip for Superconducting Qubit Readout 61

Steven Van Winckel[1], Alican Caglar[1,2], Benjamin Gys[1], Steven Brebels[1], Anton Potocnik[1], Bertrand Parvais[1], Piet Wambacq[1], Jan Craninckx[1]

[1]imec, Belgium; [2]Vrije Universiteit Brussel, Belgium

Devices & Circuits for Emerging Technologies

Date: Thursday, September 22, 2022
Time: 11:20 - 12:40
Room: Aula 1F
Chair(s): Claudio Bruschini, EPFL
Lotte Geck, Forschungszentrum Jülich

Non-Volatile Ternary Content Addressable Memory Based on Phase Change Nanoelectromechanical (NEM) Relay 65

Mohammad Ayaz Masud, Luis Hurtado, Gianluca Piazza

Carnegie Mellon University, United States

A 24V Thin-Film Ultrasonic Driver for Haptic Feedback in Metal-Oxide Thin-Film Technology Using Hybrid DLL Locking Architecture 69

Jonas Pelgrims[1], Kris Myny[1,2], Wim Dehaene[2]

[1]Katholieke Universiteit Leuven, Belgium; [2]imec, Belgium

A Monolithic SPAD-Based Random Number Generator for Cryptographic Application 73

Nicola Massari[1], Alessandro Tontini[1], Luca Parmesan[1], Matteo Perenzoni[1,2], Miloš Gruijć[3], Ingrid Verbauwhede[3], Thomas Strohm[4], Dayo Oshinubi[4], Ingo Herrmann[4], Andreas Brenneis[4]

[1]Fondazione Bruno Kessler, Italy; [2]Sony Semiconductor Solutions Corporation, Italy; [3]ITEC, Belgium; [4]Robert Bosch GmbH, Germany

Quantum-Correlated Photon-Pair Source with Integrated Feedback Control in 45 nm CMOS 77

Danielius Kramnik[1], Imbert Wang[2], J.M. Fargas Cabanillas[2], Anirudh Ramesh[3], Sidney Buchbinder[1], Panagiotis Zarkos[1], Christos Adamopoulos[1], Prem Kumar[3], M.A. Popović[2], Vladimir Stojanović[1]

[1]University of California-Berkeley, United States; [2]Boston University, United States; [3]Northwestern University, United States

Digital ML Processors

Date: Tuesday, September 20, 2022
Time: 11:10 - 12:50
Room: Aula 1E
Chair(s): Marian Verhelst, KU Leuven
Andreas Burg, EPFL

A 4.6-8.3 TOPS/W 1.2-4.9 Tops CNN-Based Computational Imaging Processor with Overlapped Stripe Inference Achieving 4K Ultra-HD 30fps 81

Yu-Chun Ding, Kai-Pin Lin, Chi-Wen Weng, Li-Wei Wang, Huan-Ching Wang, Chun-Yeh Lin, Yong-Tai Chen, Chao-Tsung Huang

National Tsing Hua University, Taiwan

A 40nm 5.6TOPS/W 239GOPS/mm² Self-Attention Processor with Sign Random Projection-Based Approximation 85

Seong Hoon Seo, Soosung Kim, Sung Jun Jung, Sangwoo Kwon, Hyunseung Lee, Jae W. Lee

Seoul National University, Korea

A 28nm 8-Bit Floating-Point Tensor Core Based CNN Training Processor with Dynamic Activation/Weight Sparsification 89

Shreyas Kolala Venkataramanaiah[1], Jian Meng[1], Han-Sok Suh[1], Injune Yeo[1], Jyotishman Saikia[1], Sai Kiran Cherupally[1], Yichi Zhang[2], Zhiru Zhang[2], Jae-Sun Seo[1]

[1]Arizona State University, United States; [2]Cornell University, United States

A Differentiable Neural Computer for Logic Reasoning with Scalable Near-Memory Computing and Sparsity Based Enhancement 93

Yuhao Ju[1], Shiyu Guo[1], Zixuan Liu[1], Tianyu Jia[2], Jie Gu[1]

[1]Northwestern University, United States; [2]Peking University, China

SIF-NPU: A 28nm 3.48 TOPS/W 0.25 TOPS/mm² CNN Accelerator with Spatially Independent Fusion for Real-Time UHD Super-Resolution 97

Sumin Lee[1], Ki-Beom Lee[1], Sunghwan Joo[2], Hong Keun Ahn[1], Junghyup Lee[1], Dohyung Kim[1], Bumsub Ham[1], Seong-Ook Jung[1]

[1]Yonsei University, Korea; [2]Samsung Electronics Co., Ltd., Korea

Resistive Memory Based In-memory Compute

Date: Tuesday, September 20, 2022
Time: 11:10 - 12:50
Room: Aula 1F
Chair(s): Manuel Le Gallo, IBM
          Cristiano Calligaro, RedCat Devices

A 40nm RRAM Compute-in-Memory Macro with Parallelism-Preserving ECC for Iso-Accuracy Voltage Scaling 101

Wantong Li, James Read, Hongwu Jiang, Shimeng Yu

Georgia Institute of Technology, United States

In-Memory Realization of In-Situ Few-Shot Continual Learning with a Dynamically Evolving Explicit Memory 105

Geethan Karunaratne[1,4], Michael Hersche[1,4], Jovin Langenegger[1,4], Giovanni Cherubini[1], Manuel Le Gallo[1], Urs Egger[1], Kevin Brew[2], Sam Choi[2], Injo Ok[2], Claire Silvestre[2], Ning Li[2], Nicole Saulnier[2], Victor Chan[2], Ishtiaq Ahsan[2], Vijay Narayanan[3], Luca Benini[4], Abu Sebastian[1], Abbas Rahimi[1]

[1]IBM Research Zürich, Switzerland; [2]IBM Research Albany, United States; [3]IBM T. J. Watson Research Center, United States; [4]ETH Zürich, Switzerland

An Embedded PCM Peripheral Unit Adding Analog MAC In-Memory Computing Feature Addressing Non-Linearity and Time Drift Compensation 109

Alessio Antolini[1], Andrea Lico[1], Eleonora Franchi Scarselli[1], Antonio Gnudi[1], Luca Perilli[1], Mattia Luigi Torres[2], Marcella Carissimi[2], Marco Pasotti[2], Roberto Antonio Canegallo[2]

[1]University of Bologna, Italy; [2]STMicroelectronics, Italy

A Highly Integrated Crosspoint Array Using Self-Rectifying FTJ for Dual-Mode Operations: CAM and PUF 113

Sehee Lim[1], Youngin Goh[2], Young Kyu Lee[1], Dong Han Ko[1], Junghyeon Hwang[2], Minki Kim[2], Yeongseok Jeong[2], Hunbeom Shin[2], Sanghun Jeon[2], Seong-Ook Jung[1]

[1]Yonsei University, Korea; [2]Korea Advanced Institute of Science and Technology, Korea

Physical Implementation of a Tunable Memristor-Based Chua's Circuit 117

Manuel Escudero[1], Sabina Spiga[1], Mauro Di Marco[2], Mauro Forti[2], Giacomo Innocenti[3], Alberto Tesi[3], Fernando Corinto[4], Stefano Brivio[1]

[1]CNR-IMM, Italy; [2]University of Siena, Italy; [3]University of Florence, Italy; [4]Politecnico di Torino, Italy

Synaptic Devices

Date: Tuesday, September 20, 2022
Time: 11:10 - 12:50
Room: Aula 1D
Chair(s): Stefan Slesazeck, NaMLab
Alessandro Spinelli, Politecnico di Milano

Ferroelectric Schottky Barrier MOSFET as Analog Synapses for Neuromorphic Computing 121

Fengben Xi[1,2], Andreas Grenmyr[1,2], Jiayuan Zhang[1,2], Yi Han[1,2], Jin-Hee Bae[1], Detlev Grützmacher[1,2], Qing-Tai Zhao[1]

[1]Peter Grünberg Institute, Germany; [2]RWTH Aachen University, Germany

Interplay Between Charge Trapping and Polarization Switching in MFDM Stacks Evidenced by Frequency-Dependent Measurements 125

Justine Barbot, Jean Coignus, Nicolas Vaxelaire, Catherine Carabasse, Olivier Glorieux, Messaoud Bedjaoui, François Aussenac, François Andrieu, Francois Triozon, Laurent Grenouillet

CEA-Leti, Université Grenoble Alpes, France

Frequency Modulation of Conductance Level in PCM Device for Neuromorphic Applications 129

Ahmed Trabelsi, Carlo Cagli, Tifenn Hirtzlin, Olga Cueto, Marie-Claire Cyrille, Elisa Vianello, Valentina Meli, Veronique Sousa, Guillaume Bourgeois, François Andrieu

CEA-Leti, France

Thermal Switching of TiO_2-Based RRAM for Parameter Extraction and Neuromorphic Engineering 133

Alessandro Milozzi[1], Daniel Reiser[2,3], Andreas Drost[2], Thomas Neuner[2,3], Marc Tornow[2,3], Daniele Ielmini[1]

[1]Politecnico di Milano, Italy; [2]Fraunhofer EMFT Research Institution for Microsystems and Solid State Technologies, Germany; [3]Technical University of Munich, Germany

Improvement of FTJ On-Current by Work Function Engineering for Massive Parallel Neuromorphic Computing 137

Suzanne Lancaster[1], Quang T. Duong[1], Erika Covi[1], Thomas Mikolajick[1,2], Stefan Slesazeck[1]

[1]NaMLab gGmbH, Germany; [2]Technische Universität Dresden, Germany

SRAM Based In-memory Compute

Date: Thursday, September 22, 2022
Time: 14:20 - 16:00
Room: Aula 1D
Chair(s): Adam Teman, Bar-Ilan University
Antoine Frappé, IEMN CNRS

A 4-Bit Mixed-Signal MAC Macro with One-Shot ADC Conversion .... N/A

Xiangxing Yang[1], Nan Sun[2]

[1]The University of Texas at Austin, United States; [2]Tsinghua University, China

A 28nm 1.644TFLOPS/W Floating-Point Computation SRAM Macro with Variable Precision for Deep Neural Network Inference and Training 145

Sangsu Jeong, Jeongwoo Park, Dongsuk Jeon

Seoul National University, Korea

All-Digital Time-Domain Compute-in-Memory Engine for Binary Neural Networks with 1.05 POPS/W Energy Efficiency 149

Jie Lou, Christian Lanius, Florian Freye, Tim Stadtmann, Tobias Gemmeke

RWTH Aachen University, Germany

A 1.23-GHz 16-Kb Programmable and Generic Processing-in-SRAM Accelerator in 65nm 153

Amitesh Sridharan[1], Shaahin Angizi[2], Sai Kiran Cherupally[1], Fan Zhang[1], Jae-Sun Seo[1], Deliang Fan[1]

[1]Arizona State University, United States; [2]New Jersey Institute of Technology, United States

A 1-to-4b 16.8-POPS/W 473-TOPS/mm² 6T-Based In-Memory Computing SRAM in 22nm FD-SOI with Multi-Bit Analog Batch-Normalization 157

Adrian Kneip, Martin Lefebvre, Julien Verecken, David Bol

Université Catholique de Louvain, Belgium

Optoelectronic Devices & Negative Capacitance FET

Date: Tuesday, September 20, 2022
Time: 15:10 - 16:10
Room: Aula 1D
Chair(s): Sara Pellegrini, STMicroelectronics
          Clara Moldovan, EPFL

Fully Integrated Si:HfO$_2$ Negative Capacitance 2D-2D WSe2/SnSe2 Subthermionic Tunnel FETs 161

Sadegh Kamaei, Ali Saeidi, Xia Liu, Carlotta Gastaldi, Clara Moldovan, Jürgen Brugger, Adrian M. Ionescu
École Polytechnique Fédérale de Lausanne, Switzerland

Electroluminescence of Si$_x$Ge$_{1-x-y}$Sn$_y$ / Ge$_{1-y}$Sn$_y$ Pin-Diodes Grown on a GeSn Buffer 165

Lukas Seidel[1], Sören Schäfer[1], Michael Oehme[1], Dan Buca[2], Giovanni Capellini[3,4], Jörg Schulze[5], Daniel Schwarz[1]
[1]University of Stuttgart, Germany; [2]Peter Grünberg Institute, Germany; [3]IHP - Leibniz Institute for High Performance Microelectronics, Germany; [4]Roma Tre University, Italy; [5]University of Erlangen-Nuremberg, Germany

GeSn-on-Si Avalanche Photodiodes for Short-Wave Infrared Detection 169

Maurice Wanitzek[1], Michael Oehme[1], Christian Spieth[1], Daniel Schwarz[1], Lukas Seidel[1], Jörg Schulze[2]
[1]University of Stuttgart, Germany; [2]University of Erlangen-Nuremberg, Germany

Sensor Interfaces

Date: Wednesday, September 21, 2022
Time: 9:20 - 10:40
Room: Aula 1E
Chair(s): Marco Grassi, University of Pavia
Georgi Radulov, TU Eindhoven

A Drift-Compensated Magnetic Spectrometer for Point-of-Care Wash-Free Immunoassays Using a Concurrent Dual-Frequency Oscillator 173

Jui-Hung Sun[1], Bill Ling[2], Md. Abdullah-Al Kaiser[1], Constantine Sideris[1]

[1]University of Southern California, United States; [2]California Institute of Technology, United States

A Full Current-Mode Timing Circuit with Dark Noise Suppression for the CERN CMS Experiment 177

Edgar Albuquerque[1], Ricardo Bugalho[1], Luis Oliveira[3], Tahereh Niknejad[1,2], Jose Silva[1,2], Alessio Boletti[2], João Varela[1,2]

[1]PETsys Electronics, Portugal; [2]Laboratório de Instrumentação e Física Experimental de Partículas, Portugal; [3]CTS-UNINOVA, Portugal

A 2.74pJ/Conversion 0.0018mm² Temperature Sensor with On-Chip Gain and Offset Correction 181

Yuting Shen, Mariska van der Struijk, Kevin Pelzers, Hanyue Li, Eugenio Cantatore, Pieter Harpe

Eindhoven University of Technology, The Netherlands

An Integrated Optical Transceiver Circuit for Power Delivery and Bi-Directional Data Communication in a Medical Catheter Device 185

Alexander Frank[1], Jens Anders[1], Joachim Burghartz[1], Bart Kootte[2], Jean Schleipen[2], Peter Jutte[2]

[1]IMS CHIPS, Germany; [2]OSYPKA AG, Germany; [3]Philips Research Eindhoven, The Netherlands

Image Sensors

Date: Wednesday, September 21, 2022
Time: 12:00 - 13:00
Room: Aula 1E
Chair(s): Robert Henderson, University of Edinburgh
Daniele Perenzoni, Sony

A Reconfigurable 224×272-Pixel Single-Photon Image Sensor for Photon Timestamping, Counting and Binary Imaging at 30.0-μm Pitch in 110nm CIS Technology 189

Leonardo Gasparini[1], Manuel Moreno García[1], Majid Zarghami[1], André Stefanov[2], Bruno Eckmann[2], Matteo Perenzoni[1,3]

[1]*Fondazione Bruno Kessler, Italy;* [2]*University of Bern, Switzerland;* [3]*Sony Seminconductor Solutions, Italy*

Statistical Measurements and Monte-Carlo Simulations of DCR in SPADs 193

Mathieu Sicre[1,2,3], Megan Agnew[4], Christel Buj[1,3], Caroline Coutier[3], Dominique Golanski[1], Rémi Helleboid[1], Bastien Mamdy[1], Isobel Nicholson[4], Sara Pellegrini[4], Denis Rideau[1], David Roy[1], Francis Calmon[2]

[1]*STMicroelectronics, France;* [2]*INSA Lyon, France;* [3]*CEA-Leti, France;* [4]*STMicroelectronics, United Kingdom*

A Scalable 64×64 Pixels Monolithic HV-CMOS Sensor for Hadron Therapy with 1ns Time Stamping Capability and In-Pixel ADC 197

Nicola Massari[1], Alessio D'Andragora[1], Matteo Perenzoni[2], Andrej Selijak[3], Carlos Chavez Barajas[4], Alan Taylor[4], Jon Taylor[4], Gianluigi Casse[4], John Pettingell[5], Ignacio Di Biase[5]

[1]*Fondazione Bruno Kessler, Italy;* [2]*Sony Seminconductor Solutions Corporation, Italy;* [3]*Jozef Stefan Institute, Slovenia;* [4]*University of Liverpool, United Kingdom;* [5]*Rutherford Cancer Centres, United Kingdom*

Biomedical Sensors

Date: Thursday, September 22, 2022
Time: 11:20 - 12:40
Room: Aula 1E
Chair(s): Mirjana Banjevic, Sensirion
Matthias Kuhl, University of Freiburg

A 32-ch Neuromodulator with Redundant Voltage Monitors Avoiding Blocking Capacitors 201

Stefan Reich[1], Markus Sporer[1], Joachim Becker[1], Stefan B. Rieger[2], Martin Schüttler[2], Maurits Ortmanns[1]

[1]University of Ulm, Germany; [2]CorTec GmbH, Germany

Fully Implantable 192×256 SPAD Sensor with Global-Shutter and Micro-LEDs for Bidirectional Subdural Optical Brain-Computer Interfaces 205

Eric H. Pollmann, Yatin Gilhotra, Heyu Yin, Kenneth L. Shepard

Columbia University, United States

A 1.8μW 5.5mm³ ADC-Less Neural Implant SoC Utilizing 13.2pJ/Sample Time-Domain Bi-Phasic Quasi-Static Brain Communication with Direct Analog to Time Conversion 209

Baibhab Chatterjee, K. Gaurav Kumar, Shulan Xiao, Gourab Barik, Krishna Jayant, Shreyas Sen

Purdue University, United States

A 50 Mb/s Full HBC TRX with Adaptive DFE and Variable-Interval 3x Oversampling CDR in 28nm CMOS Technology for a 75 Cm Body Channel Moving at 0.75 Cycle/Sec 213

Jaehyun Ko[1], Iksu Jang[1], Chanho Kim[1], Jihoon Park[1], Changjae Moon[1], Sooeun Lee[2], Byungsub Kim[1]

[1]Pohang University of Science and Technology, Korea; [2]Samsung Electronics Co., Ltd., Korea

Medical Imaging

Date: Thursday, September 22, 2022
Time: 11:20 - 12:40
Room: Aula 1D
Chair(s): Angel Rodriguez-Vasquez, Universidad de Sevilla
Michiel Pertijs, TU Delft

An Implantable Power Extraction Circuit with Integrated PMUTs for Wireless Power Delivery 217

Oi-Ying Wong[1,2], Dries Tabruyn[1,2], Veronique Rochus[1], Nick Van Helleputte[1]

[1]imec, Belgium; [2]Katholieke Universiteit Leuven, Belgium

A PMUT Transceiver Front-End with 100-V TX Driver and Low-Noise Voltage Amplifier in BCD-SOI Technology 221

Lara Novaresi[2], Piero Malcovati[1], Andrea Mazzanti[1], Edoardo Bonizzoni[1], Marco Terenzi[2], Stefano Ottaviani[2], Davide Ugo Ghisu[2], Fabio Quaglia[2], Alessandro Stuart Savoia[3]

[1]University of Pavia, Italy; [2]STMicroelectronics, Italy; [3]Roma Tre University, Italy

A Portable CMOS-Based MRI System with 67×67×83 μm^3 Image Resolution 225

Daniel Krüger[1,2], Aoyang Zhang[1], Henry Hinton[1], Victor M. Arnal[1], Yi-Qiao Song[1,4], Yiqiao Tang[4], Ka-Meng Lei[3], Jens Anders[2], Donhee Ham[1]

[1]Harvard University, United States; [2]University of Stuttgart, Germany; [3]University of Macau, China; [4]Schlumberger-Doll Research, United States

A Sub-0.01° Phase Resolution 6.8-mW fNIRS Readout Circuit Employing a Mixer-First Frequency-Domain Architecture 229

Cheng Chen[1], Zhouchen Ma[1], Yaxin Liu[1], Zhenhong Liu[1], Linfeng Zhou[1], Yan Wu[1], Liang Qi[1], Yongfu Li[1], Mohamad Sawan[2], Guoxing Wang[1], Jian Zhao[1]

[1]Shanghai Jiao Tong University, China; [2]Westlake University, China

Frequency Synthesizers

Date: Tuesday, September 20, 2022
Time: 11:10 - 12:50
Room: Aula 1A
Chair(s): Jaehyouk Choi, KAIST
Salvatore Levantino, Politecnico di Milano

A 4.7GHz Synchronized-Multi-Reference PLL with In-Band Phase Noise Lower Than Reference Phase Noise + 20logN$_{div}$ 233

Hongzhuo Liu, Wei Deng, Haikun Jia, Shiyan Sun, Qixiu Wu, Jiajie Tang, Zhihua Wang, Baoyong Chi

Tsinghua University, China

A Bang-Bang Digital PLL Covering 11.1-14.3 GHz and 14.7-18.7 GHz with Sub-40 fs RMS Jitter in 7 nm FinFET Technology 237

Staffan Ek, Patrik Karlsson, Andreas Kämpe, Roland Strandberg, Aravind Tharayil Narayanan, Martin Anderson, Hind Dafallah, Mesrop Daghbashyan, Tayebeh Ghanavati Nejad, Robert Hägglund, Nikola Ivanisevic, Robert Nilsson, Peter Nygren, Mattias Palm, Erik Säll, Sha Tao, My Chien Yee, Lars Sundström

Ericsson, Sweden

A 2.5 GHz 104 mW 57.35 dBc SFDR Non-Linear DAC-Based Direct-Digital Frequency Synthesizer in 65 nm CMOS Process 241

Dong-Hyun Yoon[1], Kwang-Hyun Baek[2], Tony Tae-Hyoung Kim[1]

[1]Nanyang Technological University, Singapore; [2]Chung-Ang University, Korea

An Inductorless Fractional-N PLL Using Harmonic-Mixer-Based Dual Feedback and High-OSR Delta-Sigma-Modulator with Phase-Domain Filtering 245

Masaru Osada, Zule Xu, Tetsuya Iizuka

The University of Tokyo, Japan

A –247.1dB FoM, –77.9dBc Reference Spur Ring-Oscillator-Based Injection-Locked Clock Multiplier with Multi-Phase-Based Calibration 249

Yeonggeun Song[1], Kyoungjoon Ha[1], Han-Gon Ko[2], Min-Seong Choo[3], Deog-Kyoon Jeong[1]

[1]Seoul National University, Korea; [2]ONE Semiconductor, Korea; [3]Hanyang University, Korea

Innovations in A/D Converters Towards Ultra Low Power

Date: Tuesday, September 20, 2022
Time: 15:10 - 16:10
Room: Aula 1A
Chair(s): Lucien Breems, NXP
          Marco Zamprogno, ST

A 590 µW, 106.6 dB SNDR, 24 kHz BW Continuous-Time Zoom ADC with a Noise-Shaping 4-Bit SAR ADC 253

Shubham Mehrotra[1,2], Efraïm Eland[1], Shoubhik Karmakar[1], Angqi Liu[1], Burak Gönen[1,3], Muhammed Bolatkale[1,2], Robert van Veldhoven[2], Kofi A. A. Makinwa[1]

[1]Delft University of Technology, The Netherlands; [2]NXP Semiconductors, The Netherlands; [3]Ethernovia, The Netherlands

A 150 mV, Sub-1 nW, 0.75%-Full-Scale INL Delta-Sigma ADC for Power-Autonomous Sensor Nodes 257

Alessandro Catania[1], Andrea Ria[2], Giuseppe Manfredini[1], Michele Dei[1], Massimo Piotto[1], Paolo Bruschi[1]

[1]University of Pisa, Italy; [2]CNR-IEIIT, Italy

A 10.4-ENOB 0.92-5.38 µW Event-Driven Level-Crossing ADC with Adaptive Clocking for Time-Sparse Edge Applications 261

Jonah Van Assche, Georges Gielen

Katholieke Universiteit Leuven, Belgium

Energy Efficient SoCs

Date: Tuesday, September 20, 2022
Time: 15:10 - 16:10
Room: Aula 1B
Chair(s): Sylvain Clerc, STMicroelectronics
Kathryn Wilcox, AMD

SmartHeaP - A High-Level Programmable, Low Power, and Mixed-Signal Hearing Aid SoC in 22nm FD-SOI 265

Jens Karrenbauer[1], Simon Klein[1], Sven Schönewald[1], Lukas Gerlach[1], Meinolf Blawat[2], Jens Benndorf[2], Holger Blume[1]

[1]Leibniz University Hannover, Germany; [2]Dream Chip Technologies GmbH, Germany

A 12nm Agile-Designed SoC for Swarm-Based Perception with Heterogeneous IP Blocks, a Reconfigurable Memory Hierarchy, and an 800MHz Multi-Plane NoC 269

Tianyu Jia[1], Paolo Mantovani[2], Maico Cassel Dos Santos[2], Davide Giri[2], Joseph Zuckerman[2], Erik Jens Loscalzo[2], Martin Cochet[3], Karthik Swaminathan[3], Gabriele Tombesi[2], Jeff Jun Zhang[1], Nandhini Chandramoorthy[3], John-David Wellman[3], Kevin Tien[3], Luca Carloni[2], Kenneth Shepard[2], David Brooks[1], Gu-Yeon Wei[1], Pradip Bose[3]

[1]Harvard University, United States; [2]Columbia University, United States; [3]IBM Research, United States

DARKSIDE: 2.6GFLOPS, 8.7mW Heterogeneous RISC-V Cluster for Extreme-Edge On-Chip DNN Inference and Training 273

Angelo Garofalo[1], Matteo Perotti[2], Luca Valente[1], Yvan Tortorella[1], Alessandro Nadalini[1], Luca Benini[1,2], Davide Rossi[1], Francesco Conti[1]

[1]University of Bologna, Italy; [2]ETH Zürich, Switzerland

Power Conversion

Date: Tuesday, September 20, 2022
Time: 16:50 - 18:30
Room: Aula 1A
Chair(s): Joseph Shor, Bar-Ilan University
Tom Van Breussegem, ICSense

A 91.6% Peak Efficiency Time-Domain-Controlled Single-Inductor Triple-Output Step-Up Converter with ±7.5 to ±12V Bipolar Output Voltages 277

Peng Cao, Danzhu Lu, Jiawei Xu, Xiaoyang Zeng, Zhiliang Hong

Fudan University, China

An Automotive-Use Dual-f_{SW}-Zone Hybrid Switching Power Converter with V_O-Jitter-Immune Spread-Spectrum Modulation 281

Jin Woong Kwak, D. Brian Ma

The University of Texas at Dallas, United States

A Galvanic-Free Secondary-Side Control Flyback Converter with Digital Adaptive On-Time Control and Direct Sequence Spread Spectrum Technique for 15.5% Error Recovery Rate Improvement 285

Wei-Cheng Huang[1], Yuan-Jin Li[1], Yi-Hsiang Kao[1], Ke-Horng Chen[1], Kuo-Lin Zheng[2], Ying-Hsi Lin[3], Shian-Ru Lin[3], Tsung-Yen Tsai[3]

[1]National Yang Ming Chiao Tung University, Taiwan; [2]Chip-GaN Power Semiconductor Corporation, Taiwan; [3]Realtek Semiconductor Corp., Taiwan

A Half-Bridge Gate-Driver for High-Efficient Boost Converter Applications with Single-Sided ZVS and an Adaptive Ringing Suppression Technique 289

Michael Hanhart, Jonas Zoche, Jan Grobe, Léon Weihs, Leo Rolff, Ralf Wunderlich, Stefan Heinen

RWTH Aachen University, Germany

A 4 to 40V Wide Input Range and Energy Re-Cycling High Power LiDAR Driver for 5% Efficiency Enhancement and 300M Long-Distance Object Detection 293

Si-Yi Li[1], Zheng-Lun Huang[1], Sheng Cheng Lee[1], Ke-Horng Chen[1], Kuo-Lin Zheng[2], Ying-Hsi Lin[3], Shian-Ru Lin[3], Tsung-Yen Tsai[3]

[1]National Yang Ming Chiao Tung University, Taiwan; [2]Chip-GaN Power Semiconductor Corporation, Taiwan; [3]Realtek Semiconductor Corp., Taiwan

RF Circuits Techniques

Date: Tuesday, September 20, 2022
Time: 16:50 - 18:30
Room: Aula 1B
Chair(s): Danilo Manstretta, Uni Pavia
          Eric Klumperink, Uni Twente

High-Voltage CMOS RF Switch with Active Biasing 297

Valentyn Solomko[1], Semen Syroiezhin[2,3], Danial Tayari[1], Jochen Essel[2], Robert Weigel[3]

[1]Infineon Technologies AG, Germany; [2]eesy-ic GmbH, Germany; [3]University of Erlangen-Nuremberg, Germany

Switching Time Acceleration for High-Voltage CMOS RF Switch 301

Semen Syroiezhin[1,3], Oguzhan Oezdamar[2], Robert Weigel[3], Valentyn Solomko[2]

[1]eesy-ic GmbH, Germany; [2]Infineon Technologies AG, Germany; [3]University of Erlangen-Nuremberg, Germany

A Low-Loss 77 GHz Sub-Sampling Passive Mixer Integrated in a 28-nm CMOS Radar Receiver 305

Alexandre Flete[1,2], Christophe Viallon[2], Philippe Cathelin[1], Thierry Parra[2]

[1]STMicroelectronics, France; [2]LAAS-CNRS, France

An FDD Auxiliary Receiver with a Highly Linear Low Noise Amplifier 309

Jin Jin[1], Simone Lecchi[2], Rinaldo Castello[2], Danilo Manstretta[2]

[1]Southeast University, China; [2]University of Pavia, Italy

A 433-MHz Wireless Burst-Chirp Modulation Transmitter with Adaptive Duty-Cycle Control and Precharge Mechanism 313

Jing-Siang Chen, Chun-Ting Chang, Yu-Te Liao

National Yang Ming Chiao Tung University, Taiwan

Novel Techniques to Boost A/D & D/A Converter Performance

Date: Wednesday, September 21, 2022
Time: 9:20 - 10:40
Room: Aula 1A
Chair(s): Mattias Palm, Ericsson
Claudio Nani, Marvell

A Pipeline ADC with Negative C-Assisted SC Amplifier Canceling Gain Error and Nonlinearity 317

Woojin Jang[1,2], Gyeong-Gu Kang[1], Yong Lim[2], Hyun-Sik Kim[1]

[1]Korea Advanced Institute of Science & Technology, Korea; [2]Samsung Electronics Co., Ltd., Korea

A 2GHz 2-Bit Continuous-Time Delta Sigma ADC with 2GHz Chopper Achieving 12nV/sqrt(Hz) 1/f Noise at 153kHz and -104.7dBc THD in 30MHz BW 321

Pierluigi Cenci[1], Hans Brekelmans[1], Shagun Bajoria[1], Marcello Ganzerli[1], Bernard Burdiek[1], Robert Rutten[1], Yihan Gao[1], Muhammed Bolatkale[1], Paul Swinkels[1], Lucien Breems[1]

[1]NXP Semiconductors, The Netherlands; [2]NXP Semiconductors, Germany

6Gs/s 8-Channel CIC SAR Ti-ADC with Neural Network Calibration 325

Evelyn Ware, Justin Correll, Seungjong Lee, Michael Flynn

University of Michigan, United States

A 56 GS/s 8-Bit 0.011 mm² 4x Delta-Interleaved Switched-Capacitor DAC in 16 nm FinFET CMOS 329

Pietro Caragiulo, Athanasios Ramkaj, Amin Arbabian, Boris Murmann

Stanford University, United States

Advanced mm-Wave 5G Circuits

Date: Wednesday, September 21, 2022
Time: 9:20 - 10:40
Room: Aula 1B
Chair(s): Kimia Ansari, Zeku
Francois Rivet, Uni Bordeaux

A 28GHz Area-Efficient CMOS Vector-Summing Phase Shifter Utilizing Phase-Inverting Type-I Poly-Phase Filter for 5G New Radio 333

Minzhe Tang, Yi Zhang, Jian Pang, Atsushi Shirane, Kenichi Okada

Tokyo Institute of Technology, Japan

A 24-30 GHz Broadband Doherty PA with a Maximum 15.37 dBm P_{avg} and 14.6% PAE_{avg} in 0.13 μm SiGe for 400 MHz BW 5G NR 337

Hao Gao[1,2], Sina Mortezazadeh Mahani[1], David Seebacher[3], Matteo Bassi[3], Gernot Hueber[1]

[1]Silicon Austria Labs, Austria; [2]Eindhoven University of Technology, The Netherlands; [3]Infineon Technologies AG, Austria

A 24-to-44 GHz Compact Linear 5G Power Amplifier with Open-Terminated Balun in 65nm Bulk CMOS 341

Kyutaek Oh[1], Hyunjin Ahn[2], Ilku Nam[1], Ockgoo Lee[1]

[1]Pusan National University, Korea; [2]Qualcomm Inc., United States

A CMOS Full-Wave Switching Rectifier with Frequency Up-Down Conversion for 5G NR Wirelessly-Powered Relay Transceivers 345

Sena Kato, Keito Yuasa, Michihiro Ide, Atsushi Shirane, Kenichi Okada

Tokyo Institute of Technology, Japan

Application Specific Accelerators

Date: Wednesday, September 21, 2022
Time: 12:00 - 13:00
Room: Aula 1A
Chair(s): Martin Brox, Micron
Christoph Studer, ETH Zurich

INTIACC: A 32-Bit Floating-Point Programmable Custom-ISA Accelerator for Solving Classes of Partial Differential Equations 349

Paul Xuanyuanliang Huang, Daniel Jang, Yannis Tsividis, Mingoo Seok

Columbia University, United States

A Scalable Bit-Serial Computing Hardware Accelerator for Solving 2D/3D Partial Differential Equations Using Finite Difference Method 353

Junjie Mu[1], Chengshuo Yu[1,2], Tony Tae-Hyoung Kim[1], Bongjin Kim[3]

[1]Nanyang Technological University, Singapore; [2]A*STAR, Singapore; [3]University of California-Santa Barbara, United States

A 283 pJ/B 240 Mb/s Floating-Point Baseband Accelerator for Massive MU-MIMO in 22FDX 357

Oscar Castañeda, Luca Benini, Christoph Studer

ETH Zürich, Switzerland

Low Power Techniques for Power Management

Date: Wednesday, September 21, 2022
Time: 12:00 - 13:00
Room: Aula 1B
Chair(s): Ravi Karadi, NXP
          Michael Lueders, Texas Instruments

A Wide-Input-Range 918MHz RF Energy Harvesting IC with Adaptive Load and Input Power Tracking Technique 361

Jing-Ren Yan, Hao-Yi Kuo, Yu-Te Liao

National Yang Ming Chiao Tung University, Taiwan

Input Nonlinear Adaptive Voltage Position Technique in the Switched-Capacitor Converter with Feedforward Compensation for 87.8% Peak Efficiency Under 8X Input Interference 365

Shu-Yung Lin[1], Sheng Cheng Lee[1], Ke-Horng Chen[1], Kuo-Lin Zheng[2], Ying-Hsi Lin[3], Shian-Ru Lin[3], Tsung-Yen Tsai[3]

[1]National Yang Ming Chiao Tung University, Taiwan; [2]Chip-GaN Power Semiconductor Corporation, Taiwan; [3]Realtek Semiconductor Corp., Taiwan

A 385mV, 270nW, Accurate Voltage Level Detector for IoT 369

Omer Nechushtan, Asaf Feldman, Joseph Shor

Bar Ilan University, Israel

Memory & On-Chip Sensing

Date: Wednesday, September 21, 2022
Time: 12:00 - 13:00
Room: Aula 1D
Chair(s): Atila Alvandpour, Linkoping University
Alexander Fish, Bar-Ilan University

An Extended Temperature Range ePCM Memory in 90-nm BCD for Smart Power Applications 373

Marcella Carissimi[1], Chantal Auricchio[1], Emanuela Calvetti[1], Laura Capecchi[1], Mattia Torres[1], Stefano Zanchi[1], Priyanka Gupta[2], Riccardo Zurla[1], Alessandro Cabrini[3], Daniele Gallinari[1], Fabio Disegni[1], Massimo Borghi[1], Elisabetta Palumbo[1], Andrea Redaelli[1], Marco Pasotti[1]

[1]STMicroelectronics, Italy; [2]STMicroelectronics, India; [3]University of Pavia, Italy

On-Chip High-Resolution Timing Characterization Circuits for Memory IPs 377

Amit Agarwal, Steven Hsu, Mark Anders, Gunjan Pandya, Ram Krishnamurthy, James Tschanz, Vivek De
Intel Corporation, United States

Capacitance-Based Voltage Regulation- and Reference-Free Temperature-to-Digital Converter Down to 0.3 V and 2.5 nW for Direct Harvesting 381

Orazio Aiello[1,2], Massimo Alioto[1]

[1]National University of Singapore, Singapore; [2]University of Genova, Italy

Frequency Generation

Date: Wednesday, September 21, 2022
Time: 14:20 - 15:40
Room: Aula 1A
Chair(s):   Akshay Visweswaran, Nokia Bell Labs
            Dmytro Cherniak, Infineon Technologies

A 24 GHz Quadrature VCO Based on Coupled PLL with -134 dBc/Hz Phase Noise at 10MHz Offset in 28 nm CMOS 385

Agata Iesurum[1], Davide Manente[1], Fabio Padovan[2], Matteo Bassi[2], Andrea Bevilacqua[1]

[1]University of Padova, Italy; [2]Infineon Technologies AG, Austria

A K-Band Gilbert-Cell Frequency Doubler with Self-Adjusted 25% LO Duty-Cycle in SiGe BiCMOS Technology 389

Lorenzo Piotto[1], Guglielmo De Filippi[1], Daniele Dal Maistro[2], Simone Erba[2], Andrea Mazzanti[1]

[1]University of Pavia, Italy; [2]Infineon Technologies AG, Austria

A 1.8-67GHz Divide-by-4 ILFD Using Area-Efficient Transformer-Based Injection-Enhancing Technique 393

Yudai Yamazaki, Jian Pang, Atsushi Shirane, Kenichi Okada

Tokyo Institute of Technology, Japan

A Fully Differential 40 MHz Switched-Capacitor Crystal Oscillator with Fast Start-Up 397

Wim Kruiskamp

Renesas Electronics Corporation, The Netherlands

Interface Circuits

Date: Wednesday, September 21, 2022
Time: 14:20 - 15:40
Room: Aula 1B
Chair(s): Piotr Kmon, AGH University of Science and Technology
Qinwen Fan, TU Delft

A 128ksps 120dB THD Low Noise Analog Front End ... 401

Gerard Mora-Puchalt[1], Gabriel Banarie[2], Pawel Czapor[2], Adrian Sherry[2], Roberto Maurino[3], Jesús Bonache[1], Italo Medina[1]

[1]Analog Devices, Inc., Spain; [2]Analog Devices, Inc., Ireland; [3]Analog Devices, Inc., Italy

A 0.01mm², 0.4V-V_{DD}, 4.5nW-Power DC-Coupled Digital Acquisition Front-End Based on Time-Multiplexed Digital Differential Amplification ... 405

Paolo Crovetti[1], Roberto Rubino[1], Pedro Toledo[1], Francesco Musolino[1], Hamilton Klimach[2], Yong Chen[3], Anna Richelli[4]

[1]Politecnico di Torino, Italy; [2]Federal University of Rio Grande do Sul, Brazil; [3]University of Macau, Macau; [4]University of Brescia, Italy

A 69dBA-730µW Silicon Microphone System with Ultra & Infra-Sound Robustness ... 409

Jose Luis Ceballos, Christopher Rogi, Fulvio Ciciotti, Cesare Buffa, Dietmar Straeussnigg, Andreas Wiesbauer

Infineon Technologies AG, Austria

A Nonuniform Sampling Lifetime Estimation Technique for Luminescent Oxygen Measurements ... 413

Ian Costanzo, Devdip Sen, John McNeill, Ulkuhan Guler

Worcester Polytechnic Institute, United States

RF & Wideband Circuits

Date: Thursday, September 22, 2022
Time: 8:30 - 9:50
Room: Aula 1A
Chair(s): Stefan Andersson, Ericsson
Jan Craninckx, IMEC

A 2.4GHz Full-Duplex Transceiver with Broadband (+120MHz), Linearity-Calibrated and Long-Delayed Self-Interference Cancellation 417

Xichen Li, Yi-Hsiang Huang, Fucheng Yin, Jacques C. Rudell

University of Washington, United States

A 3-10GHz 21.5mW/Channel RX and 8.9mW TX IR-UWB 802.15.4a/z 1T3R Transceiver 421

Elbert Bechthum, Minyoung Song, Gaurav Singh, Erwin Allebes, Charis Basetas, Pepijn Boer, Arjan Breeschoten, Stefan Cloudt, Johan Dijkhuis, Ming Ding, Sherwin Gatchalian, Yuming He, Johan van Den Heuvel, Martijn Hijdra, Paul Mateman, Bernard Meyer, Gert-Jan van Schaik, Mohieddine El Soussi, Bart Thijssen, Stefano Traferro, Evgenii Turin, Peter Vis, Nick Winkel, Peng Zhang, Yao-Hong Liu, Christian Bachmann

imec, The Netherlands

A 260mW, 6-8dB Noise Figure 1-80GHz Frequency Interleaving Receiver for Spectrum Sensing in 65nm CMOS .... N/A

Shunli Ma[1], Dong Wei[1], Jincheng Zhang[1], Tianxiang Wu[1], Yong Chen[2], Junyan Ren[1]

[1]Fudan University, China; [2]University of Macau, China

A 58 GHz Bandwidth, and Less Than 1.8% THD, Mach-Zehnder Driver, in 28 nm CMOS Technology 429

Nicola Cordioli, Danilo Manstretta, Rinaldo Castello

University of Pavia, Italy

Biomedical Interface Circuits

Date: Thursday, September 22, 2022
Time: 8:30 - 9:50
Room: Aula 1B
Chair(s): Jens Anders, University of Stuttgart
Zili Yu, IMS Chips

A 136GΩ-Input-Impedance Active Electrode for Non-Contact ECG Using Auto-Calibrated Positive Feedback and Capacitance Scaling in Femtofarad Resolution 433

Peizhuo Wang, Tianxiang Qu, Liangbo Lei, Zhiliang Hong, Jiawei Xu

Fudan University, China

A Chopper Biopotential Instrumentation Amplifier with DSL-Embedded Input Stage Achieving 109 dB CMRR and 400 mV DC Offset Tolerance 437

Surachoke Thanapitak[1], Chutham Sawigun[2]

[1]Mahidol University, Thailand; [2]imec, Belgium

A Direct Sensor Readout Circuit Using VCO-Driven Chopping with 42dB SNR at 800µVpp Input 441

Yingjie Chen[1], Marino D. Guzman[1], Beomsoo Park[1,2], Nima Maghari[1]

[1]University of Florida, United States; [2]Qualcomm Inc., United States

Mixed-Signal Compensation of Tripolar Cuff Electrode Imbalance in a Low-Noise ENG Analog Front-End 445

Rémi Dekimpe, David Bol

Université Catholique de Louvain, Belgium

Beyond 100GHz Circuits & Applications

Date: Thursday, September 22, 2022
Time: 8:30 - 9:50
Room: Aula 1E
Chair(s): Minoru Fujishima, Hiroshima Uni
Michele Caruso, Infineon

A Full D-Band Multi-Gbit RF-DAC in 90 nm SiGe BiCMOS Based on Passive Vector Aggregation 449

Tim Maiwald[1], Akshay Visweswaran[2], Klaus Aufinger[3], Robert Weigel[1]

[1]University of Erlangen-Nuremberg, Germany; [2]imec, Belgium; [3]Infineon Technologies AG, Germany

A 111-149-GHz, Compact Power-Combined Amplifier with 17.5-dBm P_{sat}, 16.5% PAE in 22-nm CMOS FD-SOI 453

Jeff Shih-Chieh Chien, James F. Buckwalter

University of California-Santa Barbara, United States

A D-Band mm-Wave Spectroscopy TX and RX in 28 nm CMOS with 15.6 dBm EIRP and 17.1 dB NF with Integrated Antennas 457

Gabriel Guimaraes, Patrick Reynaert

Katholieke Universiteit Leuven, Belgium

A Joint Space/Time Modulation Lens-Coupled 230-GHz Terahertz Source with 40°/43° 2-D Beam Steering for Fast High-Resolution Imaging/Sensing Applications 461

Hossein Jalili[1], Yuqi Liu[1], Tzu-Yuan Huang[1], Hua Wang[2]

[1]Georgia Institute of Technology, United States; [2]ETH Zürich, Switzerland

References & Filters

Date: Thursday, September 22, 2022
Time: 11:20 - 12:40
Room: Aula 1A
Chair(s): Stefano D'Amico, University of Salento
           Paras Garg, ST Microelectronics

A 956pW Switched-Capacitor Sub-Bandgap Reference with 0.44-to-3.3V Supply Range, -67dB PSRR, and 0.2% Within-Wafer Inaccuracy for Nanowatt IoT Systems .... N/A

Shuo Li, Xinjian Liu, Benton H. Calhoun

University of Virginia, United States

A 0.9-nA Temperature-Independent 565-ppm/°C Self-Biased Current Reference in 22-nm FDSOI 469

Martin Lefebvre, Denis Flandre, David Bol

Université Catholique de Louvain, Belgium

A Compact 2.5-nJ Energy/Conversion NPN-Based Temperature-to-Digital Converter with a Fully Current-Mode Processing Architecture 473

Antonio Aprile[1], Michele Folz[2], Daniele Gardino[2], Piero Malcovati[1], Edoardo Bonizzoni[1]

[1]University of Pavia, Italy; [2]TDK InvenSense, Italy

A 5.4-mW 50-MHz 29.3-dBm-IIP3 Fourth-Order Low-Pass Filter 477

Liangbo Lei, Cong Tao, Zhipeng Chen, Zhiliang Hong, Yumei Huang

Fudan University, China

mmWave Tranceivers

Date: Thursday, September 22, 2022
Time: 11:20 - 12:40
Room: Aula 1B
Chair(s): Vojkan Vidojkovic, TU Eindhoven
Gernot Hueber, Silicon Austria

A 135 GHz 32 Gb/s Direct-Digital Modulation 16-QAM Transmitter in 28 nm CMOS 481

Carl D'Heer, Patrick Reynaert

Katholieke Universiteit Leuven, Belgium

A 135 GHz 24 Gb/s Direct-Digital Demodulation 16-QAM Receiver in 28 nm CMOS 485

Carl D'Heer, Patrick Reynaert

Katholieke Universiteit Leuven, Belgium

**A Low-Power and Energy-Efficient D-Band CMOS Four-Channel Receiver with Integrated
LO Generation for Digital Beamforming Arrays** 489

Ethan Chou, Nima Baniasadi, Hesham Beshary, Meng Wei, Emily Naviasky, Lorenzo Iotti, Ali Niknejad

University of California-Berkeley, United States

A Reconfigurable Digital Beamforming V-Band Phased-Array Receiver 493

Deniz Dosluoglu[1], Kun-Da Chu[2], Diego Pena-Colaiocco[1], Ivan Zhao[3], Visvesh Sathe[1], Jacques C. Rudell[1]

[1]University of Washington, United States; [2]Apple, Inc., United States; [3]Boeing, United States

Wireline Transceivers

Date: Thursday, September 22, 2022
Time: 14:20 - 15:40
Room: Aula 1A
Chair(s): Armin Tajalli, University of Utah
Chris Rudell, University of Washington

A 112-Gb/s Single-Ended PAM-4 Transceiver Front-End for Reach Extension in Long-Reach Link 497

Xiongshi Luo[1], Xuewei You[1], Jiahan Fu[1], Zhenghao Li[1], Liping Zhong[1], Taiyang Fan[1], Zhang Qiu[1], Wenbo Xiao[1], Yong Chen[2], Quan Pan[1]

[1]Southern University of Science and Technology, China; [2]University of Macau, China

A 2×50 Gb/s Single-Ended MIMO PAM-4 Crosstalk Cancellation and Signal Reutilization Receiver in 28 nm CMOS 501

Liping Zhong, Hongzhi Wu, Weitao Wu, Wenbo Xiao, Xiongshi Luo, Dongfan Xu, Xuxu Cheng, Zhenghao Li, Taiyang Fan, Quan Pan

Southern University of Science and Technology, China

A 56-Gb/s PAM4 Transceiver with False-Lock-Aware Locking Scheme for Mueller-Müller CDR 505

Fumihiko Tachibana, Huy Cu Ngo, Go Urakawa, Takashi Toi, Mitsuyuki Ashida, Yuta Tsubouchi, Mai Nozawa, Junji Wadatsumi, Hiroyuki Kobayashi, Jun Deguchi

Kioxia Corporation, Japan

A 64Gb/s Downlink and 32Gb/s Uplink NRZ Wireline Transceiver with Supply Regulation, Background Clock Correction and Eom-Based Channel Adaptation for Mid-Reach Cellular Mobile Interface in 8nm FinFET 509

Soo-Min Lee[1], Jihoon Lim[1], Jaehyuk Jang[1], Hyoungjoong Kim[1], Kyunghwan Min[1], Woongki Min[1], Hyeonji Han[1], Gyusik Kim[1], Jaeyoung Kim[1], Chulho Kim[1], Sejun Jeon[1], Jinhoon Park[1], Hyunsu Chae[1], Sangwook Han[1], Hiep Pham[2], Xingliang Zhao[2], Qilin Gu[2], Chih-Wei Yao[2], Sangho Kim[1], Jongwoo Lee[1]

[1]Samsung Electronics Co., Ltd., Korea; [2]Samsung Electronics Co., Ltd., United States

Security

Date: Thursday, September 22, 2022
Time: 14:20 - 16:00
Room: Aula 1E
Chair(s): Makoto Ikeda, Tokyo University
Mark Anders, Intel

A 2.17-pJ/b 5b-Response Attack-Resistant Strong PUF with Enhanced Statistical Performance 513

Kunyang Liu, Gen Li, Zihan Fu, Xuanzhen Wang, Hirofumi Shinohara

Waseda University, Japan

A 183F² Gate Leakage-Based Physically Unclonable Function with Area Efficient Current Tilting-Based Masking Scheme 517

Beomsoo Park[1,2], Nima Maghari[1]

[1]University of Florida, United States; [2]Qualcomm Inc., United States

Logic-Embedded Physically Unclonable Functions for Synthesizable and Periphery-Free Implementation for Low Area and Design Cost IoT Security 521

Seonho Kim, Changyoun Im, Jongmin Lee, Soyoun Jeong, Jaerok Kim, Yoonmyung Lee

Sungkyunkwan University, Korea

Configurable Energy-Efficient Lattice-Based Post-Quantum Cryptography Processor for IoT Devices 525

Byungjun Kim, Jaehan Park, Seunghyun Moon, Kiseo Kang, Jae-Yoon Sim

Pohang University of Science and Technology, Korea

Fine-Grained Electromagnetic Side-Channel Analysis Resilient Secure AES Core with Stacked Voltage Domains and Spatio-Temporally Randomized Circuit Blocks 529

Meizhi Wang[1], Sirish Oruganti[1], Shanshan Xie[1], Raghavan Kumar[2], Sanu Mathew[2], Jaydeep P. Kulkarni[1]

[1]The University of Texas at Austin, United States; [2]Intel Research Labs, United States

Hot Research Topics in North America

Date: Tuesday, September 20, 2022
Time: 16:50 - 18:30
Room: Aula 1E
Chair(s): Chris Rudell, University of Washington

A Review on the State-of-the-Art THz FMCW Radars Implemented on Silicon .... N/A

Morteza Tavakoli Taba, S.M. Hossein Naghavi, Ehsan Afshari

University of Michigan, United States

Integrated Optical Phased Arrays on Silicon 538

Farshid Ashtiani, Firooz Aflatouni

University of Pennsylvania, United States

Full-Duplex Wireless for (Joint-) Communication and Sensing 542

Hany Abolmagd[1], Raghav Subbaraman[2], Dinesh Bharadia[2], Sudip Shekhar[1]

[1]University of British Columbia, Canada; [2]University of California-San Diego, United States

Reconfigurable Intelligent Surfaces Enabled by Silicon Chips for Secure and Robust mmWave and THz Wireless Communication 546

Kaushik Sengupta[1], Suresh Venkatesh[1], Hooman Saeidi[1], Xuyang Lu[2]

[1]Princeton University, United States; [2]University of Michigan - Shanghai Jiaotong Joint Institute, China

Hot Research Topics in Far East

Date: Thursday, September 22, 2022
Time: 14:20 - 16:00
Room: Aula 1B
Chair(s): Qiang Li, University of Electronic Science and Technology of China

A 7b 3.8GS/s 4b/Cycle SAR ADC with 16× Time-Domain Interpolation .... N/A

Dengquan Li, Yi Shen, Xin Zhao, Shubin Liu, Zhangming Zhu

Xidian University, China

A 130µW Three-Step DT Incremental ΔΣ ADC Achieving 107.6dB DR and 99.3dB SNDR with Zoom and Extended-Range Counting 554

Lairong Fang, Yijie Li, Yao Zhang, Shuwen Zhang, Xiaoyang Zeng, Zhiliang Hong, Jiawei Xu

Fudan University, China

A 0.45V 450µW 2.4GHz Flicker-Noise-Free RF Receiver Front-End Based on Switched-Capacitor TIA with 4.5dB NF and 11.5dBm OIP3 .... N/A

Chao Chen[1], Dan Huang[2], Yan Zhao[1], Yuemin Jin[1], Jun Yang[1]

[1]Southeast University, China; [2]Electronic Devices Institute, China

A 65nm 8b-Activation 8b-Weight SRAM-Based Charge-Domain Computing-in-Memory Macro Using a Fully-Parallel Analog Adder Network and a Single-ADC Interface .... N/A

Guodong Yin[1], Mufeng Zhou[1], Yiming Chen[1], Wenjun Tang[1], Zekun Yang[1], Mingyen Lee[1], Xirui Du[1], Jinshan Yue[2], Jiaxin Liu[3], Huazhong Yang[1], Yongpan Liu[1], Xueqing Li[1]

[1]Tsinghua University, China; [2]Institute of Microelectronics of the Chinese Academy of Sciences, China; [3]University of Electronic Science and Technology of China, China

A 0.016mm² Active Area 4GHz Fully Ring-Oscillator-Based Cascaded Fractional-N PLL with Burst-Mode Sampling .... N/A

Junlin Zhong[1], Xiaofeng Yang[1], Yan Zhu[1], Chi-Hang Chan[1], R.P. Martins[1,2]

[1]University of Macau, China; [2]Universidade de Lisboa, Portugal

A-SSCC Invited Papers

Date: Tuesday, September 20, 2022
Time: 16:50 - 18:30
Room: Aula 1F
Chair(s): Woogeun Rhee, Tsinghua University

FlashMAC: an Energy-Efficient Analog-Digital Hybrid Mac with Variable Latency-Aware Scheduling .... *available on Xplore*

Surin Gweon, Sanghoon Kang, Donghyeon Han, Kyoung-Rog Lee, Kwantae Kim, Hoi-Jun Yoo

Korea Advanced Institute of Science & Technology, Korea

A 33.5-37.5 GHz 4-Element Phased-Array Transceiver Front-End with High-Accuracy Low-Variation 6-Bit Resolution 360° Phase Shift and 0~31.5 dB Gain Control in 65 nm CMOS .... *available on Xplore*

Pingda Guan, Haikun Jia, Wei Deng, Zhihua Wang, Baoyong Chi

Tsinghua University, China

A 5-GHz 0.15-mm2 Collision Avoidable RFID Employing Complementary Pass-Transistor Adiabatic Logic with an Inductively Connected External Antenna .... *available on Xplore*

Saito Shibata, Reiji Miura, Yoshiki Sawabe, Kota Shiba, Atsutake Kosuge, Mototsugu Hamada, Tadahiro Kuroda

The University of Tokyo, Japan

Correlated Dual-Loop Sturdy Mash CT ΔΣ ADC with Indirect Signal Feedforward .... *available on Xplore*

Beomsoo Park[1,3], Changsok Han[2], Nima Maghari[1]

[1]University of Florida, United States; [2]Marvell Semiconductor, Inc., United States; [3]Qualcomm Inc., United States

A 1.68-23.2 Gb/S Reference-Less Half-Rate Receiver with an ISI-Tolerant Unlimited Range Frequency Detector .... *available on Xplore*

Yu-Ping Huang, Yi-Wei Chang, Wei-Zen Chen

National Yang Ming Chiao Tung University, Taiwan

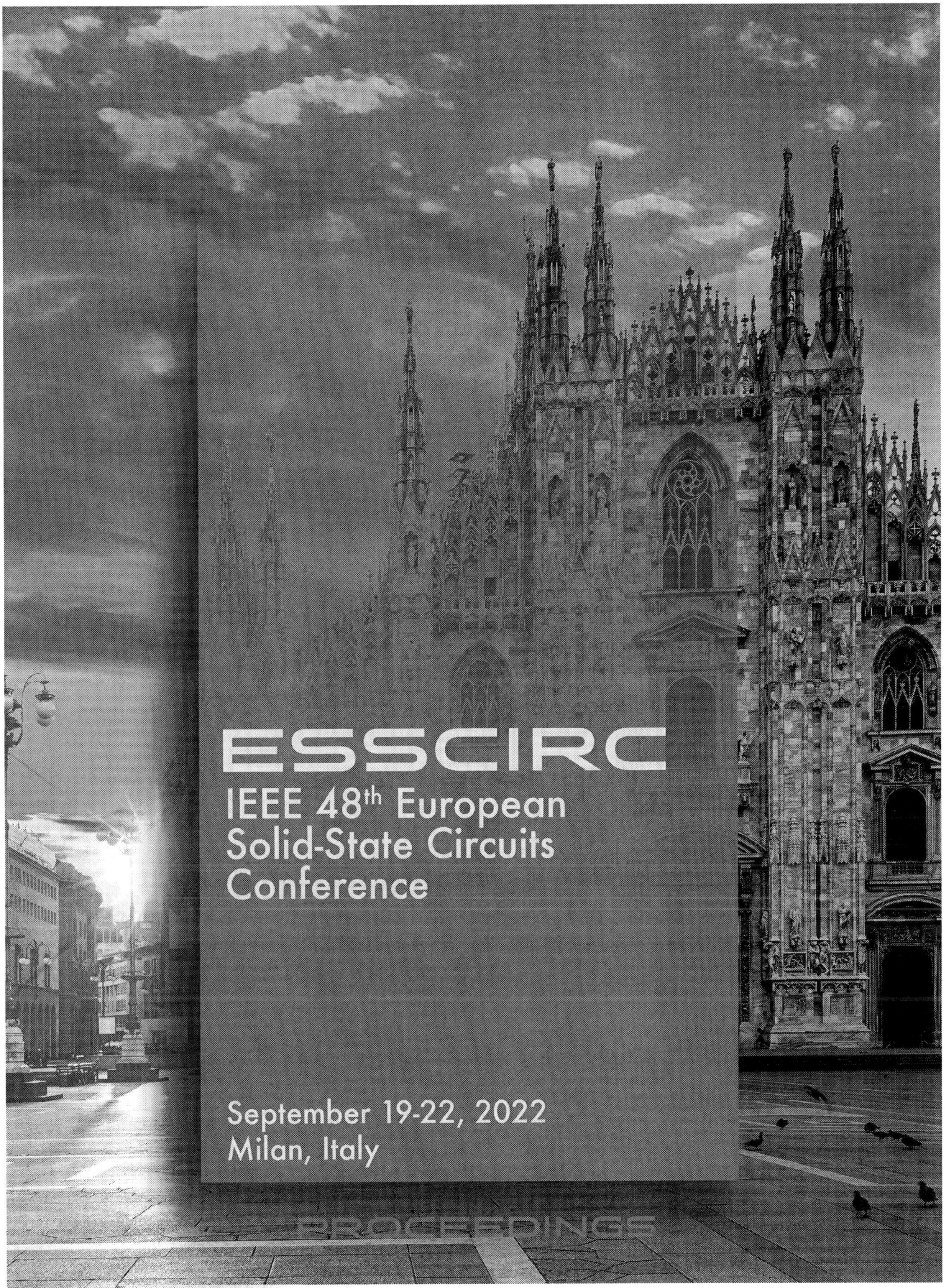

ESSCIRC

IEEE 48th European Solid-State Circuits Conference

September 19-22, 2022
Milan, Italy

PROCEEDINGS

ESSCIRC 2022 - IEEE 48th European Solid State Circuits Conference (ESSCIRC)

On Line PROCEEDINGS

ESSCIRC

IEEE 48th European Solid-State Circuits Conference

Co - Organizer

Financial Sponsorship

under the Patronage of

SPONSORS

DIAMOND SPONSORS

— SPONSORS —

GOLD SPONSORS

SONY

SYNOPSYS®

SILVER SPONSORS

 OSRAM

INVENTVM

KIOXIA

knowles

RedCat Devices

— ESSCIRC • 2022 —

SPONSORS

SILVER SPONSORS

SAMSUNG

⬡TDK

BRONZE SPONSORS

cādence®

EUROPRACTICE

GlobalFoundries™

innatera

photeon
technologies

SPRINGER NATURE

xfab

ESSCIRC · 2022

The Future of Short Reach Interconnect

Davide Tonietto
Huawei Technologies Canada Ltd.

Ottawa, Canada

davide.tonietto@huawei.com

Abstract— the unprecedented information explosion and its increasing demands on data traffic and processing are pushing a rapid and diverse evolution in short reach interconnect technologies. In the background of this race, CMOS technology is not providing the usual node over node boost in performance to help SerDes developers cope with higher bandwidth and data rates. Breakthroughs in high speed electrical interconnect and new approaches in optical interconnect, such SiPho, NPO (near package optics) and CPO (co-packaged optics) promise improved performance, energy efficiency and density. What are the specific challenges that new and emerging applications such as AI and HPC are posing on interconnect? How various interconnect technologies are going to respond to the need for die disaggregation and multi-die IC products? The focus of this paper is to provide some directions and an underlying logic to help navigate this very complex technology landscape.

Keywords—SerDes, Short Reach Interconnect, Die to Die, Chiplets, Optical Densification, NPO, CPO.

I. INTRODUCTION

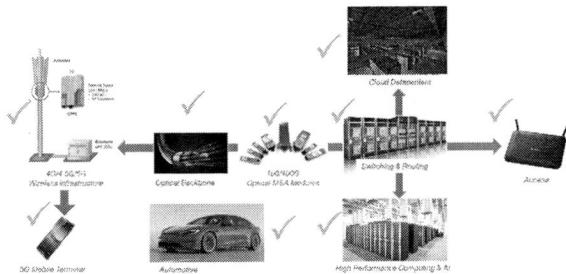

Fig. 1. : Short Reach Interconnect Landscape

SerDes are a foundation building block of the ICT industry (Fig.1.), providing the means to interconnect ICs used in every market and application, each one with different challenges. Requirements range from extreme bandwidth and performance for Ethernet applications, to extreme energy efficiency and negligible stand-by currents in the mobile space, to extreme environmental range and reliability required by wireless infrastructure, automotive or aerospace industries. This wide swath of applications and requirements is sometime referred generically as "short reach interconnect". The term really conflates a diverse range of technologies, active and passive components. To better understand "short reach interconnect" we have to think about it as a complex stack of many different technologies, used in many different combinations that together make up what constitutes "the channel". Each

technology progresses and evolves node over node, at a different pace and largely non-synchronized with the other ones. The solution for a given application and set of requirements, can present itself in multiple combinations of technologies. Each solution using different tradeoffs and optimization choices not necessarily relying on the latest "node" of a given technology.

II. THE CANARY IN THE COALMINE

Across every industry and application, energy efficiency and density are the common challenge. In this paper we will focus on high performance ASICs for Networking, Data Center and Computing, where the combined and often conflicting density and energy efficiency challenges are the most severe. In this paper we will often use the Ethernet Fabric Switch as an application example, since it is a large ASIC which typically integrates the highest number of SerDes and uses the highest ratio of SerDes vs. core logic resources among all ASICs and thus is the perfect "canary in the coalmine" to gage the impact of interconnect on the overall resources. In 2013-2014 the dominant SerDes data rate was between 10 and 15Gbps and the most common CMOS node used was 28nm. SerDes would consume approximately 10-15% of the total area and power of a 1.28Tbps Switch ASIC (Fig.2.). A bump pitch of 150um or lower for organic FcBGA substrates was already common and with the ability to rotate SerDes on the four sides of the IC die, meeting the required perimeter density was relatively straightforward. Also the escape of 128 lanes from packages sized around 50mm x 50mm and with 1mm ball pitch was not particularly challenging. Fast forward to 2022-2023 and the dominant data rate is now 112Gbps and 7nm or 5nm are the most common CMOS nodes used. Shockingly, SerDes share of resources skyrocketed to 30-40% of a 25-50T Switch. FinFET technology limitations that prevent rotation and lack of significant improvement in bump pitch dimensions on mainstream organic substrates make it very hard to fit 256 SerDes on a single die perimeter and escaping from packages as large as 100mm x 100mm is a considerable challenge As we look for answers to these challenges, in this paper we would like to use two guidelines. The first one: system constraints are not the same as ASIC constraints. This means addressing a system challenge (density, power etc.) do not necessary help mitigating the same challenges on the ASIC, in some cases, the opposite is true. Therefore in the effort to reduce the overall power of a Router or Switch system or HPC cluster or Data Center, it is worth to keep always in focus the challenges at the ASIC level. The second guideline, is that any definition of interconnect has to be strongly linked to the share

978-1-6654-8495-4/22 $31.00 © 2022 IEEE

of ASIC resources that it demands from the ASIC. So we would like to start from a new definition of Short Reach Interconnect, suitable for an age of increasingly finite ASIC resources.

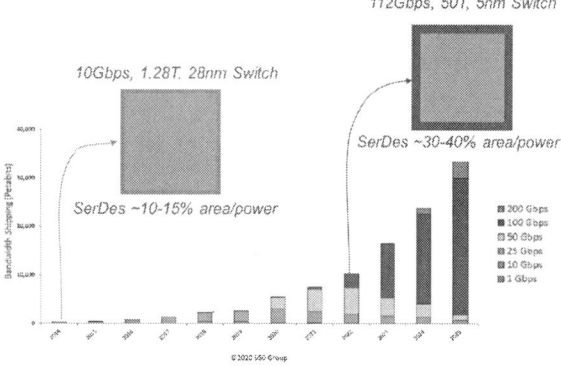

Fig. 2. Total BW Shipped vs. SerDes Resources in a Switch ASIC

III. A NEW DEFINITION OF SHORT REACH INTERCONNECT

As mentioned, "Short Reach Interconnect" is a generic term that encompasses a myriad of applications and different technologies. There are several definitions for channels and associated SerDes, based on physical reach, performance required, applications and standards [1][2]. One of the problems with a definition based on the physical reach or performance is that it depends on the physical media. For example 150m of optical fiber are "short reach" in one context, while only 20" of PCB are short reach in another. This definition largely depends on the technology used for the interconnect, without providing information on the ASIC or system resources allocation. To use an analogy: if we have to walk from Pavia to Milan (about 60Km...), we would consider that a very long reach, but if we travel by helicopter it would be a very short reach. However, we can't really afford a helicopter and even if we did, it would not fit in our garage. In other words, it is incompatible with our overall resources. So we'd like to propose a definition based on resources allocation rather than performance or physical reach, one that we can use to decide which SerDes should or should not be used inside an ASIC:

"A Short Reach Interconnect is one that satisfies data throughput and reach requirements of a host ASIC without exceeding ~25% of its total area and power"

With this definition in mind, let us now focus on SerDes energy efficiency.

IV. THE ENERGY EFFICIENCY CHALLENGE

Intuitively, efficiency gets progressively worse as the channel gets more challenging. It is interesting to observe the progression of this worsening and draw some empirical observations. For this purpose we plot, in Fig. 3, the energy

efficiency in pJ/bit as a function of channel insertion loss (IL) at Nyquist frequency in dB for 112Gbps PAM4 SerDes in 7, 5, and 3nm. These curves are based on our internal SerDes development and estimates combined with the very scarce public data available. To fill the data gaps due to lack of public data we used power scaling curves similar to [3][4].These curves should be treated as rough, empirical observations rather than a precise quantitative guideline. Energy efficiency measured in pJ/bit is estimated for a given technology and physical channel categorized following OIF definitions "XSR", "VSR" and "LR" [1][2].

A. SerDes Energy Efficiency

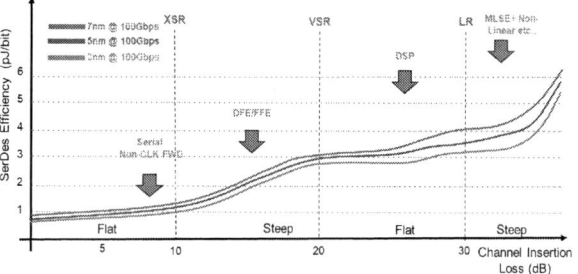

Fig. 3. Empirical Observation: Energy Efficiency vs. Channel Insertion Loss

Unsurprisingly, efficiency worsens with physical reach. That is, the harder the interconnect the less efficient the SerDes. As CMOS technology scales, the shape of the curve remains largely unchanged but efficiency improves across the whole range, with larger gains at high IL where a larger DSP benefits more from CMOS scaling. Sharp worsening in efficiency in the steep areas of the curve occurs when an additional tool or technique is deployed to deal with increased IL requiring a dramatic increase in power. As insertion loss increases beyond 35dB, the amount of energy required skyrockets due the complexity of the tools deployed to compensate channel impairments. Operating beyond this point with sufficient margin for large scale production becomes extremely inefficient. While this plot only considers insertion loss, several other impairments have a strong impact on efficiency, such as crosstalk, discontinuities and for optical links noise dispersion, non-linearity and more. Based on empirical data these curves remained similar from 50Gbps, to 112Gbps PAM4 and the same is expected for 224Gbps, as long as SerDes architectures and design methodology utilized are similar. However it seems that for the same CMOS node (i.e. 7nm) some efficiency gains at higher data rates are possible in the range between ~20 to ~30 dB, where DSP SerDes, can leverage some design efficiencies. Anywhere else faster data rates in a given node are typically less efficient mainly due to CMOS bandwidth and noise limitations, and intrinsic difficulty of generating much cleaner clocks. Also other impairments, such as crosstalk and channel discontinuities become more difficult to deal with at higher rates. We will see in the following that in order to gain efficiency while increasing data rate, CMOS scaling becomes necessary.

978-1-6654-8495-4/22 $31.00 © 2022 IEEE

B. SerDes Energy Efficiency vs. ASIC Resources

So how does this energy efficiency translate into ASIC resources? In this example we will use a 5nm, 50Tbps, 112Gbps/lane Switch Fabric ASIC, with a total expected power in the range of 600 to 800W. Using the efficiency curve as a guide, we can determine whether a certain SerDes type can fit in the "Short Reach Interconnect" definition, i.e. consume less than 25% of total power. Any SerDes below VSR and MR could fall into the "Short Reach" definition for a 600 and 800W Switch respectively, while LR and Optical SerDes could not (Fig.4.).

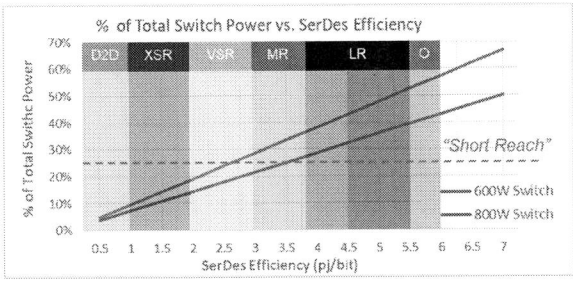

Fig. 4. : Switch resources vs. Efficiency and Type

However if we consider the forward error correction block (FEC) as part of the interconnect energy budget, as it should be but often ignored, and allocated here an estimated 1pJ/bit for an 112Gbps KP4 FEC), the scenario becomes less optimistic. Only VSR marginally qualifies as "Short Reach" for an 800W switch ASIC. Considering that a traditional Switch Fabric ASIC is equipped with "LR" SerDes to provide the maximum range of applications, this would cost between 40% and 50% of the total power. However, if we consider an XSR SerDes, even adding FEC, the Interconnect remains solidly in the "Short Reach" range for both 600 and 800W cases.

V. SERDES TYPES & ARCHITECTURES

As we proceed in the analysis it is useful to look at how the 3 main categories "LR", "VSR", "XSR" of 112Gbps PAM4 SerDes differ from each other. The following are broad-stroke descriptions of these categories while architectures and design choices vary from one developer to another. The LR usually employs a DSP based architecture, where an ADC – DAC analog front end (AFE) digitizes the bitstream and a DSP equalizes and compensates for channel impairments. Due to PAM4 signaling, FEC is necessary to achieve the target error floor. In most cases, in order to reduce development costs, the VSR is a "scaled down" version of the LR, using the same architecture but with reduced DSP capabilities. In some cases it is the same hardware, but used in a scale-down software mode [3][4]. The XSR is a significantly different SerDes, usually based on an "analog" architecture, with very limited equalization capabilities and significantly decreased requirements on performance parameters. It is striking to see in Table I, how much greater is the improvement in all efficiency metrics from VSR to XSR than it is from LR to

VSR. Also to be noted a sharp decrease in latency as we move to XSR, which is is key for certain applications. A particular category of DSP based SerDes is what is commonly referred to as "Direct Detect Optical Engine" or "Optical DSP". These SerDes are used on a wide variety of optical links ranging from few hundred meters on multimode fiber to 10-40Km on single mode fiber. They are found today in most PAM4 optical ICs inside pluggable modules and on board optical engines (OBO). They share most of the architecture and design with copper "LR" SerDes. However they need specific DSP features to compensate for optical components non-idealities. These features, combined with AFE modifications required to transmit and receive from TOSA/ROSA, drive power and area resources higher that a typical copper LR SerDes. Die to die SerDes (D2D) are an altogether separate category. This is because they comprise of a very wide variety of SerDes ranging from XSR (which OIF considers a D2D SerDes) to BoW (Bunch of Wires) a low speed, parallel SerDes proposed by ODSA [5]. As we will see the main "watershed" defining D2D SerDes is based on the type of substrate or channel they operate on.

TABLE I. 112GBPS SERDES TYPES AT A GLANCE

Parameter	SerDes Type		
	XSR/XSR+	VSR/MR	LR
Architecture	Analog	DSP	DSP
Equalization	CTLE, 3-5T FIR	CTLE, 5-15T FFE 4-5T FIR	CTLE, 1-2T DFE 25-50T FFE 5T FIR
Clock recovery	Bang-bang	Baud rate	Baud Rate
IL (db), std.-max	13	22-25	30-40
Efficiency (pJ/bit)	1-1.5	3-4	4-5
Area/lane (mm2)	0.2-0.3	0.5-0.7	0.6-0.7
Perimeter Density	0.8-1.2	0.2-0.6	0.2-0.5
Latency (ns)	3-50	100-150	100-150

VI. DSP SERDES TRENDS

At 56Gbps and even more at 112Gbps most SerDes developed to be used above 20-25dB channels are based on DSP [3][4][6][7] and we expect this trend to continue beyond 112Gbps with a notable exception to this trend in [8]. This is due to their inherent superior robustness and margin vs. conventional "analog" SerDes. They are less sensitive to PVT variations, provide significantly more equalization capabilities and thus margin and can be easily customized and scaled based on the target application. DSP SerDes became viable and competitive thanks to two simultaneous factors: PAM4 and FinFET. PAM4 made channel compensation challenges more severe, and rendered traditional discrete time equalization blocks like multi-tap DFE, much more expensive to design in the analog domain. The advent of 16nm and 7nm FiFET considerably decreased the power and area required to perform similar or better equalization in the digital domain, allowing for the first time to implement such a DSP with a lower power and area than the AFE. While DSP intrinsically scales well beyond 7nm, following Moore's law, the AFE only receives marginal benefit from deep sub-micron scaling. Still, a significant amount of area and power is used by the

DSP, therefore, it is useful to define a "natural technology node" for a given data rate, where there is a balance of 2/3 AFE and 1/3 DSP in overall resources. While this is not an absolute rule, it helps understanding whether a DSP SerDes makes sense in a given technology and data rate and brings power and area as close as possible to a conventional "analog" SerDes. If we follow this rule, 7, 5 and 3nm turn out to be the natural nodes for 56, 112G and 224Gbps respectively and we estimate that energy efficiency and density will improve node over node by ~15-20% and by 30-40% respectively (Fig.5).

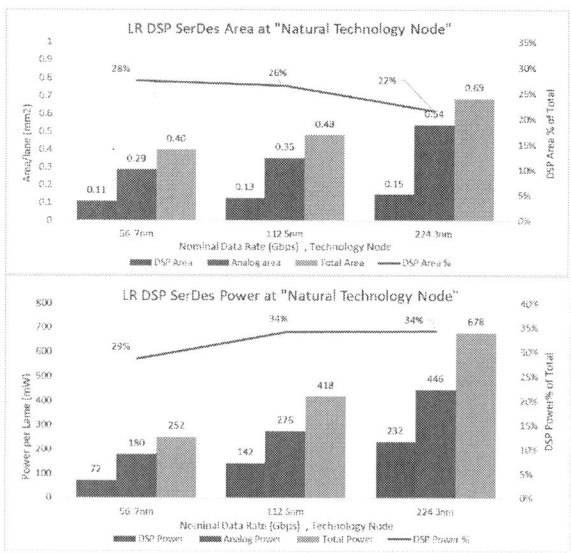

Fig. 5. : SerDes Power and Area Trends at "Natural Technology node"

Even with these improvements, the total power and area increase dramatically at 224Gbps. Also this trend assumes total channel IL to remain around 30dB, to limit the DSP complexity. It is unclear if this can be achieved with current architectures of most common large chassis used for routers, switches and HPC and could lead to sharp worsening of efficiency. Based on the trends above and on our expectations for a 100T 3nm Switch power and area, a DSP LR SerDes would consume more than 40% of the total power and area of a Switch ASIC, including all overheads. When we add FEC and D2D to connect core dies, the total resources used by interconnect could be close to 50%. This is far beyond what the new definition of "Short Reach" interconnect allows for. While DSP LR SerDes are becoming unaffordable, XSR SerDes could provide an effective alternative to LR, with ¼ of the power and less than ½ of the area, inclusive of all overheads [9][10]. XSR could increase the available space for core logic which could in turn help reduce the number of dies required for a given ASIC core.

VII. MULTI STAGE INTERCONNECT

XSR cannot cover the same channel an LR SerDes could, so a possible approach is to split it into 2 or more shorter links. While this may sound counterintuitive, when a channel

becomes extremely challenging, this solution becomes more energy efficient.

A. Backplane Example: Energy Efficiency

Let's consider two cases in an orthogonal backplane application (Fig. 6.). In the first case a single stage link of around 35dB bump to bump would require an LR SerDes at both ends. In the second case we use a repeater (a.k.a. CDR) to break up the link and transfer data from an XSR to MR SerDes.

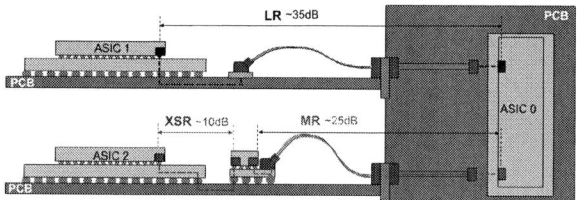

Fig. 6. : Single vs. two stage link in an orthogonal backplane example

This results into an XSR stage of around 10dB followed by an MR stage of around 25dB. Let's see what happens to the end to end energy efficiency, using the SerDes efficiency curve in Fig. 4 as a reference. We assume 5nm technology for ASICs at both ends and for the repeater IC for simplicity. In the first case an LR SerDes with an efficiency of 4.5pJ/bit at 35dB is used at both ends, and the total efficiency of this link is 9pJ/bit. In the second case the MR efficiency is 3pJ/bit at 25dB and the XSR around 1pJ/bit at 10dB. Together they create a link with 8pJ/bit efficiency. In this example the two-stage solution is better than the 1-stage in energy efficiency, because the LR SerDes works in very challenging and inefficient range of the curve, therefore replacing an LR with MR and XSR yields a better energy efficiency.

B. Backplane example: Other Benefits

As we mentioned, in order to make the right SerDes choices, it is important to understand the difference between System and ASIC constraints. The two common constraints generally are maximum current and thermal dissipation. In the single stage case all the current supplying the LR SerDes flows into the relatively small area of the ASIC package and die, generating a very high current density and a very high localized heat (Fig.7). Current availability is a considerable problem for the ASIC, given the limited number of package balls and PCB layers allocated to SerDes and ASIC thermal sinking capabilities are already strained at these levels of power consumption.

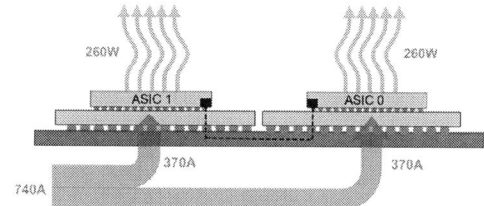

Fig. 7. : Single stage SerDes current and power distribution

978-1-6654-8495-4/22 $31.00 © 2022 IEEE

In the two-stage case (Fig.8), most of the SerDes current flows to the repeater ICs, which are distributed on a much larger surface, thus reducing current crowding and localized heat generation. System-level SerDes power and current are reduced by 12% but for the ASIC the reduction is 77%! This is an example of how an XSR SerDes as part of a two stage interconnect can provide a very significant advantage in the thermal and current consumption environment at the ASIC level and at same time some improvement at system level.

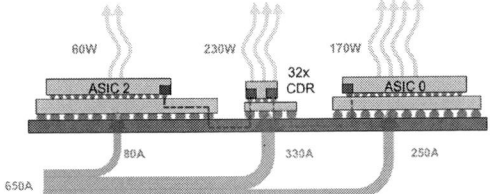

Fig. 8. : Two stage: SerDes current and power distribution

VIII. MCM AND ADVANCED MULTI DIE ASSEMBLY

Moore's law could not cope with BW growth in the last few years and forced developers to "disaggregate" cores into multiple dies often in the form of multi-chip modules (MCM). For 50T and above a few requirements are fundamental for die to die interconnect (D2D):

- D2D BW density (core) : >1Tbps/mm.
- D2D BW density (Chiplets) : >0.5Tbps/mm
- Total < 5-10% power / area of host ASIC
- Extremely low latency, ideally <<10ns

However, strategies for D2D communication vary enormously depending on the substrate used.

A. 2D and 2.5D D2D interfaces

Conventional organic substrates commonly referred as "2D" or "Low Density" are based on established technology commonly used in traditional FcBGA and they provide a bump pitch of 130-150um. The advantage of these substrates is the relatively lower cost and potentially very large maximum size; examples of organic substrates larger than 100x100mm were recently demonstrated which use large sockets to connect to PCB in order to avoid stress and planarity issues [11]. 2.5D or "High Density" substrates, use what is commonly known as silicon interposer technology like TSMC CoWoS [12] or advanced hybrid organic & silicon assembly technologies like Intel EMIB [13] or similar. These technologies provide an extremely small pitch, however for silicon interposer maximum size limited currently in mass production to ~2 standard reticles and has high loss routing layers. Advanced hybrid substrates suffer from significantly higher cost, limited supply sources and still unresolved manufacturing yield and reliability issues. Thanks to smaller bump pitch HD substrates are a much more efficient in D2D communication because they require a much lower data rate to achieve the same BW. In the side by side comparison example in Table II, both HD vs. LD use a very conventional linear, fully shielded bump grid and both achieve the required

1Tbps/mm perimeter density. LD uses differential PAM4 at 112Gbps while HD uses single ended at 10Gbps. However, HD is superior in every parameter, leading to higher energy and area efficiency, thanks to the much lower complexity of the SerDes required. Moreover, due to the native low BER floor of the low data rate NRZ, FEC is not required resulting in one order of magnitude lower latency, which is of fundamental importance for core to core interconnection.

TABLE II. LOW DENSITY VS.HIGH DENSITY D2D INTERFACE

Parameter	LD Substrate	HD Substrate
Pin pitch (um)	150	25
Signaling	PAM4, differential	NRZ, single ended
Data rate (Gbps)	112	10
Energy Efficiency (Pj/bit)	1.5 [a]	0.5
FEC required	Yes	No
Perimeter Density (Tbps/mm)	1	1
Area Density (Tbps/mm2)	0.37 [a]	2
Depth (Number of rows)	3	6
Insertion loss (dB)	2-5 @ 25Ghz	2-5 @ 5Ghz
Latency (ns)	~50 [b]	5

[a]. Not Including FEC, [b]. including ~3dB "lite" FEC

Also it is often necessary to use multiple rows to achieve the required density in both cases. However, in the LD case, to route the parallel core interface of each lane, longer and wider on die busses are required as the number of rows increases. In addition the organic substrate routing becomes more difficult and requires more layers, which adds to cost and signal integrity challenges. This leads to a progressive loss of area, perimeter and energy efficiency that usually limits the number of rows to 5 or 6 for low density substrates.

For large MCM ASICs, or ASIC core disaggregation, the most effective solutions use different substrate technologies for different goals. The superior BW efficiency of HD makes them ideal for core to core interconnect, also known as homogeneous split, using specific low data rate D2D SerDes. Lower cost, larger maximum surface achievable with LD substrates makes them best suited for chiplets, also known as heterogeneous split, using XSR SerDes.

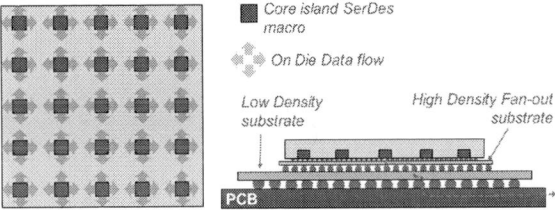

Fig. 9. SerDes Core Islands

In 2D and 2.5D D2D connectivity the major limitation is perimeter density. Also, given the constraints of FinFET technology SerDes macros cannot be rotated, so there is significant waste of area while arranging SerDes on the perimeter. Finally, as all data enters and exits from the perimeter, a significant amount of energy and area is used in moving this data toward where it is actually processed in the core. A possible way to eliminate these bottlenecks could be to use a "SerDes island" approach, where SerDes are uniformly distributed within the core and not just on the perimeter. In this way the perimeter bottleneck is removed and SerDes positioning can be extremely regular, reducing waste of area. Moreover data can enter and exit close to the location where it is processed, since high density substrates enable fan out routing from any location under the die (Fig. 9.).

B. 3D D2D interfaces

3D stacking technology dramatically increases D2D connectivity density. This technology is currently used to connect an SRAM with an ASIC die. It enables up to 100X area density improvement vs. 2.5D with pin pitches smaller than 10um. Chiplets could be now located directly under the main die and connected using reduced data rate per lane, thanks to the massive BW density enabled by the dense pitch. This would virtually eliminate the power and area overhead of interconnecting chiplets and increase the usable area in the core ASIC die (Fig. 10.).

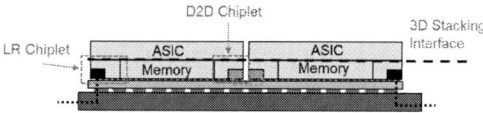

Fig. 10. 3D SerDes Chiplets

As technology progresses toward smaller nodes, cost and complexity of building SerDes IP increases, while providing little performance benefit to the analog portion of the SerDes. A way to address this issue is to use 3D stacking to partition the SerDes IP in two technology nodes. The DSP can be located inside the top die manufactured in the smallest node available, while the AFE can remain inside the chiplet developed in an older technology node. The extremely large BW requirement to connect ADC and DAC to the DSP, now residing in a separate die from AFE, is achievable using the massive BW density provided by 3D stacking (Fig. 11.).

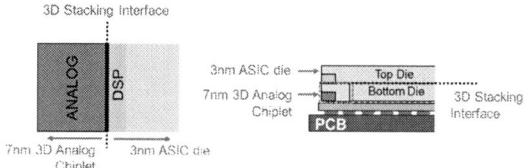

Fig. 11. 3D Hybrid SerDes Macro

IX. ESCAPE FROM PACKAGE

One of the greatest challenges at 50Tbps with data rates of 112Gbps per lane or higher, is the escape from the package substrate. Physical and signal integrity limitations are extreme due to the limited perimeter available and the large features of the ball grid escape, which create discontinuities and significantly increase crosstalk. Moreover, as PCB materials are struggling to keep up with datarate increase insertion loss and other impairments are becoming a limiting factor as well. In order to address these problems in the next few years,

flyover cables will replace conventional ball grid escape and PCB routing. By avoiding package ball grid and PCB limitations, this will provide greatly reduced discontinuities and insertion loss while providing increased total BW (Fig. 12.). Flyover cables are connected to the top of a low density organic substrate by micro sockets that fit in the form factor of the package. These sockets and cable assemblies are extremely versatile, enabling both electrical and optical flyover cables to connect [14]. Cables can be passive or active using components, such as optical engines or repeaters that reside inside the ASIC package, or can be located on the PCB in proximity of the package.

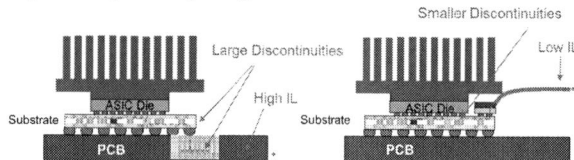

Fig. 12. Package Escape via PCB vs. Flyover Cables

X. OPTICAL DENSIFICATION

One of the most pressing questions about the future of short reach interconnect is the role of optical vs. copper and if, or how, optical can approach the efficiency and cost of copper. Optical densification is the trend that leads to a progressive increase in energy efficiency of optical engines, enabled by and combined with, an increase in in physical density. The starting point of this process is the state of the art pluggable modules based on direct detection (Fig13).

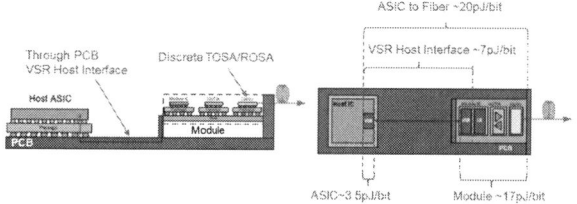

Fig. 13. State of the art 800G modules energy efficiency break down

From a few hundred meters on MMF to tens of kilometers on SMF modules flexibility and versatility are unmatchable, however their ASIC to fiber efficiency (A2F), currently at around 20pJ/bit, is still a far cry from the efficiency of copper links. If an electrical interface based on OIF "LR" 112Gbps were to be replaced with an optical one using 800G OSFP modules, the overall energy efficiency would be at least 5 times worse and the density would be more than 10X worse. Optical densification can improve energy efficiency by lowering the power consumption of the host interface, integrating optical components (TOSA/ROSA) into the optical DSP IC or finally reducing requirements on the Optical DSP inside Optical Engine. However, if we cannot reduce the application space and functionality of the optical engine, we can only benefit from the first 2 actions and total possible efficiency improvements can be only roughly 9 to 12 pJ/bit. In

Table III we summarize the possible steps from state of the art pluggable modules to NPO (near Package Optics) to 2D socketed and 2.5D soldered CPO (Co-Packaged Optics) and finally 3D CPO, indicating possible energy efficiency targets. We assume 7nm for all devices for simplicity and 112Gbps per lane or per lambda, since we believe this evolution will largely play out at this data rate.

TABLE III.   OPTICAL ENGINE TYPES AND A2F ENERGY EFFICIENCY

Type	A2F (pj/bit)	Location	Host Interface	OE
Pluggable	17	Front Panel	VSR, PCB	EIC+ Discrete TOSA/ROSA
Pluggable	15	Front Panel	VSR, Flyover	EIC w/ Integrated TIA/DRV Discrete EML
NPO	12	PCB	XSR+, Flyover	EIC w Integrated TIA/DRV +PIC, ELS
2D CPO	11	ASIC PKG[a]	XSR, LD sub.	
2.5D CPO	10	ASIC PKG[b]	D2D, HD sub.	
3D CPO	8.5	ASIC PKG[c]	D2D, 3D	

[a] Socketed assembly, [b] soldered assembly, [c] 3D stacking

Significant energy efficiency gains could be achieved by "optical densification, however, while system power is significantly reduced by optical densification, ASIC power is not (Fig. 14) With CPO, the portion of the current and power allocated to components inside the ASIC package is more than 70% compared to less than 20% with modules and NPO. This raises the question of whether CPO is attractive for 50T or even viable for 100T. For a 50T socketed CPO, 500W are allocated for optical SerDes inside the ASIC package. That is several times the amount that would fall into the "Short Reach" definition we gave earlier in this paper. There is no reasonable way to make it fit the definition without reducing significantly the SerDes DSP power, which cannot be achieved while maintaining the current Ethernet requirements.

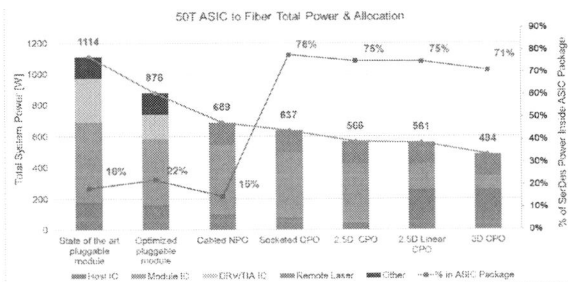

Fig. 14. 50T ASIC to Fiber Total Power and Allocation

Given this, each solution has its own advantages. Front panel pluggable modules have the best flexibility and serviceability and for most applications, and the front panel density is still good enough for 50T or 100T. NPO is the best compromise between system power and ASIC power, while CPO is the best choice for applications where system power is the main constraint. However CPO carries with it an intrinsic lack of flexibility and backward compatibility that combined with serviceability and reliability concerns will still relegate it to a niche application space for some time. The most likely outcome is a mixed approach for Ethernet short reach optics where pluggable modules, NPO and socketed CPO will coexist and dominate the market in the next 4-5 years at 112Gbps per lane. At the same time soldered CPO is viable from a pure technology perspective and we are going to see some niche product examples coming to market shortly.

XI. CLUSTERING: THE KILLER APP FOR CPO

Clustering requires massive bandwidth with low latency over very short distances of 3 to 50m. Conventional optical interconnect, based on Ethernet pluggable modules is not an ideal option due to its high power, low density and excessive latency. Copper cables on the other hand are used extensively for clustering applications, but they may not be able to keep up with bandwidth and reach requirements for very long. Moreover, size, weight and bend radius are a limitation for dense clusters. A possible solution for clustering is an XSR-like CPO which is specifically architected and standardized for these applications. It would require a very minimalistic set of features guaranteeing the lowest possible power and extremely low latency:

- Reduce the optical reach and performance requirements, thus eliminating the need for DSP
- Use of NRZ instead of PAM 4, with a native low BER floor, no FEC, to reduce complexity and latency
- Data rate lower than 50Gbps to reduce insertion loss, jitter and noise requirements
- "Direct optical drive", i.e. no retiming in the OE

This way A2F efficiency of 6pJ/bit or better is achievable: a considerable leap from Ethernet optics and very close to electrical LR efficiency. This could enable what we believe it's the "killer app" for CPO in the next 4 to 5 years.

How does this XSR-like CPO stack against copper? Beside a small decrease in efficiency and increase in system power, both still vastly better than Ethernet compliant CPO, all the other parameters are significantly better than copper. The two that stand out are latency and reach, which are the two main enablers for an effective clustering architecture especially for AI and HPC (Table IV).

TABLE IV.   COMPARISON OF "LR" COPPER VS. "XSR" OPTICAL

Parameter	"LR" Copper	"XSR" Optical
Lanes	512 Differential	1024 Single Ended
Data rate (Gbps)/ Signaling	112/ PAM4	56/NRZ
Energy Efficiency (Pj/bit)	5.5[a]	6.3[b,c]
50T System Power (W)	300[a]	361[b,c]
50T ASIC Power (W)	300[a]	200[b]
FEC	KP4 RS	no
SerDes Area (mm2)	350[a]	256[b]
Perimeter Density (Tbps/mm)	0.5	1
Latency (ns)	150[a]	5
Reach (m)	1-3	>30

a. Including FEC, b. Including ROSA, TOSA, c. Including RLS

The key issue is density: to increase efficiency this approach trades data rate for number of lanes. Compared to copper at 112Gbps it requires 2 times the number of lanes. While the bottleneck for copper is the cables themselves, for the XSR-optical it is the density of the photonic devices used and the fiber attachment unit (FAU). If we assume 100mm ASIC substrate size and 8 x 1.6T optical engines per side, when we include all the sockets mechanical and spacing overheads, we need to fit 32 fibers for data and at least 4 from RLS (Remote Laser Source), in ~3mm, i.e. a pitch of ~80um (Fig. 15). The best Mach Zhender modulators and FAU available today do not meet the beachfront density required by at least a factor of ~4 [15][16].

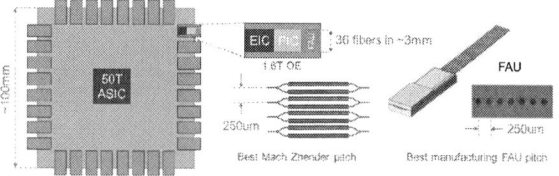

Fig. 15. Optical XSR Density Bottlenecks

Micro ring modulators, at the cost of a bit more complexity in the control circuitry and systems, can provide a considerably higher density at similar or even lower power [17]. To resolve the FAU density problem one solution would be to use very coarse WDM with at least 4 wavelengths per fiber. Another solution would be to eliminate sockets, this would significantly increase the amount of beachfront for the OEs. In conclusion it appears that CPO would be a very attractive and promising solution for a non-Ethernet clustering application especially in the AI and HPC areas.

XII. CONCLUSIONS

A. Business Model

The success of NPO, CPO and other breakthrough technologies will require and in turn accelerate a re-drawing of the Optical and IC value and supply chains. Hyperscalers developing electro-optical ASICs will need to outsource or vertical integrate to obtain the technologies required and an ecosystem of suppliers with the necessary technologies needs to emerge. Vertically integrated IC suppliers should enable ASIC developers to customize their cores while providing a scalable, off-the shelf electro optical I/O frame and packaging, using chiplet-based optical engines for connectivity. ASIC developers need to embrace a hybrid model, sourcing cores and chiplets from different suppliers, to encourage the growth of a healthy ecosystem. Standardization for chiplet communication, integration and management needs to be completed and adopted. A standard for XSR-optics suitable for clustering needs to be developed and widely adopted.

B. Technology Outlook

The future of short reach interconnect is a mix of different solutions, using different combinations of technologies to achieve different goals. Multi stage approach, with XSR replacing LR inside ASICs will be adopted, as electrical interconnect for backplane and other applications will be deployed at least until 224Gbps. However market space for longer reach copper applications will shrink progressively in the next 4-5 years. On the front panel, pluggable modules are here to stay for the foreseeable future and data rate per fiber

will increase to cope with density limitations, at the cost of ever increasing complexity of the DSP SerDes. NPO will likely replace pluggable modules in some power sensitive environments, providing the best tradeoff between system and ASIC power. CPO will appear in customized socketed ASICs developed by Hyperscalers and in some soldered CPO standard products from a few vertically integrated IC suppliers. XSR-optical, most likely CPO, will emerge as the winner in clustering applications providing the perfect mix of density, efficiency and low latency.

ACKNOWLEDGMENT

I'd like to thank Hossein Shakiba and Prof. Ali Sheikholeslami for their help in writing this paper and my colleagues of the HiLink team for providing a lot of this material and for the amazing past 11 years together.

REFERENCES

[1] "IEEE Standard for Ethernet," IEEE Std 802.3-2018 (Revision of IEEE Std 802.3-2015), pp. 1-5600, 31 Aug. 2018.

[2] N. Tracy et al., "112 Gbps Electrical Interfaces – An OIF Update on CEI-112G," Panel session, *OFC*, San Jose, 2020.

[3] M. LaCroix et al., "A 60Gb/s PAM-4 ADC-DSP Transceiver in 7nm CMOS with SNR-Based Adaptive Power Scaling Achieving 6.9pJ/b at32dB Loss," *ISSCC*, pp. 114-115, 2019.

[4] M. LaCroix et al., "A 116Gb/s DSP-Based Wireline Transceiver in 7nm CMOSAchieving 6pJ/b at 45dB Loss in PAM-4/Duo-PAM-4 and 52dB in PAM-2," *ISSCC*, pp.132-133, Feb. 2021.

[5] R. Farjadrad, M. Kuemerle and B. Vinnakota, "A Bunch-of-Wires (BoW) Interface for Interchiplet Communication," *IEEE Micro*, vol. 40, no. 1, pp. 15-24, 1 Jan.-Feb. 2020.

[6] T. Ali et al., "A 460mW 112Gb/s DSP-Based Transceiver with 38dB LossCompensation for Next-Generation Data Centers in 7nm FinFET Technology," *ISSCC*, pp.118-119, 2020.

[7] Z. Guo et al., "A 112.5Gb/s ADC-DSP-Based PAM-4 Long-Reach Transceiver with >50dB Channel Loss in 5nm FinFET," *ISSCC*, pp. 116-117, 2022.

[8] Namik Kocaman et al. , "An 182mW 1-60Gb/s Configurable PAM-4/NRZ Transceiver for Large Scale ASIC Integration in 7nm FinFET Technology," *ISSCC*, pp. 120-121, 2022.

[9] G. Gangasani et al., "A 1.6Tb/s Chiplet over XSR-MCM Channels using 113Gb/s PAM-4 Transceiver with Dynamic Receiver-Driven Adaptation of TX-FFE and Programmable Roaming Taps in 5nm CMOS," *ISSCC*, pp. 122-123, 2022.

[10] Ramy Yousry et al., " A 1.7-pJ/b 112Gbps XSR Transceiver for Intra-package Communication in 7nm FinFETtechnology," *ISSCC*, pp. 180-181, 2021.

[11] A. Chandrasekhar et al., "Server CPU Package Design Using PoINT Architecture," *ECTC*, pp. 2180-2185, 2019.

[12] P. K. Huang et al., "Wafer Level System Integration of the Fifth Generation CoWoS®-S with High Performance Si Interposer at 2500 mm2," *ECTC*, pp. 101-104, 2021.

[13] R. Mahajan et al., "Embedded Multi-die Interconnect Bridge (EMIB) – A High Density, High Bandwidth Packaging Interconnect," *ECTC*, pp. 557-565, 2016.

[14] https://www.samtec.com/solutions/flyover

[15] P. Ji, J. Gao, W. Xu, Y. Sun, W. He and H. Wu, "Electronic-Photonic Integrated Circuit Design and Crosstalk Modeling for a High Density Multi-Lane MZM Array," *ISCAS*, pp. 1-5, 2018.

[16] https://www.corning.com/microsites/coc/oem/documents/OEM-039-AEN.pdf

[17] H. Li et al., "A 3-D-Integrated Silicon Photonic Microring-Based 112-Gb/s PAM-4 Transmitter With Nonlinear Equalization and Thermal Control," *JSSC*, vol. 56, no. 1, pp. 19-29, Jan. 2021.

Semiconductors take the driver's seat - challenges and opportunities for the car of the future

Tim Gutheit
Automotive Division
Infineon Technologies AG
Neubiberg, Germany
tim.gutheit@infineon.com

Abstract—**90% of the innovations in modern vehicles are enabled by semiconductors. The trend towards autonomous driving and electrification of the drive train imposes additional requirements on electronics in the vehicle which cannot be met by using semiconductor components from the consumer domains. The talk will illustrate the main challenges on semiconductor devices in automotive use cases related to computing performance, reliability, power consumption, sensor accuracy and robustness, security and functional safety for enabling the vehicle of tomorrow. We will provide an outlook on advanced semiconductor technologies and concepts which will be instrumental to tackle these challenges successfully.**

Keywords—car electronics, automated driving, electrical drive train, automotive radar, functional safety

I. Introduction

Semiconductor devices used in automotive applications have to withstand harsh environmental conditions as e.g. ambient temperature cycles from -40°C up to 150°C, show robustness against over- and undervoltage spikes and electrical discharge, while working reliably over the entire life of the vehicle, which translates to >10,000 hrs operating lifetime. These requirements and the corresponding test methodologies are laid down in standards as e.g. AEC Q100/101 and ISO7637 [1], [2], [3]. It takes a comprehensive set of dedicated measures to fulfil these requirements. The breadth of this set spans across using specific wafer technologies and packages, special materials, design rules and device libraries, product architectures, extensive test procedures and qualification stress tests, and many more. Devices designed for consumer products and applications typically will not meet the automotive lifetime and reliability targets as shown in Fig.1.

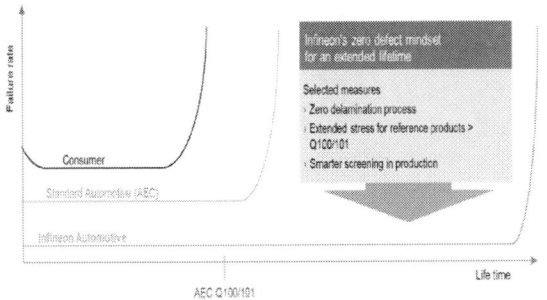

Fig. 1: *Failure rates of electronic components for automotive applications in comparison to consumer applications. Lifetime requirements according to AEC Q100/101 correspond to 10,000hrs operating lifetime in the field.*

The reliability levels which are state of the art for automotive semiconductor devices today are impressive: Whereas fail rates of devices for consumer electronics lie in the range of 200...500ppm many automotive devices reach a level of <1ppb. If cars fail in the field due to electrical faults the dominating root causes today are either a failing battery, failing connectors or broken cables, not the electronic devices themselves. As for many sub-systems in the car there is no longer a pure mechanical back-up the electro-mechanical subsystems have to fulfill the requirements of functional safety level as laid down in the ISO26262 from ASIL A..D, depending on the criticality of the respective sub-system [4].

However, with the advent of electric vehicles and automated driving functions these challenges increase further. Main drivers are

- Silicon content increase
- Functional safety, fail operational requirements
- Voltage level increase e.g. in HV battery systems
- Operating lifetime increase, mission profile change
- Higher integration levels – monolithic, heterogenuous, introducing new failure mechanisms
- High speed interfaces
- Software interference
- …

On the other hand, there are opportunities as consequence of the introduction of new vehicle electronic/electric (E/E) architectures as well: Boardnet voltage levels will be actively regulated, resulting in less spikes, reduction of complexity of wiring harness leads to less crosstalk and less unreliable connectors.

II. Examples for Innovative Electronic Systems in Road Vehicles

90% of all innovations in road vehicles are enabled by electronics, driven by progress in semiconductor technologies offering new functions and / or higher performance levels at attractive cost levels. A few examples are given in this section.

A. Sensors for Advanced Driver Assistance Systems (ADAS) and Automated Driving (AD)

ADAS systems as adaptive cruise control, traffic sign recognition, lane keeping assist, dead angle supervision, … require constant monitoring of the car's environment through a set of sensors based on various sensing principles.

Radar sensors are among the most powerful sensing elements as they provide high range and accuracy, show little sensitivity to weather or light conditions in comparison to camera-based systems. Filtering and processing of raw data remains challenging for generating robust and accurate pulse-doppler maps. Over the recent years significant progress has been made both in performance increase and in cost reduction of these automotive radar systems. The first generation 24GHz systems based on discrete III-V semiconductors have by now been substituted by 77GHz SiGe BiCMOS devices with the next integration steps in RF-CMOS technology at about to enter the market. As a result, these systems which 10 years ago were available as a high price option for premium cars only are now available in subcompact vehicles, in some cases even as standard feature. Radar systems with different aperture angles and ranges, in combination with camera and lidar systems provide a safety cocoon around the car, whose signals are increasingly used to activate safety relevant functions as automated emergency brake maneuvers. To take the next step towards higher levels of automated driving, i.e. L3+, the information from these increasing number of sensor nodes as shown in Fig. 2 need be "fused" by a safe and powerful processing unit in order to provide a comprehensive 360° model of the car's environment. The new challenges which come along with this transition are discussed in the final section of this paper.

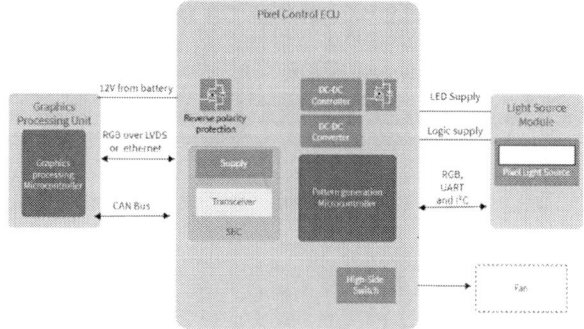

Fig. 2: *"Safety Cocoon" of sensors around car – Increase of sensor nodes for enabling advanced levels of automated driving (AD)*

B. Conversion from Internal Combustion Engine (ICE) to Electrical Drivetrain (EDT)

The replacement of the ICE is driven by legislation which enforces low levels of CO_2 emissions which cannot be achieved by conventional ICE drivetrain concepts. The conversion to electrical vehicles (EV) will go through intermediate stages as mild hybridization (MHEV), plug-in hybrid (PEV) towards full battery powered electric vehicles (BEV). Fuel cell based onboard generation of electrical power is under development as well. The new ambitious goals for decarbonization in conjunction with financial government incentives in most regions has significantly sped up the market penetration of xEV. All xEV concepts have in common that they require a battery for energy storage, chargers, and inverters for controlling the electric main motors.

The electrification of the vehicle will introduce many new functions within the car, examples are seen in Fig.3. This contributes to a significant increase of semiconductor content by approximately a factor of 2 compared to today's ICE, offering plenty of opportunities for new innovative semiconductor products.

Introduction of a high-voltage level of 400...800V for the battery, the main inverter and onboard charger brings dedicated power electronic systems into the car in the form of power modules with Silicon IGBT or SiC MOSFET devices. Minimizing the electric power dissipation losses has direct impact on range and battery cost and hence is a strong driver of pushing the roadmap for advanced power wafer technology and package concepts ahead.

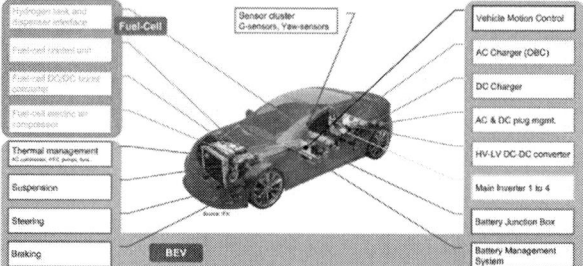

Fig. 3: *Functions with high semiconductor content which are new in xEV vehicles*

C. Innovations in Body Electronics – LED Frontlight

In parallel also in the body and comfort domain many new features are entering the market, all enabled by electronics. The most visible from the outside of the car is the conversion to LED lights from the former incandescent light bulbs. Already for those features as short circuit protection, open load detection were made possible by smart semiconductor switching devices replacing relays. The introduction of LED in the back and frontlight is more than just a visible design feature and has positive effect on road security. LED frontlight today is state-of-the-art in new vehicles, the step to matrix lights offers new features as gliding transition from low-beam to high-beam, glare protection of oncoming traffic and curve light.

A LED matrix beam system of which a system diagram is shown in Fig. 4 consists not only of the LED and their driver IC but features communication IC as CAN transceivers, microcontrollers for pattern generation and control, power supply regulation IC and many more.

The next step for this application will be the introduction of "pixel lights" where the LED array will be formed by 10k+ micro LED which can be individually controlled. These High-Density Pixel Lights will offer not only higher performance levels of the functions mentioned above but also new use cases as projection of notifications and warnings for driver or pedestrians onto the road surface.

Fig. 4: *System view of LED Pixel Frontlight [6]*

III. OUTLOOK & TRENDS, AREAS OF RESEARCH

Today most of the functions listed in the section II are managed as comfort functions (e.g. adaptive cruise control, curve light) and driver assist functions which contribute to safety (e.g. emergency brake). Still the driver remains in control and can override these systems at any time – and in critical situations is actively requested by the systems to do so on short notice. When taking the next step towards AD these functions and many more which are today mostly independent and isolated must be consolidated and controlled centrally, self-contained in a safe way, without a human driver as fallback solution.

New E/E architectures in the vehicle are required to master the complexity level which increases with the number of electronic devices and lines of software code. Whereas in the past a decentral architecture prevailed with many point-to-point connections between the individual Electronic Control Units (ECU) there is a clear trend now towards centralized and zonal architectures with safety domain controllers and compute servers hooked on an ethernet backbone. There are different architectures pursued, one representative example is depicted in Fig. 5.

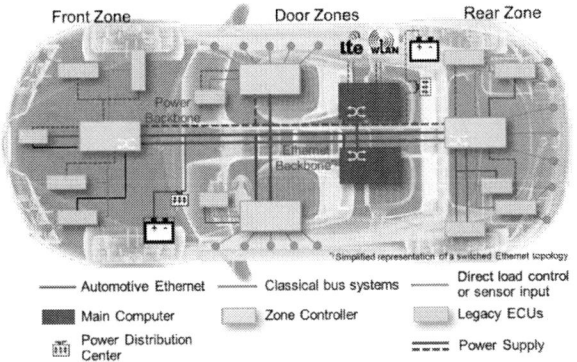

Fig. 5: *Zonal E/E architecture concept with dual power distribution backbone for dependable power for fail-operational requirements*

Computing power will increase significantly while constraints for maximum power consumption will become even more important. Less power consumption directly translates into higher range of electric cars or lower costs for the battery as one of the costliest components of xEVs. New processor architectures as for neuromorphic computing, dedicated accelerators for edge AI will provide new solution spaces. Meeting the requirements of the ISO26262 for AI based functions will continue to be challenging. The cars entering the market today already have more sub-systems which have to fulfill ASIL-B..D levels as for example the battery management, main inverter and onboard charger.

Introduction of automated driving functions will further increase requirements on functional safety, with the main challenge being the transition from fail safe functionality of critical functions of the car as powertrain, braking and steering to fail operational requirements. Implementing these requirements in a cost-effective way today is the main hurdle for introduction of true Level 3+ automated driving features. The driver is no longer a viable fall back in case of failure so the safety critical systems of the vehicle must stay operational at least until the vehicle is put to a safe stop. As a consequence, fail operational actuators must achieve random failure rates of about 10 FIT, which is a significant increase over the requirements of 500...1000 FIT for fail safe systems as braking or steering in today's vehicles. Common cause of failure as loss of power or disturbed communication must be avoided in order to fulfill the high availability level of safety critical systems. Systematic failures as software bugs must be contained with regards to their impact on safe operation. Such a fail operational concepts need significantly more effort on architectural design, and on hardware and software components themselves. Just adding a second redundant fail-operational layer is not feasible from both technical as well from commercial perspective. A very important consideration is the demand for dependable power, an aspect which is often overlooked but poses a main prerequisite for any automated driving function where there is no mechanical backup operated by the driver in case of power loss.

Another challenge poses the wireless connection of the vehicle to the internet which provides new attack vectors for compromising the safety of operation. In response to this trend the ISO/SAE21434 on Cybersecurity for Road Vehicles has been released in 2021 [5]. V2x connectivity is needed for software over the air (SOTA) updates and bug fixes, receiving and sending information on road, vehicle and driver conditions and last but not least comfort and entertainment purposes. This communication must be protected by state-of-the-art encryption schemes requiring dedicated hardware as TPM devices or secure areas on the SoC and the communication gateways. In addition, communication interfaces within the cars must be protected against manipulation or against installing of counterfeit devices and spare parts which are compromising safe operation of the vehicle. Considering the long lifetime of the vehicle of 10...20 years post quantum cryptography concepts are mandatory in order to harden the vehicle system against attacks by quantum computers which might become available over the next 10 years.

REFERENCES

[1] AEC Q100 Rev. H Base Document & Attachments, AEC Council, 2014.

[2] AEC Q101 Rev. E Base Document & Attachments, AEC Council, 2021.

[3] ISO 7637-1..5:2015, International Organization for Standardization, 2015

[4] ISO 26262:2018 „Road vehicles – Functional safety", 2018

[5] ISO/SAE 21434:2021 "Road vehicles – Cybersecurity engineering", 2021

[6] Infineon Application Note Z8F68623907, 2020

Integrated Circuits as Key Enabler
for today's Smart MEMS Sensors

Dirk Droste, Horst Symanzik,
Timo Gießelmann, Markus Ulm
Bosch Sensortec GmbH
Reutlingen, Germany

Ivano Galdi, Riccardo Campagna

Robert Bosch S.p.A. Società Unipersonale
Milano, Italy

Abstract—The requirements and capabilities of smart MEMS sensors are constantly increasing. Consequently, the challenges in development of the ASICs driving this progress increased significantly over the years. Many fields have to come together to enable this: ranging from high-precision analog mixed signal circuitry to support the ultimate goal of an "ideal" sensor transfer function with no deviation by environmental changes up to power-optimized digital logic with customized microcontrollers. Embedded software is classifying signals, correcting errors and data fusion from different domains while new process technologies are introduced, package sizes are reduced, and 3D integration is established.

To gain some comprehensive overview of these new challenges on ASIC development, in the first chapter the product portfolio of Bosch Sensortec is briefly introduced, reflecting the transition from sensors delivering calibrated raw data up to smart sensor modules with embedded algorithms, and in the second chapter, a focus on new smart MEMS sensor modules and their market requirements will be given. In the third chapter an insight about continuing optimization of performance for signal conditioning of physical and chemical sensors with respect to high signal to noise ratio, offset stability and linearity while still continuously lowering power consumption will be given. In the fourth chapter, the sequence of the concurrent increase of digital capability from edge-AI supporting structures to reduce overall power consumption of meshed sensors up to new security features for trusted domains of IoT connected devices will be discussed.

Keywords—MEMS, ASIC, smart sensors, Edge-AI

I. INTRODUCTION

Bosch Sensortec is a technology provider in sensing solutions based on microelectromechanical systems (MEMS) and dedicated to the consumer electronics world.

As a pioneer in MEMS solutions, Bosch developed the manufacturing process behind MEMS technology in 1994 that still forms the basis of today's sensor production runs [1]. Since then, Bosch has continuously broadened its technical expertise in sensing solutions and its application know-how. Starting with MEMS in application fields of Automotive Electronics, Bosch Sensortec was founded in 2005 to focus on MEMS sensor development and manufacturing dedicated to Consumer Electronics applications [2,3,4].

Bosch Sensortec now offers a broad portfolio of MEMS based sensors and solutions that enable consumer electronics devices to sense the world around them. The product portfolio includes motion sensors such as 3-axis accelerometers, gyroscopes, magnetometers and integrated 6- and 9-axis sensors, smart sensors as well as environmental sensors for measuring barometric pressure, temperature, humidity and gases. In addition, Bosch Sensortec offers optical microsystems and a comprehensive variety of software and tools.

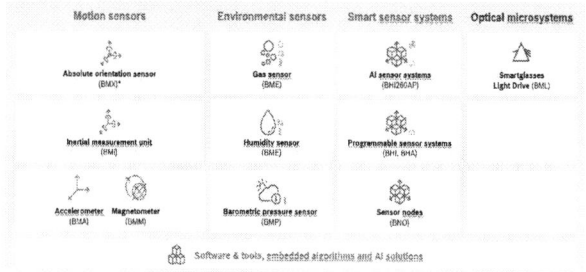

Fig. 1. Bosch Sensortec MEMS sensor and optical systems portfolio

A specific focus on Fig. 1 should be given to column "Smart Sensor Systems", as this reflects the response of Bosch Sensortec to the new market requirements for smart MEMS sensor solutions, being able to deliver higher level abstracted sensor data even with support of embedded AI, beyond just filtered and calibrated raw data.

II. SMART MEMS SENSOR MODULES AND MARKET REQUIREMENTS

Talking about smart sensors we first need to define some terms – as the wording "smart" is used in multiple meanings and context. Here, smart describes a MEMS sensor being able to process raw data with some processing unit in a free programmable algorithmic or "self-learning" way by embedded software, to recognize, mark or aggregate specific patterns of the measurand's raw data.

Analyzing the last three decades in MEMS sensors, until the first decade of this century the outputs of most MEMS sensors were mostly dedicated to providing calibrated and offset adjusted raw data or triggered interrupts of a measurand with a specific or tunable oversampling rate to the host system, and all higher-level data conversion and analytics were executed in the hosting system's CPU. Certainly, DSP-like processing structures with defined signal processing schemes and fixed or even adjustable coefficients for a flexible transfer function were already used, but just about 2010 passing, first dedicated MEMS sensors with embedded µCs were released to the market, to execute sensor signal evaluation in a module for example for significant motion, step counting or movement tracking - mostly in form of systems in package (SiP) with a readout chip and a dedicated µC as bare die, or even first monolithically integrated small CPU IPs or ASIPs.

Figure 2 as an example depicts the SiP configuration of a Smart IMU MEMS sensor module of Bosch Sensortec (BHI260AP). The readout ASIC for a 3-axis accelerometer and a 3-axis gyro MEMS is underneath both MEMS sensor dies, separated with a die film attach and the dedicated μC is placed aside; the 50 Mhz RISC μC with an ARC32 instruction set and 256 kByte of ROM and 144 kByte of RAM can run always-on sensor data processing algorithms at ultra low power consumption.

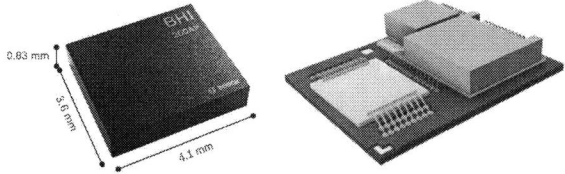

Fig. 2. BHI260AP: Smart MEMS IMU System in Package with Readout Circuit and additional μC for self learning motion tracking

One can easily imagine that integration of the μC into the readout ASIC by stepping into a smaller semiconductor technology node will achieve a significant advantage in footprint and will save power by avoding high-speed serial bus wired connections between the readout ASIC and the μC.

Figure 3 gives an overview of a smart sensor hub concept with the BHI260, as being largely utilized in current wearable or smartphone applications. The flexibility of the smart MEMS sensor to also capture data of external other sensors raises broad fusioning capabilites for new user experiences. Our clear outlook is, that any given smart MEMS sensor in future will include the "Fuser Core" as a powerful μC monolithically, but will also enable reading of supplementary sensor channels for any higher level fusion algorithms.

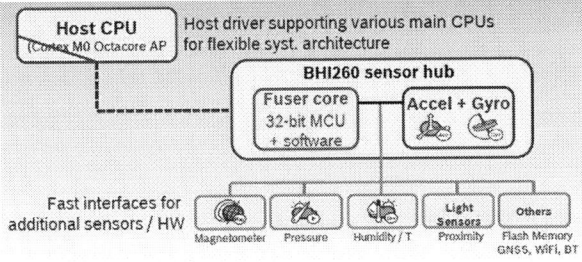

Fig. 3. General sensor hub concept for smart MEMS sensing

The ongoing structure size decrease and the simultaneous extension of these process nodes for high performance analog readout circuitry just in the last few years started to enable embedding quite powerful μC cores monolithically, to execute sophisticated signal analysis with utilization of embedded software – and to optimize the box-size of the package at the same time. It is a common prediction of most analysts, that in near future the bigger portion of total available market for MEMS sensors will be "smart" in that sense, that they will monolithically or at least per SiP integrate a quite powerful μC and a memory system being capable of applying high level signal processing, even up to edge-AI level, the latter being based on neural network architectures realized as in-memory computing. Yole's latest market report postulates that the final step for MEMS would be to "evolve from simple deterministic data collection sensors to more empathic data interpretation machines" [5].

The main challenge on this evolution is that even though everyone prefers the smartness in terms of flexible computing power, no one would like to sacrifice the classical "analog" performance indicators as signal-to-noise ratio, offset stability, linearity and power consumption. So, one has to still push ahead these KPIs while simultaneously integrating powerful μC systems on the same die - and combining this even with the simultaneous raise of mechanical stress due to lower package sizes and new 3D stacking capabilities. This will raise a new challenge. We will elaborate on this in chapter III.

It is also important to question, what the "smartness" shall be dedicated for – as a clear functional and architectural brake down of MEMS sensor systems projected to the next 10 years does not exist, due to the lack of clear requirements out of the systems; IoT systems are being vertically optimized and standards are yet to be defined. OEMs also do work on the same question: on how any future functionality, that is based on improving user experience, can be enabled by which decisions and choices for system architecture and partitioning of algorithms between application processor and an integrated processor in the sensor module. We will elaborate on this in chapter IV.

III. Optimization of Performance for Signal Conditioning

As already highlighted in the former chapter, the continuing performance optimization for signal conditioning of physical and chemical sensors is a major challenge for future MEMS sensor products, as for example discussed in [6-10].

The main KPI, which are commonly used to outline a sensor systems performance are the following three:

- signal to noise ratio
- offset stability and linearity
- power consumption

A. Signal to Noise Ratio

It is clear, that any desired SNR has always to be related to electrical power for conversion of physical measurand into an electrical value. Most literature propose to use a metrics of required electrical energy per sample conversion, which gives us a very useful figure of merit to assess power consumption and future required performances.

The diagram depicted in Figure 3 compares a Bosch device with other ones in terms of "energy per conversion" demand; the example BMA400 (a triaxial accelerometer with step counting capability based on algorithms realized with a monolithic integrated μC) can be used in different over sampling data ratios, consuming more or less power – accordingly, SNR will change by square root relation, saturating with some base current consumption. So, a very simple, but also challenging message of this plot is, that in any given architecture, optimizing SNR versus electrical energy is the main goal. But, to achieve this power optimization preserving signal conditioning performances, one needs to leave "state of the art" conversion solutions behind and make steps forward towards new architectures.

978-1-6654-8495-4/22 $31.00 © 2022 IEEE

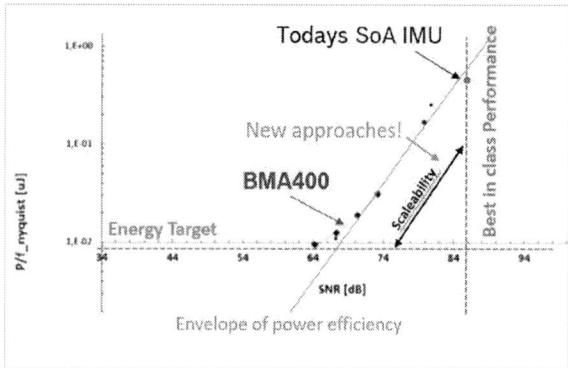

Fig. 4. Conversion Energy – to achieve SNR optimization while still reducing power consumption, new circuit approaches are mandatory

Hence for next MEMS sensor development modules in new sub micron technology, a straightforward architecture migration of existing implementations does not pay off anymore, but instead new ideas to overcome next technology limitations have to be encouraged. Lower V_{th} leaves less headroom for sophisticated g_m/I_d optimizations, especially in subthreshold domains, and upcoming parasitic effects on smaller structure sizes complicate the design phase.

Furthermore, effects of random state changes in tightened structures, commonly known as random telegraph noise (RTN) will become a new limiting factor for overall noise performance and this will give us a continuous request to not only innovate on specific MEMS sensor frontend circuitry and Analog-to-Digital converters (ADC), but also on optimized internal voltage references and power management circuities, to achieve circuitry which is robust against or suppressing RTN right at source. Being not always part of nowadays transistor models within PDKs of foundries, governing RTN will require additional effort to the design teams on elaborating effects in design phase and refine proper parasitic transistor models and circuit test benches, together with automated tests on transistor level or test structures on wafer level to validate assumptions [10].

B. Offset Stability and Linearity

Current readout-chain architectures often embed quite complex circuitries like MEMS interface power versus noise optimized analog sensor frontend and ultra-low power ADC circuitry, which exploit fully differential structures to reduce disturbances coming from power rails. Specific design techniques aiming to enhance the circuit performances like chopping or correlated double sampling, guarantee to overcome the main physical "non idealities" (offset, $1/f$ noise, RTN). But, despite all these electronic cares, mechanical parts also contribute to non-linearity and offset deviation being mostly directly included in the signal band. Hence system architectures of including sensor device in chopping and classical countermeasures like trimming at final test or a self-calibration at defined measurand "zero"-status are largely utilized.

Although the possibility to increase the overall digital content will certainly extend state of the art calibration and trimming techniques [12], "minimizing the problems at the source" still remains the best way to proceed, which likewise challenges MEMS and package designs and manufacturing.

C. Power Consumption

Modern wearable systems determine a new important KPI around MEMS sensor: power consumption, narrowing down the battery power to few mA/h. For ASIC power consumption optimization, the full set of best design guidelines and countermeasures have to be used:

- Analog: Use deep subthreshold region and apply exhaustive g_m-over-I_d optimization, use dynamic opamps, bandgaps and references wherever possible, optimize architecture versus need of robustness.

- Digital: Apply supply voltage reduction, enable power domain, clock and data path gating, apply clock tree optimization, check for manual RTL to Gate-Level optimization, apply tailored IPs (DSP, µCs)

- System: Apply multiple oscillators according to use modes, define intelligent power modes, lever technology capabilities

Power reduction is one of the most important activities in academic und industrial semiconductor community and the progress on new devices, circuits and system approaches is sustainingly huge – nevertheless, ultra low power design capabilities is one of the most important challenges in the domain of wearable sensing solutions and Bosch Sensortec is elaborating also on joint public funding research projects to keep pace to most current approaches.

As an example, for optimizing sensor readout performance while still lowering power consumption according to A, B and C we present KPIs of the MEMS pressure sensor BMP580 (fabricated in standard CMOS technology node), just being released for series mass production:

Bosch Sensortec Barometric Pressure Sensor BMP580		
Feature	Value	Unit
Measurement Range	300-1,500	hPa
Absolute Accuracy	+/- 0.3	hPa
Relatuve Accuracy	+/- 0.06	hPa
RMS Noise in pressure @ 1Hz	0.08	Pa
Current Consumption @ 1Hz	1.3	µA

Table 1: Features of MEMS pressure sensor BMP580

To enable these KPISs, all aspects of A, B and C have been applied. The concept of the capacitive bridge of the MEMS pressure sensor hereby strongly supports a very flexible approach for an offset cancellation mechanism on whole architecture by chopping. Also, the additional temperature channel of the device, which is required for compensation of temperature induced offset and sensitivity variations, is generated by a single ended reference and converted to a chopped differential signal; specific measures on circuit level have been used to achieve a reference voltage minimizing residual parasitic influences of power rails, temperature span and process variations. The general architecture of this barometric MEMS pressure sensor is depicted in Figure 5.

978-1-6654-8495-4/22 $31.00 © 2022 IEEE

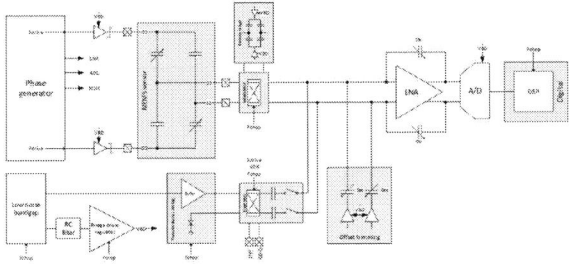

Fig. 5. Capacitive pressure sensor signal readout architecture of BMP580

As a comprehensive use case for a demonstration of the sensing capability of this capacitive MEMS pressure sensor, we measured the barometric pressure changes that results when someone walks up or down a staircase, as a classical application for indoor navigation. As the barometric pressure change over 17 cm (which is a typical stair step height) is 2.1 Pa, and to have a good measurement quality, the RMS noise has to be fairly lower and was chosen with 0.08 Pa. At the same time, the full span of barometric pressure input range is 30 hPa to 1,250 hPa, so the signal conversion architecture was defined to support a signal resolution of 24 bits.

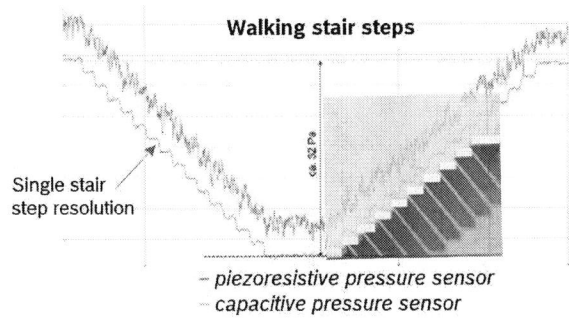

Fig. 6. Capacitive pressure sensor supporting indoor navigation

Figure 6 shows the measured barometric pressure change by this capacitive MEMS sensor used by a person walking up and down a staircase, together with measurement results of a former piezo-resistive MEMS sensor. It was feasible to realize a series production capable capacitive MEMS pressure sensor, which delivers a sufficiently precise altitude-related data to be embedded for example into indoor navigation algorithm.

Referring to roadmap considerations discussed in chapter II, any approach on a new „smart" pressure sensor will have to clearly keep and even optimize these performance indicators below, which certainly requires huge steps forward in circuit techniques and circuit architectures, as referred to in A, B and C, when switching to an advanced technology node to support required digital capability for a powerful µC system with embedded SW algorithms.

IV. INCREASE OF DIGITAL CAPABILITY

As discussed in chapter II, the digital capability for upcoming smart sensors has certainly to be increased, but still there are opposing views on concrete application use cases; system developers of next generations wearable devices are looking for even more powerful commercially available application processors, while sensor module developers claim to also being able to provide processor power at the edge. There will be a dispute on the direction of future system architectures to be implemented on any given smart application algorithm.

We did an investigation on some smart algorithms and concluded that with a given smart algorithm, the penalty to push data to CPU via standard PCB interface at appropriate sample rate and the latency requirements to wake up the system's CPU clearly flips the decision to integrate that algorithm into a dedicated µC on the same chip for signal conversion of the MEMS measurand, see Table 2.

Sensor Processor vs. Application Processor (AP)				
Use Case Example	Latency critical	Data traffic Sensor to AP	Preferred algorithm location	Rational
Wake-up gestures (e.g., wrist movement)	high	medium	Sensor	Low latency
User Interface (e.g., Multi-Tap)	High	high	Sensor	Low latency
Activity recognition	medium	medium	Sensor	Dynamic sensor control
Step count	Low	medium	Sensor	Avoid regular AP wake-up
Pedestrian dead reckoning + GNSS fusion	low	high	Distributed	Sensor calculates delta vectors. Fusion on AP
Sensor self-calibration	high	high	Sensor	Close feedback loops

Table 2: Assessment of smart algorithms on preferred aogorithm location (smart sensor module processor or application processor)

To give some comprehensive examples we measured the power consumption of a step counter algorithm (w/o power consumption of the sensor itself) for two smart MEMS acceleration sensor devices, one with SiP integrate processor (BHA260) and one with a monolithic integrated processor on readout ASIC (BMA455).

Figure 7a shows that the power consumption reduction by lowering the integration level from a SiP with dedicated processor chip to a monolithic integrated processor achieves a factor of ~10 [13]. This gives an objective statement that pushing smart algos for local sensor data processing into the readout chip largely pays off for battery lifetime of a given system. One can easily imagine that with the same approach to calculate trajectories of movements on an integrated processor, for example to support indoor navigation or pedestrian dead reckoning, a huge portion of battery power could be saved, as the GNSS/GPS engine could be kept in sleep mode for quite some while.

While this approach is beneficial for small to medium size algorithms, more demanding algorithms may not yet fit in the resource budget of the readout chip. In these cases, the SiP processor approach, or even a hybrid approach, has an advantage. However, it can be expected that, while more powerful processor systems are made available on the readout chip, the need for the SiP processor approach diminishes over time.

a) Step Counter Algorithm Power Consumption (only Algo)

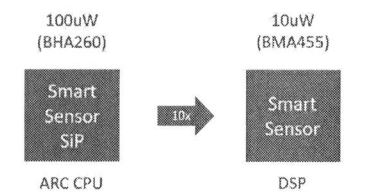

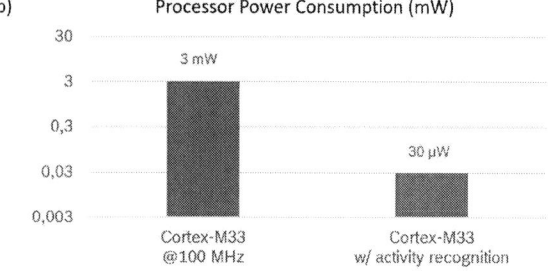

b)

Fig. 7. Power consumption of algorithms in Smart Sensors; a) SiP integrated processor versus monolithic integrated processor, b) Consumption of processor at maximum speed during activity recognition

A basic consideration of the desired algorithm has to be conducted for architectural decisions of SiP or monolithic integrated approach in general, to achieve the best optimized power consumption for required use cases of the smart sensor. Figure 7b shows the power consumption of a monolithically integrated ARM Cortex M33 CPU at maximum clock speed of 100 MHz of 3 mW. The dedicated activity recognition algorithm running with the same CPU system will not require full available performance but consumes only 30 µW. This clearly shows that there has to be a careful choice of integration of processor as SiP or as monolithically integrated device, to achieve best fit to market requirements of all KPIs including calculation capability of a smart MEMS sensor device.

But being clear in the perception of market requirements, that integration of computing power into smart MEMS sensors will be a strong, sustaining trend, the monolithic integration of a powerful µC subsystem into the readout ASIC of a MEMS sensor module is a mandatory step into the future.

V. CONCLUSION

Classical MEMS sensors evolve into edge-AI devices. Smart sensor modules with a multitude of sensors, integrated low power signal processing, for example for context awareness and specific applications, will change MEMS from the provision of raw data to meaningful data fusion output from sensor nodes for our connected world [14].

There is also a clear trend from SiP like configuration of a sensor module with a dedicated processor chip towards a

monolithic integration of an adequately powerful processor on sensor readout chip. Nevertheless, the classical Sensor-KPIs like SNR, offset stability and linearity will not step behind when sensors are becoming smarter. Further innovative circuity will be required to support enhancement of KPIs while also integrating huge digital content in shrinking semiconductor nodes. At the same time, the power consumption still needs to be lowered, which requires new approaches beyond state-of-the-art methodologies in concept definition, circuit design and device manufacturing.

As market requirements and specific application use cases for future smart sensors are hard to predict, flexible processing architectures are required.

REFERENCES

[1] F. Lärmer et al, "Bosch Deep silicon etching, improving uniformity and etch rate for advanced MEMS aplication", *Proc. IEEE MEMS 1999 Orlando USA*, pp. 211-216

[2] J. Classen et al, "Evolution of Bosch inertial measurement units for consumer electronics", *IEEE Sensors*, 2020

[3] J. Marek, "MEMS Technology- from Automotive to Consumer," *2007 IEEE 20th International Conference on Micro Electro Mechanical Systems (MEMS)*, 2007, pp. 59-60

[4] S. Finkbeiner, "MEMS for automotive and consumer electronics," *2013 Proceedings of the European Solid-State Device Research Conference (ESSDERC)*, 2013, pp. 9-14

[5] Yole, "Status of MEMS Industry 2021", p. 182

[6] G. Lammel *et al.*, "Next generation pressure sensors in surface micromachining technology," *The 13th International Conference on Solid-State Sensors, Actuators and Microsystems, Digest of Technical Papers. TRANSDUCERS'05*, 2005, pp. 35-36 Vol. 1

[7] Kulah H et al, "Noise analysis and characterization of a sigma-delta capacitive microaccelerometer", *IEEE J. Solid-State Circuits*, 2006, Vol. 41 352–61

[8] Ezekwe C D and Boser B E, "A mode-matching σδ closed-loop vibratory-gyroscope readout interface with a 0.004 ∘ /s/ √ Hz noise floor over a 50 Hz band", *IEEE Int. Solid-State Conf.—Digest of Technical Papers (IEEE)*, 2008

[9] Lobur M and Holovatyy, "A 2009 Overview and analysis of readout circuits for capacitive sensing in MEMS gyroscopes (MEMS angular velocity sensors)", *5th Int. Conf. on Perspective Technologies and Methods in MEMS Design*, 2009, pp 161–3

[10] H. -Y. Shih et al, "A CMOS MEMS Pressure Sensor for Blood Pulse and Pressure Measurement Applications," *International Symposium on Intelligent Signal Processing and Communication Systems (ISPACS)*, 2021, pp. 1-2

[11] F. M. Puglisi et al, "Random telegraph noise: Measurement, data analysis, and interpretation," *IEEE 24th International Symposium on the Physical and Failure Analysis of Integrated Circuits (IPFA)*, 2017, pp. 1-9

[12] M. V. Gheorghe, "Advanced calibration method for 3-axis MEMS accelerometers", *International Semiconductor Conference (CAS)*, 2016, pp. 81-84

[13] G. Lammel et al," Smart System Architecture for Sensors with Integrated Signal Processing and AI", *Smart Systems Integration*, 2021

[14] G. Lammel, "The future of MEMS sensors in our connected world", *MEMS* 2015.

The Next "Automation Age": How Semiconductor Technologies Are Changing Industrial Systems and Applications

Domenico Arrigo[1], Claudio Adragna[1], Vincenzo Marano[1], Rachela Pozzi[2], Fulvio Pulicelli[2], Francesco Pulvirenti[3]

STMicroelectronics, Industrial & Power Conversion Division, [1] Agrate Brianza, [2] Cornaredo, [3] Catania, Italy

{domenico.arrigo, claudio.adragna, vincenzo.marano, rachela.pozzi, fulvio.pulicelli, francesco.puvlirenti}@st.com

Abstract — **A profound transformation is making the industrial world more technically advanced and sustainable: the "Next Automation Age." The changes are driven by demand for increased safety for both personnel and equipment in factories, higher levels of intelligence in processes, and greater flexibility and efficiency.**

Newer semiconductor technologies like wide-bandgap silicon carbide and gallium nitride, combined with advanced digital control architectures, can deliver significantly higher power densities and conversion efficiency than conventional silicon technologies.

Advances in smart power BCD process technologies, with embedded phase-change memories, can facilitate transitions from analog to digital control improving power conversion and motor-control applications. Embedded on-chip, galvanic isolation extends or supersedes earlier-generation system-level protections with superior intrinsic safety and robustness. At the same time, it protects users from electric shocks while delivering exceptional power and high-speed data transfer rates across isolated barriers.

These are just some of the innovations emerging from an industrial electronics sector striving to meet the smart industry trends of the current and foreseeable sustainable future.

Keywords — *Energy efficiency, power density, wide bandgap semiconductors, digital power control, digital motor control, safety, on-chip galvanic isolation, BCD semiconductor technology.*

I. INTRODUCTION

Thanks to the confluence of technologies, we are experiencing a new type of automation. The first Automation Age was possible due to global information systems and the computerization of production: it was an epoch when machines began to perform complex tasks with little to no human interactions. Today the pervasiveness of sensors, connectivity, advanced computing, either at the edge or in the cloud, along with intelligent actuators, will accelerate the diffusion of cyber-physical systems: smart systems that include engineered interacting networks of physical and computational components [1]. A superset that includes machine learning and AI extending beyond traditional IoT, enabling what we call the Next Automation Age.

Implications are many. In factory automation this means moving to smarter, environmentally friendly, more efficient, and safer solutions with initiatives such as Industry 4.0/5.0: from preventive to predictive maintenance, increasing use of robots and cobots to improve efficiency and productivity, while preserving humans from repetitive tasks. Medical and healthcare applications are progressing to more personalized, always-on monitoring with the possibility to access medical services from the comfort of home.

Urban management is moving towards smart cities, using networks of billions of intelligent sensors and IoT nodes to improve monitoring, manage resources, assist citizens, and improve logistics with self-driven drones and vehicles. Robotics are becoming more widespread thanks to technologies developed for automatic navigation based on multiple sensors, detailed environment analysis, object recognition/detection, and tracking.

Autonomous robots are also used in agriculture (agri-robots) for farm, forestry and horticulture activities that drive productivity and maintain the ecosystem for entire populations, avoiding the dangers of manual management, thus protecting human workers from having to handle machines and chemicals. Additionally, at the same time, they could identify the state of crops and even determine if they are ready for harvesting.

The ongoing transformation needs to cope with sustainability: new societal trends are likely to shape future energy demand. Using electricity wisely is where electronic technologies and semiconductor technologies can play a leading role in reducing energy consumption.

Industry, transportation, and residential users account for approximately 45% of the total greenhouse gas emissions coming from electricity use. Overall emissions coming from electricity production and usage in industrial and automotive segments account for 67% of the global emissions [2]. These figures should give us awareness of how much we, as part of the electronic industry fraternity, can do for sustainability.

Focusing on industrial applications, semiconductor technology enables innovative trends in three main directions: power density and energy efficiency, pervasiveness of digital power control, and safety. With all of them, the target is to tailor the best cost/performance tradeoff to fit the needs of different applications, meeting the stringent requirements for environmental sustainability.

The first technology trend is incrementing power density and energy efficiency. Silicon Carbide (SiC) and Gallium Nitride (GaN) wide bandgap materials allow high switching frequencies in the range of several hundreds of kHz (see Figure 1), while maintaining efficiency well above 90%, and reduce the form factors, also thanks to the adoption of planar transformers. This enables packing more power in smaller volumes in many applications including wireless power chargers, robotic actuators, and electric vehicle charging stations. It is worth noting that a GaN HEMT is 1/10 the size and weight of a Silicon MOSFET and requires only 1/40 of the capacitance for switching.

978-1-6654-8495-4/22 $31.00 © 2022 IEEE

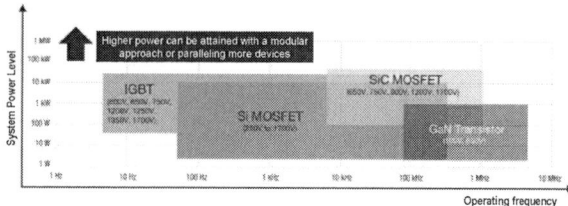

Fig. 1. Operating frequency vs. system power level for different power technologies.

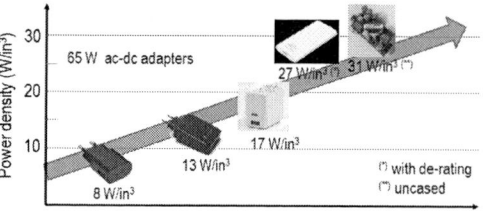

Fig. 3. Power density trajectory in chargers/adapters for portable equipment.

The second trend is power digitalization. Digital power relies on microcontrollers to run control techniques for power management [3]. This trend goes back to 1985 [4], when actuator drivers and power conversion devices were integrated on silicon with BCD (Bipolar-CMOS-DMOS) technology, as depicted in Figure 2. In 1995 for the first time, Power MOSFET, Bipolar, and CMOS components as well as non-volatile memory cells were implemented on a single chip, integrating a pulse width modulation (PWM) (up to 2A) current control loop together with an 8-bit MCU [5].

The third trend is increasing safety. The Industry 5.0 initiative, as an advancement of Industry 4.0, focuses on the interactions between man and machines in a safer working environment from a more human-centric perspective.

Extensively exposed to electrical machines and power sources, human workers require intrinsically safe interfaces with all the single machine components for protection from the risk of electrical shocks, which today is still among the first causes of death or injury in the workplace.

The evolution of semiconductor technologies plays a key role in introducing embedded galvanic isolation protection inside the chips of each single electronic modular block of a machine, guaranteeing intrinsic and modular safety at system level.

These three trends will be discussed in depth, highlighting the tools that semiconductor technologies provide to help us meet these challenges. Before going ahead, however, it must be said that the first two are tightly interconnected: a wider use of digital technologies, today essentially limited to high-end power conversion applications only (e.g., server and telecom power supplies) is instrumental in successfully reaching higher levels of energy efficiency and power density, especially in consumer and portable applications.

II. POWER DENSITY AND ENERGY EFFICIENCY

Since the beginning of the electronics era, users' needs have dictated a clear trend toward reducing the size and weight of electronic equipment across the board. Especially for portable equipment, being smaller, lighter, yet more powerful, is perceived by users as an added value.

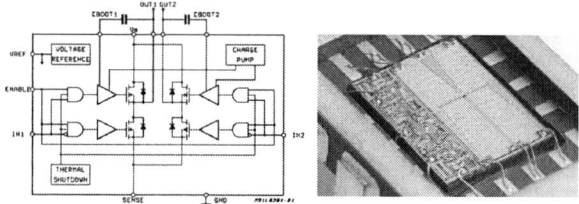

Fig. 2. First motor driver in BCD technology for industrial applications (STMicroelectronics, 1985).

Power density is one figure of merit that measures the degree of compactness of a power supply unit (PSU). Depending on the key design goal, it can be defined in different ways, typically as the ratio of its rated power to its volume (W/cm^3 or scaled equivalent units) or its weight (W/kg). Whichever definition we consider, either volumetric or gravimetric, the trend is based on "packing more power in less space", as shown in Figure 3.

It must be said, however, that the need for higher power density is not equally felt in all application sectors. For example, in industrial SMPS in automated production processes, the emphasis is on the highest robustness, reliability and safety protection functions. In other industrial sectors such as inverters for photovoltaic applications where reduction of system cost and size is at a premium, the use of wide bandgap materials (SiC, specifically) may offer a tremendous increase of gravimetric power density: from 0.7 kW/kg of a state-of-the-art silicon-based design to 1.6 kW/kg of a SiC-based design, with a comparable increase in the volumetric power density [6].

The mainstream trend in improving power density is to increase switching frequency to reduce the size of the passive components that process power (transformers, inductors, and capacitors). There is however a big drawback: essentially all power losses, which turn into heat, increase, and make it more difficult to pack the required watts in the specified volume. Quite simply: as a device gets smaller, it is more difficult to dissipate its heat (smaller volume → smaller surface). Therefore, the primary factors that limit higher switching frequency and power density are the converter power losses and the thermal performance of the system.

Power losses link power density to energy efficiency, although the latter is a broader topic concerned also with converter's efficiency requirements over a wide load range, driven by regulations and market requirements. So far, power density has roughly doubled every 10 years [7], but the fact is that power density and energy efficiency requirements clash with each other, and power designers are literally bumping into a power density barrier that might slow down this trend.

To break through this barrier, all limiting factors must be attacked in parallel; a holistic approach is required, which goes from smart architectural/topological choices to the use of appropriate active and passive components, with the addition of a good deal of innovation.

In ac-powered applications, soft-switching topologies able to work with zero-voltage switching (ZVS) at turn-on, such as the active-clamp flyback (ACF), the asymmetrical half bridge flyback (AHBF) and the LLC resonant converter, dramatically reduce frequency-related power losses. Therefore, they will be the basis of high switching frequency and high-power density designs.

978-1-6654-8495-4/22 $31.00 © 2022 IEEE

Fig. 4. IEEE Milestone plaque recognizing STMicroelectronics' BCD invention among the technologies that have advanced mankind.

In dc-dc converters used as point-of-load (POL) voltage regulators, some emerging topologies (switched capacitor converters, switched tank converters, hybrid converters) [8]-[11] are attracting attention and eroding the use of traditional topologies (e.g., buck converters).

Semiconductor technology plays a key role in the process of breaking through the power density barrier.

In particular, the contribution of BCD technology, today property of the entire microelectronics industry, starts from its invention by STMicroelectronics in mid-80s as recently recognized by the IEEE Milestone award for "Multiple Silicon Technologies on a Chip, 1985" (see Figure 4). Automotive, computer and industrial applications widely adopted this technology, which enabled chip designers to combine power, analog, and digital signal processing flexibly and reliably.

One of the benefits for applications demanding high power density resulting from the advancement of this technology is the continuous improvement of the "efficiency" of Si-based DMOS power devices with innovative internal architectures.

Figure 5 shows the evolution of the specific on-resistance figure-of-merit of BCD's power components, normalized to the 1.0 μm 3rd generation technology node (BCD3s). A 10 to 20% on-resistance reduction for each generation has been obtained through experimentation with different process and structural ingredients as reported in the figure.

The key to achieving these targets and maintaining this rate of development into the future is a deep understanding of the mechanisms underlying the physical limits [12]-[13].

As to discrete power devices, wide-bandgap (WBG) materials, namely SiC and GaN, solve most of the issues related to losses in power switches, allowing a favorable trade-off between conduction and frequency-related losses.

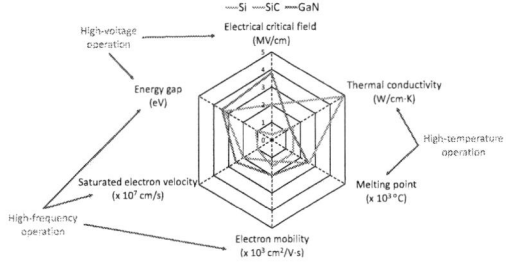

Fig. 6. Comparison of Si, SiC, and GaN for power applications.

With SiC MOSFETs addressing especially very high-voltage applications in industrial and automotive applications, GaN HEMTs (High Electron Mobility Transistor) are a nearly ideal solution in most power conversion applications with their ability to switch ten times faster than silicon MOSFETs while keeping losses acceptably low.

A quick comparison of WBG materials with Si that highlights their benefits is shown in Figure 6.

In addition to making some hard-switched topologies viable (for example, the in-fashion totem-pole Power Factor Corrector), WBG switches, when combined to soft-switching topologies, form an unrivaled pair. In fact, achieving soft switching often costs some additional conduction losses that partly offset the savings. With SiC MOSFETs and GaN HEMTs these extra losses are significantly lower than with Si MOSFETs, thus maximizing the benefits of ZVS.

Using WBG switches is not an easy task, though. This is particularly true when it comes to driving GaN HEMTs due to their limited gate-source voltage rating and their tremendous switching speed that magnifies the effect of stray inductances in both the driving path and the power path. Optimizing the circuit board layout to prevent major damage is critical more than ever.

Significant help is provided by the system-in-package (SiP) approach shown in Figure 7. Assembling the driver(s) and the GaN HEMT(s) in the same package minimizes the size of the driving path and, therefore, its stray inductance. In this way, power designers, partly relieved from the intricacies of the circuit board layout, may concentrate on the other challenges of high frequency, high power density designs.

The next step in this process is to integrate the driver(s) and the GaN HEMT(s) in the same die (System on Chip, SoC). This provides the ultimate optimization of the driving path, reducing its size to a chip-scale one.

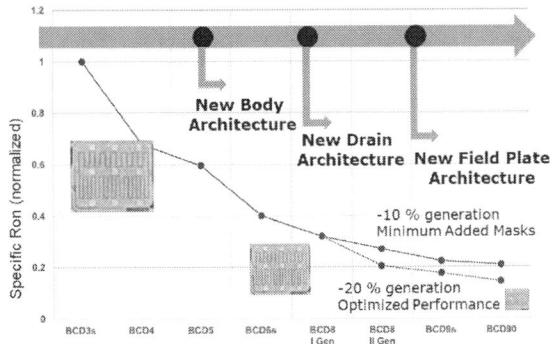

Fig. 5. Integrated power device evolution.

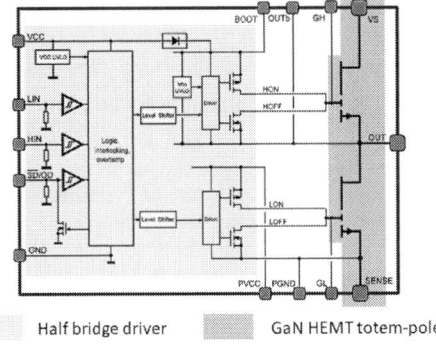

Fig. 7. MasterGaN®, a System-in-Package (SiP) embedding a high-voltage half-bridge driver and a 650 V GaN HEMT totem-pole.

This integration paves the way to an entire GaN on Si technology, a sort of GaN-based BCD technology enabling the implementation of ICs that may benefit from the speed performance of GaN in time-critical functions in all sorts of ICs, from simple controllers to advanced microprocessors.

Regardless of whether we are considering SiPs, SoCs, or discrete devices, the improvements in power density require an equally strong effort in the development of new package technologies that provide lower parasitics and better heat dissipation with reduced footprints.

This is another powerful driving force influencing package evolution that adds up to those concerning heterogenous integration, multi-chiplet approach and other trends framed in a "More than Moore" vision.

Techniques such as Direct Copper Interconnect (DCI) set the stage for the evolution toward packages with extremely low parasitics, while improving thermal performance. In fact, the main bottleneck to heat flow that package technologies need to address is the interface between semiconductors and the metal connections acting as both an electrical and thermal bridge to the printed circuit board. Recently, some SiP products (a couple of examples are shown in Figure 8) have pioneered this evolution, pushing the limits of existing technologies (die attach film, advanced frame design, and multiple die attach), and demonstrated the market potential of this approach.

These packages are ramping up now with different processes available in both flipped or traditional configurations and are already showing potential for a new wave of innovation.

New thinner and more thermally conductive die attach materials, along with solutions enabling power dissipation on both sides of the package, will enable optimal thermal management even with tight space constraints. Thermal performance and co-packaging flexibility can be improved by integrating ceramic substrate in place of the conductive exposed pad. In this way, a full system composed of dice with different substrate voltages may be housed in a single package while enhancing thermal dissipation.

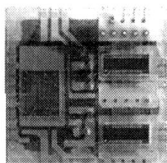 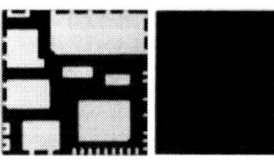

Fig. 8. Examples of customized packages for SiP assembly: driver and two 650V GaN HEMTs (left) and a complete 600V rated tri-phase power stage with gate driver (right).

III. POWER DIGITALIZATION & ENHANCED CONNECTIVITY

The evolution of Smart Industry, especially in the framework of Industry 4.0 and 5.0 initiatives, increasingly demands more connectivity, flexibility, configurability, and serviceability; all areas where semiconductor technology can help provide significant enhancements especially with the transition from analog to digital control systems.

A. Digitalization in power conversion

In power conversion, digital controllers are a consolidated reality in some application areas, such as telecom power, server power, electrical vehicles, and high-end lighting. In these applications there are communication, housekeeping, diagnostic and supervisory system requirements that need digital features to be fulfilled. It is therefore a short step to use these digital resources to control the power converter, too.

This is not the case with most consumer applications. On the one hand, these are much more cost-sensitive than those previously mentioned and digital control is generally more expensive than analog control; indeed, system requirements may not motivate power designers, traditionally using analog control, to switch to digital control, which is generally more complex, more expensive, and requires coding.

However, in recent years the cost gap between analog and digital devices is narrowing and some high-volume applications, such as ac-dc adapters and chargers, are no longer "stupid" boxes that simply turn an ac voltage into a dc voltage. This is the case, for example, of USB-PD (USB Power Delivery) compliant chargers. These devices communicate with the equipment to negotiate a power contract so they can determine how much power can be pulled from the charger. And the trend is expected to keep going in this direction. This situation has encouraged power designers to look more carefully at digital control and the benefits it may bring to power converters.

The use of digital signal processing techniques in the control loop enables the design of compensators insensitive to parameter tolerance and able to tune themselves on-the-fly to optimize converter's dynamic behavior under all operating conditions. But its benefits do not end there.

Using digital techniques, it is possible to change the way the converter works according to the operating conditions to optimize its efficiency over the widest possible range. This makes it much easier to meet energy efficiency requirements [14]-[16], which specify minimum efficiency levels under various load conditions and maximum standby power.

Digital techniques enable more sophisticated fault detection and handling algorithms using multivalued analog-to-digital converter (ADC) readings instead of comparators, diagnostic data logging, power quality and performance metrics measurements, simplified control of long timeframe processes. In many cases, system functions and converter control can be performed using a single digital processor.

To maximize the benefit of these design techniques, a "fully digital" approach is necessary, where users have the full control of the system, define the algorithms, allocate the digital and the analog resources, and write the firmware and the relevant code for debugging.

This kind of solution can use general-purpose devices such as generic microcontrollers or DSPs, and non-standard microcontrollers with peripherals specialized for the control of power conversion, with external circuitry to interface power devices and high voltage signals. Examples of this approach include high-resolution PWM generators and programmable state machines such as the SMED (State Machine Event Driven) [17].

As opposed to this, some power designers prefer a sort of "mild digital" approach, where the control algorithms and the resource allocation are fixed, and the user can adjust certain controller parameters or settings during the fine-tuning and debug phase, considerably shortening time-to-market. The devices intended for this approach look very much like analog controllers because they include all the hardware resources to connect the digital world to the converter (supply voltage regulators and gate drivers). In addition, they have a communication port to access the digital resources with a PC and a GUI (Graphical User Interface) [18] for fine-tuning parameters and monitoring functions.

Some devices of this kind integrate most of the functions required to implement a complete ac-dc converter, such as high-voltage half bridge drivers, high voltage pins able to interface with signals from the switching converter, line voltage monitoring, high voltage startup circuit, etc.

Figure 9 shows an example of a highly integrated controller that implements all the functions required to implement an isolated, USB-PD compliant external power supply for fast charging applications. It includes a microcontroller, MOSFET drivers, a complete USB-PD physical layer and a galvanically isolated communication channel that enables the microcontroller core to receive information from the primary side and control the switching of the gates on both sides of the converter. This highly integrated approach allows the power engineer to design a complete application with less effort, enjoying also the flexibility offered by the programmable digital controller.

B. Digitalization in Motion Control

Motion control is another sector where digital controllers are a consolidated reality but in full evolution.

Electric motors represent a typical example of actuators, and their use is so widespread that it is no surprise that 46% of the worldwide energy consumption is used for driving electric motors. Considering only the industrial sector, the percentage rises to 69% [19]. For this reason, efficient motor driving has a significant impact on sustainable development goals: assessments from U4E on 150 countries show a potential energy savings from electric motors of 300 TWh per annum in 2030, with a saving of 200 Mt of CO_2 emissions (equivalent to the annual electricity generated by approximately 60 coal-fired power plants with a capacity of 1,000 MW) [20].

The first step in improving efficiency is to replace motors based on brushed technologies, and ac induction motors, with more efficient brushless motors such as, for example, permanent magnet synchronous motors (PMSM). This migration, in place for several years, can reduce energy consumption up to 50% [18] and is enabled by the rapid advancement of power and digital electronics.

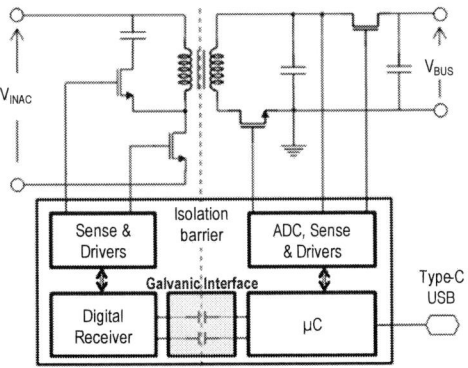

Fig. 9. Integrated ac-dc controller including USB-PD protocol and galvanically isolated communication.

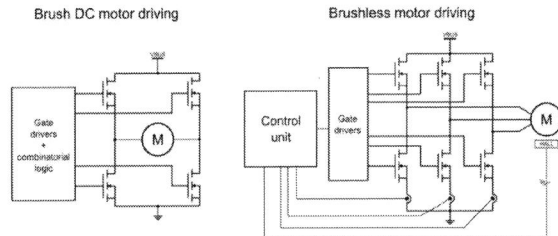

Fig. 10. Comparison between brushed dc and brushless motor driver circuitry.

In fact, a three-phase brushless motor driver is significantly more complex than a brushed one: if state-of-the-art brushed motor driving requires few power components and basic analog circuitry, like comparators and operational amplifiers, even the simplest brushless motor driver cannot be implemented without some digital controllers.

For the most advanced control solutions, driving brushless motors requires high-performance analog circuitry for sensing the motor's currents and voltages, and fast microcontrollers or DSPs able to manage complex real-time calculations, as schematically shown in Figure 10.

The basis of these advanced control algorithms is the well-established field-oriented control (FOC) strategy that dynamically controls the relation between the stator's and rotor's magnetic fields through several coordinate transformations and vector calculations, targeting the optimal displacement between the two magnetic fields for maximum efficiency.

In a direct approach, the position of the rotor's magnetic field is provided using mechanical sensors like encoders or Hall-effect based sensors, but a further level of complexity comes from its sensorless detection using model-based methods.

The advantages of FOC (Figure 11) are not limited to better efficiency and lower electric power consumption. FOC ensures better performance (e.g., silent operation and better dynamic response) and enables operating modes not achievable using other methods.

The field-weakening technique, for example, alters the motor's torque-speed curve allowing the motor to operate above the rated speed with the rated bus voltage. This result is obtained by reducing the back electro-motive force (EMF) generated by the coupling between stator coils and rotating magnetic field of motor, which opposes the supply voltage. The higher the rotating speed, the stronger the back EMF effect, which limits the ability of the driving circuitry to inject current into the motor's coil. To "weaken" the field that creates back EMF, the FOC algorithm applies a stator magnetic field in opposition to the motor one. The weakened magnetic field generates a lower back EMF, enabling higher speeds at the expense of the output torque.

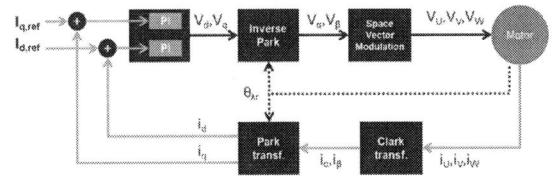

Fig. 11. Field Oriented Control algorithm.

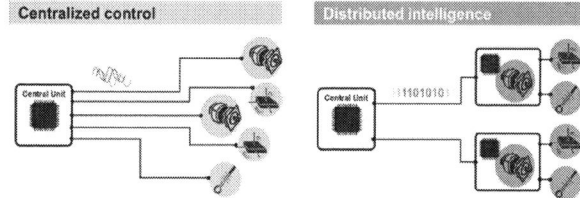

Fig. 12. Digital motors open the door to edge intelligence.

Thanks to the continuous improvement of the performance of digital electronics, FOC driving features can be easily deployed on a wider range of applications. Compared to ST's first generation of Cortex-M based MCUs (STM32F1 series), the latest high-performance series (STM32H7) offers a dual-core device with 21 times the DMIPS, 6 times the maximum clock frequency and 44% more RAM [21]-[22].

This boost in computational power leaves room for the execution of deep learning algorithms [23] combining FOC with information coming from the environment and dynamically adjust operation in a goal-oriented approach.

Fans that automatically reach an optimal balance between silence and air flow, servo motors able to react to unexpected changes [24]-[25], industrial equipment predicting a potential critical failure and promptly ask for a maintenance support: these are just few examples of how the pervasive digitalization of motor control could change the future.

The increase of digital processing capability will allow the integration of more complex functions, leading to more efficient system partitioning: from a centralized control to edge intelligence, as shown in Figure 12.

The key enabler for this target is to move towards more advanced CMOS generations. We have just delivered the first prototypes of BCD90 technology using a state-of-the-art 90 nm CMOS platform and the next generation at 40nm (BCD40) is on the way; it will be based on STMicroelectronics 40nm proprietary technology already used for our latest microcontrollers in production.

BCD90 is designed using 1.2V CMOS logic, with a density up to 500 kgate/mm^2, and a 5V second CMOS gate oxide for analog purposes and a highly efficient power MOS. While BCD40 integrates 1.1V digital CMOS, featuring 800 kgates/mm^2 density, and a 2.5V second CMOS gate oxide for best-in-class analog power components.

The second enabler for the digital processing capability increase is the availability of a new NVM (Non-Volatile Memory) concept based on a phase-changing chalcogenide composite material.

The physical storage process is the phase transition, when heated up, of the chalcogenide Germanium, Antimony, and Tellurium ternary alloy between the crystalline and the amorphous state. The two states exhibit different electrical resistivity, allowing the discrimination of "0" and "1" states of the memory cell. The interest in this kind of memory, termed PCM (Phase-Change Memory) [26], stems from certain noticeable features:

- It is integrated in the backend-of-the-line (BEOL) stage of the front-end process, with low mask increase and almost negligible impact on the other components.

- It outperforms floating gate memories, both in terms of access and speed.

- Its reliability is proven to be excellent, compatible with automotive mission profiles. This is particularly true for high-temperature data retention, required to survive product soldering in the final application board and the peak operating temperatures typical of Power ICs.

In 2018 an internal STMicroelectronics development featured a 32-bit microcontroller, monolithically integrated with an autonomous numerical torque generator for motor control, a dedicated power stage, and an analog front-end, covering smart power conversion functions from dc-dc conversion to motor drives. The device was created in a 12.7 mm^2 die in BCD technology. Future advances will rely on these evolving IC technologies as well as associated packaging solutions. These will include super-integration, wafer-to-wafer bonding, and SiP, to enable the heterogenous integration of multiple technologies such as Vertical Intelligent Power, BCD, FD-SOI CMOS, and others.

C.   Enhanced Industrial Connectivity

With the introduction of the IoT (Internet of Things) and cyber-physical systems, industrial automation is undergoing a tremendous change. This is enabled by recent advances in semiconductor technologies that allow the diffusion of digital control systems along with novel connectivity solutions.

Industrial communication networks drive smart manufacturing scenarios by real-time communication among machines and main control systems.

Operation technology networks are associated with automation and are characterized by "horizontal data flows" within the same network layer. These networks must ensure reliable, uninterrupted data transmission, even in electrically harsh environments with heavy and unpredictable noise. IT networks are usually oriented "vertically" and are designed to service end users with high transmission rates [27].

In the past, industrial communication was exclusively developed along wired protocols with the definition of proprietary fieldbus and industrial Ethernet solutions such as EtherCAT and PROFINET [28]. The need to access sensor data with the highest reliability is another requirement that IO-Link communication networks (IEC 61131-9) address particularly well. IO-Link can be used for bi-directional, point-to-point data connectivity down to the actuator and sensor level, enabling data pre-processing, sensor parameter tuning and advanced diagnostics [29]. This is achieved today thanks to a range of solutions including IO-Link transceivers and STM32 microcontrollers.

Wired communication is continuously growing and evolving from traditional fieldbus to Ethernet-based buses with the intent of connecting human machine interfaces, PLC, machines, I/Os, drives, actuators, and sensors.

Communication channels in IIoT (Industrial IoT) need to transfer several types of data, with different levels of determinism and latency. Specific industrial Ethernet protocols have therefore been deployed, such as Ethernet time-sensitive networking (TSN).

Nevertheless, the mobile use cases in smart industry such as robots and cobots, are driving the expansion of wireless networks such as 5G and Narrow Band IoT (NB-IoT) designed to meet IoT requirements, thus providing a highly flexible and scalable solution [30].

By exploiting existing cellular architecture in a cost-effective manner, NB-IoT technology has emerged as a leading solution for IoT applications suitable for long-range, ultra-reliable, low-latency communication and machine-to-machine (M2M) cellular networks for smart cities and industrial automation. However, these technologies pose new cyber-security challenges as main obstacles to introducing Industry 4.0 services for connected manufacturing in industrial factories. The advancement of semiconductor technologies that increase the computing and networking capacity in IoT devices help meet these challenges. The industrial gateway collects data from multiple wired or wireless sensor nodes and transmits them to cloud applications or a central condition monitoring system for data analysis. The industrial gateway not only acts as a bridge between different networks technologies, connecting various types of sensors directly to the Internet, but also performs rule-based analysis for Edge processing, further reducing system power consumption and network bandwidth.

IV. ENHANCED SAFETY

Industrial safety is defined by international standards such as IEC 61508, which provide tools and safety functions to assess risks and prevent danger. System safety is essential for ensuring human safety and protecting the environment. It applies to all industrial systems, including automated manufacturing, transportation, and the oil and gas industry. To protect the user from electric shock, power applications directly connected to the power grid or high-voltage lines, such as home appliances or motor drives and battery control circuits in hybrid vehicles, require isolation between the power stage and the control portion that interfaces with the user.

The galvanic isolation is a physical barrier separating the different voltages of two electric systems to prevent stray currents. Galvanic isolation guarantees safety by grounding potential differences to prevent ground loops and improve noise immunity.

There are two main categories of isolation: power and signal. In the past, ferrite core transformers were used to transfer power, while opto-couplers or transformers were both used to isolate signals.

In many cases some limitations and compromises were necessary on communication speed, power consumption, size, and reliability. For example, transformers can guarantee good communication speed and reliability, but are quite bulky; opto-couplers are smaller but are generally limited in operating speed and suffer from ageing phenomena.

In recent years, new process options in semiconductor technologies made the integration of galvanic isolation possible inside integrated circuits (IC). This allowed the coexistence in the same device of the galvanic isolation barrier with the related circuitry to implement transmission and reception of the signals, in addition to the original function that required isolation (e.g., a gate driver for power transistors).

The integrated approach brings several advantages, such as space saving, low power consumption, excellent reliability, and communication speed.

The correct choice of isolation material is a key factor to best benefit from the integration inside integrated circuits. The target is to maximize the ruggedness against surge and transients over voltages with low infant mortality (yield) and good ageing performance, while keeping short distances to minimize the power consumed to achieve the communication and obtain a high communication speed. The most used materials to integrate a galvanic isolation barrier are silicon oxide and polyimide [31].

From the point of view of device design, low-jitter and high-speed communication are required together with data integrity (immunity to voltage transients, magnetic fields) and safe fault management. Moreover, the IC package characteristics will have to meet creepage and clearance distance requirements according to the molding material used. A higher comparative tracking index (CTI) value means a lower minimum creepage distance is required.

Isolated ICs require special tests and equipment to assess isolation lifetime and later to guarantee full production testing. These are just some of the requirements needed to comply with several certification standards and regulations. For example, international standards such as UL 1577 and IEC 60747-5 define a set of tests to classify a system's isolation level.

ST's innovations in technology made significant steps in just a few years, along with the design of richer and better performing circuits taking advantage of advanced isolation characteristics. It started with basic isolation able to withstand only a few kV to today's reinforced isolation solutions able to survive more than 10kV and with excellent stability over time. The designs took advantage of different topologies to transfer signals across the isolation, such as capacitive or inductive coupling; signal transfer rates have increased to hundreds of Mbps and are destined to increase even more, in combination with more efficient power transfers and immunity to common mode transient voltages which has reached 200 V/ns.

Some examples of devices with integrated galvanic isolation are digital interfaces, analog-to-digital modulators, intelligent power switches (IPS), high-voltage gate driver for IGBTs and GaN HEMTs as well as Silicon and SiC MOSFETs. The fields of application range from those already established, such as industrial equipment and power conversion systems, to more innovative ones such as those found in hybrid and electric cars [32].

The next trend in innovation is driven by the heterogeneous integration of silicon, assembly technology and new materials enabling systems to be embedded in a single package with galvanic isolators, power components (GaN and SiC), advanced BCD smart power ICs and MCU dice.

The latest developments led to the availability of advanced system-in-package solutions, combining a galvanically isolated gate driver and two 650V 50mΩ GaN HEMTs in half bridge configuration (Figure 13).

This device is specifically designed for relatively high-power conversion applications (up to 2kW range) such as servers and industrial power supplies. In totem-pole bridgeless PFCs and resonant converters, galvanic isolation enables simpler implementations and more flexible control schemes, reducing the extra circuitry needed to interface the control and the power sections.

The QFN package (12x12 mm) housing the device has an exposed pad to simplify and improve the thermal design and avoids having to use an additional isolation layer that reduces the maximum thermal dissipation capability.

978-1-6654-8495-4/22 $31.00 © 2022 IEEE

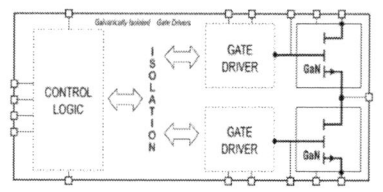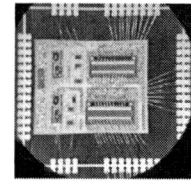

Fig. 13. Advanced system-in-package with a galvanically isolated gate driver and two 650V 50mΩ GaN HEMT in half bridge configuration.

V. CONCLUSIONS

To successfully move into the Next Automation Age, a series of distinctive capabilities are needed, blending different innovation perspectives: smart power processes, new packages, advanced architectures extracting the best figures of merit from new materials and new IPs developed in partnership with industrial leaders, with an open attitude towards the disruptive innovation coming from new game changers.

The goal is to reduce energy consumption and CO_2 emissions, optimizing the overall process and reducing operational costs by using resources more efficiently, with the objective of improving the quality of life.

In this way, between challenges and opportunities, we can imagine new horizons and bring them on the market, enabling and leading the ongoing transformation.

ACKNOWLEDGMENTS

The authors wish to thank Alberto Bianco, Michele Lauria, Enrico Poli, Domenico Ragonese, and Cristiana Scaramel for their expert support during the preparation of the paper.

REFERENCES

[1] Framework for Cyber-Physical Systems: Volume 1, Overview, National Institute of Standards and Technology (NIST), https://doi.org/10.6028/NIST.SP.1500-201.

[2] IPCC (International Panel on Climate Change), "Climate Change 2022: Mitigation of Climate Change", Apr. 2022.

[3] R. Attanasio, G. Di Caro, S. Messina, and M. Torrisi, "Digital Controller Eases Design Of Interleaved PFC For Multi-kilowatt Converters," https://www.how2power.org/article/2017/june/digital-controller-eases-design-ofinterleaved-pfc-for-multi-kilowatt-converters.php, 2017.

[4] A. Andreini, C. Contiero, and P. Galbiati, "A New Integrated Silicon Gate Technology Combining Bipolar Linear, CMOS Logic, and DMOS Power Parts," IEEE Trans. Electron Devices, vol. 33, no. 12, doi: 10.1109/T-ED.1986.22862, 1986.

[5] D. Rossi, G. Pedrazzini, G. Ricotti, and T. Cavioni, "Fully Integrated Fast Battery Charger with PWM current control and Microprocessor on Board," 1995.

[6] P. Friedrichs, "Power conversion switch technology: the who, when, where and why of using Si, SiC and GaN transistors", Industry Session presentation, Session IS04, Paper Id 3550, APEC 2020.

[7] J. W. Kolar et al. "PWM Converter Power Density Barriers", 2007 Power Conversion Conference - Nagoya, 2007, pp. P-9-P-29, doi: 10.1109/PCCON.2007.372914.

[8] M.D. Seeman, S.R. Sanders, "Analysis and optimization of switched-capacitor dc-dc converter", IEEE Transactions on Power Electronics, 23.2 (2008): 841-851.

[9] Z. C. Ye, Y. T. Lei, R. Pilawa-Podgursky, "A resonant switched capacitor based 4-to-1 bus converter achieving 2180 W/in3 power density and 98.9% peak efficiency." 2018 IEEE Applied Power Electronics Conference and Exposition (APEC). IEEE, 2018.

[10] Y. N. Chen, D. M. Giuliano, M. J Chen, "Two-stage 48v-1v hybrid switched-capacitor point-of-load converter with 24v intermediate bus."

2020 IEEE 21st Workshop on Control and Modeling for Power Electronics (COMPEL). IEEE, 2020.

[11] Y. C. Li et al., "A 98.55% efficiency switched-tank converter for data center application." IEEE Transactions on Industry Applications 54.6 (2018): 6205-6222.

[12] Riccardi, D. et al. "BCD8 from 7 to 70V: a new 0.18μm Technology Platform to Address the Evolution of Applications towards Smart Power ICs with High Logic Contents," *International Symposium on Power Semiconductor Devices and ICs (ISPSD)*, 2007.

[13] R. Roggero, G. Croce, P. Gattari, E. Castellana, A. Molfese, G. Marchesi, L. Atzeni, C. Buran, A. Paleari, G. Ballarin, S. Manzini, F. Alagi, G. Pizzo: BCD8sP "An Advanced 0.16 μm Technology Platform with State of the Art Power Devices," *International Symposium on Power Semiconductor Devices and ICs (ISPSD)*, 2013.

[14] European Commission, "Code of Conduct on Energy Efficiency of External Power Supplies - Version 5", available at https://e3p.jrc.ec.europa.eu/sites/default/files/documents/publications/code_of_conduct_for_eps_version_5_-_final.pdf.

[15] 80+ program for computer and server SMPS. Information available at https://www.clearesult.com/80plus/program-details#program-details-table.

[16] DOE Energy Conservation Standards. Information available at https://www.energy.gov/eere/buildings/standards-and-test-procedures.

[17] Loidl, J. Hajek, G. Bosisio, I. S. Bellomo, "Innovative event driven state machine (SMED) peripheral for digital control of power conversion & lighting applications with 8bit microcontrollers", PCIM Europe 2012 Proceedings, Paper #137, Page(s) 986 – 993, May 2012.

[18] "Graphical user interface for the STEVAL-PCC020V2 USB to I²C/UART interface board for STNRG011", available at https://www.st.com/en/embedded-software/stsw-stnrg011gui.html.

[19] C. U. B. Paul Waide, "Energy-Efficiency Policy Opportunities for Electric Motor-Driven Systems," International Energy Agency, 2011.

[20] United for Efficiency, "Accelerating the Global Adoption of Energy-Efficient Electric Motors and Motor Systems," UN Environment, 2017.

[21] K. S. K. Suvvala Jayaprakash, "A Comprehensive Review on Hybrid Multiport Converters for Energy Storage Devices Control and Performance of Electric Motor in EVs," International Journal of Emerging Trends in Engineering Research, vol. 8, no. 8, 2020.

[22] STMicroelectronics, "STM32H7 Series," STMicroelectronics, [Online]. Available: https://www.st.com/content/st_com/en/products/microcontrollers-microprocessors/stm32-32-bit-arm-cortex-mcus/stm32-high-performance-mcus/stm32h7-series.html.

[23] Z. H. C. C. J. X. Pengzhan Chen, "Control Strategy of Speed Servo Systems Based on Deep Reinforcement Learning, Algorithms", vol. 11, no. 5, 2018.

[24] STMicroelectronics, "STM32F1 Series," STMicroelectronics, [Online]. Available: https://www.st.com/content/st_com/en/products/microcontrollers-microprocessors/stm32-32-bit-arm-cortex-mcus/stm32-mainstream-mcus/stm32f1-series.html.

[25] Ahmat Fauzi et al, " Performance and Energy Consumption of Paddle Wheel Aerator Driven by Brushless DC Motor and AC Motor: A Preliminary Study" IOP Conf. Ser.: Earth Environ., 2021.

[26] P. Zuliani, E. Palumbo, M. Borghi, G. Dalla Libera, R. Annunziata "Engineering of Chalcogenide Materials for Embedded Applications of Phase Change Memory," *Solid-State Electronics*, vol. 111, pp. 27-31, 2015.

[27] Dr. Ovidiu Vermesan, Dr. peter Friess, "Digitising the Industry Internet of Things Connecting the Physical, Digital and Virtual Worlds", River Publishers, 2016.

[28] Martin Rostan, EtherCAT Technology Group, "Industrial Ethernet Technologies", 2014.

[29] IO-Link Consortium, "IO-Link Interface and System v1.1.3", IO-Link Community, 2019.

[30] Min Chen, Yiming Miao, Yixue Hao, Kai Hwang, "Narrow band internet of Things", 2169-3536 (c) 2017 IEEE.

[31] L.Dissado - The Incorporation of Space Charge Degradation in the Life Model for Electrical Insulating Materials 1995 – IEEE.

[32] F. Pulvirenti, G. Cantone, G. Lombardo, and M. Minieri, "Dispositivi con Isolamento Galvanico Integrato," AEIT Annual Conference, Trieste, Italy, Sep. 2014.

Reminiscing through 40 years of CMOS analog circuit design: from audio to GHz

Rinaldo Castello
University of Pavia
Pavia, Italy
rinaldo.castello@unipv.it

Invited paper

Abstract— During my research life I witness 40 years of CMOS analog design. I was lucky to be at U.C. Berkeley when all this started, and my Ph.D. with Prof. Paul Gray was part of it. I kept my focus on analog CMOS at the University of Pavia even if the migration toward digital and the dream of analog design automation decreased the interest on it. CMOS circuits moved from audio to THz as the technology scaled three orders of magnitude. This talk will show this amazing evolution with some example circuit and system primarily for RF but not only. I will look from both academia and industry presenting device, circuit and architectural innovations. I hope this talk can bring motivation to younger researchers to drive the next CMOS developments.

Keywords— CMOS, Pipe-Processing, RF, Filters, Analogn IC

I. INTRODUCTION

This presentation spans 40 years of circuit design from the beginning of my Ph.D. at Berkeley to the present as I near my retirement at the University of Pavia. During this period integrated systems complexity/performance has progressed in an astonishing way following the evolution of CMOS technology according to Moore's Law. As predicted, we have gone from few thousand transistors in the fifth order SC filter I built for my Ph.D. [1] (see fig 1) to many billions of transistors in todays most advanced processors. Factoring in the chip size difference, the 2^{20} expected element growth is almost exactly confirmed fostered by a 2^{10} linear shrinkage (from 3 μm to 3 nm) [2]. This impressive technology

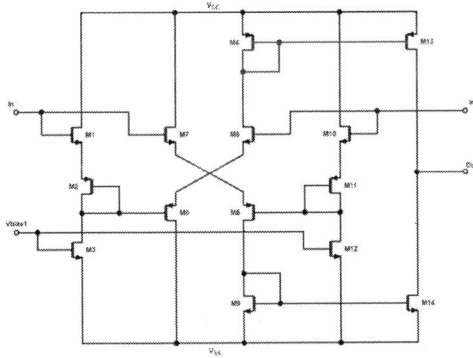

Figure 1: Class A/B sampling amplifier [1].

progresses have been accompanied and, in some sense, made possible by an equally impressive progress of the software design tools. Finally, but not less important, we have seen big progresses in systems, architectures, and circuits design. While for digital systems the progress has occurred in a

natural way, the same is not true for analog ones. This is because both technology and CAD tool have been driven by digital circuits [3]. The slight disappointment that this may bring to an analog designer is not justified if we consider the model of an integrated system as an ever-growing egg of which the analog portion is just the very thin shell.

II. THE IMPACT OF CAD TOOLS AND TECHNOLOGY ON ANALOG DESIGN

1) CAD tools have also greatly benefitted the work of analog designer

2) Analog design automation has not yet appeared on a major scale.

Regarding 1, I would like to show today Ph.D. students how I designed and verified my chip 40 years ago. We had crude simulators and layout tools, some still under development, but had no LVS or GDS tools. Figure 2 shows the enlarged plot of my SC filter that I used to search for layout errors. With the help of my fellow student Nanni Demicheli interconnects were manually checked after marking with different colors the various nodes (ground, VDD, the 4 phases of the clock etc.). After a couple of days of exhausting work, we found a killer error that I fixed saving my graduation.

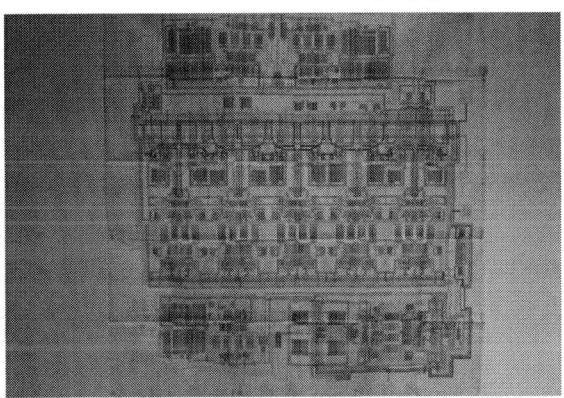

Figure 2: Enlarged plot for manual layout verification

Cad tools has made a lot easier to design and drastically increased the chance of first chip success also for analog circuits

Regarding 2, I give few reasons that might have prevented automatic analog design to become mainstream.

It is difficult to build analog systems in a hierarchical way using coded knowledge starting from building blocks. This is due to the effects occurring when interconnecting blocks that although predictable by complex models often defy intuition. As an example, we consider an amplifier capable to drive small resistors and/or large capacitors with rail-to-rail swing. In the talk I will show that for the same exact topology just connecting one side of a capacitor to a different node, as shown in Fig. 3, can completely change the circuit behavior.

For the same exact topology just connecting one side of a passive element to a different node can completely change the behavior of an amplifier

Analog libraries are more complex and costly to build than digital one. While the latter stay almost the same with scaling, the topology of analog building blocks (e.g. opamps) evolve with technology. As an example, CMOS opamps evolved very fast eventually going full circle. We started migrating the 741-like topology [4] from bipolar to CMOS and quickly move to cascode-like ones or even multiple cascode. [5]. However, the decreasing supply and gain associated with scaling pushed back to multistage. Recently multiple feedforward topologies reached unthinkable bandwidth, in one case using an inductor within the opamp [6,7]. The same circular evolution was seen in ADC where the focus went from SAR to Delta Sigma and back again.

Analog designers are jealous of their knowledge and are reluctant to put it into a tool. I saw companies offering a bonus for every topology that a designer was willing to fully document to make it re-usable but getting very poor response.

Techniques unfeasible at a given technology node become feasible in later ones. Two examples are n-path filters that were proposed 60 years ago [8] and soon discarded and are now used at RF [9] and closed loop GHz filters that have substituted gm-C in some applications [7].

My conclusion is that if technology will continue to evolve analog CAD tool will not become prevalent since their developer have to follow a moving target.

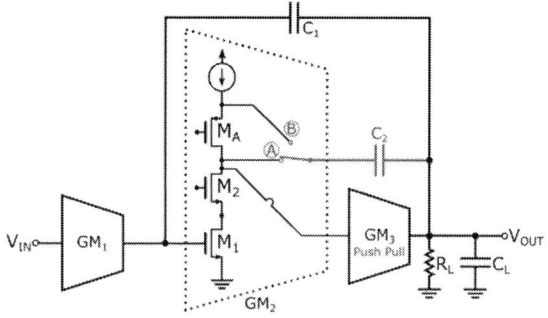

Figure 3: Multistage opamp with different feedback

Intuition, which is hard to code, remains crucial in analog design. An example is an audio amplifier used in handheld wireless terminal where intuitive understanding allowed to drastically improve performance. The improved solution was included in hundreds of millions of chips sold by ST to Nokia. The new amplifier (shown in Fig. 4) [10] was built changing the existing topology from nested Miller to double nested Miller simply opening the old layout and inserting an additional stage. The final bandwidth was a bit smaller but the distortion in the audio band more than 20 dB lower. The

intuitive way to grasp why it is so, uses human (not artificial) intelligence and will be explained in the talk.

In the early phase of Cad Tool developments, the promise of automated Analog Design reduced the number of university ad research centers perusing the topic. Also, as analog Ph.D. students we were invited to move our emphasis into software if we did not want to be out of a job [11]. I decide to still focus on analog design even when I moved to the University of Pavia, and this was a lucky choice. When the promise of Analog Automation did not materialize, the relatively small number of institutions active on the topic got a lot of interest from industry and Pavia attracted many of them.

Another topic of heated discussion I found myself involved in, was CMOS vs. BiCMOS for analog intensive systems. Many ISSCC panes and loads of articles have addressed this issue. In my view the key limitation of CMOS is 1/f noise

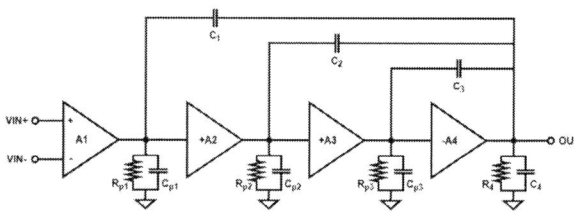

Figure 4: Doble nested Miller audio amplifier topology [10]

while the key advantage of bipolar is its much larger product fT times gm at a given I. This is because while CMOS can give either a fT competitive with the best bipolar or a gm just slightly smaller, it cannot give both at the same time. A BiCMOS trans conductor that takes advantage of this feature, shown in Fig.5, [13] was part of the first read channel ever sold by ST. Tuning of the gm is done by forcing the transistor deeper into the linear region while at the same time

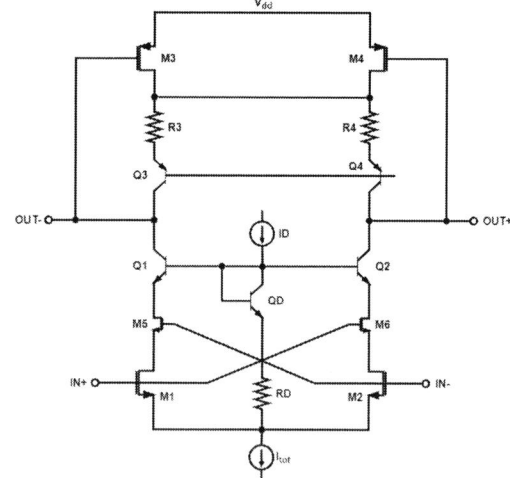

Figure 5: BiCMOS trans-conductor for disk drive [13]

preserving linearity. If the impedance at the MOS drain is very small excellent linearity is obtained, on the other hand the fT of the cascode defines the second pole of the transconductor. This calls for a cascode with both high gm and high fT for which bipolar is far superior to MOS. The excellent quality of the filter based on the above transconductor helped ST enter a new marked that grew at an

978-1-6654-8495-4/22 $31.00 © 2022 IEEE

astonishing pace. Nonetheless, even if performance where harder to meet, from the second generation on all disk drive chips (not only from ST) used CMOS.

The same migration from BiCMOS to CMOS occurred in RF transceivers [14,15] although within ST this took many product generations. In both cases the underlaying motivation was cost which, when the market goes beyond a certain size, become the ultimate decision criteria. This is almost a self-fulfilling condition. Market will grow only if cost is reduced which means unless you find a way to build your system in CMOS your market will remain a niche.

A pure CMOS implementation of a complex system is a condition to make it into a mass market product

III. THE CONCEPT OF PIPE PROCESSING

In the case of RF Transceivers, the successful migration to CMOS was due to many causes including intelligent standard definition, system design, architectural and circuit design. Among these I focus on the use of what I call Pipe Processing. This concept is shown in Figure. 6 where the signal moving

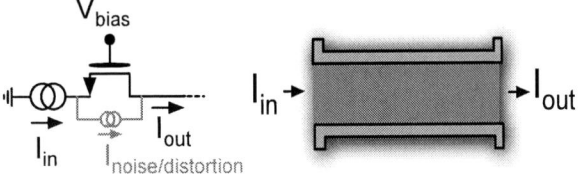

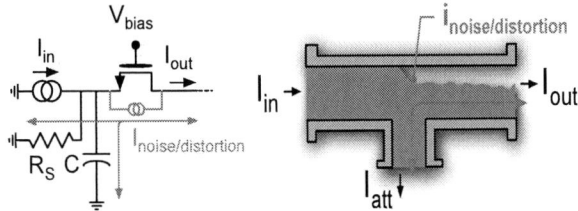

Figure 6: Hydraulic representation pipe processing.

along a processing chain is represented by the flow of water in a pipeline. The key observation is that if there is no leakage there cannot be any perturbation of the water flow which correspondingly implies no noise nor distortion can be added. Taking water to represent current, pipe processing implies lowering the input impedance of the blocks and increasing the impedance of any element shunting the nodes.

The most widely used pipe processing architecture is shown in fig.7 [16,17]. It is an RF transceiver with a current mode passive mixer connected to a base band virtual ground. This became the standard architecture within Marvell and in many other companies. Focusing on the current mode passive mixer we can see the following key features [17,18,14].

- It is suitable for a CMOS implementation.
- It benefits from technology scaling being immune to the lowering intrinsic gain while gaining from the reduction of the parasitic capacitance.
- It has a very high IIP3 (prompting the use of mixer first topologies) and adds little noise.
- It allows the implementations of high Q filters (n-path like) using its transparency.
- Its power consumption is reduced with scaling.

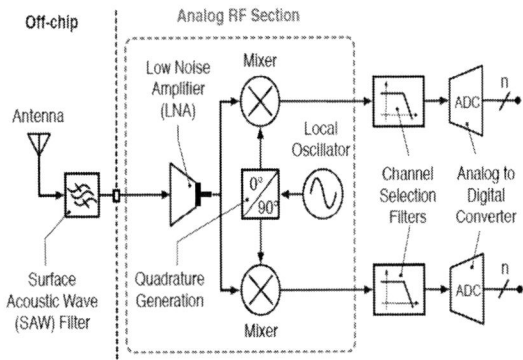

Figure 7: Typical Wireless Receiver suitable for pipe processing operation

The combination of all these advantages made the current mode passive mixer the solution of choice in virtually all integrated transceivers for many product generations. This put an end to another heated discussion on the pros and cons of Gilbert like mixers versus passive ones. The superiority of passive mixers is valid in CMOS but in my opinion remains true also in BiCMOS even if the bipolar device is very suitable for implementing a Gilbert cell [19].

The use of pipe processing via current mode passive mixers was instrumental in making CMOS RF transceivers the universally adopted solution

Unfortunately, not all the blocks in the transceiver of Fig. 7 can operate as a loss-less pipeline. The base band filter needing a shunt path to divert unwonted signals away from

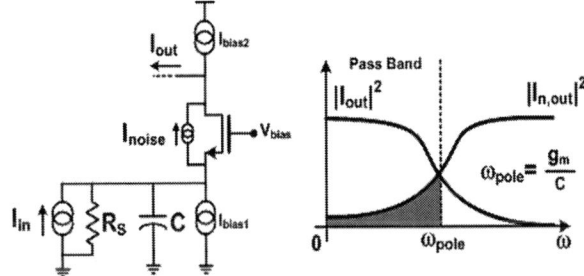

Figure 8: First order gm-C filter with in-band pipe processing operation for noise [20].

the output becomes a leaky pipe. Noise and distortion terms are therefore added to the signal. However, it is possible to minimize the amount of contamination that falls in the signal

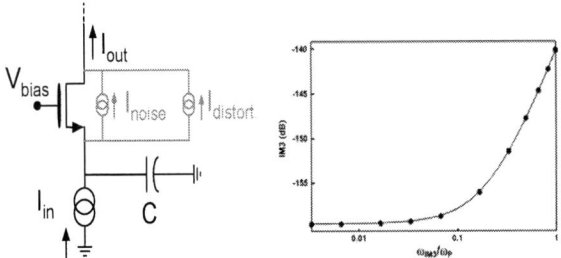

Figure 9: 1ᵗʰ order gm-C pipe filter. Simulated distortion [20]

band preserving SNDR. This is done in the circuit of Fig. 8 that we call pipe filter [20]. Simulation shows that not only the transistor noise but also the distortion components cannot enter the pipe if they fall deeply in band where the ratio between the impedance of the shunt capacitance and that at the source of the transistor is large. Such a behavior is fierily intuitive for noise considering the recirculation of the noise current shown in Fig 9a. It may not be as intuitive for the distortion, but it equally applies with respect to the frequency of the intermodulation product (not of the blocker) as shown in Fig. 9b [20]. We build more complicated pipe filters with second and third order transfer functions [21,22]. For an RF tuner done in cooperation with Marvell we implemented a pipe architecture that include the cascade of several second order pipe filter operated in the current domain as shown conceptually in Fig. 10 [23].

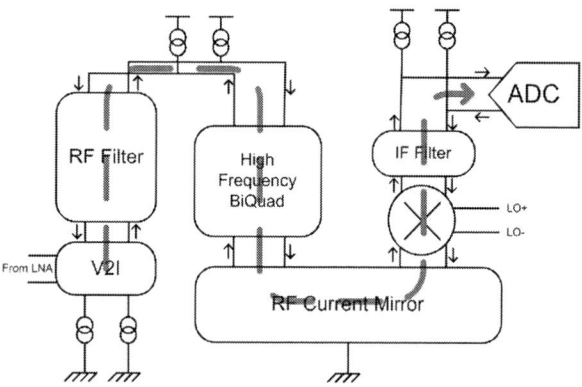

Figure 10: Silicon Tuner based in pipe processing [23].

Within the RF chain a loss-less pipe cannot be implemented also where impedance matching is performed. We addressed this issue with the un-matched LNA shown in Fig. 11 [24]. Although both NF and IIP3 improved with this technique still several issues need to be solved to make it applicable in practical cases. An important one is to ensure that all interconnects on the board are terminated at least at one side.

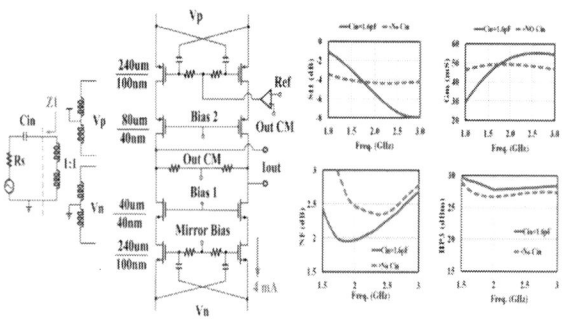

Figure 11: High performance unmatched LNA [24].

Another contribution toward the establishment of CMOS as the dominant technology for RF transceiver was the introduction of the varactor seen between the gate and the shorted source and drain terminal of an MOS transistor. The circuit, shown conceptually in Fig. 12, is the result of a cooperation with ST and was simultaneously proposed also by the group of Prof. Tom Lee at Stanford. [25-28]. The

circuit became soon widespread to the majority of RF transceiver. Unfortunately, the joint patent with ST that was issued on it did not produce the huge return we expected.

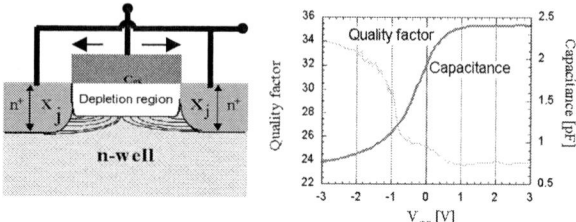

Figure 12: MOS Varactor and its V/C Characteristic [25].

IV. SAW/DUPLEXER LESS TRANSCEIVERS

Another topic that has become very hot in recent times was made possible by the availability of passive mixers i.e., n-path high-Q RF filters. The concept was known for a long time but became feasible at RF frequency with to technology scaling. The be-direction operation of a passive mixer allows to up-convert a base band low pass filter into a high-Q RF band pass filter. This has opened the possibility to reduce the

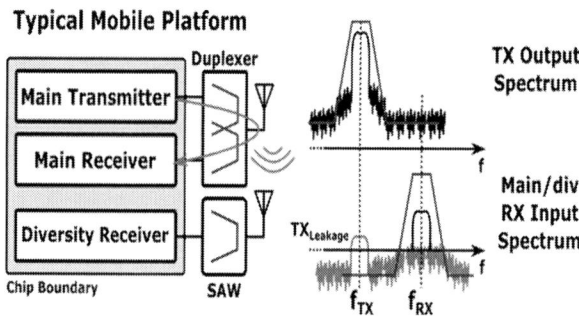

Figure 12: FDD Transceiver with main and diversity path.

number of SAW filters and duplexers that are found before FDD RF transceiver for the main and diversity channel, as shown in Fig. 10. Even more appealing is the possibility of using n-path filters to implement Full Duplex transceivers. [29-32]. We have studied both cases for the main receiver using a Hybrid transformer [33] and for the diversity receiver

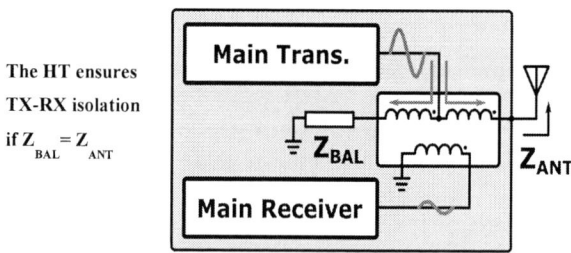

Figure 13: Duplexer-less Main Transceiver with Hybrid Transformer

using self-interference cancellation (SIC) [34,35]. As shown in fig 12 and 13. When using a hybrid transformer for the main receiver there are two main limiting effects [36,37]

1) LNA immunity to the common mode Hybrid leakage
2) the linearity of the impedance used to balance the hybrid.

978-1-6654-8495-4/22 $31.00 © 2022 IEEE   28

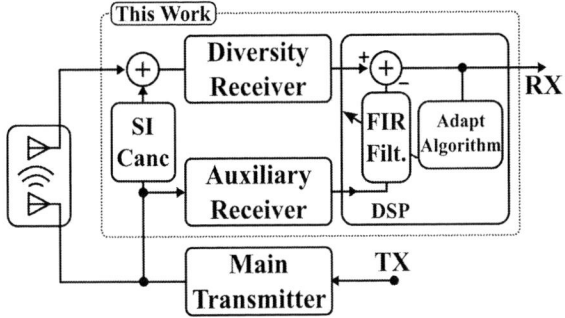

Figure 14: SAW-less Diversity receiver using SI Cancellation and Auxiliary receiver

With respect to 1) we showed that a gate-boosted common gate LNA has a better immunity to the common mode leakage compared to a CS. The result is a much higher IIP3 in the presence of the leakage for the former as shown in Fig. 15. Notice that un unmatched CG was used a discussed above.

With respect to 2) to meet the full specification of a commercial transceiver the balancing impedance (adaptive) must ensure a Hybrid IIP3 of more than 70dBm. This is almost impossible to achieve in a standard technology [38].

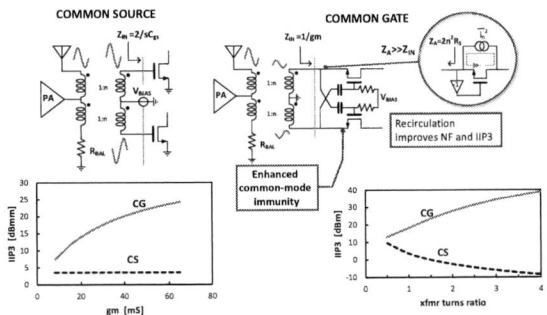

Figure 15: Linearity comparison of CS vs. boosted CG LNA in the presence of a large common mode leakage.

In addition to the use of a hybrid transformer, many other different attempts to eliminate the duplexer have appeared [30,31,39-41]. Among them integrated gyrators implemented using n-path techniques [30,31,40] or leakage cancellation for an integrated PA implemented using a DAC [41]. All these excellent works still miss the most stringent targets by more than 15 dB.

For the case of the auxiliary receiver, we combined a I/Q driven SI canceller with an auxiliary receiver that uses n-path techniques. Even with 25 dB of isolation between the two antennas our solution like all other present in literature does not satisfy the most stringent specification of the standard.

All presented solutions still have not solved a fundamental limitation associated with the reciprocal mixing produced by the phase noise of the clock.

For RF transceivers, there is a critical problem still to be solved to allow removal of the external passives i.e., reciprocal mixing due to phase noise.

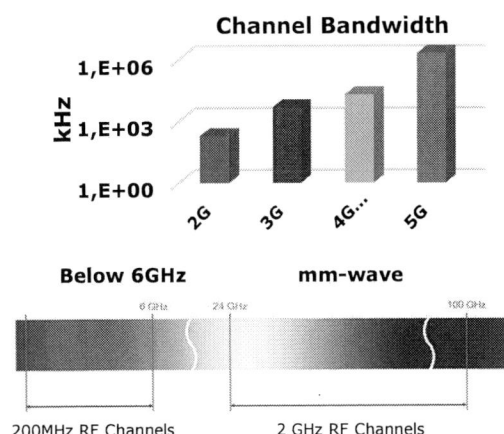

Figure 13 Spectrum and channel evolution with wireless generations.

V. VERY LARGE BANDWIDTH FILTERS

Wireless standard evolution caused proliferation of RF bands together with channel bandwidth extension [28] renewing interest on high bandwidth filters [42,43]. We designed transceivers supporting bandwidths from tens of MHz to over 1 GHz using both mixer-first and LNA first architectures [21,22]. We used both open loop and closed loop filters [7,21,22]. Beyond few hundred MHz we favored open loop solutions although we also designed a first order opamp-RC filter with a bandwidth programmable to up to 1,5 GHz [7].

A key filters specification is robustness to out of band interferers both in term of compression and IIP3. For closed loop filter we studied the effect of the opamp frequency response on IIP3[43] coming to the following conclusion. To mitigate the effect of the non-linearities in the gm and output conductance of the last stage we need to increase the loop gain at the frequency of the IM3. To mitigate the effect of the non-linearities associated with the input stage we need to increase the loop gain at the frequency of the blocker.

A summary of the conclusion of this study is given in Fig.14.

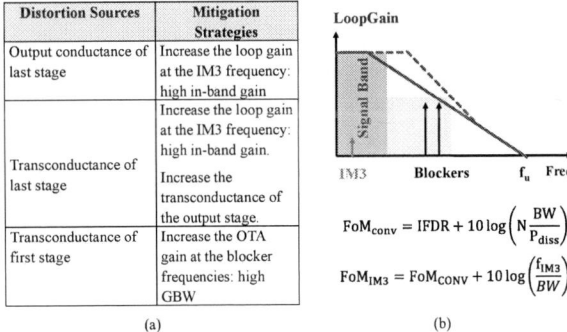

Fig. 14: TIA nonlinearity summary of the conclusions (a) and Figure of Merits definitions (b).

The first two terms in the table tend to dominate just outside the band while the last term tends to dominate far out of band. In general, better linearity requires large GBW but can be

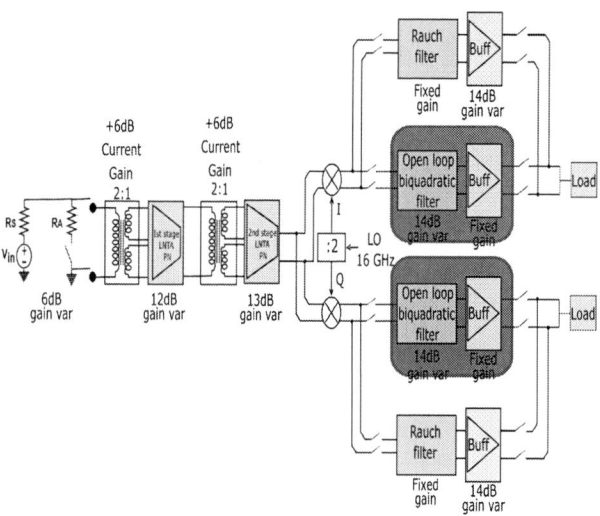

Figure 15: 5G compatible wireless receiver with bandwidth programmable from 50 MHz to 1 GHz [22].

also improved shaping the frequency response as shown in dotted lines in Figure 14b.

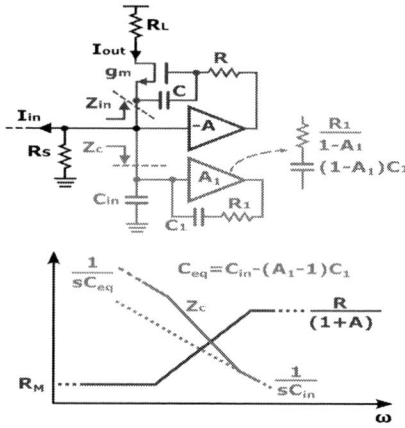

Figure 16: Conceptual diagram of a 3rd order open loop filter for a 260 MHz RF channel [21].

To process all possible bandwidth associate with the 5G standard both below and above the 6 GHz range we built the receiver shown conceptually in Fig. 15 in cooperation with Huawei. The baseband is made of two parallel paths working alternatively. The lower frequency one (up to 400 MHz) uses all close loop filters while the high frequency one (up to 1 GHz) uses an open loop third order filter followed by a first order filter/buffer also intended to drive the following stage. The open loop filter is based on the conceptual diagram shown in Fig. 16 that was used to implements the base band of a mixer first 260 MHz receiver. To extend the bandwidth to a 2 GHz RF channel first the inverting stage -A was substituted with a cross coupled diodes and second a new negative capacitor topology based on positive feedback was implemented. The resulting circuit diagram is shown in Fig. 17. 14 dB of gain control were embedded in the filter acting on the value of the top and bottom current mirrors. The closed loop filter/buffer that follows it can be programmed to have a

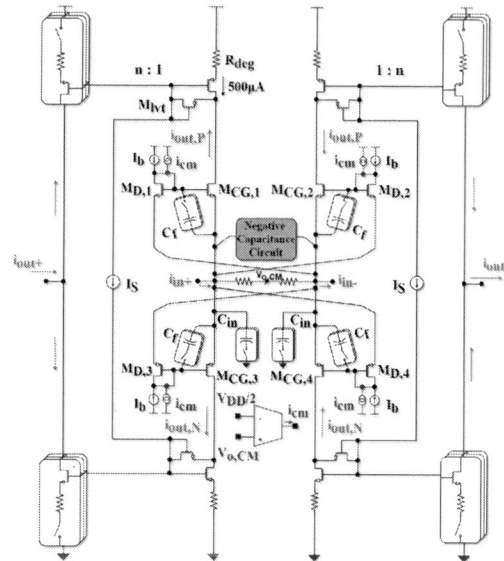

Figure 17 Schematic of 1 GHz 3rd order open loop filter [22].

bandwidth up to 1.5 GHz and uses an opamp to be described next. The most stringent specification to be achieved by the opamp is its gain at the filter band edge. This is achieved in two ways. First its unity gain bandwidth is made as wide as possible for a given power budget while ensuring closed loop stability even in the presence of PVT variations. Second the shape of its frequency response is shaped i.e., a slope of -40 dB/decade occurs from before the band edge to about 1/3 of the unity gain frequency. These features are shown in Figure 18 where the achieved values are also shown.

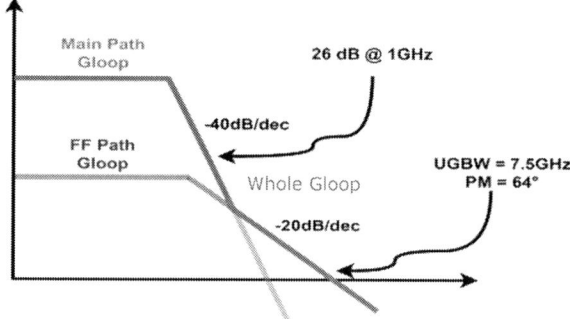

Figure 18: Frequency response of the opamp used in the closed loop filter/buffer.

The presence of multiple poles in band while ensuring closed loop stability can be classically obtained using the multiple feedforward topology shown in Figure 19 for the simple case of a three-stage main path. The number of FF paths is equal to the number of stages in the main path minus one. In this way, each FF stage bypasses two stages avoiding the creation of complex zeros in the open loop response that will produce complex conjugate poles when closing the loop. A fifth order structure with this topology was used in an 8 GHz continuous-time sigma-delta ADC [6]. We have implemented several modifications of the classical topology in cooperation with Huawei including the use of local ac feedback [42].

978-1-6654-8495-4/22 $31.00 © 2022 IEEE   30

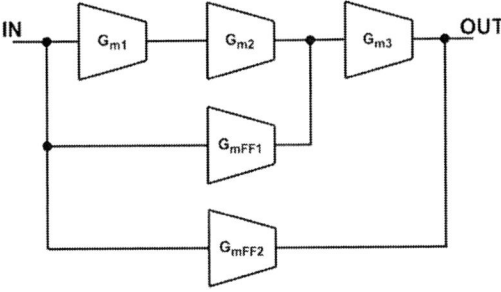

Figure 19: Classical multi-feedforward topology

The actual oapmp topology used in the GHz filter/buffer is shown in Fig. 20. It includes a three stages main path and a two-stage feed forward path. The single stage feed-forward path was eliminated to drastically reduce power. This was made possible by including in the two-stage feed forward an integrated inductor as load of its critical node where the non-dominant pole is located. This technique extends the band and improve the phase margin as shown in Figure 21. The main disadvantage of inductor peaking is size. To save area we have constructively coupled the two pairs of inductors. Since the area is proportional to the inductance, the area penalty decreases increasing the unity gain bandwidth. We expect that in future technology higher bandwidth will be achieved which will require the use of smaller inductors making their area overhead progressively less important.

Another topic of discussion is the evolution of the analog characteristics of CMOS with aggressive scaling. The move from classical planar transistor to three-dimensional structure like Fin-FET aims at increasing the density of digital circuit while preserving the Ion/Ioff ratio of digital gates. On the

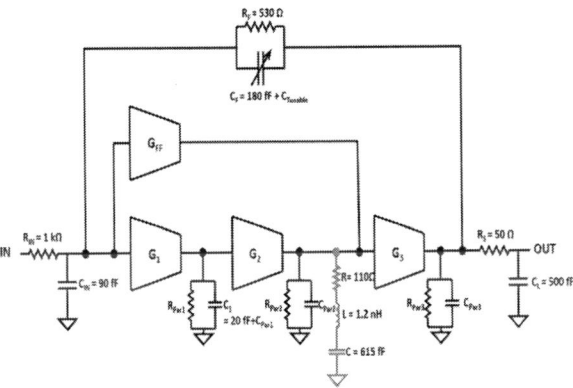

Figure 20: Topology of the feed-forward amplifier [7].

Closed loop circuits (e.g. filter) will be used to process multi GHz signal thanks to the availability of tens of GHz multiple feedforward opamp.

other hand, the evolution of analog parameters with scaling for a Fin FET is less clear. For instance, while the intrinsic gain of a minimum size device increases the same may not be true for the fT. In some papers it was claimed that transistor fT vs. technology will display a peak (corresponding to 28 nm planar CMOS) and after will start to decrease [..]. In our experience this is not the case if a custom device transistor structure is used as opposed to the standard one available in

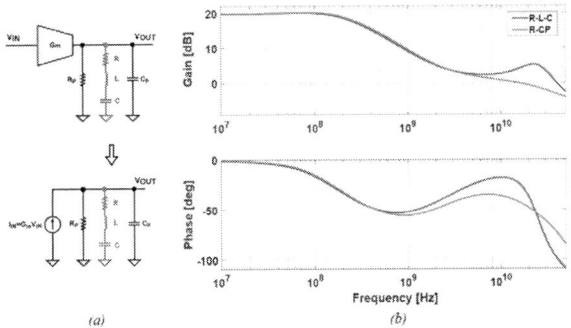

Figure 21: Effect of the integrated inductor on the amplifier open loop response [7].

the digital library. The critical point is to increase the poly pitch along the fin to reduce source access resistance. In this case a monotonically increasing fT is seen performing accurate simulation that include the effect of layout interconnect and parasitic. Our conclusion is that deeply scaled three-dimensional technology will remain competitive in applications where maximizing signal bandwidth is the primary goal like in fiber optic interfaces.

Deeply scaled down Fin-FET technologies using transistor optimized for analog performance will stay competitive even in high-speed circuit applications.

VI. CONCLUSIONS AND RECCOMANDATIONS

Having witnessed analog design evolution over the last 40 years I summarize below my conclusions

- CAD tools evolution benefit the analog designer increasing the chance of first design success
- Technology evolution benefits all aspect of analog design including the maximum achievable speed.
- Analog design automation will not become widespread as long as technology follows Moor law.
- CMOS implementation of a system is a condition to make it into a mass market product
- Analog library must constantly evolve to be useful.
- Circuit techniques previously unfeasible become feasible with technology evolution.
- Analog design is unstructured, ingenuity, creativity and intuition still define its quality.
- Topologies must match key technology feature.
- Solve the key problem to make the next big leap.

I would like also to leave the following recommendations

- Use intuitive understanding (but not oversimplify)
- Learn from other applications
- Leverage technology evolution (but stay critical)
- Broaden know-how, be visionary (but stay critical)

Most of all: always measure yourself with key principles

- You are your true judge and cannot cheat yourself
- In the long run the truth will emerge
- Teach with your example
- Credit others, value constructive criticism
- Understand other people point of view

REFERENCES

[1] R. Castello, P.R. Gray (1985). A high-performance micropower switched-capacitor filter. IEEE Journal of Solid-State Circuits pp. 1122-1132 Vol. 20.

[2] R. H. Dennard; F. H. Gaensslen ; V. L. Rideout ; E. Bassous ; A. R. LeBlanc ,"Design of Ion-implanted MOAFETS with very small physical dimensions in IEEE Journal of Solid-State Circuits, vol. 9, no 5, pp. 256-268, October 1974.

[3] Anne-Johan Annema, "Analog Circuit Performance and Process Scaling," in *IEEE Transactions on Circuits and Systems II Analog and Digital Signal Processing*, VOL. 46, NO. 6, JUNE 1999.

[4] D. Hodges, P. Gray, R. Brodersen, "Potential of MOS Technologies for Analog Integrated Circuits," Solid-State Circuits, IEEE Journal of, vol.13, no.3, pp.1770-1780, June 1978.

[5] P. W. Li, M. Chin, P.R. Gray, R. Castello, "A Ratio Independent Algorithmic Analog to Digital Converter", IEEE Journal of Solid State Circuits, vol. SC-19, pp. 828-836 , December 1984.

[6] Y. Dong et al.,"A 72 dB-DR 465 MHz-BW Continuous-Time 1-2 MASH ADC in 28 nm CMOS," in IEEE Journal of Solid-State Circuits, vol. 51, no. 12, pp. 2917-2927, Dec. 2016.

[7] N. N. Miral, D. Manstretta and R. Castello, "A 17 mW 33 dBm IB-OIP3 0.5-1.5 GHz Bandwidth TIA Based on an Inductor-Stabilized OTA," ESSCIRC 2021 - IEEE 47th European Solid State Circuits

[8] L. Franks and I. Sandberg, "An alternative approach to the realizations of network functions: The N-path filters," Bell Syst. Tech. J., pp. 1321–1350, Sep. 1960.

[9] M. Darvishi, R. Van der Zee, E. Klumperink, and B. Nauta, "A 0.1-to-1.2 GHz tunable 6th-order N-path channel-select filter with 0.6 dB passband ripple and +7 dBm blocker tolerance," in IEEE Int. Solid-State Circuits Conf. (ISSCC) Dig. Tech. Papers, 2013, pp. 171–173.

[10] S. Pernici, G. Nicollini, R. Castello. A CMOS low-distortion fully differential power amplifier with double nested Miller compensation. IEEE Journal of Solid-State Circuits pp. 758-763 Vol. 28 1993.

[12] J. Solomon, "Computer Based Design for Tomorrow Soer Chips," in IEEE International Solid-State Circuits Conference (ISSCC) Digest of Technical Papers, 1986.

[13] Alini, R.; Baschirotto, A.; Castello, R.; , "Tunable BiCMOS continuous-time filter for high-frequency applications," *Solid-State Circuits, IEEE Journal of* , vol.27, no.12, pp.1905-1915, Dec 1992

[14] A.A. Abidi, "RF CMOS comes of age," in IEEE Journal of Solid-State Circuits, vol. 39, no. 4, pp. 549-561, April. 2004.

[15] T.H. Lee, " CMOS RF': (still) no longer an Oxymoron " in 1999 IEEE Radio Frequency Integrated Circuits Symposium.

[16] Sacchi, E. et al..; , "A 15 mW, 70 kHz 1/f corner direct conversion CMOS receiver," *Custom Integrated Circuits Conference. Proceedings of the IEEE*, vol., no., pp. 459- 462, 21-24 Sept. 2003.

[17] Valla, M.; Montagna, G.; Castello, R.; Tonietto, R.; Bietti, I.; , "A 72-mW CMOS 802.11a direct conversion front-end with 3.5-dB NF and 200-kHz 1/f noise corner," *Solid-State Circuits, IEEE Journal of*, vol.40, no.4, pp. 970- 977, April 2005

[18] D. Leenaerts andW. Redman-White, "1=f noise in passive CMOS mixer for low and zero IF integrated receivers," in Proc. 27th Eur. Solid-State Circuits Conf., Villach, Austria, 2001, pp. 103–107.

[19] B. Gilbert, " A precise four-quadrant multiplier with sub nanosecond response," in IEEE Journal of Solid-State Circuits. Vol 3 no 4 1968.

[20] Pirola A., Liscidini A., Castello R., "Current-Mode, WCDMA Channel Filter With In-Band Noise Shaping". IEEE Journal of Solid-State Circuits, vol. 45, no. 6, pp. p. 1770 - 1780, June 2010.

[21] G. Pini, D. Manstretta and R. Castello, "A 260-MHz RF Bandwidth Mixer-First Receiver With Third-Order Current-Mode Filtering TIA," in IEEE Solid-State Circuits Letters, vol. 2, no. 9, pp. 183-186, Sept. 2019.

[22] K. Sohal, D. Manstretta and R. Castello, "A 2nd Order Current-Mode Filter with 14dB Variable Gain and 650MHz to 1GHz Tuning-Range in 28nm CMOS," 2021 IEEE International Symposium on Circuits and Systems (ISCAS), Daegu, Korea, 2021, pp. 1-5.

[23] Greenberg, J et al.; , "A 40MHz-to-1GHz fully integrated multistandard silicon tuner in 80nm CMOS," Solid-State Circuits Conference Digest of Technical Papers (ISSCC), 2012, vol., no., pp.162-164, 19-23 Feb. 2012

[24] E. Kargaran, D. Manstretta and R. Castello, "A 1.5–2.8 GHz current-mode LNTA achieving >25 dBm IIP3 and +8 dBm P-1dB gain compression", in Microelectronics Journal, vol. 86, no. 4, pp. 57-64, April 2019.

[25] F. Svelto, P. Erratico, S. Manzini, and R. Castello, "A metal oxide semiconductor varactor," IEEE Electron Device Lett., vol. 20, pp. 164–166, Apr. 1999.

[26] F. Svelto, S. Manzini, and R. Castello, "A Three Terminal Varactor for RF IC's in Standard CMOS Technology," IEEE TRANSACTIONS ON ELECTRON DEVICES, VOL. 47, NO. 4, April 2000.

[27] T. Soorapanth, C. P. Yue, D. R. Shaeffer, T. H. Lee, and S. S. Wong, "Analysis and optimization of accumulation-mode varactor for RF ICs," in 1998 Symp. VLSI Circuit Dig. Tech. Papers, 1998, June, pp. 32–33.

[28] P- Andreani e S. Mattisson, "On the use of MOS varactors in RF VCOs," IEEE Journal of Solid-State Circuits, 2000.

[29] "3GPP TS 38.101-2 version 16.1.0 Release 16". Retrieved 2021-01.

[30] D. Bharadia, E. McMilin, and S. Katti, "Full duplex radios," in Proc. ACM SIGCOMM Conf., New York, NY, USA, 2013, pp. 375–386.

[31] N. Reiskarimian, M. B. Dastjerdi, J. Zhou, and H. Krishnaswamy, "Highly-linear integrated magnetic-free circulator-receiver for fullduplex wireless," in IEEE Int. Solid-State Circuits Conf. (ISSCC) Dig.Tech. Papers, Feb. 2017, pp. 316–317.

[32] D. Yang, H. Yüksel, and A. Molnar, "A wideband highly integrated and widely tunable transceiver for in-band full-duplex communication" IEEE J. Solid-State Circuits, vol. 50, no. 5, pp. 1189–1202, May 2015.

[33] I. Fabiano, M. Sosio, A. Liscidini, R. Castello . SAW-less analog front-end receivers for TDD and FDD2013 IEEE International Solid-State Circuits Conference Digest of Technical Papers. 2013 IEEE International Solid-State Circuits Conference Digest of Technical Papers, pp. 82-83.

[34] E. Kargaran, S. Tijani, G. Pini, D. Manstretta and R. Castello, "Low Power Wideband Receiver with RF Self-Interference Cancellation for Full-Duplex and FDD Wireless Diversity", IEEE Radio Frequency Integrated Circuits (RFIC) Symposium 2017, Honolulu, Hawaii, 04/06/2017.

[35] D. Montanari, et al., "An FDD Wireless Diversity Receiver With Transmitter Leakage Cancellation in Transmit and Receive Bands," IEEE Journal of Solid-State Circuits, vol. 53, no. 7, pp. 1945-1959, July 2018.

[36] M. Mikhemar, H. Darabi, and A. Abidi, "A tunable integrated duplexer with 50 dB isolation in 40 nm CMOS," in IEEE Int. Solid-State Circuits Conf. (ISSCC) Dig. Tech. Papers, Feb. 2009, pp. 386–387.

[37] M. Mikhemar, H. Darabi, and A. Abidi, "An on-chip wideband and low-loss duplexer for 3G/4G CMOS radios," in Proc. IEEE Symp. VLSI Circuits, Jun. 2010, pp. 129–130.

[38] B. van Liempd *et al.*, "A +70-dBm IIP3 electrical-balance duplexer for highly integrated tunable front ends," *IEEE Trans. Microw. Theory Techn.*, vol. 64, no. 12, pp. 4274–4286, Dec. 2016.

[39] T. Zhang, C. Su, A. Najafi, and J. C. Rudell, "Wideband Dual-Injection Path Self-Interference Cancellation Architecture for Full-Duplex Transceivers" in IEEE Journal of Solid-State Circuits, vol. 53, no. 6, pp. 1563-1576, June 2018.

[40] J. Zhou, N. Reiskarimian, and H. Krishnaswamy, "Receiver with integrated magnetic-free N-path-filter-based non-reciprocal circulator and baseband self-interference cancellation for full-duplex wireless," in IEEE Int. Solid-State Circuits Conf. Dig. Tech. Papers, Jan. 2016, pp. 178–180.

[41] L. Calderin, S. Ramakrishnan, A. Puglielli, E. Alon, B. Nikolic ´, and A. M. Niknejad, "Analysis and design of integrated active cancellation transceiver for frequency division duplex," *IEEE J. Solid-State Circuits*, vol. 52, no. 8, pp. 2038–2054, Aug. 2017.

[42] Salehi M. D., Manstretta D., Castello R., "A 150 MHz TIA with Unconventional OTA Stabilization Technique via Local Active-Feed-Back" 15th Conference on Ph.D. Research in Microelectronics and Electronics (PRIME) 2019.

[43] Castello, R.; Bietti, I.; Svelto, F.; , "High-frequency analog filters in deep-submicron CMOS technology," 15th Conference, Digest of Technical Papers. ISSCC IEEE International, pp.74-75, 1999

[44] J. Lechevallier, et al. "5.5 A forward-body-bias tuned 450MHz Gm-C 3rd-order low-pass filter in 28nm UTBB FD-SOI with >1dBVp IIP3 over a 0.7-to-1V supply," 2015 IEEE International Solid-State Circuits Conference - (ISSCC) Digest of Technical Papers, 2015, pp. 1-3.

[45] G. Pini, D. Manstretta and R. Castello, "Analysis and Design of a 20-MHz Bandwidth, 50.5-dBm OOB-IIP3, and 5.4-mW TIA for SAW-Less Receivers," in IEEE Journal of Solid-State Circuits, vol. 53, no. 5, pp. 1468-1480, May 2018, doi: 10.1109/JSSC.2018.2791489.

[46] V. Subramanian et al."Planar Bulk MOSFETS Versus FinFETs: An Analog/RF Perspective," IEEE TRANSACTIONS ON ELECTRON DEVICES, VOL. 53, NO. 12, DECEMBER 2006.

From Less Batteries to Battery-Less: Enabling A Greener World through Ultra-Wide Power-Performance Adaptation down to pWs

Massimo Alioto, *Fellow, IEEE*

Abstract— **This paper presents a holistic perspective on recent advances in silicon systems for distributed and decentralized systems (e.g., IoT, AIoT), whose count is trending towards the trillions exponentially. At such unprecedented scale, batteries impose severe scaling limitations in terms of form factor, system lifetime and uptime, sensor node cost, system cost (e.g., due to battery/sensor node replacement), and a heavy environmental impact at disposal time. Accordingly, sustained growth in the number of connected devices mandates drastic battery shrinking and ultimately elimination at scale.**

Beyond the straightforward reduction in the battery utilization via long sleep modes, this paper focuses on the prominent class of miniaturized (mm-scale) battery-indifferent and battery-less systems. Recent and unfolding design techniques are presented to enable purely-harvested operation, while still maintaining sensor nodes alert/available, long lived (e.g., beyond the battery shelf life), millimeter-sized and low cost. This paper and its keynote speech companion at ESSCIRC discusses how to achieve such goals through system peak power adaptation down to the nW level, rather than conventional average. Fundamental principles are demonstrated on silicon across all major sensor node sub-systems (e.g., processing, sensor interfaces, wireless communications).

Overall, the design techniques described in this keynote paper aim to enable next-generation ubiquitous, low-cost, miniaturized yet alert sensor nodes for sustainable scale-up. From a technological viewpoint, the resulting advances aim to make scaling to the trillions economically and logistically sustainable. Even more importantly, they are crucially needed to make trillion-scale systems environmentally sustainable, addressing the gargantuan threat posed by trillions of batteries lying ahead, from production to disposal. Beyond the well-established notion of VLSI systems on a chip, sustainability in very large-scale (distributed/decentralized) systems really starts from design.

Index Terms— **Internet of Things, sustainability, circular economy, ultra-low power, battery-less systems, battery-indifferent systems, green technologies, always-on**

I. INTRODUCTION

THE increasingly data-driven nature of today's applications has led to a rapidly growing demand for distributed (e.g., IoT) and decentralized (e.g., AIoT) systems leveraging large-scale deployment of connected integrated systems, which perform physical data acquisition, processing and sensemaking on chip [1]-[4]. Such demand is fostering an exponential growth in the number of connected "edge" devices, and is trending towards the trillion scale in over a decade, as in Fig. 1 [5]. Such growth is being fueled by several applications ranging from

[1] For example, supercapacitors are well known to be unsuited for mm-scale harvested systems, due to their very high leakage [2].

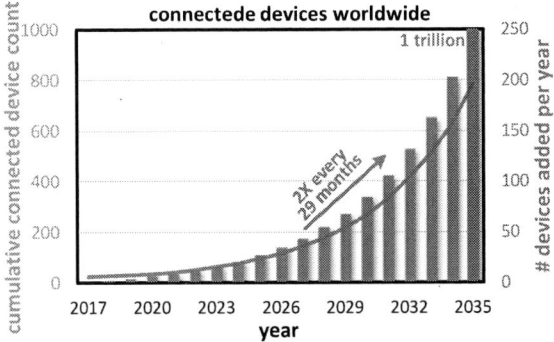

Fig. 1. Trend of IoT devices added every year and cumulative count vs. year towards the trillion scale [5].

Industry 4.0 to smart cities, buildings and homes, intelligent warehouses, smart logistics, and many others.

The semiconductor industry is not new to the scale of trillions per se, and has passed the trillion mark in terms of chips delivered worldwide at the end of 2021, reaching 1.15 trillions semiconductor units shipped during the year [6]. What is peculiar in the prospective achievement of a trillion connected devices lies in the word "connected", as their physical isolation in large-scale deployments requires full autonomy from any power or communication wired grid, and commonly from a computational perspective as well (edge computing). Edge devices indeed need to be unobtrusive (mm scale) and hence untethered, low cost (sub-$), long lived (decade-range).

A. Sustainability Challenges towards the Trillion Scale

From the above considerations, sustaining the production of a trillion devices in a decade is certainly within reach. Sustainability is instead challenged by cost, logistics, and environmental threats that the exponential trend poses on other components, and in particular batteries [1], [2]. Next-generation single-chip sensor nodes with mm-scale form factor (<<1$ cost) require batteries with comparable size (cost), although most today's sensor nodes employ much larger (and expensive) batteries in the centimeter (dollar) range [3]. These properties along with the decade-long battery shelf-life requirement make solid-state micro-batteries the only battery option, ruling out other classes of energy storage for several reasons[1] [2], [3].

Also, mm-scale battery capacity is in the μW·h range, making harvesting necessary. Unfortunately, manufacturing a trillion batteries lacks sustainability on several respects and at all stages of the battery lifecycle (Fig. 2):

- MANUFACTURING: at current extraction rates, a trillion small-sized batteries would take three years of full-capacity lithium extraction within the next decade [6], which is clearly unsustainable from a *logistical* viewpoint even in a well-functioning supply chain
- REPLACEMENT: replacing a battery to prolong the lifespan of edge devices entails a much higher cost than the node and the battery itself, making post-ownership cost dominant as publicly admitted by energy storage vendors [8], [9]. The cost of battery replacement ranges from tens of dollars to a few hundreds [10], [11]. Assuming optimistically a cost per battery replacement of 20$ and an average battery life of five years (i.e., 20% replaced every year), replacing batteries at the trillion scale would require 4 T$/year, which would correspond to the GDP of a hypothetical fourth country after US, China and Japan as of 2022. At an average time of 15 minutes per battery replacement, this would take 17 million people employed full time worldwide. This is clearly *economically* unfeasible and unsustainable.
- DISPOSAL: a trillion coin batteries disposed every five years would be sufficient to replace the stone of 500 replicas of the Colosseum in Rome (i.e., 50 Mm³). This is clearly *environmentally* unsustainable, due to the gigantic environmental impact at scale.

From the above considerations, it becomes clear that the limitations imposed by batteries are a major obstacle to the growth of the IoT due to lack of *economic*, *logistical* and *environmental* sustainability (Fig. 3). These challenges add to the scaling challenges that the semiconductor industry has been used to address over the decades, as related to cost, energy, area (density) and performance. This justifies the battery-less sensor offering and roadmap of several semiconductor companies with a 32 B$ market size in 2021, and a projected market growth beyond 100 B$ by 2026 at a robust CAGR of 27.6% [12].

B. Sustainability Challenges from a Scaling Trend Viewpoint

The battery size challenges at a given capacity are further aggravated by slower Li-ion battery scaling trends, compared to semiconductors. Indeed, energy density improvements are in the range of 2-2.5X/decade [13] and hence much slower than energy improvements in general-purpose (100X/decade) and digital signal processors (110X/decade), analog-digital circuits (17X/decade), and short-range radios (19-34X/decade) among the others [14].

From a cost perspective, prospective improvements in semiconductors and batteries are quantified by the learning

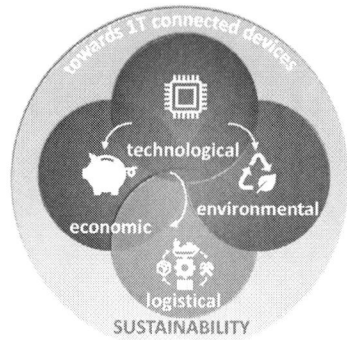

Fig. 3. Dimensions of sustainability challenges in distributed and decentralized systems, and focus of this work on advancing design to enable sustainability on all other dimensions.

curves of the related industries [15]. The learning curve of an industry expresses the historical cost reductions per unit as a function the cumulative production over a given period of time, and is typically exponential [1], [16]. The learning rate is the percentage price/unit reduction achieved when the production volume is doubled. Hence, a higher learning rate (i.e., slope of the curve) is indicative of a stronger ability to translate the accumulated knowledge and improvements in the overall design/production process into cost savings [15].

From Fig. 4a, the learning rate of the semiconductor industry over its lifespan is 55%, and is one of the very highest across industries. In contrast, from Fig. 4b the learning rate in lithium batteries is 20% [17], [18], and hence leads to much lower cost savings for a given cumulative production increase. The much faster production rate of the semiconductor industry (compare Figs. 4a-b) and the compounding effect of the learnings over production cycles determine an exponential divergence between silicon chip and battery price. Further cost scaling challenges in lithium batteries are expected due to the emergence of other less-scalable cost sources in battery manufacturing [19]. Such factors further degrade the economic sustainability of lithium batteries for edge nodes, making it a major obstacle in the scale-up towards trillions of connected devices.

This work provides a perspective on how to address the above challenges through circuit and architectural advances, as key enabler of *economic*, *logistical* and *environmental* sustainability (see Fig. 3). In the following sections, design principles and their silicon demonstrations from sub-systems to complete systems are illustrated to remove the sustainability roadblocks imposed by batteries, while still keeping the edge alert, mm-sized, low-cost and long-lived. In the following, the text is enriched with **QR codes pointing at our lab demonstration videos** showing the related silicon testchips.

II. BATTERY-INDIFFERENT AND BATTERY-LESS SYSTEMS

In general, addressing the above form factor and cost challenges is in direct contrast with the preservation of system lifetime and uptime [2] (i.e., system availability over time). Indeed, form factor and cost are immediately reduced via

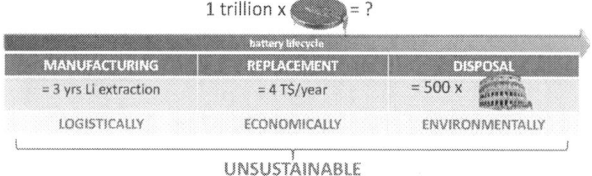

Fig. 2. Sustainability threats posed by batteries across their lifecycle.

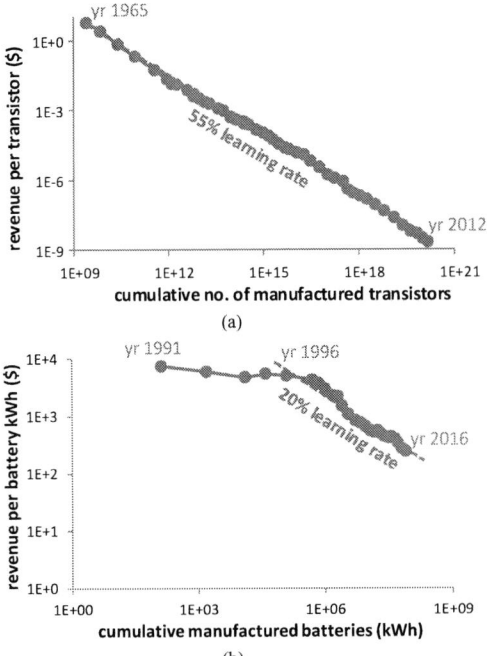

Fig. 4. Learning curve and learning rate for transistor [1], [16] and lithium battery production [18].

aggressive battery shrinking, or battery elimination as its limit case. However, the resulting more frequent battery outages (permanent in the case of battery elimination) inevitably lead to forced shutdown and missed sensing events when the instantaneous power exceeds the harvested power, thus degrading system availability over time.

Interestingly, all the above form factor, cost, lifetime and uptime goals can be simultaneously achieved by applying the following two design principles:

- **shrinking or eliminating the battery** for form factor and cost reduction
- **constantly adapting the system power to regularly fit the instantaneous harvested power**, and hence restore system availability even without battery support.

Under the above principles, system shutdowns are clearly prevented by adaptively down-scaling system power according to the available harvester power under the inevitably wide fluctuations in the environmental conditions (e.g., light intensity in a light harvester). Unfortunately, the vast majority of mm-scale harvesters typically provides a power in the nW to tens of nWs range across the inevitable fluctuations of available environmental sources [1], [2], [21]. Accordingly, the silicon system needs to exhibit an **ultra-wide power scaling down to** the aggressively low **nW-range** target [3]. At the same time, such ultra-wide power scaling needs to be free from any rigid lower bound (i.e., power floor P_{floor}) imposed by circuits or sub-systems consuming a fixed power. In other words, sustained operation at less favorable environmental conditions takes place as long as the **system power floor is pushed below the minimum harvested power** $P_{harvester,min}$, across the considered environmental conditions (see Figs. 5a-b).

From a power scheme viewpoint, the above principles certainly enable sustainable operation of **battery-less** system architectures over time, in which power fluctuations are essentially seamlessly absorbed through wide power adaptation and power floor reduction [2]. At a higher level of flexibility, the fulfilment of the same design principles enable the recently proposed class of **battery-indifferent** system architectures [2], [3], which include both a highly shrunk battery and a harvester. Battery-indifferent systems have the additional capability to operate either in a battery-powered fashion whenever the battery stores adequate charge to support the system, or in a sustained purely-harvested manner whenever the battery runs out of energy [20] (Fig. 6).

Given the diversity of considerations and goals, the above design principles for battery-indifferent and battery-less systems are described in the following for all the different sub-systems that sensor nodes typically comprise.

III. PROCESSING SUB-SYSTEM

The typical leakage power of processors and microcontrollers vastly exceeds the power delivered by harvesters under less favorable environmental conditions [2], [21]. Indeed, typical processors for IoT (e.g., ARM Cortex-M0) comprise tens of a few hundreds of logic gates [3], [22]. This requires an overall power of less than 0.1 pW/gate to keep its overall power in the nW range, which is clearly well below the leakage power of logic gates even in older technology generations (e.g., 10 pW/gate in 180nm). Accordingly, mm-sized harvesters cannot support even the processor leakage, which requires the processor to consume less than inherent

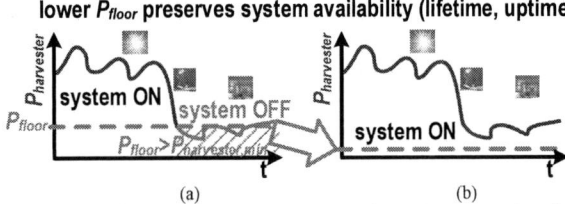

Fig. 5. a) Forced system shutdowns under pure harvesting occur when the system power floor P_{floor} becomes higher than the available harvested power $P_{harvester}$. b) System uptime is preserved by pushing P_{floor} below $P_{harvester,min}$, as requisite of battery-indifferent and battery-less systems.

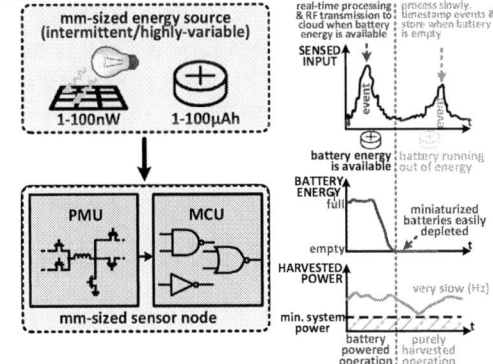

Fig. 6. Battery-indifferent systems can sustainably operate under miniaturized batteries by switching to purely-harvested mode whenever the battery runs out (frequently, in view of aggressive battery shrinkage).

transistor leakage. In other words, **sub-leakage operation** is mandated in the processing sub-system.

Sub-leakage operation was demonstrated in nW/sub-nW processors whose clock rate is restricted to the Hz range [23], [24]. Although such clock frequencies can be still useful in many applications (see Section VII), the inability to scale up performance at higher available power severely limits the applicability of a design and hence the necessary manufacturing volume and economies of scale required by low-cost edge systems. Hz-range performance inflexibility also limits the overall power efficiency, as the system becomes unable to exploit any increase in the available power [2].

From the above considerations, techniques are needed to extend the power-performance tradeoff towards the nW range (**minimum power point, MP**), while retaining the advantages of CMOS logic in terms of **minimum energy point** (MEP), **minimum clock cycle point** (MEP, achieved at nominal voltage), or any intermediate operating point. This is achieved by introducing the dual-mode logic style in Figs. 7a-c [22]. In pseudo-CMOS mode, the system voltage ΔV is set to turn the header/footer transistors M5-M6 in Fig. 7c completely on. In this mode, dual-mode logic gates operate like conventional standard CMOS logic, as desired to achieve from minimum energy to maximum performance point via conventional voltage scaling. In minimum power mode, ΔV is set to turn M5-M6 off, making the dual-mode logic gates operate like sub-leakage logic as in Fig. 7b [23], [24], and hence achieve the nW goal (minimum power point). Furthermore, setting ΔV at intermediate values covers all intermediate power-performance tradeoff from minimum power to maximum performance via the minimum energy point.

Dual-mode logic uniquely enables a very wide power-performance range [22], [25], as shown in Fig. 8. As an example, from Table I a 180-nm ARM Cortex-M0 in dual-mode logic at the minimum power mode achieves a power as low as the sub-leakage level of 595 pW, which is ~330X lower than the inherent processor leakage in standard CMOS logic [20] (i.e., its minimum power mode). Such power is on par with sub-leakage logic styles [23], [24]. The clock frequency at the minimum power point is 2.45 Hz, which is expectedly in line with other sub-leakage processors in Fig. 8. Further considerations on the applications that can fully or partially exploit such wide scalability are discussed in the final section. At the minimum energy point, the energy per cycle of 33 pJ is expectedly much lower than the energy at the minimum power point by 8.2X [22]. This confirms that the minimum

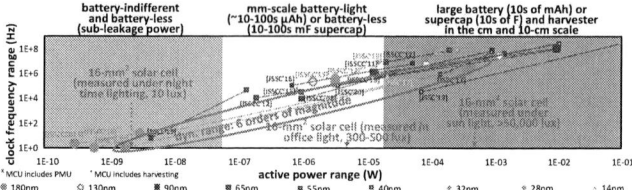

Fig. 8. Power-performance dynamic range in state-of-the-art low-power processors for the edge. Dual-mode logic reaches the widest dynamic range of six orders of magnitude [20].

power and the minimum energy point are associated with very different power-performance tradeoffs, and they cannot be achieved simultaneously. The power-performance flexibility of dual-mode logic allows to achieve either one at any point of time, as opposed to voltage scaling whose power floor is lower bounded by leakage. At the minimum cycle point, the processor has a maximum frequency of 2.8 GHz, which is again adequate to support a wide range of applications (Section VII) and in line with the majority of processors for the edge in Fig. 8. Such wide scalability allows the processor to uninterruptedly work in a battery-indifferent or battery-less fashion under a 4 mm × 4 mm commercial solar cell from sun light (minimum cycle mode) down to near-dark under lux-range light intensity (minimum power mode) [22]. Even using a low-efficiency 0.5-mm² on-chip solar cell for extreme miniaturization and low cost, the processor continues to work uninterruptedly down to 55 lux (i.e., late twilight).

In summary, dual-mode logic enables the necessary ultra-wide scaling and power floor reduction well below leakage, as needed to achieve uninterrupted battery-indifferent or battery-less processor operation under most practical environmental conditions for sustainable, mm-scale, low-cost edge systems with uncompromised lifetime and uptime (Section II).

IV. POWER MANAGEMENT

The design principle in Section II and the resulting power-performance design targets in Section III need to be supported by the system power management unit (PMU). Indeed, the latter needs to cover the entire power-performance range from the MP to the MC point, and any intermediate point from below-leakage points (MP+) via ΔV tuning to ME and above via supply voltage scaling [26] (see Fig. 9).

To achieve the above goals, the wide fluctuations of the harvested power need to be inexpensively and continuously tracked by the PMU for automatic tuning of the power-performance tradeoff. In general, the PMU needs to encompass adaptation through conventional voltage scaling (blue curve in Fig. 9) and dual-mode logic ΔV (red curve), as discussed in

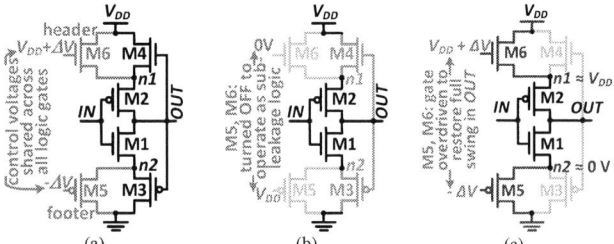

Fig. 7. a) Dual-mode inverter gate, b) its operation in minimum power mode (sub-leakage logic), c) its operation in pseudo-CMOS mode.

TABLE I. ARM CORTEX-M0 OPERATING POINTS IN DUAL-MODE LOGIC [22] (NORMALIZED TO STD CMOS LOGIC AT SAME POINT IN PARENTHESES)

	power	energy/cycle	clock frequency
min. power	595 pW (0.005X)	272 pJ (5.2X)	2.45 Hz (0.0029X)
min. energy	244 nW (0.6X)	14 pJ (1.03X)	17.4 kHz (0.58X)
min. clock cycle	420 µW	150 pJ	2.8 MHz

Section III. In addition, the PMU power floor needs to be drastically reduced as discussed in Section II. In particular, the power floor of conventional PMUs is set by the DC-DC conversion, which ultimately limits system power scalability as summarized in Fig. 10. The most direct method to remove such power contribution to the system power floor is to **suppress DC-DC conversion** altogether when it most matters, i.e. when the system is operating **around the MP point**. This means that the traditional PMU power modes need to be expanded to include an MP mode that enables **direct harvesting**, shutting down any DC-DC conversion to eliminate the related power floor.

Under direct harvesting, the system supply voltage is connected directly to either a DC harvester (e.g., solar cell, thermoelectric generator), or the rectifier of an AC harvester (e.g., vibration, RF). The following considerations focus on DC harvesters, and are immediately extended to AC harvesters by including the intermediate rectifier. As main challenge, direct harvesting exposes the system to the harvester voltage fluctuations due to environmental variations. Hence, it requires ultra-low voltage system operation so that its minimum feasible operating voltage V_{min} is lower than the harvested voltage at the worst-case condition targeted [3] (e.g., lowest light intensity for a solar cell). In other words, **ultra-low voltage operation** is not really required to reduce the system power, but to instead **suppress the extra power floor contribution of DC-DC conversion and voltage regulation**. For example, commercial solar cells for edge systems generate voltages ranging from 0.3 V and 0.6 V for light levels from near-dark to sun light [27]. This requires the system to operate uninterruptedly down to very low harvested voltages as illustrated in Fig. 11, in addition to low harvested power levels. As side benefit of dual-mode logic, V_{min} is pushed down to voltages around 0.2 V in minimum power mode [27], as relevant to lower harvested power and voltage levels. This makes dual-mode logic under direct harvesting robust against environmental fluctuations.

In addition to the above discussed direct harvesting requirement in MP mode, battery-indifferent and battery-less PMUs need to extend the power-performance scalability principles in Section II to other power modes. Consequently, setting conventional fixed bias currents in analog blocks within switching and switched-cap PMUs needs to be avoided altogether, as it would otherwise contribute to raise the power floor. As an example, this is the case of comparators, which are

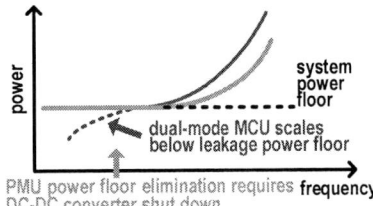

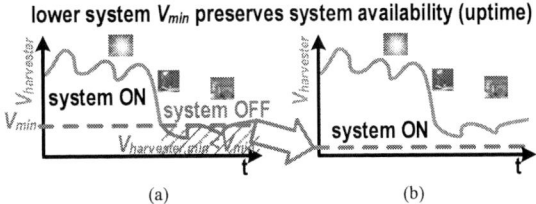

Fig. 10. PMU DC-DC conversion adds to the system power floor, and needs to be removed via direct harvesting when operating around the MP point.

Fig. 11. a) Direct harvesting exposes the system to forced system shutdowns when the harvested voltage becomes lower than the system V_{min}. b) System uptime is improved by pushing its V_{min} below $V_{harvester,min}$.

extensively used in voltage regulation loops, and zero-current switching loops [28]. Based on the processor in Section III, a switched-cap PMU to support the MC mode would require each comparator to have a bias current in the µW range, which is intolerably high in MP and MP+ modes [27]. In other words, the PMU would not be able to function in indoor environments, not to mention any practical darker light condition. A simple approach to automatically scale the power of the analog circuitry without any additional circuit and power floor burden is exemplified in Fig. 12 for a comparator. To keep the bias current proportional to the harvested power (i.e., scalable rather than contributing to the power floor), the tail transistor providing the bias current I_{bias} is directly driven by a small and independent portion of the light harvester, whose open-circuit voltage V_{OC} is proportional to $\log(P_{harvester})$. In turn, the tail transistor operating in sub-threshold generates a current I_{bias} that is proportional to $\exp(V_{OC})$, Fig. 12. Combining the logarithmic and the exponential dependences, this leads to a bias current I_{bias} that is proportional to $P_{harvester}$, as desired [27]. In the above example, keeping the bias currents scalable allows operation with an on-chip 0.85 mm² solar cell down to 55-lux light [27] (i.e., dark overcast day).

Another fundamental contribution to the power floor in circuits for harvesting is associated with conventional DC-DC conversion to assure maximum power point tracking (MPPT). The MPPT sub-system properly modulates the current absorbed from the harvester to maintain the harvester around its maximum power transfer point (e.g., solar cell voltage kept at 75% of its open-circuit voltage). To avoid such extra power floor burden around the MP point, DC-DC conversion can be suppressed altogether by directly modulating the current of the processor being powered via ΔV tuning [27] (see Fig. 7a). In other words, a feedback loop is established in MP and MP+ modes to adjust ΔV to maintain $V_{harvester} = 0.75 \cdot V_{OC}$, as required by MPPT [27]. Again, this allows to include MPPT capabilities while eliminating the power floor contribution associated with the conventional DC-DC converter for MPPT.

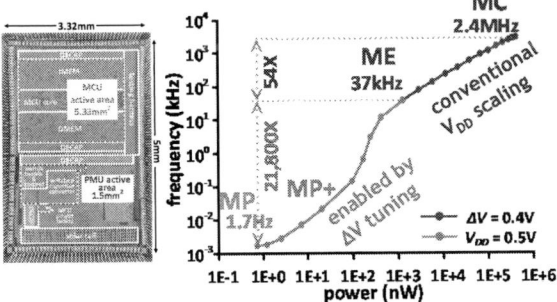

Fig. 9. Example of continuous power-performance tradeoff in battery-indifferent system with processor and integrated PMU [26].

978-1-6654-8495-4/22 $31.00 © 2022 IEEE

Overall, PMUs for battery-indifferent and battery-less systems need to 1) incorporate an extra power mode supported by direct harvesting, 2) power a system with a low V_{min} (i.e., lower than the minimum harvested voltage in the considered environmental condition range), 3) eliminate any contribution to the power floor by suppressing DC-DC converters and **making bias currents power-proportional**.

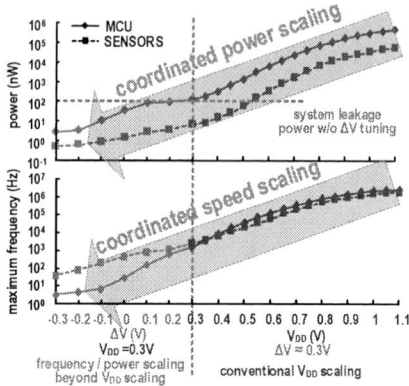

Fig. 13. Coordinated scaling of the analog sub-system and the others (e.g., processing) is necessary to avoid that [29].

V. SENSOR INTERFACES AND ALWAYS-ON CAPABILITIES UNDER PURELY-HARVESTED OPERATION

Analog circuits are subject to similar considerations in terms of its power-performance scalability and its contributions to the power floor [2], [3]. From a power-performance viewpoint, both the power consumption and the sample rate must scale by approximately the same factor as the other sub-systems (e.g., processing) across all levels of harvested power. In other words, "**coordinated power-performance scaling**" needs to be maintained in all sub-systems to make sure that

1) their performance (e.g., clock frequency in processing, sample rate in sensor interface) is kept approximately the same under any available power, and hence the system performance is not limited by any specific sub-system

2) their power remains a fixed fraction of the system power so that they does not represent any power bottleneck or power floor at the system level (**power proportionality**).

Coordinated scaling can be achieved by adopting a sensor interface architecture whose critical path delay and its dominant power contribution are defined by dual-mode logic paths, so that they match the same dependence of the processing sub-system on the harvested power. Coordinated scaling is exemplified in Fig. 13 for the sensor interface and the processing sub-system in [29]. The related multi-sensor interface supports light, capacitive, temperature and voltage-mode sensing. Focusing on the latter, a Flash ADC was adopted in [29] since its power and critical path are well-known to be associated with the comparators and the subsequent conversion logic [30]. Implementing both the comparators and the conversion logic in dual-mode logic automatically aligns the power-performance tradeoff with the processing sub-system implemented in the same logic style. From Fig. 13, the overall system scales the performance of the processing (analog) sub-system from 3.4 Hz (3.4 S/s) to 2.4 MHz (2.4 MS/s), and the power from 3.1 nW to 0.47 mW across all power modes.

Compared to other state-of-the-art complete sensor nodes including both processing and sensor interfaces in Fig. 14, coordinated scaling enables a power-performance dynamic range of 185,000. Again, the maximum performance is in line with the fastest systems, whereas the minimum energy is 3.1 nW and is very close to other sub-leakage implementations.

As shown in Fig. 15, the removal of any fixed power contribution in sensor interfaces is necessary to achieve wide power scalability and coordinated scaling. The removal of fixed bias currents translates into an appropriate choice of architectures that are digitally-intensive. At the same time, adoption of sensor interface architectures with ultra-low voltage operation (i.e., low V_{min}) are mandated in MP mode, as it relies on direct harvesting.

The above techniques enable aggressive system-level power scaling with the full sensor node being able to sense, process and detect/timestamp/store events on chip uninterruptedly across a very wide range of power levels [29]. From Fig. 16 outdoor day light (down to ~1,000 lux) is sufficient to support the system under voltage scaling, whereas the power harvested at indoor light levels pushes the system into MP+ and eventually MP mode at 1 lux, corresponding to the **moonlight level**. In other words, the edge system in [29] illustrates that it is **actually feasible to have always-on operation under purely-harvested power at any practical environmental condition**, debunking the traditional belief that batteries (or other energy storage) are needed for always-on operation.

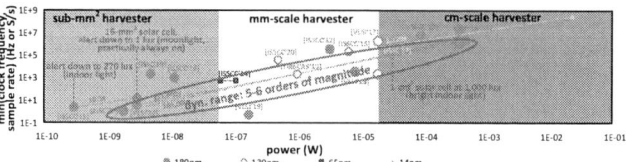

Fig. 14. Power-performance dynamic range in state-of-the-art low-power processors for the edge. Dual-mode logic reaches the widest dynamic range of six orders of magnitude [20].

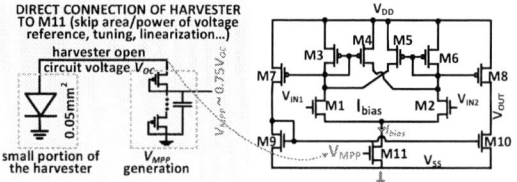

$$I_{bias} \propto \exp(V_{MPP}/nV_T) \propto \exp(\beta \log(lux)) \propto \text{harvested power}$$

Fig. 12. Direct connection of the light harvester open-circuit voltage to the gate of a tail transistor generates a bias current that is proportional to the harvested power with no extra circuitry and no additional power floor.

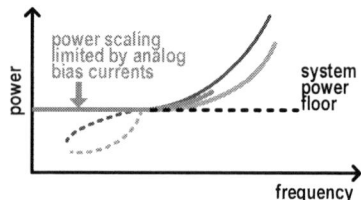

Fig. 15. Bias currents in analog circuits add to the system power floor, and they need to be removed via proper choice of the sensor interface architecture.

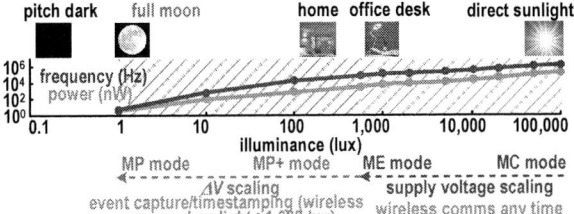

Fig. 16. Power-frequency scaling of edge system in [29] allows uninterrupted event capture down to moonlight (always-on at any practical light condition).

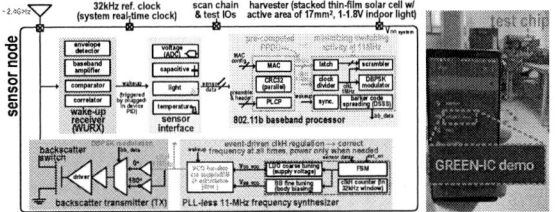

Fig. 17. Demonstration of a complete edge system with multi-sensor interface and wireless communications relying on a wake-up receiver and a PLL-less transmitter architecture with maximum peak power of 2.5 μW [32].

VI. Wireless Communications and pW-Power Components towards a Full Design Ecosystem

The full system in Section V exhibits always-on ability to uninterruptedly capture, analyze, timestamp and store events on chip any time and at any practical environmental condition. When the harvested power is sufficient to support wireless communications, the events and data stored are sent to the cloud along with their timestamp. In other words, battery-indifferent and battery-less systems can operate uninterruptedly, although at **variable data freshness** depending on available power [3].

From Fig. 16, wireless transmission at any light level above artificial indoor lights requires the **RF transmitter** to consume a power **down to the μW range**. At the same time, the RF transmitter needs to comply with existing wireless standards to enable its adoption **without disrupting the existing wireless infrastructure**. Such properties can be met by adopting 1) **backscattered radios** to suppress power-hungry RF power amplifiers and local carrier frequency synthesis [31], and 2) PLL-less transmitter architectures to suppress the PLL power, while complying with the wireless standard requirements [32].

Under the representative use case of WiFi connectivity, backscattering pushes the average RF transmission power down few tens of μW (e.g., 28 μW in [31]), and the additional **suppression of PLL pushes the peak power further down to the μW level** (e.g., ~2.5 μW in [32]). Such radios enable full compliance with the wireless standard for immediate integration with minor modifications of existing infrastructures, while backscattering a carrier with a local oscillator at a much lower frequency (e.g., in the 10-MHz range). For example, the HitchHike communication scheme [33] requires the existence of two access points within range instead of one, the occurrence of a simultaneous transmission from a third (non-backscattered) WiFi device in the neighborhoods, and the modification of the receiver to include codeword translation, while not adding any further wireless component into the system. On the other hand, Passive WiFi schemes [34] offer a different tradeoff in which no change in access point, density, receiver modification or third device with simultaneous transmission requirement is necessary. However, an additional plugged-in wireless device is necessary in the neighborhoods of a cloud of backscattered sensor nodes to handle communications and generate a sub-carrier. In contrast with an incorrect belief that unfairly favors one of the above, both approaches add minor modifications in the existing infrastructure and they are just different tradeoffs.

In general, **PLL-less backscattered radios** are based on on-chip ring oscillators to generate an intermediate frequency that

is tuned via supply and body voltage scaling up temperature event occurrence. The architecture in [32] and Fig. 17 comprises a multi-sensor interface, a 802.11b baseband processor, a PLL-less frequency synthesizer with event-driven temperature compensation and an RF switch for backscattering. The system in [32] operates uninterruptedly when powered by a 22-mm² solar cell at 500-lux light.

Finally, our research effort is progressing towards the generation of a **full design ecosystem** including fundamental building blocks with power down to pWs and aggressively scaled V_{min}. Such building blocks range from voltage [35], [36] and current references [37], amplifiers [38]-[40], wake-up oscillators [41], capacitance-to-digital converters [42], and now increasingly more sophisticated sensors such as image sensors [43]. More will come in the upcoming quarters to enable end-to-end design of battery-indifferent and battery-less systems.

VII. Remarks and Conclusion

Making the scale-up of connected devices to trillions economically, logistically and environmentally sustainable requires fundamental technological innovation at the silicon system level to dramatically shrink or eliminate batteries (Fig. 3). Next-generation alert, mm-sized, low-cost and long-lived silicon systems need to be battery-indifferent or battery-less. As basic design principles in this work, their design needs to introduce wide power scalability down to nWs, power floor removal, sub-leakage operation in digital circuits, system-level coordinated scaling, PMU enhancement with direct harvesting, and backscattering PLL-less RF architectures.

The expansion of the power-performance tradeoff towards nWs allows a wider range of applications that can exchange lower performance requirements for lower power. At the lowest performance end, Hz-range targets are typical of tasks involving events in the second scale as in environmental monitoring, smart health, smart logistics/warehousing/supply chain, structural health monitoring, smart agriculture, and several other applications [1], [2]. Applications requiring higher performance can still take full advantage of wide power scaling, and operate at the minimum power that can support the performance target. Such power flexibility is ultimately key to foster manufacturing volume and reuse across applications to ultimately make the next trillion of devices financially viable.

Acknowledgement

The author gladly acknowledges the support of the MOE2019-T2-2-189 and NRF-CRP20-2017-0003 grant,

TSMC for fabrication, Cadence via the University Program, and our Green IC group at NUS for making this possible.

REFERENCES

[1] M. Alioto (Ed.), *Enabling the Internet of Things – from Integrated Circuits to Integrated Systems*, Springer, 2017.

[2] L. Lin, S. Jain, K. Ali Ahmed, M. Alioto, *Self-Powered Sensors for Next-Gen IoT - Everywhere, Always-on and Green*, Springer, 2022 (in print).

[3] M. Alioto, "From Less Batteries to Battery-Less Integrated Systems through Ultra-Wide Power-Performance Adaptation down to pWs," *IEEE Design&Test* (invited), vol. 38, no. 5, pp. 90-133, Oct. 2021.

[4] K. L. Lueth, M. Hasan, S. Sinha, et Al. "State of IoT—Spring 2022," IoT Analytics market report, May 2022.

[5] P. Sparks, "The Route to a Trillion Devices - The Outlook for IoT Investment to 2035," White Paper, June 2017 [Online] – Available at: https://community.arm.com/cfs-file/__key/telligent-evolution-components-attachments/01-1996-00-00-00-01-30-09/Arm-_2D00_-The-route-to-a-trillion-devices-_2D00_-June-2017.pdf

[6] SIA press release, "Global Semiconductor Sales, Units Shipped Reach All-Time Highs in 2021 as Industry Ramps Up Production Amid Shortage," [Online] – Available at: https://www.semiconductors.org/global-semiconductor-sales-units-shipped-reach-all-time-highs-in-2021-as-industry-ramps-up-production-amid-shortage/

[7] R. Aitken, "How To Build A Trillion Connected Things," Semiconductor Engineering, Oct 23, 2017 [online] – Available at: https://semiengineering.com/how-to-build-a-trillion-connected-things/

[8] C. Joannin, "Metering & IoT Market & Product Manager," from SAFT website, Dec. 2020 [Online] – Available at: https://www.saftbatteries.com/energizing-iot/price-vs-value-what%E2%80%99s-behind-price-your-iot-battery

[9] "Energy Harvesting Market to rise at CAGR of 11.35% through 2030," market report by Market Research Future, Oct. 2021

[10] A. D'mello, "Energy Harvesting Brings 'Battery that Never Dies' Claim for IoT Applications," *IoTNOW*, Oct. 2020 [Online] – Available at: https://www.iot-now.com/2020/10/28/105718-energy-harvesting-brings-battery-that-never-dies-claim-for-iot-applications

[11] G. Haswell, "EMD's Energy Harvesting technology - Powering the IoT," LeadingEdgeOnly website, Ref. 00878 [Online] – Available at: https://www.leadingedgeonly.com/innovation/view/battery-free-power-for-the-iot

[12] "Battery-free Sensors Market by Sensor Type (Temperature Sensors, Humidity/Moisture Sensors, Pressure Sensors), Frequency, Industry (Automotive, Logistics, Healthcare, Food & Beverages), and Region (2021-2026)," market report by Markets&Markets, Aug. 2021.

[13] "Battery Systems for Stationary Storage," CairnERA report [Online] – Available at: http://cairnera.com/research/stationary-storage-report

[14] M. Alioto, V. De, A. Marongiu, "Energy-Quality Scalable Integrated Circuits and Systems: Continuing Energy Scaling in the Twilight of Moore's Law," *IEEE Journal on Emerging and Selected Topics in Circuits and Systems*, vol. 8, no. 4, pp. 653-678, Dec. 2018.

[15] M. Y. Jaber, *Learning Curves: Theory, Models, and Applications*, CRC Press, 2011.

[16] M. Alioto, "Editorial on TVLSI Positioning—Continuing and Accelerating an Upward Trajectory," *IEEE Trans. on VLSI Systems*, vol. 27, no. 2, pp. 253-280, Feb. 2019.

[17] M. S. Ziegler, J. E. Trancik, "Re-examining Rates of Lithium-Ion Battery Technology Improvement and Cost Decline," Energy & Environmental Science, vo. 14, no. 4, pp. 1635-1651, April 2021.

[18] M. S. Ziegler, J. E. Trancik, "Data series for lithium-ion battery technologies," Harvard Dataverse [Online] – Available at: https://dataverse.harvard.edu/dataset.xhtml?persistentId=doi:10.7910/DVN/9FEJ7C

[19] Hsieh, I-Yun Lisa, et Al., "Learning only buys you so much: Practical limits on battery price reduction," *Applied Energy*, 239, April 2019.

[20] L. Lin, S. Jain, M. Alioto, "A 595pW 14pJ/cycle Microcontroller with Dual-mode Standard Cells and Self-startup for Battery-Indifferent Distributed Sensing," in *IEEE ISSCC Dig. Tech. Papers*, Feb. 2018.

[21] M. Alioto, "Energy Harvesters for IoT: Applications and Key Aspects," short course at *VLSI Symposium 2015*, Kyoto, June 15, 2015.

[22] L. Lin, S. Jain, M. Alioto, "Sub-nW Microcontroller with Dual-Mode Logic and Self-Startup for Battery-Indifferent Sensor Nodes," *IEEE Journal of Solid-State Circuits*, vol. 56, no. 5, pp. 1618-1629, May 2021.

[23] W. Lim, I. Lee, D. Sylvester, D. Blaauw, "Batteryless Sub-nW Cortex-M0+ Processor with Dynamic Leakage-Suppression Logic," in *IEEE ISSCC Dig. Tech. Papers*, San Francisco (CA), Feb. 2015.

[24] D. Bol, R. Ambroise, D. Flandre, J. Legat, "Building Ultra-Low-Power Low-Frequency Digital Circuits with High-Speed Devices," in Proc. of *ICECS 2007*, Marrakech (Morocco), pp. 1404-1407, 2007.

[25] D. S. Truesdell, J. Breiholz, S. Kamineni, N. Liu, A. Magyar, B. H. Calhoun, "A 6–140-nW 11 Hz–8.2-kHz DVFS RISC-V Microprocessor Using Scalable Dynamic Leakage-Suppression Logic," *IEEE Solid-State Circuits Letters*, vol. 2, no. 8, pp. 57-60, Aug. 2019.

[26] L. Lin, S. Jain, M. Alioto, "Integrated Power Management and Microcontroller for Ultra-Wide Power Adaptation down to nW," VLSI Symposium 2019, Kyoto (Japan), pp. C178-179, June 2019.

[27] L. Lin, S. Jain, M. Alioto, "Integrated Power Management for Battery-Indifferent Systems with Ultra-Wide Adaptation down to nW," *IEEE Journal of Solid-State Circuits* (invited), vol. 55, no. 4, April 2020

[28] P.-H. Chen, C.-S. Wu, K.-C. Lin, "A 50 nW-to-10 mW Output Power Tri-Mode Digital Buck Converter with Self-Tracking Zero Current Detection for Photovoltaic Energy Harvesting," *IEEE Journal of Solid-State Circuits*, vol. 51, no. 2, pp. 523–532, Feb. 2016.

[29] L. Lin, S. Jain, M. Alioto, "Multi-Sensor Platform with Five-Order-of-Magnitude System Power Adaptation down to 3.1nW and Sustained Operation under Moonlight Harvesting," in Proc. of *VLSI Symposium 2020*, Honolulu (USA), June 2020.

[30] R. J. van de Plassche, *CMOS Integrated Analog-to-Digital and Digital-to-Analog Converters (2nd ed.)*, Springer, 2003.

[31] P.-H. Wang, C. Zhang, H. Yang, D. Bharadia, P.P. Mercier, "A 28µW IoT Tag That Can Communicate with Commodity WiFi Transceivers via a Single-Side-Band QPSK Backscatter Communication Technique," in *IEEE ISSCC Dig. Tech. Papers*, Feb. 2020, pp. 312-313.

[32] L. Lin, K. Ali Ahmed, P. S. Salamani, M. Alioto, "Battery-Less IoT Sensor Node with PLL-Less WiFi Backscattering Communications in a 2.5-µW Peak Power Envelope," in Proc. of *IEEE VLSI Symposium 2021*, Kyoto (Japan), June 2021.

[33] P. Zhang, D. Bharadia, K. Joshi, S. Katti, "HitchHike: Practical Backscatter Using Commodity WiFi," in Proc. of *ACM SenSys '16*, Stanford (CA), Nov. 2016.

[34] B. Kellogg, V. Talla, S. Gollakota, J. R. Smith, "Passive Wi-Fi: Bringing Low Power to Wi-Fi Transmissions," in Proc. of *13th USENIX Symposium on Networked Systems Design and Implementation (NSDI 16)*, pp. 151-164, Santa Clara (CA), Nov. 2016.

[35] L. Fassio, L. Lin, R. De Rose, M. Lanuzza, F. Crupi, M. Alioto, "Trimming-Less Voltage Reference for Highly-Uncertain Harvesting down to 0.25V, 5.4-pW," *IEEE Journal of Solid-State Circuits*, vol. 56, no. 10, pp. 3134-3144, Oct. 2021.

[36] L. Fassio, L. Lin, R. De Rose, M. Lanuzza, F. Crupi, M. Alioto, "Trimming-Less 0.2-V, 3.2-pW Voltage Reference Based on Corner-Aware Replica Combination with 1.6% Process Sensitivity, 1.4-mV Accuracy across PVT and Wafers," accepted to *IEEE ESSCIRC 2022*

[37] L. Fassio, L. Lin, R. De Rose, M. Lanuzza, F. Crupi, M. Alioto, "A 0.6-to-1.8V Trimming-Less CMOS Current Reference with Near-100% Power Utilization," *IEEE Trans. on Circuits and Systems – part II*, vol. 68, no. 9, pp. 3038-3042, Sept. 2021.

[38] P. Toledo, P. Crovetti, O. Aiello, M. Alioto, "Fully Digital Rail-to-Rail OTA with Sub-1,000 µm2 Area, 250-mV Minimum Supply and nW Power at 150-pF Load in 180nm," *IEEE Solid-State Circuits Letters*, vol. 3, pp. 474-477, Sept. 2020.

[39] O. Aiello, P. Crovetti, P. Toledo, M. Alioto, "Rail-to-Rail Dynamic Voltage Comparator Scalable down to pW-Range Power and 0.15-V Supply," *IEEE Trans. on Circuits and Systems – part II*, vol. 68, no. 7, pp. 2675-2679, July 2021.

[40] P. Toledo, P. Crovetti, O. Aiello, M. Alioto, "Design of Digital OTAs with Operation down to 0.3 V and nW Power for Direct Harvesting," *IEEE Trans. On Circuits and Systems – part I*, vol. 68, no. 9, Sept. 2021.

[41] O. Aiello, P. Crovetti, L. Lin, M. Alioto, "A pW-Power Hz-Range Oscillator Operating with a 0.3-1.8V Unregulated Supply," *IEEE Journal of Solid-State Circuits*, vol. 54, no. 5, pp. 1487-1496, May 2019.

[42] O. Aiello, P. Crovetti, M. Alioto, "Capacitance-to-Digital Converter for Operation under Uncertain Harvested Voltage down to 0.3V with No Trimming, Reference and Voltage Regulation," in *IEEE ISSCC Dig. Tech. Papers*, Feb. 2021, pp. 74-75.

[43] K. Ahmed, L. Lin, P. Salamani, M. Alioto, "Imager with Dynamic LSB Adaptation and Ratiometric Readout for Low-Bit Depth 5-µW Peak Power in Purely-Harvested Systems," IEEE VLSI Symposium, pp. 50-51, June 2022.

Innovations for Intelligent Edge

Vida Ilderem, Stefano Pellerano, Jim Tschanz, Tanay Karnik, Vivek De

Intel Corporation, Hillsboro, OR USA

(Vida.Ilderem, Stefano.Pellerano, James.Tschanz, Tanay.Karnik, Vivek.De)@intel.com

Abstract— Massive amounts of data will continue to be generated when everything becomes intelligent and connected, and where data-driven decision making becomes a necessity. Data sources are also becoming increasingly distributed and mobile. These trends have resulted in the rise of the Intelligent Edge, where compute is brought closer to the data source to address scale and latency requirements. The opportunity is to leverage the emerging confluence of compute, communication, and AI at the Edge to wisely partition complex workloads between Edge and Cloud, and to transform the way human and machines interact with the world. Consequently, industry and academia need to innovate faster than ever in key technologies to keep up with this rapidly evolving market. This keynote will cover the required technology innovations and challenges in both 1) monolithic/heterogeneous integration on the technology side, as well as (2) design and architecture on the circuits/systems side, spanning both compute and connectivity domains for a digitized world where everything can be made intelligent and connected.

I. Introduction

Our physical world's digital representation continues to grow at a rapid pace, resulting in the generation of large amounts of data of various types. We conduct more and more of our personal and professional business on the internet. For things that can't be digitized, we give them a digital representation on the internet. Everything is becoming data or has a data representation.

The proliferation of sensors and intelligent machines – and their low-latency requirements – are motivating a shift of computing to the edge of the networks, closer to devices and the data they generate. Users now expect high data rates, ubiquitous broadband connectivity, and low latency instant access. Fifth Generation (5G) wireless communication represented a significant shift for the industry where communication and computing converged to accommodate the massive influx of connected things, the speeds to deliver high resolution streaming experiences, and the responsiveness that mission critical applications demand. Therefore, 5G is the first protocol to address machine type communication along with reduced access network latency, in addition to delivering higher throughput and capacity.

The advent of IoT and the ramp of 5G resulted in the emergence of the Edge network, i.e. compute moving closer to where the data is generated and reducing the latency if required by the application. Another emerging trend is the rise of Artificial Intelligence (AI). Moving forward, we are observing a confluence of communication, computation, and AI at the network edge [1], thus heralding the era of Sixth Generation (6G) communication. Each generation of wireless technologies needs to bring a new level of experience and a set of new capabilities to enable future applications and services. 6G is no exception. This new generation establishes a solid foundation for sensing and machine intelligence requiring more energy efficient ubiquitous computing and communication in all environments. In short, the digital and network transformation that are taking place would not have been possible without innovation in semiconductor technologies. The following sections will focus on system architecture and circuit innovations required to continue to fuel this transformation.

II. System Architecture

From high-performance servers to battery-operated wireless sensor nodes, energy efficiency is more critical than ever. Endpoint devices from cameras to mobile phones to robots to cars to IoT in-field sensors are the sources of the generated data. However, the nature of the data is varied in type, size, being real time, and infused with AI that can be transported wirelessly. Therefore, to unleash the power of end nodes for intelligent "always on" devices and massive connected computing requires energy efficient solutions, starting at the system architecture level.

The type of edge node device and its computational capacity dictates the system architecture solution. An intelligent edge node with reasonable computational complexity, such as robots in a warehouse, can maximize their energy efficiency through communication-computation co-optimization by partitioning the workload between the device and the edge or cloud servers. However, set and forget sensors or perpetual sensing will require a different type of architecture due to limited access to power, leading to system level power optimization becoming essential for these battery-operated devices. We have reported an energy neutral system for agricultural application for killer moth detection and

978-1-6654-8495-4/22 $31.00 © 2022 IEEE

identification [2]. This centimeter scale edge node system was designed to be self-powered, intelligent, and secure by including an ultra-low power 14nm System on a Chip (SoC), a camera, communication unit, solar cell, micro wind turbine, photovoltaic (PV) harvester and master Power Management unit (PMU) controller, rechargeable solid-state electrolyte battery, voltage regulators, storage Flash, and multiple environmental sensors (Fig. 1). The system consumed 0.4J for 8.2 seconds of active operation and consumed only 0.2mW idle and 24mW peak power. This is an example of tailoring a system architecture to deliver results in a constrained environment. There are many other applications which would require these types of customizations through innovation in every aspect of the design.

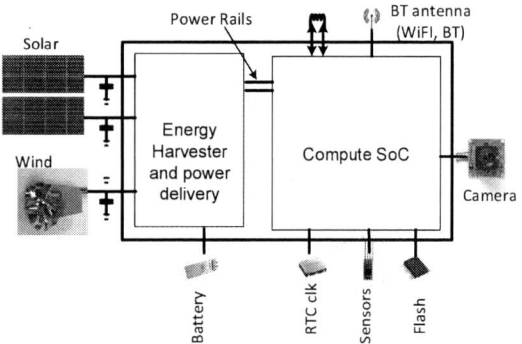

Figure 1: Energy neutral system for agricultural application for killer moth detection and identification [2]

III. RF Technologies

For wireless communication, the number of frequency bands per generation continues to increase. 5G added more spectrum for <6GHz and mmWave bands. This created a challenge for the RF frontend designs, especially for number of components/filters, area, and cost. In 6G, where computing, communication, and AI are predicted to converge, we should expect the level of complexity to increase even further via potential changes in the designs of air interface, network architecture, signal processing, and introduction of sub-THz spectrum to achieve 10x-100x platform performance improvement.

Consequently, innovation is required for frontend module (FEM) filtering technologies to reduce cost and complexity. Some candidates include highly efficient and broadband filters such as single crystal piezo materials (Lithium Niobate, Scandium doped AlN), configurable/tunable filters, and integrated multiband filters. In addition to added FEM complexity, there are going to be multiple radio access technologies available at the same time such as LTE, 5G,

mmWave, Wi-Fi, and potential sub-THz bands, where user access needs to become transparent. Therefore, virtualization of the radio access network is underway. The demand for massive MIMO, digital beamforming and larger bandwidths for higher capacity and throughput increases the complexity in radios for cellular base-stations. This complexity can be reduced by implementing a soft-radio capable of supporting different bands and standards with a single architecture. Soft-radio can be achieved through radio transceiver SoCs with multiple direct RF transceivers that are enabled by high-performance digital-to-analog converters (DACs) and analog-to-digital converters (ADCs) [3-5]. However, this architecture requires very high-speed converters to handle the direct RF sampling, and this can be enabled by advancement in IC design and process node scaling.

According to Shannon communication theory, increasing signal bandwidth is a more efficient way to increase network capacity. This is pushing wireless technology towards mmWave (30-100GHz) and sub-THz (100-300GHz) carrier frequencies, where large bands of spectrum are available. Moreover, operating at higher carrier frequencies allows for larger modulation rates with acceptable fractional bandwidth, which is key to power efficient transceiver implementations. Novel radio architectures and process scaling are enabling mmWave and sub-THz transceiver designs in CMOS technologies. Leveraging Intel 22nm FinFET technology, we have demonstrated a fully integrated D-band (110-170GHz) transmitter capable of 160Gb/s data rate at ~1pJ/b efficiency (Fig. 2) [6]. This performance is achieved while integrating all critical blocks, from bits to RF. A direct-conversion IQ digital transmitter (RF-DAC) can enable energy-efficient and wideband TX design at sub-THz frequencies, as the DAC functionality embedded in the up-conversion mixers minimizes the number of bandwidth-limiting circuit nodes in the signal path. This optimizes the fractional bandwidth used around the desired carrier frequency. Wideband PA and low-power, low-noise quadrature Local Oscillator (LO) generation at 140GHz are also key design elements to achieve 1pJ/b efficiency.

I. Power Management

Power conversion and voltage regulation play an important role in the end-to-end design efficiency, directly through losses in the voltage regulator (VR) stages and current distribution and indirectly through voltage or frequency margins that need to be added to guarantee functionality. While advanced integration has resulted in challenges in delivering large amounts of power in a small footprint, these advanced technologies also provide opportunities to move the VR closer to the final load and improve efficiency and performance.

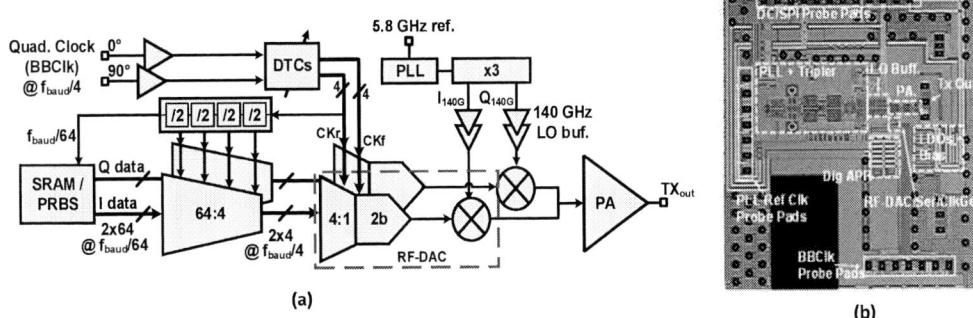

Figure 2: (a) Top-level architecture of the D-Band transmitter. (b) Die photograph. [6]

In high-performance systems, fully integrated voltage regulators (FIVRs) with package-embedded inductors can supply large currents with high efficiency and fast transient response. Implementing current-mode control allows these FIVRs to be designed in a modular way and "ganged" to support large current domains [7]. In 3D-stacked applications, this FIVR can be located in the base die, providing a direct and low-resistance current path to the load circuits directly above while utilizing through-silicon vias (TSVs) to connect to the inductors embedded in the package layers (Fig.3) [8]. In light-load applications, it is even possible to use these TSVs and a magnetic layer to build fully on-die inductor structures [9]. Heterogeneous integration also offers the possibility of employing non-CMOS technologies such as GaN to build package-integrated voltage regulators with high efficiency and with higher input voltage than supported by advanced CMOS nodes, which improves end-to-end efficiency and enables higher power density for small form factor devices [10].

linear regulators that include both analog and digital control loops provide a modular approach to balancing PSRR and energy requirements [11], while high-performance transistors in advanced process nodes enable active noise cancellation in an ultra-low-voltage linear VR [12]. Finally, digital control techniques can be leveraged in any VR topology to improve transient response and process portability and can include advanced computational control techniques that dramatically improve droop response [13]. These responsive and efficient VRs contribute to additional power reduction in the SOC by reducing voltage margins required for load-induced voltage droops.

II. Heterogenous Integration

It is widely acknowledged that Moore's Law enables new devices with higher functionality and complexity while controlling power, cost, and size. Moore predicted economics of system partitioning and optimization [14]. Implementation in more advanced CMOS nodes can enable flexible and energy-efficient systems via co-integration of digital, analog, and RF on the same chip or package-level heterogenous integration utilizing advanced interconnect solutions [15]. Chiplets and system-optimization will continue area scaling and improving performance and power efficiency, while enabling different chiplets to leverage different process nodes. This heterogeneous integration can result in yield improvement for large systems, and faster time to market with more skews, with ad-hoc, application tailored system optimizations. Continuous innovation is required for seamless integration of these chiplets with standardized interfaces, optimized thermal distribution, testability, and programmability of the final tiled system. The choice of integration, 2.5D versus 3D, is dependent on cost, complexity, and power-performance of the overall system. We are investigating methodologies for co-designing functional chiplets, interposer and package level interconnects to optimize for complex workloads [16].

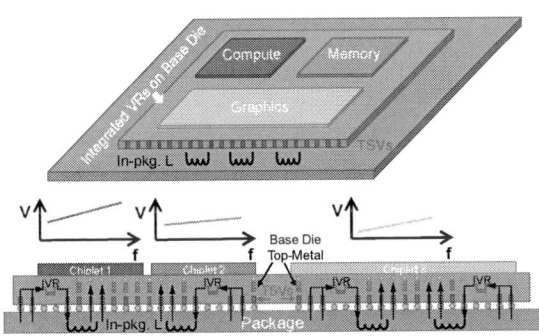

Figure 3: 3D-integrated system with IVRs in a common base-die, in-package inductors and application-specific chiplets [8].

Modern SoCs also integrate numerous special-purpose VRs to provide fine-grain dynamic voltage capability or to power sensitive mixed-signal IP blocks. To provide high power-supply rejection ratio (PSRR), these VRs often require high supply voltages which degrade energy efficiency. Dual-rail

978-1-6654-8495-4/22 $31.00 © 2022 IEEE

III. Summary

In the upcoming era of 6G, we will continue to observe massive generation of data at the network edge. These data could be infused with AI and will require vast amounts of computation and transportation. Real time data analytics applications will continue to drive system, architecture, and technology requirements. Therefore, future 6G enabling technologies can significantly benefit from process node scaling and advancements in heterogeneous integration through advanced packaging and assembly. This mix and match integration allows for greater variety and combination of processor, communication, and integrated system-level solutions, especially for scaling designs across different end-user platforms. As workloads become more complex, we must leverage Moore's Law scaling through silicon and package, and we need to innovate at the system architecture and circuit levels to achieve the best energy efficient solution.

Acknowledgements

This presentation is an accumulation of numerous innovations from our teams. We would like to thank our teams for their support and significant contribution to the state of art technologies.

References

[1] S. Talwar, *et al.*, "6G: Connectivity in the Era of Distributed Intelligence," *IEEE Communication Magazine*, Volume: 59 (11), 2021.

[2] T. Karnik, *et al.*, "A cm-scale self-powered intelligent and secure IoT edge mote featuring an ultra-low-power SoC in 14nm tri-gate CMOS," *IEEE International Solid - State Circuits Conference - (ISSCC)*, 2018, pp. 46-48.

[3] D. Gruber, *et al.*, "Direct RF sampling converters sampling rates up to 16GS/s capacitive TX RF-DAC for multiband operation," in *IEEE International Solid- State Circuits Conf. (ISSCC) Dig. Tech. Papers*, 2021.

[4] E. Thaller, *et al.*, "47.3fs subsampling ADPLL in IEEE *International Solid- State Circuits Conf. (ISSCC) Dig. Tech. Papers*, 2021.

[5] M. Clara, *et al.*, "64GS/s Calibration Reference DAC for ADC linearization," in *IEEE International Solid- State Circuits Conf. (ISSCC) Dig. Tech. Papers*, 2021.

[6] S. Callender, *et al.*, "A fully integrated 160Gb/s D-Band transmitter with 1.1 pJ/b efficiency in 22nm FinFET technology," in *IEEE International Solid- State Circuits Conf. (ISSCC) Dig. Tech. Papers*, 2022.

[7] N. Desai, *et al.*, "Peak-current-controlled ganged integrated high-frequency buck voltage regulators in 22nm CMOS for robust cross-tile current sharing," in *IEEE International Solid- State Circuits Conf. (ISSCC) Dig. Tech. Papers*, 2021.

[8] N. Desai, *et al.*, "Fully integrated voltage regulators with package-embedded inductors for heterogeneous 3D-TSV-stacked system-in-package (SIP) with 22nm CMOS active silicon interposer featuring self-trimmed, digitally

controlled on-time discontinuous conduction mode (DCM) operation," in press.

[9] H. K. Krishnamurthy, *et al.*, "A digitally controlled fully integrated voltage regulator with 3-D-TSV-based on-die solenoid inductor with a planar magnetic core for 3-D-stacked die applications in 14-nm tri-gate CMOS," *IEEE Journal of Solid-State Circuits*, 53(4), pp.1038-1048, 2017.

[10] N. Desai, *et al.*, "A 32-A, 5-V-input, 94.2% peak efficiency high-frequency power converter module featuring package-integrated low-voltage GaN NMOS power transistors," *IEEE Journal of Solid-State Circuits*, 57(4), pp.1090-1099, 2022.

[11] X. Liu, *et al.*, "A dual-rail hybrid analog/digital LDO with dynamic current steering for tunable high PSRR and high efficiency," *IEEE Symposium on VLSI Circuits*, pp. 1-2, 2020.

[12] X. Liu, *et al.*, "A 0.76 V in triode region 4A analog LDO with distributed gain enhancement and dynamic load-current tracking in Intel 4 CMOS featuring active feedforward ripple shaping and on-chip power noise analyzer," in *IEEE International Solid- State Circuits Conf. (ISSCC) Dig. Tech. Papers*, 2022.

[13] K.Z. Ahmed, *et al.*, "A variation-adaptive integrated computational digital LDO in 22-nm CMOS with fast transient response," *IEEE Journal of Solid-State Circuits*, 55(4), pp.977-987, 2020.

[14] Gordon Moore "Cramming more components onto integrated circuits," *Electronics*, 38 (8), April 19, 1965.

[15] W. Gomes, *et al.*, "8.1 Lakefield and mobility compute: A 3D stacked 10nm and 22FFL hybrid processor system in 12×12mm^2, 1mm Package-on-Package," in *IEEE International Solid- State Circuits Conference (ISSCC)*, pp. 144-146.

[16] S. Srinivasa, *et al.*, "Design methodology for scalable 2.5D/3D heterogenous tiled chiplet systems," *IEEE 23rd International Symposium on Quality Electronic Design (ISQED)*, 2022.

Coupling control in the few-electron regime of quantum dot arrays using 2-metal gate levels in CMOS technology

B. Cardoso Paz[1], V. El-Homsy[1], D. J. Niegemann[1], B. Klemt[1], E. Chanrion[1], V. Thiney[2], B. Jadot[2], P. A. Mortemousque[2], B. Bertrand[2], T. Bédécarrats[2], H. Niebojewski[2], F. Perruchot[2], S. De Franceschi[3], M. Vinet[2], M. Urdampilleta[1] and T. Meunier[1]

[1] Institut Néel, Université Grenoble Alpes, Grenoble, France.
[2] CEA-Leti, Université Grenoble Alpes, Grenoble, France.
[3] CEA-IRIG, Université Grenoble Alpes, Grenoble, France.
bruna.cardoso-paz@neel.cnrs.fr

Abstract—**Scalability is one of the biggest advantages of silicon spin qubits over other platforms, making them very promising candidates in the quest for quantum computing. In this work we approach the regime of interest for large-scale qubit integration, showing that we can deliver high electrostatic coupling control and individual tunability over an array of quantum dots (QDs). To do this we use FDSOI devices fabricated with 2-metal gate levels in an industry-compatible CMOS process. We operate them at 100mK, and in a dot-configuration where large control on tunnel barriers is leveraged. In the many-electron regime, we observe the transition of quantum dot array from single- to triple-dot configurations. Moreover, in the few-electron regime, we demonstrate the effective and in-situ modulation of the tunnel coupling between two adjacent QDs.**

Keywords—quantum dots, tunnel coupling, FDSOI

I. INTRODUCTION

The interest in CMOS-based silicon quantum dots have been recently boosted thanks to demonstrations of fidelities above 99% for two-qubit gates implemented using electron spins in semiconductor QDs [1,2]. This nails the fact that silicon spin qubits fulfill the requirements of fault-tolerance threshold, imposed by quantum error correction. Along with their long coherence time [3], silicon spin qubits have the great advantage of scalability, since they can be manufactured using advanced technologies already well established in the semiconductor industry. One of the main challenges from the process integration perspective remains the definition of well-defined quantum dots, and the fine control of charges position and displacement within the dots. For this, chemical potentials and tunnel barriers need to be separately tuned [4]. Recently, we realize the integration of 2-metal multi-gate quantum devices that could be a solution to promote the desired control over QD arrays (Fig.1). Many dot configurations are possible within such a quantum device. In a first work [5], we explored the configuration where QDs were formed under the first metal level (G-gates), at the back interface, while the second metal level (J-gates) enables control of the tunnel coupling. Due to their distance to the Si channel, the J-gates weakly modify the tunnel coupling and this possibly limits the control of Si quantum devices.

In this work, we explore a new system configuration to reach higher tunability on the tunnel barrier: the QDs are formed underneath the J-gates and the tunnel barriers are controlled by the G-gates, that are closer to the Si channel. Thanks to this, transitions from single to triple dot configurations in a quantum dot array are demonstrated in the many-electron regime. Moreover, in the few-electron regime (the regime of interest for qubit application), we could demonstrate the in-situ tuning of the tunnel coupling of more than one order of magnitude in 20mV. First, we give some details on the devices from the technological point of view and show how we use the structures as dots. Then we demonstrate coupling control over different regimes.

II. METHODOLOGY AND CHARACTERIZATION FLOW

The first step consists in exploring the QDs in the many-electron regime, which is accessed using transport measurements (current flowing through the QDs). Due to its simplicity (no need of tuning a charge detector and/or a RF setup), transport measurements can be easily integrated in a large-scale characterization flow. In this work, all experiments were carried out between 100 and 200mK using a dilution refrigerator, which have single-device test capability. But in a near future, such measurements can be done in semi-automatic wafer-level cryogenic probe-stations to evaluate the quality of the QDs and for devices screening.

The second step consists in exploring the QDs in the few-electron regime probed with charge sensing, which is the regime of interest for qubit application. When the charge occupation of the QDs is limited to the first few electrons, their size and position can be very different from those obtained in transport. Therefore, a demonstration of the tunnel coupling control in this regime is mandatory for future spin qubit measurements. The few-electron regime is demonstrated using charge sensing measurements. Since the charge sensor must be appropriately tuned to provide high sensitivity, such measurements have higher complexity in comparison to the previous one. This step will be done using

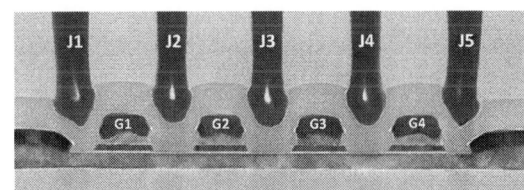

Fig. 1.   TEM image of a 4-gate linear array.

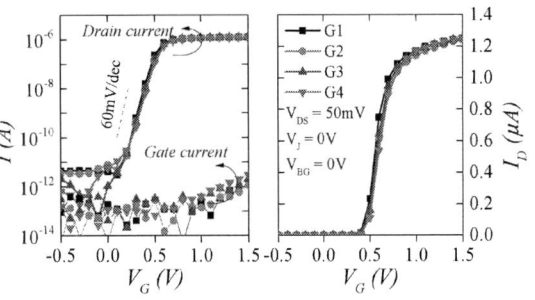

Fig. 2.   Transistor I-V characteristics measured at 300K, and V_{DS} = 50mV. Results show SS close to the theoretical limit of 60mV/dec. L = 45nm.

978-1-6654-8495-4/22 $31.00 © 2022 IEEE

two different approaches: i) single electron transistor (SET), and ii) reflectometry. One of the advantages of the reflectometry is that circuits for multiplexed qubit readout can be envisioned.

III. DEVICE CHARACTERISTICS AND QD CONFIGURATION

Devices were processed on a 300mm FDSOI wafer at CEA-Leti, following a process that has very small deviations from a standard CMOS flow [5]. The local exchange gates are integrated in a module where metallic trenches intertwine with the first gate level, whose optimized gate (TiN/poly-Si) height is only 25nm in order to improve J-gates control [5]. The distance between the J-gates and the Si qubits layer is 20nm. A pitch as small as 80nm can be achieved for both gate levels, leading to an effective gate pitch controllability of 40nm. Fig.1 shows a TEM image of a 4-gate linear QD array. At room temperature, devices behave as standard transistors (Fig.2), with smaller gate variability ($A_{VT} <$ 1mVμm [5]) than for advanced SRAM 22nm node [6].

At low temperature we explore a new configuration to define dots (chemical potential) and barriers: the dots were formed underneath the J-gates (second metal level) and the G-gates (first gate level – gate oxide thickness of 6nm) were used to control the tunnel barriers. As the J-gates control over the QDs tends to be screened by the G-gates, the configuration used in [5] presents a relatively small gate effect on the tunnel coupling. Indeed, the comparison of the extracted gate lever-arms (also known as α-factors) shows α

around 0.3 and 0.025 for the G-gates and the J-gates, respectively. This ratio of a factor ≈10 is in good agreement with electrostatic simulations. Fig.3 shows Coulomb diamonds for two distinct QDs with similar size (charging energy E_C of ≈4meV), used for the extraction of α and the gate capacitances. As a consequence of the higher coupling to the Si qubit layer, the coupling between dots could in principle be tuned with the G-gates over a larger range in this new configuration.

Additionally, the back gate was used to change the vertical position of the dots within the channel allowing to compare accumulation (charges close to the front interface) and depletion regimes (charges attracted to the back interface, where the charge noise is known to be smaller [7]).

Fig.4 shows the schematics of the bias configuration used to have a double QD at the back interface. With a back gate voltage of +25V (BOX = 145nm) and V_J = 0V, dots are formed under J-gates if V_G < 100mV. Quantum dots can only be formed under G-gates when V_G > ≈100mV, because they operate in depletion mode below 100mV. By applying 1mV to one of the reservoirs (at the drain, in this case), transport

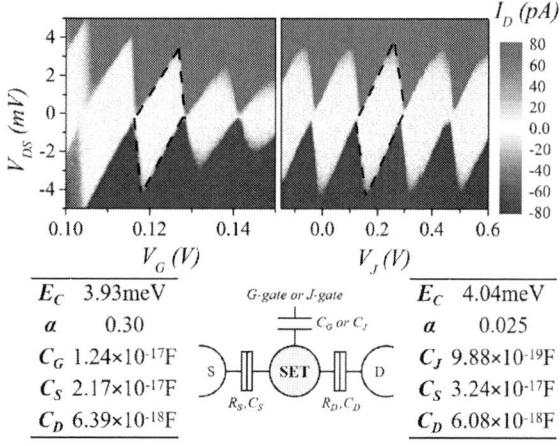

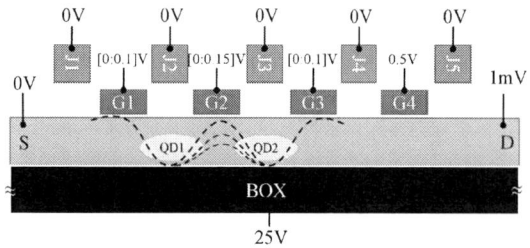

Fig. 3. Coulomb diamonds of a single QD formed under a G (left) and a J-gate (right). Extracted α-factors and capacitances for 1st and 2nd gate levels.

Fig. 4. Schematics for the double QD regime. Two QDs are formed under J2 and J3. Gate G2 is used to control the coupling exchange between QD1 and QD2. A high positive back gate value is applied to push the dots to the back interface, where the charge noise is smaller. A high positive voltage is applied to G4 in order to extend the reservoir (drain), and bring it closer to QD2.

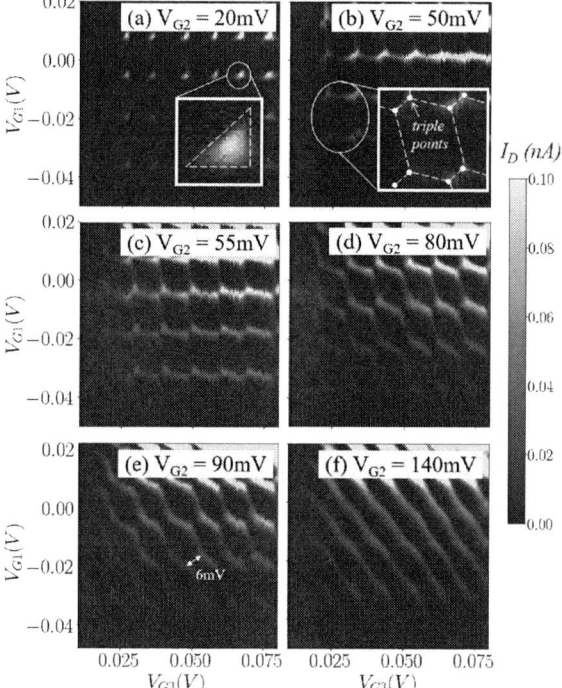

Fig. 5. Stability diagrams showing stronger coupling between two QDs, by increasing the gate voltage on the middle gate electrode (V_{G2}). V_{DS} = 1mV. The white lines were inserted as guidelines.

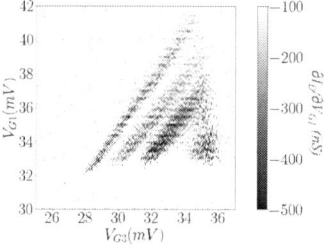

Fig. 6. Bias triangle and observation of excited states when QD1 and QD2 are weakly coupled. V_{DS} = 3mV and V_{G2} = 20mV.

can be observed from source to drain through current measurements.

IV. CONTROLLED TUNABILITY OF THE SYSTEM: FROM SINGLE TO TRIPLE QDs

Fig.5 shows the results of transport measurements as a function of the gate voltages G1 and G3, following the bias configuration presented in Fig.4. From Fig.5a to Fig.5f, only the voltage applied at G2 changes, towards more positive values, i.e. increasing the tunnel coupling between QD1 and QD2. Fig. 5 shows that we are able to tune the tunnel coupling of the two QDs in a wide range (from single to double QD regime) [8]. In Fig.5a, the weak coupling regime for double dot transport can be identified by the appearance of "bias triangles". Fig.6 shows the conductance ($\partial I_D/\partial V_{G1}$) at V_{DS} = 3mV and reveals the energy quantization of the probed dot. Indeed, when V_{DS} is increased (from 1mV in Fig.5), multiple energy levels fit in this bias window, and the alignment of the energy levels between QD1 and QD2 leads to resonances within the triangles. The dark lines in Fig.6 correspond to excited states (energy spacing ≈420µeV, similar to what has been reported in literature [9]). From Fig.5b to Fig.5e we observe the characteristic honeycomb pattern of two coupled QDs. As V_{G2} increases, the triple point gets more and more separated, indicating that the capacitive coupling between QD1 and QD2 increases. At V_{G2} = 90mV, the distance between the triple points (6mV) is comparable to the spacing of the Coulomb peaks (about 9mV). In Fig.5b (intermediate coupling regime), the co-tunneling lines (the edges of the honeycomb) become visible indicating the increase of the tunnel coupling w.r.t. Fig.5a. Then the honeycomb gradually looses its characteristic shape

indicating that tunnel coupling becomes comparable to the capacitive coupling. In Fig.5f, only diagonal lines are observed, as a result of the strong coupling between QD1 and QD2, now merged as a single large dot.

Similarly to the previously described double QD operation, the 4-gate linear array allows us to build up a triple QD, following the same approach, and then forming another QD under J4. Gates G2 and G3 are responsible for controlling the tunnel coupling between QD1-QD2 and QD2-QD3, respectively. Fig.7 shows a large stability diagram, where the current is measured as a function of both G2 and G3. Three regimes can be identified: i) three QDs can be formed when V_{G2} and V_{G3} are small; ii) when V_{G2} (or V_{G3}) is large and V_{G3} (or V_{G2}) is small, two QDs merge into a larger one and we have a double QD regime; iii) at large V_{G2} and V_{G3}, all three QDs are merged in a single large QD. Fig.7 also shows the extracted values for the E_C, where we can observe the size of the QDs for each regime. From i) to ii), the size of the QD that is not merged increases (E_C from ≈3.58meV to ≈1.76meV), because its size depends on the charge occupation.

This is the first demonstration that we can control the position and coupling of well-defined triple QDs in FDSOI technology. Therefore these results represent an important milestone on technology quality and maturity.

V. TOWARDS THE REGIME OF INTEREST FOR QUBITS: CHARGE SENSING

A. Single electron transistor - SET

One technique for charge sensing implies using a SET as a charge sensor in split-gates-like devices, as demonstrated in [10]. Fig.8 shows schematics of a 3-gate array designed to fulfil this purpose. The integration of local exchange gates is compatible with split-gate devices. The front gates in the first metal level partially wrap the silicon wire, and G2 is split intwo, allowing the definition of a SET and a QD in a split-gate geometry. To avoid coupling between them, the QDs

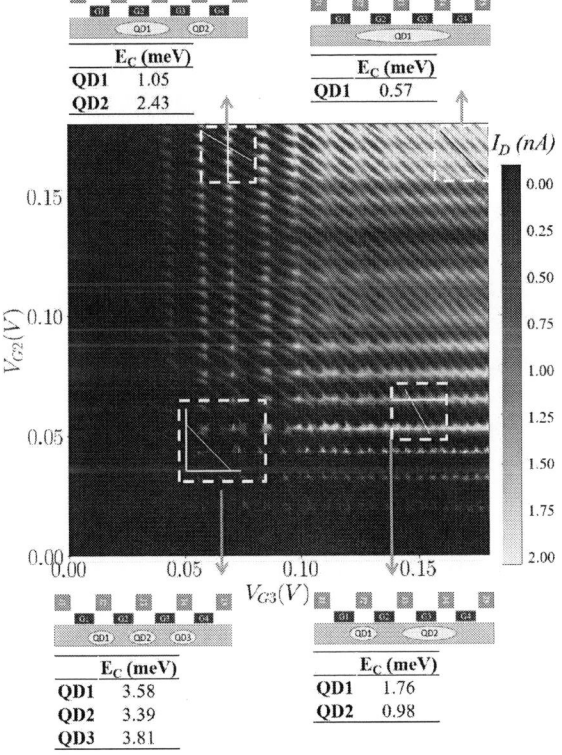

Fig. 7. G2 and G3 are used to modulate the coupling between the three QDs formed below J2, J3 and J4, at the back interface. All regimes can be identified between the transition from three weakly coupled QDs to a single large dot. Extracted E_C gives information on the size of each QD. V_{G1} and V_{G4} = 50mV.

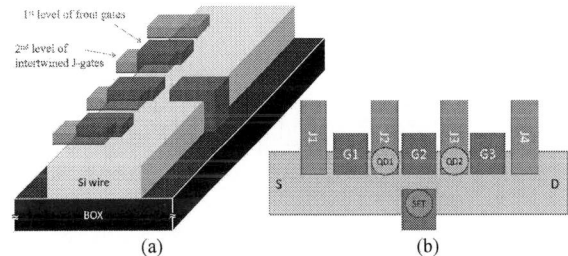

Fig. 8. Schematics of 3-gate linear array with embedded SET for charge sensing. Perspective (a) and top view (b). The SET can be used as a charge detector to count the electrons in the QDs under J2 and J3.

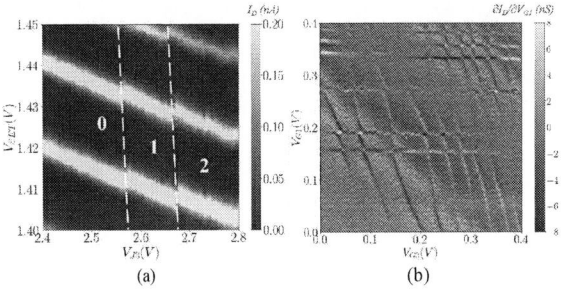

Fig. 9. Charge sensing of single (a) and double QD (b) using a SET as a charge sensor.

978-1-6654-8495-4/22 $31.00 © 2022 IEEE

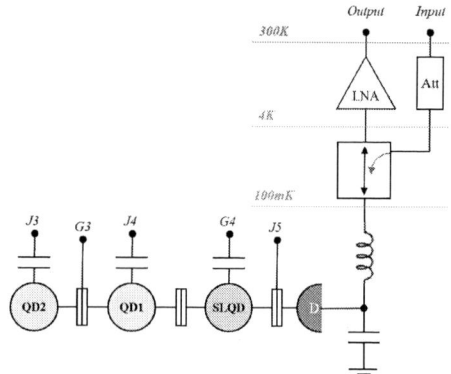

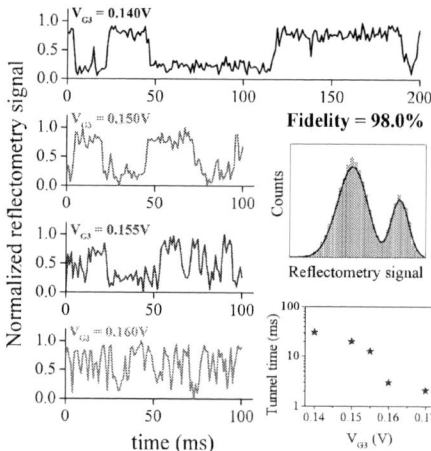

Fig. 10. Schematics of the setup used for reflectometry measurements.

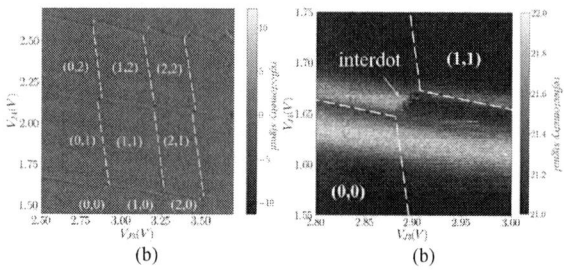

(b)   (b)

Fig. 11. (a) Stability diagram of a double QD showing the few-electron regime. (b) Stability diagram showing the interdot, where stochastic events are observed.

are formed at the front interface, i.e. $V_{BG} = 0V$.

Fig.9a shows a charge stability diagram of a QD under J3, where it is possible to count the number of electrons entering the quantum dot. Fig.9b shows the stability diagram of a double QD, under J2 and J3. In this configuration, the SET is biased at the border of a Coulomb peak to achieve the highest sensitivity. G2 is set so the dots are in the intermediate coupling. We can observe the expected honeycomb pattern, footprint of coupled dots in the few electron regime.

B. Reflectometry

Charge sensing can also be performed by probing a single lead QD using RF-reflectometry [11]. Fig.10 shows a schematic of the setup used in this work. The device is a 4-gate linear array. A QD under G4 is used as a sensor to probe a double QD (under J4 and J3), where G3 can serve to tune the coupling between the two dots. Fig.11a shows the obtained stability diagram with the respective charge occupation. In this geometry, the sensor hardly couples to the farthest QD (in this case to the QD under J3). Therefore, only the degeneracy lines associated to the dot under J4 are visible. In addition, abrupt steps along these lines are observed as a result of the change of occupation in the dot underneath J3. Fig.11b shows a zoom in the first transition, where the interdot is highlighted and we can observe stochasticity related to the tunnelling events. By sitting in a position where those stochastic events occur and measuring the signal as a function of time, it is possible to extract the characteristic tunnel time. Fig.12 shows time traces and the corresponding tunnel times for different V_{G3} values. As expected, the higher the gate voltage applied to G3 the higher the tunnel rate, since we are increasing the coupling between the two QDs. Fig.12 shows that the extracted tunnel time decreases more

Fig. 12. Time traces showing the modulation of the tunnel time with V_{G3}.

than a decade for a V_{G3} increase of 20mV. Moreover, from the histogram plot in Fig.12, we extract a charge readout fidelity as high as 98.0% for an integration time of 1ms, which is sufficient for spin qubit characterization.

VI. CONCLUSIONS

We demonstrate for the first time control over the tunnel coupling in a multiple quantum dot array using a compatible FDSOI CMOS platform. The versatility of the FDSOI technology remains an important ally to change the position of the QD in the silicon wire, since the back-gate allows to achieve both depletion and accumulation modes. The demonstration of high electrostatic control over the coupling between adjacent quantum dots is an important step towards successful manipulation of spin for qubit applications. Finally, these results provide ground for implementing different qubit designs using 2-metal gate levels.

ACKNOWLEDGMENT

This work was supported by the European Research Council (ERC) Synergy QuCube and the QLSI.

REFERENCES

[1] Xue, X. et al., Quantum logic with spin qubits crossing the surface code threshold, Nature 601, 343–347, 2022.

[2] Noiri, A. et al., Fast universal quantum gate above the fault-tolerance threshold in silicon, Nature 601, 338–342, 2022.

[3] Veldhorst, M. et al., Nature 526, 410–414, 2015., Nature 526, 410–414, 2015.

[4] De Franceschi, S. et al., SOI technology for quantum information processing, IEDM 2016.

[5] Bédécarrats, T. et al., A new FDSOI spin qubit platform with 40nm effective control pitch, IEDM 2021.

[6] Joshi, V. et al., Low-variation SRAM bitcells in 22nm FDSOI technology, VLSI 2017.

[7] Theodorou, C. G. et al., Low-frequency noise sources in advanced UTBB FD-SOI MOSFETs, TED, 61, 1161–1167, 2014.

[8] van der Wiel, W. G. et al., Electron transport through double quantum dots, Rev. Mod. Phys. 75, 1, 2002.

[9] Lim, W. H. et al., Electrostatically defined few-electron double quantum dot in silicon, Appl. Phys. Lett. 94, 173502, 2009.

[10] Chanrion, E. et al., Charge detection in an array of CMOS quantum dots, Phys. Rev. Applied 14, 024066, 2020.

[11] House, M. G. et al., High-sensitivity charge detection with a single-lead quantum dot for scalable quantum computation, Phys. Rev. Applied, 6, 044016, 2016.

978-1-6654-8495-4/22 $31.00 © 2022 IEEE

Multi-Gate FD-SOI Single Electron Transistor for hybrid SET-MOSFET quantum computing

Fabio Bersano[1], Franco De Palma[12], Fabian Oppliger[2], Floris Braakman[3], Ionut Radu[4], Pasquale Scarlino[2], Martino Poggio[3] and Adrian Mihai Ionescu[1]

[1]Nanoelectronic Devices Laboratory (NanoLab), EPFL, 1015 Lausanne, Switzerland
[2] Hybrid Quantum Circuit Laboratory (HQC), EPFL, 1015 Lausanne, Switzerland
[3] Department of Physics, University of Basel, 4056 Basel, Switzerland
[4] Soitec, Bernin, France
Email: fabio.bersano@epfl.ch, franco.depalma@epfl.ch

Abstract—In this work we explore the fabrication and experimental characterization of multi-gate FD-SOI devices that can operate both as single electron transistors (SETs) and MOSFETs. FD-SOI SET operation is achieved with electrostatically induced tunneling barriers controllable by two barrier gates (B_L and B_R) and a front gate (FG), the device operation being additionally tunable by a bottom gate (BG). Coulomb blockade current measurements at 10 mK and 4 K demonstrate the possibility of a dynamic transition from a FD-SOI MOSFET and SET operation, making this technology suitable for hybrid SET-MOSFET low power cryogenic circuits for quantum information processing. Room temperature back-gate characterization has been performed on ultra-thin film devices proving the effective reduction of charge noise for a specific bias configuration. This experimental work is a step forward towards low noise planar FD-SOI quantum dots based devices with tunable electrostatic control.

Index Terms—SET, FD-SOI, MOSFET, charge noise, cryo-CMOS, quantum computing

I. Introduction

Silicon-based spin quantum bits (qubits) are promising candidates for the implementation of a quantum processor thanks to the long coherence time of spins and scalability of the technology [1]. Their performance improved considerably over the last few years and great efforts have been made to increase gates fidelity above mK temperatures [2]–[4]. Materials and architecture have been optimized [5] to enable a fast integration of large quantum dots arrays with foundry manufacturing processes [6], [7]. However, with the growth of the number of qubits, integration of control electronics for their manipulation and readout is necessary, posing more stringent limits on power dissipation [8]. Fully depleted silicon-on-insulator (FD-SOI) transistors are known to be best suitable for ultra-low power applications compared to FinFETs [9] and their cryo-characterization has been recently reported [10]. The depletion of majority carriers combined with the threshold tunability through back-gate biasing place them as the optimal candidate for the co-integration of classical and quantum electronics [11].
Single electron transistors (SETs) can be used as charge sensors to read out the spin state of single electrons [12], [13]

This work was supported as a part of NCCR SPIN, a National Centre of Competence (or Excellence) in Research, funded by the Swiss National Science Foundation (grant number 51NF40-180604)

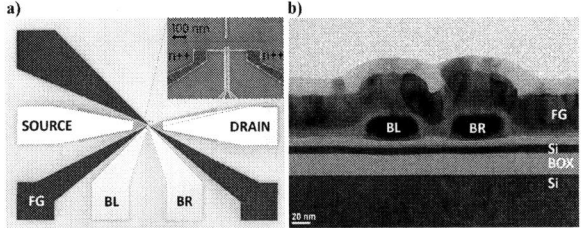

Fig. 1. a) SEM image of a fabricated ultra-thin film (7 nm) planar FD-SOI SET. b) TEM image showing the device cross-section. The same structure was fabricated on a 27 nm SOI substrate.

and their operation can be controlled by integrated CMOS circuits. Some works presented single electron and hole devices fabricated on etched SOI substrates using pattern-dependent-oxidation [14] and silicon nanowires [15]. In this work, we propose for the first time a fully planar multi-gate FD-SOI SET device based on ultra-thin silicon films whose architecture could provide both single electron and hole operation. The device can be dynamically switched to operate as a FD-SOI FET or SET through the activation of tunable tunnel barriers for the ease of integration of FD-SOI qubits with FD-SOI cryo-electronics.

II. Device fabrication

Our FD-SOI SET device has been taken as a process validation vehicle for the fabrication of fully planar (i.e., without MESA etching and raised contacts) quantum dots devices in ultra-thin silicon films. The layout of the device is shown in fig.1. A single quantum dot is electrostatically defined by two metallic gates arranged along the transistor channel (B_L and B_R) and a single top gate (front gate, F_G). The respective functions of the electrodes are to induce potential tunnel barriers and create inversion layers while tuning the chemical potential of the dot, a similar architecture was presented in [12], [13]. Separated source and drain doped regions give ohmic contacts for electrical conduction.

Two SOI substrates with different top-silicon and buried oxide (BOX) thicknesses have been processed for fabrication. The top silicon layers have been thinned down to 27 nm and

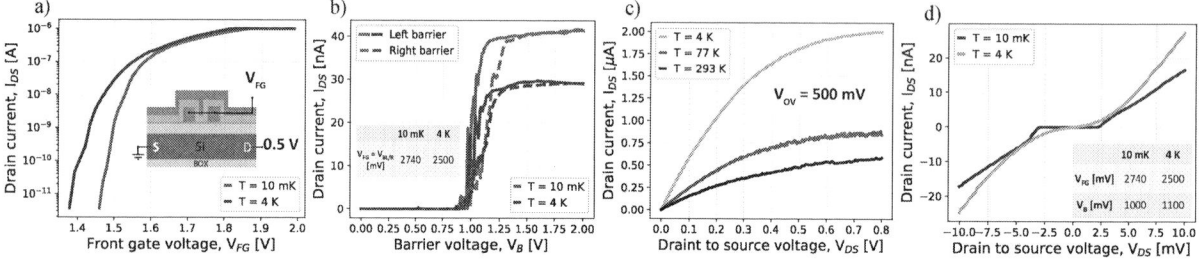

Fig. 2. Cryogenic characterization of a thin-film SET device. a) Transfer characteristic for the common top gate configuration illustrated in the inset (Al_2O_3 in green and Pd gates in grey). b) Transfer characteristic of the left and right barrier gates; a small mismatch of the pinch-off voltage was measured in the device. c) $I_{DS}(V_{DS})$ characteristics at different temperatures for a constant overdrive voltage in the common top gate configuration. d) Zoom-in of the $I_{DS}(V_{DS})$ characteristic around $V_{DS}=0$ V. For certain V_{FG} and V_B voltages a no-conduction window was measured due to Coulomb blockade.

7 nm through a few cycles of thermal dry-oxidation and wet-etching (BHF 7:1) to get fully-depleted thin and ultra-thin films. In both samples, the silicon layer was p-type with a room temperature resistivity of 10 $\Omega\cdot$cm, the estimated boron concentration is therefore 10^{15} cm^{-3}. The BOX thickness was 2 μm for the thin-film sample and 20 nm for the ultra-thin film one.

Past works proved that charge noise in MOS-quantum dots is mainly due to impurities in the oxides and at the Si/SiO$_2$ interfaces [16]. We therefore analyzed the density of charge traps in thermal gate oxides grown with different conditions through C-V characterization of MOS capacitors. A shift of the C-V curves towards ideal values was measured when performing Si dry oxidation in dichloroethane followed by rapid thermal annealing (RTA) at $300°C$ in forming gas. A gate oxide of 5 nm was grown with the same recipe. The estimated density of charge traps was $N_{eff} \approx 10^{10}$ cm^{-2}.

The source and drain contact regions have been implanted with phosphorus ions through Plasma Immersion Ion Implantation (PIII) with an energy of 1 keV and a dose ranging from $5 \cdot 10^{15}$ cm^{-2} to $1.1 \cdot 10^{16}$ cm^{-2}. Protections for bond-pads were patterned with a positive electron-beam (e-beam) resist (HSQ). A thick field oxide (Al_2O_3, 20 nm) was deposited through atomic layer deposition (ALD) and selectively etched on the active areas with H_3PO_4 50% at $60°C$. Ti/Pt contacts were patterned with standard laser lithography using LOR5A/AZ1512 photoresists, e-beam evaporation and lift-off.

All gate electrodes were patterned with single layer (PMMA950K) e-beam lithography, Ti/Pd evaporation and lift-off. The lowest e-beam dose able to resolve the mininum features size (30 nm) was selected to not damage the BOX of the SOI chips and a source-to-sample working distance of 1010 mm was used during evaporation to facilitate the lift-off. The choice of materials was mainly dictated by the ease of nanofabrication, it is indeed known that Pd has a smaller grain size compared to Al and reduced mismatch of thermal expansion coefficient to that of Si [5]. A negligible line edge roughness was measured in all the analyzed devices after lift-off. 5 nm of Al_2O_3 were deposited through ALD for a proper insulation of the two metallic layers and a RTA step was performed in forming gas at $300°C$ after each deposition.

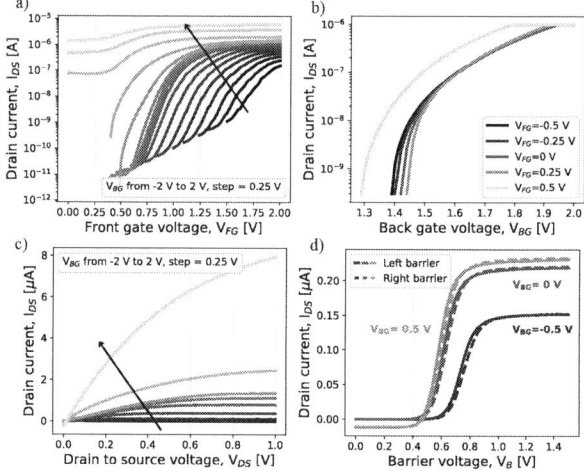

Fig. 3. Room temperature characterization of ultra-thin FD-SOI SET. a) Common top gate transfer characteristics $I_{DS}(V_{FG})$ for different back-gate voltages ($V_{FG} = V_B$, $V_{DS} = 0.5$ V). b) Back gate transfer characteristics $I_{DS}(V_{FG})$ for different front-gate voltages ($V_{DS} = 500$ mV). c) $I_{DS}(V_{DS})$ characteristics measured at $V_{FG} = 1.5$ V for different back-gate voltages. d) $I_{DS}(V_B)$ for left and right barrier gates ($V_{FG} = V_{B1/2} = 1.5$ V, $V_{DS} = 100$ mV).

Finally, a thick Al_2O_3 capping layer was deposited and wet-etched for the re-opening of contacts prior to wire bonding.

III. RESULTS

Room temperature electrical characterizations were performed with a Keithley 4200A-SCS parameter analyzer and DC cryo-measurements were taken in a LD Bluefors dilution refrigerator. We focused our analysis on the FET and SET characteristics and tested the back-gate control at room temperature on the ultra-thin film sample.

A. Front and back gated MOSFET characteristics

Transfer characteristics measured at cryogenic temperatures are shown in fig. 2. The curves prove the independent electrical control of each electrode on the drain current.

A higher subthreshold slope was measured at 10 mK in a common-top gate configuration (i.e., with same voltage applied to the barrier and front gates, fig. 2a). Higher currents were measured for decreasing temperature (fig. 2b and 2c) for similar overdrive voltages thanks to an enhancement of the electrons mobility [17]. A small mismatch in the barrier gates due to imperfect alignment of the electrodes is present, as shown in fig. 2b. For $|V_{DS}|$ voltages greater than 5 mV, no oscillations were measured whereas Coulomb blockade signatures were observed for selected V_{FG} voltages and -5 mV $\leq V_{DS} \leq 5$ mV, as evidenced by the non linearity in the plots in fig. 2b and 2d.

Room temperature measurements of the ultra-thin film devices for several back-gate bias configurations are shown in fig. 3. $I_{DS}(V_{FG})$ curves shift towards lower V_{FG} when the back-gate voltage is swept from negative to positive values until volume inversion (i.e., total inversion of the silicon film) is reached. The estimated back-gate threshold voltage is $V_{BGTH} \approx 1.45$ V and it has a dependence on V_{FG}, as shown in fig. 3b. Increasing off-currents were observed for decreasing V_{BG} and small negative values ($< |-1|$ nA) were measured for $0\,V < V_{BG} < V_{BGTH}$, as can be seen from fig. 3a and fig. 3d. This is related to the accumulation of holes and electrons at the interface between the BOX and the low doped handle wafer for positive and negative V_{BG}, respectively. This is confirmed by the poor V_{TH} control of the back-gate for $0\,V < V_{BG} < 0.8\,V$, where the substrate/BOX interface is brought from depletion to accumulation, as better illustrated in fig. 4a. In the ideal case of a metallic back-gate, the depletion capacitance does not take form and the following relation holds:

$$ V_{TH} = V_{TH0} - \frac{t_{gox}}{t_{box}} \frac{1}{1 + \frac{t_{Si}}{t_{box}} \cdot \frac{\epsilon_{ox}}{\epsilon_{Si}}} V_{BG}, \qquad (1) $$

where V_{TH0} is the front-gate threshold voltage at $V_{BG} = 0$ V and t_{gox}, t_{box} and t_{Si} the gate oxide, buried oxide and top Si thicknesses. An almost linear $V_{TH}(V_{BG})$ relation was measured even with a low doped substrate with a maximum linear coefficient (body factor γ) given for strong accumulation at the substrate/BOX interface, where the electric field in the BOX is high enough to displace the top inversion layer from the top interface, increasing the back-gate capacitance.

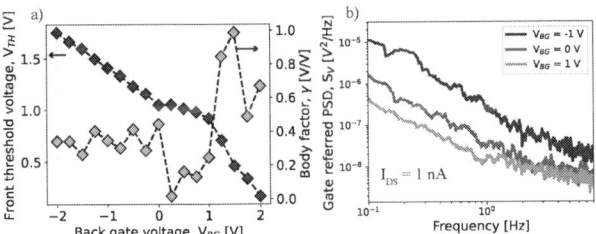

Fig. 4. a) Front gate threshold voltage dependence on back-gate voltage. The body factor is computed as the slope of the $V_{TH}(V_{BG})$ curve and it is maximized for positive V_{BG} biases. b) Frequency dependence of the gate referred power spectral density measured at room temperature. A lower $1/f$ noise is measured for a positive substrate bias.

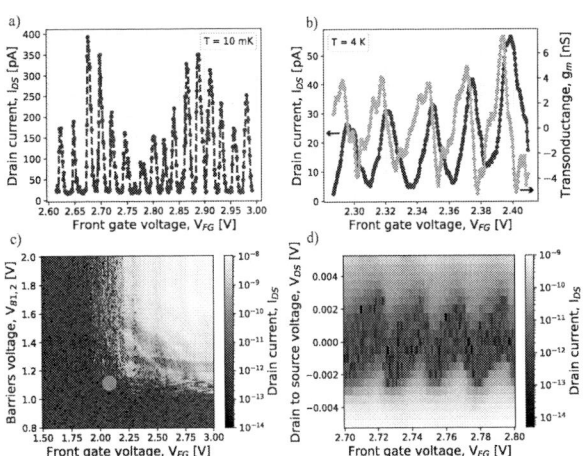

Fig. 5. a) Coulomb blockade oscillations measured at 10 mK with $V_{DS} = 1$ mV. b) Coulomb blockade oscillations measured at 4 K ($V_{DS} = 1$ mV), positive and negative values of the transistor transconductance are plotted as a function of the front gate voltage. c) Drain current measured at 10 mK for several barrier gates and front gate voltage combinations. The green dot indicates the conduction corner for which Coulomb oscillations are enabled. d) Periodic Coulomb blockade diamonds measured at 10 mK.

B. Impact of back-gate bias on drain current noise

The spectral components of the current noise in the ultra-thin film devices was analysed at room temperature for different V_{BG}. Specific front and back gate voltages were selected to give a constant drain current of 1 nA and currents were sampled for 4 min at 1 kSa/s. Three measurements for each configurations have been averaged to compensate for statistical noise from the setup. Results for the gate-referred power spectral density (PSD) are shown in fig. 4a, where

$$ S_V = \left(\frac{1}{gm^2} \right) S_I, \qquad (2) $$

being S_I the drain current PSD and $gm = dI_{DS}/dV_{FG}$ (values extracted from the curves in fig.3a). The noise is reduced for $V_{BG} > 0$ V; this effect is commonly explained in terms of displacement of inversion carriers centroid from the top Si/SiO$_2$ interface [18]. At a fixed V_{FG}, this displacement is accompanied with an enhancement of the electrons mobility and higher currents are measured for increasing back-gate bias, as shown in the curves in fig. 3c with $V_{BG} << V_{BGTH}$.

C. SET characteristics

The SET operation was tested at 10 mK and 4 K. Coulomb oscillations are visible in the $I_{DS}(V_{FG})$ plots in fig. 5a and 5b. The periodicity of the drain current peaks is 25 mV and the extracted gate-to-dot capacitance is 6.4 aF. An equivalent dot radius of 36 nm was estimated according to a disk self-capacitance model. Broader peaks were measured at 4 K, where the thermal energy is not negligible compared to charging energy of the quantum dot. Positive and negative differential conductance values (g_m) of the transistor are plotted in fig. 5b.

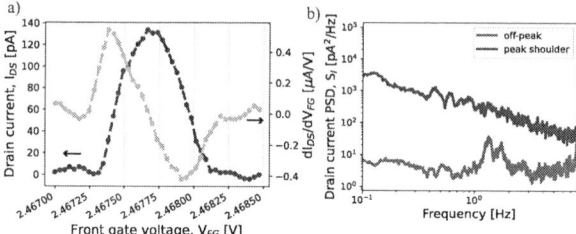

Fig. 6. a) Single current peak measured at 10 mK with $V_{DS} = 0.5$ mV and current sensitivity to front gate voltage. b) Power spectral density of the SET current measure at the points of maximum and minimum sensitivity; a peak at 1.4 Hz was measured, it is due to the pulse tube of the refrigerator.

For an accurate tuning of the device, drain current measurements were taken sweeping both front and barrier gates voltages. As shown in the color plot in fig. 5c, no conduction was enabled for barrier voltages below 1.1 V and front gate voltages lower than 2 V (at $V_{DS} = 1$ mV).

We biased the device to work at the conduction corner (green dot in fig. 5c) to analyze Coulomb blockade dependence on V_{DS} and V_{FG} with fixed potential barriers. A slight difference of $10\,mV$ was set between the two barriers to compensate for the small asymmetry in the transfer characteristics. Regular Coulomb diamonds were measured for $V_{FG} >> V_{TH}$, as shown in fig. 5d. Here we report data for a selected window of V_{FG} values for which constant diamonds dimensions were measured. For lower bias voltages, dimensions changed as a function of front-gate and barriers voltage. For a better control on the dot chemical potential a more complex architecture including a single plunger gate and lateral inversion gates should be adopted.

Different source of noise can be studied analyzing the PSD of the drain current when the SET is biased at its maximum sensitivity point. An example of an SET noise measurement is shown in fig. 6. A single peak was selected and sampled for 5 min to ensure the stability of the device and instrumentation over time. Then, the device was biased at the points of maximum and minimum sensitivity and the drain current sampled for 5 min; results are shown in fig. 6b. Typical $1/f$ noise was measured at the maximum dI_{DS}/V_{FG} point, white noise was sampled elsewhere. The noise levels are comparable to the ones reported in [19].

IV. CONCLUSION

We have reported the fabrication and characterization of four-gated FD-SOI SETs on thin and ultra-thin film substrates. Measurements were taken at room and cryogenic temperatures proving the functionality of the devices when operated both as FETs and SETs. Back-gate characterization was performed at room temperature on ultra-thin film devices and the electrical noise has been analysed. Coulomb blockade oscillations were measured at 4 K and 10 mK in SET operation and current noise measurements have been reported to prove the stability and sensitivity of the SET when operated as a charge sensor. We have found that the regime of operation could influence both sensitivity and charge noise levels as the displacement of inversion carriers centroid from the Si/SiO$_2$ interface can be dynamically controlled. This experimental work is a step forwards towards the development of optimized multi-gate FD-SOI planar quantum dot based quantum devices with tunable properties by electrostatic control.

REFERENCES

[1] D. Loss and D. P. DiVincenzo, "Quantum computation with quantum dots," *Physical Review A*, vol. 57, no. 1, p. 120, 1998.

[2] L. Petit, H. Eenink, M. Russ, Lawrie *et al.*, "Universal quantum logic in hot silicon qubits," *Nature*, vol. 580, no. 7803, pp. 355–359, 2020.

[3] L. C. Camenzind, S. Geyer, A. Fuhrer, R. J. Warburton, D. M. Zumbühl, and A. V. Kuhlmann, "A hole spin qubit in a fin field-effect transistor above 4 kelvin," *Nature Electronics*, pp. 1–6, 2022.

[4] C. H. Yang, R. Leon, J. Hwang, A. Saraiva, T. Tanttu, W. Huang *et al.*, "Operation of a silicon quantum processor unit cell above one kelvin," *Nature*, vol. 580, no. 7803, pp. 350–354, 2020.

[5] A. Saraiva, W. H. Lim, C. H. Yang, C. C. Escott, A. Laucht, and A. S. Dzurak, "Materials for silicon quantum dots and their impact on electron spin qubits," *Advanced Functional Materials*, vol. 32, no. 3, p. 2105488, 2022.

[6] F. Ansaloni, A. Chatterjee, H. Bohuslavskyi, B. Bertrand, L. Hutin, M. Vinet *et al.*, "Single-electron operations in a foundry-fabricated array of quantum dots," *Nature communications*, vol. 11, no. 1, pp. 1–7, 2020.

[7] A. Zwerver, T. Krähenmann, T. Watson, L. Lampert, H. C. George *et al.*, "Qubits made by advanced semiconductor manufacturing," *Nature Electronics*, vol. 5, no. 3, pp. 184–190, 2022.

[8] E. Charbon, F. Sebastiano, A. Vladimirescu, H. Homulle, S. Visser *et al.*, "Cryo-cmos for quantum computing," in *2016 IEEE International Electron Devices Meeting (IEDM)*. IEEE, 2016, pp. 13–5.

[9] O. Weber, "Fdsoi vs finfet: differentiating device features for ultra low power & iot applications," in *2017 IEEE International Conference on IC Design and Technology (ICICDT)*. IEEE, 2017, pp. 1–3.

[10] P. Galy, J. C. Lemyre, P. Lemieux, F. Arnaud, D. Drouin, and M. Pioro-Ladriere, "Cryogenic temperature characterization of a 28-nm fd-soi dedicated structure for advanced cmos and quantum technologies co-integration," *IEEE Journal of the Electron Devices Society*, vol. 6, pp. 594–600, 2018.

[11] H. Bohuslavskyi, S. Barraud, M. Cassé, V. Barral, B. Bertrand, L. Hutin *et al.*, "28 nm fully-depleted soi technology: cryogenic control electronics for quantum computing," in *2017 Silicon Nanoelectronics Workshop (SNW)*. IEEE, 2017, pp. 143–144.

[12] N. D. Stuyck, R. Li, S. Kubicek, F. A. Mohiyaddin, J. Jussot, B. Chan *et al.*, "An integrated silicon mos single-electron transistor charge sensor for spin-based quantum information processing," *IEEE Electron Device Letters*, vol. 41, no. 8, pp. 1253–1256, 2020.

[13] S. J. Angus, A. J. Ferguson, A. S. Dzurak, and R. G. Clark, "Gate-defined quantum dots in intrinsic silicon," *Nano letters*, vol. 7, no. 7, pp. 2051–2055, 2007.

[14] Q. Wang, Y. Chen, S. Long, J. Niu, C. Wang, R. Jia *et al.*, "Fabrication and characterization of single electron transistor on soi," *Microelectronic engineering*, vol. 84, no. 5-8, pp. 1647–1651, 2007.

[15] R. Maurand, X. Jehl, D. Kotekar-Patil, A. Corna, H. Bohuslavskyi, Laviéville *et al.*, "A cmos silicon spin qubit," *Nature communications*, vol. 7, no. 1, pp. 1–6, 2016.

[16] E. J. Connors, J. Nelson, H. Qiao, L. F. Edge, and J. M. Nichol, "Low-frequency charge noise in Si/SiGe quantum dots," *Physical Review B*, vol. 100, no. 16, p. 165305, 2019.

[17] A. Beckers, F. Jazaeri, H. Bohuslavskyi, L. Hutin, S. De Franceschi, and C. Enz, "Characterization and modeling of 28-nm fdsoi cmos technology down to cryogenic temperatures," *Solid-State Electronics*, vol. 159, pp. 106–115, 2019.

[18] C. G. Theodorou, E. G. Ioannidis, S. Haendler, N. Planes, F. Arnaud, J. Jomaah *et al.*, "Impact of front-back gate coupling on low frequency noise in 28 nm fdsoi mosfets," in *2012 Proceedings of the European Solid-State Device Research Conference (ESSDERC)*. IEEE, 2012, pp. 334–337.

[19] B. M. Freeman, J. S. Schoenfield, and H. Jiang, "Comparison of low frequency charge noise in identically patterned Si/SiO2 and Si/SiGe quantum dots," *Applied Physics Letters*, vol. 108, no. 25, p. 253108, 2016.

Cryogenic Comparator Characterization and Modeling for a Cryo-CMOS 7b 1-GSa/s SAR ADC

Gerd Kiene*, Aishwarya Gunaputi Sreenivasulu*, Ramon W.J. Overwater, Masoud Babaie, and Fabio Sebastiano

Department of Quantum & Computing Engineering & QuTech , TU Delft, The Netherlands

Email: g.kiene@tudelft.nl

Abstract—This paper reports the experimental characterization and modelling of a stand-alone StrongARM comparator at both room temperature (RT) and cryogenic temperature (4.2 K). The observed 6-dB improvement in the comparator input noise at 4.2 K is attributed to the reduction of the thermal noise and to the suppressed shot noise in the MOS transistors becoming dominant at cryogenic temperature. The proposed model is employed in the design of a loop-unrolled 2× time-interleaved 1-GSa/s 7b SAR ADC for spin-qubit readout. As predicted by the comparator model, the ADC is noise-limited at RT to a SNDR of 38.2 dB at Nyquist input, while this improves to 41.1 dB at 4.2 K, now limited by distortion, thus resulting in the state-of-the-art FoM$_W$ for cryo-CMOS ADC of 20.9 fJ/conv-step.

Index Terms—Cryo-CMOS, SAR, ADC, latching comparator, strongARM, noise measurement

I. INTRODUCTION

Quantum computers based on cryogenically cooled spin qubits in semiconductors offer a promising scaling route towards the very large number of qubits required for practical quantum applications [1]. While the core of these cryogenic computers is operating based on quantum mechanics, the algorithm execution is controlled by classical electronics typically operating at room temperature (RT). This poses a significant cabling bottleneck when increasing the qubit count, projected to grow well beyond a few thousands. To address this, a cryogenic CMOS (cryo-CMOS) electronic interface targeted to operate in the same cryostat as the qubits has been proposed [2].

The readout of semiconductor spin qubits can be performed using capacitive or resistive sensors. Their impedance can be remotely monitored by measuring the reflection coefficient at a 50-Ω line connected to the sensor via a matching network, thus achieving state-of-the-art readout speeds and allowing for frequency multiplexing of several qubit channels [3]. In such an RF readout, an RF tone weak enough to not disturb the qubit is reflected, amplified, and subsequently digitized. Due to the large bandwidth necessary for frequency multiplexing, a low-power high-speed ADC with medium resolution (6-8b) is required, thus making capacitive-DAC SAR ADCs the ideal candidates [3], [4].

The noise of such medium-resolution SAR ADCs is typically dominated by the comparator noise, which would be expected to drop by $\approx 70\times$ when reducing the temperature from RT to 4.2 K, if purely thermal in nature. However,

*These authors contributed equally to the work.

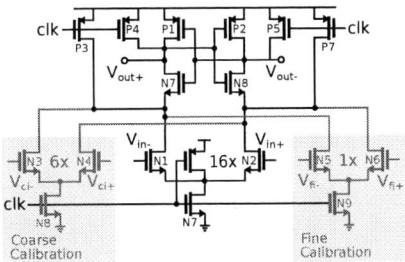

Figure 1. Comparator schematic; N3-4 (N5-6) is the coarse (fine) offset-calibration pair with input $V_{ci+,-}$ ($V_{fi+,-}$) provided by the ADC calibration circuit.

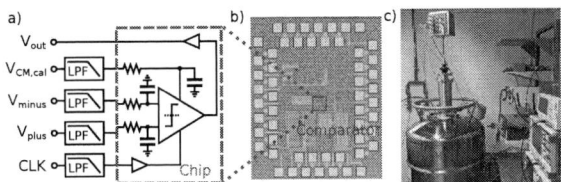

Figure 2. Test-chip for comparator characterization: a) schematic, b) chip micrograph, c) measurement setup.

characterization of various circuits at cryogenic temperatures suggests that this noise scaling is not achieved. For example, the noise figure of the LNA in [5] showed only an improvement from 2.5 dB to 0.6 dB, much less than predicted by assuming pure thermal noise. The absence of a comprehensive noise model for cryo-CMOS devices and the lack of direct noise measurements of dynamic latching circuits at cryogenic temperatures hinder the optimization of cryo-CMOS ADCs. To fill this gap, we report, for the first time, the measurement and the analysis of the noise and offset of a latching comparator at cryogenic temperatures. Based on those findings, we optimized the design of a prototype 7b 1-GSa/s SAR ADC, thus verifying the validity of the comparator characterization and modelling in a practical application.

In the following, Section II describes the proposed StrongARM comparator implementation, results and analysis. Section III covers the circuit design of the prototype ADC chip, concluding with its experimental validation.

II. DYNAMIC COMPARATOR

The StrongARM comparator is a common choice in medium-resolution ADCs for its high speed, power efficiency

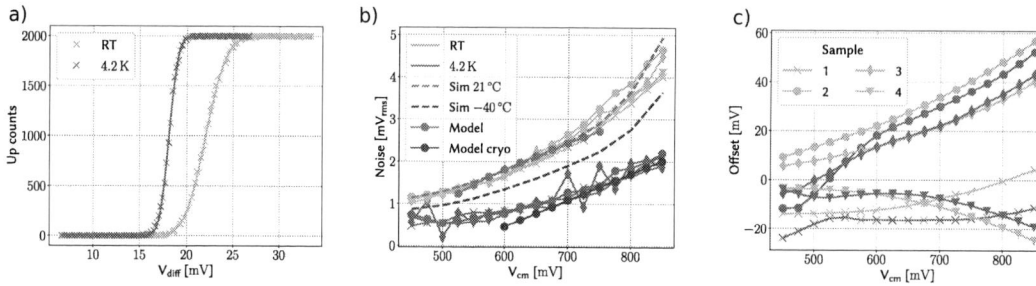

Figure 3. Comparator characterization: a) example measurement; b) noise voltage over input CM (V_{cm}) for four measured samples, simulations using the foundry device model, and the model in Eq. (5); c) measured offset over V_{cm} for four samples. Symbols for each sample are the same in b) and c).

and compactness. The proposed implementation (Fig. 1) comprises two additional coarse and fine calibration pairs to allow for an efficient foreground calibration [4]. In the following, a model for this circuit's noise and offset – relevant comparator parameters for ADC design – are derived and compared with experimental results. A specific focus is put on the behavior over input common-mode (CM), as this changes significantly due to both threshold-voltage increase at cryogenic temperature and CM variations on the capacitive DAC during a typical SAR conversion.

A. Measurements

The stand-alone comparator in Fig. 1 has been fabricated in a 40-nm CMOS process (Fig. 2b). All transistors are implemented as minimum length devices. On-chip input capacitors of 200 fF have been used to minimize the kT/C noise present at the input. A dip-stick setup employing liquid He (Fig. 2c) is used for cryogenic characterization, adopting the fixture shown in Fig. 2a. All DC input lines are filtered to reject interference coupling to the long cables in the dip-stick and are driven by 18b battery-operated RT DAC modules. We sweep the differential input voltage for a given CM and record the comparator decision statistics, from which we extract input noise and offset (Fig. 3a). On all sweeps, the CM voltage of the input and calibration pairs is the same.

When increasing the CM voltage at RT for four samples (Fig. 3b), we observe a monotonic increase of noise that is captured by the room-temperature simulation. When cooling down to 4.2 K, the transistor threshold voltage increases by about 80 mV, resulting in a shift of the curve towards a higher common-mode voltage. Compared to RT, the input noise voltage is about 2× lower and also significantly lower from simulations at −40 °C, the edge of the validity of the foundry supplied model.

The offset voltage (Fig. 3c) also shows a strong dependence on input CM, but no significant change is observed from RT to 4.2 K.

B. Model derivation

Previous analytical derivations for the noise of regenerative comparators assuming thermal-noise sources suggest that the input-referred noise power scales with absolute temperature [6], [7]:

$$\sigma_{in}^2 \propto T \tag{1}$$

However, the experimental characterization does not show the predicted large difference in noise power between RT and 4.2 K. To address the mismatch between the experimental results and the commonly accepted theory, we derive the input-referred noise by assuming the transistor noise to be a combination of thermal noise and suppressed shot noise [8].

First, we calculate the gain of the input pair during its amplification phase. This phase lasts from the positive clock edge till the input and calibration pair transistors, N1-N6, have discharged the parasitic capacitance C_p at their drains enough for the latch NMOS transistors (N7 and N8 in Fig. 1) to turn on, initiating exponential latching. The amplification time T_{amp} can be computed considering operation around common-mode as:

$$T_{amp} = \frac{2V_{th,N}C_p}{g_{m,all}V_{od}} \tag{2}$$

where $V_{th,N}$ is the NMOS device threshold voltage, $g_{m,all}$ is the combined transconductance of the input and calibration pairs, $V_{od} = V_{gs} - V_{th,N}$ is the input-pair overdrive voltage, and we assume strong-inversion operation for the input transistors. The gain of the input pair can then be written as:

$$A = \frac{g_m}{C_p}T_{amp} = \frac{2W_{in}V_{th,N}}{W_{all}V_{od}} \tag{3}$$

where W_{in} and $W_{all} = W_{in} + W_{cal}$ are the width of the input pair and the combined width of input and calibration pairs, respectively. We neglect the MOS output impedance. The power spectral density of the current noise in each half of the comparator is dominated by thermal and suppressed shot noise and it is given by:

$$S_{id} = 2qFI_{CM,all} + 4k_bT\gamma g_{m,all} \tag{4}$$

with q is the elementary charge, F the Fano factor, $I_{CM,all}$ the combined common-mode current, k_b the Boltzmann constant and T the absolute temperature. The input-referred noise in the amplification phase is then given by

$$\sigma_{in}^2 = \frac{qFW_{in}V_{od}^2}{2C_pW_{all}V_{th,N}} + \frac{2k_bT\gamma W_{in}V_{od}}{W_{all}C_pV_{th,N}} \tag{5}$$

978-1-6654-8495-4/22 $31.00 © 2022 IEEE

This expression clearly show the increase of the input-referred noise with the input common-mode $V_{cm} = V_{od} + V_{th,N}$ that we observe in the measurement.

According to our model, while thermal noise dominates at RT by contributing 84% of the total noise for $V_{cm} = 650\,\mathrm{mV}$, it becomes negligible at cryogenic temperatures. At $4.2\,\mathrm{K}$, the noise behavior is mainly determined by the shift in threshold voltage and the temperature dependence of the Fano factor. The threshold voltage (extracted from DC characterization of individual transistors) increases from $450\,\mathrm{mV}$ at RT to $530\,\mathrm{mV}$ at $4.2\,\mathrm{K}$, decreasing the overdrive voltage and therefore also lowering the noise. In practice, the $V_{th,N}$ change results in a shift of the noise curve towards higher V_{cm}.

When assuming a mix of thermal noise with $\gamma = 1$ and shot noise with $F = 0.1$ at RT, we achieve in good fitting between the foundry device model and Eq. (4) in transistor noise simulation. As shown in Fig. 3b, this also results in good match between the proposed model and the experimental data. The region of validity of our model is bounded by the subthreshold region towards lower V_{cm} and by the latch noise becoming relevant for higher V_{cm} as the input-pair gain decreases, as predicted in Eq. (3).

Using the model we can achieve a good fit for the dependency of the noise on V_{cm} even at $4.2\,\mathrm{K}$. However, here the Fano factor must be increased to $F = 0.25$ (only at $4.2\,\mathrm{K}$) to achieve an acceptable match with experimental data. Such an increase in the suppression factor could be attributed to the device behaving closer to an ideal ballistic device at lower temperatures, e.g., due to a decrease in carrier scattering in the channel. Although the shot-noise in nanometer CMOS devices have been extensively studied at RT [9], no comprehensive data on the behavior of F at cryogenic temperatures is available, thus pointing to the need for further cryogenic device characterization and physical modelling. As an outcome, the proposed model successfully captures the circuit noise behavior within the CM region, which is mainly relevant for high-speed ADC design, shown in Section III.

The comparator offset due to the input and calibration pairs can be calculated:

$$\Delta V_{in,os} = \left(\Delta V_{th,in} + \frac{\Delta \beta}{\beta} \frac{V_{od,in}}{2} \right) \sqrt{1 + \frac{W_{cal}}{W_{in}}} \quad (6)$$

where β is the device current gain, $\Delta V_{th,in}$, $\Delta \beta$ are the standard deviation of mismatch in threshold voltage and current gain of the input pair respectively. As reported in [10], the mismatch in threshold voltage shows a negligible variation at cryogenic temperature while the mismatch in β is expected to increase by 25%, thus explaining the little observed change in Fig. 3c.

III. CRYO-CMOS ADC

A. Circuit design

We verify the StrongARM comparator model in a prototype ADC tested at both RT and $4.2\,\mathrm{K}$. Similar to [4], the circuit is a $2\times$ time-interleaved loop-unrolled (LU) 7b SAR ADC. The 6b capacitive-DAC (CDAC) employs top-plate sampling

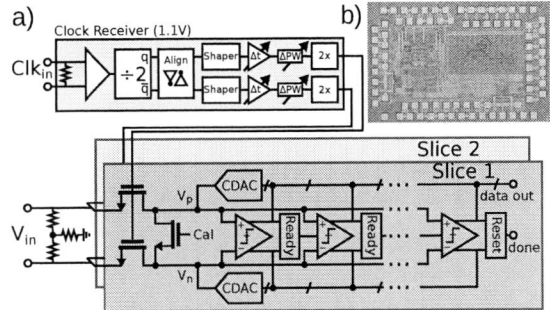

Figure 4. ADC protoype: a) block diagram; b) micrograph.

and is implemented using custom $0.5\,\mathrm{fF}$ unit cells. Since the ADC is intended for integration with an input on-chip driver in the target qubit-readout system, non-switched capacitors connected to ground are included in the CDAC to reduce the input range of the ADC to $600\,\mathrm{mV_{pp}}$, thus easing the output swing specifications of the input driver. To further facilitate the input driver, the ADC also implements a reset to V_{cm} before sampling, followed by a DAC switching similar to [4]. This provides higher V_{cm} for the comparators after the MSB decision to compensate for the increased V_{th} at cryogenic temperature.

To improve power efficiency compared to [4], the entire sampling clock chain has been implemented using thin-oxide devices. At the end of the clock chain, thin-oxide NMOS switches with $2\times$ boosted control voltage are used to sample the input signal. A pulse shaper in the clock chain tunes the duty-cycle of the sampling clock to provide more than 50% of the clock period for conversion. The $2\times$ time-interleaved ADC suffers from interleaving artifacts caused by timing mismatch. To calibrate these, a pair of digitally controlled delay lines are designed to adjust the delay of the sampling clock provided to each slice.

The StrongARM comparator measured and analyzed in Section II is used in the LU-SAR loop. The comparator design is optimized for speed and the ADC is targeted to be comparator-noise limited at RT. Since the comparator offset causes distortion for LU-SAR, the input voltages of the fine and coarse differential pair in Fig. 1 are generated by a resistive DAC to calibrate the offset, similar to [4] but with an optimized scheme with 5b (coarse) and 6b (fine) resolution. During foreground calibration, a switch with boosted static gate voltage is activated to short V_p and V_n (Fig. 4a) to calibrate every comparator at the correct V_{CM} level. Resetting the comparator tail node (drain node of N7 in Fig. 1) eliminates any signal-dependant offset, which would otherwise limit the achievable calibration accuracy, as was the case in [4].

B. Measurement

The chip is implemented in a 40-nm process, see the micrograph in Fig. 4b. Five sample dies have been bonded to the test PCBs and characterized in the same dip-stick environment as the comparator.

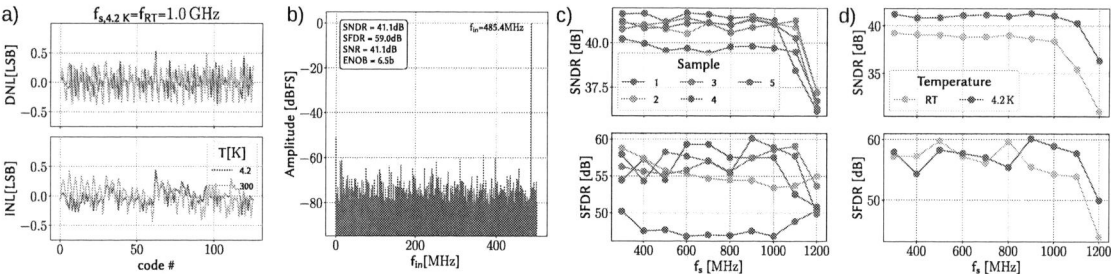

Figure 5. ADC characterization: a) INL and DNL for worst sample (sample 4) ; b) output spectrum at 1 GS/s and 4.2 K (sample 5); c) SNDR and SFDR at 4.2 K for all measured samples. SFDR does not include the TI-induced spur at $f_s/2$ d) SNDR and SFDR at RT and 4.2 K for a typical sample (sample 5).

Table I
COMPARISON TABLE

	This work		[4]		Kull, ISSCC 2013	Kull, ISSCC 2017
Temperature [K]	300	4.2	300	4.2	300	300
Architecture	TI SAR		TI SAR		SAR	PP-SAR
Max f_s[MS/s]	1000		900	1000	1300	950
Resolution [bit]	7		6-8		8	10
Technology [nm]	40		40		32	14
Supplies [V]	1.1		1.1(core), 2.5(clock)		1	0.7
Input range [Vpp]	0.6		0.7	0.7	0.5	0.5
SNDR@Nyquist [dB]	38.2^2	41.1^2	33.4	36.2	39.3	50
SFDR [dB]	>50^2	>50^2	48.4	48.5	49.6	58
Power [mW]	1.94^1		10.3^1	10.6^1	3.1	2.26
FoM$_w$[fJ/c.step]	29.2$^{1, 2}$	20.9$^{1, 2}$	260^1	200^1	28	8.9
Core area [mm²]	0.042^1		0.045^1		0.0015	0.0016

^{1}Full ADC, including clock receiver ^{2}typical sample

The comparator offset calibration based on binary search and the timing skew calibration are implemented off-chip. To optimize the linearity, manual calibration fine-tuning is done besides binary search for samples 1 and 3.

In Fig. 5a, the static performance of the worst-case sample (sample 4) is reported. Unlike [4], no signal-dependent residual offset appears and the maximum INL/DNL is below 0.55 LSB for both RT and 4.2 K. The Nyquist-rate input measurement in Fig. 5b shows an SNDR of 41.1dB. Combined with the 1.94 mW power drawn from the nominal 1.1-V supply, this results in a typical Walden FOM$_W$ of 20.9 fJ/conv·step (including the clock receiver), a $10\times$ improvement compared to [4], see Table I. All tested samples reach 1 GSa/s before dropping in SNDR value, see Fig. 5c. The dynamic performance of sample 4 is limited by the accuracy of the offset calibration. The higher sampling rates at 4.2 K are attributed to the increased speed of the comparator and the digital circuitry in the SAR loop.

Comparing the dynamic performance at RT and 4.2 K in Fig. 5d, reveals an improvement of 2-dB SNDR at 4.2 K. At 4.2 K harmonic distortion is not limiting the performance as indicated by the measured SFDR, and the SNDR is in line with the quantization-noise-limited performance of an ADC with 0.5-LSB INL. Therefore, the 6-dB improvement in comparator noise measurements at 4.2 K (Fig. 3b) makes the comparator noise negligible at cryogenic temperature.

IV. CONCLUSION

We have reported the first direct noise and offset measurements and modelling on a latching comparator at cryogenic temperatures. The gathered evidence points towards a relatively small improvement in white noise at cryogenic temperature, due to the white noise originating from a mix of thermal and shot noise. The proposed model has been applied to the design of a high-speed cryo-CMOS SAR ADC for spin-qubit readout, resulting in state-of-the-art performance compared to previously reported cryo-CMOS ADCs.

ACKNOWLEDGMENT

The authors would like to thank Intel corporation for funding and Atef Akhnoukh and Zu-yao Chang for technical support.

REFERENCES

[1] L. Vandersypen et al., "Interfacing spin qubits in quantum dots and donors—hot, dense, and coherent," npj Quantum Information, vol. 3, no. 1, pp. 1–10, 2017.
[2] B. Patra et al., "Cryo-CMOS circuits and systems for quantum computing applications," IEEE Journal of Solid-State Circuits, vol. 53, no. 1, pp. 309–321, 2017.
[3] J. Park et al., "A Fully Integrated Cryo-CMOS SoC for State Manipulation, Readout, and High-Speed Gate Pulsing of Spin Qubits," IEEE Journal of Solid-State Circuits, vol. 56, no. 11, pp. 3289–3306, 2021.
[4] G. Kiene et al., "13.4 a 1GS/s 6-to-8b 0.5 mw/Qubit Cryo-CMOS SAR ADC for Quantum Computing in 40nm CMOS," in 2021 IEEE International Solid-State Circuits Conference (ISSCC), vol. 64. IEEE, 2021, pp. 214–216.
[5] B. Prabowo et al., "A 6-to-8GHz 0.17 mW/qubit cryo-CMOS receiver for multiple spin qubit readout in 40nm CMOS technology," in 2021 IEEE International Solid-State Circuits Conference (ISSCC), vol. 64. IEEE, 2021, pp. 212–214.
[6] P. Nuzzo et al., "Noise analysis of regenerative comparators for reconfigurable ADC architectures," IEEE Transactions on Circuits and Systems I: Regular Papers, vol. 55, no. 6, pp. 1441–1454, 2008.
[7] T. Sepke et al., "Noise analysis for comparator-based circuits," IEEE Transactions on Circuits and Systems I: Regular Papers, vol. 56, no. 3, pp. 541–553, 2008.
[8] R. Navid et al., "High-frequency noise in nanoscale metal oxide semiconductor field effect transistors," Journal of applied physics, vol. 101, no. 12, p. 124501, 2007.
[9] S. Das and J. C. Bardin, "Characterization of shot noise suppression in nanometer mosfets," in 2021 IEEE MTT-S International Microwave Symposium (IMS). IEEE, 2021, pp. 892–895.
[10] P. A. t' Hart et al., "Characterization and modeling of mismatch in cryo-CMOS," IEEE Journal of the Electron Devices Society, vol. 8, pp. 263–273, 2020.

978-1-6654-8495-4/22 $31.00 © 2022 IEEE

A cryogenic SRAM based arbitrary waveform generator in 14 nm for spin qubit control

Mridula Prathapan, Peter Mueller, Christian Menolfi[§], Matthias Brändli, Marcel Kossel, Pier Andrea Francese, David Heim, Maria Vittoria Oropallo, Andrea Ruffino, Cezar Zota and Thomas Morf

IBM Zurich Research Laboratory, Rüschlikon, Switzerland. Email: mrp@zurich.ibm.com

Abstract—Realization of qubit gate sequences require coherent microwave control pulses with programmable amplitude, duration, spacing and phase. We propose an SRAM based arbitrary waveform generator for cryogenic control of spin qubits. We demonstrate in this work, the cryogenic operation of a fully programmable radio frequency arbitrary waveform generator in 14 nm FinFET technology. The waveform sequence from a control processor can be stored in an SRAM memory array, which can be programmed in real time. The waveform pattern is converted to microwave pulses by a source-series-terminated digital to analog converter. The chip is operational at 4 K, capable of generating an arbitrary envelope shape at the desired carrier frequency. Total power consumption of the AWG is 40-140 mW at 4 K, depending upon the baud rate. A wide signal band of 1-17 GHz is measured at 4 K, while multiple qubit control can be achieved using frequency division multiplexing at an average spurious free dynamic range of 40 dB. This work paves the way to optimal qubit control and closed loop feedback control, which is necessary to achieve low latency error mitigation and correction in future quantum computing systems.

Index Terms—Quantum control electronics, cryogenic CMOS, quantum computing, arbitrary waveform generator, SRAM, spin qubit systems, 14 nm bulk FinFET technology

I. INTRODUCTION

A continuous trend in the scaling of qubit numbers in quantum systems has been established over the past few years. Realization of large scale error corrected qubit systems would demand the miniaturization and integration of different components inside the cryostat. The scaling bottleneck of quantum control systems has motivated the cryo-CMOS community to explore the prospects of cryogenic circuit design at 4 K and below. The major challenge is to meet the reliability and performance requirements put forth by the future quantum computing systems, while consuming the least amount of power, limited by the available cooling power in the cryostat. Short and optimized control pulses are the key to realize fast qubit gates with improved gate fidelity. Fast gates suffer from leakage effects and additional unitary errors caused by the large bandwidth of the short control pulses [1]. Adopting techniques to increase the fidelity of short duration single qubit gates using optimized control pulses in a closed-loop fashion is crucial to build low error quantum systems [2].

Spin systems have an advantage of being able to operate at higher temperatures. CMOS compatible FinFET spin qubits operating at 1-4 K have been demonstrated by [3]. Scaling of quantum computing systems beyond 1000 qubits, irrespective

of their qubit technology, puts forth an inevitable need for intelligent sub-systems operating at proximity to the qubits. These include cryogenic control and readout circuitry with maximum affordable signal processing capabilities to ensure low-latency error correction and reduced data rate for links to room temperature, as shown in Fig. 1a. We propose an SRAM based arbitrary waveform generator, that paves the way to such systems. Advanced CMOS technology such as 14 nm bulk FinFET offers high interconnect density to support digitally assisted designs that are scalable. Control and readout circuitry implemented in advanced CMOS process nodes like 14 nm and below could be potentially co-packaged with spin qubit arrays, a well-known advantage of spin systems over other platforms. A radio frequency (RF) digital to analog converter (DAC) with wide bandwidth and sufficient spurious free dynamic range (SFDR) could be used to control multiple qubits using frequency division multiple access (FDMA). Apart from simplifying the wiring bottleneck in the cryostat, the merits of using FDMA over 1:1 qubit to control DAC ratio for large scale error corrected systems need to be evaluated.

The source series terminated (SST) transmitter proposed in [4] and its improved version [5] have a proven track of record in high speed I/O links applications. The advantages of SST concept are low power operation, high content of digital circuitry that supports scaling, and comparatively large signal swing. These concepts found their application in quantum control electronics owing to their scalability and overlapping specifications with high speed IO links, with design enhancements to achieve low power. The transmitter used in this work was designed by C. Menolfi et al., as a single ended adaptation of [4]. The main difference in architecture compared to the SST transmitter from [4], is the digital part with an SRAM based pattern generator. SRAM based architectures offer high versatility with the ability to store multiple gate sequences for qubit control. Moreover, they allow the control signals to be pre-distorted, in order to compensate the non-linearities and drifts caused by cabling. It helps to mitigate the effect of intersymbol interference (ISI) caused by cabling via feed forward equalization (FFE) at no added power cost.

II. CHIP ARCHITECTURE

The RF DAC architecture shown in Fig. 1b is an 8-bit SRAM-based single-ended SST transmitter. Four dedicated SRAM instances, each 512×16 B provide a total 32 KB on-chip memory. There is a 8×32 b wide interface between

---

[§]Presently with Cisco Systems, Inc., Thalwil, Switzerland.

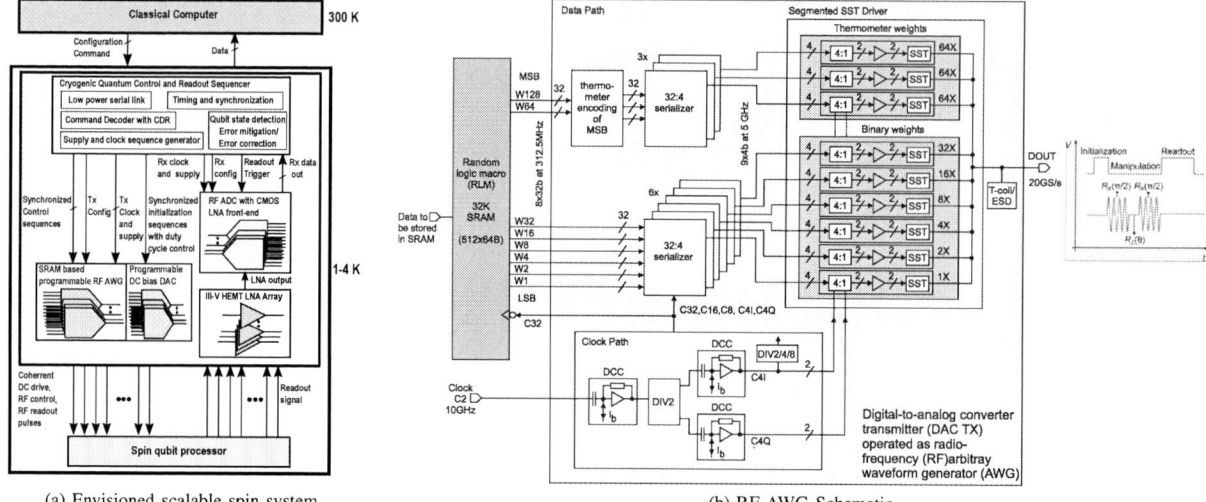

(a) Envisioned scalable spin system

(b) RF AWG Schematic

Fig. 1: Figure showing the SRAM based RF AWG architecture, together with the envisioned cryogenic qubit control system co-packed with hot spin qubits at 1-4 K. The proposed AWG can be integrated into a closed-loop feedback control system

SRAM and DAC TX. A 1/32-rate clock, C32, is used for the data capture at the DAC TX input as well as with opposite phase for the clocking of the SRAM. Because the most significant bit (MSB) is thermometer encoded to improve the linearity and also reduce the fanout spread between MSB and least-significant bit (LSB) weights in the DAC output stages, the 8 b resolution is implemented with nine 32:4 serializers that multiplex the input data to 9×4 b data streams at quarter-rate. Each quarter-rate data stream drives a specific DAC weight consisting of 4:1 multiplexer, pre-driver and SST output stage. The DAC weights are segmented into 6b binary weights between LSB and MSB-2 while the MSB and MSB-1 define the thermometer-encoded weights. The inclusion of the 4:1 multiplexer into each DAC segment increases the overall clock load but enables identical loading at the full data rate amongst all DAC weights, leading to superior timing jitter performance.

The clock path operates on an incoming half-rate clock, C2, that first goes into a duty-cycle correction (DCC) block followed by a frequency divider (DIV2) to produce quadrature clocks. Hence the DCC block in front of the DIV2 allows a quadrature-error correction (QEC). It is implemented as an AC-coupled trip-point-biased inverter with programmable bleeding current. Each quadrature clock is followed by a separate DCC block to adjust the duty cycle of the pertinent quadrature phase pairs. The quarter-rate quadrature signals C4I and C4Q are fed to the 4:1 multiplexers of the DAC segments as well as to the 32:4 serializers together with the sub-rate clocks C8, C16 and C32 that are derived from C4I via frequency division in the clock path.

The digital part consists of a pattern generator and a bidirectional serial interface as shown in Fig. 2a. The pattern generator is responsible for generating a 256 b input pattern to the RF DAC. It consists of a controller state machine, 32 KB

SRAM, data path logic with byte rotational capability and a digital serializer. The design is implemented in a 14 nm bulk CMOS FinFET technology and has an active area of 0.095 mm², including T-coil, ESD and SRAM.

III. CRYOGENIC MEASUREMENTS

The experiments described in this work were conducted at room temperature, liquid N_2 and He (300 K, 77 K, 4 K) on a dip stick set up. The chip was packaged in a miniature PCB mounted in a cryogenic dip stick shown in Fig. 2b. The bandwidth of interest is identified as 4-8 GHz, that matches both transmons and the spin qubits in general. The experiments target the control requirements of the spin system described in [3]. Fig. 2c shows measured sub-5 ns pulse sequences at 4 K, showing granular control over their amplitude, pulse duration and spacing. At 4 K the measured signal shows larger signal amplitude when compared to the same measurement at room temperature and 77 K (not shown in the figure). The effect could be attributed to a combined effect of increase in the device drain saturation current for a given applied potential [6] and reduced cable losses in the dip stick. To perform linearity measurements, the DAC was programmed to output all 256 codes, creating a DC ramp in steps at the output. 1000's of samples were averaged for each DAC input code to calculate the output voltage. The measured INL at 4 K is < 2 LSB and DNL is < 1 LSB. The ratio between MOS resistance and metal resistance does not appreciably vary over temperature, which makes the SST driver topology in this technology exceptionally linear over such broad temperature range, compared to the current steering DAC in [8].

Fig. 2d and 2e show a two tone raised cosine pulse of width 200 ns in time domain and frequency domain respectively. A power of -33 dBc IM3 at -43 dBm is calculated from Fig. 2e,

978-1-6654-8495-4/22 $31.00 © 2022 IEEE

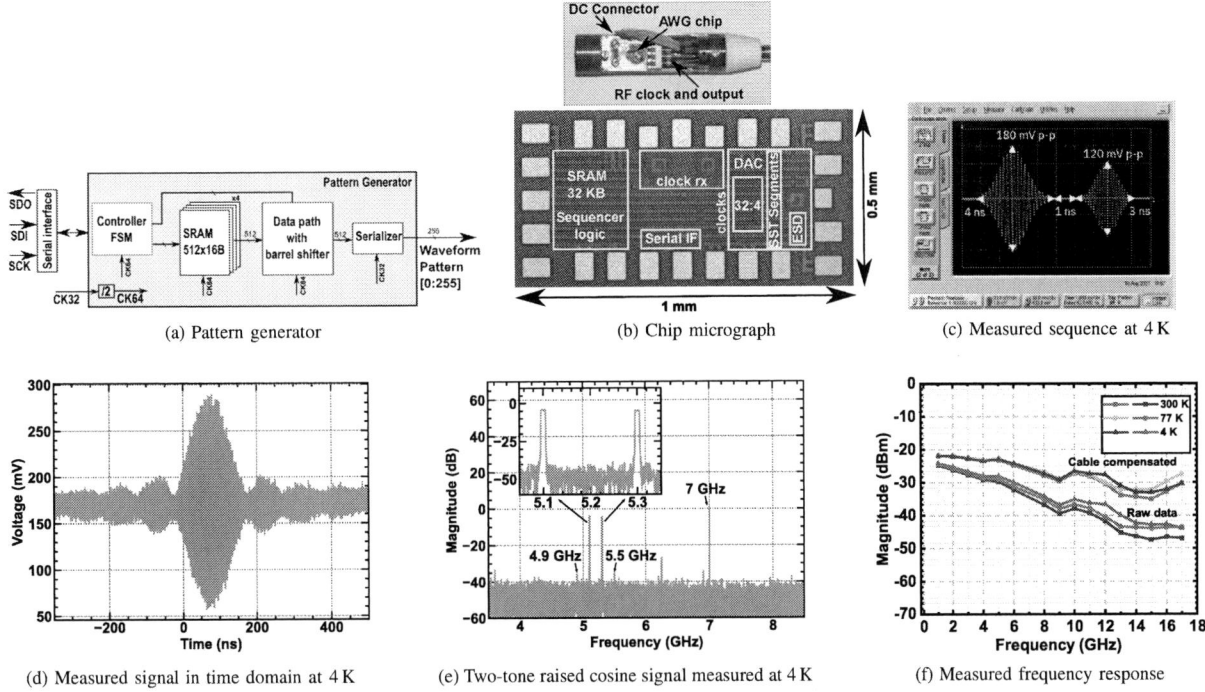

(a) Pattern generator   (b) Chip micrograph   (c) Measured sequence at 4 K

(d) Measured signal in time domain at 4 K   (e) Two-tone raised cosine signal measured at 4 K   (f) Measured frequency response

Fig. 2: (a) Simplified block level representation of the pattern generator (b) Chip Micrograph (bottom), dip stick setup (top) (c) Measured Gaussian sequences with programmable amplitude, pulse duration and spacing at 4 K (d) Time domain dual-tone raise cosine signal (e) Frequency domain shows the 3rd order intermodulation products at 4 K for 5.1 and 5.3 GHz. The magnitude is normalized to the largest amplitude corresponding to clock feedthrough at 7 GHz (f) Measured frequency response

measured at 4 K for a dual tone output of 5.1 GHz and 5.3 GHz. The tones are spaced 200 MHz apart to match the qubit spacing in [3]. A 40 dB average spurious free dynamic range is achieved over a wide bandwidth of 1 to 17 GHz, which is in accordance with the current state of the art. The total harmonic distortion is less than 2%. The operational frequency range allows the control of multiple spin qubits with different Larmor frequencies, based on the assumption that a large multiplexing ratio will be supported by the error corrected spin systems in future.

Jitter measurements were performed to analyze the drift on the control signal. Optimal linear equalizer coefficients for the cryogenic set up were calculated for a known pattern such as PRBS7. A 35% reduction in deterministic jitter is observed after feed forward equivalization (FFE) Fig. 3d. A square wave output at 5 GHz, with FFE, shows < 2 ps deterministic jitter at 4 K as shown in Table I. One of the advantages of SRAM based AWG architecture lies in its ability to adapt the FFE coefficients to equalize the changes in cryostat cabling over time. Since it is done through input data pattern, no extra power is required. Moreover, it facilitates real time calibration.

Fig. 3b shows the power measurements at 4 K. The SRAM is programmable throughout supply and frequency range shown in Fig. 3b. The power consumption could be further minimized

by dynamically lowering the power supply of the SRAM, while the data is held safe. Although the digital part of the design operates at a fraction of the reference clock, the total power consumption increases with the sampling frequency (2×reference clock), as shown in Fig. 3b. On the other hand, high frequency operation of RF up conversion architectures [7]- [9] would demand high frequency LO and mixer, adding up to their total power consumption. The digital blocks consume only 20% of the total power, where as the power consumption of [8]- [9] is largely dominated by the digital part as shown in Table II. The total power consumption is often reported as power per qubit for a given number of frequency multiplexed qubits per controller. This ratio may vary depending upon the DAC bits, linearity, target SFDR, qubit frequency spacing, etc. A total power of 2-4 mW/qubit can be estimated for the proposed controller, with a multiplexing ratio of 1:20 (drive:qubit) in our target qubit platform [3], with the qubit frequency spacing of 200 MHz within 4-8 GHz band.

TABLE I: Jitter measurements for square wave at 5 GHz

Temperature	Total jitter[a]	Random[a]	Deterministic[a]
300 K	4.80 ps	0.237 ps	1.56 ps
77 K	5.91 ps	0.239 ps	2.64 ps
4 K	4.98 ps	0.221 ps	1.95 ps

[a]Measurements were done using precision timebase scope.

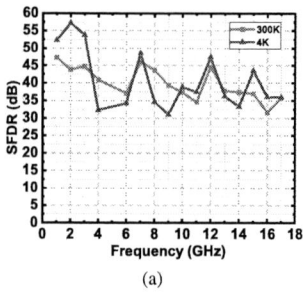 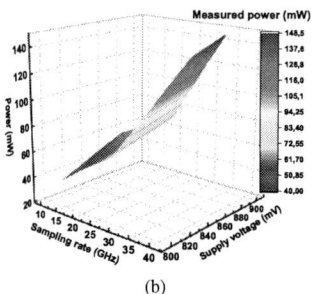 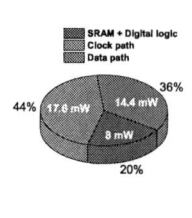 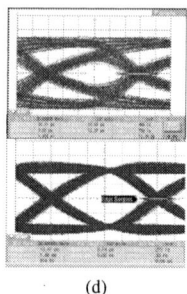

(a)   (b)   (c)   (d)

Fig. 3: (a) SFDR plots with unaccounted cable losses at 4 K (b) Power vs Sampling clock and Supply voltage at 4 K. The SRAM is programmable throughout the range. (c) Power breakdown chart at 14 GHz and 0.8 V, (d) PRBS7 eye diagram at 20 GS/s at 4 K shows that the deterministic jitter due to cabling improves by 35% by applying feed forward equalization

CONCLUSIONS

A study on the cryogenic operation of an SRAM based AWG with an RF DAC in 14 nm FinFET technology is reported. The proposed AWG can generate RF control pulses of any envelope shape with programmable amplitude, carrier frequency, pulse spacing and duration. Furthermore, it enables real time feed forward equalization to compensate drift due to changes in the cryostat cabling, at no added power cost. The

TABLE II: Comparison with current state of the art

Features	*This work*	[9]	[8]	[7]
Qubit Platforms	spin	spin	transmons, spin	transmons
Functionality	RF control	Gate pulsing, RF control, Readout	RF control	RF control
Chip area	0.5 mm²	16 mm²	4 mm²	1.6 mm²
Technology	14 nm FinFET	22 nm FinFET	22 nm FinFET	28 nm bulk
Power consumption	Total power: 40-140 mW Analog: 32-54 mW Digital: 8-16 mW	Total drive power[b]: 93-223 mW Analog: 83 mW or 5.2 mW/qubit Digital: 10-140 mW	Total power[b]: 384.4 mW Analog: 54.4 mW or 1.7 mW/qubit Digital: 330 mW	2 mW/qubit
Frequency multiplexing	yes	yes	yes	no
Drive to qubit ratio	N/A	1:16	1:32	N/A
Frequency range	1-17 GHz	11-17 GHz	2-20 GHz	4-8 GHz
IM3	−33 dBc at −43 dBm	−50 dBc at −17 dBm	−50 dBc at −18 dBm	N/A
On-chip storage	32 KB SRAM	RAM, digital memory	SRAM	Digital memory
Waveform storage	32 K points AWG	16 K points AWG	40 K points AWG	Fixed 22 points
DAC bits	8	10	10	11

[b] Calculated by multiplying the reported power/qubit by the reported number of qubits per controller for the given system

SRAM is fully functional and programmable at 4 K, within the given voltage and frequency range. The proposed digitally dominated architecture eliminates the need of local oscillators and mixers, reaping the benefits of scaling in advanced CMOS nodes, such as 14 nm or below. The measured wide signal band of 1 GHz to 17 GHz at 4 K is in accordance with the specifications of transmons and spin qubits developed thus far. Mutiple qubit control can be acheived via frequency multplexing with a target average spurious free dynamic range of 40 dB. The proposed AWG architecture supports system level integration of a cryogenic signal processor, leading to fully integrated, feedback based qubit control systems at 4 K.

ACKNOWLEDGMENT

This work was supported as a part of NCCR SPIN, a National Centre of Competence in Research, funded by the Swiss National Science Foundation (grant number 51NF40-180604). The authors thank Ralph Heller, Daniele Caimi, and BRNC for the technical support.

REFERENCES

[1] Motzoi, F., Gambetta, J. M., Rebentrost, P., Wilhelm, F. K. Simple pulses for elimination of leakage in weakly nonlinear qubits. Phys. Rev. Lett. 103, 110501 (2009).

[2] Werninghaus, M., Egger, D.J., Roy, F. et al. Leakage reduction in fast superconducting qubit gates via optimal control. npj Quantum Inf 2021.

[3] Camenzind, L. C., Geyer, S., Fuhrer, A. et al. A hole spin qubit in a fin field effect transistor above 4 kelvin. Nat Electron (2022)

[4] C. Menolfi et al., "A 112Gb/S 2.6pJ/b 8-Tap FFE PAM-4 SST TX in 14nm CMOS," 2018 ISSCC, 2018,doi: 10.1109/ISSCC.2018.8310205.

[5] M. A. Kossel et al., "8.3 An 8b DAC-Based SST TX Using Metal Gate Resistors with 1.4pJ/b Efficiency at 112Gb/s PAM-4 and 8-Tap FFE in 7nm CMOS," ISSCC, 2021, pp. 130-132

[6] A. Chabane et al., "Cryogenic Characterization and Modeling of 14 nm Bulk FinFET Technology," ESSDERC-ESSCIRC, 2021, pp. 67-70

[7] J. C. Bardin et al., "Design and Characterization of a 28-nm Bulk-CMOS Cryogenic Quantum Controller Dissipating Less Than 2 mW at 3 K," IEEE Journal of Solid-State Circuits, vol. 54, no. 11, Nov. 2019.

[8] J. P. G. Van Dijk et al., "A Scalable Cryo-CMOS Controller for the Wideband Frequency-Multiplexed Control of Spin Qubits and Transmons," in IEEE Journal of Solid-State Circuits, vol. 55, no. 11, pp. 2930-2946, Nov. 2020, doi: 10.1109/JSSC.2020.3024678.

[9] Park et al., "A Fully Integrated Cryo-CMOS SoC for State Manipulation, Readout, and High-Speed Gate Pulsing of Spin Qubits," in IEEE Journal of Solid-State Circuits, Nov. 2021, doi: 10.1109/JSSC.2021.3115988.

A 28nm 6.5-8.1GHz 1.16mW/qubit Cryo-CMOS System-on-Chip for Superconducting Qubit Readout

Steven Van Winckel[†], Alican Caglar[†]*, Benjamin Gys, Steven Brebels, Anton Potocnik, Bertrand Parvais*,
Piet Wambacq*, Jan Craninckx

imec, Leuven, Belgium

Also at Vrije Universiteit Brussel (VUB), Dept. of Electronics and Informatics (ETRO), Brussels, Belgium

alican.caglar@imec.be

Abstract—This paper presents a 28 nm cryo-CMOS system-on-chip (SoC) for the dispersive readout of superconducting qubits operating between 6.5-8.1 GHz at 4 K. The SoC includes a quadrature VCO and a full zero-IF transmitter and receiver chain, including 200 MS/s 7-bit DACs, ADCs, and digital. It can generate pulses up to 640 ns duration, and it attains a very low-power readout operation (9.8 mW) compared to its counterparts with 0.5-0.6 dB noise figure, 65-89 dB gain, and -71 dBm maximum TX output power at 4 K. Furthermore, with its duty-cycled mode, the SoC reduces the average power dissipation for the readout application.

I. INTRODUCTION

The scaling up of the number of qubits in quantum computers (QCs) obliges the adoption of more compact solutions to control and read out the state of qubits rather than using instruments at room temperature (RT) with an impractical number of interconnects between quantum environment (<1 K) and RT. In this context, the demonstration of spin and superconducting (SC) qubits co-integrated with CMOS circuits makes cryo-CMOS readout and control circuitries a promising solution for scalable QCs [1]. However, given that many qubits are required for realizing a fault-tolerant QC, and dilution refrigerators have limited cooling power at the 4 K stage, designing cryo-CMOS readout and control circuitries compliant with a stringent power budget of <1 mW/qubit becomes one of the major challenges. Readout circuitries operating in the GHz regime with high bandwidth have been proposed in the literature to lower the power consumption per qubit with simultaneous readout of multiple qubits using frequency division multiplexing [2], [3].

The dispersive readout operation for SC qubits is performed by transmitting a weak pulse (<-125 dBm) to the readout resonator of transmon qubits and reading back the reflected or transmitted signal as illustrated in Fig. 1 [4]. Since the admittance of the qubit depends on its state, the amplitude and phase of the transmitted pulse are modulated according to the qubit state. Therefore, the state of the qubit can be decoded from the phase and amplitude of the received pulse.

In this work, we propose a duty-cycled low-power (9.8 mW) readout system-on-chip (SoC) including a programmable pulse generator and receiver for the dispersive readout of SC qubits

[†]These authors contributed equally to this work.

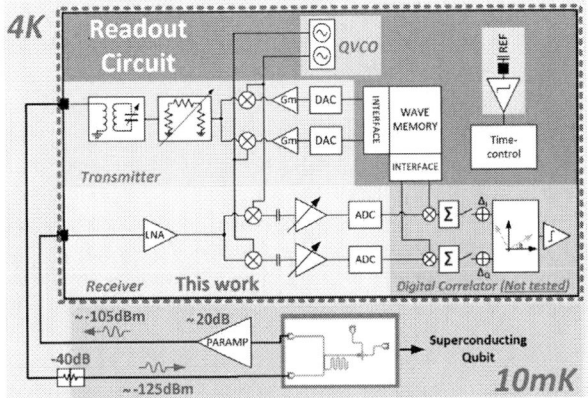

Fig. 1: Block diagram of the proposed cryo-CMOS readout SoC for SC qubits, and typical diagram of a readout system.

operating between 6.5-8.1 GHz at 4 K. In the duty-cycled operation, the whole transceiver starts working shortly before the readout operation with a fast start-up and operates only during the readout operation. In this way, it is aimed to avoid unnecessary power consumption when there is no readout operation. By the employment of the cascoded inverter-based LNA with a high-Q on-chip gate inductor [5], the receiver power consumption is significantly reduced while providing competitive noise performance compared to the state-of-the-art cryo-CMOS receivers [2], [3]. Moreover, a low-power on-chip frequency synthesis is achieved by a class-D quadrature VCO (QVCO).

II. SYSTEM-ON-CHIP ARCHITECTURE

A. System-level Overview

The proposed readout SoC comprises a 5.9-8.2 GHz QVCO and a full zero-IF TX and RX chain, including DACs, ADCs, and digital as shown in Fig. 1. A serial peripheral interface (SPI) is used to configure all the existing blocks, and a 4 Gbit/s serial data link is implemented to allow real-time off-chip monitoring of the ADCs outputs. The SoC uses a 4 GHz external reference clock to generate an on-chip clock for the 4 Gbps serial data link and a 200 MHz clock for the digital. To achieve a low-power readout operation, a time

978-1-6654-8495-4/22 $31.00 © 2022 IEEE

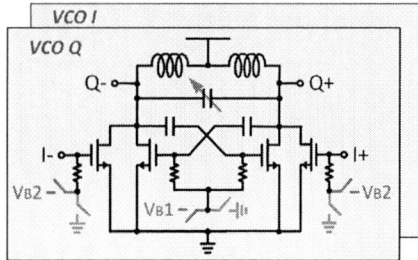

Fig. 2: Schematic of the QVCO. Switches closed during idle and readout phases are highlighted by green and red colors, respectively.

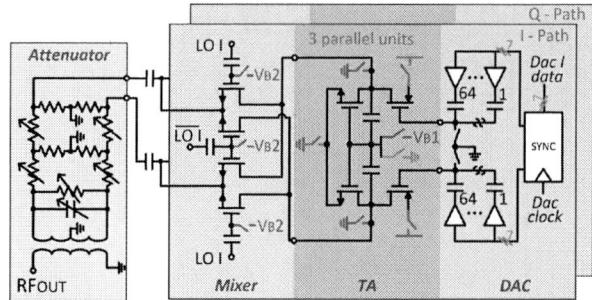

Fig. 3: Schematic of the transmitter. Switches closed during idle, biasing and readout phases are highlighted by green, blue, and red colors, respectively.

controller is implemented to run the transceiver in duty-cycled mode consisting of three phases (idle, biasing, and readout). When there is no transmission (idle), TX and RX blocks are practically not dissipating power. The fast start-up of the blocks (biasing phase) is achieved by quickly pre-charging the relevant nodes to a well-defined level shortly before the pulse transmission and putting those nodes in a high-impedance mode during the transmission (readout phase). The almost-zero leakage of transistor switches at cryo-operation enables properly keeping the intended operating points. For debugging, the receiver (including QVCO) can also operate in continuous mode.

B. Class-D Quadrature VCO

A parallel coupled 0.39 V class-D QVCO is designed as given in Fig. 2. Class-D VCOs have the capability of operating with a low VDD, which makes them appealing for low-power applications [6]. Also, thanks to its large output signal amplitude, it is implemented to directly drive the mixers instead of employment of output buffers thereby avoiding any further substantial power usage. To minimize the loss in the LC tank, an inductor with a 160 μm diameter and the simulated Q-factor of 70 at 4 K is implemented. To generate quadrature LO signals, two VCOs are coupled to each other through parallel NMOS transistors to have a high phase accuracy. The parallel coupling transistors slightly degrade the phase noise performance, however, since the phase noise in the transmitted signal is canceled by using the same LO at the down-conversion, it does not affect the readout performance. For this single-qubit readout demonstrator, quadrature accuracy (and image rejection) has not been calibrated. For quick start-up in duty-cycled mode, the gates of the cross-coupled transistors are shortly biased with a higher VDD (not illustrated in Fig. 2). During the start-up, an active circuit is enabled to enforce the oscillation to start in the desired quadrature phase sequence.

C. Pulse Generator Architecture

The TX pulse generator includes a 128-word 14-bit memory for I and Q waves (pulse duration up to 640 ns), programmable through an SPI, and custom-designed for minimal power consumption during sequential readout. Two 200 MS/s 7-bit differential charge redistribution DACs (Fig. 3) convert the stored digital I/Q baseband signal to an analog signal. The capacitor bank has a low dynamic power consumption due to the low reference voltage (0.2 V). By the same token, the capacitor drivers include only NMOS transistors. Prior to the pulse transmission, voltages are sampled and stored on each side of the DAC capacitor bank to control the input common-mode (CM) voltage of the transconductance amplifier (TA) and to perform offset compensation.

The DAC output voltages are converted into currents by the TA with 3 parallel gain units allowing gain control. To ensure a fast start-up, the CM feedback is realized by capacitance sensing of CM level and feedback through the NMOS load transistors. These capacitors are preset prior to the transmission by shorting the TA outputs to the ground and applying a biasing voltage to the NMOS gates. The output current of the TA is upconverted by a passive switching mixer. The upconverted current is transformed to voltage and attenuated through a programmable 2-stage pi-network resistive passive divider with the simulated attenuation range of 27-68 dB at 4 K. At the TX output, the differential output of the attenuator is transformed into a signal-ended signal by an 8-shaped balun to minimize the magnetic coupling from the QVCO.

D. Receiver Architecture

The receiver front-end has a cascoded inverter-based LNA with a high-Q integrated gate inductor [5]. Current re-use in the inverter topology allows achieving a decent noise figure (NF) performance with low power consumption. Moreover, the Q-factor of inductors substantially improves at 4 K. This has been leveraged using a large on-chip gate inductor to increase the passive gain and suppress the noise contribution of the active stages. The differential output of the LNA is downconverted by an IQ passive mixer driven directly by the QVCO. The gate voltage of the NMOS devices in the mixer is set before enabling the receiver to adjust the duty cycle of the switches. Both the I and Q baseband signals are amplified by an AC-coupled 4-stage variable-gain amplifier (VGA) with first-order filtering on each stage as given in Fig. 4. Each stage of the VGA has 7 parallel units, which can be (de)activated to adjust the RX gain. To quickly settle the biasing points of NMOS load devices, the capacitor connected to their gates is pre-charged via switches. The CM output voltage of the VGA units is sensed using the RC load and connected to the gate of

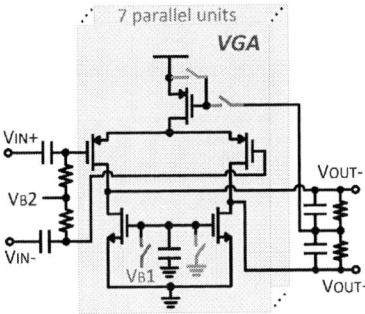

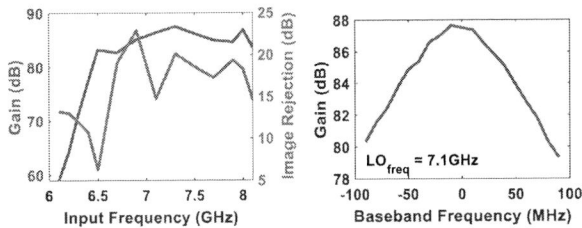

Fig. 6: Measured RX gain, image rejection characteristics across 6-8.1 GHz (left), and RX baseband gain (right) at 4 K.

Fig. 4: Schematic of the VGA's single stage. Switches closed during idle, biasing and readout phases are highlighted by green, blue, and red colors, respectively.

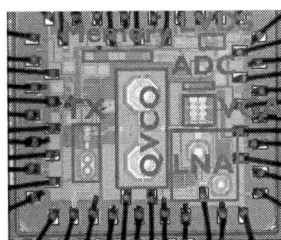

Fig. 5: Die micrograph of the 28 nm cryo-CMOS readout SoC (left) and the readout SoC wirebonded on a test PCB (right).

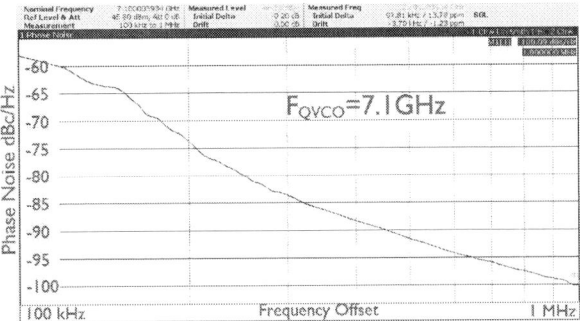

Fig. 7: Measured phase noise of the QVCO at 4 K.

the PMOS tail current source, hereby it is regulated through the current source.

To have a low-power A/D conversion, an asynchronous controlled 7-bit 200 MS/s I/Q SAR ADC with a binary capacitor bank (0.2 V reference voltage) using a switchback switching method is implemented [7]. The readout chain is completed with a digital correlator, which cross-correlates the demodulated I/Q waves with a wave pattern stored in a 128-word 14-bit memory. The calculated correlation detects the phase and amplitude variations of the transmitted pulse to discriminate the qubit state.

III. MEASUREMENT RESULTS

The proposed readout chip has been implemented in a 28 nm CMOS process. The simulations of the chip have been done using a custom cryo-model that covers basic transistor parameters at 4 K. The chip has been characterized at 4.2 K in a cryo-station. The photos of the die and test PCB are given in Fig. 5. The RF input and output of the chip are connected to 50 Ω transmission lines on the PCB, and no discrete component is used for the impedance matchings. The receiver chain performance is dominated by the LNA, and has a 0.5-0.6 dB NF, with the S11 below -10 dB between 4.2-7.4 GHz [5]. Within 6.5-8.1 GHz, the total RX gain varies 5 dB for the same baseband settings as shown in Fig. 6, and the total gain can be adjusted between 65 and 89 dB through the VGA. The baseband gain has a fourth-order filtering characteristic with 50 MHz bandwidth. The phase noise of the QVCO is -100 dBc/Hz at 1 MHz frequency offset at the 7.1 GHz

oscillation frequency (Fig. 7), and its oscillation frequency range is between 5.9-8.2 GHz (Fig. 8). The RX input P1dB and IP3 are -87.5 dBm and -77.5 dBm, respectively, obviously limited by the high baseband gain, but of no influence on system performance given the extremely low signal levels used.

Transmitted I/Q waves by the pulse generator (captured and downconverted by a vector signal analyzer (VSA)) and their demodulation by the receiver are shown in Fig. 9. Before the pulse transmission, during the start-up of the QVCO, a LO leakage is present at the TX output, which is significantly compensated with a DC offset created by the DAC during the transmission. We anticipate that the LO leakage mainly stems from the coupling between TX bondwire and QVCO. The TX output power can be programmed up to -71 dBm as given in Fig. 10. To test the readout performance of the chip, the constellation diagram of the receiver I/Q outputs has been plotted as shown in Fig. 10 by connecting the TX to RX

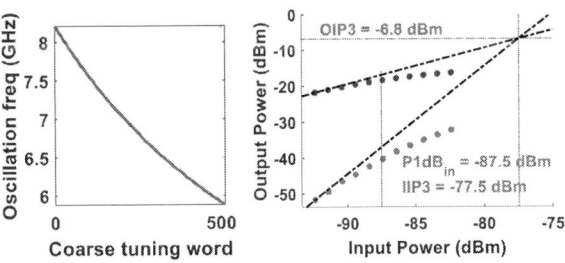

Fig. 8: Measured tuning range of the QVCO (left), measured 1-dB compression point and linearity (right) of the receiver at 4 K.

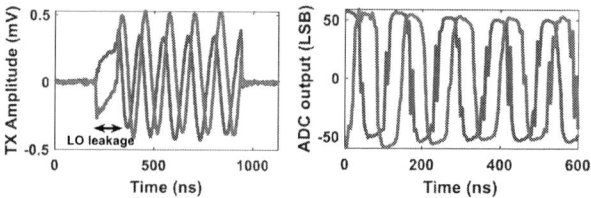

Fig. 9: Transmitted I/Q signals captured by a VSA at 4 K (left). Demodulated I/Q signals of the transmitted pulses at the receiver side at 4 K (right).

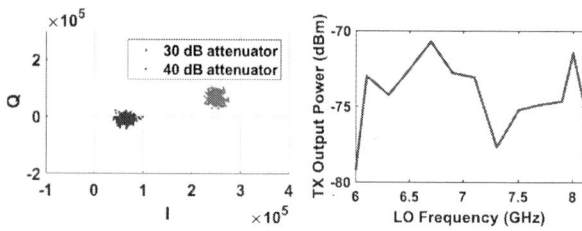

Fig. 10: Constellation diagram of I/Q signals at the receiver output at 4 K after off-chip correlation, obtained through having different attenuator values between TX and RX to imitate different states of a qubit. (left). Measured TX output power across 6-8.1 GHz at 4 K (right).

with 30 dB and 40 dB attenuators in between them to mimic the amplitude modulation induced by quantum states of qubits. The digital correlator block has not been tested, yet; therefore, the post-processing of the I/Q outputs has been done off-chip. The total power consumption of the baseband and RF blocks are 0.7 and 9.1 mW, respectively. The main contributors to the RF power dissipation are the LNA (4.2 mW) and QVCO (3 mW). With frequency division multiplexing, assuming the power consumption for the RF can be shared among 20 qubits, a 1.16 mW/qubit power consumption could be achieved.

The comparison table in Table I shows that the proposed SoC achieves a very low-power readout operation compared to the state-of-the-art with similar noise performance.

IV. CONCLUSION

This work demonstrates a 6.5-8.1 GHz full zero-IF transceiver in 28 nm CMOS working at 4 K for the dispersive readout operation of SC qubits. The low-power (9.8 mW) and low-noise (0.5-0.6 dB NF) readout operation is accomplished by taking advantage of the high passive amplification in the inverter-based LNA with a high-Q gate inductor as well as employing passive mixers and a class-D QVCO with a low VDD. Hence, for a given noise performance, the proposed readout IC achieves a much lower power consumption compared to its CMOS counterparts. Additionally, thanks to its duty-cycled mode, the average power dissipation can be lowered to prevent the temperature increase in dilution refrigerators.

ACKNOWLEDGMENT

The authors like to thank H. Suys, L. Pauwels, M. Libois, M. Mongillo, A. Grill, and R. Li for their assistance with system specifications, transistor modelling, and measurements. This work has been performed as part of imec's Industrial Affiliation Program on Quantum Computing.

TABLE I: COMPARISON WITH OTHER CRYO-CMOS QUANTUM READOUT CIRCUITRIES.

Reference	This Work	[2]	[3]
Architecture	Full Zero-IF I/Q TX and RX	Intermediate IF I/Q RX and PLL	High IF I/Q RX
CMOS technology	28 nm	40 nm	40 nm
Freq. (GHz)	6.5-8.1	5-6.5	6-8
Gain (dB)	65-89	70	58
IIP3 (dBm)	-77.5	-72	-50.8
IP1dB (dBm)	-87.5	-85	-58.4
NF (dB)	0.5-0.6	0.55	0.6
Phase noise at 1 MHz offset (dBc/Hz)	-100	-115	-
VCO tuning range	2.3 GHz	6 GHz	-
S11 (dB)	<-3.7	<-10	-
Power Diss. (mW)	9.8[1]	108	66
Area (mm^2)	2.14	2.8	0.68
Temperature (K)	4.2	3.5	4.2

[1]Excludes the power consumption used for debugging (e.g., serial link)

REFERENCES

[1] L. L. Guevel, G. Billiot, X. Jehl, S. De Franceschi, M. Zurita, Y. Thonnart, M. Vinet, M. Sanquer, R. Maurand, A. G. M. Jansen, and G. Pillonnet, "19.2 A 110mK 295µW 28nm FDSOI CMOS Quantum Integrated Circuit with a 2.8GHz Excitation and nA Current Sensing of an On-Chip Double Quantum Dot," in *2020 IEEE International Solid- State Circuits Conference - (ISSCC)*, 2020, pp. 306–308.

[2] A. Ruffino, Y. Peng, T. Y. Yang, J. Michniewicz, M. F. Gonzalez-Zalba, and E. Charbon, "13.2 A Fully-Integrated 40-nm 5-6.5 GHz Cryo-CMOS System-on-Chip with I/Q Receiver and Frequency Synthesizer for Scalable Multiplexed Readout of Quantum Dots," in *2021 IEEE International Solid- State Circuits Conference (ISSCC)*, vol. 64, 2021, pp. 210–212.

[3] B. Prabowo, G. Zheng, M. Mehrpoo, B. Patra, P. Harvey-Collard, J. Dijkema, A. Sammak, G. Scappucci, E. Charbon, F. Sebastiano, L. M. K. Vandersypen, and M. Babaie, "13.3 A 6-to-8GHz 0.17mW/Qubit Cryo-CMOS Receiver for Multiple Spin Qubit Readout in 40nm CMOS Technology," in *2021 IEEE International Solid- State Circuits Conference (ISSCC)*, vol. 64, 2021, pp. 212–214.

[4] J. C. Bardin, D. Sank, O. Naaman, and E. Jeffrey, "Quantum Computing: An Introduction for Microwave Engineers," *IEEE Microwave Magazine*, vol. 21, no. 8, pp. 24–44, 2020.

[5] A. Caglar, S. V. Winckel, S. Brebels, P. Wambacq, and J. Craninckx, "A 4.2mW 4K 6-8GHz CMOS LNA for Superconducting Qubit Readout," in *2021 IEEE Asian Solid-State Circuits Conference (A-SSCC)*, 2021, pp. 1–3.

[6] Y. Yoshihara, H. Majima, and R. Fujimoto, "A 0.171-mW, 2.4-GHz Class-D VCO with dynamic supply voltage control," in *ESSCIRC 2014 - 40th European Solid State Circuits Conference (ESSCIRC)*, 2014, pp. 339–342.

[7] G.-Y. Huang, S.-J. Chang, C.-C. Liu, and Y.-Z. Lin, "10-bit 30-MS/s SAR ADC Using a Switchback Switching Method," *IEEE Transactions on Very Large Scale Integration (VLSI) Systems*, vol. 21, no. 3, pp. 584–588, 2013.

978-1-6654-8495-4/22 $31.00 © 2022 IEEE

Non-Volatile Ternary Content Addressable Memory based on Phase Change Nanoelectromechanical (NEM) Relay

Mohammad Ayaz Masud
Electrical and Computer Engineering
Carnegie Mellon University
Pittsburgh, USA
mmasud@andrew.cmu.edu

Luis Hurtado
Electrical and Computer Engineering
Carnegie Mellon University
Pittsburgh, USA
lhurtado@andrew.cmu.edu

Gianluca Piazza
Electrical and Computer Engineering
Carnegie Mellon University
Pittsburgh, USA
piazza@ece.cmu.edu

Abstract— **We demonstrate, for the first time, a ternary content-addressable memory (TCAM) architecture based on phase change nanoelectromechanical relays (PCNRs). The non-volatility (NV), high ON-OFF ratio (10^8), and low leakage operation make PCNR an ideal candidate for high density TCAM. Additionally, PCNR devices are back-end-of-the-line (BEOL) compatible, allowing for a very small TCAM cell size of $18F^2$. A TCAM, with only 1 transistor and 2 PCNR devices (1T2P) per cell is simulated and it exhibits 133 ps search latency and 0.721 pJ energy consumption for 64 bits, making it one of the most competitive approaches for TCAM using beyond CMOS technologies.**

Keywords—Phase change material, TCAM, GeTe, NEMS

I. INTRODUCTION

The data-driven revolution in modern computational devices has reached a bottleneck where the energy-time cost of memory access is higher than the cost associated with processing. Applications that enable computation at edge devices, such as IoT sensors, suffer the most in this regard due to their limited energy resources. Architectural and device level solution is essential to mitigate this problem. One way of achieving this goal is to utilize innovative hardware designs to enable parallel access to an array of memory cells in one single search cycle. Ternary content addressable memory (TCAM) is one such circuit technology that comes into play during high frequency look-up operations [1]. It searches for a content in a 1D array and returns the address when a match is found, greatly reducing the time and energy requirement of a conventional search architecture. This architecture was originally used in network systems for operations like internet protocols (IP) look-up [2]. However, in recent years, TCAM has been employed for broader data extensive applications such as genome data analysis, natural language processing (NLP), image classification etc [3]. The on-chip circuitry also has the potential to resolve the delay associated with communication between the processor and the memory units.

CMOS-based TCAM is typically based on 16 transistor static random-access memory (16T SRAM) cells, resulting in high energy consumption and large cell area [4]. Implementation with dynamic memory (DRAM) requires frequent refresh cycles, offsetting the improvement of having only 5 transistors [5]. To solve these issues, alternative TCAM approaches utilizing several emerging devices, such as, STT-MRAM, FeFET, RRAM, electromechanical relay etc. have been studied [6-9]. The proposed approaches suffer from various limitations associated with these emerging technologies. For example, resistive switching suffers from high dynamic and static power consumption as well as long

latency while having a very low ON-OFF ratio [10]. Devices based on ferroelectric switching can provide higher ON-OFF ratio and can be used in a dense 2T TCAM architecture [11]. However, they also require high energy budget and long latency. Recently, TCAM designs have been presented with nanoelectromechanical (NEM) relay to mitigate some of these issues [12]. The ultra-high ON-OFF ratio, near-zero leakage and abrupt switching mechanism make NEM relays a suitable technology for the TCAM architecture. However, conventional NEM relay designs depend on flexure-based architecture which cannot be scaled down to sub-100 nm node sizes [13]. Moreover, most NEM relay designs are volatile and need periodic refresh cycles to hold data [14].

To solve the scalability and volatility issue of conventional NEM relays, we demonstrate a non-volatile NEM relay based on the volumetric expansion of GeTe phase change material (PCM) and implement it in a TCAM cell to minimize cell area and search energy consumption. The non-volatile nature of the relay eliminates the need for refresh cycles, while the high scalability and the low leakage enable dense TCAM architecture. In a recent work, we have presented a PCM based NEM relay architecture, called Phase Change Nanoelectromechanical Relay (PCNR), which shows very high ON-OFF ratio (10^8) and very low leakage current (30 fA) at the cost of 42 nJ writing energy [15]. Furthermore, PCNR devices are non-volatile and highly scalable. This makes PCNR an exciting new candidate for TCAM architectures. In the following sections of this paper we briefly discuss the operating principle of PCNR and demonstrate our simulation results for a 1T 2 PCNR TCAM cell. We compare the performance of this TCAM design with existing architectures.

II. DEVICE DESIGN AND OPERATION

PCMs are an exciting class of materials renowned for their rapid and reversible switching between amorphous and crystalline states. This change in nanoscale atomic arrangement alters some of the most useful physical macroscale properties, such as electrical conductivity, optical reflectance, density etc. As the material undergoes an amorphous-to-crystalline transition, its resistivity decreases by ~5 orders of magnitude and optical reflectance may increase by 10-15% [16]. Several Te-based chalcogenides, such as Ge-Sb-Te (GST) and GeTe, have been successfully used in optical and electrical data storage devices [17]. Phase change nano relay (PCNR) utilizes the density difference between two stable phases of a PCM in order to switch a relay device. Fig. 1 shows the structure of a PCNR cell. It is primarily comprised of a phase change mechanical actuator and a pair of metal contacts [18].

978-1-6654-8495-4/22 $31.00 © 2022 IEEE

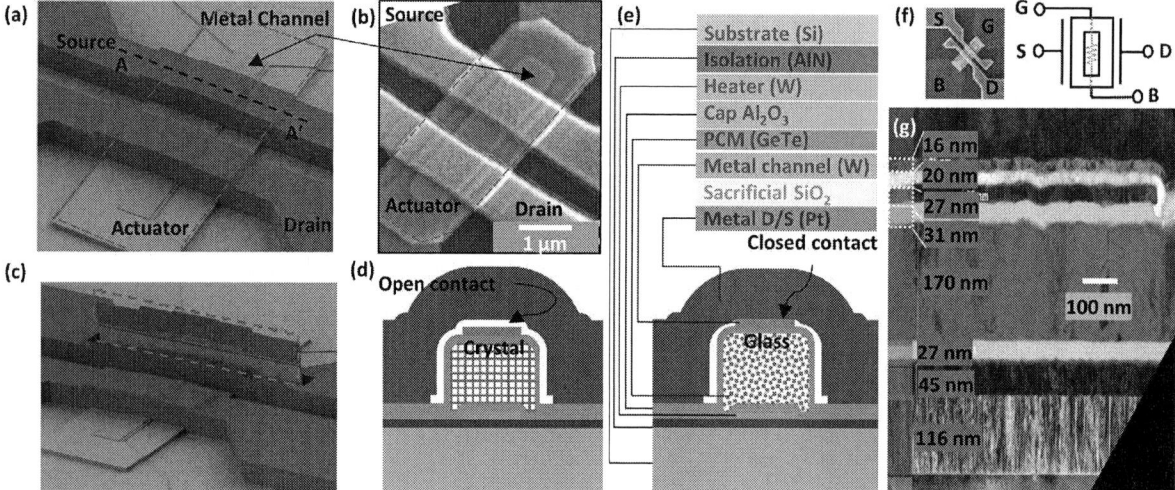

Fig. 1: Overview of PCNR device architecture and operation. (a) 3D view and (b) SEM image of a PCNR. (c) Cross-section taken along AA' line. (d)-(e) schematic cross-section with (d) open and (e) closed contact. (f) Four I/O terminals of PCNR, along with its schematic model, showing the source (S), drain (D), gate (G) and body (B). G and B connects the heater. (g) TEM image of the cross-section before release.

The key component of the phase change actuator is GeTe, which expands by 10% during its transformation from crystalline to amorphous phase [19]. The change is induced by Joule-heating from an adjacent heater layer (Gate terminal). A metal channel is then fabricated on top of the actuator. As fabricated, the channel is separated from the drain and source (D/S) metal lines by an air-gap. The air-gap is obtained by depositing a 20 nm sacrificial layer between the channel and the D/S metal layers. The fabrication process of PCNR is significantly different from conventional PCM devices, as we do not need conduction through the PCM.

In the phase-change process, a short intense thermal pulse melts the crystalline material. If the material is cooled slowly, it goes back to the original crystalline state. However, if the material is quenched fast ($>10^9$ K/s), the atoms get locked in a supercooled state [20]. This sudden quench leaves the material in an amorphous, highly resistive state. The amorphous state can be retained for more than 10 years at a temperature below 400 K [21]. To return to its crystalline phase, a longer pulse of lower intensity is applied to anneal the material at a temperature above the glass temperature which is lower than the melting point. GeTe melts at 1000 K and its glass transition happens at 500 K [22]. This transition phenomenon is also evident in atomically thin layers of PCM [23].

A. SET and RESET Operation

The amplitude and duration of the heater pulse determines the state of a PCNR cell during the write operation. SET pulse should be high enough to melt the PCM. Here we apply an 1.1 V input pulse on a 3 μm long heater to reach the melting temperature of GeTe (Fig. 2(a)). An abrupt falling edge in the SET pulse allows rapid quench. The PCM is transformed into the expanded amorphous state by the end of this heat cycle. This expansion switches the device ON by connecting the metal contact with both electrodes, writing a '1'. The RESET pulse is tuned to achieve a temperature between the glass transition temperature and the melting point of the PCM. It has a lower amplitude (0.8 V) compared to the SET pulse (Fig. 2(b)). A slower pulse is used to ensure complete crystallization of the PCM. Fig. 2(c) demonstrates the high ON-OFF ratio of PCNR. We test for leakage current for a range of read voltages. Fig. 2(d) shows that the leakage current

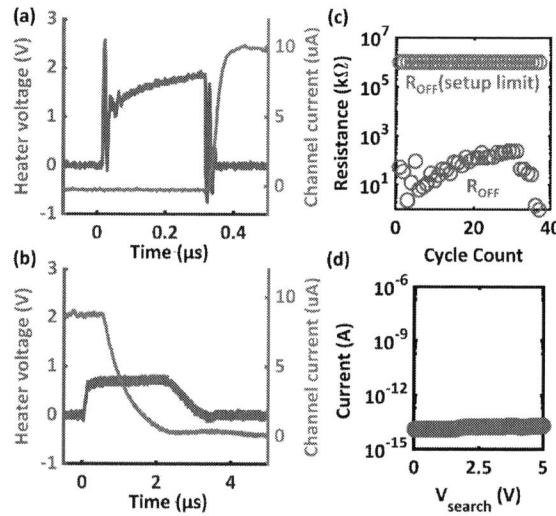

Fig. 2: Electrical characterization of a PCNR device. (a) SET and (b) RESET pulse on PCNR with 3 μm long heater. Channel current is measured from a 100 kΩ resistor in series with the drain (V_{DS}=1 V). (c) ON-OFF resistance during cycling and (d) leakage current measured in the channel for a switched OFF device.

remains unchanged even for impractically high search voltages.

B. Fabrication Process

Fig. 3 shows the process flow of the PCNR device. At first, 100 nm AlN is deposited on a Si substrate by reactive sputtering. AlN provides electrical isolation between the heater and the substrate while allowing good thermal conduction. To obtain the heater layer, 50 nm W is sputtered at 850°C and etched in a Reactive Ion Etching (RIE) tool. Next, 30 nm Al$_2$O$_3$ is deposited in an atomic layer deposition (ALD) tool at 250°C to electrically isolate the PCM from the heater. After this step, 200 nm of crystalline GeTe is deposited by co-sputtering Ge and Te at 400°C, and it is then etched by Ar$^+$ plasma in an Inductively Coupled Plasma (ICP)-RIE chamber.

978-1-6654-8495-4/22 $31.00 © 2022 IEEE

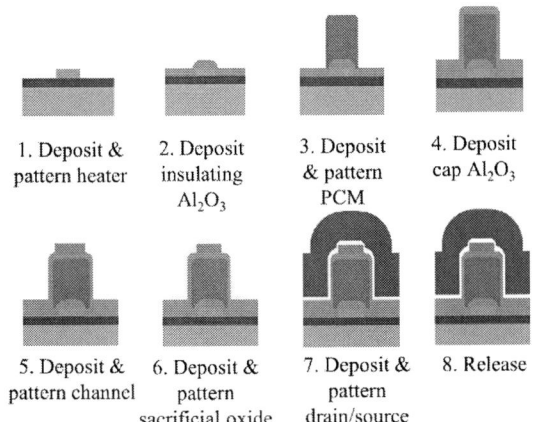

Fig. 3: BEOL compatible fabrication flow of the PCNR device.

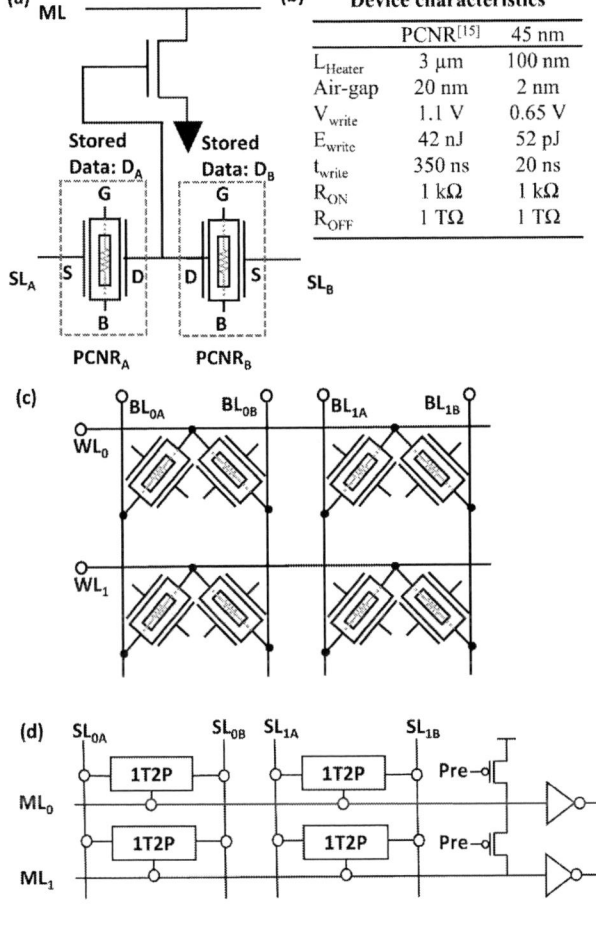

Fig. 4: PCNR TCAM architecture. (a) A TCAM cell with 1 transistor and 2 PCNR devices. (b) PCNR device characteristics obtained at 45 nm technology node. (c) Write and (d) search circuit of a simple 2×2 array. The two circuits are shown separately to help explain their independent operation. Notice that the four terminals of PCNR can be connected to the search and write circuit on a single layout.

Another 30 nm layer of Al_2O_3 is deposited by ALD to encapsulate the patterned PCM. This encapsulation layer contains the molten PCM during phase transformation. At this point, the devices are ready to be used and tested as stand-alone actuators.

Next, the channel is made by sputtering 30 nm W at room temperature. It is then patterned and etched following the same W-etch process. The airgap is formed by depositing 20 nm sacrificial SiO_2 in the ALD and patterning it in RIE. Finally, a thick layer of W is lifted-off as the D/S pair. The wafer is then etched in vapor HF to isotropically remove the sacrificial oxide.

III. PCNR TCAM ARCHITECTURE

Fig. 4 shows a PCNR TCAM cell and a simple 2x2 array during write and search operation. In the TCAM cell, PCNR devices are modeled as resistors with either low resistive state (~1kΩ) representing a logic 1 or a high resistive state (~1TΩ) representing a logic 0. The TCAM architecture assumes that both PCNR devices will not be at simultaneous low resistive state. The PCNR are BEOL compatible, leading to a cell area of $18F^2$, only 10% of the 16T TCAM cell area.

The writing portion of the PCNR devices is decoupled from the read and search operations and the cross-bar array configuration, shown in Fig. 4(c), offers minimal cell area. The non-volatile nature of the PCNR device allows writing operation at the heater crossbar array without the need of select transistors [24]. Specific word line (WL) and bit line (BL) are selected to apply a heater current on a desired device. Adjacent heaters are exposed to a fraction of the input write current due to sneak path formation. Typically, this current level is not high enough to cause inadvertent switching in adjacent cells. However, write energy increases significantly due to the undesired heat dissipation in adjacent cells. This poses a design trade-off between high cell density in selector-free architecture and low write energy consumption. Writing energy required for a single PCNR cell at 45 nm technology node is 52 pJ.

The search circuit includes the PCNR TCAM cell, pre-charging circuit, and the inverter sensing amplifier (SA). The search lines are connected to the source end of the PCNR channel. TCAMs are benchmarked based on energy consumption and delay during the search operation [25]. The search operation of the PCNR TCAM is shown in Fig. 5 with corresponding simulation set-up identical to Fig. 4(d), based on 45nm CMOS technology node.

During the search operation, the match-line (ML) is first charged to V_{DD} and is then left floating. If there is a mismatch

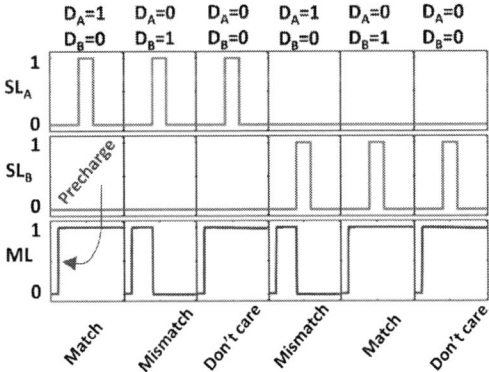

Fig. 5: Logic of the search operation in PCNR TCAM. Match line (ML) is precharged at the beginning of every search cycle. Search lines SL_A and SL_B are symmetric. For a mismatch between the SL and corresponding stored PCNR data, ML discharges within 133 ps for a 64 bit word line.

detected between the search-line (SL) input and the stored data (D) then the ML is pulled to ground via the transistor, otherwise the ML maintains the charge and results in a match. A capacitance of 0.2 fF/μm is used to represent the parasitics of the ML in the simulation setup. The worst case is denoted by discharge through a single transistor in a long array of data. This happens when only one bit has a mismatch. The search latency is calculated for the worst case of only 1-bit mismatch in a 64 bit word row and search energy is calculated based on the pre-charging of the ML for all the entries. Our simulation results exhibit 133 ps search latency and 0.721 pJ energy consumption for 64 bit word row. An overview of TCAM technology benchmark is presented in Table 1.

TABLE I.   PERFORMANCE SUMMARY OF TCAM DESIGNS

	16T [4]	2T2R [5]	2FeFET [7]	3T2N [9]	**1T2P**
Node (nm)	45	90	45	45	**45**
Non-Volatility	No	Yes	Yes	No	**Yes**
Search Voltage (V)	1	1.2	1	1	**1**
Write Voltage (V)	1	2.5	4	1	**1.1**
Search delay (ps)	582	1900	~400	106	**133**
Search energy (fJ)	1600	-	~1700	693	**721**
Area	$171F^2$	$50F^2$	$22F^2$	$32F^2$	**$18F^2$**

IV. CONCLUSION

We present the PCNR TCAM design and simulation results, exhibiting the lowest energy-delay product per unit area in comparison to available TCAM designs. The device itself offers a high ON-OFF ratio, non-volatility, and low leakage. The prototype PCNR device requires high write energy. However, scaling analysis at lower technology nodes promises pJ energy consumption during writing. Further investigation in scaling PCNR and increasing its endurance will pave the way for most energy efficient TCAM architecture.

ACKNOWLEDGMENT

The authors acknowledge the Kavcic-Moura Endowment Fund and the National Science Foundation, USA (Award 1854702) for supporting this work. The authors are grateful to the staff members of the cleanroom facility at Claire & John Bertucci Nanotechnology Laboratory in Carnegie Mellon University. We also acknowledge the use of the CMU Materials Characterization Facility at Carnegie Mellon University supported by grant MCF-677785.

REFERENCES

[1] R. Yang et al., "Ternary content-addressable memory with MoS2 transistors for massively parallel data search," *Nat Electron*, vol. 2, no. 3, pp. 108–114, Mar. 2019.

[2] R. Panigrahy and S. Sharma, "Reducing TCAM power consumption and increasing throughput," in *Proceedings 10th Symposium on High Performance Interconnects*, Stanford, CA, USA, 2002, pp. 107–112.

[3] W. Jiang, Q. Wang and V. K. Prasanna, "Beyond TCAMs: An SRAM-Based Parallel Multi-Pipeline Architecture for Terabit IP Lookup," IEEE INFOCOM 2008 - The 27th Conference on Computer Communications, 2008, pp. 1786-1794.

[4] K. Pagiamtzis and A. Sheikholeslami, "Content-Addressable Memory (CAM) Circuits and Architectures: A Tutorial and Survey," *IEEE J. Solid-State Circuits*, vol. 41, no. 3, pp. 712–727, Mar. 2006.

[5] J. Li, R. K. Montoye, M. Ishii, and L. Chang, "1 Mb 0.41 μm² 2T-2R Cell Nonvolatile TCAM With Two-Bit Encoding and Clocked Self-

Referenced Sensing," *IEEE J. Solid-State Circuits*, vol. 49, no. 4, pp. 896–907, Apr. 2014.

[6] B. Yan, Z. Li, Y. Chen, and H. Li, "RAM and TCAM designs by using STT-MRAM," in *2016 16th Non-Volatile Memory Technology Symposium (NVMTS)*, Pittsburgh, PA, USA, Oct. 2016, pp. 1–5.

[7] X. Yin, X. Chen, M. Niemier, and X. S. Hu, "Ferroelectric FETs-Based Nonvolatile Logic-in-Memory Circuits," *IEEE Trans. VLSI Syst.*, vol. 27, no. 1, pp. 159–172, Jan. 2019.

[8] L. Zheng, S. Shin, S. Lloyd, M. Gokhale, K. Kim, and S.-M. Kang, "RRAM-based TCAMs for pattern search," in *2016 IEEE International Symposium on Circuits and Systems (ISCAS)*, Montréal, QC, Canada, May 2016, pp. 1382–1385.

[9] H. Zhong et al., "DyTAN: Dynamic Ternary Content Addressable Memory Using Nanoelectromechanical Relays," *IEEE Trans. VLSI Syst.*, vol. 29, no. 11, pp. 1981–1993, Nov. 2021.

[10] M.-F. Chang et al., "17.5 A 3T1R nonvolatile TCAM using MLC ReRAM with Sub-1ns search time," in *2015 IEEE International Solid-State Circuits Conference - (ISSCC) Digest of Technical Papers*, San Francisco, CA, USA, Feb. 2015, pp. 1–3.

[11] M. Yabuuchi, M. Morimoto, Y. Tsukamoto, and S. Tanaka, "A 7nm Fin-FET 4.04-Mb/mm2 TCAM with Improved Electromigration Reliability Using Far-Side Driving Scheme and Self-Adjust Reference Match-Line Amplifier," in *2020 IEEE Symposium on VLSI Circuits*, Honolulu, HI, USA, Jun. 2020, pp. 1–2.

[12] H. Zhong, S. Cao, H. Yang, and X. Li, "Dynamic Ternary Content-Addressable Memory Is Indeed Promising: Design and Benchmarking Using Nanoelectromechanical Relays," in *2021 Design, Automation & Test in Europe Conference & Exhibition (DATE)*, Grenoble, France, Feb. 2021, pp. 1100–1103.

[13] A. Peschot, C. Qian, and T.-J. Liu, "Nanoelectromechanical Switches for Low-Power Digital Computing," *Micromachines*, vol. 6, no. 8, pp. 1046–1065, Aug. 2015.

[14] T. Qin, S. J. Bleiker, S. Rana, F. Niklaus, and D. Pamunuwa, "Performance Analysis of Nanoelectromechanical Relay-Based Field-Programmable Gate Arrays," *IEEE Access*, vol. 6, pp. 15997–16009, 2018.

[15] J. T. Best, M. A. Masud, M. P. Boer, and G. Piazza, "Phase Change Nanoelectromechanical Relay for Nonvolatile Low Leakage Switching," *Adv Elect Materials*, p. 2200085, Apr. 2022.

[16] S. Raoux, W. Wełnic, and D. Ielmini, "Phase Change Materials and Their Application to Nonvolatile Memories," *Chem. Rev.*, vol. 110, no. 1, pp. 240–267, Jan. 2010.

[17] R. Jeyasingh et al., "Ultrafast Characterization of Phase-Change Material Crystallization Properties in the Melt-Quenched Amorphous Phase," *Nano Lett.*, vol. 14, no. 6, pp. 3419–3426, Jun. 2014.

[18] J. T. Best, M. A. Masud, M. P. de Boer, and G. Piazza, "Phase Change NEMS Relay," in *2019 IEEE International Electron Devices Meeting (IEDM)*, San Francisco, CA, USA, Dec. 2019, p. 34.1.1-34.1.4.

[19] J. Best and G. Piazza, "High Work Density Gete Mechanical Phase Change Actuator," in *2019 IEEE 32nd International Conference on Micro Electro Mechanical Systems (MEMS)*, Seoul, Korea (South), Jan. 2019, pp. 962–965.

[20] J. Pries, S. Wei, M. Wuttig, and P. Lucas, "Switching between Crystallization from the Glassy and the Undercooled Liquid Phase in Phase Change Material $Ge_2Sb_2Te_5$," *Adv. Mater.*, vol. 31, no. 39, p. 1900784, Sep. 2019.

[21] A. Fantini et al., "Comparative Assessment of GST and GeTe Materials for Application to Embedded Phase-Change Memory Devices," in *2009 IEEE International Memory Workshop*, Monterey, CA, USA, May 2009, pp. 1–2.

[22] J. Pries et al., "Approaching the Glass Transition Temperature of GeTe by Crystallizing $Ge_{15}Te_{85}$," *Physica Rapid Research Ltrs*, vol. 15, no. 3, p. 2000478, Mar. 2021.

[23] F. Xiong et al., "Towards ultimate scaling limits of phase-change memory," in *2016 IEEE International Electron Devices Meeting (IEDM)*, San Francisco, CA, USA, Dec. 2016, p. 4.1.1-4.1.4.

[24] A. Chen, "Analysis of Partial Bias Schemes for the Writing of Crossbar Memory Arrays," *IEEE Trans. Electron Devices*, vol. 62, no. 9, pp. 2845–2849, Sep. 2015.

[25] B. Agrawal and T. Sherwood, "Modeling TCAM power for next generation network devices," in *2006 IEEE International Symposium on Performance Analysis of Systems and Software*, Austin, TX, USA, 2006, pp. 120–129.

978-1-6654-8495-4/22 $31.00 © 2022 IEEE

A 24V Thin-Film Ultrasonic Driver for Haptic Feedback in Metal-Oxide Thin-Film Technology using Hybrid DLL Locking Architecture

Jonas Pelgrims*, Kris Myny*† and Wim Dehaene*
*ESAT-MICAS, KU Leuven, Belgium; †Imec, Leuven, Belgium
Email: *jonas.pelgrims@esat.kuleuven.be

Abstract—This paper presents a 24V high voltage (HV) ultrasonic driver for haptic feedback, designed to drive piezo-electric micromachined ultrasonic transducers (PMUT) using a unipolar 0.8um indium-galium-zinc-oxide (IGZO) thin-film transistor technology fabricated on a polymer substrate. A new hybrid DLL architecture is proposed to overcome the IGZO technology's shortcomings and leverage the accuracy and speed capabilities of Silicon technology. The ultrasonic driver is able to drive a transducer capacitance up to 36pF with a 4bit phase resolution and a frequency of 250kHz occupying a pitch matched area per pixel of 800μmx800μm. This work presents the first fully integrated high-voltage ultrasonic driver in thin-film technology, further expanding the possible applications for thin-film technologies.

Index Terms—ultrasonic driver, thin-film, ultrasonic haptic feedback, high-voltage driver, IGZO, hybrid architecture

I. INTRODUCTION

Emerging applications of low frequency ultrasonic transducer arrays are gaining interest ranging from volumetric displays [1], ultrasonic powering to ultrasonic haptic feedback. Volumetric displays and ultrasonic haptic feedback will further expand user experiences in augmented or virtual reality. Ultrasonic haptic feedback makes use of converging ultrasonic soundwaves in a single focal point, creating modulated high acoustic pressure to realise a tactile feeling as depicted in Fig. 1. Current published and commercial systems create this acoustic output with bulk piezoelectric components in combination with large PCB systems. These PCB systems are limited in number of transducers due to either routing congestion and cost concerns. Piezoelectric Micromachined Ultrasonic Transducers (PMUT) on top of silicon CMOS drivers, as seen in biomedical imaging [2], would result in enormous array sizes for mid-air powering and haptic feedback due to the lower required oscillation frequencies and the larger number of necessary elements. This would result in very large die sizes, which is an economical less viable option and provides a large challenge on yield.

By combining thin-film large-area transducers which already show great promise [3], with a pitch-matched thin-film transistor driving circuit, a fully integrated circuit and transducer solution emerges. This opens application possibilities, such as conformal integration in a car dashboard. In

This research has been partially funded by the FWO SBO project SS004418N (Happy).

Fig. 1. Block diagram of an ultrasonic large area haptic feedback system

addition, combining driving circuits and transducers by means of large area, thin-film technologies would remove the area and resolution constraint of the transducer array.

Several circuit demonstrators based on thin-film transistors have been published in recent years targeting a multitude of domains, NFC tags for IoT communication [4], biomedical circuits [5] and even microprocessors [6]. Expanding these applications in the new domain of ultrasonic circuits is less straightforward with unipolar, flexible IGZO technology. New circuit topologies are required to deal with the nMOS only, single V_T property of the technology. In addition the IGZO transistors suffer from about 100x lower mobility compared to Si CMOS, which severely limits the operation speed of these circuits.

To overcome these limitations clever combinations of silicon CMOS and thin-film transistor circuits can be envisioned. A classical example are thin-film displays with silicon driver circuits. In this work, the accuracy of silicon circuits is leveraged to create multiple accurate delay-locked loops (DLL) for accurate phase control in driving a large-area PMUT array as illustrated in Fig. 1.

Section II elaborates on the resulting hybrid asynchronous architecture. This is followed by a description of the circuit implementation of the voltage controlled delay line (VCDL) and the ultrasonic driver pixel in section III. Finally, the measurement results of the ultrasonic driver and VCDL are discussed in section IV.

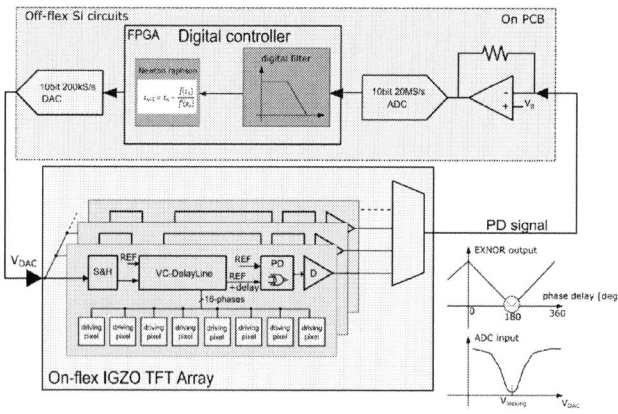

Fig. 2. Architecture diagram of the hybrid driver array

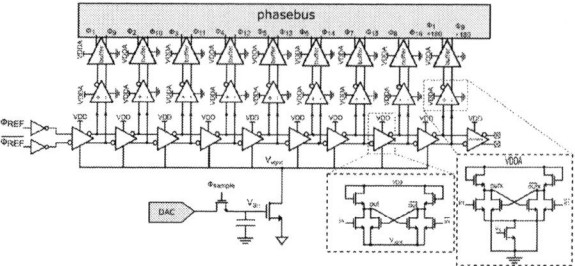

Fig. 3. Diagram and transistor schematic of the VC DLL

II. HYBRID ASYNCHRONOUS ARCHITECTURE

As the speed of operation of thin-film IGZO transistors is severely limited, a synchronous approach to create a 4bit phase resolution is not feasible. To drive a 250kHz signal with a 4bit phase resolution would require a counter operation of 4MHz. This is close to the demonstrated operating frequency of current IGZO technologies [4] and poses additional limitations in terms of area usage and power consumption.

In order to realize the necessary phase resolution, an asynchronous circuit is created by means of a delay locked loop. The implementation of this delay locked loop solely by flexible IGZO transistors would result in three main challenges. Firstly, the required loopback filter in a conventional DLL design requires a large area. Secondly creating a well balanced charge pump is a difficult task in unipolar technology due to an inherent imbalance in pull-up and pull-down strength. Thirdly flexible IGZO transistors suffer from larger process variations compared to conventional CMOS technology. This implies that the full DLL should meet the spec across very wide process corners in order to have high yield. This is even more necessary in large arrays as a small yield drop would result in a large number of dead or inaccurate driving pixels in the array.

To solve these issues we propose the hybrid architecture illustrated in Fig. 2. Here, the loop of the DLL is closed by using by using off-flex silicon circuits contributing to the locking of the DLL. To prove this concept, we have realized an evaluation PCB that emulates the silicon circuits and attach this to the flexible IGZO array. It is clear that for full deployment silicon is required to meet power and miniaturization targets. On the PCB an analog-to-digital converter (ADC) converts the phase detector (PD) signal and provides the FPGA with a 10bit 20MS/s digital data stream. In the FPGA, the loop dynamics are governed by a digital filter. A digital-to-analog convertor (DAC) closes the loop by providing an output voltage to the VCDL. Multiple DLLs can be closed with a single silicon circuit by using sample and hold (S&H) circuits and multiplexers at the output of the PD. At the DAC

output only a multiplexer is required as IGZO transistors have very low source-drain leakage current [7]. As such the IGZO transistors act as sample and hold for the VCDL input voltage. Another advantage of using the hybrid architecture, is the possible compensation of process variations of IGZO by means of the silicon circuits and further digital processing.

III. IMPLEMENTATION: IGZO SI AIDED ASYNCHRONOUS ARCHITECTURE

A. The voltage controlled delay locked loop

The unipolar circuit implementation of the VCDL is shown in Fig.3. Cross-coupled diode-load inverters have been selected for the delay elements, without resistive loads to reduce process variation impact. These inverters have a current-bleeding tail transistor to control the delay of the inverters. The bleeding transistor creates an increased ground rail which reduces the signal swing of the delay elements. This swing will be recovered by the amplifying stages driving the phase bus towards the pixels drivers.

The choices for a phase detector are limited because the speed of IGZO does not permit to create a full phase frequency detector as used in conventional DLL designs. Therefore a simple EXNOR gate is used as phase detector. In order to bring the output signal of the PD on the flex towards the silicon circuits a high-speed buffer is needed for this relatively high bandwidth signal. The possibility of voltage buffering is illustrated in Fig. 4a: a voltage buffer drives a large output bus capacitance C_{par} with a required bandwidth larger than 2MHz. As the bus capacitance for large arrays will be very substantial, reaching this bandwidth with an IGZO-based voltage buffer is infeasible. To circumvent this problem a current buffer is selected on flex, driving a silicon transimpedance amplifier. This effectively shortens the bus capacitance leading to relaxed design specifications for the current driver. The implementation of a low impedance virtual ground is not an issue on silicon. The resulting push-pull current interface with enabling transistors is shown in Fig. 4c. Either the current is pushed in the virtual ground created by the transimpedance amplifier on silicon or it is pulled to ground on flex.

B. The pulser design

To drive the PMUT transducer with a HV swing, a fully differential stacked driver is designed in such a way that the

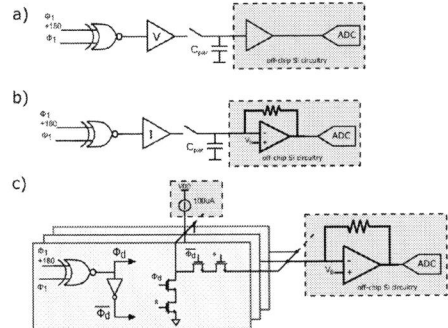

Fig. 4. Phase detector and output driving

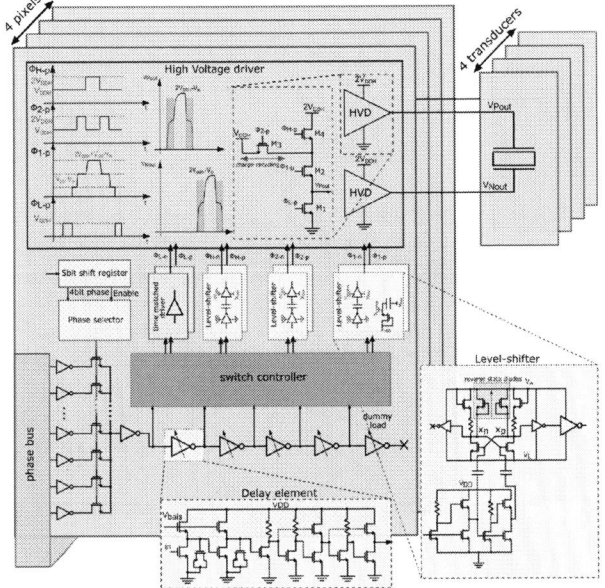

Fig. 5. Diagram of the PMUT driver pixels

IGZO transistors do not exceed the maximal voltage levels of V_{DDH}. The differential driver drives both electrodes in an alternating manner, increasing the total output voltage across the transducer with a factor of two for the a given supply voltage. To increase the driving efficiency of the PMUT transducers a single level charge recycling is also implemented at half the driving supply. For each time where both sides of the transducer needs to be charged or discharged to the full $2V_{DDH}$, the charge is recovered by shorting the output with transistor M3 to the charge recycling level V_{DDH}.

In order to drive the stacked transistors with a proper voltage swing a bootstrap supply and a unipolar capacitive level-shifter are implemented (see Fig. 5). During start-up, the voltage at nodes x_n or x_p of the capacitive unipolar level-shifter could be driven above high supply rail V_H, due to the lack of bulk diodes in a thin-film IGZO transistor. To prevent this problem, explicit diodes are added to block the reverse state. The stacked transistors in the driver need accurately timed

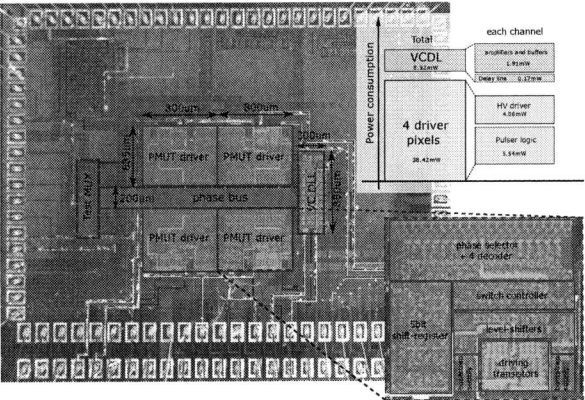

Fig. 6. Die micrograph with close-up of the driver pixel and power breakdown.

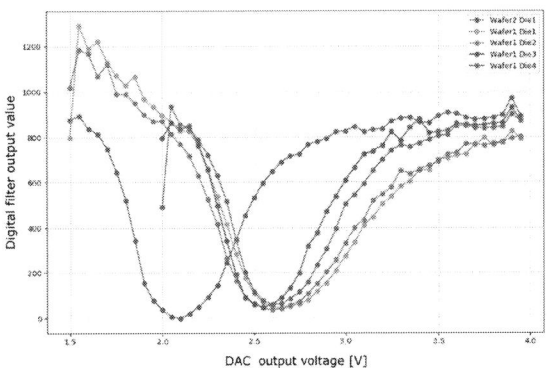

Fig. 7. Output value of the PD after the digital filter by sweeping the V_{DAC} of the VCDL for multiple samples.

driving signals as illustrated in Fig. 5. These are derived by the digital switch controller starting from the delayed versions of the input phase signal that is selected for each driver from the phase bus as described further on. The delayed versions are created by voltage controlled delay elements. To select the correct phase out of the 16 signals stemming from the phase bus of the VCDL, a 5bit shift register is implemented driving a decoder. This decoder in turn drives the pass gates to select the desirable phase. It can also disable the output pulser.

IV. MEASUREMENTS AND RESULTS

The HV driver, the partial DLL and all periphery circuits are fabricated in an 0.8 μm IGZO transistor technology on flex. Fig. 6 shows the die micrograph of a four channel HV ultrasonic driver with corresponding VCDL providing the drivers with the 4bit phase bus. A test multiplexer is added to be able to measure each phase signal in the 4bit phase bus.

By sweeping the V_{Dac} voltage of the VCDL, the output of the digital filter for different dies is shown in Fig. 7. In this curve, the output value is minimal when the VCDL reaches locking condition, i.e. the output is 180 degrees delayed

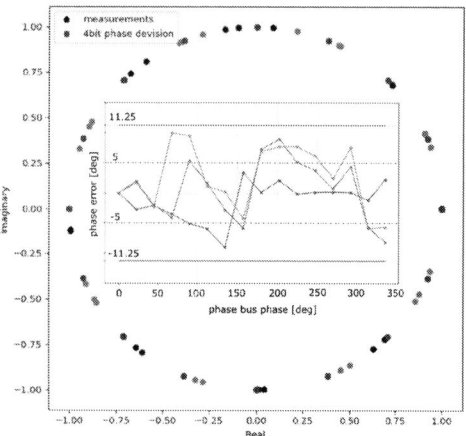

Fig. 8. Measurement results of the phase bus resolution for multiple samples after locking condition is reached.

TABLE I
MEASUREMENT RESULTS & COMPARISON

	This Work	JSSCC [8]	Nature [1]	ESSCIRC [9]
Application	Haptic feedback	imaging	Tactile display	Haptic feedback
Integration level	hybrid system on flex	silicon	PCB system	silicon
Transducer Type	PMUT	CMUT	BULK PZT	PMUT
Transducer Freq. [Hz]	250k	2.5-5M	40k	250k
Max. output voltage	24V	30V	12V	36V
Technology	800nm litho on flex IGZO	180nm BCD	PCB system FPGA	180nm BCD
Area pixel	0.8mm x 0.8mm	NaN	10.5mm x 10.5mm	0.4mm x 0.4mm
Phase resolution	4bit	4.32bit	6bit	5bit
Load cap.	36pF	40pF	0.47-1nF	240pF
Power each channel [mW]	11.68	9.85	NaN	18.65

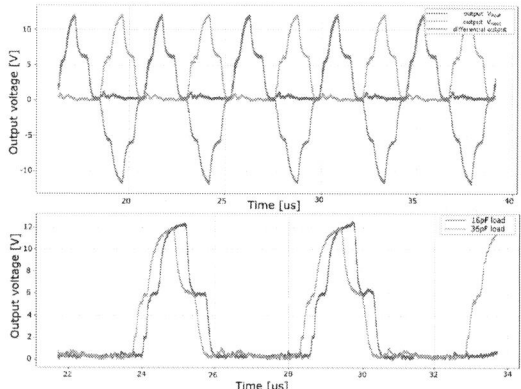

Fig. 9. Output waveform measurement of both output terminals with output load of 36pF and output waveform and a single output measurements for 36pF and 16pF load

relative to the reference clock. The phase resolution of locking in Fig. 8 demonstrates that the DLL bus meets the designed 4bit phase resolution. In addition our experiment shows that the hold and settle time of the DLL is sufficient to multiplex multiple VCDLs over a single silicon aided loop.

To characterize the output driver, a load capacitance of 16pF or 36pF is added at the output. This is equivalent to the load a typical thin-film PMUT device. The output waveform of the driver is shown in Fig. 9 driving a load of 36pF and 16pF.

In table I the thin-film ultrasonic driver for haptic feedback is compared with state-of-the-art circuits for ultrasonic driving systems targeting haptic applications. The proposed work approaches the specifications of the other circuits implemented on - costly - silicon, demonstrating the feasibility of large-area circuits on flex for mid-air ultrasound applications.

V. CONCLUSION

The first pitch-matched thin-film PMUT driving system based on IGZO transistors has been realized, capable to drive a transducer up to 36pF with 4bit phase resolution at a frequency of 250kHz and a total power consumption of 46.7mW. To accomplish this, a new hybrid silicon/flex architecture was introduced clearly combining the advantages of large-area thin-film technology driver circuits with accurate silicon filters and AD/DA convertors. The characterization results demonstrate the capabilities of flexible IGZO-based electronics for driving ultrasonic transducers. These results pave the way to further expand the possible applications of thin-film circuits to ultrasonic applications such as haptic feedback

REFERENCES

[1] Hirayama, R., Martinez Plasencia, D., Masuda, N. et al. A volumetric display for visual, tactile and audio presentation using acoustic trapping. Nature 575, 320–323 (2019).

[2] N. Sanchez et al., "An 8960-Element Ultrasound-on-Chip for Point-of-Care Ultrasound", 2021 IEEE International Solid- State Circuits Conference - (ISSCC), San Francisco,CA, USA, 2021

[3] A. Halbach et al., "Display Compatible PMUT Array for Mid-Air Haptic Feedback," 2019 20th International Conference on Solid-State Sensors, Actuators and Microsystems and Eurosensors XXXIII, Berlin, 2019

[4] Myny, K. The development of flexible integrated circuits based on thin-film transistors. Nat Electron 1, 30–39 (2018). https://doi.org/10.1038/s41928-017-0008-6 i

[5] C. Garripoli et al., "Analogue Frontend Amplifiers for Bio-Potential Measurements Manufactured With a-IGZO TFTs on Flexible Substrate," in IEEE Journal on Emerging and Selected Topics in Circuits and Systems, vol. 7, no. 1, pp. 60-70, March 20175

[6] Biggs, J., Myers, J., Kufel, J. et al. A natively flexible 32-bit Arm microprocessor. Nature 595, 532–536 (2021).

[7] H. Inoue et al., "Nonvolatile Memory With Extremely Low-Leakage Indium-Gallium-Zinc-Oxide Thin-Film Transistor," in IEEE Journal of Solid-State Circuits, vol. 47, no. 9, pp. 2258-2265, Sept. 2012

[8] K. Chen, H. Lee, A. P. Chandrakasan and C. G. Sodini, "Ultrasonic Imaging Transceiver Design for CMUT: A Three-Level 30-Vpp Pulse-Shaping Pulser With Improved Efficiency and a Noise-Optimized Receiver," in IEEE JSSCC, vol. 48, no. 11, pp. 2734-2745

[9] J. Pelgrims, K. Myny and W. Dehaene, "A 36V Ultrasonic Driver for Haptic Feedback Using Advanced Charge Recycling Achieving 0.20CV2f Power Consumption," ESSCIRC 2021 - IEEE 47th European Solid State Circuits Conference (ESSCIRC), 2021, pp. 159-16

A monolithic SPAD-based random number generator for cryptographic application

Nicola Massari[1], Alessandro Tontini[1], Luca Parmesan[1], Matteo Perenzoni[2]

[1]Fondazione Bruno Kessler (FBK)
Via Sommarive 18, Povo (TN) 38123, Italy
[2]Sony-Europe semiconductor,
Via Sommarive 18, Povo (TN) 38123, Italy

Miloš Grujić[3], Ingrid Verbauwhede[3], Thomas Strohm[4], Dayo Oshinubi[4], Ingo Herrmann[4], Andreas Brenneis[4]

[3]Imec - KU Leuven Kasteelpark Arenberg 10, B-3001
Leuven – Heverlee, Belgium
[4]Robert Bosch GmbH, Renningen, 70465 Stuttgart,
Germany

Abstract— A compact quantum random number generator based on an array of Single Photon Avalanche Diode (SPAD) is presented here. As the main feature, the proposed chip has the capability to generate random numbers without the use of an external source of light. In the present approach, SPAD devices are used as emitters and as detectors. An embedded logic allows distinguishing dark events from events coming from the photon emission. The extracted random bit sequence shows a quite uniform distribution and after being post-processed by means of an integrated circuit, used to maximize the entropy of the system, is able to pass the AIS31 test. The average bit rate is 400 kbps.

Keywords—Random number generator, SPAD, TDC

I. INTRODUCTION

The advent of Internet of Things (IoT) and the foreseen increase of number of smart devices that will be distributed in our environment, poses new technological challenges. In this scenario one of the most important issue is related to the need to protect confidential data exchanged in the network, building secure communication schemes system that prevent intruders to manipulate or control part of the system. Random number generators (RNGs) have paramount importance in defining a secure communication system and data encryption. The central role of a RNG is to generate secure keys with a high degree of unpredictability. Among different types of RNG, Quantum RNGs (QRNG), based on quantum mechanics, are considered suitable for the application thanks to the intrinsic unpredictability of the physical process. In particular, the object of this work is focused on QRNG based on photonics because of the maturity of the technology and the capability of integration.

Most examples of QRNG, found in the literature and in the market, consists of an external light source coupled to a sensor with single-photon detection capability (SPAD) and followed by a random bit extractor [1-4]. The integration of detection and logic [4] allows to realize very compact and complete devices suitable to be integrated in a miniaturized system or in a smart package [5]. Nevertheless, all these mentioned systems are still based on the assembly of an external source of light and a detector. This extra cost not only represents a limitation on the price reduction, preventing the spread of the technology in the

consumer market, but also a real bottleneck in the view of the integration of the present technology in a microprocessor.

Different works based on monolithic QRNGs have been proposed in the literature [6-10]. Authors in [6] presented first the idea to integrate both source of light and detection in the same silicon substrate. The paper shows a compact device having a central reverse-biased p-n junction as emitter surrounded by a SPAD detector. No detailed analysis on how to control this process is described while the level of integration is poor. Similar approach has been investigated in [7] where a forward-biased emitter junction was used as a source of light in an array architecture. The paper in [8] instead of using of a source of light, it relies on the spontaneous emission of dark events. Nevertheless, no direct control on the event generation is implemented while signal processing is performed by an external FPGA. The present paper proposes a monolithic solution where SPAD can be used as detector and emitter as well [9][10]. In order to control the emission of photons, two different control mechanism have been implemented: the first one influences the activity of the emitter, while the second one the emission of photons per each event. Moreover, additional logic is envisaged for event validation and distinguishing dark events from light events.

The paper is organized in the following way: in the second paragraph a description of the principle is shown followed by the chip description. Eventually, experimental results are shown for device characterization and randomness assessment.

II. THE PRINCIPLE

As described before, the core idea of the present implementation is the realization of a monolithic RNG based on SPAD devices. The principle adopted in the design, exploits the intrinsic generation of photons of a SPAD when is triggered by an avalanche event [10][11]. The implemented RNG is based on an array of identical cells. Every cell consists of three SPADs, one emitting SPAD placed in the centre and a couple of detecting SPADs on the left and right of the emitter (see Figure 1). The use of SPADs introduces the following advantages: a. good matching between the detector responsivity and the broad emission spectrum of a SPAD; b. use of a

978-1-6654-8495-4/22 $31.00 © 2022 IEEE

validation process for discarding unwanted signals thanks to the electrical monitoring of emitter events.

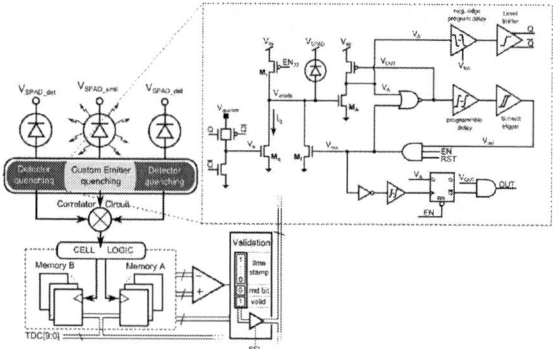

Figure 1: Basic cell of the proposed RNG and custom quenching circuit of the SPAD emitter.

Figure 1 shows a simplified schematic of a single cell. The output of the two detectors are first validated through a correlation circuit and then processed to estimate the arrival time of detected events estimated in two consecutive disjoined measurements. The comparison between these two values (t_A and t_B) gives, as the output, a single random bit ('1' if $t_A > t_B$, otherwise '0'). In case of no time stamps or time stamps with equal value, a not valid random bit is produced. At the end of the operation the cell delivers a 12b random word containing 10b time stamp (t_A), the random bit obtained after time comparison, and one extra bit for random word validation. All these operations are executed by a local embedded logic.

Two strategies have been adopted to enhance and control the photon emission of a SPAD. The first implemented strategy is based on the design of a custom quenching circuit connected to the SPAD emitter for controlling the amount of photons that can be generated during each avalanche event (see Figure 1). The key element of this circuit is represented by transistor Mq whose gate (V_q) can be connected alternatively to ground or to a reference voltage Vquench by means of a couple of switches and depending on the status of the SPAD. If enabled (EN=H), during the reset phase (RST = H), transistor Mf is activated for Tres to discharge Vanode to the ground (see time diagram of Figure 2) and node V_A changes from L to H. This voltage transition has the effect of connecting Vq to Vquench, then activating Mq. An additional circuit is used to sense the possible presence of afterpulsing. The falling edge of Vres is inverted, delayed by a time Δt and connected at the clock input of a flip-flop to sample the value of V_A immediately after the reset. In case V_A is L, because of an avalanche event occurred during the reset phase, the quenching output is masked.

In the next phase the SPAD is enabled waiting for an event. When an avalanche event occurs, a current Iq starts to flow through Mq and the SPAD junction, sustaining the avalanche current and enhancing the emission of photons. Generated photons, proportional to the current value, can be controlled by the amplitude and the duration (Tquench) of Iq respectively defined by the voltage Vquench, and Vtch. This latter sets a proper delay after the negative edge of node V_A, switching off

Mq again and SPAD enters in dead time (Tdead) until a new reset phase is executed. Under this scheme only one event is allowed per each gate time window Tw.

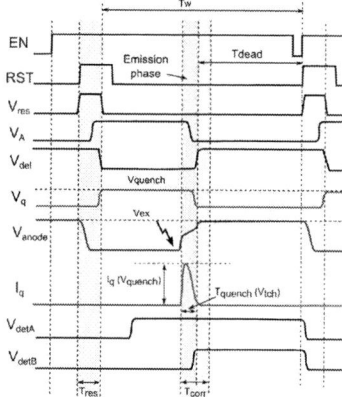

Figure 2: Typical timing diagram of the quenching circuit when working in gated mode. A correlation window Tcorr is activated when the emitter event occurs. Events outside Tcorr are discarded (VdetA) while correlated events are validated (VdetB).

The second mechanism has been used to control the emission of photons is through the implementation of an emitter based on a Perimeter Gated SPAD (PGSPAD) [12]. A PGSPAD is a SPAD device with the addition of a poly gate drawn at the perimeter of the active area. Thanks to the voltage applied to the gate (Vgate) it is possible to modulate the intensity of electric field at the guard ring and consequently also the activity of the device. This additional parameter allows us to increase the probability of having an avalanche within Tw.

Every cell has also a correlator circuit used for validating detected events. At every emitter event, the circuit generates an asynchronous correlation window with programmable width Tcorr. Only events detected within Tcorr are validated. The correlation circuit adds robustness to the random number generation by reducing the probability of spatial correlation among neighbouring cells and probability of tampering. An electrical external stimulus, injected in the emitter, can generate correlated events with low probability, being Tcorr typically in the order of few ns.

III. THE PROPOSED ARCHITECTURE

The proposed chip architecture is shown in Figure 3. As observed, the core of the device consists of two arrays of 16x8 cells for random number generation, where every cell of the array acts as an independent RNG. At every generation cycle, only a portion of the array is activated for producing random numbers. Dedicated registers allow selecting rows and columns of interest to further minimize spatial correlations. The array is configured to select at every cycle new columns, periodically scanning in this way the entire array. Only specific configurations are allowed (with 1,2, 4 or 8 columns to work in parallel). On the other hand, the scanning time defines the dead time of a single cell and, in this way, specific configuration can reduce the probability of afterpulsing. Every cell has a local memory for excluding possible noisy cell from the generation.

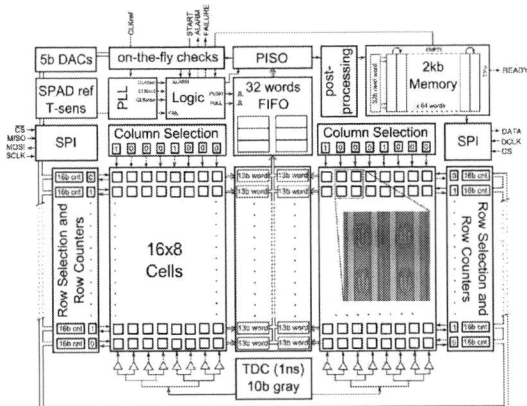

Figure 3: Chip architecture

A single common Time to Digital Converter (TDC) circuit shares a 10b-gray code running at 1GHz to the entire array for local time stamping. Core of the TDC is a VCO consisting of three stages ring oscillator made of starved NANDs, controlled by the reference voltage Vbr (see Figure 4). In the implemented scheme, a reference ring oscillator, closed in loop with an external stable clock running at 32MHz, is used for generating the reference voltage Vbr for the TDC VCO, ensuring same speed over temperature and process variation, and, at the same time, internal clocks for the chip logic and data transfer. The output of the TDC ring is connected to a 10b synchronous gray counter. All components, including the PLL filter for stability, are embedded into the chip to reduce system cost and increase the robustness of the system versus physical attacks.

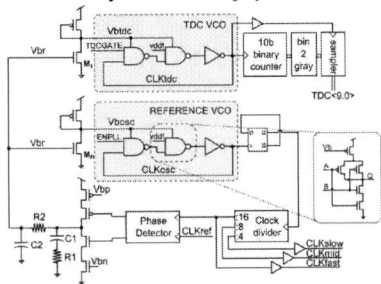

Figure 4: TDC and PLL schematic. The VCO output is connected to an asynchronous binary counter that passing through a binary to gray converter and a sampler, produces a synchronous gray code.

An SPI allows to program all internal registers for chip configuration while an embedded 5b DACs define all analogue references for the cell basing. A central logic manages the interface with the external microcontroller and controls the array scanning operation and data storage during the generation. We can distinguish different chip status. After a global reset phase, the chip is forced to stay in idle mode. In this state the chip does not produce any random numbers, lowering the power consumption, while waiting for the START signal. When START signal is asserted, the chip starts producing random bits by executing time stamp comparison. These bits are transferred from the array to a local internal memory of 2kb

to be accumulated. Before and during data generation two different checks are running in parallel with the main operation.

In the first check, the chip monitors the possible presence of total failure conditions. These conditions refer to errors in the SPAD activity (due, for example, to an unwanted or intentional erroneous bias setting) or to an exceeding of the temperature range. These checks continuously monitor not only the absolute value of SPAD activity and temperature, but also possible unexpected changes due to an extraordinary external event that can be, very likely, associated to a physical attack. The system can then detect sudden increase of SPAD activity, due to, for example, a light shining the array, or a steep decrease of temperature, due to a spry cooler, used for tampering the RNG. In case of threshold exceeding, the control block sends an alarm to the central logic which asserts the FAILURE flag stopping the generation of random number. In the presence of a total failure condition the chip need to be restarted.

The second parallel check, refers to on-the-fly tests can be applied to raw data. Implemented checks are: 1. frequency test, monitoring if the ratio between zeros and ones is close to one.; 2. correlation test, checking for possible presence of relevant patterns induced by a possible chip malfunctioning; 3. Arrival time test, monitoring that the behaviour of the TDC is always within a known range to guarantee a correct working condition. First two tests are used to verify possible relevant statistical deviation from the expected behaviour of the random sequence. If an alarm is produced, the logic of the chip asserts the ALARM flag and waits for a command to restart a new generation. In case of no alarms, if the memory is full, the chip sends the READY signal to the microcontroller to inform that the RNG has valid data to deliver. At the same time, the RNG stops generating random bits to reduce power consumption. A new generation cycle is activated when the memory is empty again (and START=H).

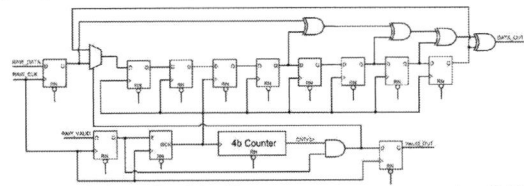

Figure 5: Schematic of the linear corrector post-processing. RAW DATA, CLK and VALID are signals delivered by the array.

The implementation of a post-processing circuit can be used for increasing the entropy per bit rate of the system in case raw bits are not compliant with the AIS-31 requirement of at least 0.997 bits of Shannon entropy [13]. We implemented three different types of information-theoretical post-processing methods: the Elias' post-processing; a linear corrector and the integer addition. The post-processing with linear corrector (see Figure 6) consists of multiplying a $k \times n$ generator matrix of a [n,k,d] code with n independent raw random bits. If the raw bits have bias of at most $b = |P(x = 1) - 0.5|$, then the min-entropy rate (and thus the Shannon entropy rate) of k post-processed bits is lower bounded by $1 - \log_{2^k}(1 + 2^{k+d} \cdot b^d)$ [14-15]. Note that unlike the ad-hoc non-information theoretically secure and non-compressing post-processing

978-1-6654-8495-4/22 $31.00 © 2022 IEEE

constructions, the linear corrector is a provably sound method to increase the entropy rate of the biased independent bits. We used the generator matrix of a [16,8,5] code for our RNG since it enables a throughput reduction of only 50 %, while the post-processed bits will have a min-entropy higher than 0.997. This post-processing is very efficiently implemented in silicone due to the cyclic structure of its generator matrix.

IV. MEASUREMENTS

We first characterize the device for proving the capability of the implemented circuit to control the emission of photons. Parameters used in this analysis are: a. Vgate: for controlling the activity of the emitter; b. Vquench and Vtch: for controlling the intensity and the duration of the emission respectively; c. Vctrl: for controlling the duration of the correlation window. Figure 6 a) shows the variation of the RNG bit rate as a function of Vquench and for different values of Vgate. The bit rate is strictly related to the capability of photons generation. In this case raw random bits are considered as result of the time stamping comparison. As observed from the plot, as soon as Vquench increases, as also the rate increases almost proportionally (please note that emission is proportional to the flowing current and not on the voltage). Moreover, is also shown that Vgate has a relevant impact on the photon generation. This plot demonstrates that the chip is able to control the emission process.

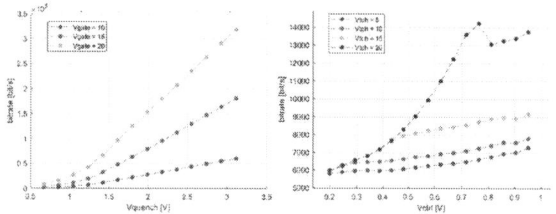

Figure 6: RNG rate a) versus Vquench and different values of Vgate (emitter activity); b) versus Vctrl (defining Tcorr) and for different values of Vtch (defining Tquench).

In Figure 6 b) the RNG bit rate is estimated as a function of the correlation window (Vctrl) and for different values of the duration of the emission (Vtch). As shown, when Tcorr > Tquench the rate does not significantly increase. This reflects on the probability of not correlated events in Tcorr. To verify the statistical quality of the RNG output, we ran the AIS-31 tests (test procedure B) and the NIST SP 800-90B estimators [16] on the random bits produced by the chip before the cryptographic post-processing. We first analyzed 25000 KB of the raw data before the linear corrector at three temperature points: -20ºC, 0ºC and 20ºC. The bias of the bits was too high to pass tests T1-T6 of the AIS-31, while on the other hand the NIST estimators assessed the min-entropy of all three sequences to above 0.93. We then ran the same tests on the random bits after the linear corrector. These bits pass all AIS-31 tests with estimated Shannon entropy higher than 0.997 and the min-entropy assessed by the NIST estimators higher than 0.94. Since the random bit extraction method implemented by this RNG is theoretically supported by an adequate stochastic model, the RNG implements the online and the total failure tests, the random bits pass the procedure B tests [13] and the bits are cryptographically post-processed in software, we can conclude that the proposed RNG is compliant with the AIS-31 methodology for RNGs.

V. CONCLUSIONS

A monolithic QRNG for cryptographic application is shown in this paper. We demonstrated the capability to control the emission of photons and we assessed the randomness produced by the device that, after embedded post-processing, is able to pass AIS-31 tests. To comply with PTG.3, the highest security level class of the AIS-31 standard, we also implemented an AES-128-based cryptographic post-processing and know-answer tests (KAT) in the microcontroller software.

ACKNOWLEDGMENT

We acknowledge the financial support from the European Commission through the Grant Agreement No. 820405 (project QRANGE). We acknowledge Lorenzo Baldessarini for doing part of the measurements.

REFERENCES

[1] Fang-Xiang Wang et al., "Robust quantum random number generator based on avalanche photodiodes", arXiv:1502.01084v3,18 Jun 2015.

[2] Simone Tisa et al., "High-Speed Quantum Random Number Generation Using CMOS Photon Counting Detectors", IEEE Journal of Selected Topics in Quantum Electronics , Vol. 21, No. 3, May/June 2015.

[3] Massari et al.: "A 16×16 pixels SPAD-based 128-Mb/s quantum random number generator with −74dB light rejection ratio and −6.7ppm/°C bias sensitivity on temperature" ISSCC 2016.

[4] Francesco Regazzoni et al., "A High Speed Integrated Quantum Random Number Generator with on-Chip Real-Time Randomness Extraction", arXiv:2102.06238v1 [quant-ph] 11 Feb 2021.

[5] https://marketing.idquantique.com/acton/attachment/11868/f-025e/1/-/-/-/-/Quantis%20QRNG%20Chip_Brochure.pdf

[6] Abbas Khanmohammadi et al., "A Monolithic Silicon Quantum Random Number Generator Based on Measurement of Photon Detection Time", IEEE Photnics Journal, Vol 7, Number 5, 7500113, October 2015.

[7] Nicola Massari et al, "A Monolithic QRNG based on an array of SPADs", Single Photon Workshop, pp153, Milan October 2019.

[8] Massimo Caccia et al., "In-silico generation of random bit streams", Nuclear Inst. and Methods in Physics Research, A 980 (2020) 164480.

[9] Fabio Acerbi et al., "Structures and Methods for Fully-Integrated Quantum Random Number Generators," in IEEE Journal of Selected Topics in Quantum Electronics, vol. 26, no. 3, pp. 1-8, May-June 2020.

[10] Fabio Averbi et al., "A Robust Quantum Random Number Generator Based on an Integrated Emitter-Photodetector Structure", IEEE Journal of Selected Topics in Quantum Electronics, March 2018.

[11] N.Massari et al., "Towards low-cost monolithic QRNGs", SPIE Photonics Europe, Strasbourg, April 2022.

[12] M. H. U. Habib et al., "Optimization of perimeter gated SPADs in a standard CMOS process" IEEE Sensors, pp. 1668-1671, 2014.

[13] W. Killmann and W. Schindler, "A proposal for: Functionality classes for random number generators," Tech. Rep., Bundesamt für Sicherheit in der Informationstechnik (BSI), Bonn, Germany, Sep. 2011.

[14] P. Lacharme,"Post-processing functions for a biased physical random number generator." In International Workshop on Fast Software Encryption, pp. 334-342. Springer, Berlin, Heidelberg, 2008.

[15] M. Grujić and I. Verbauwhede, "TROT: A Three-Edge Ring Oscillator Based True Random Number Generator With Time-to-Digital Conversion." IEEE Transactions on Circuits and Systems I (2022).

[16] M. Sönmez Turan et al., "Recommendation for the Entropy Sources Used for Random Bit Generation," Technical Report, National Institute of Standards and Technology (NIST), 2018.

Quantum-Correlated Photon-Pair Source with Integrated Feedback Control in 45 nm CMOS

D. Kramnik[1†], I. Wang[2], J. M. Fargas Cabanillas[2], A. Ramesh[3], S. Buchbinder[1], P. Zarkos[1],
C. Adamopoulos[1], P. Kumar[3], M. A. Popović[2], and V. Stojanović[1‡]

[1]Department of Electrical Engineering and Computer Sciences, UC Berkeley, Berkeley, CA 94720, USA
[2]Department of Electrical and Computer Engineering, Photonics Center, Boston University, Boston, MA 02215, USA
[3]Dept. of Electrical and Computer Eng., Northwestern University, Evanston, IL 60208, USA
Email: [†]kramnik@berkeley.edu, [‡]vlada@berkeley.edu

Abstract—Integrated photonics provides scalability needed for useful photonic quantum information processing. Many optical resonators must be aligned to the same pump wavelength to produce sources of quantum-correlated photon pairs that drive such systems, but existing solutions rely on manual alignment or offline tuning based on external photodiodes and bulky off-chip electronics, limiting scalability. Here we demonstrate feedback control of four-wave mixing (stimulated and spontaneous) in a silicon microring using circuits integrated alongside photonics in a standard 45 nm CMOS foundry process. The carrier-sweepout-generated feedback signal enables *in-situ* operation in the photon-pair generation regime, which is a key building block enabling large-scale CMOS quantum-photonic systems-on-chip.

I. INTRODUCTION

Integrated silicon photonics is an appealing quantum photonics technology platform due to its compatibility with scalable, high-fidelity CMOS manufacturing and its ability to operate at standard telecommunication wavelengths [1], [2]. Sources of indistinguishable quantum-correlated photon pairs produced by spontaneous four-wave mixing (FWM) in microring resonators are a basic building block required in many implementations of systems for quantum networking and information processing [2]. A key challenge in scaling these systems is maintaining a lock between the narrow-linewidth resonance of the microring cavity and the pump laser [3]. With higher pump powers required to maximize pair count rates, silicon microrings exhibit thermal bistability caused by self-heating from absorbed pump power [4] and can exhibit free-carrier oscillations as thermal and free-carrier dispersions take turns pushing the resonance wavelength in opposite directions [5]. These effects are substantially exacerbated by the high Q-factors needed for efficient four-wave mixing (FWM), around $10\times$ greater than those found in microring-based silicon-photonic data links [4]. Here we demonstrate the first feedback-controlled microring resonator system, with circuits implemented alongside photonics in a 45 nm SOI CMOS process, that resolves both of these issues and operates continuously during stimulated and spontaneous (photon-pair generation) four-wave mixing experiments. To our knowledge, this is the highest-Q microring wavelength-locked via an integrated thermal tuning control loop to date.

This work was funded in part by NSF EQuIP program grant #1,842,692, Packard Fellowship #2012-38222, and the Catalyst Foundation.

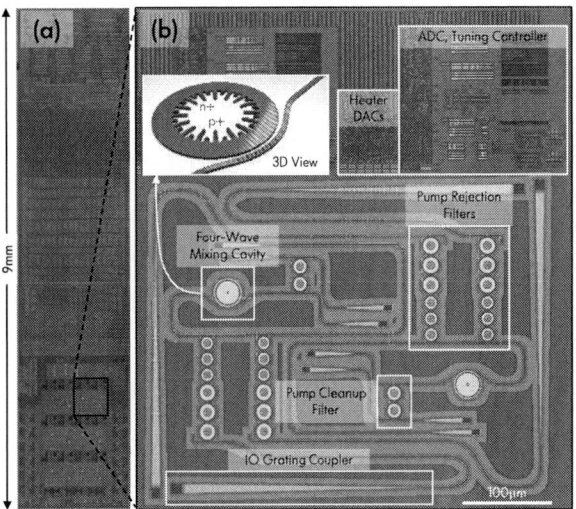

Fig. 1. (a) Die micrograph after XeF$_2$ etch to enable photonics, (b) two interleaved quantum-correlated photon-pair generator system sites with 3D render of carrier-sweepout, FWM-optimized microring structure (inset).

II. CMOS QUANTUM PHOTONICS

A. Monolithic Electronic-Photonic CMOS Process

A 2×9 mm electronic-photonic test chip with quantum photonics and integrated circuits, shown in Fig. 1(a), was fabricated in GlobalFoundries' 45RFSOI CMOS process. Twelve pairwise interleaved quantum-correlated photon-pair generator systems occupy the bottom part of the chip, with each pair consuming 0.34 mm^2 (550 μm × 620 μm) area including control circuits. The die is flip-chip packaged onto an FR-4 PCB for electrical interfacing and the Si substrate is removed using XeF$_2$ etch to enable photonics and allow optical access from the exposed back side of the chip. Optical waveguides, microring resonators and filters, and grating couplers are patterned in the transistor silicon layers to create photonic circuits alongside electronics, as shown in Fig. 1(b). We previously demonstrated passive quantum photonics in this process, including on-chip pump-rejection filters, and a manually-tuned (open-loop) photon-pair source [6], [7]. In this work, similar on-chip pump cleanup and pump rejection filters are bypassed and only the photon-pair source ring is active.

B. Microring-Based Photon-Pair Source

Four-wave mixing occurs in silicon due to its $\chi^{(3)}$ optical nonlinearity. A microring cavity resonantly enhances this effect, increasing the production rate of spontaneously-generated photon-pairs used for quantum information processing. Our system is built around a silicon microring cavity with optimized width and radius to trade off mode volume, bend loss, and integrated dispersion for optimal pair generation rate within the constraints of the CMOS platform. It contains interleaved p-i-n diode regions for carrier sweepout with geometry tailored to reduce free-carrier absorption in the area containing the optical mode, resulting in an intrinsic Q of 100,000 and loaded Q of 41,200 (FWHM = f_0/Q = 4.7 GHz) when coupled to the bus waveguide, measured at $\lambda = 1552$ nm. These all-silicon devices exhibit responsivity of 1–3 mA/W from defect-state absorption (depending on doping geometry) at -10 dBm on-chip illumination in C-band, requiring high-sensitivity readout electronics to detect the weak photocurrent signal on the order of hundreds of nA at typical pump powers (-10 dBm to 0 dBm, the high end being limited by multi-pair production). Unlike microrings used for classical optical datacom, the resonant frequency of a cavity used for four-wave mixing must be aligned precisely with the laser wavelength in order to optimize the pair generation rate, which falls off sharply with the detuning. This necessitates precise thermal tuning control that balances on the edge of instability to capture as much pump power in the resonator as possible.

C. On-Chip Tuning Circuits

Fig. 2 shows a schematic of the integrated feedback control circuits, adapted from the design previously used for read-out of photonic ultrasound and bio-sensors [8]. The microring's photocurrent is fed into an analog frontend consisting of programmable-gain inverter-based transimpedance amplifiers (TIAs) and offset-adjustment current DACs. A pseudo-differential scheme with a sense arm and replica reference arm is adopted in order to improve CMRR and reject power supply noise coupled from the digital sections. The amplified photocurrent is then digitized by a 9-bit differential SAR ADC, accumulated 512 times on-chip for averaging, and sent to a digital controller and scan chain interface to an external FPGA. A sigma-delta DAC (10-bit pulse density modulator) drives a disk-shaped heater in the center of the source ring, which has increased thermal capacitance to improve filtering of the noise-shaped spectrum, and can be set either by the on-chip digital controller or scan chain, enabling closed-loop control of the microring's resonance frequency via thermal tuning. A switching DAC is used here to increase efficiency and avoid overheating and degradation of the output transistor. The overall wavelength tuning range is approximately 1 nm (depending on HVDD), resulting in an LSB step of 980 fm (122 MHz), which is fine enough to tune within 99% of the Lorentzian peak given the linewidth of the loaded resonator. Self-heating of the microring from absorbed pump laser power pushes its resonance point to longer wavelengths as the heater

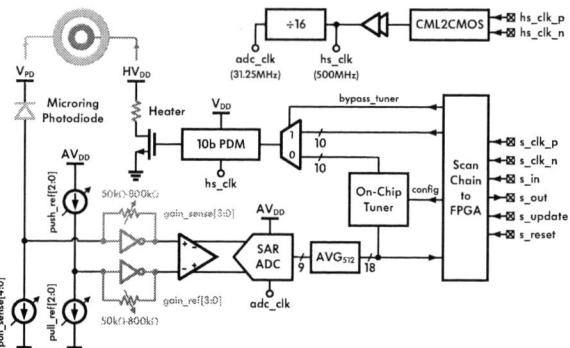

Fig. 2. Schematic diagram of readout and tuning controller circuits integrated alongside each four-wave mixing microring cavity.

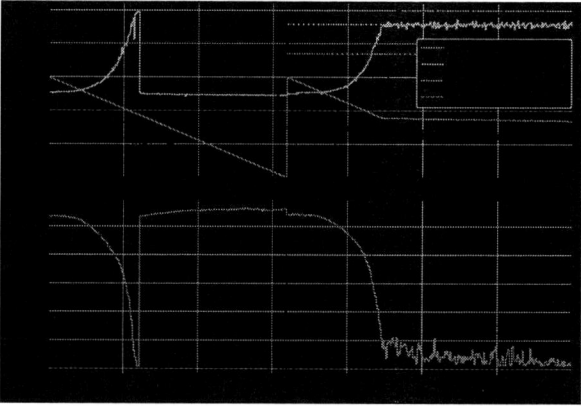

Fig. 3. Photon-pair source feedback controller waveforms and state transitions (top) and external through port monitor power (bottom), which is not used for feedback control. The slight downward slope in the through-port power is likely due to drift in the alignment of input and output coupling fibers.

is swept from hot to cold, further improving the effective resolution as the pump power is increased.

For each setting of the pump laser power and transimpedance gain (800 kΩ in the experiments reported here) the analog frontend is automatically calibrated by sweeping the offset-calibration current DACs, which speeds up experiments by eliminating manual tuning of the circuits. The ADC readings are recorded and the current DACs are set with the goal of maximizing the photocurrent sensitivity and dynamic range. The slope of the ADC readings with respect to the sense-arm pull DAC allows the photocurrent sensitivity to be calculated since the two are summed at the input of the sense arm and it is maximized when the sense arm is biased near midrail. On the other hand, the dynamic range is maximized by introducing some offset between the sense and reference arms to pull the ADC reading as low as possible, since the photocurrent can only pull it back up, but this can also start to saturate the differential input stage and degrade the photocurrent sensitivity. In practice the two effects can be traded off by maximizing a figure-of-merit defined as the product of ADC range and photocurrent sensitivity.

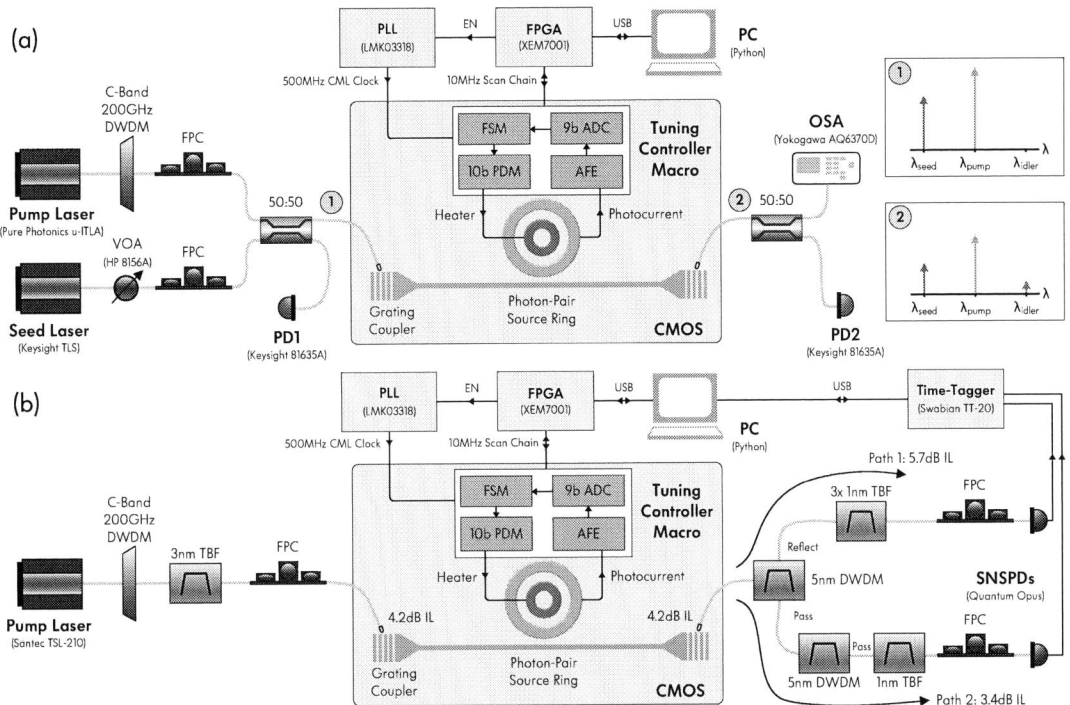

Fig. 4. (a) Stimulated four-wave mixing experimental setup and spectra, (b) spontaneous four-wave mixing experimental setup enabling simultaneous feedback control and single-photon measurements using superconducting nanowire single-photon detectors (SNSPDs).

III. CHIP MEASUREMENTS

A. Wavelength Locking

Fig. 3 shows waveforms from a feedback control experiment where the microring's resonant wavelength is locked to a fixed pump laser. After auto-calibration of the analog frontend, a 3-step finite state machine (FSM) engages and sweeps down the heater code to find the peak photocurrent (scan state), resets and finds the resonance again (seek state), and finally locks to a point near the maximum (lock state). The cavity must be returned to the hot state after sweeping past the laser to defeat the hysteresis created by self-heating [4]. In the lock state, a PI controller with integral deadband around the setpoint rejects noise from the analog section to eliminate spurious dithering of the heater DAC. A feedback controller is implemented within the digital logic on the chip with a different FSM optimized for locking with a fixed detuning between the laser and resonator, but we use off-chip control via scan chain interface to an FPGA and PC to implement our modified tuning algorithm and lock closer to the photocurrent peak (at 95 % peak current, presently limited by fiber stability). No Peltier cooler or other off-chip thermal stabilization is required. In this work the sample rate is limited to 4 Hz by the slow interface between the control FPGA and Python script used to prototype the control algorithm, but could be boosted up to 1 kHz by implementing the same control logic on the FPGA (which is then limited by the test chip's total scan chain length and 10 MHz scan clock).

B. Four-Wave Mixing

Using the dual-laser setup illustrated in Fig. 4(a) we measure $-41.2\,\mathrm{dB}$ conversion efficiency at $-5.8\,\mathrm{dBm}$ on-chip pump power in the stimulated FWM regime — Fig. 5(a,b). This is comparable to manually-aligned passive microrings in this process [9], indicating that the addition of dopings for carrier sweepout does not compromise performance (stimulated FWM efficiency is not affected by insertion losses after the resonator, making it a more reliable way to compare different resonators than off-chip photon-pair count rates). The pump laser in this case was a commercial micro-ITLA (integrated tunable laser) module, enabling future miniaturization of the system into a test-equipment-free complete module.

We also characterized quantum-correlated photon-pair generation under feedback control using the setup illustrated in Fig. 4(b). We recorded up to 300 ccps (coincidence counts per second) off-chip pair generation rate and a maximum coincidences-to-accidentals ratio (CAR) of 38 with $-1.6\,\mathrm{dBm}$ and $-5.6\,\mathrm{dBm}$ on-chip pump powers respectively, using 60% quantum efficiency in C-band superconducting nanowire single photon detectors (SNSPDs) — Fig. 5(c,d). Grating coupler insertion losses (4.2 dB per coupler) and external pump-rejection filter insertion losses (3.4 dB and 5.7 dB per channel) also limited the measured count rates. Since losing a photon in either channel of the experiment results in a missed coincidence count, the sum of insertion losses in both channels (21.9 dB) can be used to estimate the on-chip count rate, yielding a max-

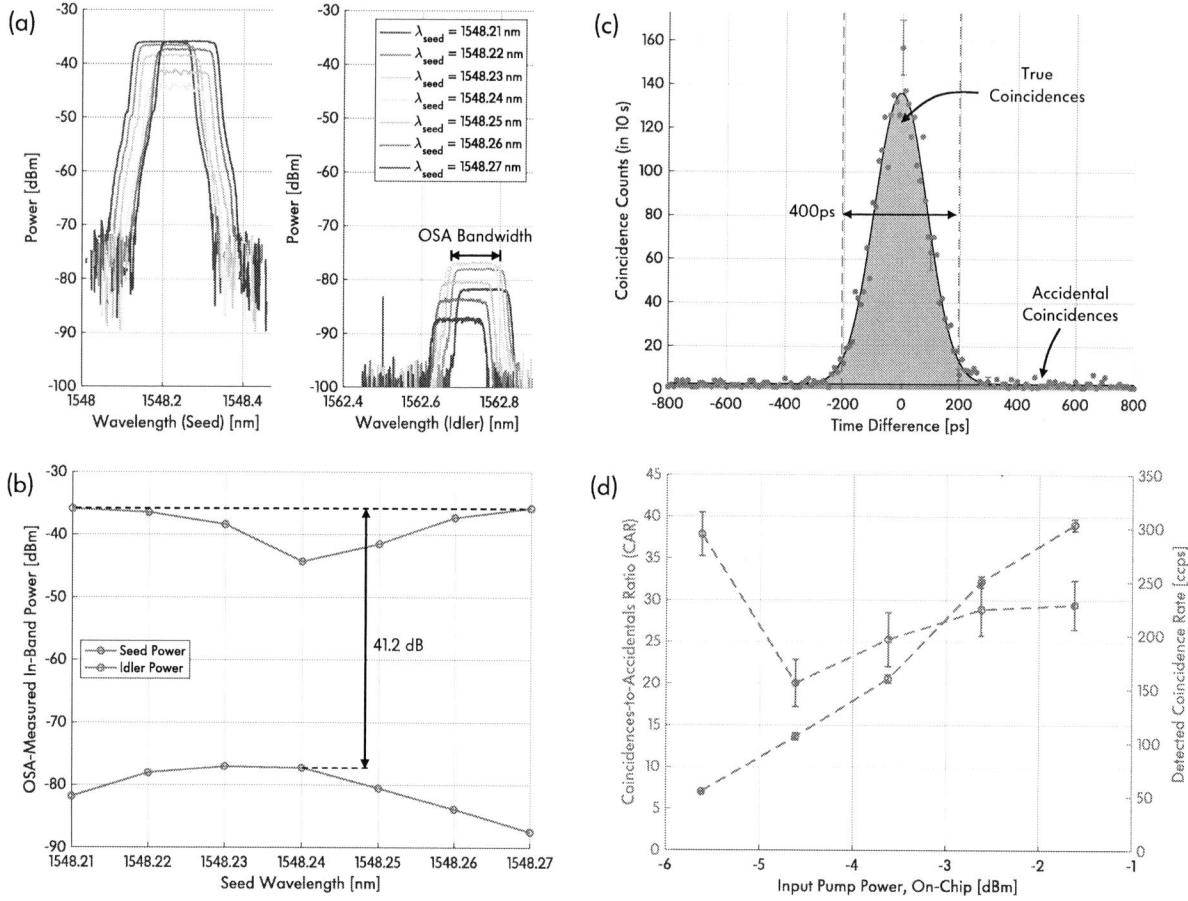

Fig. 5. Stimulated FWM through port spectra (a) and seed and idler powers and conversion efficiency (b), with $-5.8\,\mathrm{dBm}$ on-chip pump power. Spontaneous FWM coincidence histogram with $-1.6\,\mathrm{dBm}$ on-chip pump power (c) and CAR and coincidence count rate with active feedback control vs. pump power (d).

imum on-chip pair rate of 46.5 kccps. In separate experiments we have wavelength-aligned the low-loss integrated pump-rejection filters using on-chip circuits [10], paving the way for a fully-monolithic photon-pair generator system.

IV. CONCLUSION

In this work we demonstrate feedback control of a silicon four-wave mixing cavity with electronics and photonics fabricated together on the same die in a 45 nm CMOS foundry process. This represents a new step towards creating a self-contained source of entangled photons that incorporates active photonic devices and feedback control circuits onto the same CMOS chip. We utilize a standard 45 nm foundry process that can support hundreds of electronically-controlled photonic devices operating alongside millions of transistors, used previously for classical information processing [11].

ACKNOWLEDGMENT

We thank the Berkeley Wireless Research Center and Berkeley Emerging Technologies Center for support and Ayar Labs for chip fabrication.

REFERENCES

[1] Rudolph, T. "Why I am optimistic about the silicon-photonic route to quantum computing", APL Photonics. 2, 030901 (2017).

[2] Silverstone, J.W. *et al.*, "Silicon Quantum Photonics", IEEE J. Sel. Top. Quantum Electronics 22, 390-402 (2016).

[3] Carolan, J. *et al.*, "Scalable feedback control of single photon sources for photonic quantum technologies", Optica 6, 335-340 (2019).

[4] Sun, C. *et al.*, "A 45nm CMOS-SOI Monolithic Photonics Platform With Bit-Statistics-Based Resonant Microring...", IEEE JSSC (2016).

[5] Johnson, T. J. *et al.*, "Self-induced optical modulation of the transmission through a high-Q silicon microdisk resonator", Opt. Express (2006).

[6] Gentry, C. M. *et al.*, "Monolithic Source of Entangled Photons with Integrated Pump Rejection", CLEO (2018).

[7] J. Fargas-Cabanillas, D. Kramnik, A. Ramesh, *et al.*, "Tunable Source of Quantum-Correlated Photons with Integrated Pump...", FiO (2021).

[8] C. Adamopoulos, P. Zarkos, *et al.*, "Lab-on-Chip for Everyone: Introducing an Electronic-Photonic Platform...", IEEE OJ-SSCS (2021).

[9] Gentry, C. M. *et al.*, "Quantum-correlated photon pairs generated in a commercial 45nm...", Optica 2, 1065-1071 (2015).

[10] Fargas Cabanillas, J. M. *et al.*, "Monolithically Integrated High-Order Vernier Filters for Electronic-Photonic...", CLEO (2022).

[11] C. Sun, M.T. Wade, Y. Lee, *et al.* "Single-chip microprocessor that communicates directly using light", Nature 528, 534–538 (2015).

A 4.6-8.3 TOPS/W 1.2-4.9 TOPS CNN-based Computational Imaging Processor with Overlapped Stripe Inference Achieving 4K Ultra-HD 30fps

Yu-Chun Ding, Kai-Pin Lin, Chi-Wen Weng, Li-Wei Wang, Huan-Ching Wang,
Chun-Yeh Lin, Yong-Tai Chen, and Chao-Tsung Huang
Department of Electrical Engineering, National Tsing Hua University, Hsinchu, Taiwan

Abstract—In this paper, we present an energy-efficient accelerator chip which supports high-quality CNN-based computational imaging applications at 4K Ultra-HD 30fps. To address the huge requirement of DRAM bandwidth and computing energy, an overlapped stripe inference flow and a structure-sparse CONV3x3 engine are proposed respectively. The former reduces DRAM bandwidth to 0.81-1.74 GB/s when supporting high-quality CNN inference with 16 to 29 layers at 4K Ultra-HD 30fps. The latter reduces computing complexity by 40% without noticeable quality degradation, e.g. 0.02-0.03 dB of PSNR drop. More specifically, it uses only 4.9 intrinsic TOPS of computing capability at 200 MHz to approach the quality of dense models which demand up to 8.2 TOPS. In addition, a coarse-grained reconfigurable datapath is designed to support diverse applications including super-resolution, denoising, and style transfer with high hardware efficiency. Fabricated in 40nm CMOS, this chip achieves 4.6-8.3 TOP/W of energy efficiency for high-quality computational imaging applications. We also implement an FPGA-aided system to demonstrate real-time processing for the diverse applications supported by the fabricated chip.

I. INTRODUCTION

Convolutional neural networks (CNNs) have demonstrated superior quality for image restoration tasks such as super-resolution (SR) and denoising (DN). They can even provide some novel applications, like style transfer (ST), which are difficult to achieve by traditional methods. Therefore, CNNs have great potential to revolutionize the image pipeline on cameras and displays.

However, high-resolution feature maps in computational imaging CNNs could demand enormous external memory access (EMA) and thus make it hard to be supported in real time on the accelerators optimized for classification CNNs which have low-resolution input and gradually-downsampled features. Previous SR CNN accelerator chips [1], [2] reduce EMA through layer-fusion inference. Consecutive convolution layers are fused together to reduce the off-chip transaction of high-resolution intermediate feature maps. As a result, these works can support the lightweight and shallow FSRCNN [3] in real time for CNN-based SR. For the 8-layer FSRCNN SRx2 inference at merely Full-HD (FHD) output, in [1] 15.7 MB of EMA per frame is demanded for using small on-chip SRAM of size 77KB to store intermediate features. On the other hand, a larger SRAM of size 1056KB is used in [2] to reduce off-chip access. Additional memory cost including off-chip access and on-chip storage will be demanded to further enlarge

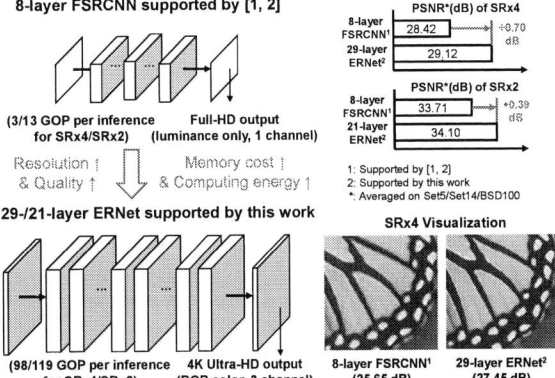

Fig. 1. Comparison of image quality and computing complexity for shallow FSRCNN and deep ERNets. This work supports the 29- and 21-layer CNN inference for SRx4 and SRx2, respectively.

image resolution and increase model layers to enhance image quality as shown in Fig. 1. Besides, the dramatic computing requirement for the deep models at ultra-high resolution, e.g. 98-119 GOP for one 4K Ultra-HD (UHD) frame output in Fig. 1, also poses a limitation on these previous works for real-time acceleration. Moreover, computational imaging can provide much more applications than only SR, but this diversity is not demonstrated in these works as well.

In this paper, we present an energy-efficient accelerator chip to support high-quality and diverse-function CNN inference at 4K UHD 30fps for computational imaging applications. It addresses design challenges with three primary contributions: 1) a memory-efficient overlapped stripe inference (OSI) flow which reduces off-chip and on-chip memory costs simultaneously; 2) a structure-sparse CONV3x3 engine for energy-efficient and high-quality inference; 3) a coarse-grained reconfigurable datapath (CGRD) to support multiple imaging applications with high hardware efficiency. The detail will be introduced in the following.

II. OVERLAPPED STRIPE INFERENCE FLOW

Fig. 2 shows our proposed OSI flow and the comparison to previous methods [1], [2], [4], [5] for reducing memory cost including SRAM size and EMA. In [1], [5], consecutive

978-1-6654-8495-4/22 $31.00 © 2022 IEEE

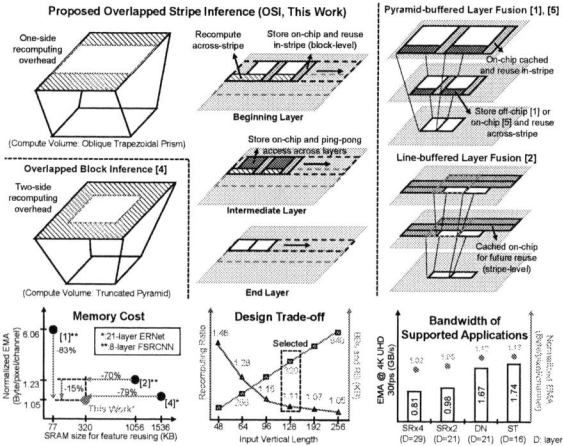

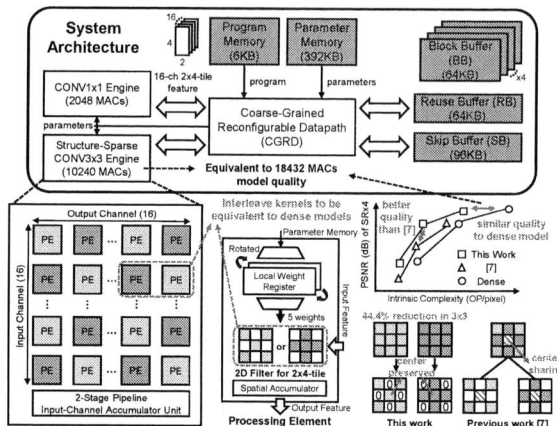

Fig. 2. Overlapped stripe inference flow for memory cost reduction including EMA and SRAM size.

Fig. 3. System architecture with structure-sparse CONV3x3 engine. The CONV3x3 engine adopts a 3-stage pipeline and consists of 256 PEs. Each PE serves 2D filtering for one 2x4-tile in one cycle, and output from PEs will be accumulated in two cycles in the 2-stage input-channel accumulator unit.

convolution layers are calculated at once until the final output patch is generated. The overlapped features between adjacent output will be cached all on-chip [5] to eliminate off-chip access, or the across-stripe ones are selectively stored in off-chip [1] to reduce SRAM footprint. The sizes of pyramid-shaped overlapped features dramatically increase in deep models and demand unfordable memory cost. In [2], a line-based fusion approach is adopted to avoid the growing memory cost of feature pyramid by only caching a few lines in each fused layer. However, it still requires a large on-chip memory of size 1056KB for storing the high-resolution across-stripe features. On the other hand, [4] performs an overlapped block inference flow which recomputes across-block features, instead of reusing them in either off-chip or on-chip memory. Although a hardware-oriented CNN structure, ERNet [6], is devised for alleviating recomputing overhead, [4] still needs to deploy as large as 1536KB of on-chip block buffers for deep models.

In this work, we propose to recompute across-stripe features to save EMA and reuse in-stripe ones in small-size SRAM. The OSI flow can shrink the demands of EMA and SRAM size at the same time. The one-side recomputing leads to smaller computing overheads and enables asymmetric block shapes. We preserve longer block length on the recomputed side to reduce the overhead and shrink the reused side to save SRAM size. In each block, the convolution will be performed in a layer-by-layer manner, and four block buffers (BBs) are used to cache the intermediate features between layers. The reuse buffer (RB) is devised for feature reusing among the adjacent blocks in the same stripe. As a result, we use as small as 320KB of BBs and RB, and only 0.81-1.74 GB/s of EMA for inference of high-quality SR/DN/ST models (21-to-29-layer ERNets and 16-layer ResNet) at 4K UHD 30fps. Compared to [1] and [2] which support the shallow 8-layer FSRCNN, this work reduces normalized EMA, in terms of bytes per output color component, by 83% and 15% respectively.

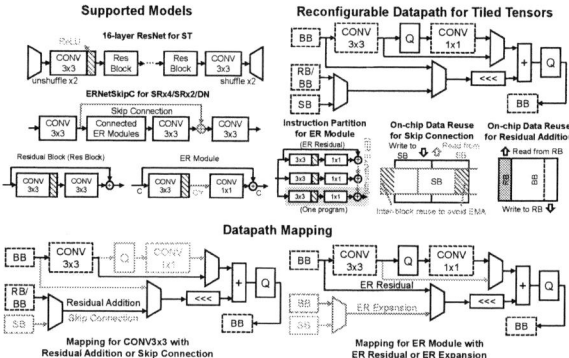

Fig. 4. Coarse-grained reconfigurable datapath to address application diversity. Both ERNet and ResNet can be supported in this work for high-quality SR, DN, and ST.

III. SYSTEM ARCHITECTURE

Fig. 3 shows the system architecture of this chip. There are three computing engines for CONV3x3, CONV1x1, and datapath, respectively. We keep the whole model parameters in the global parameter memory to avoid EMA for parameter re-transmission across blocks. In the following, we will introduce the structure-sparse CONV3x3 engine and the coarse-grained reconfigurable datapath in details.

A. Structure-sparse CONV3x3 Engine

Huge computing energy is demanded for CNN inference, which poses a challenge to resource-constrained edge devices. Weight pruning can greatly reduce the computing cost but the considerable circuit overheads for irregularity may degrade its complexity saving. Thus we design a highly-parallel CONV3x3 engine with structural sparsity in kernels for low-overhead complexity saving as shown in Fig. 3. We reduce the complexity in a regular representation which directly zeros

978-1-6654-8495-4/22 $31.00 © 2022 IEEE

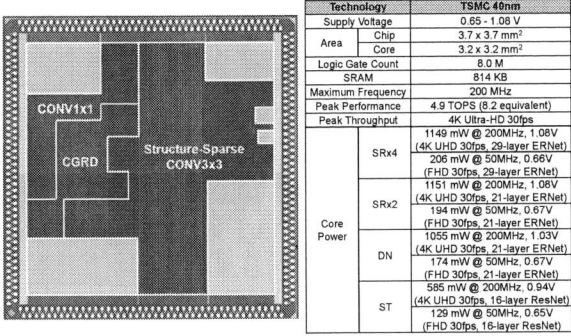

Technology		TSMC 40nm
Supply Voltage		0.65 - 1.08 V
Area	Chip	3.7 x 3.7 mm²
	Core	3.2 x 3.2 mm²
Logic Gate Count		8.0 M
SRAM		814 KB
Maximum Frequency		200 MHz
Peak Performance		4.9 TOPS (8.2 equivalent)
Peak Throughput		4K Ultra-HD 30fps
Core Power	SRx4	1149 mW @ 200MHz, 1.08V (4K UHD 30fps, 29-layer ERNet) 206 mW @ 50MHz, 0.66V (FHD 30fps, 29-layer ERNet)
	SRx2	1151 mW @ 200MHz, 1.08V (4K UHD 30fps, 21-layer ERNet) 194 mW @ 50MHz, 0.67V (FHD 30fps, 21-layer ERNet)
	DN	1055 mW @ 200MHz, 1.03V (4K UHD 30fps, 21-layer ERNet) 174 mW @ 50MHz, 0.67V (FHD 30fps, 21-layer ERNet)
	ST	585 mW @ 200MHz, 0.94V (4K UHD 30fps, 16-layer ResNet) 129 mW @ 50MHz, 0.65V (FHD 30fps, 16-layer ResNet)

Fig. 5. Chip micrograph and performance summary. Main processing engines are highlighted and light gray regions represent on-chip SRAMs.

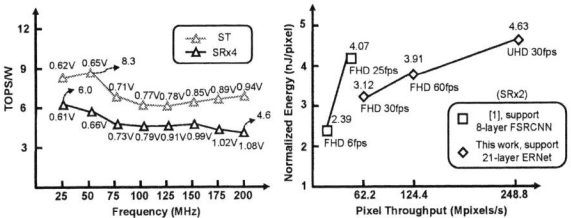

Fig. 6. Measurement results (at room temperature).

out some values in the kernels. More specifically, the kernel patterns are interleaved to maintain an equivalent receptive field compared to a corresponding dense model. This structure-sparse CONV3x3 engine reduces 44.4% of intrinsic complexity compared to a naïve dense one while achieving similar image quality by 0.02-0.03 dB of PSNR drop. [7] applies a similar approach but shares informative center positions to save more complexity and support reconfigurability in kernel sparsity; however, the center sharing causes significant image quality degradation by 0.08-0.11 dB of PSNR. As a result, our CONV3x3 engine achieves better complexity-quality tradeoff compared to a dense alternative and [7]. It consists of 10240 intrinsic MACs which are equivalent to 18432 MACs for dense models. Along with the dense CONV1x1 engine, this work can deliver up to 8.2 equivalent TOPS at 200 MHz with only 4.9 intrinsic TOPS, which achieves 40% of complexity reduction for energy-efficient and high-quality inference.

To optimize weight reuse for block-based inference, we adopt a similar weight-stationary strategy in [4] for the CONV3x3 and 1x1 engines. The parameters of one convolution layer are stored in the local weight registers and reused for the whole block. These local weight registers are distributedly close to their corresponding MACs to enhance energy efficiency. Meanwhile, the parameters of the next layer will be decoded from parameter memory and sent to the other set of local weight registers. These two register sets follow a ping-pong scheme and rotate for every layer. In this way, we can use low-cost register access and avoid frequent SRAM access for energy-efficient weight reuse.

B. Coarse-grained Reconfigurable Datapath

To support diverse applications at high throughput, the processor needs to maintain high utilization in the highly-parallel CONV engines while addressing a wide variety of model structures. As shown in Fig. 4, the CGRD is carefully optimized to support ER module, residual block and skip connection, with different channel number and model depth. The basic processing granularity is as coarse as a 16-channel 2x4-tile 3D tensor for efficient data movement. The accumulation unit is shared for both inter- and intra-instruction accumulation

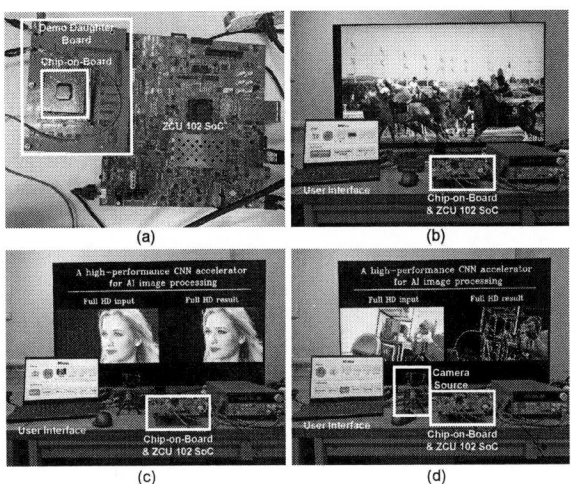

Fig. 7. FPGA-aided chip demo system setup. The chip is connected to Xilinx ZCU 102 SoC through the PCB daughter board as shown in (a). It performs diverse CNN-based computational imaging applications on input images from the real-time camera or the off-line video, including SR, DN, and ST as shown in (b), (c), and (d) respectively.

paths for area saving. The cross-layer addition paths including residual and skip connection are completed without EMA for high arithmetic intensity. Through modifying the instructions stored in the program memory, the CGRD can be reconfigured to form a variety of computing modules, in particular for supporting ERNet and ResNet.

IV. IMPLEMENTATION SUMMARY

Chip specifications and micrograph are summarized in Fig. 5, and measurement results are shown in Fig. 6. The core size is 3.2x3.2 mm² in 40nm CMOS, comprising 8.0M logic gates and 814KB of on-chip SRAM. The core power was measured on well-trained CNN models and averaged on two images from the benchmark dataset, Set5, for four different CNN applications. Operating at 200MHz, the chip achieves 4.9 TOPS of CNN acceleration for high-performance computational imaging at 4K UHD 30fps. It dissipates 585mW on the 16-layer ResNet for ST and 1055-1151mW for the ERNets for SRx4, SRx2 and DN. The power consumption of ST is smaller due to its higher activation sparsity.

The normalized energy for generating one pixel is presented on the right of Fig. 6: 3.12-4.63 nJ/pixel for completing a

Comparison	VLSIC 2019 [1]	VLSIC 2021 [2]	This Work
Process	65nm	12 nm	40nm
Die Area (mm²)	16.0	7.0	13.7
Maximum Frequency (MHz)	200	930	200
SRAM (KB)	572 (Feature: 77)	1780 (Feature: 1056)	814 (Feature: 416)
MAC Numbers	1152	2048	12288
Precision (W/A)	8/16	8/8	8/8
Peak TOPS	0.5	3.8	Intrinsic: 4.9 (Equivalent: 8.2)
Supported Applications	SRx2, SRx4	SRx2	SRx2, SRx4, DN, ST
Peak Pixel Throughput	FHD 25fps (SRx2) FHD 60fps (SRx4)	N/A	4K UHD 30fps (SRx2, SRx4, DN, ST)
Normalized EMA (Byte/pixel/channel)	3.14-6.06 (8-layer FSRCNN)	1.23 (8-layer FSRCNN)	1.02-1.17 (16-to-29-layer ResNet/ERNets)
Energy Efficiency (TOPS/W)	1.1-1.9	5.0-15.8	Intrinsic: 4.6-8.3 (Equivalent: 7.6-15.0)
Area Efficiency (GOPS/mm²)	28.8	542.9	Intrinsic: 358.8 (Equivalent: 598.0)

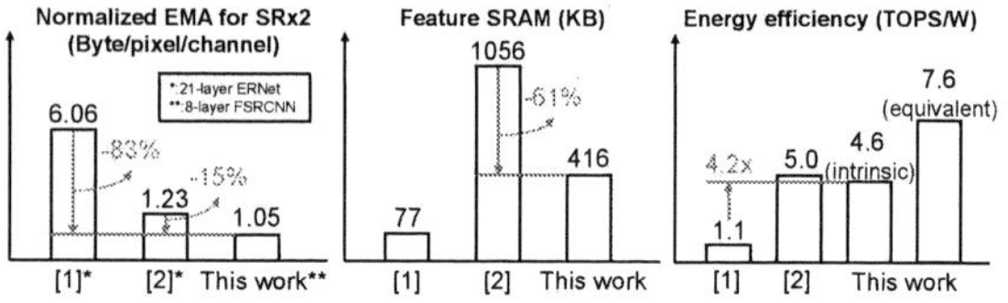

Fig. 8. Performance comparison.

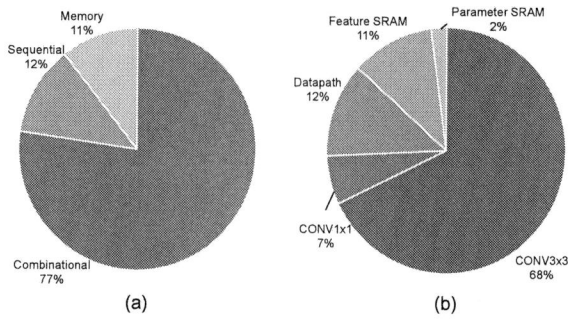

Fig. 9. Relative power consumption in terms of (a) circuit types and (b) main engines from post-layout simulation for SRx2.

21-layer ERNet at FHD to 4K UHD 30fps. For comparison, [1] demands 4.07 nJ/pixel for the 8-layer FSRCNN at only FHD 25fps. We also scaled the working frequency and voltage to measure energy efficiency for different computing performance metrics. For example, for ST the peak energy efficiency is 8.3 TOPS/W at 0.65V and 50MHz, and for SRx4 it is 6.0 TOPS/W at 0.61V and 25MHz. The deep ERNets supported by this work show 0.39-0.70 dB of PSNR gain over the FSRCNN supported by [1], [2] as mentioned in Fig. 1. This work can support these deeper models at 4K UHD 30fps while [1] only supports FSRCNN at up FHD 60fps. The FPGA-aided demo system of the fabricated chip is illustrated in Fig. 7.

Fig. 8 shows the comparison with state-of-the-art SR CNN accelerator chips [1], [2]. This work supports multiple computational imaging CNN applications (SR/DN/ST) at 4K UHD 30fps. Even through supporting deeper models, this work achieves 83% and 15% EMA reduction compared to [1] and [2], respectively. Fabricated in 40nm CMOS process, this work shows 4.2x energy efficiency improvement compared to the 65nm design [1], and provides comparable energy efficiency compared to the 12nm design [2].

Power consumption breakdown from post-layout simulation is shown in Fig. 9. The combinational circuits contribute about 77% of power consumption for the highly-parallel convolution. The sequential circuits occupy about 12%, consisting of the local weight registers in the CONV engines, the pipelined registers in the datapath, and the clock tree. The SRAMs consume the rest 11%. The CONV engines contribute up to 75% while the parameter SRAM occupies only 2% of power. These results demonstrate that most of the power is consumed in the combinational logic of CONV engines thanks to the energy-efficient data movement of parameters and features and the highly-parallel acceleration.

V. CONCLUSION

In this work, we present a memory-efficient, energy-efficient, and diverse-function accelerator chip for high-quality CNN-based computational imaging applications at 4K UHD 30fps. It addresses the huge requirement of EMA and computing energy with the proposed OSI flow and the structure-sparse CONV3x3 engine, respectively. The former reduces the EMA to 0.81-1.74 GB/s for high-quality CNN inference with 16 to 29 layers at 4K UHD 30fps. The latter reduces the computing complexity by 40% without noticeable quality degradation, e.g. 0.02-0.03 dB of PSNR drop. More specifically, it approaches the quality of 8.2 TOPS dense models with only 4.9 TOPS of computing capability at 200MHz. In addition, the proposed CGRD can support multiple image applications with high hardware efficiency. As a result, this 40nm fabricated chip can achieve 4.6-8.3 TOPS/W of energy efficiency for high-quality CNN-based SR, DN and ST at the edge.

ACKNOWLEDGMENT

This work was supported by MOST 109-2218-E-007-034. We would like to acknowledge chip fabrication support provided by TSRI. We also thank NCHC for providing computational and storage resources.

REFERENCES

[1] J. Lee *et al.*, "A Full HD 60 fps CNN Super Resolution Processor with Selective Caching based Layer Fusion for Mobile Devices," in *2019 Symposium on VLSI Circuits*.

[2] K. Goetschalckx *et al.*, "DepFiN: A 12nm, 3.8TOPs depth-first CNN processor for high res. image processing," in *2021 Symposium on VLSI Circuits*.

[3] C. Dong *et al.*, "Accelerating the Super-Resolution Convolutional Neural Network," in *European Conference on Computer Vision (ECCV) 2016*.

[4] C.-T. Huang *et al.*, "eCNN: A Block-Based and Highly-Parallel CNN Accelerator for Edge Inference," in *2019 52th Annual IEEE/ACM International Symposium on Microarchitecture (MICRO)*.

[5] M. Alwani *et al.*, "Fused-layer CNN accelerators," in *2016 49th Annual IEEE/ACM International Symposium on Microarchitecture (MICRO)*.

[6] C.-T. Huang, "ERNet Family: Hardware-Oriented CNN Models for Computational Imaging Using Block-Based Inference," in *2020 IEEE International Conference on Acoustics, Speech and Signal Processing (ICASSP)*.

[7] C.-W. Weng *et al.*, "A Quality-Oriented Reconfigurable Convolution Engine Using Cross-Shaped Sparse Kernels for Highly-Parallel CNN Acceleration," in *2021 IEEE 3rd International Conference on Artificial Intelligence Circuits and Systems (AICAS)*.

A 40nm 5.6TOPS/W 239GOPS/mm^2 Self-Attention Processor with Sign Random Projection-based Approximation

Seong Hoon Seo*, Soosung Kim*, Sung Jun Jung, Sangwoo Kwon, Hyunseung Lee, Jae W. Lee

Seoul National University, Seoul 08826, Republic of Korea

{andyseo247, soosungkim, miguel92, kwonsw055, hs_lee, jaewlee}@snu.ac.kr

Abstract—Transformer architecture is one of the most remarkable recent breakthroughs in neural networks, achieving state-of-the-art (SOTA) performance on various natural language processing (NLP) and computer vision tasks. *Self-attention* is the key enabling operation for transformer-based models. However, its quadratic computational complexity to the sequence length makes this operation the major performance bottleneck for those models. Thus, we propose a novel self-attention accelerator that skips most of the computation by utilizing an approximate candidate selection algorithm. Implemented in a 40nm CMOS technology, our 5.64 mm^2 chip operates at 100-600 MHz consuming 48.3-685 mW to achieve the energy and area efficiency of 0.354-5.61 TOPS/W and 239 GOPS/mm^2, respectively.

I. INTRODUCTION

Since the advent of the transformer architecture [1], the self-attention mechanism has become one of the most important building blocks in neural networks (NN). From the representative NLP models such as BERT [2] and GPT-3 [3] to computer vision [4] and even multi-modal models [5], the self-attention mechanism has widely been adopted in various domains, and its importance will continue to grow.

The self-attention mechanism is an operation that computes the relations among input entities. Input entities may vary depending on the context. For instance, an entity is a word or token in NLP, while it may be an image patch in vision classification. Fig. 1 is a visualization of one self-attention operation within BERT [6]. The darker the connected line is, the stronger the relationship between the two words is. In the example, the adverb "where" strongly attends to "in" and "Milan", while the pronoun "it" attends to "conference".

Convolutional and recurrent operations, which are the dominant primitives for vision and NLP, respectively, only capture the relationship among local, adjacent entities [7]. Thus, they cannot directly compute the global dependencies. Self-attention mechanism overcomes this limitation and directly captures the long-range dependencies among distant entities.

$$\text{Self-attention}(Q, K, V) = \text{Softmax}(\frac{QK^T}{\sqrt{d}})V \quad (1)$$

Equation (1) shows the formal definition of the self-attention operation. It takes $n \times d$ query (Q), key (K), and value (V) matrices as inputs, which are linear projections of the input entities. n is the sequence length of the input entities whose attention will be computed, and d is the dimension of projected vectors for each entity. QK^T is often divided by $\sqrt{d}$ as a

* These authors contributed equally to this work.

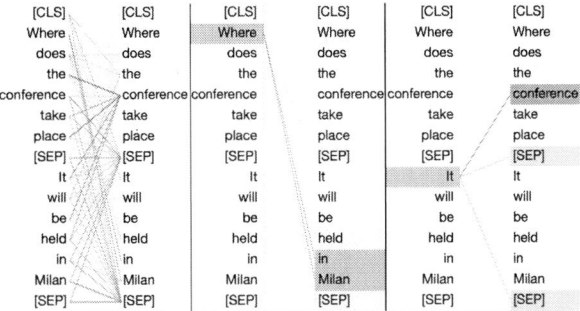

Fig. 1. Visualization of the self-attention mechanism using BERT [6]

scaling factor for stable gradients [1]. The softmax function ($(\sigma(V)_i = \frac{e^{V_i}}{\sum_{j=1}^{n} e^{V_j}}$) returns a weight, which is then multiplied by V to output a weighted sum of value vectors.

Although the self-attention operation is mainly composed of matrix multiplications that benefit from the parallelism of GPUs and NPUs, it is computationally expensive. The amount of required computation quadratically increases as the length of the input entities (i.e., n) increases, as the self-attention mechanism has a complexity of $O(n^2d)$. This often limits the maximum length of the input sequence to 512 or less for BERT. Therefore, input sentences must be split into multiple sub-sequences, preventing the model from capturing the global relationship between input entities in different sub-sequences.

A recent study shows that the self-attention mechanism can be *approximated* with negligible accuracy loss by estimating the angular distance between a query-key vector pair based on sign random projection (SRP) [8]. From the observation that the softmax function maps most of the values to near-zero except for a few large values, Ham et al. [8] proposes an approximation algorithm that prunes most query-key pairs that do not contribute to the final output.

We elaborate on this idea, presenting a 40nm implementation of a self-attention accelerator with SRP-based approximation. The implemented chip is fully integrated with a popular NN framework for end-to-end inference of NLP models. Our chip achieves a peak energy efficiency of 5.61 TOPS/W, 202× higher than a high-performance GPU (NVIDIA GV100), and the SOTA technology-normalized area efficiency.

The remaining paper is organized as follows. Section II overviews the SRP-based approximation in [8], followed by implementation details of the chip in Section III. Section IV presents experimental results, and Section V concludes.

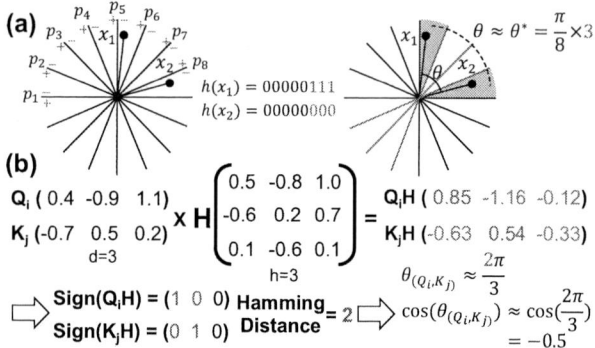

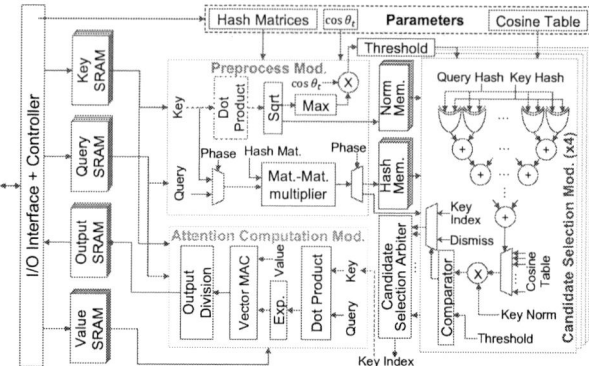

Fig. 2. (a) Sign Random Projection (b) Computation steps of SRP

Fig. 3. Architecture of the self-attention accelerator

II. SIGN RANDOM PROJECTION-BASED APPROXIMATION

This section summarizes the approximation algorithm that utilizes sign random projection (SRP) to estimate the dot-product similarity [8]. The first stage of the self-attention mechanism is the multiplication of query (Q) and transposed key (K^T) matrix. This matrix multiplication computes the dot-product similarity between each pair of d-dimensional query vector (Q_i) and key vector (K_j), where ith query/key vector is a linear projection of the ith input entity. In other words, $QK^T_{(i,j)} = Q_i \cdot K_j$.

The softmax function sharpens the difference between values to project most values to near-zero except for a few large ones, which also make near-zero the multiplication of such values by the value matrix (V) (i.e., Softmax(QK^T)$_{(i,j)} \cdot V_{(j,d)}$), to barely affect the final outcome. Based on this observation, Ham et al. [8] proposes a method to identify and skip (Q_i, K_j) pairs whose softmax output will be near-zero, significantly reducing the amount of computation without loss of accuracy.

Based on $(Q_i \cdot K_j)/||Q_i|| = ||K_j||cos\theta$, the scheme approximates the magnitude of the dot-product similarity by computing the norm of key vectors and the estimated angle in advance. As illustrated in Fig. 2(a), SRP estimates the angle between two vectors with the Hamming distance between hashed bit-vectors. Binary hashing is performed using the relative location against each hyperplane (p_i). Fig. 2(b) shows how SRP is computed. Each input vector is first multiplied with a $d \times h$ hash matrix (H), where each column vector corresponds to a different hyperplane. The output is then transformed into a h-bit vector based on the sign of each value. The Hamming distance of two bit-vectors estimates the angle.

Using the estimated angle, The query-normalized similarity ($||K_j||cos\theta$) can be estimated, which is then compared against the product of $\max_{1 \le j \le n} ||K_j||$ and pre-trained cosine threshold $cos\theta_t$. Query-key vector pairs whose estimated similarity is below this threshold are skipped. $cos\theta_t$ is computed for each value of the configurable hyperparameter p using the training dataset. p controls the degree of approximation, with a higher value resulting in a more aggressive approximation. For each p, we find $cos\theta_t$ that selects query-key pairs whose softmax output is greater than p/n. We discuss how the choice of p affects the accuracy and latency in Section IV.

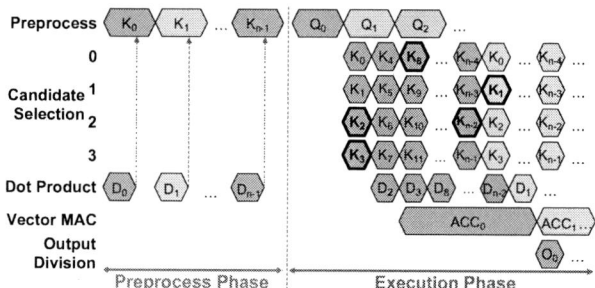

Fig. 4. Execution flow of the self-attention accelerator

III. IMPLEMENTATION DETAILS

A. Overview

Fig. 3 depicts the overall architecture of the implemented accelerator. It consists of one preprocess module, four parallel candidate selection modules, one attention computation module, and a controller in charge of state transition and off-chip communication. We envision our design to be integrated into existing processors (e.g., GPU or NPU) such that the accelerator can directly access the input matrices from the device's scratchpad memory. However, for validation and measurement of our accelerator, the prototype chip includes 192 KB SRAM for input and output matrices.

Execution Flow. The execution flow is depicted in Fig. 4. During the preprocess phase, the hash and the norm of key vectors are computed by the preprocess module. During the execution phase, the same module computes the hash of each query vector. Candidate selection modules consume the query hash and the preprocessed values to deliver the indices of selected query-key pairs (e.g., (Q_0, K_2), (Q_0, K_3) in Fig. 4) to the attention computation module, which in turn performs the self-attention operation in a pipelined manner.

Number Representations. The proposed accelerator uses INT12 and custom FP17 with a 6-bit mantissa. For dot-product similarity, INT12 is precise enough to prevent the loss of end-to-end accuracy of the model. For exponent and its multiplication with the value matrix, custom FP17 with 10-bit exponent and 6-bit mantissa is used. This prevents the overflow of the softmax output's denominator (i.e., $\sum_{j=1}^{n} e^{V_j}$) and allows us to decompose the softmax function into the

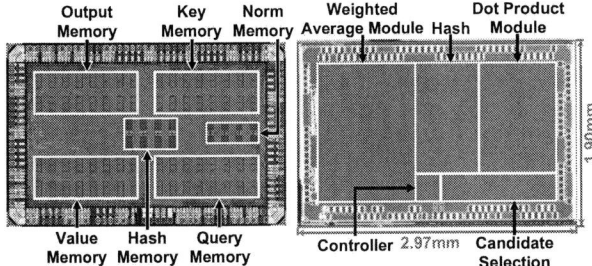

Fig. 5. Chip layout (left) and micrograph (right)

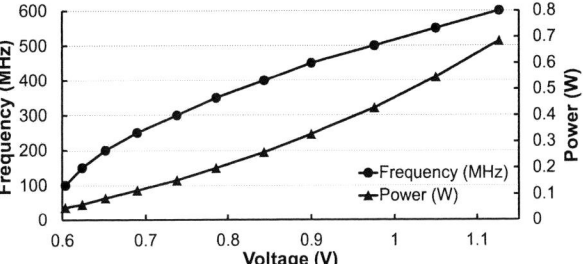

Fig. 6. Voltage-frequency scaling

accumulation stage and division stage, hiding the computation bottleneck of exponent accumulation required before division.

B. Hardware Details

Preprocess Module. For each input vector, the preprocess module computes its h-bit hash value using SRP. We choose $h=d=64$ for our implementation. Naïvely multiplying a $d \times h$ matrix incurs too much overhead for preprocessing. Instead, we use the algorithm in [8] to decompose one large matrix multiplication into three smaller multiplications. In $h=d=64$, this reduces the required number of multiply-and-accumulate (MAC) operations per hashing from $64^2=4096$ to $3 \cdot 4^4=768$.

During the preprocess phase, the module also computes the L2 norm of each key vector, keeping track of the maximum key norm. As mentioned in Section II, the maximum key norm is multiplied by a pre-trained cosine threshold to generate a similarity threshold. Norm is computed by taking the dot product of the input vector itself and taking the square root of the product. Since the attention computation module is idle during this phase, the preprocess module "borrows" its dot-product unit, maximizing resource utilization.

Candidate Selection Module. Four candidate selection modules operate in parallel during the execution phase. Query hash is passed from the preprocess module, and the key hashes are retrieved from the hash memory. The module first computes the Hamming distance between the query and key hashes. A lookup table maps the result to the corresponding cosine value, which is multiplied by the key norm to compute query-normalized similarity. Finally, the module compares the similarity against the threshold and passes the indices of selected query-key pairs to the attention computation module.

Attention Computation Module. Attention computation module is composed of a dot-product unit and a weighted average sub-module. The former computes the dot product of query-key vector pairs selected by candidate selection modules. The latter consists of an exponent, a vectorized MAC, and an output division unit. These units are pipelined to compute the weighted sum of value vectors. First, the exponent unit returns the exponent of the dot product (e^{D_i}), internally accumulating each exponent. Second, the vectorized MAC unit multiplies the exponent with the value vector, then accumulates the output. Once accumulation for a single query vector is complete, the output division unit divides the accumulated vector with the sum of exponents ($\sum e^{D_i}$). As shown in Fig. 4, performing division at the end of the pipeline prevents unnecessary stalls waiting for the sum of exponents.

IV. MEASUREMENT RESULTS

The prototype chip shown in Fig. 5 is fabricated in 40nm CMOS technology with an area of 5.64 mm^2. As shown in Fig. 6, the chip operates at a maximum frequency of 600 MHz with 1.126 V.

Methodology. We measure the performance of the chip on BERT-large model [2] using Stanford Question Answering Dataset (SQuAD 2.0) [9]. To compare the performance with existing processors, we run the same workload on NVIDIA GV100 GPU [10], Jetson TX2 GPU [11], and Google TPUv3 [12]. The performance of these processors is measured using FP16/BF16, advantageous for them compared to using FP32.

We develop an evaluation environment that allows the end-to-end inference of BERT on the existing ML framework (PyTorch) with offloading of the self-attention mechanism to the chip. Fig. 7 depicts the details of the offloading mechanism. During inference, the self-attention operations are exported to the C++ offloader, which exposes an API for the end-user. The offloader forwards the input matrices to the PCIe-connected FPGA, which relays them to the chip. Once the chip completes the computation, the FPGA receives the output matrix from the chip, relaying it back to the offloader.

Performance. The chip achieves peak energy and area efficiency of 5.61 TOPS/W and 239 GOPS/mm^2 as shown in Table I. The chip is up to 92.5× more energy-efficient than TPUv3, which has the highest TOPS/W among the three. Fig. 8 shows the effect of approximation on performance and accuracy. As the degree of approximation increases, the chip aggressively skips more query-key vector pairs, directly leading to throughput improvement. Moderate approximation of $p=2$ achieves 3.4× speedup over no approximation ($p=0$), with negligible (sub-1%) accuracy loss compared to the baseline

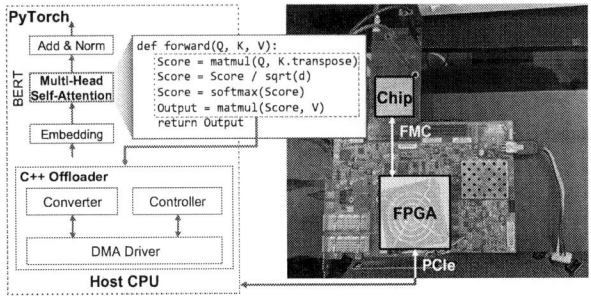

Fig. 7. End-to-end inference system based on PyTorch

TABLE I. COMPARISON TABLE

	GV100 GPU [10]	Jetson TX2 GPU [11]	TPU v3 (v3-8) [12]	ISSCC 2022 [13][a]	This Work
Technology (nm)	12	16	16	28	40
Die Area (mm²)	815	31.4[b]	700	6.82	5.64
Frequency (MHz)	1627	1300	940	50 - 510	100 - 600
Power (W)	250	8.69	1800	0.012 - 0.273	0.0483 - 0.685
ML Algorithm	BERT-large	BERT-large	BERT-large	GPT-2, ViT, Swin-Transformer	BERT-large
ML Dataset	SQuAD 2.0	SQuAD 2.0	SQuAD 2.0	wikiText-2, ImageNet	SQuAD 2.0
Peak Performance (TFLOPS, TOPS)	6.95	0.14	109	1.92[c]	1.35[d]
Energy Efficiency (TFLOPS/W, TOPS/W)	0.0278 (0.00250)[e]	0.0163 (0.00260)[e]	0.0607 (0.00971)[e]	2.60 - 17.2[c] (1.27 - 8.41[c])[e]	0.354 - 5.61[d]
Area Efficiency (GFLOPS/mm², GOPS/mm²)	8.53 (0.768)[e]	4.50 (0.720)[e]	156 (25.0)[e]	282[c] (138[c])[e]	239[d]

a) The effect of computations outside the self-attention layers are excluded as explained in Section IV   b) Die area is approximated with the number of shader cores
c) 90% output sparsity   d) 88% output sparsity   e) Assumes 40nm implementation and energy is proportional to (Technology)² [14]

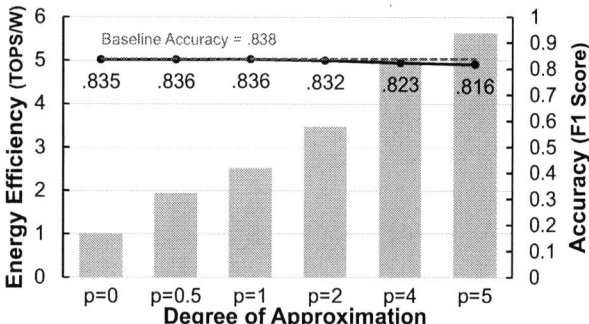

Fig. 8. Effect of approximation on performance and accuracy

accuracy of GPU using FP32 (0.838 in Fig. 8). If further accuracy loss can be traded in for increased performance, a more aggressive approximation ($p=5$) can lead to over $5.5\times$ speedup at the cost of a small (some 2%) accuracy loss.

Comparison with transformer accelerator by Wang et al. [13]. Table I compares our prototype chip with a recently proposed transformer accelerator [13]. For fair comparison, dense feed-forward layers are removed as they are out of scope of our design and also lower the transformer accelerator's TOPS and TOPS/W. Peak TOPS and TOPS/W are measured at 88% output sparsity ($p=5$) for our chip and 90% for the accelerator by Wang et al. In contrast, minimum TOPS and TOPS/W are measured with no approximation in both cases.

Our chip results in 0.354-5.61 TOPS/W, 239 GOPS/mm². By comparison, the transformer accelerator achieved 1.27-8.41 TOPS/W, 138 GOPS/mm² on average, after normalizing the technology based on the simple scaling model presented by Biswas and Chandrakasan [14], where both energy and area are proportional to the square of technology. Our self-attention accelerator achieves comparable energy efficiency and higher area efficiency against the transformer accelerator due to the absolute computation reduction of our candidate selection mechanism.

V. CONCLUSION

This paper presents a self-attention accelerator with an approximate candidate selection algorithm. This chip is fab-

ricated in 40nm CMOS with an area of 5.64 mm². Our chip selects query-key vector pairs that are estimated to be highly related, skipping the computation for the non-selected pairs. Thanks to the efficient candidate selection algorithm, the chip achieves $92.5\times$ higher peak TOPS/W than TPUv3 and $1.73\times$ higher technology-normalized area efficiency than the state-of-the-art transformer accelerator.

ACKNOWLEDGMENTS

This work was supported by a National Research Foundation of Korea (NRF) grant (NRF-2020R1A2C3010663) and an Institute of Information & Communications Technology Planning & Evaluation (IITP) grant (2021-0-00105), funded by the Korea government (MSIT). The EDA tool was supported by the IC Design Education Center (IDEC), Korea. Jae W. Lee is the corresponding author.

REFERENCES

[1] A. Vaswani et al., "Attention is All you Need," in NeurIPS, 2017.
[2] J. Devlin et al., "BERT: Pre-training of Deep Bidirectional Transformers for Language Understanding," in NAACL, 2019.
[3] T. Brown et al., "Language Models are Few-Shot Learners," in NeurIPS, 2020.
[4] A. Dosovitskiy et al., "An Image is Worth 16x16 Words: Transformers for Image Recognition at Scale," in ICLR, 2021.
[5] W. Kim et al., "ViLT: Vision-and-Language Transformer Without Convolution or Region Supervision," in ICML, 2021.
[6] J. Vig, "A Multiscale Visualization of Attention in the Transformer Model," in ACL, 2019.
[7] X. Wang et al., "Non-Local Neural Networks," in CVPR, 2018.
[8] T. J. Ham et al., "ELSA: Hardware-Software Co-design for Efficient, Lightweight Self-Attention Mechanism in Neural Networks," in ISCA, 2021.
[9] P. Rajpurkar et al., "Know What You Don't Know: Unanswerable Questions for SQuAD," in ACL, 2018.
[10] "Quadro GV100," https://www.nvidia.com/en-us/design-visualization/store/, NVIDIA, 2018.
[11] "Jetson TX2 Module," https://developer.nvidia.com/embedded/jetson-tx2, NVIDIA, 2017.
[12] N. P. Jouppi et al., "Ten Lessons From Three Generations Shaped Google's TPUv4i," in ISCA, 2021.
[13] Y. Wang et al., "A 28nm 27.5TOPS/W Approximate-Computing-Based Transformer Processor with Asymptotic Sparsity Speculating and Out-of-Order Computing," in ISSCC, 2022.
[14] A. Biswas et al., "CONV-SRAM: an energy-efficient SRAM with in-memory dot-product computation for low-power convolutional neural networks," in JSSC, vol. 54, no. 1, 2019.

A 28nm 8-bit Floating-Point Tensor Core based CNN Training Processor with Dynamic Activation/Weight Sparsification

Shreyas Kolala Venkataramanaiah*, Jian Meng*, Han-Sok Suh*, Injune Yeo*, Jyotishman Saikia*,
Sai Kiran Cherupally*, Yichi Zhang[†], Zhiru Zhang[†], and Jae-sun Seo*
*Arizona State University, USA [†]Cornell University, USA

Abstract—**We present an 8-bit floating-point (FP8) training processor which implements (1) highly parallel tensor cores (fused multiply-add trees) that maintain high utilization throughout forward propagation (FP), backward propagation (BP), and weight update (WU) phases of the training process, (2) hardware-efficient channel gating for dynamic output activation sparsity, (3) dynamic weight sparsity based on group Lasso, and (4) gradient skipping based on FP prediction error. We develop a custom ISA to flexibly support different CNN topologies and training parameters. The 28nm prototype chip demonstrates large improvements in FLOPs reduction (7.3×), energy efficiency (16.4 TFLOPS/W), and overall training latency speedup (4.7×), for both supervised and self-supervised training tasks.**

Index Terms—**Convolutional neural networks, deep neural network training, structured sparsity, hardware accelerator**

I. INTRODUCTION

Training deep convolutional neural networks (CNNs) requires a large amount of memory and iterative computation, which necessitates speedup and energy reduction for both cloud and edge devices. Several prior works reported CNN training processor designs, but some of them did not exploit sparsity during training [1], [2]. Bit-slice input/output sparsity is exploited in [3], but did not consider weight sparsity during training. In [4], the sparse channels are randomly selected to simplify the hardware design, but resulted in training accuracy loss. A global threshold is used in [5] to generate the element-wise sparse masks without sorting, but the non-structured sparse elements are not skippable.

In this work, we present a new sparse CNN training accelerator that exploits structured activation sparsity, structured weight sparsity, and gradient skipping to dynamically reduce the unimportant operations and achieve high speedup. We developed both hardware-efficient sparse training algorithms and the custom sparse training accelerator with programmable instructions. The 28nm prototype chip demonstrates large improvements in FLOPs reduction (7.3×), energy efficiency (16.4 TFLOPS/W), and overall training latency speedup (4.7×) across supervised and self-supervised training tasks.

II. HARDWARE-EFFICIENT SPARSE TRAINING ALGORITHM

A. Dynamic Structured Weight Sparsification

We generate structured weight sparsity dynamically during CNN training via group Lasso [6], where the unimportant weight groups get penalized without any sorting. Suppose the weight groups in DNN layer l is $W_{l,g}$. The Lasso penalty term

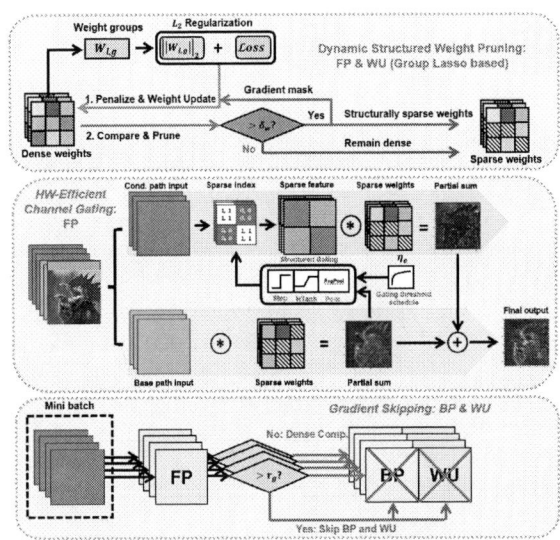

Fig. 1: Proposed efficient sparse training techniques.

in the loss function ($\hat{\mathcal{L}}$ in Eq. (1)) is the summation of group L_2 norm of $W_{l,g}$ over all groups (G_l) and layers (L).

$$\hat{\mathcal{L}} = \mathcal{L} + \lambda \sum_{l=1}^{L} \sum_{g=1}^{G_l} ||W_{l,g}||_2, \tag{1}$$

$$w_{i,g}^* = \underbrace{(1 - \eta \frac{\lambda}{||W_{l,g}||_2})w_{i,g}}_{GL_{scale}} - \underbrace{\eta \frac{\partial \mathcal{L}}{\partial \nabla w_{i,g}}}_{\text{Normal gradient}} \tag{2}$$

In Eq. (2), $\nabla w_{i,g}$ is the gradient of the weight element $w_{i,g} \in W_{l,g}$. The weight $w_{i,g}^*$ is updated by the normal gradient and also scaled down by the regularization scaler GL_{scale}. Since L_2 norm of the weight group is pre-computed, GL_{scale} can be easily implemented in hardware. Smaller weights are gently scaled down, while larger weights are penalized harder. As a result, every $|w_{i,g}| \in W_{l,g}$ decreases simultaneously, so the element-wise threshold can easily sparsify the entire group. The resultant structured sparse masks are applied to both FP and WU. Following the hardware design, we choose K_l (# of output channels) × C_l (# of input channels) = 8×8 as the sparse group size for efficient hardware mapping.

B. Structured Activation Skipping

We also exploit the activation sparsity by skipping the unsalient features during FP. In CGNet [7], activation sparsity

978-1-6654-8495-4/22 $31.00 © 2022 IEEE

was employed during CNN inference without any auxiliary salience predictors. The input/output channels are divided into base and conditional paths, and the gated base path's feature score determined the skipping of conditional path computation. However, the learnable threshold and the non-linear gating function are expensive for hardware implementation.

To elevate the hardware compatibility, we propose structured CGNet (SCG), including (a) low-precision threshold schedule, (b) non-linearity relaxation, and (c) structured feature map sparsification. The learning pattern of the gating threshold is generalized as a gradually increased low-precision scheduler, and the non-linear Sigmoid gating function [7] is simplified as a scaled and shifted HardTanh function:

$$\mathcal{G}(y_{base}^*) = \underset{min=0,max=1}{\text{HardTanh}}\left(0.25 \times (y_{base}^* + 2) - 0.5\right) \quad (3)$$

We generate structured sparsity of feature maps by computing the *group salience* scores of the base path output features via 3-D average pooling and HardTanh (Fig. 1). The group salience scores determine the structured computation skipping of the conditional path. Table I shows the negligible accuracy degradation of the highly hardware-compatible SCGNet.

TABLE I: High hardware compatibility and negligible accuracy drop of the SCGNet with ResNet-18 model on CIFAR-10.

Method	Gating Func.	Threshold	Cond. Path Sparsity	Granularity	Acc.
Baseline	-	-	0.0%	-	94.78%
CGNet [7]	Sigmoid	Learnable	70.21%	Element-wise	94.45%
SCGNet	**HardTanh**	**Scheduled**	68.71%	Structured	94.41%

C. Gradient Skipping

Inputs with high confidence during FP will have minimal weight update in the training process, thus can be skipped from the BP and WU phases [8]. As shown in Fig. 1, we exclude inputs with high softmax confidence from the BP and WU phases. Combined with the structured weight/gradient sparsity, the proposed scheme achieves high energy efficiency in both gradient accumulation and the gradient itself.

Putting together the structured weight sparsity, structured activation skipping and gradient skipping, ResNet-18 FP8 training result for CIFAR-10 dataset is shown in Fig. 2.

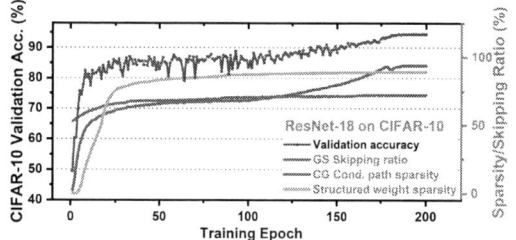

Fig. 2: Validation accuracy, sparsity, and skipping ratio of ResNet-18 FP8 training on CIFAR-10 dataset.

III. CHIP ARCHITECTURE AND OPERATION

Fig. 3 shows the overall architecture of the sparse training processor, featuring four sparse compute cores (SpCC), a central core (CC), a global controller, and external I/O interface.

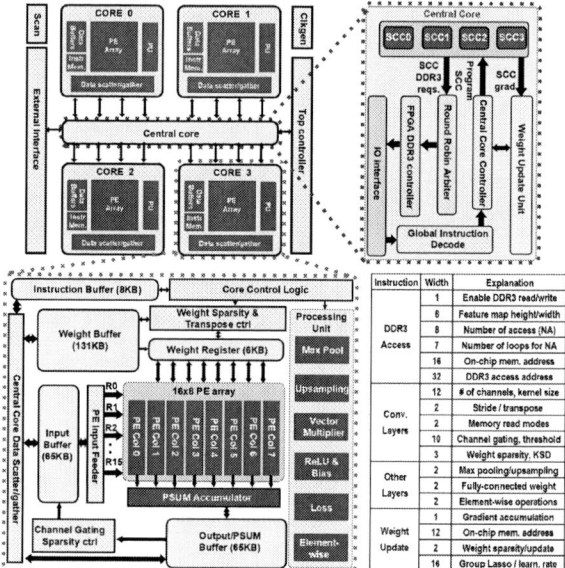

Fig. 3: Overall chip architecture and custom ISA.

A. Sparse Compute Cores (SpCC) and Central Core (CC)

Each SpCC is a programmable core that can be configured individually using the custom ISA (Fig. 3). The ISA is flexible to support all operations for CNN training, facilitates off-chip memory accesses during training, and controls all proposed sparsity/dataflow optimizations. The CC synchronizes all four SpCCs, accumulates gradients from SpCCs in WU phase, and processes off-chip DDR3 access requests using a round-robin arbiter. The SpCC includes (1) a 16×8 PE array for MAC operations, (2) SRAMs to store activations, weights, instructions, and sparsity masks, (3) vector processing units for non-MAC operations, (4) sparsity controllers to exploit dynamic structured sparsity of activations/weights, and (5) data scatter/gather units to store SRAM data in the required format.

During WU, the CC performs (1) weight gradient (WG) accumulation, (2) structured weight sparsity generation for new weights, and (3) weight update at the end of the batch based on stochastic gradient descent. After all four SpCCs complete the WG computations, a gradient accumulation instruction is dispatched to the CC controller. The CC controller loads the weight update memory with the previous WGs and reads all computed WGs from the SpCCs. The WG accumulator obtains all gradients and accumulates them using an FP16 adder tree.

B. PE Array

Each SpCC has a PE array that consists of eight PE columns with 16 PEs and a PE load balancer (Fig. 4). The PE column shares the weights obtained from the weight register, and the PE row shares the input activations during FP and BP. Each PE is a configurable dot-product engine with eight FP8 (1-5-2) multipliers and one FP16 (1-5-10) adder. The multiplier products are aligned by matching the exponents and shifting the mantissa. The aligned mantissa products are sent to the

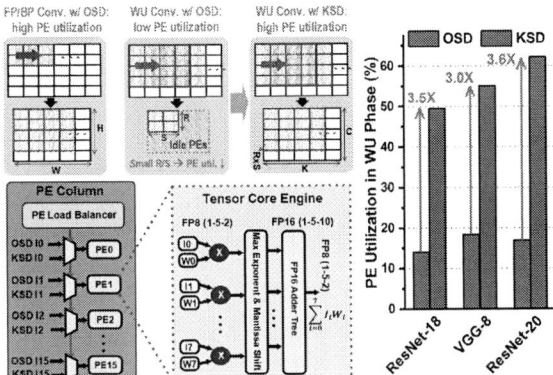

Fig. 4: PE array supporting OSD (FP/BP) and KSD (WU).

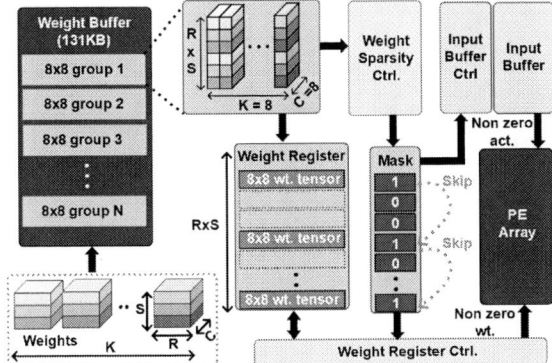

Fig. 5: Convolution with dynamic structured weight sparsity.

8-way FP16 adder tree, whose output is normalized and quantized back to FP8 precision (nearest neighbor rounding).

The PE array supports standard, transposed, and weight update convolutions required for training. During FP and BP, it follows output stationary dataflow (OSD) and computes 16 output pixels of 8 output channels in parallel with high PE utilization. However, OSD leads to low PE utilization for weight gradient computation during WU, because the typically small kernel size (e.g. 3×3) makes it difficult for the 4-D tensor weight gradient $[K, C, R, S]$ (K/C: number of output/input channels, $R\times S$: kernel size) to efficiently use the PE array. To overcome this, we propose a new kernel stationary dataflow (KSD) during WU, where each PE computes one output kernel ($R\times S$) and the PE array computes 8/16 output/input channels (K/C) in parallel. The PE load balancer unit dynamically switches from OSD to KSD during WU phase, improving the PE utilization (Fig. 4 (right)).

C. Structured Weight Sparsity

Fig. 5 shows the FP/BP operation with structured weight sparsity (WS) that is generated in. $8\times8\times1\times1$ ($K\times C\times R\times S$) groups. WS controller selects a weight group, stores it in the weight register, and compares the weights with a deterministic threshold to generate the sparsity mask. Only the non-zero weight groups are executed in the PE array.

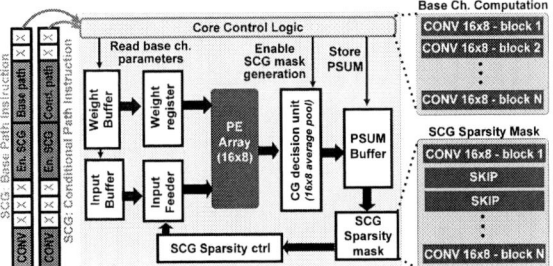

Fig. 6: Proposed SCG dataflow for dynamic activation sparsity.

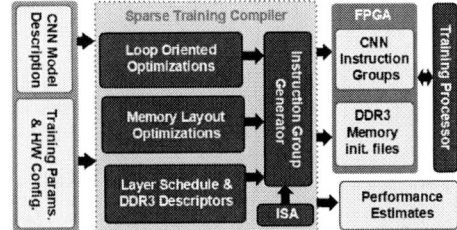

Fig. 7: Compiler for the sparse training processor.

D. Structured Channel Gating (SCG) Dataflow

Fig. 6 describes the SCG dataflow. We first compute all outputs of a given CNN layer using the base input channels in the SCG-enabled convolutions. Upon a base path instruction, the SpCC reads the parameters required to compute only the base path from input/weight memory, and enables the SCG decision unit to compute a SCG sparsity mask, a 16×8 average pooling module, and a comparator. Structured activation sparsity is achieved by performing 16×8 average pooling instead of element-wise comparison. The SpCC reads the sparsity mask for the conditional path (CP) and enables structured output activation skipping. The CP outputs are accumulated with base channel outputs stored in the partial sum memory. Since the 16×8 block structure exactly matches the PE array size, the SCG sparsity controller efficiently eliminates input/weight memory accesses as well as the computations associated with the skipped 16×8 CP blocks, largely reducing latency/power.

E. Sparse Training Compiler

Fig. 7 illustrates the sparse training compiler (STC), which takes CNN models from TensorFlow or PyTorch frameworks as inputs, extracts the layer-wise details, and generates an instruction snippet deployed on the accelerator. STC also takes (1) training parameters such as SCG threshold, learning rate, group Lasso parameters, etc. and (2) hardware configuration details such as SpCC selection, enable/disable the sparsity memory layout, etc. as inputs. Based on these inputs and the given on-chip memory and PE array size of the accelerator, STC performs loop optimizations including unrolling the convolution loops and dividing the convolution into smaller tiles.

IV. MEASUREMENT RESULTS

The prototype chip was fabricated in 28nm CMOS (Fig. 8(a)). Fig. 8(b) shows power/energy measurements of

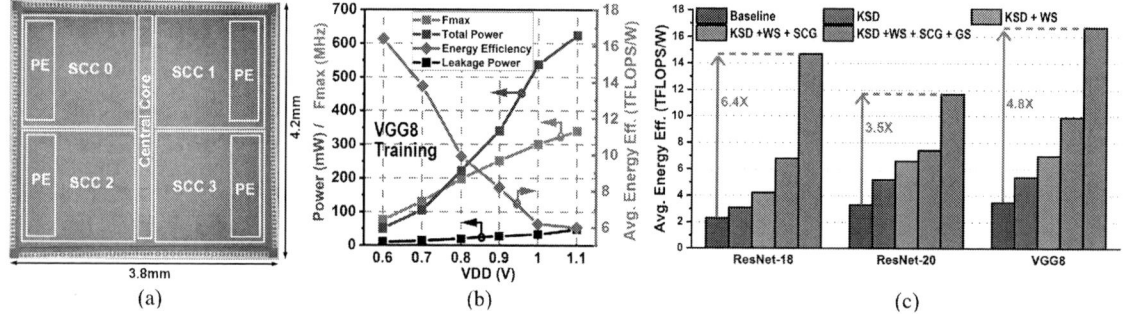

Fig. 8: (a) Chip micrograph. (b) Power/energy measurements with voltage and frequency scaling. (c) Average energy-efficiency improvement breakdown with the proposed techniques on ResNet-18/ResNet-20/VGG8 models.

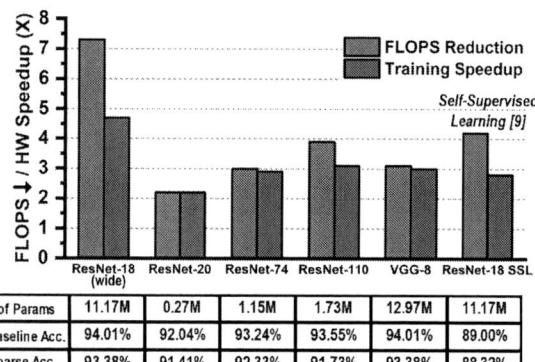

Fig. 9: FLOPS, speedup, and accuracy for training of CNNs.

	ResNet-18 (wide)	ResNet-20	ResNet-74	ResNet-110	VGG-8	ResNet-18 SSL
# of Params	11.17M	0.27M	1.15M	1.73M	12.97M	11.17M
Baseline Acc.	94.01%	92.04%	93.24%	93.55%	94.01%	89.00%
Sparse Acc.	93.38%	91.41%	92.33%	91.73%	93.38%	88.32%

the prototype chip with voltage and frequency scaling. Including the skipped operations, throughput of 3.76 TFLOPS is achieved at 1.1V, and average energy-efficiency of 16.4 TFLOPS/W is achieved at 0.6V for VGG8 training. The proposed architecture efficiently exploits the structured sparsity (CG, WS) and improves PE utilization (KSD) during training achieving ~6.4× improvement in energy-efficiency (Fig. 8(c)) compared to the baseline of training dense CNNs without activation/weight sparsity, KSD and GS.

We trained various CNNs for both supervised and self-supervised learning [9] tasks with our chip programmed using custom ISA. As shown in Fig. 9, most of the FLOPs reduction (up to 7.3×) results in corresponding training speedup (up to 4.7×) with minimal accuracy degradation. Table II shows the comparison to prior works, where our work achieves higher average energy-efficiency for actual DNN training, including and excluding the skipped operations. Our training speedup is ~2.7× higher than that of the state-of-the-art work [5].

V. CONCLUSION

In this work, we present an energy-efficient 8-bit floating-point programmable training processor with a custom ISA. The proposed KSD/WS/CG/GS schemes collectively achieved a high amount of hardware-efficient sparsity and computation skipping (up to 7.3×). The 28nm training processor was evaluated across various CNNs, and achieved high energy-efficiency and 4.7× speedup compared to training dense CNNs.

TABLE II: Comparison with prior works.

		[3]	[4]	[5]	This Work
Technology		28nm	65nm	28nm	28nm
Precision		Dyn. FXP	FP8/16	BFP8/16	FP8/16
Sparsity Support	I/O	Element (bit-slice)	Fine / coarse	Element / channel	Structured (channel gating)
	W	N/A	Fine / coarse	Element / kernel	Structured (group Lasso)
Supply (V)		0.58-1.04	0.78-1.1	0.58-1.1	0.6-1.1
Area (mm²)		12.96	16.0	20.96	16.4
Freq. (MHz)		2-250	50-200	40-440	75-340
On-Chip SRAM (MB)		0.55	0.34	0.63	1.25
Throughput (TFLOPS)		3.6-8.13	0.61[1]-18.0[2]	0.9[1]-58.7[2]	3.76[4]
Power (mW)		1.9-500	0.49-425	23-363	51.1-623.7
Energy-Efficiency (TFLOPS/W or TOPS/W)	Dense[1]	3.3-7.5 (12b-8b)	3.1	4.3	3.5 @VGG8 2.3 @ResNet-18
	Ideal[2] (90% I, 90% W)	N/A	146.5	276.5	N/A
	Peak-Skipped[3]	14.5 @ResNet-9 15.2 @ResNet-18	10.6 @AlexNet	N/A	37.6 @VGG8 64.7 @ResNet-18
	Avg.-Skipped[4]	10.1 @ResNet-9 10.7 @ResNet-18	~6.8 @AlexNet	N/A	16.4 @VGG8 14.4 @ResNet-18
	Avg.-Executed[5]	N/A	N/A	N/A	5.4 @VGG8 3.1 @ResNet-18
Training Speedup		N/A	N/A	1.76X	4.7X

[1] sparsity = 0%, [2] input/output/weight sparsity = 90% (not relevant for any specific DNN training)
[3] peak energy-efficiency including skipped operations during sparse DNN training
[4] average energy-efficiency including skipped operations throughout sparse DNN training
[4] average energy-efficiency for actual executed operations throughout sparse DNN training

VI. ACKNOWLEDGEMENTS

This work is supported in part by NSF and JUMP C-BRIC, a SRC program sponsored by DARPA.

REFERENCES

[1] A. Agrawal *et al.*, "A 7nm 4-core AI chip with 25.6 TFLOPS hybrid FP8 training, 102.4 TOPS INT4 inference and workload-aware throttling," in *IEEE ISSCC*, 2021.

[2] J. Park *et al.*, "A 40nm 4.81 TFLOPS/W 8b floating-point training processor for non-sparse neural networks using shared exponent bias and 24-Way fused multiply-add tree," in *IEEE ISSCC*, 2021.

[3] D. Han *et al.*, "HNPU: An adaptive DNN training processor utilizing stochastic dynamic fixed-point and active bit-precision searching," *IEEE Journal of Solid-State Circuits*, vol. 56, no. 9, pp. 2858–2869, 2021.

[4] S. Kim *et al.*, "A 146.52 TOPS/W deep-neural-network learning processor with stochastic coarse-fine pruning and adaptive input/output/weight skipping," in *Symp. on VLSI Circuits*, 2020.

[5] Y. Wang *et al.*, "A 28nm 276.55 TFLOPS/W sparse deep-neural-network training processor with implicit redundancy speculation and batch normalization reformulation," in *Symp. on VLSI Circuits*, 2021.

[6] W. Wen *et al.*, "Learning structured sparsity in deep neural networks," in *NeurIPS*, 2016.

[7] W. Hua *et al.*, "Channel gating neural networks," in *NeurIPS*, 2019.

[8] D. Shin *et al.*, "Prediction confidence based low complexity gradient computation for accelerating DNN training," in *ACM/IEEE DAC*, 2020.

[9] T. Chen *et al.*, "A simple framework for contrastive learning of visual representations," in *ICML*, 2020.

978-1-6654-8495-4/22 $31.00 © 2022 IEEE

A Differentiable Neural Computer for Logic Reasoning with Scalable Near-Memory Computing and Sparsity Based Enhancement

Yuhao Ju
Electrical and Computer Engineering
Northwestern University
Evanston, IL, United States
Yuhaoju2017@u.northwestern.edu

Shiyu Guo
Electrical and Computer Engineering
Northwestern University
Evanston, IL, United States
ShiyuGuo2021@u.northwestern.edu

Zixuan Liu
Electrical and Computer Engineering
Northwestern University
Evanston, IL, United States
ZixuanLiu2021@u.northwestern.edu

Tianyu Jia
School of Integrated Circuits
Peking University
Beijing, China
tianyuj@pku.edu.cn

Jie Gu
Electrical and Computer Engineering
Northwestern University
Evanston, IL, United States
jgu@northwestern.edu

Abstract—Logic reasoning represents a new class of artificial intelligence. This work presents the first hardware implementation of the Differentiable Neural Computer accelerator based on brain inspired "working memory" concept for reasoning tasks. A special near-memory computing architecture is developed achieving high scalability and over 90% utilization of computing resources. Sparsity based enhancements such as zero skipping, data compression are applied with 30% speedup of the computing latency. A 65nm test chip was fabricated with demonstrations on a variety of logic reasoning tasks showing 700X and 46X speedup compared with CPU and GPU and up to 1.28TOPS/W power efficiency.

I. INTRODUCTION

Despite the recent success in image and voice recognition applications, a missing capability from the current deep learning based artificial intelligence (AI) is realizing human like logic reasoning. Fig. 1 shows several common cognitive reasoning tasks such as deductive/abstract/sequential reasoning, algorithm deduction, graphic traverse, etc. where sequential relationships are being inferred from context of graphs or texts. While exhaustive or sophisticated heuristic search algorithms are traditionally used to solve such problems, applying deep neural network (DNN) to reasoning tasks allows a differentiable solution, e.g. learning through back-propagation without human intervention. However, existing CNN or LSTM architectures suffer from limited memory space due to the entanglement of computing and memory elements leading to poor performance in long sequential reasoning tasks. Recently, models of differentiable neural computer (DNC) or Memory-augmented Neural Network (MANN) were developed for reasoning tasks [1-2]. As shown in Fig. 1, DNC incorporates content memory operations through special "read/write heads" to infer logical information from content memory contents overcoming limited memory space issues of CNN or LSTM. Such a capability resembles human brain's "working memory" which uses an "attention" based controller to access vocal or visual memory of the brain [3]. This work implemented an end-to-end logical inference processor based on DNC algorithm with offline trained models [4]. As highlighted in Fig. 1, the challenges of ASIC acceleration of DNC include (1) large amount of memory access from the attention mechanism with 10.6X more memory request than conventional CNN, (2) highly sparse input and memory

This work is supported in part by NSF grant CCF-2008906.

contents and (3) complex model with eight operating phases making the ASIC acceleration very challenging. In this work, for the first time, an ASIC logic reasoning processor was designed to accelerate cognitive reasoning tasks with 700X/46X improvement over commercial CPU/GPU. The contributions include (1) A scalable near-memory architecture is developed to overcome the memory bandwidth challenges of the algorithm; (2) Special input zero skipping and data compression techniques are applied to exploit sparsity of the data; (3) Efficient transpose multiplication is introduced to avoid large data exchange among computing tiles; (4) Reconfigurable MACs are designed to support the eight operating phases with above 90% PE utilization rate for the challenging mapping of the software model.

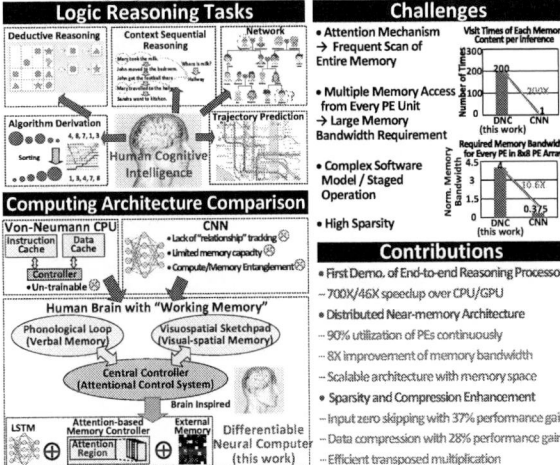

Fig. 1. Logic reasoning tasks with different computing architectures and main contributions of this work.

II. ARCHITECTURE AND ALGORITHM

Fig. 2 shows the top-level DNC algorithm. A LSTM serves as a central controller which preprocesses sequential input data and manages the access of various memory banks through memory controllers. The memory controllers include "write head" and "read head" which realize an "attention" mechanism to select a region of "focus" from the large content memory, similar to human brain's memory retrieval mechanism. For "read head", the "attention" includes cosine similarity and matrix/dot operations where the entire contents of content memory are scanned for "related" information. A special "attention memory" is used to keep the "linkage" information, i.e. logical/sequential relationship among the contents of content memory. For "write head", a usage memory is added to keep track of the content memory usage for efficient recall, e.g. allocation of new memory for incoming information. Each iteration passes through a sequence of control/update/attention/recall/result operations and after hundreds of iterations, the final result is obtained from a fully connected network (FCN).

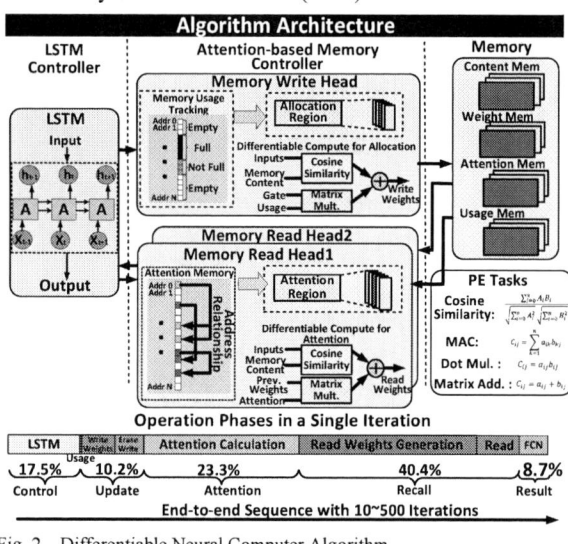

Fig. 2. Differentiable Neural Computer Algorithm.

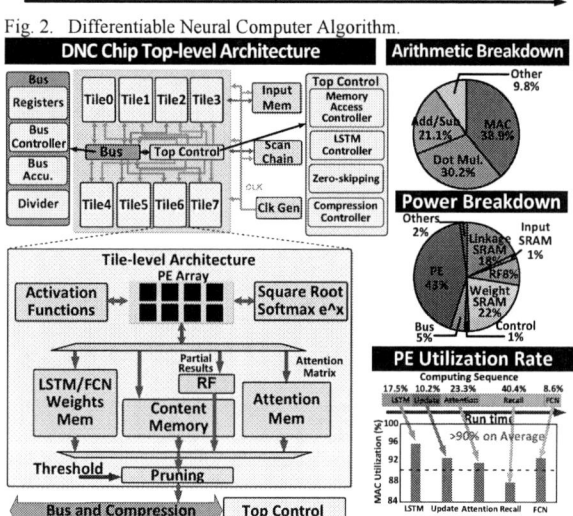

Fig. 3. DNC Chip Top-level Architecture, PE utilization, arithmetic and power breakdown.

As shown in the arithmetic operation breakdown in Fig. 3, besides extensive memory operation, DNC needs to support a variety of operations using PE units including MAC operations (38.9%) from LSTM and FCN, dot and vector multiplication (30.2%) and matrix addition/subtraction (21.1%) for memory similarity calculation. This leads to challenges in utilization of PE and high memory bandwidth required from memory banks, i.e. weight memory for LSTM/FCN, main content memory, usage memory and attention memory. To overcome the challenges, as shown in Fig. 3, a distributed near-memory computing (NMC) architecture is developed where the chip is divided into computational tiles connected by a global bus. Each tile embeds a small processing element (PE) array with 8 MACs, distributed memories, a register file and special function modules, e.g. SoftMax. For fitting into a small chip budget, 8 tiles were implemented and can be proportionally scaled up.

Fig. 3 also shows the PE utilization of this design with over 90% on average. Power breakdown is also shown in Fig. 3 with 43% from PEs and 50% from memory.

III. NEAR-MEMORY COMPUTING ARRAY

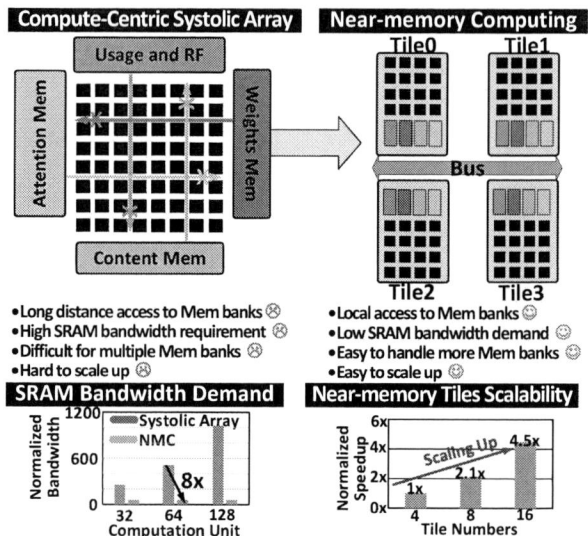

Fig. 4. Comparison between conventional systolic array and near-memory computing architectures.

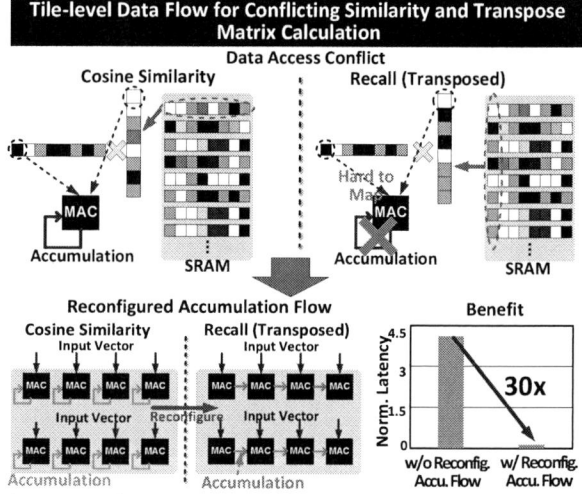

Fig. 5. Reconfigured dataflow for tile-level conflicts.

Fig. 4 compares the proposed NMC architecture with a conventional systolic array (SA). The required access from many different types of memory in DNC causes low efficiency, data collision, large travel distance and poor scalability from SA. NMC allows data to stay locally broadcasting only processed data with 8X reduction of memory bandwidth. In addition, NMC is scalable in throughput with computing tiles in contrast with SA. Optimization of dataflow for different computing phases, e.g. LSTM, attention, etc. are performed at tile level. As shown in Fig. 5, a data mapping conflict between similarity and recall operation with transposed matrix calculation is observed. A reconfigurable flow is used to pass accumulation results in different directions with significant latency enhancement.

IV. ZERO-SKIPPING AND DATA COMPRESSION

Fig. 6 shows a reconfigurable MAC unit which was developed to deal with a variety of operations including MAC, dot multiplication and addition/subtraction. In addition,

978-1-6654-8495-4/22 $31.00 © 2022 IEEE

extensive clock gating and a configurable hybrid precision of 8 bits (LSTM) and 16 bits (Read/Write Head) are used for PE array with minor accuracy loss (3~4% from 32 bits).

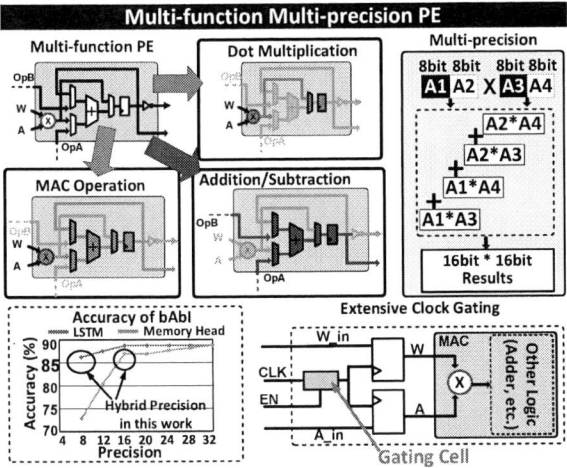

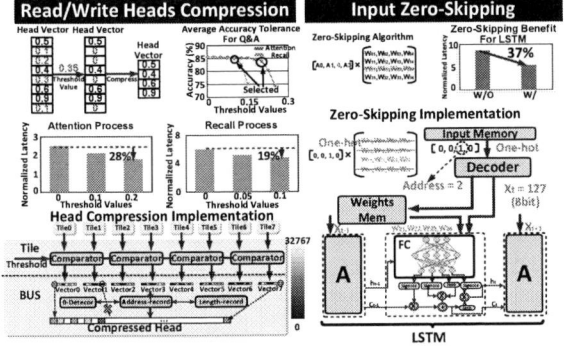

Fig. 6. PE Reconfiguration, multi-precision, clock-gating.

Fig. 7. Data compression and zero-skipping techniques used in this work leveraging sparsity of DNC accelerator.

Fig. 7 shows the sparsity and compression techniques used in this work. In write/read head operations, non-zero weights are compressed in a global bus before sending to each computing tile with preset threshold to prune the write/read weights to enhance sparsity. The data compression technique results in 28% speedup for the attention calculation with negligible overhead and minimal accuracy loss. Due to high sparsity of incoming data stream, input zero-skipping with associated detection and decoder logic as shown in Fig. 7 is also implemented to skip large amount of related MAC operations and weight loading in LSTM for frequent one-hot input. As a result, 96% of FCN operations or 37% of total LSTM operations are being bypassed.

V. MEASUREMENT RESULTS AND DEMONSTRATION CASES

A 65nm test chip was fabricated running at 350MHz at nominal 1V supply. Different reasoning tasks using DNC models trained offline was sent into the chip for evaluation with end-to-end operations. Fig. 8 to 10 show detailed description of four examples of reasoning tasks implemented in the test chip including copy task, finding family relationship based on family tree, graph traversal task for traversing London underground stations within a given number of steps and context-based Q&A using bAbI database [5]. Attention mechanism from attention memory is highlighted to show the sequential relationship discovered by the chip. Performance comparison with CPU and GPU and accuracy comparison with floating point model are also shown.

As in the copy task in Fig. 8, DNC receives a sequence of vectors as input data and generate the same vector pattern in as the output. The attention memory is used to store the relationship, i.e. the sequence, between the different addresses of the content memory. For example, the large value in the coordinate (1,2) of the attention memory represents the address 1 and 2 (blue circle) are highly related in relationship enabling "copying" the sequence of the vectors. As shown in Fig. 9 for the family tree example, relationships for immediate family, i.e. father, mother, son, daughter are encoded as vectors sent into DNC to build the family tree graph. The attention memory represents family relationship. As an example, "Amy, David, Father" and "Mary, Amy, Mother" are inputs that include family relationships to be recorded by the attention memory. By performing inference, DNC can generate any relationship between 2 people in the family tree. In this task, the DNC accelerator can achieve about 693X speedup than CPU (Ryzen 5 2600X) with 4% accuracy loss.

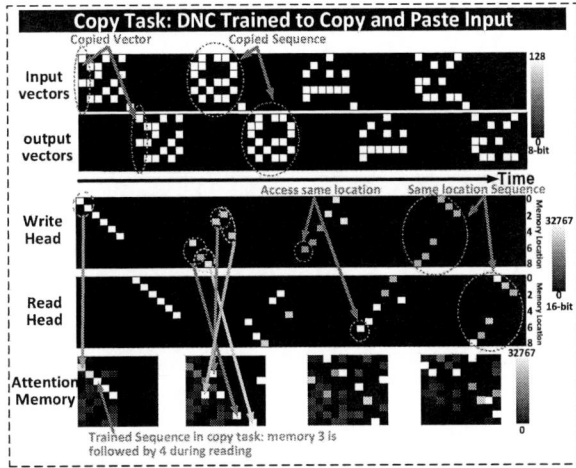

Fig. 8. Detailed demonstration of copy task.

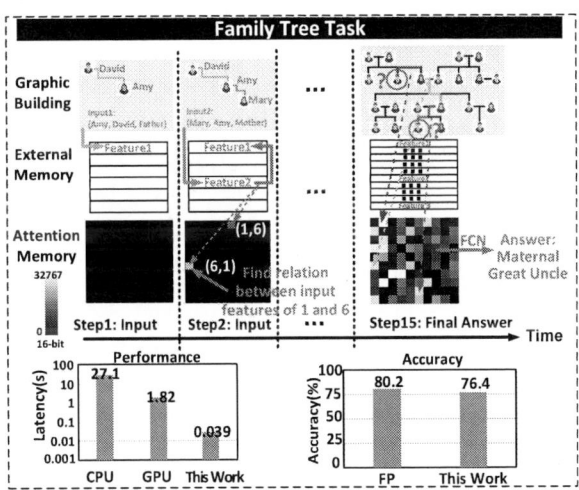

Fig. 9. Detailed demonstration of family tree task.

Finding the logic relationship between two objects in the sequential context or graph is another important application of DNC. As the Q&A task shown in Fig. 10, DNC receives the text contents and is able to give the answer about the questions on the relationship of the objects. The speedup compared with CPU is around 709X. The London underground traversal task is also shown in Fig. 10, DNC can find the traversal path back to the starting station after receiving the information from

978-1-6654-8495-4/22 $31.00 © 2022 IEEE

London underground map in advance. The speedup compared with CPU is 697X with about 4% accuracy degradation.

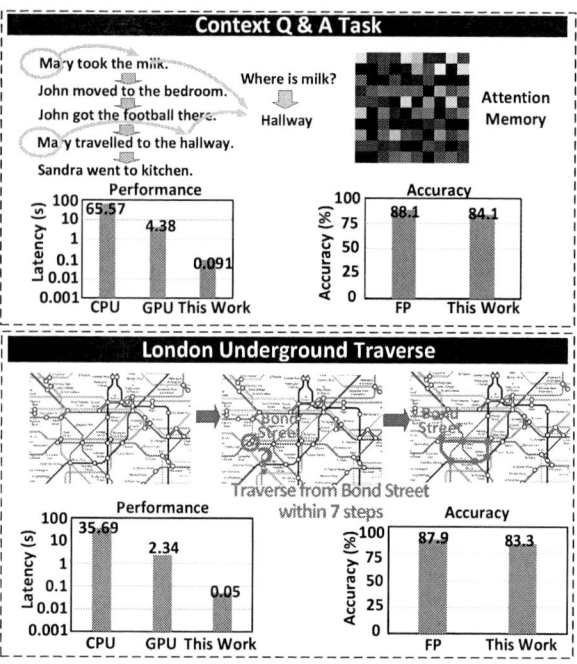

Fig. 10. Details of Q&A task and London underground traversal task.

In total, eight different logic reasoning tasks spanning across diversified jobs including sorting, copying, repeated copy, recall, sort, context Q&A, graphic traversal and shortest path, were tested to verify the functionality and performance against commercial CPU and GPU. Fig. 11 shows the measurement results on power and latency. An average of above 90% utilization has been observed among computing phases. The test chip achieved over 640X and 46X speedup over CPU and GPU processors across the eight test cases. End-to-end speedup of 30% was also achieved from the applied sparsity enhancement techniques.

The comparison table was shown in Fig. 12. As this work is the first implementation of a reasoning processor, comparison was made mainly to prior DNN accelerators specially for LSTM/FCN in similar technology. A maximum efficiency of 1.28TOPS/W is observed for 8-bit LSTM. Compared with a prior simulation-based work using a related but different computation model of MANN [6], a 21X improvement of efficiency is observed from this work. Fig. 13 shows the chip micrograph.

VI. CONCLUSION

A 65nm test chip using Differentiable Neural Computer model was implemented to perform logic reasoning tasks for the first time. A special NMC architecture was developed rendering lower requirement of SRAM bandwidth with better scalability. Input zero-skipping and data compression techniques are applied to achieve 28% reduction on attention calculation and 37% reduction of LSTM operations. Eight different logic reasoning tasks are demonstrated using the test chip. 700X and 46X speedup compared with commercial CPU and GPU are observed on the logic reasoning tasks with a power efficiency of up to 1.28TOPS/W and over 90% utilization of PE units in all the eight computing phases.

REFERENCES

[1] J. Rae, et al., "Scaling Memory-Augmented Neural Networks with Sparse Reads and Writes," NIPS, 2016.

[2] T. Munkhdalai, et al., "Meta Networks", ICML, 2017.

[3] A. Baddeley, et al., "Working Memory: Theories, Models, and Controversies," Annual Review of Psychology, Jan. 2012.

[4] Graves, A., Wayne, G., Reynolds, M. et al., "Hybrid computing using a neural network with dynamic ex-ternal memory," Nature, 2016.

[5] Facebook bAbI, https://research.fb.com/downloads/babi

[6] J. Stevens, et al., "Manna: An Accelerator for Memory-Augmented Neural Networks," MICRO, pp. 794–806, 2019.

[7] D. Shin, J. Lee, J. Lee and H. -J. Yoo, "14.2 DNPU: An 8.1TOPS/W reconfigurable CNN-RNN processor for general-purpose deep neural networks," ISSCC, 2017.

[8] Y. -H. Chen, T. Krishna, J. Emer and V. Sze, "14.5 Eyeriss: An energy-efficient reconfigurable accelerator for deep convolutional neural networks," ISSCC, 2016.

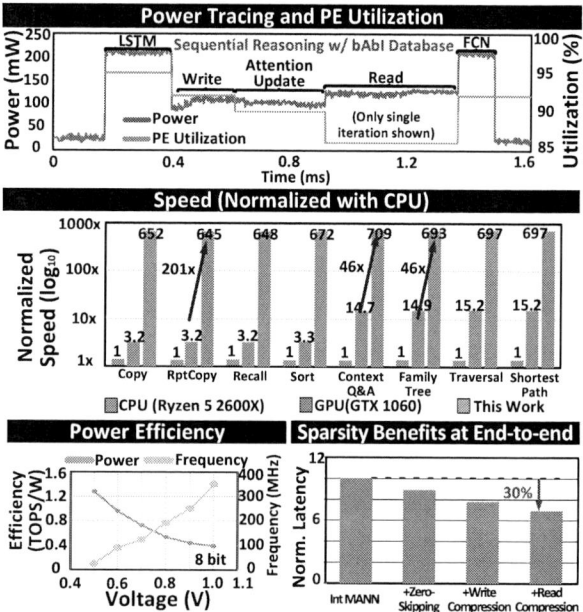

Fig. 11. Measurement results

TABLE I.   COMPARISON TABLE

	MICRO2019[6]	DNPU[7]	Eyeriss[8]	This Work
Core	MANN	CNN,FC,LSTM	CNN	DNC
Num. of PE	3*256	768(16bit)	168	64
Process(nm)	15nm Nangate Open Cell Library	65nm	65nm	65nm
Area(mm2)	40	16	12.25	7.75
Supply Vdd	-	1.1V	1.0V	1.0V
Power	16W (TDP)	279mW	278mW	230mW
Freq. (MHz)	500	200	200	350
Data Type	FP32	INT1~16	INT16	INT8(LSTM) INT16(Read/Write Head)
Memory	39.8MB	-	181KB	200KB
Power Efficiency	18GOPS/W (Simulated Results)	3.9TOPS/W (4b) 1.0TOPS/W (16b)	0.241TOPS/W (1V,16b)	389.6GOPS/W (1V,8b) 1.28GOPS/W (0.5V,8b)

Fig. 12. Comparison Table with prior work.

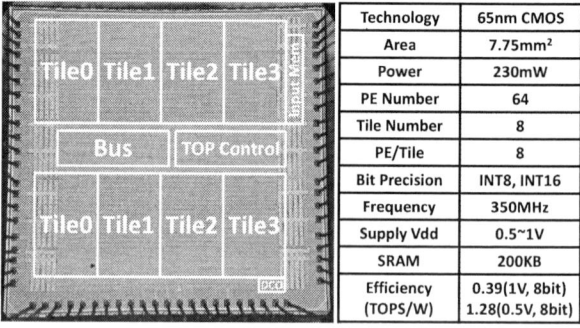

Technology	65nm CMOS
Area	7.75mm²
Power	230mW
PE Number	64
Tile Number	8
PE/Tile	8
Bit Precision	INT8, INT16
Frequency	350MHz
Supply Vdd	0.5~1V
SRAM	200KB
Efficiency (TOPS/W)	0.39(1V, 8bit) 1.28(0.5V, 8bit)

Fig. 13. Micrograph of the test chip.

SIF-NPU: A 28nm 3.48 TOPS/W 0.25 TOPS/mm^2 CNN Accelerator with Spatially Independent Fusion for Real-Time UHD Super-Resolution

Sumin Lee[1], Ki-Beom Lee[1], Sunghwan Joo[2], Hong Keun Ahn[1], Junghyup Lee[1], Dohyung Kim[1],

Bumsub Ham[1], and Seong-Ook Jung[1]

[1]*School of Electrical and Electronics Engineering, Yonsei University*, Seoul 03722, Republic of Korea

[2]*Samsung Electronics*, Hwasung-si, Gyeonggi-do 18448, Republic of Korea

Abstract—This paper proposes a convolutional neural network (CNN)-based super-resolution accelerator for up-scaling to ultra-HD (UHD) resolution in real-time in edge devices. A novel error-compensated bit quantization is adopted to reduce bit depth in the SR task. Spatially independent layer fusion is exploited to satisfy high throughput requirements at UHD resolution by increasing parallelism. Burst operation with write mask in the dual-port SRAM increases the process element utilization by allowing the concurrent multi-access without exploiting additional memory. The accelerator is implemented in the 28nm technology and shows at least 4.3 times higher FoM (TOPS/mm^2 × TOPS/W) of 0.87 than the state-of-art CNN accelerators. The implemented accelerator supports up-scaling up to 96 frames-per-seconds in UHD resolution.

Keywords—AI accelerator, Bit quantization, Convolutional neural network, Dataflow, Super-resolution

I. INTRODUCTION

Since convolutional neural network (CNN) has shown remarkable results in the image processing field of computer vision, it has been employed in many advanced applications such as classification, object detection, and super-resolution (SR). Especially, CNN-based SR has demonstrated superior visualization to conventional interpolation algorithms such as bicubic interpolation at the expense of leveraging massive parameters and computational intensity [1] [2]. However, there are challenges in implementing a CNN-based SR accelerator operating in real-time on resource-limited edge devices.

Contrary to classification, intermediate feature maps in the SR are not shrunken by the pooling layer, as shown in Fig. 1 (a). Since the intermediate feature maps generate excessive external memory access (EMA), many prior works focus on reducing the EMA by fusing some adjacent layers in the neural network. However, the previous methods have the problems such as the requirement of large cache memory and degradation in process element (PE) utilization. Furthermore, since the SR is more sensitive to the quantization error due to the regression model, 14 times higher bit-depth integer representation than the classification task is mandatory to maintain performance [3] [4]. Moreover, modern CNN-based SR accelerators support up to full-HD (FHD, 1080p) resolution in real-time operation [5] [6]. As shown in Fig. 1 (b), since the up-scaling to ultra-HD (UHD, 2160p) resolution with CNN-based SR requires four times higher system throughput than the FHD resolution, minimizing propagation bottleneck within a small cache size is demanded for real-time operation of edge devices.

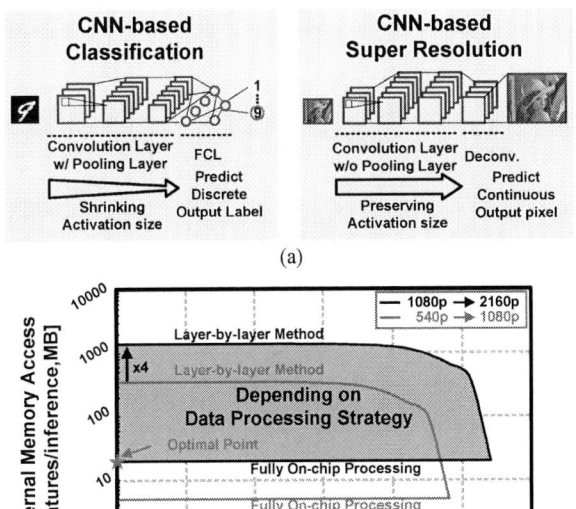

(a)

(b)

Fig. 1. (a) Task differences between classification and super-resolution (b) External memory access vs. on-chip cache size for full-HD and ultra-HD resolution in FSRCNN with 16-bit features

This paper proposes high throughput and low-power CNN-based SR accelerator for up-scaling to UHD resolution in real-time with a small EMA and cache size. The proposed accelerator is implemented in a 28nm technology and supports up to UHD resolution with a high throughput of 96 frames per second (fps). To the best of our knowledge, the implemented accelerator supports the real-time operation of CNN-based SR in UHD resolution for the first time.

II. PROPOSED CNN-BASED SR ACCELERATOR

A. Error-Compensated Bit Quantization

Since classification is the task of predicting discrete class label output, it is tolerant to noise such as bit quantization error. Prior work proves that acceptable accuracy for the classification task is achieved with the binary representation for both activation and weight [3]. However, since the SR is one of the regression tasks for predicting continuous quantity output, it is very sensitive to quantization error. A previous study shows that a minimum 14-bit depth integer is required

978-1-6654-8495-4/22 $31.00 © 2022 IEEE

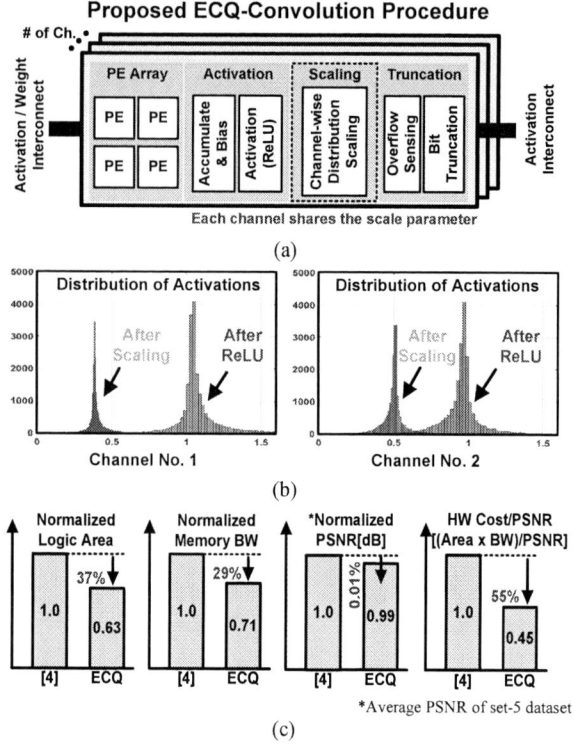

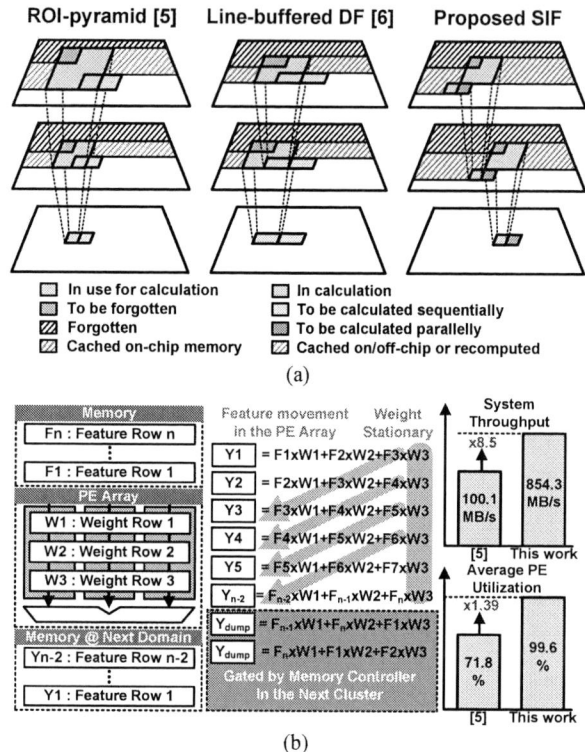

Fig. 2. (a) Proposed error-compensated bit-quantization (ECQ)-based convolution process (b) Scaling effect of distribution in *woman* of set-5. The graphs are shown in channel no.1 and no.2, respectively (c) Decreases of hardware overhead with an insignificant PSNR degradation

Fig. 3. (a) Previous layer fusion approaches and proposed spatially-independent fusion approach (b) Activation processing procedures in each layer and throughput comparison with a prior work

to maintain the peak signal-to-noise ratio (PSNR) compared to single-precision floating-point representation [4].

Thus, the error-compensated bit quantization (ECQ) is proposed to resolve the problem. In the ECQ, the scaling process is added between activation function and truncation, as shown in Fig. 2 (a). In the scaling process, a channel-wise single scaling parameter is multiplied with the feature maps after the ReLU function to adjust the distribution of activations. For instance, the distributions of activations are shown in Fig. 2 (b) when inferencing *woman* image in the set-5 dataset by adopting the scaling parameter. The purpose of adopting the scaling parameter is to maintain a similar distribution scale after bit quantization. In the quantization process, since the bit truncation with upper and lower bounds is leveraged, the distribution scale before quantization differs from that after quantization. Thus, the scale parameters compensate for the distribution errors with post-processing to represent the values exceeded in fixed points, leading to the reduction in quantization errors.

In this work, the signed 9-bit and 8-bit depth integers are used for activations and parameters that include weight, bias, and scaling, respectively. Although the scale parameters are added, the computation load required for adopting ECQ is negligible because a single scale parameter is in charge of all weights in a channel. Compared to the accelerator in [4], the accelerator with ECQ achieves a 37% and 29% reduction in logic area and bandwidth, respectively, with an insignificant PSNR degradation of 0.01% in the set-5 dataset. As a result, since the hardware overhead for maintaining PSNR is decreased by about 55%, the ECQ is suitable for CNN-based SR accelerators in resource-limited edge devices.

B. Spatially Independent Layer Fusion

Fig. 3 (a) shows three different layer fusion approaches [5] [6]. The intermediate feature maps in CNN generate huge EMA due to the limitation of on-chip cache capacity, resulting in the bottleneck of accelerating inference. Thus, the efficient data flow that determines data processing order is important to accelerate inference by reducing dispensable EMA. Previous works resolve the EMA problem by fusing some adjacent layers in CNN. The region-of-influence (ROI) pyramid is one of the fusing methods [5]. In the ROI pyramid, the features in the receptive field are processed first and then propagated to the next layers. However, the ROI pyramid is applied only to a few layers due to the limitation of on-chip cache size because the receptive field is extended by kernels [7].

The line-buffered depth-first (DF) in [6] improves the cache size problem by partitioning the receptive fields in the fused layers. Since the partitioning in the DF method maximizes the propagation while discarding un-reusable features, the DF method preserves the required on-chip cache size when fusing layers. However, the DF method degrades parallelism because the PEs in the fused layers should be stalled until the relevant processes are completed. Although the DF method enlarges the output patch size to improve parallelism, increasing the output patch size generates duplicated receptive field degrading EMA.

The proposed spatially independent fusion (SIF) resolves the throughput limitation without increasing output patch size by hierarchical pipelining the whole network. In the SIF, each ROI is applied only between two adjacent layers while the spatially separated ROIs in the network are connected

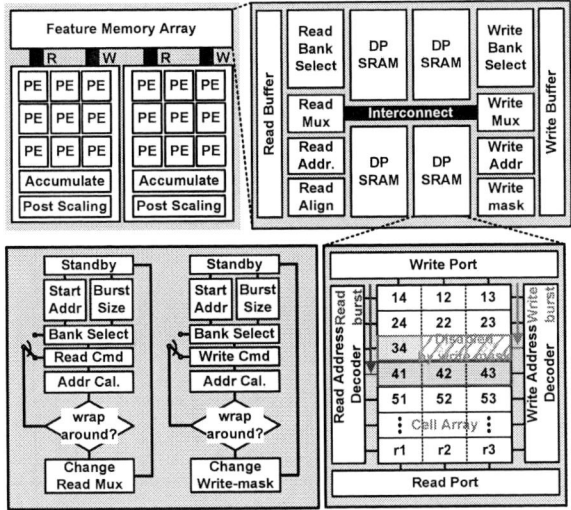

Fig. 4. Feature memory configuration including the write-mask operation and flowchart of read and write operation

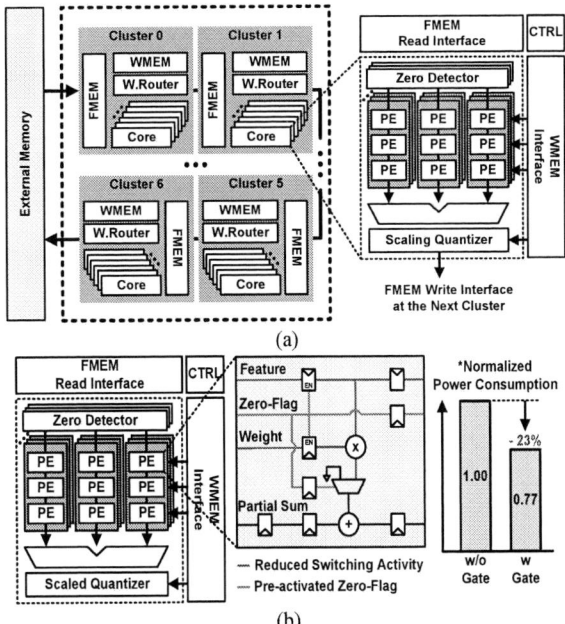

Fig. 5. (a) Overall architecture of the proposed CNN-based SR accelerator (b) pre-aware sparse-gating scheme and comparison of power consumption

consecutively for elaborating the pipelining. Since the SIF disconnects the fused ROIs spatially, PEs can operate independently without waiting for the complete process of adjacent ROIs. In the SIF, each ROI in the layer processes the line features sequentially, as shown in Fig. 3 (b). The connection of separated ROIs is implemented by controlling the memory access sequences described in the following subsection. Since the SIF increases throughput by the elaborated pipelining in the inference, the data flow becomes suitable for consistent high throughput systems such as video streaming services. The SIF achieves a throughput of 854.3 MB/s, higher about 8.5 times than the fusion method in [5].

C. Burst transfer with write mask in dual-port SRAM

The purpose of the on-chip cache memory is to deliver feature maps to the PE array with minimum latency. At the same time, the generated feature maps from the previous layers are written temporarily in the feature memory for reducing the dispensable EMA. In accelerating inference, enabling concurrent multi-access is required in the memory because PEs in the adjacent layers process simultaneously. However, since the typical single-port SRAM (SPSRAM, 6T) allows only single access to the cell array, prior work focuses on hiding the multi-access latency by leveraging the additional SPSRAM [8]. Although the approach improves PE utilization, data-centralization within the same cell array cannot be allowed. The distributed data complicates the data management and requires duplicated access to the memory.

In this work, the dual-port SRAM (DPSRAM, 8T) with a write mask is adopted to resolve the problem. Fig. 4 shows the feature memory configuration using the DPSRAM array with operating flowcharts. The DPSRAM has independent ports allowing concurrent multi-access within the same cell array. In a single-channel convolution of CNN, the write port requires smaller bandwidth than the read port due to multiply-accumulate operation. Thus, the write mask is adopted to balance the bandwidth while preserving reusable data. When the write address is wrap-around, the position of the write mask is swapped. Since the column order changes within the same cell array, the read multiplexers are required to align the data in ascending order. The burst transfer in DPSRAM operates with a different starting address at each

layer for implementing the SIF. The starting address and burst size are dependent on the network size and target image resolution, respectively. The memory scheduling for the SIF achieves 99.6% PE utilization without any duplicated memory access, 27.8% higher than that in the accelerator in [5].

D. Overall architecture of the proposed CNN-based SR acclerators

Fig. 5 shows the overall architecture of the proposed CNN-based SR accelerator. The accelerator has a series of clusters composed of feature and weight memories (FMEM, WMEM), routers, and multiple cores. Each core has zero detectors that process sparsity for low power and a 3D shape PE array with a scaling quantizer for ECQ. A total of 1,500 PEs are deployed in the overall chip. The zero detectors prevent the propagation of the sparsity generated by ReLU at the previous clusters. Before dispensable switching activity occurs in the PE array, the zero detectors are pre-aware of the sparsity and gate the propagation path, leading to the average 23% power reduction in the set-5 dataset. The feature memory using DPSRAMs connects two adjacent clusters, allowing concurrent multi-access including burst transfer in both ports.

III. MEASUREMENT RESULTS AND COMPARISON

Fig. 6. shows the chip specification and photograph of the proposed CNN-based SR accelerator. The accelerator is implemented in the 28nm logic CMOS technology. The core occupies the area of 2.43 mm² with 47.3 KB DPSRAM. The accelerator operates up to 200 MHz at 1.1 V with an area efficiency of 0.25 TOPS/mm². The power consumption is 179.3 mW at the maximum frequency and supply voltage. The energy efficiency is measured as 3.48 TOPS/W and 3.37 TOPS/W at 50 MHz and 200 MHz, respectively. Table I summarizes the specifications of state-of-art CNN accelerators. The figure of merit (FoM) is defined as the area-energy efficiency product to compare the accelerators

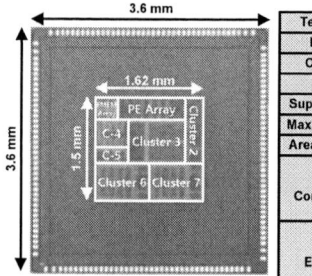

Technology	Samsung 28nm Logic CMOS
Die Area	3.6 x 3.6 mm²
Core Area	1.62 x 1.5 mm²
SRAM	47.3 KB
Supply Voltage	0.9 V ~ 1.1 V
Max. Frequency	200 MHz
Area Efficiency	0.25 TOPS/mm²
Power Consumption	43.4 mW (@ 50MHz, 1.0V)
	179.3 mW (@ 200MHz, 1.1V)
Energy Efficiency	3.48 TOPS/W (@ 50MHz, 1.0V)
	3.37 TOPS/W (@ 200MHz, 1.1V)

Fig. 6. Chip specification and photograph of proposed CNN-based SR accelerator

TABLE I.   COMPARISON FOR THE STATE-OF-ART CNN ACCELERATOR

	VLSI 2019[5]	VLSI 2021[6]	ISSCC 2020[9]	This Work
Technology	65 nm	12 nm	7 nm	28 nm
Core Area	16 mm²	4.5 mm²	3.0 mm²	2.4 mm²
Max. Frequency	200 MHz	930 MHz	880 MHz	200 MHz
Fusion Method	ROI	128 DF	N.A	SIF
SRAM	572 KB	1780 KB	2176 KB	47.3 KB
Precision (W/A)	8 / 16	8 / 8	8 / 8	8 / 9
Peak TOPs	0.23	3.81	3.60	0.60
Area Efficiency [TOPS/mm²]	0.01	0.84	1.18	0.25
Normalized Area Efficiency[a]	0.08	0.16	0.07	0.25
Energy Efficiency [TOPS/W]	1.9	5.0	6.83	3.48
Normalized Energy Efficiency[a]	1.07	1.25	2.26	3.48
FoM[ab]	0.09	0.20	0.16	0.87

[a] Normalized to the 28nm process
[b] FoM (Figure of merit) is defined as TOPS/W x TOPS/mm²

comprehensively. For a fair comparison, technology normalization is applied to all accelerators when extracting FoMs. The proposed CNN accelerator achieves at least 4.3 times higher the FoM of 0.87 compared to the state-of-art CNN accelerators.

The SR performance is shown in Fig. 7. The CNN-based SR accelerator achieves about 4 dB higher PSNR than the conventional bicubic interpolation at the *butterfly* image in the set-5 dataset. In addition, the throughput of 96 fps is achieved with a maximum frequency of 200 MHz. When inferencing UHD resolution, the EMA of 11.12 MB inevitably occurs in the input and output layers to transfer or receive feature maps. It is noted that the EMA in the internal layers is not required in this work. In addition, the proposed accelerator supports various resolutions below UHD resolution with an up-scaling factor of two.

IV. CONCLUSION

This paper proposes the CNN-based SR accelerator to UHD resolution in real-time operation. The scaling parameter in the ECQ is adopted to reduce bit depth significantly in SR task while maintaining PSNR. In addition, the SIF is exploited to increase throughput by elaborating the pipelining in CNN. Finally, burst transfer with a write mask increases the PE utilization by allowing concurrent multi-

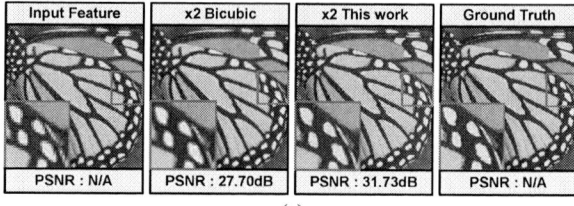

SR Performance Summary				
Support Scale	x2 Up-scaling	DataSet	PSNR	SSIM
Output Resolution	~ UHD (2160p)	Set-5	36.23 dB	0.95
Throughput	~ 96 fps @ 200MHz	Set-14	31.97 dB	0.91
EMA (I/O Layer)	~ 11.12 MB @ 9bit	B100	31.08 dB	0.89
EMA (Int. Layer)	0 MB (On-chip)	Urban 100	28.93 dB	0.89

(b)

Fig. 7. (a) Comparison of up-scaling performance between CNN-based SR and bicubic interpolation at the *butterfly* image in the set-5 dataset (b) SR performance summary

access without additional memory. The implemented accelerator using 28nm technology achieves the highest FoM compared to the state-of-art accelerators and supports up-scaling to UHD resolution with a throughput of 96 fps.

ACKNOWLEDGMENT

This research was supported by the Samsung Research Funding & Incubation Center for Future Technology (SRFC-IT1802-06). The chip fabrication and EDA tool were supported by the IC Design Education Center (IDEC), Korea.

REFERENCES

[1] C. Dong, C. C. Loy, K. He, and X. Tang, "Learning a Deep Convolutional Network for Image Super-Resolution," in *Computer Vision – ECCV 2014*, Cham, D. Fleet, T. Pajdla, B. Schiele, and T. Tuytelaars, Eds., 2014: Springer International Publishing, pp. 184-199.

[2] C. Dong, C. C. Loy, and X. Tang, "Accelerating the super-resolution convolutional neural network," in *European conference on computer vision*, 2016: Springer, pp. 391-407.

[3] J. Kim *et al.*, "Area-Efficient and Variation-Tolerant In-Memory BNN Computing using 6T SRAM Array," in *2019 Symposium on VLSI Circuits*, 9-14 June 2019, pp. C118-C119.

[4] Y. Kim, J. Choi, and M. Kim, "A Real-Time Convolutional Neural Network for Super-Resolution on FPGA With Applications to 4K UHD 60 fps Video Services," *IEEE Transactions on Circuits and Systems for Video Technology*, vol. 29, no. 8, pp. 2521-2534, 2019.

[5] J. Lee, D. Shin, J. Lee, J. Lee, S. Kang, and H. J. Yoo, "A Full HD 60 fps CNN Super Resolution Processor with Selective Caching based Layer Fusion for Mobile Devices," in *2019 Symposium on VLSI Circuits*, 9-14 June 2019, pp. C302-C303.

[6] K. Goetschalckx and M. Verhelst, "DepFiN: A 12nm, 3.8TOPs depth-first CNN processor for high res. image processing," in *2021 Symposium on VLSI Circuits*, 13-19 June 2021, pp. 1-2.

[7] D. Im, D. Han, S. Choi, S. Kang, and H. J. Yoo, "DT-CNN: An Energy-Efficient Dilated and Transposed Convolutional Neural Network Processor for Region of Interest Based Image Segmentation," *IEEE Transactions on Circuits and Systems I: Regular Papers*, vol. 67, no. 10, pp. 3471-3483, 2020.

[8] H. Y. Shen, Y. C. Lee, T. W. Tong, and C. H. Yang, "4.7 A 91mW 90fps Super-Resolution Processor for Full HD Images," in *2021 IEEE International Solid- State Circuits Conference (ISSCC)*, 13-22 Feb. 2021, vol. 64, pp. 66-68.

[9] C. Lin *et al.*, "7.1 A 3.4-to-13.3TOPS/W 3.6TOPS Dual-Core Deep-Learning Accelerator for Versatile AI Applications in 7nm 5G Smartphone SoC," in *2020 IEEE International Solid- State Circuits Conference - (ISSCC)*, 16-20 Feb. 2020, pp. 134-136.

A 40nm RRAM Compute-in-Memory Macro with Parallelism-Preserving ECC for Iso-Accuracy Voltage Scaling

Wantong Li, James Read, Hongwu Jiang, and Shimeng Yu

School of Electrical and Computer Engineering,
Georgia Institute of Technology,
Atlanta, GA 30332, USA
E-mail: shimeng.yu@ece.gatech.edu

Abstract — **Compute-in-memory (CIM) employing resistive random access memory (RRAM) has been widely investigated as an attractive candidate to accelerate the heavy multiply-and-accumulate (MAC) workloads in deep neural networks (DNNs) inference. Supply voltage (VDD) scaling for compute engines is a popular technique to allow edge devices to toggle between high-performance and low-power modes. While prior CIM works have examined VDD scaling, they have not explored its effects on hardware errors and inference accuracy. In this work, we design and validate an RRAM-based CIM macro with a novel error correction code (ECC), called MAC-ECC, that can be reconfigured to correct errors arising from scaled VDD while preserving the parallelism of CIM. This enables RRAM-CIM to perform iso-accuracy inference across different operation modes. We design specialized hardware to implement the MAC-ECC decoder and insert it into the existing compute pipeline without throughput overhead. Additionally, we conduct measurements to characterize the effect of VDD scaling on errors in CIM. The macro is taped-out in TSMC N40 RRAM process, and for 1×1b MAC operations on DenseNet-40 network it achieves 59.1 TOPS/W and 70.9 GOPS/mm² at VDD of 0.7V, and 43.0 TOPS/W and 112.5 GOPS/mm² at VDD of 1.0V. The design maintains <1% accuracy loss on the CIFAR-10 dataset across the tested VDDs.**

Keywords — ***RRAM, non-volatile memory, compute-in-memory, error correction code, voltage scaling***

I. INTRODUCTION

In recent years, deep neural networks (DNNs) have provided remarkable performance gains in fields spanning from computer vision to language processing. Compute-in-memory (CIM) has been widely explored as a promising paradigm to efficiently process the extensive workloads of multiply-and-accumulate (MAC) operations in the inference of DNNs. Among memory types in CIM design for edge intelligence, resistive random-access memory (RRAM) is an attractive candidate due to its high density, low leakage power during standby, and non-volatile storage of DNN weights [1]. Edge devices often can toggle between different modes such as high-performance and low-power modes, and voltage scaling is a popular approach to dynamically scale the supply voltage (VDD) according to the performance requirements. Unlike VDD scaling for digital compute mostly affecting its speed, effects of VDD scaling may be more detrimental to analog compute. However, for many applications, the inference accuracy cannot be severely deteriorated regardless of the device operation mode. Some prior RRAM-based CIM prototypes [2] have examined VDD scaling, but hardware errors due to the scaling and resulting effects to inference accuracy have not been fully explored.

In this work, we introduce an RRAM-based CIM macro equipped with *in-situ* error correction code (ECC) to enable iso-accuracy inference across different VDDs. The novel error correction scheme [3], termed MAC-ECC, is used for single-error correction double-error detection (SECDED) for CIM-based MAC unit. MAC-ECC does not protect individual memory cells, but instead protects the partial sum while preserving the row parallelism of CIM. This is because conventional ECC reads out data and parity from the memory array row by row and then performs syndrome calculation at the array edge, whereas the proposed MAC-ECC performs parallel read-out from multiple rows. A few prior works have proposed other methods to protect MAC results. An analog ECC [4] is proposed to correct MAC errors in the memory array, but the technique is insensitive to small errors and does not consider the peripheral sensing errors. Belief propagation in LDPC decoding proposed by [5] takes multiple iterations to converge, complicating its integration into CIM pipeline without incurring throughput penalty. The major contributions of this work are as follows:

- We demonstrate the first RRAM-based CIM macro capable of iso-accuracy inference across VDD scaling through utilizing MAC-ECC. The macro is flexible to be reconfigured into high-performance mode or low-power mode, by adjusting VDD and the corresponding MAC-ECC scheme.

- We present the silicon demonstration of the parallelism-preserving MAC-ECC scheme for CIM application. The correction capability of this code is both analyzed in simulation and showcased with chip testing involving analog computations. Specialized hardware that implements the MAC-ECC decoder is designed to reduce overhead and avoid throughput penalty.

- We characterize the effect of VDD scaling on voltage-mode CIM through chip measurement, and present the accuracy degradations to the CIFAR-10 dataset with and without applying MAC-ECC.

II. PROPOSED RRAM-CIM MACRO WITH MAC-ECC

A. RRAM-CIM Macro Architecture

To validate the MAC-ECC functionality and characterize the effects of VDD scaling on CIM, we designed and fabricated an RRAM-based CIM macro with embedded 1T1R RRAM in TSMC 40 nm process. Fig. 1 shows the top-level schematic of the RRAM-CIM macro with *in-situ* MAC-ECC. The RRAM array is divided into data and parity sections, which store DNN weights and MAC-ECC parity respectively. The weight significance is encoded to different data columns. Each row of activation is shared by both data and parity, and input precision is encoded by multiple

---

This work is supported by NSF-CCF-1903951, and ASCENT, one of the SRC/DARPA JUMP centers.

compute cycles. On-chip write-verify [6] is employed to tighten the resistance distribution of each cell state. During compute, the digital input vector asserts multiple wordlines (WLs) simultaneously to enable current flow in the bitlines (BLs). The total current represents the MAC result and is converted to a voltage signal through resistive divider between a pull-up PMOS and the activated RRAM cells. A 3-bit flash analog-to-digital converter (ADC) consisting of 7 voltage-mode sense amplifiers (VSAs) then quantizes the voltage, and the output goes through digital shift-add and is accumulated to outputs from previously processed rows. The parity columns have their own set of ADCs, so their outputs can be read out in parallel with the data columns.

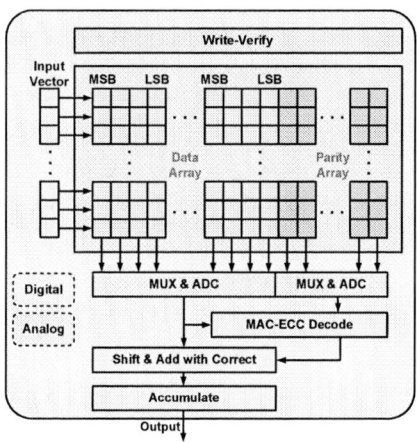

Fig. 1. Schematic of the implemented RRAM-CIM macro with in-situ MAC-ECC.

B. In-situ MAC-ECC for CIM

MAC-ECC belongs to the class of linear block code, and corrects errors based on the arithmetic distance instead of the Hamming distance. For instance, Hamming code will define the number *100* to deviate by 3 bits from *011*, whereas MAC-ECC will assign arithmetic context to the numbers as binary and recognize that they arithmetically differ by 1. This property helps to preserve the parallelism of CIM, since the values in CIM memory cells are accumulated arithmetically along the columns. Fig. 2 presents an overview of the MAC-ECC encoding and decoding procedures. In step 1 of MAC-ECC, the parity array is encoded as a function of the data array and a fixed error locator vector. For inference applications, the parity array can be encoded off-chip and programmed to the array before the chip's deployment. In step 2, both data and parity columns are sensed by ADCs during compute. In step 3, the error locator and output vector with possible errors are used to calculate the syndrome through their element-wise

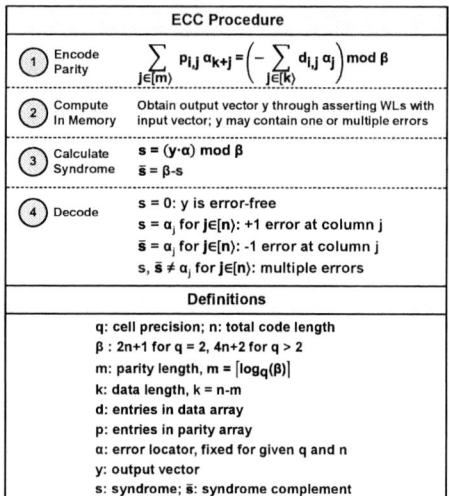

Fig. 2. Procedure for MAC-ECC encoding and decoding.

multiplication, and the result goes through a modulo function. Finally in step 4, the syndrome is decoded by matching its value to one of the elements in the error locator. The decoded error location and sign information is sent to the shift-add module for error correction.

Fig. 3a illustrates the pipelined operations among column sensing with 4-bit weight, digital compute, and MAC-ECC. Steps to decode MAC-ECC are entirely hidden in the existing compute pipeline to avoid throughput penalty. Data column sensing starts with the most significant bit (MSB) columns, and only the MSB columns are equipped with MAC-ECC parity in this example. After sensing the MSB data and its parity columns, the output is sent to both the shift-add block and the MAC-ECC decoder. The decoding process completes when *col2* is ready for shift-add. In case of 1-bit errors in the MSB columns, the errors are corrected while *col2* is being shift-added. Since a 1-bit error in the MSB column of a 4-bit weight contributes an effective error of 8, the shift-add result is added or subtracted by 8 depending on the error sign. Fig. 3b shows the schematic of the RRAM array with peripheral flash ADC that consists of VSAs. Fig. 3c shows the hardware implementation for the MAC-ECC decoder. We use look-up tables (LUTs) to store the constants α and β, which only occupy tens of Bytes. To achieve 1-cycle syndrome calculation with low overhead, we design a customized modulo block that consists of 6-stage successive comparators and subtractors. This module is reconfigurable for multiple MAC-ECC schemes and can support up to a total code length of 31. Similar stages can be attached to extend the supported code length. Parallel comparators are designed to locate the error based on the

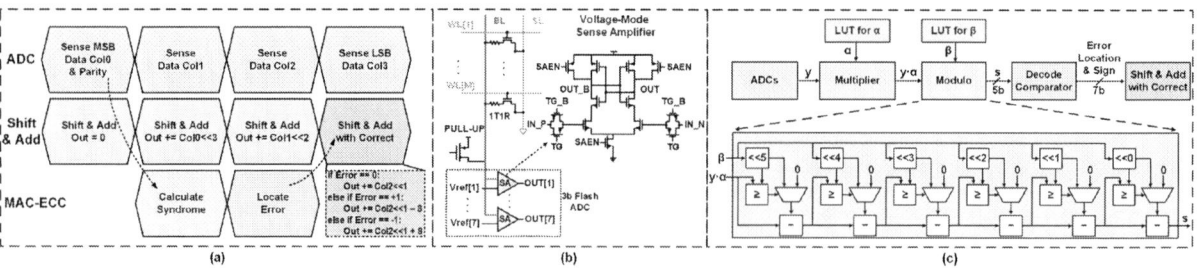

Fig. 3. (a) Compute pipeline with MAC-ECC decoding; (b) schematic of 1T1R array and peripheral 3-bit flash ADC consisting of voltage-mode sense amplifiers; (c) MAC-ECC decoder implementation, highlighting the customized modulo block.

978-1-6654-8495-4/22 $31.00 © 2022 IEEE

syndrome value. Since the modulo and decode comparators only output data of bitwidths 5 and 7 respectively, we choose to use their outputs for pipelining to decrease the number of registers for immediate data.

III. EVALUATION

A. Analyses of DNN Workload and MAC-ECC

In this work, we use DenseNet-40 [7] as the target workload for CIM-based inference. After quantized to 4-bit weights and 8-bit activations, DenseNet-40 retains an appreciable baseline accuracy of 90.5% for the CIFAR-10 dataset while only containing 1M parameters, making it an attractive network for inference in lightweight edge devices. With its fewer number of parameters and densely connected layers, however, DenseNet-40 is less error-tolerant than other networks that can achieve similar accuracies in CIFAR-10. In the inference simulation of DenseNet-40, we study its level of fault tolerance by injecting errors to the MAC partial sums contributed from each weight bit. We define the bit error rate (BER) as the total arithmetic errors divided by the number of activated RRAM cells. As shown in Fig. 4, we inject the same set of errors to different weight bits, ranging from injecting to all the weight bits at once to injecting only to the least significant bit (LSB). We observe that leaving the MSB intact can substantially preserve the inference accuracy, which leads us to only protect the MSB columns with MAC-ECC in this work.

With software simulation to reconstruct the MAC-ECC procedure and using it to protect the outputs of simulated MAC operations that receive customizable errors, we evaluate the correction capabilities of MAC-ECC schemes under various raw BERs. The notation for a MAC-ECC scheme follows (n, k), where n is the total code length and k is the data length. Fig. 5 shows the correction capabilities for $(13, 8)$, $(22, 16)$, and $(30, 24)$ MAC-ECC schemes, which protect 8, 16, and 24 data columns respectively. The BER reduction metric represents how much errors can be reduced from the raw BER. At low error regime (<0.1% raw BER), a

weak MAC-ECC such as $(30, 24)$ can provide at least $2\times$ error reduction. However, at high error regime (>0.5% raw BER), a stronger scheme such as $(13, 8)$ is required to obtain similar level of protection.

B. Measurement Results

To understand the effect of voltage scaling on RRAM-CIM that employs voltage-mode sensing, we measure the BL voltages corresponding to each ADC output partial sum at varying VDDs. In this experiment, we scale the VDDs for the RRAM array, ADCs, and digital modules together. As observed in Fig. 6, the ADC sense margins are diminished with reducing VDD. Because thermal noise is independent of VDD, the reduced sense margins make the ADC outputs more error-prone at lower VDDs. Overall, scaling VDD to 0.7V decreases the sense margins by 52.4% on average compared to operating at 1.0V. In addition, BL voltages of the larger partial sums with VDD at 0.7V sit very close to the threshold voltage of the sense amplifier input, so further voltage scaling becomes infeasible in this design.

To characterize the error level of our RRAM-CIM chip, we first program the RRAM array with representative weight patterns from DenseNet-40 and equip the MSB columns with $(16, 10)$ MAC-ECC, with an example segment shown in Fig. 7. We then provide test vectors and collect the ADC outputs at varying VDDs, with and without activating MAC-ECC. As shown in Fig 8, more errors expectedly arise at lower VDDs due to reduced sense margins. The collected ADC error distribution before MAC-ECC represents the raw errors in the chip. The ADC outputs after applying MAC-ECC involve inter-column dynamics since only 1-bit errors can be corrected among all the protected columns. The ADC error distributions after correction are used to validate our software simulation framework, in which errors sampled from measured initial ADC error distribution are injected to partial sums and the MAC-ECC is recreated in software to correct some of the injected errors during network inference. We find that the measured and simulated ADC error distributions after applying MAC-ECC indeed closely match.

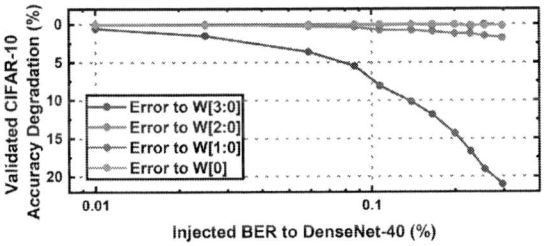

Fig. 4. Simulated CIFAR-10 accuracy degration by injecting BERs from all weight bits to only the LSB.

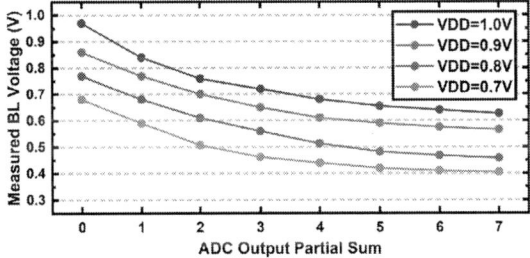

Fig. 6. Measured BL voltages for each ADC output partial sum at VDDs from 0.7V to 1.0V.

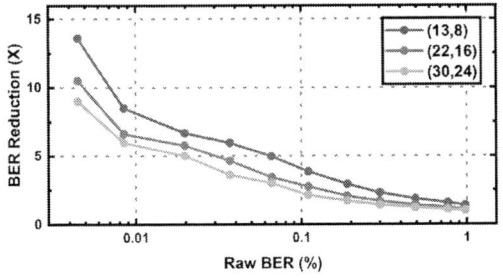

Fig. 5. Simulated correction capabilitites of several MAC-ECC schemes.

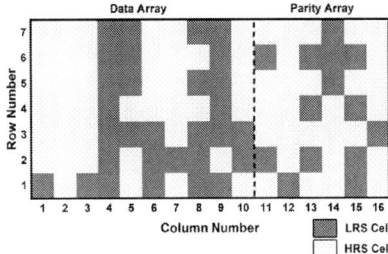

Fig. 7. Example segment of data and parity arrays representative of DenseNet-40 weight patterns, used for collecting ADC error distributions before and after applying $(16, 10)$ MAC-ECC.

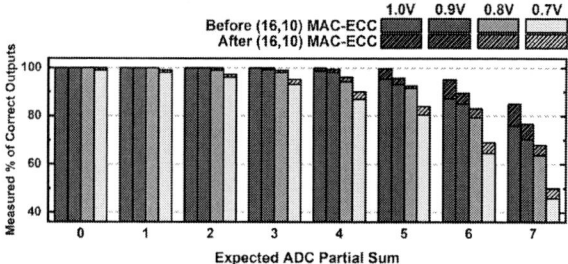

Fig. 8. Measured vs. expected ADC outputs at varying VDDs, before and after applying (*16, 10*) MAC-ECC.

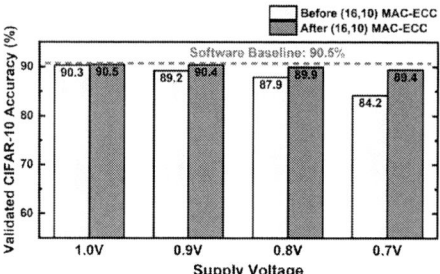

Fig. 9. Simulated CIFAR-10 accuracy before and after MAC-ECC.

C. Evaluation of Iso-Accuracy Inference Engine

With the validated inference simulation framework, we first evaluate the CIFAR-10 accuracy from DenseNet-40 before and after error correction with (*16, 10*) MAC-ECC and at varying VDDs. As shown in Fig. 9, at VDD of 1V the accuracy hardly degrades from the software baseline, and using a stronger (*16, 10*) MAC-ECC helps to fully recover the accuracy. At VDD of 0.7V, the accuracy before correction suffers 6.3% loss and can be recovered to 1.1% loss after applying (*16, 10*) MAC-ECC. The results indicate that a weaker MAC-ECC may be used for higher VDD to retain similar amount of accuracy loss as when a stronger MAC-ECC is used for lower VDD. Thus, we can apply MAC-ECC of different strengths to achieve iso-accuracy across different VDDs. Standard inference benchmarks such as MLPerf [8] require networks for mobile applications to exhibit at most 1% accuracy loss from baseline. We follow this standard to keep CIFAR-10 accuracy loss after correction around 1% or less. Under each VDD condition from 0.7V to 1.0V, we use simulation to find the least costly MAC-ECC scheme needed to keep the accuracy within the 1% loss target. Table I summarizes the least costly MAC-ECC scheme for each VDD, along with the associated macro performances and MAC-ECC hardware overheads. Note that the energy and area for the parity columns are considered in the metric calculations, but the CIM computations for parity are not counted into the total number of operations performed. The results show that at higher VDDs, the macro can operate at higher frequency at the expense of reduced energy efficiency. A higher VDD also reduces the MAC-ECC strength requirement to achieve <1% accuracy loss,

Table I. Iso-Accuracy Performance Summary with VDD Scaling

VDD	Least costly MAC-ECC for <1% loss	Frequency	Energy Efficiency (TOPS/W)	Compute Efficiency (GOPS/mm²)	Energy Overhead	Area Overhead
1V	No ECC	115MHz	43.0	112.5	0%	0%
0.9V	(31, 25)	100MHz	46.2	93.1	3.73%	3.1%
0.8V	(25, 19)	90MHz	52.4	82.8	4.95%	4.11%
0.7V	(16, 10)	80MHz	59.1	70.9	7.48%	6.06%

Fig. 10. Prototype die photo and chip summary.

Technology	TSMC 40nm w/ RRAM
Array size	128 × 128b
Rows activated simultaneously	7
Operating voltage	0.7V to 1.0V
Clock frequency	80MHz to 115MHz
Compute efficiency 1×1b MAC (GOPS/mm²)	70.9 (at 0.7V) to 112.5 (at 1.0V)
Energy efficiency 1×1b MAC (TOPS/W)	59.1 (at 0.7V) to 43.0 (at 1.0V)

thus lessening the hardware overhead. At lower VDDs, stronger MAC-ECC schemes are necessary and the macro operates slower, but energy efficiency improves despite the slightly increased MAC-ECC overhead. Overall, operating at 1V provides ~60% increase in throughput compared to at 0.7V, whereas operating at 0.7V allows for ~40% improvement in energy efficiency compared to at 1V. This flexibility allows the RRAM-CIM macro to operate either in high-performance mode where high throughput is desired, or in low-power mode where energy saving is critical. For both modes the network maintains low accuracy loss with the help of reconfigurable MAC-ECC. Fig. 10 shows the prototype die photo and the chip summary.

IV. Conclusion

In this work, we design an integrated RRAM-CIM macro with *in-situ* MAC-ECC, which corrects errors in ADC outputs and preserves CIM parallelism. With reconfigurable MAC-ECC to correct additional errors resulting from supply voltage scaling, this work presents the first RRAM-based CIM macro capable of iso-accuracy DNN inference at VDD from 0.7V to 1.0V. Operated at 0.7V the macro achieves 59.1 TOPS/W and 70.9 GOPS/mm², and at 1.0V the design achieves 43.0 TOPS/W and 112.5 GOPS/mm². Therefore, we improve the versatility of RRAM-CIM to be deployed in various edge applications with minimal accuracy loss.

Acknowledgment

The authors thank TSMC for offering N40 RRAM shuttle.

References

[1] E. Esmanhotto *et al.*, "High-density 3D monolithically integrated multiple 1T1R multi-level-cell for neural networks," *IEEE International Electron Devices Meeting (IEDM)*, 2020.

[2] J.-M. Hung *et al.*, "An 8-Mb DC-current-free binary-to-8b precision ReRAM nonvolatile computing-in-memory macro using time-space-readout with 1286.4-21.6TOPS/W for edge-AI devices," *IEEE International Solid- State Circuits Conference (ISSCC)*, 2022.

[3] R. M. Roth, "Fault-tolerant dot-product engines," *IEEE Transactions on Information Theory*, vol. 65, no. 4, pp. 2046-2057, 2019.

[4] C. Li *et al.*, "Analog error correcting codes for defect tolerant matrix multiplication in crossbars," *IEEE International Electron Devices Meeting (IEDM)*, 2020.

[5] Q. Lou *et al.*, "Embedding error correction into crossbars for reliable matrix vector multiplication using emerging devices," *ACM/IEEE International Symposium on Low Power Electronics and Design (ISLPED)*, 2020.

[6] W. Li *et al.*, "A 40nm RRAM compute-in-memory macro featuring on-chip write-verify and offset-cancelling ADC references," *IEEE European Solid-State Circuits Conference (ESSCIRC)*, 2021.

[7] G. Huang *et al.*, "Densely connected convolutional networks," *IEEE Conference on Computer Vision and Pattern Recognition (CVPR)*, 2017.

[8] V. J. Reddi *et al.*, "MLPerf inference benchmark," *International ACM/IEEE Symposium on Computer Architecture (ISCA)*, 2020.

978-1-6654-8495-4/22 $31.00 © 2022 IEEE

In-memory Realization of In-situ Few-shot Continual Learning with a Dynamically Evolving Explicit Memory

G. Karunaratne[*§], M. Hersche[*§], J. Langenegger[*§], G. Cherubini[*], M. Le Gallo[*], U. Egger[*],
K. Brew[†], S. Choi[†], I. Ok[†], C. Silvestre[†], N. Li[†], N. Saulnier[†], V. Chan[†], I. Ahsan[†],
V. Narayanan[‡], L. Benini[§], A. Sebastian[*], A. Rahimi[*]

[*]IBM Research, Zürich, Switzerland [†]IBM Research, Albany, NY, USA
[‡]IBM T. J. Watson Research Center, NY, USA [§]ETH Zürich, Zürich, Switzerland

Abstract—Continually learning new classes from few training examples without forgetting previous old classes demands a flexible architecture with an inevitably growing portion of storage, in which new examples and classes can be incrementally stored and efficiently retrieved. One viable architectural solution is to tightly couple a stationary deep neural network to a dynamically evolving explicit memory (EM). As the centerpiece of this architecture, we propose an EM unit that leverages energy-efficient in-memory compute (IMC) cores during the course of continual learning operations. We demonstrate for the first time how the EM unit can physically superpose multiple training examples, expand to accommodate unseen classes, and perform similarity search during inference, using operations on an IMC core based on phase-change memory (PCM). Specifically, the physical superposition of few encoded training examples is realized via in-situ progressive crystallization of PCM devices. The classification accuracy achieved on the IMC core remains within a range of 1.28%–2.5% compared to that of the state-of-the-art full-precision baseline software model on both the CIFAR-100 and miniImageNet datasets when continually learning 40 novel classes (from only five examples per class) on top of 60 old classes.

Index Terms—In-memory Computing, Continual Learning, Few-shot Learning, Hyperdimensional Computing, Non-volatile Memory Devices

I. INTRODUCTION

Few-shot continual learning (FSCL), aka few-shot class-incremental learning [1], [2], requires a learner to incrementally learn new classes from very few training examples, without forgetting the previously learned classes. The learner is exposed to a series of sessions, whereby each session introduces distinct unseen classes by providing only a few training examples per class. From these few examples, the learner should quickly and incrementally learn novel classes without forgetting the previously learned old classes. After learning novel classes in each session, the learner is evaluated on several query samples from all the classes, to which it was exposed so far (i.e., the union of the classes from the previous and the current sessions). FSCL is a very challenging research problem that could impose significant additional compute and memory costs on the learner.

Very recently, a low-cost solution was proposed that avoids expensive gradient-based computations for learning unseen classes during the course of FSCL [3]. This solution is inspired by a robust few-shot learner [4] that brings together deep neural networks with hyperdimensional computing [5], [6] to be able to represent raw images with high-dimensional holographic binary or bipolar vectors. Inspired by this powerful combination, the learner in [3] is composed of a *frozen* controller and a *dynamically evolving* explicit memory (EM) for FSCL. The controller is a deep convolutional network (including a final fully connected layer) that is interfaced with the EM unit to dynamically store or retrieve the acquired knowledge about the classes. The controller interacts with the EM through write and read operations using d-dimensional holographic vectors. Although the controller remains stationary, the EM dynamically updates its contents by new examples, and grows its size by storing new classes. Hence, it is desirable to have an EM unit with dynamically evolving contents that retains the acquired knowledge about already-seen classes in both compressed and nonvolatile manner.

II. PROPOSED EXPLICIT MEMORY UNIT FOR FSCL

Fig. 1(a)(b) illustrates the main stages of FSCL involved in [3]: a pretraining and metalearning stage, and an inference stage. The first stage is all done in software. The goal of this rather elaborate stage is to train the controller (here, a ResNet-12) to be able to generate d-dimensional[1] quasi-orthogonal real-valued vectors for different classes. By the end of this stage, the controller has learned how to assign 256-dimensional quasi-orthogonal, and thus dissimilar, vectors to novel classes in the EM, which allows the controller to remain stationary afterwards. This pretraining and metalearning stage is based on just the first session's (S1) training samples. Based on the datasets used in our experiments, as explained in Section IV-A, the first session contains 60 classes, each including 500 samples, out of the total 100 classes.

---

[1]As seen later in Section III, d is fixed to 256 so that the vectors occupy an entire column of a PCM crossbar array.

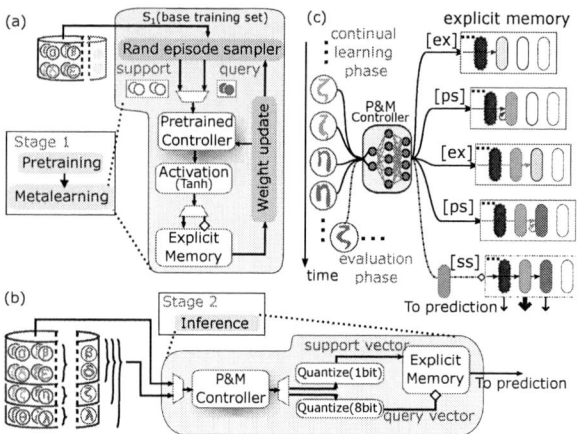

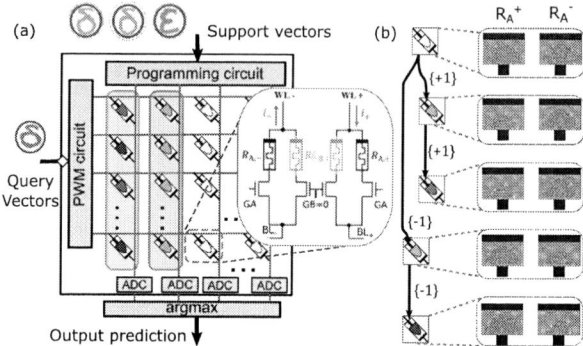

Fig. 1. **(a)** Stage 1 (Pretraining and Metalearning), the controller weights are updated over a series of episodes. **(b)** In Stage 2 (Inference), the pretrained and metalearned (P&M) controller generates support and query vectors over a series of sessions. **(c)** EM operations during inference stage can be divided into two phases. Phase 1 continual learning consists of: [ex]=expansion (allocating a new column in the array to store the first support vector from a previously unseen class); [ps]=physical superposition (few support examples are accumulated on the relevant class memory). Phase 2 evaluation phase consists of: [ss]=similarity search between a query vector and class vectors on the array at that point in time. Black and white slots in EM correspond to already filled and empty slots, respectively.

Next, the inference stage is composed of two phases, as shown in Fig. 1(c): a continual learning phase of novel classes from very few training/support examples per class (in our results, 5 training examples, hence 5-shot), and a query evaluation phase in which we evaluate the accuracy over a batch of query examples containing a number of samples (100 in the datasets we used) per each class encountered so far. We quantize the controller's output vector elements to 1-bit, hence the controller generates bipolar support vectors (of training examples) to be written, or accumulated, in the EM unit during the continual learning phase. For the query vectors (of query samples) during the evaluation phase, we quantize the controller's output vector elements to 8-bit, so an analog input corresponding to the 8-bit query vector element is applied in each row of the crossbar array. This performs a similarity search between query vector and the class vectors. The resulting vector received via the columns of the crossbar array is used to classify the query sample. The class corresponding to the maximum element of the resulting vector is then taken as the predicted class.

For the inference stage, the EM is implemented on an IMC core with a unit-cell array comprising PCM devices (see Fig. 2). In every continual learning phase: (i) when the first example of a new class appears, the EM is expanded by choosing a fully reset column of unit-cells and, based on the sign of the bipolar support vector element, the unit-cell conductance is increased or decreased by the application of single SET pulses; ii) when an example from a previously seen class appears, a similar update is performed on the column of unit cells corresponding to that particular class.

Fig. 2. **(a)** Schematic illustration of the EM implemented on a cross-bar array of unit-cells. Each unit-cell comprises 4 PCM devices (only 2 used for this experiment) organized in a differential configuration. During continual learning, the expansion and physical superposition operations are performed by applying SET pulses to a column of devices. Similarity search is realized by applying read voltage pulses with varying pulse width and subsequently digitizing the accumulated current using ADCs. **(b)** Illustration of the partial crystallization of the differential pair of devices under different +1/-1 pulse sequences during superposition operation.

We exploit the in-situ accumulation via progressive crystallization of PCM to realize physical superposition of few related support vectors on a single class vector. This means, as shown in Fig. 2(b), starting from a fully reset pair of PCM devices, fine grained SET pulses are applied on either the positive or the negative device depending on whether the type of accumulation is incremental or decremental based on the input bipolar support vector element. This creates a multibit analog EM, whereby the size of the EM is set just by the number of classes, as opposed to the product of the number of classes and the number of training examples per class [ ]. The nonvolatility of the resulting analog states preserves the acquired knowledge about already-seen classes. Finally, during the evaluation phase, the frequent similarity search between the 8-bit query vector (of a query image) and the analog class vectors are computed in-memory by exploiting Kirchhoff's circuit laws, and the result is directly used for classification.

Moreover, given that the controller requires no parameter updates after the first stage of pretraining and metalearning, it can be treated as a deep neural network with stationary weights that can also benefit from an IMC-based implementation, as demonstrated for example in [ ].

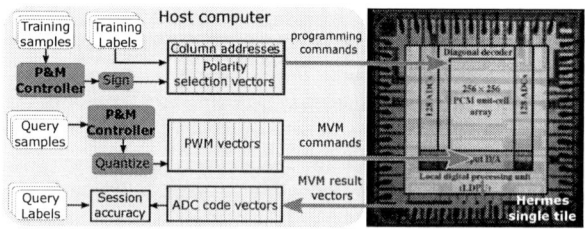

Fig. 3. Experimental Setup. On the right: the micrograph of the Hermes chip. The chip is accessed using a sequence of programming and MVM commands prepared and sent by the host computer via a FPGA-based interface.

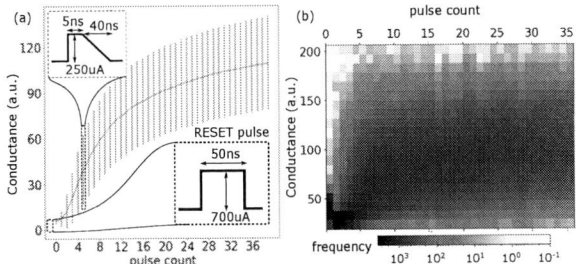

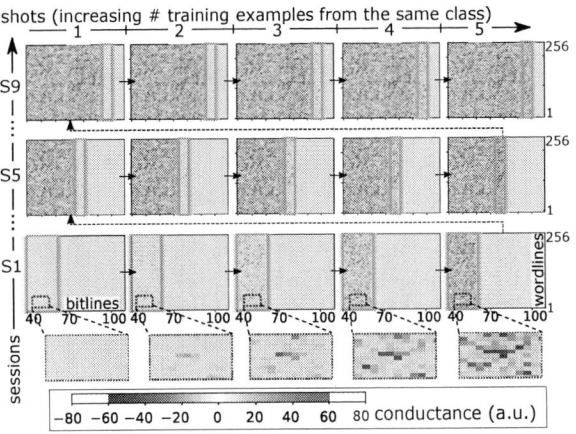

Fig. 4. **(a)** Progressive crystallization of PCM devices in the IMC core. The conductance distribution corresponding to 65,536 devices is shown with the successive application of SET pulses after an initial RESET. The blue curve denotes the mean conductance and the red bars indicate one standard deviation. **(b)** A 2-D illustration of the conductance distribution for the same experiment. It can be seen that the dynamic range (number of SET pules needed to span the conductance range) is fully adequate, because of the very small number (5) of training examples available during FSCL.

III. EXPERIMENTAL SETUP

The experiments are carried out on the Hermes chip with a 256x256 unit-cell array of PCM devices organized in a differential configuration [ ] (see Fig. 3), accessed by a host computer via FPGA (Field Programmable Gate Array)-based interface. The bipolar support vectors are sent as SET pulses to the corresponding column of 256 unit cells of the initially fully RESET array. Based on the +1/-1 vector elements, the conductance of the positive/negative polarity devices are increased. The evolution of the conductance distribution of individual PCM devices as a function of the number of applied SET pulses on initially RESET devices is shown in Fig. 4.

For similarity search in the evaluation phase, a 4-quadrant matrix-vector multiply (MVM) is performed between the 8-bit query vector and the set of analog class vectors stored in the unit-cell array.

IV. EXPERIMENTAL RESULTS

A. Datasets used for evaluation

For the accuracy evaluation, we use the CIFAR-100 and the miniImageNet dataset, restructured to comply with the FSCL setting [ ]–[ ]. Both datasets contain natural images of 100 classes in total, which are divided into a first session (S1) containing 60 classes with 500 training and 100 query examples per class, and eight novel sessions (S2–S9) with 5 novel classes introduced in each session containing 5 support examples and 100 query examples per class. The updated class vectors occupy 256 rows and 60 columns (classes) on the array in S1, evolving to 100 columns (classes) in the last S9. The conductance evolution of unit cells of the crossbar array corresponding to CIFAR-100 is shown in Fig. 5.

B. Classification accuracy

The accuracy obtained for each dataset with our EM on the IMC hardware and various full-precision software baselines are illustrated in Fig. 6. Compared to the full-precision software baseline in [ ], the accuracy degradation with our EM realization is at worst 2.5% and at best 1.28% across

Fig. 5. A 2-D map of the array conductance (of the unit-cells with differential PCM devices) for the FSCL experiment on CIFAR-100 dataset. With each new session, more columns of the array are selected corresponding to the new classes in that session. And with each training example per class, the corresponding class vectors are updated with the application of SET pulses. The unit-cells activated for the update at each panel are highlighted in green. A close up view of the conductance evolution in a 10x10 region of the crossbar is shown in the bottom row.

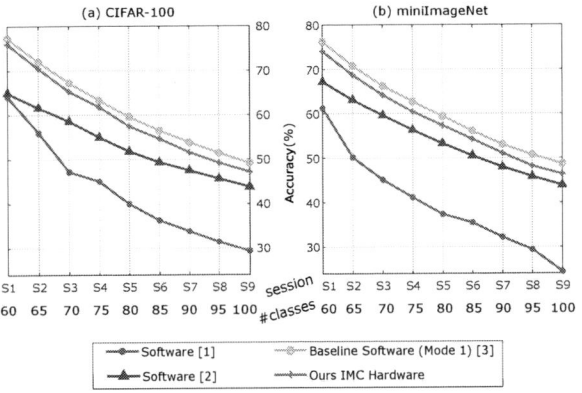

Fig. 6. FSCL classification accuracy for IMC vs. FP32 software implementations for **(a)** CIFAR-100 and **(b)** miniImageNet. As is expected in FSCL, the accuracy progressively reduces with increasing number of sessions, as the learner is progressively exposed to more classes. However, the IMC accuracy is still within 2.5% of the full-precision software baseline [ ] for all sessions. Notably, our IMC hardware accuracy surpasses other full-precision software methods [ ], [ ], for all sessions across the two datasets.

all sessions. This still makes the accuracy of our EM on the IMC hardware from 2.5% to 11.01% higher than the other best performing full-precision software methods reported in [ ], [ ], considering all sessions across CIFAR-100 and miniImageNet datasets.

C. Energy estimation

The energy consumption during incremental class vector updates is estimated using the programming parameters, which include peak pulse current of 150 uA, flat pulse duration 5 ns, trailing edge pulse duration 40 ns and source voltage 2.34 V. These parameters yield an energy expense of 8.78 pJ per PCM device during one programming cycle. Considering the vector dimension of 256, the total programming time and energy spent during an incremental update of one class vector are 11.5 us and 2.25 nJ, respectively. Given that 25 (5-shots from 5 classes) class vectors are updated during all subsequent sessions (S2–S9), the time and energy spent on updating the class vectors in these sessions are estimated to be 57.6 us and 56.2 nJ, respectively.

In comparison, the time and energy spent on similarity search of a single query (including digital to analog conversion, PCM read and analog to digital conversion) are estimated to be 520 ns and 7.74 nJ, respectively, during the last session. This leads to a total similarity search time and energy of 5.2 ms and 77.3 uJ respectively, as in this session we evaluate 10,000 queries (100 queries per class) in total, compared to just 25 updates of the class vectors. The limited number of updates and the application of SET pulses with low energy ensure that the durability of PCM is not significantly affected [9], [10].

TABLE I
COMPARISON WITH THE RELATED WORKS

	Kazemi et al. [ ]	Li et al. [ ], [ ]	Karunaratne et al. [ ]	This work
Few-shot learning	✓	✓	✓	✓
Continual learning				✓
In-situ accumulation				✓
Holographic rep. in EM			✓	✓
Analog multibit EM				✓
Truly $\mathcal{O}(1)$ search^a		✓		✓
Query vector dim. (d)	128–192	128	512	256
Number of classes	≤20	32	**≤100**	**≤100**
Similarity search energy^b	-	17.5 pJ	25.6 pJ	19.1 pJ
Programming energy^c	-	99 pJ	6240 pJ	**8.78 pJ**
Dataset(s)	Omniglot	Omniglot	Omniglot	miniImageNet & CIFAR100

^{a} Actual crossbar (no emulation based on individual memory devices) performing all dot product operations in parallel in-memory
^{b} Core energy normalized to one class vector of length 256 ($d = 256$)
^{c} per vector element

D. Comparison

We compare features of our work against the related works in Table I. Our work shares few-shot learning capability with [ ], [ ]–[ ], holographic representation of vectors with [ ], and truly $\mathcal{O}(1)$ similarity search capability with [ ], [ ]. However, our work is the first to demonstrate continual learning capability using the in-situ accumulation property. This makes our EM unit the fist multibit analog IMC core, whereas in [ ], [ ] storage and queries use binary vectors. Furthermore, our EM unit can handle up to 100 class vectors for the natural image datasets, making it the largest truly $\mathcal{O}(1)$ similarity search engine with analog states to date.

In Table I, we also compare energy saving we gain by programming the support vectors using the in-situ accumulation instead of programming them from scratch in the novel bit lines. Programming with the in-situ accumulation is at least 4.7× energy efficient than the normal programming, because the in-situ accumulation requires a SET pulse of

shorter duration. Our similarity search energy remains on par with other works.

V. CONCLUSION

We present a hardware based on an in-memory compute core consisting of PCM devices to implement the EM operations required in FSCL. We demonstrate for the first time how support vectors are accumulated in-situ using the progressive crystallization property of PCM. The proposed approach leads to physically superposed representations and enhanced energy savings, while retaining the accuracy within 2.5% from the full-precision software baseline.

ACKNOWLEDGEMENTS

This work is supported by the IBM Research AI Hardware Center, and the Center for Computational Innovation at Rensselaer Polytechnic Institute for computational resources on the AiMOS Supercomputer.

REFERENCES

[1] X. Tao et al., "Few-shot class-incremental learning," in *Proceedings of the IEEE/CVF Conference on Computer Vision and Pattern Recognition (CVPR)*, 2020. 1, 3

[2] G. Shi et al., "Overcoming catastrophic forgetting in incremental few-shot learning by finding flat minima," *Advances in Neural Information Processing Systems*, vol. 34, 2021. 1, 3

[3] M. Hersche et al., "Constrained Few-shot Class-incremental Learning," in *Conference on Computer Vision and Pattern Recognition (CVPR)*, 2022. 1, 3

[4] G. Karunaratne et al., "Robust high-dimensional memory-augmented neural networks," *Nature Communications*, vol. 12, no. 1, pp. 1–12, 2021. 1, 2, 4

[5] P. Kanerva, "Hyperdimensional Computing: An Introduction to Computing in Distributed Representation with High-Dimensional Random Vectors," *Cognitive Computation*, vol. 1, no. 2, pp. 139–159, 2009. 1

[6] R. W. Gayler, "Vector symbolic architectures answer Jackendoff's challenges for cognitive neuroscience," in *Proceedings of the Joint International Conference on Cognitive Science. ICCS/ASCS*, 2003, pp. 133–138. 1

[7] V. Joshi et al., "Accurate deep neural network inference using computational phase-change memory," *Nature Communications*, vol. 11, no. 1, pp. 1–13, 2020. 2

[8] R. Khaddam-Aljameh et al., "HERMES-core—a 1.59-TOPS/mm^2 PCM on 14-nm CMOS in-memory compute core using 300-ps/LSB linearized CCO-based ADCs," *IEEE Journal of Solid-State Circuits*, vol. 57, no. 4, pp. 1027–1038, 2022. 3

[9] T. Tuma et al., "Stochastic phase-change neurons," *Nature Nanotechnology*, vol. 11, no. 8, pp. 693–699, 2016. 4

[10] M. Le Gallo et al., "An overview of phase-change memory device physics," *Journal of Physics D: Applied Physics*, vol. 53, no. 21, p. 213002, 2020. 4

[11] A. Kazemi et al., "In-memory nearest neighbor search with FeFET multi-bit content-addressable memories," in *2021 Design, Automation Test in Europe Conference Exhibition (DATE)*, 2021, pp. 1084–1089. 4

[12] H. Li et al., "One-shot learning with memory-augmented neural networks using a 64-kbit, 118 GOPS/W RRAM-based non-volatile associative memory," in *2021 Symposium on VLSI Technology*. IEEE, 2021, pp. 1–2. 4

[13] H. Li et al., "SAPIENS: A 64-kb RRAM-based non-volatile associative memory for one-shot learning and inference at the edge," *IEEE Transactions on Electron Devices*, vol. 68, no. 12, pp. 6637–6643, 2021. 4

978-1-6654-8495-4/22 $31.00 © 2022 IEEE

An embedded PCM Peripheral Unit adding Analog MAC In-Memory Computing Feature addressing Non-linearity and Time Drift Compensation

Alessio Antolini[1*], Andrea Lico[1], Eleonora Franchi Scarselli[1], Antonio Gnudi[1], Luca Perilli[1],
Mattia Luigi Torres[2], Marcella Carissimi[2], Marco Pasotti[2], Roberto Antonio Canegallo[2]
[1]DEI – ARCES, University of Bologna, viale Carlo Pepoli 3/2, 40123 Bologna, Italy.
[2]STMicroelectronics, via Camillo Olivetti 2, 20864 Agrate Brianza, Italy.
*Correspondence: alessio.antolini2@unibo.it

Abstract—This paper presents an integrated peripheral unit interfaced to an embedded Phase-change Memory (ePCM) macrocell, with the aim of adding Analog In-memory Computing (AIMC) feature without any modifications to the internal structure of the memory array. The testchip has been designed and manufactured in a 90-nm STMicroelectronics CMOS technology. The unit allows the execution of signed Multiply and Accumulate (MAC) operations at the edge of the memory array exploiting the physical characteristics of memory devices. I-V characteristic non-linearity and transconductance time drift of PCM cells are overcome through a regulated bitline readout circuitry with time-coded inputs, along with a drift compensation technique based on a conductance ratio. Measurements results show 1-σ accuracy of 95.56% in MAC operations, whose decrease in time is roughly negligible at room temperature and is less than 1% after 24 hours at 85°C bake.

Keywords—Analog in-memory computing (AIMC), drift compensation, multiply and accumulate operation (MAC), Phase-change memory (PCM).

I. INTRODUCTION

In the field of Beyond-von Neumann architectures, Analog In-memory Computing (AIMC) has been proposed as a valid strategy to reduce the amount of energy consumption and latency due to internal data transfers. The aim of AIMC is performing computations within the memory unit, typically leveraging the physical features of the memory devices [1]. Among resistive non-volatile memories (NVMs), Phase-change Memory (PCM) is a promising technology in this field, thanks to the intrinsic capability of its memory cells to store multilevel data [2]. A relevant AIMC task is the Multiply and Accumulate (MAC) operation, which is a kernel of the Matrix Vector Multiplication (MVM). MAC can be implemented on PCM arrays mapping the matrix coefficients on memory cells conductances, but, despite recent advances, computation accuracy is deeply affected by PCM devices I-V non-linearity and conductance time drift [2],[3].

This paper presents a peripheral unit interfaced with a 128-kB embedded PCM (ePCM) array in a 90-nm STMicroelectronics CMOS technology, with the purpose of executing one-step MAC operations with both signed inputs and coefficients. The developed testchip is mainly intended to demonstrate a readout technique for non-linearity and time drift compensation, which differs from solutions based on empirical models and post-processing compensation [4],[5],[6]. Moreover, the unit is conceived to avoid any

changes of the internal structure of the memory. This paper is organized as follows: Section II describes the implementation of the proposed unit; Section III illustrates experimental results; Section IV concludes the paper.

II. AIMC UNIT IMPLEMENTATION

A. Testchip structure and interface to the ePCM array

Figure 1 shows a simplified schematic of the 128-kB ePCM array architecture [7] and AIMC unit interface. The AIMC unit is directly connected to the main bitlines (MBL) and during AIMC computation standard ePCM read and program operations are disabled. To perform MAC tasks, the AIMC unit sets the voltage of each MBL and reads the current of the cells belonging to the addressed word line (WL). Unlike other works [5],[6],[8], where MVM is performed in a single step, the proposed solution implements a MVM with multiple consecutive MACs; this requires a sequential activation of different WLs, but prevents the row decoder from being modified, so that the ePCM can be employed as a binary memory as well.

B. MAC computation architecture

The proposed architecture, shown in Fig. 2, is designed to perform a single signed MAC operation $Z = \boldsymbol{w} \cdot \boldsymbol{x} = \sum_i w_i x_i$. The input array $\boldsymbol{x} = [x_1, ..., x_n]$, where x_i are 5-bit signed data, is stored in the control unit. A set of DACs converts the 4-bit absolute value of each x_i to analog value V_i, while the sign bits $x_{i,sign}$ are directly connected to the readout circuit. Each element w_i of the array $\boldsymbol{w} = [w_1, ..., w_n]$ is expressed through the absolute value of its conductance g_i and its sign g_{si}, which are stored in two different PCM cells.

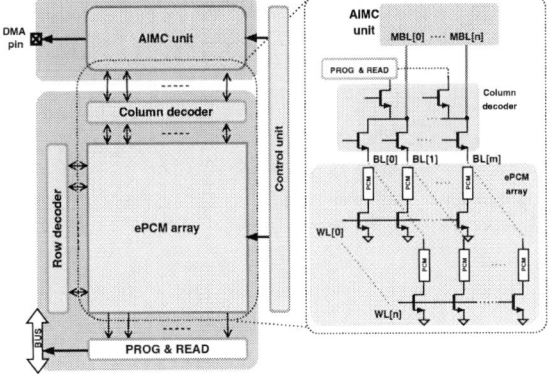

Fig. 1. Left: Block diagram. Right: Simplified schematic of the array architecture and column decoder. DMA pin is used to access various internal analog signals.

This project has received funding from the ECSEL Joint Undertaking (JU) under grant agreement No 101007321. The JU receives support from the European Union's Horizon 2020 research and innovation programme and France, Belgium, Czech Republic, Germany, Italy, Sweden, Switzerland, Turkey.

978-1-6654-8495-4/22 $31.00 © 2022 IEEE

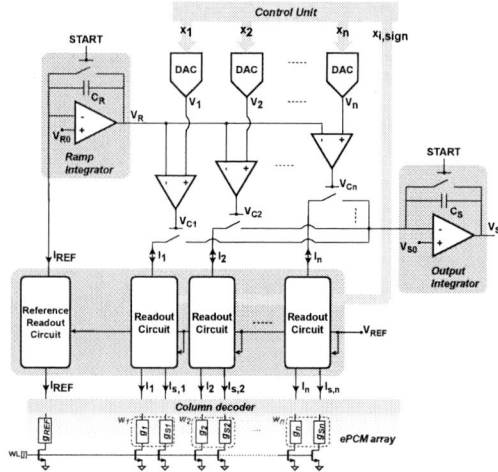

Fig. 2. Block diagram of the proposed MAC architecture.

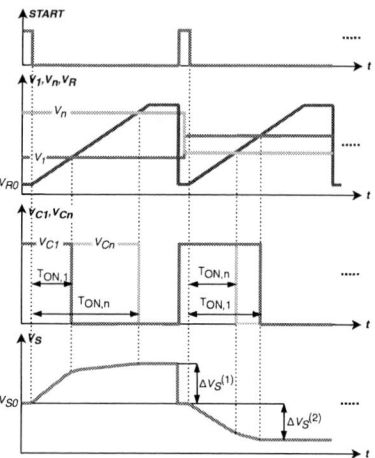

Fig 3. Sketch of waveforms showing two consecutive MAC operations. The first one represents a positive MAC, while the second a negative one.

The Reference Readout Circuit sets the read voltage V_{REF} across the reference conductance g_{REF}. According to Fig. 2 and 3, when the START signal switches to logic low, a current $I_{REF} = g_{REF}V_{REF}$ is integrated on capacitance C_R, generating a ramp signal V_R starting from V_{R0}:

$$V_R(t) = \frac{I_{REF}}{C_R}t + V_{R0}. \quad (1)$$

The same reference read voltage V_{REF} is applied across each weight cell g_i through n Readout Circuits. Each current $I_i = g_iV_{REF}$ is then sourced to or sunk from the output integrator circuit according to the sign of the product w_ix_i, which is obtained combining the corresponding sign cell g_{Si} value and $x_{i,sign}$ sign bit, as described in Paragraph D. Current I_i is then integrated on capacitance C_S for a time window T_{ON_i}, which begins at the START falling edge and ends when the output $V_{C,i}$ of the i-th comparator switches to logic low, i.e., when $V_R(t) = V_i$. According to (1):

$$T_{ON_i} = \frac{(V_i - V_{R0})\,C_R}{I_{REF}} = \frac{(V_i - V_{R0})\,C_R}{g_{REF}V_{REF}}, \quad (2)$$

is the time-coded version of V_i. Summing all the n currents $\pm I_i$, the output variation $\Delta V_S = V_S - V_{S0}$ is:

$$\Delta V_S = \sum_{i=1}^{n}\left[\pm\frac{I_iT_{ON,i}}{C_S}\right] = \frac{C_R}{C_S}\sum_{i=1}^{n}\left[\pm\frac{g_i}{g_{REF}}(V_i - V_{R0})\right]. \quad (3)$$

Considering $V_i - V_{R0}$ and g_i/g_{REF} as the absolute values of x_i and w_i, respectively, $\Delta V_S = (C_R/C_S)\sum_i w_ix_i = (C_R/C_S)Z$ is therefore proportional to the signed MAC operation Z.

C. Drift compensation

The drift of a generic cell conductance $g(t)$ has been shown to follow the power law $g(t) = g_0(t/t_0)^{-\alpha}$ [9], where g_0 is the conductance at arbitrary initial time t_0, and α is the drift coefficient, which is positive and cell-to-cell variable. The MAC result is proportional to the conductance ratio g_i/g_{REF}, and combining the drift model of $g(t)$ with (3), the MAC operation evaluated at time t_1 after t_0 becomes:

$$\Delta V_S(t_1) = \frac{C_R}{C_S}\sum_{i=1}^{n}\left[\pm\frac{g_{0,i}}{g_{0,REF}}\left(\frac{t_1}{t_0}\right)^{-(\alpha_i - \alpha_{REF})}(V_i - V_{R0})\right], \quad (4)$$

where $g_{0,i}$ and $g_{0,REF}$ are the weight and reference cells conductance at t_0, respectively. Therefore, each resulting drift coefficient is reduced to $\alpha_i - \alpha_{REF}$, and drift is partially compensated. In other words, the slope of ramp $V_R(t)$ decreases accordingly to the reference cell conductance drift; this leads to an increase in integration time T_{ON_i}, which compensates for the drift-induced drop of weight cells currents. Moreover, the adoption of time-coded inputs T_{ON_i}, along with cells being read at fixed voltage, addresses cells I-V characteristic non-linearity issue.

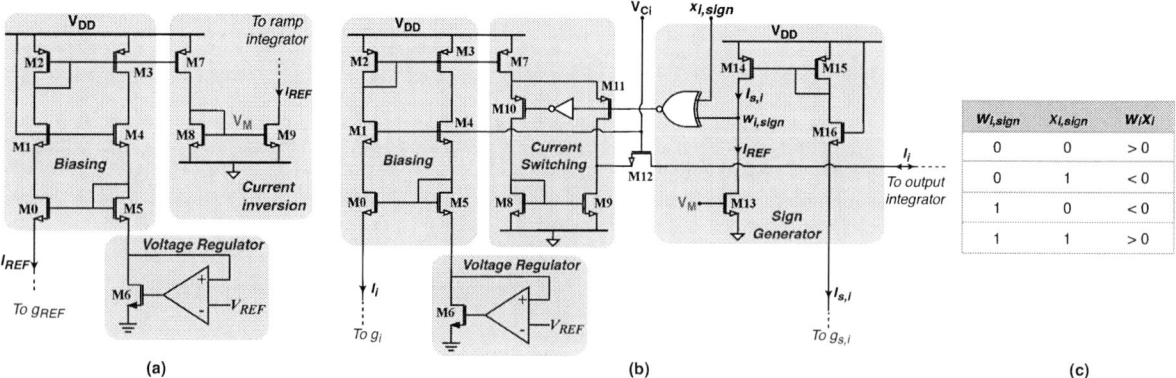

Fig. 4. (a) Schematic of reference readout circuit. (b) Schematic of the cells readout circuit, with sign generation system. (c) Table summarizing the management of signed inputs and coefficients.

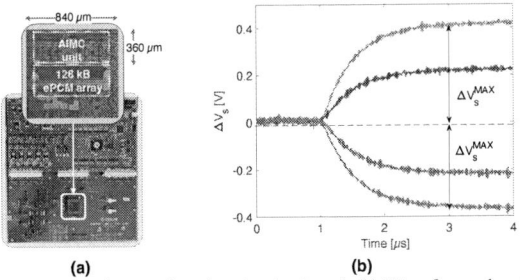

Fig 5. (a) Die micrograph and evaluation board. (b) Waveforms showing four different MAC operations.

D. Reference and Readout circuit with sign management

The detailed schematic of the Reference Readout Circuit is shown in Fig 4.a. The biasing circuit along with the voltage regulator allows to read the reference cell at a fixed reference voltage level V_{REF}. The reference voltage V_{REF} applied to the source terminal of transistor M5 is generated using a voltage regulator circuit, composed of an operational amplifier and transistor M6. Feedback from the source of transistor M5 is provided to the non-inverting input of the amplifier, while the inverting input is connected to V_{REF}, which is generated from a band-gap circuit. Current mirrors M5-M0 and M2-M3 provide voltage feedback that forces the gate-to-source voltages of transistors M0 and M5 to be equal. Thus, neglecting voltage drop through column decoder shown in Fig. 1 right, reference voltage V_{REF} is applied to the reference cell too. The current mirroring ratio of both current mirrors is 10:1, which is the same used between transistors M2 and M7 to provide I_{REF} to the ramp integrator.

Fig. 4.b shows one of the n Readout Circuits. The biasing and voltage regulator circuits are equal to those previously described and apply V_{REF} to the selected cells. Moreover, the n biasing circuits share a single voltage regulator. The current switching and sign generator circuits allow to manage both the input and the weight signs. Sign cell current I_{Si}, which is the bit line current from the weight bit cell g_{Si}, is mirrored by current mirror M14-M15 and compared to reference current I_{REF}, mirrored from reference readout circuit to M13 by means of voltage V_M. The result of the current comparison is a logic signal $w_{i,sign}$ that represents the sign of w_i. When $I_{Si} < I_{REF}$, this being indicative of a positive sign, $w_{i,sign}$ voltage value is logic low. Whereas, when $I_{Si} > I_{REF}$ (i.e., a negative sign), $w_{i,sign}$ is logic high. $w_{i,sign}$ is then combined with the sign bit $x_{i,sign}$ to produce a control signal applied to transistors M10 and M11 of the current switching circuit. Thus, the sign of the multiplication, as summarized in the Table of Fig. 4.c, determines the direction of current I_i applied to the integrator.

III. CHARACTERIZATION RESULTS

The testchip includes a single 128-kB ePCM array interfaced with the AIMC unit, and it is mainly intended to validate the proposed drift compensation technique. In this first prototype, the dimension of the input and coefficient arrays is $n = 12$. Circuits have been designed with $V_{DD} = 1.2$ V and $V_{REF} = 0.3$ V, leading to PCM cells currents ranging from hundreds of nA to 10 µA. The minimum time required to perform a single MAC operation is 150 ns, and depends on the reference conductance value, as shown in (2), while the maximum output voltage ΔV_S^{MAX} is ±400 mV. According to the intention of this work, this Section presents the AIMC

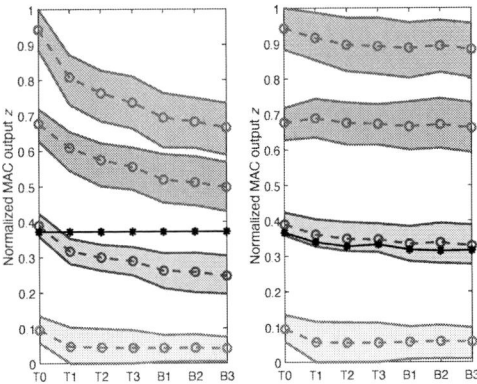

Fig. 6. Measured normalized outputs z after programming (T0), T1 = 1 day, T2 = 4 days and T3 = 7 days at room temperature, and then after B1 = 1 hour, B2 = 5 hours and B3 = 24 hours bake at 85°C. Left: constant reference current (black line); right: PCM reference current (black line). Dashed lines identify the mean measured values, while areas borders identify the minimum and the maximum.

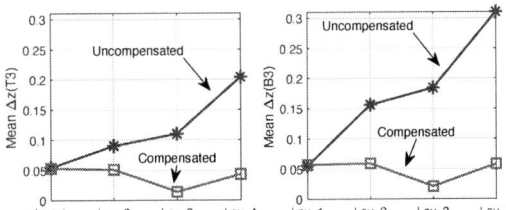

Fig. 7. Normalized mean drift error Δz as a function of the PCM cells levels, in T3 (left) and B3 (right).

performances in terms of MAC accuracy. The result of the generic operation was obtained measuring on DMA pin the output voltage ΔV_S, as defined in (3). Data were digitally converted using a 16-bit ADC available on a dedicated evaluation board. Examples of measured ΔV_S are reported in Fig. 5.b. In the following Paragraphs, results are related to the normalized MAC output z, defined as:

$$z \doteq \frac{Z}{Z^{MAX}} = \frac{\boldsymbol{w} \cdot \boldsymbol{x}}{\max\{\boldsymbol{w} \cdot \boldsymbol{x}\}}, \qquad (5)$$

which is obtained measuring $\Delta V_S / \Delta V_S^{MAX}$.

A. Evaluation of conductance time drift compensation

The drift compensation technique on individual cells has been tested evaluating a set of normalized MAC outputs depending on a single weight w_i. To this purpose, 960 PCM cells, belonging to 80 different WLs, have been programmed with a dedicated iterative algorithm [10] with four different conductance levels. Then, in accordance with (5), all but the i-th input x_i have been set to 0, whereas x_i was chosen equal to the maximum x^{MAX}; then, the normalized MAC output $z = w_i/w^{MAX}$ was measured. This operation has been repeated for all the 960 cells after T1 = 1 day, T2 = 4 days, and T3 = 7 days from initial time T0. Then, the testchip has been baked at 85°C in a controlled climate chamber to accelerate cells drift phenomena [11]. Measures have been repeated after B0 = 1 hour, B1 = 5 hours and B3 = 24 hours bake.

All measurements have been performed first with a ramp signal V_R generated with an external source constant in time, thus expecting no drift compensation (Fig. 6-left); then, V_R was generated by a PCM reference cell g_{REF} (Fig. 6-right). As expected, in the first case, normalized outputs z tend to

978-1-6654-8495-4/22 $31.00 ©2022 IEEE

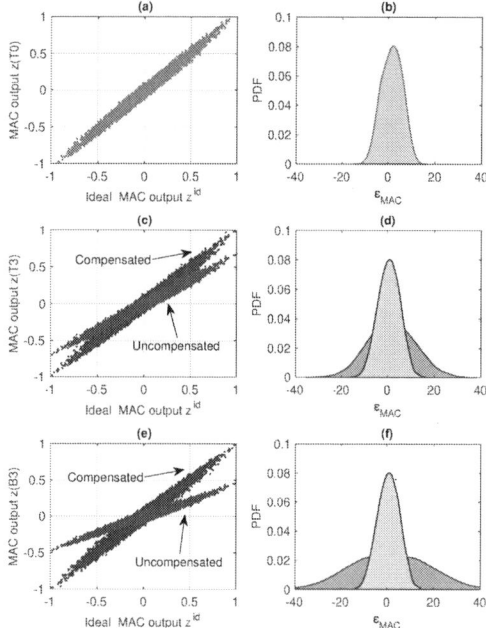

Fig. 8. Experimental results of MAC operations and PDFs of corresponding MAC error: (a-b) after programming, (c-d) after T3 = 7 days, and (e-f) after 24 hours bake at 85°C, with constant reference current (purple data) and with PCM reference (blue data).

decrease in time under the effect of cells conductance drift, which becomes even stronger after the bake. On the contrary, in the second case, the compensation mechanism successfully reduces the drop of results in time, in agreement with (4), keeping the four considered output levels widely separated. Nonetheless, the spread of MAC operations belonging to the same level is unaffected by drift compensation, as it is related to programming precision and to PCM cell-to-cell drift variability. In this example, g_{REF} was programmed to the second conductance level. Similar results were obtained when g_{REF} was programmed with the third or the fourth level. To quantify the effect of compensation, the normalized drift errors after 7 days at room temperature $\Delta z(T3) \doteq z(T0) - z(T3)$, and after 24-hours bake $\Delta z(B3) \doteq z(T0) - z(B3)$ have been calculated for each multiplication in both uncompensated and compensated case. Results of Fig. 7 show that the proposed technique keeps mean drift error under 6%, even after bake.

B. MAC accuracy

To ultimately evaluate the MAC accuracy, PCM cells of several WLs have been randomly programmed with the same conductance levels of the previous test, along with RESET state. Moreover, weights were chosen half with positive sign and half with negative sign. The employed value of g_{REF} was the same as in the previous test. Once a WL had been fully programmed, a set of 10000 different input vectors $\boldsymbol{x}$ of length $n = 12$ was applied to measure the normalized MAC outputs z. In Fig. 8.a the whole set of outputs immediately after programming is reported with dots as a function of the output $\boldsymbol{z}^{id}$; the latter is defined considering the nominal programming values of $\boldsymbol{x}$ and target values of $\boldsymbol{w}$. The distribution of error $\varepsilon_{MAC} = 100(\boldsymbol{z}^{id} - \boldsymbol{z})$ is plotted in Fig. 8.b. The error values lie between −12.17% and +13.9%, and its standard deviation $\sigma(\varepsilon_{MAC})$ is 4.44%. Thus, the AIMC unit mean accuracy, defined as $100 - \sigma(\varepsilon_{MAC})$, is equal to 95.56%. Evidently, this performance is limited by the programming algorithm finite tolerance and the circuit non-idealities.

The same operations have been repeated after T3 = 7 days, both with a constant reference current and with PCM variable reference current. Results are reported in Fig. 8.c, where all the measured normalized MAC outputs z are plotted in the two cases, together with the distributions of the corresponding MAC error (Fig. 8.d). In the uncompensated case, ε_{MAC} varies between −30.2% and +32.81%, with a standard deviation equal to 10.58%, while ε_{MAC} range is reduced to −16.02 and +15.52% and $\sigma(\varepsilon_{MAC})$ to 4.66%, when using PCM reference. The same test was finally carried out after B3 = 24 hours bake at 85°C and results are reported in Fig. 8.e and 8.f. In this case, ε_{MAC} varies from −42.42% to 41.8%, with $\sigma(\varepsilon_{MAC}) = 17.71\%$ with no compensation, while, when the PCM reference is used, ε_{MAC} and $\sigma(\varepsilon_{MAC})$ in B3 are similar to the T3 case. It is evident that the compensation feature of the AIMC unit keeps the computation error fairly invariant in both cases. Accordingly, the MAC accuracy in T3 is equal to 95.34%, and to 94.97% in B3, with a decrease of 0.22% and 0.59% only with respect to the initial one.

IV. CONCLUSION

In this paper a peripheral unit adding analog in-memory multiply and accumulate (MAC) computing function to an embedded phase-change memory (ePCM) macrocell has been presented. The unit exploits an innovative readout scheme to address non-linearity of I-V characteristic and time drift of cells conductances. The unit is conceived to operate with signed inputs and coefficients and does not require any modification to the internal structure of the ePCM. MAC operations are performed with a 1-σ accuracy of 95.56%, which is not significantly affected in time by drift effects, even after 24-hours bake at 85°C.

V. ACKNOWLEDGMENT

The Authors wish to thank Chantal Auricchio and Laura Capecchi from STMicroelectronics Italy for their fundamental contribution to the testchip design and development.

VI. REFERENCES

[1] W. Haensch, T. Gokmen, and R. Puri, "The Next Generation of Deep Learning Hardware: Analog Computing," *Proc. IEEE*, vol. 107 , 2019.

[2] G. W. Burr, B. N. Kurdi, J. C. Scott, C. H. Lam, K. Gopalakrishnan, and R. S. Shenoy, "Overview of candidate device technologies for storage-class memory," *IBM J. Res. Dev.*, vol. 52, no. 4–5, 2008.

[3] G. W. Burr *et al.*, "Neuromorphic computing using non-volatile memory," *Adv. Phys. X*, vol. 2, no. 1, Jan. 2017.

[4] C. Paolino *et al.*, "Compressed Sensing by Phase Change Memories: coping with encoder non-linearities,", *ISCAS* 2021.

[5] V. Joshi *et al.*, "Accurate deep neural network inference using computational phase-change memory," *Nat. Commun.*, vol. 11, 2020.

[6] X. Sun *et al.*, "PCM-Based Analog Compute-In-Memory: Impact of Device Non-Idealities on Inference Accuracy," *IEEE Trans. Electron Devices*, vol. 68, no. 11, 2021.

[7] M. Carissimi *et al.*, "2-Mb Embedded Phase Change Memory with 16-ns Read Access Time and 5-Mb/s Write Throughput in 90-nm BCD Technology for Automotive Applications," *ESSCIRC 2019 - IEEE 45th Eur. Solid State Circuits Conf.*, 2019.

[8] R. Khaddam-Aljameh *et al.*, "HERMES Core-A 14nm CMOS and PCM-based In-Memory Compute Core using an array of 300ps/LSB Linearized CCO-based ADCs and local digital processing," *IEEE Symp. VLSI Circuits, Dig. Tech. Pap.*, vol. 2021-June, 2021.

[9] D. Ielmini, A. L. Lacaita, and D. Mantegazza, "Recovery and drift dynamics of resistance and threshold voltages in phase-change memories," *IEEE Trans. Electron Devices*, vol. 54, no. 2, 2007.

[10] A. Antolini *et al.*, "Characterization and programming algorithm of phase change memory cells for analog in-memory computing," *Materials (Basel).*, vol. 14, no. 7, 2021.

[11] F. G. Volpe, A. Cabrini, M. Pasotti, and G. Torelli, "Drift induced rigid current shift in Ge-Rich GST phase change memories in low resistance state," *2019 26th IEEE Int. Conf. Electron. Circuits Syst. ICECS*, 2019.

978-1-6654-8495-4/22 $31.00 © 2022 IEEE

A Highly Integrated Crosspoint Array Using Self-rectifying FTJ for Dual-mode Operations: CAM and PUF

Sehee Lim[1*], Youngin Goh[2*], Young Kyu Lee[1], Dong Han Ko[1], Junghyeon Hwang[2], Minki Kim[2], Yeongseok Jeong[2], Hunbeom Shin[2], Sanghun Jeon[2], and Seong-Ook Jung[1]

School of Electrical and Electronics Engineering, Yonsei University, Seoul, Korea[1];
School of Electrical Engineering, Korea Advanced Institute of Science and Technology (KAIST), Daejeon, Korea[2].
*Equal contribution. Corresponding email: sjung@yonsei.ac.kr, jeonsh@kaist.ac.kr

Abstract—This study proposes a self-rectifying ferroelectric tunnel junction (SR-FTJ) crosspoint array which affords dual-mode operation as content-addressable memory (CAM) and physically unclonable function (PUF). The PUF mode operation of the proposed SR-FTJ crosspoint array does not require any temporary transfer of stored CAM data to buffer, which prevents additional area overhead or potential data security threats. Experimental measurements verify the feasibility of the dual-mode operation. Further simulations using the SR-FTJ model reflecting measured characteristics show that the proposed SR-FTJ crosspoint array reduces area by at least 88.1 % up to 97.4 % compared to the previous structures that implement individual CAM and PUF. In addition, the proposed SR-FTJ crosspoint array achieves the lowest search energy (2.05 fJ/search/bit) and the highest randomness (Hamming weight of 0.5000) among the structures.

Keywords—*Content-addressable memory, crosspoint array, dual-mode operation, physically unclonable function, self-rectifying ferroelectric tunnel junction.*

I. INTRODUCTION

As the fourth industrial revolution has drastically increased data transfer among Internet of Things (IoT) devices, existing von Neumann architecture encountered speed and energy limits derived from memory bottlenecks. In-memory-computing has been developed to resolve memory bottlenecks by processing data entirely in memory without transferring data to a processing unit. Content-addressable memory (CAM) is one of the in-memory-computing techniques specialized for high-speed searching for cache controllers or routers. The recent development of data processing has explored new CAM applications such as neuromorphic associative memory, pattern matching, and reconfigurable computing [1], however, there exist security issues derived from unconstrained access by the public [2]. This security issue requires IoT devices to perform additional cryptographic functions such as a physically unclonable function (PUF). Since IoT devices process large amounts of data in limited hardware, high area efficiency is demanded in circuit designing for IoT devices. Although there have been several attempts to improve the density of CAMs [3], [4] and PUFs [5], [6] using emerging memory devices, they focused on reducing the cell area of CAM or PUF individually, leading to no significant increase in the area efficiency.

This paper proposes a self-rectifying ferroelectric tunnel junction (SR-FTJ) crosspoint array with dual-mode operations for the first time. The proposed SR-FTJ crosspoint array can satisfy the stringent area requirements of IoT devices by implementing multiple functions in an array, different from previous CAMs [3], [4] and PUFs [5], [6], [7]

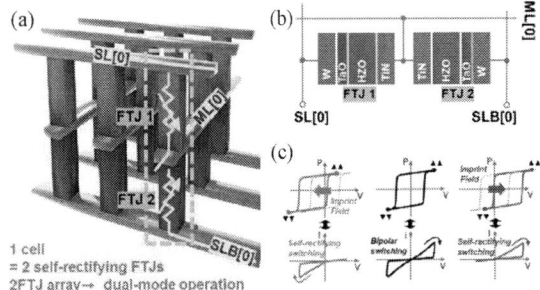

Fig. 1. (a) 3D schematic diagram of 2 SR-FTJ cell based crosspoint array. (b) Device structure of 1 cell in which 2 SR-FTJs are connected in series. (c) Different from conventional FTJ (in black), imprinted FTJs (in red and blue) show self-rectifying switching.

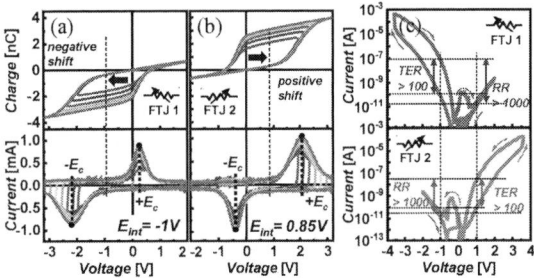

Fig. 2. Measured charge-voltage (Q-V) hysteresis curve and corresponding current-voltage (I-V) curve with a different voltage range of (a) FTJ 1 and (b) FTJ 2. Asymmetry hysteresis loops are obtained due to internal bias. (c) Measured DC I-V curve of FTJ 1 and 2 showing a self-rectifying switching with opposite rectifying direction.

that have an insuperable limit on improving area efficiency due to the individual implementation. The experimental results verify the feasibility of the dual-mode operations, and the performance evaluation using the simulation results shows area reduction by at least 88.1 % up to 97.4 %, the lowest search energy of 2.05 fJ/search/bit, and the highest randomness with Hamming weight of 0.5000 compared to previous CAMs and PUFs.

II. PROPOSED SR-FTJ CROSSPOINT ARRAY

A. SR-FTJ Characteristics and Array Composition

HfO$_2$-based FTJ has been developed as a promising nonvolatile memory device with high CMOS compatibility, scalability, low power, and non-destructive readout. The FTJ consists of an ultra-thin ferroelectric layer sandwiched between electrodes, which is 2-terminal based metal-ferroelectric-metal structure. In FTJ, the polarization reversal

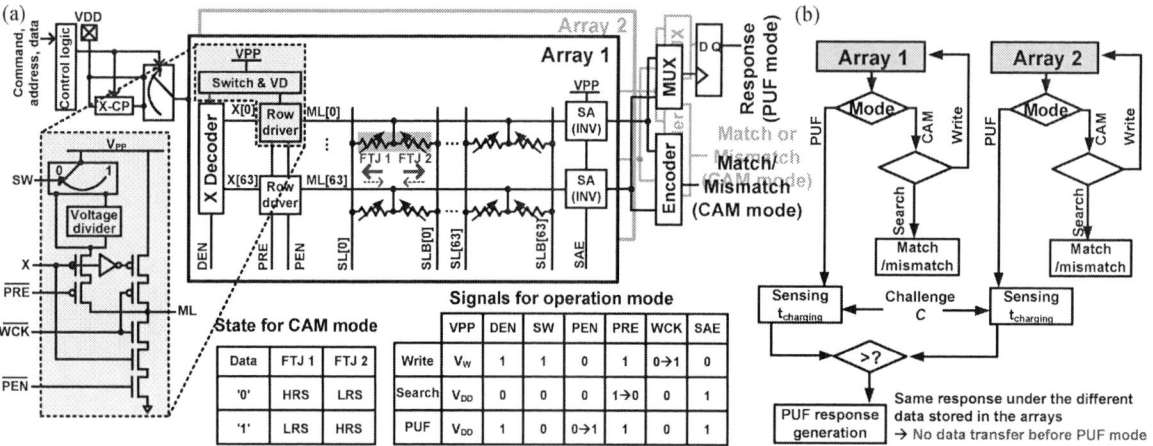

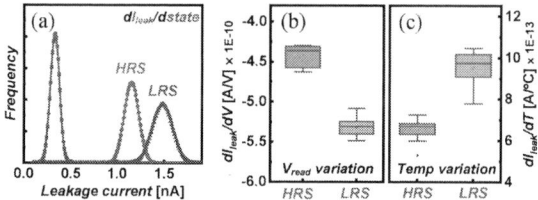

Fig. 3. (a) Overall structure of the proposed SR-FTJ crosspoint array and the peripheral circuits with the SR-FTJ state and signal information for CAM/PUF modes. A cell indicated in blue has two SR-FTJs, and on/off current and leakage current (I_{leak}) flowing through the two SR-FTJs are expressed in thick and dotted arrows, repectively. (b) Operational flowchart for showing the concept of the dual-mode operation.

caused by an electric field applied through two electrodes modulates the potential barrier, inducing a tunneling electroresistance (TER) effect. The FTJ-based crosspoint arrays with $4F^2$ memory cell size have an advantage with high-density integration compared to 3-terminal devices-based array. However, this crosspoint array indispensably requires FTJ with rectifying properties to suppress the sneak current paths. In addition, for dual-mode operation as CAM and PUF, self-rectifying properties in FTJ are necessary.

This paper proposes a 2 SR-FTJ-based cell that stacks W/TaO/HZO/TiN and TiN/HZO/TaO/W for dual-mode operation. Fig. 1(a) shows a 3D schematic diagram of the proposed SR-FTJ crosspoint array with 2 SR-FTJ based cells, search-line (SL), search-line bar (SLB), and match-line (ML). Two symmetrically 3D stacked SR-FTJs (FTJ 1 and FTJ 2) in a cell show opposite rectifying directions as described in Fig. 1(b). In this stack, the TaO layer which acts as a positive fixed charge layer generates a large imprint field, inducing a highly shifted hysteresis curve and intrinsic rectifying resistive switching properties [8]. As shown in Fig. 1(c), the P-V curve is shifted in a negative direction when TaO is inserted at a top interface (FTJ 1), and the P-V curve is shifted in a positive direction when TaO is inserted at a bottom interface (FTJ 2).

Fig. 2(a) and (b) show measured charge-voltage (Q-V) loops and corresponding current-voltage (I-V) curves of FTJ 1 and FTJ 2. It is verified that the Q-V and I-V curves of FTJ 1 are horizontally shifted in a negative direction, while those of FTJ 2 are shifted in a positive direction. As a result, FTJ 1 and 2 represent self-rectifying switching behavior (on/off ratio >100, rectifying ratio >1000) with opposite rectifying directions (Fig. 2(c)).

B. CAM Mode Operation

Fig. 3(a) shows the overall structure of the proposed SR-FTJ crosspoint array with its operational information. The proposed SR-FTJ crosspoint array performs the dual-mode operation, as shown in Fig. 3(b): 1) writes data and searches the stored data in CAM mode and 2) generates reliable cryptographic keys regardless of the stored CAM data in PUF mode. FTJ 1 and FTJ 2 in a cell storing CAM data of '0' ('1') are in high resistance state (HRS) (low resistance state (LRS)) and LRS (HRS), respectively. The peak-to-peak

Fig. 4. (a) $dI_{leak}/dstate$ and I_{leak} of measured SR-FTJs in HRS and LRS. Box plot of (b) dI_{leak}/dV and (c) dI_{leak}/dT of measured SR-FTJs in HRS and LRS. The measurements of SR-FTJs in HRS and LRS are indicated as red and blue, respectively.

voltage (V_{PP}) used for the write operation is the write voltage (V_W). The write operation is composed of two steps: At the 1st step, the ML of the selected row is driven to 0. SL and SLB are driven to V_W (0) and 0 (V_W), respectively, to write '0' ('1'). Then, the voltage difference between two terminals of SR-FTJs connected to SL or SLB of V_W becomes $-V_W$, writing the SR-FTJs into HRS. At the 2nd step that drives the ML of the selected row to V_W, the rest of the SR-FTJs in the selected row are written into LRS by the voltage difference of $+V_W$. The MLs of the unselected rows are fixed to $V_W/2$ during the write operation to prevent write disturbance.

Before the search operation, all MLs are precharged to the supply voltage (V_{DD}). Then, SL and SLB are driven to 0 (V_{DD}) and V_{DD} (0) to search '0' ('1'), respectively. If there is any mismatched bit, SR-FTJs in LRS are positively biased, which turns on the SR-FTJs to discharge ML. In the case of a match, only I_{leak} flows through all SR-FTJs in LRS due to the negative voltage bias, while other SR-FTJs are turned off under the positive voltage bias because they are in HRS. Thus, the ML voltage is maintained as V_{DD}, which is sensed by an encoder to output the address of the row storing matched data.

C. PUF Mode Operation

PUF is a cryptographic function that generates a random output (response) according to an input (challenge) using intrinsic device-to-device (D2D) variations. With a challenge that selects a row each in two SR-FTJ crosspoint arrays, the MLs of the selected rows are pre-discharged to 0. Driving all SL/SLB to V_{DD} induces leakage current (I_{leak}) of all SR-FTJs in the selected row, which gradually charges the selected ML. At that time, MLs of the unselected rows are driven to V_{DD} to

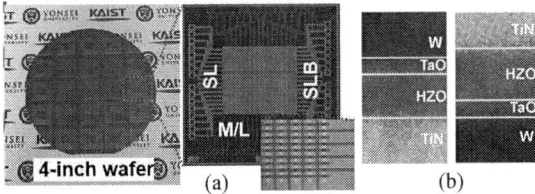

(a)

(b)

Fig. 5. (a) 4-inch wafer-level demonstration and optical microscope image of 16 × 16 SR-FTJ crosspoint array. (b) HRTEM image of W/TaO/HZO/TiN stacked FTJ 1 and TiN/HZO/TaO/W stacked FTJ 2.

avoid unnecessary power consumption. The charging time ($t_{charging}$) depends on the intrinsic D2D variations of the sum of I_{leak} and ML capacitance in a row. The digital response is determined by comparing the mismatch of $t_{charging}$ between the two selected rows. If $t_{charging}$ of the selected row in array 1 is faster (slower) than that in array 2, the response is 0 (1).

Different from on/off current having a significant difference according to the state of SR-FTJ, I_{leak} slightly increases as the SR-FTJ switches from HRS to LRS ($dI_{leak}/dstate$) with a negligible D2D variation, as shown in Fig. 4(a). Suppose that the sum of I_{leak} of all SR-FTJs in a row is $\sum I_{leak}$, and all SR-FTJs show the same $dI_{leak}/dstate$. Since the number of SR-FTJs in HRS and LRS in a row is always the same regardless of the stored CAM data, the increase and decrease in $dI_{leak}/dstate$ caused by SR-FTJs switching from HRS to LRS and from LRS to HRS, respectively, are canceled out even if the stored CAM data changes. Thus, $\sum I_{leak}$ remains the same irrelevant to the stored CAM data, enabling the dual-mode operation without any data transfer before PUF mode.

I_{leak} is also a suitable PUF variation source due to its robustness under wide environmental variations. The changes in I_{leak} according to voltage and temperature (VT) (dI_{leak}/dV and dI_{leak}/dT) show insignificant D2D variations under wide VT variations, as shown in Fig. 4(b) and (c). The constant changes in I_{leak} prevent the mismatch of $\sum I_{leak}$ between two selected rows from being flipped, leading to a low bit error rate (BER) under wide VT variations. The low BER is important for PUFs which must generate constant responses over a wide range of environmental conditions for successful authentication. To further improve stability, several error-prone challenge-response pairs (CRPs) are discarded by iteratively generating responses while changing the stored CAM data during the enrollment phase.

III. SR-FTJ CROSSPOINT ARRAY EVALUATION

A. Evaluation Settings

1) Experimental Settings: To verify the functionality of the proposed SR-FTJ crosspoint array, we fabricated a highly scalable SR-FTJ crosspoint array on a 4-inch wafer, as shown in Fig. 5(a). Fig. 5(b) shows the TEM image of the fabricated SR-FTJ. In this structure, a 100 nm-thick W bottom electrode with a low thermal expansion coefficient acts as a tensile stressor for effective crystallization. A 1 nm-thick TaO is introduced as a positive fixed charge layer to generate a built-in field. A 5.5 nm-thick $Hf_{0.7}Zr_{0.3}O_2$ is prepared as a tunneling barrier and a 100 nm-thick TiN top electrode is employed as a capping layer. The SR-FTJs are programmed into LVT by applying 4 V for 10 ns and erased into HVT by applying –4 V for 10 ns.

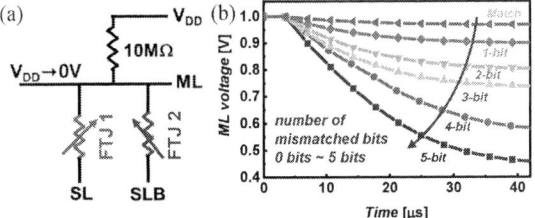

Fig. 6. (a) Operation of a 2 SR-FTJ based cell. (b) The ML voltage according to the number of mismatched bits.

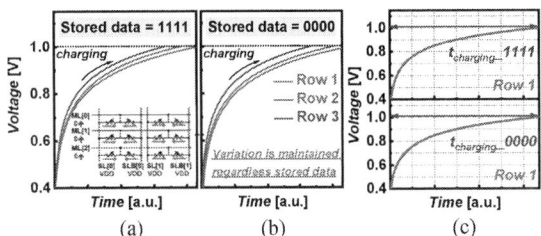

Fig. 7. Waveform of fabricated 4 × 3 SR-FTJ crosspoint array in PUF mode with all rows storing (a) 1111 and (b) 0000. (c) Comparison of $t_{charging}$ between 1111 and 0000.

2) Simulation Settings: For an extensive analysis of a large-sized array, we performed Monte-Carlo simulations using industry-compatible 180 nm technology with an SR-FTJ model reflecting the mean and sigma value of measured characteristics such as I_{on}, I_{off}, I_{leak}, dI_{leak}/dV, dI_{leak}/dT, and $dI_{leak}/dstate$. Note that our current experimental set-up can afford the fabrication and measurement of SR-FTJs sizing at least 30 × 30 μm² so that the characteristics of the smaller-sized SR-FTJs for the simulations are estimated through the scaling factor obtained by measuring the fabricated SR-FTJs in different sizes. The simulations are based on 64 × 64 arrays with V_W of 4 V. Parasitic resistance and capacitance on the metal lines are included in the simulations.

B. Experimental Demonstration

With the fabricated SR-FTJ crosspoint array based on 2 SR-FTJ cells shown in Fig. 6(a), the search operation is performed with varying the number of mismatched bits. Fig. 6(b) shows ML voltage remains as V_{DD} at the match and is discharged at the mismatch. ML discharging rate increases with the number of mismatched bits due to the increased discharge current.

Fig. 7(a) and (b) show the ML waveform of the fabricated 4 × 3 SR-FTJ crosspoint array with all rows storing 1111 and 0000, respectively. Rows 1, 2, and 3 show different $t_{charging}$ due to the intrinsic D2D variations of I_{leak} in SR-FTJs composing each row, and $t_{charging}$ is entirely consistent regardless of the stored CAM data, as shown in Fig. 7(c). The consistency of $t_{charging}$ demonstrates the feasibility of the proposed SR-FTJ crosspoint array for PUF applications without impinging on the stored CAM data.

C. Simulation Results

We evaluate the uniqueness and stability of the proposed SR-FTJ crosspoint array using inter Hamming distance (HD_{inter}) and BER, respectively. Fig. 8(a) shows the mean HD_{inter} of the proposed SR-FTJ crosspoint array, which is close to the ideal 50 %. Fig. 8(b) shows BER under VT variations based on the response generated at the nominal

978-1-6654-8495-4/22 $31.00 © 2022 IEEE

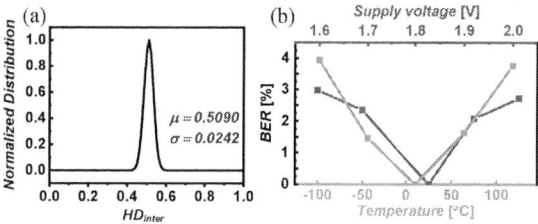

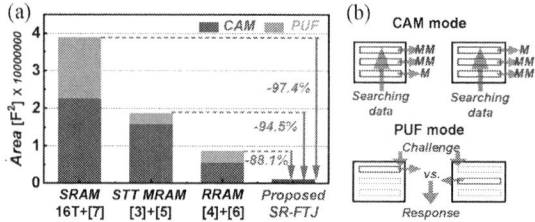

Fig. 8 (a) HD$_{inter}$ and (b) BER under supply voltage and temperature variations of the proposed SR-FTJ crosspoint array.

Fig. 9 (a) Comparison of area based on the estimation when the bits of stored data and generated responses of the CAMs and PUFs are the same with those in ten 64 x 64 SR-FTJ crosspoint arrays considering discarded CRPs. (b) The activated rows of the proposed SR-FTJ crosspoint arrays for CAM/PUF mode operations. M and MM refers to match and mismatch, respectively.

condition with a V$_{DD}$ of 1.8 V and a temperature of 25 °C.

Fig. 9(a) compares the area when the number of stored and generated bits of the previous CAMs and PUFs is the same as that of ten 64 x 64 SR-FTJ crosspoint arrays. The compared structures are composed of CAMs and PUFs using the same memory devices: SRAM, spin-transfer torque magnetoresistive RAM (STT MRAM), and resistive RAM (RRAM). Owing to the distinctive feature that an array is used for multiple purposes as CAM data storage and PUF response generation, as shown in Fig. 9(b), the proposed SR-FTJ crosspoint array reduces area by at least 88.1 % up to 97.4 % compared to the previous structures.

Table I summarizes the comparison of the proposed SR-FTJ crosspoint array with previous CAMs and PUFs. The proposed SR-FTJ crosspoint array shows the smallest cell area and area/CRP due to the SR-FTJs stacked in a 3D direction without any diode or selector. The smallest cell area leads to the lowest search energy due to the small capacitance of signal lines. Hamming weight of the proposed SR-FTJ crosspoint array is closest to the ideal 50 %, implying the highest randomness. Although the BER of the proposed SR-FTJ crosspoint array is worse than the previous nonvolatile memory-based PUFs, the BER can be lowered further by using majority voting or error correction code which are widely used techniques for PUF stability.

IV. CONCLUSION

This paper proposes an SR-FTJ crosspoint array targeted at the hardware-restricted IoT devices which require both a fast address lookup and a high level of security. We stacked 2 SR-FTJs vertically in a crosspoint array for ultra-high density. In addition, we exploited the leakage current of SR-FTJs as a PUF variation source based on the measurement results of the fabricated SR-FTJs. The constant difference in the leakage current according to the state of SR-FTJ enables the dual-mode operation without disturbing the stored CAM

TABLE I. PERFORMANCE COMPARISON

		SRAM 16T+[7]	STT MRAM [3]+[5]	RRAM [4]+[6]	This work
CAM	Dual-mode	N	N	N	Y
	Nonvolatility	N	Y	Y	Y
	Cell area [F²]	553.09	387.65	138.27	**25**
	Search energy* [fJ/search/bit]	13.53	-	9.60	**2.05**
PUF	Area/CRP [F²]	338–970	172.07	142.25–179.17	**50**
	HD$_{inter}$	0.481–0.495	-	0.4999	0.5090
	Hamming weight	-	0.498	0.5001	**0.5000**
	Condition	0.5–0.9V 0–80°C	0.8–1.2V -25–75°C	1.4–2.2V 2.5–5.5V -25–75°C	**1.6V–2.0V -100–125°C**
	BER [%]	3.17 (native)	0.04 (worst T)	0 (worst VT)	3.94 (worst VT)
Normalized Area**		x37.82	x18.29	x8.43	**x1**

*Energy is normalized based on industry-compatible 180nm technology.
**Area is normalized considering discarded CRPs in the proposed SR-FTJ crosspoint array.

data. Comprehensive evaluations with experimental and simulation results demonstrate that the proposed SR-FTJ crosspoint array is a promising solution for IoT devices.

ACKNOWLEDGMENT

This work was supported by National Research Foundation of Korea (NRF) grant funded by the Korea government (MSIT) (No. 2020M3F3A2A01081918). The EDA tool was supported by the IC Design Education Center (IDEC), Korea.

REFERENCES

[1] R. Karam, R. Puri, S. Ghosh, and S. Bhunia, "Emerging trends in design and applications of memory-based computing and content-addressable memories," *Proceedings of the IEEE*, vol. 103, issue. 8, pp. 1311-1330, Aug. 2015.

[2] B. Liu, C. Yang, H. Li, Y. Chen, Q. Wu, and M. Barnell, "Security of neuromorphic systems: challenges and solutions," *IEEE International Symposium on Circuits and Systems (ISCAS)*, May, 2016.

[3] S. Matsunaga, S. Miura, H. Honjou, K. Kinoshita, S. Ikeda, T. Endoh, H. Ohno, and T. Hanyu, "A 3.14 µm² 4T-2MTJ-cell fully parallel TCAM based on nonvolatile logic-in-memory architecture," *Symposium on VLSI Circuits*, Jun. 2012.

[4] C.-C. Lin, J.-Y. Hung, W.-Z. Lin, C.-P Lo, Y.-N. Chiang, H.-J. Tsai, G.-H. Yang, Y.-C. King, C. J. Lin, T.-F. Chen, and M.-F. Chang, "A 256b-wordlength ReRAM-based TCAM with 1ns search-time and 14x improvement in wordlength-energyefficiency-density product using 2.5T1R cell," *IEEE International Solid-State Circuits Conference (ISSCC)*, Jan.-Feb. 2016.

[5] S. Lim, B. Song, and S.-O. Jung, "Highly independent MTJ-based PUF system using diode-connected transistor and two-step postprocessing for improved response stability," *IEEE Transactions on Information Forensics and Security*, vol. 15, pp. 2798-2807, Mar. 2020.

[6] B. Lin, Y. Pang, B. Gao, J. Tang, D. Wu, T.-W. Chang, W.-E. Lin, X. Sun, S. Yu, M.-F. Chang, H. Qian, and H. Wu, "A highly reliable RRAM physically unclonable function utilizing post-process randomness source," *IEEE Journal of Solid-State Circtuis*, vol. 56, no. 5, pp. 1641-1650, Jan. 2021.

[7] S. Jeloka, K. Yang, M. Orshansky, D. Sylvester, and D. Blaauw, "A sequence dependent challenge-response PUF using 28nm SRAM 6T bit cell," *Symposium on VLSI Circuits*, Jun. 2017.

[8] Y. Goh, J. Hwang, M. Kim, M. Jung, S. Lim, S.-O. Jung, and S. Jeon, "High performance and self-rectifying hafnia-based ferroelectric tunnel junction for neuromorphic computing and TCAM applications," *IEEE International Electron Devices Meeting (IEDM)*, Dec. 2021.

[9] K. Ni, et al., "Ferroelectric Ternary Content-Addressable Memory for One-Shot Learning," *Nature Electronics*, vol. 2, pp. 521-529, Nov. 2019.

Physical Implementation of a Tunable Memristor-based Chua's Circuit

Manuel Escudero[*], Sabina Spiga[*], Mauro di Marco[†], Mauro Forti[†], Giacomo Innocenti[‡],
Alberto Tesi[‡], Fernando Corinto[§] and Stefano Brivio[*]

[*]CNR—IMM, Unit of Agrate Brianza, Agrate Brianza, 20864, Italy
[†]Università degli Studi di Siena, Siena, 53100, Italy
[‡]Università degli Studi di Firenze, Florence, 50121, Italy
[§]Politecnico di Torino, Turin, 10129, Italy

Abstract—Nonlinearity is a central feature in demanding computing applications that aim to deal with tasks such as optimization or classification. Furthermore, the consensus is that nonlinearity should not be only exploited at the algorithm level, but also at the physical level by finding devices that incorporate desired nonlinear features to physically implement energy, area and/or time efficient computing applications. Chaotic oscillators are one type of system powered by nonlinearity, which can be used for computing purposes. In this work we present a physical implementation of a tunable Chua's circuit in which the nonlinear part is based on a nonvolatile memristive device. Device characterization and circuit analysis serve as guidelines to design the circuit and results prove the possibility to tune the circuit oscillatory response by electrically programming the device.

Index Terms—Memristor, oscillators, chaos, Chua's circuit, nonlinear systems

I. Introduction

The continuous growth in terms of performance and energy efficiency of von-Neumann centralized computer architectures is reaching its limit, which is especially challenging for tasks that deal with huge amounts of data, such as classification or optimization. New paradigms are taking advantage of different forms of analog nonlinear dynamics as an engine for computation, mainly with a two-fold approach: (i) dynamic features operating as working memory, usually referred as local computation or in-memory computing and aiming for massive parallel processing, and (ii) computational performance improvement due to the complexity of nonlinear systems [1].

The focus of this work lies on the Chua's circuit, an autonomous oscillator that exhibits a plethora of behaviors, including chaotic oscillations [2], which can be used as the basis for novel nonlinear computing systems, like oscillator-based computing [3] and reservoir computing [4]. For this circuit to show such complex oscillations, a nonlinear element is required, which usually is built using operational amplifiers or diodes [5]. However, these solutions do not offer configurability and technological scalability at the same time.

The possibility of exploiting memristor nonlinearity of their I-V characteristics in the Chua's circuit has already been considered, e.g. [6]. Memristors are two terminal devices with nonlinear conduction mechanisms holding an internal state, e.g. its resistance, that depends on its previous history. Mem-

ristive devices are attractive due to their highly scalable, non-volatile memory properties and different dynamical features they can provide. However, to our best knowledge, there has yet not provided experimental proof of a Chua's circuit implementation using real memristive devices.

In this paper, a physical hardware implementation of a memristor-based Chua's circuit is demonstrated. Thanks to the tunable nonlinearity of the I-V characteristics, we demonstrate bifurcations and chaos, observed from experimental circuit responses. Our results are achieved by carefully matching both the requirements of the circuit and of the device characteristics.

II. Tunable Chua's Circuit Enabled by Memristor Nonlinearity

A. Memristor characteristics

The nonvolatile memristive devices used in this work are based on a Pt/HfO$_2$/TiN stack prepared as already discussed in [7]. The devices are measured through a B1500 Keysight semiconductor parameter analyzer. Fig. 1 reports a representative switching operation. A high negative voltage produces the creation of a first filament in a fresh device (not shown) [7], which initiates the bipolar switching with RESET (SET) operations at positive (negative) voltages. A compliance current at 1 mA is required to prevent devices breakdown. RESET operation partially dissolves the formed filament (high resistance state, HRS) and SET operations reinstate it (low resistance state, LRS) under a compliance current control. The resistance states are retained for more than 10 years and can be programmed for thousands of cycles [8] [9]. The HRS resistance value can be controlled by the maximum voltage applied during the RESET sweep (V$_{STOP}$). We notice that the higher the HRS resistance the higher the SET transition voltage (V$_{SET}$, vertical transition) in the negative polarity.

The HRS current-voltage ($i_M - v_M$) characteristics is nonlinear and is well described by a 5th order polynomial from V$_{SET}$ to V$_{STOP}$, as

$$i_M = p_1 v_M + p_2 v_M^2 + ... + p_5 v_M^5 \qquad (1)$$

B. Proposed Chua's circuit

Fig. 2(a) depicts the original Chua's circuit [2], containing different linear passive elements as well as the nonlinear block

978-1-6654-8495-4/22 $31.00 © 2022 IEEE

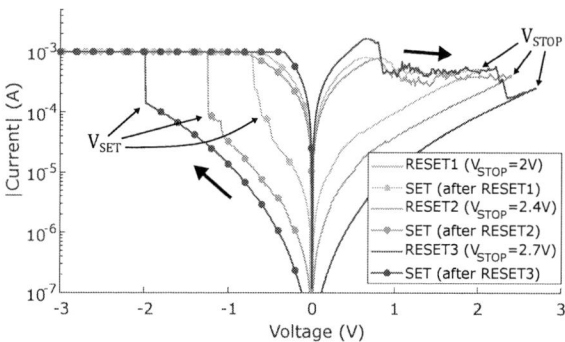

Fig. 1. Pt/HfO$_2$/TiN device switching operation, displaying three RESET/SET cycles using V_{STOP} = 2, 2.4 and 2.7 V.

with a piecewise linear voltage-dependent current i_R of Fig. 2(b). The state of this circuit is defined by the three state variables v_1, v_2 and i_L. Three equilibrium points exist, labelled as P_0, P_+ and P_-, which for the variable v_1 are defined as solutions of

$$i_R(v_1) + Gv_1 = 0 \qquad (2)$$

or graphically found at the crossing of i_R with the load line $-G$, as shown in 2(b). Note that given such i_R, P_+ and P_- only exist if

$$di_R/dv(0) < -G \qquad (3)$$

where $di_R/dv(0)$ is the nonlinear block conductance at P_0.

The circuit shows a variety of trajectories, i.e. the evolution of the state variables through time, on the basis of the values of the circuit parameters and stability of the equilibrium points: periodic and nonperiodic oscillations around P_+ or P_-, or nonperiodic oscillations around both P_+ and P_- (double-scroll, shown in Fig. 2(c)).

The memristor-based Chua's circuit in our study is shown in Fig. 2(d), where the only difference is found in the nonlinear block. This block is composed of two elements connected in parallel: the nonvolatile device, from now on referred to as memristor, and an operational amplifier implementing a negative resistor $-R_N$. These elements contribute with a current i_M from (1) and $i_N = -G_N v_R$, as depicted in Fig. 2(e). Summing both currents in the parallel association, the total current i_R is depicted in Fig. 2(f). Smooth nonlinear functions with similar characteristics to i_R have been already confirmed to show chaotic attractors when used in the Chua's circuit [10].

The complete memristor-based Chua's circuit can be described as a dynamical system using the state equations

$$\begin{aligned} \frac{dv_1}{dt} &= \frac{1}{C_1}\left((v_2 - v_1)G - i_R(v_1)\right) \\ \frac{dv_2}{dt} &= \frac{1}{C_2}\left((v_1 - v_2)G + i_L\right) \\ \frac{di_L}{dt} &= -\frac{v_2}{L} \end{aligned} \qquad (4)$$

where the different parameters are reported in Fig. 2(d). As in the original Chua's circuit, three equilibrium points exist given i_R shown in 2(f) if (3) is satisfied (in this case

$di_R/dv(0) = p_1 - G_N$); defined by (2), P_0 is at $v_1 = 0$ V, while P_+ and P_- depend on the memristor i-v characteristics, G and G_N. The circuit trajectory is expected to be similar to the one in Fig. 2(c). It is worth noting that this implementation does not only benefit from the memristor nonlinear characteristics but the nonlinear block can be electrically tuned by programming the memristor to different HRS, as suggested in Fig. 2(e) and 2(f). In this case, it is expected that if the memristor resistance state is decreased, P_+ and P_- shift towards P_0, as shown in Fig. 2(f).

III. CIRCUIT DESIGN CONSIDERING PHYSICAL MEMRISTOR CONSTRAINTS

In this section, we present a design approach that matches device features and circuit requirements.

The circuit needs a static nonlinear characteristics, that the memristor can provide as long as v_1 (the voltage across the memristor) is not extended below $-V_{SET}$ or above V_{STOP}. From Fig. 1 it is clear that $|V_{SET}| < V_{STOP}$, thus V_{SET} is the most limiting value. For this reason, V_{SET} has been characterized by programming a single device using a progressive RESET procedure, in which successive RESET operations increasing the voltage are applied until reaching a target V_{STOP}. After each progressive RESET, a SET operation is carried out to capture V_{SET}. A single device is programmed with progressive RESET with V_{STOP} values from 1.5 to 2.7 V and repeated 10 times to account for the intrinsic memristor cycle-to-cycle variability. Results in fig. 3(a) show how $|V_{SET}|$ generally increases with the used V_{STOP}. Moreover, the memristor resistance state increases with V_{STOP}, where R_{PROG} is the memristor resistance measured at 0.1 V after applying the progressive RESET. Each programmed state is fitted with (1) and the fitting parameters p_i are depicted in 3(b) as a function of R_{PROG}. It is worth noting that the linear term parameter p_1 is a good estimator of the memristor conductance at low voltage. As R_{PROG} increases, p_1 becomes comparable with the nonlinear terms parameters. As a result, it is convenient to program the device to higher resistance states to achieve both high $|V_{SET}|$ and nonlinear characteristic.

Next, the circuit is designed to ensure operation inside the voltage range supported by the memristor and at the same time to exhibit oscillatory behavior. Specifically, we look for the conditions to reproduce a double-scroll attractor analogous to the scenario presented in [2]. Oscillations are guaranteed as long as the three equilibrium points in the system are unstable. At P_0, the instability is guaranteed if the trace of the Jacobian matrix of (4) evaluated at this point is forced to zero, which occurs only if

$$G_N = G + \frac{C_1}{C_2}G + p_1 \qquad (5)$$

Note that (5) implies (3). Additionally, it can be easily observed that P_+ and P_- are also unstable when (5) is used, again by analyzing the Jacobian matrix trace evaluated at P_+ and P_-. In conclusion, G_N calculated from (5) guarantees the three equilibrium points existence and their instability.

978-1-6654-8495-4/22 $31.00 © 2022 IEEE

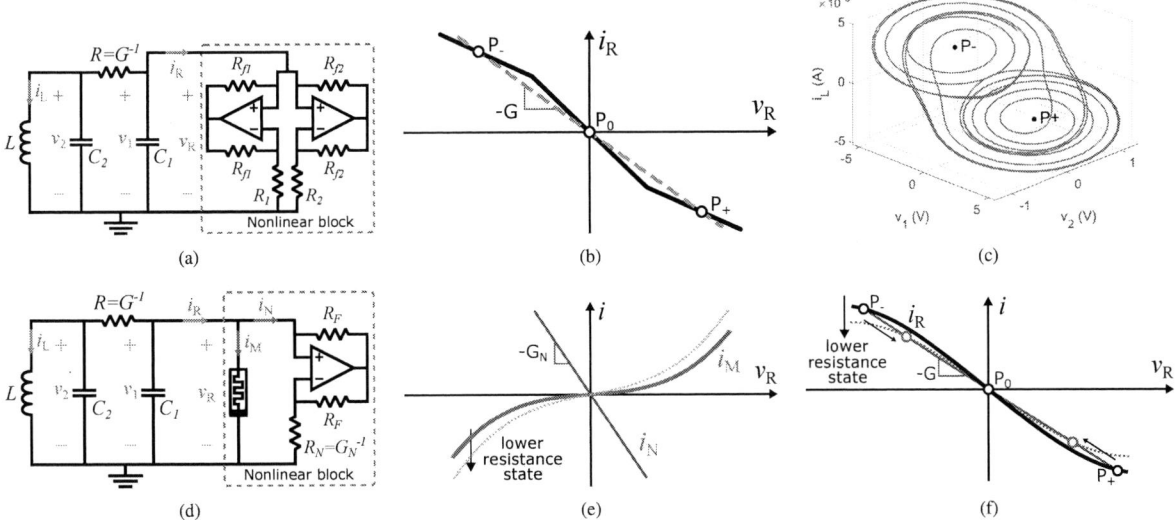

Fig. 2. Original Chua's circuit (a) scheme, (b) detailed nonlinear block i-v characteristic and (c) example of a trajectory of a double-scroll attractor in the state space. Proposed memristor-based Chua's circuit (d) scheme, (e) memristor i-v characteristics at two different HRS extracted from measurement data (in blue) and the negative impedance converter (in red), and (f) complete nonlinear block i-v characteristic for both resistance states.

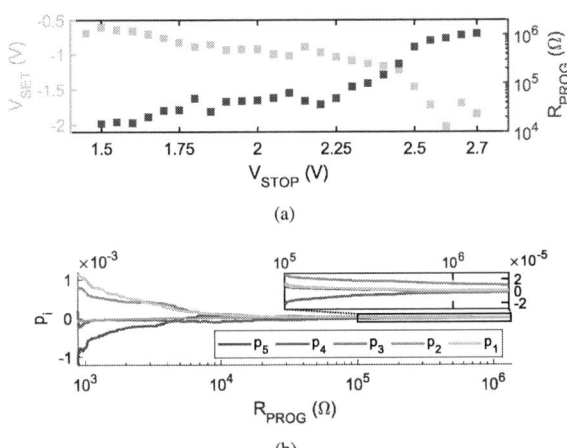

(a)

(b)

Fig. 3. (a) Average V_{SET} and resistance measured at 0.1 V R_{PROG} after performing a progressive reset operation according to different used V_{STOP}. (b) Average fitting parameters for different R_{PROG}.

Now, by combining (2) and (5),

$$G = \frac{C_2}{C_1} \left[\frac{i_R(v_{eq})}{v_{eq}} - p_1 \right] \qquad (6)$$

where v_{eq} is the absolute value of the v_1 coordinate for P_+ and P_-. This last expression is particularly useful, because it allows to design G (and G_N) by fixing P_+ and P_- position, which are in turn related with the amplitude of the trajectory in v_1 dimension.

TABLE I

TABLE I. MEMRISTOR I-V CHARACTERISTICS FITTING PARAMETERS ($V_{STOP} = 2.6$ V) USED IN THE DESIGN AND IMPEDANCES CALCULATED

p_1	$1.91 \cdot 10^{-6}$	p_2	$3.11 \cdot 10^{-7}$	p_3	$1.91 \cdot 10^{-5}$
p_4	$-5.20 \cdot 10^{-6}$	p_5	$1.77 \cdot 10^{-6}$	v_{eq}	0.9 V
R	7.643 kΩ	R_N	6.856 kΩ	L	410 mH
C_1	10 nF	C_2	100 nF		

Finally, C_1, C_2 and L values influence the attractor exhibited by the circuit. Guided by the bifurcation analysis in [2], a double-scroll attractor can be achieved by fixing the adimensional parameters of Chua's circuit to $\alpha = 10$ and $\beta = 14.22$, where

$$\alpha = \frac{C_2}{C_1} \quad , \quad \beta = \frac{R^2 C_2}{L} \qquad (7)$$

Following the previous device physical constraints and previous exposed conditions, the Chua's circuit has been designed by programming a memristor with $V_{STOP} = 2.6$ V. More especifically, component values C_2, L, G and G_N are calculated with (5)-(7) using the fitting parameters of the programmed memristor i-v characteristics, fixing P_+ and P_- at $v_{eq} = 0.9$ V (lower than $|V_{SET}|$ at $V_{STOP} = 2.6$ V, as shown in Fig. 3(a)) and choosing C_1 to 10 nF. Both fitting parameters and circuit impedances values used are shown in Table I.

IV. PHYSICAL IMPLEMENTATION AND EXPERIMENTAL VALIDATION

The previously designed Chua's circuit has been built, connected to the memristor and the state variables are captured.

978-1-6654-8495-4/22 $31.00 © 2022 IEEE

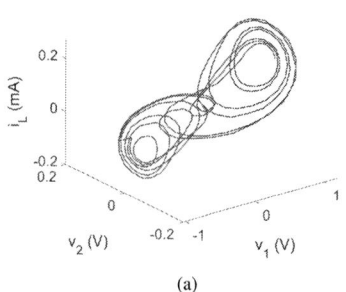
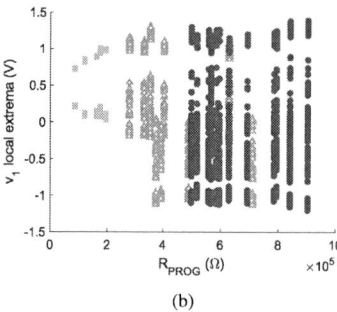
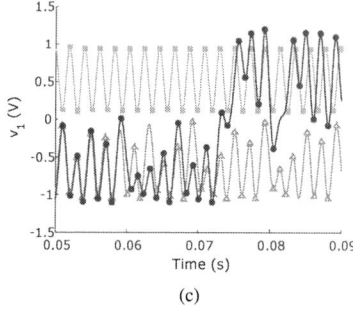

(a)   (b)   (c)

Fig. 4. Experimental validation results of the tunable memristor-based Chua's circuit. (a) Trajectory obtained after designing the circuit to show a double-scroll attractor. (b) All trajectories obtained after each programming operation; only the local extrema of v_1 are shown, coloured according to the types of identified trajectories: double-scroll attractor in red, single-scroll attractor in green and periodical oscillation in blue. (c) Examples of the v_1 temporal evolution for each kind of identified trajectory.

Fig. 4(a) shows the resulting trajectory, which is indeed the double-scroll attractor, with $P+$ and $P-$ located approximately at $v_1 = \pm 0.9$ V, validating our design procedure. Notice the asymmetry of the trajectory, probably due an slight asymmetry in the memristor i-v characteristics or non-idealities of the circuit elements.

Fig. 4(b) reports an experimental bifurcation plot as a function of the R_{PROG} demonstrating that the memristor is properly providing the nonlinearity required for chaotic behaviour and that it allows tuning the dynamics from periodic trajectory (blue squares), single-scroll attractor (green triangles) to double-scroll attractor (red circles). The bifurcation plot only depicts the local extrema of v_1, which tells us qualitatively the type of obtained oscillation. For a better understanding of the results, an example of each kind of oscillatory trajectory is also included in Fig. 4(c). Note that both the single-scroll attractor and the periodic trajectories oscillate either around P_+ or P_-, which depends only on the initial circuit conditions, not controlled during our experiments.

It is worth mentioning that, generally, the voltage spanning of v_1, which is the voltage drop on the memristor, is wider for higher R_{PROG} in agreement with the displacement of the equilibrium points P_+ and P_- depicted in Fig. 2(f). Such growth of v_1 spanning range with R_{PROG} does not lead to unintentional memristor state perturbation as also V_{SET} grows with R_{PROG}. Such combination helps the circuit design.

Although the bifurcation plot in Fig. 4(b) shows that for some R_{PROG} bands it is expected to obtain a certain type of trajectory, we have found few single-scroll attractor among resistance states in which double-scroll attractors are dominant. The same observation is valid for the general trend of decreasing amplitude with R_{PROG}. The main cause is the the device intrinsic cycle-to-cycle variability, that hinders repeatable trajectories.

V. CONCLUSIONS

We successfully implemented physically a Chua's circuit powered by the nonlinearity of the nonvolatile Pt/HfO$_2$/TiN device. Moreover, we showed that circuit tunability is possible using nonvolatile resistive switching devices, and different kinds of trajectories can be obtained after changing the memristor state. All these achievements have been possible after a careful design based on the dynamic system analysis and the constraints observed from the device characterization. By using a dedicated nonlinear programmable device for this circuit, the circuit complexity is alleviated and configurability features are provided at the same time. This offers an attractive alternative for implementing scalable chaotic oscillators. For real applications, the programming operation may need to control precisely the memristor state in order to guarantee a robust tuning. Therefore, program and verify techniques can be considered for this purpose.

ACKNOWLEDGMENT

This work is partially supported by the PRIN-MIUR project COSMO (Prot. 2017LSCR4K).

REFERENCES

[1] B. Kia, J. F. Lindner, and W. L. Ditto, "Nonlinear dynamics as an engine of computation," *Philos. Trans. R. Soc., A*, vol. 375, no. 2088, 2017.
[2] M. P. Kennedy, "Three Steps to Chaos—Part II: A Chua's Circuit Primer," *IEEE Trans. Circuits Syst. I*, vol. 40, no. 10, pp. 657–674, 1993.
[3] G. Csaba and W. Porod, "Coupled oscillators for computing: A review and perspective," *Appl. Phys. Rev.*, vol. 7, no. 1, 2020.
[4] G. Tanaka *et al.*, "Recent advances in physical reservoir computing: A review," *Neural Networks*, vol. 115, pp. 100–123, 2019.
[5] L. Fortuna, M. Frasca, and M. G. Xibilia, *Chua's Circuit Implementations: Yesterday Today and Tomorrow*. Singapore: World Scientific, 2009.
[6] M. Itoh and L. O. Chua, "Memristor oscillators," *Int. J. Bifurcation Chaos Appl. Sci. Eng.*, vol. 18, no. 11, pp. 3183–3206, 2008.
[7] S. Brivio, J. Frascaroli, and S. Spiga, "Role of metal-oxide interfaces in the multiple resistance switching regimes of Pt/HfO$_2$/TiN devices," *Appl. Phys. Lett.*, vol. 107, no. 2, p. 023504, jul 2015.
[8] ——, "Role of Al doping in the filament disruption in HfO$_2$ resistance switches," *Nanotechnology*, vol. 28, no. 39, p. 395202, sep 2017.
[9] J. Frascaroli, F. G. Volpe, S. Brivio, and S. Spiga, "Effect of Al doping on the retention behavior of HfO2 resistive switching memories," *Microelectron. Eng.*, vol. 147, pp. 104–107, nov 2015.
[10] A. I. Khibnik, D. Roose, and L. O. Chua, "On Periodic Orbits and Homoclinic Bifurcations in Chua's Circuit with a Smooth Nonlinearity," *Int. J. Bifurcation Chaos Appl. Sci. Eng.*, vol. 03, no. 02, pp. 363–384, apr 1993.

978-1-6654-8495-4/22 $31.00 © 2022 IEEE

Ferroelectric Schottky Barrier MOSFET as Analog Synapses for Neuromorphic Computing

Fengben Xi[*,1,2], Andreas Grenmyr[1,3], Jiayuan Zhang[1,3], Yi Han[1,2], Jin Hee Bae[1], Detlev Grützmacher[1,2], and Qing-Tai Zhao[*,1]

[1]*Peter-Grünberg-Institute (PGI 9) and JARA-FIT, Forschungszentrum Jülich, 52428 Jülich, Germany*
[2]*Faculty of Mathematics, Computer Science and Natural Sciences, RWTH Aachen University, 52074, Aachen, Germany*
[3]*Faculty of Electrical Engineering and Information Technology, RWTH Aachen University, 52074, Aachen, Germany*
*E-mail: f.xi@fz-juelich.de; q.zhao@fz-juelich.de

Abstract—In this paper, artificial synapses based on ferroelectric Schottky barrier MOSFETs (FE-SBFETs) are presented. The FE-SBFETs are fabricated with Si doped HfO_2 ferroelectric layers scaling down to a gate length of 40 nm and using single crystalline $NiSi_2$ contacts on silicon-on-insulator (SOI) substrates. The ferroelectric polarization switching dynamics gradually modulate the Schottky barriers, thus programming the device conductance by applying stimulus on the gate to imitate the short- and long-term plasticity of biological synapse, including excitatory/inhibitory post-synaptic current (EPSC/IPSC), paired-pulse facilitation (PPF) and long-term potentiation (LTP) and long-term depression (LTD) behaviors. Based on a multilayer perceptron artificial neural networks, a high recognition accuracy (83.6%) is achieved for handwritten digits. These findings demonstrate FE-SBFET has high potential as an ideal synaptic component for the future intelligent neuromorphic network.

Keywords—Artificial synapse, ferroelectric polarization, Si:HfO2, FE-SBFET, neuromorphic application

I. INTRODUCTION

The brain-inspired neuromorphic computing has emerged as an alluring promising technology for the post-Moore era [1][2]. The emerging Non-volatile memories based neuromorphic devices, such as memristor [3], magnetic random-access memory (MRAM) [4] and phase change memory (PCM) [5] allow to analogue weight storage to complete cognitive tasks more efficiently. However, the integration, stability and power consumption are still bottlenecks for these devices in the application of neuromorphic system [6].

Recently ferroelectric MOSFETs (FeFETs) [7] based on doped HfO_2 are being evaluated showing very promising synaptic functions with high performance, cost-effective integration, CMOS compatibility and low power consumption [8]. In our previous work [9][10], we have reported analog synapse based on the long channel ferroelectric SBFET (FE-SBFET). Compared with traditional FeFETs, the Ni silicide source/drain Schottky barrier (SB) MOSFET (SBFET) structure shows relatively low temperature processing and simple fabrication process [11]–[15].

In this work, a synapse based on the FE-SBFET scaling down to 40 nm is presented. The ferroelectric polarization switching dynamics gradually modulate the channel potential, thus programming the device conductance by applying stimulus on the gate to imitate the short- and long-

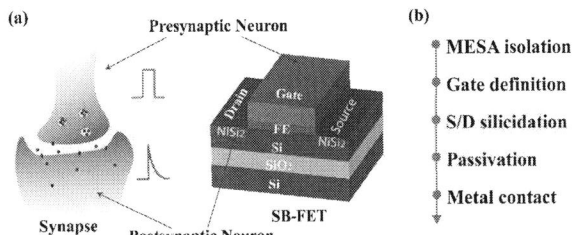

Fig. 1. (a) Schematic diagram of the FE-SBFET analog synapse, where the gate and drain are used as the pre- and post- synaptic terminals. (b) Key fabrication process flow for the FE-SBFET.

term plasticity of biological synapse. The fabricated analog synapse based on FE-SBFETs employing ferroelectric Si:HfO_2 film and atomic crystalline $NiSi_2$ SB contacts on silicon-on-insulator (SOI) showing excitatory/inhibitory post-synaptic current (EPSC/IPSC), paired-pulse facilitation (PPF) and long term potentiation and depression (LTP/LTD) characteristics. A high learning accuracy of 83.6% is also achieved based on a multilayer perceptron artificial neural networks.

II. DEVICE FABRICATION

As shown in Fig. 1(a), the gate and drain of fabricated devices are connected with the pre- and post- synaptic neuron to mimic the synapse behaviors based on ferroelectric partial polarization switching dynamics in Si:HfO_2 thin films. The channel is modulated by the gradually ferroelectric polarization, thus changing the conductance of the device. An increase of the number of pulses will strengthen the domain polarization of ferroelectric layers, resulting in higher channel electric field. Thus, the applied gate stimulus gradually program the conductance of the device. Fig. 1(b) shows the key fabrication process flow for FE-SBFETs, which were fabricated on an SOI substrate with a 25 nm thick top p-type Si (~10^{16} B/cm^3) layer and a buried SiO_2 layer (BOX) with the thickness of 145 nm. After mesa definition and RCA cleaning, a 10 nm thick Si:HfO_2 layer and a 40 nm thick TiN layer were deposited by atomic layer deposition (ALD) and sputtering, respectively. Then a 30 s rapid thermal annealing at 800 °C with N_2 was performed to crystallize the Si:HfO_2 into ferroelectric phase. The gate with a length of 40 nm was defined by E-beam lithography and reactive ion etching. Then, single crystalline $NiSi_2$ layers were formed at source/drain regions using conventional self-aligned silicidation process with 3 nm Ni at 800 °C in forming gas atmosphere [16].

978-1-6654-8495-4/22 $31.00 © 2022 IEEE

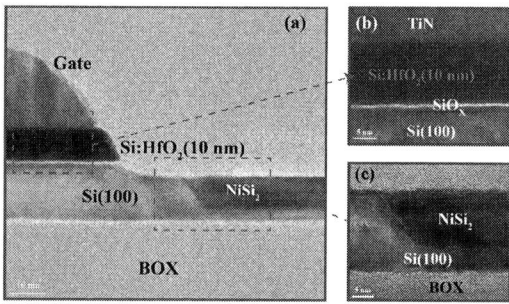

Fig. 2. (a) The TEM image of SB-FeFET gate stack. (b) TEM images of gate and channel regions of FeFETs with 10 nm ferro-layer and 1.6nm-thick IL. (c) TEM images of source/drain regions of FeFETs with 10-nm-thick NiSi₂ layers, showing atomic contact with Si.

Fig. 2(a) shows a cross-sectional TEM image of a FE-SBFET. Due to shadow effects of the gate during the 3 nm Ni sputtering deposition, there is a gap region between the gate and source and drain which induce lower on current showing below. Fig. 2(b) focuses on the gate and channel regions. A 10 nm ferroelectric layer and a thin SiO_x interfacial layer (IL) are observed at the interface between $Si:HfO_2$ and Si. It is clearly visible that a 10 nm thick $NiSi_2$ with atomic flat $NiSi_2$ /Si interface in Fig. 2(c) is formed offering superior properties of uniform and stable SB contacts. Meanwhile, a metal-ferroelectric-metal (MFM) ferroelectric capacitor with the $TiN/Si:HfO_2/TiN$ stack was also fabricated as reference. Finally, the sample was passivated by SiO_2 and metallized with Al after via opening.

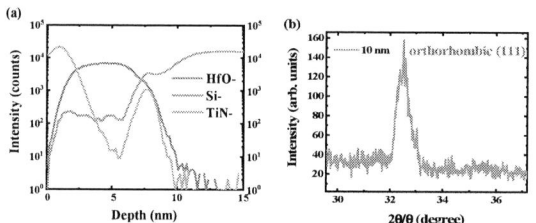

Fig. 3. (a) ToF-SIMS results of $TiN/Si:HfO_2/Si$ capacitor structure. The Si doping is clearly shown in the profile. (b) X-ray diffraction (XRD) result measured for 10 nm thick $Si:HfO_2$, and shows a distinguishable orthorhombic peak.

III. RESULTS AND DISCUSSION

A. Ferroelectric material characteristics

Fig. 3a shows the element distributions measured with time-of-flight secondary ion mass spectrometry (ToF-SIMS) on $TiN/Si:HfO_2/Si$ with a 10 nm 3.8 mol% Si doped HfO_2 ferroelectric layer, showing a uniform Si distribution within the HfO_2 layer. The X-ray diffraction (XRD) (Fig. 3(b)) shows a clear orthorhombic peak.

The ferroelectricity of $Si:HfO_2$ is characterized by a TiN MFM capacitor with a sandwich structure of TiN/10 nm $Si:HfO_2/$. Fig. 4(a) shows the typical polarization–electric field (P-E) ferroelectric hysteresis loops of the MFM capacitor. The ferroelectric polarization gradually increases by upgrading the electric field from 1 to 6 MV/cm², showing the controllable polarization characteristics of the $Si:HfO_2$ ferroelectric thin film. The wake-up within 10^3 cycles and fatigue measurement over 10^6 cycles are presented in Fig. 4(b). As indicated in Fig. 4(c), the 10 nm thick $Si:HfO_2$ layer

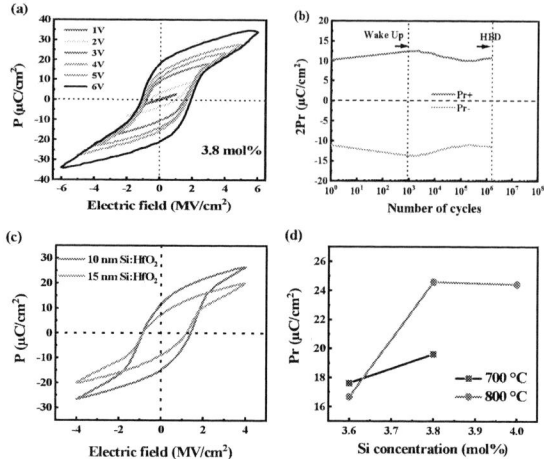

Fig. 4. (a) P-E characteristics of a 10 nm thick $Si:HfO_2$ layer measured with a $TiN/Si:HfO_2/TiN$ MFM structure by sweeping the electric field with various ranges. (b) Wake-up and endurance measurements for $Si:HfO_2$ within 10^6 cycles. (HBD stands for hard break down). (c) P-E value of 10 and 15 nm thick $Si:HfO_2$ with 3.8% doped Si. (d) Remnant polarization of $Si:HfO_2$ layer with different doping concentrations annealed at 700 °C or 800 °C.

shows higher polarization than the 15 nm thick $Si:HfO_2$ layer with the same Si doping. Figure 4(d) shows the dependence of ferroelectricity in $Si:HfO_2$ on different doping concentrations from 3.6 to 4 mol% and annealing temperatures at 700 °C or 800 °C. The polarization becomes stronger when the annealing temperature and the Si doping concentration increases to 800 °C and 3.8 mol% Si, respectively. This enhancement is supposed to be due to the occurrence of more orthorhombic phases inside the $Si:HfO_2$ [17]. Considering a relative stable device operation, the 10 nm film with 3.8 mol% Si at 800 °C is utilized for the further device fabrication.

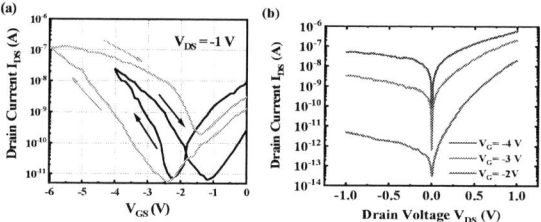

Fig. 5. (a) I_{DS}-V_{GS} transfer characteristics at V_{DS} = -1 V for an FE-SBFET, showing a clockwise hysteresis for p-FET and ambipolar switching behavior. (b) I_{DS}-V_{DS} output curves of the corresponding device with varying V_{GS} from -2 to -4 V.

B. I-V characteristics of FE-SBFETs

The drain current (I_{DS}) and gate voltage (V_{GS}) transfer characteristics of the FE-SBFET were shown in Fig. 5(a) with different dual sweeping of V_{GS} in forward and reverse directions at a drain voltage (V_{DS}) of -1 V, showing a large clockwise hysteresis for the p-FET branch. The memory window changed from 1.0 to 2.1 V when the sweeping voltage increased for -4 to -6 V. Higher gate sweeping voltage shows bigger memory window due to stronger ferroelectric polarization switching. The ambipolar behavior is a typical feature for SB-MOSFETs. The device depicts an I_{ON}/I_{OFF} ratio of 1.6×10^4. The I_{DS}-V_{DS} curve for the corresponding FE-SBFET shows also the typical SBFET characteristic (Fig. 5(b)).

978-1-6654-8495-4/22 $31.00 © 2022 IEEE

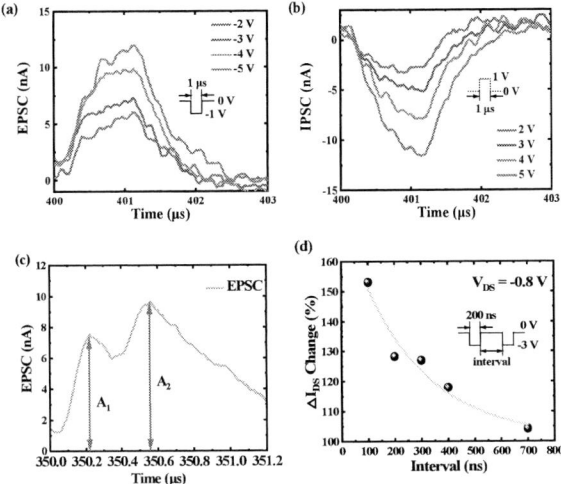

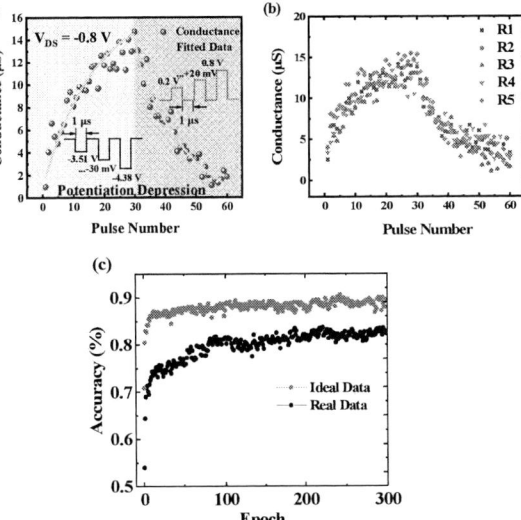

Fig. 6. (a) EPSC, and (b) IPSC generated by a sequence of gate spikes with an identical pulse duration time (1 μs) and various potentiating pulse amplitudes V_{AM} ranging from -2 (2) to -5 (5)V, both measured at V_{DS} = -1 V. (c) EPSC responses triggered by a pair of input pulses (-1 V, 1 μs) with a time interval of 200 ns at V_{AM} = -3 V and V_{DS} = -0.8 V. A_1 and A_2 represent the EPSC generated by the first and second presynaptic spike pulses, respectively. (d) PPF characteristics for an FE-SBFET synapse by measuring the synaptic strength change as a function of the pulse interval, showing a biological synapse behavior with a single exponential decay. measured at V_{DS} = -1 V

Fig. 7. Emulation of biological synaptic behaviors and repetitive channel conductance modulation. (a) Measured synapse channel conductance at V_{DS} = -0.8 V as a function of pulse numbers, showing potentiation and depression characteristics. (b) Exemplary 5 cycles of potentiation and depression after 10k repeated pulse measurements. (c) Pattern recognition simulation accuracy acquired with the ideal and real data from a. The high learning accuracy of 83.6% and 90.6% are achieved with real and ideal data.

C. Characteristics of synapse based on FE-SBFETs

To demonstrate synaptic functionalities of the FE-SBFET based analog synapse, the gate and drain terminals are used to connect the pre- and post- synaptic neurons, as shown in Fig.1. After that, the post-synaptic currents triggered by presynaptic spikes from the gate terminal correspond to EPSC/IPSC showing the short-term synaptic plasticity (STP) of synapses, which is the most fundamental synapse property to measure the synaptic strength variations. As illustrated in Fig. 6(a), the EPSC response of an FE-SBFET based analog synapse was characterized by measuring the transient drain currents while a voltage pulse stimulus on the gate, which were generated by single presynaptic spike with different pulse amplitudes ranging from -2 to -5 V at V_{DS} = -1 V and a pulse width = 1 μs. Similarly, the IPSC responses are shown in Fig. 6(b).

In the nervous system, there is also another important short-term synaptic plasticity-paired-pulse facilitation which is related to decode the temporary information and measure the synaptic strength change. As shown in Fig. 6(c), the PPF characteristics of a FE-SBFET synapse was measured by the EPSC variation as a function of presynaptic spike time interval when a pair of input pulses (200 ns and -3 V) were introduced. The PPF index is defined as:

$$PPF \ index = A_2 / A_1 \cdot 100\% \tag{1}$$

where A_1 and A_2 are the amplitude of the EPSC corresponding to the first and second pre-synaptic stimulus. The PPF index is plotted in Fig. 6(d) showing a short-term synaptic behavior by fitting PPF with the following equation:

$$PPF \ index = C_0 + C_1 \cdot e^{-t/\tau_0} \tag{2}$$

where C_0 corresponding to the background signal (1.01), t the interval time between the input pulses, C_1 (0.74) related to the initial facilitation magnitude, and τ_0 is the decay time constant (249 ns). The PPF index decreases from 153.1% to 104.2% with increasing the interval time between two presynaptic spikes from 100 to 700 ns.

In addition, long-term potentiation and depression are essential memory features of nervous synapses, which reflect the long-term synaptic plasticity between two neurons. The channel conductance resembles the information transmission in biological synapses. By applying an external stimulus on the gate, the synapse weight is tuned by modulating the connection strength similar to the synaptic plasticity between neurons. The continuous and repetitive channel conductance modulation of the FE-SBFET based analog synapse by applying a series pulse is exhibited in Fig. 7, showing the long-term synaptic plasticity. Fig. 7(a) shows the measured conductance of a FE-SBFET analog synapse as a function of pulse numbers for pulse waveforms with 1 μs width, 1 μs interval and a non-identical amplitude, where the nonlinearity α_P = 0.10, α_D = 0.12, ON/OFF ratio = 15 and asymmetry β = -0.43 are extracted according to [15] by fitting the experimental data with 30 programming states. The results clearly indicate synaptic characteristics with LTP/LTD. Fig. 7(b) shows exemplarily 5 cycles of potentiation/depression after 10k pulse measurements. The corresponding scheme of the applied synaptic spikes is illustrated in the insets of Fig. 7(a).

In order to show further neuromorphic properties of the FE-SBFET based analog synapse, a Multilayer Perceptron (MLP), consisting of 400 input nodes, 100 hidden nodes, and

10 output nodes was developed to perform the neural network simulation. The nodes are fully connected with different weights representing different conductive states of the artificial synapse. Each output node is predicting the probability that the input contains a 20×20 pixels image of a certain number between 0-9 from the MNIST dataset. Based on the back propagation iterations, after each iteration, the weights between the nodes are updated by a certain value extracted from the LTD/LTP, which is decided by the difference between the predicted and true output value. The response of each applied LTD/LTP pulse to each weight follows the LTD/LTP behavior of our devices. The hidden nodes add all the input nodes multiplied with each corresponding weight. The sum is then used as a variable in the activation function of the hidden node. The value of the activation function from the hidden node then multiplies by the weights in the second weight layer. In the final step of the forward propagation operation, the output node sums up all products from all the hidden nodes and the connected weights. Therefore, this simulation is related to the neuromorphic performance of our devices by comparing the achieved accuracy of the predictions for the output nodes.

Parameters that affect the neuromorphic performance of our device include the ON/OFF ratio, the number of conductive states, and the linearity of the LTD/LTP update behavior, the conductance and the cycle variations. For the result from this simulation, as shown in Fig. 7(c), the accuracy for the real device increases rapidly for the first epochs and stabilizes after 100 epochs and the highest accuracy is 83.6 %. For the ideal device where the linearity of the device is ideal shows better performance and the accuracy can reach 90.6%. For each epoch, the MLP trains by updating the weights for 4000 images and then test the accuracy at 10000 images without changing the weights. For this simulation, Adam optimizer was the used as the optimization algorithm. The simulation was developed on NeuroSim+ platform [19].

IV. CONCLUSIONS

We demonstrated a three terminal ferroelectric Schottky barrier FET with single crystalline $NiSi_2$ contacts and a $Si:HfO_2$ based 40 nm ferroelectric gate to implement synaptic properties. The fabricated artificial synapse exhibits the short- and long-term plasticity, including excitatory/inhibitory post-synaptic current, paired-pulse facilitation, and long term potentiation and depression. A high recognition accuracy (83.6%) with a multilayer perceptron artificial neural networks is achieved. Further optimization of the device structure, for example using FDSOI and improved ferroelectricity will be investigated to improve the performance.

ACKNOWLEDGMENT

This work was partially supported by the Federal Ministry of Education and Research (BMBF, Germany) in the project NEUROTEC (16ME0398K).

REFERENCES

[1] N. K. Upadhyay, H. Jiang, Z. Wang, S. Asapu, Q. Xia, and J. Joshua Yang, "Emerging Memory Devices for Neuromorphic Computing," *Adv. Mater. Technol.*, vol. 4, no. 4, pp. 1–13, 2019.

[2] X. Wang *et al.*, "MoS^2 Synaptic Transistor With Tunable Weight Profile," *IEEE Trans. Electron Devices*, vol. 65, no. 8, pp. 3543–3547, Aug. 2018.

[3] S. Choi, J. Yang, and G. Wang, "Emerging Memristive Artificial Synapses and Neurons for Energy-Efficient Neuromorphic Computing," *Adv. Mater.*, vol. 32, no. 51, pp. 1–26, 2020.

[4] D. W. Kim *et al.*, "Double MgO-Based Perpendicular Magnetic Tunnel Junction for Artificial Neuron," *Front. Neurosci.*, vol. 14, no. April, pp. 1–9, 2020.

[5] I. Boybat *et al.*, "Neuromorphic computing with multi-memristive synapses," *Nat. Commun.*, vol. 9, no. 1, pp. 1–12, 2018.

[6] M. K. Kim, Y. Park, I. J. Kim, and J. S. Lee, "Emerging Materials for Neuromorphic Devices and Systems," *iScience*, vol. 23, no. 12, p. 101846, 2020.

[7] H. Mulaosmanovic *et al.*, "Novel ferroelectric FET based synapse for neuromorphic systems," in *Symposium on VLSI Technology*, 2017, pp. T176–T177.

[8] J. Luo *et al.*, "Capacitor-less Stochastic Leaky-FeFET Neuron of Both Excitatory and Inhibitory Connections for SNN with Reduced Hardware Cost," in *2019 IEEE International Electron Devices Meeting (IEDM)*, 2019, vol. December, pp. 6.4.1-6.4.4.

[9] F. Xi *et al.*, "Artificial Synapses Based on Ferroelectric Schottky Barrier Field-Effect Transistors for Neuromorphic Applications," *ACS Appl. Mater. Interfaces*, vol. 13, no. 27, pp. 32005–32012, 2021.

[10] F. Xi, Y. Han, A. Tiedemann, D. Grutzmacher, and Q.-T. Zhao, "4-Terminal Ferroelectric Schottky Barrier Field Effect Transistors as Artificial Synapses," in *IEEE 51st European Solid-State Device Research Conference (ESSDERC)*, 2021, pp. 291–294.

[11] M. Zhang, J. Knoch, Q. T. Zhao, U. Breuer, and S. Mantl, "Impact of dopant segregation on fully depleted Schottky-barrier SOI-MOSFETs," *Solid. State. Electron.*, vol. 50, no. 4, pp. 594–600, Apr. 2006.

[12] C. Urban *et al.*, "Radio-Frequency Study of Dopant-Segregated n-Type SB-MOSFETs on Thin-Body SOI," *IEEE Electron Device Lett.*, vol. 31, no. 6, pp. 537–539, Jun. 2010.

[13] L. Knoll, Q. T. Zhao, R. Luptak, S. Trellenkamp, K. K. Bourdelle, and S. Mantl, "20 nm Gate length Schottky MOSFETs with ultra-thin NiSi/epitaxial NiSi 2 source/drain," *Solid. State. Electron.*, vol. 71, pp. 88–92, 2012.

[14] V. Sessi *et al.*, "A Silicon Nanowire Ferroelectric Field-Effect Transistor," *Adv. Electron. Mater.*, vol. 6, no. 4, p. 1901244, Apr. 2020.

[15] V. Sessi, H. Mulaosmanovic, R. Hentschel, S. Pregl, T. Mikolajick, and W. M. Weber, "Junction Tuning by Ferroelectric Switching in Silicon Nanowire Schottky-Barrier Field Effect Transistors," in *2018 IEEE 18th International Conference on Nanotechnology (IEEE-NANO)*, 2018, vol. 2018-July, pp. 1–4.

[16] L. Knoll, Q. T. Zhao, S. Habicht, C. Urban, B. Ghyselen, and S. Mantl, "Ultrathin Ni Silicides With Low Contact Resistance on Strained and Unstrained Silicon," *IEEE Electron Device Lett.*, vol. 31, no. 4, pp. 350–352, Apr. 2010.

[17] T. S. Böscke, J. Müller, D. Bräuhaus, U. Schröder, and U. Böttger, "Ferroelectricity in hafnium oxide thin films," *Appl. Phys. Lett.*, vol. 99, no. 10, pp. 0–3, 2011.

[18] M. Jerry *et al.*, "Ferroelectric FET analog synapse for acceleration of deep neural network training," *Tech. Dig. - Int. Electron Devices Meet. IEDM*, vol. 6, no. c, pp. 6.2.1-6.2.4, 2018.

[19] P.-Y. Chen, X. Peng, and S. Yu, "NeuroSim+: An integrated device-to-algorithm framework for benchmarking synaptic devices and array architectures," in *2017 IEEE International Electron Devices Meeting (IEDM)*, 2017, pp. 6.1.1-6.1.4.

Interplay between charge trapping and polarization switching in MFDM stacks evidenced by frequency-dependent measurements

Justine Barbot, Jean Coignus, Nicolas Vaxelaire, Catherine Carabasse, Olivier Glorieux,
Messaoud Bedjaoui, François Aussenac, François Andrieu, François Triozon and Laurent Grenouillet

CEA, LETI, Univ. Grenoble Alpes, Grenoble, France

Abstract—**Experimental analysis of polarization switching in metal-ferroelectric-metal and metal-ferroelectric-dielectric-metal junctions is reported. Combined GIXRD and electrical analyses demonstrate that insertion of Al₂O₃ dielectric layer boosts the ferroelectric polarization. Ferroelectric switching measurements at various frequencies show that the injection and trapping of charges into the ferroelectric-dielectric stack have a large influence on the polarization switching.**

Index Terms—**Alumina, ferroelectric devices, hafnium oxide, memory, switching frequency, trapping, tunnel junction.**

I. INTRODUCTION

Ferroelectric dielectric bilayer stacks have been extensively studied for Ferroelectric Tunnel Junctions (FTJs), as their asymmetry boosts the difference between ON and OFF currents [1]. Ferroelectric properties of these junctions are typically determined via the measurement of transient IV characteristics at one frequency. From this method, inconsistent ferroelectric performance has been reported on Metal/Ferroelectric/Dielectric/Metal junctions (MFDM). Indeed, dielectric layer is suspected to either enhance or reduce the remanent polarization. In details, a first group of papers supports that the insertion of dielectric layer changes the crystallographic structure of ferroelectric film, boosting the polarization [2], [3]. A second group claims that dielectric layer degrades the remanent polarization by reducing the electric field on the ferroelectric layer [4], [5]. A third group affirms that charge injection through leaky dielectric layer can result in the same or greater remanent polarization in metal-ferroelectric-metal stack (MFM) than in metal-ferroelectric-dielectric-metal stack (MFDM) [6]–[10].

Therefore, there is an interest to investigate more deeply the ferroelectric properties of MFM and MFDM. This is the purpose of this paper which reports a comparison of HfO₂:Si/Al₂O₃ (HSO/Al₂O₃) bi-layer stacks with reference HfO₂:Si (HSO) one. In details, this article examines each of the three claims quoted above. Firstly, any modification of crystal structure due to the insertion of Al₂O₃ will be investigated through grazing incidence X-ray diffraction (GIXRD). Secondly, impact of Al₂O₃ layer on the remanent polarization will be studied through Positive-Up-Negative-Down (PUND) characteristics at one frequency. Thirdly, charge injection in dielectric layer of MFDM stacks will be covered through PUND characteristics acquired from 10 Hz to 10^5 Hz.

II. EXPERIMENT

A. Samples

In this study, HSO reference ferroelectric capacitors and HSO/Al₂O₃ bi-layer stacks were integrated between two metal lines using 130 nm node Back-End of Line (BEOL)

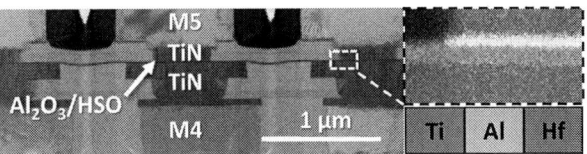

Fig. 1: Left: STEM cross section of a 0.109 µm² TiN/HSO(10 nm)/Al₂O₃ (2.5 nm)/TiN scaled FTJ integrated in 130 nm BEOL between metal lines; Right: EDX close-up view of the scaled FTJ showing uniformity of both dielectric and ferroelectric layers.

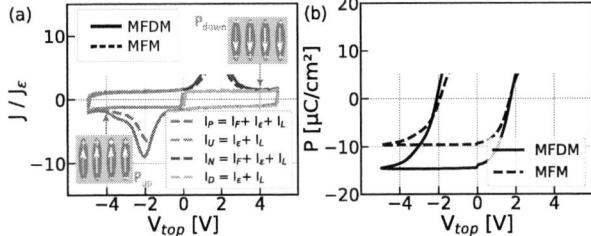

Fig. 2: Comparison of PUND results on MFDM and MFM devices after wake-up set at 1000 triangular cycles (± 5 V, 100 kHz): (a) Normalized current density-voltage characteristics with respect to the dielectric current density (J_ε); (b) corrected polarization-voltage hysteresis loop. PUNDs were acquired at ± 5 V, 10 kHz without any delay between each consecutive pulse. Here, MFDM and MFM refer to TiN/HSO(10 nm)/Al₂O₃(1 nm)/TiN and TiN/HSO(10 nm)/TiN respectively. Colors on graph (a) refers to the colors used on the electrical pulse scheme of Fig. 6.

process flow and design rules. The TiN bottom electrode was reactively sputtered by physical vapour deposition (PVD) at 350 °C and planarized by chemical mechanical polishing (CMP) to reach a 0.18 nm RMS surface roughness. The 10 nm-thick HfO₂ layer was deposited at 300 °C by Atomic Layer Deposition (ALD) with HfCl₄ and H₂O precursors in ASM Polygon Pulsar Chamber. The HfO₂ film was then Si-doped by ion implantation [11]. On the top of HSO layers, Al₂O₃ film was deposited at 300 °C with trimethylaluminum and H₂O, targeting thicknesses of 1, 1.5, 2 and 2.5 nm. The bi-layer stack was then coated by a 100 nm-thick TiN layer. The selective etching of the TiN top electrode defines the 330 nm diameter capacitors arranged in parallel, leading to equivalent area ranging from 50 µm² to 400 µm². A cross-section of the MFDM stack between metal lines is shown in Fig. 1.

B. Electrical characterization

The ferroelectric switching was studied using the Positive-Up-Negative-Down (PUND) method, which relies on time-resolved transient current measurements to investigate the ferroelectric domain switching. Five triangular pulses were applied to the top electrode of the capacitors to collect the

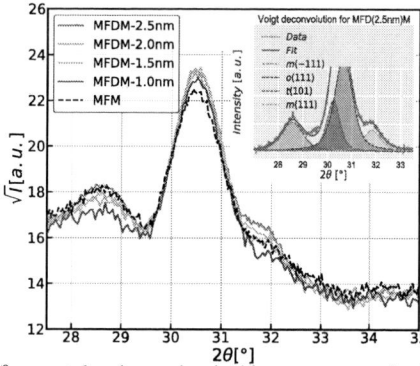

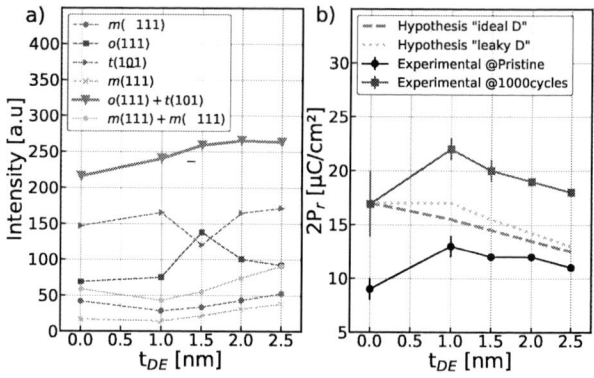

Fig. 3: 2θ scans taken in grazing-incidence geometry for film stacks composed of 10 nm HSO with and without Al_2O_3 layer. Inset: Voigt deconvolution of the 2θ for TiN/HSO(10nm)/Al_2O_3(2.5nm)/TiN, dashed peaks indicate the expected peak positions for the monoclinic (m-), polar orthorhombic (o-) and tetragonal (t-) phases.

joint contribution of the leakage (I_L), dielectric (I_ε) and switching currents (I_F) (see Fig. 2(a)). I_ε was ascertained on the positive current plateau of U and D pulses at 2 V. I_F was then extracted from the non-switchable contributions ($I_\varepsilon + I_L$) by a current subtraction: I_P-I_U and I_N-I_D. Integration of I_F over time yielded the P-V hysteresis loop (see Fig. 2(b)), from which the remanent polarization ($2P_r$) is extracted at null bias. The PUND method is by far the most appropriate way to ascertain I_F and then $2P_r$ without experimental artefact from the set-up, and leakage contributions [12].

C. Grazing-incidence X-ray diffraction measurements

Grazing-incidence X-ray diffraction (GIXRD) measurements were carried out on a Panalytical Empyrean X-ray 2-circle goniometer. 2θ scans were acquired in the grazing incidence geometry, at an incidence angle of 0.5°, in order to minimize the signal coming from the underlying substrate.

III. RESULTS

A. Impact of the Al_2O_3 layer on the crystallographic structure of HSO thin film

Firstly, the effect of Al_2O_3 layer on crystallographic structure of HSO is investigated by a comparison between MFDM and MFM stacks. For this comparison, GIXRD and PUND measurements were performed.

GIXRD provides information on the crystallographic parameters of HSO films. A superposition of the GIXRD data from the five studied samples is shown in Fig. 3 and is indexed for orthorhombic (o-), monoclinic (m-), tetragonal (t-) phases. Given the process temperatures, the Al_2O_3 layers are amorphous. Ferroelectricity in HSO is known to be correlated to the polar orthorhombic (o-) phase [13]. o- and t- peak intensities of MFDM samples are greater than the MFM reference sample, while the background remains the same. Among the four MFDM samples, peak intensities increase as Al_2O_3 thickness increases. Deconvolution using Voigt profile was used to extract the different phase contributions. Fig. 4(a) shows the specific intensity of each phase. The intensities of o(111)+t(101) phase increases up to 1.5 nm of Al_2O_3 after it levels out. One could remark that the increase of o- and t- peak intensities with alumina thickness is less smooth when the peaks are taken separately. This is due to the structural similarities of the o- and t- phases, the Voigt deconvolution cannot clearly distinguish the o-

Fig. 4: (a) Intensity of each crystal phase of HSO films for each Al_2O_3 thickness extracted via Voigt deconvolution; (b) median $2P_r$ on 200 mm wafer (70 capacitors) at pristine and after 1000 wake-up cycles ($\pm$ 5 V, 100 kHz) extracted by a PUND ($\pm$ 5 V, 10 kHz) for all studied stacks. Hypothesis "ideal D" schematically illustrates the evolution of $2P_r$ for ideal MFDM, while hypothesis "leaky D" illustrates leaky dielectric in MFDM stack. In both hypothesis, the crystal structure of the ferroelectric is assumed to be the same as in MFM.

phase from the t- phase of HSO [14]. Overall the GIXRD results confirm a modification of the texture or/and phase proportion of HSO layer by the thickening of the Al_2O_3 layer. These crystallographic or/and micro-structural modifications can boost the ferroelectric response, thus promoting $2P_r$. The increase of o+t intensity in function of t_{DE} on Fig. 4(a) suggests that $2P_r$ increases with the thickening of the Al_2O_3 layer. These suggestions will be verified hereafter through electrical measurements.

B. Remanent polarization evolution with Al_2O_3 thickness

Standard PUND measurements at $\pm$ 5 V, 10 kHz were carried out to quantify the ferroelectric response of pristine (i.e. fresh) and woken-up samples (i.e. after 1000 cycles at $\pm$ 5 V, 100 kHz) samples. Remanent polarization $2P_r$ as a function of Al_2O_3 thickness (t_{DE}) is depicted in Fig 4(b), and shows two distinct trends: a $2P_r$ increase with the insertion of Al_2O_3 layer (MFM vs MFDM), followed by a continuous $2P_r$ decrease with Al_2O_3 thickening. Morphological modifications of HSO highlighted by GIXRD measurements could explain the origin of $2P_r$ boost between MFM and MFDM. However, it is clearly not sufficient to explain the decline of $2P_r(t_{DE})$ curve. As both wake-up and PUND voltages were always set at $\pm$ 5 V, one may attribute this $2P_r$ decline to the voltage drop through the interfacial dielectric layer. Indeed, MFDM stack configuration could reduce the electric field applied to the ferroelectric layer. However, as this decline is found at pristine state, and as voltages higher than twice the coercive voltages are applied to the stacks ($|V_c| > 2.5$ V for 2.5 nm of Al_2O_3), other electrical-based hypotheses must be studied. Fig 4. (b) presents two of these hypotheses. For ease of comprehension, these hypotheses will omit the impact of Al_2O_3 layers on HSO morphology. The first hypothesis is "ideal D" where the the Al_2O_3 layers are perfectly insulating (ideal MFDM). In this stack, the thicker the Al_2O_3 layer, the smaller $2P_r$. Indeed, dielectric layer inhibits the compensation of ferroelectric polarization by free carriers in the metal. Uncompensation yields to an electric field opposite to the polarization as shown by blue arrow in

Fig. 5 (b). This resulting depolarization field degrades the polarization [15], leading to lower $2P_r$ in MFDM than MFM stack. The second hypothesis, more realistic, is "leaky D". As depicted in [8]–[10], charge injection through D layer either modelled by Fowler Nordheim or trap-assisted transport can compensate and stabilize the ferroelectric polarization. This leads to similar $2P_r$ in MFDM and MFM reference samples as long as sufficient charges are injected (Fig. 5 (c)). When Al_2O_3 layer gets sufficiently thick, the amount of injected charges is reduced, resulting in $2P_r$ drop. When Al_2O_3 layer gets sufficiently thick, it becomes insulating, and $2P_r$ values converge to the $2P_r$ values of "ideal D" hypothesis.

As a result of the conduction properties of Al_2O_3 (i.e. quite leaky dielectric), the hypothesis "leaky D" reflects the experimental $2P_r$-t_{DE} decline.

To summarize, PUND results support the existing literature idea that the insertion of Al_2O_3 boosts the $2P_r$ by structural modification of HSO layer. To this day, the precise origin, either a preferred texture or a greater proportion of the o/t phase, is not completely established. In addition, impact of the depolarization field modulated by charge injection on 2Pr, should not be discredited. Charge injection will be enlightened in the following thanks to PUND measurements carried out over a wide range of frequencies.

C. Charge trapping evidenced by frequency-dependent measurement

To highlight the interplay between charge trapping across the dielectric and ferroelectric properties, the PUND frequency was varied from 10 Hz up to 5.10^5 Hz. For each frequency, 10 PUND periods (5000 data points) were collected to average the current noise at high frequencies. The maximum voltages were set to 3 V and 4 V for MFM and MFDM stacks, respectively. These selected voltages were close to the saturation voltage (i.e. $V_{max} \sim 2V_c$, where V_c is the coercive voltage) and prevented breakdown at low frequency. For each stack, the median frequency response of five capacitors was studied. The global voltage pulse scheme (shown in Fig. 6) experienced by the capacitor was adapted to minimize wake-up and fatigue-related modulation of $2P_r$ [16], [17]. To address this purpose, prior to measurements, each capacitor overcame a wake-up through 1000 bipolar triangular cycles. To avoid data interdependencies, devices were electrically reinitialized between each complete PUND measurement.

The corresponding P-V curves of MFM sample are shown in Fig. 7, from which $2P_r$(f) of Fig. 8 is extracted. The $2P_r$ evolution with frequency can be divided in two stages separated at the inflection point 10^5 Hz, as depicted in Fig. 8. For frequency above 10^5 Hz, the drop of $2P_r$ is accompanied with an increases of V_c as observed on P-V hysteresis in Fig. 7. These variations go with slight lags of dielectric current. This indicates that a RC delay affects the measure of ferroelectric properties. Given the small studied surface, such delay mostly comes from parasitic factor in the sensing circuit [18]. Thereupon, the study of switching dynamics will be now limited to frequency below 10^5 Hz.

From 10 Hz to 10^5 Hz, $2P_r$ decreases slightly from 17 to 15 µC/cm² (slope -0.65 µC.s/cm²) on MFM stack. This behaviour is explained in the light of ferroelectric switching kinetics [3], [19]. From 10 Hz to 10^5 Hz, the frequency increase makes defect migration harder within ferroelectric.

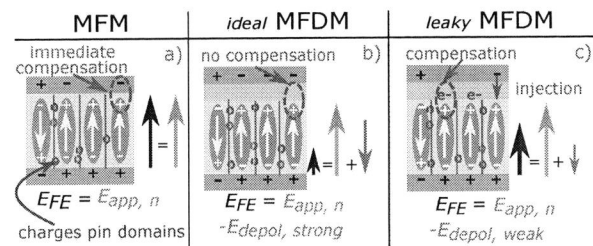

Fig. 5: Compensation in MFM, ideal MFDM and leaky MFDM stacks. By applying a bias, an external electric field E_{app} (pink arrow) is applied to the stack. E_{FE} (black arrow) is the total electric field perceived by the ferroelectric layer. In case of MFDM stacks (b) and (c), depolarization fields E_{depol} (blue arrow) counterbalances E_{app}. In ideal MFDM (b), E_{depol} is so strong that polarization is less stable than in MFM, leading to lower measured $2P_r$. In leaky MFDM (c), electron injection (e-) at D/F interface reduces the effect of E_{depol}, allowing to reach the same $2P_r$ as MFM stack. Charge accumulated at domain walls (green dots) can also affect polarization by pinning domains.

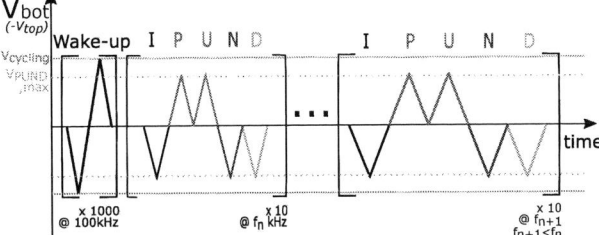

Fig. 6: Total voltage pulse scheme used on each capacitor to study the dynamic hysteresis of MFM and MFDM stacks. $V_{cycling}$ and $V_{PUND,max}$ were adapted to the total thickness of the stack, such as $V_{cycling} \geq V_{PUND,max}$. $V_{cycling}$ was set to 4V and 5V for MFM and MFDM devices respectively. $V_{PUND,max}$ was set to 3V, 4V for MFM and MFDM devices respectively.

Consequently, ferroelectric domains are progressively harder to unpin and remnant polarization decreases [20], [21].

The MFDM P-V curves at various frequencies are shown in Fig. 7. From 10 Hz to 10^5 Hz, the $2P_r$ of 1.5 nm Al_2O_3 MFDM stack decreases from 30 to 25 µC/cm² (slope -1.27 µC.s/cm²) (see Fig. 8). Steeper slopes are observed on all MFDM samples compared to the reference MFM one. These slopes are modulated by the dielectric thickness. Steep slopes for MFDM stacks suggest an additional contribution to the time-dependent ferroelectricity exhibited in MFM stack. In MFDM, this additional contribution is attributed to the reduction of injected charges in the D layer or at D/F interface with increased frequency. Charge injection relies on tunneling mechanisms and frequency dependent trap-assisted transport. As frequency is increased, ratio of the occupied traps declines. Indeed, when the voltage swing is carried out too rapidly, the trap capture/emission does not follow the electric field dynamics. In this condition, the quantity of trapped charges remains small, which limits the compensation of polarization charges and consequently decreases the as-measured $2P_r$. It is worth reminding that charge injection mechanism has exponential dependence on dielectric thickness. Less charges are injected when the dielectric layer is thicker. This last point explains the decrease of the absolute slope of $2P_r$(log(f)) curve with thicker dielectric thickness.

IV. CONCLUSION

As shown by GIXRD and PUND measurements, the insertion of Al_2O_3 layer modulates the ferroelectric performances

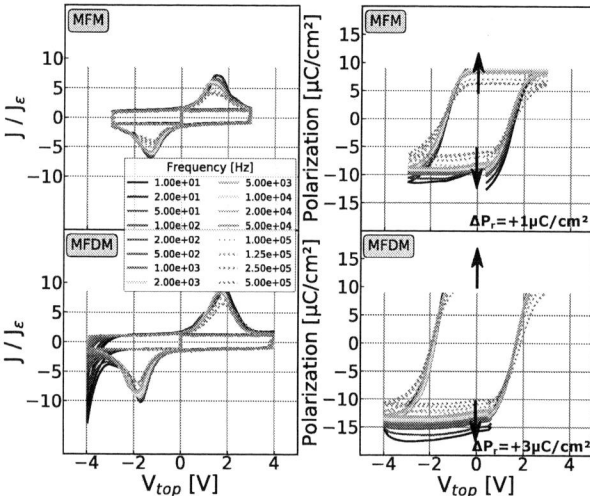

Fig. 7: PUND at various frequencies obtained during the voltage pulse scheme of Fig. 6. Left: Normalized current density-voltage characteristics with respect to the dielectric current density (J_ϵ); Right: Corrected polarization-voltage hysteresis loop. MFDM and MFM refer to TiN/HSO(10 nm)/Al$_2$O$_3$(1 nm)/TiN and TiN/HSO(10 nm)/TiN respectively. Measurement above 10^5 Hz (dotted lines) are affected by RC delays.

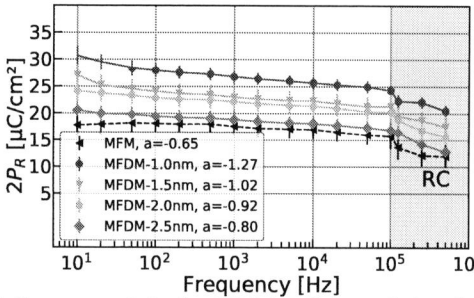

Fig. 8: Remanent polarization ($2P_r$) for all the studied stacks versus frequency. Each data point corresponds to the median $2P_r$ obtained on 5 capacitors. Errors bars are standard deviations. Above 10^5 Hz, measurements are affected by RC delay. a represents the slope of $2P_r(\log(f))$ expressed in µC.s/cm² for each Al$_2$O$_3$ thicknesses.

(i.e. remanent polarization) by two ways. Firstly, insertion of Al$_2$O$_3$ boosts the ferroelectric response of ferroelectric film. This boost relies on texture or phase proportion modification. Secondly, thickening of Al$_2$O$_3$ reduces the measured remanent polarization, by increasing the depolarization field. As viewed, this field is tuned by the conduction properties of Al$_2$O$_3$. The thinner the Al$_2$O$_3$ layer is, the more injected charges screen the depolarization field. The injection and trapping of charges into the dielectric layer in MFDM stacks was demonstrated through PUND study at various frequencies. In the MFM reference devices, remanent polarization evolves slightly with frequency, due to depinning/pinning of domains. In MFDM devices, the remanent polarization decreases by almost 5 µC/cm² when the frequency is increased from 10 Hz up to 5.10^5 Hz. This supports the idea that the exchange of trapped charges in the dielectric layer modulates the ferroelectric switching behavior.

Remanent polarization can be enhanced by favoring the amount of injected charge through the dielectric layer. In this respect, stacks with thin dielectric layers are promising candidates for high remanent polarization, leading to FTJs with high ON/OFF ratio. However, dielectric layer will have to be optimized in respect with device endurance.

Acknowledgment: This work was supported by European Union through the BeFerroSynaptic project (GA:871737). The authors wish to thank Beferrosynaptic's partners for fruitful discussions.

REFERENCES

[1] B. Max, M. Hoffmann, S. Slesazeck, and T. Mikolajick, "Ferroelectric tunnel junction based on ferroelectric-dielectric HfZrO$_2$/Al$_2$O$_3$ capacitor stacks," *Conf. ESSDERC*, pp. 142–145, 2018.

[2] V. Gaddam, D. Das, and S. Jeon, "Ferroelectricity Enhancement in HfZrO$_2$ Capacitors by Incorporating Ta$_2$O$_5$ Dielectric Seed Layers," *Conf. EDTM*, pp. 1–3, Apr. 2020.

[3] J. Wang, M. Qin, M. Zeng, X. Gao, and G. Zhou, "Excellent ferroelectric properties of Hf$_{0.5}$Zr$_{0.5}$O$_2$ thin films induced by Al$_2$O$_3$ dielectric layer," *IEEE Electron Device Lett.*, vol. 40, pp. 1937–1940, Dec. 2019.

[4] A. K. Tagantsev, M. Landivar, E. Colla, and N. Setter, "Identification of passive layer in ferroelectric thin films from their switching parameters," *J. Appl. Phys.*, vol. 78, pp. 2623–2630, Aug. 1995.

[5] A. Shekhawat, G. Walters, N. Yang, J. Guo, T. Nishida, and S. Moghaddam, "Data retention and low voltage operation of Al$_2$O$_3$/Hf$_{0.5}$Zr$_{0.5}$O$_2$ based ferroelectric tunnel junctions," *Nanotechnology*, vol. 31, p. 39LT01, Sept. 2020.

[6] H. J. Lee, M. H. Park, G. H. Kim, J. Y. Seok, Y. J. Kim, and C. S. Hwang, "Polarization switching and discharging behaviors in serially connected ferroelectric Pt/Pb(Zr,Ti)O$_3$/Pt and paraelectric capacitors," *J. Appl. Phys.*, vol. 109, p. 114113, June 2011.

[7] H. Ju Lee, M. Hyuk Park, Y. Jin Kim, C. Seong Hwang, J. Hwan Kim, and H. Funakubo, "Improved ferroelectric property of very thin Mn-doped BiFeO$_3$ films by an inlaid Al$_2$O$_3$ tunnel switch," *J. Appl. Phys.*, vol. 110, p. 074111, Oct. 2011.

[8] X. Wang, X. Sun, Y. Zhang, L. Zhou, J. Xiang, X. Ma, and H. Yang, "Impact of charges at ferroelectric/interlayer interface on depolarization field of ferroelectric FET with metal/ferroelectric/interlayer/Si gate-stack," *IEEE Transactions on Electron Devices*, vol. 67, pp. 4500–4506, Oct. 2020.

[9] R. Fontanini, M. Massarotto, R. Specogna, F. Driussi, M. Loghi, and D. Esseni, "Modelling and design of FTJs as high reading-impedance synaptic devices: depolarization field, read current and 4-bit synaptic weight resolution," *IEEE J. Electron Devices Soc.*, p. 4, 2021.

[10] H. W. Park, S. D. Hyun, I. S. Lee, S. H. Lee, Y. B. Lee, and M. Oh, "Polarizing and depolarizing charge injection through a thin dielectric layer in a ferroelectric–dielectric bilayer," *Nanoscale*, vol. 13, no. 4, pp. 2556–2572, 2021.

[11] T. Francois, J. Coignus, L. Grenouillet, J. Barnes, N. Vaxelaire, and J. Ferrand, "Ferroelectric HfO$_2$ for memory applications: impact of Si doping technique and bias pulse engineering on switching performance," *Conf. IMW*, pp. 1–4, May 2019.

[12] I. Fina, L. Fàbrega, E. Langenberg, X. Martí, F. Sánchez, and M. Varela, "Nonferroelectric contributions to the hysteresis cycles in manganite thin films: A comparative study of measurement techniques," *J. Appl. Phys.*, vol. 109, p. 074105, Apr. 2011.

[13] M. Hyuk Park, H. Joon Kim, Y. Jin Kim, W. Lee, T. Moon, and C. Seong Hwang, "Evolution of phases and ferroelectric properties of thin Hf$_{0.5}$Zr$_{0.5}$O$_2$ films according to the thickness and annealing temperature," *Appl. Phys. Lett.*, vol. 102, no. 24, p. 242905, 2013.

[14] Y. J. Kim, M. H. Park, W. Jeon, H. J. Kim, T. Moon, and Y. H. Lee, "Interfacial charge-induced polarization switching in Al$_2$O$_3$/Pb(Zr,Ti)O$_3$ bi-layer," *J. Appl. Phys.*, vol. 118, p. 224106, Dec. 2015.

[15] B. Max, T. Mikolajick, M. Hoffmann, S. Slesazeck, and T. Mikolajick, "Retention characteristics of Hf$_{0.5}$Zr$_{0.5}$O$_2$-Based ferroelectric tunnel junctions," *Conf. IMW*, pp. 1–4, May 2019.

[16] T. Schenk, E. Yurchuk, S. Mueller, U. Schroeder, S. Starschich, U. Böttger, and T. Mikolajick, "About the deformation of ferroelectric hystereses," *Appl. Phys. Rev.*, vol. 1, p. 041103, Dec. 2014.

[17] D. Zhou, J. Xu, Q. Li, Y. Guan, F. Cao, and X. Dong, "Wake-up effects in Si-doped hafnium oxide ferroelectric thin films," *Appl. Phys. Lett.*, vol. 103, p. 192904, Nov. 2013.

[18] Z. Gao, S. Lyu, and H. Lyu, "Frequency dependence on polarization switching measurement in ferroelectric capacitors," *J. Semicond.*, p. 6, 2022.

[19] Y. Zhang, X. L. Zhong, J. B. Wang, H. J. Song, Y. Ma, and Y. C. Zhou, "The electrical and switching properties of a metal-ferroelectric (Bi$_{3.15}$Nd$_{0.85}$Ti$_3$O$_{12}$)-insulator (Y$_2$O$_3$-stabilized ZrO$_2$)-silicon diode," *Appl. Phys. Lett.*, vol. 97, p. 103501, Sept. 2010.

[20] S. S. Fields, S. W. Smith, S. T. Jaszewski, T. Mimura, D. A. Dickie, and G. Esteves, "Wake-up and fatigue mechanisms in ferroelectric Hf$_{0.5}$Zr$_{0.5}$O$_2$ films with symmetric RuO$_2$ electrodes," *J. Appl. Phys.*, vol. 130, p. 134101, Oct. 2021.

[21] X. Li, C. Li, Z. Xu, Y. Li, Y. Yang, and H. Hu, "Ferroelectric properties and polarization fatigue of La:HfO$_2$ Thin-film capacitors," *ppss (RRL)*, vol. 15, p. 2000481, Apr. 2021.

Frequency modulation of conductance level in PCM device for neuromorphic applications

A. Trabelsi*, C. Cagli, T. Hirtzlin, O. Cueto, M. C. Cyrille, E. Vianello, V. Meli, V. Sousa, G. Bourgeois, F. Andrieu

CEA-Leti, Minatec Campus, 17 Rue de Martyrs, 38054 Grenoble, France

*E-mail: ahmed.trabelsi@cea.fr

Abstract—In this study we report for the first time the control of conductance level in PCM cells by means of a frequency modulation of progressive SET pulses. We show that by applying a train of progressive SET pulses, the conductance increases gradually and eventually saturates to a value G_{sat}. The latter can be tailored by changing the duty cycle of the pulse train. We propose a simple physics-based model to explain this effect. First, we simulated the thermal condition in the active region and showed that the increase of conductance was due to nucleation of a spherical hollow region around a central core, where re-amorphization takes place during programming. Based on this we illustrate that the frequency modulation can be analytically described by an equilibrium equation where the increase of conductance is balanced by the intrinsic chalcogenide resistance drift. The model is in very good agreement with data and shows that a fine tuning of the PCM device can be achieved. A frequency blind modulation of the programming pulse is believed to be much easier to be implemented in neuromorphic circuits as synaptic device [1][2].

Keywords—phase change device, neuromorphic, modelling, conductance modulation.

INTRODUCTION

Phase-change memory (PCM) is the most mature emerging non-volatile memory technology; it has been studied since the 60's and it is now-a-day particularly suitable for sub-40nm nodes in embedded applications [3]. More recently PCM has also been extensively investigated for applications other than data or code storage. The use of PCMs for learning neural networks has attracted in fact a lot of interest [4]. Mostly, the learning technique is based on the backpropagation gradient algorithm which has many weaknesses, in particular the lack of locality and the need for very precise updates [5]. Other promising algorithms have also been studied recently, notably by introducing a third parameter (3-factor learning rule and e-prop) [6]. Here we report for the first time a technique to control the cell conductance by the modulation of the frequency of a programming SET pulse train. The tailoring of the PCM conductance is indeed essential for neuromorphic applications. It has already been shown that conductance can be modified by applying the same programming pulse several times [7], but the conductance control via frequency modulation is a novel approach disclosed here for the first time. This frequency-dependent modulation of conductance opens a new pathway to on-chip learning with PCM devices, where PCM devices act as a synaptic weight.

In this paper we first show that the conductivity of a PCM device can be tailored by modifying the frequency of a SET pulse train in a reliable way. Next, we show, thanks to electro-

thermal simulations, that the inner-most part of the GST mushroom is re-amorphized after each pulse and based on this evidence we propose a simple analytical model that links the pulse frequency to the PCM conductance. This model is based on the inter-play between the pulse induced increase of conductance and the intrinsic drift in the amorphous PCM phase that counters it. By modulating the pulse frequency, the equilibrium between these two effects is shifted towards higher or lower conductance.

Experiment

To study the conductance modulation in PCM, we first built a 16kbit 1T1R NOR array on 300mm FDSOI wafer as illustrated in Fig. 1. Bitlines (BL) and wordlines (WL) are decoded by MUX which connect external pads to the selector gates and the top and bottom electrodes (TE and BE respectively). As selector we used a standard logic 28nm FDSOI thin oxide NMOS. The PCM cell is depicted in Fig. 1. It is integrated between metal 5 and 6 and the active material is $Ge_2Se_2Te_5$ (or GST 225 for short), connected to the lower W-plug by a 5nm wide TiN heater.

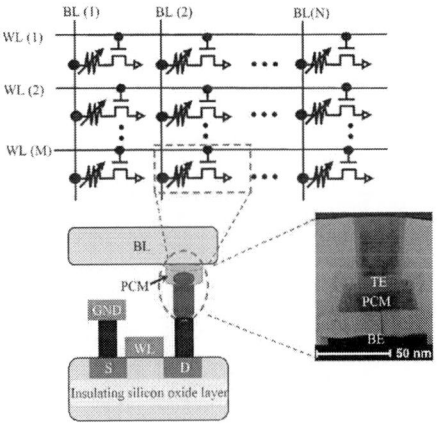

Figure 1: Schematic of the PCM 16kbit array.

The digital signals to decode the cell addresses are provided by an external microcontroller (Arduino Mega), while the analog signals to program and read the PCM devices are provided by an Agilent B1530 equipped with 4 PMUs. The experiment we performed is described in Fig. 2. After an initial RESET operation (I_{RESET}=557μA 20ns/500ns/20ns rise, width and fall times respectively), a train of successive voltage pulses is applied, while the current is monitored by the PMU

978-1-6654-8495-4/22 $31.00 © 2022 IEEE

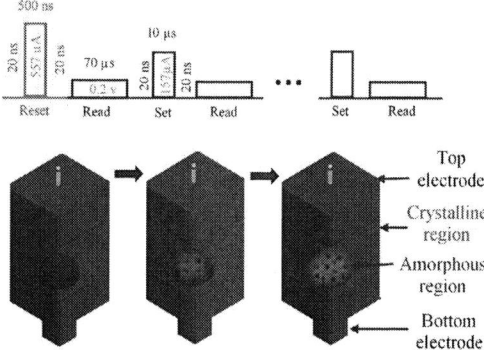

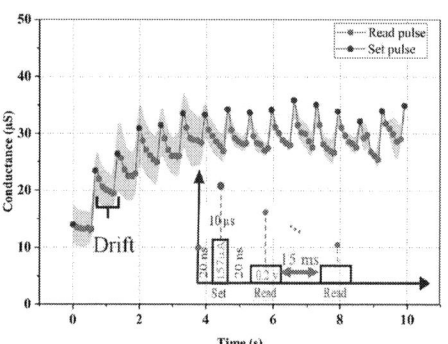

Figure 2: Up: Schematic of the applied pulse train. Down: Starting from fully amorphous state, PCM is gradually crystallized using voltage pulses with the same amplitude.

(I_{SET}=156µA 20ns/10µs/20ns). After each pulse the conductivity is dynamically measured at low field (I_{read}=0.2V).

When multiple readings between each SET pulse are collected, the typical outcome is shown in Fig. 3. Each SET pulse increases the conductivity, while drift lowers it. Conductivity increase and drift are always counteracting each other, and this leads eventually to a conductivity saturation as was also previously reported by other authors [8] [9] [10].

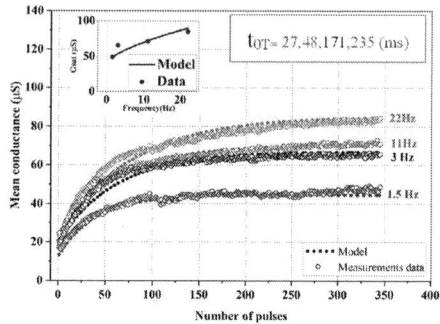

Figure 4: Average on 20 devices of the conductance increase during the experiment depicted in Fig. 2. Data are fitted with ∆G =1.1µS and the t_{0T} values reported in the box. Inset: G_{sat} vs pulse applied frequency.

DISCUSSION AND MODELLING

To understand and model the increase of G_{sat} vs frequency we first conducted a multi-physics simulation as described in Fig 5. The simulation model couples trap assisted tunneling with the heat equation and nucleation/growth physics. A detailed description of the simulation tool can be found in [11]. As Fig. 5 shows, when the SET pulse is applied the inner core of the amorphous dome melts down as the temperature is higher than the melting point. As the pulse has a steep falling edge (20ns) the inner core is once again amorphized. Differently, in a roughly spherical hollow region around the inner core, the temperature stays below the melting point. It is in this region that nucleation occurs and GST crystals form leading to an increase of conductivity. Each successive pulse re-amorphizes the inner core and increases the crystal volume around it. Simulations show that the increase of conductivity is directly connected to the increase of crystal volume which improves the current percolation path. In this simulation however the drift physics is not implemented yet and consequently no decrease of conductivity can be observed.

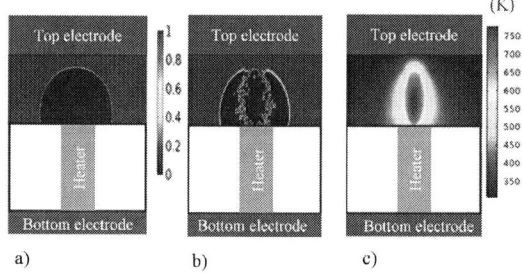

Figure 5: Multi-physics simulation for a progressive crystallization phenomenon. A) Initial fully RESET state. B) The mushroom inner core is melt-quench at each SET pulse. C) Temperature profile during a 150µA electrical SET pulse

Nonetheless based on this result we can surely assume that the conductivity drift after each SET pulse must be systematically identical pulse after pulse, and this allows us to write the drift equation in a recursive way. The drift after pulse n+1 can thus be described by:

$$G_{n+1}(t) = G_n \left(\frac{t}{t_{0T}} \right)^{-\nu} \quad (1)$$

Figure 3: Evolution of the average conductance and standard deviation for 20 devices. Drift is measured by reading the cell conductance five times.

It is important to stress that drift is intrinsically present as the GST mushroom is only partly crystallized, while the remaining portion is kept amorphous and, as we will discuss shortly, it is re-melted and re-amorphized at each SET pulse. We repeated this experiment by modifying the time between successive SET pulses. Data are reported in Fig. 4. In this case the conductivity after each pulse was only read once, to increase the pulse frequency range. We thus applied a train of 350 pulses with frequencies from 1.5Hz up to 22 Hz. Each curve in Fig.4 represents the average over 20 different cells. As can been seen, the conductivity saturates at different values (G_{sat}) depending on the applied frequency. The inset of Fig. 4 shows the saturation values as a function of the frequency. By increasing the frequency, a higher G_{sat} can be obtained.

Where the drift pre-factor t_{0T} depends on the period T. Clearly, after each pulse we have:

$$G_1 = (1 - \gamma)\, G_0 + \Delta G$$

$$G_2 = (1 - \gamma)\, G_1 + \Delta G$$

$$\dots$$

$$G_n = (1 - \gamma)\, G_{n-1} + \Delta G \quad (2)$$

Where ΔG is the conductivity increase due to the additional crystallization as described earlier and γ is defined as:

$$(1 - \gamma) := \left(\frac{t_{0T}}{T}\right)^\nu \quad (3).$$

Eq. 3 is derived from Eq. 1 noting that the pulse period T is the time the cell is allowed to drift between two consecutive pulses.

At saturation the increase of conductance is balanced by the drift occurred over a period, so that $G_n = G_{n+1} =: G_{sat}$. From Eq. 2 we have:

$$G_{sat} = (1 - \gamma)\, G_{sat} + \Delta G$$

And thus:

$$\gamma G_{sat} = \Delta G \quad (4).$$

If we assume that ΔG is not dependent on the period T nor on the cell conductance, Eq. 4 allows to describe G(t) for different frequencies. We measured the drift coefficient ν running drift experiments at room temperature (see Fig. 6) and extracted the coefficients with the well-known drift equation as shown in the inset of Fig. 6. As data show, ν is constant in a relatively large range of resistances and has a mean value of 0.025.

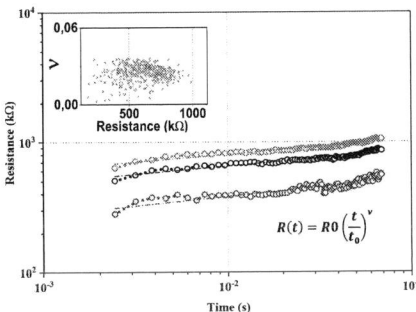

Figure 6: Drift measurements at room temperature. The drift coefficient was measured for 4000 devices and averaged to 0.025

This value is then used in Eq. 2 to fit data as shown in Fig. 4. $\Delta G = 1.1\,\mu S$ was extracted as a fit parameter and is consistent with the increase of conductance reported in Fig. 4. In the same way t_{0T} were extracted for each period T. The model proves to be in good agreement with data and this shows that Eq. 4 provides a good analytical tool to map the pulse frequency to the saturation conductance level. The inset of Fig 4 reports the model expectation of G_{sat}, where t_{0T} have been described as a power law function of the period T. This allows to program the cell conductance by using a frequency modulation approach where no pulse amplitude modulation is required, nor is any program/verify routine. By looking at Eq. 4, one can easily notice that the saturation conductance does not depend on G_0. This might look strange at first sight

but can be understood remembering that the saturation level corresponds to the G level at which drift and conductance increase cancel out, thus logically it cannot be linked to G_0. To verify this, we conducted an additional experiment where multiple cells are initially prepared at different G_0 levels and are then programmed with the same pulse train at the same pulse frequency of 22 Hz. The results in Fig. 7 show that all four curves (average of 20 cells), initially programmed with G_0 ranging between 18 to 43 μS, saturate at the same $G_{sat} \approx 80\,\mu S$.

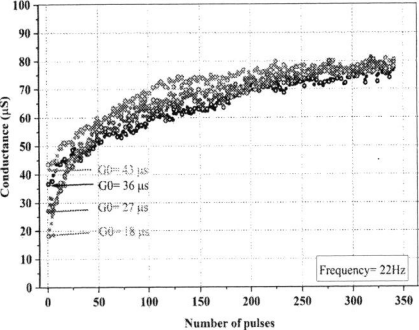

Figure 7: starting form a different conductance values (G_0) and applying SET pulses of the same frequency the same G_{sat} is achieved. Shown data report the average over 20 devices.

These results strongly support the claim that the frequency modulation of the saturation conductance can be described in good approximation by the simple balance between drift and partial crystallization. This equilibrium easily compensates the variability in the progressive crystallization (see Fig. 3 and 4) that comes from device-to-device variability plus the intrinsic nucleation variability.

CONCLUSION

In this study we investigated the modulation of conductance in PCM devices exploiting a train of identical SET pulses. We show that the conductance can be adjusted changing the frequency of the applied pulses. We demonstrate that this mechanism is fully understood as an equilibrium between nucleation and drift. Multiphysics simulations show that each pulse causes the re-amorphization of a central GST dome core, while a progressive nucleation takes place in its surroundings. This allows us to reset the drift equation after each pulse, and lead to a conductance saturation at which the increase of conductance is counter balanced by a drift induced decrease of the same amount. By adjusting the pulse frequency, the equilibrium point is moved allowing the fine-tuning of the final conductance. The latter does not depend on the initial conductance value, but only on the applied frequency. This technique can be used in neuromorphic inspired synaptic circuit, where, for instance, the synaptic weight is stored in the PCM conductance value by pulsing an identical stimulus at a calibrated frequency. This approach would allow the storage of the synaptic weight without the need of any program/verify routines nor the need to tailor the programming pulse, thus significantly reduce the circuit complexity.

978-1-6654-8495-4/22 $31.00 © 2022 IEEE

ACKNOWLEDGMENT

A.T was supported by the CEA NUMERICS program, which has received funding from the European Union's Horizon 2020 research and innovation program under the Marie Sklodowska-Curie grant agreement No 800945.

T.H was supported under the MeM-Scales EU Horizon 2020 research and innovation program under the grant agreement 871371.

REFERENCES

[1] Nandakumar, S. R., Irem Boybat, Manuel Le Gallo, Evangelos Eleftheriou, Abu Sebastian, and Bipin Rajendran. "Experimental Demonstration of Supervised Learning in Spiking Neural Networks with Phase-Change Memory Synapses." Scientific Reports 10, no. 1 (December 2020): 8080.

[2] Sarwat, Syed Ghazi, Benedikt Kersting, Timoleon Moraitis, Vara Prasad Jonnalagadda, and Abu Sebastian. "Phase-Change Memtransistive Synapses for Mixed-Plasticity Neural Computations."

[3] Arnaud, F., P. Ferreira, F. Piazza, A. Gandolfo, P. Zuliani, P. Mattavelli, E. Gomiero, et al. "High Density Embedded PCM Cell in 28nm FDSOI Technology for Automotive Micro-Controller Applications." In 2020 IEEE International Electron Devices Meeting (IEDM), 24.2.1-24.2.4, 2020.

[4] Ambrogio, Stefano, Nicola Ciocchini, Mario Laudato, Valerio Milo, Agostino Pirovano, Paolo Fantini, and Daniele Ielmini. "Unsupervised Learning by Spike Timing Dependent Plasticity in Phase Change Memory (PCM) Synapses." Frontiers in Neuroscience 10 (2016).

[5] Nandakumar, S. R., Manuel Le Gallo, Irem Boybat, Bipin Rajendran, Abu Sebastian, and Evangelos Eleftheriou. "A Phase-Change Memory Model for Neuromorphic Computing." Journal of Applied Physics 124, no. 15 (October 21, 2018): 152135.

[6] Demirağ, Yiğit, Filippo Moro, Thomas Dalgaty, Gabriele Navarro, Charlotte Frenkel, Giacomo Indiveri, Elisa Vianello, and Melika Payvand. « PCM-Trace: Scalable Synaptic Eligibility Traces with Resistivity Drift of Phase-Change Materials ».

[7] Sebastian, Abu, Manuel Le Gallo, and Evangelos Eleftheriou. "Computational Phase-Change Memory: Beyond von Neumann Computing." Journal of Physics D: Applied Physics 52, no. 44 (August 2019): 443002.

[8] Suri, Manan, Daniele Garbin, Olivier Bichler, Damien Querlioz, Dominique Vuillaume, Christian Gamrat, and Barbara DeSalvo. « Impact of PCM Resistance-Drift in Neuromorphic Systems and Drift-Mitigation Strategy ».

[9] Papandreou, N., A. Sebastian, A. Pantazi, M. Breitwisch, C. Lam, H. Pozidis, and E. Eleftheriou. « Drift-resilient cell-state metric for multilevel phase-change memory ». In 2011 International Electron Devices Meeting, 3.5.1-3.5.4, 2011.

[10] Zhang, Wei, and Evan Ma. "Unveiling the Structural Origin to Control Resistance Drift in Phase-Change Memory Materials." Materials Today 41 (December 1, 2020): 156–76.

[11] O. Cueto, A. Trabelsi, C.Cagli, M.C Cyrille submitted to IEEE conf. SISPAD 2022.

Thermal switching of TiO$_2$-based RRAM for parameter extraction and neuromorphic engineering

Alessandro Milozzi[1], Daniel Reiser[2,3], Andreas Drost[2], Thomas Neuner[2,3], Marc Tornow[2,3] and Daniele Ielmini[1].

(1) Dipartimento di Elettronica, Informazione e Bioingegneria (DEIB) - Politecnico di Milano, Milan, Italy
(2) Fraunhofer EMFT Research Institution for Microsystems and Solid State Technologies, Munich, Germany
(3) Molecular Electronics, Technical University of Munich, Garching, Germany

Abstract—Recently, resistive switching random access memory (RRAM) has gained maturity for storage class memory and in-memory computing. For these applications, an improved control of the switching phenomena can lead to higher data density and computing accuracy, thus paving the way for RRAM-based artificial intelligence (AI) accelerators for edge computing. This work presents a study of thermally-induced switching in TiO$_2$-based RRAM devices. Thermal switching is explained by defect rediffusion controlled by the activation energy for defect migration in TiO$_2$. Experiments and simulations support thermal switching as a tool for parameter extraction in RRAM, as well as for novel neuromorphic cognitive functions for brain-inspired computing.

Index Terms—RRAM, device modeling, neuromorphic engineering, embedded memory, neural network.

I. INTRODUCTION

Among the emerging memory technologies, resistive switching random access memory (RRAM) is one of the most promising thanks to the back-end integration, low power and good scalability in crosspoint arrays [1]. RRAM relies on the atomic migration and diffusion of ionized defects [2], which in turn is responsible for the stochastic variation and time-dependent fluctuation of resistive states [3]. Variations result in memory bit errors and poor inference accuracy in neural networks [4], thus they should be mitigated by advanced materials engineering and better programming algorithms. In this perspective, novel characterization tools may provide a deeper understanding of the switching phenomena, thus supporting a knowledge-based modeling and design of RRAM.

In this work, we study thermal switching in TiO$_2$-based RRAM devices [5]. After programming the device in the high resistance state (HRS) by electrical reset, we observe thermally-induced set process at elevated temperature around 150°C. This is interpreted as the thermally-activated rediffusion of defects across the depleted gap in the HRS. The dependence on electrode materials is explained by a different activation energy of diffusion, which is confirmed by the Kissinger technique to extract the activation energy. Thermal switching thus emerges as a powerful tool for the characterization of physical parameters of RRAM. Finally, we illustrate novel brain-inspired homeostatic plasticity which could support the development of neuromorphic systems.

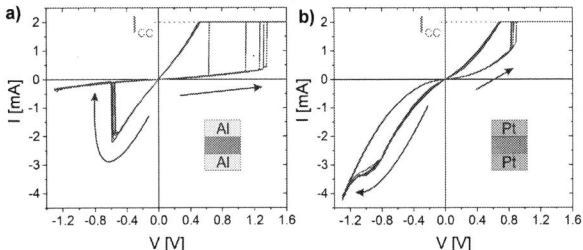

Fig. 1: Electrical characteristics of RRAM devices. (a) Measured I-V curve for Al/TiO$_2$/Al RRAM. (b) Measured I-V curve for Pt/TiO$_2$/Pt RRAM.

II. RRAM CHARACTERIZATION

Fig. 1 show the electrical characteristics of RRAM devices with Al/TiO$_2$/Al stack (a) and Pt/TiO$_2$/Pt stack (b). The voltage was applied at the top electrode (TE), while the bottom electrode (BE) was grounded. The RRAM devices were initially formed at positive bias with a current compliance I$_{cc}$ = 4 mA. Both RRAM devices show stable bipolar switching behavior with set and reset processes taking place at positive and negative voltage, respectively. The shape of the switching characteristics depends on the electrode, with the Al/TiO$_2$/Al stack showing digital (abrupt) switching, while the Pt/TiO$_2$/Pt

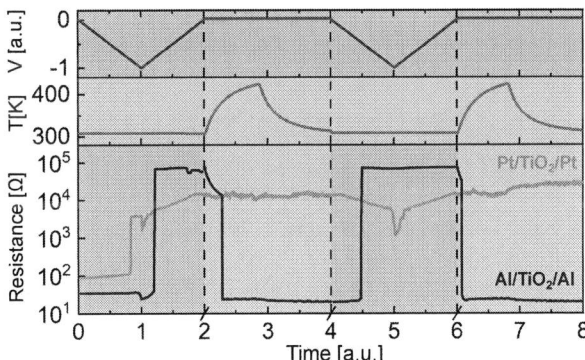

Fig. 2: Thermal switching of RRAM devices. First, a negative voltage ramp for electrical reset (top), then the temperature is increased (centre) which induces a thermal set transition in Al/TiO$_2$/Al RRAM (bottom). No thermal switching is seen in Pt/TiO$_2$/Pt RRAM

978-1-6654-8495-4/22 $31.00 © 2022 IEEE

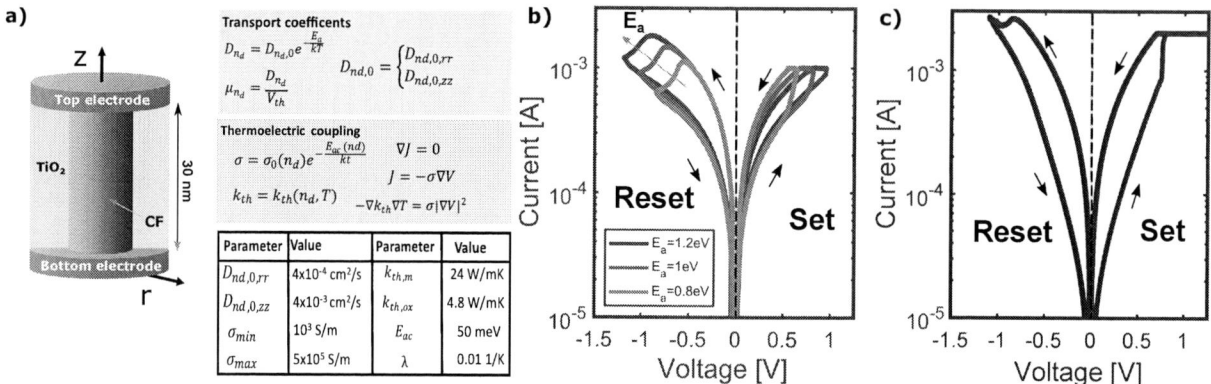

Fig. 3: Numerical model. (a) Geometry of the simulated structure including top/bottom electrodes and oxide layer with CF in cylindrical coordinates. The table reports the main equations and physical parameters. (b) calculated I-V curves for increasing E_a, controlling the set and reset voltages. (c) Experimental I-V curves for the $Al/TiO_2/Al$ RRAM device.

stack shows analog-type switching especially along the reset transition. The two devices also show markedly different reset voltage V_{reset}, with V_{reset} = 0.6 V and V_{reset} = 0.9 V for $Al/TiO_2/Al$ and $Pt/TiO_2/Pt$, respectively.

Fig. 2 shows the thermal switching experiments, indicating the applied voltage V (top), the ambient temperature T (centre) and the device resistance measured at V_{read} = 25 mV (bottom). For $Al/TiO_2/Al$ stack, the device was initially programmed in HRS by an electrical reset operation with negative voltage, then T was increased to 150°C, which induced set transition to low resistance state (LRS) [5]. Note that the read voltage of 25 mV is negligible, thus the reset transition must be entirely attributed to the increase of T. The same experiment was executed with the $Pt/TiO_2/Pt$ stack (see Fig. 2), although no thermally-induced set transition was seen. No thermal switching was visible after electrical set transition.

III. THERMAL SWITCHING MODEL

To explain the thermal switching in the $Al/TiO_2/Al$ RRAM device, we carried out simulations by our finite-element method (FEM) numerical model of RRAM switching [6]. The model relies on the drift-diffusion mechanisms for the resistive switching, where ionized defects, such as oxygen vacancies and metallic impurities, migrate in response to the applied electric field and the concentration gradient. Fig. 3a shows a sketch of the simulated structure, consisting of a conductive filament (CF) connecting the TE and BE with cylindrical symmetry. The CF, arising from the forming operation, consists of a localized region with a high concentration of defects. Defect migration is described by the drift-diffusion equation given by:

$$\frac{\partial n_d}{\partial t} = \nabla(D_n \nabla n_d - \mu F n_d) \qquad (1)$$

where n_d is the concentration of defects, D_n is the ionic diffusivity, μ is the ionic mobility and F is the local electric field. The diffusivity and mobility are linked by the Einstein rule and are thermally activated according to the Arrhenius law with an activation energy E_a. Electrical and thermal conductivities depend on the defect concentration and temperature to reflect the insulator-metal transition within the oxide layer [6]. Anisotropic diffusion was assumed with the vertical diffusivity D_{zz} being larger than the radial diffusivity D_{rr}, as a result of the localized structural modification due the forming and switching. Fig. 3a reports the relevant model parameters used in the simulations.

A. Simulations of electrical switching

Fig. 3b shows the calculated I-V curves with the numerical model of Fig. 3a. A select transistor was connected in series to the RRAM device to limit the current during the set transition as in the experiments in Fig. 1. The simulations results agree well with the measured characteristics of the $Al/TiO_2/Al$ RRAM device in Fig. 3c. To assess the impact of E_a on the set/reset voltages, simulations were carried out at increasing E_a, namely E_a = 0.8 eV, 1 eV and 1.2 eV. Both V_{reset} and the set voltage V_{set} increase at increasing E_a, as a result of the higher barrier for ionic migration, thus requiring a higher field and a higher Joule heating to induce the transition. The higher V_{reset} in the $Pt/TiO_2/Pt$ stack, compared with the $Al/TiO_2/Al$ RRAM, can thus be explained by a higher E_a for defects migration.

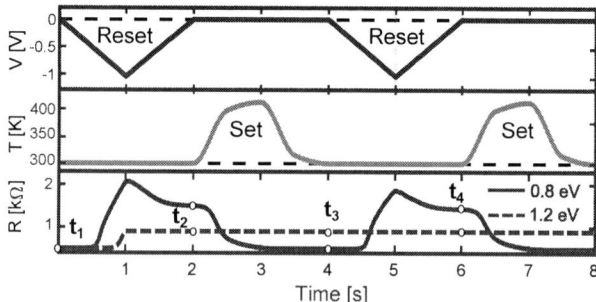

Fig. 4: Simulation of thermal switching. The reset operation is induced by the negative voltage ramp (top), while the temperature increase (centre) causes a set transition (bottom) for E_a = 0.8 eV but not for E_a = 1.2 eV, in agreement with the behaviors of $Al/TiO_2/Al$ and $Pt/TiO_2/Pt$, respectively, in Fig. 2.

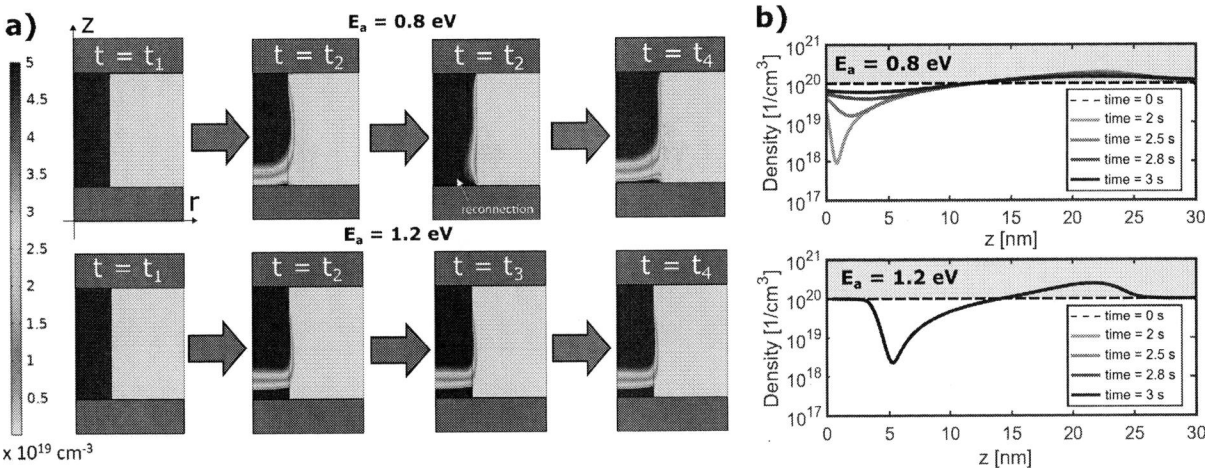

Fig. 5: Thermal reset process. (a) Contour plot of simulation results for the defect concentration at increasing times for $E_a = 0.8$ eV (top) and $E_a = 1.2$ eV (bottom). (b) Defect distribution along the CF axis ($r = 0$) as a function of z for 0.8 eV (top) and 1.2 eV (bottom). Defect rediffusion causes the collapse of the CF for $E_a = 0.8$ eV, which explains the thermal reset process in Fig. 2.

B. Simulations of thermal switching

The thermal switching effect was simulated by the same numerical model of Fig. 3. Fig. 4 shows the applied voltage (top), temperature (center) and the calculated resistance (bottom) as a function of time during the thermal annealing at 150°C for $E_a = 0.8$ eV and 1.2 eV. First, a negative voltage sweep is applied to prepare the device in the HRS by electrical reset operation, then T is increased to a maximum value of 150°C to induce thermal switching. This operation was simulated two times to assess the repeatability of the thermal switching process. Starting from the HRS, the resistance drops to the LRS for $E_a = 0.8$ eV, while no thermal reset is seen for the larger $E_a = 1.2$ eV. This is in agreement with the behavior of Al/TiO$_2$/Al and Pt/TiO$_2$/Pt, respectively, in Fig. 2. Fig. 5a shows the 2D defect concentration profiles at increasing time during annealing, while Fig. 5b shows the integrated defect concentration as a function of the vertical axis. The thermally-induced set transition for $E_a = 0.8$ eV can be explained by the rediffusion of defects across the depleted gap, namely the region with low defect concentration for $z < 5$ nm, close to the BE interface, which is responsible for the high resistance of the HRS. Due to the large concentration gradient dn/dz in the gap region, defects diffuse thus resulting in the steep drop of resistance. For $E_a = 1.2$ eV, the annealing temperature and time are not sufficient to stimulate diffusion, thus preventing the thermal reset process. The activation energy thus appears as a key parameter controlling electrical and thermal switching processes in RRAM.

C. Kissinger study of thermal switching

To validate our interpretation of thermal switching as due to thermally-driven diffusion, we used an adapted Kissinger technique for the extraction of activation energy in our devices [7,8]. In the Kissinger analysis, V_{reset} is monitored as a function of the voltage sweep rate. Due to the strong thermal activation of reset transition, V_{reset} marks the point where the transition temperature T_P is reached within the CF. For increasing sweep rate, T_P increases, thus evidencing the activation energy E_a of migration [7,8].

Fig. 6a shows the I-V curve at V<0 for Al/TiO$_2$/Al RRAM and Pt/TiO$_2$/Pt RRAM, respectively. The reset transition indicates that $|V_{reset}|$ increases at increasing sweep rate, as expected from the thermal activation of reset. Note that the critical temperature cannot be directly measured, thus, the usual Kissinger plot of the heating rate $\beta = dT/dt$ divided by T_P^2

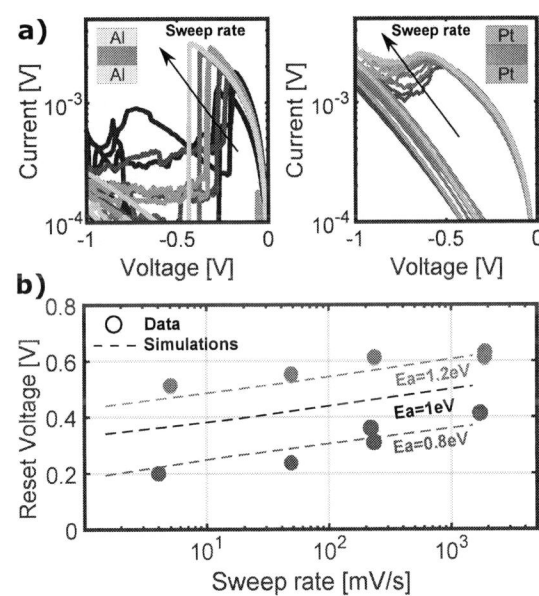

Fig. 6: Kissinger analysis. (a) Measured IV curves at increasing sweep rate for Al/TiO$_2$/Al (left) and Pt/TiO$_2$/Pt (right). (b) Measured V_{reset} as a function of the sweep rate, compared to simulation results for increasing E_a

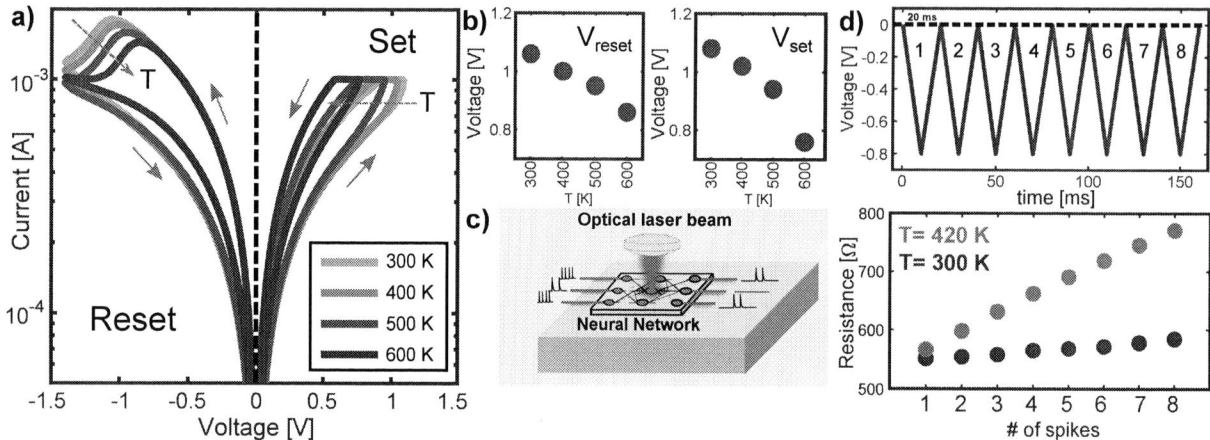

Fig. 7: Homeostatic intrinsic plasticity with RRAM. (a) Simulated I-V curves for increasing T, indicating a decrease of set and reset voltages. (b) Calculated set and reset voltages as a function of T. (c) Sketch of a neuromorphic computing system where an optical laser induces local heating, thus facilitating RRAM-neuron integration. (d). Applied voltage spikes and calculated R for T = 300 K and T = 420 K, evidencing T-dependent neuron integration.

as a function of 1/kT is not possible in this case [7,8]. Fig. 6b reports instead the measured V_{reset} as a function of the sweep rate. V_{reset} was also computed from the numerical model assuming various values of E_a, namely E_a = 0.8 eV, 1 eV and 1.2 eV. The simulation results agree well with the measured characteristic of Al/TiO$_2$/Al RRAM and Pt/TiO$_2$/Pt RRAM for E_a = 0.8 eV and E_a = 1.2 eV, respectively. Our model can thus fully account for the experimental characteristics of both thermal switching (Fig. 2) and electrical switching (Fig. 1) of the two RRAM stacks with just one set of parameters and different E_as.

IV. HOMEOSTATIC PLASTICITY

The capability of temperature-induced change of the device resistance paves the way for novel bioinspired neuromorphic functions. In particular, homeostatic plasticity mechanisms act on specific regions of the brain on a global basis to modify the learning properties of neurons and synapses [9]. In homeostatic intrinsic plasticity, the release of particular neurotransmitters such as dopamine changes the neuron threshold of a particular area of the brain, thus facilitating the neural firing activity. In our devices, this effect can be emulated adopting a RRAM device as an integrate and fire neuron [10] and inducing a local increase of temperature, which decreases the threshold of the RRAM-based neuron.

This is illustrated in Fig. 7a, showing the calculated I-V curves of a RRAM device with E_a = 0.8 eV for increasing temperature. Fig. 7b shows the calculated V_{set} and V_{reset} as a function of T: As T increases, both V_{set} and V_{reset} decrease, as a result of the cooperation of thermally- and electrically-induced switching. Fig. 7c shows a sketch of an homeostatic plasticity behavior of the RRAM neuron, where an optical laser beam can induce a local or global heating, thus facilitating spike integration and fire. Fig. 7d shows the computed RRAM conductance in response to a train of spikes for T = 25°C and T = 150°C: in the latter case, the

RRAM neuron integration is accelerated because of the local heating. These results pave the way for mixed optical/electrical neuromorphic systems thanks to the cooperation of thermally- and electrically-induced switching in RRAM devices.

V. CONCLUSIONS

In this work, we presented experiments on thermally-induced switching in Al/TiO$_2$/Al and Pt/TiO$_2$/Pt RRAM devices. We explained the phenomenon as a rediffusion of defects in TiO$_2$ across the depleted gap of HRS. The intepretation is supported by detailed numerical simulations and experimental Kissinger analysis. Thermal switching can thus be used as a new tool for the extraction of RRAM parameters, as well as for mixed optical/electrical neuromorphic systems.

ACKNOWLEDGMENT

This work received funding from the European Union's Horizon 2020 research and innovation program (grant agreement no. 899559) and it was partly supported by BMBF (funding codes: 16FMD01K, 16FMD02, 16FMD03). We greatfully acknowledge S. Altmannshofer, C. Hochreiter, J. Stuprich, D. Linke, M. Wackerle, U. Seidelmann, C.-H. Vu, D. Linke (all Fraunhofer EMFT, Munich) for technical support and I. Eisele (EMFT) for helpful discussions.

REFERENCES

[1] D. Ielmini and H.S Philip Wong, Nature Electronics 1.6, 333-343 (2018).
[2] D. Ielmini, Semiconductor Science and Technology 31.6, 063002 (2016).
[3] S. Ambrogio et al., 2013 IEEE International Electron Devices Meeting, 31.5.1-31.5.4 (2013).
[4] V. Milo et al., APL Materials 7, (2019).
[5] D. Reiser et al., 2020 IEEE 20th International Conference on Nanotechnology 342-347 (2020).
[6] S. Larentis et al., IEEE Trans. Electron Devices 9, 2468-2475 (2012).
[7] S. Yu and H.S. Philip Wong, IEEE Electron Device Lett. 12, 1455-1457 (2010).
[8] C. Cagli et al., IEEE Trans. Electron Devices 8, 1712-1720 (2009).
[9] G. Turrigiano, Cold Spring Harbor perspectives in biology 4.1, a005736 (2012).
[10] S. Lashkare et al., IEEE Electron Device Lett. 39, 484-487 (2018).

Improvement of FTJ on-current by work function engineering for massive parallel neuromorphic computing

Suzanne Lancaster
NaMLab gGmbH,
Dresden, Germany
suzanne.lancaster@namlab.com

Quang T. Duong
NaMLab gGmbH,
Dresden, Germany
quang.duong@namlab.com

Erika Covi
NaMLab gGmbH
Dresden, Germany,
erika.covi@namlab.com

Thomas Mikolajick
NaMLab gGmbH & IHM TUD
Dresden, Germany
thomas.mikolajick@namlab.com

Stefan Slesazeck
NaMLab gGmbH
Dresden, Germany
stefan.slesazeck@namlab.com

Abstract— HfO$_2$-based ferroelectric tunnel junctions (FTJs) exhibit attractive properties for adoption in neuromorphic applications. The combination of ultra-low-power multi-level switching capability together with the low on-current density suggests the application in circuits for massive parallel computation. In this work, we discuss one example circuit of a differential synaptic cell featuring multiple parallel connected FTJ devices. Moreover, from the circuit requirements we deduce that the absolute difference in currents I$_{on}$ − I$_{off}$ is a more critical figure of merit than the tunneling electroresistance ratio (TER). Based on this, we discuss the potential of FTJ device optimization by means of electrode work function engineering in bilayer HZO/Al$_2$O$_3$ FTJs.

Keywords — ferroelectric tunneling junction, FTJ, work function engineering, differential synaptic pair.

I. INTRODUCTION

In non-von-Neumann architectures the latency and power consumption required for computational tasks is reduced by repealing the separation between memory and computing engines. Instead, data is processed where it is stored [1]. Especially in power-efficient edge computing, the ability for normally-off computing requires the utilization of non-volatile memory devices, thus enabling the system to be powered on and off frequently to process data only when needed or to cope with frequent power interruptions. In recent years the ferroelectric tunnel junction (FTJ) has emerged as an interesting non-volatile memory device featuring non-destructive readout [2], high-speed write operation [3], low voltage operation, and high endurance cycling [4]. The concept of an FTJ was first introduced by Esaki et al. [5]. The operating principle is based on a difference between potential barrier height and width at a metal-ferroelectric interface depending on the direction of the spontaneous polarization (P). This polarization can be switched using an external electric field. According to Simmons [6], the tunneling current density is inversely proportional to the potential barrier, with

$$J \sim \exp(-\sqrt{\varphi}), \qquad (1)$$

where φ is the potential barrier height above the Fermi level. Therefore, two distinct resistance states could be generated depending on the direction of the polarization. The difference in conductance of the two states is called the tunneling electroresistance ratio (TER), given by TER=(G$_{on}$-G$_{off}$)/G$_{off}$, where G$_{off}$ is the conductance for the P$_{down}$ state and G$_{on}$ is the current for the P$_{up}$ state [7].

Recently, a differential synaptic cell featuring two FTJ devices has been proposed [8] as is depicted in Fig. 1 a. The circuit concept is based on a combination of a current normalizing differential synaptic weighting element [9] and a 2T1C read-out circuit to cope with the small on-currents of small-scaled FTJ devices [10]. In this circuit the readout is performed in two steps by first pre-charging the gates of T$_3$ and T$_4$ via the access transistors to bit-line (BL) while already applying the readout voltage difference between plate-line (PL) and BL, and in a second phase the access transistors are switched off in order to develop a voltage difference ΔV between nodes n$_1$ and n$_2$ by charging the gates of T$_3$ and T$_4$ via the FTJ devices C$_1$ and C$_2$.

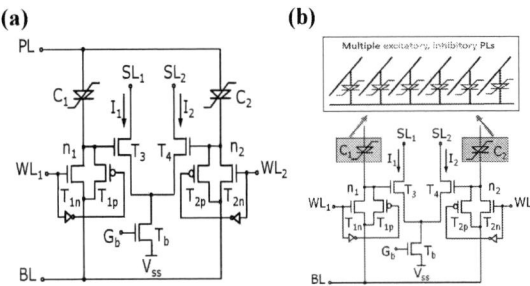

Fig. 1. (a) Differential synaptic cell using two FTJ devices. © 2021 IEEE. Reprinted, with permission, from [8]; (b) differential synaptic cell using multiple parallel FTJ devices.

As has been discussed already in [8] and [11], the voltage difference ΔV depends mainly on the FTJ current density J$_{FTJ}$, the FTJ self-capacitance per area C$_{0,FTJ}$ and the signal development time t$_{read}$ according to the equation:

$$\Delta V = (J_{FTJ}\ t_{read})\ /\ C_{0,FTJ} \qquad (2)$$

Moreover, it has been shown by means of circuit simulation using calibrated FTJ models, that despite the very low on-current densities of the used FTJ devices in the range of <1 pA/µm² this synaptic weighting cell can be operated successfully at biologically plausible time scales with read times in the 100 µs range and typical values of ΔV in the range of 100 mV. In this simplest version of the synaptic cell 7 transistors are used. The voltage signals to operate the cell such as PL, BL, word-line (WL), and its complement voltages are provided from outside and the respective transistors are not shown. Hence, when looking from a

978-1-6654-8495-4/22 $31.00 © 2022 IEEE

system level perspective on this circuit concept it turns out that the number of transistors per synaptic weighting element is large, which makes the design not very efficient, both in terms of silicon area as well as power consumption.

II. PARALLEL FTJs IN DIFFERENTIAL SYNAPTIC CELL

Fig. 1 b depicts a differential synaptic cell that uses multiple parallel connected FTJs. The idea of this concept is to integrate multiple FTJ devices in a cross-bar array on top of the CMOS circuitry. Thus, multiple FTJ devices could be read in parallel, thereby greatly reducing the CMOS circuitry overhead. This configuration also allows the implementation of multi-level cells by parallel operation of multiple single-level cells that can eventually reduce the programming circuit overhead. Another interesting aspect arising for a large number, n, of FTJs is that the overall node capacitance connected to n_1 and n_2 increases by a factor of n with respect to the capacitance of the individual FTJ. That enables to perform program and erase operations of single FTJs without the need to connect the nodes n_1 and n_2 actively to the BL, but rather simply using the capacitive voltage divider between one active and (n - 1) passive FTJs by activating the respective PLs. Hence, the access transistor would be used for pre-charge during read operation only, which allows to get rid of the pFET access transistors (T_{1p}, T_{2p}) and removes the need for generation of complementary WL signals from the circuit in Fig. 1a.

Due to the increased capacitance of the nodes n_1 and n_2 by a factor n, the readout of single FTJs in a kind of single-bit access will not be possible anymore since according to (2) the ΔV scales with $1/n$. However, the case is different for the parallel readout as one would envisage for typical non-von Neumann architectures: the dot product can be realized by applying voltage pulses to individual PLs, the pulse time representing a first analogue value, and weighting this input signal by the conductivity of the individual FTJs, finally accumulating the resulting charge on n_1 and n_2. Another implementation could be using just digital representation of data in large vectors, which would be interesting for hyper dimensional computing (HDC) concepts [12].

Interestingly, in this parallel configuration of multiple FTJs the maximum voltage difference ΔV that develops during read phase at the gates of T_3 and T_4 would not be altered compared to the case of a single FTJ, since both current density as well as capacitance would scale linearly with the number of FTJs:

$$\Delta V = (J_{FTJ}\ t_{read})\ /\ C_{0,FTJ} = (\mathbf{n}\ J_{FTJ}\ t_{read})\ /\ (\mathbf{n}\ C_{0,FTJ}). \quad (3)$$

Hence, after readout ΔV represents the weighted sum of all inputs, and in addition the natural limit of ΔV that is given by (3) realizes a window function that, depending on the interpretation of the resulting difference in currents I_1 and I_2 could be further interpreted as a linear approximation of a tanh or sigmoid activation function.

In summary, the proposed differential cell featuring multiple parallel FTJs for realization of multiple weighting elements offers many advantages. However, in order to make the readout less susceptible to noise, for a given FTJ technology ΔV can be only increased in the trade-off against increasing the readout time t_{read}. Hence, for device engineering one of

the most important targets is still to increase the attainable current density J_{FTJ} while keeping the capacitance per area $C0_{FTJ}$ low. In addition, the above argumentation highlights that in our readout concept the absolute current difference between I_{ON} and I_{OFF} is a much more important figure of merit than the oft-discussed TER.

III. ELECTRODE WORK FUNCTION ENGINEERING

A. Sample preparation

According to (1) the current density of the FTJ can be increased by lowering the effective potential barrier at the injecting electrode. In the case of our double layer FTJ [7] injection takes place from the top electrode which is in contact with the Al_2O_3 layer. This can be attained by the reduction of the work-function by replacing the TiN TE (work function of 4.7 – 5.2 eV [13]) by either Al (work function of 4.0 – 4.2 eV) or by doping Al into the TiN TE.

In our experiments, three different stacks have been prepared for direct comparison: TiN/$Hf_{0.5}Zr_{0.5}O_2$/Al_2O_3/TiN (referred to as *standard stack*), TiN/$Hf_{0.5}Zr_{0.5}O_2$/Al_2O_3/Al (referred to as *Al top electrode stack*), and TiN/$Hf_{0.5}Zr_{0.5}O_2$/Al_2O_3/[TiN/TiAl/TiN] (referred to as *Al-doped TiN top electrode stack*).

At first, starting with a mid- doped Si wafer, 30 nm of tungsten (W) is deposited by physical vapor deposition (PVD) at room temperature. Next, a 10 nm TiN bottom electrode is deposited at room temperature by sputtering under high vacuum conditions. The ferroelectric film $Hf_{0.5}Zr_{0.5}O_2$ (HZO) with a nominal thickness of 10 nm is grown by atomic layer deposition (ALD) at 280 °C with a 1:1 ratio of HfO_2 and ZrO_2 with ozone as oxidant. Similarly, a 2 nm Al_2O_3 layer is also deposited by ALD at 280 °C, using TMA as precursor and ozone as the oxidant source.

From this point, three different top electrodes have been deposited to complete the structure. The first sample utilizes Al top electrode with a thickness of approximately 25 nm. After aluminium oxide deposition, and before TE deposition, the anneal process was performed, followed by the e-beam evaporation of Al using a shadow mask to form a final capacitor structure.

The second sample contains a TiN top electrode (TE) that has been deposited with a nominal thickness of 10 nm. In the last samples, the laminate structure TiN/TiAl/TiN is deposited by PVD, with three different nominal thicknesses of TiAl, that is 2.5, 5.0, and 7.5 nm in order to verify the effect of Al contents towards current density and ferroelectric properties. In both cases, the stacks undergo a crystallization anneal after TE deposition at 500 °C for 20 seconds in N2 atmosphere. A Ti/Pt dot was patterned using a shadow mask during e-beam evaporation, followed by the standard cleaning solution (SC-1, ammonium hydroxide and hydrogen peroxide) to etch the TiN electrode and define the capacitor area.

The final devices used for electrical characterization have diameters of 200 μm. The polarization-voltage (P-V) hysteresis and cycling endurance were measured with a ferroelectric device tester (Aixacct TF3000). A Keithley

978-1-6654-8495-4/22 $31.00 © 2022 IEEE

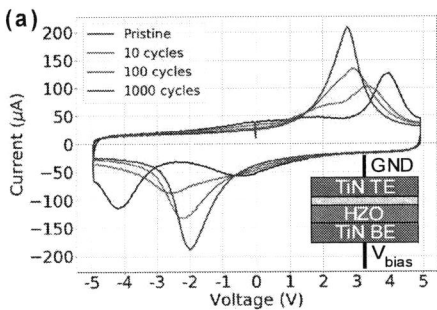

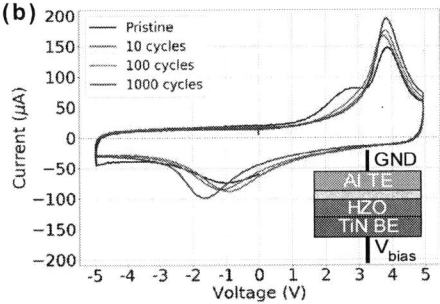

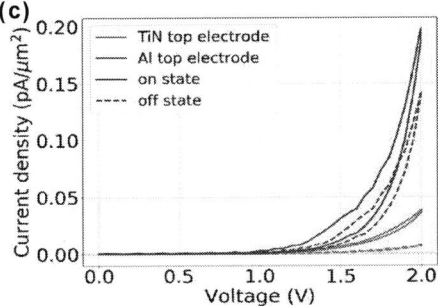

Fig.2. (a) I-V characteristics of TiN/Hf0.5Zr0.5O2/Al2O3/TiN sample (b) I-V characteristics of TiN/Hf0.5Zr0.5O2/Al2O3/Al measured at +/- 5 V, 1 kHz (cycling at 5 V, 10 kHz). Positive voltages indicate the P_{up} state, and negative voltages indicate P_{down}. (c) Comparison of the current-voltage response of devices with different electrodes in the P_{up} and P_{down} state.

4200S was used to analyze the resistive switching behavior by obtaining the DC current-voltage response of the FTJ.

B. Al top electrode

In a first experiment the TiN TE was replaced by Al. Fig. 2 shows the comparison of the measured transient IV-switching characteristic and the DC-IV I_{ON}/I_{OFF} currents of the FTJ devices. A wake-up effect can be observed clearly in the standard stack, merging two switching peaks into one peak [14]. In contrast, this effect is diminished in the Al top electrode stack, suggesting a more even distribution of oxygen vacancies directly after annealing [15]. Another feature is the shift of the hysteresis to the right in the Al top electrode sample, which means that a large voltage is needed to switch to the positive state (P_{up}), and the coercive voltage is smaller on the negative side (P_{down}), which can be explained due to the internal field caused by the work function difference between top and bottom electrodes [16]. This internal field may also be the cause for the redistribution of charges prior to electric field cycling.

In the standard process (TiN top electrode), after 1000 cycles, the switching voltages are -1.6 V and 2.3 V, with a

remanent polarization $P_r = -22.61 \ \mu C/cm^2$ and $+21.66 \ \mu C/cm^2$, whereas, in the Al top electrode sample, these values are -1.5 V and 3.6 V, $P_r = -22.1 \ \mu C/cm^2$ and $P_r = +15.94 \ \mu C/cm^2$, respectively (as shown in Fig. 2). The overall larger coercive field of the Al-TE sample can be explained by an increase of the Al_2O_3 interlayer thickness by 0.9 nm due to partial oxidation of the Al-electrode during Al evaporation directly on Al_2O_3 and subsequent annealing, which was revealed by CV and XRR measurements (not shown). Despite the thicker tunneling barrier, when comparing with the DC-IV response of the standard stack we observed an increase in tunneling current by a factor of 4, with approximately $0.2 \ pA/\mu m^2$ at 2 V readout for on-state, revealing the successful increase of the on-current by lowering the electrode work function. However, from reliability investigations we could identify that further oxidation of the top electrode is a critical degradation issue in the Al-TE based FTJ devices, strongly affecting the device's stability over time. Additionally, the strong built-in field impacts the stability of the two distinct resistance states and we calculated that due to fast polarization loss, only ~15% of switched domains contribute to the measured I_{ON}, meaning that the theoretical maximum I_{OFF} could be much larger than shown here. In order to improve the stability of the devices, work-function engineered TEs were manufactured by doping Al into a TiN TE [17].

C. Al-doped TiN top electrode

The typical current-voltage (I-V) switching characteristic is shown in Fig. 3 a for three different TiAl insertion layers. Increasing the Al content in the top electrode increases the polarization P_r value, whereas the coercive voltage V_c, is decreased. The increase in P_r can be explained by a larger orthorhombic phase fraction as was revealed by measuring the grazing-incidence X-ray diffraction spectra of the HZO films (not shown). Furthermore, it is evidence that the work

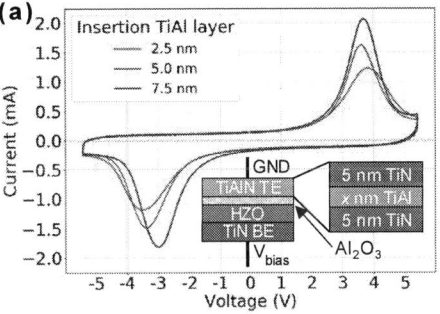

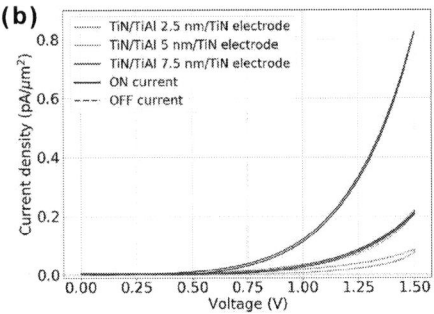

Fig. 3. (a) I-V switching characteristics measured at +/- 5.5 V, 1 kHz (waking up after 1e4 cycles at 5 V, 10 kHz) and (b) DC-IV on-current characteristic of the Al-doped TE samples.

978-1-6654-8495-4/22 $31.00 ©2022 IEEE   139

function of the TiN decreases with a thicker insertion TiAl layer, resulting in a higher work function difference between the bottom and top electrode. Therefore, a higher effective electric field is applied on the HZO film (in this case, the thickness of HZO and Al_2O_3 remain unchanged), thus lowering the coercive voltage and increasing the emnant polarization P_r. Additionally, more TiAl may lead to a lower electrode resistivity. The resistivities measured on a 4-point probe are 1.05E-1, 1.55E-5, 8.4E-6 Ω.m for 2.5, 5.0, and 7.5 nm TiAl insertion layer thickness, respectively.

The double-sweep I-V curves of three devices are shown in Fig. 3 b for the forward and backward sweep direction. With increasing TiAl thickness, the ON current rises due to the smaller work function, resulting in smaller potential barriers and a larger tunneling current. Obviously, the OFF current also increases. However, since in the differential synaptic circuit the decisive difference voltage $\Delta V_1 - \Delta V_2$ of the nodes n_1 and n_2 depends on the difference $I_{ON} - I_{OFF}$ rather than the current ratio I_{ON} / I_{OFF}, this effect is not detrimental. The $I_{ON} - I_{OFF}$ difference increased from <0.1 pA/μm^2 for the standard sample (see Fig. 2) to 0.6 pA/μm^2 for the TE with a 7.5 nm TiAl interlayer.

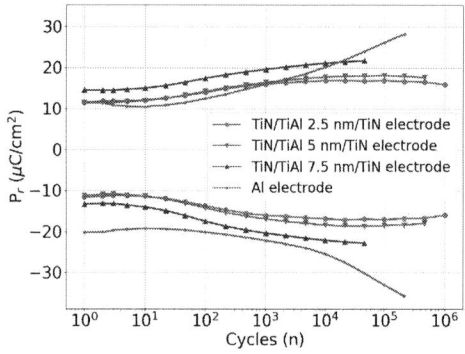

Fig. 4. Endurance cycling of the FTJ devices with Al-TE and TiN/TiAl/TiN TEs, cycled at +/- 5.5 V, 100 kHz; measured at 10 kHz, +/- 5.5 V.

Finally, the endurance cycling measurements of the Al-doped TE FTJ samples (see Fig. 4) revealed a much better reliability of the device compared to the Al-top electrode. In the case of Al-only, a high leakage after switching at cycles above 10^4 obscures the polarization switching and could lead to unstable device behavior, whereas the P_r for devices with TiAlN electrodes is seen to stabilize.

IV. CONCLUSION

In this paper we have explored the possibility of using multiple FTJs in a differential synaptic cell for massive parallel neuromorphic computing. By analyzing the circuit considerations, we showed that due to the simultaneous scaling of both current density and self-capacitance, the voltage difference ΔV between the On- and Off-states is not improved by adding additional devices. From this, we conclude that the most important figure of merit for characterizing such multi-FTJ cells is the difference in current $\Delta I = I_{ON} - I_{OFF}$, with $\Delta V \propto \Delta J_{FTJ}$.

To achieve a large ΔI, work function engineering of the TE can be applied. We have investigated Al, with a smaller work function than the standard TiN electrodes. However, pure Al as a TE leads to problems in reliability. We have shown that by instead doping Al into the stable TiN electrodes, ΔI can be increased from <0.1 pA/μm^2 for the standard sample to 0.6 pA/μm^2.

ACKNOWLEDGMENT

This work was supported by European Union through the BeFerroSynaptic project, Grant 871737 and by the DFG through the Memristec priority program, in the scope of the project ReLoFemRis (SL 305/2-1).

REFERENCES

[1] G. Indiveri, S.-C. Liu. Memory and information processing in neuromorphic systems. *Proceedings of the IEEE*, 2015, 103. Jg., Nr. 8, S. 1379-1397.

[2] Garcia, V., Fusil, S., Bouzehouane, K. *et al.* Giant tunnel electroresistance for non-destructive readout of ferroelectric states. *Nature* **460,** 81–84 (2009). https://doi.org/10.1038/nature08128

[3] Chanthbouala, A., Crassous, A., Garcia, V. *et al.* Solid-state memories based on ferroelectric tunnel junctions. *Nature Nanotech* **7,** 101–104 (2012). https://doi.org/10.1038/nnano.2011.213

[4] Si Joon Kim, Jaidah Mohan, Harrison Sejoon Kim, Jaebeom Lee, Chadwin D. Young, Luigi Colombo, Scott R. Summerfelt, Tamer San, Jiyoung Kim, Low-voltage operation and high endurance of 5-nm ferroelectric Hf0.5Zr0.5O2 capacitors, 2018, Applied Physics Letters, https://aip.scitation.org/doi/abs/10.1063/1.5052012

[5] L. Esaki, B. B. Laibowitz, and P. J. Stiles, "Polar switch," IBM Tech. Discl. Bull., vol. 13, no. 8 p. 2161, 1971.

[6] John G. Simmons, Electric Tunnel Effect between Dissimilar Electrodes Separated by a Thin Insulating Film, 1963, Journal of Applied Physics, https://aip.scitation.org/doi/abs/10.1063/1.1729774

[7] Max, B., Hoffmann, M., Slesazeck, S. and Mikolajick, T., 2018, September. Ferroelectric tunnel junctions based on ferroelectric-dielectric Hf 0.5 Zr 0.5. O 2/Al 2 O 3 capacitor stacks. In *2018 48th European Solid-State Device Research Conference (ESSDERC)* (pp. 142-145). IEEE.

[8] Covi, Erika, et al. "Ferroelectric tunneling junctions for edge computing." *2021 IEEE International Symposium on Circuits and Systems (ISCAS).* IEEE, 2021.

[9] M. Ye. Zhuravlev, Y. Wang, S. Maekawa, and E. Y. Tsymbal, "Tunneling electroresistance in ferroelectric tunnel junctions with a composite barrier", Appl. Phys. Lett. 95, 052902 (2009)

[10] M. Payvand, M. V. Nair, L. K. Müller, and G. Indiveri, A neuromorphic systems approach to in-memory computing with non-ideal memristive devices: From mitigation to exploitation. *Faraday Discussions, 213*, 487-510, 2019.

[11] S. Slesazeck, et al., A 2TnC ferroelectric memory gain cell suitable for compute-in-memory and neuromorphic application. 2019 IEEE International Electron Devices Meeting (IEDM). IEEE, 2019. S. 38.6. 1-38.6. 4.

[12] Khan, A.A., Ollivier, S., Longofono, S., Hempel, G., Castrillon, J. and Jones, A.K., 2021. Brain-inspired Cognition in Next Generation Racetrack Memories. *arXiv preprint arXiv:2111.02246.*

[13] J. Hölzl & F. K. Schulte, "Work function of metals", Solid Surface Physics, Springer 2006.https://doi.org/10.1007/BFb0048919

[14] Jiang, Pengfei, et al. "Wake-Up Effect in HfO2-Based Ferroelectric Films." *Advanced Electronic Materials* 7.1 (2021): 2000728.

[15] Schenk, T., Hoffmann, M., Ocker, J., Pešić, M., Mikolajick, T. and Schroeder, U., 2015. Complex internal bias fields in ferroelectric hafnium oxide. ACS applied materials & interfaces, 7(36), pp.20224-20233.

[16] Pešić, Milan, et al. "Built-in bias generation in anti-ferroelectric stacks: Methods and device applications." *IEEE Journal of the Electron Devices Society* 6 (2018): 1019-1025.

[17] Chen, Y.F., Hsu, L.W., Hu, C.W., Lai, G.T. and Wu, Y.H., 2021. Enhanced Tunneling Electro-Resistance Ratio for Ferroelectric Tunnel Junctions by Engineering Metal Work Function. *IEEE Electron Device Letters.*

Gap in pagination due to withheld paper.

Pages 141-144

A 28nm 1.644TFLOPS/W Floating-Point Computation SRAM Macro with Variable Precision for Deep Neural Network Inference and Training

Sangsu Jeong, Jeongwoo Park, Dongsuk Jeon
Graduate School of Convergence Science and Technology
Seoul National University, Seoul, Korea
djeon1@snu.ac.kr

Abstract—**This paper presents a digital compute-in-memory (CIM) macro for accelerating deep neural networks. The macro provides high-precision computation required for training deep neural networks and running state-of-the-art models by supporting floating-point MAC operations. Additionally, the design supports variable computation precision, enabling optimized processing for different models and tasks. The design achieves 1.644TFLOPS/W energy efficiency and 57.9GFLOPS/mm² computation density while supporting a wide range of floating-point data formats and computation precisions.**

Keywords—Compute-in-memory, Neural Network, Floating-point, Variable Precision

I. INTRODUCTION

Deep neural network (DNN) has achieved unprecedented success in artificial intelligence applications due to its superior performance to conventional algorithms. Various hardware accelerators for DNN have been extensively studied, but the need for moving a large amount of data still remains a challenge in improving throughput and efficiency. Compute-in-memory (CIM) mitigates this issue by integrating computation units into an existing memory block, which significantly reduces data movement overheads. While prior CIM designs demonstrated remarkable energy efficiency with various computation techniques such as current-domain [1-4], charge-domain [5-6], and digital-domain computations [7], most of them only support fixed-point operations with limited precision. However, floating-point (FP) data formats are now widely adopted both in inference and training for complex models. The required precision in inference significantly varies across neural network models and tasks (Fig. 1(a)), suggesting that general-purpose CIM macro should support variable precision. In addition, using optimal computation precision could substantially accelerate deep neural network training. In simulations, combining FP8 and FP32 computations for training achieved a **68.8%** reduction in BitFLOPs, a metric for evaluating bit-level computation costs by normalizing operation count to data precision (Fig. 1(b)). These challenges motivate an efficient processing unit that supports floating-point operations in reconfigurable precision for neural network processors.

However, it is cumbersome to implement floating-point MAC operations entirely in an SRAM-based CIM macro. For instance, the design in [8] only processes exponent computations inside memory and offloads mantissa operations to external processing units. The CIM macro presented in [9] performs fixed-point operations internally, requiring external conversion of weights and input activations to fixed-point values. Another design in [10] realizes floating-point arithmetic operations without external processing, but it only supports row-wise operations and hence requires a large

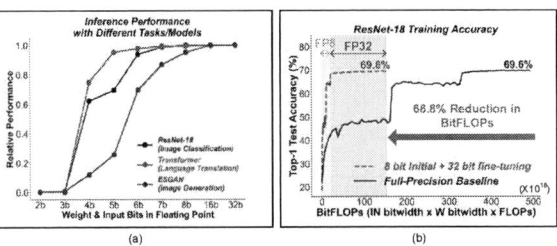

Fig. 1. (a) Optimal precision for inference varies with tasks and models. (b) Using optimal precision in each stage greatly accelerates training process.

number of cycles for accumulation, making it unsuitable for deep learning applications.

In this paper, we propose an SRAM macro that fully implements floating-point MAC operations with variable precision and input data formats. The key features of our design are as follows:

- Full implementation of floating-point MAC operations without external processing blocks
- Direct support for existing floating-point data formats without additional data manipulations
- Reconfigurable computation precision for optimal inference and training performance
- Low-latency MAC operations through Multi-way Windowed MAC (MWMAC) scheme

Our design achieves 1.644TFLOPS/W max efficiency and supports a wide range of floating-point number formats with variable multiplication and accumulation precisions, making it an attractive solution for deep learning applications.

II. OVERALL COMPUTATION FLOW

Fig. 2 depicts the computation flow of the proposed SRAM macro. In floating-point MAC operations, the multiplication results (products) have a very large dynamic

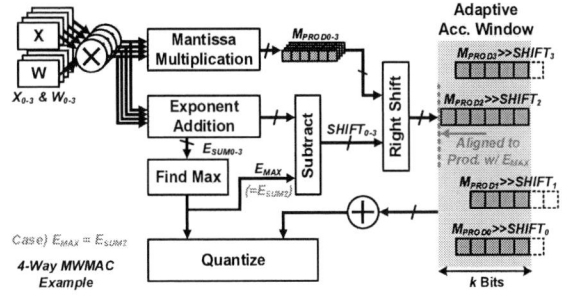

Fig. 2. Computation flow of the proposed MWMAC scheme.

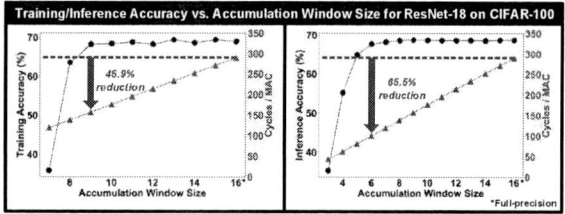

Fig. 3. Performance improvements due to accumulation window size optimization.

range; hence, accumulating those products incurs a large processing overhead when implemented in hardware. Our MWMAC scheme efficiently realizes this accumulation process using an adaptive accumulation window, as shown in Fig. 2. The macro first takes multiple input-weight pairs and generates their products by calculating exponent sums (E_{SUM}) and mantissa products (M_{PROD}). The macro obtains the largest exponent sum (E_{MAX}), and the starting point of the accumulation window is aligned to the MSB of the product whose E_{SUM} is equal to E_{MAX}. Then, the bits in the accumulation window are used for accumulation. We calculate each M_{PROD} only down to the bits equivalent to the accumulation window size. This noticeably reduces multiplication latency with negligible accuracy degradation as lower bits do not affect accumulation results if we ignore carry bits. Since MWMAC aligns all M_{PRODS} with respect to the accumulation window at once by finding E_{MAX}, it avoids the cost of alignment, which is the most time- and power-consuming process in floating-point operations [10]. This accumulation scheme allows for high-precision accumulation in contrast to the scheme presented in [9], where a large amount of information is lost during initial conversion and bit truncation of weights and input activations before multiplication.

Since only those bits in the adaptive accumulation window are used for accumulation, the accumulation window size determines both accumulation and multiplication precisions, as well as dictates processing throughput. Fig. 3 displays how the training and inference performance of ResNet-18 varies with the accumulation window size (k) for inputs and weights in bfloat16. Using optimal accumulation window size significantly reduces the latency by 45.9% and 65.5% for training and inference, respectively, while achieving similar accuracy to the full-precision baseline (k=16).

III. ARCHITECTURE AND IMPLEMENTATION

A. Overall Architecture

The overall architecture and the bit cell structure of the proposed CIM macro are depicted in Fig. 4. The macro consists of 256×64 weight storage 8T SRAM (WS-SRAM), 256×32 mantissa product 8T SRAM (MP-SRAM), and 256×9 exponent sum 10T CAM (ES-CAM), along with a digital near-memory computing block and an adder tree. In our design, the entire word of weights is loaded at once vertically through BL and BLB, as in typical SRAM, for easier integration with other blocks, while RBL and ML are used for row-wise computation. MWMAC operation begins by multiplying mantissas and adding exponents of the weight and input, and the results are written in MP-SRAM and ES-CAM, respectively. The location of the accumulation window is determined by finding the largest exponent sum (E_{MAX}). Starting from the LSB in the accumulation window, bits with the same weight in the products are combined in the adder tree that receives 256 1-bit inputs at once, producing the final accumulation results. The accumulation result is quantized back into the desired floating-point format using E_{MAX}.

B. Bit-serial Floating-Point Multiplication

The proposed design implements floating-point multiplications in a bit-serial manner to support an arbitrary data format (Fig. 5). During bit-serial multiplication, all rows are processed simultaneously to maximize the performance. The weights in WS-SRAM are read horizontally through RBL while the inputs are driven to ML. The near-memory computing block drives WBL/WBLB to the calculated write-back value, such as the sum of activation bit, weight bit, and carry in. By activating WWL, the write-back value is written

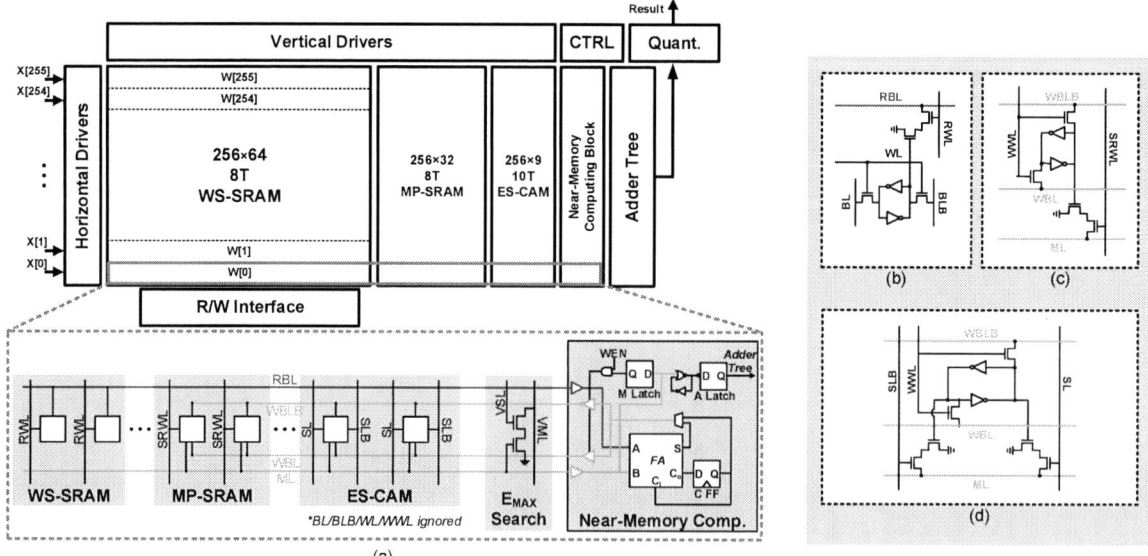

Fig. 4. (a) Arhitecture of the proposed CIM macro and bit cell structures of (b) 8T WS-SRAM, (c) 8T MP-SRAM, and (d) 10T ES-CAM

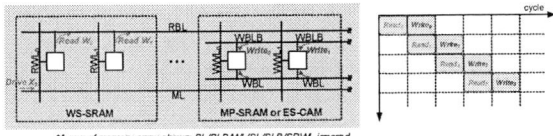

Fig. 5. Pipelined bit-serial operations

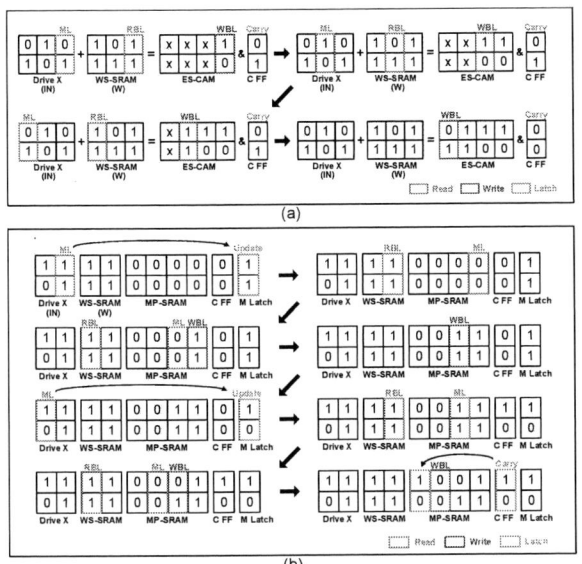

Fig. 6. Bit-serial operation examples: (a) 4-bit exponent addition and (b) 2-bit mantissa multiplication.

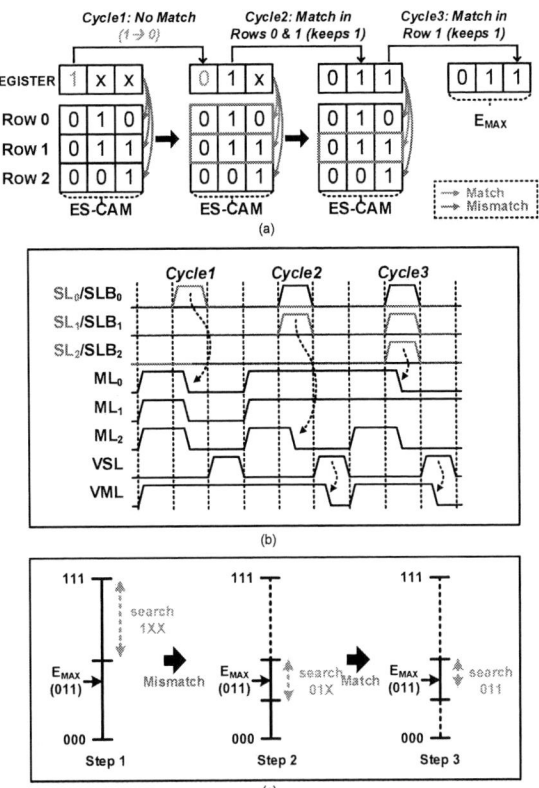

Fig. 7. (a) Computation flow E_{MAX} search and (b) Timing diagram of E_{MAX} search. (c) Similarity with binary search

in ES-CAM or MP-SRAM. Since horizontal read and write lines are isolated from each other, bit-serial read and write could be pipelined, which not only improves throughput but also reduces power consumption due to lower parasitic capacitance on each line, compared to the design in [11] adopting 10T bit cells for bit-serial operations.

Fig. 6 displays the flow of bit-serial multiplication, which consists of exponent addition and mantissa multiplication. During the mantissa multiplication, the M latch in the near-memory computing block holds the weight bit as a multiplier. If the M latch holds 1, the row would be updated by write-back values; otherwise, the row should not be updated, resulting in a half-select issue. To address this, when the M latch holds 0, WBL/WBLB are precharged. After the exponent addition and mantissa multiplication, the XOR operation required for obtaining the sign bit is realized by resetting the C flip-flop and writing back S into the last cell of MP-SRAM.

C. Maximum Exponent Sum (E_{MAX}) Search

After multiplying each pair of input and weight, the search for E_{MAX} follows (Fig. 7). Moving from MSB to LSB by one bit at a time, ML and VML are precharged and SL/SLB lines are set to a search word. SL=1/SLB=0 and SL=0/SLB=1 are used for 1 and 0, respectively, whereas SL=0/SLB=0 represents a don't care. If ML remains high after SL/SLB are activated, it suggests a match. VSL is pulled up after evaluating ML, and VML will be discharged if at least one row matches. If there is a match, the current bit of the register keeps 1; otherwise, it is reset to 0. At the end of this sequential process, the register stores the E_{MAX} in the products. This process is similar to binary search since each step determines one bit of E_{MAX}, reducing the search range by half

successively. This scheme efficiently implements E_{MAX} search without an additional comparison tree adopted in [9].

D. Window Search and Accumulation

Fig. 8 describes the accumulation step. In this step, SL/SLB and SRWL are activated to find E_{SUM} and M_{PROD} simultaneously. A match occurs when the E_{SUM} equals the search word and the corresponding M_{PROD} bit is 1. If the

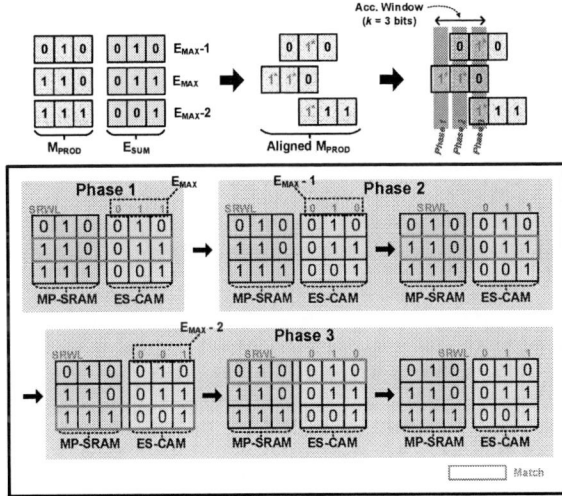

Fig. 8. Computation flow of window search and accumulation.

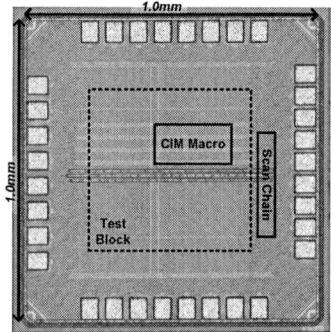

Technology	28nm
Die Size	$1.0 \times 1.0 mm^2$
Macro Size	$0.0439 mm^2$
Bitcell Size	$0.390 \times 1.745 \mu m^2$
Memory Size	16kb
Supply Voltage	0.5 - 0.9V
Frequency	53 - 403MHz
Data Format	FP with Variable Precision
Peak Energy Efficiency	1.644 - 0.099 TFLOPS/W
Peak Area Efficiency	57.9 - 4.84 GFLOPS/mm^2

Fig. 9. Chip micrograph and summary

accumulation window size is k, the accumulation step is divided into k phases. In each phase, the bits having the same weight (i.e., the bits on the same column in the accumulation window) are combined. This is achieved by incrementing the search word of ES-CAM by one and moving the search bit location of MP-SRAM toward LSB by one bit in each cycle. The search result is temporarily stored in the A latch in the near-memory computing block. At the end of the phase, the values of A latches are combined in the adder tree to generate a partial sum. The values of positive and negative A latches are separately processed in even and odd cycles, respectively, to minimize the adder tree area. This does not increase latency since it runs in parallel with the next search operation. A shift-and-add circuit accumulates the fixed-point partial sums generated in each cycle. The final accumulation result is sent to the quantization unit with E_{MAX} to generate the final MWMAC result in a desired floating-point format.

IV. MEASUREMENT RESULTS

The design is fabricated in a 28nm CMOS process, and Fig. 9 shows a chip micrograph and design summary. The macro operates with a clock frequency of 53-403MHz at 0.5-0.9V. The design is confirmed to work with all floating-point data formats between FP8 and FP32 and supports a wide range of computation precision ($k = 6$ to 29). It achieves a peak energy efficiency of 1.644TFLOPS/W for FP8 and 0.099TFLOPS/W for FP32. The max computation density is 57.9GFLOPS/mm^2 and 4.84GFLOPS/mm^2 for FP8 and FP32, respectively.

Table I shows comparisons to recently presented floating-point CIM macros. Our design implements entire floating-point MAC operations without external processing circuits [8] or input/weight conversion and alignment blocks [9], and achieves significantly higher efficiency than a standalone floating-point CIM macro in [10]. In addition, the design directly supports various input data formats and reconfigurable computation precision, making it suitable for general-purpose deep learning processors. For instance, recent GPUs support multiple floating-point formats, such as TensorFloat-32 (1-8-10) and BrainFloat-16 (1-8-7); our design can be seamlessly integrated into those accelerators as a cache memory by supporting an arbitrary data format.

V. CONCLUSION

In this work, we propose an SRAM CIM macro efficiently implementing entire floating-point MAC operations. Our design supports a wide range of floating-point data formats with variable computation precision to optimally accelerate

TABLE I. COMPARISONS TO PRIOR FLOATING-POINT CIM MACROS.

		[8] VLSI'21	[9] ISSCC'22	[10] JSSC'20	This work
Technology		28nm	28nm	28nm	28nm
Unit Macro Size		64Kb	64Kb	32Kb	16Kb
Supply Voltage		0.68-1.1V	0.6-1.0V	0.6-1.1V	0.5-0.9V
Precision Reconfigurability	Data	X	O	O	O
	Acc.	X	X	X	O
Usability to Cache		X	X	O	O
Supported Floating-point Format		Bfloat16	Bfloat16, FP32	FP32	Variable (FP8-FP32)
Operations Supported in Memory		Exponent Computation	Fixed-point MAC	Vector Operations	MAC
Peak Energy Efficiency		1.43 TFLOPS/W[1] (0.76V)	46.2, 5.9 TFLOPS/W (0.6V, BF16, FP32)	0.048[3], 0.0067[4] TFLOPS/W (1.1V)	1.644-0.099 TFLOPS/W (0.5V, FP8-FP32)
Peak Compute Density		27.4 GFLOPS/mm^2 [1, 2] (1.1V)	1.15, 0.15 TFLOPS/mm^2 (1.0V, BF16, FP32)	1.19[3], 0.17[4] GFLOPS/mm^2 (1.1V)	57.9-4.84 GFLOPS/mm^2 (0.9V, FP8-FP32)

1) w/o sparsity, 2) estimated from figure, 3) vector multiplication, 4) vector addition

training and inference for state-of-the-art DNN models while achieving high energy efficiency of 1.644TFLOPS/W.

ACKNOWLEDGMENT

This work was supported by the National Research Foundation of Korea (Grant No. NRF-2022R1C1C1006880) and Samsung Electronics. The EDA tool was supported by the IC Design Education Center.

REFERENCES

[1] J. Zhang et al., "In-Memory Computation of a Machine-Learning Classifier in a Standard 6T SRAM Array," *IEEE J. Solid-State Circuits*, vol. 52, no. 4, pp. 915-924, Apr. 2017.

[2] X. Si et al., "A Twin-8T SRAM Computation-in-Memory Unit-Macro for Multibit CNN-Based AI Edge Processors," *IEEE J. Solid-State Circuits*, vol. 55, no. 1, pp. 189-202, Jan. 2020.

[3] X. Si et al., "A Local Computing Cell and 6T SRAM-Based Computing-in-Memory Macro With 8-b MAC Operation for Edge AI Chips," *IEEE J. Solid-State Circuits*, vol. 56, no. 9, pp. 2817-2831, Sept. 2021.

[4] J. -W. Su et al., "Two-Way Transpose Multibit 6T SRAM Computing-in-Memory Macro for Inference-Training AI Edge Chips," *IEEE J. Solid-State Circuits*, vol. 57, no. 2, pp. 609-624, Feb. 2022.

[5] H. Valavi, et al., "A 64-Tile 2.4-Mb In-Memory-Computing CNN Accelerator Employing Charge-Domain Compute," *IEEE J. Solid-State Circuits*, vol. 54, no. 6, pp. 1789-1799, Jun. 2019.

[6] Z. Chen et al., "CAP-RAM: A Charge-Domain In-Memory Computing 6T-SRAM for Accurate and Precision-Programmable CNN Inference," *IEEE J. Solid-State Circuits*, vol. 56, no. 6, pp. 1924-1935, Jun. 2021.

[7] Y. -D. Chih et al., "An 89TOPS/W and 16.3TOPS/mm^2 All-Digital SRAM-Based Full-Precision Compute-In Memory Macro in 22nm for Machine-Learning Edge Applications," in *IEEE Int. Solid-State Circuits Conf. (ISSCC) Dig. Tech. Papers*, Feb. 2021, pp. 252-253.

[8] J. Lee et al., "A 13.7 TFLOPS/W Floating-point DNN Processor using Heterogeneous Computing Architecture with Exponent-Computing-in-Memory," in *Proc. IEEE Symp. VLSI Circuits*, Jun. 2021, pp. 1-2.

[9] F. Tu et al., "A 28nm 29.2TFLOPS/W BF16 and 36.5TOPS/W INT8 Reconfigurable Digital CIM Processor with Unified FP/INT Pipeline and Bitwise In-Memory Booth Multiplication for Cloud Deep Learning Acceleration," in *IEEE Int. Solid-State Circuits Conf. (ISSCC) Dig. Tech. Papers*, Feb. 2022, pp. 254-255.

[10] J. Wang et al., "A 28-nm Compute SRAM With Bit-Serial Logic/Arithmetic Operations for Programmable In-Memory Vector Computing," *IEEE J. Solid-State Circuits*, vol. 55, no. 1, pp. 76-86, Jan. 2020.

[11] K. C. Akyel et al., "DRC²: Dynamically Reconfigurable Computing Circuit based on memory architecture," in *Proc. IEEE Int. Conf. Rebooting Comput. (ICRC)*, Oct. 2016, pp. 1-8.

All-Digital Time-Domain Compute-in-Memory Engine for Binary Neural Networks With 1.05 POPS/W Energy Efficiency

Jie Lou, Christian Lanius, Florian Freye, Tim Stadtmann and Tobias Gemmeke
Chair of Integrated Digital Systems and Circuit Design, RWTH Aachen University, Germany
Email: lou@ids.rwth-aachen.de

Abstract—**This paper presents an all-digital time-domain compute-in-memory (TDCIM) engine for binary neural networks (BNNs), which is based on commercial standard cells facilitating technology mapping. The proposed TDCIM engine exploits energy-efficient computing principles, supports data reuse and employs double-edge triggered operation. Time domain wave-pipelining technique is also introduced to improve throughput by 1.5x while preserving accuracy. We use Structured Data-Path (SDP) placement and custom routing flow during place and route (P&R) to reduce systematic variations. The measured arrival time of different MAC results is sufficiently bounded to preserve accuracy across PVT variations. Fabricated in a 22nm process, the proposed BNN engine can achieve an energy efficiency of 1.05 POPS/W at 0.5V matching the accuracy of the PyTorch baseline of 99.14% on the MNIST dataset.**

Index Terms—**time-domain, compute-in-memory, binary neural network, commercial standard cells, double-edge operation, wave-pipelining**

I. INTRODUCTION

Convolutional neural networks (CNNs) can provide high accuracy in a variety of applications such as image classification and object localization. However, CNNs with high-precision weights and activations also consist of a large number of multiply-and-accumulate (MAC) computations and memory accesses [1], [2], making them challenging to use on energy-constrained devices. BNN with weights and activations quantized into +1 and -1, which drastically simplifies multiplications into XNOR logic, can be used to reduce computational complexity and memory requirements [1]–[5]. The compute-in-memory (CIM) micro-architecture is also being widely explored to reduce memory accesses and improve energy efficiency. Time domain (TD) computing has attracted attention due to its inherent analog signal processing properties and compatibility with digital circuits. It encodes numeric values as relative distances of discrete arrival times rather than voltage level, and takes advantage of the inherent delay characteristics of the circuit itself to achieve energy-efficient operation. Moreover, its error probability decreases exponentially with higher bit position. However, CIM related research is mainly implemented by analog addition of currents or charges [2], [3] and TD implementation most frequently use custom logic [5], [6], both making voltage and technology scaling demanding.

In this paper, we present an energy efficient TDCIM engine for BNN realized in the 22nm fdSOI technology featuring (1)

weight-stationary reuse and activation-broadcasting for convolution layers to reduce energy expenditure of memory accesses and data movements; (2) TD computing realizing MAC operations (TD-MAC), batch normalization and binarization; (3) double-edge triggered operation for higher energy efficiency and throughput; (4) overlapping time-domain wave-pipelining to improve 1.5x throughput while preserving accuracy; (5) all-digital design flow based on commercial standard cells suitable for technology scaling without analog re-design effort (capacitors, ADCs/DACs).

II. PROPOSED TIME DOMAIN DESIGN

A. Time-Domain Architecture

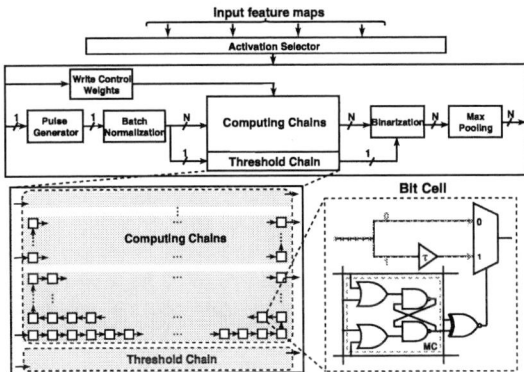

Fig. 1: Proposed TDCIM overall architecture with details on bit cell schematic.

Figure 1 shows the overall architecture and bit cell schematic of the proposed TDCIM macro. It mainly consists of a controller to write weights, a pulse generator, and the TD-MAC array with multiple computing chains. During initialization the weights are written into the memory cells (MC). As shown in bottom-right, the OAI-gate based SR-latch constitutes the MC [7], which is used to store the 1b weight of the BNN. The bitwise multiplication is performed with an XNR2 gate and controls an adjustable delay. If the result of dot product is +1, the signal will pass through an additional delay gate (DLY) producing a longer delay, while a -1 will pass the edge through the common MUX, only. The proposed bit cell can operate on both rising and falling transitions, which increases throughput by 2x without any additional hardware effort. Thereby, the dominant static power is effectively cut by two. To optimize the aspect ratio of the TD-MAC array, the bit cells of each computing chain are arranged in a snake-like pattern.

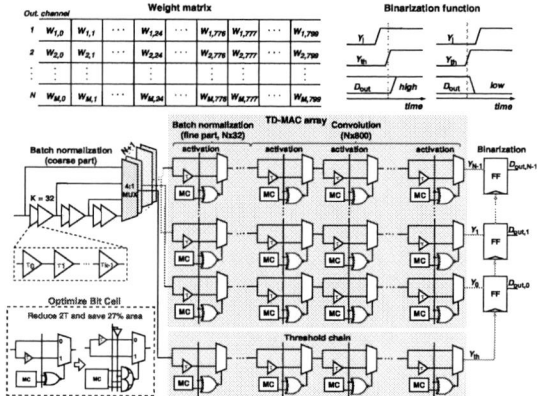

Fig. 2: Array data reuse principle and implementation of TD-MACs, Batch normalization and Binarization in time domain.

Figure 2 shows the unrolled architecture of the TDCIM engine. For the convolution layer, the weights of one filter of size RxRxC (here 5x5x32) are loaded into the distributed memory of one computing chain. Hence, N computing chains process N filters in parallel. The input activations are distributed vertically to all computing chains, while TD-MAC computations run from left to right. For BNN architectures, batch normalization (BN) can be realized as offsets to the inputs [8]. To reduce area, power and timing uncertainty, the BN is divided into a shared coarse and a fine part per delay chain, as shown on the left of Fig. 2. The shared delay chain provides 3 coarse delay steps of 32 units each, which is tapped by multiplexers to the corresponding delay chain. The fine part comprises 32 bit-cells each with a step of 1 unit. Therefore, the range of supported BN offset values is 0-128. Furthermore, a common offset can be provided by the threshold chain. As shown on the right of Fig. 2, the binarization function is realized by using a D-flip-flop (FF) connecting the accumulated delay time Y_i to the D input and the threshold delay time Y_{th} to the clock pin, respectively. The circuit generates high (+1) or low (-1) output values as function of relative arrival times Y_i and Y_{th}. This binarization step implicitly realizes the time-to-digital conversion (TDC) required in TD computing. For this proposed architecture, the memory weights and BN offsets need to be updated only after all the corresponding binarization results have been computed, while the input activations can be reused by all computing chains, thus reducing data movement and lowering memory fetch power. Moreover, the bit cell can be further optimized by replacing the XNR2 gate with an AOI22 gate which saves two transistors and 27% area.

B. Time-Domain Wave-pipelining Technique

The throughput in TD is improved by adopting wave-pipelining technique (Fig. 3). The baseline (1) corresponds to no wave-pipelining with the launch of the subsequent edge being postponed until the latest possible arrival time of the current one with the timing requirement shown in Eq. 1, where N is the number of bit cells in a computing chain. Adopting wave-pipelining, the timing requirement is relaxed by considering the minimal delay time of an edge as visualized with the dashed line marked (2). At the same time, this creates new timing requirements on the activation signals not to interfere with other edges in flight (Eq. 2). To limit control overhead, the delay chains are split into groups of simultaneously switching activation signals according to the ratio α of the delay unit required for the convolution computation, and the number of bit cells corresponding to the two groups is N_{grp1} and N_{grp2}. Further speed-up of this TD wave-pipelining is achieved by tolerating collisions of traveling edges outside the narrow sampling windows. This overlapped wave-pipelining is marked with (3).

$$T \geq N(\tau_{\text{MUX}} + \tau_{\text{DLY}}) \tag{1}$$

$$N_{\text{grp1}} > N(1 - \frac{\alpha_{\min}}{\alpha_{\max}})\frac{1}{\frac{1}{\alpha_{\max}}\frac{\tau_{\text{MUX}}}{\tau_{\text{DLY}}} + 1}$$

$$T_{\text{act1}} \geq N_{\text{grp1}}(\tau_{\text{MUX}} + \tau_{\text{DLY}})$$

$$T_{\text{act2}} \geq N_{\text{grp2}}(\tau_{\text{MUX}} + \tau_{\text{DLY}}) \tag{2}$$

$$T_{\text{act2}} \leq N_{\text{grp1}}\tau_{\text{MUX}}$$

$$T \geq N\tau_{\text{MUX}} + \frac{N}{2}\tau_{\text{DLY}} \tag{3}$$

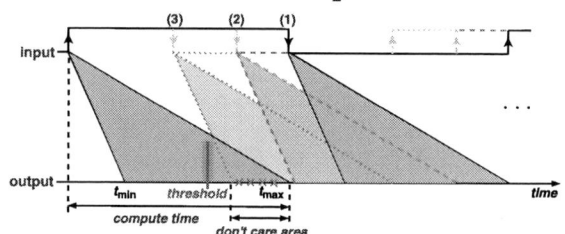

Fig. 3: Time-domain wave-pipelining technique.

C. SDP Placement and Custom Routing

Timing uncertainty sets a lower bound on delay per unit and thereby limits performance and energy efficiency. To provide enough margin for PVT variations and supply noise it is important to eliminate unnecessary uncertainties as much as possible. During the P&R phase, systematic variations in delay times of the physical design are basically eliminated by adopting a SDP placement and custom routing flow (Fig. 4). The latter uses wiring templates replicated by shifting and rotating within each delay chain thereby minimizing variations in parasitic wire loading. Figure 5 shows the delay distributions for the cases without SDP (w/o SDP), with SDP (w/ SDP) and with custom routing (w/ CR). The delay distribution w/ SDP is similar to the case w/o SDP, but the standard deviation per bit cell including extracted wire loading is significantly improved. By employing custom routing, the spread in delay times is further reduced eliminating overlaps which leads to error free operation after tuning of the threshold timing.

Fig. 4: SDP Placement and custom routing of bit cell.

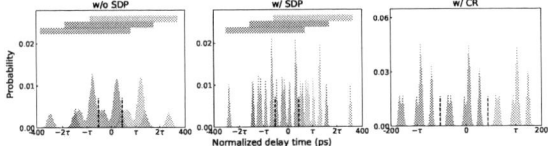

Fig. 5: Delay distribution comparison of w/o SDP, w/ SDP and w/ CR (base on post-layout simulation).

III. Measurement Results

Four delay cells (DLY40, DLY60, DLY80 and DLY100) are available in the foundry provided library. DLY60 was chosen during sign-off providing a reasonable trade-off between reliable operation and performance. A TD-MAC array including periphery and test circuitry for at-speed validation was taped-out. Matching the BNN specification for MNIST [9], it consists of 8 computing chains and a threshold chain, each composed of 832 bit cells to accommodate different layers of the network.

A. Threshold Chain Measurement Setup

The threshold delay was measured by controlling the number of delay cells passing through the threshold chain and activating a feedback path incorporating an inverter, thus generating oscillators of different frequencies. Measured threshold delay variations per stage follow a Gaussian distribution (Fig. 6a and 6b). In order to perform a high resolution measurement of the computing delay time, the threshold chain was set to multiple combinations with the same MAC result 400 to cover the entire possible delay region (Fig. 6c). Process variations lead to small deviations in the delay time which can then be sorted to achieve a timing reference with a resolution of 20ps (Fig. 6d).

B. Measured Computing Delay

The computing chain with specific MAC results are randomly configured and swept with selected threshold combinations to obtain the trend of its accuracy with threshold delay, which is a cumulative distribution function (CDF). The probability distribution function (PDF), which is the computing delay time, can be derived from the CDF. Fig. 7a-7c show the measured arrival time distribution before and after calibration for a single MAC result 400, multiple MAC results 397-402 and across 10 dies at 0.8V, respectively. After calibration, the difference in distribution of each computing chain is reduced, thus reducing the overall computation error. The computation error is essentially induced by the delay variation of the delay cell. The underlying trend of σ/μ with voltage is shown in Fig. 8a. When the voltage is reduced, a

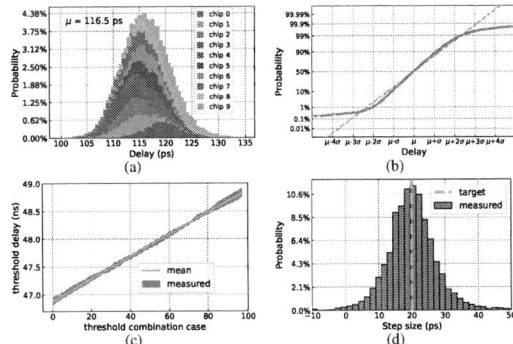

Fig. 6: Threshold chain measurement: (a) threshold cell delay across 10 dies; (b) normal probability plot; (c) selected threshold combinations; (d) achieved 20ps threshold resolution.

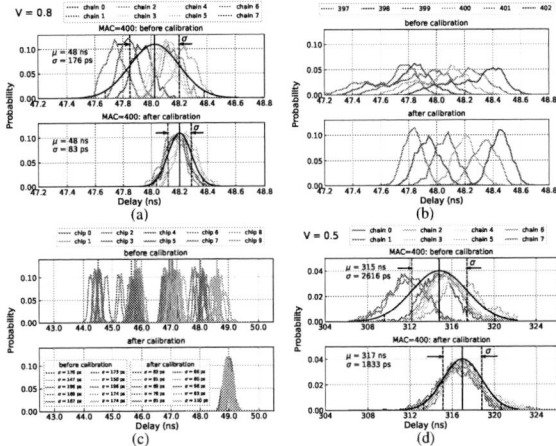

Fig. 7: Measured computing delay before and after calibration: (a) MAC = 400 at 0.8V; (b) MAC = 397-402 at 0.8V; (c) MAC = 400 at 0.8V across 10 dies; (d) MAC = 400 at 0.5V.

smearing effect on the edge timing can be observed and bears a higher error rate. The MAC result of 400 at 0.5V is shown in Fig. 7d.

C. Measured Performance

Figure 8b-8d highlights the performance results with different delay cells and different computation modes at 50% toggle rate. The overlapped wave-pipelining mode can achieve 1.5x better throughput compared to the normal mode of operation. The measured energy efficiency of the TD-MAC array is 452.6 TOPS/W with 99.20/% accuracy at 0.8V in overlapped wave-pipelining mode. This increases to 1.05 POPS/W when the voltage is reduced to 0.5V, leading to 0.167 µJ/image while still achieving 99.14% accuracy of the binary reference model. Considering the use of the DLY40, performance jumps to 1.51 POPS/W at 0.5V. Training with noise can improve the error tolerance of the neural network [10]. Errors due to increased random variations at supply voltages below 0.6V,

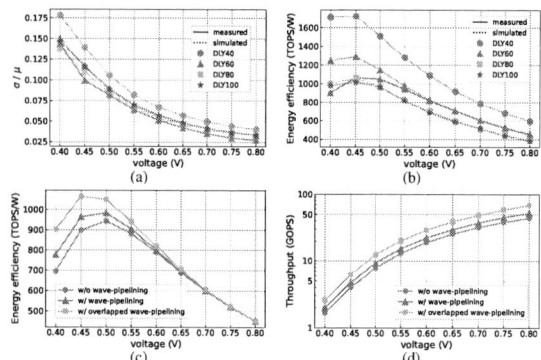

Fig. 8: Impact of voltage scaling on: (a) σ/μ for different delay cells; (b) energy efficiency for different delay cells; (c) energy efficiency for different computation mode; (d) throughput for different computation mode.

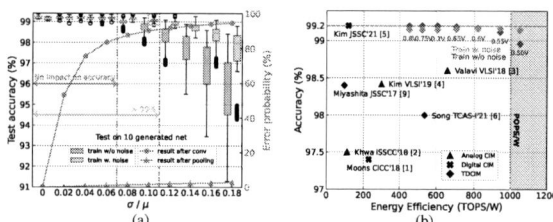

Fig. 9: (a) Test accuracy with variation. (b) performance comparison with prior work.

corresponding to σ/μ greater than 0.7 for DLY40, can be partially compensated by adding random noise in the computation during the inference phases of the training. Figure 9b and Table I show the unscaled comparison with previous BNN engines, where the proposed TDCIM can achieve POPS/W and outperform prior work in terms of energy efficiency and accuracy. The die micrograph and power breakdown are shown in Fig. 10

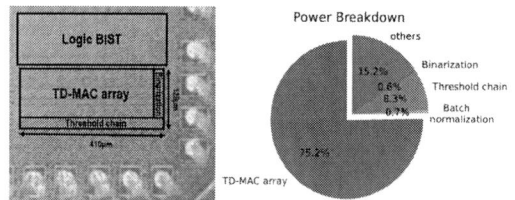

Fig. 10: Die micrograph and power breakdown.

IV. CONCLUSION

In this paper, we present a TDCIM macro based on 22nm commercial standard cells, which is optimized for BNN and suitable for technology scaling. The TD-MAC array not only enables energy-efficient operation, but also allows data reuse and parallel computation. The adopted double-edge triggered operation and wave-pipelining technique can effectively im-

TABLE I: Comparison with State-of-the-art Work

	ISSCC'18 [2]	VLSI'19 [3]	JSSC'21 [4]	TCAS-I'21 [5]	This Work
Technology	65 nm	28 nm	65 nm	40 nm	22 nm
Circuit Type	Analog CIM	Analog CIM	Digital CIM	TDCIM	TDCIM
Memory Type	SRAM	SRAM	SRAM	SRAM	Standard Cell
Array Size	4kB	16kB	16kB	8kB	6.25kB
Bit Cell Area (μm^2)	0.525	0.362	10.53	4.97	3.06
Supply Voltage (V)	1 & 0.8	1.0	0.6-0.8	0.9	0.5...0.8
Network Type	Binary Network				
Accuracy (%)	97.5	98.42	99.2	98	99.14...99.20
Throughput (GOPS)	N/A	615.6	567	5.76	12.33...68.86
Energy Efficiency (TOPS/W)	111.6	300.3	117.3	537-716	1050.6...452.6

prove energy efficiency and throughput. By using SDP placement and custom routing, the TD-MAC array can perform high precision computing under PVT variations. The proposed design can surpass POPS/W at 0.5V and still maintain good accuracy on the MNIST dataset.

ACKNOWLEDGMENT

This work was funded by the German BMBF project NEUROTEC under Grant no. 16ES1134.

REFERENCES

[1] B. Moons et al., "Binareye: An always-on energy-accuracy-scalable binary CNN processor with all memory on chip in 28nm CMOS," in IEEE Custom Integr. Circuits Conf. (CICC), 2018, pp. 1–4.

[2] W.-S. Khwa et al., "A 65nm 4Kb algorithm-dependent computing-in-memory SRAM unit-macro with 2.3ns and 55.8TOPS/W fully parallel product-sum operation for binary DNN edge processors," in IEEE Int. Solid-State Circuits Conf. (ISSCC), 2018, pp. 496–498.

[3] J. Kim et al., "Area-efficient and variation-tolerant in-memory BNN computing using 6T SRAM array," in IEEE Symp. on VLSI Circuits, 2019, pp. 118–119.

[4] H. Kim, et al., "Colonnade: A reconfigurable SRAM-based digital bit-serial compute-in-memory macro for processing neural networks," IEEE J. Solid-State Circuits, vol. 56, no. 7, pp. 2221–2233, 2021.

[5] J. Song et al., "TD-SRAM: Time-domain-based in-memory computing macro for binary neural networks," IEEE Trans. Circuits Syst.I, Reg. Papers, vol. 68, no. 8, pp. 3377–3387, 2021.

[6] P.-C. Wu et al., "A 28nm 1Mb time-domain computing-in-memory 6T-SRAM macro with a 6.6ns latency, 1241GOPS and 37.01TOPS/W for 8b-MAC operations for edge-AI devices," in IEEE Int. Solid-State Circuits Conf. (ISSCC), 2022, pp. 190–192.

[7] X. Fan et al., "Synthesizable memory arrays based on logic gates for subthreshold operation in IoT," IEEE Trans. Circuits Syst.I, Reg. Papers, vol. 66, no. 3, pp. 941–954, 2019.

[8] D. Miyashita et al., "A neuromorphic chip optimized for deep learning and CMOS technology with time-domain analog and digital mixed-signal processing," IEEE J. Solid-State Circuits, vol. 52, no. 10, pp. 2679–2689, 2017.

[9] T. Stadtmann et al., "From quantitative analysis to synthesis of efficient binary neural networks," in IEEE Int. Conf. Mach. Learn. Appl. (ICMLA), 2020, pp. 93–100.

[10] S. Buschjäger et al., "Margin-maximization in binarized neural networks for optimizing bit error tolerance," in Des. Automat. Test Europe Conf. Exhib. (DATE), 2021, pp. 673–678.

A 1.23-GHz 16-kb Programmable and Generic Processing-in-SRAM Accelerator in 65nm

Amitesh Sridharan[1§], Shaahin Angizi[2§], Sai Kiran Cherupally[1], Fan Zhang[1], Jae-sun Seo[1], Deliang Fan[1]

[1]Arizona State University, USA, [2]New Jersey Institute of Technology, USA §both authors contribute equally

Abstract—We present a generic and programmable Processing-in-SRAM (PSRAM) accelerator chip design based on an 8T-SRAM array to accommodate a complete set of Boolean logic operations (e.g., NOR/NAND/XOR, both 2- and 3-input), majority, and full adder, for the first time, all in a single cycle. PSRAM provides the programmability required for in-memory computing platforms that could be used for various applications such as parallel vector operation, neural networks, and data encryption. The prototype design is implemented in a SRAM macro with size of 16 kb, demonstrating one of the fastest programmable in-memory computing system to date operating at 1.23 GHz. The 65nm prototype chip achieves system-level peak throughput of 1.2 TOPS, and energy-efficiency of 34.98 TOPS/W at 1.2V.

Index Terms—In-memory computing (IMC), SRAM, programmability.

I. INTRODUCTION

Traditional von-Neumann computing architectures, such as CPUs and GPUs, demonstrate limitations in memory bandwidth and energy efficiency. However, their high demand lies in their programmability and flexible functionality. Such platforms execute a wide spectrum of bit-wise logic and arithmetic operations. In this regard, recent application-specific processing-in-memory (PIM) designs suffer from the major challenge that their performance is intrinsically limited to one specific type of algorithm or application domain, which means that such PIM platforms cannot keep pace with rapidly evolving software algorithms [1]. To overcome this limitation, state-of-the-art generic and programmable PIM architectures (e.g., [2]) exploit alternatives to conventional bit-parallel algorithms. For example, it is possible to realize arithmetic operations using bit-serial algorithms. However, it comes at a cost of high latency and more intermediate data write-back if multiple computing cycles are needed for basic in-memory Boolean logic functions [1]–[5].

To address this challenge, we propose a programmable processing-in-SRAM accelerator (PSRAM) that combines the PIM computation efficacy with the programmability. More importantly, for the first time, PSRAM realizes a complete set of Boolean operations (both 2- and 3-input), majority, and full adder in only one single memory cycle, demonstrating one of the fastest in-memory computing macros to date operating at 1.23 GHz. Furthermore, our one-cycle in-memory Boolean logic design also eliminates the redundant intermediate data write-back operations for 3-input logic and full adder that typically need multiple cycles with extra latency and energy in prior multi-cycle in-memory logic designs [1]–[5]. We demonstrate PSRAM for three applications: bulk bitwise vector operations, low-precision deep learning acceleration, and the Advanced Encryption Standard (AES) computation.

II. PROPOSED PSRAM DESIGN

The proposed PSRAM leverages the charge-sharing feature of the 8T SRAM cell on Read Bit-Line (RBL) and elevates it to implement 2-input and 3-input Boolean logic between two or three selected rows in a single memory read cycle. The key idea comes from the observation that certain discharge rate of the precharged RBL is determined by the data value stored in the simultaneously selected memory cells attached to the same bit-line. For instance, by activating three memory rows via Read Word Lines (RWL), e.g., RWL0-RWL2 (Fig. 1), if $S_{0,0}$, $S_{1,0}$, and $S_{2,0}$ memory cells all store '0's, then the read access transistors (T8) remain OFF, and the RBL precharged voltage does not discharge. On the other side, if all cells store '1's, the RBL voltage will rapidly discharge through T8s. Similarly, based on different combinations of the values stored in those memory cells, the discharged voltage value will be different if sampled at a preset frequency and VDD, which could be sensed by our follow up 'logic-SA' design to implement different logic functions through selecting different voltage references. Theoretically, there will be four different voltage levels based on all possible combinations of three memory cell data in the same bit-line. In our design, to yield a sufficiently large sense margin, as shown in Monte Carlo simulations (Fig. 5), the read path transistor (T7 and T8) size is designed to be $3\times$ as shown in Fig. 1.

To implement a programmable logic function, a new re-configurable logic-SA is designed as in Fig. 1. It consists of three sub-SAs with voltage references (i.e., $V_{Ref1} < V_{Ref2} < V_{Ref3}$), each dedicated to distinct logic functions. In this way, by activating three memory rows (i.e., input operand vectors) at the same time, each sub-SA performs a neat voltage comparison between the reference voltage and the discharged RBL voltages (w.r.t. different discharge rate corresponding to stored memory cell data), which respectively generates (N)OR3, (MAJ)MIN, and (N)AND3 logic output (complementary SA), and more importantly, at the same time.

A novel single-cycle in-SRAM XOR3 (full adder's Sum) logic is developed through an interesting observation as shown in the bottom-right truth table of Fig. 1. When the majority function (MAJ) output (green box in the truth table) is '0', the corresponding XOR3's output is the same as the OR3's output. When the majority function output is '1', XOR3's output can be achieved through AND3 as highlighted by the purple box. Based on our last paragraph description, our logic-SA could simultaneously get the OR3, MAJ and AND3 logic outputs, then we propose to design the XOR3 logic through a two-transistor 2:1 multiplexer (with MAJ output as the

978-1-6654-8495-4/22 $31.00 © 2022 IEEE

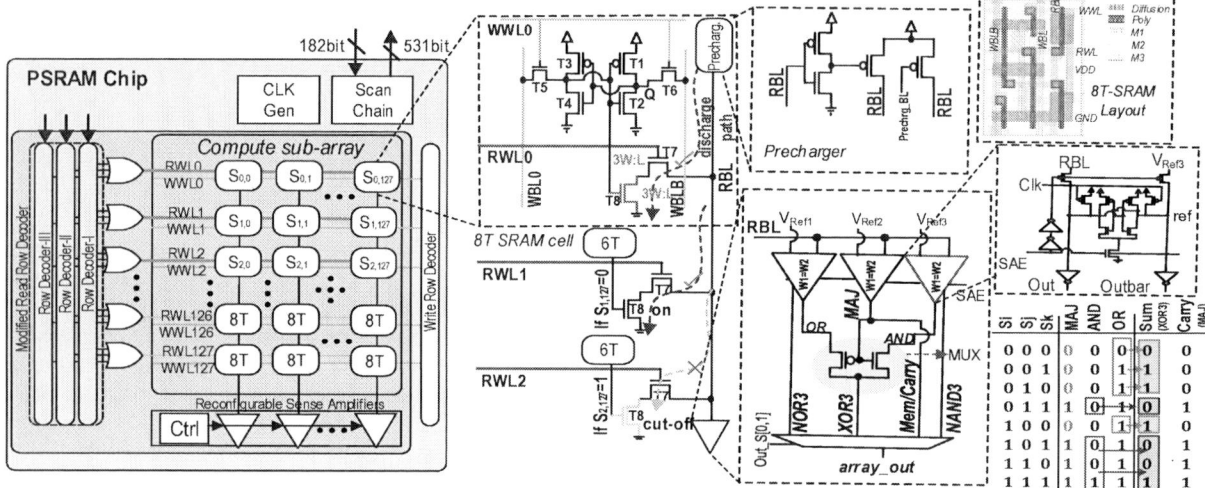

Fig. 1: PSRAM chip with 8T SRAM cell as the operand memory and the proposed single-cycle logic-SA design.

selector) circuit highlighted in the proposed reconfigurable logic-SA. The Boolean logic of in-memory XOR3 can be given as $XOR3 = MAJ(S_i, S_j, S_k).AND(S_i, S_j, S_k) + MIN(S_i, S_j, S_k).OR(S_i, S_j, S_k)$. In this way, assuming three vector operands are pre-stored in the memory, parallel in-memory full adder logic can be implemented for the first time in a single memory cycle, where MAJ and XOR3 outputs generate the carry-out and Sum signals, respectively. The two-input bit-wise operations will be readily implemented by initializing one row to '0'/'1'. All in-memory logic simulations are first shown in Fig. 2, showing corresponding functionality.

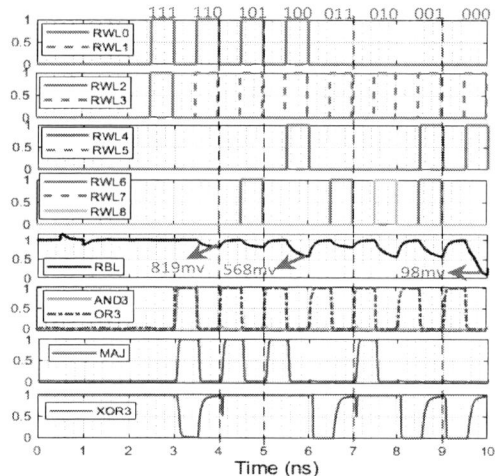

Fig. 2: In-memory logic simulation waveforms.

III. MEASUREMENT RESULTS

A. Performance Measurement

We prototyped the PSRAM macro (128×128) in TSMC 65nm CMOS (Fig. 3). The macro has a 16-kb capacity and occupies 0.17 mm² (with decoder) in the chip floorplan. The

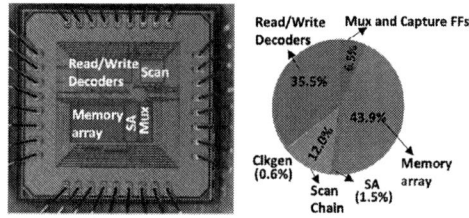

Fig. 3: Die micrograph and core area breakdown.

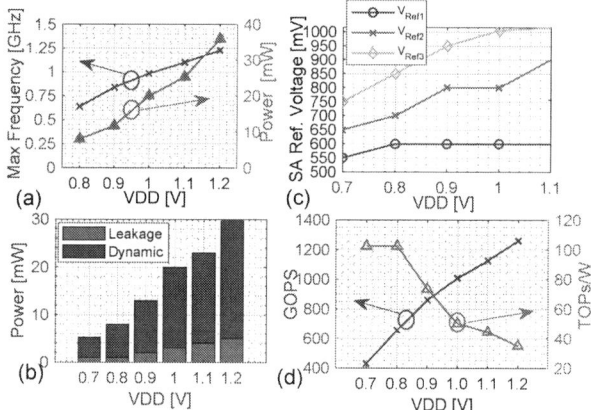

Fig. 4: (a) Frequency scaling over different VDDs, (b) Static and dynamic power consumption, (c) V_{Ref} scaling over different VDDs, and (d) throughput scaling over different VDDs.

bit-cell has an area of 4.56 μm^2 (1080 F^2 when scaled according to feature size), which is designed using logic rules. For efficient integration, the logic-SAs are pitch matched w.r.t. the column and occupy 3.4% of the array size (0.082 mm²). The complete core area breakdown is shown in Fig. 3. The PSRAM macro consumes 36 pJ (includes power consumed by all components on the die) and takes 813 ps to generate 512 outputs of the complete 3 input logic set (AND3, XOR3, OR3, MAJ). This represents a peak throughput of $2 \times 128 \times 4/813ps$ = 1259.52 GOPs at 1.2V supply and a compute density of 583.12 GOPS/mm^2. PSRAM achieves a significant speedup of 4-157$\times$ when compared to state-of-the-art in-memory computing works [1]–[4]. We report the maximum frequency, power consumption and throughput w.r.t. different VDDs in Fig. 4.

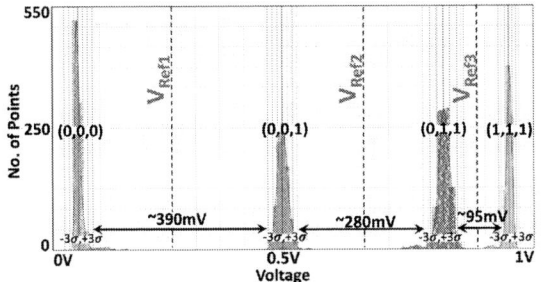

Fig. 5: Monte-Carlo simulation.

TABLE I: Reference voltage ranges measured on chip.

VDD/V_{Ref}	V_{Ref1}(V)	V_{Ref2}(V)	V_{Ref3}(V)
0.7V @ 0.42Ghz	509m-546m	603m-647m	658m-693m
0.8V @ 0.64Ghz	452m-616m	620m-733m	745m-780m
0.9V @ 0.84Ghz	414m-661m	669m-750m	829m-889m
1.0V @ 0.984Ghz	503m-711m	735m-902m	908m-995m
1.1V @ 1.1Ghz	550m-754m	760m-994m	999m-1.083
1.2V @ 1.23Ghz	554m-790m	815m-1.08	1.09-1.16

B. SA reference voltage(V_{Ref}) analysis

The RBL sense margins are first tested through post-layout Monte Carlo simulations in Cadence Spectre for the four possible sensing voltages, as shown in Fig. 5, where the sensing margin is reported considering both process (inter-die) and mismatch variations (intra-die) for core VDD (1.0 V) at 1 GHz. During the chip measurements, off-chip voltage references are provided (V_{Ref}) to the SAs. To conduct the V_{Ref} variation analysis on chip, we test all 128 bit-lines, 100 times, for all possible bit value combinations in memory. 10 chips are tested and we report all the reference voltage ranges at different VDDs and the corresponding maximum frequencies with zero logic errors in Table I. It is found that at lower voltages the maximum operating frequency is limited by the reduction of V_{Ref} ranges. A higher VDD also yields a larger sensing margin.

IV. APPLICATIONS

• Case Study I: Bulk Bitwise Boolean Vector Operations. The PSRAM could be leveraged to implement bulk bitwise Boolean logic operations efficiently between vectors stored in the same memory sub-array. This can lead to efficient re-use of the internal memory bandwidth. Table II compares the latency for a set of vector operations of interest, implemented by three generic PIM designs. We achieve the best performance of each design, where input vectors A($a_0 a_1 ...$) B($b_0 b_1 ...$) and C($c_0 c_1 ...$) are stored in separate rows of the memory. We draw two conclusions from Table II. Firstly, our PSRAM is the only design that supports a full-set of Boolean logic (both 2-input and 3-input) and integer operations. Second, due to the

TABLE II: Latency comparison of vector Boolean logic operations supported by PSRAM and prior accelerators.

Parameters	JSSC'18 [4]	JSSC'20 [2]	PSRAM
Capacity (KB)	8	16	2
Technology (nm)	40	28	65
Frequency (GHz)	0.029	0.475	1.23
NOT (ns / # of Cycle)	34.72 / 1	2.1 / 1	0.81 / 1
NAND2 (ns / # of Cycle)	34.72 / 1	2.1 / 1	0.81 / 1
NAND3 (ns / # of Cycle)	69.44 / 2	4.2 / 2	0.81 / 1
NOR2 (ns / # of Cycle)	34.72 / 1	2.1 / 1	0.81 / 1
NOR3 (ns / # of Cycle)	69.44 / 2	4.2 / 2	0.81 / 1
X(N)OR2 (ns / # of Cycle)	34.72 / 1	2.1 / 1	0.81 / 1
XOR3 (ns / # of Cycle)	69.44 / 2	4.2 / 2	0.81 / 1
Majority (ns / # of Cycle)	n/a	n/a	0.81 / 1
FULL-ADD (ns / # of Cycle)	69.44 / 2	4.2 / 2	0.81 / 1
FULL-SUB (ns / # of Cycle)	69.44 / 2	4.2 / 2	1.62 / 2
ADD-RCA (4-bit) (ns # of Cycle)	n/a	n/a	3.24 / 4
ADD-CSA (4-bit) (ns # of Cycle)	n/a	n/a	4.05 / 5
ADD-Serial* (4-bit) (ns)	173.6	10.5	4.05
SUB-Serial† (4-bit) (ns)	312.48	18.9	7.29
MULT-Serial‡ (4-bit) (ns)	1180.48	71.4	27.54
MULT-Serial (8-bit) (ns)	3541.44	214.2	82.62

*N+1 cycles, †2N+1 cycles, ‡N^2+ 5N-2 cycles

complexity of some operations (e.g., ADD/SUB/MULT), they cannot be implemented in a time-efficient manner by the prior designs [2], [4], while PSRAM outperforms all prior works in latency.

• Case Study II: Binary-Weight Neural Networks. We also implement the binary-weight neural network (BWNN) with various weight configurations for AlexNet and report the energy, latency and other performance metrics in Table III and Fig. 7. The general HW/SW framework developed for BWNN consists of image and kernel banks, and PSRAM sub-arrays. Weights and activation are constantly quantized to 1-bit and q-bit using the same method as [6], respectively, and then mapped to the parallel PSRAM sub-arrays. The top-1 accuracy after quantization on ImageNet dataset is reported in Fig. 7. For hardware mapping, considering n-activated PSRAM chips with the size of 128×128 (Fig. 6), each sub-array can handle the parallel ADD/SUB (multiply-and-accumulate operations are converted to ADD/SUB in BWNNs) of up to 128 elements of m-bit ($2m \leq 128$) and so accelerator could process $n \times 128$ elements simultaneously within computational sub-arrays to maximize the throughput. The memory sub-array data mapping for PSRAM is depicted in Fig. 6. We reserve four rows for Carry results initialized by zero and up to 32 rows for Sum results. Every pair of corresponding elements to be added together is aligned in the same bit-line. Herein, channel 1 (Ch1) and Ch2 should be aligned in the same sub-array. With m=32-bit, Ch1 elements occupy the first 32 rows of the sub-array followed by Ch2 in the next 32 rows.

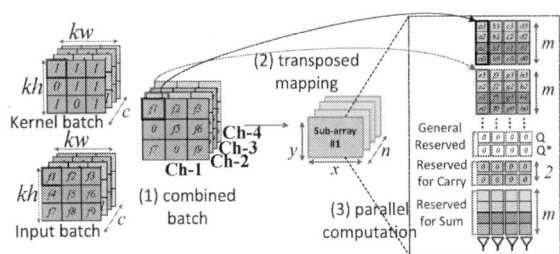

Fig. 6: BWNN hardware mapping.

The addition algorithm starts bit-by-bit from the LSBs of

TABLE III: Comparison with state-of-the-art SRAM based PIM accelerators.

Reference	PSRAM	BWNN Accelerators		Generic Accelerators	
		JSSC'19 [1]	JSSC'19 [3]	JSSC'20 [2]	JSSC'18 [4]
Technology	65nm	65nm	65nm	28nm	40nm
Bit cell Density	8T	10T	8T	8T Transposable	10T
Supply Voltage	0.8-1.2V	0.8-1.2V	0.68-1.2V	0.6 − 1.1V	0.5-0.9V
Max Frequency	1230MHz (1.2V)	5MHz	100MHz	475MHz (1.1V)	28.8MHz (0.7V)
SRAM Macro Size	2KB	2KB	4.8KB	16 KB	8KB
Performance (GOPS)	1259.52	8	295	32.7	14.7
Performance per unit area (GOPS/mm^2)	583.12	126	23.4	27.3	70
Energy-Efficiency (TOPS/W)	34.98	40.3	20.6	5.27 (add) 0.55 (mult.)	31.28
Reconfigurable	Programmable	N/A	N/A	Programmable	N/A

[1] We assume 2 operations (OPs) per NAND3/XOR3/X(N)OR3/NOR3 (cascaded logic), similar to MAC (1 mult. + 1 add).

the two words and continues towards MSBs. For evaluation, a 7-layer BWNN is adopted with distinct weight configurations of <W:I>: <1:1>, <1:2>, <1:8>. Our evaluation result reported in Fig. 7 shows that PSRAM can process AlexNet on average with 35 mJ energy per inference and ~0.5 ms latency. The process energy and latency include the amount required by multiple PSRAM chips working as a whole entity. More detailed performance comparison with other recent SRAM based PIM designs are reported in Table III.

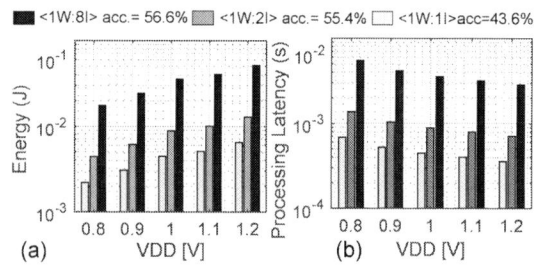

Fig. 7: (a) PSRAM energy consumption and (b) processing time for running the AlexNet (ImageNet dataset).

• **Case Study III: Data Encryption.** We further take the Advanced Encryption Standard (AES) data encryption algorithm as the third case-study. To facilitate working with input data (with a standard input length of 128 bits), each input byte data is distributed into 8-bit such that eight PSRAM sub-arrays are filled by 4×4 bit-matrices [7]. After mapping, PSRAM supports the required AES bulk bit-wise operations to accelerate each transformations inside the memory. As shown in Fig. 8, all AES transformations are mainly based on (N)AND and XOR operations that are fully supported in PSRAM. In SubBytes, MixColumns, and AddRoundKey stages, parallel in-memory XOR2 and (N)AND2 operations contribute to more than 90% of the operations. In ShiftRows stage, state matrix will undergo a cyclical shift operation by

a certain offset. We use the 128-bit AES software implementation as the baseline from [4], a 350nm ASIC [8], and a 40nm ASIC [4] designs for comparison. Table IV shows that PSRAM achieves the highest speed-up over baseline. This mainly comes from the massively-parallel and high throughput XOR operation supported in PSRAM.

TABLE IV: 128-bit AES performance.

Platforms	#Cycles	Freq. (MHz)	Time (μS) (Norm.)	Energy (nJ) (Norm.)
Baseline [4]	6358	24	265 (1x)	64.2 (1x)
ASIC [8]	5429	0.847	6410 (24x)	10259 (160x)
Recryptor [4]	726	28.8	25.2 (0.1x)	7.05 (0.11x)
PSRAM	718	1230	0.58 (0.002x)	19.21(0.3x)

V. CONCLUSION

In this work, we present a programmable PSRAM chip design in TSMC 65nm CMOS technology. For the first time, the PSRAM could execute a complete set of Boolean logic vector operations (i.e., NOR/NAND/XOR, both 2- and 3-input), majority, and full adder, all in a single memory cycle. We also demonstrate three case studies leveraging our PSRAM design, including parallel vector operation, neural networks, data encryption, etc. PSRAM paves a new path towards the generic, programmable and fast in-SRAM computing.

ACKNOWLEDGEMENTS

This work is supported in part by NSF grants 2003749 and 2144751, by SRC and DARPA.

REFERENCES

[1] A. Biswas et al., "Conv-sram: An energy-efficient sram with in-memory dot-product computation for low-power convolutional neural networks," *IEEE JSSC*, 2018.

[2] J. Wang et al., "A 28-nm compute SRAM with bit-serial logic/arithmetic operations for programmable in-memory vector computing," *IEEE JSSC*, 2020.

[3] H. Valavi et al., "A 64-tile 2.4-mb in-memory-computing cnn accelerator employing charge-domain compute," *IEEE JSSC*, 2019.

[4] Y. Zhang et al., "Recryptor: A reconfigurable cryptographic cortex-m0 processor with in-memory and near-memory computing for iot security," *IEEE JSSC*, 2018.

[5] J. Yue et al., "14.3 a 65nm computing-in-memory-based cnn processor with 2.9-to-35.8 tops/w system energy efficiency using dynamic-sparsity performance-scaling architecture and energy-efficient inter/intra-macro data reuse," in *IEEE ISSCC*, 2020.

[6] J. Faraone et al., "Syq: Learning symmetric quantization for efficient deep neural networks," in *CVPR*, 2018.

[7] S. Mathew et al., "53 Gbps native $GF(2^4)^2$ composite-field AES-encrypt/decrypt accelerator for content-protection in 45nm high-performance microprocessors," in *IEEE Symp. on VLSI*, 2014.

[8] M. Hutter et al., "A cryptographic processor for low-resource devices: Canning ecdsa and aes like sardines," in *IFIP*, 2011.

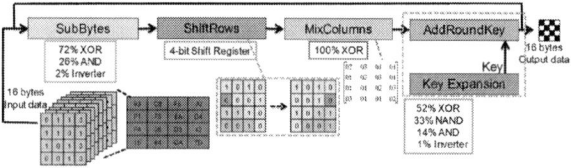

Fig. 8: AES block diagram with the gate utilization.

A 1-to-4b 16.8-POPS/W 473-TOPS/mm^2 6T-based In-Memory Computing SRAM in 22nm FD-SOI with Multi-Bit Analog Batch-Normalization

Adrian Kneip, Martin Lefebvre, Julien Verecken and David Bol

ICTEAM Institute, Université catholique de Louvain, Louvain-la-Neuve, Belgium

{adrian.kneip, martin.lefebvre, julien.verecken, david.bol}@uclouvain.be

Abstract—**Computing in-memory (CIM) is rapidly becoming an enticing solution to accelerate convolutional neural networks (CNNs) at the edge. Yet, low-precision current-based CIM-SRAMs face severe SNR degradation due to numerous analog non-idealities and high quantization noise when performing analog-to-digital conversion prior to digital batch-normalization (DBN). In this paper, we propose a dual-supply 1-to-4b CIM-SRAM macro in 22nm FD-SOI using 6T foundry bitcells, co-designed with a CIM-aware CNN training framework to overcome these challenges. The macro includes a multi-bit analog BN (ABN) unit combined with self-calibrating dual-phase sense-amplifiers (SCDP-SAs). Measurement results show peak 1b-normalized power and area efficiencies of 16.8POPS/W and 473TOPS/mm^2 at 0.4/0.8V supply and 100MHz, surpassing existing low-precision designs.**

Index Terms—**Compute in-memory (CIM), SRAM, Analog batch-normalization (ABN), ADC, Convolutional neural networks (CNNs), 22nm FD-SOI.**

I. INTRODUCTION

NOWADAYS, edge devices face ever-growing processing and communication challenges to execute data-driven AI algorithms, such as convolutional neural networks (CNNs) [1]. The compute in-memory (CIM) paradigm has recently proven to be a valuable solution by performing highly efficient analog dot-product (DP) operations inside SRAM arrays [2], [3]. However, CIM-SRAMs still face several design challenges, as depicted in Fig. 1(a). First, CIM-SRAMs based on high-density 6T bitcells from the foundry suffer from exacerbated variability in deeply scaled bulk CMOS technology, leading to accuracy loss [4]. Novel architectures based on charge [3] or digital [5] computing help to reduce this loss, yet at the expense of area and/or energy efficiency. Alternatively, the FD-SOI technology allows further downscaling at constant accuracy. Second, most conventional CIM-SRAMs focus on DP acceleration and perform remaining CNN operations in the digital domain, after ADC quantization of the DP results [6], [7]. In the case of (digital) batch-normalization (DBN), ADCs with less than 8b precision do not provide enough BN input data diversity and hinder the learning of DBN parameters, resulting in up to 20% accuracy loss on MNIST as shown in Fig. 1(a). The accuracy drop can be reduced by pre-training the DBN weights with full-precision (32b)

This work is supported by the Fonds de la Recherche Scientifique - FNRS, Belgium, under CDR Grant J.0014.20 and A. Kneip's Research Fellowship.

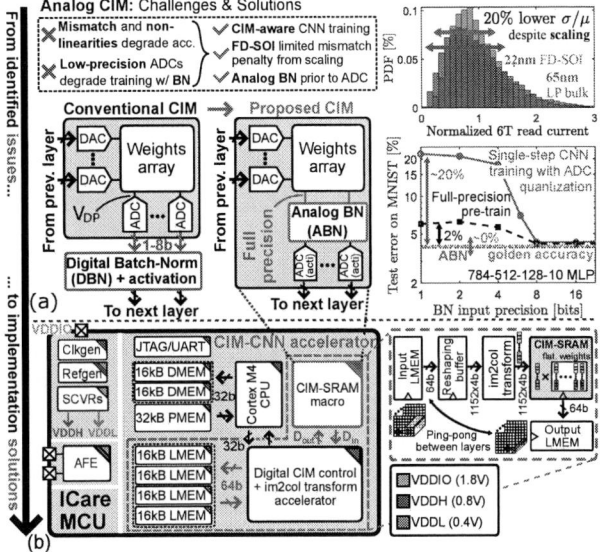

Fig. 1. (a) Design challenges in analog CIM-SRAMs and possible solutions. (b) Architecture of the ICare MCU, embedding a custom CIM-SRAM macro surrounded by a massively parallel digital controller, with im2col acceleration capability and tightly-coupled local memories (LMEMs).

ADCs, then fine-tuning with the actual low-precision (1-8b) ADC quantization. Still, a 2% accuracy loss remains below the 6b ADC precision compared to golden accuracy, obtained for digital hardware with 32b BN inputs. This puts significant pressure on the ADC, requiring a 6-to-8b design and making it a bottleneck with respect to throughput and energy efficiency [6]. Instead, performing BN in the analog domain prior to ADC quantization allows to bypass the precision bottleneck, thereby recovering golden accuracy with compact 1-to-4b ADCs. Yet, analog BN (ABN) has only been proposed for binary CIM-SRAMs to this day [3].

In this work, we propose a mixed-signal CIM-CNN accelerator based on a 1152×512 dual-supply CIM-SRAM macro in 22nm FD-SOI using high-density 6T-SRAM bitcells and featuring a custom multi-bit ABN unit. The accelerator is embedded within the ICare microcontroller unit (MCU) shown in Fig. 1(b) and features a digital controller with a 1-to-4b configurable datapath and four 16kB local SRAM memories (LMEMs). The controller uses highly parallel pipelined oper-

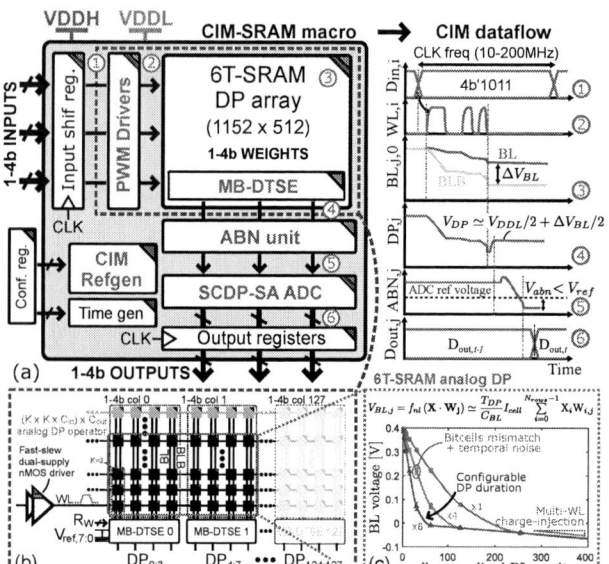

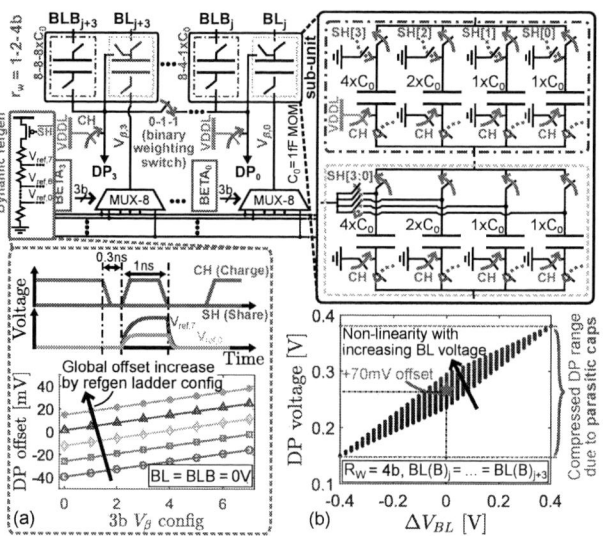

Fig. 2. (a) Top-level architecture of the CIM-SRAM macro, with qualitative depiction of a CIM operation. (b) Zoom on the analog 6T current-based DP array, relying upon differential BL discharge and DTSE conversion. (c) One-sided post-layout transfer function of the DP operation. Non-linearity and variability are included in a custom CIM-aware training framework.

Fig. 3. Architecture of the MB-DTSE unit, consisting of four adjacent DTSE sub-units with tunable total integration capacitance. The chosen weight's precision defines the total capacitance value to perform inter-column binary weighting during the share (SH) phase. (a) MB-DTSE supports DP voltage offsetting with 3b local V_β and global ladder configurability. (b) Post-layout transfer characteristic of the MB-DTSE unit, suffering from non-linearity and dynamic range reduction due to parasitic coupling.

ation to minimize data transfer latency and handles image-to-column (im2col) transforms. Altogether, the CIM-CNN accelerator achieves a peak throughput of 7.4TOPS at 100MHz and a total energy efficiency of 2.8POPS/W, including the controller and LMEM access energy. Considered alone, the CIM-SRAM macro reaches peak 1b-normalized area and energy efficiencies of 473TOPS/mm^2 and 16.8POPS/W, respectively. Finally, a custom CIM-aware CNN training framework is developed to account for analog non-idealities.

The rest of the paper focuses on the CIM-SRAM macro and is organized as follows. Section II presents the overall CIM-SRAM architecture, highlighting its dataflow. Then, Section III focuses on the co-design of the ABN and ADC units within the macro, while Section IV presents measurement results.

II. CIM MACRO ARCHITECTURE

The CIM-SRAM macro architecture is depicted in Fig. 2(a) with its different power domains. CNN weights are flattened on a row-wise basis to perform column-parallel analog DP operations. For convolutional layers (CONV-2D), the macro supports a kernel size of 3×3 and enables data reuse thanks to an input shift register, limiting the amount of LMEM transfers. Moreover, using 0.8V V_{DDH} and 0.4V V_{DDL} voltages respectively for the control signals and the highly parallel datapath reduces the DP energy without significant speed degradation.

As the analog DP operation starts, 4b digital inputs are sequentially applied to the wordlines (WLs) with binary pulse-width-modulated (PWM) encoding. Depending on the weights stored in each column, the V_{DDL}-precharged bitline (BL) then discharges over time using pull-down currents, obtaining the transfer function shown in Fig. 2(c) by time integration. The

configurable DP duration T_{DP} not only allows to compensate for PVT variations but also to tune the BL dynamic range to fit the expected distribution of BL results, shaped at train time. Therefore, we add T_{DP} as a custom learning parameter during CNN training to maximize information on the BL. Nevertheless, allowing a full BL swing makes bitcells subject to data corruption events as the BL goes close to 0V [4]. To avoid such events, we underdrive WLs to V_{DDL} with bitcells supplied at V_{DDH}. While this increases the sensitivity to mismatch, Fig. 2(c) shows that the relative 1-σ deviation of the BL voltage stays below an acceptable 10% for the longest T_{DP} configuration in the considered technology.

Once the analog DP finishes, the resulting voltage difference $\Delta V_{BL} = V_{BL} - V_{BLB}$ undergoes a multi-bit differential-to-single-ended (MB-DTSE) conversion before entering the ABN stage and being digitized by the ADC. The MB-DTSE unit shown in Fig. 3(a) combines ideas from [8] and [9] to simultaneously perform DTSE conversion and binary capacitive weighting between blocks of four adjacent columns. It operates in two phases. In the *charge phase*, which takes place during the analog DP operation, DTSE sub-units integrate their connected BL and BLB voltages on 8fF MoM capacitors. In the ensuing *share phase*, the outputs of the sub-units are shorted to perform binary-weighted charge-sharing, obtaining the single-ended DP result. The weights resolution r_w controls the sharing between DP outputs as well as the capacitance in each sub-unit such that the ideal DP voltage yields

$$V_{DP} = \frac{V_{DDL}}{2} + \sum_{i=0}^{r_w-1} \frac{2^i}{2^{r_w}-1} \times \frac{\Delta V_{BL,i}}{2}. \qquad (1)$$

In practice, parasitic capacitances compress the effective DP

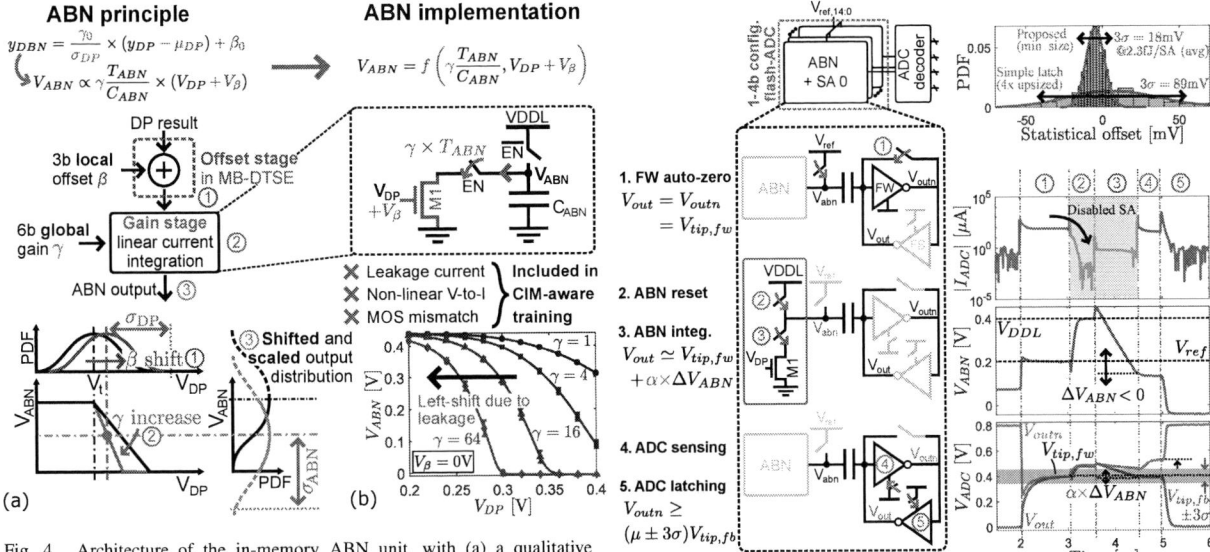

Fig. 4. Architecture of the in-memory ABN unit, with (a) a qualitative depiction of its working principle, with consecutive 3b local shift and 6b global scaling stages, (b) the implementation of the current integration stage, controlled by the shifted DP voltage. Its post-layout transfer function on the bottom-right highlights the main non-idealities to be considered at train time.

Fig. 5. The 1-to-4b flash ADC, co-integrated with the ABN unit. Self-calibrating dual phase sense-amplifiers (SCDS-SAs) achieve a 18mV 3-σ input offset with a sub-ns 2.3fJ/SA decision. Disabling SAs during ABN operations avoids high short-circuit current while keeping the AZ information.

dynamic range and shift the result based on the DP node voltage during the charge phase. Interestingly, we can take advantage of this shift to align the limited DP range with the desired ABN input voltage range by precharging the output node to V_{DDL}. Still, parasitics also introduce additional dependence on V_{BL} in Eq. (1), leading to the non-linearity observed in Fig. 3(c). As the integration capacitance cannot be further increased due to the pitch-matching constraint, we accept this non-linearity and model it in training, thereby limiting the accuracy degradation on MNIST below 1%.

Furthermore, the MB-DTSE unit implements a configurable DP offset, as shown in Fig. 3(b). We control the offset value by applying a configurable V_β voltage on the bottom plate of the BL integration capacitance during the *share* phase. A 3b pitch-matched latch stores the offset configuration per column and selects one of eight reference voltages V_{ref} generated by a resistive ladder as V_β. The ladder is only enabled for the duration of the *share* phase, which avoids significant power overhead during the rest of CIM operations. Moreover, a five-level configuration provides PVT compensation as well as the ability to offset the DP globally.

III. MULTI-BIT ABN-ADC CO-DESIGN

The ABN unit intends to shift and scale the distribution of DP voltages to match the behavior of DBN. As qualitatively depicted in Fig. 4(a), we apply these operations in two steps. First, we use the configurable offset of the MB-DTSE to apply a 3b V_β shift per output channel. Then, this shifted DP voltage is used to control the discharge rate of a current integration unit. There, we apply the gain factor γ by globally tuning the integration time on 6b, changing the slope of the ABN response. While ideally linear, this integration stage suffers from MOS transistor non-idealities, altering its transfer function as

seen in Fig. 4(b). Namely, M_1 subthreshold leakage limits the maximum ABN gain by transforming the linear scaling effect into a shift of the ABN transfer function towards lower V_{DP} values at high gain γ. Besides, such architecture also faces intrinsic non-linearity and column-to-column mismatch. These effects are critical and have to be considered during the off-line CNN training, as shown later in Section IV.

A dense 1-to-4b flash ADC based on custom self-calibrating dual-phase sense-amplifiers (SCDP-SAs) is connected to the ABN output, requiring four ABN+SA units per column. The 4b output precision is enabled by combining the results of four adjacent columns within the ADC decoder. Similar to the MB-DTSE unit, a resistive ladder dynamically generates the required SA decision voltages V_{ref} for the target output precision. A 2fF auto-zero (AZ) MoM capacitance C_{az} connects the ABN node to the SCDP-SA forward (FW) and feedback (FB) inverters, which are respectively driven by separate control signals. The SCDP-SA combines the advantages of the conventional latch and single-ended AZ topologies, achieving a sub-ns 2.3fJ/SA decision with minimum-sized devices and a 3-σ input-referred offset deviation of 18mV, which is $3.5\times$ lower than a conventional latch built with $4\times$ upsized transistors. The SCDP-SA operates in five phases, as described in Fig. 5. First, the FB inverter is disabled prior to ABN operations to perform FW offset compensation, setting V_{out} to the FW inverter's tipping point. During phases 2 and 3, the ABN operation takes place and both inverters are disabled. In this way, V_{out} follows changes in V_{ABN} through C_{az} coupling, with a reduction factor α related to parasitic charge-sharing. Moreover, disabling inverters during ABN avoids to suffer from significant short-circuit current. After ABN operations, the FW inverter is first enabled to sense the $\alpha \times \Delta V_{ABN}$

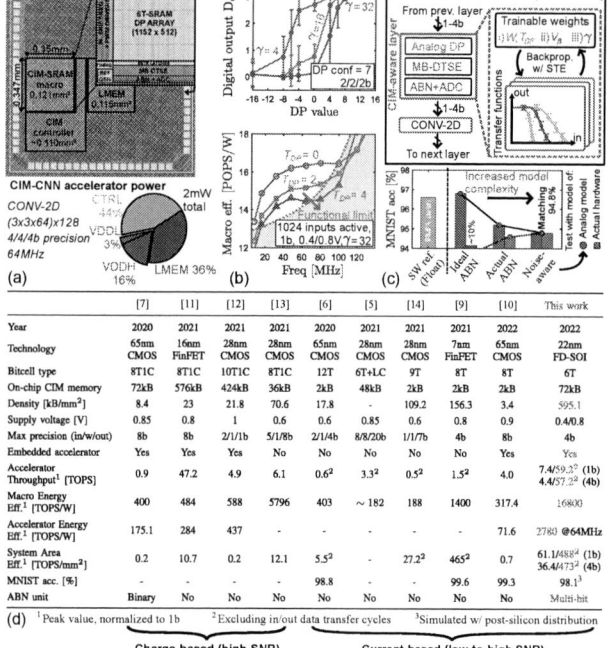

	[7]	[11]	[12]	[13]	[6]	[5]	[14]	[9]	[10]	This work
Year	2020	2021	2021	2021	2020	2021	2021	2021	2022	2022
Technology	65nm CMOS	16nm FinFET	28nm CMOS	28nm CMOS	65nm CMOS	28nm CMOS	28nm CMOS	7nm FinFET	65nm CMOS	22nm FD-SOI
Bitcell type	8T1C	8T1C	10T1C	8T1C	12T	6T+LC	9T	8T	8T	6T
On-chip CIM memory	72kB	576kB	424kB	36kB	2kB	48kB	2kB	2kB	2kB	72kB
Density [kB/mm²]	8.4	23	21.8	70.6	17.8	-	109.2	156.3	3.4	595.1
Supply voltage [V]	0.85	0.8	1	0.6	0.6	0.85	0.6	0.8	0.9	0.4/0.8
Max precision (in/w/out)	8b	8b	2/1/1b	5/1/8b	2/1/4b	8/8/20b	1/1/7b	4b	8b	4b
Embedded accelerator	Yes	Yes	Yes	No	No	No	No	No	Yes	Yes
Accelerator Throughput[1] [TOPS]	0.9	47.2	4.9	6.1	0.6[2]	3.3[2]	0.5[2]	1.5[2]	4.0	7.4/59.3[2] (1b) 4.4/57.2[2] (4b)
Macro Energy Eff.[1] [TOPS/W]	400	484	588	5796	403	~182	188	1400	317.4	16800
Accelerator Energy Eff.[1] [TOPS/W]	175.1	284	437	-	-	-	-	-	71.6	2780 @64MHz
System Area Eff.[1] [TOPS/mm²]	0.2	10.7	0.2	12.1	5.5[2]	-	27.2[2]	465[2]	0.7	61.1/488[4] (1b) 36.4/473[2] (4b)
MNIST acc. [%]	-	-	-	-	98.8	-	-	99.6	99.3	98.1[3]
ABN unit	Binary	No	No	No	No	No	No	No	No	Multi-bit

(d) [1] Peak value, normalized to 1b   [2] Excluding in/out data transfer cycles   [3] Simulated w/ post-silicon distribution

Charge-based (high-SNR)   Current-based (low-to-high SNR)

Fig. 6. (a) Chip microphotograph, highlighting the CIM-CNN accelerator and showcasing the layout of the CIM-SRAM macro. (b) Measured transfer function and energy efficiency under different conditions, at 0.4/0.8V. (c) CIM-aware training framework, with simulated accuracy improvements after each modelling step. (d) Comparison to the state of the art.

change, such that its output voltage V_{outn} resolves in the correct direction. If the sensed $\alpha \times \Delta V_{ABN}$ is large enough, the change in V_{outn} will overcome the FB inverter's offset deadband and ensure correct decision in the latching phase. As only a few mV is required on $\alpha \times \Delta V_{ABN}$ to overcome the FB deadband, the SCDP-SA offset is much lower than that of a conventional latch.

IV. MEASUREMENT RESULTS

The proposed CIM-SRAM macro has been fabricated in the ICare 22nm FD-SOI MCU and occupies a total area of 0.121mm² with a 67% array efficiency, as shown in Fig. 6(a). Thanks to its 6T current-based DP architecture and ultra-low voltage datapath, the CIM-SRAM macro achieves 1b-normalized peak energy and area efficiencies of 16.8POPS/W and 473TOPS/mm² at 0.4/0.8V and 100MHz with all inputs activated. Importantly, system-level efficiency remains limited by the in/out data transfer through the controller and LMEM accesses, calling for a future holistic optimization of the full CIM-CNN accelerator. In order to limit the accuracy loss due to analog non-idealities, we developed the *ModelCIM* CIM-aware CNN training framework from Fig. 6(c), which relies on statistical hardware models and look-up tables.

Compared to the state of the art in Fig. 6(d), this work achieves the densest design (relative to technology) on top of the highest 1b-normalized energy and area efficiency to date. Furthermore, to the best of our knowledge, this is

the first design including an in-memory multi-bit ABN unit. However, this architecture targets low-precision computing and can therefore not properly map more complex tasks with limited memory footprint (e.g., ImageNet), contrary to [10], [11]. Moreover, capacitance-based binary weighted precision and column-parallel flash ADCs are hardly scalable above the 4b precision. These points highlight the well-known SNR-efficiency trade-off in CIM designs, the present work focusing on improving the efficiency boundary at low SNR.

V. CONCLUSION

In this work, we proposed a novel 72kB CIM-SRAM macro in 22nm FD-SOI with 6T foundry current-based dot-product (DP) operators and a dual-supply strategy enabling ultra-low power datapath. The macro features a multi-bit analog batch-normalization (ABN) unit to overcome the ADC quantization bottleneck in low-precision CIM-based CNNs. Dense 4b flash ADCs based on self-calibrating dual-phase sense-amplifiers (SCDP-SAs) showcase high resilience to input-referred offset with only 2.3fJ per SA decision. Measurement results showcase 1b-normalized peak macro area and energy efficiencies of 473TOPS/mm² and 16.8POPS/W respectively, surpassing previous low-precision CIM-SRAM designs to date.

REFERENCES

[1] M. Horowitz, "Computing's energy problem (and what we can do about it)," in *IEEE ISSCC*, 2014, pp. 10–14.

[2] J. Zhang *et al.*, "A Machine-Learning Classifier Implemented in a Standard 6T SRAM Array," in *IEEE VLSI*, 2016, pp. 1–2.

[3] H. Valavi *et al.*, "A 64-Tile 2.4-Mb In-Memory-Computing CNN Accelerator Employing Charge-Domain Compute," *IEEE JSSC*, vol. 54, no. 6, pp. 1789–1799, 2019.

[4] A. Kneip *et al.*, "Impact of Analog Non-Idealities on the Design Space of 6T-SRAM Current-Domain Dot-Product Operators for In-Memory Computing," *IEEE TCAS-I*, vol. 68, no. 5, pp. 1931–1944, 2021.

[5] J.-W. Su *et al.*, "A 28nm 384kb 6T-SRAM Computation-in-Memory Macro with 8b Precision for AI Edge Chips," in *IEEE ISSCC*, 2021, pp. 250–251.

[6] S. Yin *et al.*, "XNOR-SRAM: In-Memory Computing SRAM Macro for Binary/Ternary Deep Neural Networks," *IEEE JSSC*, vol. 55, no. 6, pp. 1733–1743, 2020.

[7] H. Jia *et al.*, "A Programmable Heterogeneous Microprocessor Based on Bit-Scalable In-Memory Computing," *IEEE JSSC*, vol. 55, no. 9, pp. 2609–2621, 2020.

[8] M. Lefebvre *et al.*, "A 0.2-to-3.6TOPS/W Programmable Convolutional Imager SoC with In-Sensor Current-Domain Ternary-Weighted MAC Operations for Feature Extraction and Region-of-Interest Detection," in *IEEE ISSCC*, 2021, pp. 118–120.

[9] M. Sinangil *et al.*, "A 7-nm Compute-in-Memory SRAM Macro Supporting Multi-Bit Input, Weight and Output and Achieving 351 TOPS/W and 372.4 GOPS," *IEEE JSSC*, vol. 56, no. 1, pp. 188–198, 2021.

[10] L. Yue *et al.*, "STICKER-IM: A 65 nm Computing-in-Memory NN Processor Using Block-Wise Sparsity Optimization and Inter/Intra-Macro Data Reuse," *IEEE JSSC (Early Access)*, pp. 1–14, 2022.

[11] J. Lee *et al.*, "A Programmable Neural-Network Inference Accelerator Based on Scalable In-Memory Computing," in *IEEE ISSCC*, 2021, pp. 236–238.

[12] S. Yin *et al.*, "PIMCA: A 3.4-Mb Programmable In-Memory Computing Accelerator in 28nm for On-Chip DNN Inference ," in *IEEE VLSI*, 2021, pp. 1–2.

[13] J. Lee *et al.*, "Fully Row/Column-Parallel In-memory Computing SRAM Macro employing Capacitor-based Mixed-signal Computation with 5-b Inputs," in *IEEE VLSI*, 2021, pp. 1–2.

[14] V. Rajanna *et al.*, "SRAM with In-Memory Inference and 90Activity Reduction for Always-On Sensing with 109 TOPS/mm2 and 749-1,459 TOPS/W in 28nm," in *IEEE ESSCIRC*, 2021, pp. 127–130.

Fully integrated Si:HfO₂ Negative Capacitance 2D-2D WSe₂/SnSe₂ Subthermionic Tunnel FETs

Sadegh Kamaei[1], Ali Saeidi[1], Xia Liu[2], Carlotta Gastaldi[1], Clara Moldovan[1], Jürgen Brugger[2], and Adrian M. Ionescu[1].

[1]EPFL, Nanoelectronic Devices Laboratory (NanoLab), 1015, Lausanne, Switzerland

[2]EPFL, LMIS1, 1015, Lausanne, Switzerland

Email: Sadegh.kamaeibahmaei@epfl.ch, Adrian.ionescu@epfl.ch

Abstract— **We report the first experimental demonstration and performance characterization of** *a fully integrated negative capacitance (NC) WSe₂/SnSe₂ p-type Tunnel FETs (TFETs)*, **validating the use of NC as a technology booster to achieve a significantly improved sub-thermionic electronic switch. A WSe₂/SnSe₂ TFET with sub-60 mV/dec subthreshold slope (SS) is employed as the baseline TFET and characterized by using internal metal as a gate. The universal boosting impact of an NC effect of silicon-doped HfO₂ on digital and analog performances of 2D/2D TFETs is reported. A sub-30 mV/dec point SS and 50 mV/dec average swing over 2.5 decades of current with $I_{on}/I_{off} > 10^4$ at $V_D = 500$ mV are reported. Moreover, the low-slope region and I_{60} figures of merit are extended by 1.5 orders of magnitude due to the internal voltage amplification of the NC. The NC area of the polarization characteristic of Si:HfO₂ is extracted for all the investigated drain voltages based on the experimental data measured in DC mode. Importantly, the supply voltage is reduced by 0.3 V by NC to achieve the same output current, I_{on}. Our results fully demonstrate the combined merits of band-to-band-tunneling in 2D/2D WSe₂/SnSe₂ heterostructure with NC as a universal performance booster for 2D Tunnel FETs, offering to future 2D platforms a path towards improved energy efficiency.**

Keywords—WSe₂/SnSe₂, 2D/2D TFETs, Negative capacitance, NC 2D/2D TFETs.

I. INTRODUCTION

Tunnel FETs with quantum-mechanical band-to-band tunneling (BTBT) of carriers have emerged as an alternative structure aiming at achieving sub-60 mV/dec SS and addressing the power dissipation issue [1]. Two-dimensional (2D) materials are undoubtedly attractive for TFET devices because of the unique and layer-dependent physical properties, variety of electrical band gap, and potential for abrupt van der Waals (vdW) heterojunctions [2]. Among various 2D/2D heterojunctions used for the realization of TFETs, WSe₂/SnSe₂ configuration has been reported as a promising material system for achieving BTBT and sub-60 mV/dec SS due to its broken band gap alignment [3-6]. Additionally, the base material of this 2D system, WSe₂, recently presents as a TMDC that appears even more suitable than MoS₂ as a 2D platform for the development of future complementary 2D FET technology [7]. However, achieving SS values below 60 mV is very challenging, and the steep slope region is limited to one decade of current [3, 4]. Apart from using the BTBT of carriers, it has been proposed that the integration of the NC [8] into the gate stack of TFETs would be beneficial for enhancing the BTBT probability due to the internal voltage amplification [9-10]. In this context, we have combined the advantages of tunneling transport of charge in an integrated 2D/2D platform with the differential voltage amplification effect of a NC gate stack to demonstrate *the first reported NC 2D/2D TFET*. CMOS compatible poly-crystalline Si:HfO₂ ferroelectric is employed as the NC booster. An improved SS down to 24 mV/dec, an average SS of 50 mV/dec over almost 2.5 decades of drain current with an $I_{on}/I_{off} > 10^4$, and an enhanced transconductance, g_m, up to 9 times of the base values, highlight that NC can enhance both digital and analog performances of 2D/2D TFETs.

II. DEVICE FABRICATION

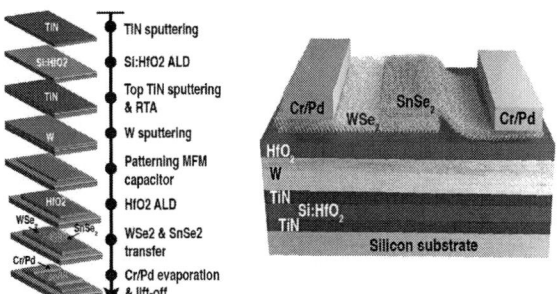

Fig. 1. 3D schematic view of the realized NC WSe₂/SnSe₂ TFET and summary of the fabrication process flow.

The experimental configuration of the NC WSe₂/SnSe₂ TFET is shown in Fig. 1, together with the summary of the proposed process flow. The ferroelectric stack consists of a 10-nm-thick titanium-nitride (TiN) layer deposited by sputtering on a silicon substrate, a 10-nm-thick Si:HfO₂ deposited by atomic layer deposition (ALD) that serves as a ferroelectric layer and top TiN sputtered layer. Rapid thermal annealing (RTA) at 600 °C for 2 minutes was performed to activate the ferroelectricity of Si:HfO₂. Then, 20-nm-thick tungsten (W) was sputtered as an internal metal gate (IMG). Subsequently, the top W/TiN layers were patterned by conventional photolithography on the AZ1512 layer and ion beam etching (IBE), to define the MFM capacitors. The NC gate structure was completed by depositing 10-nm-thick HfO₂ as a linear dielectric. Then, WSe₂ flakes obtained by mechanical exfoliation of a commercial WSe₂ bulk, are transferred to the substrate. SnSe₂ flakes were transferred deterministically on selected WSe₂ flakes, using a PDMS stamp. The present devices are fabricated on 6-nm-thick WSe₂ flakes and relatively thick SnSe₂ flakes. The source and drain contacts were deposited on the flakes with an electron-beam lithography

---

This work was financially supported by the European Research Council (ERC) under the ERC Advanced Grants Milli-Tech (695459) and ERC-2016-ADG, Project "MEMS 4.0" (742685).

978-1-6654-8495-4/22 $31.00 © 2022 IEEE

(EBL) step performed on a PMMA/MMA bilayer resist followed by evaporation and lift-off of a Cr (6 nm)/Pd (60 nm) stack. Finally, the HfO_2 was locally etched by wet etching process to access the W contact. Fig. 2a shows a cross-sectional high-resolution TEM image of a representative fabricated device, confirming a sharp interface between WSe_2 and $SnSe_2$ flakes. An optical image of the complete device is shown in Fig. 2b. The structure of the Si:HfO_2 capacitor and the atomic crystal configuration of the film are depicted in Fig. 2c (top). The piezoelectric force microscopy (PFM) maps reveal a sharp and coherent polarization domain pattern, which is expected for good quality polycrystalline ferroelectric layers (Fig. 2c bottom). The P-V and C-V characteristics of the Si:HfO_2 capacitor are also measured and shown in Fig. 2d.

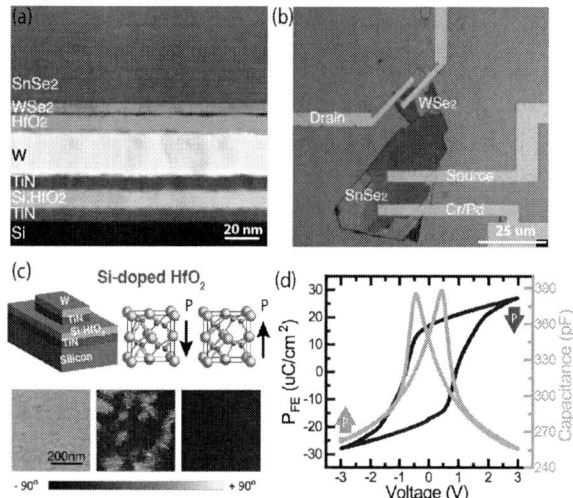

Fig. 2. a) High-resolution TEM image of the cross-section of the device, showing WSe_2/$SnSe_2$ heterojunction, ferroelectric capacitor, internal metal gate and the HfO_2 as a linear dielectric. b) Optical image of a representative device. c) Schematic diagram of the MFM capacitor and crystal structure of the Si:HfO_2 (top), together with PFM maps of domain switching in a 10-nm-thick Si:HfO_2 ferroelectric capacitor (bottom). d) P-V and C-V curves obtained from the MFM structure with an RTA of 600 ^{0}C for 2 mins.

III. RESULTS

All the electrical measurements were performed at ambient conditions and room temperature using an HP 4156C semiconductor parameter analyzer. In this section, we demonstrate based on the *first ever fully monolithically integrated device* the enhancement of the NC effect of the ferroelectric as a performance booster for 2D/2D TFETs. The average width of the heterojunction area is 1.5 µm and the WSe_2 thickness is 6 nm.

A. *2D/2D WSe_2/$SnSe_2$ TFET*

First, we characterized the WSe_2/$SnSe_2$ TFET at room temperature. Fig. 3a shows the transfer characteristic of the device at different drain voltages when the WSe_2 is biased as the drain and the IMG as the gate electrode. The device exhibits

a p-type behavior with an ON/OFF current ratio larger than 10^4. The output characteristic (I_D-V_D) curves shown in Fig. 3b exhibit a clear gate tunable negative differential resistance (NDR) region, a fingerprint of the tunneling mechanism, for all internal gate voltages. The peak to valley current ratio of the NDR region exceeds 9 for the negative gate biases, showing a robust BTBT current in the heterojunction. The gate leakage current of the device was measured to be less than 10 pA.

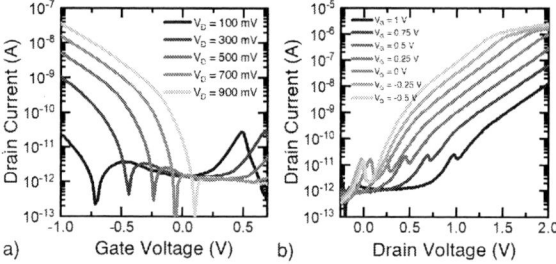

Fig. 3. a) Transfer characteristic of the WSe_2/$SnSe_2$ TFET under different top gate biases and, b) the output characteristic of the WSe_2/$SnSe_2$ TFET at multiple gate voltages with evident NDR, confirming the BTBT as the main carrier transportation mechanism.

The extracted SS as a function of the drain current is also illustrated in Fig. 4. The TFET achieves a sub-thermionic swing for all drain voltages. A minimum SS of 35 mV/dec at lower bias and sub-60 mV/dec SS over 1.5 decades of drain current at higher bias are obtained, values superior to the state-of-the-art of 2D/2D TFETs [3-6]. The presented results confirm that the BTBT between the $SnSe_2$ conduction band and the WSe_2 valance band is the main carrier injection mechanism of the current, similar to [3, 6].

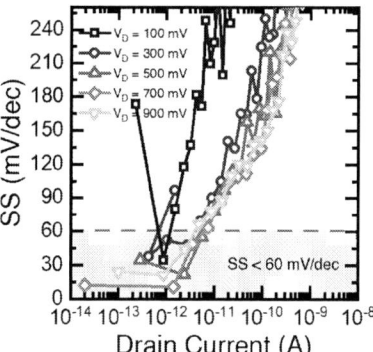

Fig. 4. The extracted SS of the WSe_2/$SnSe_2$ TFET (room temperature) at different drain biases. Sub-60 mV/dec SS shows the BTBT of carriers in our device.

B. *NC 2D/2D WSe_2/$SnSe_2$ TFET*

The transfer characteristics of the NC TFET and its baseline transistor at $V_D = 0.5$ V are plotted in Fig. 5a, showing a clear performance improvement. The forward sweep of the gate voltage is considered for the NC 2D/2D TFET reported data. In

978-1-6654-8495-4/22 $31.00 © 2022 IEEE

order to decouple the effect of the threshold voltage (V_{TH}) variation, curves are plotted with respect to the effective gate voltage: $V_{gs\_eff} = V_{gs} - V_{TH}$. The gate voltage was swept in the +/- 4 V range. The NC is partially stabilized in our devices (over a defined region of operation), leading to a low but non-negligible hysteretic behavior. The gate leakage is recorded and systematically verified to be extremely low (less than 10 pA) compared to I_D and cannot impact the reported effects. The SS of the two devices are also illustrated in Fig. 5b. The I_{60} figure of merit, representing the maximum current below which the TFET still sustains a sub-60 mV/dec swing, is improved by more than one decade. The NC 2D/2D TFET exhibits an SS_{min} of 24 mV/dec and an average SS of 50 mV/dec over almost 2.5 decades of the drain current. The internal gain of the NC, dV_{int}/dV_g, is extracted by measuring the internal node and plotted in Fig. 5c. A differential voltage amplification up to 2.3 is observed in the subthreshold region of the NC TFET characteristic. This affects the surface potential to increase faster than the applied gate voltage in the noted region and enhances the tunneling probability. As a result, the SS is significantly improved. Considering the conservation of the electric field, the polarization characteristic of the Si:HfO$_2$ is extracted (Fig. 5d). The ferroelectric performs two separate NC regions, demonstrating a zig-zag polarization characteristic. This is mainly due to the fact that the polycrystalline Si:HfO$_2$ is showing two main polarization domains and a multi-domain ferroelectric capacitor in steady states cannot hold more than one negative capacitance domain at a time [11]. The transconductance dependence on the gate bias is shown in Fig. 6, showing a steeper transition and an enhancement up to 9 times of its original value.

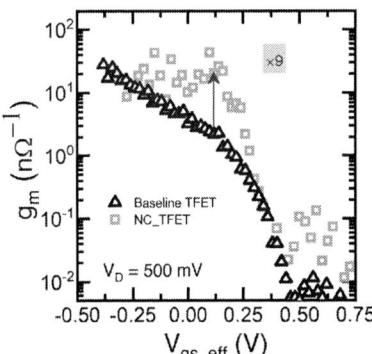

Fig. 6. Direct comparison between transconductance characteristic, g_m, of the baseline TFET and the NC TFET at $V_D = 500$ mV. The NC device exhibits a steeper transition and an improvement larger than 9 times.

C. Impact of drain bias on the NC 2D/2D TFET characteristic.

To study the impact of V_D, the NC 2D/2D TFET is measured at different drain biases. Figs. 7a and 7b show the V_D dependency of the I_D-V_G curves and SS of the device. Besides, the general effect of V_D on the increase of on-current and off-current, the low-slope region of the device is extended for higher V_D values. In fact, the boosting effect in a NC transistor can be dramatically controlled by V_{DS} electric field, since the charge and MOS capacitance vary with V_{DS} [10]. In a negative capacitance transistor, the ferroelectric polarization charge density and the channel charge density should match. Thereby, the operation point, which is the intersection of the transistor charge line and the PV characteristic of the ferroelectric, is dependent on the drain voltage. This means that by varying V_{DS}, the operation point of the NC TFET changes and, hence, the boosting effect of NC. The internal voltage amplification shows the degradation of the NC gain at a lower V_D (see Fig. 7c). The extracted P-V for all the investigated drain voltages are depicted in Fig. 7d, showing a clear negative capacitance area during the operation of the device. As can be seen, the NC region limits by decreasing V_D, which is consistent with the reported internal gain in Fig. 7c.

D. Benchmarking

Table 1 summarizes the electrical performance of our device compared to the state-of-the-art 2D/2D TFET. The NC 2D/2D TFET proposed in this work *clearly outperforms all previous 2D/2D heterojunction devices in terms of SS$_{min}$, low-slope region and I_{60},* while maintaining a relatively high ON/OFF current ratio (>10^4). Additionally, Fig. 8 benchmarks state-of-the-art best p-type TFETs (TFETs that show SS near or <60 mV/decade) [12] in terms of SS vs I_D per width. Our device structure shows superior performance while offering an easy and low cost manufacturing process.

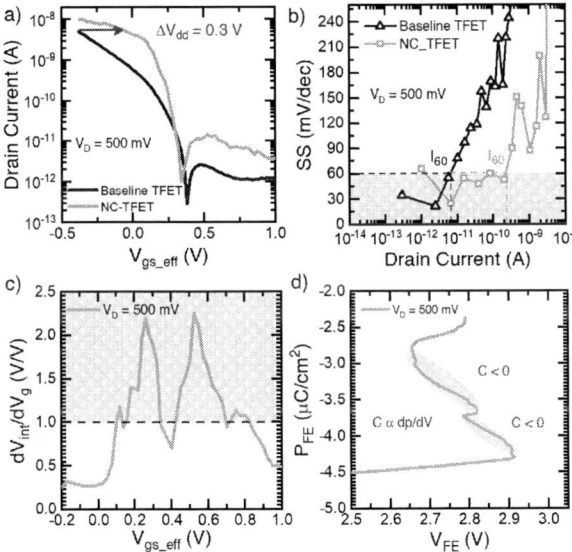

Fig. 5. Direct comparison between (a) I_d-V_g curves and, (b) SS of both the NC 2D/2D TFET and its baseline transistor at $V_D = 500$, confirming a significant improvement in the supply voltage, low-slope region, and I_{60}. (c) An internal voltage amplification, dV_{int}/dV_g, higher than one is observed. (d) The extracted P-V plot with a clear S-shaped confirms the existence of the NC effect.

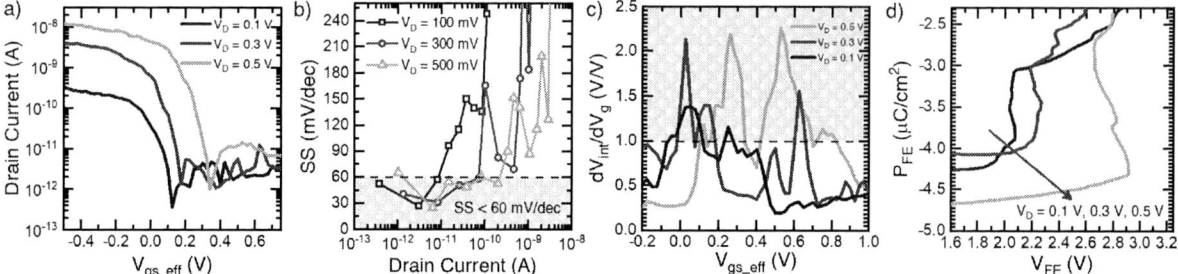

Fig. 7. a) Transfer curves of a NC 2D/2D TFET at the investigated drain biases. The impact of V_D variation on the SS as a function of I_D (b), internal voltage amplification (c), and extracted S-shaped P-V of the Si:HfO$_2$ ferroelectric capacitor (d), characteristics. Although, the NC TFET still shows an improvement compared to its baseline, the NC effect is degraded at lower V_D.

Table 1. Performance comparison of our NC TFET with recently reported 2D/2D TFETs. Our device exhibits the smallest reported subthreshold slope.

Material system	Sub-60 region	SS$_{min}$ mV/dec	I$_{60}$ (μA/μm)	I$_{on}$/I$_{off}$	Ref
WSe$_2$/SnSe$_2$	1 dec.	35	4×10^{-5}	10^5 ($V_D = 0.5$ V)	[2]
WSe$_2$/SnSe$_2$	~ 1 dec.	37	3×10^{-5}	10^5 ($V_D = 0.5$ V)	[3]
MoTe$_2$/MoS$_2$	1 dec.	34	~ 10^{-5}	10^6 ($V_D = -0.6$ V)	[12]
BP/MoS$_2$	-	65	-	10^4 ($V_D = -0.8$ V)	[13]
WSe$_2$/SnSe$_2$	-	90	-	10^6 ($V_D = 0.5$ V)	[4]
NC + WSe$_2$/SnSe$_2$	2.3 dec.	24	2×10^{-4}	$> 10^4$ ($V_D = 0.5$ V)	here

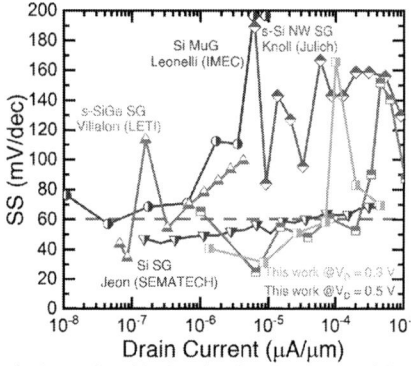

Fig. 8. Benchmarking based on best p-type state-of-the-art TFETs. Our 2D/2D device shows promising performance compared to the other technologies.

IV. CONCLUSION

In summary, we have reported for the first time a full experimental study on a 2D platform that integrates two steep slope concepts in a single NC WSe$_2$/SnSe$_2$ TFET, demonstrating the value of such an approach in terms of performance enhancement over any previous reported 2D tunnel FET devices. The point sub-thermionic swing of 24 mV/dec and average swings of 50 mV/dec over about 2.5

decades of current with relatively high ON/OFF current ratio ($>10^4$) are demonstrated. More importantly, the I$_{60}$ figures of merit are boosted up to 1.5 orders of magnitude due to the voltage amplification of the NC effect. The experimental results of this work offer a significant contribution to the performance engineering of 2D/2D energy efficient switches.

REFERENCES

[1] Ionescu, A.M. and Riel, H., "Tunnel field-effect transistors as energy-efficient electronic switches," *nature*, vol. 479, no. 7373, pp.329-337, 2011.

[2] Kim, S., et al., "Thickness-controlled black phosphorus tunnel field-effect transistor for low-power switches," *Nature nanotechnology*, vol. 15, no. 3, pp.203-206, 2020.

[3] Oliva, N., et al., "WSe$_2$/SnSe$_2$ vdW heterojunction Tunnel FET with subthermionic characteristic and MOSFET co-integrated on same WSe$_2$ flake," *npj 2D Materials and Applications*, vol. 4, no. 1, pp.1-8, 2020.

[4] Yan, X., et al., "Tunable SnSe$_2$/WSe$_2$ heterostructure tunneling field effect transistor," *Small*, vol. 13, no. 34, p.1701478, 2017.

[5] Fan, S., et al., "Tunable negative differential resistance in van der Waals heterostructures at room temperature by tailoring the interface," *ACS nano*, vol. 13, no. 7, pp.8193-8201, 2019.

[6] Kamaei, S., et al., "Gate energy efficiency and negative capacitance in ferroelectric 2D/2D TFET from cryogenic to high temperatures," *npj 2D Materials and Applications*, vol. 5, no. 1, pp.1-10, 2021.

[7] Chiang, C.C., et al., "Air-Stable P-doping in Record High-Performance Monolayer WSe$_2$ Devices," *IEEE Electron Device Letters*, 2021.

[8] Salahuddin, S. and Datta, S., "Use of negative capacitance to provide voltage amplification for low power nanoscale devices," *Nano letters*, vol. 8, no. 2, pp.405-410, 2008.

[9] Lee, M.H., et al., "Ferroelectric negative capacitance hetero-tunnel field-effect-transistors with internal voltage amplification," In *2013 IEEE International Electron Devices Meeting* (pp. 4-5). IEEE 2013.

[10] Saeidi, et al., "Nanowire tunnel FET with simultaneously reduced subthermionic subthreshold swing and off-current due to negative capacitance and voltage pinning effects," *Nano letters*, vol. 20, no. 5, pp.3255-3262, 2020.

[11] Saeidi, A., et al., "Negative capacitance as universal digital and analog performance booster for complementary MOS transistors," *Scientific reports*, vol. 9, no. 1, pp.1-9, 2019.

[12] Lu, H. and Seabaugh, A., "Tunnel field-effect transistors: State-of-the-art," *IEEE Journal of the Electron Devices Society*, 2(4), pp.44-49, 2014.

[13] Koo, B., et al., "Vertical-tunneling field-effect transistor based on MoTe$_2$/MoS$_2$ 2D–2D heterojunction," *Journal of Physics D: Applied Physics*, vol. 51, no. (47), p.475101, 2018.

[14] Xu, J. et al., "Tunneling field effect transistor integrated with black phosphorus-MoS$_2$ junction and ion gel dielectric," *Applied Physics Letters*, vol. 110, no. 3, p.033103, 2017.

Electroluminescence of $Si_xGe_{1-x-y}Sn_y$ / $Ge_{1-y}Sn_y$ pin-Diodes Grown on a GeSn Buffer

Lukas Seidel
Institute for Semiconductor Engineering
University of Stuttgart
Stuttgart, Germany
lukas.seidel@iht.uni-stuttgart.de

Sören Schäfer
Institute for Semiconductor Engineering
University of Stuttgart
Stuttgart, Germany
st155402@stud.uni-stuttgart.de

Michael Oehme
Institute for Semiconductor Engineering
University of Stuttgart
Stuttgart, Germany
michael.oehme@iht.uni-stuttgart.de

Dan Buca
Peter-Gruenberg-Institute 9
Forschungszentrum Jülich
Jülich, Germany
d.m.buca@fz-juelich.de

Giovanni Capellini
IHP – Leibniz-Institut für innovative Mikroelektronik, Germany
Dip. Scienze, Universitá Roma Tre, Italy
capellini@ihp-microelectronics.com

Jörg Schulze
Chair of Electron Devices
Friedrich-Alexander-University
Erlangen, Germany
joerg.schulze@fau.de

Daniel Schwarz
Institute for Semiconductor Engineering
University of Stuttgart
Stuttgart, Germany
daniel.schwarz@iht.uni-stuttgart.de

Abstract—We present the growth, fabrication, and characterization of a $Ge_{1-y}Sn_y$ pin-diode and a $Si_xGe_{1-x-y}Sn_y$ / $Ge_{1-y}Sn_y$ pin-diode. The pin-diodes are grown by molecular beam epitaxy on a partially relaxed $Ge_{1-y}Sn_y$ buffer grown by reduce-pressure chemical vapor deposition. The analysis of the crystal shows that the $Ge_{1-y}Sn_y$ pin-diode is lattice-matched grown and the $Si_xGe_{1-x-y}Sn_y$ / $Ge_{1-y}Sn_y$ pin-diode is pseudomorphic grown with respect to the buffer. Temperature-dependent direct current measurements reveal a threshold voltage shift from 0.3 V to 0.55 V and a series resistance that shows metallic behavior. Furthermore, by comparing the electroluminescence spectra at 13.4 K and 293 K we observe a 10 times higher signal for the $Ge_{1-y}Sn_y$ pin-diode and a 3 times higher signal for the $Si_xGe_{1-x-y}Sn_y$ / $Ge_{1-y}Sn_y$ pin-diode at cryogenic temperatures. The peak energies at an injection current density of 2.5 kA/cm^2 are 575 meV and 610 meV, respectively.

Keywords—*SiGeSn, GeSn, GeSn Substrate, Low Temperature, Electroluminescence*

I. INTRODUCTION

In the last years, there was a keen interest in building a photonic integrated circuit (PIC) on a Si platform [1]. PICs can be used for many different applications, e. g. on- and inter-chip communication, sensors, or photonic neural networks [2].

Main elements of a PIC are emitters, waveguides, modulators, and detectors. From all the development of emitters is the most challenging to fabricate and to implement on a Si platform. The reason lies in the indirect nature of the group IV elements Si and Ge. Thus, a bipolar current in a pn-junction formed out of those materials recombines rather non-radiative than radiative.

For building an efficient light emitter a direct bandgap material is essential. Approaches to achieve this are hybrid integration of III-V materials on Si [3], using tunneling to generate direct electrons [4], and tuning the bandgap of Ge by the incorporation of Sn [5]. Recent advances in the last approach resulted in a proof-of-concept for a laser on a Si platform [6].

The key part of this approach is the negative direct bandgap of α-Sn. By forming the binary alloy semiconductor $Ge_{1-y}Sn_y$, the direct, as well as the indirect bandgap, are decreasing with increasing Sn content. The fact that the energy of the direct bandgap decreases faster than the indirect bandgap leads to a transition of $Ge_{1-y}Sn_y$ from an indirect to a direct bandgap semiconductor at a Sn concentration of 6.5 % [7]. One bottleneck is the lattice mismatch between Ge and $Ge_{1-y}Sn_y$, which results either in compressively strained layers or in defects during the relaxed growth of $Ge_{1-y}Sn_y$ on Ge. One solution to overcome this is to grow a relaxed $Ge_{1-y}Sn_y$ buffer and consecutively grow relaxed $Ge_{1-y}Sn_y$ active layers on top of it. Additionally, by adding Si into $Ge_{1-y}Sn_y$ and creating a ternary alloy $Si_xGe_{1-x-y}Sn_y$, the bandgap can be decoupled from the lattice constant [5]. With this, it is possible to grow lattice-matched $Si_xGe_{1-x-y}Sn_y$ cladding layers that form barriers for electron- and hole-confinement in a pseudomorphic grown $Ge_{1-y}Sn_y$ recombination region.

In this paper, we present the growth and fabrication of a $Ge_{1-y}Sn_y$ pin-diode and a $Si_xGe_{1-x-y}Sn_y$ / $Ge_{1-y}Sn_y$ pin-diode on a $Ge_{1-y}Sn_y$ buffer. With high-resolution X-ray diffraction (HR-XRD) we analyze the grown layers and characterize the performance of the diodes with temperature-dependent, static direct current (DC) and electroluminescence (EL) measurements. We discuss the utilization of such a material system as amplification material for laser applications.

II. METHODS

A. Sample description and growth by RP-CVD and MBE

The two samples are grown on a (100) p$^-$-Si substrate. They differ in the composition of the p- and n-regions of the active layers (see Fig. 1). The growth process is divided into two sequences. First, a thick slightly tensile strained Ge-layer followed by a ~500 nm thick partially relaxed $Ge_{1-y}Sn_y$ buffer layer with a Sn concentration of $y = 8 \%$ are grown by reduced-pressure chemical vapor deposition (RP-CVD).

Before the consecutive growth of the active layers by molecular beam epitaxy (MBE), the wafer was cleaned in hydrofluoric acid (HF) and baked at $T_S = 300 °C$ for $t = 30 min$ [8]. The MBE growth starts with a 300 nm thick

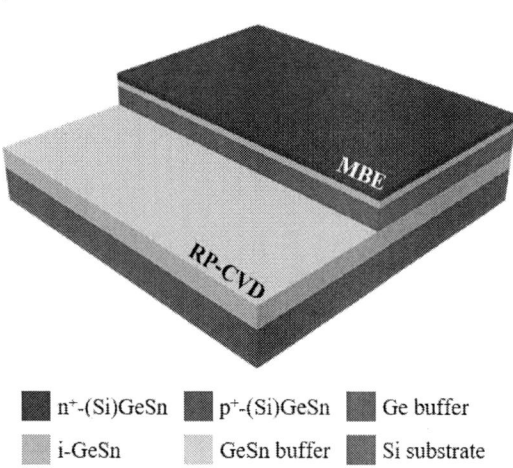

n⁺-(Si)GeSn ▪ p⁺-(Si)GeSn ▪ Ge buffer

i-GeSn ▪ GeSn buffer ▪ Si substrate

Fig. 1. Schematic layer stack of pin-diodes grown by MBE on a $Ge_{1-y}Sn_y$ buffer grown by CVD.

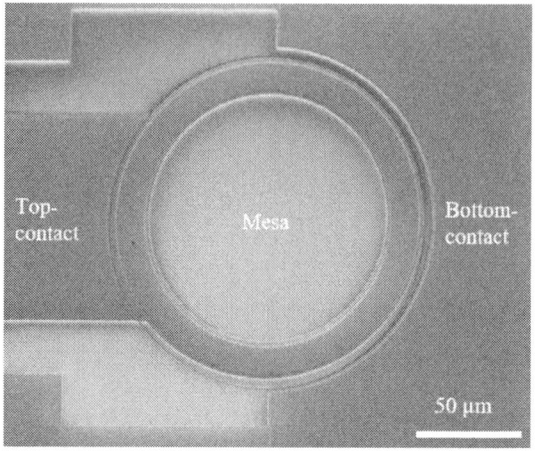

Fig. 2. Top view SEM image of a fabricated pin-diode with 80 µm radius. The mesa and the top- and bottom contacts are highlighted in false colors.

B-doped p⁺-layer with a nominal doping concentration of $N_A = 5 \cdot 10^{19} \ cm^{-3}$. After the p⁺-layer a 100 nm thick intrinsic $Ge_{1-y}Sn_y$-layer followed by a 100 nm thick Sb-doped n⁺-layer with a doping concentration of $N_D = 5 \cdot 10^{19} \ cm^{-3}$ are grown.

The p+- and n+-layer of sample A are strain-relaxed $Ge_{1-y}Sn_y$-layers with a nominal Sn concentration of $y = 8.3 \ \%$. Sample B is a hetero-diode with intrinsic $Ge_{1-y}Sn_y$ embedded into doped $Si_xGe_{1-x-y}Sn_y$ layers. The $Ge_{1-y}Sn_y$-layer contains nominal $y = 5.7 \ \%$ Sn and the $Si_xGe_{1-x-y}Sn_y$ layer $x = 14.1 \ \%$ and $y = 9.5 \ \%$, respectively. The motivation for embedding the $Ge_{1-y}Sn_y$-layer into the $Si_xGe_{1-x-y}Sn_y$ -layers is to confine the electrons and holes that are injected in the recombination region. The composition of the $Si_xGe_{1-x-y}Sn_y$-layers of sample B is chosen to lattice-match the $Ge_{1-y}Sn_y$ buffer and the Sn concentration for the embedded layer results from a trade-off between direct bandgap material (lower bound), tensile strain (upper bound) and good crystal quality (lower Sn concentration).

B. Fabrication process with a low thermal budget

The cylindrically shaped devices are fabricated in a single mesa process. As illustrated in the scanning electron microscopy (SEM) image in Fig. 2 the final devices have a ring-shaped top contact for electro-optical operation. To avoid Sn segregation all process steps are kept below a temperature of 250 °C.

Patterning for the mesa, the oxide windows, and the metallization is done with contact lithography using the i-line of a Hg-vapor lamp. Structuring of the mesa and the metallization is done via inductive-coupled plasma reactive ion etching (ICP-RIE) in an HBr plasma. To reduce the maximum sample temperature the metallization is etched in multiple steps. Due to lower power, this is not necessary for etching the mesa. In a trade-off between oxide quality and thermal budget, the 300 nm thick SiO_2 passivation layer is deposited at $T_S = 250 \ °C$ via plasma-enhanced CVD. The oxide windows were etched in an RIE system with CHF_3 followed by an etch in buffered HF. For the metallization, a combination of 50 nm Ti and 2 µm Al is used. The Ti acts as a diffusion barrier to avoid Al spiking at high injection currents.

C. Static DC and EL measurement setup

The measurements shown in sections III.B and III.C are taken in a He-cooled closed-cycle cryostat. Both devices are mounted and bonded on a printed circuit board which is electrically connected via coaxial cables to a vacuum feedthrough connector. The injection of the current during the EL measurements, as well as the DC measurements, are done with a Keithley 2601A.

DC measurements are taken in the range of -1 V to 1 V in steps of 10 mV. The EL spectra are measured with an integration time of 100 ms. Each spectrum for one constant injection current represents the mean spectrum of 2,000 measurements.

The emitted light of the devices is picked up from the top with a 300 µm fiber which can be aligned in all three axes. Outside of the cryostat, the fiber is connected to a round-to-linear fiber, which is connected to a NIRQuest 512-2.5 spectrometer. A relative calibration for the system is done with a Tungsten Halogen Light Source.

III. RESULTS AND DISCUSSION

A. Crystal analysis by HR-XRD

Fig. 3 shows the space map (SM) around the asymmetrical (-2-24) Bragg reflex taken by HR-XRD. For both samples, the peaks of the Ge-buffer and the partially relaxed $Ge_{1-y}Sn_y$ buffer are visible. Additionally, sample B shows two peaks that we assign to the $Si_xGe_{1-x-y}Sn_y$ and the intrinsic $Ge_{1-y}Sn_y$ layers.

As indicated by the relaxation line, the Ge-buffer is slightly tensile strained ($\varepsilon_\parallel = 0.1 \ \%$) with respect to the Si substrate. In Fig. 3a the high intensity of the $Ge_{1-y}Sn_y$ buffer confirms the lattice-matched growth of the $Ge_{1-y}Sn_y$ pin-diode of sample A. From the out-of-plane and in-plane lattice parameters we can calculate the relaxed lattice parameter a_0 and with Vegard's law we calculate the Sn concentration of the $Ge_{1-y}Sn_y$ layers of sample A to $y = 7.7 \ \%$ with a compressive strain of $\varepsilon_\parallel = -0.12 \ \%$ [9]. From the pseudomorphic line in Fig. 3b we see that the $Si_xGe_{1-x-y}Sn_y$ and $Ge_{1-y}Sn_y$ layers are both tensile strained. Similar calculations for the intrinsic $Ge_{1-y}Sn_y$ layer of sample B lead to a Sn concentration of $y = 5.7 \ \%$ and a tensile strain of

978-1-6654-8495-4/22 $31.00 © 2022 IEEE

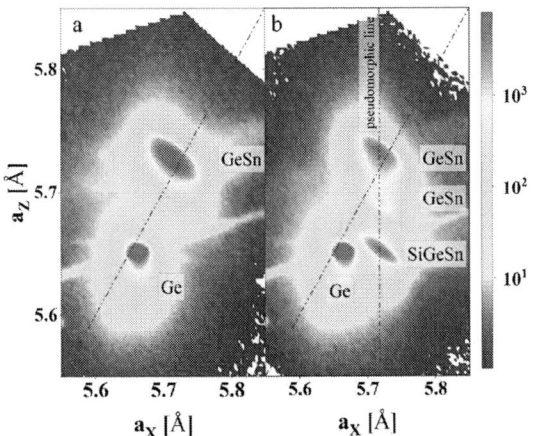

Fig. 3. SM image of (a) the $Ge_{1-y}Sn_y$ pin-diode and (b) the $Si_xGe_{1-x-y}Sn_y$ / $Ge_{1-y}Sn_y$ pin-diode. For both samples the relaxation line and in (b) the pseudomorphic line are shown.

$\varepsilon_\| = 0.23$ %. The $Ge_{1-y}Sn_y$ buffer has a higher Sn concentration ($y = 8.45$ %) and a compressive strain of $\varepsilon_\| = -0.17$ %.

B. Temperature-dependent DC characteristic

The temperature-dependent DC characteristics of devices with a radius of 80 µm for both samples are shown in Fig. 4. Measurements are taken in a range from 293 K down to 13.4 K in steps of 50 K.

We observe a high reverse current density at a voltage of -1 V and attribute this to high intrinsic charge carrier generation in the 100 nm thick intrinsic region of the diodes. By reducing the temperature, the current density in the reverse direction drops down.

At small voltages and low temperatures, we see a deviation from a normal pin-diode behavior. As explained in [10] band-to-band-tunneling from the n^+-layer into the parasitic p-doped intrinsic $Ge_{1-y}Sn_y$ layer could lead to such a behavior. Further investigations of smaller devices are required to proof this statement.

In the forward direction, the threshold voltage U_{th} shifts to higher voltages for decreasing temperature. We determine U_{th} by fitting the DC characteristics in the linear regime of the forward current and extrapolating to the x-intercept. With U_{th} we can calculate the temperature-dependent series resistance R_{Series} by using the following formula.

$$R_{Series} = (1\,V - U_{th}) / I(1\,V) \qquad (1)$$

Here, $I(1\,V)$ is the measured absolute current at 1 V. The temperature-dependent threshold voltage, as well as the series resistance, are given in Tab. I. An interesting fact is the decreasing series resistance with decreasing temperature. This behavior is normally known from metals and states that the semiconductor itself is not the limiting factor for the current, but some metallic material. Further investigations are necessary to figure out the origin for this behavior.

TABLE I.   TEMPERATURE-DEPENDENT THRESHOLD VOLTAGE AND SERIES RESISTANCE

T (K)	U_{th} (mV)		R_{Series} (Ω)	
	A	B	A	B
293	288	303	3.05	4.38
250	349	342	2.83	4.12
200	405	401	2.64	3.90
150	458	454	2.42	3.70
100	499	497	2.23	3.47
50	536	543	2.09	3.28
13.4	550	565	2.04	3.15

C. Temperature-dependent EL characteristic

In Fig. 5 the EL spectra of both samples are shown for two temperatures (13.4 K and 293 K) and five different injection current densities ranging from 0.5 kA/cm² to 2.5 kA/cm² in equidistant steps. The spectra are normalized to the maximum intensity I_{max} for each sample.

From the values of I_{max}, one can see that the emission intensity at the maximum injection current density of sample A is nearly four times higher than the intensity of sample B. This is attributed to a more direct $Ge_{1-y}Sn_y$ layer in sample A, due to a higher Sn concentration and thus to a higher energetic separation between the Γ- and L-valley.

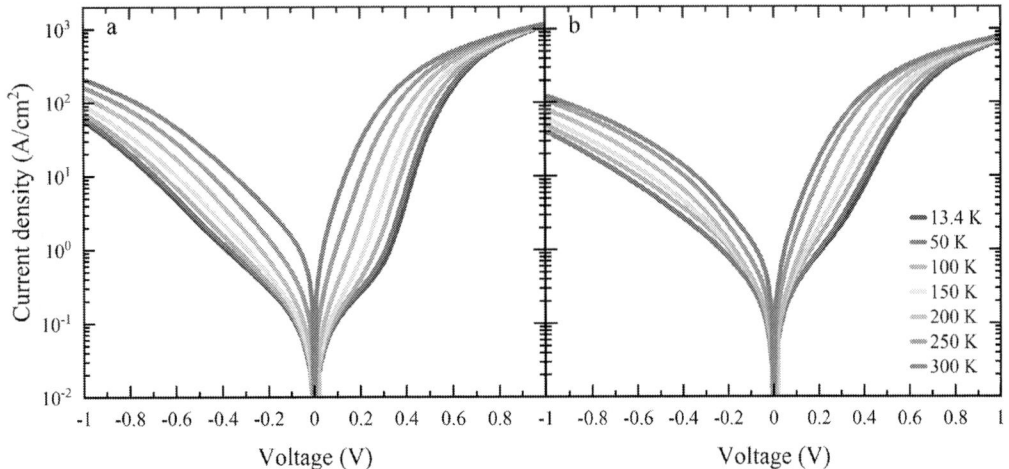

Fig. 4. Temperature dependent DC characteristics of (a) a $Ge_{1-y}Sn_y$ pin-diode and (b) a $Si_xGe_{1-x-y}Sn_y$ / $Ge_{1-y}Sn_y$ pin-diode with a radius of 80 µm.

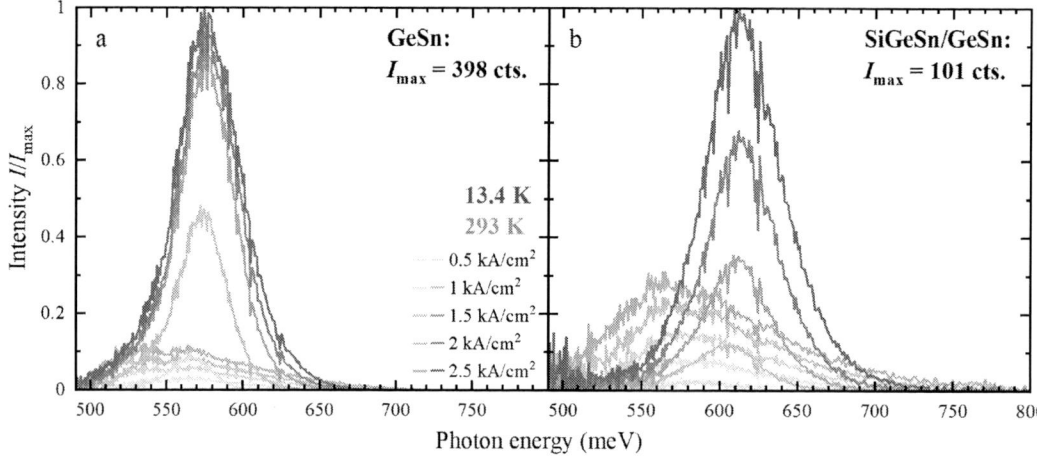

Fig. 5. EL spectra of (a) a $Ge_{1-y}Sn_y$ pin-diode and (b) a $Si_xGe_{1-x-y}Sn_y$ / $Ge_{1-y}Sn_y$ pin-diode with a radius of 80 µm at different injection current densities. The spectra are normalized to the maximum intensity of each device.

However, both samples show a higher intensity at 13.4 K compared to the intensity at 293 K and the peak energy shifts accordingly with the expected bandgap temperature indicating a direct semiconductor. It has been shown that a low Sn concentration and tensile strain is a way to realize efficient direct bandgap alloys [11]. Even the slightly tensile strained $Ge_{1-y}Sn_y$ layer of sample B behaves direct [12].

The evaluation of the peak energies as a function of the injection current density shows different behaviors for the different temperatures. At 293 K the heating of the devices results in a lowering of the bandgap and a shift to lower energies, respectively. We attribute the opposite shift of the spectra at 13.4 K to higher energetic carriers at higher injection current densities. The peak energies at maximum injection current densities are 575 meV for sample A and 610 meV for sample B, respectively.

IV. CONCLUSION

We showed the successful growth and fabrication of $Ge_{1-y}Sn_y$ and $Si_xGe_{1-x-y}Sn_y$ / $Ge_{1-y}Sn_y$ pin-diodes on a $Ge_{1-y}Sn_y$ buffer. From the SM we see the lattice matched growth of the $Ge_{1-y}Sn_y$ diode and pseudomorphic growth of the $Si_xGe_{1-x-y}Sn_y$ / $Ge_{1-y}Sn_y$ diode. The resulting Sn concentrations of the intrinsic region are 7.7 % with slight compressive strain for $Ge_{1-y}Sn_y$ diode and 5.7 % with slight tensile strain for $Si_xGe_{1-x-y}Sn_y$ / $Ge_{1-y}Sn_y$ diode.

By analyzing the temperature-dependent DC characteristics, we can determine a threshold voltage shift from 288 mV at 293 K to 550 mV at 13.4 K for the $Ge_{1-y}Sn_y$ diode and from 303 mV at 293 K to 565 mV at 13.4 K. We calculate a low series resistance of 3.05 Ω for sample A and 4.38 Ω for sample B. Furthermore, the series resistance is decreasing with decreasing temperature and thus behaves like a metal. This is a very good basis for laser operation at high temperatures because parasitic heating of the device by the series resistance is very small.

The EL-characteristic proves the direct material with increasing intensities at decreasing temperature. The $Ge_{1-y}Sn_y$ pin-diode with a higher Sn concentration shows a four times higher intensity than the $Si_xGe_{1-x-y}Sn_y$ / $Ge_{1-y}Sn_y$ pin-diode. At 13.4 K the diodes have a peak energy of 575 meV for sample A and 610 meV for sample B.

ACKNOWLEDGMENT

This research received partial funding from the Deutsche Forschungsgemeinschaft, DFG (OE 520/7-2) and the German Federal Ministry of Education (BMBF) under the ForMikro project FKZ: 16ES1076.

REFERENCES

[1] S. Y. Siew et al., "Review of Silicon Photonics Technology and Platform Development," *J. Lightwave Technol.*, vol. 39, no. 13, pp. 4374–4389, 2021, doi: 10.1109/JLT.2021.3066203.

[2] A. N. Tait et al., "Neuromorphic photonic networks using silicon photonic weight banks," *Sci Rep*, vol. 7, no. 1, p. 7430, 2017, doi: 10.1038/s41598-017-07754-z.

[3] G.-H. Duan et al., "Hybrid III--V on Silicon Lasers for Photonic Integrated Circuits on Silicon," *IEEE J. Select. Topics Quantum Electron.*, vol. 20, no. 4, pp. 158–170, 2014, doi: 10.1109/JSTQE.2013.2296752.

[4] R. Koerner et al., "The Zener-Emitter: A novel superluminescent Ge optical waveguide-amplifier with 4.7 dB gain at 92 mA based on free-carrier modulation by direct Zener tunneling monolithically integrated on Si," in *2016 IEEE International Electron Devices*, 22.5.1-22.5.4.

[5] N. von den Driesch et al., "Advanced GeSn/SiGeSn Group IV Heterostructure Lasers," *Advanced Science*, vol. 5, no. 6, p. 1700955, 2018, doi: 10.1002/advs.201700955.

[6] Y. Zhou et al., "Electrically injected GeSn lasers on Si operating up to 100 K," *Optica, OPTICA*, vol. 7, no. 8, p. 924, 2020, doi: 10.1364/OPTICA.395687.

[7] S. Gupta, B. Magyari-Köpe, Y. Nishi, and K. C. Saraswat, "Achieving direct band gap in germanium through integration of Sn alloying and external strain," *Journal of Applied Physics*, vol. 113, no. 7, p. 73707, 2013, doi: 10.1063/1.4792649.

[8] D. Schwarz et al., "MBE-Grown Ge0.92Sn0.08 Diode on RPCVD-Grown Partially Relaxed Virtual Ge0.92Sn0.08 Substrate," in *2019 42nd International Convention on Information and Communication Technology, Electronics and Microelectronics (MIPRO)*, 2019, pp. 50–54.

[9] D. Schwarz, H. S. Funk, M. Oehme, and J. Schulze, "Alloy Stability of Ge1−xSnx with Sn Concentrations up to 17% Utilizing Low-Temperature Molecular Beam Epitaxy," *Journal of Elec Materi*, vol. 49, no. 9, pp. 5154–5160, 2020, doi: 10.1007/s11664-020-08188-6.

[10] C. Schulte-Braucks et al., "Negative differential resistance in direct bandgap GeSn p-i-n structures," *Appl. Phys. Lett.*, vol. 107, no. 4, p. 42101, 2015, doi: 10.1063/1.4927622.

[11] A. Elbaz et al., "Ultra-low-threshold continuous-wave and pulsed lasing in tensile-strained GeSn alloys," (in En;en), *Nat. Photonics*, vol. 14, no. 6, pp. 375–382, 2020, doi: 10.1038/s41566-020-0601-5.

[12] N. von den Driesch et al., "SiGeSn Ternaries for Efficient Group IV Heterostructure Light Emitters," *Small (Weinheim an der Bergstrasse, Germany)*, vol. 13, no. 16, 2017, doi: 10.1002/smll.201603321.

GeSn-on-Si Avalanche Photodiodes for Short-Wave Infrared Detection

Maurice Wanitzek
Institute of Semiconductor Engineering
University of Stuttgart
Stuttgart, Germany
maurice.wanitzek@iht.uni-stuttgart.de

Michael Oehme
Institute of Semiconductor Engineering
University of Stuttgart
Stuttgart, Germany
michael.oehme@iht.uni-stuttgart.de

Christian Spieth
Institute of Semiconductor Engineering
University of Stuttgart
Stuttgart, Germany
st156759@stud.uni-stuttgart.de

Daniel Schwarz
Institute of Semiconductor Engineering
University of Stuttgart
Stuttgart, Germany
daniel.schwarz@iht.uni-stuttgart.de

Lukas Seidel
Institute of Semiconductor Engineering
University of Stuttgart
Stuttgart, Germany
lukas.seidel@iht.uni-stuttgart.de

Jörg Schulze
Chair of Electron Devices
Friedrich-Alexander-University
Erlangen-Nuremberg
Erlangen, Germany
joerg.schulze@fau.de

Abstract— In this work, we report the growth, fabrication, and characterization of GeSn-on-Si Avalanche Photodiodes with a Sn concentration of up to 2.2 %. The $Ge_{1-x}Sn_x$ absorption layer was grown at a very low temperature of 200 °C by using molecular beam epitaxy. We show record low dark currents limited by a perimeter leakage path, only and therefore not influenced by the Sn concentration, while the absorption for a wavelength of 1,550 nm increases significantly.

Keywords—GeSn, Avalanche Photodiode, Short-Wave Infrared, Lidar

I. Introduction

In recent years, applications for detectors operating in the short-wave infrared (SWIR) region have increased significantly. For example, eye-safe imaging, medical diagnostics, and fiber-optic telecommunications, etc. [1, 2, 3].

Eye-safe detectors are especially interesting for light detection and ranging (lidar), since the maximum range of the current state-of-the-art Si-based lidar is limited by the legally prescribed laser power requirements. Si-based lidar use Si single-photon avalanche diodes (SPAD) as receivers mainly using wavelengths of 905 nm. A solution to this limit is the shift of the working wavelength into the SWIR region since in this wavelength range orders of magnitude more light intensity can be emitted. The wavelength of 1,550 nm for optical telecommunications is ideal here, as there are several cost-effective solutions for the emitter side. In addition to materials from group III/V, the material Ge is also suitable for detecting light with a wavelength of 1,550 nm.

Ge as an absorption material has already been widely investigated for example as detectors in infrared cameras, as avalanche photodiodes (APD) in high data rate communication systems, or as SPADs for various applications [4, 5, 6]. However, the direct bandgap of Ge at room temperature is exactly 0.8 eV or 1,550 nm. This is also shown by the absorption spectrum in Fig. 1. As soon as a Ge device is cooled, e.g. to reduce the dark current, the direct band gap shifts to smaller wavelengths or higher energies, respectively. This significantly reduces the sensitivity of Ge detectors at 1,550 nm. That becomes a problem, especially in Ge-based SPADs in which the absorption at the wavelength of 1,550 nm is greatly decreased since they as of now need cryogenic cooling to function correctly [6]. Implementing Sn inside the Ge crystal will reduce the bandgap of the material and

AI-SEE is a co-labelled PENTA and EURIPIDES² project endorsed by EUREKA. National Funding Authorities: Austrian Research Promotion Agency (FFG), Business Finland, Federal Ministry of Education and Research (BMBF) and National Research Council of Canada Industrial Research Assistance Program (NRC-IRAP).

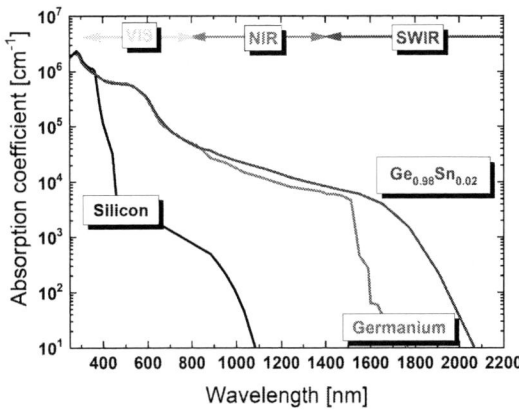

Fig. 1: Absorption coefficients of Si, Ge, and $Ge_{0.98}Sn_{0.02}$ as a function of the wavelength.

therefore increase the absorption for wavelengths of 1,550 nm and above [7]. A vast improvement compared to Ge can already be achieved with small Sn concentrations, as shown in Fig. 1. $Ge_{1-x}Sn_x$ as an absorption material will therefore allow good absorption at 1,550 nm even at cryogenic temperatures.

II. Growth and Fabrication

A. Low-temperature growth of GeSn on Si substrates

The overall design of the GeSn-on-Si APD is the well-known separate absorption, charge and multiplication structure. An illustration of this layer stack with the internal electrical field can be seen in Fig. 2.

The GeSn-on-Si APDs are grown using a solid source 6"-molecular beam epitaxy (MBE) system with a base pressure of $p < 10^{-10}$ mbar. While Ge and Si are deposited via an electron beam evaporator, Sn and B are deposited via effusion cells. For the samples, we use an As-doped n^{++} Si wafer with (100) orientation as a substrate, which will be utilized as the bottom contact. The growth starts with the deposition of the 500 nm thick i-Si multiplication layer at a growth temperature of 600 °C. The background doping of the i-Si is n-type in the order of 10^{15} cm^{-3}, confirmed with Hall measurements. After that, a p-type doped 100 nm thick Si layer is deposited at 450 °C which serves as the charge layer. With the charge layer, one can tune the internal electrical field strengths, so that the electrical field is not too high in the low bandgap

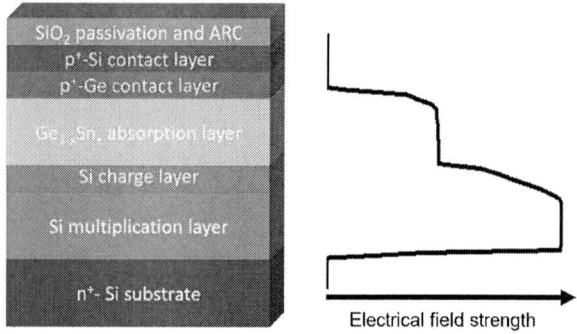

Fig. 3: Schematic illustration of the GeSn-on-Si APD with the internal electrical field.

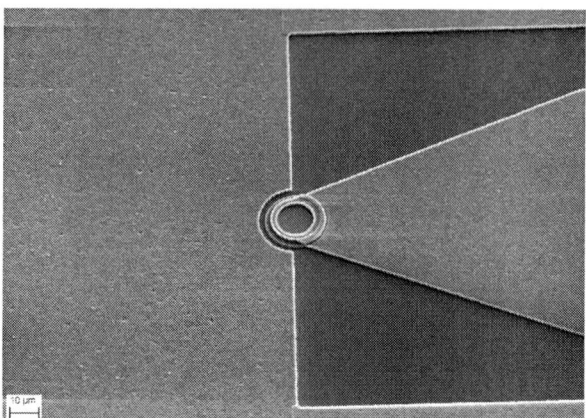

Fig. 2: A top view SEM image of a finished device is shown, which represents a diode with a 20 µm diameter.

absorption material. The nominal doping of the charge layer is $8 \cdot 10^{16}$ cm^{-3}. The 300 nm thick Ge$_{1-x}$Sn$_x$ absorption layer is deposited at a very low temperature of 200 °C.

After that, the substrate temperature is increased again and a 100 nm thick heavily p-doped Ge layer is deposited at 330 °C followed by a 100 nm thick heavily p-doped Si layer deposited at 450 °C. These layers form a hetero contact and allow easy processing later on.

B. Sample Variations

Using the same growth process, four different GeSn-on-Si APDs with varying Sn concentrations of 0%, 0.9%, 1.2%, and 2.2% in the absorption layer were grown for this experiment. The samples are named as can be seen in Tab. 1.

C. Fabrication

Following the MBE growth, the wafers are diced in squares and fabricated using a single mesa process. The main fabrication steps are:

- Mesa etching using an inductively coupled plasma reactive ion etching (ICP-RIE) with HBr
- Passivation with SiO$_2$ in a plasma-enhanced chemical vapor deposition system at 390 °C using TEOS as a precursor
- Contact hole opening with reactive ion etching using CHF$_3$
- Contact metal deposition using sputtering of Al
- Al contact structuring using ICP-RIE with HBr

A more detailed description of the fabrication process can be found elsewhere [8]. A finished GeSn-on-Si APD can be seen in the scanning electron microscope (SEM) image in Fig. 3, showing a mesa diode with a 20 µm diameter and its metal contacts. The top metal has a ring structure to allow illumination from the front side.

D. Material and layer analysis

After the growth process, all samples were characterized by high-resolution x-ray diffraction. Space map scans of the asymmetric (-2-24) diffraction were performed to give information about Sn concentration, crystallinity, material quality, and strain. In Fig. 4, the space maps of the GeSn-on-Si APDs with a Sn concentration up to 2.2 % are illustrated.

Samples with a Sn concentration > 0 % show a Ge$_{1-x}$Sn$_x$ peak from the Ge$_{1-x}$Sn$_x$ absorption layer and an additional Ge

TABLE I. Sample variations with their corresponding Sn concentration.

Sample	A	B	C	D
Sn con-centration	0 %	0.9 %	1.2 %	2.2 %
a_{perp}	5,656 Å	5,665 Å	5,673 Å	5,681 Å
a_{par}	5,650 Å	5,664 Å	5,661 Å	5,669 Å

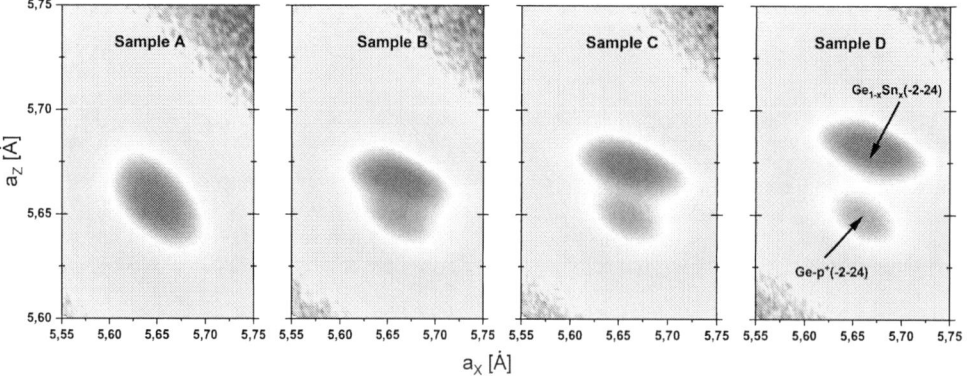

Fig. 4: Space map plots for the samples showing the Ge and Ge$_{1-x}$Sn$_x$ peaks.

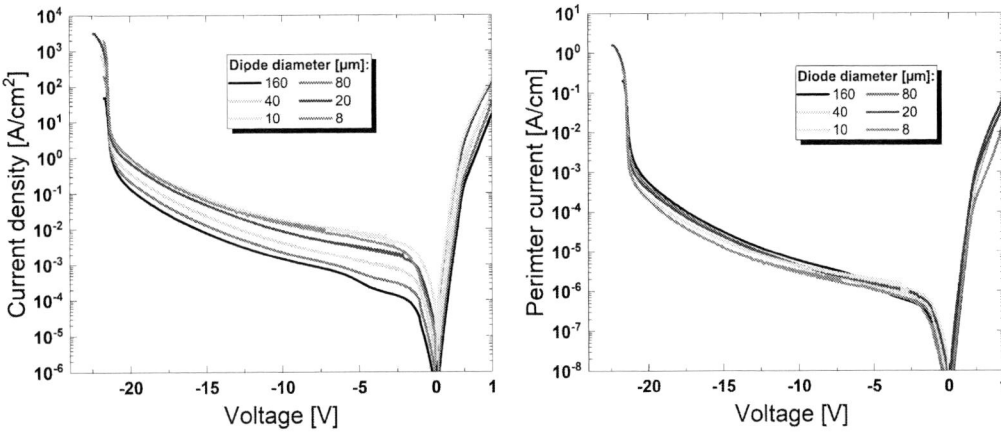

Fig. 5: Area and perimeter normalized I(V) curves of the GeSn-on-Si APD with 2.2 % Sn concentration for varying diode diameters.

peak coming from the top p^+-Ge contact layer. This peak increasingly overlaps with the $Ge_{1-x}Sn_x$ absorption layer peak for decreasing Sn concentration until, as in sample A, one cannot differentiate between the peaks anymore.

The strain in the different absorption layers is slightly compressive and measured to be between 0.007 % and 0.12 %. With additional space scans after the fabrication process, we could not find any differences. This proves that especially the thermal budget of the fabrication process did not influence the material layers meaningfully.

III. CHARACTERIZATION

Following the growth and fabrication process, we performed electrical and electrooptical characterizations.

A. Electrical Characterization

All samples were electrically characterized via a 4-point measurement. Mesa diodes with ranging diameters between 8 and 160 μm were measured to investigate the volume or perimeter proportional behavior of the leakage current. Fig. 5 exemplary shows the measured I(V) curves nominalized to the area and perimeter.

The breakdown voltage of -21.4 V is independent of the diode diameter with slight variations over the whole sample due to the flux distribution of the electron beam evaporators. All GeSn-on-Si APDs as well as the Ge reference APD show a peripheral proportionality in the dark current characteristics. Additional improvements in the mesa passivation for example with atomic layer deposition of Al_2O_3 are promising to further reduce the dark current. In Fig. 6, the dark currents of the different samples are compared for 160 μm and 20 μm diameter diodes.

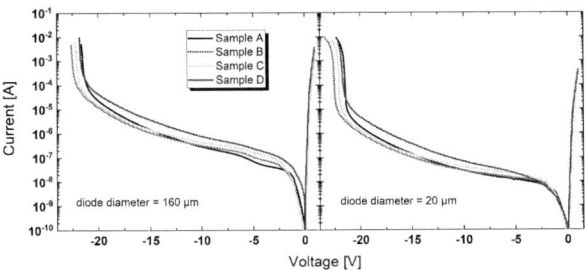

Fig. 6: Dark current characteristic of the samples for 160 μm and 80 μm diode diameter.

Overall, the dark currents of the samples are smaller than what has been published before [9, 10]. It can be seen that there is no significant increase in dark current for an increasing Sn concentration in the absorption layer. This proves that the low-temperature growth of the $Ge_{1-x}Sn_x$ is not degrading the dark current. Which is in contrast to what is to be expected in simple GeSn pin diodes [11]. This is due to the layer structure of the GeSn-on-Si-APD which consists of an internal blocking Si diode.

Since the breakdown voltage of the device is mostly defined by the Si multiplication layer thickness no influence of Sn concentration can be found. Nevertheless, the breakdown voltage varies between -21.4 V and -22.4 V across the samples which is caused by growth rate fluctuations.

B. Electrooptical Characterization

The electrooptical measurements were performed using a laser at a wavelength of 1,550 nm. For that, we illuminated the mesa diodes through a single-mode fiber and measured the current as a function of reverse voltage. In Fig. 7 the resulting photocurrents at a wavelength of 1,550 nm are illustrated.

One can see that the photocurrent of the GeSn-on-Si APDs only becomes significant after a certain reverse voltage is reached. This voltage is called the punch-through voltage and describes the voltage at which the electrical field penetrates through the p-doped charge layer into the $Ge_{1-x}Sn_x$ absorption layer. For all samples this happens at approximately -5.3 V. Since the absorption layer is only slightly doped, the electrical field expands rapidly which corresponds to the sharp rise in photocurrent. The photocurrent will only increase slowly for increasing reverse voltages until the multiplication of the APD becomes significant and the photocurrent rises again.

At a wavelength of 1,550 nm, the photocurrent increases significantly for increasing Sn concentration in the absorption layer. This is caused by the lowering of the bandgap of the absorption layer. Over a wide range of bias voltages, a 4-fold increase in photocurrent is achieved with a Sn concentration of 2.2 % compared to the Ge reference APD.

We measured the photocurrent as a function of optical power and then calculated the responsivity in dependence on the applied voltage. These responsivities are illustrated in Fig. 8.

978-1-6654-8495-4/22 $31.00 © 2022 IEEE

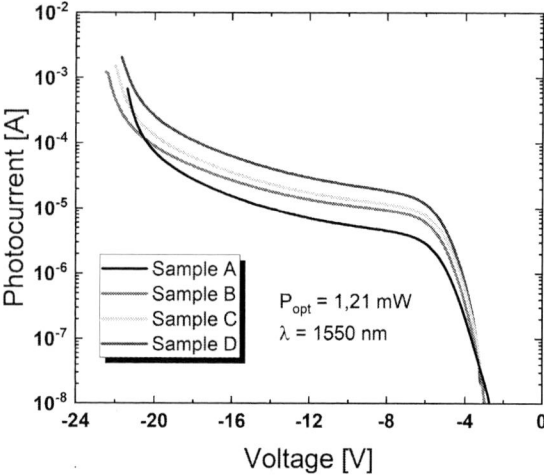

Fig. 7: Photocurrent as a function of reverse voltage at a wavelength of 1,550 nm.

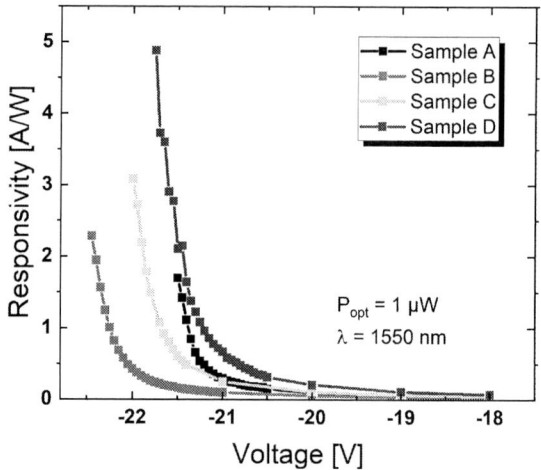

Fig. 8: Responsivity as a function of reverse voltage at a wavelength of 1,550 nm and a power of 1 μW.

One can see, that the optical responsivities for a wavelength of 1,550 nm increase drastically for increasing Sn concentration. At a wavelength of 1,550 nm and a power of 1 μW, the maximum responsivity of 4.88 A/W is achieved, which is the highest achieved responsivity with GeSn APDs to the knowledge of the authors.

IV. CONCLUSION

We successfully demonstrated the use of low-temperature $Ge_{1-x}Sn_x$ epitaxy at 200 °C for GeSn-on-Si APDs. This will allow the $Ge_{1-x}Sn_x$ to be grown in the backend-of-line to further enable the commercial use of $Ge_{1-x}Sn_x$ as an absorber material. The samples show an overall low dark current which is dominated by perimeter leakage and therefore the further reduction of dark current with improved fabrication processes is promising. The GeSn-on-Si APD with 2.2 % Sn shows a 4-fold increased photocurrent compared to a Ge-on-Si APD at a wavelength of 1,550 nm. The responsivities increasingly improve for increasing Sn concentration in the $Ge_{1-x}Sn_x$ absorption layer. A responsivity of 4.88 A/W is achieved with a Sn concentration of 2.2 %.

ACKNOWLEDGMENT

The research leading to these results is part of the AI-SEE project, which is a co-labelled PENTA and EURIPIDES² project endorsed by EUREKA. Co-funding is provided by the following National Funding Authorities: Austrian Research Promotion Agency (FFG), Business Finland, Federal Ministry of Education and Research (BMBF) and National Research Council of Canada Industrial Research Assistance Program (NRC-IRAP).

REFERENCES

[1] M. P. Hansen and D. S. Malchow, "Overview of SWIR detectors, cameras, and applications, " Proc. SPIE 6939, Thermosense XXX, 69390I, 2008.

[2] R. Soref, D. Buca, and S. Yu, "Group IV photonics. driving integrated optoelectronics," Optics&Photonics News Vol. 27, No. 1, pp. 32-39, 2016.

[3] E. Kasper, M. Kittler, M. Oehme and T. Arguirov, "Germanium tin: silicon photonics toward the mid-infrared," Photon. Res. Vol. 1, No. 2, pp. 69-76, 2013.

[4] M. Oehme et al, "Backside illuminated Ge-on-Si NIR camera," IEEE Sensors Journal Vol. 21, No. 17, 2021.

[5] Y. Kang et al, "Monolithic germanium/silicon avalanche photodiodes with 340 GHz gain-bandwidth product," Nat. Photon. 3, pp. 59-63, 2009.

[6] P. Vines et al, "High performance planar germanium-on-silicon single-photon avalanche diode detectors," Nat. Commun. 10, 1086, 2019.

[7] G. He and H. A. Atwater, "Interband transitions in Sn_xGe_{1-x} alloys," Phys. Rev. Lett., Vol. 79, No. 10, pp. 1937-1940, 1997.

[8] M. Wanitzek, M. Oehme, D. Schwarz, K. Guguieva and J. Schulze, "Ge-on-Si avalanche photodiodes for LIDAR applications", 2020 43rd International Convention on Information, Communication and Electronic Technology (MIPRO), pp. 8-12, 2020.

[9] D. Zhang et al, "GeSn on Si avalanche photodiodes for short wave infrared detection," Proc. SPIE 10846, Optical Sensing and Imaging Technologies and Applications, 108461B, 2018.

[10] Y. Dong et al, "Germanium-Tin on Si avalanche photodiode: device design and technology demonstration," IEEE Transactions on Electron Devices, Vol. 62, No. 1, pp. 128-135, 2015.

[11] M. Oehme et al, "GeSn p-i-n detectors integrated on Si with up to 4% Sn," Appl. Phys. Lett. 101, 141110, 2012.

A Drift-Compensated Magnetic Spectrometer for Point-of-Care Wash-Free Immunoassays using a Concurrent Dual-Frequency Oscillator

Jui-Hung Sun*, Bill Ling†, Md. Abdullah-Al Kaiser*, Constantine Sideris*

Email: {juihungs, mdabdull, csideris}@usc.edu, blling@caltech.edu

*Department of Electrical and Computer Engineering, University of Southern California, Los Angeles, CA, USA

†Department of Chemical Engineering, California Institute of Technology, Pasadena, CA, USA

Abstract—A 2x2 magnetic spectrometer array using a concurrent dual-frequency transformer-based oscillator in 65nm CMOS is presented. Concurrent dual-frequency operation allows compensation of the drift of the free-running sensing oscillator without reconfiguring to switch the frequency of oscillation, which enables wash-free magnetic label assays and single-site multiplexed spectroscopy. The spectrometer achieves a sensitivity of 0.7ppm consuming only 3.1mW per cell over a wide frequency range of 1.2-1.65 / 2.9-4 GHz. A biotin-streptavidin immunoassay using iron oxide magnetic nanoparticle labels is performed without any washing steps, demonstrating high sensitivity and viability for point-of-care diagnostics.

Keywords—spectrometer, concurrent oscillator, magnetic nanoparticle, immunoassay, biosensor, diagnostics.

I. INTRODUCTION

Rapid, low-cost, and accessible point-of-care diagnostics have gained prominence due to their potential for quick and accurate detection of disease. Bioassays using magnetic labels are of particular interest since they can provide distinct advantages over the fluorescent labels used in much commercially available analysis equipment. Magnetic nanoparticle (MNP) labels do not suffer from signal-bleaching or significant background. Furthermore, MNPs can be detected by inexpensive and/or portable biosensor platforms, eliminating the bulky and expensive optics required for fluorescent assays. Several magnetic biosensor implementations in CMOS have been demonstrated, encompassing GMR [1], Hall effect [2], and LC-resonance shift [3, 4] sensing modalities. However, these sensors require complex surface chemistry functionalization and thorough washing steps to remove unbound MNPs from the sensing surface, restricting viability for point-of-care use.

Recently, proof-of-concept wash-free MNP-based immunoassay experiments were demonstrated which measure changes in MNP relaxation due to bound target biomolecules [5-7]. When target proteins bind to multiple MNPs coated with polyclonal antibodies, aggregate MNP clusters are formed. Although the total amount of magnetic material is unchanged, the clusters exhibit different magnetic relaxation dynamics versus free unbound MNPs, because dipolar interactions due to nearby magnetic domains affect the Néel relaxation times and clustering changes the Brownian motion [8]. Such changes in relaxation times directly translate to variations in the frequency-domain magnetic susceptibility of the MNPs, which can be

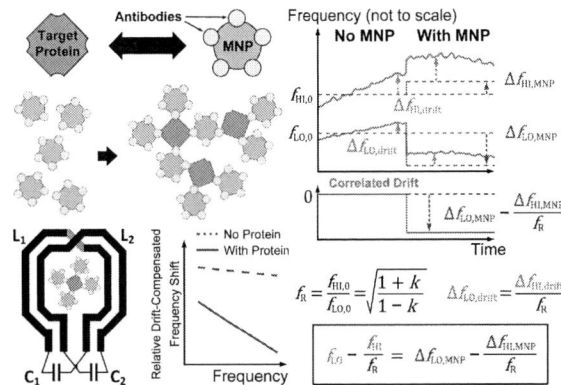

Fig 1. Concept of LC-resonance-shift magnetic immunoassay and drift compensation using a dual-frequency transformer-based oscillator.

measured by using frequency-sweep spectroscopy. LC-resonance shift-based sensing is particularly well suited for this application, as highly sensitive, low-power and low-cost LC-shift sensors can be implemented in standard CMOS processes without requiring any post-fabrication processing or modification. The presence of MNPs is sensed through induced frequency shifts in on-chip LC oscillators, which do not require any external biasing or excitation magnetic fields. Although such an approach is highly sensitive to magnetic materials, its sensitivity can be significantly limited by the drift of the free-running oscillators, caused by temperature variations and other slow-varying noise sources. A reference oscillator may be used to mitigate some of the drift, but at the cost of additional chip area, power consumption, and noise. Furthermore, a reference may not be viable for microfluidics-free, open-well based assays where the whole chip surface may be covered with liquid sample, making isolating the reference from samples difficult.

In this work, we present the first realization of a fully integrated CMOS biosensor that can perform wash-free magnetic immunoassays using frequency-sweep spectroscopy. The sensor achieves very high sensitivity and frequency stability by using a concurrent dual-frequency (CDF) oscillator and a self-referenced drift cancellation scheme. The magnetic sensing signal is encoded in the difference between the two oscillation frequencies, while the drift affects both frequencies in the same direction and appears in their sum.

978-1-6654-8495-4/22 $31.00 © 2022 IEEE

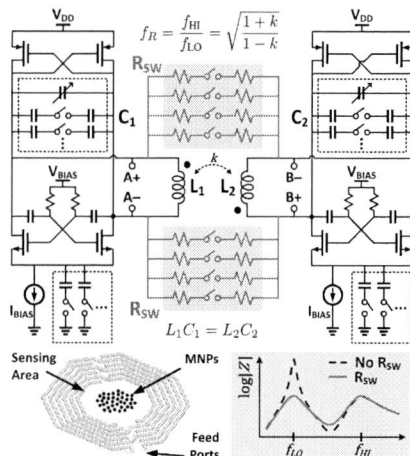

$$f_R = \frac{f_{HI}}{f_{LO}} = \sqrt{\frac{1+k}{1-k}}$$

$$L_1 C_1 = L_2 C_2$$

Fig 2. Concurrent dual-frequency transformer-based oscillator.

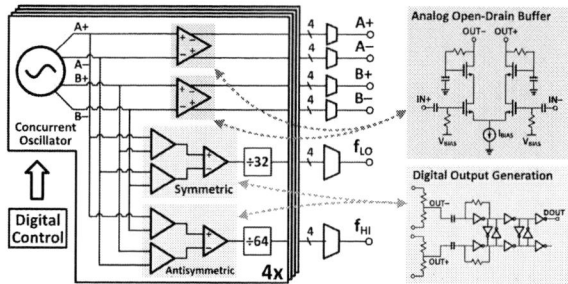

Fig 3. Magnetic spectrometer array system block diagram.

II. SENSOR OPERATION

To realize self-referenced drift cancellation, the transformer-based resonant tank approach introduced in [4] is adapted to the proposed CDF oscillator. The 4th order transformer tank has two resonance frequencies, f_{LO} and f_{HI}. With a fully symmetric tank, the ratio $f_R = f_{HI} / f_{LO}$ between the two frequencies depends only on the transformer coupling factor k and not the absolute inductance and capacitance values. This significantly reduces the impact of process and temperature variations, and thus shifts in frequency caused by temperature fluctuations and other slow-varying noise sources will be highly correlated between f_{LO} and f_{HI}. The absolute values of f_{LO} and f_{HI} are chosen such that f_{LO} lies below the typical resonance frequency of MNPs, while f_{HI} lies above it. Therefore, when MNPs are introduced, f_{LO} shifts downward while f_{HI} either remains largely unchanged or shifts upwards (in the opposite direction of f_{LO}), since the MNPs have a smaller (usually negative) susceptibility at f_{HI}. Since the shifts in f_{LO} and f_{HI} due to drift are correlated, the drift will appear in the sum of the measured f_{LO} and f_{HI} signals. By taking the difference between the measured f_{LO} and reconstructed f_{LO} (using the measured f_{HI}), the drift-induced shifts are cancelled while the MNP-induced shifts are unaffected or amplified if the MNP susceptibility at f_{HI} is either small or negative, respectively. Thus, an accurate measurement of the magnetic content in the sample without the drift can be obtained.

A. Concurrent Operation

With a single-frequency transformer-based oscillator as in [4], this drift compensation method is limited by the need to periodically switch the sensing oscillator between f_{LO} and f_{HI}. Sequential sensing and reconstruction lead to a loss of correlation where noise and drift that vary faster than the switching window will not be cancelled. Also, the startup transients caused by reconfiguring the frequency may introduce additional thermal fluctuations, which further exacerbate the loss of correlation and leads to a reduction in sensitivity and frequency stability. CDF operation allows reconstruction of the frequency data using fully concurrent operation without requiring any switching, thus eliminating the loss of correlation.

B. Wash-Free Magnetic Immunoassay

To perform a frequency-sweep magnetic immunoassay, a biosample is mixed with antibody-coated MNPs and applied directly to the sensing transformer. As the resonance frequencies of the CDF oscillator are swept by tuning the tank capacitance, the drift-compensated frequency shift relative to f_{LO} varies (usually decreasing with f_{LO}, as the MNP susceptibility typically decreases with frequency). The presence of target biomolecules will be indicated by a different relative frequency shift curve versus a control sample prepared without the target analyte, due to the difference in relaxation time constants of the aggregated versus randomly dispersed MNPs. The magnetic immunoassay and CDF drift compensation method are illustrated in Fig. 1.

III. SPECTROMETER IMPLEMENTATION

A. CDF Oscillator

The CDF transformer-based oscillator is shown in Fig. 2. To allow stable concurrent oscillation, the impedances associated with f_{LO} and f_{HI} must be approximately equal. Previous transformer-based concurrent oscillator implementations [9, 10] typically rely on asymmetry between the two sides of the transformer to equalize the impedances. However, symmetry is necessary for the tracking between f_{LO} and f_{HI} to be independent of absolute component values. In a symmetric oscillator, startup would naturally favor f_{LO} since the tank impedance is significantly greater at f_{LO} than at f_{HI}. To make concurrent oscillation possible, the Q associated with f_{LO} must be adjusted to bring the f_{LO} and f_{HI} peaks close in magnitude. This is accomplished with a pair of switched resistor banks connected anti-symmetrically across the transformer nodes.

Furthermore, the 5th order nonlinearity of the transconductors in the oscillator must satisfy a particular condition for both resonance frequencies to have the same steady-state closed-loop gain and thus support concurrent oscillation [9]. In order to achieve the required nonlinearity with a fully symmetric oscillator and with conventional CMOS devices, the NMOS cross-coupled pairs are operated in class-C. Using a lowered gate bias and tunable tail capacitance, the conduction angle is lowered to allow stable concurrent oscillation over a wide frequency range. Matched switched-capacitor banks and varactors in the oscillator permit tuning f_{LO} and f_{HI} together, enabling frequency-sweep spectroscopy.

B. Sensor Output

Sensor readout can be obtained either via wideband open-drain analog buffers on each side of the transformer or via

978-1-6654-8495-4/22 $31.00 © 2022 IEEE   174

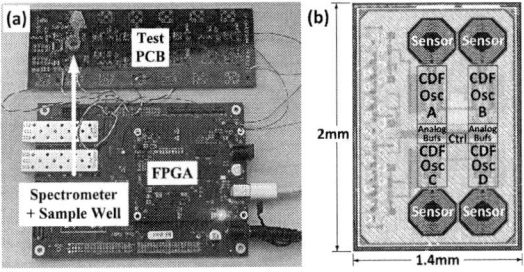

Fig. 4. (a) Experiment setup with spectrometer chip in sample well and FPGA; (b) Chip micrograph.

divided-down digital outputs. The analog open-drain outputs for all sensor cells in the spectrometer are multiplexed together, allowing output selection by enabling the tail current source of the respective buffer. The digital outputs are generated by extracting the common and differential mode voltages from each sensor cell CDF oscillator by combining each pair of transformer nodes in phase (f_{LO}) and out of phase (f_{HI}) respectively using resistors. This approach is highly area-efficient for extracting f_{LO} and f_{HI}, in contrast to on-chip filters. The outputs are then passed through digital buffers and on-chip 64x (f_{HI}) and 32x (f_{LO}) dividers, allowing frequency data to be obtained via simple external frequency-counting circuitry. A block diagram of the spectrometer is shown in Fig. 3.

IV. MEASUREMENT RESULTS

We implement a prototype 2x2 spectrometer array in a bulk 65nm CMOS process. Each sensor cell uses a 1.2V supply with adjustable bias current from 2.6mA to 10mA, allowing operation with as little as 3.1mW of power. The bias current is tuned in discrete steps by using a bank of binary-weighted switched resistors. The sensing coils are 1:1 transformers with k = 0.77 and 3.36nH of inductance per coil. The frequency tuning range is 1.2 to 1.65GHz for f_{LO} and 2.9 to 4GHz for f_{HI}. All circuits and passives in each sensor cell are interleaved and laid out in a common-centroid arrangement to maintain the accuracy of the tracking between f_{LO} and f_{HI}. Each cell has a sensing area of 390x360μm², sufficient for DNA, cell, and protein assays.

The proof-of-concept portable spectrometer system is comprised of the spectrometer IC mounted as chip-on-board on a PCB interfaced with an FPGA. Samples can be deposited on the chip surface in a sample well mounted directly on top of the IC. The experimental setup and die micrograph of the IC are shown in Fig. 4. The measured phase noise under concurrent oscillation was −115dBc/Hz and −111dBc/Hz at 1MHz offset for 1.22GHz and 2.90GHz, respectively. Measured CDF oscillation time-domain waveforms and the spectrum using the lowest capacitor bank setting are shown in Fig. 5.

A. Drift Compensation

To verify the long-term tracking and reconstruction capabilities of the spectrometer, we perform a 5.5 hour test with one cell in concurrent oscillation. Fig. 6 shows measured and reconstructed f_{LO} data at the lowest and highest-frequency capacitor bank settings, indicating excellent agreement between the reconstructed and measured frequencies. The measured noise floor of the drift-compensated signal is 1.1kHz over the full test, which is 52x smaller than the noise floor without

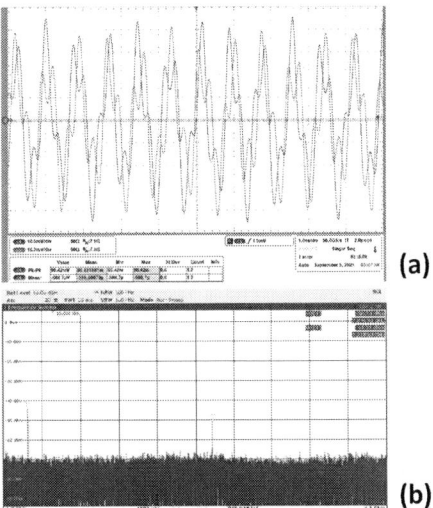

Fig. 5. (a) Measured CDF oscillation waveforms at the lowest capacitor bank setting; (b) Corresponding measured spectrum.

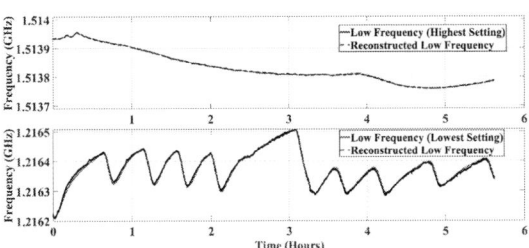

Fig. 6. Long-term drift cancellation over a 5.5-hour period for the highest and lowest capacitor bank settings and varactor at 0V.

compensation and corresponds to a detection limit of 0.7ppm.

Fig. 7 shows measured and reconstructed f_{LO} data and the drift-compensated transient response due to the introduction of 10nm iron oxide MNPs. The relative frequency shifts for 10 and 20nm MNPs (2μL volume, 1mg/mL each) are plotted in Fig. 8. The different relative frequency shift spectra of the two samples indicate the single-site multiplex sensing capability of the spectrometer. A shift of 100ppm at the lowest frequency for the 10nm MNPs corresponds to an estimated detection limit of 14ng of MNPs based on the measured sensor noise floor of 0.7ppm.

B. Wash-Free Magnetic Immunoassay

We perform a biotin-streptavidin immunoassay to demonstrate viability for single-step wash-free detection. Streptavidin is a tetrameric protein with a high affinity for binding to biotin. When streptavidin is mixed into a solution of biotinylated MNPs, aggregate clusters are formed, leading to enhanced dipole-dipole interactions and altered AC susceptibility when compared to free unbound MNPs without streptavidin. Control (0.2mg/mL 10nm MNPs) and experiment (0.2mg/mL 10nm MNPs + 0.5mg/mL streptavidin) samples consisting of 20μL total volume each and buffered in 1X PBS are prepared and left to incubate at room temperature for 30min. 5μL of each sample is directly applied to the sensing cells without any additional preparation. The measured drift-

TABLE I
COMPARISON WITH STATE-OF-THE-ART CMOS MAGNETIC SENSORS

Reference	[1]	[2]	[3]	[4]	This Work
Sensor Type	GMR	Hall	LC (inductive)	LC (transformer)	LC (transformer)
Technology	180 nm	180 nm	65 nm	65 nm	65 nm
Power (min.) [mW]	1.39*	330	1.8	3	3.1
Frequency [GHz]	N/A	N/A	1.1 - 3.3	1.4 / 3.7	1.2–1.65, 2.9–4
Frequency Shift Sensitivity [ppm]	N/A	N/A	N/R	0.3 ppm	0.7 ppm
Sensing Area / Cell [µm²]	Off-chip	320x500†	260x260	250x250	390x360
External Magnetic Field Required	Yes	No	No	No	No
Surface Chemistry / Washes Required	Yes	Yes	Yes	Yes	No
Drift Compensation	No	No	No	Yes	Yes
Frequency-sweep Spectroscopy	No	No	Yes	No	Yes

* Excluding sensor bias
† Area per sensor sub-bank; 4 total

N/A : Not Applicable
N/R : Not Reported

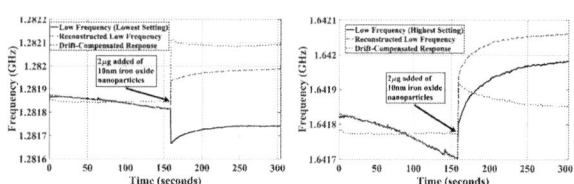

Fig. 7. Time-domain measured and reconstructed f_{LO} and drift-compensated response to the addition of 10nm sized iron oxide MNPs.

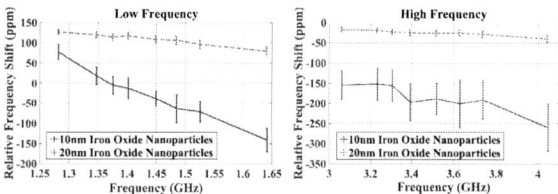

Fig. 8. Frequency shift spectra of 10nm and 20nm MNPs. Error bars show one standard deviation around the mean across 5 measurements.

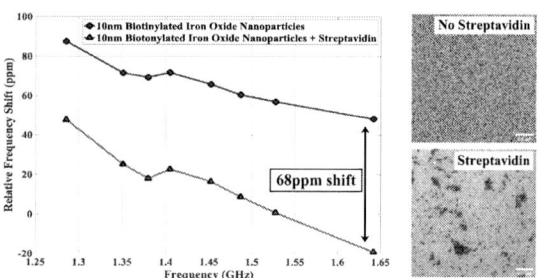

Fig. 9. Biotin-streptavidin immunoassay; measured frequency-shift spectrum using 10nm biotinylated MNP with and without streptavidin; TEM micrographs of each sample.

compensated frequency shift versus f_{LO} for the control (MNP only) and experiment (MNP and streptavidin) samples over 1.3 to 1.65GHz are plotted in Fig. 9, along with TEM micrographs of each sample. A large relative frequency shift of 68ppm is obtained, verifying that the sensor can be used for wash-free protein detection.

V. CONCLUSION

A reference-free drift-compensated magnetic spectrometer in 65nm CMOS enabling wash free immunoassays has been presented. The CDF oscillator topology is introduced, which allows simultaneous sensing and drift compensation over a wide frequency range for magnetic spectroscopy. A comparison of the spectrometer with prior work is shown in Table I. The presented work achieves, to the best of the authors' knowledge, the first demonstration of a wash-free immunoassay using a fully integrated CMOS magnetic biosensor. The viability of the spectrometer for portable point-of-care applications is shown by the achieved high sensitivity with low power consumption.

ACKNOWLEDGMENT

This work was supported by the Office of Naval Research (N00014-21-1-4005).

REFERENCES

[1] X. Zhou, M. Sveiven and D. A. Hall, "11.4 A Fast-Readout Mismatch-Insensitive Magnetoresistive Biosensor Front-End Achieving Sub-ppm Sensitivity," *ISSCC Dig. Tech. Papers*, pp. 196-198, Feb. 2019.

[2] S. Gambini et al., "A CMOS 10kpixel baseline-free magnetic bead detector with column-parallel readout for miniaturized immunoassays," *ISSCC Dig. Tech. Papers*, pp. 126-128, Feb. 2012.

[3] C. Sideris and A. Hajimiri, "An Integrated Magnetic Spectrometer for Multiplexed Biosensing," *ISSCC Dig. Tech. Papers*, pp. 300-301, Feb. 2013.

[4] C. Sideris, P. P. Khial and A. Hajimiri, "Design and Implementation of Reference-Free Drift-Cancelling CMOS Magnetic Sensors for Biosensing Applications," *IEEE J. Solid-State Circuits*, vol. 53, no. 11, pp. 3065-3075, Nov. 2018.

[5] K. Wu et al., "A Portable Magnetic Particle Spectrometer for Future Rapid and Wash-Free Bioassays," *ACS Appl. Mater. Interfaces*, vol. 13, no. 7, pp. 7966–7976, Feb. 2021.

[6] V. K. Chugh et al., "Magnetic Particle Spectroscopy-Based Handheld Device for Wash-Free, Easy-to-Use, and Solution-Phase Immunoassay Applications," *Proc. 2020 Design. Med. Devices Conf.*, Apr. 2020.

[7] K. Wu et al., "One-Step, Wash-free, Nanoparticle Clustering-Based Magnetic Particle Spectroscopy Bioassay Method for Detection of SARS-CoV-2 Spike and Nucleocapsid Proteins in the Liquid Phase," *ACS Appl. Mater. Interfaces*, vol. 13, no. 37, pp. 44136–44146, Sep. 2021.

[8] N. J. Darton, A. Ionescu, and J. Llandro, *Magnetic Nanoparticles in Biosensing and Medicine*. Cambridge: Cambridge University Press, 2019.

[9] A. Li and H. C. Luong, "A reconfigurable 4.7–6.6GHz and 8.5–10.7GHz concurrent and dual-band oscillator in 65nm CMOS," *IEEE Radio Freq. Int. Circuits Symp.*, pp. 523-526, June 2012.

[10] S. Oh, K.-S. Seo and J. Oh, "Low Phase Noise Concurrent Dual-Band (5/7 GHz) CMOS VCO Using Gate Feedback on Nonuniformly Wound Transformer," in *IEEE Microwave and Wireless Comp. Lett.*, vol. 31, no. 2, pp. 177-180, Feb. 2021.

A Full Current-Mode Timing Circuit with Dark Noise Suppression for the CERN CMS Experiment

E. Albuquerque[1], R. Bugalho[1], L. B. Oliveira[3], T. Niknejad[1,2], J.C. Silva[1,2], Alessio Boletti[2] and J. Varela[1,2]

(1) PETsys Electronics, Oeiras, Portugal
(2) LIP, Lisbon, Portugal
(3) CTS-UNINOVA, DEEC, FCT NOVA, Caparica, Portugal
email: efmalbuquerque@gmail.com, l.oliveira@fct.unl.pt, joao.varela@cern.ch

Abstract— **In this paper we present an analog circuit for the new MIP Timing Detector of the CMS experiment at CERN, featuring, for the first time, a silicon implementation of the Differential Leading Edge Discriminating technique to suppress SiPM dark noise. This technique also stabilizes the baseline, leading to a time resolution of 25 ps at beginning of life and 55 ps at end of life while dissipating less than 4 mW. The full analog front-end ASIC has 32 channels and has been designed in a CMOS 130 nm technology with a total die area of 8.5 x 5.2 mm2. The radiation tolerance of this design has been confirmed by radiation tests.**

Keywords—Radiation detectors, Analog front-end, SiPM dark current, Minimum Ionizing Particles Timing Detector.

I. INTRODUCTION

Radiation detectors are widely used in particle physics experiments at the European Laboratory for Particle Physics (CERN) located near Geneva, Switzerland. Currently, the main accelerator at CERN is the LHC (Large Hadron Collider). A broad range of detector technologies are used for particle tracking: silicon detectors, scintillating materials coupled to photo-sensors for energy measurements, and various types of ionization gas detectors. Pulses generated by these sensors are processed and digitized by full custom integrated electronic circuits [1]. The two largest experiments at the LHC are ATLAS and CMS, which have several millions of individual radiation sensors coupled to analog front-end (AFE) channels embedded in the detector structure [2]. The AFEs are exposed to very high radiation doses (up to several hundred Mrad) and particle fluence (up to 10^{15} n$_{eq}$/cm^2), which require radiation tolerant circuits.

The work described in this paper was developed in the framework of the Compact Muon Solenoid (CMS) experiment [3]. The existing detectors are expected to remain in operation until 2025, when the LHC will undergo a major upgrade to increase the proton beams luminosity by a factor of ten in the High-Luminosity LHC (HL-LHC). To match the new challenging beam conditions offered by the HL-LHC, a large portion of the CMS detector will be replaced, and, for the first time, a new timing detector will be featured [4] with an improved time resolution under 50 picosecond (average over the detector lifetime). These are unprecedent requirements for a large particle physics experiment operating in a high-luminosity proton collider, improving the present performance by one order of magnitude [5].

In this paper we present the AFE time measurement electronics developed for the new CMS timing detector. The AFE is the main block of the front-end ASIC, which also features TDCs, an ADC per channel and a digital block for data filtering and transmission. Preliminary results on this ASIC [6, 7] are completed here with a full description of the circuits used in the timing branch.

The long exposure to high radiation levels with damaging effects to the crystalline structure of the Silicon Photo-Multipliers (SiPMs) generates dark noise that strongly degrades the AFE time resolution through its lifetime. In order to attenuate the effect of dark noise, the DLED (Differential Leading Edge Discrimination) technique has been applied [8]. This technique reduces the effect of dark noise and mitigates baseline fluctuations and pulse pile-up, since the output pulse is bipolar. The DLED technique is implemented here in silicon for the first time, and we obtain a measured time resolution of 25 ps at the beginning of life (BoL) and 55 ps at the end of life (EoL), complying with a stringent specification of less than 50 ps time resolution average, over 10 years lifetime. The analog circuit must be able to handle Minimum Ionizing Particles (MIP) at a rate of 2.5 M hit/s per channel and low energy particles (<1 MeV) at a rate of up to 5 M hits/s per channel. The circuit is also required to be resistant to a Total Ionization Dose of 3 Mrad.

In section II we review the state of the art. In section III we present the timing branch of the AFE and its key blocks. In section IV we present the measurement results, including radiation tests, and finally, in section V we draw the conclusions.

II. STATE OF THE ART

A. SiPM pulses and statistics

The AFE is an event driven system, i.e., the current pulses from the SiPMs trigger all the events in the system. These current pulses are generated when a particle crosses a crystal bar and a number of optical photons are emitted. The pulse amplitude is proportional to the number of photons yielded by the particle interaction with the crystal [9].

Due to exposure to high radiation levels, the SiPM gain and the photoelectron (p.e.) yield are reduced, hence, the current pulse amplitude is reduced over time. The damage to the crystalline structure also yields spontaneous small current pulses, known as Dark Counts (the average rate of counts without any incident light is known as Dark Count Rate (DCR)) [10]. At BoL the DCR is negligible (<1 MHz), but its value is expected to increase with aging of the SiPMs, up to a rate of about 30 GHz at EoL, despite the SiPMs being operated at -45°C.

At BoL, the signal-to-noise ratio and timing resolution are determined by noise in the electronic circuits, referred to here as electronics noise. At EoL, the signal-to-noise ratio and timing resolution are deteriorated, due to radiation, by dark noise which is considerably higher than electronics noise and by the reduction of the SiPM current pulse amplitude by a factor of 4.

This project was funded by Fundação para a Ciência e Tecnologia, Portugal developed in the frame of the Collaboration Agreement LIP-CERN KN436/EP and the Development Contract LIP-PETsys of July 15, 2019, and within the projects: foRESTER PCIF/SSI/0102/2017 and ROBUST EXPL/EEI-EEE/0776/2021.

B. ASICs for Radiation Detectors

The sensor technology based on LYSO crystals and SiPMs adopted here for the Barrel Timing Layer (BTL) has been used in the last ten years in medical imaging positron emission tomography (PET). Table I shows a comparison of the main features of several ASICs developed for PET applications. None of these ASICs is radiation tolerant.

Table I – Comparison of ASICs for PET (adapted from [11])

ASIC	Timestamp Digitization	Power /mW/ channel	Time Resol. /ps	Crystal Height /mm	Ref.
FlexToT	External	11	123	5	[12]
STiC3	TDC on ASIC	25	240	15	[13]
PETA4	TDC on ASIC	<40	460	25	[14]
TRIROC	TDC on ASIC	10	432.7	10	[15]
TOFPET2	TDC on ASIC	3.6-7.2	119	3	[16]

The time resolution is the main figure of merit of the BTL detector and is determined by spurious noise that appear throughout the analog signal processing chain and have a stochastic behaviour, namely: photo-statistics, SiPM dark noise, electronics noise, and time discretization. In BTL, before irradiation, the required time resolution is 25 ps [5], and the contributions (added quadratically) are 20 ps from photo-statistics, 10 ps from electronics noise, and 10 ps from time discretization.

Considering the parameters that set the SiPM current pulse amplitude (photon yield, light collection efficiency, SiPM photon detection efficiency and gain) we find that pulses in BTL are five times smaller than in PET. Hence, to have the required time resolution, the AFE electronics noise must be improved by one order of magnitude and the remaining noise contributions must be halved with respect to PET. These are very challenging targets.

Throughout the BTL life, due to radiation, dark noise increases dramatically, and, at the EoL it determines the time resolution. In order to reach the 50 ps average time resolution, dark noise must be reduced by a factor of at least 2, hence, the AFE must have some kind of dark noise suppression.

C. Dark noise suppression

Considering the different types of hits on the SiPM and their statistics as well as dark noise, discriminating and time tagging relevant events with a time resolution better than 60 ps is impossible without noise reduction. The challenge relies on the fact that relevant events that must be time-tagged and dark counts have the same shape and frequency content.

For timing measurements, instead of signal-to-noise ratio, we must maximize slew-rate to noise ratio. A particularly well-suited technique is DLED [8], shown in Fig. 1, which with appropriate delay settings, preserves the rising edge of the pulse till the point of maximum slew-rate, and suppress the rest of the pulse (including the decaying tail). This considerably reduces dark noise power, while preserving the slew-rate of the rising edge, hence, improving the slew-rate to noise ratio and timing resolution.

The DLED block output pulse can be made arbitrarily shorter than the input pulse from the SiPM, by appropriately setting the delay value. Moreover, the stream of pulses due to MIP and lower energy events at the output of the DLED block have a considerably lower amplitude and are bipolar, whereas the input pulses from the SiPM are unipolar (the amplitude reduction has no impact on later Leading Edge Discrimination since the SiPM current pulses are very large). Thus, baseline drifts are dramatically reduced, improving threshold accuracy and timing precision.

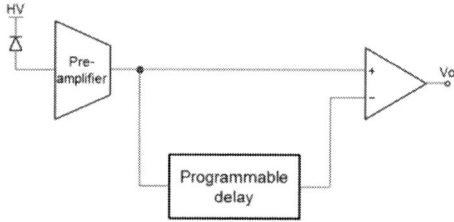

Fig. 1 - DLED block diagram.

III. ARCHITECTURE AND IMPLEMENTATION

The current in the preamplifier is mirrored in two branches for timing and energy measurements; in this paper we will focus only on the timing branch of the AFE. The circuit is designed to be robust to Process, Voltage and Temperature (PVT) variations as well as radiation tolerant. Radiation characterization of the target technology done at CERN has showed that robustness to process corners also implies radiation tolerance.

A. Pre-amplifier Circuit

We have found that the best suited preamplifier for the SiPM sensing device is the simple Common-Gate stage (Fig. 2), since it provides a low input impedance to sink the current from the SIPM, and the output current can be mirrored in different branches for timing and energy measurements. The pre-amplifier is a current buffer between the SiPM and the following signal processing stages.

We have also investigated Regulated Common-Gate amplifiers [17], which have the advantage of a higher gain and a lower input impedance than the simple Common-Gate amplifier (an order of magnitude lower); however, the design of the regulation amplifiers proved to be very challenging in terms of stability, noise contribution and saturation margins through corners. Furthermore, in regulated common-gate pre-amplifiers the input impedance varies in the frequency band of interest, impacting the input signal.

B. Implementation of the DLED Circuit.

The DLED technique seems appropriate to improve the EoL timing resolution, but, finding a suitable circuit implementation with affordable power consumption, reasonable area, wide enough bandwidth, high yield and low channel to channel spread, is a very challenging task.

Since the SiPM signal is a current, a full current-mode approach with current preamplifier and current-mode DLED block followed by a current comparator seems a natural solution. We have developed this solution with just four low impedance nodes in the signal path from SiPM input to comparator output, which leads to a very large bandwidth and achieves the target EoL timing resolution below 60 ps. When a current pulse appears at the input of the preamplifier Iin (Fig. 2), it is mirrored through two different paths: a) direct path through mirrors MN-D/MN2 and again through PMOS mirrors MP1/MP2; b) Invert and delay path through mirrors MN-D/MN1. At the output node, Io (I1-I2) is the DLED output current.

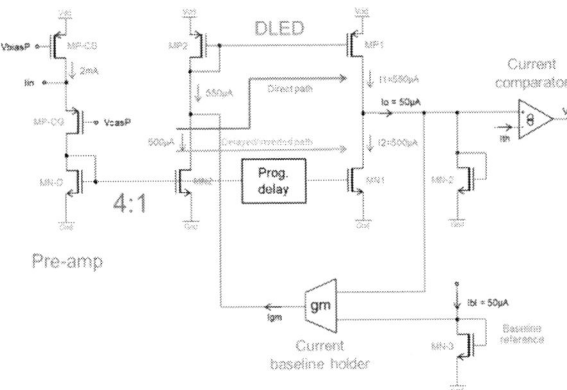

Fig. 2. Full AFE including pre-amplifer, DLED, baseline holder and comparator.

Ideally, if the delay is close to zero, the current pulse at the output of the DLED block is fully cancelled. As the delay increases towards the peaking time, the output current pulse is a replica of the input current pulse till the point where the delay ends and it will start decaying due to the subtraction of the inverted and delayed pulse. Hence, as the delay increases toward the peaking time the pulse amplitude will increase, and for very large delays (higher than the pulse width), the output pulse will be a replica of the input pulse. The delay controls the time at which the subtraction branch is effective and starts to cancel the pulse from the direct branch, so it also determines the output pulse width.

An ideal delay line cannot be implemented in a CMOS process, so we have used the best possible alternative, a cascade of RC sections. Since resistors and capacitors have high value spreads (up to 25%), to overcome this problem, instead of a fixed delay, we have designed a variable delay line with seven taps that can be programmed to have a maximum delay of 750 ps. The ability to vary the delay value provides an additional degree of freedom to fine tune the pulse shape, since increasing the delay value increases the output pulse amplitude.

In order to have a stable and predictable baseline, a baseline holding feedback loop is used (Fig. 2). This is achieved by sinking/sourcing a current of the node connecting MP2/MN2 with a transconductor, such that Io is always in the vicinity of the reference current value of 50 μA. The simulated baseline value spread through the 16 corners is only 2.18 μA.

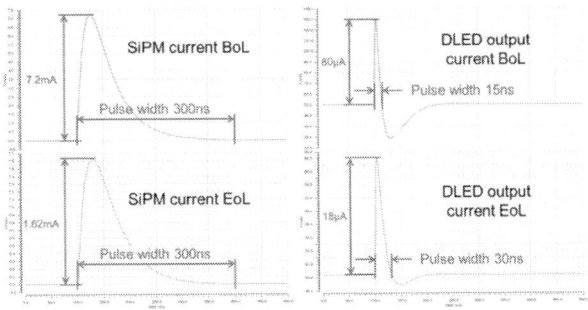

Fig. 3. Current-mode signal processing in the AFE timing branch.

The baseline holding amplifier is designed with a single low frequency pole in the mHz region; hence, there are no stability issues, since all the other poles are at very high frequencies (hundreds of MHz). This is very important, since the dynamics of the baseline holder will not modify the pulse shape. Fig. 3 shows the SiPM current at the input of the preamplifier and the current exiting the DLED block in BoL and EoL conditions. Due to the highly non-linear behaviour of the DLED block for large signals, the unconditional stability and robustness of the loop has been thoroughly checked with large 30 k p.e. pulses (direct and reverse) and dark noise current. To trigger the timing information retrieval process of an incoming pulse, the current comparator from [18] with positive feedback and a low output impedance is used.

IV. MEASUREMENT RESULTS

A. Experimental Setup

Sensor modules with 16 LYSO crystal bars, glued at both ends to linear arrays of 16 SiPMs are used in the measurements. The sensor module used in these measurements has a slit on the external wrapping allowing the scintillation crystal to be excited with a UV laser (375 nm) that emulates the MIP energy. The laser power is adjusted to obtain a given number of photoelectrons per pulse. EoL performance is measured on un-irradiated SiPMs, and dark noise is emulated with blue LED light. The HDR2 SiPM from manufacturer HPK (one of the candidates for use the BTL detector) is used in the present AFE characterization. Fig. 4 shows the test board with two Time-of-Flight at High-Rate (TOFHiR) chips (left) and the TOFHiR layout (right); the die area is 8.5 x 5.2 mm².

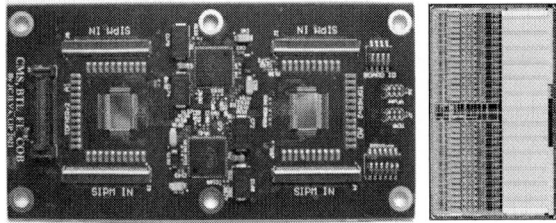

Fig. 4. TOFHiR test board (left) and layout (right).

B. Time resolution of MIP equivalent pulses

In BTL, the timing of a MIP particle is obtained by averaging the two measurements in a single LYSO bar. The bar time resolution may be derived from the Coincidence Time Resolution (CTR) of the two channels in the crystal bar (σ_{bar}=CTR/2). For the measurements in BoL conditions we used an UV laser pulse tuned to generate a LYSO pulse with 9500 photoelectrons with the SiPMs operated at an overvoltage of 3.5 V (SiPM gain 3.8×10^5) [9].

In order to emulate the large dark noise at the EoL, we used background LED blue light. The calibration of the LED light is done by measuring the SiPM current as a function of the LED voltage. The SiPM current is converted into equivalent DCR by taking into account the SiPM gain at the operating over-voltage. We measure the channel time resolution of laser pulses as anticipated at EoL conditions using the method described previously, while illuminating the SiPMs with a blue LED light.

Fig. 5 shows the BoL time resolution versus the comparator threshold for the optimal setting of the delay in the DLED circuit. We measured a time resolution of 25 ps, at the optimum threshold, which is in agreement with the simulations. The same figure also shows the time resolution for EoL pulses with 6000 p.e., SiPM gain of 1.5×10^5, and

978-1-6654-8495-4/22 $31.00 © 2022 IEEE

equivalent DCR of 30 GHz. In this measurement, the SiPM is operated with an over-voltage of 1.5 V as anticipated at EoL to limit dark noise. For the optimum setting of the delay (720 ps), we obtain a time resolution of 55 ps. These results match simulations and fully comply with BTL specifications.

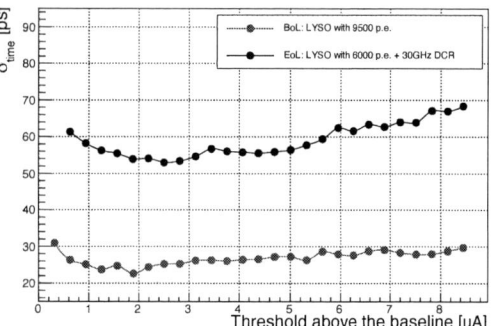

Fig. 5. AFE time resolution in BoL and EoL conditions versus the comparator threshold.

C. Radiation Tests

The tolerance of the AFE to Total Ionization Dose (TID) was tested at the X-ray irradiation facility at CERN. The maximum expected dose in the barrel MIP Timing Detector is 3 Mrad. However, we have irradiated the ASICs up to 7 Mrad in steps of 0.5 and 1.0 Mrad, which is a factor 2.3 above the requirement to account for uncertainties in the estimation of TID (irradiations were done at -25°C). Overall we have observed minor or negligible effects on the AFE. Fig. 6 shows the measured EoL time resolution versus irradiation dose, and within the measurement accuracy no significant variation were observed.

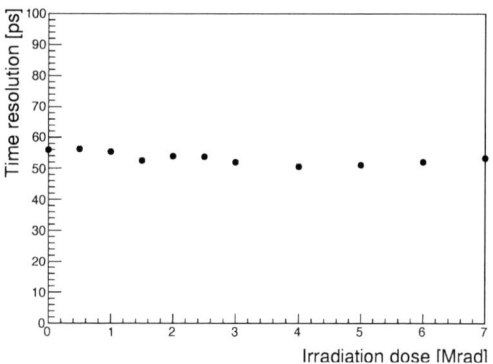

Fig. 6. Measured EoL time resolution versus irradiation dose.

TABLE II. Timing branch performance key parameters.

Number of channels	32
Voltage supply (V)	1.2
Technology (nm)	130
Radiation tolerance (up to 7 MRad)	Yes
Dark noise suppression	Yes
Max MIP rate/ch (MHz)	2.5
Power consumption/channel (mW)	4
BoL (ps)	25
EoL (ps)	55

V. Conclusions

In this paper a full current-mode radiation tolerant AFE timing branch is presented for the new CMS barrel MIP Timing Detector at CERN. The proposed circuit achieves a time resolution of 25 ps at the beginning of life, and 55 ps at

the end of life, while dissipating less than 4 mW. This performance is achieved under stringent EoL conditions, featuring reduced current pulse amplitude by a factor of four and very large dark noise, which proves the effectiveness of the dark noise suppression by the DLED circuit. The feasibility of this critical block of the CMS timing detector has been proved and the required performance has been achieved thanks to the full current-mode DLED implementation described in this paper.

Acknowledgments

We thank the MTD project for all the support provided to the AFE development (specially for the MTD sensor team for providing the sensor modules used in the measurements). We also thank the coordinators of the CERN X-ray irradiation facility for making available this facility and the CMS collaboration for creating the conditions that made this work possible.

References

[1] A. Rivetti, *"CMOS: Front-End Electronics for Radiation Sensors"*, ISBN 9781138827387, Published July 27, 2017 by CRC Press

[2] W. Snoeys et al., "Integrated circuits for particle physics experiments," in *IEEE Journal of Solid-State Circuits*, vol. 35, no. 12, pp. 2018-2030, Dec. 2000.

[3] S. Chatrchyan *et al.* (CMS Collaboration) (2008), "The CMS Experiment at the CERN LHC", *JINST* 3 (2008) S08004.

[4] The CMS Collaboration, *"Technical Proposal for the Phase II Upgrade of the Compact Muon Solenoid"*, CMS-TDR-15-02, ISBN 9789290834168, cds.cern.ch/record/2020886

[5] The CMS Collaboration, *"Technical Design Report of the MIP Timing Detector for the CMS Phase-2 Upgrade"*, 15 Mar 2019, CERN-LHCC-2019-003; CMS-TDR-020.

[6] E. Albuquerque *et al.*, "The TOFHIR2 readout ASIC of the CMS Barrel MIP Timing Detector", 2020 *IEEE Nuclear Science Symposium and Medical Imaging Conference* Nov. 2020.

[7] E. Albuquerque *et al.*, "Results with the TOFHIR2X version of the front-end ASIC of the CMS Barrel Timing Layer", *2021 IEEE Nuclear Science Symposium and Medical Imaging Conference*, Nov. 2021.

[8] A. Gola, C. Piemonte and A. Tarolli, "Analog Circuit for Timing Measurements With Large Area SiPMs Coupled to LYSO Crystals," in *IEEE Transactions on Nuclear Science*, vol. 60, no. 2, pp. 1296-1302, April 2013.

[9] S. Gundacker and A. Heering, "The silicon photomultiplier: fundamentals and applications of a modern solid-state photon detector", 2020 *Phys. Med. Biol.* 65 17TR01.

[10] E. Garutti, Yu. Musienko, *"Radiation damage of SiPMs"*, *Nuclear Instruments and Methods in Physics Research Section A*, Volume 926, 11 May 2019, Pages 69-84.

[11] V. Nadig et al.,, "Investigation of the Power Consumption of the PETsys TOFPET2 ASIC", in *IEEE Transactions on Radiation and Plasma Medical Sciences,* vol. 4, no. 3, pp. 378-388, May 2020.

[12] J. M. Cela et al., "A Compact Detector Module Design Based on FlexToT ASICs for Time-of-Flight PET-MR", *IEEE Transactions on Radiation and Plasma Medical Sciences*, vol. 2, no. 6, pp. 549–553, Nov. 2018.

[13] V. Stankova et al., "STIC3 - silicon photomultiplier timing chip with picosecond resolution", *Nuclear Instruments and Methods in Physics Research A*, vol. 787, pp. 284– 287, 2015.

[14] I. Sacco et al., "PETA4: a multi-channel TDC/ADC ASIC for SiPM readout", *Journal of Instrumentation*, vol. 8, 2013.

[15] S. Ahmad et al., "TRIROC: a multi-channel SiPM readout ASIC for PET/PET-TOF application", *IEEE Transactions on Nuclear Science*, vol. 62, no. 3, pp. 664–668, 2015.

[16] R. Bugalho et al., "Experimental characterization of the TOFPET2 ASIC", 2019 *JINST* 14 P03029.

[17] M. M. Silva and L. B. Oliveira, "Regulated Common-Gate Transimpedance Amplifier Designed to Operate With a Silicon Photo-Multiplier at the Input," in *IEEE Transactions on Circuits and Systems I: Regular Papers*, vol. 61, no. 3, pp. 725-735, March 2014.

[18] H. Traff, "Novel Approach to HighSpeed CMOS Current Comparators", *IEEE Electronic Letters*, vol.28, No.3, 1992.

A 2.74pJ/conversion 0.0018mm^2 Temperature Sensor with On-chip Gain and Offset Correction

Yuting Shen, Mariska van der Struijk, Kevin Pelzers, Hanyue Li, Eugenio Cantatore and Pieter Harpe
Integrated Circuits Group, Eindhoven University of Technology, Eindhoven, The Netherlands
Email:y.shen@tue.nl

Abstract—**This paper presents a dynamic temperature sensor in 65nm CMOS with on-chip analog gain and offset correction for low power systems. By shifting the reset phase of the N-bit ADC, offset correction with a range of $\pm 2^{(N-1)}$LSB is realized. Fine tuning capacitors are introduced to improve the offset correction accuracy to 0.5LSB. By adding programmable parasitic capacitors, gain errors up to 6.3% can be compensated. Thanks to the proposed analog correction techniques, the gain errors are reduced to 0.73% and the offsets are reduced to 0.5LSB. This sensor consumes 2.74pJ per conversion and only occupies an area of 0.0018 mm^2 including the extra correction techniques. It has an RMS resolution of 0.47K, leading to a FoM of 0.6 pJ·K^2.**

Index Terms—**analog correction, gain, offset, on-chip, temperature sensor**

I. INTRODUCTION

Nowadays, temperature sensors are used in a wide range of applications such as Internet-of-the-Things (IoT), biomedical and environmental monitoring systems. For these applications, moderate resolution, low power and small area are desired.

One challenge for temperature sensors is that they are sensitive to mismatch and process variations, which could result in significant offset and gain variations from chip to chip. Hence, correction needs to be done individually. While the power consumption of temperature sensors can be reduced already to pW-level [1], the gain and offset corrections are still done off-chip in most scientific publications. When integrated on-chip, they may consume more power and area than the sensor itself [2]. This is especially true for low-to-medium resolution pW-level sensor interfaces.

To address the above issue, [3], [4] use a Wheatstone bridge as the sensing element and add trimming resistors to compensate for the spread of the resistors. [5] uses an accurate electrothermal filter based sensor to calibrate an inaccurate but efficient Wien-bridge based sensor. However, these sensors aim for high precision, and their energy per conversion (6.45 to 790nJ) and area (0.12 to 0.43mm^2) are relatively large. For very small, low power sensors (e.g [6] with 2.18pJ/conversion and 0.0017mm^2), it is much more challenging to efficiently implement on-chip correction.

The goal of this work it to implement on-chip gain and offset correction for temperature sensors used in low power systems. A dynamic resistive temperature sensor with integrated analog correction techniques is proposed. By shifting

This work with project number 16594 is financed by the Dutch Research Council (NWO).

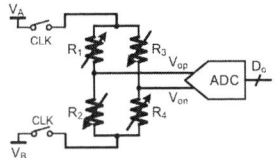

Fig. 1: A simplified model of a dynamic resistive temperature sensor.

the reset phase of the readout ADC and adding tunable capacitors, the gain and offset errors of the temperature sensor are corrected with limited area and power cost. The final sensor dissipates 2.74pJ per conversion and only occupies an area of 0.0018mm^2 including the on-chip gain and offset correction.

This paper is organized as follows. Section II reviews the architecture of a resistive sensor and analyses its gain and offset errors. Section III presents the proposed analog gain and offset correction techniques. Section IV introduces the circuit implementation details. Measured results are shown in Section V and conclusions are drawn in Section VI.

II. REVIEW OF A RESISTIVE TEMPERATURE SENSOR

A. Architecture

Fig. 1 shows a simplified model of a resistive temperature sensor which consists of a Wheatstone bridge sensing front-end and an ADC back-end [7]. The differential output of the resistive bridge can be calculated as:

$$V_{out} = V_{op} - V_{on} = \left(\frac{R_2(1+\alpha_2\Delta T)}{R_1(1+\alpha_1\Delta T) + R_2(1+\alpha_2\Delta T)} - \frac{R_4(1+\alpha_1\Delta T)}{R_4(1+\alpha_1\Delta T) + R_3(1+\alpha_2\Delta T)} \right)(V_A - V_B) \tag{1}$$

Here, resistor 1 and 4 are resistors with a temperature coefficient α_1 and a resistance of R_1 and R_4. Resistor 2 and 3 are resistors with a temperature coefficient α_2 and a resistance of R_2 and R_3. When $R_1 = R_2 = R_3 = R_4$, V_{out} only depends on the resistor temperature coefficients and the temperature.

An ADC is used to read out the output of the resistor bridge. SAR ADCs are often chosen thanks to their high efficiency at moderate resolutions [7]. The overall gain (code/temperature) of the sensor can then be calculated as:

$$gain = gain_{Whb} \times gain_{ADC} \tag{2}$$

And the overall offset of the sensor equals to:

$$offset = offset_{Whb} + offset_{ADC} \tag{3}$$

978-1-6654-8495-4/22 $31.00 © 2022 IEEE

Fig. 1 shows a basic example to read out temperatures from a resistive sensing front-end. Moreover, to mitigate 1/f noise, a system-level correlated double sampling (CDS) could be introduced by swapping the supply voltages of the resistive bridge (V_A,V_B) [7], as will be shown later in Fig. 5.

B. Analysis of gain and offset errors

Due to process variations and random errors, resistors suffer from mismatch. This results in gain and offset errors for the output of the resistive bridge [8].

Assume R_1, R_2, R_3, R_4 have relative errors of e_1, e_2, e_3, e_4, respectively, then (1) can be rewritten as:

$$V_{out} = \frac{A}{B}(V_A - V_B) \qquad (4)$$

where,

$$A = (e_2 + e_3 - e_1 - e_4) + 2(\alpha_2 - \alpha_1)\Delta T + (e_2\alpha_2 + e_3\alpha_2 - e_1\alpha_1 - e_4\alpha_1)\Delta T + H.O.T \qquad (5)$$

$$B = 4 + 2(e_2 + e_3 + e_1 + e_4) + 2(e_2\alpha_2 + e_3\alpha_2 + e_1\alpha_1 + e_4\alpha_1)\Delta T + H.O.T \qquad (6)$$

since both e and α are relatively small, this is approximately:

$$V_{out} \approx \left(Off + \frac{1}{2}(\alpha_2 - \alpha_1)\Delta T + Gain_{err}\Delta T \right)(V_A - V_B) \qquad (7)$$

where,

$$Off = \frac{1}{4}(e_2 + e_3 - e_1 - e_4) \qquad (8)$$

$$Gain_{err} = \frac{1}{4}(e_2\alpha_2 + e_3\alpha_2 - e_1\alpha_1 - e_4\alpha_1) \qquad (9)$$

The offset (Off) is directly related to the resistance error. Assume that α is in the order of $2e^{-3}$ and the temperature range is about 140^oC, when e_1 is in the order of 10%, this could result in a 3σ error similar to the full-scale range. The gain error ($Gain_{err}$) is related to the derivative of the resistance versus temperature. The gain variation will be a secondary effect as (9) shows, as both e and α are small.

The temperature coefficients also suffer from random variation. Assume that α_1 has a relative error of $e_{\alpha 1}$, then the actual value of R_1 equals to:

$$R_{1,real} = R_1(1 + e_1)\left(1 + \alpha_1(1 + e_{\alpha 1})\Delta T \right) \qquad (10)$$

$e_{\alpha 1}$ only affects R_1 in the order of $e_1 \times \alpha_1 \times e_{\alpha 1}$. Since α and e_α are small, the random variation of α will be less important than the random variation of the resistance.

Besides resistive bridge imperfections, there are also gain and offset errors from the ADC which affect the final output as equations (2) and (3) show.

III. PROPOSED GAIN AND OFFSET CORRECTION

To correct the gain and offset errors for low power sensors, on-chip digital solutions could easily take several times the power and the area of the sensor itself. Thus, low-cost analog methods are preferred. According to (2) and (3), either $gain_{WhB}$ and $Offset_{WhB}$ can be corrected directly [3], [4] or $gain_{ADC}$ and $Offset_{ADC}$ can be tuned to compensate for the gain and offset errors caused by the resistive bridge. In this

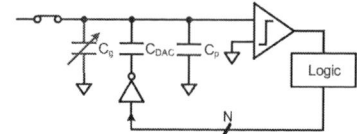

Fig. 2: Diagram of gain correction in the SAR ADC.

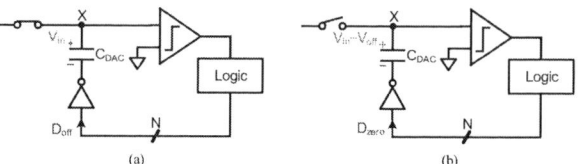

Fig. 3: Operation of offset correction (a) tracking phase (b) reset phase.

work, gain and offset correction techniques based on the later option are proposed. Compared to tuning the resistive bridge itself, tuning the ADC's gain and offset is more area efficient thanks to the ability to re-use most of the existing DAC capacitors, and the area efficiency of capacitors in general. On top of this, implementing gain and offset correction in the resistive bridge is more cumbersome. Firstly, because the resistors are relatively large components. Secondly, because adding trimming resistors also adds parasitics, which is disadvantageous in a dynamic structure [7]. Thirdly, because the extra switches in the bridge could cause additional non-linearity.

A. Gain correction

Fig. 2 shows a diagram of the proposed analog gain correction. With a tunable capacitor C_g, the ADC's gain can be calculated from the original ADC gain as:

$$gain_{ADC,C_g} = \frac{C_{DAC} + C_p + C_g}{C_{DAC} + C_p} \times gain_{ADC} \qquad (11)$$

Depending on the estimation of the front-end gain variation, the tunable range of C_g can be chosen accordingly.

B. Offset correction

As analyzed in Section II, the offset error caused by the resistive bridge mismatch has a wide spread. To realize such a wide range offset correction, the following offset correction technique is proposed (Fig. 3), which is similar to the comparator offset calibration technique proposed in [9] but now being used to correct the sensor's front-end variation with a full code range. Compared to the conventional SAR ADC operation, the DAC reset phase is shifted in time, and the DAC is preset to an offset value during tracking (Fig. 4).

During the tracking phase, the DAC control code is set to D_{off}, which corresponds to the offset voltage to be compensated. Then, after V_{in} is sampled at the top plates of the DAC capacitors, the bottom plates of the DAC capacitors are reset to 0 by changing the digital control code from D_{off} to D_{zero}. The charge redistribution that takes place effectively subtracts the offset voltage from the input signal. Because node X is floating, the voltage at the comparator input is now equal to $V_{in} - V_{off}$. In this way, an offset correction with a range of $\pm 2^{(N-1)}$LSB can be realized. Here, N stands for the number of bits of the ADC.

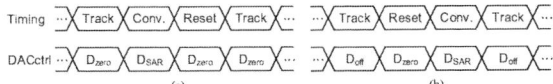

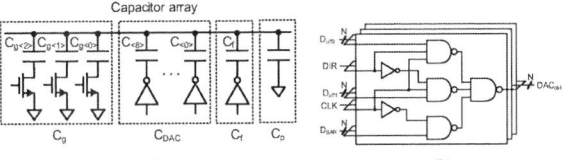

Fig. 4: Timing diagram and DAC control signals of (a) conventional SAR conversion (b) SAR conversion with the proposed offset correction.

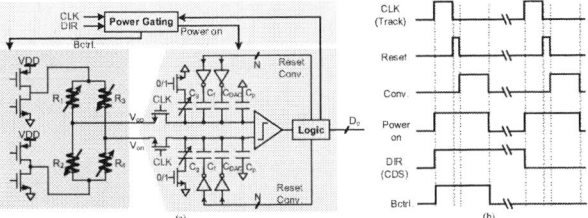

Fig. 6: Capacitor array (a) and logic cell of the offset correction (b).

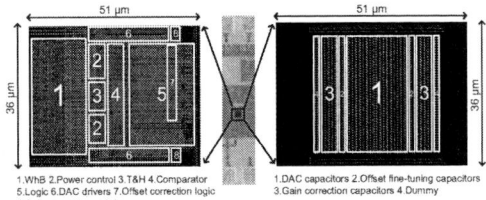

Fig. 5: Architecture of the sensor (a) and timing diagram (b).

Fig. 7: Die photo and layout view (left: up to Metal 5; right: Metal 6 and 7).

IV. CIRCUIT IMPLEMENTATION

A. Architecture

Fig. 5(a) shows the architecture of the dynamic resistive sensor, which is based on the design in [6]. Here, resistor 1 and 4 are N-type diffusion resistors with a positive temperature coefficient, while resistor 2 and 3 are P-type polysilicon resistors with a negative temperature coefficient. The nominal value of all resistors is 100 $k\Omega$. This is selected as a trade-off between sensitivity and small area.

A 9-bit asynchronous charge redistribution SAR ADC is used to read out the resistive bridge. NMOS transistors are used as sampling switches. A 2-stage dynamic comparator is used. The DAC cells are implemented as unit-length capacitors placed on top of the active circuits to minimize area cost. The nominal value of the total DAC capacitance is 300fF.

Besides the sensing front-end and the ADC back-end, an automatic power gating control block is applied to maximize the stand-by time and to minimize the leakage power of the system [6]. Once the conversion is done, a latched signal will be generated and the power supply will be switched off.

The power gating system and the sampling switches of the ADC are driven by a 1V supply voltage (VDDH). The rest of the circuit is driven by a 0.6V supply (VDD).

Fig. 5(b) shows the timing diagram of the sensor. The sampling clock (CLK) and the direction signal (DIR) are provided externally. The other clock signals are generated internally and included in the power consumption. The DIR signal controls the polarity of the system to implement CDS.

B. Gain and offset correction

The capacitor array implementation of this work is shown in Fig. 6(a). Except for the normal DAC capacitors and parasitic capacitance, a 3-bit programmable capacitor C_g is designed to calibrate the gain of the sensor, which can achieve gain correction in a range of 6.3% with a step size of 0.8%. The offset can be corrected in a range of $\pm2^{(N-1)}$LSB. To enable a step size of 0.5LSB, one additional fine tuning capacitor C_f (with a value of 0.5LSB) is added.

Fig. 6(b) shows the design of the logic cell for the proposed offset correction technique. When the CLK signal is high (tracking phase), the DAC drivers are controlled by the pre-set values. Depending on the direction signal (DIR), the

corresponding pre-set signal will be chosen. Thus, the bottom plate of the DAC is pre-set to an offset value during the tracking phase. When the CLK signal is low, the DAC follows the output from the SAR logic. The bottom plate of the DAC is firstly switched back to the reset mode and then a normal SAR conversion is performed.

V. MEASURED RESULTS

This work is fabricated in 65nm CMOS and occupies an area of 0.0018 mm^2 (Fig. 7). Supplies of 1V and 0.6V are used. 20 samples are measured over a temperature range from -20 to 120 °C. The on-chip gain and offset correction coefficients are set based on 2 measured temperature points (0 and 100 °C).

Fig. 8 shows the output codes for 20 samples when the proposed analog correction is enabled and disabled. The measured gain and offset distribution are summarized in Fig. 9. When the on-chip correction is disabled, the measured gain has a variation of 6.47% while the offset value varies from 12 to 35LSB. With correction, the gain variation is reduced to 0.73%. The offset values are calibrated to within 0.5LSB from the desired value.

Fig. 10 shows the temperature error of the sensor with off-chip digital and on-chip analog correction. After digitally removing the systematic non-linearity with a fixed 5th order polynomial batch correction, the residual temperature error can be observed in Fig. 11. As can be seen, the analog correction performs at least as well as digital correction and may have a benefit due to better cancellation of ADC mismatch thanks to CDS [10]. This sensor (with on-chip gain/offset correction and off-chip systematic distortion correction) has an inaccuracy of -0.3/0.4 °C, resulting in a relative inaccuracy of 0.5%.

The RMS resolution is 0.471K and 0.473K at room temperature for this sensor with and without correction, respectively. They are similar because the noise contribution is almost the same for both cases.

This temperature sensor consumes 2.74pJ per conversion with the proposed analog correction techniques and 2.62pJ without correction. For both gain and offset correction, the extra capacitors are placed on the top of the existing circuitry. The extra logic only occupies an area of 180μm^2 (Fig. 7).

978-1-6654-8495-4/22 $31.00 © 2022 IEEE

TABLE I: Performance summary and comparison

	[3]	[4]	[5]	[6]	[7]	This work
Sensor type	Res	Res	Res	Res	Res	Res
Technology [nm]	180	180	180	65	65	65
Area [mm^2]	0.12	0.43	0.2	0.0017	0.084	0.0018
Supply voltage [V]	1.8	1.5	1.8	1/0.6	1/0.6	1/0.6
Temp. range [^{o}C]	-55 to 125	-45 to 125	-55 to 125	-20 to 120	-10 to 120	-20 to 120
Samples	20	14	20	16	10	20
Inaccuracy (Trimming points) [K]	±0.14(2^{bc})	±0.4(2^{bc})	±0.03(2^{bc})	1.2 (2^{ac})	-0.34/0.29 (2^{ac})	-0.3/0.4 (2^{ac})
Resolution [K]	0.16m	0.01	0.45m	0.53	0.38	0.47
Power [μW]	79	64.5	66	0.108	0.571	0.137
Conversion time [s]	10m	100μ	10m	20.6μ	10μ	20μ
Conversion energy [pJ]	790000	6450	660000	2.18	5.71	2.74
Resolution FoM[1] [pJ·K^2]	0.02	0.65	0.13	0.61	0.82	0.60

^{a} Maximum values ^{b} 3σ values ^{c} With nonlinearity removal
[1] $FoM = Energy/Conversion \cdot (Resolution)^2$

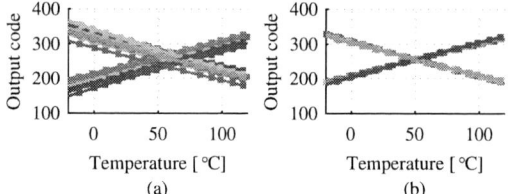

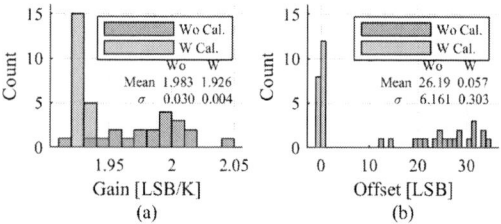

Fig. 8: Output code when analog correction is disabled (a) and enabled (b).

Fig. 9: Distribution of gain (a) and offset (b), without and with analog correction.

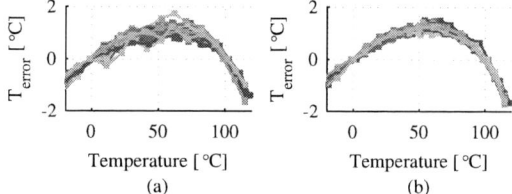

Fig. 10: Temperature error with off-chip digital gain/offset correction (a), and with on-chip analog correction (b).

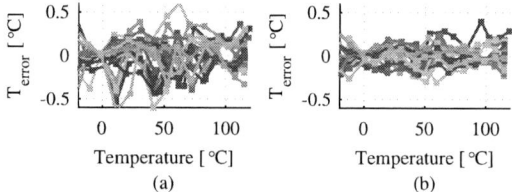

Fig. 11: Temperature error after fixed systematic error removal with off-chip digital gain/offset correction (a), and with on-chip analog correction (b).

Overall, the proposed on-chip gain and offset correction is very area and power efficient.

Table I summarizes the performance of this work and compares it to state-of-the-art designs. Compared to prior art with on-chip correction techniques [3]–[5], this work achieves a competitive resolution FoM with much lower conversion energy and a 67× to 239× smaller area. However, these references aim for much higher precision at higher power levels. Compared to [6], which used off-chip gain and offset calibration, this work integrates these calibrations on-chip at the cost of 11% extra area and 25% extra energy, while achieving better accuracy, resolution, and FoM.

VI. CONCLUSION

This work presents an ultra-low power temperature sensor with on-chip gain and offset correction. Thanks to the proposed efficient analog correction techniques, this sensor achieves a full code range offset correction with a step of 0.5LSB and a 6.3% gain correction with a step of 0.8%. The overall design consumes 2.74pJ conversion energy, uses 0.0018mm^2 area, and achieves an RMS resolution of 0.47K. This results in a resolution FoM of 0.6 pJ·K^2. These results show that even for very small and low power sensors, gain and offset correction can be done on-chip with limited overhead.

ACKNOWLEDGMENT

The authors would like to thank Haoming Xin for his contributions to the design.

REFERENCES

[1] K.A.A. Makinwa, "Smart Temperature Sensor Survey," [Online]. Available: http://ei.ewi.tudelft.nl/docs/TSensor_survey.xls
[2] M. Perrott et al., "A temperature-to-digital converter for a MEMS-based programmable oscillator with better than ±0.5ppm frequency stability," ISSCC, 2012.
[3] S. Pan and K. A. A. Makinwa, "A Wheatstone Bridge Temperature Sensor with a Resolution FoM of 20fJ·K2," ISSCC, 2019.
[4] C.-H. Weng et al., "A CMOS Thermistor-Embedded Continuous-Time Delta-Sigma Temperature Sensor With a Resolution FoM of 0.65 pJ ^{o}C^2," in IEEE JSSC, vol. 50, no. 11, pp. 2491-2500, Nov. 2015.
[5] S. Pan et al., "A Hybrid Thermal-Diffusivity/Resistor-Based Temperature Sensor with a Self-Calibrated Inaccuracy of ±0.25° C(3 σ) from -55°C to 125°C," ISSCC, 2021.
[6] K. Pelzers et al., "A 2.18-pJ/conversion, 1656-μm^2 Temperature Sensor With a 0.61-pJ·K² FoM and 52-pW Stand-By Power," in IEEE SSCL, vol. 3, pp. 82-85, 2020.
[7] H. Xin et al., "A 0.34-571nW All-Dynamic Versatile Sensor Interface for Temperature, Capacitance, and Resistance Sensing," ESSCIRC 2019.
[8] Pan, S., Makinwa, K.A.A. (2022). Wheatstone Bridge–Based Temperature Sensors. In: Resistor-based Temperature Sensors in CMOS Technology. Analog Circuits and Signal Processing. Springer, Cham.
[9] X. Peng et al., "A Novel Comparator Offset Calibration Technique for SAR ADCs," EDSSC, 2018.
[10] H. Xin, "Low Power Versatile All-Dynamic Sensor Interfaces," Ph.D dissertation, Technische Universiteit Eindhoven, 2020.

An Integrated Optical Transceiver Circuit for Power Delivery and Bi-directional Data Communication in a Medical Catheter Device

Alexander Frank, Jens Anders and
Joachim Burghartz
IMS CHIPS
Institut für Mikroelektronik
Stuttgart, Germany
frank@ims-chips.de

Bart Kootte
OSYPKA AG
Rheinfelden, Germany

Jean Schleipen and Peter Jutte
Philips Group Innovation, Research
Eindhoven, The Netherlands

Abstract—In this paper[*], the realization of an optical transceiver circuit (OTC) integrated into a customized catheter system is presented. The electronics at the distal end of the catheter is located far away from the external bed-side unit and is connected by an optical link to control its functions. By means of light only, the optical link simultaneously delivers power and establishes a bi-directional data communication. The optical link consists of just a few components, i.e. a multi-mode fiber, a blue LED, an external unit and the optical transceiver circuit. The LED is located at the catheter tip and is used by the OTC to operate as an optical transceiver and energy harvester. Several circuit blocks are integrated into the OTC, to provide a regulated voltage of 1.8 V at a maximum current of 2.1 mA. The OTC establishes a communication with a speed of up to 15.6 kBits/s for receiving and 1.35 MBits/s for transmitting data. Because of the small area and only less components to set up the link at the sensor side, the optical link is very suitable for application to the catheter system presented in this paper. The concept can also be applied to other biomedical or industrial sensor systems, where conventional approaches, using electrical wiring, are unpractical.

Keywords—Bi-directional optical interface, transceiver, biomedical applications, LED, energy harvesting, MRI compatible

I. INTRODUCTION

An ongoing trend in biomedical data acquisition is to read-out and process signals as close as possible at the sensor side using an integrated circuit. This avoids long distance sensor signal transmission, resulting in better signal quality and in some cases also helps to reduce the cable-count. A prominent use case is given by biomedical catheter applications, where electrophysiological signals are captured for health monitoring or treatment. These systems are built up to be as compact as possible and to have an integrated sensor in the catheter tip to detect electrophysiological signals. An integrated circuit is also placed in the catheter tip to directly process the sensor signals. The application requires to have a bi-directional data transmission, to send either digitized data inside-out or configuration settings outside-in to adjust the data acquisition. More importantly, power needs to be delivered to the catheter tip. Conventional methods by hard-wiring cables to the catheter tip are becoming more and more challenging, as the catheter is miniaturized and the cable count is limited. Radio frequency (RF) transmission is less practical due to high losses in human tissue. Here, we present a novel concept to simultaneously establish a bi-directional communication and power transfer between a user and a

compact catheter system, containing minimum discrete components to allow for a compact form factor. To achieve this, we have designed an optical transceiver chip (OTC) embedded in the catheter tip. The OTC is connected to an optical link, which consists of a blue LED, a multimode fiber and the bed-side electro-optical unit. The optical link simultaneously delivers power and transmits data by means of light only. Furthermore, the used thin multimode fiber can be easily integrated into a catheter and offers low loss of signal quality and optical power. In addition, the optical link could also be used in Magneto Resonance Imaging (MRI), where long wires are not allowed to avoid high voltage induction.

The paper is organized as follows: In Section II, the optical link is described in detail. Next, in Sections III and IV, the circuit implementation of the OTC, chip layout and system integration are described. Finally, in Section V, the experimental results are discussed.

II. OPTICAL LINK

The optical link is used to establish both power and data transfer between a bed-side electro-optical unit and the electronics inside a catheter tip (Figure 1). Both parts are located several meters away and physically connected by a multimode fiber, which guides optical light. The electronics inside the catheter tip consists of a blue LED, an optical transceiver circuit (OTC) integrated into a customized ASIC, and capacitors. Inside the electro-optic unit, a semiconductor laser generates light with a wavelength of 405 nm and optics are mounted to couple the laser light into the fiber. The light is passed to the distal end of the fiber and illuminates a blue LED, which operates as a power receiver as well as a data transmitter. The power delivery function is achieved by relying on the photovoltaic effect, which converts the laser

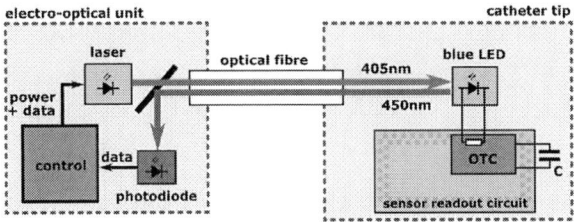

Figure 1: Schematic of the optical link and its required components. The optical transceiver stage is integrated into a sensor readout circuit, which requires power and an interface for communication. An electro-optical unit is used to drive the laser and to read-out the received data.

---

[*]This work is part of the POSITION-II project funded by the ECSEL Joint Undertaking (Ecsel-783132-Position-II-2017-IA) and BMBF (16ESE0304).

power P_{laser} into electrical energy. The generated current applies a voltage at the LED and supplies the OTC with power. Meanwhile, the bi-directional data transmission function is realized with the help of the physical effect called photo-induced electroluminescence (PEL) [1]. It relies on the fact that, if the resistance of a connected load is high enough, the LED will generate a voltage higher than given by the bandgap of the semiconductor and emits light without an external current source back into the fiber. This can be used to operate the LED as a data transmitter by simply modulating the resistance of a connected load and, thus, emitting light back into the fiber. The wavelength of the back-emitted light must be distinguishable from the laser light. The blue InGaN LED is a suitable candidate. It provides on one hand a high bandgap of 2.8 eV, at which a theoretically voltage of 2.8 V is achievable, and on the other hand an emission wavelength of 450 nm [2]. The back-emitted light of the LED is separated from the laser light using a dichroic mirror and directed to a silicon photodiode to convert the modulated light into an electrical signal. A transimpedance amplifier and a FPGA are used afterwards to post-process the data [3].

A bi-directional data communication can be established by using PEL and the photovoltaic effect. Data are sent from the external electro-optical unit to the OTC by modulating the laser power P_{laser}. As a consequence, the voltage drop V_{LED} across the LED will change and will be detectable by a resistor divider n_1. A logic "1" is transmitted, if P_{laser} is higher than a reference and vice versa (Figure 2 (a)). In order to send back data from the catheter tip to the electro-optical unit, the LED's current $I_{LED,min}$ needs to be modulated, thereby providing more or less power to emit light back into the fiber. This can be done by sourcing an additional current $I_{load,TX}$, thereby ensuring a constant load current I_{OTC}. In this manner, a logical "1" is sent by sourcing a higher current 2 x $I_{load,TX}$ and a logical "0" is sent by sourcing 1 x $I_{load,TX}$ (Figure 2 (b)).

III. CIRCUIT IMPLEMENTATION

An overall schematic of the OTC is illustrated in Figure 3. The schematic consists of all required on-chip units and off-chip components, such as the blue LED, the capacitors and the customized sensor readout circuit. With the help of the off-chip components, the OTC can be used to deliver power and to establish data transfer to the sensor readout circuit simultaneously. The latter, in turn, is implemented closely to the OTC on the same ASIC and requires a stabilized voltage and enough current to run properly. In addition, the customized sensor readout circuit uses the power-on-reset (POR) function of the OTC to ensure a correct start-up sequence.

The internal circuit parts of the OTC built up with a low drop-out linear regulator (LDO), a POR, a biasing circuit, a logic block to manage transmission, a diode D_1, a programmable current source CS_0, and a circuit RX to detect the received data. As described in Section II, the blue LED is used to act both, as an energy harvester and transceiver. Therefore, the LED's anode is connected to the power input of the OTC and passed to the circuit RX, D_1 and CS_0. Because of the maximum voltage level is 2.8 V, these circuit blocks and the LDO require to be built up with MOS devices operating at a voltage higher than 2.8 V, i.e., 3 V MOS devices are used, while the others as well as the customized

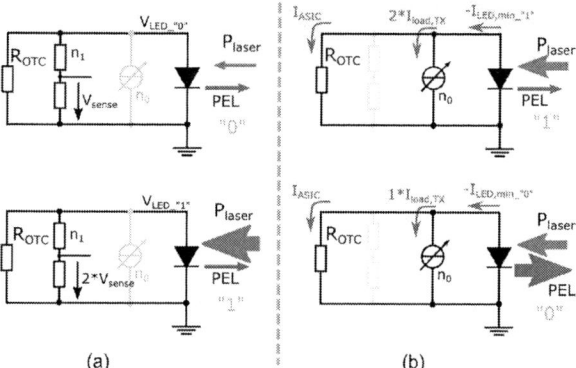

Figure 2: Implementation of the optical bi-directional communication. Data are received by changing the laser power and measuring the voltage drop V_{sense} (a). Data are transmitted using n_0 to change the total current $I_{LED,min}$ by $I_{load,TX}$ (b).

circuit is using 1.8 V MOS devices. In the following sub-sections, each circuit block and their related functions are discussed in detail.

1) Energy Harvesting and Voltage regulation with LDO
An integrated linear low drop-out regulator, diode D_1 and capacitors are implemented to supply the sensor readout circuit. In this example, the sensor readout circuit requires a stabilized voltage of 1.8 V with a maximum current consumption of 2.1 mA, which needs to be provided even during data transmission. Because of the fact, that data are transmitted by turning off and on the laser light, meaning a varying input voltage V_{LED} from 0 V to 2.8 V, the OTC's voltage regulation circuit has to accommodate with a huge input voltage range. A suitable way to ensure a stable output voltage of 1.8 V was found by the following topology: If the laser light is turned off for a short time, the diode D_1 is used to prevent current flow back to the LED. In addition, an external capacitor $C_{2,ext} = 4.7$ μF is added to buffer the input voltage $V_{IN,LDO}$ and ensures stable operation condition above the minimum designed input voltage of 2.0 V. If the laser light is turned on, the voltage V_{LED} of the LED requires to be high enough despite of the loss, because of the diode's forward voltage. A large PMOS transistor in diode configuration is chosen to minimize that voltage. The topology of the LDO is built up adopting a conventional approach consisting of an error amplifier, PMOS transistor, a bandgap circuit and an external capacitor $C_{1,ext} = 4.7$ μF. The latter is used to stabilize and to buffer the output voltage of 1.8 V

2) Data transmission proximal-to-distal
The transmitted data of the electro-optical unit are received by using the circuit block RX of the OTC. As described in section II, data are sent by turning on and off the laser light. The voltage drop of V_{LED} is sensed by the circuit RX and further processed by the digital core. To do so, a programmable resistor network n_1 is implemented to sense V_{LED}, which divides the voltage V_{LED} to a lower voltage V_{sense} in order to fit to the 1.8 V MOS requirements. A Schmitt Trigger comparator ST_1 compares V_{sense} with a fixed reference voltage V_{ref}. In case the laser is turned off, V_{LED} is lower than V_{ref}, which results to $V_{RX} = 0$ V (or logical "0"). And vice

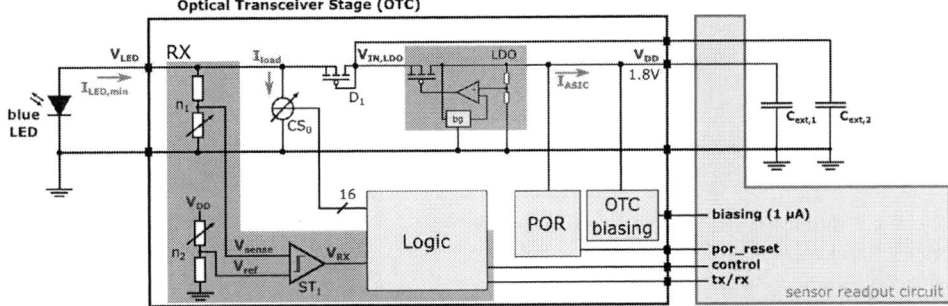

Figure 3: Schematic of the optical transceiver stage (OTC). A blue LED is connected to its input to deliver power and transmit data. Several circuit block inside the OTC are used to process data transmission and ensures stable power to a sensor readout circuit.

versa, if V_{LED} is higher than V_{ref}, the laser is turned on and thus $V_{RX} = 1.8$ V ("1"). The decision result V_{RX} is passed to the logic unit of the OTC, where data are processed further. The RX block is designed to receive data at maximum rate of 15.6 kBits/s. To adjust sensitivity and common-mode voltage of ST_1, the voltage divider n_1 and n_2 can be controlled digitally.

3) Data transmission distal-to-proximal
Data are transmitted to the electro-optical unit by changing the current flow I_{load} of source CS_0. The amount of current requires to be changeable during run-time and determines the amount of back-emitted light into the fiber. Therefore, a programmable current source functionality is implemented at CS_0, where the current I_{load} can be changed stepwise from 0.13 mA to 1.92 mA (4-bit resolution). CS_0 is built up using a conventional current mirror structure consisting of a binary weighted NMOS devices. This is because of the benefit, that a current source and the resulting current I_{load} isn't affected by its input voltage. This allows to source a constant current I_{load} even if the voltage V_{LED} is varying. CS_0 is designed to set the current with a speed of up 1.35 MBits/s.

4) Clock recovery implementation
The OTC has integrated a logic block, which controls data transmission and sets configuration registers of internal circuit parts. Since the optical interface provides no electrical connection between the electro-optical unit and OTC, both parts are not clock synchronized and just assuming their transmitter's default frequency. Because of process variations, the OTC's clock can differ in the range of +/- 30 %. That means, that both parts require techniques to recover the clock and data of a received input signal. For the OTC, this is solved by oversampling signal V_{RX} of up to 16 times and minimize the influence of process variations. A speed of up to 15.6 kBits/s can be achieved, if using a clock frequency of 1.35 MHz and PWM modulation. Same applies for the electro-optical unit, where the transmitted data of the OTC are oversampled by 100 MHz to recover the clock and data.

IV. CIRCUIT LAYOUT AND SYSTEM INTEGRATION

The OTC is integrated into an ASIC placed in a catheter tip for sensing electrophysiology signals inside the atrium of a human heart. Figure 4 shows the chip micrograph with the OTC block highlighted. The ASIC is realized in a 180 nm CMOS technology and occupies an area of 0.07 mm². The pads of the OTC are separated in such a way, that two

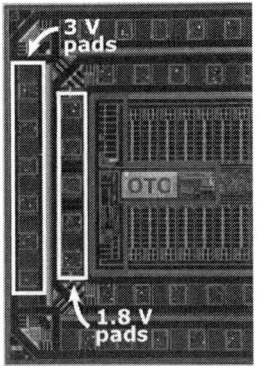

Figure 4: Micrograph of the OTC. The circuit is connected to 3.0 V and 1.8 V pad rings. In addition, the OTC is designed to fit to the area requirements of the customized circuit.

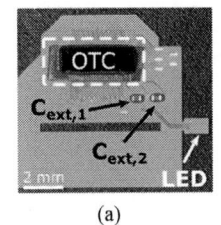 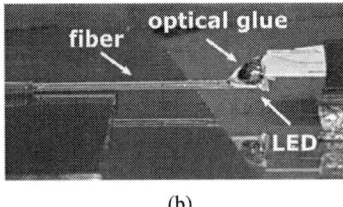

(a)   (b)

Figure 5: (a) The customized ASIC with integrated OTC is placed on top of a liquid polymere substrate. The capacitors and LED are solderd close to the ASIC (LED is placed on the bottom of the LCP). (b) The fiber is glued to the LED.

different types of pads with different voltage regimes are used, either 3 V and 1.8 V. Moreover, the pitch of the pads is chosen to ensure enough space for reliable gold bumping connection.

The overall system ASIC is placed on top of a liquid crystal polymer (LCP) substrate (Figure 5 (a)). Other components as the LED and capacitors are soldered on the bottom and top side of the LCP as closes as possible to the ASIC. The LCP area of the LED is folded in such a way, that the fiber can be glued horizontally (Figure 5 (b)).

V. MEASUREMENT SETUP AND RESULTS

The measurement setup is built up to test the performance of the OTC and to show the functionality of the overall optical link. The setup consists of an electro-optical unit, a PC to control the measurement sequence and an oscilloscope to record data (see Figure 6). As can be seen in Figure 7, the optical power of the laser is ramped up to 25 mW (58 mW electrical power) and then held constant to give time to start-up the ASIC (blue curve). If enough power is delivered to the LED, the OTC begins to regulate to a supply voltage of $V_{DD} = 1.8$ V (green curve). As the next step, the customized circuit automatically sends identifier packets by using the OTC's transmission unit. The electro-optical unit is configured to detect these identifier packets in order to recognize a successfully start-up of the OTC (red curve). Next, the OTC is ready to receive data. This can be done between two identifier packets to avoid interferences. In this example, data are sent to the OTC to change internal settings of the customized circuit (see command "set register" and "set current" in Figure 7). Finally, a command is sent, which starts to perform measurement and the transmission of data with up to 1.35 Mbit/s (called, burst mode; see red curve). At the carried-out measurement, I_{load} was modulated between 0.13 mA (logical "0") and 1.17 mA (logical "1") to achieve sufficient back-emitted optical power for a high signal amplitude at the photodiode. A summary of selected performance values is given in Table 1.

Table 1: Performance summary of the OTC

	This work
Technology	180 nm CMOS
Core Supply Voltage	3 V, 1.8 V
Circuit Area (without pad area)	0.07 mm²
Type of Energy Harvesting	optical
Required optical laser power	25 mW
Minimum input voltage	2.0 V
Supply output for load	1.8 V
Maximum current for load	2.1 mA
Speed of transmission	< 1.35 Mbit/s
Speed of receiving	< 15.6 kBit/s
On-Chip LDO	yes
Power-on-Reset	yes
Off-chip capacitors	2 x 4.7 µF

VI. CONCLUSION

In this paper, an optical transceiver circuit (OTC) integrated into a customized ASIC residing in a catheter, is presented. It uses a single optical fiber acting as optical link to provide power and a communication interface simultaneously. A blue LED is placed closed to the OTC and a fiber is glued to the LED to transfer data and power by means of light only. The concept is verfied by measurement results. The optical link as well as the OTC are perfectly suited for applications, where less space inside a catheter is available and a bi-directional data transmission interface are required. Further investigations will cover circuit techniques to eliminate the use of external capacitors and tests inside an MRI system.

Acknowledgment

The authors would like to thank Cor Scherjon from IMS CHIPS, Thorsten Göttsche, Benjamin Burg from OSYPKA AG, Martin van der Mark, Martin Pekař, Anneke van Dusschoten from Philips MEMS & Micro Devices.

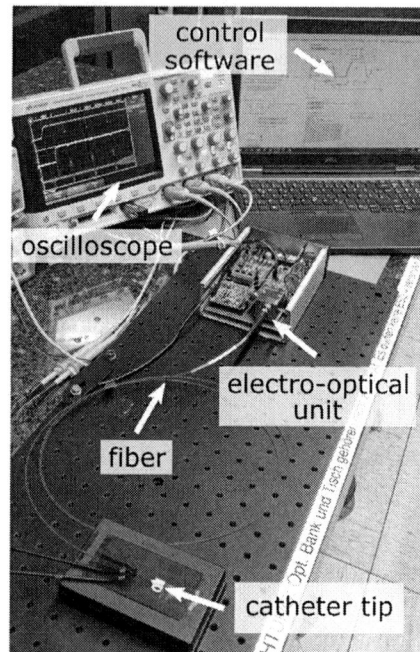

Figure 6: Setup to perform optical measurements with the OTC.

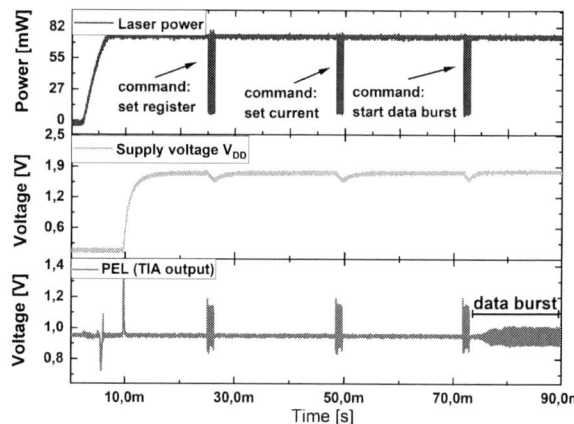

Figure 7: Measurement results of a bi-directional communication using the OTC and optical link.

References

[1] M. F. Schubert, Q. Dai, J. Xu, J. K. Kim und E. F. Schubert, „Electroluminescence induced by photoluminescence excitation in GaInN/GaN light-emitting diodes," *Applied Physics Letters*, Bd. 95, p. 191105, November 2009.

[2] M. B. van der Mark, A. van Dusschoten und M. Pekar, „All-optical power and data transfer in catheters using an efficient LED," in *Optical Fibers and Sensors for Medical Diagnostics and Treatment Applications XV*, 2015.

[3] A. Frank, B. Kootte, T. Göttsche, P. Jutte, J. Schleipen, V. Henneken, M. van der Mark, E. Bihler, P. Dijkstra, J. Anders und J. Burghartz, „A 96-Channel Electrophysiology Catheter with Integrated Read-Out ASIC and Optical Link," Mexico, 2021.

A reconfigurable 224×272-pixel single-photon image sensor for photon timestamping, counting and binary imaging at 30.0-μm pitch in 110nm CIS technology

Leonardo Gasparini
Center for Sensors & Devices
Fondazione Bruno Kessler
Trento, Italy
gasparini@fbk.eu

Manuel Moreno García
Center for Sensors & Devices
Fondazione Bruno Kessler
Trento, Italy
Now at Sony Semiconductor Solutions

Majid Zarghami
Center for Sensors & Devices
Fondazione Bruno Kessler
Trento, Italy
Now at Sony Semiconductor Solutions

André Stefanov
Institute of Applied Physics
University of Bern
Bern, Switzerland
andre.stefanov@iap.unibe.ch

Bruno Eckmann
Institute of Applied Physics
University of Bern
Bern, Switzerland
Now at METAS

Matteo Perenzoni
Center for Sensors & Devices
Fondazione Bruno Kessler
Trento, Italy
Now at Sony Semiconductor Solutions

Abstract—**A 60k-pixel SPAD-based, multi-functional image sensor is designed and fabricated in a 110-nm CMOS image sensor technology. The 30.0-μm pixel with 12.9% fill-factor can operate in multiple working modes to count photons with a 7-bit counter, to timestamp them with an in-pixel time-to-digital converter down to 184ps resolution and 8-bit depth, or to generate a fast sequence of 7 gated binary images. Thanks to its reconfigurability, up to 3kHz frame rate and less than 80 mW power consumption, the sensor can be applied to a large spectrum of applications: here its use in a Fluorescence Lifetime Imaging Microscopy setup, in a quantum physics experiment, and in the acquisition of a high dynamic range scene are demonstrated.**

Keywords—*CMOS SPAD, imager, TDC, quantum optics and binary imaging.*

I. INTRODUCTION

IMAGERS based on Single-Photon Avalanche Diodes (SPAD) in CMOS technology combine single-photon sensitivity, sub-nanosecond timing and in-pixel or on-chip processing capabilities, e.g., to count photons, timestamp them or perform autonomous event detection. These unique advantages have been exploited in the past in applications like fluorescence lifetime imaging microscopy (FLIM), as in the pioneering work in [1] where a large pixel with small fill-factor implements a time-to-digital converter (TDC) in a 130nm technology, and more recently in [2] where the use of a 40nm process enabled an impressive scaling, but substantially with similar functionalities. This and other biomedical applications have been addressed [3] exploiting similar architectures. Quantum physics [4], 3D ranging with photon hit control strategy [5] or real-time laser peak search [6], and high-dynamic range (HDR) quanta imaging [7] have also been explored, on the one hand realizing more and more specialized and targeted implementations, on the other hand restricting the possible fields of application and thus narrowing the effective value of the imager. However, the different pixel topologies often share some common features that can be flexibly exploited to realize multiple functions. Additionally, the intrinsically excellent timing resolution of SPADs calls for integrated time-to-digital

converters (TDC), a typically area-consuming circuit which is difficult to optimize. With this objective, this work presents a multi-functional CMOS imager based on SPADs, introduced in [8], implementing compact and optimized programmable in-pixel circuitry, including an area-efficient TDC implementation. The pixel array can work in multiple modes: pixel-wise fine (sub-nanosecond resolution) or coarse ($\geq$ 10 ns) timestamping, photon counting or fast binary imaging.

II. SENSOR ARCHITECTURE

The imager finds its strength in the pixel ability to change its pixel functionality thanks to highly flexible reconfigurable connections of the in-pixel memory elements. Figure 1 shows the pixel schematic, which consists of three parts: the SPAD and its front-end, a trigger generation and processing logic, and the single-wire readout. The implemented concept reuses a bank of flip-flops in the processing logic either as a counter for a TDC based on a ring-oscillator or global clock, as a photon counter for the SPAD pulses, and as a shift register for bit-plane imaging and readout.

A. Pixel general operation

Within the *quenching and front-end* circuits, a programmable SRAM memory keeps noisy SPADs off through M0. The SPAD can be either synchronously charged through M1, using a global, fast (10 ns) pulse, or passively quenched through R_{poly} and M2.

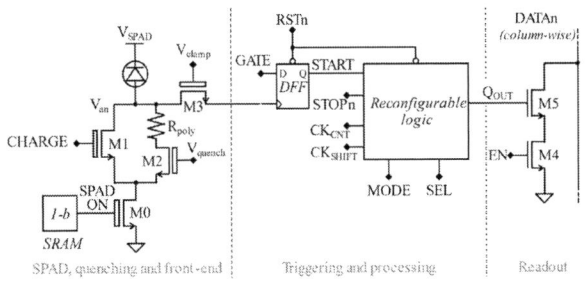

Figure 1. Pixel block diagram

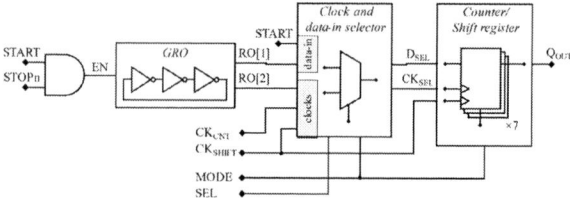

Figure 2. Block diagram of the in-pixel reconfigurable logic

TABLE 1 PIXEL WORKING MODES ACCORDING TO MODE AND SEL.

MODE / SEL	7-bit register	Pixel working mode
0 / 0	Counter	Fine timestamping
0 / 1	Counter	Coarse timestamping / Counting
1 / 0	Shift register	Binary imaging
1 / 1	Shift register	Readout

M3 clamps the voltage step following the SPAD avalanche to the digital logic domain voltage (1.2V), thus operating as an electrical interface between the front-end and the *triggering and processing* logic of Figure 1. The DFF operates as triggering logic, rising START only if a photon is detected while GATE = H and feeding it to the reconfigurable logic to perform the required task. On the right side, the *readout* of the pixel DATA is implemented through a column-shared serial bus in an open drain configuration by means of M4 and M5.

B. In-pixel reconfigurable logic

The reconfigurable logic block in Figure 2 contains (i) a regulated gated ring oscillator (GRO), (ii) a selector for the CK_{SEL} and D_{SEL} signals, and (iii) a set of 7 DFFs with re-routable inputs (DFF_{config}), to be configured through MODE and SEL as detailed in TABLE 1. The reconfigurability is achieved by means of multiplexers distributed in the *CK and data-in selector* and in front of the DFF_{config}'s as shown in Figure 3 and Figure 4, respectively.

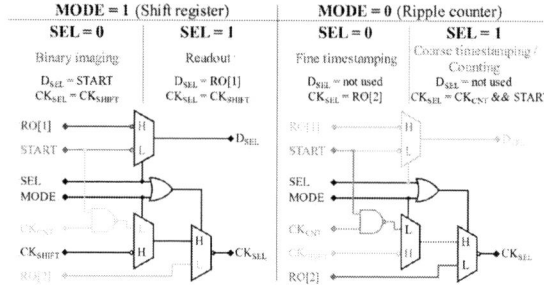

Figure 3. Schematic diagram and configurations of the *Clock and data-in selector*. Signals not used in each configuration are greyed out.

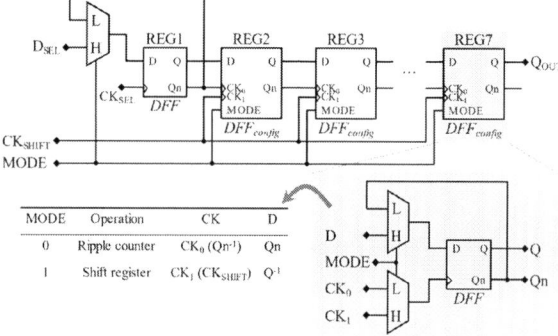

Figure 4. Schematic diagram of the 7-bit *Counter/Shift register* and of the DFF_{config}. Q^{-1} and Qn^{-1} are the outputs of the previous DFF.

MODE = H, as shown in Figure 3(left), activates the shift register operation connecting the CK_{SHIFT} signal to the clock input of all the DFFs and the output of each DFF to the input of the following one, while SEL selects the input of the first DFF. This operating mode is used with SEL = L to generate binary images at rates up to 100 MHz inserting the value of START into the shift register, and with SEL = H to serially read out data (with row-wise distribution of CK_{SHIFT}) and reset the pixels by inserting 0's (the GRO under reset).

MODE = L configures the register as a ripple counter. From Figure 3(right), with SEL = H, the counter is clocked by the AND between the globally distributed CK_{CNT} and START. In this configuration, the photon counting operation is achieved by opening multiple observation windows and pulsing CK_{CNT} at the end of each observation, while the coarse timestamping functionality is achieved providing a train (up to 127) of pulses on CK_{CNT} within a single observation. Alternatively, SEL = L configures the sensor in fine timestamping mode: the counter clock is connected to the GRO, which is started when a photon is detected and stopped by the global STOP.

The GRO implementation is shown in Figure 5. With 12 transistors, this is the most compact GRO implementation, to the best of our knowledge. It is based on 3 inverters in series, the last one (INV2) being tri-state (M8 and M9 operate as switches), and a pass transistor (TG), with one and only one between INV2 and TG being active at time. M6 regulates the GRO supply voltage. Under reset (RSTn = L, EN = L), INV2 is tri-stated and the GRO behaves as an inverter-based latch enforcing RO[0:2]="011" through M7 and TG. After disabling of the reset, a photon detected during the observation sets EN = H, which opens TG and activates INV2, starting the GRO oscillation. RO[2] clocks the counter. When the STOP signal is activated, EN goes back to L and the GRO operates as a static memory that preserves the value of RO[1]. During the readout phase, the value of RO[1] is shifted in the register, thus providing an additional 8th, least significant bit to the timestamp.

C. Sensor architecture and readout

The overall sensor architecture in Figure 6 is completed by the logic to program the in-pixel SRAM cells, the drivers for the control signals, and the readout logic. Skew of control signals introduced by the large RC network of the pixels is reduced by means of left+right and top+bottom buffer trees, each covering half of the array. Readout is performed through a raster scan of the array with a zero-suppression mechanism that exploits the sparsity of data of several CMOS SPAD imaging applications. Pixels are set in readout mode as per TABLE 1, configuring the DFFs as a shift register.

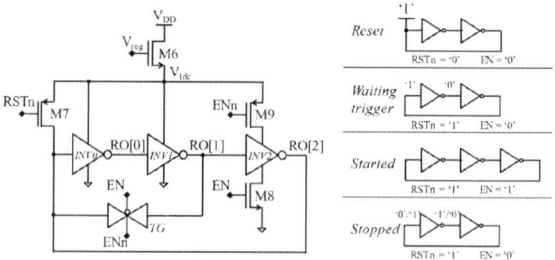

Figure 5. GRO schematic diagram and working principle.

978-1-6654-8495-4/22 $31.00 © 2022 IEEE

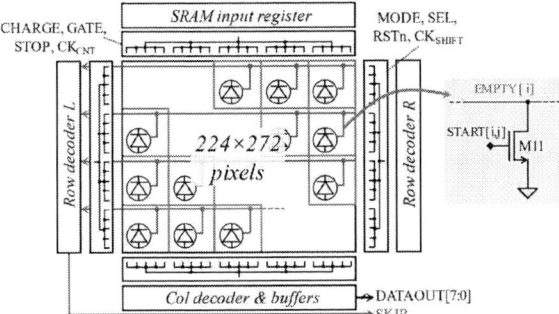

Figure 6. Top-level block diagram.

While the first CK_{SHIFT} period enables transferring any TDC LSB which is latched into the GRO, during the following cycles the same operation enables reset of the pixel by resetting the GRO and letting its value propagate into the register. The row decoders select one row at a time (the logic is duplicated left and right as the CK_{SHIFT} signal is routed row-wise from both sides of the array). Here, the zero-suppression mechanism notifies on SKIP the absence of triggered pixels in the currently selected row [4]. For this purpose, a pull-down transistor in each pixel discharges a common row line if a detection has occurred within the GATE signal. In presence of at least one triggered pixel, row data are transferred to the column buffers and then outside the chip.

III. SENSOR CHARACTERIZATION

The 7.5×9.1 mm^2, front-side illuminated chip has been fabricated in a 110-nm 1P4M CMOS Image Sensor technology, with a pixel pitch of 30 µm. Figure 7 shows the chip micrograph, the layout of the pixel and its area breakdown. The pixel contains a 115.83-µm^2 p+/deep n well SPAD [9] achieving a 12.9% fill factor. The dead-area ring around the diode that separates it from the low-voltage electronics occupies 50% of the area. The remaining part is mostly taken by the SPAD front-end and the GRO (16%), and by the 7-bit register (21%).

The SPAD device, with a breakdown voltage of 18.8 V, exhibits a median DCR of 100 Hz at 3.5 V excess bias, with 80% of the devices below 200 Hz. The fine TDC has a resolution of 183.8 ps and its linearity has been evaluated by means of a code-density test using weak illumination, showing a peak-to-peak DNL of less than 0.3 LSB a peak-to-peak INL of about 1 LSB (see Figure 9). Illuminating the array with an 80-ps FWHM laser, the single-shot precision of the pixel has been evaluated to 261 ps FWHM.

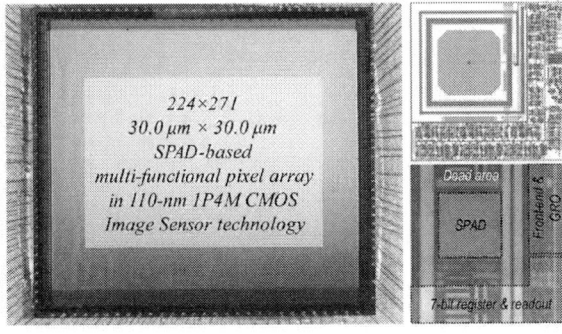

Figure 7. Micrograph of the chip (left). Pixel area breakdown (right).

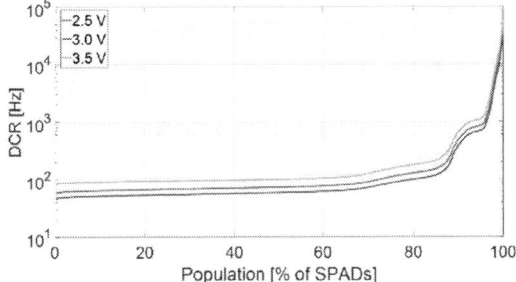

Figure 8. SPAD DCR population of a chip, at multiple excess biases.

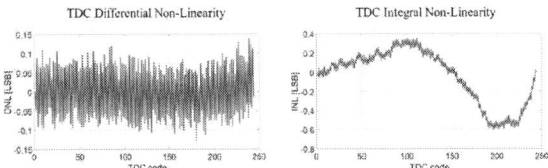

Figure 9. TDC DNL and INL for a representative pixel.

In terms of power consumption, the worst-case scenario occurs when the sensor operates in fine-timestamping mode, due to the GRO. In this case, both power consumption and effective frame rate depend on how much data are sparse. We measured three configurations: with <0.001%, 0.2% and 2% pixel triggering rate and it consumes 36 mW at 3000 fps (dominated by IOs, row skipping highly effective), 13.4 mW at 700 fps and 78 mW at 290 fps (dominated by GROs, no row skipping), respectively.

The chip has been tested in multiple contexts: in a low-cost, compact, wide-field FLIM setup, in a quantum optics testbed, and for the acquisition of an HDR scene. Figure 10 shows the intensity and fluorescence decay times of a standard sample of Convallaria Majalis acquired using the photon counting and fine timestamping modes, respectively, without excitation filter. A color-coded map of decay times is modulated according to the pixel intensity to obtain a combined image. The right plot compares the histograms of the photon arrival times of two portions of the sample characterized by different decay times.

Figure 11 demonstrates the sensor ability to discriminate momentum entangled photon pairs generated in a non-linear crystal by means of spontaneous parametric down-conversion [10] and acquired in the far field. The pairs are generated simultaneously (< 1 ps apart), at the same position (few micrometers apart), have the same wavelength (810 nm, in our case) and exit the crystal along a cone in opposite directions. In the far-field, the first-order correlation function $G^{(1)}(x,y)$ corresponds to the light intensity on the sensor and shows the direction of propagation of the photons. For the non-collinear configuration, $G^{(1)}$ is expected to show a ring-like distribution. Temporal coincidences allow to observe spatial correlations between the photons of a pair on that ring. They are indicated by the second-order correlation function $G^{(2)}(\rho+)$ that corresponds to the distribution of the center of gravity of the pairs. Because of their strong momentum correlations, a small spot centered in 0 is expected. Measurements, corrected for accidental coincidences mostly occurring among uncorrelated photons, confirm the expectations.

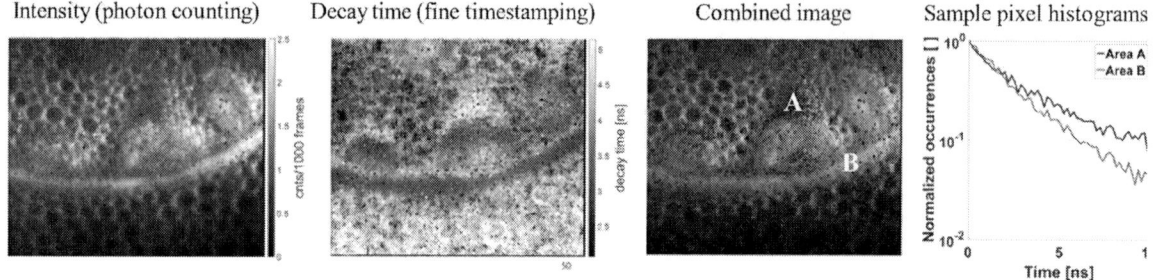

Figure 10. Images of a sample of Convallaria Majalis in a FLIM setup: intensity (left), lifetime (mid-left), combined (mid-right) and histograms (right).

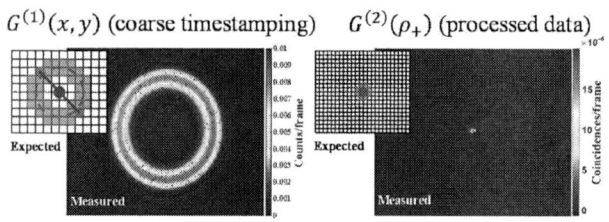

Figure 11. Expected and measured first- and second-order correlation functions of a source of momentum-entangled bi-photons in the far-field.

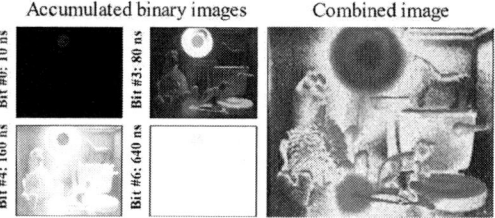

Figure 12. HDR image generated using the sensor in binary imaging mode with an exponentially increasing gate time.

Figure 12 uses the binary imaging mode to observe an HDR scene. Each of the 7 bits of the in-pixel register is associated to a different observation time (acquired in the same exposure with one readout), in this case doubling at every step from 10 ns up to 640 ns, to cope with multiple levels of intensity. The acquisition of the 7-bitplane image is repeated 256 times to externally build 7 8-bit images, which are then combined to obtain the final HDR image.

IV. CONCLUSIONS

A 224×272-pixel single-photon image sensor has been realized in 110 nm CIS technology, offering a broad range of functionalities: photon timestamping with a fine (184 ps) or coarse ($\geq$ 10 ns) resolution, photon counting or high-speed binary imaging. This is achieved integrating in a 30-μm pixel a ring oscillator and a block of logic that can be reconfigured in real time to obtain the desired pixel functionality. The imager has been demonstrated on a

FLIM setup, a quantum optics experiment and the acquisition of a scene characterized by a high dynamic range. This demonstrates the sensor flexibility and its potential for a broad range of applications at a low cost, considering the FSI, non-3D-stacked process technology. TABLE 2 summarizes and compares the sensor characteristics with the state of the art.

REFERENCES

[1] C. Veerappan, *et al.*, "A 160×128 single-photon image sensor with on-pixel 55ps 10b time-to-digital converter," in Proc. of the ISSCC, San Francisco, CA, USA, 2011, pp. 312–314.

[2] R. K. Henderson, *et al.*, "A 192 x 128 time correlated spad image sensor in 40-nm cmos technology," IEEE J. Solid-State Circuits, vol. 54, no. 7, pp. 1907–1916, July 2019.

[3] C. Bruschini, *et al.* Single-photon avalanche diode imagers in biophotonics: review and outlook. Light Sci Appl 8, 87 (2019), doi: 10.1038/s41377-019-0191-5

[4] L. Gasparini *et al.*, "A 32×32-pixel time-resolved single-photon image sensor with 44.64μm pitch and 19.48% fill-factor with on-chip row/frame skipping features reaching 800kHz observation rate for quantum physics applications," in Proc. of the ISSCC, San Francisco, CA, USA, 2018, pp. 98-100.

[5] E. Manuzzato, *et al.*, "A 64×64-Pixel Flash LiDAR SPAD Imager with Distributed Pixel-to-Pixel Correlation for Background Rejection, Tunable Automatic Pixel Sensitivity and First-Last Event Detection in Proc. of the ISSCC, San Francisco, CA, USA, 2022, pp. 96-98.

[6] S. Park, *et al.*, "An 80×60 Flash LiDAR Sensor with In-Pixel Histogramming TDC Based on Quaternary Search and Time-Gated Δ-Intensity Phase Detection for 45m Detectable Range and Background Light Cancellation," in Proc. of the ISSCC, San Francisco, CA, USA, 2022, pp. 98-100.

[7] Y. Ota *et al.*, "A 0.37W 143dB-Dynamic-Range 1Mpixel Backside-Illuminated Charge-Focusing SPAD Image Sensor with Pixel-Wise Exposure Control and Adaptive Clocked Recharging," in Proc. of the ISSCC, San Francisco, CA, USA, 2022, pp. 94-96.

[8] L. Gasparini, *et al.*, "Scalable, Multi-functional CMOS SPAD arrays for Scientific Imaging", presented at the 2020 International SPAD Sensor Workshop.

[9] M. Moreno-Garcia, *et al.*, "Low-Noise Single Photon Avalanche Diodes in a 110nm CIS Technology," in Proc. of the ESSDERC, 2018, pp. 94-97.

[10] C. Couteau, "Spontaneous parametric down-conversion," Contemp. Phys., vol. 59, no. 3, pp.1-14, Aug 2018.

TABLE 2: IN-PIXEL PROCESSING SENSORS COMPARISON

Reference	This work	[2]	[1]
Year	2022	2019	2011
Architecture	TDC fine/coarse Counting Binary img	TDC fine Counting	TDC fine Counting
Process [nm]	110 CIS	40	130
Pixel			
Pitch [μm]	30	18.4 × 9.2	50
SPAD size [μm]/ FF []	11.0 / 12.9%	5.4 / 13%	6 / 1.2%
DCR [Hz] (@ V_{ex} [V])	100 (3.5)	25 (1.5)	50 (0.73)
TDC			
Area [μm²]	340	84.6	2244
(Area/L_{min}^2 [μm²/nm²])	(0.028)	(0.053)	(0.133)
Depth [bits]/Range [ns]	8 / 45	12 / 163	10 / 55
Time resolution [ps]	184	36	55
DNL/INL p-p [LSB]	0.26 / 1.07	0.9 / 5.64	0.6 / 4.0
Precision, FWHM [ps]	261	208	170
Chip			
Array size [pix]	224×271	192×128	160×128
Size [mm²]	7.5×9.1	3.2×2.4	11.0×12.3
Max frame rate [fps]	$\leq$ 3 k	18.6 k	250 k
Power [mW]	$\leq$ 78	31	550

Statistical measurements and Monte-Carlo simulations of DCR in SPADs

Mathieu Sicre[1,2,3,*], Megan Agnew[4], Christel Buj[1,3], Caroline Coutier[3], Dominique Golanski[1], Rémi Helleboid[1], Bastien Mamdy[1], Isobel Nicholson[4], Sara Pellegrini[4], Denis Rideau[1], David Roy[1] and Francis Calmon[2]

*Email : mathieu.sicre@st.com

[1]TR&D, STMicroelectronics, Crolles, France
[2]INL, UMR CNRS 5270, Université de Lyon, INSA Lyon, Villeurbanne, France
[3]CEA, LETI, Univ. Grenoble Alpes, Grenoble, France
[4]Imaging Division, STMicroelectronics, Edinburgh, U.K

Abstract—**Dark Count Rate (DCR) in 3D-stacked CMOS Single-Photon Avalanche Diode (SPAD) is investigated by means of measurements and simulations at various temperatures and voltages. This study strengthens previous hypotheses on the roles of depleted regions and interfaces in DCR generation by examining device architectures. A nonradiative multiphonon-assisted trapping model (NRM) is used to calculate the carrier capture/emission rate of defect sites. Systematic comparison between measurement and Empirical Monte-Carlo (EMC) simulation, accounting for the stochastic diffusion of carriers in computation of the avalanche breakdown probability (P$_t$), was performed to investigate trap and avalanche positions within the device. This simulation unveils device regions contributing to different DCR dynamics by de-embedding carrier avalanche localizations. Based on this simulation methodology, the DCR statistical distribution induced by local device-to-device process variation is covered by randomly setting the defect positions and sizes within the device.**

Keywords—*Single-Photon Avalanche Diode (SPAD), avalanche breakdown probability, Dark Count Rate (DCR), Technology Computer Aided Design (TCAD), defects, thermal activation energy, Empirical Monte-Carlo, statistical analysis*

I. INTRODUCTION

Silicon-based Single-Photon Avalanche Diodes (SPADs) are required for obstacle and facial recognition [1], optical wireless communication [2] and quantum random number generation (QRNG) [3]. To fulfil the requirements for most applications, the SPAD sensor must have a low noise floor, in other words a low Dark Count Rate (DCR). The multi-step processes with various ion implantations together with a limited thermal budget may worsen the defect density, further increasing the DCR noise. Using numerical simulations that couple Technology Computer Aided Design (TCAD) software with in-house company solver, a significant effort has been made in the semiconductor industry to model device operation including these noise aspects. The results of these simulations are used as input parameters for the production line to reduce the noise and reach high-quality standards. Even with this collective effort, few traps may remain active. Carriers generated from these local defects may lead to avalanche breakdown interfering with the photo-generated signal. These parasitic events drive the DCR level. To identify the origin of these carriers and their Recombination-Generation (R-G) rate, DCR has been modeled in the literature [4][5]. These papers mainly focused on carrier generation in the depleted regions. Our previous study [6] reinforced these electric field-dependent hypotheses achieved by modulating the architecture of the p-n junction (Fig. 1a (1)

and (2)). To capture P$_t$ outside the multiplication region, the simulation was realized along the electric field lines according to the McIntyre model [7]. Our previous results emphasized the critical role of interfaces by varying the manufacturing processes at the side (3) and bottom (4) interfaces.

The previous modeling was limited to predict the median DCR trend and did not consider the statistical DCR fluctuation reported on Fig. 1 (b) due to the local device-to-device process variation. This local process variation can be inferred to the dispersion of defect localization and size [8]. These assumptions will be verified hereafter through temperature and field dependent measurement and simulation on numerous SPAD samples. To simulate this statistical dispersion, defect position and size are randomly set within Region of Interest (RoI) reported on Fig. 1a.

An Empirical Monte-Carlo method (EMC) [9][10] was used to link trap generated carrier position and avalanche localization within the device considering the stochastic property of carriers' diffusion and defect distribution. Based on this methodology, the carrier avalanches, either multiplying in the center or in the peripheral parts, are divided into regions contributing to DCR computation. To go beyond the previously used conventional SRH generation-recombination framework [6][11], the considered carrier generation mechanism from these defects is a nonradiative multiphonon-assisted trapping model (NMP) [12]. The R-G rate coupled with the resulting P$_t$ at each structure mesh point results in the DCR revealing the potential parasitic noise sources. Statistical conclusions are drawn about experimental DCR distribution by stochastically simulating the dynamics of a large sample of individual carriers in the device.

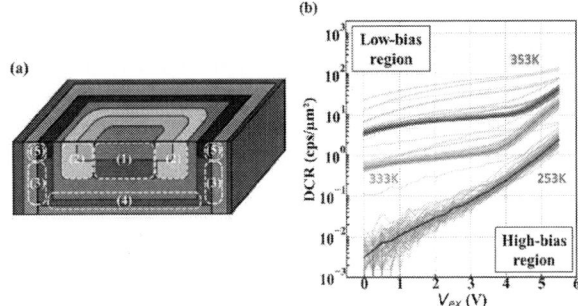

Figure 1. SPAD architecture with Regions of Interest (RoI): (1) High-field avalanche region, (2) Low-field peripheral ring region, (3) Side interface, (4) Bottom interface and (5) Near-contact region. Contacts on top of regions (1) and (5) are not shown here. (b) Measured DCR as a function of V$_{ex}$ at 253K (blue), 333K (orange) and 353K (red). The bold lines are the medians.

II. EXPERIMENTAL FRAMEWORK

To assess DCR contributions, SPAD architectures have been engineered to induce significant change in the electric field distribution within the p-n junction. The studied SPADs are based on a silicon n-on-p structure [13][14] (Fig. 1a). The multiplication region is composed of highly doped n- and p-wells (1), while a p-epitaxial lightly doped region allows a separate large photon collection volume. Lightly n- and p-doped guard rings (2) prevent lateral breakdown at high biasing voltage. An optical stack (4) at the bottom surface increases photons back side transmission to the silicon. Deep Trench Isolation (DTI) (3) isolates devices to avoid crosstalk. Ohmic contacts (5) are yielded by heavily doped implantation at the top. The Design of Experiment (DoE) is composed of four different SPAD architectures, in which variation of wells (1) and rings (2) are performed without changing the process parameters:

TABLE I.   SPAD ARCHITECTURE VARIATIONS

Architecture name	Variation localization
SPAD 1 & 2	Wells (1)
SPAD 3 & 4	Rings (2)

To capture local process variation impact on the defect distribution, measurements of median DCR values (at 253K, 333K and 353K) are performed on 64 identical SPADs at reversed excess bias above breakdown voltage (V_{ex}) for each device architecture. All devices have identical passive quench circuit. The DCR is normalized per unit photo-sensitive area (cps/μm^2).

III. BREAKDOWN PROBABILITY SIMULATION

To determine the avalanche breakdown probability P_t, the dopant profile and the electrostatic distribution were simulated considering each process step. The defined processes were simulated thanks to Sentaurus Process tool [15]. The electrical outputs were computed by solving the Poisson equation and current continuity equations coupled with drift-diffusion transport model in Sentaurus Device module [16]. The avalanche breakdown probability P_t was computed using either the McIntyre model [7] presented in [6][17], or with an Empirical Monte-Carlo method (EMC) [9][10]. These models will be compared hereafter through study of simulated breakdown probability maps to identify potential avalanche-leading defect sites.

In the McIntyre model, the carrier trajectories inside the device are approximated by the electric field streamlines. Starting from this assumption, the non-linear McIntyre model [7] was solved and coupled with the impact-ionization coefficients $\alpha_{n,p}(F,T)$ calculated from the Chynoweth law [18] along these field lines to simulate P_t.

In the EMC method, the stochastic transport and avalanche multiplication probability are simulated starting from carriers generated at each device mesh point. The carrier drift along the electric field lines is considered as previously mimicked from the McIntyre model. In addition, the random-walk formulation of carrier diffusion is added. To determine the carrier path, the carrier displacement is computed by solving drift-diffusion equation [9][10] multiplied by a random number drawn from a gaussian distribution at each time step. To consider the carrier drift velocity saturation in the high electric-field of the multiplication region, the extended Canali model [19] was used. In parallel, the ionization integrals [7] are computed for each particle along time. These integrals are compared to a random number drawn from a uniform distribution defined at each time step to consider the stochastic multiplication process. When the ionization integrals become higher than this reference number, a multiplication occurs generating an electron-hole pair by impact-ionization in the depletion region. This procedure is repeated until all carriers recombine, are collected at the contacts, or reach the avalanche breakdown threshold set at 100 carriers to compute P_t. The simulation was run 1500 times at each structure mesh point with one initial particle each time. The simulation convergence has been studied and a typical time step of 10fs has been found to be a good compromise between simulation time and accuracy.

In Fig. 2 and 3, the breakdown probabilities simulated from the McIntyre model to the EMC method are compared (only half of the device is shown) at low- and high-biases. These applied voltages activate different potential avalanche areas within the device, either in central or peripheral parts. As can be seen, the near contact region (right-top corner which is zoom in the inset) exhibits significant difference: from zero avalanche breakdown probability for the McIntyre model to up to 5% for the EMC method in this region. This is due to an electrostatic potential barrier at the device top right-hand side which cuts the direction of the electric field lines represented on Fig. 2 and 3. In the McIntyre model, the carrier trajectories follow the electric field streamlines. Therefore, this electrostatic potential barrier stops the simulation path, and thus disables the computation of P_t. In the EMC method, the carrier diffusion enables to overcome this electrostatic barrier. Consequently, carriers have two new possible paths: they can diffuse and be collected to the contacts without avalanche triggering leading to P_t reduction (A) or reach an avalanche active region resulting in P_t increase (B).

To conclude, the EMC approach unveils new carrier paths by simulating the random movement and multiplication of carriers. If a carrier is generated within these regions, it will add several path choices leading to different DCR levels in opposition with carrier trajectory constrained by electric field streamlines. Taking advantage of this carrier feature, the avalanche triggering localization will be studied based on the EMC method in the following.

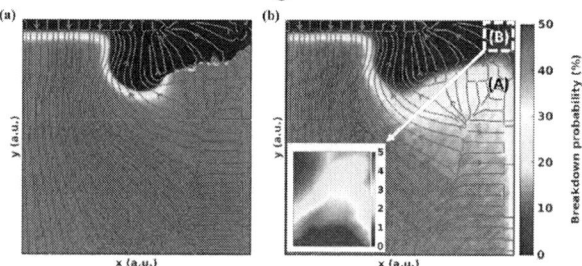

Figure 2. Breakdown Probability with (a) McIntyre model compared to the (b) Empirical Monte Carlo method superimposed over multiple electric field lines simulated at Vex=2V and T=253K. Inset: Zoom on the near-contact region (dashed box) with a refined scale. (A) shows a P_t decrease due to carrier diffusion to contacts and (B) a P_t increase owing to carrier diffusion overcoming electrostatic potential barrier.

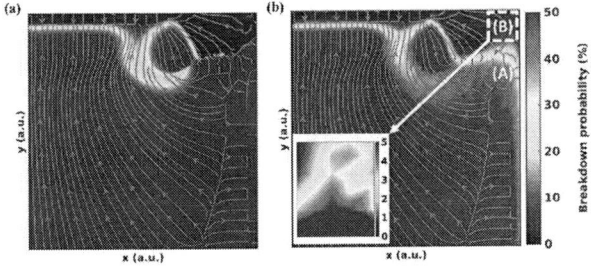

Figure 3. Same as Fig. 2 simulated at Vex=5V and T=253K.

To correlate localization of carrier generation to avalanche triggering, we arbitrary split the breakdown probability P_t in two areas at the edge of the peripheral region (from (1) to (2) on Fig. 1a). This allows to follow the path of a carrier to the avalanche triggering, either in the central part (Fig. 4a) or in the peripheral part (Fig. 4b). The EMC method predicts that some carriers generated from the side interface have a probability to trigger peripheral avalanches contributing to the DCR high-bias dynamic.

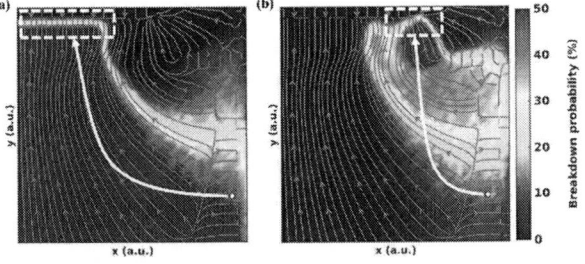

Figure 4. Breakdown Probability computed with Empirical Monte-Carlo superimposed over multiple electric field lines at Vex=5V and at 253K. The breakdown probability contributions are arbitrary split at the edge of the peripheral region (from (1) to (2) on Fig. 1a). Each color point represented a starting position (such as empty circle) considering carrier whose ending in (white arrow): (a) the central region and (b) the peripheral region (represented by dashed white rectangles).

IV. DARK COUNT RATE DESCRIPTION

A. Previous results based on median DCR

With the avalanche breakdown probability simulated at each device point coupled with the R-G rate, the defect distribution within the device can be determined based on the comparison between DCR measurements and simulations. The previous DCR study [6] were performed with the McIntyre model [7]. In this previous study, replicating the defect distribution on the dopant impurity distribution based on the Scharfetter model [20] was discredited to reproduce DCR trends. However, setting defects within RoI depicted on Fig. 1 allows to accurately replicate DCR over a broad range of voltages and temperatures. From these results, the low-bias region (Fig . 1b) is explained on the scope of BTBT within the main junction at low temperature and diffusion from the interfaces at higher temperature. The high-bias region (Fig. 1b) is related to the breakdown of the lateral junction at high bias voltage (Fig. 3 and Fig. 4b).

The defect distribution defined constant for all architectures allows to reconstruct the median DCR trends (bold lines on Fig. 1b) [6]. However, this average value does not capture the intrinsic population (from 1st to 3rd quartiles) and the extrinsic population (outliers). The local process-induced statistical fluctuation will be studied based on the EMC approach together with a randomization of defect

positions and sizes (from a single-defect to a large cluster of a hundred defects) within the RoI reported in Fig. 1.

B. Study of recombination-generation mechanisms

With the randomization of defect positions and sizes together with EMC approach and the NRM emission R-G rate, the same temperature-voltage dependence must be found. The experimental results of temperature and field dependent measurements have been used as inputs for simulations. This procedure enables the extraction of thermal activation energy E_a to identify the combination of carrier recombination-generation mechanisms within RoI (e.g. within the depleted part or at the interfaces). These effective thermal activation energies refer to the amount of energy required to carry a carrier. The NRM model was set with mid-gap traps with nonradiative capture cross section computed at 10^{-15}cm^2 [12] to compare with previous modeling results [6]. From the experimental Arrhenius' plot of median DCR, two main activation energies can be extracted from Fig. 6 in both voltage regimes. This temperature-voltage dependence will be explained based on simulation results.

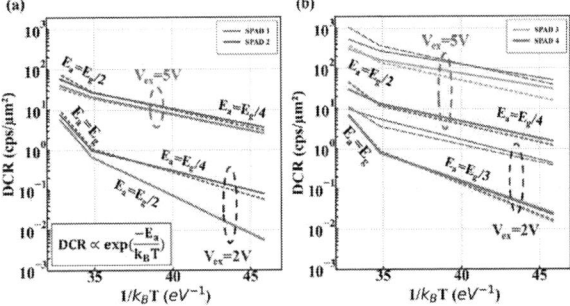

Figure 6. DCR as a function of $1/k_BT$ measured (bold lines) and simulated (dashed lines) at V_{ex}=2V and 5V for architecture variations of: (a) wells (SPAD 1 and 2) and (b) rings (SPAD 3 and 4). The thermal activation energy E_a is equivalent to the slope of the line following the Arrhenius law $\ln(DCR)$ vs $1/k_BT$ with k_B the Boltzmann constant. The activation energies extracted from simulated P_t (-0.08eV at V_{ex} = 2V and -0.03eV at V_{ex}=5V) were removed to the presented DCR activation energy. These results highlight either diffusion ($E_a=E_g$) or NRM and tunneling generation ($E_g/4<E_a<E_g/2$) mechanisms (with E_g the silicon energy gap). Typical outlier behavior is shown in grey dashed lines.

According to Fig. 6, the low-bias region is driven by carrier diffusion from interfaces ($E_a=E_g$) at high temperature, and by tunneling ($E_g/4<E_a<E_g/2$) at low temperature. The tunneling part is a combination between NRM and BTBT generation of carriers. For instance, the SPAD 2 which has the highest BTBT contribution at low temperature (Fig. 7a) has an activation energy E_a of $E_g/4$. Conversely, the SPAD 1 has an activation energy of $E_g/2$ closer to the NRM generation from mid-gap traps. The high-bias region is driven by field-enhanced trap generation in the periphery resulting in a lower activation energy ($E_g/4<E_a<E_g/2$).

C. Statistical analysis of Dark Count Rate

With the generation mechanism well-calibrated on experimental results, the local process-induced statistical fluctuation can be appreciated on Fig. 7 and 8. The device architectural constraints allow to verify our assumptions based on different electric field distributions. On one hand, for the intrinsic population inside the low-bias region, defect clusters randomly set along the side interface (4) together within the near-contact region (5) and within the multiplication region (1) catch this statistical dispersion (Fig

8a). For the high-bias region, only defects randomly set at the edge of the wells and the rings, where the defects concentration is supposed to be high considering the multiple implant steps, reproduce this trend (Fig. 8b). On the other hand, the extrinsic population is replicated with defect randomly set within the wells (1) and the rings (2).

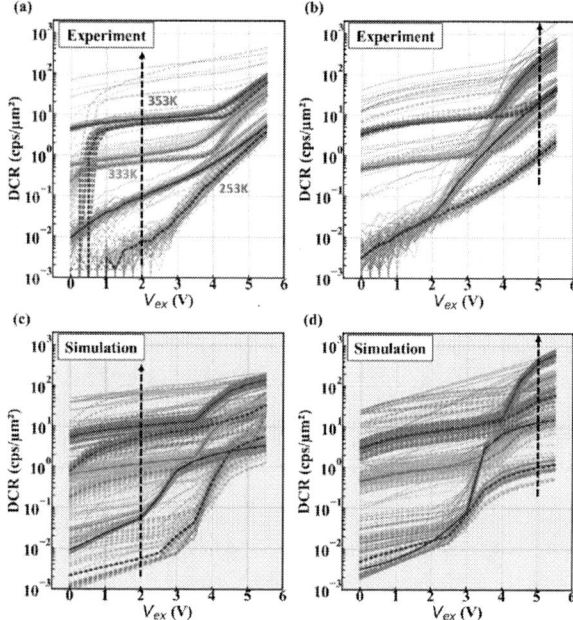

Figure 7. Measured and simulated DCR as a function of V_{ex} at 253K (blue), 333K (orange) and 353K (red): (a)(c) wells (SPAD 1 and 2) and (b)(d) rings (SPAD 3 and 4) from reference to variation (arrow direction). The bold lines are the medians. (a) and (b) are experiments and (c) and (d) are simulations.

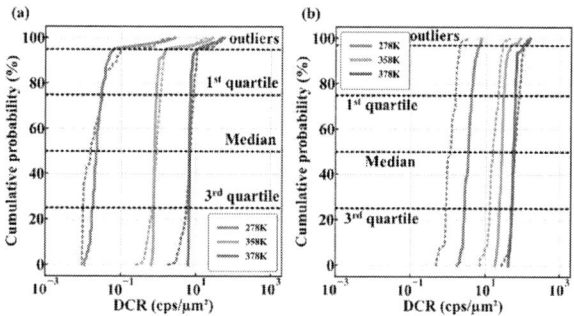

Figure 8. Cumulative probability as a function of DCR simulated for SPAD 4 at (a) $V_{ex}=2V$ and (b) $V_{ex}=5V$. Simulation are in dashed line and experiment in bold line.

V. CONCLUSION

This study reports on a threefold simulation and measurement methodology namely voltage, temperature, and device architectural constrain to strengthen assumptions on defect distribution and R-G mechanisms responsible for DCR in SPADs. An Empirical Monte-Carlo method was used to simulate the carrier trajectories from the generation centers considering stochastic transport and avalanche multiplication probability to ascertain trap positions within the device. This simulation approach unveils new potential avalanche-leading carrier paths beyond McIntyre modeling which is constrained by the device electric field distribution. The splitting of the breakdown probability into central and peripheral active avalanche areas sheds light on the side interface contribution

to the high-bias region. The local device-to-device process variation is reproduced by means of randomization of defect positions and sizes based on the EMC method.

ACKNOWLEDGMENT

The authors gratefully acknowledge VIZTA project (https://www.vizta-ecsel.eu/) for funding this study (ECSEL JU Agreement No 826600).

REFERENCES

[1] F. Piron, D. Morrison, M. R. Yuce and J. -M. Redouté, "A Review of Single-Photon Avalanche Diode Time-of-Flight Imaging Sensor Arrays," in IEEE Sensors Journal, vol. 21, no. 11, pp. 12654-12666, 1 June1, 2021.

[2] E. Sarbazi, M. Safari and H. Haas, "Statistical Modeling of Single-Photon Avalanche Diode Receivers for Optical Wireless Communications," in IEEE Transactions on Communications, vol. 66, no. 9, pp. 4043-4058, Sept. 2018.

[3] F. Acerbi et al., "Structures and Methods for Fully-Integrated Quantum Random Number Generators," in IEEE Journal of Selected Topics in Quantum Electronics, vol. 26, no. 3, pp. 1-8, May-June 2020.

[4] A. Panglosse et al., "Dark Count Rate Modeling in Single-Photon Avalanche Diodes," in IEEE Transactions on Circuits and Systems I: Regular Papers, vol. 67, no. 5, pp. 1507-1515, May 2020.

[5] Y. Xu, P. Xiang, X. Xie and Y. Huang, "A new modeling and simulation method for important statistical performance prediction of single photon avalanche diode detectors," Semiconductor Science and Technology, 31(6), 9, June 2016.

[6] M. Sicre et al., "Dark Count Rate in Single-Photon Avalanche Diodes: Characterization and Modeling study," IEEE 47th European Solid State Circuits Conference (ESSCIRC), 2021, pp. 143-146.

[7] R. J. McIntyre, "On the avalanche initiation probability of avalanche diodes above the breakdown voltage," in IEEE Transactions on Electron Devices, vol. 20, no. 7, pp. 637-641, July 1973.

[8] A. Jay et al., "Clusters of Defects as a Possible Origin of Random Telegraph Signal in Imager Devices: a DFT based Study," International Conference on Simulation of Semiconductor Processes and Devices (SISPAD), 2021, pp. 128-132

[9] C. Jacoboni and P. Lugli, "The Monte Carlo Method for Semiconductor Device Simulation," Computational Microelectronics, Wien: Springer-Verlag, 1989.

[10] R. Helleboid et al., "A new Monte Carlo method for electronic transport and avalanche simulation in single-photon avalanche diodes," unpublished.

[11] W. Shockley and W. Read, "Statistics of the Recombinations of Holes and Electrons," Physical Review, vol. 87, no. 5, pp. 835-842, 1952.

[12] D. Garetto, et al., "Analysis of defect capture cross sections using non-radiative multiphonon-assisted trapping model," Solid-state Electronics 71, 2011.

[13] Michael Hofbauer, Bernhard Steindl and Horst Zimmermann, "Temperature Dependence of Dark Count Rate and After Pulsing of a Single-Photon Avalanche Diode with an Integrated Active Quenching Circuit in 0.35 μm CMOS", Journal of Sensors, 2018.

[14] N. Moussy and J.L. Ouvrier-Buffet, "SPAD-TYPE PHOTODIODE," U.S. Patent 9,780,247, B2, issued October 3, 2017.

[15] Sentaurus Process User Guide, Version P-2019.03, March 2019.

[16] Sentaurus Device User Guide, Version P-2019.03, March 2019.

[17] R. Helleboid et al., "Comprehensive modeling and characterization of Photon Detection Efficiency and Jitter in advanced SPAD devices," IEEE 51st European Solid-State Device Research Conference (ESSDERC), 2021, pp. 271-274.

[18] A. G. Chynoweth, "Ionization Rates for Electrons and Holes in Silicon," Physical Review, vol. 109, no. 5, pp. 1537–1540, 1958.

[19] C. Canali et al., "Electron and Hole Drift Velocity Measurements in Silicon and Their Empirical Relation to Electric Field and Temperature," IEEE Transactions on Electron Devices, vol. ED-22, no. 11, pp. 1045–1047, 1975.

[20] J. G. Fossum, "Computer-Aided Numerical Analysis of Silicon Solar Cells," Solid-State Electronics, vol. 19, no. 4, pp. 269–277, 1976.

A scalable 64×64 pixels monolithic HV-CMOS sensor for hadron therapy with 1ns time stamping capability and in-pixel ADC

Nicola Massari[1], Alessio D'Andragora[1], Matteo Perenzoni[2], Andrey Selijak[3]

[1]Fondazione Bruno Kessler (FBK)
[2]Sony Europe semiconductor, Via Sommarive 18, Povo (TN) 38123, Italy
[3]Jozef Stefan Institute, Jamova cesta 39, 1000 Ljubljana, Slovenia

Carlos Chavez Barajas[4], Alan Taylor[4], Jon Taylor[4], Gianluigi Casse[4], John Pettingell[5], Ignacio Di Biase[5]

[4]University of Liverpool, Physics Dep., Oxford St, Liverpool L69 7ZE, United Kingdom
[5]Rutherford Cancer Centres, Celtic Springs, Spooner Cl, Newport NP10 8FZ, United Kingdom

Abstract— **In this paper we present an array of 64x64 pixels, designed for proton beam source characterization. The configurable pixel allows to measure the energy or the arrival time of detected particles in a short time window, or it can count events for a pre-defined observation time. The adopted solution uses a signal gating followed by a sample and hold and in-pixel single slope ADC to avoid pile-up effect when the flux of protons is high. In order to guarantee a fine time resolution (1ns) across the entire array, a scalable architecture has been conceived for possible extending the resolution of the sensor with no major modifications. The chip has been electrically characterized and tested with optical stimulus and using a beta source.**

Keywords—HV-CMOS, particle detector, hadron therapy

I. INTRODUCTION

Hadron therapy is a promising technique for treating tumour cells. The great advantage lies on the property of accelerated particles, such as carbon ions or protons, to release almost all their energy at a specific depth inside the body (Bragg peak) depending on their energy. Nevertheless, due to the interaction with the crossing materials, a portion of these particles deviates from the main focus of the beam hitting healthy organs potentially damaging them. In recent years, different techniques have been proposed to monitor the process during the treatment to precisely control the volume of interaction of the focused beam within the body [1-3]. These techniques range from implantable detectors [1], to external scintillator-based particle detectors [2] or ionoacoustic detectors [3], based on secondary products originating from the interactions of particles with the human tissue. Main target of the monitoring is to find the exact point of the Bragg peak during the treatment. All these solutions are difficult to be applied in a real scenario because of the complexity of the measurement or because they are too invasive. In the present work we propose a concept of a new sensor based on a monolithic High Voltage (HV) CMOS detector and conceived for a system for the proton beam characterization before the treatment. The idea of the system is shown in Figure 1. It consists of a water equivalent human phantom made by a tank of water. The box has an aperture with

a transparent window for injecting the particle beam into the phantom. The detection system consists of two sensing layers fixed to a mechanical moving arm that places the detectors at different distances with respect to the source. At every distance the system evaluates the arrival time of particles and their energy, to precisely measure the Bragg peak point. Time information allows the system to reconstruct the trajectory of particles by correlating the time stamp and the position of hits in the two sensing layers.

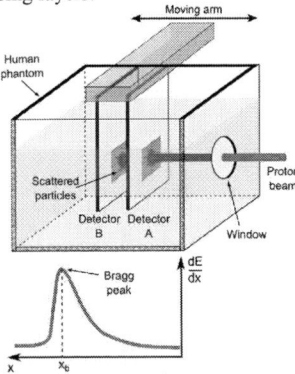

Figure 1: Description of the system for proton source characterization: a water equivalent human phantom is used in order to estimate the position at which the Bragg peak is observed depending on the distance of the sensing layers to the source and on the tracking of particles crossing the detectors.

II. THE CHIP ARCHITECTURE

When an ionizing particle interacts with silicon, it generates a certain amount of charge depending on its energy. Figure 2 shows the principle of the monolithic HV-CMOS approach to efficiently collect these charges [4]. The sensing diode, a deep-nwell/psub, is inversely biased by applying to the substrate a negative voltage (-HV). In the depletion region, further extended when using a high resistivity substrate, the electric field E allows to fast collect the generated charges by drift. As observed in the cross section, CMOS circuits are embedded

978-1-6654-8495-4/22 $31.00 © 2022 IEEE

within the same deep-nwell used for sensing. In particular, the psub layer electrically separates nwells from the deep-nwell. This layer originates also a parasitic Cpsub, greater than Cdnsub, that mainly affects the noise of the readout circuit of the pixel. Moreover, the combination of layer allows to define two distinct grounds (vss_ana and vss_dig) physically separating analogue to digital circuits.

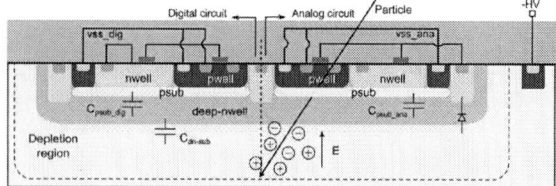

Figure 2: Cross section of the pixel and working principle for charge collection. CMOS circuit is embedded into the deep-nwell

Particle detectors typically use a continuous time Charge Amplifier (CA) followed by an event discriminator to detect particles. The most used technique for measuring the energy released by a crossing particle is the Time-over-Threshold (ToT) approach [4].

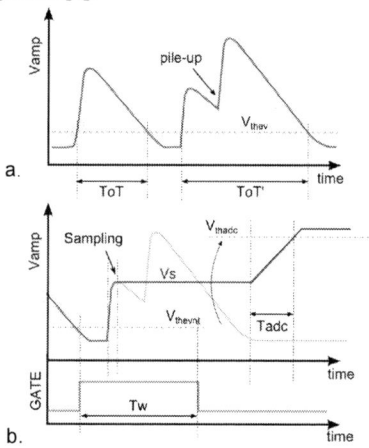

Figure 3: a) Typical ToT approach and possible pile-up effect and consequent error in the energy measurement (ToT'). b) The proposed solution, thanks to the sampling of the signal peak and the gating technique, avoids the pile-up to affect the measurement.

Figure 3 shows the principle. The ToT conversion measures the time needed for discharging the feedback capacitance of the CA after an event has been detected. This compact solution achieves 5-6b resolution at most [5-6]. A longer discharging time increases the energy resolution, assuming same time reference, but, on the other hand, it reduces the maximum event rate. Therefore, in case of high particle rates, the ToT approach can suffer of the pile-up effect (see Figure 3a) with consequent distortion of the energy spectrum (see the increase of ToT' value in Figure 3a). Due to the high rate of protons measured at the centre of the beam (in the order of 10^7-10^8 events/s/pixel), an alternative solution has to be used. The proposed approach combines two techniques for avoiding signal distortion: a.

signal gating; b. data sampling and conversion. Signal gating is used to synchronize the two layers allowing the time stamp correlation in a specific observation window Tw. All particles detected before and after Tw will ignored by the sensor. The process for data sampling is instead shown in Figure 3b and is used for avoiding signal pile-up. The idea is to sample and hold the peak of the integrated charge and perform the ADC conversion as soon as the fast gating window is closed.

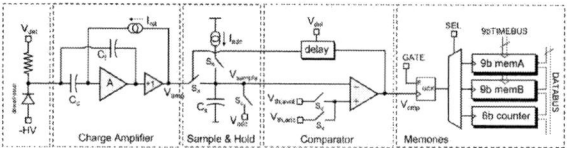

Figure 4: Schematic of the main pixel

The Figure 4 shows the proposed pixel schematic. The first stage of the pixel is a CA, ac-coupled with the detector (a deep-nwell/psub diode) in order to be insensitive to possible leakage current [4]. A comparator, with 5b programmable threshold for offset compensation, discriminates the presence of a hitting particle. A global GATE signal defines a temporal window Tw where the output of the comparator is enabled. In order to avoid false triggering due to possible previous detections, the circuit processes only the first rising edge of the comparator after GATE is asserted. When a true event is detected within Tw, the sampling of node Vamp in Cs is activated. A voltage controlled analogue delay is introduced in the loop to sample the signal close to its peak. During the A/D conversion, the sampled value Vsample is charged with a fixed current and compared by the same comparator used for event discrimination but with a different threshold Vthadc (see Figure 3b). The width of Vcmp (T_{ADC}), measured with a digital counter, will be inversely proportional to the value of Vs. The maximum conversion time is equal to 512 ns, but can be flexibly reduced in case of lower energy resolution.

Both A/D conversion and time stamping operations need the time information. The solution adopted in the present architecture for providing time information for all pixels in a scalable solution is shown in Figure 6.

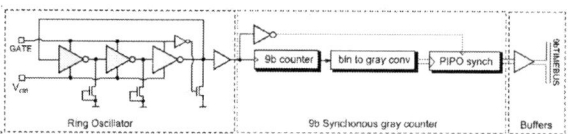

Figure 5: Schematic of the master pixel

The detector is composed by an array of 4x4 identical clusters each consisting of 16x16 pixels for a total number of 64x64 cells. Among all elements of a cluster there is a specific pixel, called Master Pixel that, blind to particle detection, is responsible for synchronizing the measurement and for providing the time reference (TIMEBUS) for all pixels belonging to the cluster. As shown in Figure 5, the master pixel consists of a ring oscillator connected to a synchronous 9b gray counter and followed by a buffer network for propagating the

978-1-6654-8495-4/22 $31.00 © 2022 IEEE

TIMEBUS to all pixels. The ring oscillator is a 3-stages single-ended current controlled ring working at the nominal frequency of 1GHz and having 4-bit SRAM for ring speed calibration. The speed of the ring is adjusted through an external voltage control in an open loop configuration. The ring works in a start/stop mode, being enabled only during Tw (~ 512 ns). Specifically, it is activated only during the time of arrival measurement and during the A/D conversion. This allows to reduce power consumption and possible injection of noise during particle detection. The counter, to ensure the synchronization with the main clock, has a sampler (register of 9 flip-flops) sensitive to the falling edge of the clock. Similar to the energy measurement, also the timing measurement needs to work in gating mode. This essential feature allows in fact to guarantee the correlation of the time information between the two sensing layers.

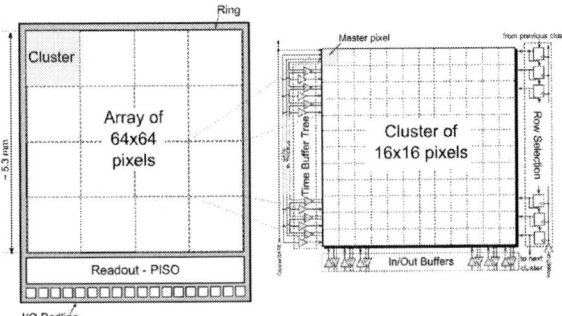

Figure 6: Architecture of the sensor. The array of 64x64 pixels is divided in 4x4 clusters each one containing 16x16 pixels. Among these, the master pixel, blind to the radiation, is responsible for signal synchronization (definition of the time gate window) and for delivering the time information (TIMEBUS) to all pixels of the cluster.

Figure 7: Microphotograph of the chip assembled in the PCB

An additional 6b counter is also implemented mainly for detector alignment and testing purpose. The array can work in three different modes: energy, timing and counting mode. As observed in Figure 4, every main pixel has two memories, allowing maximum two time-stamps or two energy values detected in two disjoined Tw or, eventually, one time-stamp and one energy value related to a single detection.

III. MEASUREMENTS

The chip, fabricated in a standard CMOS 0.15μm technology, has been assembled in a custom PCB (see Figure 7) and fully characterized and tested with a radioactive source. The pixel shows a sensitivity of about 3.8 μV/e- and a noise level of about 1.5 ke- mainly due to the value of capacitance C_{psub} not minimized in this design. The detector was first calibrated

compensating the comparator offset and the baseline variation of the CA (Figure 8) and possible speed variations of 16 ring oscillators (Figure 9). As observed, both the spread of the offset (from σ = 16.5 mV to 1.5 mV) and the ring speed (from 0.6% to 0.15% of speed variation) were improved after calibration.

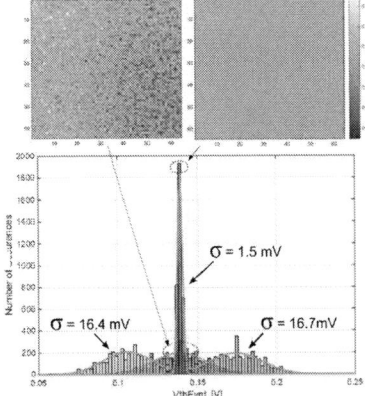

Figure 8: Compensation of the pixel offset. The histogram (bottom plot) shows the distribution of pixel offset before the and after the calibration. The two additional plots, at the right and left of the central curves, are the offset distribution measured at respectively maximum (+15) and minimum (-15) value of the calibration code. Two sampled images demonstrate the effect of the procedure

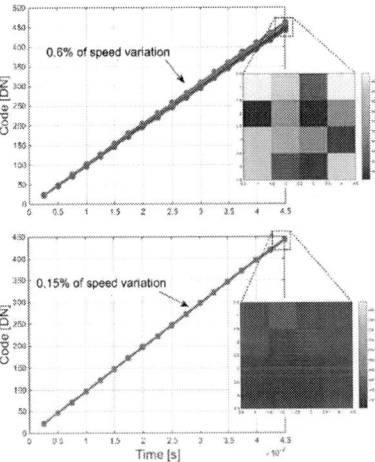

Figure 9: Calibration of the 16 ring oscillators and plotted characteristic with final values sampled at the end of the characteristic.

Figure 10, 11 and 12 show experimental results obtained by the sensor when stimulated by a β source. In this case the source, a ^{90}Sr having a nominal activity of 37MBq, was placed at about 8-9 mm from the detector. The first image (see Figure 10) shows the increase of the detection rate as a function of the reverse bias of the sensing diode. The increase of the electric field and the wider extension of the depletion region improve the efficiency of charge collection. Figure 11 represents a heat map obtained by placing, between the source and the detector, a metal plate with a small hole in the middle, shielding the majority of the array. Figure 12a shows the energy spectrum obtained by exposing the detector to the β without any shield.

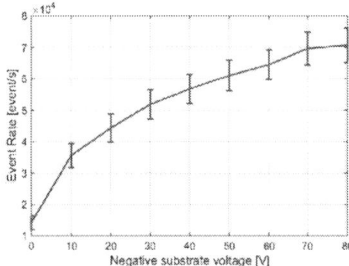

Figure 10: Values of particle rates as a function of the bias applied to the substrate.

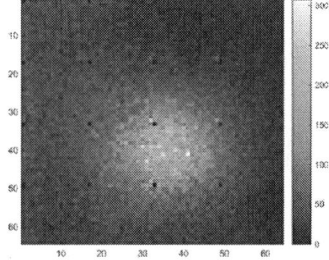

Figure 11: heat map obtained using a β source and a metal shield with a hole. The bias voltage of the substrate was VHV = -70V and the total integration time was of 2.3 min.

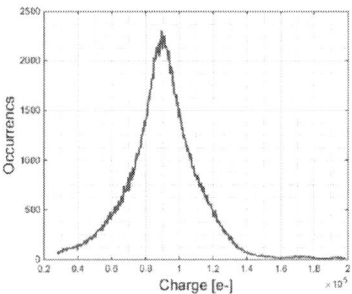

Figure 12: Energy measurement of β source.

To demonstrate the time stamping capability of the array, an external laser was synchronized with the GATE signal in order to deliver a single impulse during Tw. The delay between the synchronization and the effective pulse was changed and the time stamp was plotted accordingly (see Figure 13). Table I summarizes main parameters and achieved results in comparison with other sensor architectures. The overall power measured consumption is about 170mW, corresponding to about 41.5 µW/pix which slightly higher than previous works but still lower than system requirements (<1W/cm²).

II. CONCLUSIONS

The design of a HV-CMOS detector for hadron therapy is presented in this paper. The chip architecture is designed to be scalable for providing a time stamp with a resolution of 1ns. The detector can work in three different modes: energy, time and particle heat mode. A preliminary characterization of the chip shows that the calibration is working while the response to the β source is demonstrated.

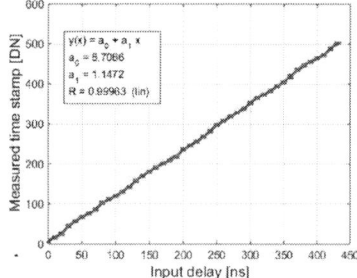

Figure 13: Time measurement using an external laser triggered by the time gate signal. A resolution of about 1ns is observed. Every point is the result of an average of 1000 measurements (error bars are also included).

Table I: chip performance summary and comparison

	Present work	Velopix [5]	TimePix3 [6]	Monopix [4]
Technology	CMOS 150nm	CMOS 130nm	CMOS 130nm	CMOS 150nm
Detector	Monolithic	Hybrid	Hybrid	Monolithic
Detector resol	64x64	256x256	256x256	129x36
Functionality*	ToA, En, PC	ToA, En	ToA, En,PC	ToA,PC
Pixel pitch µm	83	55	55	50x250
Energy resol	8b	6b	13b (iToT)	-
Counter	6b	-	10b	1b
Time resol	9b - 1 ns	25 ns	1.56 ns	8b – 25ns
Power cons	170mW** @1.8V	<1.5W @1.2V		36µW/pix

* ToA= Time of Arrival; En = energy; PC = Particles Counting   ** Measured at 1.5 kfps

REFERENCES

[1] K.Lee et al., "A 64×64 Implantable Real-Time Single-Charged-Particle Radiation Detector for Cancer Therapy", pp., ISSCC 2020.

[2] M.Riva et al., "Acoustic Analog Front End for Proton Range Detection in Hadron Therapy", IEEE Trans Biomed Circuits Syst, Vol.12, pp. 954-962, Aug 2018.

[3] E. Manuzzato et al., "A 16 × 8 Digital-SiPM Array With Distributed Trigger Generator for Low SNR Particle Tracking,", pp. 75-78, ESSCIRC 2019.

[4] I.Caicedo et al., "The Monopix chips: Depleted monolithic active pixel sensors with a column-drain read-out architecture for the ATLAS Inner Tracker upgrade", JINST 2018.

[5] T. Poikela et al., "The VeloPix ASIC", JINST 2017.

[6] T.Poikela et al., "Timepix3: a 65K channel hybrid pixel readout chip with simultaneous ToA/ToT and sparse readout", JINST 2014.

A 32-ch Neuromodulator with redundant Voltage Monitors avoiding Blocking Capacitors

Stefan Reich[1], Markus Sporer[1], Joachim Becker[1], Stefan B. Rieger[2], Martin Schüttler[2] and Maurits Ortmanns[1]

[1]*University of Ulm, Ulm, Germany;* [2]*CorTec GmbH, Freiburg, Germany*

Email: stefan.reich@uni-ulm.de

Abstract—Neural recording and modulation has evolved rapidly in recent years. Closed-loop neuromodulation systems have been successfully demonstrated for the treatment of Parkinson's disease and epilepsy. Chronically implanted medical devices, requiring compliance to rigorous safety regulations, employ numerous safety measures to protect the patient. One such measure to prevent direct current from being applied to the tissue in case of a system failure is typically the usage of external blocking capacitors between the electrodes and the neuromodulator. These capacitors can cause significant magnetic resonance imaging (MRI) magnetic susceptibility artifacts that appear as a shading effect. This paper presents some of the challenges which arise when evolving neuromodulation hardware from benchtop prototypes to biomedical systems for chronic implantation in humans. We propose a novel safety measure to mitigate the MRI shading issue while still complying to the relevant safety regulations. Furthermore, we propose several additional features that improve the flexibility and usability of a 32-channel neuromodulation platform. The neuromodulator was fabricated in a $180\,\text{nm}$ HV CMOS process and realizes a fully digital-to-neural interface.

Index Terms—Neuromodulation, Blocking capacitors, MRI artifacts, Electrode Protection

I. INTRODUCTION

Closed-loop neuromodulation, i.e., neural stimulation adapted with parameters extracted from neurally recorded biomarkers, has shown great promise in the development of novel therapeutic treatment for various neurological ailments, such as Parkinson's disease and epilepsy.

State of the art (SoA) neuro-recording hardware has evolved rapidly, exhibiting a large variety of architectures ranging from the still popular AC-coupled front-ends [1], [2], over chopper-stabilized recorders [3], [4] to direct-digitization front-ends [5], [6], which incorporate the noise-critical first amplifier or Gm-stage into $\Delta\Sigma$- or VCO-ADC loops. The latter allows for very power- and area-efficient designs, which partially take advantage of technology scaling. While fine-line CMOS is beneficial for the design of the neural recorder, it increases the system cost (neuromodulators are a low-volume market) and also has a limiting impact on the capabilities of the neural stimulator, as high voltage (HV) compliance is not generally available in smaller CMOS nodes. Stacking multiple LV-devices to achieve HV-compliance has been demonstrated [7], but the drawbacks of limited stimulator current, comparably large area consumption and reliability risks, especially at start-up, can outweigh the benefits.

This work was funded by BMBF grant MR:Implant, and by DFG grant EI 1078/1-1.

Fig. 1. Neuromodulation platform from [2] with encapsulation. Blocking capacitor footprints are visible towards the edge of the PCB in the middle, highlighting area consumption. Courtesy: CorTec GmbH.

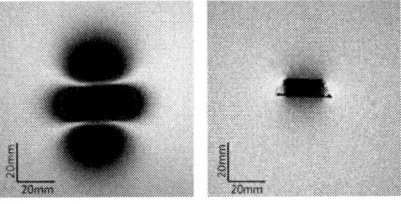

Fig. 2. MRI shading of the ceramically excapsulated neuromodulator (cf. Fig. 1) due to ferro-magnetic constituents in blocking capacitors (left), and reference image without blocking capacitors (right). Courtesy CorTec GmbH.

SoA bidirectional neural interfaces are rarely optimized for flexibility but rather target specific applications: In [8], a HV-compliant stimulator is combined with a large-bandwidth (BW) recorder, but noise performance and maximum stimulation current $I_{\text{stim,max}}$ is limited. The design in [9] combines HV-stimulation with arbitrary stimulation waveforms, but the recorder does not allow for spike (action potentials - APs) detection. Finally, the implementation in [10] allows for local field potentials (LFPs) and APs recording and $I_{\text{stim,max}} = 4\,\text{mA}$, but the HV-compliance is limited to $5\,\text{V}$, and only square wave stimulation is supported. In contrast, [2] presented a neuromodulation platform featuring serial digital control, HV-compliance of up to $18\,\text{V}$ with stimulation currents of up to $I_{\text{stim,max}} = 10\,\text{mA}$, arbitrary waveform current- and voltage-controlled stimulation (CCS, CVS), as well as a full-band recorder with various gain and BW settings, state of the art noise performance and on-chip digitization.

When transferring such academically developed integrated circuits (IC) for neuromodulation into implantable medical devices, cf. Fig. 1, additional factors need to be considered other than just the standalone IC performance. These include system size, functionality (use case flexibility) and system safety ensuring conformity with relevant safety regulations.

Arguably the most important safety consideration is electrode safety. During stimulation, tissue damage can be avoided by maintaining a relationship between charge density and the charge per phase delivered by the electrode. By applying active or passive charge balanced biphasic stimulation pulses, corrosion, dissolution, and tissue damage can be avoided ensuring safe stimulation. Furthermore, electrode damage can be identified with an electrode impedance estimation and potentially harmful failure currents can be detected with monitoring circuitry.

In [11] a fault current detector was introduced to monitor unwanted stimulation currents e.g., by broken output devices; to minimize area consumption, the current monitor was only applied to the counter electrode (as common return path). A shortcoming of many prior techniques is that the safety monitoring circuit is part of the neurointerface IC and is functionally dependent upon it, therefore not guaranteeing first-fault-safety. First-fault-safety is the requirement that the system must not fail in scenarios where a single, isolated error occurs or a single component fails resulting in an unacceptable risk to the patient, cf., ISO-14708-1.

Such safety, e.g., against faulty stimulator, is typically achieved by using external blocking capacitors in the signal path between electrode and neuromodulator, preventing DC current into the tissue. This however requires a large number of discrete components, which consume large area, cf. Fig. 1, and add a shadow in MRI due to ferro-magnetic constituents, as is shown in Fig. 2. Ceramic capacitors made from non-magnetic materials exist but are infeasible due to their low capacitance per volume. This is a significant drawback in a clinical scenario, as MRI is a standard diagnostic procedure and the brain area and tissue directly in contact with the implant are of major interest.

In a prior implementation [2], a 32ch neuromodulator included safety features such as electrode impedance estimation, passive charge balancing and a digital interface with a strict protocol to detect communication errors. However, as there was no independent first-fault-detection mechanism at the electrodes, coupling capacitors were still needed on system level, cf. Fig. 1. This paper presents an improved version of that neuromodulator ASIC, where several additional safety and regulation-compliance features are incorporated alongside functional improvements. A novel and first-fault-prove voltage monitoring system is implemented, which allows to achieve regulation-compliance without blocking capacitors. Furthermore, the ASIC is optimized for flexibility and compatibility with commercial hardware for easier system integration.

The paper is organized as follows: The improvements to the neuromodulator are motivated and detailed in Section II. Measurement results are shown in Section III, whereas Section IV concludes the paper.

II. CIRCUIT LEVEL IMPLEMENTATION

The neuromodulator ASIC features 32 bidirectional front-end channels, each of which is realized as indicated in Fig. 3 and digitally controlled by a global state machine

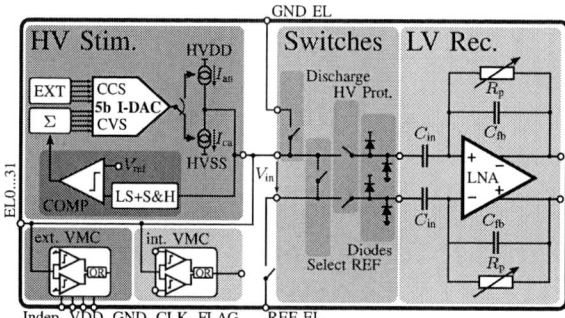

Fig. 3. Simplified schematic of a bidirectional channel, depicting the first stage of the recorder (green), the HV stimulator unit (red), the HV switches (orange) and the voltage monitoring circuits (purple). The connectivity of the GND (discharging) and REF (rec. reference) electrodes are depicted as well.

programmable through SPI communication. Two 16-bit I-$\Delta\Sigma$-ADCs digitize the recorded neural signals and provide a $10.24\,\mathrm{Mbit\,s^{-1}}$ data stream via SPI. Thus, the neuromodulator can be fully controlled by a microcontroller unit (MCU) and serves as digital-neural interface. Every front-end can be randomly accessed as recording or stimulation site.

A. Stimulation

The stimulation front-end is based on a 5-bit current-DAC with HV push-pull output stage. It can be reconfigured into CCS or CVS mode, the latter by embedding the I-DAC and HV output into a DSM based loop [2]. The new stimulation module is improved for increased dynamic range by rescaling of the current mirror gain options, now featuring $66\,\mathrm{dB}$ of dynamic range with stimulation currents from $5\,\mu\mathrm{A}$ to $10\,\mathrm{mA}$. The highly flexible waveform generation from [2], which is based on the state-machine approach of [12] with sequential execution of stimulation commands, is improved to feature programmable loops within the command table, thereby allowing repetitive execution of waveforms and commands.

B. Recording

The recording front-end has sophisticated programmability [2], allowing to record either LFPs, APs or both bands with selectable gain and BW settings. However, the ADC sampling frequency was disadvantageously constant at $20\,\mathrm{kHz}$, which significantly exceeds the required data rate in LFP-only recording. This increased the default processing load on the MCU, which needed to downsample the data in order to avoid excess power consumption in the Bluetooth link.

The presented neuromodulator features a data rate reduction module allowing to gate the digital data stream in compliance with the SPI-interface. The module can be programmed such that only every n-th sample per channel is transmitted. Additionally, individual channels can be deactivated entirely. Propagating only every n-th sample may appear simplistic, since averaging would allow to still benefit from disregarded samples noise-wise. This is however only true for white noise, and not for the correlated 1/f noise which dominates in the LFP

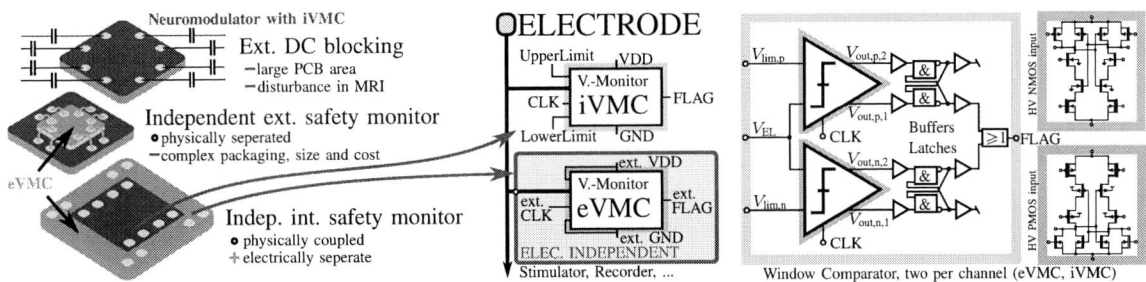

Fig. 4. Blocking-capacitor-less first-fault-safety feature: Idea of monolithic integration of a physically coupled but electrically independent voltage monitoring circuit (VMC). Electrode voltage is monitored by two independent window comparator based VMC: iVMC has adjustable limits (2-bit) and is powered and controlled by the neuromodulator ASIC. eVMC has separate power and control signals and is electrically independent from the other circuits.

domain. Since downsampling is infeasible in the white noise dominated AP-domain due to BW requirements, there is no significant benefit in averaging, which was thus omitted.

C. Front-end Circuitry

The front-end circuitry was modified from [2] such that the reference electrode (REF-EL) is no longer used as a charge sink for the passive discharge of electrodes post stimulation, as this may lead to artifacts and charging of REF-EL. Instead, a ground electrode (GND-EL) is introduced, which is used as discharge destination. This allows for a cleaner reference in the recording. An internal HV-switch matrix allows to disconnect the dedicated REF-EL, and to instead use any of the electrodes in the electrode array as a reference. This feature is frequently required in practical application, since the characteristics of implanted electrodes are unknown prior to implantation and can vary over time, thus making it disadvantageous to define a fixed reference a priori.

D. First-Fault-Safety without DC Blocking Capacitors

Two electrically independent instances of voltage monitoring circuits (VMCs) are added to each electrode. These are based on window comparators, which are implemented via two low-kickback-noise StrongARM latches for minimal clock feed-through to the electrodes and to the recording front-end (see Fig. 4, right). The StrongARM latches have HV-MOS transistors as input devices, such that they can withstand the electrode voltages during stimulation. The comparator, which monitors the upper limit $V_{\text{lim,p}}$, is built with an NMOS input device, while the lower window limitation $V_{\text{lim,n}}$ is monitored with a HV-PMOS based comparator. Both comparator implementations are designed for a simulated PVT variation of $<30\,\text{mV}$ (3σ). The two VMC instances are distinctly different: There is an internally powered iVMC and an externally powered eVMC. The iVMC is the more sophisticated instance as its window range can be digitally programmed: The available ranges are [$\pm 375\,\text{mV}$; $\pm 750\,\text{mV}$; $\pm 1.125\,\text{V}$; $\pm 1.5\,\text{V}$] around the body potential. Any electrode exiting this voltage window triggers an interrupt flag, such that the MCU can react quickly and query which electrode has triggered the alert. Disadvantageously, this monitoring system

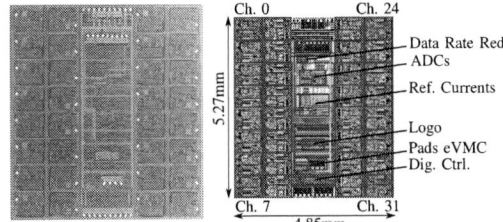

Fig. 5. Chip photograph and annotated layout overview.

is only operational while the ASIC functions as intended. As the stimulator ASIC could be faulty, stimulating DC currents and not providing iVMC feedback, coupling capacitors would still be necessary for first-fault safety (cf. Fig. 4, left/top). An alternative, would be to introduce a physically and electrically separated, second monitoring device, e.g., stacked to the first (cf. Fig. 4, left/middle), which disadvantageously would increase packaging complexity. As a solution we propose to realize an external second monitor, eVMC, which is physically combined in the same substrate, but electrically separated (cf. Fig. 4, left/bottom). It has an independent supply and ground connection, alongside an extra external trigger input and a single output flag, which is asserted when one or more of the electrodes are outside of the window region. The separate supplies allow eVMC to function as an independent circuit in the same substrate, coupled to the electrodes using the same pads as the neuromodulator. Thus, the two VMCs add two independent layers of safety against first-fault conditions, thereby achieving regulation compliance without requiring external blocking capacitors.

Both VMCs react to an event by asserting a flag and thereby triggering an interrupt in the MCU. It is the MCUs task to determine whether this event indicates a hardware error or was caused by stimulation. This also allows built-in self-test (BIST) of both VMCs after implantation.

III. MEASUREMENT RESULTS

The design was fabricated in a 180 nm HV-CMOS process. The chip photo and layout overview are given in Fig. 5. The chip consumes 6.5 mW of power and an area of 4.85 mm ×

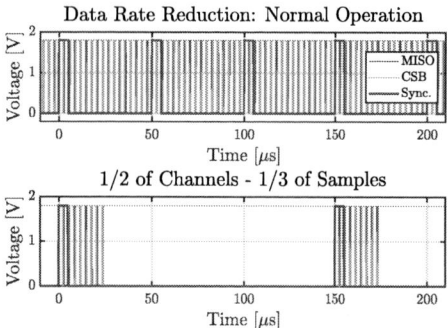

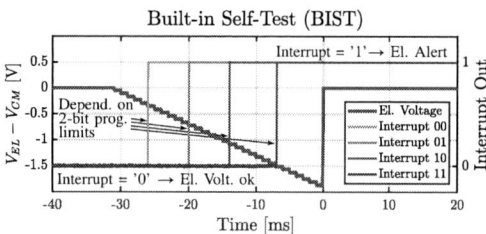

Fig. 6. Measurement results of the data rate reduction module: Unaltered vs. reduced data rate (half of the channels, every third sample).

Fig. 7. Measurement of BIST. Electrode voltage is stepped by stimulation, triggering an interrupt via iVMC once 2-bit programmed window limits are exceeded.

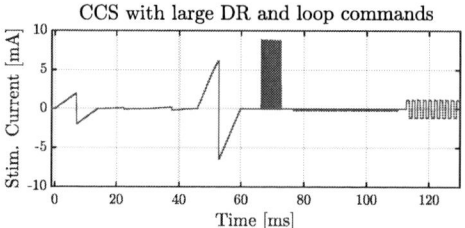

Fig. 8. Exemplary, arbitrary stimulation waveforms using loops commands.

being recorded. The presented blocking-capacitor-less first-fault safety method achieves regulation compliance without external capacitors, thereby avoiding large area consumption and shading in MRI.

5.27 mm. Measurement data of the data rate reduction module is depicted in Fig. 6: The top half shows the normal operation at full data rate, while the bottom half shows the same signals in a setting where half the channels are deactivated, and only every third sample of the remaining channels is propagated, thus reducing the data rate by 6x in this example. Fig. 7 shows the electrode voltage V_{EL}, starting at body potential V_{CM}, which is then stepped down via stimulation. Depending on the programmed window setting of the iVMC, the corresponding interrupt flag is asserted once the electrode voltage exceeds the defined range around its default potential. The same procedure can be used as BIST. In Fig. 8, the loop-function in the stimulation commands and the improved dynamic range are demonstrated by various arbitrary current stimulation (CCS) waveforms into a resistive load. (In-vitro) recording and other functionality are unaltered against the original neuromodulator platform in [2] and therefore not repeated here.

IV. SUMMARY AND CONCLUSION

We presented functional and safety enhancements to a 32-ch neuromodulation ASIC, improving flexibility and usability in a clinical environment. The digital control was modified to comply with the SPI standard in order to avoid a complex custom interface. The novel data rate reduction module allows the adaption of the data stream to various applications, reducing both the number of channels and the samples per second

REFERENCES

[1] R. Harrison and C. Charles, "A low-power low-noise CMOS amplifier for neural recording applications," *IEEE Journal of Solid-State Circuits*, vol. 38, no. 6, pp. 958–965, 2003.

[2] S. Reich, M. Sporer, M. Haas, J. Becker, M. Schüttler, and M. Ortmanns, "A High-Voltage Compliance, 32-Channel Digitally Interfaced Neuromodulation System on Chip," *IEEE Journal of Solid-State Circuits*, vol. 56, no. 8, pp. 2476–2487, 2021.

[3] T. Denison, K. Consoer, W. Santa, A.-T. Avestruz, J. Cooley, and A. Kelly, "A 2 μW 100 nV/rtHz Chopper-Stabilized Instrumentation Amplifier for Chronic Measurement of Neural Field Potentials," *IEEE Journal of Solid-State Circuits*, vol. 42, no. 12, pp. 2934–2945, 2007.

[4] S. Reich, M. Sporer, and M. Ortmanns, "A Chopped Neural Front-End Featuring Input Impedance Boosting With Suppressed Offset-Induced Charge Transfer," *IEEE Transactions on Biomedical Circuits and Systems*, vol. 15, no. 3, pp. 402–411, 2021.

[5] H. Chandrakumar and D. Marković, "A 15.2-ENOB 5-kHz BW 4.5-μ W Chopped CT $\Delta\Sigma$ -ADC for Artifact-Tolerant Neural Recording Front Ends," *IEEE Journal of Solid-State Circuits*, vol. 53, no. 12, pp. 3470–3483, 2018.

[6] C. Pochet, J. Huang, P. Mercier, and D. Hall, "A 174.7-dB FoM, 2nd-order VCO-based ExG-to-Digital Front-End Using a Multi-phase Gated-Inverted Ring Oscillator Quantizer," *IEEE Transactions on Biomedical Circuits and Systems*, pp. 1–1, 2021.

[7] J. P. Uehlin, W. A. Smith, V. R. Pamula, E. P. Pepin, S. Perlmutter, V. Sathe, and J. C. Rudell, "A Single-Chip Bidirectional Neural Interface With High-Voltage Stimulation and Adaptive Artifact Cancellation in Standard CMOS," *IEEE Journal of Solid-State Circuits*, vol. 55, no. 7, pp. 1749–1761, 2020.

[8] Y.-K. Lo, C.-W. Chang, Y.-C. Kuan, S. Culaclii, B. Kim, K. Chen, P. Gad, V. R. Edgerton, and W. Liu, "22.2 A 176-channel 0.5cm3 0.7g wireless implant for motor function recovery after spinal cord injury," in *2016 IEEE International Solid-State Circuits Conference (ISSCC)*, 2016, pp. 382–383.

[9] B. C. Johnson, S. Gambini, I. Izyumin, A. Moin, A. Zhou, G. Alexandrov, S. R. Santacruz, J. M. Rabaey, J. M. Carmena, and R. Muller, "An implantable 700μW 64-channel neuromodulation IC for simultaneous recording and stimulation with rapid artifact recovery," in *2017 Symposium on VLSI Circuits*, 2017, pp. C48–C49.

[10] X. Liu, M. Zhang, A. G. Richardson, T. H. Lucas, and J. Van der Spiegel, "Design of a Closed-Loop, Bidirectional Brain Machine Interface System With Energy Efficient Neural Feature Extraction and PID Control," *IEEE Transactions on Biomedical Circuits and Systems*, vol. 11, no. 4, pp. 729–742, 2017.

[11] M. Ortmanns, A. Rocke, M. Gehrke, and H.-J. Tiedtke, "A 232-Channel Epiretinal Stimulator ASIC," *IEEE Journal of Solid-State Circuits*, vol. 42, no. 12, pp. 2946–2959, 2007.

[12] E. Noorsal, K. Sooksood, H. Xu, R. Hornig, J. Becker, and M. Ortmanns, "A neural stimulator frontend with high-voltage compliance and programmable pulse shape for epiretinal implants," *IEEE Journal of Solid-State Circuits*, vol. 47, no. 1, pp. 244–256, 2012.

Fully Implantable 192×256 SPAD Sensor with Global-Shutter and Micro-LEDs for Bidirectional Subdural Optical Brain-Computer Interfaces

Eric H. Pollmann, Yatin Gilhotra, Heyu Yin, and Kenneth L. Shepard
Columbia University, New York, NY, USA
Email: ehp2121@columbia.edu, yg2753@columbia.edu, hy2693@columbia.edu, shepard@ee.columbia.edu

Abstract— We demonstrate a fully implantable optoelectronic neural interface device featuring an array of single-photon avalanche photodiode (SPAD) detectors with a global shutter and monolithically integrated with an array of flip-chip bonded micro-LEDs (μLED) for fluorescence excitation and optogenetic stimulation. The device is integrated with optical filters and a lens-less computation mask to create a 200-μm-thick implantable device. To enable the global shutter, an area-efficient 10b roll-over counter is used in-pixel. With a phase unwrapping algorithm, these counters can be used in a "modulo" fashion providing high dynamic range extension. Our SPAD sensor architecture achieves a better noise·power figure-of-merit (FoM) than comparable photodiode image sensors.

Keywords—optoelectronic implants, single photon avalanche diodes, optogenetics, heterogeneous co-integration, brain computer interfaces.

I. INTRODUCTION

Neural interfaces play a significant role in exploring and understanding the brain. Despite the primacy of electrophysiological neural implants, optical neural interfaces have drawn recent interest because of the ability to introduce cell-type specificity into recording and actuation. Current optical interfaces rely primarily on microscopes, which despite work on miniaturization, are volumetrically inefficient [1], physically occupying more space than the volume they image, making them impractical as implanted devices. The key to reaching the limits of miniaturization, for both electrophysiological and optoelectronic implants, is heterogenous integration onto a CMOS integrated chip. In this work, we extend upon our previous implantable optoelectronic platform [2] by significantly improving upon the sensitivity, dynamic range, and power consumption of the SPAD image sensor.

II. SYSTEM DESIGN

This subdural implant design is shown in Fig. 1. The device consists of an array of 256×192 monolithic SPADs with a global shutter for low-light-intensity imaging and dual color (blue and red) flip-chip bonded μLEDs as light sources for fluorescence excitation and optogenetic stimulation, respectively. Implementing global-shutter integration, more than 41.6dB improvement in sensitivity can be achieved compared to rolling-shutter SPAD imagers [2]. Optical co-integration of the fluorescence sources, filters, and computational mask and thinning of the CMOS chip brings the total device thickness to

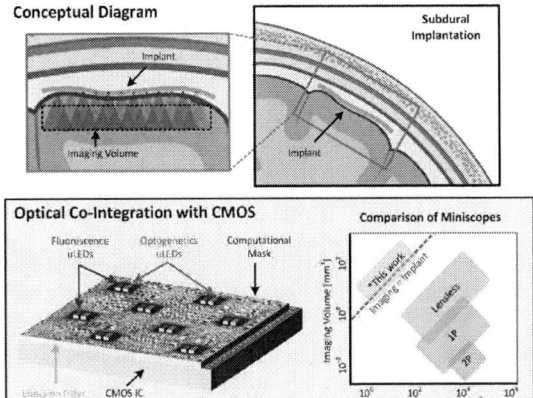

Fig. 1: Conceptual illustration of the fully-integrated neural implant.

less than 200 μm, making the implant volume of the imager less than the imaging volume of the tissue.

The imager controller is shown in Fig. 2. The chip's frame rate is 200fps, when reading the entire array, or 400fps, when reading half the array (128x192); the latter allows for the imaging of fast state-of-the-art voltage indicators [3]. The 100MHz RefClk supports the data transmission to meet the desired frame rate. The digital core operates on a divide by 8 DigClk to save power. Each pixel consists of two passively quenched SPADs whose outputs are combined into an in-pixel 10b counter. The data is read off the imager in a rolling column readout structure where the pixel data is buffered with up to 16 repeater cells throughout the array and loaded into a data serializer for streaming off chip. The *Col* signal allows for the data to reach the end of the array before the *Load* signal latches the data into the data serializer. Once the data is loaded, it is serialized off the chip at ~100Mbps.

The details of a single pixel are also shown in Fig. 2. The active diameter of each SPAD is 6μm; n-well sharing between the two adjacent chamfered square SPADs, as shown in the SPAD cross section, results in a fill-factor (FF) of 11.5%. The quenching circuit is passive-quench and passive-reset (PQRC) with dead-time determined by the VQ bias on the quench transistor. An off-pixel 3b digital to analog converter (DAC) provides the global VQ bias from 440mV to 1.06V in steps of 88mV. It uses the conventional resistor-divider topology for high linearity and consumes 800uW of power. To bring the

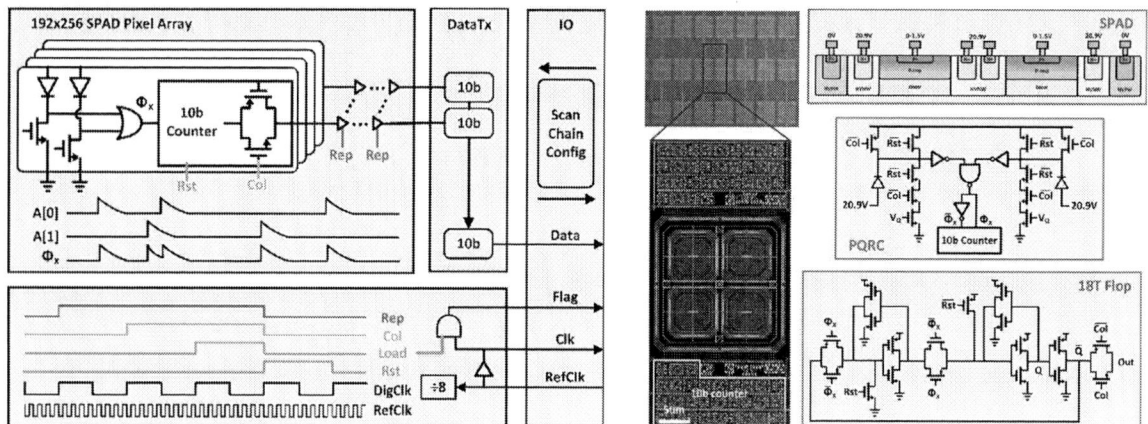

Fig. 2: Chip Block Diagram of SPAD pixel array, reset and readout waveforms, and data transmission. Schematic of SPAD, quenching circuit, and area-optimized flip-flop for 10b counter.

counter in-pixel, a divide-by-2 element was designed that uses an area-efficient custom 18T flip-flop circuit. The flip-flop is a positive-edge triggered, cross-coupled latch-based primary and replica design. The timing parameters of flip-flop support >2GHz photon arrival rate while consuming <3mW of power for the entire array at highest activity rate.

The μLED driver is designed to be compatible with various light emitting sources. The ANODE contact is global and be set above the turn-on voltage for a given light source. For example, blue OLEDs [4] and VCSELs [5] have relatively high turn ON voltages ~5-6V yet provide additional benefits such as high pixel density or narrow linewidth, respectively. The CATHODE contact has pixel-level control and can be modulated between ground and VDD.

III. SYSTEM VERIFICATION

The CMOS die shown in Fig. 3 is fabricated in a 0.13μm HV process. The SPADs are arrayed in blocks of 16×16 pixels with a pitch of 25μm. 12.5% of the array is removed for μLED bond pads and drivers. Commercially available, LA UB08FP2 (blue) and LA UR08FP2 (red) μLEDs, for fluorescence excitation and optogenetic stimulation, respectively, are flip chip bonded to the arrayed bond pads and are under filled with black epoxy (Epotek 320LV) to prevent light leakage. The spectrum of the blue μLED, conditioned by an interference excitation filter deposited

directly on the μLED surface, overlaps with commonly available fluorescence indicators.

The co-integration of the necessary optical components for fluorescence imaging as aware as the optical characterization are shown in Fig. 4. There are 24 blue and 24 red μLEDs arrayed across the imager and controlled via the scan chain. The measured IV-curves and optoelectronic efficiencies show the ideal operating point for each of the μLEDs. In addition to the excitation filter directly deposited onto the μLEDs, we co-

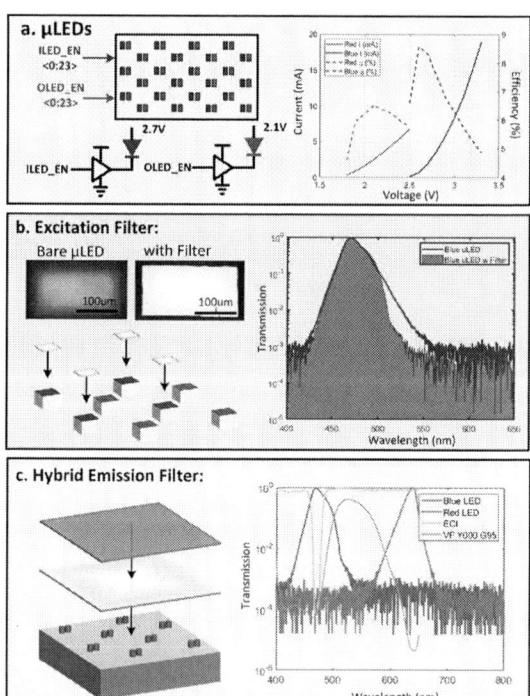

Fig. 4: a) Optical Characterization of μLEDs, b) Excitation Filter, and c) Hybrid Emission Filter.

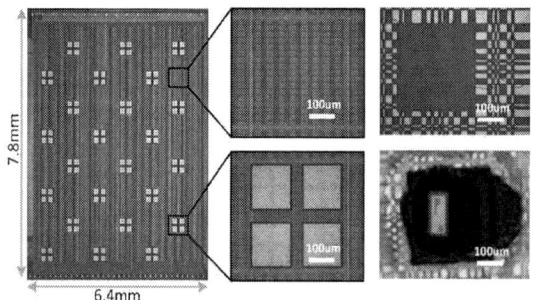

Fig 3. Die photo, SPAD array, and μLED pads fully packaged.

978-1-6654-8495-4/22 $31.00 © 2022 IEEE

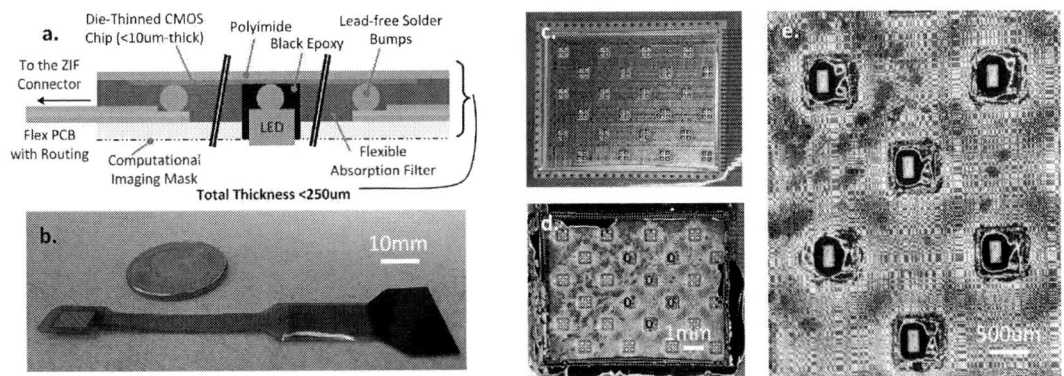

Fig. 5: a) Cross-section of packaging stack up with <250µm thickness, b) Chip packaged with flexible PCB, c-d) Chip before and after epoxy underfill of the absorption filter, e) magnified image showing µLEDs and computational mask.

optimized a hybrid absorption/interference filter, which is deposited and packaged directly on the surface of the imager to reject the 470-nm blue excitation light in favor of 520-nm green emission with an optical density (OD) exceeding six.

The interference filter provides OD3 rejection over narrow spectrum and low angles of incidence (AOI), and the custom absorption filter provides additional OD3 rejection over a broader spectrum and all AOIs. The custom absorption filter is made by mixing Valifast Yellow 3150, Valifast Green 1501,

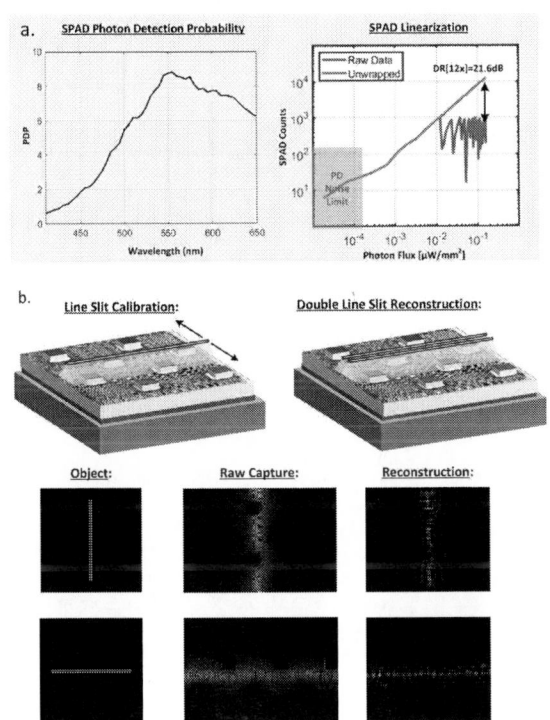

Fig. 6: a) SPAD photon detection probability and linearization with dynamic range extension, b) Computational mask calibration and double line slit reconstruction.

Cyclopentanone, and KMPR photoresist. The SPAD peak photon detection probability (PDP) is 9% at 550nm and the dark count rate is 17Hz at an excess bias voltage of 1.5V. Fig. 5 shows the final flexible packaging of the chip. The CMOS substrate is die-thinned as to become mechanically flexible and reduce its volumetric footprint for insertion underneath the dura.

The dynamic range of the image sensor determines the maximum and minimum signal intensities that can be recorded. Fluorescence signal is dependent on the absorption cross section and quantum efficiency of the fluorophore, the size and expression in the neuron, and the incident light flux. Any excitation light leakage or background fluorescence reduces the dynamic range. To enable the high-dynamic range required in fluorescence imaging, the chip operates as a modulo sensor [6], with counter wrapping at high intensity levels as shown in Fig. 6. The dynamic range of the sensor can be extended by a factor of 12 by this technique for an improvement from 60.2dB to 81.8dB. Furthermore, we leverage computational imaging techniques to obtain large field-of-view (FoV), high-resolution images without the use of focusing lenses. We employ the Texas Two-Step model for calibration and reconstruction of fluorescent point scenes [2, 7]. The double line slit reconstruction demonstrates a 50µm lateral resolution.

Finally, the inherent low-noise performance of SPADs, due to low dark-count rate (17Hz) and no read noise, enable low-power sensing which is important for implantable applications to limit tissue heating. Since the SPAD is an activity-based high-gain sensor, power consumption scales with light intensity, in contrast to photodiode imagers which rely on fixed amplification overhead. Our SPAD image sensor consumes ~18mW from VDD and ~160uW from the SPAD voltage in the dark. As the light intensity increases, the activity factor on VDD increases in the in-pixel counters and data transmission blocks. For medium light intensities, our chip consumes ~20mW from VDD and ~500uW from the SPAD voltage. Then, as the light levels increase and the counters start to rollover, VDD saturates around ~21mW and only the SPAD power consumption increases with light. Additionally, we measured the power draw from the cable drivers to be around 16mW. Future

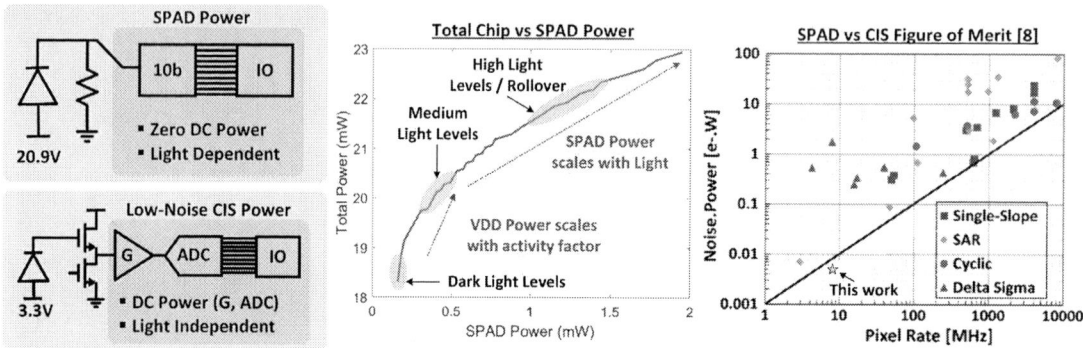

Fig. 7: Power consumption of the SPAD image sensor is light dependent while the low-noise CIS sensor has fixed overhead. VDD power scales with activity factor and saturates at medium light levels whereas SPAD power continues scaling with light. Comparison graph of CIS noise·power figure-of-merit [8].

implementations can use a more power-efficient data transmission module to further reduce power on VDD.

We compare our SPAD imager performance against a CIS noise-power figure-of-merit for low-light scenarios, commonly found in biological environments. The SPAD sensor pixel rate is defined as the number of pixels × the frame rate which is $192 \cdot 256 \cdot 200 fps$. The SPAD noise·power FoM is $\sqrt{DCR \times t_{int}} \cdot P \approx 0.087$ and it shows comparable or better FoM performance than low-noise CMOS image sensors [8] shown in Fig. 7. Comparison with current state-of-the-art [9, 10] is shown in Table 1.

	VLSI'14 [9]	VLSI'17 [10]	TBCaS'21 [2]	This Work
CMOS Technology	180nm	180nm	130nm	130nm
Pixel Pitch	35	55	30	25
FF/PDP (%)	14.4/2.7	28/5*(QE)	5/12	11.5/9
Dark Count (Hz)	200,000*	27fA	26	17
SPAD Array Size	72x60	80x36	160x160	256x192
Chief Ray Angle	10	18	70	70
Field of View (mm2)	2.1x2.5	4.7x2.25	5.4x5.4	5.1x6.8
Resolution	1000*	220	60	50
Frame-rate (fps)	N/R	20	125	200/400
Power (mW)**	83.8	3.5*	40	18 (VDD)
Data-rate (Mb/s)	N/R	N/A	40	100
Integrated Light Source	No	No	Yes	Yes

* Estimated from figures/data, ** Illumination power not included

Table 1. Comparison table with state-of-the-art.

IV. CONCLUSION

In this work, we improved upon the previous work with greater sensitivity, better spatial and temporal resolution, and larger pixel array. By achieving global shutter integration, we see a great improvement in sensitivity, allowing for low-power imaging of dim biological scenes. Future implementations can improve upon the resolution of the imager advanced technology nodes as well as utilize wireless power and data telemetry to achieve a truly implantable form factor.

ACKNOWLEDGMENT

This work was supported in part by DARPA under Contract N66001-17-C-4012 and by the National Science Foundation under Grant 1706207.

REFERENCES

[1] K. K. Ghosh *et al.*, "Miniaturized integration of a fluorescence microscope," *Nature Methods,* vol. 8, no. 10, pp. 871-878, 2011.

[2] S. Moazeni *et al.*, "A Mechanically Flexible, Implantable Neural Interface for Computational Imaging and Optogenetic Stimulation Over 5.4×5.4mm2 FoV," *IEEE Transactions on Biomedical Circuits and Systems,* vol. 15, no. 6, pp. 1295-1305, 2021.

[3] M. Z. Lin and M. J. Schnitzer, "Genetically encoded indicators of neuronal activity," *Nature Neuroscience,* vol. 19, no. 9, pp. 1142-1153, 2016.

[4] A. Morton *et al.*, "Photostimulation for In Vitro Optogenetics with High-Power Blue Organic Light-Emitting Diodes," *Advanced Biosystems,* vol. 3, no. 3, p. 1800290, 2019.

[5] R. T. ElAfandy, J.-H. Kang, B. Li, T. K. Kim, J. S. Kwak, and J. Han, "Room-temperature operation of c-plane GaN vertical cavity surface emitting laser on conductive nanoporous distributed Bragg reflector," *Applied Physics Letters,* vol. 117, no. 1, p. 011101, 2020.

[6] H. Zhao, B. Shi, C. Fernandez-Cull, S. K. Yeung, and R. Raskar, "Unbounded High Dynamic Range Photography Using a Modulo Camera," in *2015 IEEE International Conference on Computational Photography (ICCP)*, 2015, pp. 1-10

[7] J. K. Adams *et al.*, "Single-frame 3D fluorescence microscopy with ultraminiature lensless FlatScope," *Science Advances,* vol. 3, no. 12, p. e1701548, 2017.

[8] S. Kawahito, "Column-Parallel ADCs for CMOS Image Sensors and Their FoM-Based Evaluations," *IEICE Transaction on Electronics,* vol. E101-C, no. 7, pp. 444-456, 2018.

[9] C. Lee, B. Johnson, and A. Molnar, "An on-chip 72×60 angle-sensitive single photon image sensor array for lens-less time-resolved 3-D fluorescence lifetime imaging," in *2014 Symposium on VLSI Circuits Digest of Technical Papers*, 2014, pp. 1-2

[10] E. P. Papageorgiou, B. E. Boser, and M. Anwar, "Chip-scale fluorescence imager for in vivo microscopic cancer detection," in *2017 Symposium on VLSI Circuits*, 2017, pp. C106-C107

A 1.8μW 5.5mm³ ADC-less Neural Implant SoC utilizing 13.2pJ/Sample Time-domain Bi-phasic Quasi-static Brain Communication with Direct Analog to Time Conversion

Baibhab Chatterjee, K Gaurav Kumar, Shulan Xiao, Gourab Barik, Krishna Jayant and Shreyas Sen

Purdue University, West Lafayette, Indiana 47907, USA. Email: {bchatte, gauravk, xiao208, gbarik, kjayant, shreyas}@purdue.edu

Abstract—**Untethered miniaturized wireless neural sensor nodes with data transmission and energy harvesting capabilities call for circuit and system-level innovations to enable ultra-low energy deep implants for brain-machine interfaces.** *Realizing that the energy and size constraints of a neural implant motivate highly asymmetric system design* (a small, low-power sensor and transmitter at the implant, with a relatively higher power receiver at a body-worn hub), we present Time-Domain Bi-Phasic Quasi-static Brain Communication (TD-BPQBC), offloading the burden of analog to digital conversion (ADC) and digital signal processing (DSP) to the receiver. The input analog signal is converted to time-domain pulse-width modulated (PWM) waveforms, and transmitted using the recently developed BPQBC method for reducing communication power in implants. The overall SoC consumes only 1.8μW power while sensing and communicating at 800kSps. The transmitter energy efficiency is only 1.1pJ/b, which is >30X better than the state-of-the-art, enabling a fully-electrical, energy-harvested, and connected in-brain sensor/stimulator node.

Keywords—**Wireless, neural, implant, ADC-less, BPQBC, analog to time conversion (ATC), time to digital conversion (TDC).**

I. INTRODUCTION

A. Background and Related Work

In recent years, several miniaturized wireless neural sensor nodes [1]-[9] have been demonstrated for data communication and powering in brain-machine interfaces (BMIc). Most of them use RF, Inductive, Optical (OP), Ultrasound (US) and Magneto-Electric (ME) techniques, which either suffer from significant tissue absorption (RF/Inductive methods), or transduction losses (OP, US and ME techniques). The high amount of tissue absorption in traditional radiative RF requires large transmission power for system-level operation. For example, [9] uses a transmit power of 0.5W, which exceeds ICNIRP safety guidelines [10] by ~10X. US [1] or OP [3], [8] telemetry are safer, at the cost of significant loss (up to 110 dB loss as shown in [1]) due to scattering and skull absorption, requiring a sub-cranial interrogator which needs to be surgically placed, and reduces end-to-end efficiency. Magneto-Electric (ME) [7] methods exhibit extremely low body-absorption, but has large transduction loss (0.1mT magnetic field requirement in [7] which is equivalent to ~300kV/m electric field for iso-energy-density), lowering the end-to-end efficiency. Capacitive body-coupled powering and communication techniques [11]-[12] have been shown to work well with wearable devices, but the same techniques are not proven conclusively for implants.

Bi-Phasic Quasistatic Brain Communication (BPQBC) [4],[6] has recently been shown as a promising alternative for powering and data transmission in such implants which utilizes fully-electrical quasi-static signaling to transmit signals from a brain implant to an external hub placed on the skull/skin, with low end-to-end channel loss (<60dB for 50mm channel length, avoiding any challenge due to transduction loss) and ≈52pJ/b effective energy

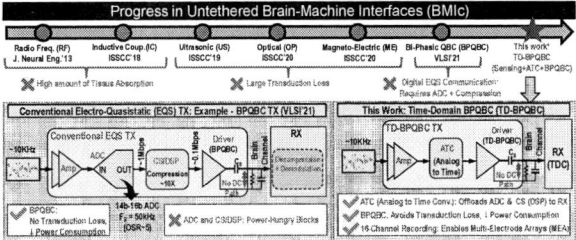

Fig. 1. Progress in Untethered BMIc, leading to Time-domain BPQBC (TD-BPQBC): data transmission from a brain implant to a body-worn hub, with analog information being encoded in pulse-width of a PWM signal.

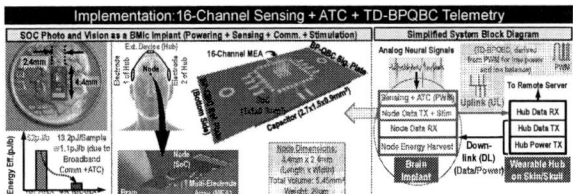

Fig. 2. Details of the node implementation and performance summary.

efficiency due to digital data compression. However, [4] assumes that the communication is narrowband and an ADC precedes the BPQBC transmitter (TX), generating the digital bits from neural signals as shown in Fig. 1 (bottom left). A compressive Sensing (CS) DSP core helps in reducing the number of bits to be transmitted, thereby reducing the communication burden. Both the ADC and the DSP are power-hungry, and thus the TX needs to be heavily duty cycled (0.1% as shown in [4], without even considering the ADC power). This duty cycle limitation results from the requirement of operating the implants with harvested energy, which is usually in the range of a few μW when powered from wearable sources, within safety limits for human-centric applications [10].

B. Our Contribution

Realizing that the energy and size constraints of a neural implant motivate highly asymmetric system design, we present, for the first time, Time-Domain BPQBC (TD-BPQBC), which offloads the burden of ADC and DSP to the body-worn receiver (RX), while transmitting the input analog signal by converting it to time-domain using an analog to time converter (ATC), as shown in Fig. 1 (bottom right). The ATC converts an analog neural signal to a pulse-width modulated (PWM) wave, which is conducive toward digital-friendly BPQBC techniques due to its 2-level signaling, leading to ultra-low-power. In this work, we implemented a 5.5 mm³ sensor and a 1.1pJ/b TX at the implant that sends the TD-

978-1-6654-8495-4/22 $31.00 © 2022 IEEE

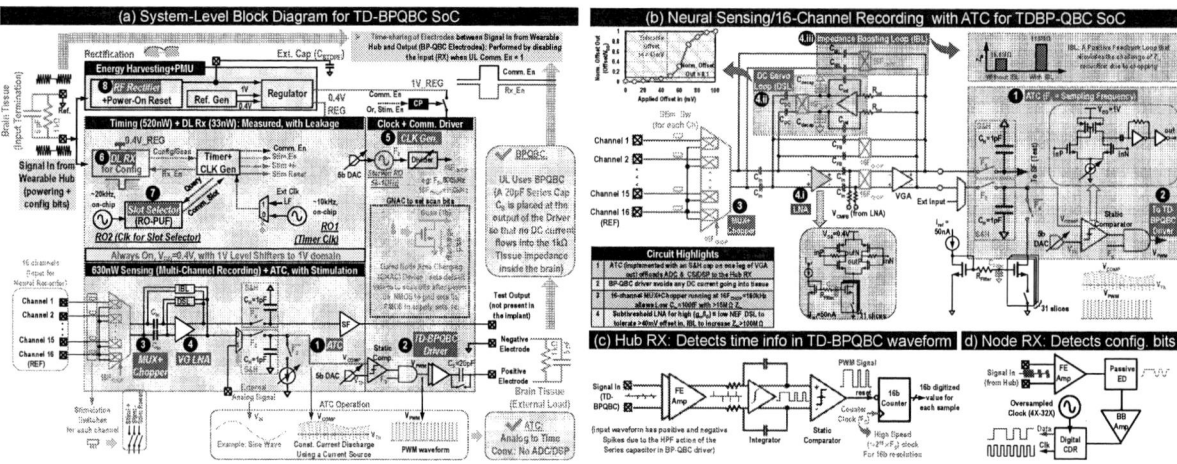

Fig. 3. (a) System Diagram of TD-BPQBC SoC, with 1) a 120nW ATC (analog to time conversion), 2) a 9.8μW TD-BPQBC driver, 3) combined 16MUX+Chopper front end, 4) a variable gain (40-60dB) LNA, along with auxiliary circuits: 5) variable frequency clock generator, 6) downlink RX, 7) Communication slot selector preventing collision, 8) an RF rectifier for energy harvesting, with power-on reset; (b) Circuit Diagram and Implementation Highlights of the 16-channel Neural Recording Front-end with ATC and BPQBC; (c) Hub UL RX and (d) BMIc node DL RX design.

BPQBC waveform to a relatively higher power receiver at a body-worn hub, forming an uplink (UL). The downlink (DL) path is from the wearable hub to the node, which is used to transfer power to the node and for configuring the implant.

Hence, the *key contributions* of this work are as follows: (1) This is the *first implementation of time-domain fully-electrical brain communication*, showing that energy efficiencies of up to 1.1pJ/b is possible (>30X improvement over literature); (2) Analysis of *offloading the burden of ADC and DSP to the UL RX* is presented, in terms of the trade-off in achievable signal to distortion ratio (SNDR) vs. sampling frequency at the ATC and the resolution of the time to digital converter (TDC) at the UL RX, showing that a high sampling frequency (~800kHz) as well as a moderate to high TDC bit-resolution (>12 bits) is required for < 2dB degradation in SNDR as compared to the input; (3) Development of a 5.5mm³ Neural SoC for simultaneous powering, multi-channel sensing, stimulation and communication. Together, these key contributions lead to the *first fully electrical, energy-harvested, and connected in-brain sensor/stimulator node, with orders of magnitude lower powering requirement* compared to other brain communication modalities that often require field transduction (conversion between electrical energy and optical/ultrasound/magnetic energies, resulting in higher end-to-end system loss).

II. OPERATING PRINCIPLES AND CIRCUIT DESIGN

As shown in Fig. 2, the implementation involves a brain implant/node with 1) 16-channel sensing, 2) ATC, 3) an UL TD-BPQBC data TX, 4) bi-phasic current stimulation (500nA-1mA), 5) a DL TD-BPQBC data RX and 6) energy harvesting. The node sends the TD-BPQBC signals through the brain tissue, and communicates with a hub (placed on the skin/ skull) containing 1) an UL data RX and 2) a DL power and data TX. The use of duty cycling with a variable communication slot selector ensures that two nodes are unlikely to communicate to the hub at the same time (similar to [4]), and hence broadband communication can be implemented. This improves the 52pJ/b energy efficiency in BPQBC to 13.2pJ/Sample in TD-BPQBC, while the ATC improves it further to 1.1pJ/b (>45x improvement vs. [4]) by detecting 12b/sample in the UL hub RX.

The 65nm TD-BPQBC SoC implementation is shown in Fig. 3(a), featuring 1) an 120nW ATC for converting the analog information to time, 2) a 9.8μW TD-BPQBC driver to reduce power consumption and maintain ion balance, 3) a combined MUX+Chopping front end for 16 input channels, 4) a variable gain (40-60dB) low-noise amplifier (LNA), along with auxiliary circuits for clocking, DL RX, communication slot selection and powering. These features are described in detail in Fig. 3(b). The 16-channel neural recording front-end, interfacing with 16 on-chip electrodes (please see die photo as shown in Fig. 4), utilizes a combined MUX+Chopping stage with 16X higher clock frequency (16F$_{CHOP}$) than that is required for a single channel (F$_{CHOP}$). The use of a higher chopping frequency through the use of combined MUX+Chopper ensures that the succeeding LNA stage can be designed with small (W/L) transistors and small AC-coupling capacitors (C$_{in}$≈100fF), which keeps the input impedance (Z$_{in}$) of the chopping front-end to acceptable values (>15MΩ) even without any impedance boosting loop (IBL). Similar to [13], a separate positive feedback IBL increases the Z$_{in}$ to > 118MΩ by providing the required charging currents during chopping. A DC servo loop (DSL) helps in tolerating up to 45mV of DC electrode offset. The LNA is designed with subthreshold transistors to 1) reduce the power consumption and 2) design at a better (g$_m$/I$_D$) point for a better noise efficiency factor (NEF). The 120nW ATC is implemented by 1) passing the output of the amplifier through a sample and hold (S&H) with a

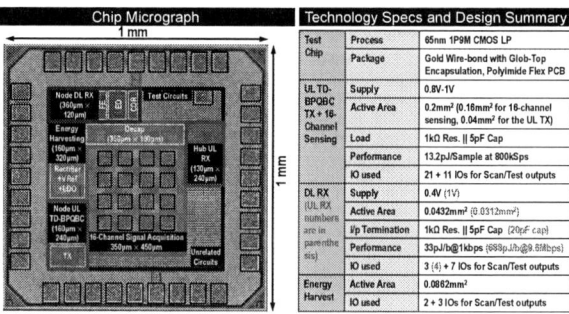

Fig. 4. Die micrograph of the SoC and Design Summary.

	Chip Micrograph	Technology Specs and Design Summary	
Test Chip	Process	65nm 1P9M CMOS LP	
	Package	Gold Wire-bond with Glob-Top Encapsulation, Polyimide Flex PCB	
UL TO-BPQBC TX + 16-Channel Sensing	Supply	0.8V-1V	
	Active Area	0.2mm² [0.16mm² for 16-channel sensing, 0.04mm² for the UL TX]	
	Load	1kΩ Res. ‖ 5pF Cap	
	Performance	13.2pJ/Sample at 800kSps	
	IO used	21 + 11 IOs for Scan/Test outputs	
DL RX (UL RX numbers are in parenthesis)	Supply	0.4V (1V)	
	Active Area	0.0432mm² [0.0312mm²]	
	I/p Termination	1kΩ Res. ‖ 5pF Cap (20pF cap)	
	Performance	33pJ/1kbps [669p.J/b@9.6Mbps]	
	IO used	3 (4) + 7 IOs for Scan/Test outputs	
Energy Harvest	Active Area	0.0862mm²	
	IO used	2 + 3 IOs for Scan/Test outputs	

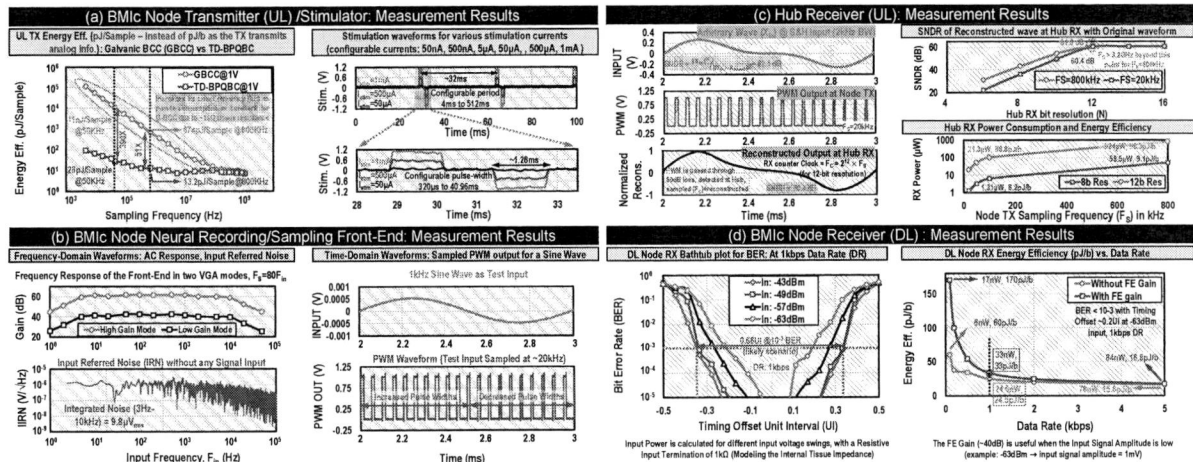

Fig. 5. (a) Measurement for TD-BPQBC UL TX and stimulator: 51X better energy efficiency than GBCC @800kSps; (b) Measurement for front-end, with 1) 40-60dB gain, 2) 9.8μVrms noise, 3) time-domain waveforms; (c) Measurement for UL Hub RX: reconstructed output with < 2dB degradation in SNDR w.r.t. original wave if Hub RX bit res. ≥ 12-bit; (d) DL Node RX performance in terms of BER and Energy Efficiency, showing 33pJ/b efficiency at 1kbps

sampling frequency of $F_S \approx 800$kHz, 2) discharging the S&H capacitor (C_H) during the hold mode through a constant current discharge, and 3) compare the linear discharging waveforms with a constant voltage using a static comparator, thereby generating a PWM waveform, which is provided to the TD-BPQBC driver. By sharing the hold mode with the discharge cycle, this method requires only one capacitor for ATC ($C_H \approx 1$pF). Similar to BPQBC [4], the TD-BPQBC driver at the UL TX blocks DC current paths into the brain tissue by use of a 20pF series capacitor (C_S), which is large enough to ensure that $\approx 80\%$ of the driver AC voltage is available at the ≈ 5pF load as presented by the brain tissue (obtained through FEM simulations in HFSS using the NEVA EM Human v2.2 model), but is still small enough (100μm × 100μm) for on-chip implementations. Fig. 3(c)-(d) show the UL hub RX and DL node RX topologies. The UL RX uses an integrator to compensate for the high-pass filtered (differentiated) TD-BPQBC waveform due to C_S, and utilizes a 16b counter with a high-speed clock ($F_C = 2^N \times F_S$) to get N-bit resolution for each sample at the UL RX. Please note that the ATC's current and capacitances will determine the full scale at the UL RX, while the comparator in the UL TX may introduce a fixed offset for the PWM signal, which is taken care of using a 5-bit digital offset compensation circuit. The DL RX detects configuration bits from the DL hub TX to program the node.

III. MEASUREMENT RESULTS

Fig. 5(a) shows the measured results for the UL TX and the stimulator (3 different current configs), while Fig. 5(b) presents the results for the recording/sampling front-end. Compared to traditional Galvanic Body Channel Communication (GBCC), TD-BPQBC shows improved energy efficiencies (51X better at 800kHz and 390X better at 50kHz) as the DC current path to the tissue is blocked by C_S. The recording front end has 40-60dB programmable gain and an integrated input referred noise of 9.8μVrms. Fig. 5(c)-(d) show the performance of the UL hub RX and DL node RX. The UL hub RX receives reduced-amplitude PWM signals (with ~50dB loss representing the body channel) with analog information residing on the edge locations. Since the implant is ADC-less, the overall performance of the ATC system is measured by a calculating the SNDR (which is a ratio of the L2 norms of the signal to noise and distortion), both at the input of the

ATC and at the reconstructed analog waveform at the RX. The moderate aliasing distortion present in the reconstructed signal in Fig. 5(c) (left-hand-side) can be removed by using a low-pass filter with sharp roll-off. For $F_S = 20$kHz, the SNDR degrades from 62.1dB at the input to 60.4dB at the reconstructed waveform at the RX, when 12-bit resolution is used at the UL RX (N=12, $F_C = 2^{12} \times F_S$). However, the SNDR at the reconstructed waveform depends on both F_S and RX bit-resolution (N). A higher F_S results in a better SNDR because of a higher oversampling ratio. Similarly, a higher N will result in a better SNDR, till the point it is saturated with the input SNDR (~62.1dB). As shown in Fig. 5(c) (right-hand-side), by using a higher $F_S = 800$kHz, reconstructed SNDR becomes 61.6dB for N = 12. Please note that with $F_S = 800$kHz, SNDR beyond N > 12 was not measured as $F_C > 3.2$GHz at this point. At $F_S = 800$kHz, the energy efficiency for the UL RX is 9.1pJ/b for N=8 and 96.3pJ/b for N=12. The degradation in energy efficiency from N=8 to N=12 is expected, as we only achieve a linearly increasing bit-resolution at the cost of an exponentially increasing F_C at the UL RX, which results in an overall degradation in pJ/b. However, in terms of the UL TX, this asymmetric TD-BPQBC link enables the 1.8μW Neural Node SoC with 5% duty cycling. Out of the 1.8μW, only 0.49μW is consumed in the duty-cycled TX. Without any duty cycling, the node will consume 11.1μW.

With 5% duty cycling, and N=12 at the UL RX, this results in an UL TX energy efficiency of 1.1pJ/b, which is a > 45x improvement over [4]. As shown in Fig. 5(d), the bit error rate (BER) for the DL RX at 1kbps data rate (DR) with 31-bit-PRBS shows a timing margin of >0.66UI for a BER of 10^{-3} at -43dBm input (10mV$_{pp}$, calculated with 1kΩ R_{in} || 5pF C_{in}, i.e. the effective termination in the brain tissue [4]). The energy consumption of the DL RX is 33pJ/b with a front-end (FE) on-chip amplifier, and 24.5pJ/b without the FE amplifier at 1kbps.

The Full-System Block Diagram for the TD-BPQBC SoC is shown in Fig. 6(a). The 5.5mm³ node is placed on the brain of a C57BL/6J Mouse, adhering to the overseeing Animal Care and Use Committee guidelines. Two RX electrodes (signal+ref) are placed on the surface of the head. The Node communicates the signal through the Brain Tissue, and is received at the hub where an N-bit counter/TDC converts the PWM waveform to digital bits.

978-1-6654-8495-4/22 $31.00 © 2022 IEEE

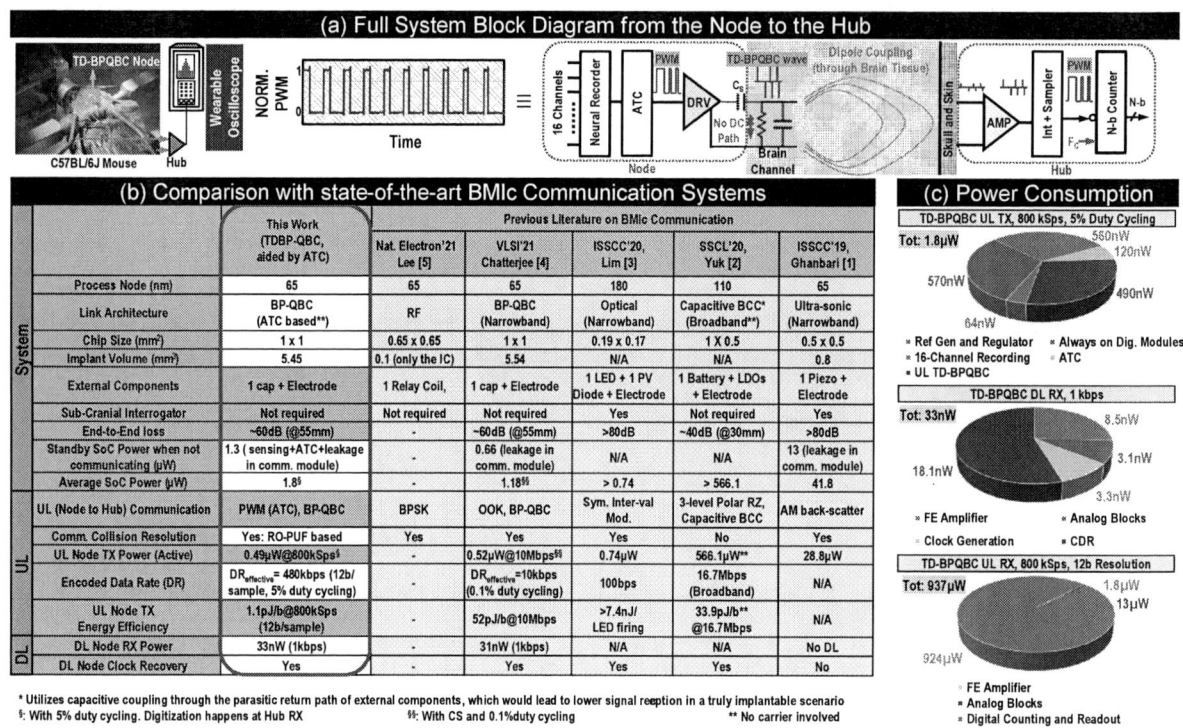

Fig. 6. (a) Vision of the full system block diagram from the Node to the Hub; (b) state-of-the-art BMIc telemetry SoC comparison, showing TD-BPQBC's >30X better energy efficient UL TX as compared to [2],[4]; (c) Power Consumption Summary for UL TX (in the Node), DL RX (in the Node) and UL RX (in the Hub).

IV. COMPARISON WITH STATE-OF-THE-ART

Fig. 6(b) compares the TD-BPQBC SoC performance with state-of-the-art BMIc telemetry systems, exhibiting the best energy efficiency (30X improvement vs. [2], and >45X vs. [4]), while preserving the benefits of BPQBC, namely, 1) max. channel length (up to 55mm) and 2) no transduction loss due to fully-electrical quasi-static signaling with information embedded in the time-domain. Although optical and ultrasonic techniques can result in small implant sizes, the low end-to-end system loss of the TD-BPQBC method would allow room for further miniaturization.

V. CONCLUSIONS AND FUTURE WORK

The implemented SoC is the first *ADC-less TD-BPQBC link* for simultaneous powering, 16-channel sensing, data communication and stimulation. The combination of ATC with TD-BPQBC results in ~1.1pJ/b energy efficiencies at the UL TX, while a UL RX with TDC bit resolution > 12 can reconstruct the analog signal with < 2dB degradation in the SNDR, when a sampling frequency of 800kHz is used in the ATC of the UL TX. The trade-offs involving power, RX bit-resolution, sampling frequency and SNDR are analyzed. The presented SoC is the first fully electrical, energy-harvested, and connected in-brain sensor/stimulator node, with orders of magnitude lower powering requirement compared to other brain communication modalities that often require field transduction. This will potentially motivate future ADC-less asymmetric network architectures for energy-efficient brain implants. As future work, body-coupled galvanic and bi-phasic modalities of power transfer, and further miniaturization for such mm-scale implants will be explored in further detail.

REFERENCES

[1] M. Ghanbari *et al.*, "A 0.8mm³ Ultrasonic Implantable Wireless Neural Recording System with Linear AM Backscattering," *ISSCC*, 2019.

[2] B. Yuk *et al.*, "An Implantable Body Channel Comm. System With 3.7-pJ/b Reception and 34-pJ/b Transmission Efficiencies," *SSCL* 2020.

[3] J. Lim *et al.*, "A 0.19×0.17mm² Wireless Neural Recording IC for Motor Prediction with NIR-Based Power and Data Telemetry," *ISSCC*, 2020.

[4] B. Chatterjee *et al.*, "A 1.15µW 5.54mm³ Implant with a Bidirectional Neural Sensor and Stimulator SoC with Bi-Phasic Quasi-static Brain Communication achieving 6kbps-10Mbps Uplink with Compressive Sensing and RO-PUF based Collision Avoidance," *VLSI*, 2021.

[5] J. Lee *et al.*, "Neural Recording and Stimulation using Wireless Networks of Microimplants," *Nature Electronics*, 2021.

[6] B. Chatterjee *et al.*, "Bi-Phasic Quasistatic Brain Communication for Fully Untethered Connected Brain Implants," *preprint arXiv:2205.08540*

[7] Z. Yu *et al.*, "An 8.2 mm³ Implantable Neurostimulator with Magneto-electric Power and Data Transfer," *ISSCC*, 2020.

[8] S. Lee *et al.*, "A 330µm×90µm Opto-Electronically Integrated Wireless System-on-Chip for Recording of Neural Activities," *ISSCC*, 2018.

[9] J. Lee *et al.*, "An Implantable Wireless Network of Distributed Microscale Sensors for Neural Applications," *Int. IEEE/EMBS Conf. on Neural Engineering (NER)*, 2019.

[10] "ICNIRP Guidelines for Limiting Exposure to Time-Varying Electric, Magnetic and Electromagnetic Fields," [Online]. Available: https://www.icnirp.org/cms/upload/publications/ICNIRPemfgdl.pdf. [Accessed: Jan-12-2022].

[11] J. Li *et al.*, "Human-Body-Coupled Power-Delivery & Ambient-Energy-Harvesting ICs for a Full-Body-Area Power Sustainability," *ISSCC*, 2020.

[12] B. Chatterjee *et al.*, "A 65nm 63.3µW 15Mbps TRX with Switched-Capacitor Adiabatic Signaling and Combinatorial-Pulse-Position Modulation for Body-Worn Video-Sensing AR Nodes," *ISSCC*, 2022.

[13] S. Mondal and D. A. Hall, "A 13.9-nA ECG Amplifier Achieving 0.86/0.99 NEF/PEF Using AC-Coupled OTA-Stacking," *JSSC*, 2020.

A 50 Mb/s Full HBC TRX with Adaptive DFE and Variable-Interval 3x Oversampling CDR in 28nm CMOS Technology for A 75 cm Body Channel Moving at 0.75 Cycle/sec

Jaehyun Ko[1], Iksu Jang[1], Chanho Kim[1], Jihoon Park[1], Changjae Moon[1], Sooeun Lee[2] and Byungsub Kim[1]

[1]Department of Eletrical Engineering, Pohang University of Science and Technology (POSTECH), Pohang, Korea,
[2]Samsung Electronics, Hwaseong, Korea,
E-mail: byungsub@postech.ac.kr

Abstract—This paper reports a 50 Mb/s full human body communication (HBC) transceiver (TRX) that communicates through a 75 cm body channel moving at speed of 0.75 cycle/sec. For the first time, adaptive decision feedback equalization (DFE) and a clock and data recovery (CDR) were integrated in a HBC TRX to adapt to a moving body channel in real time. In the proposed design, the transmitter (TX) consists of inverter-based driver that transmits non-return-to-zero (NRZ) signal to human body. The receiver (RX) utilized a 10-tap adaptive DFE to compensate for time-varying inter symbol interferences (ISIs) of moving body channel. Also, for clock synchronization, variable-interval 3x oversampling CDR (VI-3x-CDR) was adopted to overcome wide and asymmetric data-dependent jitter (DDJ). A test chip was fabricated in 28 nm CMOS technology. Compared with the prior arts that were measured with moving body channel, the data rate and the area were improved by 10x and 3x, respectively.

Keywords—Human body communication (HBC), Body channel communication (BCC), Decision feedback equalization (DFE), Wireless body area network (WBAN).

I. INTRODUCTION

As the number of connectivity around human body is enormously increasing for medical and entertainment purposes, energy efficient (long battery time) and high-speed (high quality of data) wearable and implantable devices are increasingly demanded. For those devices, early developed IEEE standards (802.15 task group) were established for the general connection of personal device at short distance: ZigBee (802.15.4), Bluetooth (802.15.1) and Ultra-wide band (UWB, 802.15.3). However, the power efficiency and data rate of these schemes were limited because the RF signal that broadcasts to air around the body suffers huge channel loss. To overcome such fundamental limitations, human body communication (HBC) which utilizes human body as communication medium is getting increasing attention because of the low channel loss of a human body channel [1], [4]-[7] as a promising solution for energy-efficient and high-speed wearable and implantable

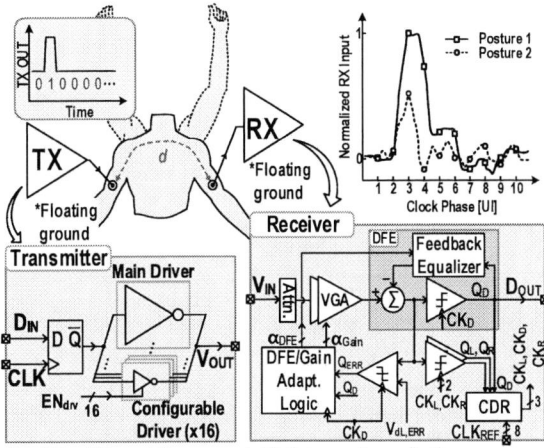

Fig. 1. An overall block diagram of the proposed HBC transceiver and SBRs measured with two different postures.

devices. In the past years, many researcher have focus on increasing the maximum speed of HBC TRX. The data rate of state-of-the-art HBC TRXs have reached over tens of Mb/s [1],[7]. However the achievable speed of HBC drastically drops with human motion because the electrical response of the body channel varies a lot with posture change. There have been some prior works that attempted to address this problem [4]-[6]. Reference [4] presented the first magnetic field HBC (mHBC) that achieved maximum data rate of 5 Mb/s while walking. As the inductances of coils at TX and RX vary with posture change, a frequencylocked-loop was adopted to maintain LC resonance frequency by tuning capacitance. However, channel variation by inductor mis-alignment by posture change and large coil area are still in issue. In electric field HBC (eHBC), prior works [5]-[6] tried to mitigate channel variation due to variable ground effect in which the capacitively-coupled return path varies.

The eHBC channel gain is determined by both the body path and the return path. P-OFDM HBC TRX [5] compensated for the channel variation with reconfigurable gain LNA by

978-1-6654-8495-4/22 $31.00 © 2022 IEEE

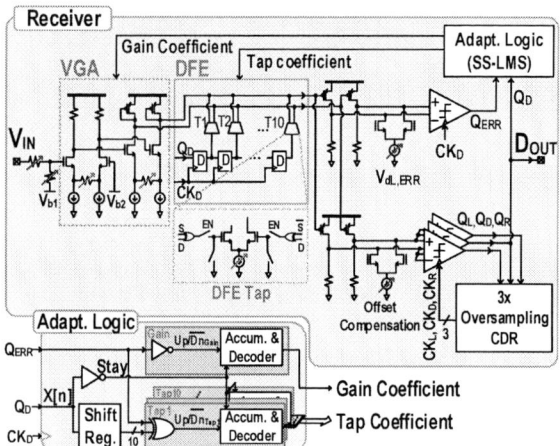

Fig. 2. A schematic diagram of the RX DFE datapath.

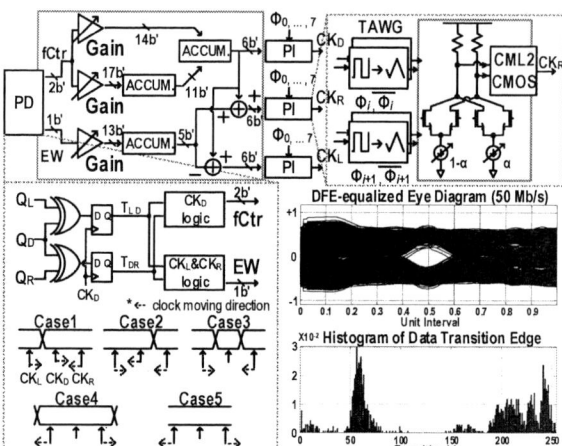

Fig. 3. A block diagram of the CDR and relevant simulations (an eye diagram and a jitter histogram).

continuously monitoring RC time constant of channel, which is associated with the channel status. However it only achieved the maximum data rate of 2 Mb/s while shaking head. It also occupied large area because of a huge OFDM modulation blocks. The work reported in [6] adopted resonance-matching technique in which the channel gain is automatically adjusted by changing inductance. It achieved the maximum data rate of 4 Mb/s, and showed that the channel gain was maintained for various distances between TX and RX. However, a large inductance is a big problem in implementation. Therefore, so far, no HBC TRX faster than 5 Mb/s has been demonstrated with a moving human body.

In this paper, we report a 50 Mb/s full human body communication (HBC) transceiver (TRX) that communicates through a 75 cm body channel moving at speed of 0.75 cycle/sec. To maintain high data rate with a quickly moving body channel, adaptive decision feedback equalization (DFE) and a clock and data recovery (CDR) were integrated in a HBC TRX for the first time. In experiment, the proposed TRX achieved 50 Mb/s with a body channel moving at the speed of 0.75 cycle/sec. This paper organized as follow. In section II, the proposed circuit design is explained. In section III, the measurement setup and the results are presented, and section IV discusses the results.

II. THE PROPOSED DESIGN

A. Overall architecture

The proposed HBC TRX consists of an inverter based TX and an adaptive DFE RX with a variable gain amplifier (VGA) and a variable-interval 3x oversampling CDR (VI-3x-CDR) (Fig. 1). The TX driver consists of a main driver (a large inverter) and 16 configurable driver (small inverters) that can be digitally controlled to adjust driving strength. By using this driver, baseband NRZ signal is transmitted through a body. In the RX, DFE cancels intersymbol interference (ISI) (Fig. 1) caused by the channel distortion which is the main obstacle in tens-of-Mb/s HBC [1]. Because two single-bit responses (SBRs) measured with two different postures greatly differ

from each other (Fig. 1), DFE/CDR adaptation is necessary to body movement. For VGA/DFE adaptation, the sign-sign least mean square (SS-LMS) algorithm [3], which is widely used in high-speed wireline, was adopted for its good performance and efficiency. For CDR adaptation, VI-3x-CDR [2] was adopted for reliable operation. The finite-state machines (FSMs) for DFE/CDR adaptation were synthesized for the full-rate clock frequencies up to 50 MHz. Because human motion is much slower, the FSMs can continuously track time-varying characteristics of a moving body channel.

B. Receiver DFE datapath circuit

The RX employs a full-rate 10-tap current-summing DFE adopted from [1] (Fig. 2). In HBC, the channel gain varies a lot depending on the channel distance and the data rate. Also, the input common-mode voltage may fluctuate. Under such circumstances, keeping the linearity of the analog circuits is challenging. The input gain is controlled by a two-stage VGA and a programmable voltage-dividing termination to keep the linearity of the circuits. Also, a multi-stage topology using current mirrors provides a wide linear operation range. The 10 DFE taps cancel 10 post-cursor ISIs. The two output amplifiers of the DFE summer have threshold control knobs. One cancels the datapath offset for data sampling (DFE output: $Q_D=D_{OUT}$) and edge sampling (left: Q_L, right: Q_R). The other one subtracts the desired signal level ($V_{dL,ERR}$) for error sampling (Q_{ERR}). Q_D, Q_L, and Q_R are used in CDR adaptation [2]. DFE adaptation uses Q_D and Q_{ERR}. In the adaptation logic, update steps and cycles can be reconfigured to adjust adaptation bandwidth depending on channel circumstances.

C. Variable-interval 3x oversampling CDR

VI-3x-CDR [2] was adopted and modified for the proposed CDR based on phase interpolators (PIs). Because the SBR's ISI tail is large and longer than the DFE tap count (Fig. 1), the ISIs left after the DFE subtraction are not negligible, causing large and complex data-dependent jitter. Fig. 3 shows simulation results after the DFE compensation. The eye width is only 0.2 UI. The jitter histogram is widely dispersed in an

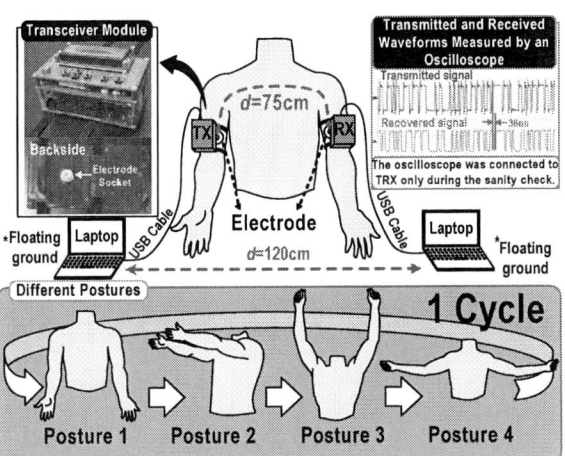

Fig. 4. Measurement setup and postures.

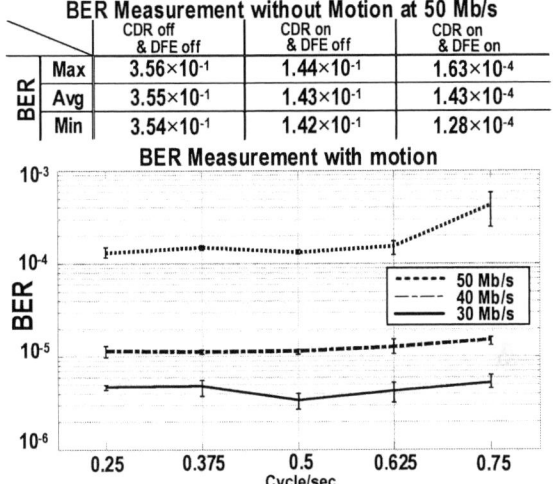

BER Measurement without Motion at 50 Mb/s

		CDR off & DFE off	CDR on & DFE off	CDR on & DFE on
BER	Max	3.56×10^{-1}	1.44×10^{-1}	1.63×10^{-4}
	Avg	3.55×10^{-1}	1.43×10^{-1}	1.43×10^{-4}
	Min	3.54×10^{-1}	1.42×10^{-1}	1.28×10^{-4}

Fig. 5. BER measurements with a 75-cm body channel.

asymmetric irregular shape that has many peaks, and is also severely skewed from the typical edge locations (0.5 UI apart from the center of the eye diagram). Therefore, a typical CDR cannot work with this body channel because it usually locates the data clock (CK_D) 0.5 UI apart from the average phase of the jitter distribution. On the other hand, the VI-3x-CDR adjusts the edge clocks (CK_L and CL_R) to track the edges of the data eye rather than tracking the jitter density while placing the data clock (CK_D) in the middle of the two edge clocks [2]. Therefore, the VI-3x-CDR samples the center of the eye irrespective of the magnitude and shape of the jitter [2]. Fig. 3 conceptually describes the proposed CDR behavior. If only one transition occurs (Case 1: $Q_L \neq Q_D = Q_R$, or Case 2: $Q_L = Q_D \neq Q_R$), the CDR tries to increase (Case 1) or to decrease (Case 2) the CK_D phase while trying to reduce the phase difference between CK_L and CK_R. If two transitions occur (Case 3: $Q_L \neq Q_D \neq Q_R$), the CDR tries to reduce the phase difference between CK_L and CK_R without changing CK_D. If no transition

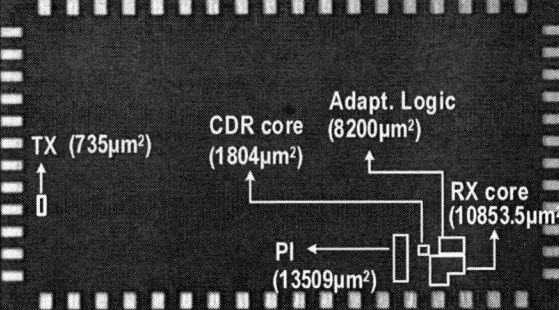

Fig. 6. A Microphotograph of the chip.

is detected five times consecutively (Case 4 and Case 5: $Q_L = Q_D = Q_R$), the CDR tries to increase the phase difference between CK_L and CK_R without changing CK_D. For this adaptation, phase detectors (PDs) generate two control signals: 1) fCtr to control CK_D phase and 2) EW to control the phase difference between CK_L and CK_R. fCtr and EW are fed to digital filters to produce the phase codes, from which three PIs generate CK_D, CK_L, and CK_R using 8-phase clocks ($\Phi_0 \ldots \Phi_7$) to sample Q_D, Q_L, and Q_R, respectively. The PI adopted typical current mode topology with resistive load, and tri-angular wave generators (TAWG) which consists of digitally configurable current-starving inverters and a MOS capacitor bank were placed before the interpolation to keep phase linearity at wide range of clock speed. Except PIs and samplers, the CDR was fully synthesized.

III. EXPERIMENT RESULTS

A. Measurement setup

The TRX was fabricated with an on-chip bit error rate tester (BERT) circuit in 28 nm CMOS (Fig. 6). Two test modules of the TX and the RX were configured and separately powered by two different battery-powered laptops with floating grounds to avoid ground coupling through the power/USB cables (Fig. 4). An electrode on the backside of a module directly contacts the body without a lead wire (Fig. 4). The test modules were tied to upper arms with straps for stable contact during the test. The TX and the RX communicated through a 75 cm body channel. The waveforms of the transmitted and received bit patterns were measured with an oscilloscope for a sanity check before the quantitative bit error rate (BER) measurement. The measured bit patterns were identical (Fig. 4). The oscilloscope was used only during the sanity check, and then disconnected from the TRX during the BER measurement to avoid ground coupling via power cables.

B. Experiment results

BER was measured multiple times by the on-chip BERT with and without body channel movement in various conditions. The BER measurements at 50 Mb/s with a motionless 75-cm body channel are reported in Fig. 5. When either DFE or CDR was disabled, the BER remained higher than 10^{-1}. When both DFE and CDR were enabled, the BER dropped to 1.43×10^{-4}, verifying that DFE and CDR play critical

TABLE I
PERFORMANCE SUMMARY AND A COMPARISION TABLE

		This work	ISSCC'19 [4]	JSSC'17 [5]	ASSCC'19 [6]	VLSI'17 [1]	VLSI'19 [7]
Process		28 nm	65 nm	65 nm	180 nm	65 nm	180 nm
Supply voltage		TX: 1.2 V, RX: 1 V	0.6 V	1.1 V	0.8 V	N/A	1 V
HBC Technique		Capacitive	mHBC	Capacitive	Capacitive	Capacitive	Galvanic
Communication Scheme		NRZ	OOK	8 P-OFDM BPSK	OOK	NRZ	Bipolar RZ
Data Rate		30 - 50 Mb/s	5 Mb/s	200 Kb/s -2 Mb/s	4 Mb/s	150 Mb/s	100 Mb/s
BER		$< 10^{-3}$	No BER measurement with moving body channel	$< 10^{-7}$	$< 10^{-5}$	$< 10^{-6}$	$< 10^{-9}$
Channel Distance		75 cm	100 cm	10 cm	150 cm	20 cm	A few cm
Human Motion While Testing		0.75 cycle/sec of pre-defined motion	While walking	While shaking head	Changing ground path distance from 2 cm to 20 cm	No motion	No motion
Adaptation		Yes	Yes	Yes	Yes	No	No
Area [mm²]		0.036	0.12	0.542	2.8	0.00558	N/A
Energy Efficiency [pJ/bit]	TX	33.72	7.15	435	30	3.3	4.75
	RX	68.84	4.7	550	41	13.3	26.8
	Total	102.56	11.85	985	71	16.6	31.55

roles in overcoming the channel distortion and in timing recovery, respectively. The BER measurements with a 75-cm moving channel in various conditions are plotted in the min-max chart (Fig. 5). For precise measurement, the human subject repeated the cycle of four postures in Fig. 4 while the movement speed was adjusted to the beat of a metronome. At the fastest motion speed of 0.75 cycle/sec, the worst BERs were measured 6.36×10^{-6}, 1.63×10^{-5}, and 5.84×10^{-4} at the data rates of 30, 40, and 50 Mb/s, respectively (Fig. 5). In all test conditions (data rate: 30-50 Mb/s, motion speed: 0.25-0.75 cycle/sec), the measured BER remained smaller than 10^{-3}, verifying that DFE and CDR successfully adapted to the channel.

IV. DISCUSSION

The proposed HBC TRX achieved the maximum data rate of 50 Mb/s through a 75 cm body channel moving at speed of 0.75 cycle/sec for the first time while consuming 5.1 mW (102.5 pJ/b) and occupying only 0.036mm². Compared with the prior state-of-the-arts that were measured with moving body channels [4]-[6], the proposed TRX improved the maximum data rate by 10x at costs of 3x smaller area and reasonable energy consumption. Comparing the results at similar BERs reported in [5] and [6], our TRX achieved much faster data rates of 30 Mb/s and 40 Mb/s at costs of slightly higher BERs of 6.36×10^{-6} and 1.63×10^{-5}, respectively. The prior works that were not measured with moving body channels achieved much faster speeds of 100 Mb/s [7] and 150 Mb/s [1]. However, comparing our work with them is not fair because the achievable maximum data rate greatly decreases with a human motion. It is also noticeable that the proposed TRX has an advantage in area efficiency with technology scaling because it has a large portion of digital circuits and does not require bulky analog elements like inductors or analog filters.

ACKNOWLEDGMENT

This work was supported in part by COMPA grant funded by the Korea government (MIST) (No. 20211100); in part by the MIST, Korea, under the ITRC support program (IITP-2021-0-02052) supervised by the IITP; in part by the BK21 FOUR Project of NRF for the Dept. of EE, POSTECH; in part by Samsung Electronics Co., Ltd (Cluster Academia Collaboration Program, POSTECH Samsung Semiconductor Education Program). Authors also appreciate IDEC for tool support and Samsung Electronics for chip fabrication.

REFERENCES

[1] J.-H. Lee, K. Kim, M. Choi, J.-Y. Sim, H.-J. Park, and B. Kim, "A 16.6-pJ/b 150-mb/s body channel communication transceiver with decision feedback equalization improving> 200% area efficiency," in 2017 Symposium on VLSI Circuits. IEEE, 2017, pp. C62–C63.

[2] S.-H. Lee et al., "A 5-Gb/s 0.25-μm CMOS jitter-tolerant variable-interval oversampling clock/data recovery circuit," IEEE J. Solid-State Circuits, vol. 37, no. 12, pp. 1822–1830, Dec. 2002.

[3] V. Stojanovic et al., "Autonomous dual-mode (PAM2/4) serial link transceiver with adaptive equalization and data recovery," IEEE J. Solid-State Circuits, vol. 40, no. 4, pp. 1012–1026, Apr. 2005.

[4] J. Park and P. P. Mercier, "A Sub-10-pJ/bit 5-Mb/s Magnetic Human Body Communication Transceiver," IEEE J. Solid-State Circuits, vol. 54, no. 11, pp. 3031-3042, Nov. 2019.

[5] W. Saadeh, M. A. B. Altaf, H. Alsuradi, and J. Yoo, "A 1.1-mW ground effect-resilient body-coupled communication transceiver with pseudo OFDM for head and body area network," IEEE J. Solid-State Circuits, vol. 52, no. 10, pp. 2690–2702, Oct. 2017.

[6] J. Zhao et al., "A 4-Mbps 41-pJ/bit on-off keying transceiver for body-channel communication with enhanced auto loss compensation technique," in Proc. IEEE Asian Solid-State Circuits Conf. (A-SSCC), Macau, Macao, Nov. 2019, pp. 173–176.

[7] Y. Jeon et al., "A 100Mb/s Galvanically-Coupled Body-Channel-Communication Transceiver with 4.75pJ/b TX and 26.8 pJ/b RX for Bionic Arms," 2019 Symposium on VLSI Circuits, 2019, pp. C292-C293, doi: 10.23919/VLSIC.2019.8778040.

An Implantable Power Extraction Circuit with Integrated PMUTs for Wireless Power Delivery

Oi-Ying Wong[1,2], Dries Tabruyn[1,2], Veronique Rochus[1], Nick Van Helleputte[1]

[1] imec, Leuven, Belgium. [2] KU Leuven, Leuven, Belgium.

sylvia.wong@imec.be, dries.tabruyn@kuleuven.be, Veronique.Rochus@imec.be, Nick.VanHelleputte@imec.be

Abstract — **A power-extraction IC (PEC) capable of extracting and boosting ultrasound energy to power ultra-low-power medical implants is presented. This work focusses on circuits that can convert extremely low levels of incident energy as would typically be found in deep-tissue miniaturized implants. The PEC chip is designed to extract energy from a piezoelectric micromachined ultrasonic transducer (PMUT) array operating at 2.5MHz. It can self-start with a ~48mVp AC input voltages. The PEC was designed and implemented in Silterra 0.13μm PMUTs-on-CMOS process, which makes ultrasonic wireless power delivery (UWPD) using monolithically integrated PMUTs possible. The PEC harvests ultrasound energy and charges an on-chip 4.5nF capacitor to ~1V. From the on-chip capacitor, it can then deliver an average power of 7.9μW for 88μs to a load. Using an off-chip 0.1μF storage capacitor, this can extend to 1.7ms. The chip also included an experimental 5.4mm×5.4mm PMUT array. The fully integrated chip was tested in-vitro and could power a load at 10kPa (3.6mW/cm²) received ultrasound pressure.**

Keywords—ultrasonic wireless power delivery, low power, PMUT, low voltage, deep implants

I. INTRODUCTION

Recent years have seen an increased research effort in ultra-low-power microwatt-level sensing circuitry (temperature, strain, biopotentials, tissue-impedances, ...) for medical implants. This makes wirelessly-powered, highly integrated and miniaturized implantable medical devices feasible. The absence of batteries and complex cabling and integration yields compact, durable, sustainable and safer implants. Among different wireless power delivery (WPD) methods, ultrasound (US) holds a lot of promise for extremely small and deep-tissue implants [1]. Conventional UWPD designs [2-4] mostly use discrete bulk piezoelectric (PE) crystals. Reducing the thickness of such transducers, to achieve ultra-thin and conformal implants, will increase their resonance frequency. However, this also means an increased attenuation loss in human tissue, which is not preferred for power transfer. Monolithically integrated PMUTs are an interesting alternative that can result in very thin solutions capable of operating at the MHz range. The use of PMUT for UWPD has been proposed and investigated in previous works, but no CMOS PEC able to self-start from very low incident power levels has been shown to date.

While PMUTs are interesting, they do suffer from worse power conversion efficiency (POE) than bulk piezo. Combined with the ambition to be able to power deep-tissue implants, the incident energy and PMUT output voltages can be extremely low, necessitating the need for a dedicated CMOS PEC. The main design challenges are to start up and maintain operation at very low input voltage and power levels, as well as performing power conversion with large mismatch between the input and the required output voltage

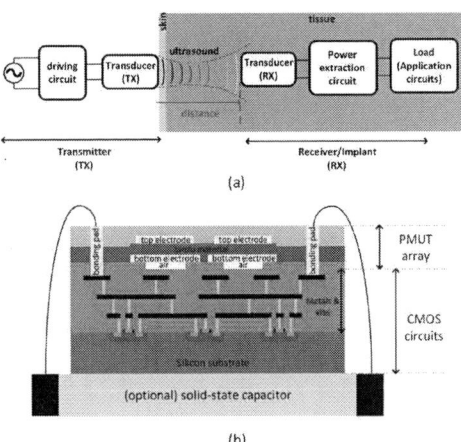

Fig. 1 (a) A typical UWPD system, and (b) illustration on the structure of the harvester on-chip RX in this work.

and power. Research on ultra-low-voltage [5] or ultra-low-power [6] converters and fully-integrated low voltage (LV) start-up circuits [7] has drawn much attention in wireless powering because of the need for smaller, self-powered autonomous devices able to operate with weak power sources. However, ref. [5-7] only work with a DC source. The PEC in [5] requires a bulk inductor or large number of external components to start up and operate. The fully-integrated converter in [6] and the ultra-low-voltage start-up circuit in [7] have very low driving capability that can only power very high impedance loads (>1Mohm) [7]. PECs with an AC input can be found in various kinetic and RF energy harvesting systems. However, these PECs are usually designed to operate with a high voltage amplitude because the input voltage is higher or boosted by matching circuits.

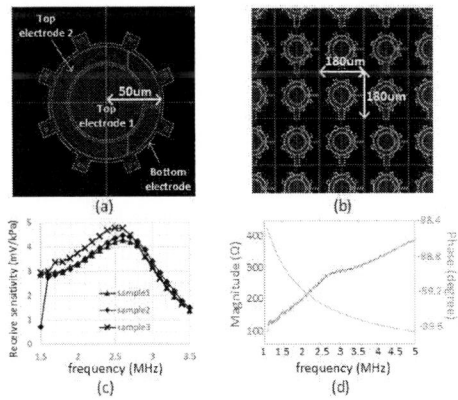

Fig. 2 (a) A PMUT device, (b) PMUT array, and the measurement results of the (c) receive sensitivity and (d) output impedance with frequencies of a 30×30 PMUT array sample.

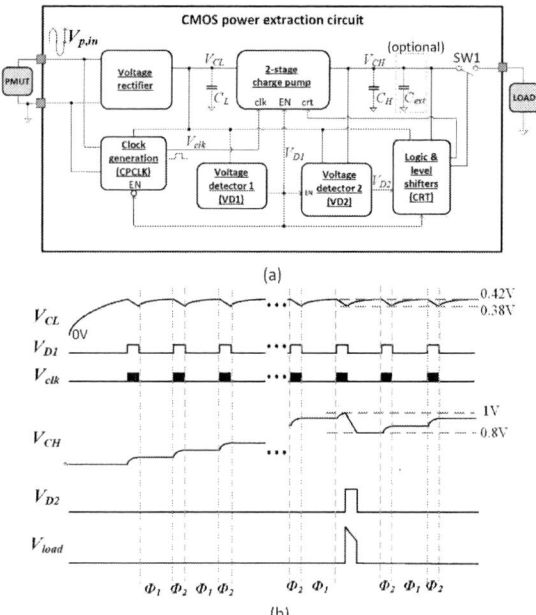

(a)

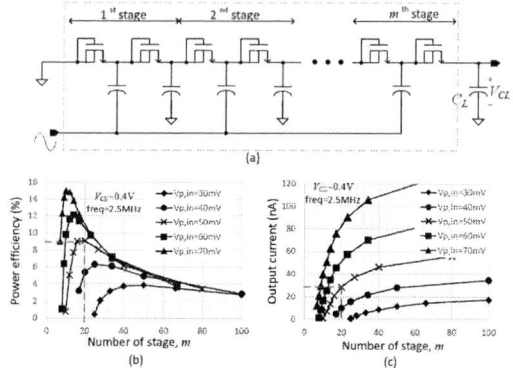

(b)

Fig. 3 (a) Proposed power extraction circuit and (b) its operation.

In this work, a CMOS low power (LP) PEC is presented capable of start-up from ~48mVp AC input using monolithic integrated PMUT for deep implants. This work can convert much lower incident US power compared to other PECs for UWPD.

II. SYSTEM ARCHITECTURE

The whole UWPD system (see Fig. 1a) consists of a transmitter (TX) outside the body where electrical energy is converted into acoustic energy and a receiver (RX) inside the body (i.e. the implant) which does the reverse. This work focusses specifically on the power conversion circuits of the RX capable of driving heavily duty-cycled microwatt power sensing circuits that could eventually leverage the same ultrasound transducers to backscatter the recorded signals. The structure of the proposed RX is shown in Fig. 1b. A PMUT array is stacked on top of the CMOS chip using a MEMS process. An optional solid state storage capacitor can be placed underneath for applications requiring higher energy per measurement/communication cycle. Lower frequencies will result in less attenuation loss (~0.6dB/cm/MHz in human tissue) but require larger cavity sizes which are difficult to process reliably. In this work a 5.4mm × 5.4mm PMUT array was designed to resonate at 2.5MHz (see Fig. 2c). Hence, within the FDA limit of 720mW/cm², only 3.2% of the acoustic power (~25kPa) would reach a 10cm deep implant under perfect focus and alignment. Misalignment and shadowing due to bones will result in even lower received pressures, meaning the RX should be able to extract energy from very weak acoustic pressures.

A. PMUT array

The RX transducer is built using PMUTs device from a commercial PMUT-on-CMOS process from Silterra (see Fig. 2a). The resonance frequency depends on the material stack and cavity width. The transducer elements were designed to maximize receive sensitivity at 2.5MHz. The transducer

element cavity width was optimized for the lowest possible frequency using a piezolayer of Scandium 9.5% doped Aluminum Nitride. The transducer elements cannot be too large due to processing rules regarding the cavity. To ensure sufficient energy can be captured a 30×30 array of parallel-connected elements is used. Fig. 2c and 2d show the measured receive sensitivity (in mV/kPa) and impedance of the 5.4mm × 5.4mm PMUT array at different frequencies. The PMUT array shows the highest sensitivity at 2.5MHz. It is expected that the sensitivity in Fig. 2c is slightly underestimated as parasitic capacitances present in the measurement setup were not corrected for. The output impedance of the PMUT array is basically a capacitance of ~0.34nF. Therefore, from Fig. 2c and 2d, it can be estimated that under an US pressure of 12.5kPa, the open-circuit voltage is around 53mV with a maximum available power of 3.7µW at 2.5MHz. Obviously, a smaller array could be used if less power or less depth is needed.

B. Power extraction circuit (PEC)

The PEC is designed to self-start with input voltages of ~50mVp. To maintain operation with very low incident power levels, the active power consumption of the PEC is <130nW. To achieve this, the conversion happens in 2 cascaded stages (see fig. 3a). First (phase Φ_I) a voltage rectifier (VR), implemented as a passive Dickson charge pump, charges the on-chip 850pF C_L to around 0.4V. Once V_{CL} reaches around 0.4V (phase Φ_2), the active 2-stage charge pump (CP) can more efficiently boost to above 1V and charges the on-chip 4.5nF C_H, or optionally a larger external storage capacitor. The 0.4V V_{CL} threshold is chosen to minimize leakage and standby power.

The detailed operation of the proposed PEC is illustrated in Fig. 3b. The architecture is designed to minimize the standby currents of all circuits as much as possible. The VR is always on to charge C_L and its output voltage is monitored by a voltage detector, VD1. VD1 output goes HIGH when V_{CL} rises above 0.42V and goes LOW when V_{CL} drops below 0.38V. This signal enables the CP and its control blocks including clock generation circuitry (CPCLK) and a second output voltage monitoring circuit (VD2). So as soon as V_{CL} is charged to the right value, a bit of energy is pumped from C_L to C_H. When V_{CH} is charged to above 1V, the output of VD2 will go HIGH allowing the load, including a supply regulation circuit if needed, to be driven by C_H via the pass transistor (SW1). When V_{CH} drops below 0.8V, the load will

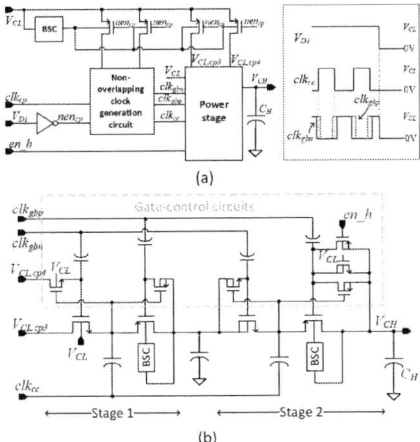

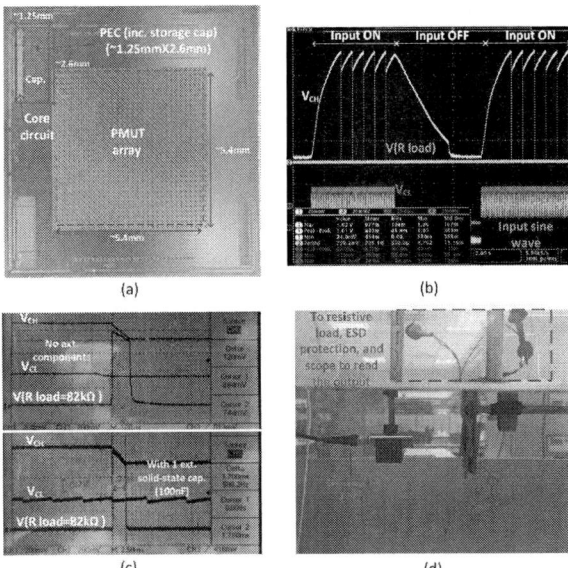

Fig. 5 (a) Architecture of the 2-stage charge pump and (b) the schematic and control signals for the power stage of the charge pump.

be disconnected so C_H can be charged up again. The on-chip capacitor is sized to be able to deliver around 7.9µW during 88µs to the load, which could suffice for a simple measurement and backscattering the result. If more energy is needed, an external capacitor can be sized accordingly. Except VR and VD1, all other circuits are disabled when C_L does not have enough energy. Different from conventional PECs, the momentarily output power can be boosted to a level much higher than that available at the input. This method allows to continuously scavenge the ultrasound energy and trigger a measurement as soon as sufficient energy is available.

III. CIRCUIT IMPLEMENTATION AND SIMULATION RESULTS

The first rectifier is implemented as a fully-integrated passive voltage multiplier shown in Fig. 4a. To ensure self-start, low-Vth transistors are used. The standby currents of all the circuitry connected to V_{CL} are minimized and simulated to be less than 8.8nA (see Fig. 6b-c). A drawback of such an architecture is of course the large reverse current

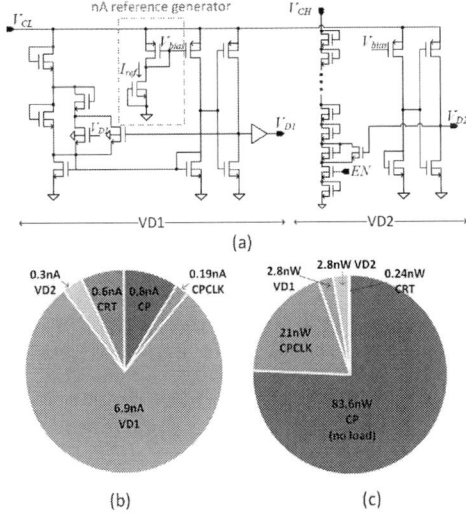

Fig. 6 (a) Schematics of the two voltage detectors, and simulation results on the (b) current consumptions of various circuit blocks in Φ_1, and (c) power consumptions on various circuit blocks in Φ_2.

Fig. 7 (a) chip photo of the PEC, (b) output waveforms on a 82kΩ load (with/without external capacitor), with sine wave generator input, (c) transient response of the PEC, with sine wave generator input, and (d) acoustic measurement setup for the PMUT-on-CMOS chip.

which becomes larger for larger input/output voltages. This architecture is hence not optimal to achieve the best efficiency for reasonably large input/output voltages. However, the whole design is optimized for weak input energy levels anyway and the VR only operates up to 0.42V output voltage. Fig. 4b-c give the simulated power efficiencies and the corresponding output currents of the VR with different number of stages and peak input AC voltage $V_{p,in}$, for a fixed transistor size and V_{CL}=0.4V. If 20 stages are used, the VR can achieve a peak POE of 9% at only $V_{p,in}$=50mV. In these conditions 29nA of output current can be given.

The schematic of the active CP is shown in Fig. 5. Since it can operate on a reasonably large 0.4V input voltage, an active 2-stage Dickson CP is chosen to maximize efficiency. Each charge transfer switch (CTS) is controlled by a dedicated gate-control circuit to reduce the shoot-through currents and making its gate control less dependent on V_{CH}. To prevent the reverse current from C_H and reduce the standby current, the bulks of those transistors connected to C_L and C_H are biased at higher voltages during Φ_1 by bootstrapping circuits (BSC, in Fig. 5). With a voltage of 0.4V from V_{CL} and the duty-cycled operation, the CP achieves a peak POE of 81.6%. The CP operates with a clock at half the input AC frequency, which is generated by CPCLK. It uses a class-B stage to generate the clock directly from the input AC signal.

The circuit architecture for VD1 and VD2 in this work is shown in Fig. 6a. It consists of a current comparator to compare a nA reference current with a current derived from V_{CL}/V_{CH}. Fig. 6c-d plot the simulated current and power consumptions of the circuit blocks in the proposed PEC at V_{CL}=0.4V during Φ_1 and Φ_2.

IV. MEASUREMENT RESULT

The PEC was implemented in Silterra 0.13µm PMUT-on CMOS process (Fig. 7a). The electrical performance of the

978-1-6654-8495-4/22 $31.00 © 2022 IEEE   219

TABLE I. COMPARISON WITH STATE-OF-THE-ART

	UWPD systems			LV LP converters in other energy harvesting circuits			this work
	[2]	[3]	[4]	[5]	[6]	[7]&	
Transducer	bulk piezo	bulk piezo	bulk piezo	TEG	Solar cell	TEG	PMUT
US frequency (MHz)	1.78	2	0.79	-	-	-	2.5
US transducer size (mm) [volume (mm³)]	0.75×0.75×0.75 [0.42]	0.75×0.75×0.75 [0.42]	1×1×1.8 [1.8]	-	-	-	5.4×5.4×0.01 [0.29]
US Intensity (mW/cm²)	-	135	40	-	-	-	3.6 (10kPa)
Input source	AC	AC	AC	DC	DC	DC	AC
Min. startup (peak) voltage (V)	>2.1	>1.2	>2	0.011	0.14	0.05	0.048
Total power consumption (µW)	37.7	140	-	1.725	~0.1	~0.297	0.29*
Output power (µW)	25	-	-	1.38 @V_{in}=10.5mV	~0.04@ V_{in}=0.25V	0.005 @V_{in}=55mV	7.9 for 88µs/ 1700µs + avg:0.0013 @$V_{p,in}$=60mV
(PEC) output voltage (V)	1	1.2	1.9	<1V	4	0.84	0.74-0.9
Regulation	Y	Y	Y	N	N	N	N
No. of ext. components	0	0	1 cap.	5 ind. & 2 cap.	0	0	0/1 cap. (optional)
Technology (nm)	65	65	180	130	180	180	130

* maximum available power with source resistance of 1000Ω; +with optional external capacitor of 100nF; &only the performance of the LV startup circuit in ref. [7]

PEC was first characterized by connecting a resistor Rs in series with a 2.5MHz sine wave generator (peak output voltage of $V_{p,scr}$). Two samples were characterized with Rs=200Ω, emulating the large on-chip PMUT test array. These two samples can self-start and maintain their operation at $V_{p,scr}$ of 44mV and 45mV, respectively. Therefore, it is expected that the PEC can self-start with an US pressure below 25kPa as explained in Section IIA. To investigate the minimum required PMUT array output power (i.e. for even smaller arrays) at which the PEC can operate, the chip was also tested with a higher Rs (lower output power from the PMUT array, $P_{av,max}$). The two samples can self-start and operate with R_s=1000Ω for $V_{p,scr}$=47 and 48mV, respectively. This implies that only about 280-288nW of array output power is required to drive the PEC, which theoretically would allow to reduce the PMUT array by a factor 4. A transient measurement of the chip is shown in Fig. 7b. The PEC can function correctly up to $V_{p,scr}$ of at least 100mV for frequencies of 1-3MHz. The PEC can provide an output power of 7.9µW for 88µs without the use of any external component to a simulated load of around 82kΩ. This can extend to 1.7ms with an additional 100nF solid-state capacitor (see Fig. 7c). The feasibility of the PMUT-on-CMOS integration for USPD was verified with a single chip with both the PEC and a 30×30 PMUT array on it. The PMUT array was powered by a commercial US probe (2.25MHz Olympus) in water (see Fig. 7d). The chip can successfully start-up and operate at minimum measured $V_{p,in}$ of ~40mV. Based on the measured receive sensitivities of three PMUT array samples by a hydrophone (see Fig. 2c), the average received US pressure in the in-vitro tests is expected to be around 10kPa.

Different from other UWPD systems [2-4], this work is about UWPD with monolithic PMUT. The average output power and POE are lower than the conventional ones, but this work can deliver microwatts of power from much lower US power levels. Also thanks to the on-chip boost circuits, microwatt-level circuits can be powered in a duty-cycled fashion from sub-microwatt received power. Although a large PMUT array is demonstrated in this work, thanks to the low input power levels of 280nW, smaller arrays can be implemented. Table I includes also a comparison with other energy harvesting circuits for completion. [5-7] only operate from a DC source which is not applicable to US. The others either require a large number of inductors [5], operate above 100mV [6] and/or have low output power levels [7]. These works do perform better with a large input or over a larger input range, though none can operate from extremely weak US power sources.

V. CONCLUSION

A LV LP PEC able to use a PMUT array to extract power from a weak US source was implemented and verified. It can be used to power heavily duty-cycled microwatt deep implants.

The authors would like to thank the TD-MEMS team from Silterra Malaysia Sdn. Bhd. for their help in manufacturing the monolithic CMOS+PMUT devices.

REFERENCES

[1] G. Barbruni et al., "Miniaturised wireless power transfer systems for neurostimulation: a review," IEEE Trans. Biomed. Circuits Syst., vol. 14, no. 6, pp. 1160-1178, Dec. 2020.

[2] M. Ghanbari et al., "A sub-mm3 ultrasonic free-floating implant for multi-mote neural recording," IEEE J. Solid-State Circuits, vol. 54, no. 11, pp. 3017-3030, Sept. 2019..

[3] S. Sonmezoglu and M. M. Maharbiz, "A 4.5mm³ deep-tissue ultrasonic implantable luminescence oxygen sensor," in Proc. IEEE Int. Solid- State Circuits Conf., 2020, pp. 454-456.

[4] M. J. Weber et al., "A miniaturized single-transducer implantable pressure sensor with time-multiplexed ultrasonic data and power links," IEEE J. Solid-State Circuits, vol. 53, no. 4, pp. 1089-1101, Apr. 2018.

[5] R. L. Radin et al., "A 7.5-mV-input boost converter for thermal energy harvesting with 11-mV self-startup," IEEE Trans. Circuits Syst. II: Express Briefs, vol. 67, no. 8, pp. 1379-1383, Aug. 2020.

[6] W. Jung et al., "An ultra-low power fully integrated energy harvester based on self-oscillating switched-capacitor voltage doubler," IEEE J. Solid-State Circuits, vol. 49, no. 12, pp. 2800-2811, Dec. 2014.

[7] S. Bose, T. Anand and M. L. Johnston, "Integrated cold start of a boost converter at 57 mV using cross-coupled complementary charge pumps and ultra-low-voltage ring oscillator," IEEE J. of Solid-State Circuits, vol. 54, no. 10, pp. 2867-2878, Oct. 2019.

A PMUT Transceiver Front-End with 100-V TX Driver and Low-Noise Voltage Amplifier in BCD-SOI Technology

Lara Novaresi*[†], Piero Malcovati*, Andrea Mazzanti*, Edoardo Bonizzoni*, Marco Terenzi[†],
Stefano Ottaviani[†], Davide Ugo Ghisu[†], Fabio Quaglia[†], Alessandro Stuart Savoia[‡]

*Department of Electrical, Computer and Biomedical Engineering, University of Pavia, Italy

[†]STMicroelectronics, Cornaredo (MI), Italy

[‡]Department of Industrial, Electronic and Mechanical Engineering, Roma Tre University, Italy

Abstract—This paper presents an analog transceiver for PMUT-based portable ultrasound medical imaging probes, working in the 2–4 MHz frequency range. The transceiver, fabricated in a 160-nm BCD-SOI technology, delivers a three-level train of pulses with amplitude up to ±50 V to the ultrasound transducer and collects the back-scattered echoes with a receiver chain, consisting of a low-noise voltage amplifier with programmable gain and a buffer. The circuit, connected to a PMUT, achieves a RX sensitivity of 50 mV/kPa at minimum gain and an input-referred noise spectral density of 13 nV/$\sqrt{Hz}$ at 2.3 MHz, consuming 5.4 mW. The peak RX sensitivity, obtained with acoustic measurements in a water tank, is 50 mV/kPa at 2.3 MHz.

I. Introduction

Currently, clinical specialists heavily rely on ultrasound technology (echography) for diagnostics guidance and monitoring in several medical applications, such as abdominal, cardiac and cerebrovascular examination. Echography represents a powerful alternative to X-ray imaging (despite some limitations regarding penetration depth), thanks to its convenient features, such as being a real-time sensing solution with little clinical invasiveness to the patient. Echography relies on the propagation of high-frequency acoustic waves inside the human body and on the subsequent sensing of their echoes, in order to produce a dynamic image.

The manufacturing process of ultrasound transducers for echographic probes is recently steering towards MEMS integration technologies, at the expense of traditional bulk transducers. In particular, the so-called micromachined ultrasound transducers (MUTs) represent a promising solution in this field, due to their low cost, high design flexibility and reproducibility, making them very attractive for portable and handheld ultrasound imaging systems. In particular, piezoelectric MUTs (PMUTs), which are based on the flexural motion of a micromachined membrane coupled with a thin piezoelectric film, are excellent candidates for echographic applications, thanks to their large transmission efficiency.

For exploiting the potential of these transducers, dedicated integrated interface circuits are required [1]. This paper presents an integrated transceiver, realized in BCD-SOI technology, specifically tailored for a PMUT-based imaging system. The device, whose block diagram is shown in Fig. 1,

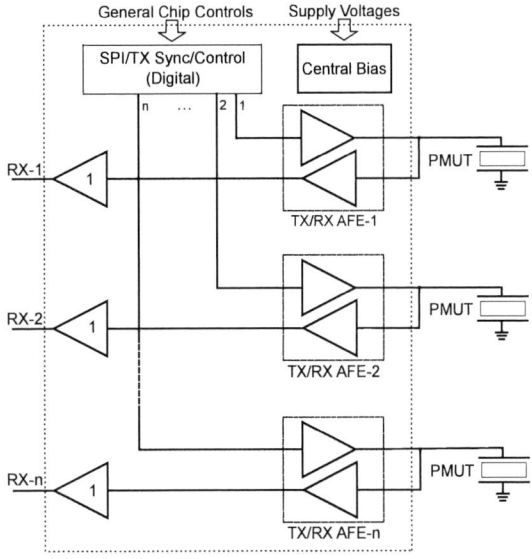

Figure 1. Block diagram of the integrated transceiver for PMUTs

consists of a 1-D array of PMUTs, each connected to an analog front-end (AFE) circuit, including a three-level transmitter (TX), a low-noise receiver (RX) and a TX/RX switch. The top terminal of each PMUT is connected to the corresponding AFE circuit for both TX and RX, while the bottom terminal is connected to a fixed bias voltage (typically −40 V). The shape of the TX pulses for the different channels can be controlled independently with digital signals generated on-chip. The circuit includes also a bias block, common to all channels, and a serial register to load the TX pulse waveforms and RX settings (e. g. gain programming). The maximum supply voltage for the TX path is ±50 V, while the supply voltage of the RX chain is 1.8 V. The paper is organized as follows. Section II describes in detail the circuit implementation of all the building blocks of the transceiver, while Section III reports the achieved experimental results. Finally, conclusions are drawn in Section IV.

978-1-6654-8495-4/22 $31.00 © 2022 IEEE

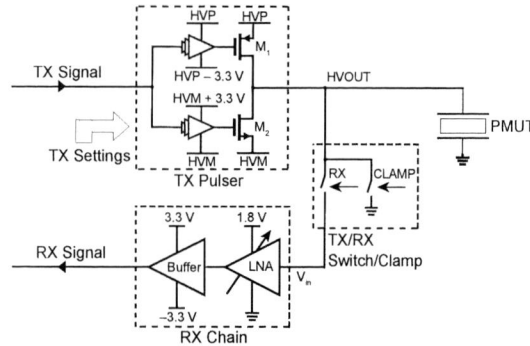

Figure 2. Block diagram of the AFE circuit

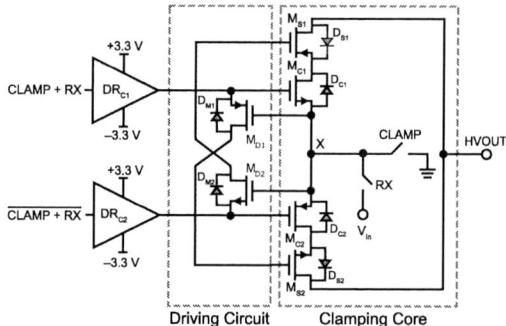

Figure 3. Schematic of the TX/RX Switch/Clamp

II. ANALOG FRONT-END CIRCUIT

In order to properly design the AFE circuit, a suitable equivalent circuit of the PMUTs composing the array of the imaging probe is required, since the features of the transducers heavily affect the specification of the AFE circuit blocks. Therefore, the PMUTs have been simulated, using the finite-element method (FEM), to extract a model compatible with circuit simulators [2]. This model allows estimating the TX and RX sensitivity of each element of the array, as well as the frequency response and the equivalent output capacitance C_p of the PMUTs. Capacitance C_p represents the capacitive load that the TX path must drive and the source impedance for the RX chain which amplifies the echoes.

The considered PMUTs feature a resonance frequency of 2.7 MHz and a −3-dB RX bandwidth ranging from 2 MHz to 4 MHz with a peak sensitivity equal to 3.5 mV/kPa. Moreover, C_p is approximately equal to 2 nF. The equivalent in-band noise pressure generated by the PMUT in RX mode is equal to $116\,\mu Pa/\sqrt{Hz}$ (corresponding to 164 mPa integrated over the −3-dB bandwidth of the PMUT), which is basically negligible, considering that the expected sound pressure of the echoes for this application is varying between 10 Pa and 10 kPa. As a consequence, the noise limitation at system level will be given mostly by the contribution of the AFE circuit.

The key challenges for the design of a transceiver for the considered PMUTs are the limited amount of power that can be dissipated in the integrated circuit, the harmonic distortion constraints that have to be respected both in the TX and RX phases, as well as the signal-to-noise ratio (SNR) and dynamic range specifications of the RX chain, which are critical aspects to avoid compromising the image quality [1].

The proposed transceiver has been designed in a BCD-SOI technology featuring 160-nm high-speed CMOS transistors, 0.35-μm I/O CMOS transistors and high voltage (HV) DMOS transistors, withstanding maximum V_{GS} and V_{DS} of 3.3 V and 100 V, respectively. The block diagram of the complete AFE circuit is shown in Fig. 2. The circuit consists of three main blocks: the TX Pulser, the TX/RX Switch/Clamp and the RX Chain.

A. TX Pulser

The TX Pulser consists of power transistors M_1 and M_2, which provide two-level pulses to the PMUTs, with amplitude up to ±50 V, defined by the power supply voltages *HVP* and *HVM*. Transistors M_1 and M_2 are driven by a tapered chain of buffers realized with 0.35-μm I/O low-voltage devices. The TX Pulser is implemented with a modular architecture, which allows selecting from 1/4 to 4/4 of the maximum current driving capability (4 A) and, hence, regulating the pulse rise and fall times, as well as the peak current absorbed by the integrated circuit, which depends on the number of PMUTs actuated at the same time. The maximum time allowed for the rising and falling edges of the pulses is 35 ns. The duration and delay of the levels in the pulse train can be programmed independently through the digital interface. Considering that the duty cycle of the TX phase is low (transmission occurs in short bursts), the TX Pulser peak current does not affect significantly the average power consumption of the complete integrated transceiver.

B. TX/RX Switch/Clamp

In order to implement three-level pulses in transmission and switch between TX and RX phases, it is necessary to either connect the top plate of the PMUTs to ground, to the output of the TX Pulser or to the input of the RX Chain. This function is realized by the TX/RX Switch/Clamp block, whose schematic is shown in Fig. 3. Simple single-transistor switches cannot be used in this case, since it would be impossible to keep them all off when node *HVOUT* switches from *HVP* to *HVM* or viceversa. In the adopted circuit, the Clamping Core block consists of two HV clamping transistors, M_{C1} and M_{C2}, driven at low voltage, and two HV switching-off transistor, M_{S1} and M_{S2}, driven at high-voltage, connected in series. The body diodes D_{S1}-D_{C1} and D_{C1}-D_{C2} are connected in anti-series to prevent current flow when the transistors are off. The Driving Circuit block is realized with two cross-coupled high-voltage driving transistors, M_{D1} and M_{D2}, connected between the gates of the clamping and switching-off transistors. In this way, the Driving Circuit block controls the switching on or off of M_{S1} and M_{S2}, while M_{C1} and M_{C2} are driven at low voltage directly by the input driver blocks DR_{C1} and DR_{C2}. The output terminal

978-1-6654-8495-4/22 $31.00 © 2022 IEEE

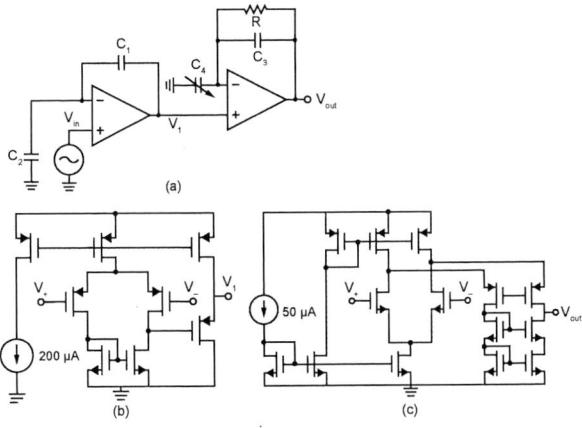

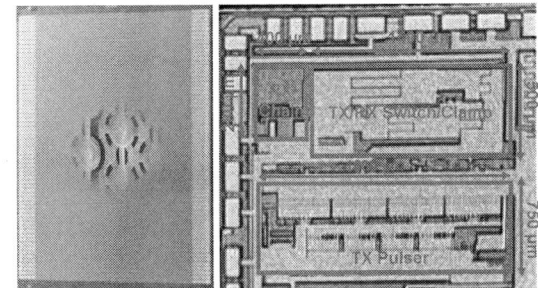

Figure 5. Micrograph of the PMUT array and of the proposed transceiver

Figure 4. Schematic of the LNA (a), of the operational amplifier used in the first stage (b) and of the operational amplifier used in the second stage (c)

HVOUT is, therefore, connected to node X through M_{S1} and M_{S2} when control signal *CLAMP + RX* is high. Node X can, then, be connected to ground to generate three-level pulses (*CLAMP* high) or to the RX chain input in the RX phase (*RX* high) through standard low-voltage switches. In these cases, both M_1 and M_2 in the TX pulser are off. When the TX Pulser is active and either M_1 or M_2 are on (*CLAMP + RX* low), node X is disconnected from the output terminal *HVOUT*.

C. RX Chain

The RX Chain consists of a Low-Noise Amplifier (LNA), realized with two cascaded gain stages, followed by a buffer for driving the output load. The LNA, whose block diagram is shown in Fig. 4a, is implemented as a Voltage Amplifier (VA) with high input impedance. A VA has been chosen instead of a transimpedance stage [3], since in the presence of a large MUT output capacitance (C_p), as in the case of PMUTs, the VA is by far more power efficient. Indeed, achieving the required bandwidth and noise specifications with a transimpedance stage, in the presence of a large value of C_p, requires an impractically large input transistor transconductance and, hence, power consumption, which is not the case for the VA, whose noise and bandwidth performance are in first approximation not affected by C_p. The gain of the VA with capacitive feedback, given by $1 + C_2/C_1$, is also not affected by C_p.

The key aspect in the design of the RX Chain is the trade-off between noise and power consumption. The required LNA input-referred noise can be obtained by imposing the condition SNR $\geq$ 0 dB in the PMUT bandwidth with the minimum detectable input pressure ($P_{S,min}$), in order to guarantee that a clear image can be created. To compute the SNR, the input-referred noise power spectral density (PSD) of the LNA ($v_{n,in}^2$) is integrated over the −3-dB bandwidth of the PMUT (2–4 MHz) obtaining

$$V_n = \sqrt{\int_{2\,\text{MHz}}^{4\,\text{MHz}} v_{n,in}^2\, df} \leq \mu_S P_{S,min}. \tag{1}$$

where μ_S is the PMUT RX sensitivity. Considering a minimum detectable input pressure $P_{S,min} = 10\,\text{Pa}$ and adding some margin, we obtain $v_{n,in} \leq 14\,\text{nV}/\sqrt{Hz}$ and, hence, assuming

$$v_{n,in}^2 \approx \frac{8kT}{3g_m}, \tag{2}$$

the transconductance g_m of the LNA input transistors can be sized appropriately. The required value of g_m has to be obtained with the minimum possible power consumption, while respecting the power budget of the entire ultrasonic probe (which comprises 64 channels), which, for safety reasons, has to be lower than 2 W.

The gain of the two-stage LNA can be programmed with 3 bits from 23 dB to 33 dB, by acting on capacitor C_4, as shown in Fig. 4a, in order to accommodate the eventual variability of the PMUT sensitivity, without exceeding the output voltage swing of the LNA. At maximum gain, the 1-dB compression point of the LNA has to be larger than 10 kPa (maximum sound pressure), while the total harmonic distortion in the above conditions has to be lower than −30 dB.

The operational amplifier used for the first stage of the LNA (Fig. 4b) consists of a PMOS differential pair followed by a level-shifter, while the second-stage of the LNA is realized with a folded-cascode operational amplifier with NMOS input pair (Fig. 4c). Finally, the output of the LNA is connected to a buffer with power supply ±3.3 V to drive the output load.

III. EXPERIMENTAL RESULTS

The proposed integrated transceiver for PMUTs has been fabricated in a 160-nm BCD-SOI technology. The micrograph of a single channel is shown in Fig. 5 (a chip contains 4 channels and 16 chips are used for the entire ultrasonic probe). The area of the transceiver is 1.8 mm×1.8 mm.

Fig. 6 shows the measured output signal of the TX path loaded by an equivalent circuit of the transducer, consisting of a 240-Ω resistor in parallel to a 1.8-nF capacitor, with *HVP* = 30 V, *HVM* = −30 V and maximum current capability. The slope of the rising and falling edges is 1.5 V/ns. The unwanted second harmonic in the TX pulses is 30 dB below the fundamental. The transfer function of the RX chain for the different gain configurations is shown in Fig. 7a. The achieved −3-dB bandwidth ranges from 650 kHz to 40 MHz. The input referred noise with maximum gain is 13 nV/$\sqrt{Hz}$,

978-1-6654-8495-4/22 $31.00 © 2022 IEEE

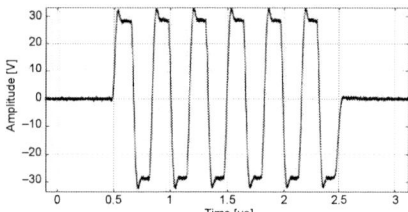

Figure 6. Measured output of the TX path loaded by the equivalent circuit of the transducer

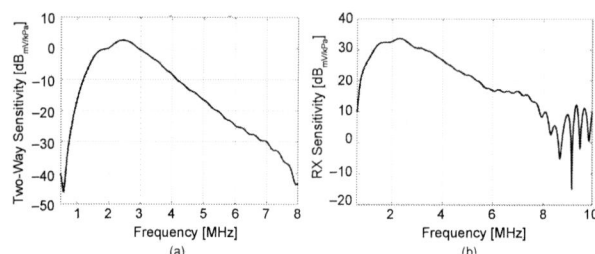

Figure 8. Measured RX sensitivity (a) and two-way sensitivity (b) of the transceiver connected to the PMUT in a water tank

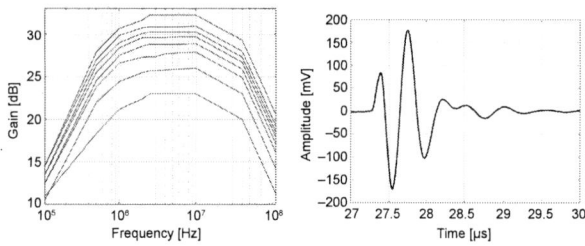

Figure 7. Transfer function of RX chain for different gain settings (a) and pulse echo signal acquired with the actual transducer in a water tank (b)

Table I
PERFORMANCE SUMMARY OF THE RECEIVER AND
COMPARISON WITH THE STATE OF THE ART

Parameter	This Work	[5]	[6]	[7]
Process [nm]	165	180	180	180
Transducer	PMUT	—	PZT	PZT
f_{in} [MHz]	2.5	5	7.5	13
RX Bandwidth [MHz]	2–4	13	—	20
LNA Power Consumption [mW]	5.4	48.31	2	6
RX Front-End Input Noise [nV/$\sqrt{Hz}$]	13 @ 3 MHz	5 @ 5 MHz	37	14 @ 15 MHz
FoM [nV $\sqrt{W}$/$\sqrt{Hz}$]	0.955	1.1	1.65	1.084

as expected, with a power consumption of 5.4 mW. The measured input-referred 1-dB compression point is 35 mV, which corresponds to the expected PMUT output voltage at the maximum expected sound pressure (10 kPa). The measured equivalent on resistance of the TX/RX Switch/Clamp is 7.9 Ω.

TX pressure and pulse-echo measurements on the complete transceiver connected to the PMUT have been carried out in a water tank. The RX frequency response was estimated by processing the results of two measurements achieved by exciting four adjacent PMUT array elements biased at 30 V with 5-V, 50-ns unipolar pulses. The TX pressure has been first acquired using a hydrophone placed on axis at 40 mm from the transducer surface. The echo signal, generated by a stainless steel planar reflector placed at 20 mm from the transducer surface, was then acquired by the RX chain of the transceiver configured with a minimum gain (23 dB), as shown in Fig. 7b. The RX sensitivity, shown in Fig. 8b, was finally computed by dividing the pulse-echo frequency response, shown in Fig. 8a, obtained by performing a FFT of the time domain received signal, by the TX frequency response. The achieved peak RX sensitivity value is 50 mV/kPa at 2.3 MHz.

Tab. I summarizes the measured performances of the receiver and compares them with the state of the art. Considering the low-is-better Figure of Merit (FoM), defined as

$$\text{FoM} = v_{n,in}\ \sqrt{P}, \qquad (3)$$

where P is the power consumption, the proposed receiver turns out to be significantly more efficient with respect to the other reported devices.

IV. CONCLUSIONS

This paper presented an integrated transceiver for a PMUT based imaging probe to be employed in medical ultrasound

diagnostic. The circuit can deliver a three-level square wave with amplitude up to ±50 V to the transducers and receive the generated echoes in the sound pressure range from 10 Pa to 10 kPa with 3.7-mPa/$\sqrt{Hz}$ input-referred noise spectral density in the 2–4 MHz band, thus ensuring good imaging quality, while consuming only 5.4 mW (FoM < 1 nV $\sqrt{W}$/$\sqrt{Hz}$). The peak RX sensitivity, obtained with acoustic measurements in a water tank, is 50 mV/kPa at 2.3 MHz.

ACKNOWLEDGMENTS

This work is part of the Moore4Medical project funded by the ECSEL Joint Undertaking under grant number H2020-ECSEL-2019-IA-876190.

REFERENCES

[1] M. Sautto et al., "A CMUT Transceiver Front-End with 100-V TX Driver and 1-mW Low-Noise Capacitive Feedback RX Amplifier in BCD-SOI Technology", *Proc. ESSCIRC '14*, pp. 407-410, 2014.

[2] A. S. Savoia et al., "Design, Fabrication, Characterization, and System Integration of a 1-D PMUT Array for Medical Ultrasound Imaging", *IEEE IUS*, pp. 1-3, 2021.

[3] L. Novaresi et al., "Acquisition RX Chain for PMUT-Based Highly Integrated Ultrasound Imaging Systems", *Proc. PRIME '22*, in press.

[4] S. Rossi et al., US Patent 8,749,099, 2014.

[5] T. Kim et al., "A CMOS Analog Front-End for Driving a High-Speed SAR ADC in Low-Power Ultrasound Imaging Systems", *Proc. IEEE ISOCC '16*, pp. 163-168, 2016.

[6] E. Kang et al., "A Reconfigurable 24 × 40 Element Transceiver ASIC for Compact 3D Medical Ultrasound Probes", in *Proc. ESSCIRC '17*, pp. 211-214, 2017.

[7] D. M. van Willigen et al., "A Transceiver ASIC for a Single-Cable 64-Element Intra-Vascular Ultrasound Probe", *IEEE J. Solid-State Circ.*, vol. 56, no. 10, pp. 3157-3166, 2021.

A Portable CMOS-based MRI System with $67 \times 67 \times 83\,\mu m^3$ Image Resolution

Daniel Krüger[1,2], Aoyang Zhang[1], Henry Hinton[1], Victor M. Arnal[1], Yi-Qiao Song[1,4], Yiqiao Tang[4], Ka-Meng Lei[3], Jens Anders[2], and Donhee Ham[1]

[1]Harvard University, Cambridge, MA, USA. [2]University of Stuttgart, Stuttgart, Germany. [3]University of Macau, Macau, China. [4]Schlumberger-Doll Research, Cambridge, MA, USA.
Email: dkrueger@seas.harvard.edu

Abstract—This paper presents the smallest-ever, portable CMOS-based magnetic resonance imaging (MRI) system, comprising a fully-integrated and digitally-assisted CMOS RF transceiver IC and a permanent magnet with $B_0 \approx 0.51\,T$ with the corresponding f_0 for the ^{1}H proton spins being 21.8 MHz. The system has a volume less than a shoebox, weighs less than 10 kg, and provides an imaging volume of $6 \times 6 \times 6\,mm^3$ with an image resolution of $67 \times 67 \times 83\,\mu m^3$. In addition to MRI, the presented system can also perform multi-dimensional nuclear magnetic resonance (NMR) relaxometry and high-resolution NMR spectroscopy, combining all state-of-the-art NMR modalities in a single, portable device.

Keywords—MRI, NMR, CMOS, imaging, on-chip, portable.

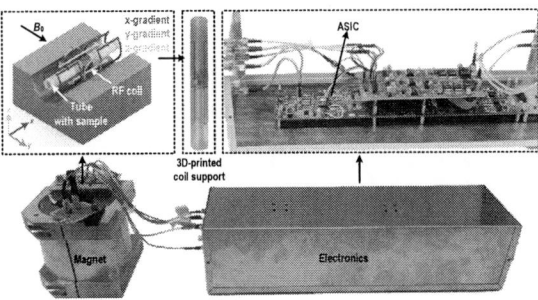

Fig. 1: Portable, CMOS-based MRI system.

I. INTRODUCTION

Magnetic Resonance Imaging (MRI) revolutionized medicine with its power to elucidate the tissue structure details and blood flow activities inside the body non-invasively. It is based on nuclear magnetic resonance (NMR), where a nuclear spin (e.g., ^{1}H proton spin) in a static magnetic field (B_0) manifests resonance at an RF frequency of $f_0 = \gamma B_0 / 2\pi$, with γ being the gyromagnetic ratio of the nuclear spin. An ensemble of nuclear spins in a material (e.g., water or fat) excited by an RF magnetic field at f_0 undergoes precession at f_0, with relaxation times depending on the material composition. MRI is the spatial mapping of local proton densities and relaxation times across the sample, where the mapping is achieved by using magnetic field gradients.

Commercial MRI systems [1], which use superconducting magnets, cryogenic support, and discrete RF electronics are bulky, heavy, costly, and thus available only at dedicated facilities. Their miniaturization into a portable platform–by replacing the superconducting magnet with a permanent magnet and by integrating the RF electronics into a CMOS chip–would enable on-site and on-demand imaging of *ex vivo* biological tissues as well as artificial organoids, small organisms, and organic materials, and can thus further greatly impact medicine and biological/physical sciences all the more. However, while NMR systems without imaging capability have been miniaturized in recent years [2-6], MRI system miniaturization has proven challenging. Although MRI is based on NMR, its miniaturization is more challenging, as it involves a complex engineering due to the additional magnetic field gradients (to choose slices to image) that must work in synchrony with RF magnetic fields (to manipulate the nuclear spins). While a small MRI system was recently reported [7], its magnetic field gradient manipulation, the most essential component of MRI, was not performed on-chip.

Here, we report the smallest-ever portable MRI system consisting of a fully-integrated and digitally-assisted CMOS RF transceiver IC and a 0.51-T permanent magnet (Fig. 1, bottom). With the system occupying a volume of less than 5 liters and weighing less than 10 kg, compared to conventional MRI systems, especially preclinical ones, where samples at the molecular or tissue level (e.g., biopsies) are analyzed, the reported system achieves a volume and weight reduction of 2700 times and 480 times (c.f. [1]), respectively, and provides an imaging volume of $6 \times 6 \times 6\,mm^3$ with an image resolution of $67 \times 67 \times 83\,\mu m^3$. Moreover, the presented system not only enables MRI, but also multidimensional NMR relaxometry and high-resolution NMR spectroscopy in a single, portable device.

The paper is organized as follows: In Section II we give an overview of the proposed MRI system. Section III describes the utilized NMR transceiver chip architecture with its individual building blocks. In Section IV we discuss the electrical characterization of the chip and provide

978-1-6654-8495-4/22 $31.00 © 2022 IEEE

measurement results obtained with the presented MRI system. Section V concludes the paper with a brief summary.

II. SYSTEM OVERVIEW

Fig. 1 shows the proposed MRI system. It consists of three main parts, a CMOS integrated RF transceiver circuit, a permanent magnet with $B_0 \sim 0.51$ T with the corresponding f_0 for the ^{1}H proton spins being 21.8 MHz, as well as a set of NMR/MRI and gradient coils. The RF coil, wound around a capillary sample tube, is connected to the CMOS transceiver to inject RF magnetic field pulse sequences that excite the nuclear spins, and to monitor the resulting nuclear spin motions. In addition, the Golay coil pair (red), the Maxwell coil pair (blue), and the Saddle coil pair (yellow) generate the magnetic field gradients necessary for spatial encoding. These three gradient coils are formed around 3D-printed support structures and connected to the field gradient control of the CMOS chip.

Despite the advances in form factor and weight, permanent magnets suffer from lower magnetic field strengths and higher magnetic field inhomogeneities, which significantly lower the MRI signal and therefore impede high resolution MRI. To ameliorate the signal-to-noise (SNR) of the image and realize fast image acquisition, we have designed an application-specific integrated circuit (ASIC), whose die micrograph is shown in Fig. 2.

III. TRANSCEIVER ASIC

Fig. 3 shows the block diagram of the CMOS RF transceiver ASIC, highlighting the main building blocks, namely, a low-noise RF receiver, an RF transmitter, a 3D-gradient control, and a digital/mixed-signal controller (pulse programmer digital logic, shift-register memory banks, and delay-locked loop (DLL) based multi-phase generator).

To preserve the already weak MRI signal, it is most important to reduce the receiver noise. Furthermore, to allow for a broad variety of experiments, a large dynamic range is desirable. In order to minimize the SNR degradation and to increase the dynamic range, the front-end of the RF receiver consists of a fully-differential, two-stage, low-noise amplifier (LNA) followed by a digitally controlled (5-bit) attenuator. Further improvement of the receiver noise figure by 3 dB is achieved by using quadrature mixers for frequency down-

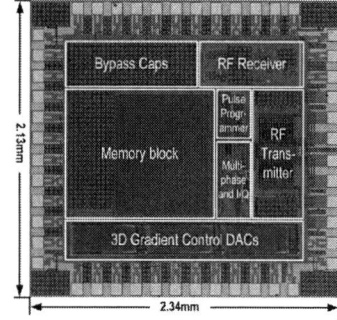

Fig. 2: Annotated micrograph of the transceiver ASIC.

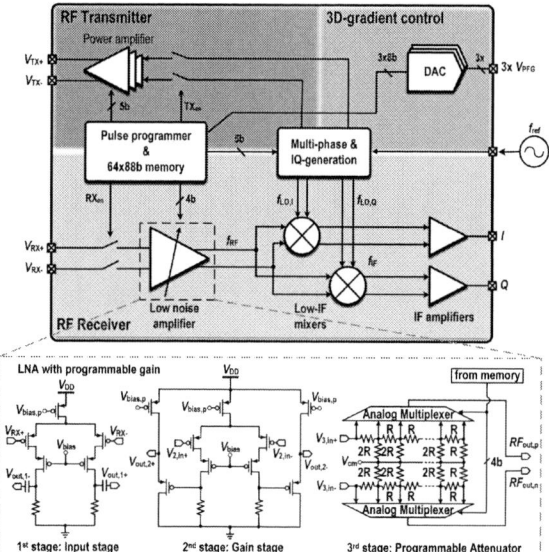

Fig. 3: Block and transistor level diagram of the transceiver ASIC.

conversion. Driven by quadrature local oscillators they produce two intermediate frequency signals, which are subsequently low pass filtered and amplified to produce the two quadrature outputs, I and Q, which jointly contain information on the nuclear spin motion. With an overall, tunable receiver gain of 30 to 100 dB, the receiver dynamic range can be optimally adjusted to the NMR application at hand.

The RF transmitter sends out RF pulse sequences with fine phase and amplitude control in order to manipulate the nuclear spin motion with high precision. The core is a class-D^{-1} power amplifier (Fig. 4, left), which is designed for wide amplitude tuning range, efficiency enhancement, and low AM-AM distortion. It incorporates 32 equally weighted unit amplifiers that run in parallel. Each unit amplifier stacks two NMOS-

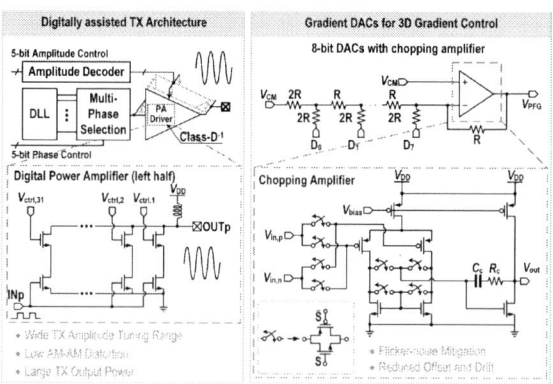

Fig. 4. Block and transistor level diagrams of the digitally assisted transmitter architecture and gradient DACs.

978-1-6654-8495-4/22 $31.00 © 2022 IEEE

Digital control logic (w/o config register)

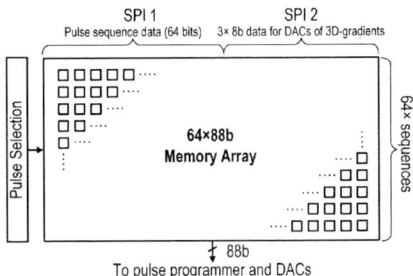

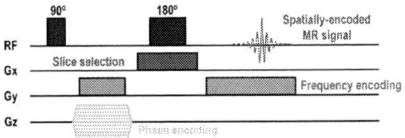

Fig. 5. Digital control logic and utilized MRI sequence.

transistors. The bottom transistor functions as a current source, driven by the phase-modulated RF signal, provided by the on-chip DLL, which produces any one of the 32 different RF phases equally spaced between 0 and 2π, with a phase resolution of $\pi/16$. The cascode transistor acts as a switch, enabled by a 5-bit binary to 31-bit thermometer decoder and controlled by the digital control logic. This enables an RF output amplitude modulation with 31 differing amplitude and 32 phase values, allowing for a broad range of RF pulse sequences.

The most important feature of the presented chip is its 3D-gradient field control (Fig. 4, right), which is essential for MRI. The 3D-gradient control incorporates three 8-bit digital-to-analog converters (DACs) to perform a 3D static magnetic field gradient control in the x, y, and z directions. Each DAC is of the R-2R topology and uses a chopper-stabilized op-amp in a negative feedback loop to mitigate offset, low-frequency drifts and $1/f$ noise for high fidelity. To synchronize relative timings between the field gradients and RF excitation signals, the gradient code for each DAC is stored in the on-chip memory bank, and linked to the corresponding pulse sequence address in the digital control logic (pulse programmer).

This digital control logic, whose simplified function is shown in Fig. 5, manages the entire operation of the chip. It not only controls the phase and amplitude of the RF transmitter (as discussed earlier), but also stores up to 64 sets of digital codes, representing specific pulse sequences, greatly improving the user-friendliness of the presented system. An additional 64-bit configuration register in the memory block is to control common signals and reduces the number of I/O-pads. Setting and refreshing of the memory is done through a standard serial peripheral interface (SPI).

IV. MEASUREMENT RESULTS

A. ASIC Characterization

The NMR/MRI transceiver ASIC, fabricated in a standard $0.18\,\mu m$ CMOS technology, occupies $4.98\,mm^2$ (Fig. 2). The measured receiver voltage gain ranges from 30 to 100 dB, which by choosing the maximum gain, leads to a measured input-referred voltage noise density at room temperature of $0.78 nV/\sqrt{Hz}$. The input-referred noise has been obtained by measuring the output voltage noise (differential input shorted) and dividing the results by the previously measured voltage gain.

In addition, we have characterized the RF transmitter. Fig. 6, top left shows the measurement results for the PA. For an output frequency of 21.8 MHz ($f_0 = 42.6$ MHz $\cdot B_0$), and a precision 1-Ω load resistor the measured output current of the PA is tunable from $25\,mA_{pp}$ to $290\,mA_{pp}$. With a power-matched load ($4.9\,\Omega$), the PA delivers a maximum output power of 220 mW, which certifies an output power tuning range in excess of 14 dB.

Furthermore, for the 3D gradient-control DACs we have measured DNLs less than +/- 0.2 LSB, sufficiently linear for high-resolution imaging (Fig. 6, top left).

B. NMR/MRI Results

Before acquiring the MR images, we first performed NMR spectroscopy (Fig. 6, bottom left) and relaxometry (Fig. 6, top right) experiments.

Fig. 6, bottom left, shows the measured ^{1}H NMR free induction decay (FID) from ethanol, and its corresponding NMR spectrum. Resolving the chemical shifts mandated a static magnetic field resolution better than 1 ppm, which we achieved by utilizing the gradient coils for first-order shimming.

Fig. 6, top right, shows the measured 2D correlations of spin-lattice relaxation time (T_1) and spin-spin relaxation time

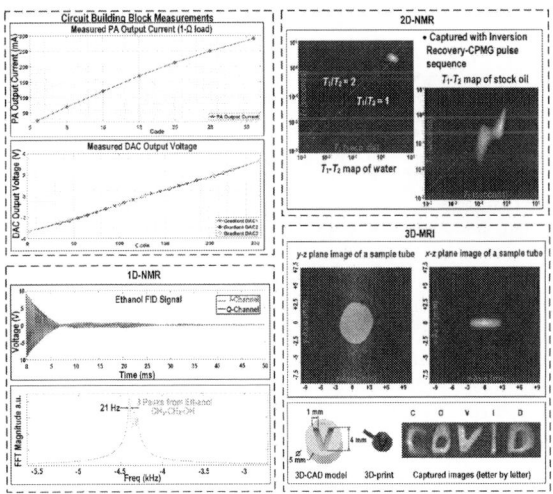

Fig. 6. Measurement results. PA and DAC output, 1D-, and 2D-NMR, as well as 3D-MRI.

TABLE I. Comparison with state-of-the-art.

Specifications	This work	[3] H. Bürkle ESSCIRC'21	[4] S. Hong VLSI'20	[5] K. Lei ISSCC'16	[6] N.Sun ISSCC'10	[7] K. Lei Anal. Chem.'20	[8] J.Handwerker Nat. Methods'20
System Performance							
Functionality	**NMR Relaxometry NMR Spectroscopy MRI**	NMR Relaxometry	NMR Relaxometry	NMR Relaxometry	NMR Relaxometry	NMR Relaxometry NMR Spectroscopy MRI	NMR Relaxometry NMR Spectroscopy MRI
Magnet	**Permanent (0.51 T)**	Permanent (1.1 T)	Permanent (0.5 T)	Permanent (0.46 T)	Permanent (0.5 T)	Permanent (0.51 T)	Superconducting (14.1 T)
Imaging	**On-Chip**	N/A	N/A	N/A	N/A	Off-Chip	Off-Chip
ASIC Performance							
Technology	**0.18 μm CMOS**	0.35 μm HV-CMOS	0.18 μm CMOS	0.18 μm CMOS	0.18 μm CMOS	0.18 μm CMOS	0.18 μm CMOS
Frequency Range [MHz]	**10-60**	0.1-200	21	20	21	10-40	600
RX IRN [nV/√Hz]	**0.78**	1.1	0.95	N/A	0.93	0.82	1.26
RX Gain [dB]	**30-100**	N/A	65-125	N/A	N/A	34-100	N/A
TX P_{out} [mW]	**220**	500	84	16	80	182	N/A
On-chip AM/PM	**Yes**	No	No	No	Yes	Yes	No
Area (mm²)	**5**	5.8	1.8	7.6	11.3	4	1.8

(T_2) for water and for stock oil. The more complex T_1-T_2 data for the oil are due to the composition of the coil with various weight hydrocarbon molecules.

For MRI, the three gradient coil pairs are used for slice selection, and phase and frequency encoding, respectively. The Golay coil pair (Fig.1, red coils) generates a field gradient in the x direction, the Maxwell coil pair (Fig. 1, blue coils) generates a field gradient in the y direction, and the Saddle coil pair (Fig. 1, yellow coils) generates a field gradient along the z direction. With the 0 to 3.3 V output voltage range of each DAC, within the magnet bore (diameter: 26 mm), the gradient coils are optimized for a field-of-view of 6 mm, and a field gradient up to 200 mT/m. Fig. 6, bottom right, shows an MRI image of the y-z cross-section of a tube containing water, and an MRI image of the x-z cross-section of the same water-filled tube. Furthermore, as also shown in the figure, we have imaged 5 different alphabet phantoms (1 mm deep trenches filled with water) on the x-z planes. The achieved spatial imaging resolution is as high as $67 \times 67 \times 83$ μm³.

Table I compares this work, the smallest MRI system consisting of a fully-integrated and digitally-assisted CMOS RF transceiver IC, with the recent CMOS-based magnetic resonance systems (mostly limited to NMR [3-6], and the two MRI works using either off-chip gradient control [7] and/or a superconducting magnet [8]).

V. Conclusion

In this paper we have presented the smallest MRI system combining a fully-integrated and digitally-assisted transceiver ASIC and a permanent magnet. For the first time, the gradient control has been implemented on-chip, greatly reducing the complexity of the off-chip electronics and the required programming efforts. In addition to MRI, the system can also perform multi-dimensional NMR relaxometry and high-resolution NMR spectroscopy. Combining all state-of-the-art NMR modalities in a single, portable device, this system

presents an important step towards on-site and on-demand imaging of e.g. *ex vivo* biological tissues and can thus further greatly impact medicine and biological/physical sciences.

Acknowledgment

We thank Dr. Isik Kizilyalli, program director of the Advanced Research Projects Agency-Energy (ARPA-E), for the support of this research under contract DE-AR0001063. This work was also funded in part by the University of Macau (MYRG2020-00132-IME).

References

[1] "Ultra high field MRI: BioSpec," *Bruker* [Online]. Available: https://www.bruker.com/en/products-and-solutions/preclinical-imaging/mri/biospec-ultra-high-field-mri.html [Accessed: 4/18/22].

[2] Jens Anders, Frederik Dreyer, Daniel Krüger, Ilai Schwartz, Martin B. Plenio, Fedor Jelezko, "Progress in miniaturization and low-field nuclear magnetic resonance," *Journal of Magnetic Resonance*, Volume 322, 2021.

[3] H. Bürkle, T. Klotz, R. Krapf and J. Anders, "A 0.1 MHz to 200 MHz high-voltage CMOS transceiver for portable NMR systems with a maximum output current of 2.0 App," *ESSCIRC 2021 - IEEE 47th European Solid State Circuits Conference(ESSCIRC)*,2021, pp.327-330.

[4] S. Hong and N. Sun, "A Portable NMR System with 50-kHz IF, 10-us Dead Time, and Frequency Tracking," *2020 IEEE Symposium on VLSI Circuits*, 2020, pp. 1-2.

[5] K.-M. Lei *et al.*, "28.1 A handheld 50pM-sensitivity micro-NMR CMOS platform with B-field stabilization for multi-type biological/chemical assays," *2016 IEEE International Solid-State Circuits Conference (ISSCC)*, 2016, pp. 474-475.

[6] N. Sun *et al.*, "Palm NMR and one-chip NMR," *2010 IEEE International Solid-State Circuits Conference - (ISSCC)*, 2010, pp. 488-489.

[7] K.-M. Lei, Dongwan Ha, Yi-Qiao Song, Robert M. Westervelt, Rui Martins, Pui-In Mak, and Donhee Ham, *Analytical Chemistry 2020* 92 (2), 2112-2120.

[8] Handwerker, J., Pérez-Rodas, M., Beyerlein, M. *et al.* A CMOS NMR needle for probing brain physiology with high spatial and temporal resolution. *Nat Methods* **17,** 64–67 (2020).

A Sub-0.01° Phase Resolution 6.8-mW fNIRS Readout Circuit Employing a Mixer-First Frequency-Domain Architecture

Cheng Chen[1,2], Zhouchen Ma[2], Yaxin Liu[2], Zhenhong Liu[2], Linfeng Zhou[2], Yan Wu[2],
Liang Qi[2], Yongfu Li[2], Mohamad Sawan[3], Guoxing Wang[1,2], Jian Zhao[2]

[1]National Key Laboratory of Science and Technology on Micro/Nano Fabrication, Shanghai Jiao Tong University, China.
[2]Department of Micro/Nano Electronics, Shanghai Jiao Tong University, China.
[3]Westlake University, China. (Email: zhaojianycc@sjtu.edu.cn)

Abstract—**This paper presents a frequency-domain functional near-infrared spectroscopy readout circuit for absolute value measurement of tissue optical characteristics. A mixer-first analog front-end (AFE) structure and a 1-bit Σ-Δ phase-to-digital converter (PDC) are proposed to lower both the required circuit bandwidth and the laser modulation frequency to save the power while maintain high resolution. The IC achieves sub-0.01° phase resolution and 6.8-mW power consumption. 9 optical solid phantoms are produced to evaluate the chip. Compared to a self-built high-precision measurement platform which combines a network analyzer with an APD module, the maximum measuring errors of absorption coefficient and reduced scattering coefficient are 14.2% and 16.5%, respectively.**

Index Terms—*frequency domain near-infrared spectroscopy, mixer-first AFE structure, phase-to-digital converter*

I. INTRODUCTION

Functional near infrared spectroscopy (fNIRS)-based optical imaging is a noninvasive and non-ionizing neuroimaging technology applied in functional brain imaging, brain-computer interface and psychology studies, etc [1]. Generally, fNIRS has a better spatial resolution than electroencephalogram (EEG) techniques, a lower cost than MRI [2]. Depending on natural advantage of artifact insensitivity, fNIRS is promising to realize nature-scenario operation, which is crucial in many wearable applications [3].

Continuous-wave (CW) scheme is a common and simple implementation in fNIRS, which injects light with constant intensity into the brain, and detects the amplitude change of the outgoing light to obtain the relative change of the absorption coefficient ($\Delta\mu_a$). Since light source does not require modulation and the bandwidth of the readout circuit only needs to cover the near-DC range, the energy efficiency of the CW-fNIRS is well optimized in the previous state-of-the-art work, reaching the μW level [4]–[6]. However, CW-fNIRS only measures light amplitude attenuation in the tissue, resulting in insensitive to any information about the neural activity encoded in the time-of-flight of photons through the brain [7]. Therefore, CW-fNIRS cannot detect the absolute value of absorption coefficient (μ_a) and the reduced scattering coefficient (μ_s'). For this reason, frequency-domain (FD) fNIRS modulates the optical source to radio frequency, and

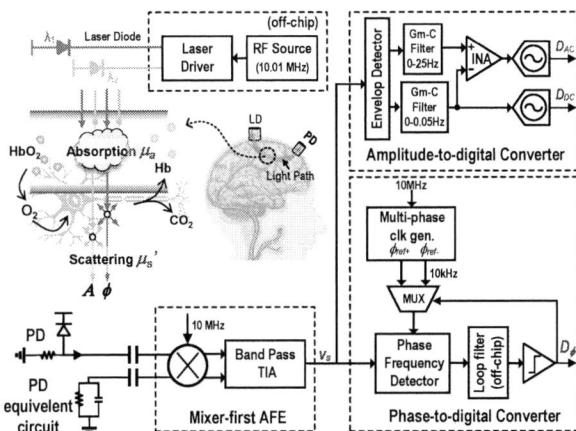

Fig. 1. Block diagram of the FD-fNIRS circuit.

detects the amplitude (A) and phase (Φ) of the outgoing light simultaneously to detect absolute value of μ_a and μ_s' [8]. Moreover, it has proved that fNIRS imaging with phase information (FD-fNIRS) can achieve higher spatial resolution than that with only amplitude information (CW-fNIRS) [1], and more robust to motion artifact [7]. It has potential to achieve a non-invasive multi-modal high-resolution functional brain imaging.

However, the structure of FD-fNIRS system is complicated, which increases hardware cost and volume [7]. It is inconvenient for brain research in nature-scenario, especially for those requiring multiple subjects like hyper-scanning research [9]. FD-fNIRS integrated circuits are designed to miniaturize the system, and cut down the power consumption [2] [10], where heterodyning mixers are employed after the trans-impedance amplifier (TIA) to amplify the time delay, and thus mitigate the requirement of phase detectors. However, the existing FD-fNIRS technique encounters a power-accuracy dilemma, a higher modulation frequency (>80 MHz) will mitigate the required resolution of the phase detector or PDC. However it will significantly increases the power consumption of the front-end TIA. Therefore, the key design issue is to reduce the power overhead of front-end TIA and improve the precision

978-1-6654-8495-4/22 $31.00 © 2022 IEEE

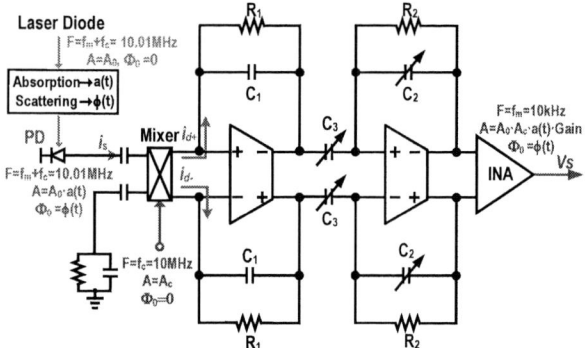

Fig. 2. Schematic of mixer-first AFE.

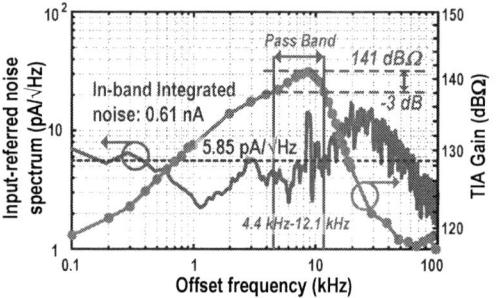

Fig. 3. Amplitude-frequency response and input-referred noise spectrum of AFE.

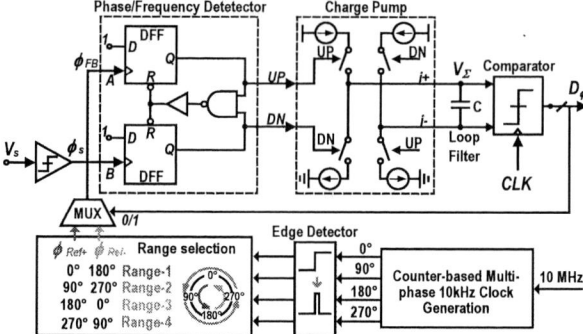

Fig. 4. Schematic of the proposed 1-bit Σ-Δ PDC.

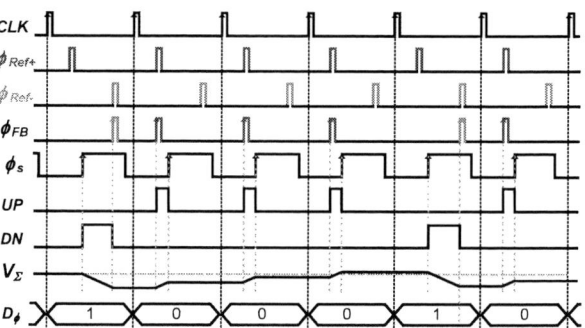

Fig. 5. The example timing diagram of the proposed 1-bit Σ-Δ PDC.

of PDC.

To improve the energy efficiency of FD-fNIRS chip, this work has the following features: 1) A mixer-first AFE structure mitigates the required bandwidth of TIA to save power dramatically and improve trans-impedance gain; 2) An on-chip Σ-Δ PDC is proposed to lower the laser modulation frequency to 10 MHz while achieving a 0.0096° phase resolution.

Section II introduces both the system and circuit-level implementation of the proposed FD-fNIRS chip. Section III shows the measurement results and performance comparison with the state-of-the-art works. Finally, section IV concludes this article.

II. PROPOSED FD-fNIRS READOUT CIRCUIT

Fig. 1 shows the proposed FD-fNIRS circuit block diagram, a 10.01-MHz modulated light from laser diode injects into the brain. When the neurons located on the light-path are active, the μ_a and μ'_s of the active tissue will change and result in amplitude and phase responses of the received photodiode (PD) light current. This current is demodulated and amplified by a mixer-first AFE with a 10-MHz demodulating clock. The heterodyned low-side-band 10-kHz signal (V_s) maintains the amplitude and phase information, which is fed into both a PDC and an amplitude-to-digital converter, respectively for the subsequent μ_a and μ'_s detection. Both converters are employed with a noise shaping architecture to improve the in-band resolution.

Fig. 2 shows the AFE circuit, which consists of a mixer and a band-pass TIA. Utilizing the input signal current property, a passive mixer is used to down-convert the signal frequency first from 10.01 MHz to 10 kHz to save the power of TIA. The band-pass TIA consists of an integrator, followed by a differentiator and INA to form an inherent band-pass filter with 10-kHz center frequency to filter out the upper side harmonics after demodulation. Another benefit is that the trans-capacitance integrator in the first stage reduces the resistive current noise of TIA. The measured performance of AFE is showns in Fig. 3, where the amplitude gain is 141 dBΩ and the input referred current noise is 5.85 pA/$\sqrt{\text{Hz}}$.

Fig. 4 shows the proposed PDC with 1st-order phase-domain Σ-Δ architecture. Fig. 5 shows the example of the proposed PDC operation. The comparator converts the input signal V_s into a digital signal ϕ_s firstly. The tri-state phase/frequency detector (PFD) and charge pump converts phase difference between ϕ_s and the phase feedback signal ϕ_{FB} to charging/discharging current ($i+$, $i-$), whose pulse width is proportional to the phase difference. A integrator capacitor is connected after the charge pump as the loop filter. Then, a comparator quantizes the integral output voltage V_Σ into a 1-bit digital code D_ϕ to alternatively connect one of the two reference phase signals ϕ_{ref+} and ϕ_{ref-} to the other input of the PFD. Four reference phase signals are generated by a counter-based clock generator. ϕ_{ref+} and ϕ_{ref-} are manually selected from range-1 to range-4. It is noted that the phase of *CLK* is equal to ϕ_{ref-}+90°. The 1-bit stream (D_ϕ) will convert to a digital phase output through an off-chip decimation filter. Benefit from the 1-bit phase feedback and

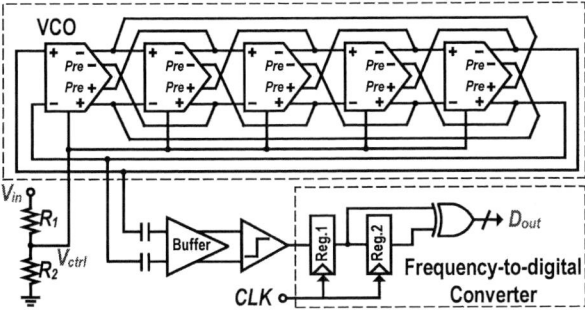

Fig. 6. Schematic of the VCO-based ADC.

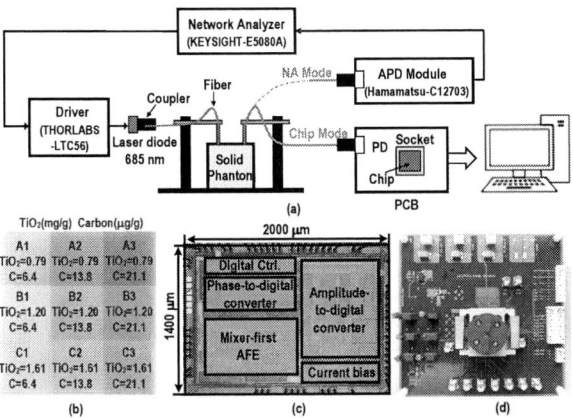

Fig. 7. (a) Optical characteristic measurement setup with proposed chip and instrument-based reference system, (b) 9 optical solid phantoms with different doping concentrations, (c) die photo, (d) testing PCB.

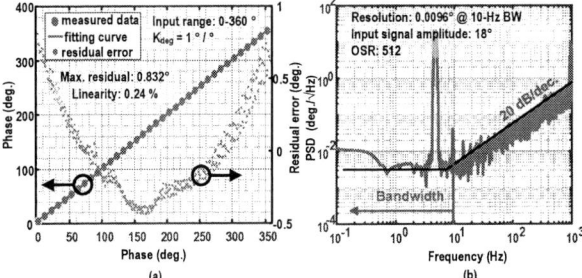

Fig. 8. Measured (a) linearity and (b) noise of PDC.

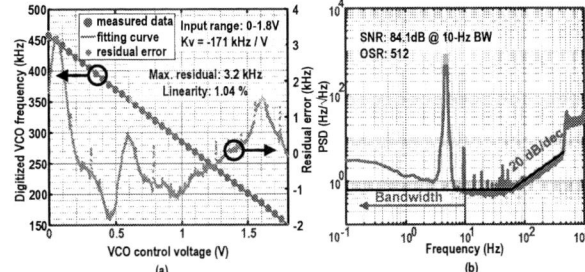

Fig. 9. Measured (a) linearity and (b) noise of VCO-based ADC

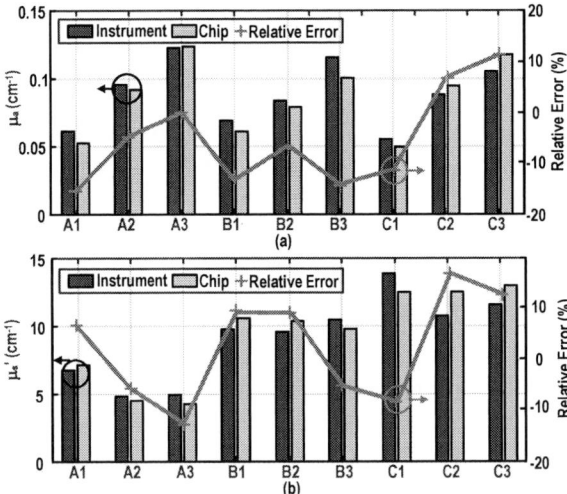

Fig. 10. Measured absolute value of (a) μ_a and (b) μ'_s comparison between the proposed FD-fNIRS chip and instrument-based reference system.

noise shaping, the proposed PDC achieves higher resolution and linearity, compared with conventional open-loop PDC [11].

In the amplitude-to-digital converter, the envelope detector [12] demodulates the amplitude information of the fNIRS signal. Two Gm-C filters and instrumentation amplifier (INA) amplify AC (>0.05 Hz) and DC (<0.05 Hz) of amplitude signals independently because of their small perfusion index (PI, AC/DC ratio). Two open-loop VCO-based ADCs digitize the amplified signals respectively. Fig. 6 shows the VCO-based ADC circuit. A linearity-enhanced VCO with large input range (0-1.8V) is employed, which offers proportional negative feedback to limit the original VCO-control voltage (V_{ctrl}). Frequency-to-digital converter with 2-DFF and 1-XOR quantizes the VCO's frequency with a 1st-order quantization noise shaping.

III. MEASUREMENT RESULTS AND PERFORMANCE COMPARISON

The readout circuit was implemented in a 0.18-μm CMOS technology and occupies 2.8 mm^2. The die photo and testing PCB are shown in Fig. 7 (c) and (d), respectively. The measured performance of PDC is shown in Fig. 8. The maximum residual error of PDC is 0.24% within $0° \sim 360°$ input range. The measured phase noise density is $0.003°/\sqrt{Hz}$ and the phase error is $0.0096°$ in 10-Hz bandwidth by applying a

signal with sinusoidal varying phase shift at the input of PDC. The measured linearity and noise of the VCO-based ADC is shown in Fig. 9. It achieves 1% linearity within 0~1.8V input voltage range, and the signal-to-noise ratio (SNR) is 84.1 dB.

To demonstrate the IC's capability of absolute tissue optical characteristics measurement, 9 resin based optical solid phantoms with different μ_a and μ'_s shown in Fig. 7 (b) are molded with different doping concentrations of nano-carbon and titanium oxide (TiO$_2$) powders, which adjust the μ_a and μ'_s of the phantoms, respectively. Fig. 7 (a) shows optical measurement setup. Network Analyzer (Keysight-E5080A) controls laser diode controller kit (Thorlabs-LTC56) to generate 10.01-MHz AC current to drive a 685-nm laser diode (HL6750MG), while DC bias current is self-generated by the current controller in

TABLE I
PERFORMANCE SUMMARY COMPARED WITH THE STATE-OF-THE-ART WORKS.

Parameter	JSSC'2018 [4]	ISSCC'2018 [5]	ISSCC'2021 [6]	TbioCAS'2017 [10]	TbioCAS'2018 [2]	This Work
NIRS Technology	CW	CW	CW	FD	FD	FD
Laser Modulation Frequency	DC	DC	DC	80MHz	80MHz	**10MHz**
Technology	65nm	180nm	180nm	130nm	130nm	180nm
Power Supply	1.2/3.3V	1.2/3.3V	1.2/3.3V	1.8/1.2V	1.5V	1.8V
Power Consumption	$30\mu W$	$132\mu W$	$28\mu W$	18mW (AFE)	30mW (AFE)	**6.8mW**
AFE Maximum Gain	235dBΩ	100dBΩ	114dBΩ	100dBΩ	100dBΩ	**141dBΩ**
AFE Dynamic Range	60dB	87dB	134dB	60dB	33dB	60dB
AFE Input Referred Noise	N.A.	$200pA/\sqrt{Hz}$	$10pA/\sqrt{Hz}$	$\sim 8pA/\sqrt{Hz}$ *	$3.2pA/\sqrt{Hz}$ **	$5.85pA/\sqrt{Hz}$
Amplitude Linearity	N.A.	N.A.	N.A.	$\sim 2\%$*	$\sim 3\%$*	1%
Amplitude SNR	74dB	87dB	95.96dB	98dB (off-chip)	98dB (off-chip)	84.1dB
PDC Linearity				$\sim 1.50\%$*	$\sim 1\%$*	**0.24%**
PDC Resolution		Unable to measure the phase Φ		0.2° (off-chip)	0.2° (off-chip)	**0.0096°**
PDC Bandwidth		and the absolute value of μ_a and μ_s'		5Hz	10Hz	10Hz
μ_a Maximum Error				30%	23%	**14.2%**
μ_s' Maximum Error				30%	23%	**16.5%**

* Estimated from the measurement results in the publications. ** Simulated results.

the kit. The laser is irradiated on the surface of the phantom vertically by fiber coupling. The outgoing light is similarly coupled to the PD, which is welded on the back of the chip testing PCB. For comparison, a reference measuring system is established, which substitutes the chip with a low noise off-the-shelf APD module (Hamamatsu-C12703). All the 9 solid phantoms are measured alternately using the reference system and our chip. The results in Fig. 10 show that the maximum error for μ_a and μ_s' of our chip are 14.2% and 16.5%, compared with the reference system.

Table I presents the performance summary and comparison with state-of-the-art fNIRS readout circuits. The CW-fNIRS chips cannot measure the absolute value of μ_a and μ_s' [4]–[6]. Compared with existing FD-fNIRS ICs which only provide analogue output [2] [10], this work has both digital output for amplitude and phase responses of optical current. Thanks to the mixer-first AFE structure and high-resolution Σ-Δ PDC, phase resolution is improved by 20× and the modulation frequency is reduced by 8×, while power consumption is saved by 2.6~4.4×.

IV. CONCLUSION

In this work, we propose a mixer-first FD-fNIRS readout circuit to measure absolute value of tissue optical characteristics. A mixer-first AFE structure is adopted to save the power consumption of TIA dramatically and increase tran-impedance gain. Furthermore, an on-chip 1-bit Σ-Δ PDC with 0.0096° phase resolution and 0.24% phase linearity is proposed for higher phase resolution and lower laser modulation frequency. Fabricated in 180-nm technology and occupying 2.8 mm^2, the FD-fNIRS chip achieves 14.2% and 16.5% maximum error of μ_a and μ_s', while only consumes 6.8-mW power under a 1.8-V supply.

REFERENCES

[1] M. Doulgerakis, A. T. Eggebrecht, and H. Dehghani, "High-density functional diffuse optical tomography based on frequency-domain mea-

surements improves image quality and spatial resolution," *Neurophotonics*, vol. 6, no. 3, pp. 1–14, 2019.

[2] Y. Miao and V. J. Koomson, "A CMOS-Based Bidirectional Brain Machine Interface System With Integrated fdNIRS and tDCS for Closed-Loop Brain Stimulation," *IEEE Transactions on Biomedical Circuits and Systems*, vol. 12, no. 3, pp. 554–563, 2018.

[3] C. Chen, Z. Ma, Z. Liu, L. Zhou, G. Wang, Y. Li, and J. Zhao, "An Energy-Efficient Wearable Functional Near-infrared Spectroscopy System Employing Dual-level Adaptive Sampling Technique," *IEEE Transactions on Biomedical Circuits and Systems*, pp. 1–10, 2022.

[4] U. Ha, J. Lee, M. Kim, T. Roh, S. Choi, and H.-J. Yoo, "An EEG-NIRS Multimodal SoC for Accurate Anesthesia Depth Monitoring," *IEEE Journal of Solid-State Circuits*, vol. 53, no. 6, pp. 1830–1843, 2018.

[5] J. Xu, M. Konijnenburg, B. Lukita, S. Song, H. Ha, R. van Wegberg, E. Sheikhi, M. Mazzillo, G. Fallica, W. De Raedt, C. Van Hoof, and N. Van Helleputte, "A 665 μW silicon photomultiplier-based NIRS/EEG/EIT monitoring asic for wearable functional brain imaging," in *2018 IEEE International Solid - State Circuits Conference - (ISSCC)*, 2018, pp. 294–296.

[6] Q. Lin, S. Song, R. V. Wegberg, M. Konijnenburg, D. Biswas, C. V. Hoof, F. Tavernier, and N. V. Helleputte, "A 28 μW 134dB DR 2nd-Order Noise-Shaping Slope Light-to-Digital Converter for Chest PPG Monitoring," in *2021 IEEE International Solid- State Circuits Conference (ISSCC)*, vol. 64, 2021, pp. 390–392.

[7] J. J. Wathen, M. J. Fitch, V. R. Pagán, G. W. Milsap, E. G. McDowell, L. Spietz, Z. E. Markow, J. W. Trobaugh, E. J. Richter, A. T. Eggebrecht, J. P. Culver, D. W. Blodgett, and S. M. Hendrickson, "A 32-channel frequency-domain fNIRS system based on silicon photomultiplier receivers," in *Optical Techniques in Neurosurgery, Neurophotonics, and Optogenetics*, vol. 11629, 2021, pp. 62–79.

[8] S. Fantini and A. Sassaroli, "Frequency-Domain Techniques for Cerebral and Functional Near-Infrared Spectroscopy," *Frontiers in Neuroscience*, vol. 14, pp. 1–18, 2020.

[9] Y. Pan, S. Dikker, P. Goldstein, Y. Zhu, C. Yang, and Y. Hu, "Instructor-learner brain coupling discriminates between instructional approaches and predicts learning," *NeuroImage*, vol. 211, pp. 1–13, 2020.

[10] C. C. Sthalekar, Y. Miao, and V. J. Koomson, "Optical Characterization of Tissue Phantoms Using a Silicon Integrated fdNIRS System on Chip," *IEEE Transactions on Biomedical Circuits and Systems*, vol. 11, no. 2, pp. 279–286, 2017.

[11] R. Yun and V. M. Joyner, "A Monolithically Integrated Phase-Sensitive Optical Sensor for Frequency-Domain NIR Spectroscopy," *IEEE Sensors Journal*, vol. 10, no. 7, pp. 1234–1242, 2010.

[12] B. Lin, Z. Ma, M. Atef, L. Ying, and G. Wang, "Low-power high-sensitivity photoplethysmography sensor for wearable health monitoring system," *IEEE Sensors Journal*, vol. 21, no. 14, pp. 16 141–16 151, 2021.

978-1-6654-8495-4/22 $31.00 © 2022 IEEE

A 4.7GHz Synchronized-Multi-Reference PLL with In-Band Phase Noise Lower than Reference Phase Noise + 20logN$_{div}$

Hongzhuo Liu, Wei Deng, Haikun Jia, Shiyan Sun, Qixiu Wu,
Jiajie Tang, Zhihua Wang, and Baoyong Chi

School of Integrated Circuits, BNRist, Tsinghua University, China

Email: wdeng@tsinghua.edu.cn

Abstract—This paper presents a PLL using synchronized multiple references with in-band phase noise (PN) lower than the conventional theoretical limit of reference phase noise + 20logN$_{div}$. The proposed technique provides design flexibility to alleviate the jitter-power trade-off in PLL. Realized in 65-nm CMOS, the prototype achieves in-band PN 1.7-dB lower than the conventional theoretical limit, with four synchronized references. The proposed PLL can be configured in a single reference mode achieving 0.54-ps integrated jitter. In the contrast, the proposed PLL using four synchronized references achieves 0.35-ps integrated jitter.

Keywords—phase-locked loops (PLL), phase noise (PN), reference clock

I. INTRODUCTION

The advancing of various applications requires lower clock jitter, demanding better PN performance of PLLs [1]. The out-of-band PN is determined by the VCO, in which there is a well-known trade-off relation between PN and power consumption. The in-band PN is mainly determined by the reference, usually a crystal oscillator (XO) without the possibility to trade off PN against power. To optimize the overall PN, the PN contributions of the reference and of the VCO are often designed to be approximately equal. As analyzed in [1], the bottleneck of reference PN results in a stringent jitter-power trade-off in PLL. In order to provide flexibility and scalability in reference PN reduction, this paper introduces a multi-reference phase averaging technique, using synchronized voltage-controlled XO (VCXO) as references.

The proposed technique can be used to replace a low-PN but expensive reference oscillator by multiple poor-PN but cheap reference oscillators, and it is particularly suitable for the MEMS oscillators that have poor PN performance but may well be used in quantities. MEMS oscillators are manufactured in batches using silicon process, and have many superiorities over the quartz oscillators, such as smaller size, slower aging, better reliability, and packaging compatibility with IC, as illustrated in Fig. 1. However, the poor PN of the existing MEMS oscillators limits the achievable PLL in-band PN and the overall jitter, to which this paper provides a solution. In this paper, VCXO means voltage-controlled MEMS oscillator.

Manufacturing	Packaging	Reliability	Phase Noise
MEMS✓ Quartz✗	Quartz / MEMS	MEMS✓ Quartz✗	PN / MEMS / Quartz / t
☺ Silicon process; Batch production.	☺ Smaller size; Compatible with IC.	☺ Slower aging; Temperature and vibration stable.	☹ Poor PN.

Fig. 1 MEMS oscillator compared with quartz oscillator.

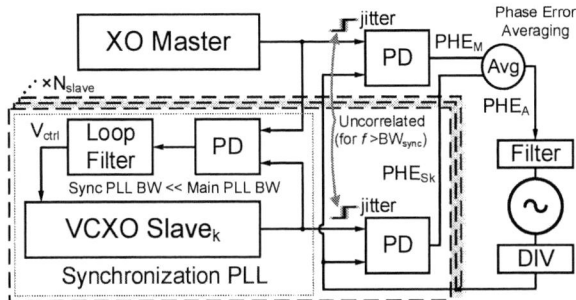

Fig. 2: Conceptual diagram of the proposed synchronized-multi-reference PLL.

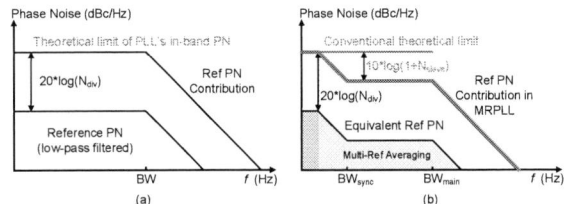

Fig. 3: Reference PN contribution of (a) conventional PLL, and (b) multi-reference PLL.

II. THE PROPOSED MULTI-REFERENCE PLL

A. Principles of the Multi-Reference Technique

The idea is to utilize the fact that different references are independent PN sources, and their PNs are uncorrelated. By averaging multiple references in the phase domain, the Multi-Reference PLL (MRPLL) could obtain an equivalent reference with lower PN, and consequently reduce the in-band PN of the PLL. The key question in implementing the proposed multi-reference technique is to synchronize the references. Therefore, the MRPLL uses VCXOs instead of XOs as the references. There are additional loops to phase-lock the VCXOs.

Fig. 2 shows a conceptual diagram of the MRPLL. There is one master XO and N$_{slave}$ slave VCXOs, and every slave is phase-locked by the master with a synchronization PLL. The bandwidth of the synchronization PLL (BW$_{sync}$) is much less than the main PLL (BW$_{main}$). At frequencies lower than BW$_{sync}$, the master VCXO dominates the overall PN, which is similar to a conventional PLL. For BW$_{sync}$ < f < BW$_{main}$, the PN of VCXO master and slaves are uncorrelated and each of the references equally contributes to the overall PN. The MRPLL averages the phase errors (PHE) produced by the phase detectors of all references, therefore the equivalent reference PN could be reduced by 10*log(1+N$_{slave}$) dB, as shown in Fig. 3.

978-1-6654-8495-4/22 $31.00 © 2022 IEEE

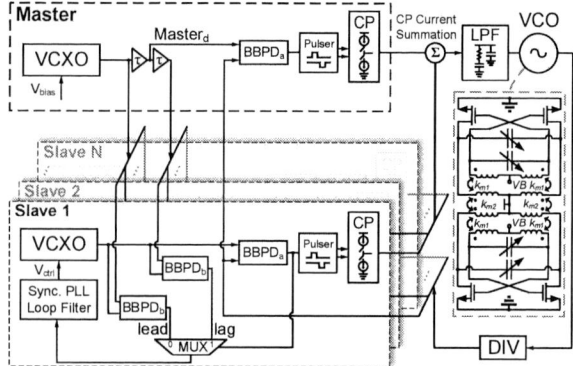

Fig. 4: Block diagram of the proposed MRPLL.

B. PLL Architecture

This work uses bang-bang phase detector (BBPD) for its simplicity. Due to its highly nonlinear nature, the valid input range of BBPD is very small [2]-[3]. Therefore, the relatively large mismatch of input offsets of different BBPDs may saturate the BBPDs. Fig. 4 shows the detailed block diagram of the proposed MRPLL. The BBPD used for the main PLL and the synchronization PLL are denoted as $BBPD_a$ and $BBPD_b$, respectively. There are 2 delays (τ), 2 $BBPD_b$s and a MUX in the VCXO synchronization mechanism for facilitating the locking procedure. The master VCXO is delayed by τ ($\tau << T_{ref}$) and sent into the master $BBPD_a$. This delayed signal is named 'Master$_d$'. The output of a $BBPD_a$ determines the polarity of a pulse generator, which controls a charge pump (CP). The CP currents are summed up to accomplish PHE averaging.

C. Synchronization Method

When the VCXO master and slaves are not synchronized, and the VCO is not locked, the output of slave $BBPD_a$ are uncertain, almost random. The MUX selects one of the two $BBPD_b$s randomly, and equivalently it is one $BBPD_b$ with an uncertain input time offset ε ($\varepsilon = -\tau$ or τ). Such a $BBPD_b$ with an uncertain offset behaves like a "noisy" BBPD, and the equivalent circuit is shown in Fig. 5 (a). It is adequate to initially pull the slave VCXOs. Simulations on behavioral models showed that this mechanism works with τ up to 5% T_{ref}.

When the VCXOs are synchronized, the time difference between the rising edges of a slave and Master$_d$ is less than τ. The 'lead' signal in Fig. 4 is constantly 1, and the 'lag' signal is constantly 0. The MUX works as an inverter, inverting the output of a slave $BBPD_a$ and send it to the synchronization loop filter, as shown in Fig. 5 (b). The synchronization loop filter is a proportional-integral filter on PCB.

The locking procedure of the proposed MRPLL has 4 steps as shown in Fig. 6. First, the VCXO slaves are pulled toward the VCXO master. Second, all the slaves are locked within an interval ($-\tau,\tau$) around Master$_d$, and the VCO starts to be pulled toward this cluster of references. Third, all the slaves and the VCO are coarsely locked within a narrow interval and they are tracking Master$_d$, The length of the interval is decreasing, until it is less than the valid input range of $BBPD_a$. Fourth, slave VCXOs and VCO are strictly phase-locked by Master$_d$ and the MRPLL goes into the LTI model. With the help of this coarse-fine synchronization mechanism, the MRPLL is immune to a sudden perturbation.

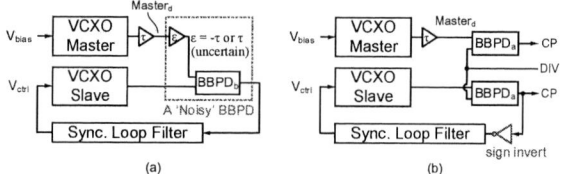

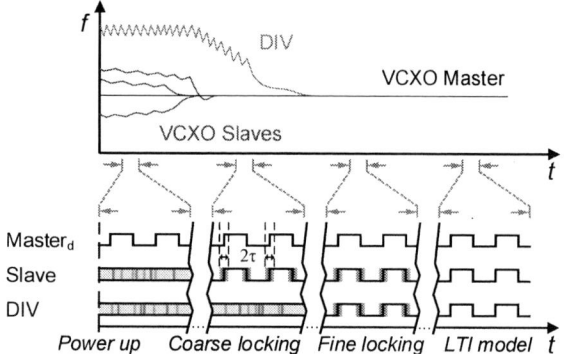

Fig. 5: Equivalent circuit when the VCXO master and the VCXO slaves are: (a) not synchronized, (b) synchronized.

Fig. 6: Synchronization procedure of the VCXOs and the VCO

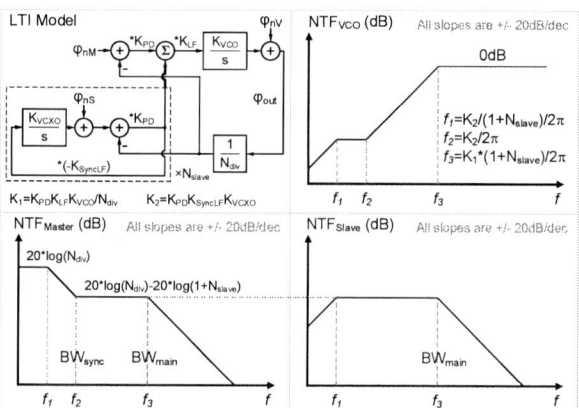

Fig. 7: LTI model and noise transfer functions of the VCO, the VCXO master, and the VCXO slave.

D. LTI Model

Fig. 7 shows the LTI model of the proposed MRPLL and the noise transfer functions (NTF) of the VCO, the VCXO master, and the VCXO slave. The LTI model is corresponding to the equivalent circuit in Fig. 5 (b). All the loop filters are approximated as only proportional gain K. It is assumed that the loop gain of the main PLL (K_1) is much larger than the loop gain of the synchronization PLL (K_2) in deriving these formulas. Note that BW_{main} would increase linearly with the number of references if the loop gain were constant, which is an intuitive result.

For $f_2 < f < f_3$, the NTF_{Master} and NTF_{Slave} are equally $20*\log(N_{div}) - 20*\log(1+N_{slave})$ dB. Assuming that the PN of a VCXO is x dBc/Hz, then the overall reference PN contribution is $x + 20*\log(N_{div}) - 10*\log(1+N_{slave})$ dBc/Hz. There is a $10*\log(1+N_{slave})$ dB reduction in the equivalent reference PN, offering the opportunity to reduce PLL's in-band PN by using more VCXOs.

978-1-6654-8495-4/22 $31.00 © 2022 IEEE

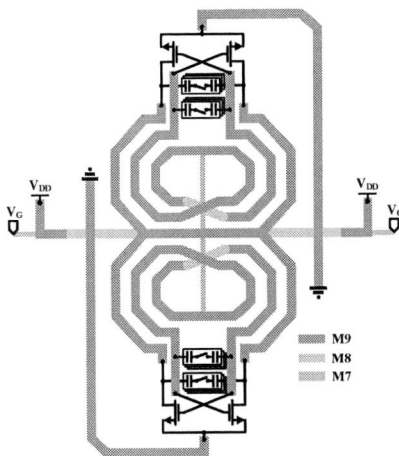

Fig. 8: Schematic of the proposed dual-core Class-F VCO with tail filtering.

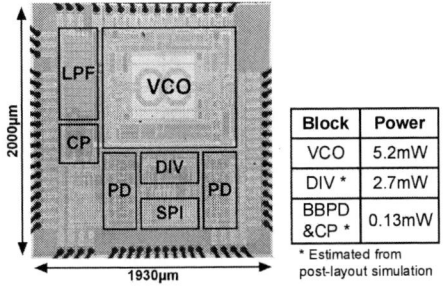

Fig. 9: Die micrograph and power consumption.

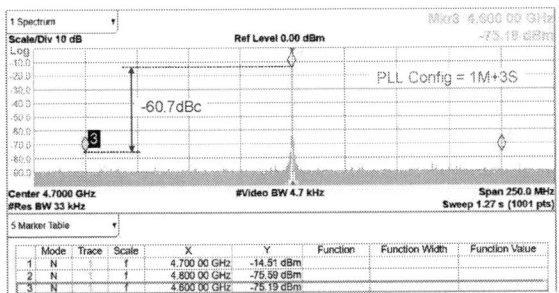

Fig. 10: Measured spectrum.

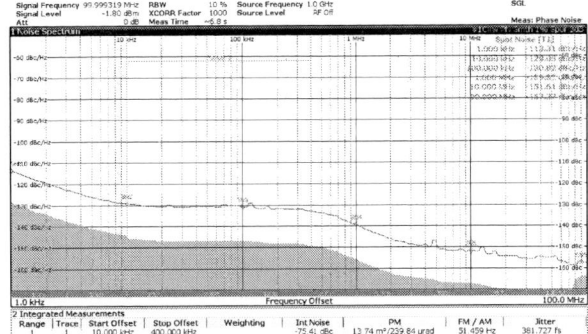

Fig. 11: Measured phase noise of the MEMS VCXO.

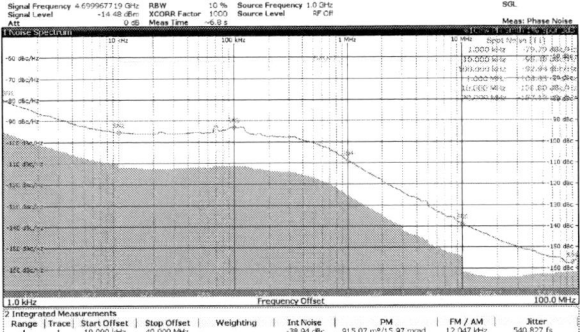

Fig. 12: Measured phase noise of the PLL with 1 reference (1M).

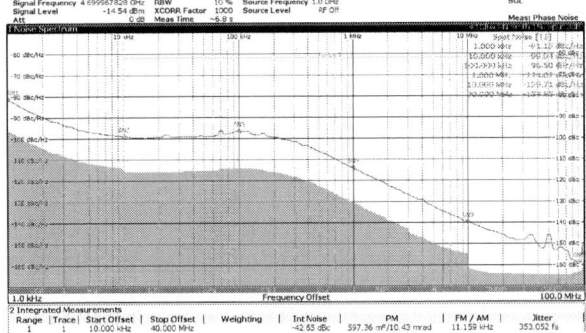

Fig. 13: Measured phase noise of the PLL with 4 references (1M+3S).

E. Circuit Implementation

Fig. 8 shows the schematic of the VCO. Different from the dual-core inductor-coupling VCO in [4], each core in the proposed VCO is an independent Class-F topology with a high-k transformer, enabling that the first and third harmonic frequency change simultaneously when tuning the two capacitor banks at gate and drain, which achieves one-dimensional tuning and reduces the tuning complexity of the Class-F VCO. The PN performance of the VCO is further improved by implementing the tail filtering at the common source of the cross-coupled transistors. The bias and power are fed through the center tap of the transformer. The VCO covers the frequency range of 4.6-to-5.8 GHz.

III. EXPERIMENTAL RESULTS

A. Measurement Results

The chip was fabricated in TSMC 65nm CMOS process. Fig.9 shows the die micrograph and power consumption. Fig. 10 shows the measured spectrum.

Fig. 11 shows the measured PN of the VCXO, which is a commercial product of voltage-controlled MEMS oscillator. There are spurs in the VCXO spectrum, which causes spurs in the closed-loop PLL spectrum. The VCXO has 0.38-ps jitter integrated from 10 kHz to 400 kHz. The VCXO frequency is 100 MHz and the PLL output frequency is 4.7 GHz. PLLs using different numbers of VCXOs are tested, as shown in Fig. 12 (1 master VCXO) and Fig. 13 (1 master VCXO and 3 slave VCXOs). The PLL bandwidth is around 200 kHz. The MRPLL with 4 VCXOs achieves 0.35-ps jitter integrated from 10 kHz to 40 MHz.

Fig. 14 shows a comparison of the PN (10 kHz – 100 kHz) with and without the multi-reference technique. The theoretical limit in a conventional PLL is calculated as VCXO PN + 20*log(47) dB. With the slave VCXOs removed, the PLL becomes a conventional single-reference PLL, and the PN is higher than the theoretical limit. With 4 VCXOs, the MRPLL achieves in-band PN lower than the conventional theoretical limit, proving the feasibility of the proposed multi-reference technique.

TABLE I. COMPARISON WITH PRIOR WORK

	This work		[5] CICC'17	[6] JSSC'16	[7] JSSC'16
	Conventional 1-Ref	Proposed 4-Ref			
Technology (nm)	65		65	65	40
Power (mW)	8.0	8.4	5.86	4.2	3.8
Reference Freq. (MHz)	100		49.15	100	55
Output (GHz)	4.7		2.36	2.2	3.96
RMS jitter (ps)	0.54	0.35	0.63	0.38	0.64
	(10k-40M)		(1k-100M)	(10k-40M)	(1k-100M)
In-band PN (dBc/Hz) @ Offset Freq. (Hz)	-95.6	-98.9	-123.5	-112	-94.4*
	@60k		@300k	@300k	@50k
Reference-induced theoretical limit of in-band PN (dBc/Hz)	-97.2		-126.4	-115.5*	-97.0*
	@60k		@300k	@300k	@50k
Difference between in-band PN and the theoretical limit (dB)	1.6	-1.7	2.9	3.5	2.6
Jitter FoM** (dB)	-236	-240	-236	-242	-238

* Estimated from figures

** Jitter FoM = 20*log(Jitter/1sec) + 10*log(Power/1mW)

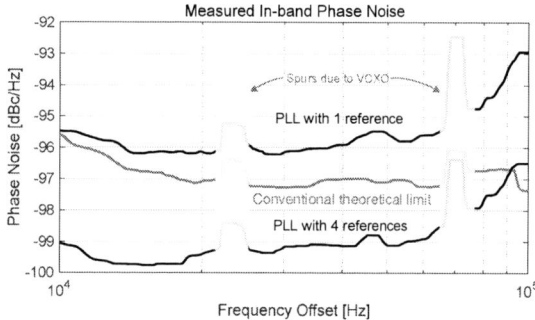

Fig. 14: Comparison of the in-band PN of the PLL with/without the multi-reference technique.

B. Comparison with Prior Work

Most of the literatures tend to choose a reference clock with lowest available PN, in order to minimize the PLL's overall jitter. However, there is a trade-off between reference's PN performance and its cost. Table I summarizes some of the previously reported PLLs that used a reference clock with plain PN performance. Using similar poor-PN reference, this work achieves similar FoM and nearly-half integrated jitter compared with [7], which would be impossible if it were not for the multi-reference technique.

For the conventional single-reference PLLs, there is a reference-induced theoretical limit of in-band PN, and none of the prior work could have in-band PN lower than that limit. This work achieves in-band PN 1.7dB lower than the former theoretical limit.

IV. CONCLUSION

This paper presents a new PLL architecture with multiple references to reduce PLL's in-band PN. The prototype 4.7GHz MRPLL achieves in-band PN 1.7dB lower than the former theoretical limit of reference PN + 20*log(N_{div}) dB, which demonstrates the effectiveness of the proposed technique.

The proposed multi-reference technique enables the trade-off between PLL's in-band PN and the number of references, providing additional design flexibility to alleviate the performance bottleneck imposed by the reference PN.

This architecture has good scalability, and there may be various types of MRPLL in the future.

ACKNOWLEDGMENT

This work was partly supported by the National Key R&D Program of China under Grant No. 2020YFB1807300, the Shenzhen Science and Technology Program (No. JCYJ20200109113601723 and No. JCYJ20210324115610028), the Tsinghua-Samsung Joint Research Project, and the Beijing Innovation Center for Future Chips (ICFC), Tsinghua University.

REFERENCES

[1] B. Razavi, "Jitter-Power Trade-Offs in PLLs," in IEEE Transactions on Circuits and Systems I: Regular Papers, vol. 68, no. 4, pp. 1381-1387, April 2021, doi: 10.1109/TCSI.2021.3057580.

[2] L. Avallone, M. Mercandelli, A. Santiccioli, M. P. Kennedy, S. Levantino and C. Samori, "A Comprehensive Phase Noise Analysis of Bang-Bang Digital PLLs," in IEEE Transactions on Circuits and Systems I: Regular Papers, vol. 68, no. 7, pp. 2775-2786, July 2021, doi: 10.1109/TCSI.2021.3072344.

[3] T. Seong, Y. Lee, S. Yoo and J. Choi, "A 320-fs RMS Jitter and – 75-dBc Reference-Spur Ring-DCO-Based Digital PLL Using an Optimal-Threshold TDC," in IEEE Journal of Solid-State Circuits, vol. 54, no. 9, pp. 2501-2512, Sept. 2019, doi: 10.1109/JSSC.2019.2918940.

[4] W. Deng et al., "A Self-Adapted Two-Point Modulation Type-II Digital PLL for Fast Chirp Rate and Wide Chirp-Bandwidth FMCW Signal Generation," in IEEE Journal of Solid-State Circuits, vol. 57, no. 4, pp. 1162-1174, April 2022

[5] S. S. Nagam and P. R. Kinget, "A −236.3dB FoM sub-sampling low-jitter supply-robust ring-oscillator PLL for clocking applications with feed-forward noise-cancellation," 2017 IEEE Custom Integrated Circuits Conference (CICC), 2017, pp. 1-4, doi: 10.1109/CICC.2017.7993677.

[6] T. Siriburanon et al., "A 2.2 GHz -242 dB-FOM 4.2 mW ADC-PLL Using Digital Sub-Sampling Architecture," in IEEE Journal of Solid-State Circuits, vol. 51, no. 6, pp. 1385-1397, June 2016, doi: 10.1109/JSSC.2016.2546304.

[7] T. -K. Kuan and S. -I. Liu, "A Bang Bang Phase-Locked Loop Using Automatic Loop Gain Control and Loop Latency Reduction Techniques," in IEEE Journal of Solid-State Circuits, vol. 51, no. 4, pp. 821-831, April 2016, doi: 10.1109/JSSC.2016.2519391.

A Bang-Bang Digital PLL Covering 11.1-14.3 GHz and 14.7-18.7 GHz with sub-40 fs RMS Jitter in 7 nm FinFET Technology

Staffan Ek, Patrik Karlsson, Andreas Kämpe, Roland Strandberg, Aravind Tharayil Narayanan, Martin Anderson,
Hind Dafallah, Mesrop Daghbashyan, Tayebeh Ghanavati Nejad, Robert Hägglund, Nikola Ivanisevic,
Robert Nilsson, Peter Nygren, Mattias Palm, Erik Säll, Sha Tao, My-Chien Yee, Lars Sundström
Ericsson, Sweden
staffan.ek@ericsson.com

Abstract—This paper presents an integer-N bang-bang digital PLL for synthesis of a high purity clock targeting output frequencies of 12 and 16 GHz using a 500 MHz reference. The PLL uses a self-resetting differential comparator-based BBPD with low hysteresis and a dual DCO architecture for lowest phase noise at respective output frequency. The PLL is implemented in a 7 nm FinFET process with an area of 0.18 mm^2 and achieves <40 fs RMS jitter integrated between 1 kHz and 100 MHz with a phase noise of -118.6 dBc/Hz at 1 MHz offset, while consuming 85.4 mW. The jitter varies less than 1.5dB across a -40 C to +85 C ambient temperature range.

I. INTRODUCTION

Cellular communication is governed by stringent requirements on spectral efficiency, manifested by challenging transmitter emission and receiver response requirements. Frequency synthesizers used for RF local oscillator and wideband data converter clock generation in transceivers are therefore critical for the fulfilment of these requirements. This is particularly true for base stations where emissions at low MHz offsets are predominately allocated to the transmit chains and the RF power amplifiers, hence putting further constraints on the synthesizers.

This paper presents the design and measurements of an integer-N PLL for generating 12 GHz and 16 GHz sample clocks from a 500 MHz reference with focus on minimized RMS jitter, using a 7 nm FinFET technology. A digital architecture is chosen due to its merits in terms of process node scalability, design portability, and programmability, but also as it ensures good phase tracking capabilities in a multi-transceiver system with several PLLs.

II. PLL ARCHITECTURE

A. Overview

The performance of integrated digital PLLs progressively gets closer and closer to the analog counterparts, in part attributed to continued technology scaling but also because of improved understanding of the architectural options leading to only a few preferred options [1] [2]. Yet, the best reported jitter figures are still obtained from pure analog PLLs, and sampling PLLs in particular [3]. Unfortunately, analog loop filter designs suffer from poor phase tracking over temperature and tuning making, e.g., co-integration of

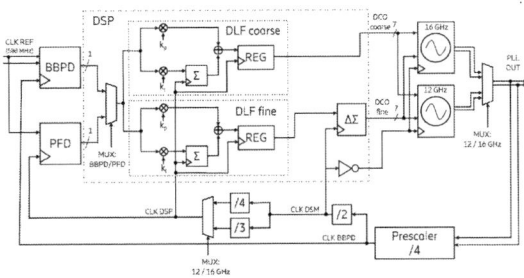

Fig. 1. Bang-Bang digital PLL architecture.

several PLL instances clocking different transceiver branches in multi-antenna systems challenging. The reference source assumed for this work is used as a yardstick for the phase noise (PN) budget up to the 1 MHz mark. The contribution from the reference can be reduced by lowering the PLL bandwidth below that point. This however rapidly increases the phase noise requirement on the RF-oscillator (RFO) to the point where it will consume a disproportional amount of power, and ultimately render the 7 nm FinFET technology inadequate. Instead, the strategy employed here is to ensure that all other noise contributions are well below the reference noise. This is a departure from the common approach to balance the contributions from the reference and RFO and, hence, the design will not be primarily optimized for best FoM.

Clearly the trend with digital loop filter PLLs replacing charge-pump analog dittos is not always well motivated as a means to reach state of the art spectral purity clock or LO signals. New noise sources (quantization) and not the least non-linearities need to be carefully managed through calibrations to even get close to the performance of an analog counterpart. While the preference for digital centric PLLs are often backed up by claims about reduced design effort and faster time-to-market, the complexity and reported design efforts often indicate the opposite.

The analog PLL in [4] is based on similar frequency plan as in this work and achieves excellent RMS jitter performance. But inspecting the path from the reference clock input to the output of the loop filter reveals a range of noise contributors that require careful attention to optimize the overall noise performance.

978-1-6654-8495-4/22 $31.00 © 2022 IEEE

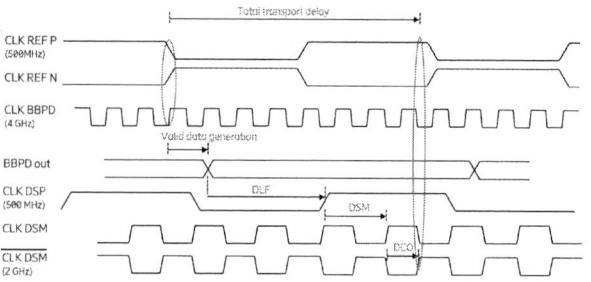

Fig. 2. Timing diagram

The strategy used in this work is to digitize the phase error as early as possible, both to ensure that reference noise will dominate and to facilitate a short design time. The phase error digitizer in an integer-N PLL obviously does not need a large range. Further, to reach state of the art performance, the RMS jitter of the phase error to be digitized should be minimal. Simulations of the RMS jitter level at the input of the digitizer show values lower or in line with state of the art time-to-digital converter (TDC) resolution [1]. Operating a TDC under such conditions means that more than two digitized levels seldom will be used, and hence phase error digitization based on a bang-bang phase detector (BBPD) will suffice. A digital BBPD PLL is also a good option to secure robust output phase tracking, a much needed property in multi-transceiver systems like advanced antenna system base stations.

B. Bang-Bang digital PLL

The non-linear characteristics of the BBPD have implications, one of which is the transport delay through the signal processing part that needs to be small to avoid limit-cycles degrading performance [5]. A look-ahead loop filter is used in [6] to minimize the delay, down to only about 1/3 of a reference clock cycle. However, transient system simulations performed in this work show negligible performance degradation using a transport delay of 1 instead of 0 reference cycles. Hence, considering the 7 nm process capabilities, a datapath without internal pipeline registers is used. The overall structure of the PLL is shown in Fig. 1 and the corresponding timing diagram for processing of detected phase errors is shown in Fig. 2. The total transport delay is approx. 1 reference cycle.

III. CIRCUIT DESIGN

A. DCO

One DCO is designed for each respective target frequency of 12 GHz and 16 GHz. The DCOs are based on an architecture optimized for low phase noise to keep its contribution around the 1 MHz offset well below the reference clock phase noise. Firstly, it is supplied with 1.8 V to enable high voltage swing together with thick oxide devices. Secondly, the DCOs are implemented as dual cores to increase current in the inductor. As the environment is expected to be hostile when integrating different functionality on the same die, 8-shaped inductors are chosen for improved magnetic coupling immunity. The two cores are each driving one respective side of the 8-shaped inductor, thereby avoiding metal crossings that otherwise would

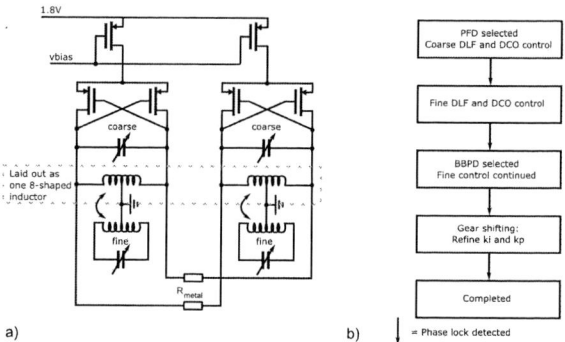

Fig. 3. a) Architecture of DCO. b) PLL start-up state flow.

degrade the inductor Q-value. To secure in-phase operation of the two seemingly separate oscillators cores, they are weakly coupled through metal resistors as shown in Fig. 3. PMOS devices are used in the active part for lower noise and MoM-type coarse tuning capacitors are used for high Q. The use of MoM capacitors, characterized by comparably larger process spread over MOS type capacitors, is one main reason for covering the two output frequencies using two separate DCOs instead of one wideband DCO. The coarse tuning bank is segmented with 4 binary LSBs and 3 thermometer coded MSBs for improved linearity. Coarse bank switching can have a detrimental effect on DCO behaviour. When performing initial frequency calibration and phase locking using the coarse tuning bank, the technique proposed in [7] is adopted. An additional switch, controlled by a pulse generated each time a coarse tuning value is changed, suppresses amplitude modulation of the DCO.

The fine tuning is implemented with MOS-devices coupled to the main oscillator cores through a secondary winding. A small mutual coupling enables the use of larger devices for better matching while maintaining a small frequency step. The down-transformed voltage swing in the secondary winding also allows use of fast switching thin oxide devices. The fine tuning bank has 128 levels and is controlled through a full binary to thermometer converter with row and column decoders to decrease logic depth. Finally, all DCO cores have amplitude detectors for amplitude calibration.

B. Frequency dividers and BBPD

The usage of a BBPD comes with an input dependent phase error quantization noise level. The larger the jitter at the input, the more quantization noise is added at the output. Although dividers are acting as samplers folding down high frequency noise contributions regardless of architecture, with a BBPD the frequency divider out-of-band noise components have a larger impact on overall PLL performance when compared with a linear phase detector. In other words, the noise floor level of the dividers is crucial, and generally a maximized slew rate throughout the chain should be aimed for. The approach chosen though is a compromise between phase noise performance and interference immunity. The first two divide-by-2 stages (prescaler) generating the sampling clock for the BBPD are implemented in current mode logic for power supply and common mode rejection. The prescaler output samples the reference clock through a differential comparator-based BBPD. The BBPD is self-resetting during the low half period of the sample clock minimizing hysteresis.

C. DSP unit

The PLL digital signal processing (DSP) unit, synthesized from RTL, contains the main digital loop filter (DLF) functionality implementing proportional and integral paths to make the PLL a type-II second order system. The high precision loop filter gets the input data bit generated from full custom logic that selects valid data from the stream of samples generated by the over-sampled BBPD. The data bit is processed and the result latched with the PLL output divided to reference frequency. The loop filter output data stream is fed to a second order delta-sigma modulator clocked at 1/8 the PLL output frequency. The binary to thermometer decoder output in the DCO is then latched at an opposite edge of the DSM sample clock to allow for timing variation in the interface to analog domain. The DSP also includes supporting circuitry, such as a finite state machine (FSM) for startup procedure control, a snapshot function to read out internal states and a RAM to record sequences of specific nodes in the DSP unit. The start-up default procedure traverses a set of states where either a lock detector output or a timer triggers state switching. Additionally, large flexibility is provided to allow the FSM to be stopped in any state e.g. at coarse tuning to do DCO amplitude calibration. The different states are shown in Fig. 3b).

IV. MEASUREMENT RESULTS

The total area of the PLL core circuitry measures $0.18\,\text{mm}^2$ with $0.13\,\text{mm}^2$ and $0.02\,\text{mm}^2$ allocated by the DCOs and decoupling, respectively. The power consumption is 85.4 mW distributed according to the pie chart in Fig. 8b. It includes DCO buffers for both cores, where only one buffer is used in the feedback loop. Fig. 8a shows the layout with an inset of a partial chip photo provided by the chip manufacturer (naked chips where not made available for subsequent micro photography). The DCO tuning ranges are 11.1-14.3 GHz and 14.7-18.7 GHz, respectively, covering the targeted frequencies with margin. The centering is slightly higher than expected partly due to a package ground plane placed close to the DCO inductors not taken into account during the design phase.

The input dependent equivalent gain of the BBPD increases loop gain uncertainty compared to when using a linear phase detector. The main parameter affecting the gain is the jitter present at the BBPD input where the gain is inversely proportional to the jitter [8]. The BBPD jitter may vary when different reference clock generators are used, temperature conditions changes, or the DCO is operated with backed off DCO current settings. Hence there is a need for loop gain calibration and the approach chosen is described in [9]. The loop calibration is implemented in MATLAB, guided by BBPD output sequences recorded in the on-chip RAM, thus completely in digital domain. An example showing how the algorithm works towards an optimized PLL RMS-jitter

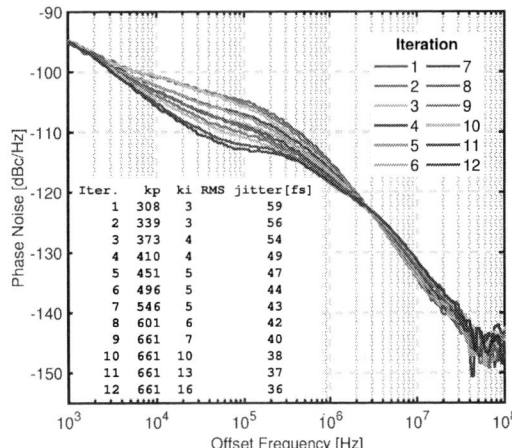

Fig. 4. Measured PLL phase noise during loop-gain calibration. $f_{out} = 15.729\,\text{GHz}$

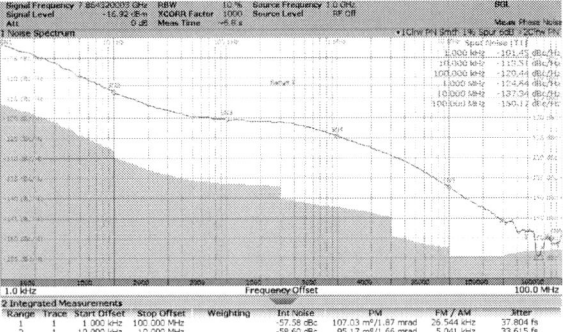

Fig. 5. Measured PLL phase noise at the divide-by-2 output. $f_{out} = 15.729\,\text{GHz}$

is illustrated in Fig. 4. A linear search is performed, initially looking for k_p (proportional gain) with an average auto-correlation coefficient of the BBPD sequence slightly above 0. The final adjustments towards 0 is then done increasing k_i (integral gain). Notably, the gain parameters may be adjusted without introducing any settling transients (verified through RAM-recordings) and therefore the algorithm may be safely run in the background while operating a transceiver. The phase noise for a 15.729 GHz clock measured with a Rohde-Schwarz FSWP at the first divide-by-2 output is shown in Fig. 5. The 491.52 MHz reference clock is generated by a Rohde-Schwarz SMA100B with 100 kHz and 1 MHz offset phase noise of -149.8 and -156.1 dBc/Hz respectively.

Integrating a PLL together with a transceiver system for cellular base station may mean operation under varying ambient temperatures. For that reason the performance was verified for a range of temperatures with results in Fig. 6. Both RMS jitter and phase noise at 1 MHz offset is kept reasonably constant across all temperature conditions.

Reference spur levels are increased when using a high reference clock amplitude for optimal jitter performance. One reason is an I/O with tight connection between reference clock input and test output. A comparison between a level generating

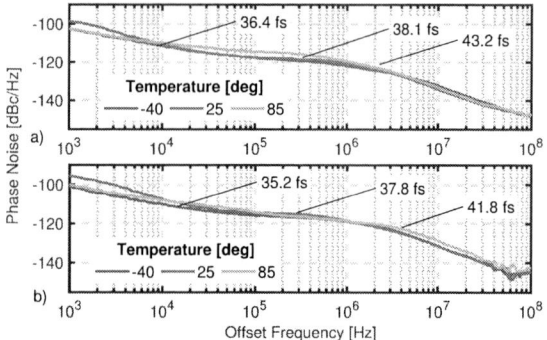

Fig. 6. Measured PLL phase noise over temperature. RMS jitter figures 1 kHz to 100 MHz indicated. a) f_{out} = 11.796 GHz. b) f_{out} = 15.729 GHz.

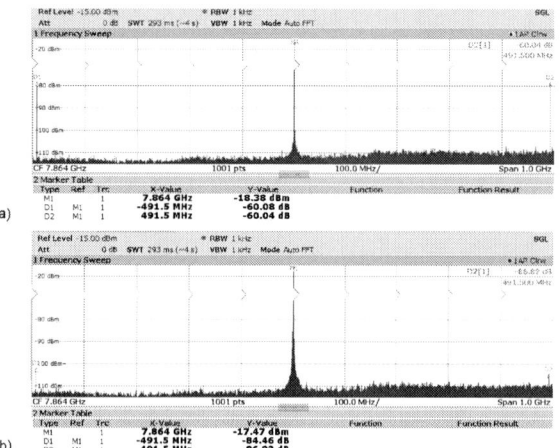

Fig. 7. Measured PLL output spectrum at the divide-by-2 output. f_{out} = 15.729 GHz. a) High reference level. b) Minimal reference level.

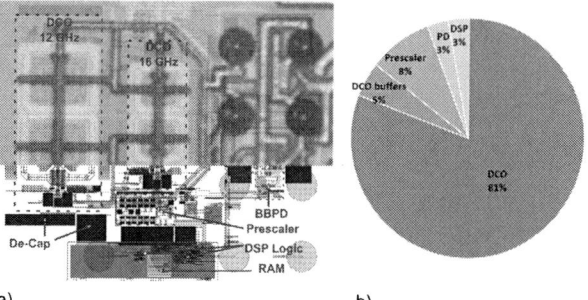

Fig. 8. a) PLL layout with chip-photo overlay. b) Power chart.

TABLE I
COMPARISON WITH PREVIOUSLY PUBLISHED PLLS WITH SIMILAR OUTPUT FREQUENCY

Parameter	This Work	[1] ISSCC 2021	[2] JSSC 2021	[3] VLSI 2021	[4] ISSCC 2018	[6] JSCC 2019
Tech. [nm]	**7**	16	28	28	16	7
Type	**Digital**	Digital	Digital	Analog	Analog	Digital
PD	**BB**	TDC	BB	Sampling	PFD	BB
f_{min} [GHz]	**11.1**(1)	12.1	12.9	19	7.4	13.6
f_{max} [GHz]	**18.7**(1)	16.6	15.1	19	14	15.7
f_{REF} [MHz]	**491.52**	245.76	250	250	500	350
Jitter[fs](2)	**37.8**	49.9	69.5	20.3	57.7(3)	143
PN [dbc/Hz](4)						
@100 kHz	**-114.4**	-112.9	-110.3	-124.6	-111.9	-103.0
@1 MHz	**-118.6**	-115.1	-113.3	-128.5	-115.1	-107.7
Ref.Spur [dBc](4)	**-54.0**	-75.1	-72.8	-67.6	<-51.5	-
P [mW]	**85.4**	56	10.8	12	45	40
FoM_j [dB](5)	**-249.1**	-248.6	-252.8	-263	-248.2	-240.8

(1) 400 MHz not covered mid-range (2) RMS 1 kHz to 100 MHz
(3) Extracted from published PN plot (4) Normalized to 15.789 GHz
(5) $FoM_j = 10\log((\text{Jitter/s}))^2(\text{Power}/1\text{mW})$

optimal phase noise and a minimal level for functionality is shown in Fig. 7. The minimal level yields a 25 dB improvement down to -84.5 dBc proving that the reference has a leakage path into parts of the RF portion of the PLL.

The key performance parameters of the PLL are compared with other similar PLLs in Table I. This work has resulted in the lowest RMS jitter reported for a fully integrated digital PLL with an FoM on-par with recently published work. The competitive phase noise @1 MHz offset also enables an efficient transmit chain fulfilling 3GPP emissions requirements for sub-6GHz equipment.

V. CONCLUSIONS

An 11.1 to 18.7 GHz Bang-Bang digital PLL with 37.8 fs RMS jitter for a 15.729 GHz clock has successfully been designed. The measurements demonstrate stable properties over a wide range of temperature conditions using a background calibration loop to optimize digital loop filter parameters. This proves that the bang-bang digital PLL is a natural choice for high spectral purity integer-N clock synthesizers in the frequency range, minimizing analog design complexity and enabling reuse.

REFERENCES

[1] E. Thaller et al., "A 32.6 K-Band 12.1-to-16.6GHz Subsampling ADPLL with 47.3fs Jitter Based on a Stochastic Flash TDC and Coupled Dual-Core DCO in 16nm FinFET CMOS," in *2021 IEEE International Solid- State Circuits Conference (ISSCC)*, vol. 64, 2021, pp. 451–453.

[2] S.M. Dartizio et al., "A 12.9-to-15.1-GHz Digital PLL Based on a Bang-Bang Phase Detector With Adaptively Optimized Noise Shaping," *IEEE Journal of Solid-State Circuits*, 2021.

[3] Y. Zhao and B. Razavi, "A 19-GHz PLL with 20.3-fs Jitter," in *2021 Symposium on VLSI Circuits*, 2021.

[4] D. Turker et al., "A 7.4-to-14GHz PLL with 54fs rms jitter in 16nm FinFET for integrated RF-data-converter SoCs," in *2018 IEEE International Solid - State Circuits Conference - (ISSCC)*, 2018, pp. 378–380.

[5] M. Zanuso et al., "Noise Analysis and Minimization in Bang-Bang Digital PLLs," *IEEE Transactions on Circuits and Systems II: Express Briefs*, vol. 56, no. 11, pp. 835–839, 2009.

[6] D. Pfaff et al., "A 14-GHz Bang-Bang Digital PLL with sub-150fs Integrated Jitter for Wireline Applications in 7nm FinFET," in *2019 IEEE Custom Integrated Circuits Conference (CICC)*, vol. 55, no. 3, 2020, pp. 580–591.

[7] P. Andreani et al., "A TX VCO for WCDMA/EDGE in 90 nm RF CMOS," *IEEE Journal of Solid-State Circuits*, vol. 46, no. 7, pp. 1618–1626, 2011.

[8] N.D. Dalt, "Markov Chains-Based Derivation of the Phase Detector Gain in Bang-Bang PLLs," *IEEE Transactions on Circuits and Systems II: Express Briefs*, vol. 53, no. 11, pp. 1195–1199, 2006.

[9] S. Jang et al., "An Optimum Loop Gain Tracking All-Digital PLL Using Autocorrelation of Bang–Bang Phase-Frequency Detection," *IEEE Transactions on Circuits and Systems II: Express Briefs*, vol. 62, no. 9, pp. 836–840, 2015.

A 2.5 GHz 104 mW 57.35 dBc SFDR Non-linear DAC-based Direct-Digital Frequency Synthesizer in 65 nm CMOS Process

Dong-Hyun Yoon[1], Kwang-Hyun Baek[2], and Tony Tae-Hyoung Kim[1]

[1]Nanyang Technological University, Singapore, Singapore, E-mail: thkim@ntu.edu.sg,
[2]Chung-Ang University, Seoul, Korea

Abstract—This paper presents a direct-digital frequency synthesizer (DDS) with dynamic performance enhancement techniques. First, a fixed-weight decoder with an auxiliary DAC is proposed to remove the truncation spur of the phase accumulator. Second, the proposed tri-state decoding scheme reduces the number of current sources for reducing timing mismatches and capacitances. Finally, a fine current source reusing technique is developed to reduce the number of current sources and power consumption. The proposed DDS is fabricated in 65 nm CMOS technology. The worst SFDR is 57.35 dBc at 2.5 GHz with power consumption of 104 mW. The measured figure of merit is 18,124 GHz·$2^{(SFDR/6)}$/W.

Keywords—Direct-digital frequency synthesizer, spurious-free dynamic range, signal-to-noise and distortion ratio

I. INTRODUCTION

Direct-digital frequency synthesizers (DDSs) are the most attractive solutions for modern wireless communications because of their fast frequency switching, sub-hertz resolution, and wide frequency range characteristics [1]-[5]. Fig.1(a) illustrates the concept of the basic DDS and the limitation of each block. DDS consists of a phase accumulator (PACC), phase-to-amplitude (P2A), and digital-to-analog converter (DAC). PACC accumulates phase using frequency control word (FCW). P2A converts the phase information to sinusoidal amplitude. Finally, DAC generates the sinusoidal output. Many studies have improved spurious-free dynamic range (SFDR) and signal-to-noise and distortion ratio (SNDR) performances. Even though the NLDAC segment decoder [2][4] achieves high dynamic performance with low power dissipation, dynamic performances are still limited by the truncation spur of PACC and the non-linearity of DAC. To improve the DAC linearity, thermometer decoding is used to ensure monotonicity and prevent glitches. However, many current sources are required, which causes timing mismatches and large capacitances. Therefore, the output spectral purity is degraded, especially at high frequencies [6]. In addition, dithering has been presented to remove the truncation spur of PACC [5]. However, because it dithers not only truncation spur but also the fundamental frequency, it degrades SNDR performance.

This paper proposes three techniques to overcome the limitations of PACC and DAC: fixed-weighted decoding (FWD), tri-state current source, and fine current reusing. The FWD scheme with auxiliary DAC removes the truncation spur in PACC. The proposed tri-state current source and fine current reusing reduce the number of current sources and DAC drivers, improving the DAC linearity. Fine current reusing is proposed to reduce power consumption by replacing coarse current sources with unused fine current sources.

II. PROPOSED DDS

Fig. 1(b) depicts the block diagram of the proposed DDS. The proposed DDS utilizes the segmented NLDAC architecture. It is composed of PACC, P2A decoder, and NLDAC. PACC has a 32-bit frequency control word (FCW) for fine frequency tuning steps and 12-bit output. P2A converts the phase information to sinusoidal amplitude based on the decoding table in [2]. Finally, NLDAC generates the

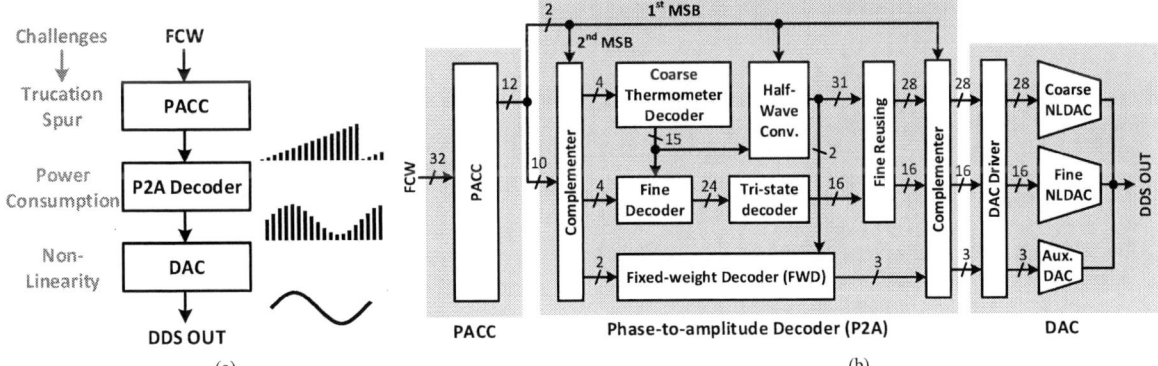

Fig. 1. (a) Concept of DDS and challenges and (b) detailed block diagram of the proposed DDS.

This work was supported by Samsung Electronics under Grant NTU REF 2019-1528.

978-1-6654-8495-4/22 $31.00 © 2022 IEEE

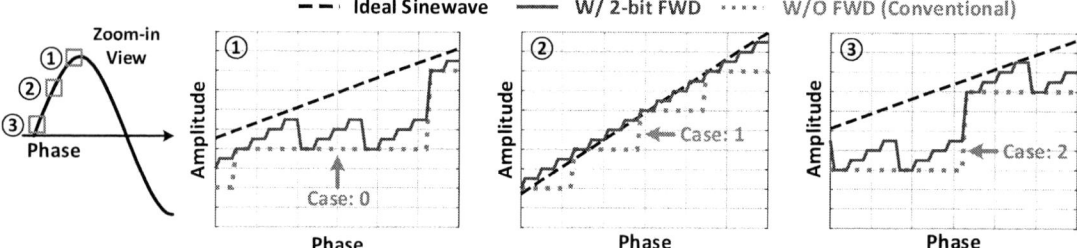

Fig. 2. Waveforms of ideal sine wave, conventional DDS and proposed 2-bit FWD with auxiliary DAC at different phases.

analog sine waveform. The proposed fixed-weighted decoding (FWD), fine reusing, and tri-state decoder are placed in the P2A. The proposed blocks improve spectral purity in the digital domain and the dynamic performance of analog blocks by reducing the parasitic components and the number of current sources. Also, the auxiliary DAC does not require an additional design consideration. The details will be discussed in II. A.

A. Fixed-Weighted Decoder

PACC requires many input bits (FCW) for high-frequency resolution. The PACC output is truncated to reduce the power and circuit complexity of P2A and DAC. However, truncation severely degrades the dynamic performance. The dithering technique in [5] converts the truncation spur into noise. However, it also spreads the fundamental frequency. Thus, dithering has a trade-off between the SFDR and SNDR. Also, if more PACC output bits are used, truncation spur can be reduced. However, the P2A decoder consumes large power for sinusoidal amplitude conversion depending on the phase. Thus, it requires almost twice power for PACC and P2A when increasing the PACC output by 1-bit. FWD with auxiliary DAC is proposed to remove the truncation spur without excessive power dissipation. Fig.2 shows the waveforms of the ideal sinewave, the conventional DDS, and the proposed 2-bit FWD with auxiliary DAC. The P2A decoder's output follows the ideal sinewave depending on the predefined decoding table to implement the sine amplitude. On the other hand, the proposed FWD always shows equal amplitude. The FWD output exhibits a sawtooth waveform regardless of phase. Note that the proposed FWD removes the truncation spur by reducing the error between the ideal sinewave and the DDS output without implementing a sine amplitude. Therefore, the errors are significantly reduced depending on the FWD's number of bits. Fig. 3 presents the simulated results of 1/2/3-bit FWD over the auxiliary DAC's static errors. The 2-bit FWD improves the SNDR and SFDR by 1.7 and 8.7 dB, respectively. When the auxiliary DAC static error is ±50 % (±0.5 LSB), SNDR and SFDR degradations are only 0.2 and 3.5 dB. It indicates the auxiliary DAC does not require excessive overhead despite the additional bits. This work employs the 2-bit FWD after considering the overhead and the performance improvement. Moreover, although the proposed DDS has 9 (main DAC) + 2 (auxiliary DAC) bit amplitude resolution, it has an equal mismatch requirement of 9-bit DDS. Monte Carlo simulation with 1k samples was conducted to verify the robustness of the mismatches. In this work, the I_σ/I_{unit} of the main DAC is 1.6 %. Since the I_σ/I_{unit} is proportional to

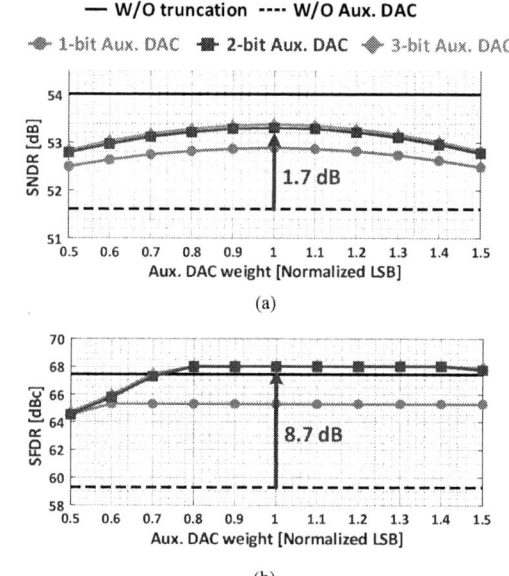

Fig. 3. (a) SNDR and (b) SFDR simulation results over auxiliary DAC static error.

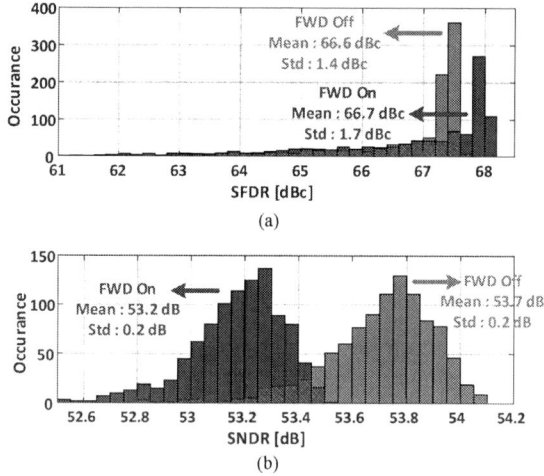

Fig. 4. Monte Carlo simulation results of (a) SNDR and (b) SFDR of with/without 2-bit FWD.

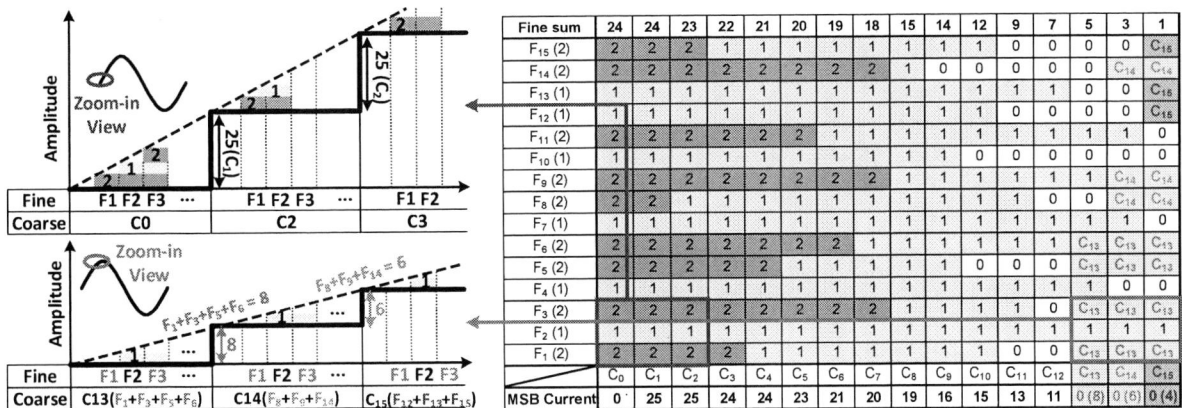

Fine sum	24	24	23	22	21	20	19	18	15	14	12	9	7	5	3	1
F_{15} (2)	2	2	2	1	1	1	1	1	1	1	0	0	0	0	0	C_{16}
F_{14} (2)	2	2	2	2	2	2	2	1	0	0	0	0	0	0	C_{14}	C_{14}
F_{13} (1)	1	1	1	1	1	1	1	1	1	1	1	1	0	0	0	C_{16}
F_{12} (1)	1	1	1	1	1	1	1	1	1	1	1	0	0	0	0	C_{16}
F_{11} (2)	2	2	2	2	2	2	1	1	1	1	1	1	1	1	0	0
F_{10} (1)	1	1	1	1	1	1	1	1	1	0	0	0	0	0	0	0
F_9 (2)	2	2	2	2	2	2	2	1	1	1	1	1	1	C_{14}	C_{14}	
F_8 (2)	2	2	1	1	1	1	1	1	1	1	1	0	0	C_{14}	C_{14}	
F_7 (1)	1	1	1	1	1	1	1	1	1	1	1	1	1	1	0	
F_6 (2)	2	2	2	2	2	2	1	1	1	1	1	C_{13}	C_{13}	C_{13}		
F_5 (2)	2	2	2	2	1	1	1	1	1	0	0	C_{13}	C_{13}	C_{13}		
F_4 (1)	1	1	1	1	1	1	1	1	1	1	1	1	0	0	0	
F_3 (2)	2	2	2	2	2	2	1	1	1	1	0	C_{13}	C_{13}	C_{13}		
F_2 (1)	1	1	1	1	1	1	1	1	1	1	1	1	1	1	1	
F_1 (2)	2	2	1	1	1	1	1	1	1	1	0	C_{13}	C_{13}	C_{15}		
	C_0	C_1	C_2	C_3	C_4	C_5	C_6	C_7	C_8	C_9	C_{10}	C_{11}	C_{12}	C_{13}	C_{14}	C_{15}
MSB Current	0	25	25	24	24	23	21	20	19	16	15	13	11	0 (8)	0 (6)	0 (4)

Fig. 5. Segmented decoding table with fine current source reusing.

the area of the current source, the I_o/I_{unit} of 2-bit auxiliary DAC is 0.032 % (The I_{unit} of 2-bit auxiliary DAC and area are the 1/4 of the main DAC). Fig. 4 shows the SFDR and SNDR simulation results. The difference of standard derivation (Std.)/Mean is only 0.4%.

B. Fine Current Source Reusing

Designing high-frequency DAC is another critical challenge in DDS. The thermometer decoding is an essential scheme for the high performance of DAC for the monotonicity and the low glitches. However, the thermometer scheme requires more bits than the binary scheme. Even though the 4:4 segmented decoder reduces the number of bits, many current sources and DAC drivers are still required. The third harmonic of DAC is expressed by $[R_L N/4|Z_O|]^2$, where R_L, N, and Z_o are the load resistance, the number of current sources, and the current source's output impedance, respectively. The capacitance is the dominant component of Z_o, especially at a high frequency. Therefore, the thermometer scheme degrades the spectral purity by increasing N. It also causes timing mismatches, leading to higher non-linearity. To address this issue, this paper proposes techniques for reducing the number of current sources. Fig. 5 shows the 4:4 segmented decoding table for quarter sinewave. As shown in the left-top zoom-in view of Fig. 5, the coarse amplitude is simply implemented using sine-weighted current sources. On the other hand, the fine amplitude uses different weights depending on the phase. Because the amplitudes rarely change around the top of a sinewave, some fine current sources are wasted. For example, $F_{1,3,5,6,8,10,12-15}$ are always zero in C_{13}. Since the current of C_{13} and $F_{1,3,5,6}$ are equal to eight, $F_{1,3,5,6}$ can replace C_{13}. The proposed fine current reusing scheme replaces $C_{13,15,16}$ with the fine current sources. This removes three coarse current sources and decreases the current source array power by 3.2 %. In addition, because the common centroid layout is required for current source matching, the layout complexity is also reduced.

C. Tri-state Current Source

The fine decoding table has three levels (0/1/2). Conventional DDS uses two current sources to implement three levels, as shown in Fig. 5 [2][4]. Therefore, many current

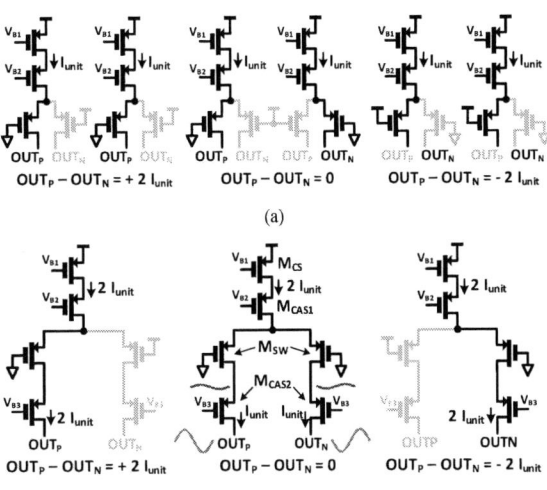

Fig. 6. Operation principle of (a) conventional current source and (b) proposed tri-state decoding scheme.

sources are inevitable. However, the proposed tri-state current source requires only one current source to generate three levels, as shown in Fig. 6. As a result, the number of fine current sources decreases from 24 to 16. The proposed current source's output voltages (OUT_P/OUT_N) vary depending on the phase. When both switches are turned on, they operate in the deep triode region. Thus, the output current is the function of $V_{DS,MSW}$. Because the output voltages affect $V_{DS,MSW}$, each switch delivers a different current to the output node. Therefore, the bottom cascode transistor (M_{CAS2}) is added to keep $V_{DS,MSW}$. As a result, both switches deliver an equal current regardless of output voltage. Furthermore, the bottom cascode transistors have been frequently used for a high-frequency DAC [7]. First, M_{CAS2} increases the output resistance of the current source. Also, the parasitic capacitances of M_{SW}, M_{CAS1}, M_{CS} are divided by the intrinsic gain ($g_m r_o$) of M_{CAS2}. Since the capacitance is the dominant factor at the high frequency, M_{CAS2} also prevents the harmonic of DAC.

978-1-6654-8495-4/22 $31.00 © 2022 IEEE

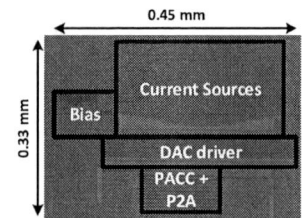

Fig. 7. Test chip die photo.

Fig. 8. Measured SFDR/SNDR performance over output frequency.

TABLE I: STATE-OF-ART DDS PERFORMANCE COMPARISON

Parameter	Unit	Proposed	[1]	[2]	[3]	[4]
Process	-	CMOS 65 nm	CMOS 65 nm	CMOS 55 nm	SiGe 350 nm	CMOS 90 nm
FCW Width	Bit	32	24	32	8	24
Amplitude Resolution	Bit	9 + 2	10	9	9	11
Frequency	GHz	2.5	7	2	5	1.3
Worst SFDR	dBc	57.35	32	55.1	45.7	52
Power	mW	104	601	130	460	350
Efficiency	mW/ GHz	41.6	85.9	65.0	92.0	269.2
Diff out Ampl.	Vpp	0.6	1.2	0.6	N/A	0.8
Area	mm²	0.15	1.15	0.10	2.10	0.9
FoM₁	-	18124	470	8944	2133	1509
FoM₂	-	120826	408	89443	1016	1677

$FoM_1 = 2^{SFDR_{worst}/6} f_{CLK}/power$ / $FoM_2 = FoM_1/area$

IV. CONCLUSION

This paper presents high dynamic performance DDS with three techniques. First, the FWD with auxiliary DAC removes the truncation spur of PACC without SNDR degradation. Also, the tri-state current source and fine current reusing reduce the number of current sources, improving the DAC linearity. As a result, the proposed DDS achieves SFDR of 57.35 dBc with 104 mW at the 2.5 GHz clock frequency. Furthermore, it shows the best FoMs among the recent DDS papers.

III. MEASUREMENT RESULTS

The proposed DDS is fabricated in a 65 nm CMOS with a 0.15 mm2, as shown in Fig. 7. All measurements have been performed under a 2.5 GHz clock and 600 mV output differential peak-to-peak amplitude. Fig. 7 also shows the measured SFDR and SNDR over the oautput frequency. In terms of SNDR, FWD removes the truncation spur over the entire frequency range. On average, it improves SNDR by 0.7 dB. The best and worst SNDR are 54.1 and 44.3 dB, respectively. The proposed FWD shows better SFDR improvement at lower frequencies. Fig. 8 presents the power spectral density of the best and worst SFDR. The truncation causes dominant spur at 154.1 MHz output because the DAC non-linearity is not critical at low frequency. As a result, the proposed FWD improves SFDR by 10.1 dB, from 59.7 to 69.8 dBc. At 818.7 MHz, the major harmonic is not truncation spur but the third harmonic of DAC. Therefore, FWD improves SFDR marginally even though it removes the truncation spur. However, the proposed tri-state current source and the fine current reusing improve SFDR by saving some current sources and DAC drivers. Table I shows the comparison with the prior state-of-the-art DDSs. The proposed DDS achieves worst SFDR of 57.35 dBc with power consumption of 104 mW. FoM₁ and FoM₂ are 18,124 and 120,826, respectively, showing the best performance in Table I.

REFERENCES

[1] A. M. Alonson, et al., "A 12.8-ns-Latency DDFS MMIC With Frequency, Phase, and Amplitude Modulations in 65-nm CMOS," *IEEE JSSC*, vol. 53, no. 10, pp.2840-2849, Oct. 2018

[2] T. Yoo, et al., "A 2 GHz 130 mW Direct-Digital Frequency Synthesizer With a Nonlinear DAC in 55 nm CMOS," *IEEE JSSC*, vol.49, no. 12, pp. 2976-2989, Dec. 2014

[3] C. -Y. Yang, et al., "A 5-GHz Direct Digital Frequency Synthesizer Using an Analog-Sine-Mapping Technique in 0.35- m SiGe BiCMOS," *IEEE JSSC*, vol. 46, no.9, pp. 2064-2072, Sep. 2011

[4] H. C. Yeoh, et al., "A 1.3-GHz 350-mW Hybrid Direct Digital Frequency Synthesizer in 90-nm CMOS ," *IEEE JSSC*, vol.45, no.9, pp. 1845-1855, Sep. 2010

[5] J. -M, Choi, et al., "Design and Analysis of Low Power and High SFDR Direct Digital Frequency Synthesizer," *IEEE Access*, vol.8, pp. 67581 – 67590, Apr. 2020

[6] L. Duncan, et al., " A 10-bit DC-20-GHz Multiple-Return-to-Zero DAC With >48-dB SFDR," *IEEE JSSC*, vol. 52, no. 12, pp. 3262-3275, Dec. 2017.

[7] C. -H. Lin, et al., "A 12 bit 2.9 GS/s DAC With IM3 < - 60 dBc Beyond 1 GHz in 65 nm CMOS," *IEEE JSSC*, vol. 44, no. 12, pp. 3285-3293, Dec. 2009

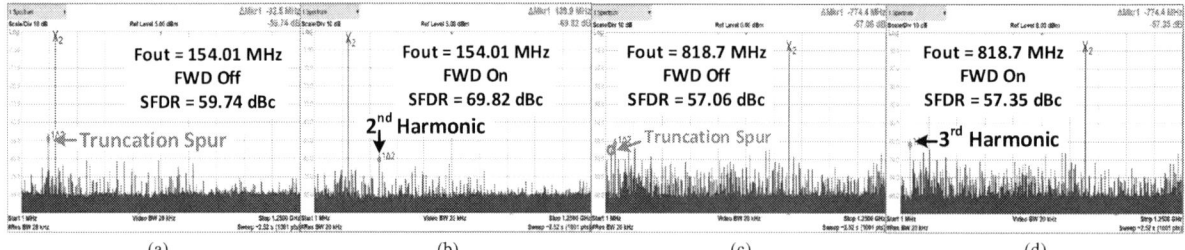

Fig. 9. Spectral density at (a),(b) 154.01 / (c),(d) 818.7 MHz (a),(c) with / (b),(d) without fixed-weighted decoder.

An Inductorless Fractional-N PLL Using Harmonic-Mixer-Based Dual Feedback and High-OSR Delta-Sigma-Modulator with Phase-Domain Filtering

Masaru Osada[1], Zule Xu[2]*, and Tetsuya Iizuka[1,2]
Email: osada@silicon.u-tokyo.ac.jp, {xuzule, iizuka}@vdec.u-tokyo.ac.jp
1, Department of Electrical Engineering and Information Systems, The University of Tokyo, Tokyo, Japan
2, Systems Design Lab, School of Engineering, The University of Tokyo, Tokyo, Japan

Abstract—An inductorless Harmonic-Mixer (HM) based fractional-N PLL is proposed. It simultaneously achieves Delta-Sigma-Modulator (DSM) noise suppression and a wide loop bandwidth by employing a high-OSR DSM and nested-PLL-based phase-domain lowpass filtering inside of the dual-feedback architecture. A 2.8-3.5 GHz prototype implemented in 65-nm CMOS achieves a –227.6dB FoM with an 8 MHz bandwidth, with no calibration circuitry and with a compact layout containing no inductors.

Index Terms—Phase-Locked-Loop, Fractional-N, Ring-VCO, Harmonic-Mixer, Nested-PLL, high-OSR DSM

I. Introduction

Inductorless PLLs using ring VCOs are attractive for their useful properties such as small area, wide tuning range, multi-phase output, and little coupling with other circuits. However, applying ring VCOs to fractional-N PLLs has proven difficult because of the conflicting bandwidth requirements posed by the suppression of DSM quantization noise and that of the ring VCO phase noise. An approach seen in many prior works is to cancel out the DSM noise using a Digital-to-Time-Converter (DTC) [1]–[3] for example. However, such solutions suffer from increased settling time and design complexity that are required to calibrate the gain and linearity of the noise cancellation circuit. Also, small imperfections in the design and calibration can degrade the output spectrum greatly [1], meaning that these solutions may not be robust to PVT variations and can be difficult to implement.

Recently, calibration-free fractional-N PLLs that use Sample-and-Hold (S/H) based HMs to suppress the noise have been proposed [4]–[8]. These architectures take advantage of the unity feedback gain achieved by frequency subtraction instead of frequency division seen in conventional architectures, to avoid the amplification of the DSM noise. The solution is also robust since it relies on the inherent property of frequency subtraction rather than the performance of a particular circuit block like a DTC. While very promising, the DSM noise suppression achieved by the avoidance of noise amplification is not quite enough to allow the wide loop bandwidth required

*Zule Xu is currently with IMEC Netherlands.

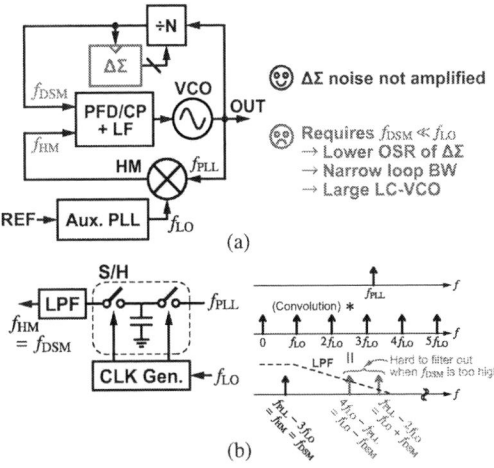

Fig. 1. (a) The dual-feedback PLL [5], [6] and (b) the frequency limitation of the DSM due to the S/H operation in the HM

to suppress the phase noise of ring VCOs to a satisfactory level. The existing works on HM-based PLLs [4]–[8] all resort to LC-VCOs, and the loop bandwidths are around 1MHz.

Increasing the DSM frequency would allow us to further suppress the noise, but this is not a practical solution in HM-based PLLs. For instance, in the dual-feedback architecture shown in Fig. 1(a), the DSM frequency is equal to the HM output frequency, and it must be kept sufficiently lower than the HM LO frequency. Fig 1(b) illustrates an example of a HM implementation based on a S/H circuit with its input and output frequency components. Through the sampling action, the PLL output frequency f_{PLL} is down-converted by a harmonic of f_{LO} (the 3rd harmonic in this case) to f_{HM}, which is equal to $f_{\mathrm{PLL}} - 3f_{\mathrm{LO}}$ here [9]. A large HM output frequency $f_{\mathrm{HM}} = f_{\mathrm{DSM}}$ relative to the HM LO frequency would make it difficult to filter out unwanted tones such as $f_{\mathrm{LO}} \pm f_{\mathrm{DSM}}$ shown in red in Fig. 1(b) at the HM output, since they would be close to the desired signal at f_{HM}. The operation speed of the DSM itself can also be a bottleneck, especially if we require a large

978-1-6654-8495-4/22 $31.00 © 2022 IEEE

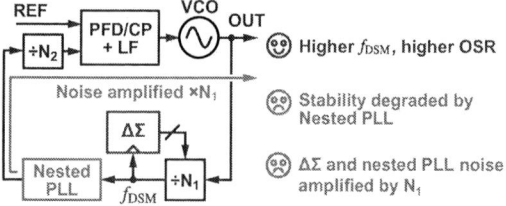

Fig. 2. The high-OSR DSM + nested PLL configuration with its stability issue and noise contribution

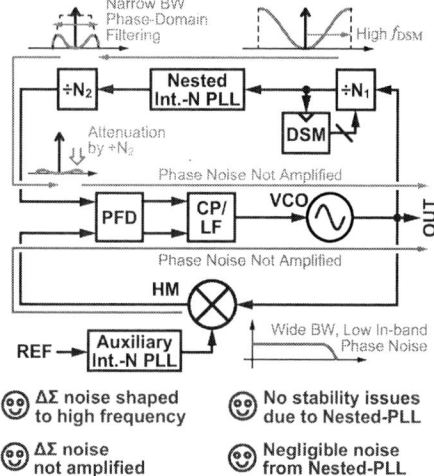

Fig. 3. The proposed architecture and its noise suppression mechanism

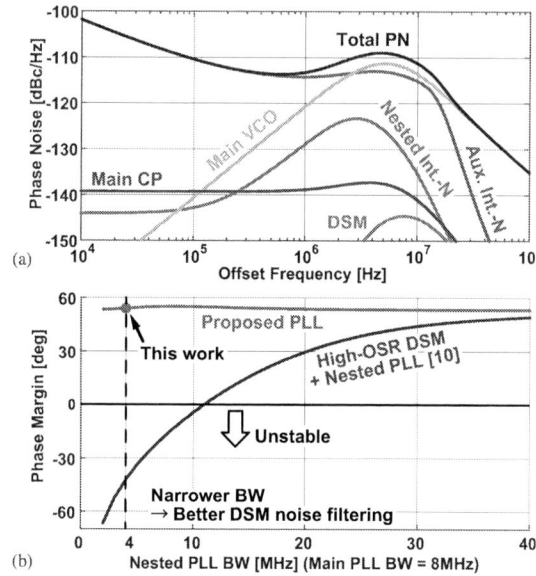

Fig. 4. (a) The noise contribution breakdown of the proposed architecture, and (b) the stability comparison between the proposed architecture and [10]

bit size for the DSM or if we are using a mature process node where the operation speed of the transistors is limited.

II. PROPOSED ARCHITECTURE

To overcome the insufficient DSM noise suppression in HM-based PLLs and allow a wider loop bandwidth for the suppression of ring VCO phase noise, we employ a scheme where we split the feedback divider into two parts and insert a nested PLL in between. Originally proposed in [10] and shown in Fig. 2, splitting the feedback divider and clocking the DSM with the intermediate frequency allows us to increase the noise shaping frequency, while the nested PLL serves as a phase-domain anti-aliasing filter to avoid the potential noise folding that can occur due to the phase sampling in the second divider. The proposed architecture shown in Fig. 3 uses this scheme inside of the HM-based dual-feedback PLL, resulting in strong DSM noise suppression by increasing the noise shaping frequency while also avoiding its amplification.

The proposed architecture also solves issues that are present in the configuration in Fig. 2 [10]. Firstly, in [10], the noise contributed by the nested PLL is amplified at the output due to the PLL transfer function, potentially resulting in large noise/power overhead from the nested PLL. In the proposed architecture, however, the unity-DC-gain transfer function of the main PLL [5], [6] and the attenuation by the second divider

results in negligible noise contribution from the nested PLL. As shown in red in Fig. 4(a), the expected noise contribution from the nested integer-N PLL in our design is negligible compared to that of the dominant noise sources, indicating a low overhead. Secondly, in Fig. 2 [10], the open-loop transfer function of the main PLL is greatly affected by the nested PLL, forcing the bandwidth of the nested PLL to be much wider than that of the main PLL in order to maintain a sufficient phase margin. This in turn means that not much phase noise filtering is achieved i.e., the nested PLL serves solely as an anti-aliasing filter and the filtering of the DSM noise must be done by the main PLL, limiting its bandwidth. As shown in the blue line in Fig. 4(b), the nested PLL bandwidth must be much larger than the main PLL bandwidth of 8MHz to maintain a positive phase margin for the overall open-loop gain. According to [10], the nested PLL bandwidth must be at least 2.5 times larger than that of the main PLL when we consider the PVT variation of the loop characteristic. In the proposed architecture, however, the feedback gain of the main PLL is dominated by the HM based path, meaning that the nested PLL has a negligible effect on the overall open-loop transfer function and stability, and its bandwidth can be made narrow to effectively filter out the DSM noise. As shown in the red line in Fig. 4(b), the nested PLL bandwidth has little effect on the overall open-loop gain. Our work employs a bandwidth of 4MHz for the nested PLL, which is only 0.5 times that of the main PLL, without suffering from any stability issues. This leads to aggressive filtering of the DSM noise, resulting in negligible noise contribution from the DSM as shown in green in Fig. 4(a).

In summary, the proposed architecture provides great suppression of the DSM noise by avoiding noise amplification,

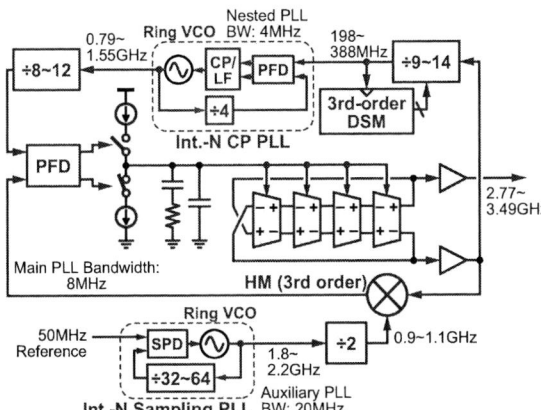

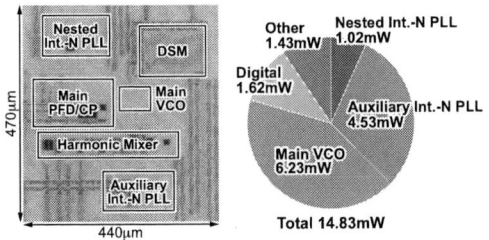

Fig. 5. Implementation details of the proposed PLL

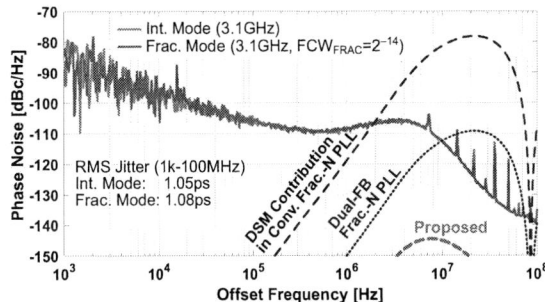

Fig. 7. Measured phase noise of the overall PLL along with calculated DSM noise contribution in conventional fractional-N PLL, dual-feedback PLL [5], [6], and proposed PLL

Fig. 6. Chip micrograph and power breakdown

increasing the noise shaping frequency, and with aggressive noise filtering by the narrow-bandwidth nested PLL. All of this is done with a low noise/power overhead, high robustness to stability issues, and without relying on complex calibration schemes.

III. MEASUREMENT RESULTS

A 2.8-3.5GHz fractional-N synthesizer based on the proposed architecture was fabricated in 65-nm CMOS as a proof-of-concept. The implementation details are shown in Fig. 5. The main and nested PLLs use type-II charge-pump based architectures for ease of implementation, while the auxiliary integer-N PLL uses a Sampling Phase Detector (SPD) based architecture to extend the loop bandwidth close to $0.5f_{REF}$. Note that the main PLL bandwidth can be further extended by using an SPD based architecture as well, though a Frequency Locked Loop (FLL) would be required due to the large frequency variation at the HM output during the locking transient. All of the VCOs are ring-based and the layout contains no inductors. Fig. 6 shows the chip micrograph and power breakdown. The contribution from the nested PLL is small compared to that of the other blocks, meaning our solution is realized with little overhead with respect to the dual-feedback PLL [5], [6].

The measured phase noise spectra of the entire synthesizer in integer and fractional modes are shown in Fig. 7, along with the DSM noise contribution in the proposed architecture, and the predicted DSM contribution if we had used a conventional fractional-N PLL or a simple dual-feedback PLL [5]–[7] with the same bandwidth for the main PLL (8MHz) and same PFD/CP frequency (85MHz). In the case of the conventional PLL shown with the dashed line, the combination of noise amplification, lack of filtering, and low DSM frequency means that the DSM noise significantly degrades the out-of-band phase noise. For the dual-feedback PLL shown with the dotted line, the DSM noise is reduced due to the avoidance of noise amplification, but it still degrades the out-of-band phase noise because of the remaining issues of low DSM frequency and lack of filtering. As shown with the red dashed line, the proposed architecture further suppresses the DSM noise by addressing all of these issues, and the negligible difference in the measured phase noise between the integer and fractional modes demonstrates the validity of its concept.

Fig. 8 shows examples of spectra with fractional and reference spurs, while Fig. 9 shows a plot of the fractional spurs for different fractional values of the FCW. The worst-case fractional spur is –53dBc, while the reference spur is –67dBc. The tones at 7MHz and its harmonics in Fig. 8(b) are caused by the power supply coupling with the digital circuitry and can be mitigated by more careful layout. Finally, the performance comparison with recent ring-based fractional-N synthesizers in Table I shows that the proposed architecture achieves state-of-the-art performance among calibration-free works in terms of jitter and FoM by realizing simultaneously a wide loop BW and DSM noise suppression. We can also see that the core area is comparable to that of prior art, meaning that area overhead is not an issue in our work even with the use of 3 PLLs, thanks to the fact that the main, auxiliary, and nested PLLs all use ring VCOs. It should be noted that although some of the works employing DTC-based architectures with calibration show better performance than our work, this is largely due to the fact that our design and architectural choice is based on achieving sure operation and ease of implementation for the proof-of-concept, and is not completely optimized for jitter and power consumption. For instance, an SPD based architecture with a dedicated FLL can be used for the main PLL, allowing us to further increase the loop bandwidth and reduce the noise/power contribution from the main VCO.

978-1-6654-8495-4/22 $31.00 © 2022 IEEE   247

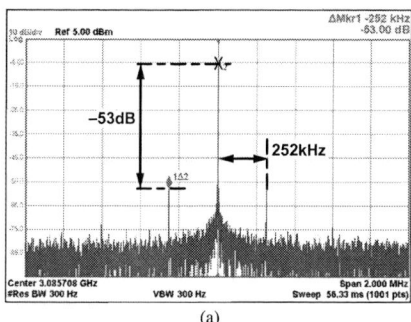

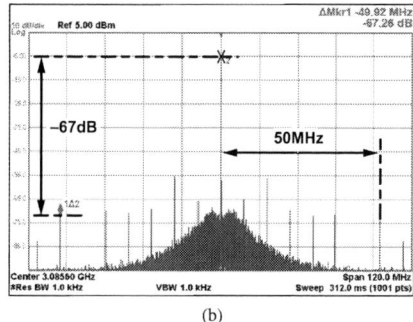

Fig. 8. Examples of PLL output spectra containing (a) fractional and (b) reference spurs

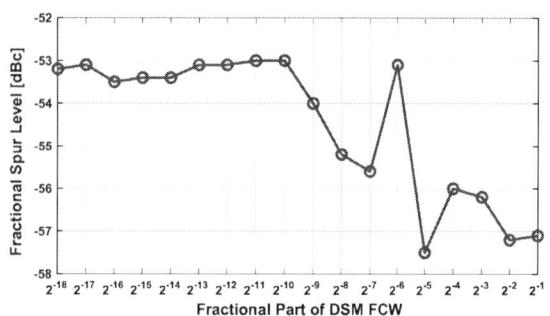

Fig. 9. A plot of the fractional spur levels for different fractional FCWs

IV. CONCLUSION

This work proposes a calibration-free, inductorless, fractional-N PLL that uses the HM-based dual-feedback architecture and a phase-domain filter based on a nested-PLL. The DSM noise is suppressed to a negligible level through avoidance of noise amplification, an increase in the DSM shaping frequency, and the narrow bandwidth filtering by the nested PLL that is allowed by the inherent stability of the proposed architecture.

ACKNOWLEDGMENT

This work is supported in part by the Institute for AI and Beyond, the University of Tokyo and in part by JSPS KAKENHI Grant Numbers JP20K14786, JP21H03406, and JP21J21917.

TABLE I
PERFORMANCE COMPARISON

Method	This Work	[2] Park ISSCC '21	[3] Zhang ISSCC '21	[11] Zhang JSSC '20	[12] Kong JSSC '18
	Dual-Feedback + High-OSR DSM + Nested PLL Filter	DTC w/ NL cancellation + PDS-DSM	DTC w/ TP Calibration	DSM + space-time averaging	DSM + Time-Domain FIR
Calibration Required?	No	Yes	Yes	No	No
Ref. Freq. [MHz]	50	100	50	50	22.6
Output Freq. [GHz]	3.1	5.3	1.5	2.4	2.4
BW [MHz]	8	20*	10*	2.8	5.7
RMS Jitter [fs] (Integ. Range)	1079 (1k-100M)	365 (10k-30M)	1670 (10k-10M)	2260 (1k-100M)	1500 (10k-50M)
Ref. Spur [dBc]	−67	−77	−44	−67	−60
Frac. Spur [dBc]	−53	−63	−60	−47	−41
Power [mW]	14.83	9.27	11.95	4.85	10
FoM** [dB]	−227.6	−239.1	−224.8	−226.1	−226.5
Core Area [mm²]	0.112	0.146	0.18	0.086	0.096
Process Node [nm]	65	65	65	40	45
* Estimated from PN spectrum **FoM = $10\log_{10}[$ (Jitter/1s)^2 (Power/1mW)]					

REFERENCES

[1] S. Levantino, G. Marucci, G. Marzin, A. Fenaroli, C. Samori and A. L. Lacaita, "A 1.7 GHz Fractional-N Frequency Synthesizer Based on a Multiplying Delay-Locked Loop," in IEEE Journal of Solid-State Circuits, vol. 50, no. 11, pp. 2678-2691, Nov. 2015, doi: 10.1109/JSSC.2015.2473667.

[2] H. Park, C. Hwang, T. Seong, Y. Lee and J. Choi, "32.1 A 365fsrms-Jitter and -63dBc-Fractional Spur 5.3GHz-Ring-DCO-Based Fractional-N DPLL Using a DTC Second/Third- Order Nonlinearity Cancelation and a Probability-Density-Shaping ΔΣM," 2021 IEEE International Solid- State Circuits Conference (ISSCC), 2021, pp. 442-444, doi: 10.1109/ISSCC42613.2021.9365798.

[3] Q. Zhang, S. Su, C. -R. Ho and M. S. -W. Chen, "29.4 A Fractional-N Digital MDLL with Background Two-Point DTC Calibration Achieving -60dBc Fractional Spur," 2021 IEEE International Solid- State Circuits Conference (ISSCC), 2021, pp. 410-412, doi: 10.1109/ISSCC42613.2021.9365819.

[4] D. Yang et al., "16.6 A Calibration-Free Triple-Loop Bang-Bang PLL Achieving 131fsrms Jitter and-70dBc Fractional Spurs," 2019 IEEE International Solid- State Circuits Conference - (ISSCC), 2019, pp. 266-268, doi: 10.1109/ISSCC.2019.8662494.

[5] M. Osada, Z. Xu and T. Iizuka, "A 3.2-to-3.8 GHz Harmonic-Mixer-Based Dual-Feedback Fractional-N PLL Achieving -65 dBc In-Band Fractional Spur," in IEEE Solid-State Circuits Letters, vol. 3, pp. 534-537, 2020, doi: 10.1109/LSSC.2020.3037311.

[6] M. Osada, Z. Xu and T. Iizuka, "A 3.2-to-3.8GHz Calibration-Free Harmonic-Mixer-Based Dual-Feedback Fractional-N PLL Achieving −66dBc Worst-Case In-Band Fractional Spur," 2020 IEEE Symposium on VLSI Circuits, 2020, pp. 1-2, doi: 10.1109/VLSICircuits18222.2020.9162799.

[7] D. Yang, D. Murphy, H. Darabi, A. Behzad, R. Ruby and R. Parker, "An FBAR Driven -261dB FOM Fractional-N PLL," 2021 IEEE Radio Frequency Integrated Circuits Symposium (RFIC), 2021, pp. 147-150, doi: 10.1109/RFIC51843.2021.9490409.

[8] D. Yang et al., "A Sub-100MHz Reference-Driven 25-to-28GHz Fractional-N PLL with -250dB FoM," 2022 IEEE International Solid-State Circuits Conference (ISSCC), 2022, pp. 384-386.

[9] T. Iizuka and A. A. Abidi, "FET-R-C Circuits: A Unified Treatment—Part I: Signal Transfer Characteristics of a Single-Path," in IEEE Transactions on Circuits and Systems I: Regular Papers, vol. 63, no. 9, pp. 1325-1336, Sept. 2016.

[10] P. Park, D. Park and S. Cho, "A 2.4 GHz Fractional-N Frequency Synthesizer With High-OSR ΔΣ Modulator and Nested PLL," in IEEE Journal of Solid-State Circuits, vol. 47, no. 10, pp. 2433-2443, Oct. 2012, doi: 10.1109/JSSC.2012.2209809.

[11] Y. Zhang et al., "A Fractional-N PLL With Space–Time Averaging for Quantization Noise Reduction," in IEEE Journal of Solid-State Circuits, vol. 55, no. 3, pp. 602-614, March 2020, doi: 10.1109/JSSC.2019.2950154.

[12] L. Kong and B. Razavi, "A 2.4-GHz RF Fractional- N Synthesizer With BW = $0.25f_{REF}$," in IEEE Journal of Solid-State Circuits, vol. 53, no. 6, pp. 1707-1718, June 2018, doi: 10.1109/JSSC.2018.2796544.

A −247.1dB FoM, −77.9dBc Reference Spur Ring-Oscillator-Based Injection-Locked Clock Multiplier with Multi-Phase-Based Calibration

Yeonggeun Song[1], Kyoungjoon Ha[1], Han-Gon Ko[2], Min-Seong Choo[3] and Deog-Kyoon Jeong[1]

[1]Department of Electrical and Computer Engineering, Seoul National University, Seoul, Korea
[2]ONE Semiconductor, Suwon, Korea
[3]Division of Electrical Engineering, Hanyang University, Ansan, Korea
Email: ygsong@isdl.snu.ac.kr, dkjeong@snu.ac.kr

Abstract—This paper presents a ring-oscillator (RO)-based injection-locked clock multiplier (ILCM) with a new background calibration technique that utilizes a multi-phase generation capability of the RO. By detecting phase changes before and after the injection pulse, both a frequency error and an injection path offset are calibrated. The frequency calibrator operates at the injection rate with high bandwidth, which contributes to further suppressing flicker noise of the RO and producing much lower RMS jitter. The path offset calibrator operating at the pulse-gating rate makes the ILCM converge to the state with a minimum reference spur. For a low-power implementation, a sub-sampling bang-bang phase detector is employed for each calibration loop and all of the loops operate at the reference clock rate. Fabricated in 28-nm CMOS, the proposed ILCM achieves 143.6-fs RMS jitter with a −77.9-dBc reference spur and consumes 9.4 mW at the 4.8-GHz operation, which translates to a FoM of −247.1 dB.

Keywords—*injection-locked clock multiplier (ILCM), multi-phase calibration, two-point calibration, spur reduction*

I. Introduction

A phase-realignment mechanism of a ring-oscillator (RO)-based injection-locked clock multiplier (ILCM) offers strong suppression of the RO-induced noise due to its high bandwidth nature. To make the most of the injection effects, a free-running frequency of the RO must be precisely tuned with the target frequency. However, the delay difference between the frequency calibrator and the injection path, i.e., path offset (PO), causes a remaining frequency error (FE), which degrades the reference spur performance. Therefore, various background calibration techniques have been proposed [1-5]. Fig. 1 depicts the conceptual block diagrams of the ILCM's calibration. In the conventional pulse-gating ILCM [3-4], the injection pulse (IP) is periodically gated and the FE is detected. However, it is restricted in a trade-off between the achievable injection effects and the bandwidth of the FE calibration. The ILCM adopting a replica delay cell (RDC) [2], [5] can detect the FE without the degradation in the bandwidth of the frequency calibration, but it requires a larger power budget since the RDC and the phase detector operate at the oscillation frequency.

In this paper, an ILCM with a multi-phase-based calibration (MPC) is proposed. The FE is detected by examining the deviation in the pre-injection phases, ϕ_{PRE}, at every injection rate and the PO is identified by the direction of the injection effects, i.e., injection pushing or injection pulling, observed in the post-injection phases, ϕ_{POST}, when the IP is gated. Since the FE is calibrated at every injection rate rather than pulse-gating rate and all calibration loops operate at a reference clock rate using a sub-sampling bang-bang phase detector (SS-BBPD), the proposed ILCM offers a

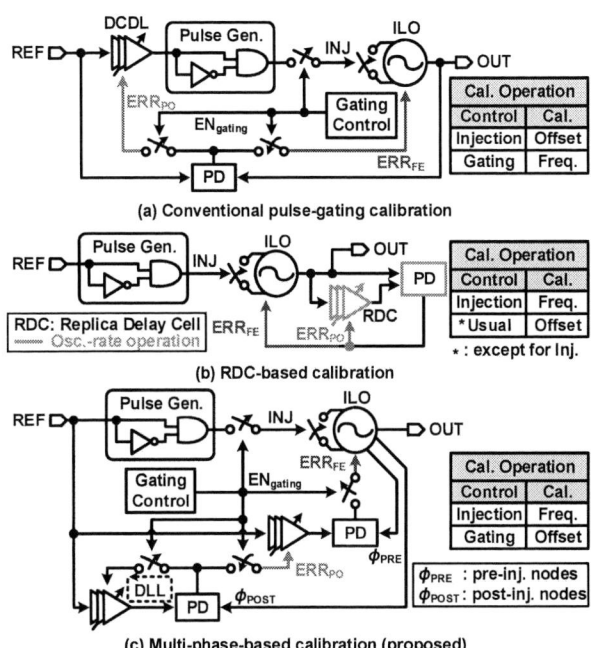

Fig. 1. Conceptual diagrams of the ILCM's calibration methodology.

low-power and high-bandwidth frequency calibration simultaneously.

II. Proposed Multi-Phase-Based Calibration

A. Proposed ILCM Architecture

Fig. 2 shows the overall architecture of the proposed ILCM. A conventional bang-bang phase-locked loop structure is used for an initial locking and disabled later in the ILCM mode. The IP, S_{INJ}, shorts differential oscillating nodes, ϕ_{INJ} ($\phi_{0°}$ and $\phi_{180°}$), and the frequency is tuned by a digitally-controlled resistor (DCR). The pre- and post-injection nodes, ϕ_{PRE} ($\phi_{315°}$ and $\phi_{135°}$) and ϕ_{POST} ($\phi_{45°}$ and $\phi_{225°}$), are sub-sampled by S_{PRE} and S_{POST} with SS-BBPDs to calibrate the FE and PO, respectively. S_{PRE} and S_{POST} are the delayed signals of a reference clock, S_{REF}, by t_{PRE} and t_{POST}, respectively. To explore whether the S_{INJ} is aligned with the rising or falling edge of $\phi_{0°}$, ϕ_{INJ} is sub-sampled by S_{POST}. Since all of S_{INJ}, S_{PRE} and S_{POST} are synchronous to S_{REF}, the required delay range of each digitally-controlled delay line (DCDL) can be reduced. The gating control block adjusts the pulse-gating rate through a gating-rate control word (GRCW) and affects the calibration bandwidth. The injection strength, β, is varied by 4-bit binary weighted injection switches. For precise calibration, 20-bit delta-sigma

978-1-6654-8495-4/22 $31.00 © 2022 IEEE

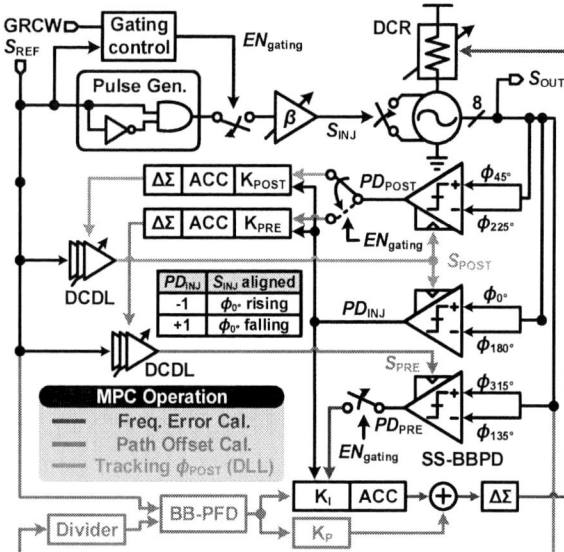

Fig. 2. Overall architecture of the proposed ILCM with the MPC.

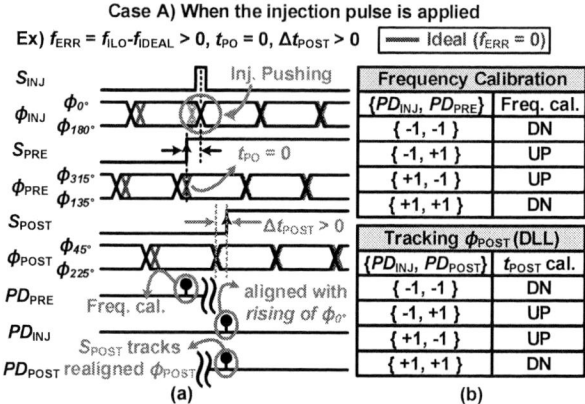

Fig. 3. (a) Timing diagram and (b) decision tables of the proposed MPC operation, when the injection pulse is applied.

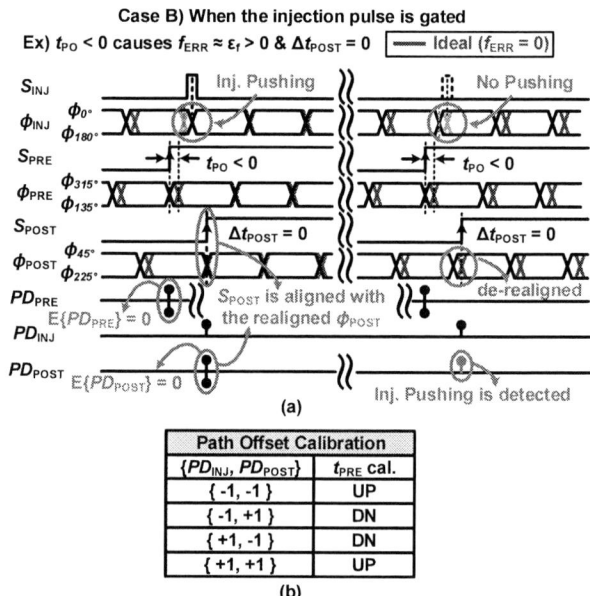

Fig. 4. (a) Timing diagram and (b) decision tables of the proposed MPC operation, when the injection pulse is gated.

modulators ($\Delta\Sigma$s) are employed in each calibration loop to obtain the finer frequency and delay resolution.

B. Proposed MPC Operation

The proposed MPC operation is described in two steps. The first is when the IP is applied while the other is when the IP is gated. When the IP is enabled, both the FE calibration (blue line in Fig. 2) and a delay-locked loop (DLL) that S_{POST} tracks the realigned ϕ_{POST} (green line in Fig. 2) are executed. On the other hand, when the IP is disabled, the PO is calibrated (red line in Fig. 2). For an unambiguous description of the MPC, some variables are defined as follows. The FE, f_{ERR}, is defined as an oscillation frequency error from the target frequency, f_{IDEAL}, ($f_{ERR} = f_{ILO} - f_{IDEAL}$). And the PO, t_{PO}, is the time difference of the rising edge of S_{PRE} from its ideal position ahead of S_{INJ} by $T_{IDEAL}/8$, where T_{IDEAL} is a period corresponding to f_{IDEAL}. Δt_{POST} is the time difference of the rising edge of S_{POST} from its ideal position later than S_{INJ} by $T_{ILO}/8$, where T_{ILO} is a period corresponding to f_{ILO}. The detailed MPC operation is illustrated with example cases via timing diagrams and corresponding decision tables as shown in Figs. 3 and 4.

Fig. 3 depicts the FE calibration and the DLL behavior for the case of $f_{ERR} > 0$, $t_{PO} = 0$ and $\Delta t_{POST} > 0$, when the IP is enabled. S_{INJ} realigns ϕ_{INJ} with the injection pushing and PD_{INJ} indicates that S_{INJ} is aligned with the rising edge of $\phi_{0°}$. The transition of ϕ_{PRE} leads the rising edge of S_{PRE} and the corresponding PD_{PRE} denotes the polarity of f_{ERR}. In the similar fashion, the transition of ϕ_{POST} leads the rising edge of S_{POST} and the corresponding PD_{POST} denotes the polarity of Δt_{POST}. Consequently, the FE is calibrated and S_{POST} tracks the realigned transition of ϕ_{POST}, according to the decision tables shown in Fig. 3(b).

Fig. 4(a) illustrates how the PO is calibrated. Before the IP is gated, the MPC converges into the steady state where both expectations of PD_{PRE} and PD_{POST} are zero, as explained in Fig. 3. Meanwhile, even with a small negative t_{PO}, a small positive FE (ε_f) corresponding to t_{PO} remains and an imperfect locked condition is reached. In this case, the phase realignment generated for every injection occurs in the

same direction, resulting in a poor reference spur performance. However, when the IP is gated, the phase realignment does not take place. Thus, S_{POST} samples the de-realigned ϕ_{POST} and the direction of the injection effect, i.e., injection pushing or injection pulling, can be identified by PD_{POST}. Therefore, the PO is calibrated according to the decision table shown in Fig. 4(b) and the proposed MPC locks to the point with the minimum reference spur.

C. Circuit Implementation

The circuit description of the injection-locked oscillator (ILO) is shown in Fig. 5(a). The ILO is composed of a 4-stage differential RO and injection switches. The IP is placed at only one of four switches and the rest are dummies with disabled switches. The injection strength, β, is controlled by 4-bit binary weighted switches, of which the unit size is W. The phase domain response curve in the vicinity of the locking point is plotted and β versus injection switch size is simulated as shown in Fig. 5(b). As the size of the injection switch becomes larger, β also increases, but its rate of increase gradually diminishes.

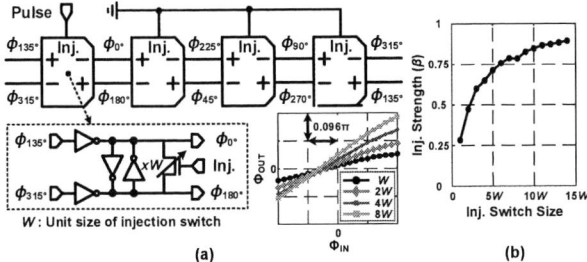

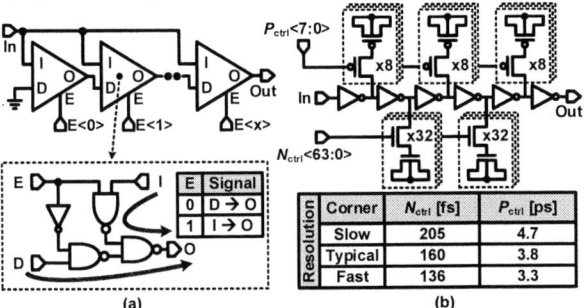

Fig. 5. (a) Circuit description of the ILO with its PDR curve and (b) injection strength curve.

Fig. 6. Circuit description of (a) DCDL coarse and (b) DCDL fine.

The DCDL consists of coarse- and fine- tuning cells as described in Fig. 6. Each coarse cell is a NAND-based DCDL and has a resolution of two NAND propagation delays. The fine-tuning DCDL adjusts its delay through an 8-bit P_{ctrl} and a 64-bit N_{ctrl} by changing load capacitance, according to the number of the activated MOSFETs. P_{ctrl} compensates for the narrow range of N_{ctrl}. If N_{ctrl} is stuck to either the minimum or the maximum value, P_{ctrl} is activated one by one. The simulated resolutions of the fine-tuning DCDL are shown in the bottom of Fig. 6(b).

III. EXPERIMENTAL RESULTS

The proposed ILCM is fabricated in the 28-nm CMOS technology and occupies an active die area of 0.062 mm² as shown in Fig. 5(a). The supply domains of the ILCM are divided into three parts. One is for the ILO (VDD$_{ILO}$), another is for the frequency calibrator and the other is for the rest (VDD$_{ANA}$). The power breakdown of each supply domain is shown in Fig. 5(b). The ILCM consumes 9.36 mW with the 4.8-GHz output clock. The measured phase noise and spectrum are shown in Fig. 8. The measured RMS jitter is 143.6 fs integrated from 10 kHz to 40 MHz and the reference spur of −77.89 dBc is achieved. Fig. 9 depicts the phase noise plots of the ILCM with and without the MPC. The ILCM without the MPC is measured at the same locked condition as the ILCM with the MPC, and all calibration loops are disabled. The phase noise at the low-frequency offset of the ILCM without the MPC is higher than that of the ILCM with the MPC. It indicates that continuous frequency calibration further suppresses flicker noise of the RO, which helps to achieve the much lower RMS jitter [5]. The measured clock eye diagram with its time interval error (TIE) histogram is shown in Fig. 10. The symmetric and Gaussian-shaped TIE histogram represents that the ILCM is successfully injection-locked with the MPC. To further verify the MPC operation, the integrated RMS jitter and the reference spur with supply (VDD$_{ILO}$) variations in five different chips are measured as shown in Fig. 11. Without

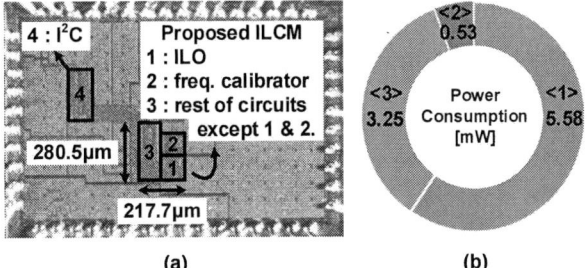

Fig. 7. (a) Die photomicrograph and (b) power breakdown at 4.8GHz operation

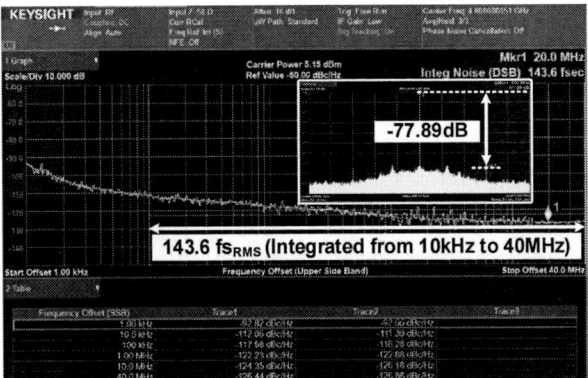

Fig. 8. Measured phase noise and spectrum of the output clock.

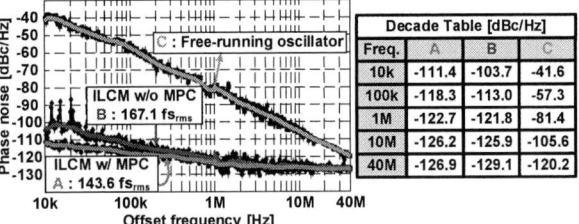

Fig. 9. Measured phase noises with and without the MPC.

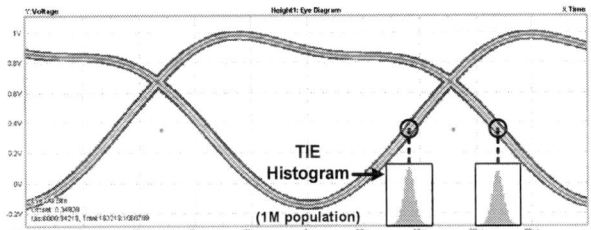

Fig. 10. Measured clock eye with its jitter histogram.

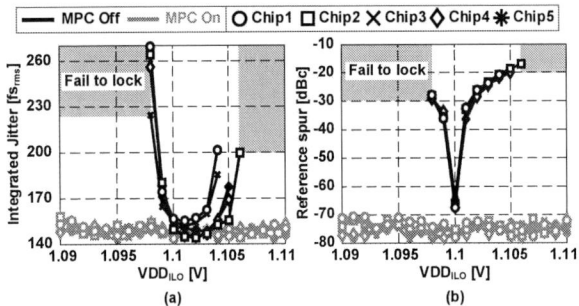

Fig. 11. Measured (a) integrated jitter and (b) reference spur with supply (VDD$_{ILO}$) variations in five different sample chips.

978-1-6654-8495-4/22 $31.00 © 2022 IEEE

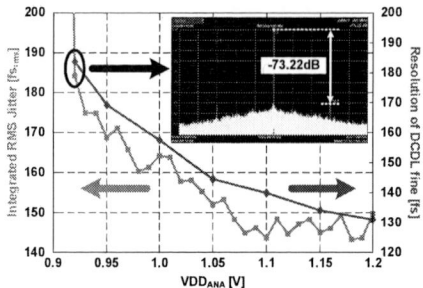

Fig. 12. Measured integrated jitter and resolution of DCDL fine cell with supply (VDD$_{ANA}$) variations.

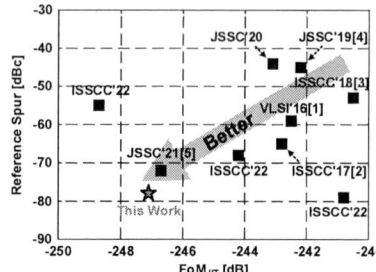

Fig. 13. Benchmark of performances of this work and state-of-the-art RO-based ILCMs.

the MPC, both the integrated jitter and the reference spur rapidly deteriorate when the supply voltage deviates from the central value of 1.1 V. Furthermore, the ILCM without the MPC fails to lock, if the supply voltage deviation becomes larger. However, with the MPC, the ILCM maintains robust performance with the consistently low integrated jitter and the low reference spur in multiple samples. Fig. 12 shows the measured integrated jitter and the resolution of the fine-tuning DCDL with supply (VDD$_{ANA}$) variations. Since the proposed MPC structure is comprised of the three digitally-controlled calibration loops, the performance has a strong dependency on the resolution of the DCDL. The two lines in Fig. 12 show that both the integrated jitter and the resolution of the fine-tuning DCDL increase as VDD$_{ANA}$ decreases. The RMS jitter of 184.3 fs and the resolution of 183.7 fs are measured at 0.92 V. When it comes to the reference spur, relatively consistent performance is obtained from 1.2-V to 0.92-V VDD$_{ANA}$ variations and the spectrum measured at 0.92 V is plotted in Fig. 12. In this case, the reference spur of −73.22 dBc is measured. Table I compares this design against the state-of-the-art RO-based ILCMs. Since the proposed MPC performs the low-power and high-bandwidth frequency calibration, the ILCM exhibits the best FoM along with the lowest reference spur.

IV. CONCLUSION

In this work, we present a low-jitter, low-reference-spur RO-based ILCM. With the MPC that utilizes a multi-phase generation capability of the RO, the ILCM successfully compensates for a frequency error and an injection path offset. Thanks to the high-bandwidth frequency calibration, the ILCM further suppresses flicker noise of the RO and achieves much lower RMS jitter. In addition, by identifying the direction of the applied injection, the path offset is calibrated and the ILCM reaches the minimum reference spur point in the steady state. For a low-power implementation, the SS-BBPDs are used in all calibration loops. The ILCM achieves 143.6-fs RMS jitter with a −77.9-dBc reference spur and consumes 9.4 mW at the 4.8-GHz operation, which translates to a FoM of −247.1 dB.

REFERENCE

[1] Y. Lee, H. Yoon, M. Kim and J. Choi, "A PVT-robust −59-dBc reference spur and 450-fs$_{RMS}$ jitter injection-locked clock multiplier using a voltage-domain period-calibrating loop," in *Proc. Symp. VLSI Circuits*, Jun. 2016, pp. C190-C191.

[2] S. Kim et al., "A 2.5GHz injection-locked ADPLL with 197fsrms integrated jitter and −65dBc reference spur using time-division dual calibration," *IEEE ISSCC Dig. Tech. Papers*, Feb. 2017, pp. 494-495.

[3] K. M. Megawer et al., "A 5GHz 370fsrms 6.5mW clock multiplier using a crystal-oscillator frequency quadrupler in 65nm CMOS," *IEEE ISSCC Dig. Tech. Papers*, Feb. 2018, pp. 392-394.

[4] A. Elkholy, D. Coombs, R. K. Nandwana, A. Elmallah and P. K. Hanumolu, "A 2.5–5.75-GHz ring-based injection-locked clock multiplier with background-calibrated reference frequency doubler," in *IEEE J. Solid-State Circuits*, vol. 54, no. 7, pp. 2049-2058, July 2019.

[5] S. Yoo et al., "A low-jitter and low-reference-spur ring-VCO-based injection-locked clock multiplier using a triple-point background calibrator," in *IEEE J. Solid-State Circuits*, vol. 56, no. 1, pp. 298-309, Jan. 2021.

TABLE I. PERFORMANCE COMPARISON WITH THE STATE-OF-THE-ART RO-BASED ILCMs

	This work	VLSI'16 [1] Y. Lee	ISSCC'17 [2] S. Kim	ISSCC'18 [3] K. M. Megawer	JSSC'19 [4] A. Elkholy	JSSC'21 [5] S. Yoo
Tech. [nm]	28	40	65	65	65	65
Ref. Freq. [MHz]	300	180	156.25	54	125	100
Output. Freq. [GHz]	4.8	1.44	2.5	4.752	5	2.4
Mult. Factor (N)	16	8	16	88	40	24
Inte. Jitter, σ_{RMS} [fs] (range) [MHz]	143.6 (0.01 - 40)	450 (0.001 - 40)	198 (0.01 - 40)	370 (0.01 − 30)	337 (0.01 - 40)	140 (0.01 - 30)
Ref. Spur [dBc]	−77.9	−59	−65	−53	−45	−72
Power [mW]	9.36	2.8	13.5	6.5	5.3	11.0
Power Eff. [mW/GHz]	1.95	1.94	5.40	1.37	1.06	4.58
Freq. cal. Methodology (cal. BW / cal. Power)	Multi-Phase (High / Low)	Period-Charging (Low / Low)	Replica Delay (High / High)	Pulse-Gating (Low / Low)	Pulse-Gating (Low / Low)	Replica Delay (High / High)
Area [mm^2]	0.062	0.061	0.064	0.16	0.09	0.055
*FoM$_{JITTER}$ [dB]	−247.1	−242.5	−242.8	−240.5	−242.2	−246.7

$$*\text{FoM}_{JITTER} = 10 \cdot \log\left[\left(\frac{\sigma_{RMS}}{1s}\right)^2 \cdot \left(\frac{Power}{1mW}\right)\right]$$

A 590 μW, 106.6 dB SNDR, 24 kHz BW Continuous-Time Zoom ADC with a Noise-Shaping 4-bit SAR ADC

Shubham Mehrotra[1,2], Efraïm Eland[1], Shoubhik Karmakar[1], Angqi Liu[1], Burak Gönen[1,3], Muhammed Bolatkale[1,2], Robert van Veldhoven[2], and Kofi A. A. Makinwa[1]

[1]Delft University of Technology, Delft, [2]NXP Semiconductors, Eindhoven, [3]Ethernovia, Zeist, The Netherlands
Email: shubham.mehrotra@nxp.com

Abstract—**This paper presents a continuous-time zoom ADC for audio applications. It combines a 4-bit noise-shaping coarse SAR ADC and a fine delta-sigma modulator with a tail-resistor linearized OTA for improved linearity, energy efficiency, and handling of out-of-band interferers compared to previous designs. In 160 nm CMOS, the prototype chip occupies 0.36 mm², achieves 107.2 dB SNR, 106.6 dB SNDR, and 107.3 dB dynamic range in a 24 kHz bandwidth while consuming 590 μW from a 1.8 V supply. This translates into a Schreier figure-of-merit (FoMS) of 183.4 dB and a FoMSNDR of 182.7 dB.**

Keywords—*A/D conversion, audio analog to digital converter (ADC), continuous-time delta-sigma ADC, dynamic zoom ADC, low-power circuits, noise-shaping SAR ADC, high linearity operational transconductance amplifier (OTA)*

I. INTRODUCTION

ADCs for audio applications such as battery-powered CODECs often require high linearity, dynamic range (DR), and energy efficiency. Zoom ADCs meet these requirements by using a low-power coarse SAR ADC to update the references of a fine delta-sigma modulator (ΔΣM). However, previous discrete-time (DT) zoom ADCs [1, 2] place stringent requirements on the linearity of the input and reference drivers needed to charge their sampling capacitors with large signal-dependent peak currents. The resistive input of a continuous-time (CT) zoom ADC [3] is easier to drive but a power-hungry OTA is needed to ensure that its first integrator achieves sufficiently high linearity. Another limit on the linearity of zoom ADCs is the non-unity STF of their fine ΔΣMs which causes in-band leakage of the SAR ADC's quantization noise (also referred to as "fuzz"). This can be mitigated, at the expense of complexity, by adding a digital filter after the SAR ADC [2], or by using a feedforward path in the fine ΔΣM to enforce a unity STF [1]. However, the effectiveness of both techniques is limited by analog spread.

To address these limitations, this work presents a CT zoom ADC that employs passive noise-shaping in its coarse SAR ADC to robustly and efficiently shape its fuzz out of band (OOB). Furthermore, the 1ˢᵗ integrator of the fine ΔΣM employs a tail-resistor linearized (TRL) OTA, allowing it to be designed solely for noise, rather than for linearity, thus reducing power dissipation. Together, these techniques result in high energy efficiency (FOMSNDR = 182.7 dB), and the lowest in-band noise and distortion spectral density (NDSD = −150.4 dBFS/Hz) amongst state-of-the-art audio ADCs.

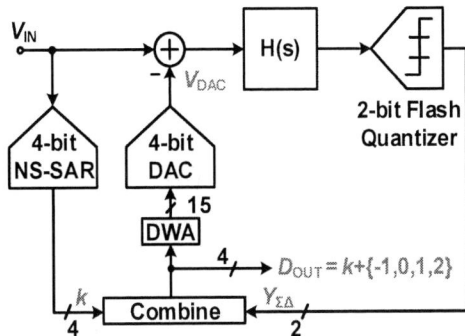

Fig. 1. Proposed CT zoom ADC architecture.

II. PROPOSED CT ZOOM ADC

A. Architecture

Fig. 1 shows the simplified architecture of the proposed zoom ADC. It consists of a 4-bit asynchronous SAR ADC and a 2-bit CT ΔΣM running concurrently at a sampling frequency (F_s) = 5.12 MHz. The SAR output (k) updates the references of ΔΣ DAC with 1 LSB of over-ranging such that $V_{REF-} = (k-1).V_{LSB} < V_{IN} < (k+2).V_{LSB} = V_{REF+}$, where V_{LSB} is the quantization step-size of the fine ΔΣ DAC. As in [1], the resulting 3 LSB range can then be well matched to the steps of the 2-bit ΔΣM.

In a zoom ADC, OOB interferer tracking is limited by the conversion speed of the SAR ADC and F_s, making the ΔΣM susceptible to overload in the presence of fast OOB interferers. Although these interferer signals can be easily removed by using a low pass filter at the input, a sharp cut-off filter is undesirable in a CT ΔΣM that should benefit from relaxed anti-aliasing filter requirements. For improved robustness to OOB interferers, this work uses a 4-bit SAR ADC as a coarse quantizer rather than the 5-bit SAR of previous designs [1, 2, 3], thereby relaxing the anti-overloading filter requirements. However, this introduces more quantization noise and fuzz in the zoom ADC's output spectrum. Apart from limiting its in-band linearity, the presence of OOB fuzz increases the requirements on the digital decimation filter. In this work, this is addressed by implementing noise-shaping in the 4-bit SAR ADC, which effectively decorrelates and shapes the quantization noise of the SAR ADC OOB, and thus reduces the fuzz significantly.

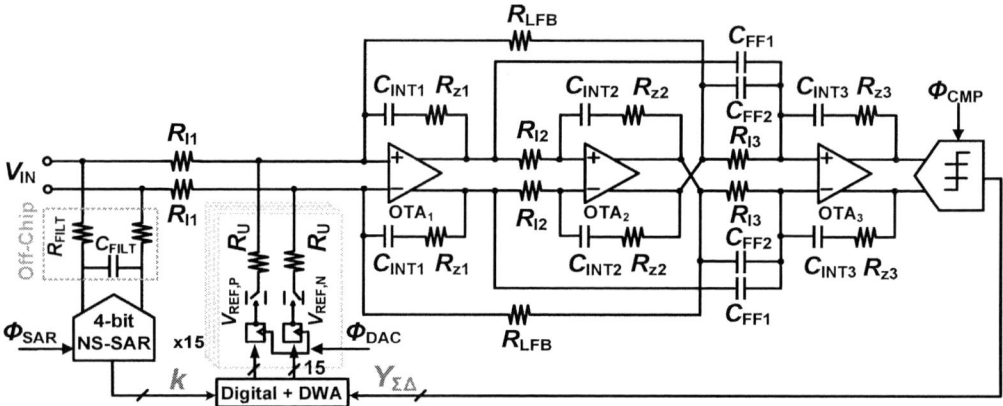

Fig. 2. Simplified schematic of the proposed CT zoom ADC.

To maintain energy efficiency, the noise shaping (NS) in the SAR ADC is realized with a passive integrator [4].

B. Circuit Implementation

Fig. 2 shows a simplified schematic of the proposed zoom ADC. The $\Delta\Sigma$M employs a 3rd order CIFF loop filter. In contrast to [3], the loop filter employs a local feedback loop (via R_{LFB}) which creates an NTF notch for better noise shaping. Furthermore, the linearity of OTA$_1$ is improved by the tail-resistor linearization technique [5]. The resulting OTA (Fig. 3) is >17x more linear than the one used in [3], allowing it to be biased at half the tail current ($I_{tail} = 56$ μA, $R_{tail} = 167$ Ω) to meet noise requirements while achieving a DC gain of 60 dB over PVT. The improved linearity also facilitates the use of noise shaping in the SAR ADC, despite this producing ~3x more virtual ground swing than a conventional 4-bit SAR (Fig. 4). OTA$_1$ is chopped at F_s, which mitigates its $1/f$ noise, and also improves its CMRR and PSRR. As in [3], OTA$_{2,3}$ are implemented as current-reuse amplifiers, while the feedback DAC consists of a 4-bit unary NRZ R-DAC. The 2-bit flash quantizer has a conversion time of <8 ns over PVT, which even with the added delay of the DWA logic, is well within the fixed loop delay of $T_s/8 = 24$ ns.

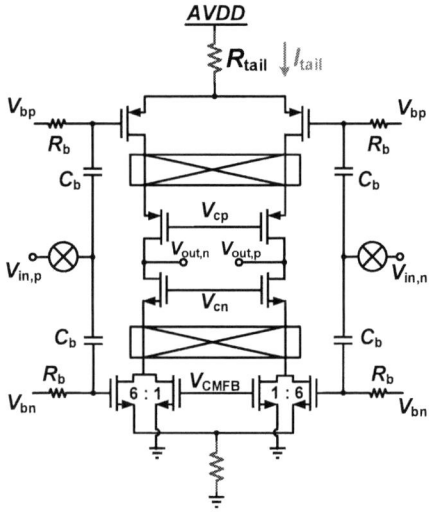

Fig. 3. Tail-Resistor-Linearized OTA$_1$.

As shown in Fig. 5, the SAR ADC consists of a 4-bit C-DAC ($C_u = 1.8$ fF), a comparator, asynchronous digital logic, and a passive noise shaper. At the end of the SAR conversion, the residue signal is sampled on one of two pairs of capacitors ($C_{1R(P,N)}$ or $C_{2R(P,N)}$). These are then used in a ping-pong scheme to sample the residue during a conversion cycle ($C_{1R(P,N)}$ in ϕ_{NS1} and $C_{2R(P,N)}$ in ϕ_{NS2}) and then transfer it to a pair of integration capacitors (C_{IP} and C_{IN}) during the next cycle [4]. The aggressiveness of this passive noise-shaping scheme is determined by the ratio $C_{(1,2)RP}/C_{IP}$, which is set to 0.6 to ensure that the output swing of the SAR ADC does not exceed the range of the fine $\Delta\Sigma$M.

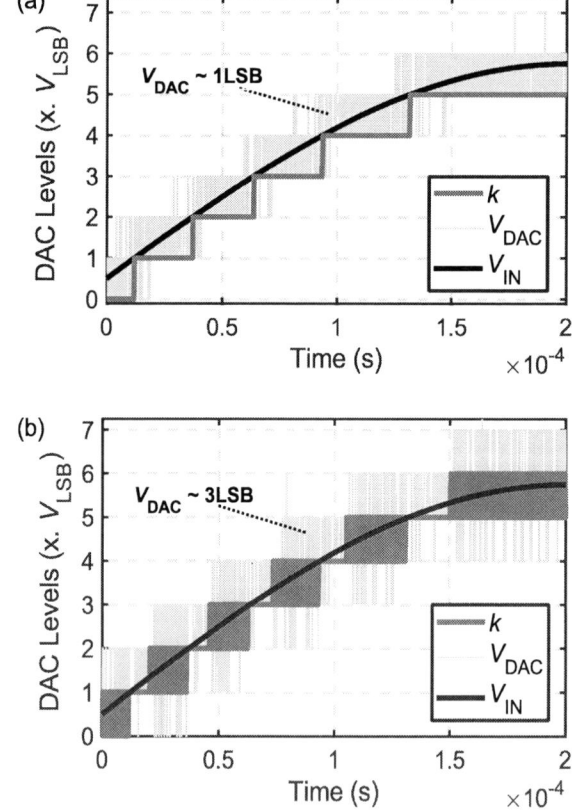

Fig. 4. SAR ADC (k) and $\Delta\Sigma$ DAC (V_{DAC}) output (a) w/o NS in SAR ADC (b) w/i NS in SAR ADC.

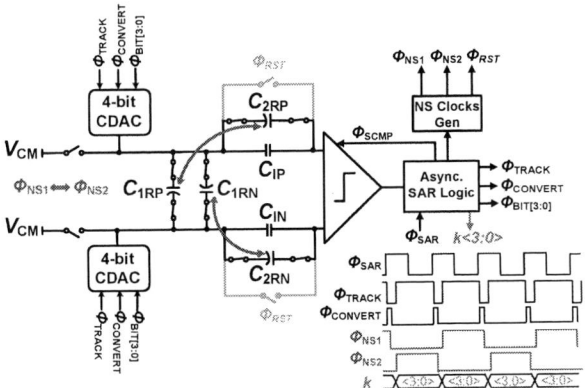

Fig. 5. Simplified schematic of the 4-bit NS-SAR ADC.

III. MEASUREMENT RESULTS

The fabricated prototype chip occupies an active area of 0.36 mm² (Fig. 6) and draws a current of 327.7 μA from a 1.8 V supply. The analog, digital, and reference blocks account for 40.1%, 26.4%, and 33.5% of the total power consumption, respectively. For a 1 kHz, –0.35 dBFS input signal, the measured SNDR and THD of the 4-bit SAR ADC with NS "OFF" is 30.3 dB and –30.4 dB, respectively, (Fig. 7). With NS "ON", its in-band quantization noise is suppressed by about 9.5 dB, while its SNDR and THD improve by 14.5 dB and 14.6 dB, respectively. With NS "OFF", the measured peak SNR, SNDR, and THD of the zoom ADC are 107.2 dB, 105.4 dB, and –110.5 dB, respectively (Fig. 8). Turning NS "ON" effectively suppresses in-band fuzz and improves the SNDR, and THD of the zoom ADC by 1.2 dB and 5.5 dB, respectively. The tone at $F_s/2$ with NS "ON" comes from the SAR ADC as it toggles predominately between 2 levels while resolving the residue. The design achieves a DR of 107.3 dB (Fig. 9) with NS "ON", which is mainly limited by residual DWA tones. Fig. 10 shows the integrated in-band noise of the ADC (20 Hz to 24 kHz) when a –1.5 dBFS input signal is swept over frequency. The integrated noise is unaffected for input frequencies < 95 kHz, which is higher than that reported for previous zoom ADCs. Over the entire audio band, the measured CMRR and PSRR of the ADC are greater than > 66 dB (Fig. 11). For -6 dBFS input signals around F_s, the measured alias rejection of the zoom ADC is > 80 dB in the audio band (Fig. 12). As in [3], a relaxed 800 kHz RC low-pass filter (R_{FILT} =20 kΩ, C_{FILT} =5 pF) is enough to suppress the effects of aliasing in the SAR ADC itself.

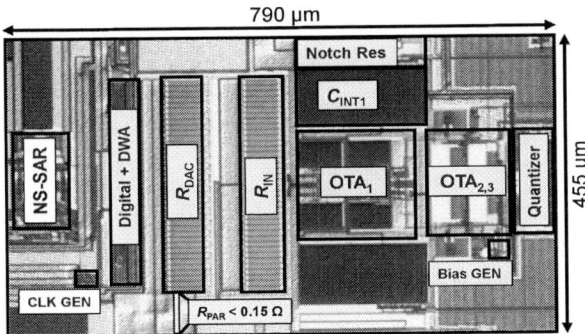

Fig. 6. Die Micrograph of the CT zoom ADC.

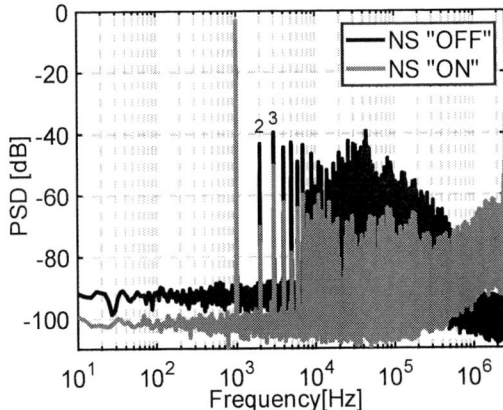

Fig. 7. Measured PSD of SAR ADC.

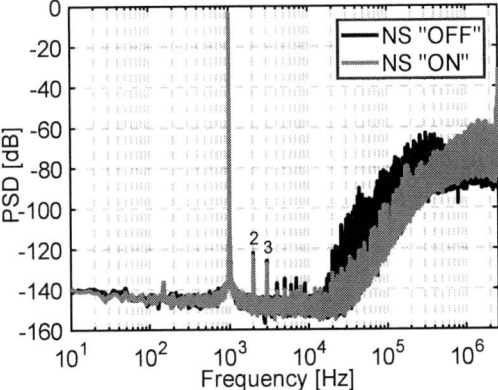

Fig. 8. Measured PSD of zoom ADC (2^{21} points, 8 averages, Blackman-Harris window).

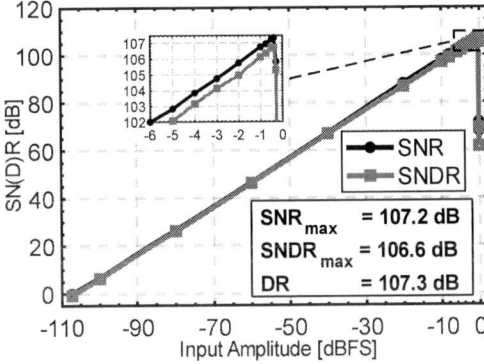

Fig. 9. Measured SN(D)R of zoom ADC across input amplitude.

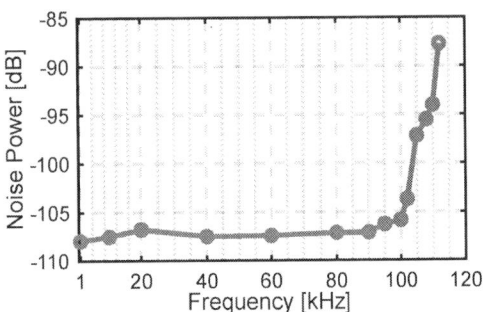

Fig. 10. Total in-band noise power across F_{IN} for a -1.5 dBFS input signal.

Table I. Performance summary and comparison

	This Work	ISSCC'22 [9]	ISSCC'21 [6]	ISSCC'21 [7]	JSSC'21 [1]	ISSCC'20 [8]	JSSC'20 [3]
Architecture	**CT Zoom**	DT PPD	CT	CT FIR DAC	DT Zoom	CT 3-level	CT Zoom
Tech (nm)	**160**	180	28	65	160	65	160
Area (mm²)	**0.36**	0.03	0.07	0.39	0.27	0.28	0.27
Supply (V)	**1.8**	1.8/1.1	1.8/1.0	1.2	1.8	1.2	1.8
Power (µW)	**590**	203.5	116	139	440	134	618
F_s (MHz)	**5.12**	5.80	6.144	7.2	3.5	8	5.12
BW (kHz)	**24**	20	24	24	20	24	20
SNR$_{max}$ (dB)	**107.2**	106.7	100.7	102.0	107.5	101.0	108.1
SNDR$_{max}$ (dB)	**106.6**	105.4	100.6	100.9	106.5	99.4	106.4
DR (dB)	**107.3**	108.8	104.4	104.8	109.8	103.5	108.5
NDSD⁺(dBFS/Hz)	**-150.4**	-148.4	-144.4	-144.7	-149.5	-143.2	-149.4
FoM$_{SNDR}$* (dB)	**182.7**	185.3	183.7	183.3	183.1	181.9	181.5
FoM$_S$ (dB)**	**183.4**	188.7	187.5	187.2	186.4	186.0	183.6

⁺Noise+Distortion Spectral Density (NDSD) = $-(SNDR_{max} + 10 \log_{10}(BW))dBFS/Hz$
*FoM$_{SNDR}$ = SNDR + 10log$_{10}$(BW/Power)   **FoM$_S$ = DR + 10log$_{10}$(BW/Power)

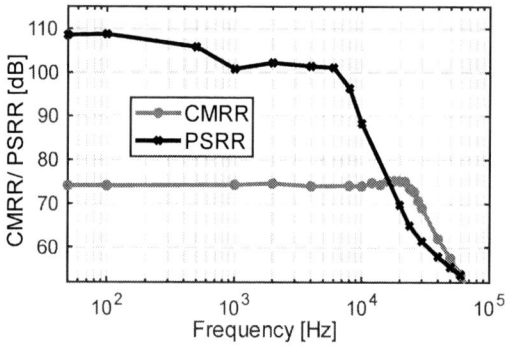

Fig. 11. Measured CMRR & PSRR across input/supply frequency.

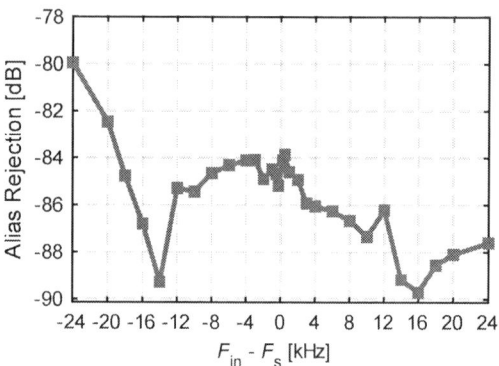

Fig. 12. Alias rejection of the CT zoom ADC (input: -6 dBFS).

Table I summarizes the performance of the designed CT zoom ADC and compares it with that of other state-of-the-art audio ADCs. When compared to the previous CT zoom ADC [3], this design achieves higher energy efficiency (FoM$_{SNDR}$) over a somewhat wider bandwidth (24 kHz versus 20 kHz), demonstrating the effectiveness of the techniques used in this work. The ADCs in [1, 9] have better FoM but are DT and require high-power input buffers to drive their switched-capacitor front ends, resulting in overall higher system power. The ADCs in [6, 7, 8] have higher energy efficiency but are realized in more advanced technologies resulting in significantly lower digital power and area, especially for clock generation and the DWA logic. It should be noted that the designs in [6, 7, 8] also target significantly lower (~6 dB) SNDR$_{max}$ specifications.

IV. CONCLUSIONS

This paper describes a CT zoom ADC for audio applications. Through the use of analog techniques such as passive noise shaping in the coarse SAR ADC and tail-resistor-linearization in the OTA$_1$, high linearity is achieved without compromising the energy efficiency. The proposed ADC achieves a competitive Schreier figure-of-merit (FoM$_S$) of 183.4 dB and FoM$_{SNDR}$ of 182.7 dB, together with the lowest in-band noise and distortion spectral density (NDSD = −150.4 dBFS/Hz) amongst state-of-the-art audio ADCs.

REFERENCES

[1] E. Eland, S. Karmakar, B. Gönen, R. van Veldhoven and K. A. A. Makinwa, "A 440-µW, 109.8-dB DR, 106.5-dB SNDR Discrete-Time Zoom ADC With a 20-kHz BW," in *IEEE Journal of Solid-State Circuits*, vol. 56, no. 4, pp. 1207-1215, April 2021

[2] S. Karmakar, B. Gönen, F. Sebastiano, R. van Veldhoven and K. A. A. Makinwa, "A 280 µW Dynamic Zoom ADC With 120 dB DR and 118 dB SNDR in 1 kHz BW," in *IEEE Journal of Solid-State Circuits*, vol. 53, no. 12, pp. 3497-3507, Dec. 2018

[3] B. Gönen, S. Karmakar, R. van Veldhoven and K. A. A. Makinwa, "A Continuous-Time Zoom ADC for Low-Power Audio Applications," in *IEEE Journal of Solid-State Circuits*, vol. 55, no. 4, pp. 1023-1031, April 2020

[4] Y. -Z. Lin, C. -Y. Lin, S. -C. Tsou, C. -H. Tsai and C. -H. Lu, "20.2 A 40MHz-BW 320MS/s Passive Noise-Shaping SAR ADC With Passive Signal-Residue Summation in 14nm FinFET," *2019 IEEE International Solid- State Circuits Conference - (ISSCC)*, 2019, pp. 330-332

[5] S. Pan and K. A. A. Makinwa, "A 10 fJ·K² Wheatstone Bridge Temperature Sensor With a Tail-Resistor-Linearized OTA," in *IEEE Journal of Solid-State Circuits*, vol. 56, no. 2, pp. 501-510, Feb. 2021, doi: 10.1109/JSSC.2020.3018164.

[6] C. Lo *et al.*, "10.1 A 116µ W 104.4dB-DR 100.6dB-SNDR CT ΔΣ Audio ADC Using Tri-Level Current-Steering DAC with Gate-Leakage Compensated Off-Transistor-Based Bias Noise Filter," *2021 IEEE International Solid- State Circuits Conference (ISSCC)*, 2021, pp. 164-166, doi: 10.1109/ISSCC42613.2021.9365807.

[7] S. Mondal, O. Ghadami and D. A. Hall, "10.2 A 139 µ W 104.8dB-DR 24kHz-BW CT ΔΣM with Chopped AC-Coupled OTA-Stacking and FIR DACs," *2021 IEEE International Solid- State Circuits Conference (ISSCC)*, 2021, pp. 166-168

[8] M. Jang, C. Lee and Y. Chae, "9.2 a 134µw 24khz-bw 103.5db-dr ct ΔΣ modulator with chopped negative-r and tri-level fir dac," *2020 IEEE International Solid- State Circuits Conference - (ISSCC)*, 2020, pp. 1-3

[9] C. Y. Lee and U. -K. Moon, "A 0.0375mm² 203.5µW 108.8dB DR DT Single-Loop DSM Audio ADC Using a Single-Ended Ring-Amplifier-Based Integrator in 180nm CMOS," *2022 IEEE International Solid-State Circuits Conference (ISSCC)*, 2022, pp. 412-41

A 150 mV, Sub-1 nW, 0.75%-Full-Scale INL Delta-Sigma ADC for Power-Autonomous Sensor Nodes

A. Catania[1]*, A. Ria[2], G. Manfredini[1], M.Dei[1], M. Piotto[1], and P. Bruschi[1]

[1]Department of Information Engineering, University of Pisa, Italy
[2] IEIIT Istitute of Italian National Research Council (CNR), Pisa, Italy
*alessandro.catania@unipi.it

Abstract— Measured performances of a Delta Sigma ADC prototype operating down to 150 mV supply show 600 pW consumption while being able to convert dc signals with INL<0.75% of input full scale. The modulator shows an SNDR of 59.3 dB over a bandwidth of 0.3 Hz and 17 Hz at a supply voltage of 0.15 V and 0.3 V, respectively. Dc characterizations over a set of 5 samples between 0.18 V and 0.3 V supply range show consistent and repeatable offset (<0.51% of full scale) and integral non-linearity (<1.45% of full scale). Such performances candidate the proposed ADC for its use in ultra-low voltage, low sampling rate, power-autonomous sensor nodes.

Keywords— Ultra-low voltage, Ultra-low power, Delta-Sigma ADC, Inverter-Like, Energy-harvester powered

I. INTRODUCTION

An increasing number of emerging applications is requiring the development of extremely miniaturized sensor nodes, capable of monitoring critical physical and chemical quantities relying only on the power provided by energy harvesters. Among the latter, great interest is received by enzymatic biofuel-cells [1,2], especially for wearable and implantable devices. These sources pose great challenges to the designers, due to their low output voltages, often of the order of only a few hundred mV. Use of boost dc-dc converters is made difficult by their low efficiency when applied to Ultra-Low Voltage (ULV) and Ultra-Low Power (ULP) harvesters. Then, it is desired to develop complete sensor systems capable of operating with an as small as possible supply voltage. Prototypes of CMOS digital standard-cells capable of operating with sub-100 mV supply voltage have been demonstrated [3], while analog circuits are lagging behind, with only a few examples capable of overcoming the 200 mV barrier. To fill this gap, the development of ULV CMOS analog cells is therefore of paramount importance.

A key block that is required in most sensor systems is the Analog to Digital Converter (ADC). Interestingly, data logging of biometric parameters (e.g. temperature, humidity, concentration of analytes in body fluids) involves very slow conversion rates, often below one sample per second. In these cases, the usual trade-offs between bandwidth, resolution and power consumptions are less relevant and the actual challenge is obtaining an adequate dc accuracy even with extremely low supply voltages and power consumptions [4]. Several examples of ULV ADCs have been presented in the last decade. The functionality of SAR ADCs for supply voltages equal [5] or even lower [6] than 200 mV have been demonstrated. For the quoted relaxed sampling rate constraints, single-bit Discrete Time (DT) Delta Sigma ($\Delta\Sigma$) ADCs represent a viable alternative to SAR ADCs, offering advantages in terms of area occupation due to their intrinsic insensitivity to device mismatch. Examples of DT-$\Delta\Sigma$ ADCs compatible with supply voltages from 300 mV to 250 mV have been proposed [7-10]. Lower supply-voltage limits have

been reported for $\Delta\Sigma$ ADCs where the integrator function is accomplished in the phase-domain by a Voltage Controlled Oscillator (VCO) [11, 12]. A major drawback of open-loop VCO-based ADCs is low control over gain and offset errors, which strongly depend on the accuracy of the VCO voltage-to-frequency conversion law. Considering classical DT-$\Delta\Sigma$ architectures, reduction of the supply voltage (V_{dd}) impacts on the amplifier's gain-bandwidth product and output range. Inverter-like amplifiers become the mandatory solution when V_{dd} gets lower than the MOSFET's threshold voltage [10]. Notice that for supply-voltages lower than nearly 200 mV, the interval of output voltages where both devices of the inverters operate in saturation region vanishes, leading to available voltage gains in the order of a few units. In addition, stabilization of the output common mode voltage of Fully-Differential (FD) amplifiers formed only by standard CMOS inverters is problematic. The popular Nauta's six-inverter solution [13] was found to be highly inefficient in terms of output range [14].

In this work we propose a second-order FD DT $\Delta\Sigma$ ADC, based on switched capacitor (SC) integrators that exploits recently introduced solutions for mitigation of the aforementioned problems occurring at extremely low dc voltages. A 9-inverter topology [15], which, differently from the Nauta's solution, does not degrade the original inverter output swing, is used for the fully differential amplifiers. Acceptable dc performances are maintained even at supply voltages below 200 mV thanks to the adoption of a two-stage SC integrator [16] that boosts the dc gain of the amplifiers, while reducing the offset of the latter through Correlated Double Sampling (CDS). The proposed ADC architecture was originally proposed in [17], where the potentiality of application in ULV conditions was supported by simulations. This work describes for the first time the results of measurements performed on a prototype designed using the UMC 0.18 µm CMOS process.

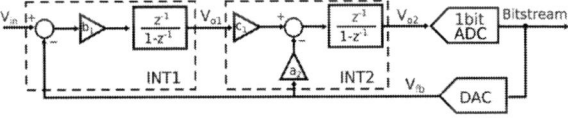

Fig. 1. Block diagram of a 2nd order, single-bit quantizer, Discrete-Time $\Delta\Sigma$ modulator.

II. CIRCUIT DESCRIPTION

The topology chosen for the proposed ULV $\Delta\Sigma$ modulator is a DT 2nd order Cascade of Integrator FeedBack (CIFB) with a single-bit quantizer (Fig.1). This standard topology was adopted for its good compromise between resolution, linearity and design complexity. Moreover, a fully-differential architecture was preferred due to its intrinsic robustness to common-mode disturbs and its doubled signal range. The schematic view of the whole $\Delta\Sigma$ modulator is represented in Fig.2, including the time diagram of the clock phases.

This project has received funding from the European Union's Horizon 2020 research and innovation programme under the Marie Skłodowska-Curie grant agreement No. 893544.

978-1-6654-8495-4/22 $31.00 © 2022 IEEE

The first integrator INT1 represents the most critical block in terms of noise, offset and dc gain requirements, as it was extensively analysed in [17]. For this reason, the switched-capacitor integrator presented in [16] was employed. The coefficient b_1 (Fig.1) is realized by the capacitive ratio C_S/C_F, while the other capacitors C_T and C_H are needed to reach the low dc-gain sensitivity and the dynamic offset cancellation technique, which are peculiar of this integrator topology. These characteristics are essential when we are working with ultra-low supply voltages and inverter-like amplifiers, characterized by intrinsic low dc gain and large sensitivity to device mismatch. The fully-differential amplifiers A_1 and A_2 are implemented exploiting the completely inverter-like topology presented in [15] and depicted in Fig.3-a. Differently from other fully-differential inverter-like amplifiers as [13], it maximizes the differential output range, reaching at the same time a good common-mode stabilization. Thanks to the Common-Mode Stabilization Loop made by Inv_{3-9}, the amplifier shows a common-mode gain lower than one, which allows stabilization of the common-mode signals of the integrator. Inverters Inv_{1-9}, employed in the fully-differential amplifier, are depicted in Fig.3-b. The body terminal of the pMOS is connected to a potential close to ground in order to lower its threshold voltage and improve the inverter speed. The body terminal is connected to the nMOS M_d (shared by all the inverters in the modulator) to limit the current flowing through the forward-biased body-source and body-drain junctions of the pMOS. However, at ultra-low supply voltages (e.g. lower than 0.5 V), the current flowing through the two junctions is negligible and the pMOS body terminal is biased with a potential very close to ground.

The performance of the second integrator INT2 is not as critical as the first one, as discussed in [17]. For this reason, we did not choose the same topology of INT1 but we opted for a standard parasitic-insensitive switched-capacitor topology, which employs only one fully-differential amplifier. Capacitive ratios C_{S2A}/C_{F2} and C_{S2B}/C_{F2} implement modulator coefficients c_1 and a_2, respectively. The fully-differential amplifier A_3 is identical to A_1 and A_2. The constant bias voltage V_{inv} used in INT1 and INT2 is generated by a single input-output connected inverter (Inv_0 in Fig.2).

The 1-bit ADC is a latched comparator, whose schematic view is shown in Fig.4. It works on the two clock phases, pre-amplifying the input signal during phase 2 (Inv_{10-11}) and deciding during the following phase 1 (Inv_{12-13}). Each clocked inverter is formed by a CMOS inverter and two switches to disable it during the off-phase. The 1-bit DAC provides the differential feedback signal to INT1 and INT2. The reference voltage corresponds to the supply voltage V_{dd}, thus the DAC simply consists of two cascaded inverters used to buffer the output differential signal of the comparator and its schematic view is omitted for simplicity.

Proper non-overlapped control signals, whose timing is depicted in Fig.2, drive all modulator switches. Excluding the switches in the 1-bit ADC, which are implemented as pass-transistors, the rest of the switches are complementary pass-gates. The high and low voltage levels of the control signals are shifted to $2V_{dd}$ and $-V_{dd}$, respectively, by means of a clock-boosting circuit [9], reducing the on-resistance of both the pMOS and the nMOS switches, which is particularly critical in ULV design.

III. EXPERIMENTAL RESULTS

The proposed $\Delta\Sigma$ modulator was fabricated with the 0.18 µm UMC CMOS process, using the 1.8 V MOSFET core devices. Sizing of all the inverters employed in the fully-differential amplifiers A_1, A_2 and A_3 (which are nominally identical), as well as sizing of the inverters in the ADC and in the DAC and of all the pass-gates, is reported in Table I. Values of the capacitors employed in INT1 and INT2 are also shown in Table I. Fig.5 shows a composition of an optical micrograph of the $\Delta\Sigma$ modulator and its superimposed layout in order to display devices and interconnections otherwise hidden below the planarization dummies. The geometrical dimensions of the clock boosting circuit and the $\Delta\Sigma$ modulator are indicated in the picture; the overall area occupation is 0.088 mm².

The oversampling clock signal was generated by a digital finite state machine, starting from the on-chip oscillator frequency and successively divided by means of programmable counters, which allow fine tuning of the oversampling frequency. The clock signal was level-shifted from the digital supply level (1.8 V) to the supply voltage V_{dd} of the $\Delta\Sigma$ modulator. At the output of the modulator, the bitstream was level-shifted from V_{dd} to 1.8 V. Both the clock and the bitstream waveforms were acquired by means of a Rohde & Schwarz RTB2004 oscilloscope and transferred to a personal computer, where the produced bit sequence was recovered and processed by programs written with the Python language. The software was used to estimate the bitstream spectra and implement a 3rd order CIC filter with a decimation factor (=OSR) of 64. The dc input differential signals were provided by a two-channel Source Measure Unit (SMU) Keysight B20902B. Sinusoidal and step stimuli were provided with an Agilent 33220A Arbitrary Waveform Generator.

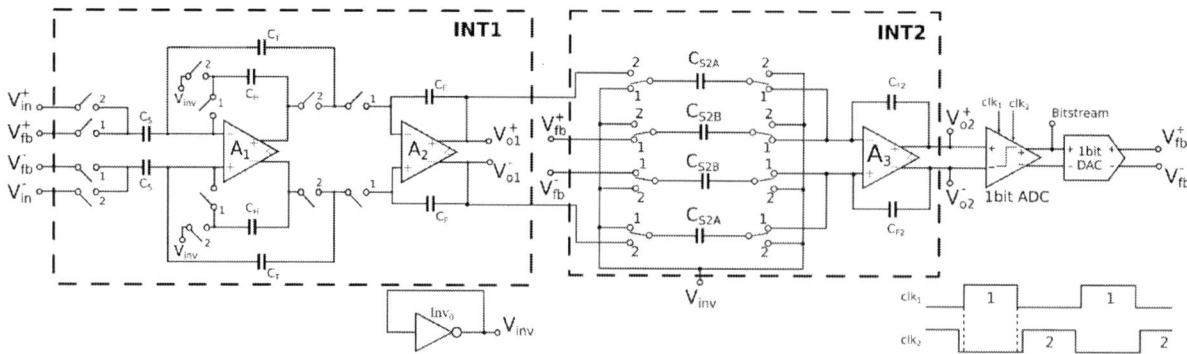

Fig. 2. Schematic view of the 2nd order, single-bit quantizer, fully-differential $\Delta\Sigma$ modulator.

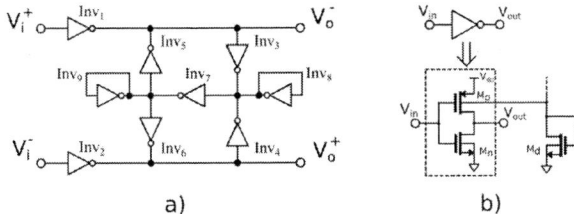

a)

b)

Fig. 3. Schematic view of the fully-differential inverter-based amplifier (a) and of the CMOS inverter with body-bias of Mp (b).

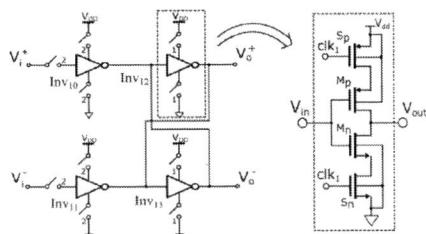

Fig. 4. Schematic view of the inverter-like 1-bit ADC, with the detail of the clocked inverter.

TABLE I
SIZES OF MOSFETS AND CAPACITORS

	L_{nMOS}	W_{nMOS}	L_{pMOS}	W_{pMOS}
Inv0	1 μm	5 μm	1 μm	5 μm
Inv1,2	1 μm	10 μm	1 μm	10 μm
Inv3,4	1 μm	2 μm	1 μm	2 μm
Inv5,6	1 μm	5 μm	1 μm	5 μm
Inv7,8,9	1 μm	500 nm	1 μm	500 nm
Inv10,11,12,13 (ADC 1-bit)	180 nm	2 μm	180 nm	6 μm
DAC inverters	180 nm	12.5 μm	180 nm	12.5 μm
Pass Gates	180 nm	960 nm	180 nm	1.92 um
C_S	C_T, C_H, C_{F2}	C_F	C_{S2A}	C_{S2B}
500 fF	1 pF	4 pF	490 fF	125 fF

Dc characterization was performed at different supply voltages, from 0.15 V to 0.3 V. The reference voltage of the ADC is V_{dd}, so the maximum input differential range is [-V_{dd}, V_{dd}]. All the results are normalized to the full-scale range of the ADC (FS=V_{dd}) to facilitate comparison of measurements performed at different supply voltages. Fig.6 shows the Integral Non-Linearity (INL) as a function of the differential dc input voltage for a fixed common mode dc input equal to V_{dd}/2, at different supply voltages (0.15 V, 0.18 V, 0.2 V and 0.3 V). The INL is lower than 1.5% of FS for a wide range of differential input voltages (from -90% to 90% of FS), for all the tested supply voltages. The current consumption I_{sup} and the oversampling frequency f_{ovs} at each supply voltage are listed in the inset of Fig.6. The dc operating point of the inverter-like amplifiers is sensitive to the supply voltage and their bias currents depend exponentially on the V_{dd} variations due to the weak inversion bias region. Thus, also the bandwidth of the amplifiers and, consequently, the maximum oversampling frequency of the ADC increase at higher supply voltages. It is worth noting the very low current consumption at V_{dd}=0.15 V resulting in a power consumption of only 600 pW.

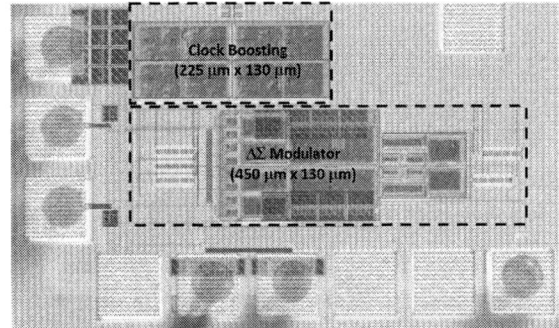

Fig. 5. Optical micrograph (plus superimposed layout) of the ΔΣ modulator and the clock boosting circuit.

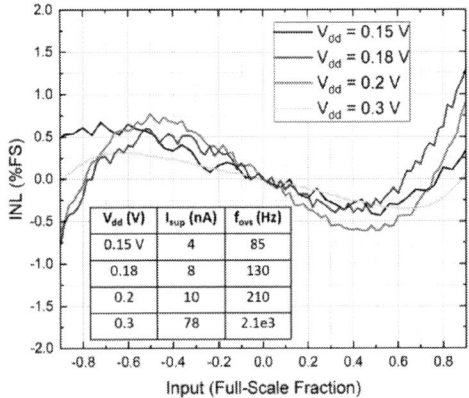

V_{dd} (V)	I_{sup} (nA)	f_{ovs} (Hz)
0.15 V	4	85
0.18	8	130
0.2	10	210
0.3	78	2.1e3

Fig. 6. INL vs differential input voltage at different supply voltages. The table in the inset shows the current consumption and the oversampling frequency for each V_{dd}

Fig. 7 shows the step response of the ΔΣ modulator at the output of the software CIC filter, with a supply voltage of 0.15 V, an input common mode of V_{dd}/2 and an oversampling frequency of 85 Hz. Comparing the input differential square waveform and the output data, a significant offset can be recognized. This is mainly due to the low dc gain of the inverter-like amplifiers at this extremely low supply voltage, which causes a degradation of the effectiveness of the CDS technique [16]. It is worth noting also the settling time of the output waveform, which coincides with the typical three decimated data settling time of the employed 3rd order CIC filter. Finally, the inset shows repeated acquisitions after the settling time, from which an rms equivalent noise of 354 ppm of FS was estimated.

Fig.8 shows the spectrum of the modulator bitstream for two different supply voltages, 0.15 V and 0.3 V. Different oversampling and stimulus frequencies were set in the two testing conditions, as well as different oversampling ratios (128 at V_{dd}=0.15 V, 64 at V_{dd}=0.3 V) and different amplitudes. The ratio between the input tone amplitude and the converter full-scale was kept constant. In both operating conditions, the modulator shows a Signal to Noise And Distortion ratio (SINAD) of 59.3 dB, while the Total Harmonic Distortion (THD) is close to -60 dB or lower, which represent a promising result of these preliminary analysis.

978-1-6654-8495-4/22 $31.00 © 2022 IEEE

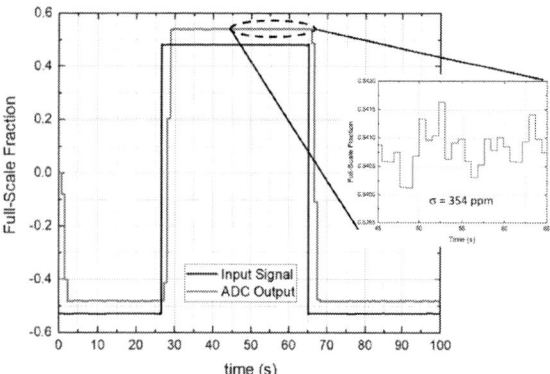

Fig. 7. Step response of the $\Delta\Sigma$ modulator supplied at $V_{dd}=0.15$ V (FS=0.15 V). The output bitstream was processed by means of Python 3rd order CIC filter with a decimation factor of 64.

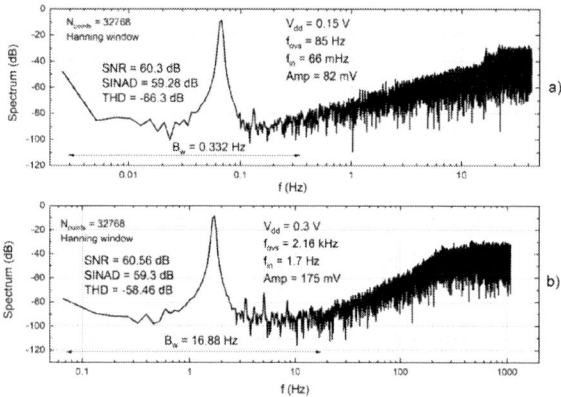

Fig. 8. Spectra of the output bitstream at $V_{dd} = 0.15$ V a) and $V_{dd} = 0.3$ V b).

Dc characterizations over a set of 5 samples at three different supply voltages (0.18 V, 0.2 V and 0.3 V) were performed and the average value μ and the standard deviation σ are reported in Table II. The standard deviation of the modulator offset is similar at 0.18 V and 0.2 V, while it significantly decreases at the highest supply voltage, due to the intrinsic increase of the amplifier dc gain. A similar trend can be recognized also in the maximum INL (measured over a differential input range from -90% to +90% of FS).

TABLE II
Dc Measurements on 5 Samples at Different VDD Values

V_{dd} (V)	I_{sup} (nA)		Offset (%FS)		INL-MAX (%FS)	
	μ	σ	μ	σ	μ	σ
0.18	9.9	1.2	-0.51	0.7	1.4	0.4
0.2	14.2	2.36	-0.06	0.83	1.45	0.55
0.3	108.6	17	-0.16	0.19	0.43	0.13

CONCLUSIONS

The proposed $\Delta\Sigma$ modulator is capable of working with supply voltage as low as 150 mV and a power consumption of only 600 pW. These performances are made possible by means of the high-gain, offset-free, switched-capacitor

integrator and the high output swing, fully-differential, inverter-like amplifier. It shows a maximum INL of 0.75% of full-scale at $V_{dd}=0.15$ V and a SINAD = 59.3 dB over a bandwidth of 0.3 Hz. Despite of the very low conversion rate, the proposed ADC is perfectly suitable for data logging in power-autonomous sensor nodes (as wearable and implantable devices), where the signal bandwidth of several biometric parameters is typically below 1 Hz.

REFERENCES

[1] A.F. Yeknami, X. Wang, I. Jeerapan, S. Imani, A. Nikoofard, J. Wang, P. Mercier, "A 0.3-V CMOS Biofuel-Cell-Powered Wireless Glucose/Lactate Biosensing System", IEEE J. Solid State Circuits, vol. 53, pp. 3126-3139, November 2018.

[2] C. Gozales-Solino, M. Di Lorenzo, "Enzymatic Fuel Cells: Towards Self-Powered Implantable and Wearable Diagnostics", Biosensors, vol. 8, pp. 1-18, January 2018.

[3] N. Lotze and Y. Manoli, , "Ultra-Sub-Threshold Operation of Always-On Digital Circuits for IoT Applications by Use of Schmitt Trigger Gates", IEEE Trans. Circ. Sys. I, vol. 64, pp. 2920-2933, Nov.ember2017.

[4] M. Dei, J. Aymerich, M. Piotto, P. Bruschi, F. del Campo, and F. Serra-Graells, "CMOS Interfaces for Internet-of-Wearables Electrochemical Sensors: Trends and Challenges," MDPI Electronics, vol. 8, no. 2, p. 150, Jan. 2019.

[5] A. Petrie, W.Kinnison, Y. Song, K. Layton, S.W. Chiang, "A 0.2-V 10-bit 5-kHz SAR ADC with Dynamic Bulk Biasing and Ultra-Low-Supply-Voltage Comparator", proc. of IEEE CICC conference, Boston, MA, 22-25 March 2020.

[6] X. Zhou, Q. Li, "A 160mV 670nW 8-bit SAR ADC in 0.13μm CMOS", proc. of IEEE CICC, San Jose, CA, USA 9-12 September 2012.

[7] L. Lv, X. Zhou, Z. Quiao, Q. Li, "Inverter-Based Subthreshold Amplifier Techniques and Their Application in 0.3-V $\Delta\Sigma$-Modulators", IEEE J. Solid State Circuits, vol. 54, pp. 1436-1445, May 2019.

[8] F. Michel and M. S. J. Steyaert, "A 250 mV 7.5 μW 61dB SNDR SC $\Delta\Sigma$ modulator using near threshold votage biased inverter amplfiers in 130 nm CMOS", IEEE J. Solid-State Circuits, vol. 47, no. 3,pp. 709–721, Mar. 2012

[9] A. Catania, L. Benvenuti, A. Ria, G. Manfredini, M.Piotto, P. Bruschi, "A 2 nW 0.25 V 140 dB-FOM Inverter-Based First Order $\Delta\Sigma$ Modulator, IEEE Trans. Circ. Sys. II, vol. 7, pp. 1514-1518, Sept. 2020.

[10] L. Lv, X. Zhou, Z. Qiao and Q. Li, "Inverter-Based Subthreshold Amplifier Techniques and Their Application in 0.3-V $\Delta\Sigma$-Modulators," IEEE J. Solid State Circuits, vol. 54, no. 5, pp. 1436-1445, May 2019.

[11] U. Wismar, D. Wisland and P. Andreani, "A 0.2V, 7.5 μW, 20 kHz $\Sigma\Delta$ modulator with 69 dB SNR in 90 nm CMOS", in proc. of ESSCIRC 2007, Munich, Germany, 11-13 September 2007.

[12] V. Nguyen, F. Schembari and R. B. Staszewski, "A Deep-Subthreshold Variation-Aware 0.2-V Open-Loop VCO-Based ADC," IEEE J. Solid State Circuits, Early Access, Oct. 2021

[13] B. Nauta, "A CMOS transconductance-C filter technique for very high frequencies," IEEE J. Solid State Circuits, vol. 27, no. 2, pp. 142-153, Feb. 1992.

[14] L. H. Rodovalho, C. R. Rodrigues and O. Aiello, "CMOS inverter linearization technique with active source degeneration", in proc. of 2021 IEEE Nordic Circuits and Systems Conference (NorCAS), Oslo, Norway, 26-27 October 2021.

[15] G. Manfredini, A. Catania, L. Benvenuti, M. Cicalini, M. Piotto, and P. Bruschi, "Ultra-Low-Voltage Inverter-Based Amplifier with Novel Common-Mode Stabilization Loop," MDPI Electronics, vol. 9, no. 6, p. 1019, Jun. 2020.

[16] P. Bruschi, A. Catania, S. Del Cesta and M. Piotto, "A Two-Stage Switched-Capacitor Integrator for High Gain Inverter-Like Architectures," IEEE Trans. Circ. Sys II, vol. 67, no. 2, pp. 210-214, Feb. 2020.

[17] L. Benvenuti, A. Catania, G. Manfredini, A. Ria, M. Piotto, and P. Bruschi, "Design Strategies and Architectures for Ultra-Low-Voltage Delta-Sigma ADCs," MDPI Electronics, vol. 10, no. 10, p. 1156, May 2021.

A 10.4-ENOB 0.92-5.38 μW Event-Driven Level-Crossing ADC with Adaptive Clocking for Time-Sparse Edge Applications

Jonah Van Assche, Georges Gielen
ESAT-MICAS, KU Leuven, Belgium
Email: jonah.vanassche@kuleuven.be

Abstract—This paper proposes a novel event-driven level-crossing ADC (LCADC) that highly improves power efficiency and accuracy compared to existing LCADC implementations. For applications with sparse signals such as ECGs, neural action potentials, etc., such LCADC can have a large data reduction at the ADC output. The new LCADC topology uses clocked comparators and event-driven adaptive clocking to overcome the high power consumption and signal-dependent distortion of regular asynchronous LCADCs. Due to the introduced clock, the ADC can seamlessly be integrated with any type of digital processing circuit, event-driven or conventional. Fabricated in a 40nm CMOS technology, it obtains 10.4 ENOB (this is the highest reported accuracy in literature). The ADC shows dynamic power consumption: it consumes 5.38 μW for a 15 kHz full-scale sine wave, while it consumes only 0.92 μW for a typical ECG signal. The peak Walden FOM of the ADC is 138 fJ/conv, an improvement of 35% in power efficiency beyond the state of the art. For an ECG application, the IC achieves a data reduction of 30%, clearly indicating that the LCADC can achieve a large power reduction at system level.

Index Terms—Level-Crossing ADC, Event-Based Sensor Interface, Edge Computing

I. INTRODUCTION

In many sensing applications like IoT nodes, wearable healthcare devices, etc., local on-chip data processing in the edge is increasingly used [1], [2], as it offers shorter latencies and reduces the overall system power consumption, by reducing the amount of data to be transmitted off-chip or even avoiding transmission completely. However, integrating extra on-chip data processing can compromise the low-power requirement for sensor front-ends, since the processsing circuits can potentially increase the power consumption of the system [1]. A way of reducing the data (and hence the processing/transmission power), is to employ so called event-driven processing [1], [2], [4]. For sparse burst-like signals, such as ECGs or neural action potentials, the relevant information of the signal only occurs sporadically in time. Sampling these signals with traditional fixed-clock sensing readouts (see Fig. 1a) is inefficient since this does not leverage the sparse information rate that these (bio)signals have. In contrast, when the sensor uses event-driven (or level-crossing) sampling, the sparse signal characteristics are fully exploited by sampling only when the signal change crosses a certain threshold level (see Fig. 1b) [4], [5], inherently leading to an event-driven, reduced output data stream to be processed/transmitted. Subse-

quent digital circuits could also work event-driven (e.g. spiking neural networks), such that the power consumption of both the ADC and the entire system application can significantly be reduced [1]–[4].

The level-crossing ADCs (LCADCs) reported in literature [1], [2], [5]–[7], however, do not achieve good power efficiencies and do not achieve a high ENOB (which is a requirement in sensing applications). The continuous-time, always-on comparators in traditional LCADC architectures limit their performance both in accuracy (by introducing signal-dependent distortion) as well as in power consumption [3], [5], [7]. Previous LCADCs were also using two DACs for the reference window [1], [5]–[7], which takes up a large area, and which consumes a substantial amount of power. Moreover, due to the asynchronous nature of the LCADC, integration with regularly clocked discrete-time DSPs is not straight forward, and retiming is required [7]. These disadvantages limit the usability of existing LCADC in a full system implementation. This paper describes the design of a novel LCADC topology (see Fig. 2) that uses an adaptive clocking strategy to overcome the limitations of previous asynchronous LCADCs by means of following innovations:

- power-efficient and highly accurate clocked comparators are introduced to reduce the power;
- an on-chip clock generator generates multiple sample clocks for the ADC to enable adaptive clocking;
- event logic dynamically adjusts the sample clock of the ADC to the signal activity limiting the power.

Fabricated in a 40nm CMOS technology, the designed prototype chip achieves obtains an ENOB of 10.4, which is to the author's knowledge the highest reported in literature. The

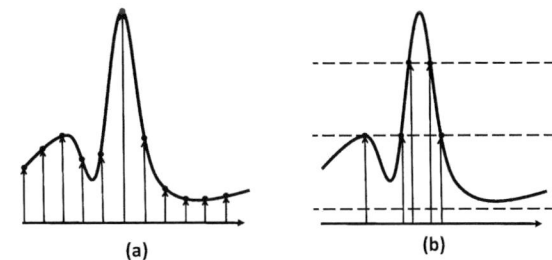

Fig. 1: (a) Regular Nyquist sampling and (b) event-driven level-crossing sampling.

978-1-6654-8495-4/22 $31.00 © 2022 IEEE

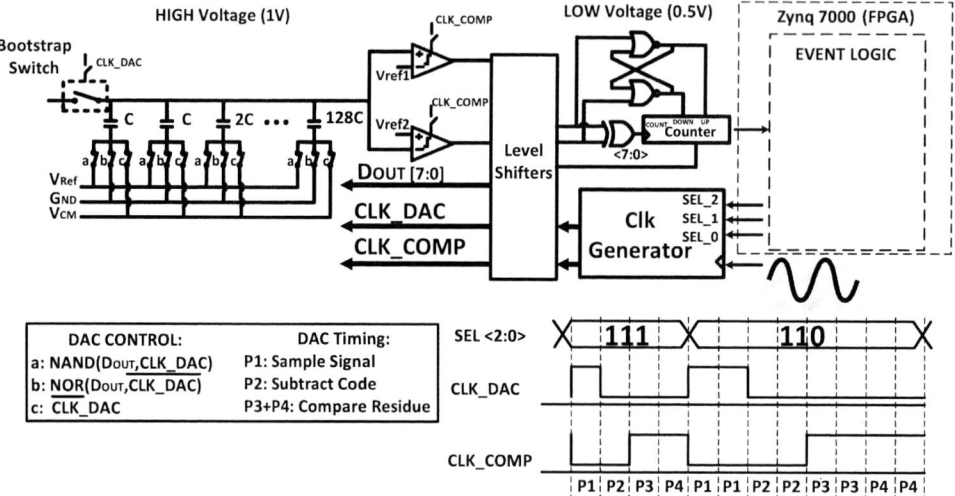

Fig. 2: Overview of the proposed level-crossing ADC and its timing diagram.

ADC also achieves a best-in-class 138 fJ/conv peak Walden FOM (an improvement of 35% compared to existing works) and has a dynamic power consumption ranging from 0.92 μW for an ECG signal to 5.38 μW for a full-scale sine wave of 15 kHz. At the system level, the IC reduces the output data stream by 30% compared to a Nyquist-rate ADC for ECG monitoring applications. The paper is structured as follows. Section II discusses the new LCADC architecture and how it improves on existing implementations. Section III shows the measurement results obtained from the prototype IC and compares it to other designs. Section IV concludes the paper.

II. ARCHITECTURE OF THE LCADC

A. ADC Operation

The new LCADC circuit architecture and timing diagram are shown in Fig. 2. The ADC algorithm works as follows. First, a signal is sampled on the capacitive DAC during a period P1 ($= T_{CLK}/4$); the previous sample is then subtracted during period P2 by the capacitive DAC, which results in a residue voltage smaller than 1 LSB. This residue is then compared to two reference values during half a clock period (P3 and P4). The counter at the output is updated when the residue voltage crosses one of the reference values (either it counts 1 LSB up or down). The counter value is then sent to the event logic, that counts the amount of periods in which the ADC output remains constant. After the ADC output has not changed for a (programmable) amount of clock periods, the clock frequency of the ADC is slowed down by means of the SEL[2:0] signal. The on-chip clock generator (see Fig. 3) takes this signal as input and outputs the appropriate clock signals for the comparators and DAC, as illustrated in the timing diagram of Fig. 2.

B. Topology Improvements

To improve the power efficiency and accuracy of the LCADC topology towards higher resolutions, our design (see Fig. 2) uses power-efficient discrete-time clocked comparators

instead of continuous-time comparators that have to draw a constant current. These clocked comparators are notably more power efficient, and, due to their clocked nature, don't introduce signal-dependent distortion.

If an LCADC misses a level crossing, distortion is introduced, the so called slope overload [3], [5]. In our system, the clock frequency required to avoid slope overload is given by:

$$f_{\text{CLK}} = \frac{\frac{\partial V_{\text{in}}}{\partial t}}{\text{LSB}} \qquad (1)$$

where $\frac{\frac{\partial V_{\text{in}}}{\partial t}}{\text{LSB}}$ is the maximum slew rate of the input signal. For the prototype design, a target quantizer resolution of 8 bit and a bandwidth of 15 kHz (which is sufficient for most biosignals) were chosen. This results in a required system clock of 25 MHz.

Due to the clocked nature of the comparators, the LCADC output is automatically synchronized, avoiding the need for an extra TDC [7] and allowing interfacing the ADC directly with a discrete-time DSP. However, with a fixed-rate clock, the ADC would not scale its power consumption with the signal activity, losing its advantage for sparse signals. Therefore, we use an adaptive clocking strategy: an on-chip clock generator generates multiple clock frequencies from one input clock (see Fig. 3), and a clock frequency is selected dynamically by means of a 3-bit control signal SEL[2:0], generated by event

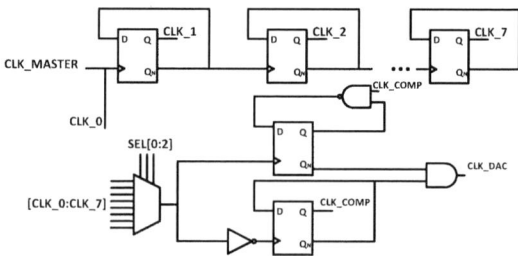

Fig. 3: Clock generator block of the LCADC.

logic that monitors the signal activity (see Fig. 2). Since the power consumption of all ADC building blocks scales with the clock frequency, this adaptive clocking strategy leads to significant power savings for time-sparse input signals.

Traditional level-crossing ADCs lack a clock or sampling stage, which means that they do not introduce quantization errors and that they are not limited by kT/C noise, such that they can achieve superior in-band SNDR performance compared to a Nyquist-rate ADC with equal quantizer resolution [4], [5]. The proposed architecture introduces quantization; however, due to the high clock frequency that is required (see equation 1), the quantization noise is distributed over a large frequency range. The in-band SNDR will therefore be much larger compared to that of a conventional Nyquist-rate ADC. Previous LCADCs needed two DACs to monitor if the input signal crosses a reference level (one DAC for the upper threshold and one for the lower). Our design uses only one DAC, that subtracts the previous code from each new sample, resulting in a residue voltage that is always bound between -1 LSB and 1 LSB. The residue is then compared to two fixed reference values to monitor potential level crossings. This "residue quantization" [6] thus only requires one DAC, saving area and power.

III. MEASUREMENT RESULTS

A prototype chip with 8-bit LCADC and 15 kHz signal bandwidth (SBW) has been fabricated in a 40nm CMOS technology. Fig. 4 shows the die photo and bonded chip. The total chip area is $1.5 \times 1.5 \text{mm}^2$, with 0.01mm^2 of active area. Mismatch in the DAC has been calibrated by means of a one-time foreground calibration with a 10-kHz sine wave. Fig. 5 shows the measured spectrum for a sine input of 10 kHz, resulting in a SNDR in the 15 kHz signal band of 62.5 dB. Since the SBW is much smaller than the sample rate, the in-band quantization noise is lower compared to that of a classical 8b Nyquist ADC; hence, a higher SNDR is reached. Fig. 6 shows the measured SNDR in the 15-kHz signal band for different input frequencies, before and after calibration. The peak SNDR of 10.4 ENOB (64.4 dB) is reached at 15 kHz. Fig. 7 shows the power consumption versus the input sine frequency, at several fixed clock frequencies (in this mode, the SEL control signal is kept constant) and in event-driven mode (when the event logic adapts the ADC clock

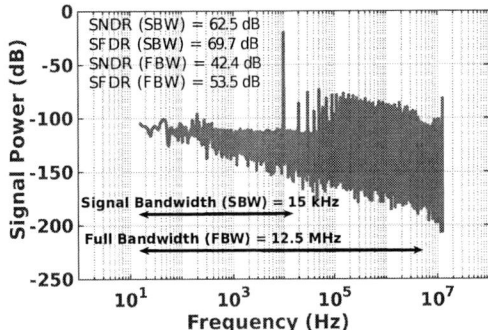

Fig. 5: 15000000-point FFT for a 10-kHz input sine.

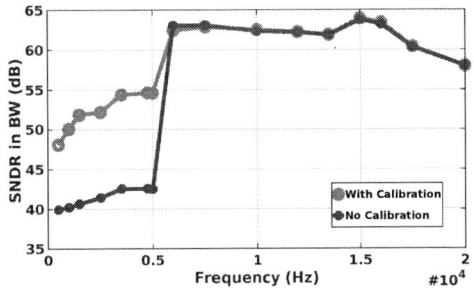

Fig. 6: SNDR in the 15-kHz signal band for different input frequencies.

frequency based on the signal activity). In fixed mode, the power consumption varies with the applied clock frequency. Once a clock frequency is chosen (in function of the slew rate of the input signal, see equation 1), the power consumption varies only weakly with the input frequency. As the ADC only generates a few discrete clock frequencies, it works inefficiently in the fixed mode. In contrast, in the event-driven mode, the power consumption varies from 0.92 μW for an ECG (see Fig. 9) to 5.38 μW for a 15 kHz full-scale sine wave (excluding I/O and clock buffers). This highlights the activity-based power consumption of the ADC. To illustrate the possible power savings in subsequent digital blocks, Fig. 9 plots the power of the I/O buffers, showing a reduction of 3x for ECG. Clearly, the I/O buffers consume significantly less thanks to the data-dependent output of the LCADC, which achieves a 30% reduction in data compared to a Nyquist-rate

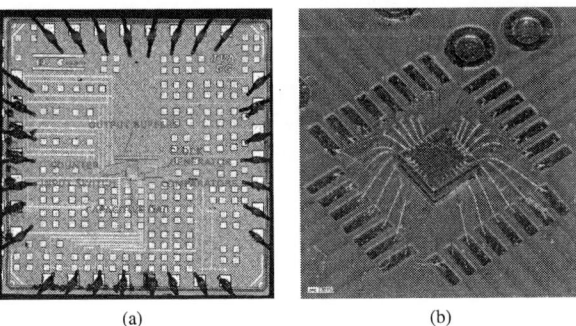

Fig. 4: (a) Die photograph and (b) chip bonded to PCB.

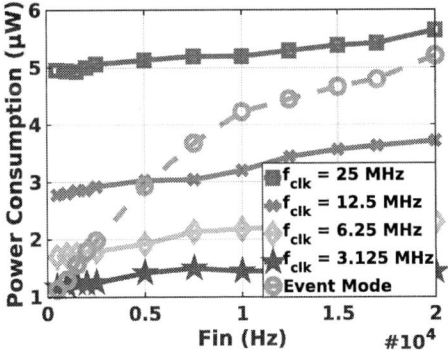

Fig. 7: Measured power consumption as a function of the sine wave input frequency in fixed and event-driven modes.

978-1-6654-8495-4/22 $31.00 © 2022 IEEE

TABLE I: COMPARISON WITH STATE-OF-THE-ART LEVEL-CROSSING ADCS.

	Weltin-Wu JSSC13 [5]	Wang JSSC20 [6]	Wu JSSC17 [7]	He A-SSCC21 [1]	Wang JSSC21 [2]	This work
Application	Sensors	IOT	RF	ECG	Wake-Up Circuit	Biomedical Sensing
Topology	AR LCS	AR LCS	LCS	LCS	LCS	LCS
Technology (nm)	130	28	65	40	180	40
Supply Voltage (V)	0.8	/	1	0.9/1/1.1	0.6	0.5/1
ADC Resolution	8b	7b	4b	/	5b	8b
Discrete DSP Compatible?	No	Yes	Yes	No	Yes	Yes
Area (mm^2)	0.36	0.0126	0.3	0.06**	/	0.012
Power Consumption	3-8.5 μW	205 μW	30 mW	7 -17 μW	75 nW- 1μW	5.38 μW
Bandwidth	20 kHz	1.42 MHz	19 MHz	1000 Hz	80 kHz	15 kHz
ENOB (Bits)	7.5-8.7	8.6	9.7	4.7-9.6	/	10.4
SNDR (dB)	47-54	53.53	59.9	30-59.5	/	64.4
SFDR (dB)	58.3 (1 kHz)	58	56	72	/	81.8
Walden FOM (fJ/conv)*	210-880	186.1	977	109000-11000	195***	138

*FOM = Power/2*$f_{Nyquist}$*2^{ENOB}
**Estimated from die photo.
***Assuming 5b ENOB (was not reported).

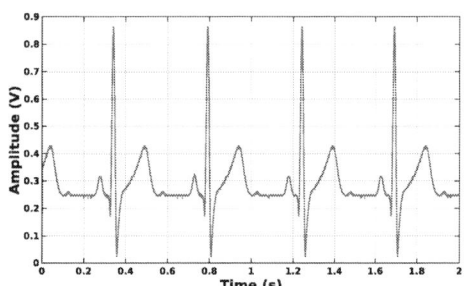

Fig. 8: Sampled ECG signal.

ADC for ECG signals.

Table I shows a comparison of achieved performances of recently published LCADCs. Our design highly outperforms the state of the art, the achieved ENOB of 10.4 is to our knowledge the highest reported in literature for LCADCs. The chip also reaches a best-in-class peak Walden FOM of 138 fJ/conv, which is an improvement of 35% compared to the best implementation till now.

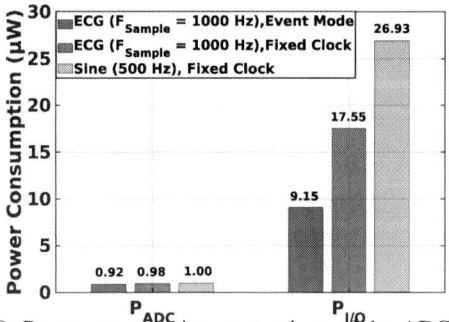

Fig. 9: Power consumption comparison of the ADC and I/O buffers for an ECG signal and a 500-Hz sine wave.

IV. CONCLUSION

This paper has presented a novel event-driven level-crossing ADC (LCADC). By replacing the continuous-time comparators by clocked discrete-time comparators and using an adaptive clocking strategy, the LCADC achieves a much higher power efficiency and accuracy compared to state-of-the-art LCADCs. The prototype chip in 40nm CMOS achieves a record ENOB of 10.4 bits and a peak Walden FOM of 138 fJ/conv.

The system-level benefits of the topology have also been illustrated: a data reduction for ECG applications is noted. Digital circuits, both regularly clocked and event-driven (e.g. spiking neural networks) as used in edge computing, can greatly benefit from the proposed architecture.

ACKNOWLEDGMENT

This work was supported by the KU Leuven internal research grant 'Time is Money'. The authors would like to thank EUROPRACTICE MPW and design tool support for tape-out sponsoring.

REFERENCES

[1] Y. He et al., "A 28.2 C Neuromorphic Sensing System Featuring SNN-based Near-sensor Computation and Event-Driven Body-Channel Communication for Insertable Cardiac Monitoring," 2021 IEEE Asian Solid-State Circuits Conference (A-SSCC), 2021, pp. 1-3, doi: 10.1109/A-SSCC53895.2021.9634787.

[2] Z. Wang et al., "A Software-Defined Always-On System With 57-75-nW Wake-Up Function Using Asynchronous Clock-Free Pipelined Event-Driven Architecture and Time-Shielding Level-Crossing ADC," in IEEE Journal of Solid-State Circuits, vol. 56, no. 9, pp. 2804-2816, Sept. 2021, doi: 10.1109/JSSC.2021.3074636.

[3] J. Van Assche and G. Gielen, "Power Efficiency Comparison of Event-Driven and Fixed-Rate Signal Conversion and Compression for Biomedical Applications," in IEEE Transactions on Biomedical Circuits and Systems, vol. 14, no. 4, pp. 746-756, Aug. 2020, doi: 10.1109/TB-CAS.2020.3009027.

[4] Y. Tsividis, "Event-Driven Data Acquisition and Digital Signal ProcessingA Tutorial," in IEEE Transactions on Circuits and Systems II: Express Briefs, vol. 57, no. 8, pp. 577-581, Aug. 2010, doi: 10.1109/TC-SII.2010.2056012.

[5] C. Weltin-Wu and Y. Tsividis, "An Event-driven Clockless Level-Crossing ADC With Signal-Dependent Adaptive Resolution," in IEEE Journal of Solid-State Circuits, vol. 48, no. 9, pp. 2180-2190, Sept. 2013, doi: 10.1109/JSSC.2013.2262738.

[6] H. Wang, F. Schembari and R. B. Staszewski, "An Event-Driven Quasi-Level-Crossing Delta Modulator Based on Residue Quantization," in IEEE Journal of Solid-State Circuits, vol. 55, no. 2, pp. 298-311, Feb. 2020, doi: 10.1109/JSSC.2019.2950175.

[7] T. Wu, C. Ho and M. S. Chen, "A Flash-Based Non-Uniform Sampling ADC With Hybrid Quantization Enabling Digital Anti-Aliasing Filter," in IEEE Journal of Solid-State Circuits, vol. 52, no. 9, pp. 2335-2349, Sept. 2017, doi: 10.1109/JSSC.2017.2718671.

SmartHeaP - A High-level Programmable, Low Power, and Mixed-Signal Hearing Aid SoC in 22nm FD-SOI

Jens Karrenbauer*, Simon Klein*, Sven Schönewald*, Lukas Gerlach*, Meinolf Blawat[†],
Jens Benndorf[†], Holger Blume*
* IMS, Leibniz University Hannover
{karrenbauer, klein, schoenewald, gerlach, blume}@ims.uni-hannover.de
[†] Dream Chip Technologies GmbH
{meinolf.blawat, jens.benndorf}@dreamchip.de

Abstract—To handle the advances in hearing aid algorithms, the need for high-level programmable but low-power hardware architectures arises. Therefore, this paper presents the Smart Hearing Aid Processor (SmartHeaP), a mixed-signal system on chip (SoC) fabricated in 22 nm fully-depleted silicon-on-insulator (FD-SOI) with an adaptive body biasing (ABB) unit and a total die size of 7.36 mm^2. The proposed SoC consists of two application-specific instruction set processor (ASIP) architectures: firstly, a Cadence Tensilica Fusion G6 instruction set architecture, extended with custom instructions for audio processing, and secondly, a Cadence Tensilica LX7 for wireless interfacing, e.g., Bluetooth Low Energy. Furthermore, an analog front-end and digital audio interfaces are added. The large local memory of 2 MB and a high-level software environment enables memory-intensive algorithms to be deployed quickly. Typical hearing aid algorithms in a real-time setup are used to evaluate the power consumption of the SoC at different operating frequencies. At 50 MHz, a mean power consumption of less than 2.2 mW was measured, resulting in an efficiency of 34.8 μW/MHz.

Index Terms—22nm FD-SOI, ABB, ASIC, ASIP, hearing aid, instruction extension, low power, system on chip (SoC), Tensilica

I. INTRODUCTION AND RELATED WORK

Even though hearing aids are well-known and have already been on the market for a long time, the supply for patients is still inadequate [1]. For better acceptance, hearing aids are becoming smaller. However, these designs also leave less space for batteries, additional chips beside the processors, and other peripheral devices. In addition, new and even more complex algorithms are researched to improve the benefit for hearing impaired persons. One approach is to identify, locate, and separate speakers in complex and noisy surroundings with binaural algorithms [2]. For further enhancement of the signal-to-noise ratio, neural networks, can be used to denoise the audio signals [3]. Nevertheless, these advances also tend to have higher memory requirements, which most current hearing aid architectures can not provide. Consequently, a need for flexible low-power architectures including all required memory and peripherals while retaining a small footprint arises.

The main focus of hearing aid researches are application-specific instruction set processor (ASIP) architectures [3]–[10].

These architectures, equipped with custom hardware accelerators, achieve a power consumption of only a few milliwatts while still allowing for a certain degree of programmability. Therefore, some of them are fabricated in small and energy-efficient technologies like a 28nm low power CMOS; others already include an analog front-end or support programming via assembly. A detailed comparison can be found in Table I.

The Smart Hearing Aid Processor (SmartHeaP) aims to combine these benefits into one hearing aid system on chip (SoC). Using the advanced technologies of 22 nm, integrated with an analog front-end and interfaces for binaural signal processing, the two ASIPs embedded on the SoC can operate in various hearing aid scenarios while consuming only a few milliwatts. Furthermore, with the large local memories and the high-level programmability, new algorithms can quickly be deployed on the enhanced audio ASIP, while the second ASIP can be used for binaural connectivity.

The paper is organized as follows: In Section II the general design and the optimizations of the ASIPs of the SmartHeaP SoC are described in detail, followed by an evaluation of the implementation results of the fabricated SoC in Section III. Finally, Section IV concludes this paper.

II. SMARTHEAP SOC

To achieve a highly flexible, computational powerful, and energy-efficient ASIP, the SmartHeaP SoC was manufactured in the 22 nm fully-depleted silicon-on-insulator (FD-SOI) technology denoted as 22 FDX by GLOBAL-FOUNDARIES [11]. An adaptive body biasing (ABB) block is included in the SoC to achieve high energy efficiency. The transistors can operate at lower voltages with this IP while remaining at the same frequency as corresponding cells with zero body biasing [12]. Fig. 1 shows a simplified block diagram of the SoC.

A. Analog Front-end

The SoC includes an analog front-end to reduce the components on the hearing aid and, therefore, the device's size. It consists of two 20-bit ADC and one 20-bit DAC and enables

---

978-1-6654-8495-4/22 $31.00 © 2022 IEEE

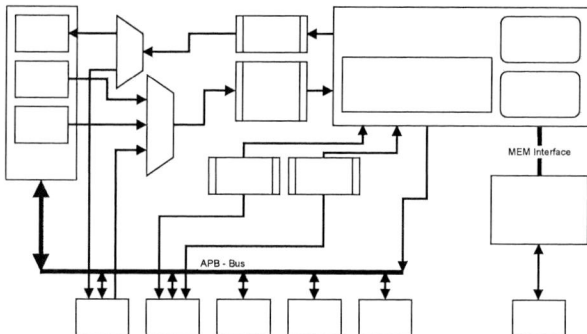

Fig. 1: Block diagram of the SmartHeaP SoC.

the direct connection of two microphones and one speaker to the SoC. A 20-bit wide and 128 words long FIFO memory is attached to each analog interface for block-wise processing of the received audio data. Instead of the analog front-end, the inputs of the queues can be multiplexed to a digital inter-IC sound (I^2S) interface. To receive data from a wireless transmission such as near-field magnetic induction (NFMI) or other microphones of a second hearing aid, an additional standardized two-way I^2S interface is connected to two 32-bit wide and 256 words long FIFOs.

B. Processing Architecture

The cores of the SmartHeaP SoC are both application-specific instruction-set processor (ASIP) architectures based on the Cadence Tensilica LX7 family [13]. While one is optimized for audio processing, the second ASIP is designated for Bluetooth low energy (BLE) communications. Both omit a floating-point unit to save energy and chip area. For the audio processing ASIP, the Tensilica Fusion G6 instruction set architecture (ISA) [14] is added to the ISA of the LX7.

The BLE core is based on the LX7 RISC architecture using a 5 stage pipeline. A 32-bit multiplier and a 32-bit integer divider are added to the base ISA. In addition, 128 KB of instruction RAM and 256kB data RAM are included in the system to handle a complete Bluetooth stack. To exchange data between the two cores a direct memory interface is used. With this interface both cores can simultaneously read and write from/into the data memory of the other core without additional cycles or a direct memory access (DMA) unit.

In comparison to the BLE core, the audio core is more complex. The core has four issue slots, supports very long instruction words (VLIW), and packed single instruction multiple data (SIMD) operations. It operates on native 32-bit wide general-purpose registers while containing special 256-bit wide vector registers. Therefore, two 256-bit wide load and store units are included in the first and second issue slot of the architecture.

C. Instruction Extensions

Besides the standard ISA extensions of the Fusion G6, like hardware loops or bit manipulating operations, several

newly designed fixed-point custom instructions were added to enable fast and energy-efficient digital signal processing. These instructions, based on [15], include an additional register type, combining 32-bit real and complex-valued samples into one. In addition, different special hardware-aware units, like a complex multiply and accumulate (CMAC) [16] or complex add and sub units, are added to operate with these registers. They allow the optimization of computations in the frequency domain or Fourier transformations themselves, which are crucial in today's hearing aid applications.

Trigonometric and exponential functions are another intense computational task for fixed-point architectures. Nevertheless, they often appear in the audio domain, e.g., for scaling different sound pressure levels in a compressor. The coordinate rotation digital computer (CORDIC) algorithm is a well-known method to compute them. A hardware unit calculating one iteration of the CORDIC kernel in one cycle is included in the ASIP. With software intrinsics during run-time the unit can be configured to operate in hyperbolic or vectoring mode and use one of three phase types (circular, linear, and hyperbolic). Thirty-two 32-bit wide lookup registers for storing the angular tables of the CORDIC were added to increase the benefits of the unit in repeated iterations like block processing of audio data. Once initialized, they can be reused for several CORDIC iterations. Two of these CORDIC units were added in two different issue slots of the architecture, including the corresponding new standalone and SIMD instructions to operate with them.

D. Memory Subsystem and Programmability

The audio core of the SmartHeaP SoC is equipped with 256 KB of instruction memory to handle the control flow of even complex hearing aid algorithms. Because of the recent development of computing and memory demanding algorithms [17], 2 MB of data memory separated into two memory banks were added to the audio core. This large amount of local memory enables low access times without a cache and eliminates the corresponding additional wait cycles through cache misses. Furthermore, it is possible to deploy memory-intensive hearing aid algorithms or even a complex chain of different algorithms.

With advances in neural networks, their use also extends to the hearing aid domain, e.g., with better noise reduction or more advanced algorithms for speaker separation. Even applying the most recent compression techniques, neural networks still need at least several hundred kB [18], with increasing demand depending on the network complexity. The large local memory of the SmartHeaP SoC allows the development and evaluation of these networks directly on a hearing aid SoC.

The SmartHeaP SoC comes with a complete programming environment, including a hardware extensions aware compiler toolchains for C/C++ (GCC, Clang) automatically generated by the Tensilica ecosystem. Furthermore, the already presented framework in [19] was adapted so that Matlab fixed-point code can easily be ported onto the SoC while using hardware-aware intrinsics. These advances drastically the reduce hearing aid

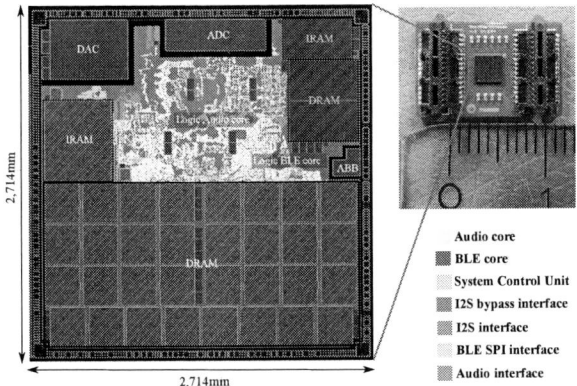

Fig. 2: Layout and photo of the fabricated SmartHeaP SoC

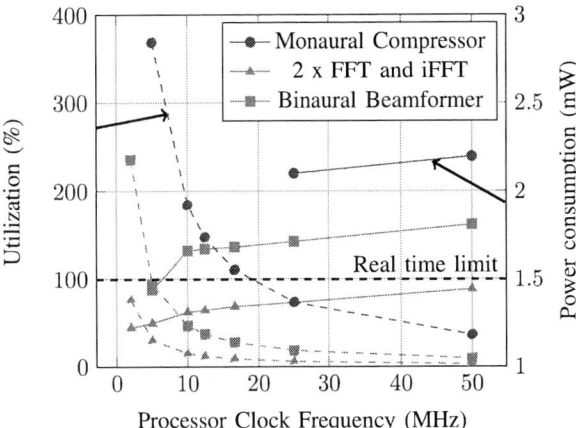

Fig. 4: Measured processor utilization and power consumption vs. frequency during a 4 ms time period. The utilization is is displayed by dashed lines, the power consumption by solid lines.

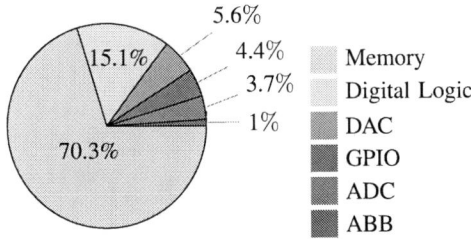

Fig. 3: Area distribution of the SoC

algorithms' design and maintenance time compared to other recently published hearing aid processors [4], [10] without this support.

III. IMPLEMENTATION RESULT

Fig. 2 depicts the SoC-Layout of the fabricated chip. The chip area is 7.36 mm^2, while the audio core's data memories occupy around 70 % of the chip, as seen in Fig. 3. It is divided into two banks of 16 instances per bank to keep the critical path and the power consumption of the memories low [7]. In total, the chip deploys four different voltage domains. For the digital core logic, 0.5V is used, and 0.65V supplies the memories. Furthermore, the analog part of the chip uses a 0.8V domain, and the chip-IO is supplied by 1.8V.

In contrast to IoT devices, hearing aids are always-on and need constant computing power instead of short peaks of high computational loads. Therefore, the goal of the presented SoC is to reduce the overall power consumption with a lower maximal clock speed compared to recently published low-power IoT devices (e.g. [20]) in the same technology. Thus, no power gating and only clock gating with a maximum operating frequency of 50 MHz are used. With this design goal more transistors with less leakage were used.

Different hearing aid algorithms are deployed on the SoC to evaluate the real power consumption and utilization: two 256-point FFT and iFFT assuming two microphone inputs, a binaural MVDR Beamformer [21] using four microphone inputs to steer the beam between four quadrants, and a

monaural compressor. The MVDR Beamformer and monaural compressor work in the frequency domain, thus also internally computing a 256-point FFT and iFFT. Applying overlap-add and 50 % zero padding, 64 audio samples are required to compute one FFT frame. The 64 samples were streamed into the system via one I2S interface and sampled with 16 kHz, thus resulting in 4ms processing time per frame. This time is also assumed as a real-time constraint for the processor utilization in Fig. 4. To keep the results shown in Fig. 4 comparable to the related works presented in Table I, the mean power consumption during the 4 ms time-frame of the memories and the digital parts of the SoC are depicted. The presented values are measured at the power supply of the 0.5V and 0.65V domains.

The evaluations were done for various clock frequencies ranging from 2 MHz to 50 MHz. The results show that the binaural beamforming can be performed in real-time with a power consumption of 1.45 mW @ 5 MHz. At 50 MHz an efficiency of 34.8 μW/MHz is reached. It appears that the power consumption changes only slightly with the clock frequency. This observation is explained by the low utilization at higher frequencies, where the idle power dominates the power consumption. As intended by the always-on design, the idle power is small compared to the dynamic power of the algorithm, and the power consumption is mainly driven by the complexity of the algorithm, not the clock frequency. This is confirmed by Fig. 4, showing that the FFT and iFFT @ 50MHz consume only 1.42 mW, which is less than the monaural compressor with 2.2 mW, while both algorithms change by less than 0.2 mW over the various clock frequencies. Furthermore, due to the low utilization, this evaluation also shows the potential of the proposed SoC to deploy more complex or even a chain of algorithms.

TABLE I
COMPARISON WITH RELATED WORKS.

	[9]	[10]	[3], [7]	[4]	[6]	[8]	this Work
Year	2018	2018	2018	2019	2020	2020	2022
Analog Front-End	Mixed Signal	16-bit ADC/DAC	12-bit ADC	-	-	-	20-bit ADC/DAC
Programmability	C	Assembly	Programmable	Assembly	Programmable	Programmable	C/C++
Technology	65 nm	130 nm	28 LP CMOS	40 nm	40 nm	40 nm	22 FDX
Supply	1.18 V	0.9 V	0.55 V	1.1 V	0.6 V	0.7 V	0.5 V Core 0,65 V Mem
Power	1.25 mW	1.1 mW	4 mW	0.6 mW @ 10 MHz	2.17 mW @ 5 MHz	1,5 mW	1.45 mW @5 MHz
Max freq.	15.36 MHZ	-	200 MHz	50 MHz	20 MHz	10.5 MHz	50 MHz
Size	30 mm^2 (Package)	9.3 mm^2	9 mm^2	3.6 mm^2	4.2 mm^2	0.3 mm^2	7.365 mm^2
Memory	256 KB EEPROM on SoC Package	-	16 - 320 KB per core	56 KB data +65 KB instr	327 KB	16 KB data +16 KB instr	2,5 MB data +375 KB instr
Architecture	CoolFlux DSP + ARM Cortex M3 + 2 co-processors	RISC based ASIP	Cortex M0 + Stream Micro Processor (SuP)	4 KAVUAKA + 10 co-processors	4 processing clusters	Open RISC 1200	Fusion G6 + LX7 + ISA Extension

IV. CONCLUSION AND FUTURE WORK

This paper presents the SmartHeaP SoC, a mixed-signal chip produced in 22FDX. It is designed to achieve the low power requirements of a hearing aid while being high-level programmable. The integrated analog front-end and the different audio interfaces can reduce the need for additional peripherals. Furthermore, the SoC includes two Tensilica ASIPs, one equipped with custom ISA extensions for trigonometric computations and operations in the frequency domain. The total die size is 7.36 mm^2, while the 2 MB data memory occupies 70 % of the chip to deploy memory-intensive algorithms. In a real-time hearing aid scenario with a monaural compressor and a streamed audio signal a power consumption of 1.45 mW @ 5 MHz is measured. With the high-level software environment, the presented SoC can be used for further evaluation of hearing aid algorithms.

ACKNOWLEDGMENT

This work was funded by the Federal Ministry of Education and Research of Germany (BMBF) - project number 16ES0759. We also like to thank the partners for the excellent cooperation, especially Hörzentrum Oldenburg for the algorithmic support, Cadence Design Systems for providing the Tensilica and other IPs and support related EDA tools, SoC and architecture design methodology, and GlobalFoundries for the fabricated 22 FDX silicon.

REFERENCES

[1] N. S. Reed, E. Garcia-Morales, and A. Willink, "Trends in hearing aid ownership among older adults in the united states from 2011 to 2018," *JAMA Internal Medicine*, vol. 181, no. 3, pp. 383–385, 2021.

[2] C. Seifert et al., "Real-time implementation of a gmm-based binaural localization algorithm on a vliw-simd processor," in *2017 IEEE International Conference on Multimedia and Expo (ICME)*, 2017, pp. 145–150.

[3] Y. Pu et al., "An ultra-low-power 28nm cmos dual-die asic platform for smart hearables," in *2018 IEEE Biomedical Circuits and Systems Conference (BioCAS)*, 2018, pp. 1–4.

[4] L. Gerlach, G. Paya-Vaya, and H. Blume, "Kavuaka: A low power application specific hearing aid processor," in *2019 IFIP/IEEE 27th International Conference on Very Large Scale Integration (VLSI-SoC)*. IEEE, 06.10.2019 - 09.10.2019, pp. 99–104.

[5] S.-W. Kim, M.-J. Kim, and J.-S. Kim, "High–performance dsp platform for digital hearing aid soc with flexible noise estimation," *IET Circuits, Devices & Systems*, vol. 13, no. 5, pp. 717–722, 2019.

[6] Y.-C. Lee, T.-S. Chi, and C.-H. Yang, "A 2.17-mw acoustic dsp processor with cnn-fft accelerators for intelligent hearing assistive devices," *IEEE Journal of Solid-State Circuits*, vol. 55, no. 8, pp. 2247–2258, 2020.

[7] Y. Pu et al., "A 9-mm 2 ultra-low-power highly integrated 28-nm cmos soc for internet of things," *IEEE Journal of Solid-State Circuits*, vol. 53, no. 3, pp. 936–948, 2018.

[8] Y.-J. Lin et al., "A 1.5 mw programmable acoustic signal processor for hearing assistive devices with speech intelligibility enhancement," *IEEE Transactions on Circuits and Systems I: Regular Papers*, vol. 67, no. 12, pp. 4984–4993, 2020.

[9] Semiconductor Components Industries, LLC:, "Wireless-enabled audio processor for hearing aids," 2018.

[10] C. Chen and L. Chen, "A 79-db snr 1.1-mw fully integrated hearing aid soc," *Circuits, Systems, and Signal Processing*, vol. 38, no. 7, pp. 2893–2909, 2019.

[11] R. Carter et al., "22nm fdsoi technology for emerging mobile, internet-of-things, and rf applications," in *2016 IEEE International Electron Devices Meeting (IEDM)*, 2016, pp. 2.2.1–2.2.4.

[12] S. Hoppner et al., "How to achieve world-leading energy efficiency using 22fdx with adaptive body biasing on an arm cortex-m4 iot soc," in *ESSDERC 2019 - 49th European Solid-State Device Research Conference (ESSDERC)*, 2019, pp. 66–69.

[13] Cadence Design Systems, Inc., "Xtensa lx7 processor," 2016.

[14] ——, "Tensilica fusion g dsp family," 2017.

[15] N. Werner, G. Payá-Vayá, and H. Blume, "Case study: Using the xtensa lx4 configurable processor for hearing aid applications," *ICT Open*, November 2013.

[16] M. Hemnani et al., "Hardware optimization of complex multiplication scheme for dsp application," in *2015 International Conference on Computer, Communication and Control (IC4)*. IEEE, 2015, pp. 1–4.

[17] L. Gerlach, G. Payá-Vayá, and H. Blume, "A survey on application specific processor architectures for digital hearing aids," *Journal of Signal Processing Systems*, vol. 14, no. 4, p. 1218, 2021.

[18] S. C. Klein, J. Kantic, and H. Blume, "Fixed point analysis workflow for efficient design of convolutional neural networks in hearing aids," *Current Directions in Biomedical Engineering*, vol. 7, no. 2, pp. 787–790, 2021.

[19] J. Karrenbauer et al., "Design space exploration framework for tensilica-based digital audio processors in hearing aids," in *2020 9th International Conference on Modern Circuits and Systems Technologies (MOCAST)*, 2020, pp. 1–6.

[20] P. D. Schiavone et al., "Quentin: an ultra-low-power pulpissimo soc in 22nm fdx," in *2018 IEEE SOI-3D-Subthreshold Microelectronics Technology Unified Conference (S3S)*. IEEE, 2018, pp. 1–3.

[21] K. Adiloğlu et al., "A binaural steering beamformer system for enhancing a moving speech source," *Trends in hearing*, vol. 19, 2015.

A 12nm Agile-Designed SoC for Swarm-Based Perception with Heterogeneous IP Blocks, a Reconfigurable Memory Hierarchy, and an 800MHz Multi-Plane NoC

Tianyu Jia[1,*], Paolo Mantovani[2,*], Maico Cassel dos Santos[2,*], Davide Giri[2], Joseph Zuckerman[2], Erik Jens Loscalzo[2], Martin Cochet[3], Karthik Swaminathan[3], Gabriele Tombesi[2], Jeff Jun Zhang[1], Nandhini Chandramoorthy[3], John-David Wellman[3], Kevin Tien[3], Luca Carloni[2], Kenneth Shepard[2], David Brooks[1], Gu-Yeon Wei[1], Pradip Bose[3]

[1]Harvard University, Cambridge, MA, [2]Columbia University, New York, NY, [3]IBM Research, Yorktown Heights, NY

* These authors have equal contributions. Email: pbose@us.ibm.com

Abstract— **This paper presents an agile-designed domain-specific SoC in 12nm CMOS for the emerging application domain of swarm-based perception. Featuring a heterogeneous tile-based architecture, the SoC was designed with an agile methodology using open-source processors and accelerators, interconnected by a multi-plane NoC. A reconfigurable memory hierarchy and a CS-GALS clocking scheme allow the SoC to run at a variety of performance/power operating points. Compared to a high-end FPGA, the presented SoC achieves 7× performance and 62× efficiency gains for the target application domain.**

I. INTRODUCTION

The slowdown of CMOS scaling and limited effectiveness of parallelism via homogeneous multi-core processors have pushed modern computing systems toward heterogeneous SoC architectures. Heterogeneous architectures deliver superior energy-efficient performance by combining general-purpose processors with fixed-function accelerators. Heterogeneity, however, increases the complexity of the design and verification process. Open-source hardware (OSH) addresses this complexity challenge by promoting design reuse [1]. This work focuses on the emerging application domain of vehicular swarm perception (Fig. 1), which expands the vehicle perceptive field by sharing neighbors' sensor data through wireless V2V (vehicle-to-vehicle) communication and reduces false predictions [2-3]. Compared to the computations for the autonomous driving of a single vehicle [4], which include CNNs for object detection and general-purpose computing for decision making, swarm-based perception additionally relies on FFT and Viterbi decoding for sensor signal processing and wireless communication. Hence, this leads to the design of an SoC architecture with a highly heterogeneous architecture.

This paper presents a domain-specific SoC with a tile-based architecture for the target application domain. We designed the SoC with an agile design methodology that promotes the reuse of existing OSH IP blocks and simplifies the development of new ones. Fig. 1 shows its main steps: 1) the SoC components are selected from a library of reusable OSH IPs based on extensive workload analysis; 2) the tile sockets seamlessly integrate the OSH IPs, and the generation of the full SoC RTL is automated based on parameterized configurations; 3) a

hierarchical physical design strategy leverages the modularity of the tile-based architecture and clocking scheme. Compared to agile design approaches for homogeneous multi-core chips [5], our methodology mitigates the complexity of heterogeneous SoC design by decoupling the design and integration of the heterogeneous IPs. Our approach scales up for the development of SoCs with larger and more heterogeneous arrays of tiles.

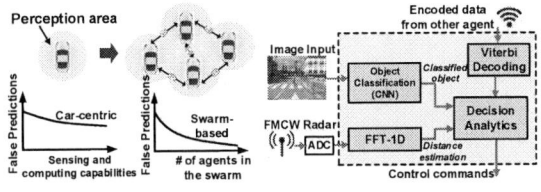

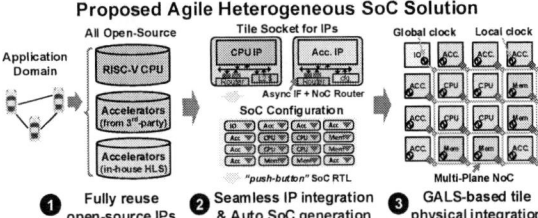

Fig. 1 Swarm-based perception in autonomous vehicle with its key computation kernels, and the agile-designed heterogeneous SoC with a tile-based architecture.

II. SoC ARCHITECTURE AND AGILE DESIGN METHODOLOGY

Fig. 2 shows the overall SoC architecture comprising an array of 4×4 tiles connected by a 2D-mesh multi-plane network-on-chip (NoC). One of the four RISC-V CPU cores [6] acts as the host and boots the Linux operating system. To support parallel processing of camera and sensor data inputs, three NVDLA DNN inference accelerators [7] and three FFT accelerators are deployed to perform object detection and distance estimation tasks. One Viterbi accelerator is deployed to decode the incoming vehicle messages. Both the FFT and Viterbi accelerators are designed in-house using high-level synthesis (HLS) [8]. For higher modularity, each IP block is encapsulated within a tile socket, which connects it to a local bus

978-1-6654-8495-4/22 $31.00 © 2022 IEEE

(e.g., AXI4) as a master and is connected to the NoC via asynchronous interfaces. The tile socket also implements system-level services, including services specific to the particular type of tile: e.g. DMA and configuration registers for an accelerator tile. The NoC, tile sockets, and distributed reconfigurable memory hierarchy are extended from ESP, an open-source SoC platform [9].

The last-level cache (LLC) is partitioned into four memory tiles, each containing a 64-bit wide off-chip memory link. Combined, the four off-chip links support the real-time workload bandwidth requirements. Any subset of the memory tiles can be selected at runtime, and the memory hierarchy is reconfigurable to support different cache-coherence modes. The test-chip prototype relies on a modular FPGA system to connect each memory tile via FMC connectors to a 2GB DDR3 card, which stores image and sensor data for swarm perception. There is one IO tile in the SoC containing ROM and peripheral IO, such as Ethernet and UART.

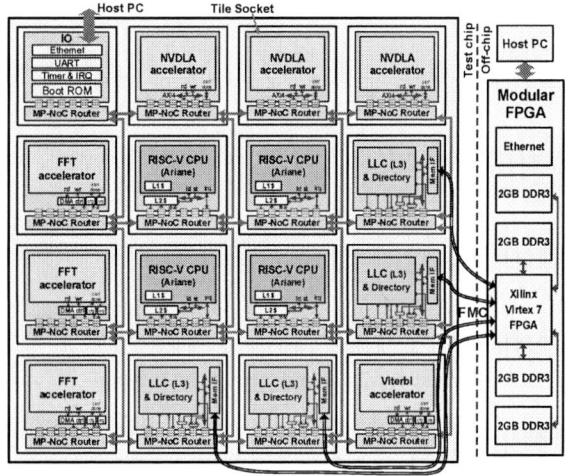

Fig. 2 The architecture of the domain-specific heterogeneous SoC.

Fig. 3 illustrates the details of the agile design methodology, which leverages the modularity of the tile-based architecture. Driven by the analysis of the target swarm-based perception application[1], the key computation kernels, e.g. CNN, FFT, Viterbi decode, are identified. Existing OSH IPs are evaluated and selected for these key computations. Reusing existing OSH IP blocks can significantly reduce the SoC design cycle. Each IP is seamlessly integrated into the SoC by the tile socket, which decouples its design from the rest of the system, thus simplifying the integration of heterogeneous blocks. Instanced from the open-source ESP SoC platform [9], the full SoC RTL, including the NoC and system-level services, is automatically generated. Co-generation of corresponding testbenches is also provided to enable rapid evaluation of the SoC performance and architecture optimizations by FPGA emulation.

During physical design, the inherent regularity of the tile-based architecture decouples each tile from its location in the top-level floorplan, i.e., the same tile can be replicated to meet

[1] Target workload Mini-ERA: https://github.com/IBM/mini-era

workload requirements. A hierarchical timing-closure flow is adopted for independent timing signoff between the local clock frequency of each tile and the global NoC frequency. The physical design of all tiles is conducted in parallel, while the global NoC timing is closed later based on the interface logic model (ILM) timing models. Such timing closure flow allows flexible reuse or respin of pre-existing IPs, further trimming design time. The entire SoC exclusively uses synthesizable designs to avoid any manual layout effort.

Thanks to our agile SoC design methodologies, the proposed SoC was designed in 4 months by less than 10 full-time designers. This design cycle is considerably shorter than the 7 month cycle of a prior agile-designed multi-core processor [5] and was achieved despite the additional challenges posed by the higher heterogeneity of the SoC architecture.

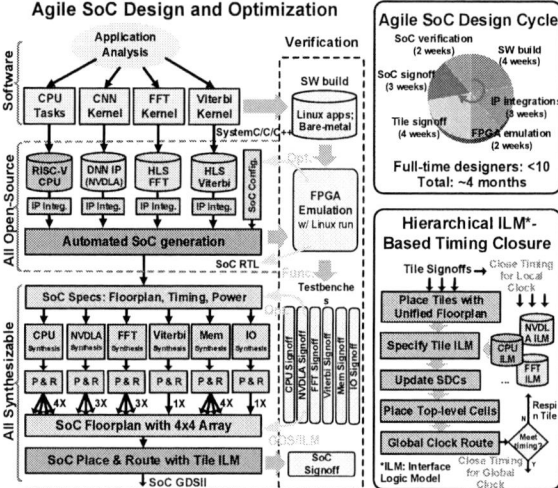

Fig. 3 Agile SoC design/optimization and ILM-based timing closure.

III. SYSTEM-LEVEL SERVICES AND DYNAMIC RECONFIGURATION

The tile socket provides each IP with system-level services such as DMA access and local reconfigurability. To support data exchange among heterogeneous tiles, the mesh NoC has six physical planes, as shown in Fig. 4. Planes 1-3 provide coherence channels between CPUs, accelerators, and LLC partitions. Planes 4-5 support DMA access for the accelerators. Plane 6 is dedicated to interrupts and memory-mapped IO and registers. Asynchronous buffers in the tile sockets connect the NoC routers in the six planes to the logic inside each tile.

In the SoC, the accelerators can communicate with the memory hierarchy via three dynamically configurable cache-coherence modes [10]: non-coherent DMA (accelerator bypasses the cache hierarchy and accesses main memory directly), LLC-coherent DMA (memory requests are sent directly to the LLC and coherence is enforced by software), and the coherent DMA (memory requests are sent directly to the LLC and the hardware maintains full coherence). Each mode supports different degrees of hardware coherence and offers distinct benefits depending on the active workloads, system-level contention, and accelerator properties.

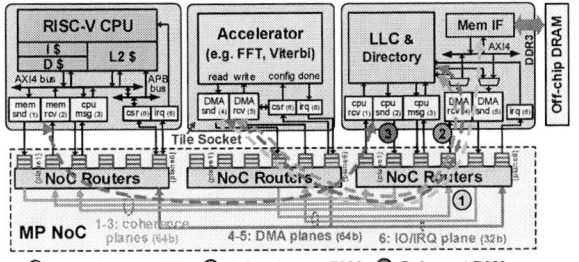

Fig. 4 The multi-plane NoC interconnection for heterogeneous tiles and three reconfigurable cache coherence modes.

The SoC implements a communication synchronous GALS (CS-GALS) clocking strategy, as shown in Fig. 5. Each IP tile is synchronous to a local clock (*clk_tile*) with a local power supply. The NoC, which sits in a global power domain, is synchronously driven by a chip-wide global clock (*clk_noc*). During the timing closure, only the clock skews between neighboring tiles need to be constrained, which relaxes traditional timing-closure constraints. To support the heterogeneity of the IP blocks, the frequency of each tile, as well as the NoC, can be adjusted dynamically and independently of one another, making them globally asynchronous. The NoC router consists of crossbar switches routed to four neighbors synchronously and to the local tile via asynchronous interfaces. Together with look-ahead router design, the CS-GALS enables single-hop-per-cycle throughput for each plane, which outperforms the asynchronous NoC strategy in prior tile-based SoCs [11, 13-14].

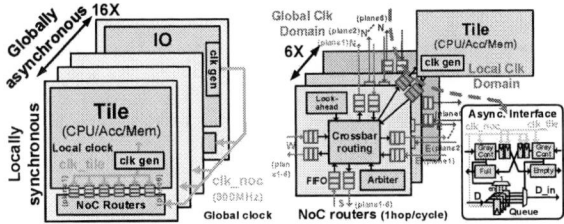

Fig. 5 CS-GALS clocking and NoC router w/ asynchronous interface.

IV. MEASUREMENT RESULTS

The presented domain-specific SoC is fabricated using 12nm FinFET technology. Fig. 6 shows the die photo and its test setup with a modular FPGA system. In the SoC, each tile occupies an identical 1×1mm area for flexible placement and consistent timing closure between the NoC routers. The IP designs occupy more than 90% of the area in each tile, while the NoC logic is placed on the periphery for straight tile-to-tile routing. Only two accelerators (FFT, Viterbi) slightly underutilize tile area, and the entire design incurs less than 15% overhead compared to a bespoke design with custom sizes for each tile. The total active area of the SoC is 21.6mm².

The chip is assembled on a flip-chip package containing 18 power domains. During testing, the test board is connected to the FPGA motherboard through 3 FMC connectors. The FPGA test system is modular, with the flexibility of utilizing different daughter cards. An Ethernet link provides a debug interface from a PC for accessing memory-mapped regions of the SoC.

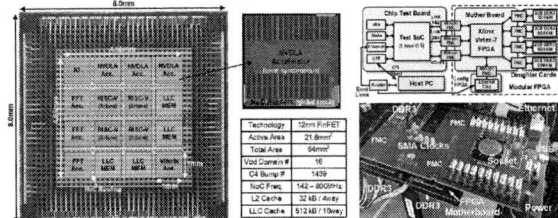

Fig. 6 Die photo and the test setup with modular FPGA system.

We first evaluate the performance and benefits of each tile. As shown in Fig. 7, the operating frequencies of each accelerator were measured across a range of supply voltages, from 0.5V to 1V. We run the CPU with a minimum supply voltage of 0.7V for stable operation of the operating system. We present the benchmark measurements at the nominal 0.8V. The deployment of accelerators significantly improves the performance, i.e. workload latency, compared to standalone CPU operations. At 0.8V, offloading computation to an FFT accelerator achieves 71× and 233× latency and energy reductions, respectively. Similarly, offloading the Viterbi decode kernel to its dedicated accelerator obtains a 20× latency and 56× energy improvement.

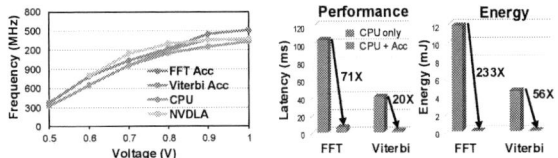

Fig. 7 (Left) V/F scaling for each tile, (Right) Benefits of offloading tasks to dedicated accelerators.

The benefit of the reconfigurable memory hierarchy is evaluated across different workload sizes, as shown in Fig. 8. When the accelerator and CPU share an LLC partition (e.g., when using one memory tile to save energy), the non-coherent DMA mode performs best by avoiding LLC contention, as it accesses DRAM directly. When the accelerator owns its own dedicated LLC partition, there are significant performance benefits from the coherent-DMA and LLC-coherent DMA modes, in which the accelerator performs DMA directly to the LLC and potentially avoids off-chip DRAM access.

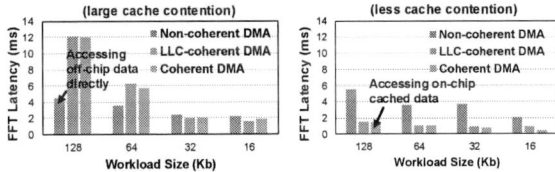

Fig. 8 (Left) Accelerator share one LLC partition with the CPU, (Right) Accelerator owns its dedicated LLC partition.

The accelerator performance is highly correlated with the memory bandwidth, which varies at runtime depending on the SoC operations. The proposed tile-based architecture simplifies the dynamic provisioning of the available four LLC partitions and corresponding off-chip memory links to meet workload demands, e.g. by scaling them up to match the parallel execution of accelerators for performance improvement. As illustrated in Fig. 9, the LLC memory partitions and the corresponding off-chip links are scaled together with the FFT accelerator parallelism to avoid a memory bottleneck. The CS-GALS approach also allows each accelerator to run at its optimal frequency, independently from the rest of the SoC. For example, when the workload is memory bound, the frequency of the FFT accelerator can be reduced from the maximum 1.2GHz to 470MHz while maintaining similar workload latency, thus achieving a 2× energy reduction.

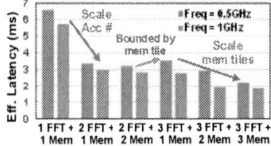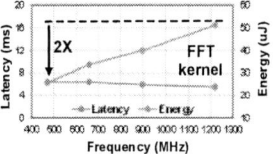

Fig. 9 (Left) Accelerator performance with scalable LLC partition and memory links, (Right) Scale frequency under CS-GALS.

Altogether, for the swarm perception workload Mini-ERA, our SoC achieves a 7× performance and 62× energy improvement compared to an implementation of the same design on a high-end Xilinx Virtex UltraScale XCVU440 FPGA, as shown in Fig. 10. For this workload, the SoC consumes 1.36W at 0.8V, with 7.2% attributable to the NoC. The measured NoC frequency reaches 800MHz at 0.8V.

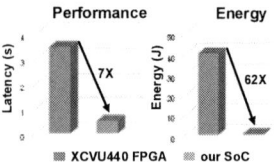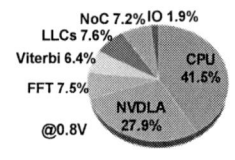

Fig. 10 (Left) Benefits for the target swarm perception application, (Right) SoC power breakdown.

Fig. 11 compares this work to prior tile-based chip designs, which feature *homogeneous* arrays of either processors [11-12] or accelerators [13]. In contrast, the proposed SoC contains a variety of *heterogeneous* OSH IPs and achieves much higher heterogeneity than prior open-source SoCs [15]. The NoC enables data transfers between the tiles with a maximum 281Gb/s throughput, while supporting reconfigurability of the memory hierarchy for a range of performance/power operating points. The scalability of the proposed SoC architecture and the associated methodology benefit engineering productivity when designing SoCs with even larger tile arrays.

	JSSC'17 [11]	VLSI'19 [12]	JSSC'20 [13]	This work
Process	32nm	16nm	16nm	12nm
Area	57.41 mm²	15.25 mm²	6 mm²	21.6 mm²
Design Target	CPU Processor	CPU Processor	DNN Accelerator	Domain-Specific SoC
Tile Array Size	1000	496	16	16
Tile Property	Homogeneous	Homogeneous	Homogeneous	Heterogeneous
Tile(s)	16b RISC CPU	32b RISC-V CPU	MAC Array	64b RISC-V, NVDLA, FFT, Viterbi, Mem, IO
GALS Clocking	Yes	No	Yes	Yes
Tile-to-Tile Clock	Asynchronous	Synchronous	Asynchronous	Synchronous
Clock Domains	2012	3	20	17
NoC planes x BW	1x16b	1x32b	1x64b	5x64b + 1x32b
Tile-to-Tile BW	45.5 Gb/s	32 Gb/s	70 Gb/s	281 Gb/s
NoC Power in SoC	~7%	Not reported	10%	7.2%
Frequency	115 - 1770 MHz	10 – 1400 MHz	161 – 2001 MHz	165 -1520MHz
Voltage	0.56 – 1.10V	0.6 - 0.98 V	0.41 – 1.2 V	0.5 - 1.0V
Power	1.3 - 39.6 W	7.47 W	30mW – 4.16 W	240mW* – 1.83 W

* Memory links and NoC at 0.8V

Fig. 11 The comparison table with other tile-based designs.

V. CONCLUSION

We presented an SoC with a tile-based architecture for the application domain of swarm-based perception. We developed the SoC with an agile design methodology that simplifies the reuse of OSH IPs. The heterogeneous IPs are integrated with tile sockets, which enable system-level services, and are interconnected by a multi-plane NoC. The CS-GALS clocking and reconfigurable memory hierarchy allow flexible performance tuning based on workload demands. The SoC delivers 7× performance and 62× efficiency gains compared to a high-end FPGA implementation.

ACKNOWLEGEMENT

This research was developed with funding from DARPA. The views, opinions and/or other findings expressed are those of the authors and should not be interpreted as representing the official views or policies of the Department of Defense or the U.S. Government. Distribution Statement A. Approved for public release: distribution unlimited.

REFERENCES

[1] D. Daly, "EE5: Is an open-source hardware revolution on the horizon?," *IEEE International Solid-State Circuits Conference (ISSCC)*, pp. 555-557, Feb. 2020.

[2] T. Wang, et al., "V2VNet: vehicle-to-vehicle communication for joint perception and prediction", *European Conference on Computer Vision (ECCV)*, pp. 605-621, 2020.

[3] E. Sisbot, A. Vega, et al., "Multi-vehicle map fusion using GNU radio", *Proceedings of the GNU Radio Conference*, vol. 4, no. 1, 2019.

[4] K. Matsubara, et al., "A 12nm autonomous-driving processor with 60.4TOPS, 13.8TOPS/W CNN executed by task-separated ASIL D control", *IEEE International Solid-State Circuits Conference (ISSCC)*, pp. 56-57, Feb. 2021.

[5] C. Schmidt, et al., "An eight-core 1.44GHz RISC-V vector machine in 16nm FinFET", *IEEE International Solid-State Circuits Conference (ISSCC)*, pp. 58-59, Feb. 2021.

[6] F. Zaruba and L. Benini, "The cost of application-class processing: energy and performance analysis of a Linux-ready 1.7-GHz 64-Bit RISC-V core in 22-nm FDSOI technology", *IEEE Transactions on VLSI Systems (TVLSI)*, vol. 27, no. 11, pp. 2629-2640, Nov. 2019.

[7] Nvidia, NVIDIA Deep Learning Accelerator (NVDLA). http://nvdla.org/primer.html, 2018.

[8] B. Khailany, et al., "A modular digital VLSI flow for high-productivity SoC design", *Design Automation Conference (DAC)*, 2018.

[9] P. Mantovani, et al., "Agile SoC development with open ESP", *IEEE International Conference on Computer-Aided Design (ICCAD)*, 2020.

[10] J. Zuckerman, et al., "Cohmeleon: learning-based orchestration of accelerator coherence in heterogeneous SoCs", *IEEE International Symposium on Microarchitecture (MICRO)*, pp. 350-365, Oct. 2021.

[11] B. Bohnenstiehl, et al., "KiloCore: A 32-nm 1000-processor computational array", *IEEE Journal of Solid-State Circuits (JSSC)*, vol. 52, no. 4, pp. 891-902, Apr. 2017.

[12] A. Rovinski, et al., "A 1.4 GHz 695 Giga Risc-V inst/s 496-core manycore processor with mesh on-chip network and an all-digital synthesized PLL in 16nm CMOS", *IEEE Symposium on VLSI Circuits (VLSI)*, pp. C30-C31, Jun. 2019.

[13] B. Zimmer, et al., "A 0.32–128 TOPS, scalable multi-chip-module-based deep neural network inference accelerator with ground-referenced signaling in 16nm", *IEEE Journal of Solid-State Circuits (JSSC)*, vol. 55, no. 4, pp. 920-932, Apr. 2020.

[14] M. Fojtik, et al., "A fine-grained GALS SoC with pausible adaptive clocking in 16 nm FinFET", *IEEE International Symposium on Asynchronous Circuits and Systems (ASYNC)*, pp. 27-35, May 2019.

[15] A. Gonzalez, et al., "A 16mm² 106.1 GOPS/W heterogeneous RISC-V multi-core multi-accelerator SoC in low-power 22nm FinFET", *IEEE European Solid State Circuits Conference (ESSCIRC)*, Sep. 2021.

DARKSIDE: 2.6GFLOPS, 8.7mW Heterogeneous RISC-V Cluster for Extreme-Edge On-Chip DNN Inference and Training

Angelo Garofalo[*], Matteo Perotti[†], Luca Valente[*], Yvan Tortorella[*],
Alessandro Nadalini[*], Luca Benini[*†], Davide Rossi[*], and Francesco Conti[*]
[*]Department of Electrical, Electronic and Information Engineering (DEI), University of Bologna, Italy
[†]Integrated Systems Laboratory (IIS), ETH Zürich, Switzerland

Abstract—Extreme-edge applications using Deep Learning (DL) have strict requirements in terms of latency, throughput, accuracy, and flexibility. Heterogeneous clusters are promising architectural solutions that combine the programmability of DSP-enhanced cores with the performance and efficiency boost of specialized accelerators. We present DARKSIDE, a System-on-Chip with a heterogeneous cluster of 8 RISC-V cores enhanced with 2-b to 32-b mixed-precision integer arithmetic. To further speed-up key compute-intensive Deep Neural Network (DNN) kernels, the cluster is enriched with three specialized digital accelerators: an accelerator for low-data-reuse depthwise convolution kernels (up to 30 MAC/cycle); a minimal overhead datamover to marshal 1-b to 32-b data on-the-fly; a 16-b floating point Tensor Product Engine (TPE) for tiled matrix-multiplication acceleration. DARKSIDE is implemented in 65nm CMOS technology. The cluster achieves a peak integer performance of 65 GOPS and a peak efficiency of 835 GOPS/W when working on 2-b integer DNN kernels. When targeting floating-point tensor operations, the TPE provides up to 18.2 GFLOPS of performance or 300 GFLOPS/W of efficiency – enough to enable on-chip floating-point training at competitive speed coupled with ultra-low power quantized inference.

Index Terms—Extreme-Edge; Heterogeneous Cluster; Tensor Product Engine; Ultra-Low-Power AI

I. INTRODUCTION

Recently, we have seen a significant push toward *TinyML*: the deployment of Machine Learning (ML) and Deep Learning (DL) algorithms, such as Deep Neural Networks (DNNs), on devices operating at the extreme edge of the Internet-of-Things (IoT). Emerging TinyML applications require advanced capabilities, including not only on-chip inference, but also network tuning or partial training [1] under an active power budget of a few milliwatts as well as a very tight silicon area budget. Heterogeneous computing is a promising solution to deliver TinyML's required performance, efficiency and flexibility while coping with strict power and cost constraints [2]: multiple programmable cores provide flexible and efficient execution for generic parallel kernels, while specialized hardware accelerators provide extra performance and efficiency boost on key kernels that dominate the computational workload.

Low- and mixed-precision ($\leq$8-bit) DNNs, derived from quantization-aware training [3], are one of the critical kernels to accelerate in emerging extreme-edge devices. Several digital [4], [5] and mixed-signal [6] ASICs have been proposed, but these highly specialized units are inflexible and must still be coupled with processors. An alternative approach is introducing hardware-accelerated extensions for low-precision arithmetic at the ISA level. Previous works [7], [8] show that extending the RISC-V ISA with 2b-32b uniform SIMD sum-of-dot-product, fused MAC-load operations, and status-based mixed-precision instructions can achieve ASIC-like efficiency on DNN workloads without compromising programmability.

However, some kernels are intrinsically challenging to parallelize. For example, depthwise convolution layers have a small footprint in terms of weights but are not computationally efficient [9] due to their poor intrinsic data reuse. Likewise, data marshaling operations (e.g., low-bitwidth transpose) commonly used in DNNs are inefficient as they heavily rely on sub-byte swap operations. Other kernels require a performance boost that cannot be achieved without using a massive number of cores, which would blow up the SoC's area unacceptably. In particular, on-chip learning requires floating-point (FP) arithmetic, and the workload is 10-100$\times$ larger than inference [1]. While recent work has shown that FP16 and even FP8 can be used [10], [11], the performance requirements are very high and require fixed-function acceleration to fit within a TinyML compatible area budget.

In this work, we present DARKSIDE, a PULP-based [2], [12] heterogeneous computing SoC that targets emerging TinyML inference and on-chip training applications. We introduce four main innovations in DARKSIDE: *1)* RISC-V cores with advanced low-bitwidth mixed-precision integer computing capabilities (RVNN); three dedicated hardware accelerators: *2)* a Depth-Wise convolution Engine (DWE), *3)* a low-overhead DataMover for marshaling operations, and *4)* a low-power Tensor Product Engine (TPE) for efficient FP16 matrix multiplications. The cores and accelerators are tightly integrated into a shared-L1 cluster to enable advanced hardware/software cooperation. The chip has been fabricated in TSMC 65nm technology and shows peak integer performance (2-bit) of 65 GOPS with an efficiency of 835 GOPS/W. On TPE-accelerated FP16 workloads, it achieves up to 18.2 GFLOPS and a peak efficiency of 300 GFLOPS/W at 2.6 GFLOPS, achieving peak performance and efficiency similar to 8-bit integer operations.

II. SoC OVERVIEW

Fig. 1a shows the architecture of the DARKSIDE cluster. It features 8 RISC-V-based cores and three cluster-coupled specialized accelerators to boost the performance and the efficiency of integer and floating-point DNN compute-intensive kernels. Each accelerator is integrated using the PULP Hardware Processing Engine (HWPE) interface, which exposes

978-1-6654-8495-4/22 $31.00 © 2022 IEEE

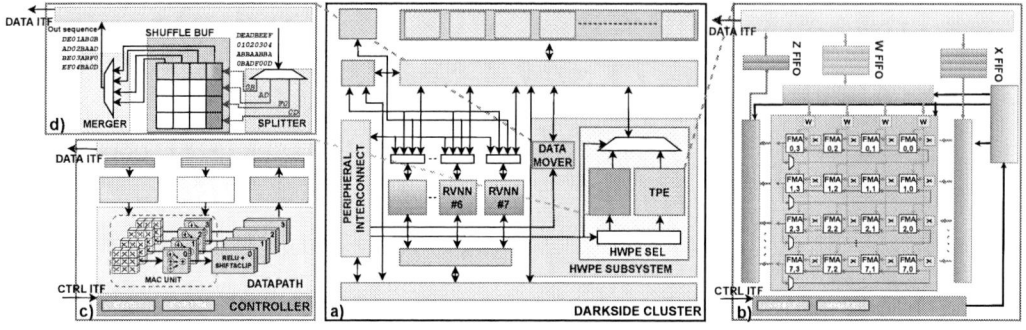

Fig. 1. a) Overview of the DARKSIDE cluster and details of the specialized accelerators: b) TPE, c) DWE, d) DataMover.

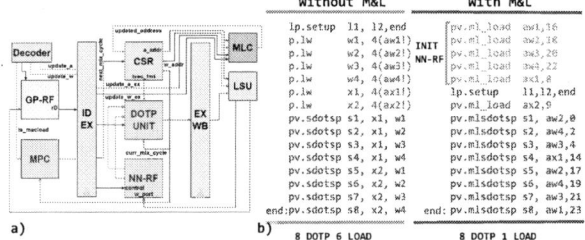

Fig. 2. a) Pipeline extension in the RVNN core to support the M&L instruction. b) Example of *MatMul* kernel.

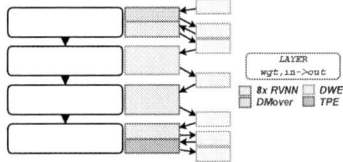

Fig. 3. Example of DARKSIDE heterogeneous operation on a DNN with 3 quantized layers and a final trainable fully-connected layer.

to the rest of the cluster a data transfer port and a control port to program the accelerator with memory-mapped control registers. In each HWPE, specialized streamers move data between the accelerators and the L1 memory through the data port, converting the memory accesses into coherent streams to feed the accelerator's datapath.

The cores and the accelerators share 128kB of L1 scratchpad memory through a single-cycle latency Heterogeneous Cluster Interconnect (HCI), leveraging a request/grant protocol. The cluster also features a two-level hierarchical instruction cache, a DMA controller for data transfers between L1 and higher-level memories, and an Event Unit block that manages fine-grained parallel thread dispatching and clock gating of idle cores while they wait on a synchronization barrier. In the proposed SoC, the cluster resides in a dedicated power and clock domain and is surrounded by other IPs integrated into a different power and clock domain, namely the *Fabric* domain, which serves as a programmable testbench in the context of this work. The Fabric domain features a controlling RISC-V processor, 256kB of L2 memory containing also the cluster code, a standard set of peripherals, and FLLs for clock generation.

1) Status-Based Fused MAC-Load SIMD operations: The RVNN cores of the DARKSIDE cluster extend the 32-bit

RISC-V processor presented in [13], which implements the RV32IMC Instruction Set Architecture (ISA) plus custom 2-bit to 32-bit mixed-precision SIMD instructions, including *dot-product (dotp)* based operations, supported through a status-based execution. Our key enhancement is a fused MAC-load (M&L) operation to reduce the load overheads within the innermost loops of compute-intensive kernels. The key added value of this new instruction extension of this approach is shown in Fig. 2b. From the micro-architectural viewpoint, we extend the datapath of the processor to fuse one load and one *dotp*-based SIMD operation in one single-cycle latency instruction, as shown in Fig. 2a. The M&L instructions fetch the source operands of the *dotp* from a dedicated Register File, NN-RF, while the accumulators reside in the GP-RF. When a *dotp* instruction is executed, one of the two source operands can be updated with new data fetched from memory through the load branch without the need for an explicit load instruction. This extension implies a gate count increase of the core of just 8.3% compared to Dustin's baseline, while it enables an instruction/cycle improvement of 57% on the 8-bit MatMul kernel, achieving up to 91% MAC utilization (compared with 58% of the baseline).

2) Tensor Product Engine: The Tensor Product Engine (TPE) accelerates matrix multiplications (MatMuls) of the kind $Z = X \cdot W$, using the IEEE 754 binary-16 representation (FP16 in the following). The datapath is a semi-systolic array of 32 FP16 Fused Multiply-Add (FMA) units [14]. The architecture is organized in 8 rows, each of 4 columns. The four FMAs of each row are cascaded; each one passes the intermediate result as input to the unit to its left. The left-most FMA feeds back the computed partial product as accumulation input to the right-most FMA of the same row, as shown in Fig. 1b, to internally perform the computation of matrices larger than the size of the array. To meet the timing requirements, each FMA features three internal pipeline registers. To maximize throughput, the **X**-matrix elements of each FMA are held steady for the number of cycles necessary to the FMAs of one row to compute the partial results. On the other hand, **W**-matrix operands are streamed-in at each cycle and broadcasted to all the FMAs of the same column. During the computation, **X** and **W** data are continuously reloaded. The reload has no computational overhead, resulting in an overall 98.8% FMA utilization (31.6 MAC/cycle) even when the **X** and **W** dimensions are larger than the array size. Each row of FMAs computes 16 elements of a row of the **Z**-matrix, that

978-1-6654-8495-4/22 $31.00 © 2022 IEEE

are stored in the memory only at the end of the computation, maximizing internal data reuse.

3) Depth-Wise Convolution Engine: The Depth-Wise Convolution Engine (DWE) can process the low-reuse depth-wise component of the depth-wise + point-wise kernels often used in recent neural networks on 8-bit signed input tensors and weights, leaving the much better-parallelizable point-wise kernels to M&L-accelerated software. The DWE, as shown in Fig. 1c, employs a weight-stationary data flow and targets 3×3 depth-wise layers, the one most commonly encountered in DNNs. Overall, the DWE includes 36 MAC units to implement 3×3 kernels on 4 channels at the same time. The *weights* buffer hosts $16\times 3\times3$ kernels that are kept stationary during the whole DWE operation to maximize data reuse. The *input* buffer hosts 3×3 8-bit pixels over 16 channels that are consumed in 4 cycles of operation of the MAC units. Every 4 cycles of operation, the streamer uses 3 cycles to load the 16-channel pixels to the right with respect to the ones currently stored in the input buffer to implement the spatial sliding window mechanism over the input image. The accumulation of intermediate results is performed over $16\times$ 32-bit registers, accessed 4 at a time in the 4-cycle operation loop. Afterward, non-linear activation functions and ancillary operations such as shifting and clipping are applied to re-quantize the results to 8-bit precision. In the last cycle of the 4-cycle operation loop, the 16-channel 8-bit pixel is streamed out of the accelerator.

4) DataMover: On-the-fly data marshaling operations can drastically reduce the performance of DNN workloads, both for inference and training. The DARKSIDE cluster includes a DataMover unit that can be used to perform transposition operations on DNN 3-dimensional tensors stored in L1 memory, with configurable data chunk size d =32-b, 16-b, 8-b, 4-b, 2-b or 1-b. The DataMover splits incoming streams from memory into chunks of size d. The chunks are buffered internally into a Shuffle Buffer composed of $32\times$ 32-bit registers with the output (transposed) format; only $32/d$ registers are used, depending on the data size d configuration (e.g., 4 registers for 8-bit chunks). After $32/d$ input stream transactions, all the full 32-bit transposed words are ready for stream-out using a 3D strided address generator to address memory. The DataMover then continues operations by streaming in a new set of $32/d$ words to transpose.

A. HW/SW Cooperation

The DARKSIDE cluster can be used to implement complex AI behavior by means of cooperation between the RVNN cores and the HWPEs. Fig. 3 shows an example of a DNN with 3 mixed-precision quantized layers using all the cluster blocks and communicating by means of software-managed buffers allocated on the shared TCDM memory.

III. MEASUREMENTS AND BENCHMARKING

Fig. 4 shows the chip micrograph of DARKSIDE. The SoC is implemented with TSMC 65nm technology. The die area, including the Fabric domain, is 12 mm², while the cluster area is 4.84 mm², partitioned as shown in Fig. 4. Fig. 5 reports the maximum operating frequency and the power consumption of the cluster over the 0.75V to 1.2V voltage range. The power is measured on the silicon prototype, running integer and floating-point compute-intensive kernels (MatMuls). Linearly

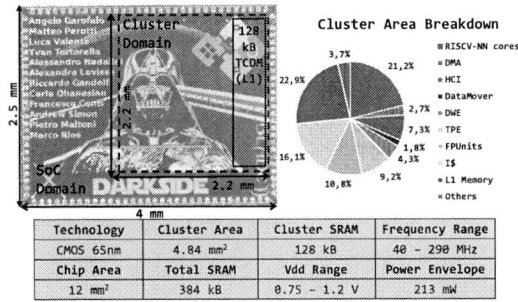

Technology	Cluster Area	Cluster SRAM	Frequency Range
CMOS 65nm	4.84 mm²	128 kB	40 - 290 MHz
Chip Area	Total SRAM	Vdd Range	Power Envelope
12 mm²	384 kB	0.75 - 1.2 V	213 mW

Fig. 4. Chip micrograph and specifications. Area breakdown of the DARK-SIDE cluster.

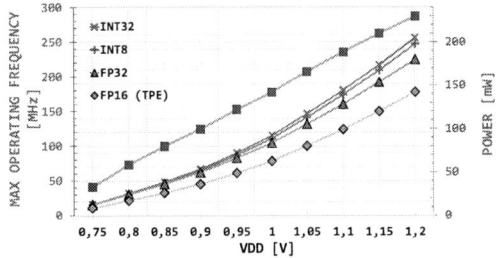

Fig. 5. Voltage sweep vs. max frequency vs. power consumption.

increasing with the voltage, we reach the highest operating frequency of 290 MHz at 1.2 V.

Fig. 6 shows the performance and the efficiency achieved by the cluster on integer and floating-point MatMuls. The measurements are taken at the maximum operating frequency, sweeping the supply voltage from 0.75V to 1.2V. On ML workloads (i.e., integer 8-/4-/2-bits) deployed on RVNN cores, DARKSIDE delivers up to 65 GOPS with a peak efficiency of 835 GOPS/W. The TPE boosts the performance and the efficiency of FP16 MatMuls to 18.2 GFLOPS and 300 GFLOPS/W, respectively, $17.7\times$ and $21.8\times$ higher compared to a software scalar execution on the 4 FPUs of the DARKSIDE cluster.

Fig. 7 compares the proposed SoC with fully programmable IoT end-nodes [2], [13], [15], [16], while Fig. 8 highlights the performance gains achieved by the new extensions and accelerators whose design is the key contribution of this work. Compared with more traditional single-core IoT end-nodes [15], [16], DARKSIDE delivers significantly better performance and efficiency by exploiting a mix of software parallelism, mixed-precision M&L ISA extensions, and cluster-coupled domain-specific accelerators. Compared to Vega [2] featuring a similar 8-cores cluster, on an 8-bit matrix multiplication workload DARKSIDE delivers slightly better integer performance (17 GOPS vs. 15.2 GOPS) despite the much less scaled technology node used for implementation (65nm vs. 22nm), thanks to the M&L support on the ISA. Furthermore, larger performance and efficiency gains are achieved thanks to the native support for sub-byte operations, as shown in Fig. 7 and Fig. 8. Compared to Dustin [13], featuring a cluster with 16 processors with mixed-precision extensions implemented in the same technology node, the proposed cluster achieves 1.13x better performance with half of the cores thanks to the M&L

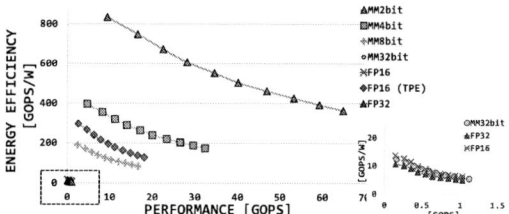
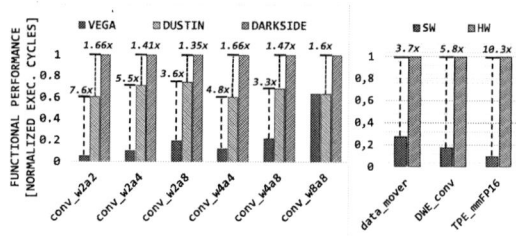

Fig. 6. Performance and Energy Efficiency of the DARKSIDE Cluster. The measurements are performed at the max frequency, sweeping the supply voltage between 0.75V and 1.2V.

Fig. 8. Left: Normalized single-core performance of quantized DNN kernels. Darkside's core performance is compared with Dustin's core [13] and Vega's core [2]. Right: performance improvement of execution on cluster-coupled accelerators over software (8-cores).

	SleepRunner [15]	SamurAI [16]	Vega [2]	Dustin [13]	This Work
Technology	CMOS 28nm FD-SOI	CMOS 28nm FD-SOI	CMOS 22nm FD-SOI	CMOS 65nm	CMOS 65nm
Die Area	0.68 mm2	4.5 mm2	12 mm2	10 mm2	12 mm2
Application	IoT GP	IoT GP + DNN	IoT GP + NSAA+DNN	IoT GP + DNN + QNNs	IoT GP + NSAA + DNN + QNNs
CPU	CM0S	1x RISCY	10 x RISCY	16 x RISCY MPIC CORES	8 x RISCY-NN
ISA	Thumb-2 subset	RVC32IMFXpulp	RVC32IMFXpulp	RVC32IMCXumpic	RVC32IMFXpulpNN2
Int Precision (bits)	32	8, 16, 32	8, 16, 32	2, 4, 8, 16, 32 (plus mixed-precision)	2, 4, 8, 16, 32 (plus mixed-precision)
FP Precision	--	--	FP32, FP16, bfloat	--	FP32, FP16
Supply Voltage	0.4 - 0.8 V	0.45 - 0.9 V	0.5 - 0.8 V	0.8 - 1.2 V	0.75 - 1.2 V
Max Frequency	80 MHz	350 MHz	450 MHz	205 MHz	290 MHz
Power Envelope	320 μW	96 mW	49.4 mW	156 mW	213 mW
Best Integer Performance	31 MOPS (32b)	1.5 GOPS (8b)	15.6 GOPS (8b)	15 GOPS (8b) 30 GOPS (4b) 58 GOPS (2b)	17 GOPS (8b) 32 GOPS (4b) 65 GOPS (2b)
Best Integer Efficiency	97 MOPS/mW (32b)	230 GOPS/W (8b)	614 GOPS/W (8b)	303 GOPS/W (8b) 570 GOPS/W (4b) 1152 GOPS/W (2b)	191 GOPS/W (8b) 396 GOPS/W (4b) 835 GOPS/W (2b)
Best FP32 Performance	--	--	2 GFLOPS	--	1.03 GFLOPS
Best FP32 Efficiency	--	--	79 GFLOPS/W	--	12 GFLOPS/W
Best FP16 Performance	--	--	3.3 GFLOPS	--	18.2 GFLOPS
Best FP16 Efficiency	--	--	129 GFLOPS/W @ 1.27 GFLOPS	--	300 GFLOPS/W @ 2.6 GFLOPS

Fig. 7. Comparison of DARKSIDE with the State-of-the-Art.

extension. Note that the cores are only 8.3% larger than those instantiated in Dustin. On the other hand, the energy efficiency of Dustin is higher than DARKSIDE thanks to the power reduction achieved thanks to the Vector Lockstep Execution Mode (VLEM). On 16-bit floating-point matrix-multiplication workloads (FP16), DARKSIDE surpasses Vega both in terms of peak performance and efficiency by more than 6× and 2×, respectively, thanks to the boost in performance and efficiency delivered by the TPE. TPE improves FP16 *MatMuls* performance by 10.3× with respect to software execution on 8-cores (Fig. 8). As a final remark, we observe 5.8× of speed-up when 8-bit depth-wise kernels are offloaded to the dedicated DWE in DARKSIDE with respect to software execution. On the same trail, a small accelerator like the DataMover can save more than 3.7× execution cycles in data marshaling operations compared to the same task offloaded to the cores.

IV. CONCLUSION

We presented DARKSIDE, a low-power heterogeneous compute cluster for extreme-edge DNN inference and on-chip training. The cluster features 8 RISC-V cores, enhanced with 2-bit to 32-bit mixed-precision integer SIMD instructions and fused mac-load operations. It also includes specialized accelerators to boost the performance of integer depth-wise convolutions, reduce the latency of data marshaling operations, and enhance the performance and efficiency of FP16 kernels. The proposed SoC, implemented in TSMC 65nm technology, can achieve up to 65 GOPS peak performance on ML workloads, with 835 GOPS/W of energy efficiency.

On FP16 kernels offloaded to the TPU, the SoC achieves 18.2 GFLOPS with 300 GFLOPS/W, surpassing the efficiency and performance of state-of-the-art SoCs implemented in much more scaled and expensive technology nodes.

ACKNOWLEDGEMENT

This work was partially supported through the H2020 WiPLASH (g.a. 863337) EU Project. The authors thank R. Gandolfi, P. Maltoni, and the ETH Design Zentrum.

REFERENCES

[1] L. Ravaglia *et al.*, "A TinyML Platform for On-Device Continual Learning With Quantized Latent Replays," *IEEE Journal on Emerging and Selected Topics in Circuits and Systems*, vol. 11, no. 4, pp. 789–802, 2021.

[2] D. Rossi *et al.*, "Vega: A Ten-Core SoC for IoT Endnodes With DNN Acceleration and Cognitive Wake-Up From MRAM-Based State-Retentive Sleep Mode," *IEEE Journal of Solid-State Circuits*, vol. 57, no. 1, pp. 127–139, 2021.

[3] I. Hubara *et al.*, "Quantized neural networks: Training neural networks with low precision weights and activations," *The Journal of Machine Learning Research*, vol. 18, no. 1, pp. 6869–6898, 2017.

[4] J. Lee *et al.*, "UNPU: A 50.6 TOPS/W unified deep neural network accelerator with 1b-to-16b fully-variable weight bit-precision," in *2018 IEEE International Solid-State Circuits Conference-(ISSCC)*. IEEE, 2018, pp. 218–220.

[5] Y.-H. Chen *et al.*, "Eyeriss v2: A flexible accelerator for emerging deep neural networks on mobile devices," *IEEE Journal on Emerging and Selected Topics in Circuits and Systems*, vol. 9, no. 2, pp. 292–308, 2019.

[6] C. Zhou *et al.*, "AnalogNets: ML-HW Co-Design of Noise-robust TinyML Models and Always-On Analog Compute-in-Memory Accelerator," *arXiv preprint arXiv:2111.06503*, 2021.

[7] A. Garofalo *et al.*, "XpulpNN: Enabling Energy Efficient and Flexible Inference of Quantized Neural Networks on RISC-V Based IoT End Nodes," *IEEE Transactions on Emerging Topics in Computing*, vol. 9, no. 3, pp. 1489–1505, 2021.

[8] A. Ottavi *et al.*, "A mixed-precision RISC-V processor for extreme-edge DNN inference," in *2020 IEEE Computer Society Annual Symposium on VLSI (ISVLSI)*. IEEE, 2020, pp. 512–517.

[9] A. Garofalo *et al.*, "A Heterogeneous In-Memory Computing Cluster For Flexible End-to-End Inference of Real-World Deep Neural Networks," *arXiv preprint arXiv:2201.01089*, 2022.

[10] A. Rodriguez *et al.*, "Lower numerical precision deep learning inference and training," *Intel White Paper*, vol. 3, pp. 1–19, 2018.

[11] X. Sun *et al.*, "Hybrid 8-bit floating point (HFP8) training and inference for deep neural networks," *Advances in neural information processing systems*, vol. 32, 2019.

[12] G. Lucas and G. Kurtz, "Star Wars Episode IV, A New Hope," 1977.

[13] G. Ottavi *et al.*, "Dustin: A 16-Cores Parallel Ultra-Low-Power Cluster with 2b-to-32b Fully Flexible Bit-Precision and Vector Lockstep Execution Mode," *arXiv preprint arXiv:2201.08656*, 2022.

[14] S. Mach *et al.*, "FPnew: An Open-Source Multi-Format Floating-Point Unit Architecture for Energy-Proportional Transprecision Computin," *IEEE Transactions on Very Large Scale Integration (VLSI) Systems*, vol. 29, no. 4, pp. 774-78, 2020.

[15] D. Bol *et al.*, "SleepRunner: A 28-nm FDSOI ULP Cortex-M0 MCU With ULL SRAM and UFBR PVT Compensation for 2.6–3.6-μW/DMIPS 40–80-MHz Active Mode and 131-nW/kB Fully Retentive Deep-Sleep Mode," *IEEE Journal of Solid-State Circuits*, vol. 56, no. 7, pp. 2256–2269, 2021.

[16] Miro-Panades *et al.*, "SamurAI: a 1.7MOPS-36GOPS Adaptive Versatile IoT Node with 15,000x Peak-to-Idle Power Reduction, 207ns Wake-up Time and 1.3TOPS/W ML Efficiency," in *2020 IEEE Symposium on VLSI Circuits*. IEEE, 2020, pp. 1–2.

A 91.6% Peak Efficiency Time-Domain-Controlled Single-Inductor Triple-Output Step-Up Converter with ±7.5 to ±12V Bipolar Output Voltages

Peng Cao, Danzhu Lu, Jiawei Xu, Xiaoyang Zeng and Zhiliang Hong

State Key Laboratory of ASIC and System, Fudan University, Shanghai, China

Email: {pengcao, lvdanzhu, jwxu, xyzeng, zlhong}@fudan.edu.cn

Abstract—This paper presents a time-domain-controlled (TDC) single-inductor triple-output (SITO) step-up converter with enhanced load transient response. The proposed converter can convert 2.7 to 5V input voltage into a pair of bipolar high output voltages ranging from ±7.5 to ±12V and a 5V positive output voltage with a maximum output power of 9.6W. By using the time-domain controller and an ADC-assisted fast settling loop, the converter exhibits a fast transient response with low undershoot/overshoot voltages. The proposed converter was fabricated in a 180nm BCD process. With a 2MHz switching frequency, the converter achieves a peak conversion efficiency of 91.6% at an output power of 2.25W.

Keywords—step-up converter, single-inductor, bipolar-output voltages, time-domain-controlled, fast transient response

I. INTRODUCTION

Bipolar supply voltages are widely used in OLED displays, radio/audio amplifiers, data acquisition, audio amplifier and industry test instruments. These devices typically operate with bipolar high voltages up to ±12V. On the other hand, to accommodate a portable or wearable form factor, these devices are often powered by lithium-ion batteries with a limited output voltage between 2.7V and 4.2V. Hence, designing a high-efficiency bipolar step-up DC-DC converter with a wide output voltage range is of great importance.

Conventional designs rely on two separate converters, i.e. a boost and an inverting buck-boost converters, to generate bipolar output voltages [1], [2]. This structure requires two inductors and two control stages, resulting in size and cost penalties. This issue can be solved by converting one of the positive output voltage of a traditional single-inductor multi-output (SIMO) converter into the negative output voltage with an extra capacitor [3], but the additional switching capacitor topology reduces the conversion efficiency. A simultaneous energy transferring (SET) single-inductor bipolar-output (SIBO) structure [4] is proposed to achieve high conversion efficiency and small output voltage ripple at the same time, but its output voltages are regulated at different magnitudes. Moreover, one shortcoming of the previous bipolar output converters is that their output voltages are usually limited to ±6V with less than 4W output power [3]–[5]. In addition, another bottleneck of the prior art bipolar-output converters is their slow transient response due to narrow loop bandwidth [3]–[6]. This considerably hampers their use in high voltage applications. In

summary, there are three key challenges to be tackled for the state-of-the-art converters with bipolar output voltages: 1) to achieve high conversion efficiency, 2) to enhance the transient response; 3) to reduce the number and size of off-chip components and increase system integration level.

To mitigate these issues, this paper proposes a time-domain-controlled SITO step-up converter with bipolar output voltages and an ADC assisted fast-settling loop. The operating principle of the SITO converter and circuits implementations are depicted in Section II and Section III, respectively. Section IV provides the measurement results. Section V concludes the paper.

II. TIME-DOMAIN-CONTROLLED SITO CONVERTER

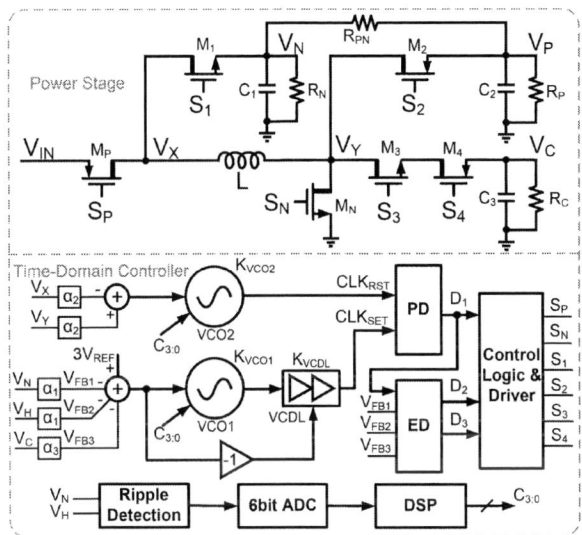

Fig. 1: Architecture of the proposed SITO converter.

In Fig. 1, the proposed SITO converter consists of two main building blocks, i.e. a SITO power stage and a time-domain controller. The power stage can generate three output voltage levels, including a pair of bipolar output voltage V_N, V_P between ±7.5V and ±12V and a 5V control voltage V_C. To withstand high output voltages, HVDMOS transistors are used for the power switches. Note that transistors M_3 and M_4 are connected in series to facilitate unidirectional switching and provide HV isolation.

In the control stage, traditional step-up converter usually needs a nanofarad level external capacitor to realize Proportional-Integral (PI) compensation [1]–[3], which increases the size and cost of the converter. Inspired by [6], we utilize a time-domain controller to implement a compact converter with enhanced transient response, where the traditional error amplifier and PI compensation are replaced by voltage-controlled oscillators (VCOs) and voltage-controlled delay lines (VCDLs). The PWM signal D_1 is obtained after comparing the phase differences between CLK_{SET} and CLK_{RST} in a phase detector (PD). D_2 and D_3 are generated in an energy distribution (ED) block, together with D_1 to modulate the charging time for three output voltages.

The timing diagram and operation of the SITO converter is illustrated in Fig. 2. In one switching period, the inductor is first energized by turning on the switch S_P and S_N, and then the inductor current discharges to three outputs V_N, V_P and V_C by turning on the switch S_1, S_2 and S_3 in sequence. To perform stability analysis, an equivalent small signal model of the SITO converter is depicted in Fig. 3. The phase to inductor current gain $G_{i\Phi}$, phase to output voltage gain $G_{v\Phi}$ and loop gain G_{loop} can be derived from (1)-(3), where $\omega_c = \alpha_2 V_{out,eq} K_{VCO2}/2\pi$, K_{VCO1} and K_{VCO2} are the voltage-to-frequency gain of the two VCOs, K_{VCDL} is the voltage-to-delay gain of VCDL, D is the duty cycle, α_1 and α_2 are the feedback coefficient, $V_{out,eq}$ is the equivalent output voltage of SITO. From (2) and (3), the control-to-output transfer function

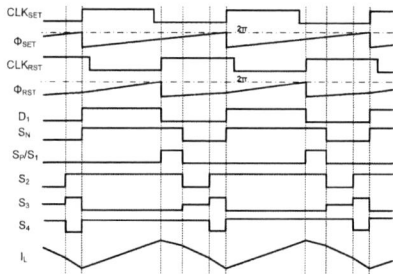

Fig. 2: Timing diagram and operation of the SITO converter.

$$G_{i\phi} = \frac{i_L}{\Phi_{SET}} = \frac{\frac{1}{2\pi}G_{id}}{1+\frac{1}{2\pi}\alpha_2 L K_{VCO2}G_{id}} \approx \frac{1}{\alpha_2 L K_{VCO2}} \times \frac{1}{1+\frac{s}{\omega_c}} \quad (1)$$

$$G_{v\phi} = -\alpha_1 G_{i\phi}G_{vi} = \frac{\alpha_1(1-D)R}{2\alpha_2 L K_{VCO2}} \times \frac{1-\frac{sL}{(1-D)^2R}}{\left(1+\frac{s}{\omega_c}\right)\left(1+s\frac{RC}{2}\right)} \quad (2)$$

$$G_{loop} = -\alpha_1 G_{i\phi}G_{vi}\left(\frac{K_{VCO1}}{s}+K_{VCDL}\right) \quad (3)$$
$$= \frac{\alpha_1(1-D)R}{2\alpha_2 L K_{VCO2}} \times \frac{\left(1-\frac{sL}{(1-D)^2R}\right)\left(\frac{K_{VCO1}}{s}+K_{VCDL}\right)}{\left(1+\frac{s}{\omega_c}\right)\left(1+s\frac{RC}{2}\right)}$$

and loop gain of the TDC SITO is similar to that of a traditional current-mode boost converter. According to control theory, the SITO system stability can be guaranteed by properly setting K_{VCO1}, K_{VCO2} and K_{VCDL}.

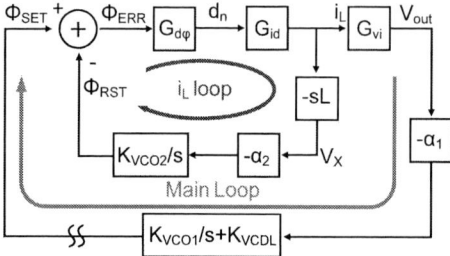

Fig. 3: Small signal model of time-domain-controlled SITO converter.

III. CIRCUIT IMPLEMENTATIONS

A. VCO and VCDL with ADC Assisted Fast-Settling Loop

Fig. 4 presents the fully differential VCO and VCDL with an ADC-assisted loop for transient response enhancement. The converter inductor current I_L is sensed by transconductance amplifier G_{M1} and the output voltage errors are detected by $G_{M2} - G_{M4}$. The clock signals $CLK1$ and $CLK2$ are generated in two CCOs driven by $G_{M1} - G_{M4}$. The succeeding VDCL block provides the time-delay compensation and generates signals CLK_{SET} and CLK_{RST}. To improve the transient response, an ADC-assisted fast settling loop is introduced in the top of Fig. 4. It includes a ripple detector to sense the output ripple voltage. The AC component of V_N, is extracted and converted to an absolute value (i.e. $|V_{AC}|$) by the amplifier. Then $|V_{AC}|$ is digitized by a 6b ADC and further processed by the DSP. The 4b output signals $C\langle 3:0\rangle$ of the DSP controls the variable current source I_{fc1} and I_{fc2} to enable the fast loop-settling by directly modulating the frequency of the VCO in case of a transient response.

B. PD with Duty Limit

In Fig. 5, PD is used to generate the PWM signal D_1 of the proposed TDC SITO converter. The operating waveforms of PD is illustrated underneath its schematic. To achieve a fast transient response, a cycle-slip detector is also used in this design [7]. In practice, the duty of the PWM signal for the boost converter is often set below 90% to ensure stability. Therefore, a duty cycle limiting block is added to the traditional PD block [7] to guarantee the 50ns minimum inductor discharging time. The maximum duty cycle of the PWM signal D_1 is then limited to 90% with a 2MHz switching frequency.

C. Time-Domain Energy Distribution Block

Fig. 6 illustrates the proposed time-domain output energy distribution block. Two transconductance amplifiers with the RC loads serve as the PI compensation for the energy distribution loop. Their output voltages V_{C1} and V_{C2} are used to modify the delay time between D_1 and D_2 and between

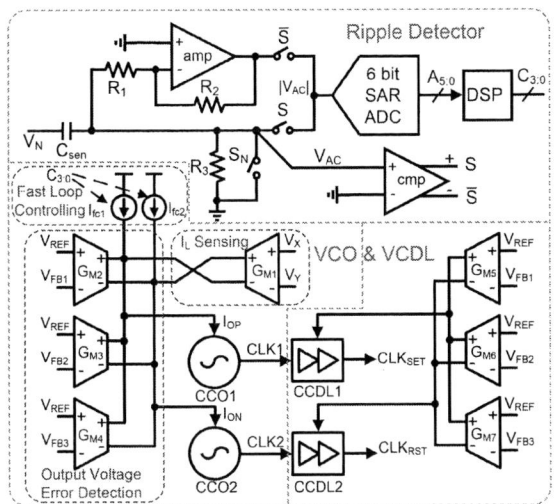

Fig. 4: Proposed VCO and VCDL with ADC-assisted fast-settling loop.

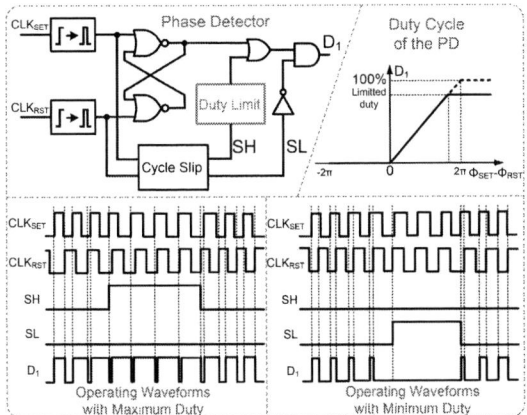

Fig. 5: Proposed PD with its operating waveforms.

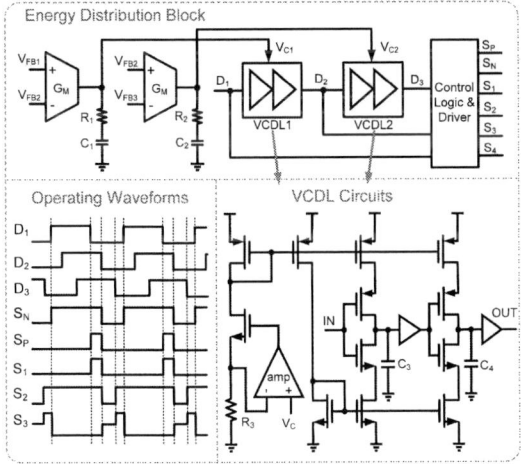

Fig. 6: Output Energy distribution block.

D_2 and D_3, respectively. These delays determine the active periods of the inductor current that charges the load to three output voltages. The operating waveforms and detailed VCDL circuits are given in the bottom-right of Fig. 6. The control voltage V_C is first converted into a current equal to V_C/R_3, and then it is mirrored to charge the capacitor C_3 and C_4 to generate the delay signals. The delay time of VCDL is inversely proportional to the control voltage V_C.

IV. MEASUREMENT RESULTS

The proposed SITO converter was implemented in a 180nm BCD process. As shown in Fig. 7, the total chip area is $4.6 \times 2.65\text{mm}^2$ while the controller occupies 0.37mm^2. The converter operates with a switching frequency of 2MHz. The input voltage range is 2.7V to 5V and the output voltages is between ±7.5V and ±12V with a maximum output current of 400mA. Fig. 8 shows the SITO's steady state operating waveforms with the bipolar output voltages of $V_P=-V_N=10\text{V}$ and load currents of $I_P=-I_N=100\text{mA}$, respectively. The output ripple voltages are 35mV for V_N, 38mV for V_P and 50mV for V_C. Load transient response waveforms are given in Fig. 9. When the load current steps between 100mA and 200mA, the undershoot and overshoot voltages of the three output are all below 160mV with a recovery time of less than $100\mu s$. Fig. 10 shows the measured conversion efficiency versus load current at multiple input voltages and output voltages. The proposed SITO converter achieves a peak conversion efficiency of 91.6% for 5V input, ±7.5V output and load currents of

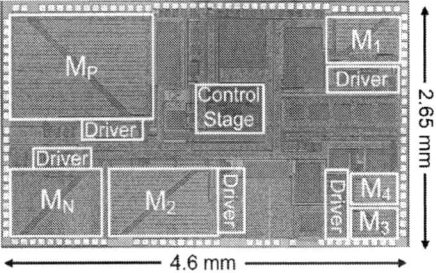

Fig. 7: Chip micrograph.

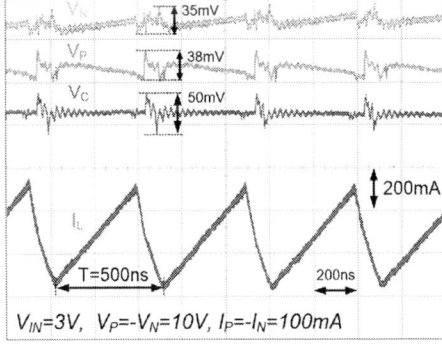

Fig. 8: Measured steady state waveforms.

$I_P=-I_N$=150mA.

Table I compares the performance of this work with the state-of-the-art works. This design features the highest output voltage range up to ±12V among [3]–[6] with the maximum output power of 9.6W, and a highly competitive conversion efficiency of 91.6%. In addition, the proposed SITO converter shows much better load transient response performance than previous works.

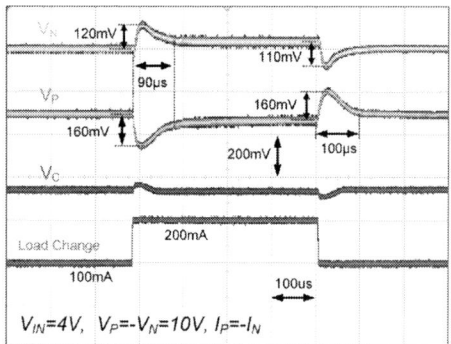

Fig. 9: Measured load transient response waveforms.

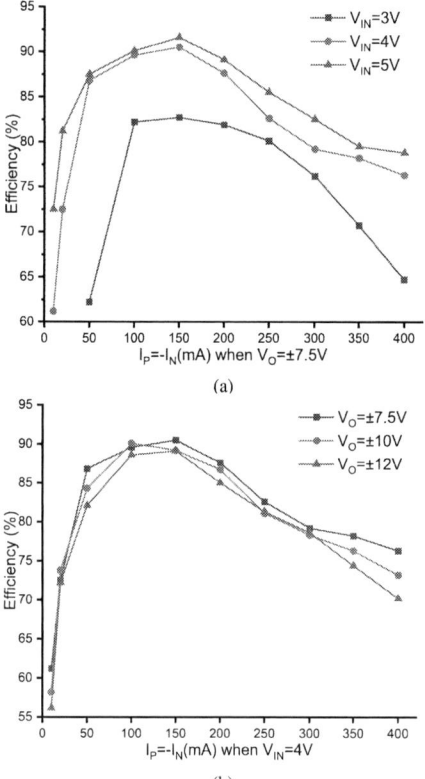

Fig. 10: Measured conversion efficiency versus load current with (a) different input voltages; (b) different output voltages.

TABLE I: Comparison with state-of-the-art works

	Texas Instruments [1]	JSSC2009 C.-S. Chae [3]	ISSCC2020 S.-W. Hong [4]	JSSC2021 F. Mao [5]	CICC2011 Y.-H. Lee [6]	This work
Process	N/A	0.5μm Bi-CMOS	0.5μm CMOS	0.35μm CMOS	0.25μm CMOS	0.18μm BCD
Topology	Boost/Buck-Boost	SIBO	SETSIBO	Hybrid SIBO	SIBO	TDC SITO
Input Voltage	2.7-5.5V	2.7-4.5V	3.7V	2.7-4.5V	2.7-3.6V	2.7-5V
Output Voltage	≤±15V	4.58V, -6.24V	4.6V, -4.9V	5.3V, -4.7V	±10V	±7.5-±12V, 5V
Switching Frequency	1.38 MHz	1 MHz	1.4 MHz	1 MHz	0.9 MHz	2 MHz
Inductors	2×(3.3~6.8μH)	4.7μH	10μH	10μH	10μH	2.2μH
Output Ripple	<28mV	<15mV	<28mV	<30mV	N/A	<50mV
Peak Efficiency	89%@+10V 81%@-10V	82.3%	94.1%	89.3%	90%	91.6%@±7.5V 90.1%@±10V 89.1%@±12V
Maximum Output Power	9W@+15V 6W@-15V	1.1W	2.85W	3.5W	N/A	9.6W
Chip Area	N/A	4.1 mm²	2.8 mm²	3.68 mm²	4.35 mm²	12.2 mm²
Power Density	N/A	0.27W/mm²	1.02W/mm²	0.95W/mm²	N/A	0.79W/mm²
Load Transient Recovery Time/ Shoot Voltage @Load Step*	~400us <280mV @20↔60mA	~100us <400mV @10↔80mA	~50us <430mV @30↔300mA	~1ms <500mV @30↔350mA	~120us <200mV @20↔40mA	100us <160mV @100↔200mA

* Estimated from the paper

V. CONCLUSION

In this paper, we present a time-domain-controlled SITO step-up converter. The converter can generate a pair of bipolar high output voltages up to ±12V and a positive output voltage of 5V with the maximum output power of 9.6W. The input voltage range is from 2.7V to 5V, allowing the converter to be powered from lithium-ion batteries or from fixed 3.3V or 5V supplies. The converter achieves a peak efficiency of 91.6% with a 2.2μH inductor. With the ADC-assisted fast-settling loop, the converter exhibits a fast transient response with low undershoot/overshoot voltages during the load step process.

ACKNOWLEDGMENT

The authors would like to thank Analog Devices Inc. for funding this project.

REFERENCES

[1] Texas Instruments, "TPS65131, Positive and Negative Output DC-DC Converter," Available: https://www.ti.com/product/TPS65131, 2016.

[2] Analog Devices, "ADP5070, 1A/0.6A, DC-to-DC Switching Regulator with Independent Positive and Negative Outputs," Available: https://www.analog.com/en/products/adp5070, 2015.

[3] C. -S. Chae, H. -P. Le, K. -C. Lee, G. -H. Cho and G. -H. Cho, "A Single-Inductor Step-Up DC-DC Switching Converter With Bipolar Outputs for Active Matrix OLED Mobile Display Panels," *IEEE Journal of Solid-State Circuits*, vol. 44, no. 2, pp. 509-524, Feb. 2009.

[4] S. Hong, "A 1.46mm² Simultaneous Energy-Transferring Single-Inductor Bipolar-Output Converter with a Flying Capacitor for Highly Efficient AMOLED Display in 0.5μm CMOS," *IEEE International Solid- State Circuits Conference - (ISSCC)*, Feb. 2020, pp. 200-202.

[5] F. Mao et al., "A Hybrid Single-Inductor Bipolar-Output DC-DC Converter With Floating Negative Output for AMOLED Displays," *IEEE Journal of Solid-State Circuits*, vol. 56, no. 9, pp. 2760-2769, Sep. 2021.

[6] Y. -H. Lee et al., "A near-zero cross-regulation single-inductor bipolar-output (SIBO) converter with an active-energy-correlation control for driving cholesteric-LCD," *IEEE Custom Integrated Circuits Conference (CICC)*, 2011, pp. 1-4.

[7] J. Kang, J. Park, M. Jeong and C. Yoo, "A Time-Domain-Controlled Current-Mode Buck Converter With Wide Output Voltage Range," *IEEE Journal of Solid-State Circuits*, vol. 54, no. 3, pp. 865-873, March 2019.

An Automotive-Use Dual-f_{SW}-Zone Hybrid Switching Power Converter with V_O-Jitter-Immune Spread-Spectrum Modulation

Jin Woong Kwak and D. Brian Ma
Integrated Power System Laboratory (IPSL)
The University of Texas at Dallas, Richardson, TX 75080, USA
Email: jinwoong.kwak@utdallas.edu

Abstract—To break challenging design trade-offs between power switch on-times, electromagnetic interference (EMI) suppression and output voltage V_O regulation in highly demanded high step-down power converters, this paper introduces a new hybrid converter topology with a dual- switching frequency (f_{SW})-zone operation. It relaxes on-times of high-side power switches at high f_{SW} and retains independent V_O regulation with no jittering under spread-spectrum modulations (SSMs). The design work is validated by an IC prototype, which is fabricated on a 180nm HV BCD process using a die area of 1.84mm². It achieves a high step-down conversion from 24V to 1V over a load range of 100mW to 2.7W. Compared to classic half-bridge topology, it relaxes the on-time of high-side switch by about 12 times from 10.4ns to 125ns at 4MHz. Compatible with all major SSM techniques, it reduces peak EMI by 18.25 dB and 20.75 dB, with fixed-rate SSM (FR-SSM) and random SSM (RSSM), respectively. In both cases, the converter experiences no V_O jittering with V_O ripples below 50mV.

Keywords—*Automotive-use switching power converters, high step-down power converters, on-time limitation, EMI suppression, spread-spectrum modulation, V_O jittering*

I. INTRODUCTION & BACKGROUND

The proliferation of modern electric vehicles has imposed ever-increasing demands on automotive electronics integration within stringently limited vehicle space. In response to such demands, automotive power circuit modules must achieve high step-down voltage conversion ratios to reduce module numbers for high power density. Meanwhile, powering critical electronic systems that are highly sensitive to EMI-caused failures requires effective measures to minimize EMI emission. However, in order to achieve high step-down voltage conversion, the on-time (T_{ON}) of a power converter usually has to be significantly shortened (Fig. 1(a)). For instance, for a classical half-bridge converter at a switching frequency (f_{SW}) of 4MHz, the T_{ON} is 75.8ns to achieve a 3.3V/1V conversion. However, if a 24V/1V conversion should be accomplished, the T_{ON} must be reduced to about 10ns, which becomes comparable or even shorter than the delays of the feedback loop (t_{fb}) and the gate driver (t_{gd}). Given that these delay elements can be highly sensitive to I/O voltage and load conditions as well as temperature and process variations, regulation accuracy and stability of the converter would be severely compromised. The situation further deteriorates as f_{SW} increases, which is regarded as the most effective way to minimize the power passives sizes and improve the power density [1].

On the other hand, compared to low voltage portable electronics, the much higher levels of voltages and currents in automotive electronics account for significantly larger dv/dt and di/dt switching actions in power converters, resulting in severely

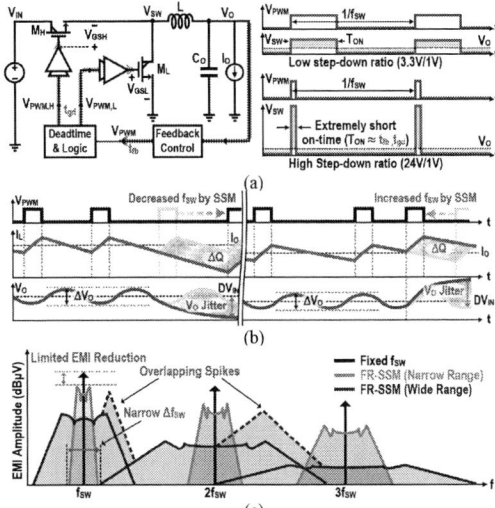

Fig. 1. High step-down power converter design trade-offs: (a) extremely short T_{ON} at high f_{SW}, (b) V_O jittering under SSM, and (c) EMI spectrum overlaps.

elevated EMI emission. Recently, adaptive gate driving techniques [2, 3] are reported. By adjusting the gate charging current to slow down certain period of turn-on process of power switches, the slew rates of switching current di/dt can be reduced to help suppress EMI. However, such an approach degrades efficiency due to elevated switching loss. In addition, it requires a sophisticated Miller Plateau tracking circuit loop, which is difficult to implement at high f_{SW}. On the other hand, the SSM techniques reported in [1, 4] reduce EMI effectively by dithering the f_{SW} of a power converter over a modulation range Δf_{SW}. However, because the f_{SW} dithering can cause instant variation on switching cycles, which a feedback loop may not be fast enough to respond, substantial V_O ripples and V_O jittering effect can often be observed (Fig. 1(b)). Meanwhile, the SSM techniques become less effective when Δf_{SW} increases due to potential spectrum overlaps as depicted in Fig. 1(c).

To break the above design trade-offs, this paper introduces a newly developed, dual-f_{SW}-zone, hybrid step-down power converter, which not only relaxes T_{ON}, but also accomplishes simultaneous SSM-based EMI reduction and jitter-immune V_O regulation. The remainder of this paper is organized as follows. In Section II, we first introduce the development of the power converter along with its operation scheme and design benefits. V_O-jitter-immune SSM scheme and its implementation are then addressed in Section III. Section IV details the system implementation and the experimental results. Finally, Section V concludes this work.

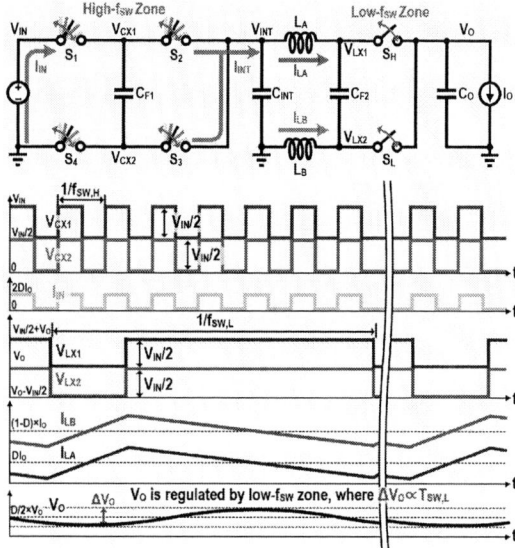

Fig. 2 Topology and operation of the dual-f_{SW}-zone power converter.

II. DUAL-f_{SW}-ZONE TOPOLOGY & OPERATION

As shown in Fig. 2, a dual-f_{SW}-zone, high step-down, hybrid power converter is constructed with two distinct parts. At the front-end, a switched capacitor (SC) power stage creates an internal power rail $V_{INT}(=V_{IN}/2)$ across C_{INT}, which reduces the voltage stress of the following power devices by 50%. While the front-end SC stage handles high input voltage but low current, the following inductive stage is designed to deliver high current to the output V_O. At reduced voltage rating, it can be achieved more efficiently with lower voltage rated devices for S_H and S_L. Moreover, the T_{ON} of the latter inductive stage is doubled, enabling high f_{SW} operation for high power density. In addition to serving as power passives for DC/DC conversion, the C_{INT}, L_A, and L_B naturally divide the converter into two zones, namely high-f_{SW} and low-f_{SW} zones. The dual-f_{SW}-zone operation successfully breaks the trade-offs between T_{ON}, V_O jitter, and EMI noise in such a high step-down converter. In the high-f_{SW} zone, all switches $S_1 \sim S_4$ in the SC frond-end operate with a duty ratio of 50%. The switching nodes V_{CX1} are V_{CX2} serve as key indicators of faster switching actions in the timing diagram. With the fixed duty ratio of 50%, despite of high f_{SW} operation, the T_{ON}s of $S_1 \sim S_4$ are much relaxed. On the other hand, the inductive stage employs much smaller duty ratios on S_H and S_L to execute high step-down conversion. However, because it operates at a much lower f_{SW}, the on-time T_{ON}s of S_H and S_L are also significantly relaxed. Hence, no switches in the topology requires extreme short T_{ON}. On the other hand, if the SSM is implemented in the SC front-end for EMI reduction, it makes little impact on the inductive stage, which can operate at a fixed f_{SW} because of the separated f_{SW}-zones, making the topology fully immune to V_O jittering under SSM operation.

The operation of the proposed topology highlights two unique features beneficial for high step-down voltage conversion. First, as depicted in Fig. 3(a), throughout the steady state, C_{INT} achieves a bidirectional lowpass filtering on both I_{IN} and ΔI_{LA}. The pulsating I_{IN} is lowpass-filtered by the C_{INT}, averaged to V_{INT}, then delivered to the inductive stage, isolating

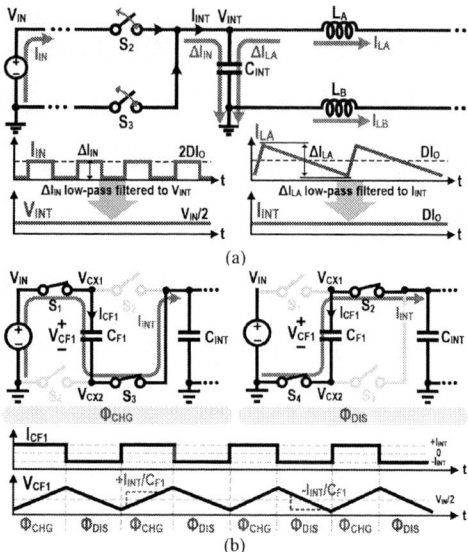

Fig. 3 Secondary benefits of the dual-f_{SW}-zone topology: (a) bidirectional lowpass filtering by C_{INT}, (b) soft-charging and discharging of C_{F1} by I_{INT}.

the two f_{SW}-zones while the current ripple ΔI_{LA} is filtered by C_{INT} to I_{INT}, minimizing the RMS currents carried by $S_1 \sim S_4$. This feature enables the reduction on both EMI and conduction power losses. Second, the proposed topology achieves a soft-charging and discharging on C_{F1} through I_{INT} which is lowpass-filtered inductive load I_{LA}, as shown in Fig. 3(b). This eliminates the hard-charging induced loss under high V_{IN}, further improving the power efficiency. Both features make the proposed topology suitable for achieving both high step-down and low EMI.

III. V_O-JITTER IMMUNE SSM SCHEME

A V_O-jitter-immune SSM scheme is developed for optimal operation of EMI reduction and V_O regulation. As shown in Fig. 4(a), when a conventional half-bridge converter operates in SSM, the f_{SW} is continuously modulated by the f_{SW} modulator, which can be reflected on both V_{SW} and the inductor current I_L. As a result, the conventional solution inherently suffers from a large jitter at V_O under SSM operation. In contrast, Fig. 4(b) shows the SSM operation of the proposed dual-f_{SW}-zone topology, where the $f_{SW,H}$ of the high-f_{SW}-zone is continuously modulated within a range of $\Delta f_{SW,H}$ while the $f_{SW,L}$ of the low-f_{SW} zone remains constant. Thanks to the 50% duty ratio switching operation of the high-f_{SW}-zone, $T_{ON,H}$ is not confined by the conversion ratio, and $f_{SW,H}$ can be set at a higher frequency than $f_{SW,L}$ to re-position the fundamental frequency of EMI from the f_0 of the conventional topology to the much higher $f_{0,H}$. With EMI spectra positioned at higher frequencies, wider $\Delta f_{SW,H}$ can be employed for the SSM to achieve a larger peak EMI attenuation without inducing spectrum overlaps. On the other hand, the natural frequency zone isolation of the topology achieves V_O-jitter immunity automatically. The dithered $f_{SW,H}$ in the front-end is effectively filtered by C_{INT}, leaving the V_O regulation to the low-f_{SW}-zone stage operating at a fixed $f_{SW,L}$. With the V_O-jitter immunity, a much smaller output capacitor C_O can be employed, which compensates the footprint overhead of C_{INT}. Compared to the prior SSM schemes [1, 4], the proposed design achieves higher $f_{SW,H}$ and wider $\Delta f_{SW,H}$ with

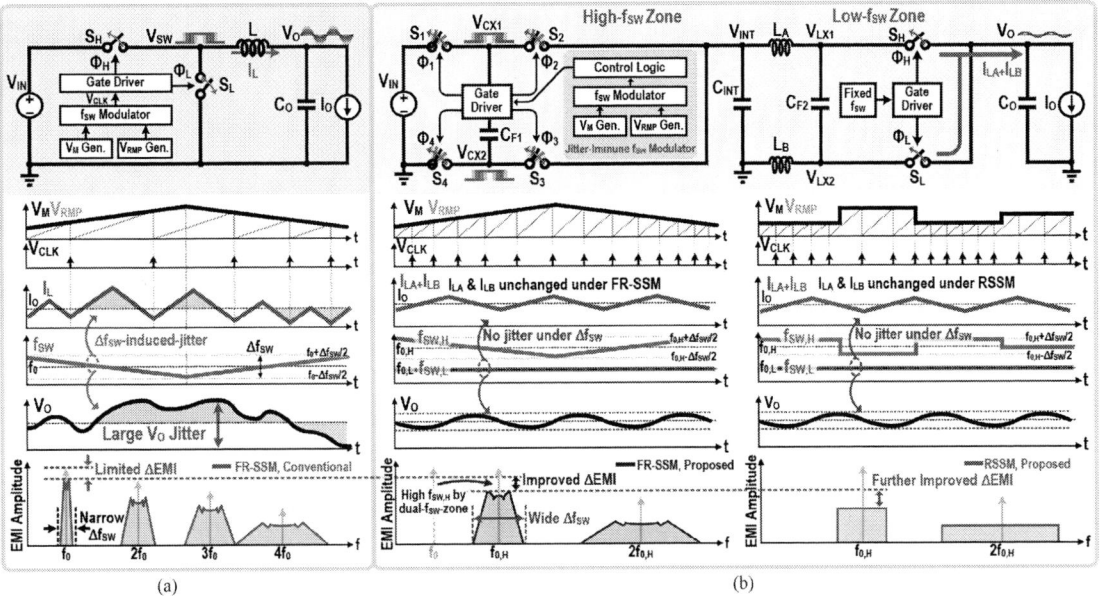

Fig. 4. SSM operations on (a) conventional half-bridge buck converter, (b) proposed dual-f_{SW}-zone converter.

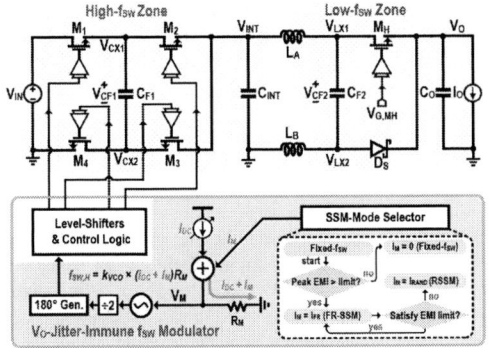

Fig. 5 System block diagram of the dual-f_{SW}-zone converter.

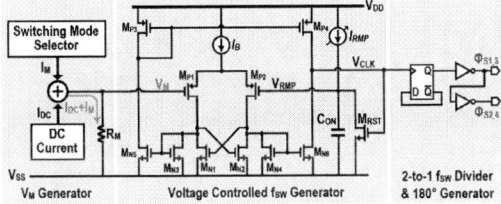

Fig.6 Circuit implementation of the V_O-jitter-immune f_{SW} modulator.

relaxed on-times, facilitating a more significant EMI suppression in high step-down conversion.

IV. SYSTEM IMPLEMENTATION & EXPERIMENTAL RESULTS

The system block diagram of the proposed dual-f_{SW}-zone converter is illustrated in Fig. 5. The switching operation scheme is determined according to a SSM-mode selector, which selects the SSM mode according to the targeted EMI suppression. Fig. 6 details the circuit schematic of the V_O-jitter-immune f_{SW} modulator. The modulation signal V_M is first selected from the SSM-mode selector according to the targeted EMI suppression

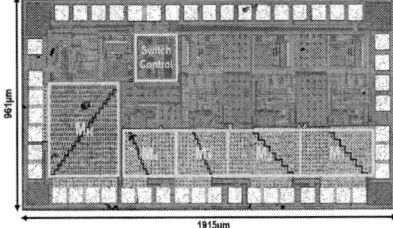

Fig. 7 Chip micrograph.

and is compared to a ramp control signal V_{RMP}. At the trip point, it triggers V_{CLK}, which is then used to generate $\Phi_{S1} \sim \Phi_{S4}$ for the high-f_{SW} zone. According to the selection of V_M, the high-f_{SW} zone can operate with either fixed $f_{SW,H}$, fixed-rate SSM (FR-SSM), or random SSM (RSSM). Note that under all of switching modes, only $f_{SW,H}$ is modulated, whereas $f_{SW,L}$ stays constant, providing V_O-jitter-immune regulation to the output.

Fig. 7 shows the chip micrograph. The proposed converter was fabricated in a 180nm BCD process, with a die area of 1.84mm^2 including $M_1 \sim M_4$, M_H and gate drivers. It accommodates a V_{IN} ranging from 12V to 24V and offers a regulated V_O programmable from 1V to 1.8V. It delivers a maximum I_O of 1.5A, with a peak efficiency of 84.0%.

To validate the steady-state operation and the soft-charging and discharging on C_{F1}, Fig. 8 shows the measured switching nodes and V_{CF1}. At a V_{IN} of 24V, measured key switching nodes reveal valid and steady switching actions with reduced voltage swings of $V_{IN}/2$ (=12V), verifying lower voltage stress on each power device in Fig. 8(a). Furthermore, the smoothly ramped transitions of V_{CF1} in Fig. 8(b) validate the soft-charging and discharging on C_{F1}, which implies much reduced capacitor charge loss at high V_{IN}.

Fig. 9 showcases a major benefit of the dual-f_{SW}-zone operation. In Fig. 9(a), all switches operate at a uniformed f_{SW},

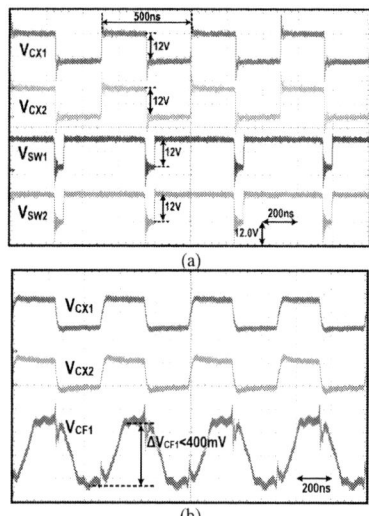

(a)

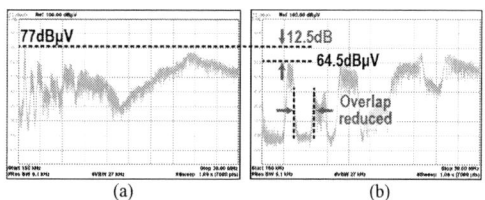

(b)

Fig. 8(a) Measured switching nodes in steady state, and (b) soft-charging and discharging on the flying capacitor C_{F1}.

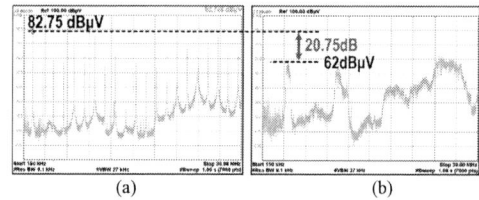

(a)   (b)

Fig. 9 Compared to FR-SSM with uniformed f_{SW} operation to all switches in (a), dual-f_{SW}-zone operation in (b) reduces both peak EMI and spectrum overlaps.

(a)   (b)

Fig. 10 Compared to a fixed-f_{SW} operation in (a), the RSSM operation at $f_{SW,H}$=4MHz±0.4MHz reduced the peak EMI by 20.75dB.

modeling conventional power converters incapable of the dual-f_{SW}-zone operation. A FR-SSM is applied in the test, with a f_{SW} of 1MHz and a modulation ratio $\Delta f_{SW}/f_{SW}$ of ±10%. In comparison, the same $\Delta f_{SW}/f_{SW}$ of ±10% is applied the proposed dual-f_{SW}-zone converter in Fig. 9(b), However, thanks to the dual-f_{SW}-zone operation, the 50% duty ratio at the SC stage relaxes the T_{ON}s of input switches, allowing a higher frequency operation at $f_{SW,H}$=4MHz. Meanwhile, the switches in the low-f_{SW}-zone remain at 1MHz ($f_{SW,L}$) for high step-down conversion. The measurement reveals that a higher $f_{SW,H}$ reduces spectrum overlap between neighboring harmonics. It also achieves an additional 12.5dB peak EMI reduction.

Fig. 10 compares the fixed-$f_{SW,H}$ operation at 1MHz in Fig. 10(a) with the RSSM scheme with $f_{SW,H}$ at 4MHz±10% in Fig. 10(b). Thanks to the natural 50% duty operation of the high-f_{SW}-zone, the $f_{SW,H}$ can be increased from 1MHz to 4MHz while T_{ON}

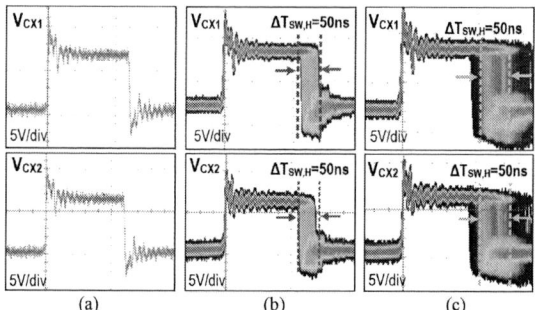

(a)   (b)   (c)

Fig. 11 Measured switching nodes V_{CX1} and V_{CX2}, verifying the adaptation of the converter to (a) fixed-f_{SW}, (b) FR-SSM, and (c) RSSM operation.

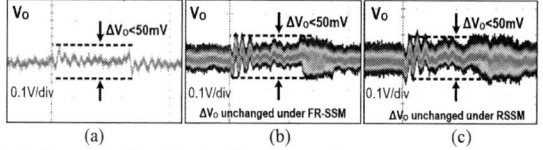

(a)   (b)   (c)

Fig. 12 V_O retains <50mV ripple without observable jittering, when the converter operates (a) at a fixed-f_{SW}, (b) under FR-SSM, and (c) under RSSM.

is retained at 125ns. Along with the RSSM applied to high-f_{SW}-zone, it reduces the peak EMI by 20.75dB, which is comparable to [1, 2].

Fig. 11 shows the measured V_M, $V_{CX1,2}$, and V_O with a fixed f_{SW}, a FR-SSM and a RSSM, respectively. The modulated $f_{SW,H}$ at 2MHz±10% is reflected on the switching nodes V_{CX1} and V_{CX2}. As shown in Fig. 12, V_O ripples stay below 50mV in all cases, validating the V_O jitter immunity and the V_O regulation independence of any SSMs.

V. CONCLUSIONS

In this work, a dual-f_{SW}-zone power converter with V_O-jitter-immune spread-spectrum modulation scheme is designed and demonstrated. With the proposed topology, the switching frequencies in the dual operation zones are independent of each other. By only modulating the switching frequency in the high-f_{SW}-zone, the EMI suppression no longer faces any SSM-related design complications, achieving no jittering at V_O and low EMI. A prototype of this work achieves 20.75dB peak EMI reduction without V_O jitter while delivering a 12~24V/1V conversion with a peak efficiency of 84.0% over a load range of 100mW to 2.7W.

REFERENCES

[1] D. Yan, D. B. Ma., "An Automotive-Use Battery-to-Load GaN-Based Switching Power Converter With Anti-Aliasing MR-SSM and In-Cycle Adaptive ZVS Techniques", *IEEE J. Solid-State Circuits*. pp. 1186-1196, Apr. 2021.

[2] Y. Chen and D. B. Ma, "A 10-MHz Closed-Loop EMI-Regulated GaN Switching Power Converter Using Emulated Miller Plateau Tracking and Adaptive Strength Gate Driving," *IEEE J. Solid-State Circuits*, pp. 531-540, Feb. 2021.

[3] C. Yang, et al., "Design and Characterization of a 10-MHz GaN Gate Driver Using On-Chip Feed-Forward Gaussian Switching Regulation for EMI Reduction", *IEEE J. Solid-State Circuits*, pp. 3521-3532, Nov. 2021.

[4] Y. Chen and D. B. Ma, "EMI-Regulated GaN-Based Switching Power Converter With Markov Continuous Random Spread-Spectrum Modulation and One-Cycle on-Time Rebalancing," *IEEE J. Solid-State Circuits*, pp. 3306-3315, Dec. 2019.

978-1-6654-8495-4/22 $31.00 © 2022 IEEE

A Galvanic-Free Secondary-Side Control Flyback Converter with Digital Adaptive On-Time Control and Direct Sequence Spread Spectrum Technique for 15.5% Error Recovery Rate Improvement

Wei-Cheng Huang[1], Yuan-Jin Li[1], Yi-Hsiang Kao[1], Ke-Horng Chen[1], Kuo-Lin Zheng[2], Ying-Hsi Lin[3], Shian-Ru Lin[3], Tsung-Yen Tsai[3]

[1]National Yang Ming Chiao Tung University, [2]Chip-GaN Power Semiconductor Corporation, Hsinchu, Taiwan [3]Realtek Semiconductor Corp, Hsinchu, Taiwan

Abstract—**For today's convenient lightweight AC power adapters, the proposed galvanic-free secondary-side controlled flyback converter can reduce power modules, since the bulky pulse transformer and photo-coupler can be eliminated by using the main transformer to send the feedback control signal. The proposed direct sequence spread spectrum (DSSS) can have good encryption on the system side and improve the bit error recovery rate from 0% to 15.5%, which can tolerate large noise interference. In addition, digital adaptive on-time control is adopted to improve steady-state efficiency, with light-load and heavy-load efficiencies reaching 92% and 93.8%, respectively.**

Keywords—*direct sequence spread spectrum (DSSS), flyback converter, photo-coupler*

I. INTRODUCTION

Modern secondary-side AC-DC conversion techniques [1] can be tightly controlled through the Universal Serial Bus (USB) Type C port in the host system for high efficiency and flexibility [2-4]. In Fig. 1(a), the secondary side of conventional flyback converters uses pulse transformer [2] or photo-coupler to transmit digital or analog feedback control signals, which can be easily affected by the attack of hackers and noise interference. The soft-switching flyback converter [3] uses the main transformer working in discontinuous conduction mode (DCM) to transmit negative current, which is susceptible to interference due to large parasitic capacitance and magnetic leakage of the transformer. However, it fails to transmit the control signal in the continuous conduction mode (CCM) due to the existence of dc inductor current. Any interference may cause failure in controllers and cause large damage to the loading USB-C system.

In Fig. 1(b), under heavy loads, any missing control signal that occurs at the primary side will cause the system to shut down. At light loads, any accidental transmission of signals caused by hackers or noise will cause serious overvoltage faults at the host system. Hence, it is necessary to include encryptions on the system side to improve the noise immunity and security level in Internet-of-Things (IoT) applications. Thus, based on the transmission in CCM and DCM by the main transformer, this paper proposes the direct sequence spread spectrum (DSSS) technique to enhance security, reduce power module, and enhance overall efficiency.

The organization of this paper is as follows. Section II shows the galvanic-free secondary-side control flyback converter with the proposed DSSS technique to ensure security and noise immunity, which the received signal is proved for correctness. Circuit implementations are shown in Section III. Experimental results are shown in Section IV to prove this work. Finally, conclusions are made in Section V.

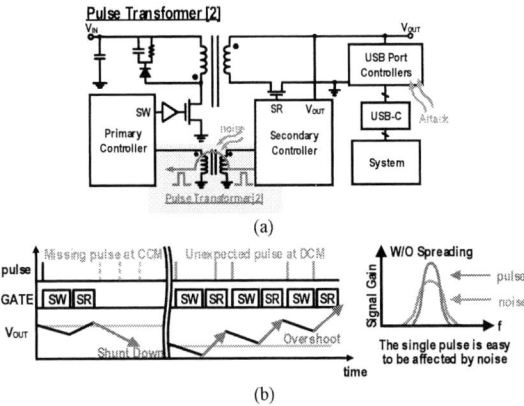

Fig. 1. (a) Transmission error due to noise (or attack from hackers) and magnetic leakage of transformer. (b) Regulation failure due to noise.

II. PROPOSED GALVANIC-FREE SECONDARY-SIDE CONTROL FLYBACK CONVERTER

The proposed galvanic-free flyback converter in Fig. 2 can remove bulky and expensive pulse transformer, which consists of three parts: (1) request energy, (2) store energy, (3) release energy. Using deadtime periods to send control signals, the control signals fed back from the secondary side can be transmitted to the primary side through the main transformer to request energy without interfering with the normal operation, which works correctly in DCM and CCM.

A. DSSS technique

Since high DC drive current in CCM will affect the transmission of feedback control from the secondary side, this paper adopts the DSSS technique to encrypt the primary side control signal to avoid interference. The loaded control signal (AC coupled signal) can avoid the transformer saturation. The DSSS technique consists of an SS encryption circuit and an SS filter circuit, which is in the secondary side and the primary side, respectively. In Fig. 3(a), when the output V_{OUT} is lower than the reference voltage, the SS encryption circuit couples the encrypted energy request signal to the primary side through the main transformer during the SS periods, which is marked as "1". After the SS filter circuit receives the signal AUX from the auxiliary side, the specific spreading code (SS Code) is used to

978-1-6654-8495-4/22 $31.00 © 2022 IEEE

decode it into the original data REC, which leads to the flyback converter entering the energy storage period, which is marked as the switching (SW) periods. Then, the primary side current Ip increases to store energy. However, to effectively control the energy transferring from the primary side, the steady state efficiency enhancement (SSEE) technique is used.

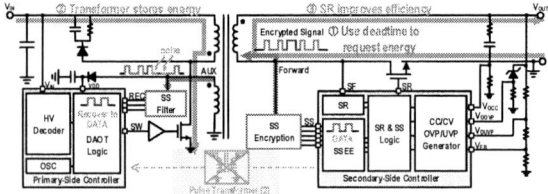

Fig. 2. Architecture of the proposed galvanic-free flyback converter with direct sequence spread spectrum (DSSS) technique.

B. SSEE technique

The digital adaptive on-time control (DAOT) adaptively adjusts the energy based on V_{IN} in each switching cycle to increase stability. At light loads in DCM, the proposed SSEE technique is used to transmit the load information packet in DATA$_{TX}$ to the primary side controller (PSC), to linearly adjust the on-time T_{on} according to the REC for high efficiency. Once the energy request can be correctly determined and stored by the DAOT, the flyback converter enters the secondary regulation (SR) periods. The secondary side current Is decreases from its peak value to zero to regulate V_{OUT} in Fig. 3(a).

In time domain, the recovery can be achieved successfully even under large interferences, as shown in Fig. 3(a). In the frequency domain, since the transmission signal is spread by the spread spectrum, the frequency range of the transmission signal and noise is well distinguished. Thus, even under large interference, the control signal can be completely restored at the primary side, as shown in Fig. 3(b).

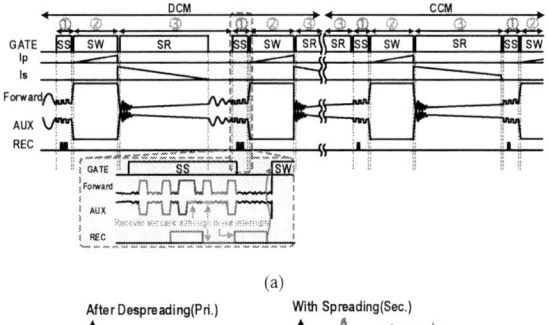

(a)

(b)

Fig. 3. (a) Waveforms of DCM and CCM operation. (b) The pulse spectrogram compared with the de-spreading encrypted signal spectrogram.

III. CIRCUIT IMPLEMENTATIONS

The timing diagram and the relationship between the SS encryption and the SS filter are shown in Fig. 4(a). The SS encryption sends the control signals through the forward pin while the SS filter receives the control signal from the AUX pin.

A. DSSS technique

The SS encryption circuit is shown in Fig. 4(b). When the feedback V_{FB} is less than the V_{REF}, the signal SS is set high and the SS_B is set low to turn on the coupling path. Thus, the signal DATA can be transmitted from the secondary side to the primary side through the forward pin. However, the main transformer has large parasitic capacitance, copper loss (coil resistance heating), and unavoidable magnetic leakage. The transmission signal via the main transformer will be distorted. Therefore, the signal DATA needs to be spread. The DATA and SS Code are mixed into the encrypted signal DATA$_{TX}$ through the exclusive OR operation.

Hence, the SS filter circuit at the primary side in Fig. 4(c) can decode the AUX signal to the REC signal. The SS filter circuit can first filter out most noise when the LC oscillation goes through a high-pass filter (C_{HP} and R_{HP}). To ensure confidentiality, both the SS encryption circuit and the SS filter circuit in the communication will have the same SS Code. The SS filter circuit receives 4 continuous results in the shift register to do exclusive OR operation with the SS Code. That is, all the results of the exclusive OR operation are added together to generate the signal ADD. The REC signal restores AUX to DATA by evaluating whether the value of ADD is greater than 2. Thus, any interference can be reduced to ensure the correct control from the secondary side.

The mathematical proof of the recovery signal REC equals DATA is shown in (1), which shows the advantage of the DSSS technique.

$$
\begin{aligned}
&\text{DATA}_{TX} = \text{SSCode(s)} \oplus \text{DATA(s)} \\
&\text{REC} = \text{SSCode(p)} \oplus \text{DATA}_{TX} \\
&\because \text{SSCode(p)} = \text{SSCode(s)} \\
&\therefore \text{REC} = \text{SSCode(p)} \oplus \text{SSCode(s)} \oplus \text{DATA(s)} \\
&\qquad = 0 \oplus \text{DATA(s)} = \text{DATA(s)}
\end{aligned}
\tag{1}
$$

The advantage of this technology is that the errors caused by noise interference during the transmission process can be detected and corrected for accurate control, as shown in the timing diagram in Fig. 4(a). The spread signal DATA$_{TX}$ is equivalent to white noise in the frequency spectrum, as show in Fig. 4(d) and is not susceptible to interference.

B. SSEE technique

The slave PSC controller in Fig. 5(a) waits for a digital turn-on command from the secondary side controller (SSC) in Fig. 5(b). When V_{OUT} drops and energy is needed, the SR control will first be turned off, then the encrypted signal will be transmitted to the PSC. After the data is restored at the primary side, the power switch SW is controlled by the programmable on-time T_{on} from the data of the SSC controller. In DCM control, a DAOT control is proposed, which controls the T_{on} to expand at heavy load and reduce at light load for high efficiency.

Therefore, the proposed SSEE technique in Fig. 5(c) is not only embedded in the DSSS technique for high security, but also changes T_{on} in DCM for higher steady-state efficiency.

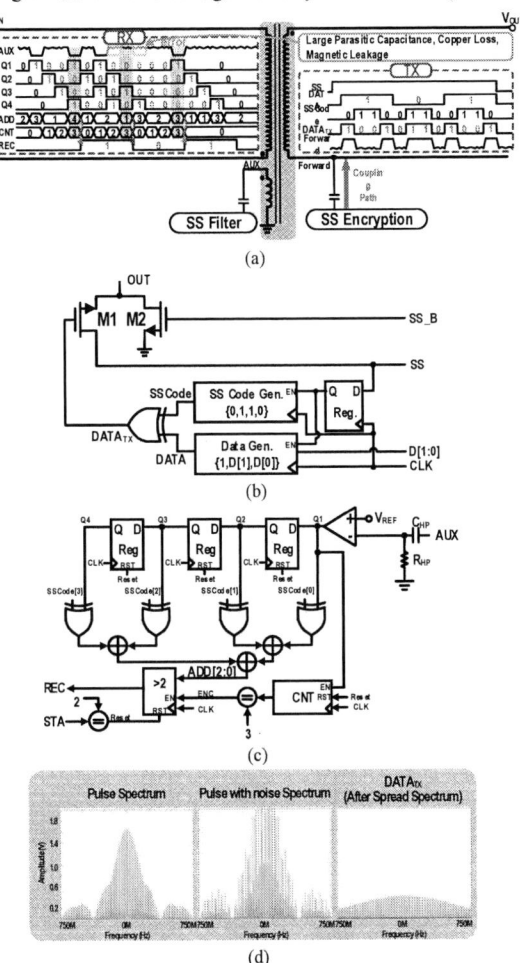

(a)

(b)

(c)

(d)

Fig. 4. (a) SS filter and SS encryption circuit of the proposed DSSS technique and the transmission signal waveforms (b) The SS encryption circuit in the secondary side. (c) The SS filter circuit in the primary side. (d) The spectrum of the $DATA_{TX}$ proves the theoretical spread spectrum model.

The discharge time of the output capacitor is Dc in Fig. 5(d), also expressed as Q, which is the time period when the SR turns off to set the signal CV high. If Q is larger than 48, it indicates that too much energy is supplied from the PSC controller. Thus, the PSC controller will decrease T_{on} to reduce the delivery energy when it receives the REC '$111_{(2)}$' from the SSC controller. If the value of Q is between 32 and 16, the PSC controller linearly increases T_{on} to deliver more energy, so that Dc is shrunk to a balanced range to improve efficiency. However, CCM does not have the characteristics related to T_{on} and the load. Therefore, D is set to 0 when Q is less than 16 to avoid T_{on} change in CCM or boundary. The PSC controller will not adjust T_{on} after receiving REC=0, but will adjust T_{on} based

on V_{IN}, thereby equalizing the delivery energy of each transmission.

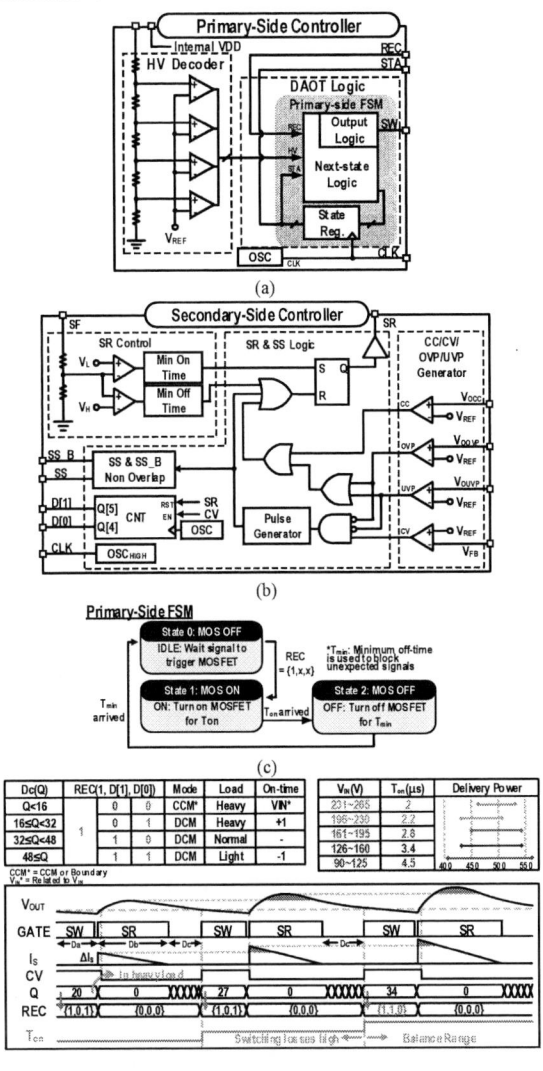

(a)

(b)

(c)

(d)

Fig. 5. (a) The PSC controller. (b) The SSC controller. (c) Proposed SSEE technique. (d) Operation waveforms.

IV. EXPERIMENATL RESULTS CIRCUIT

The proposed PSC and SSC controllers are fabricated in a 0.18μm BCD process, of which the silicon areas are 0.95 ×1.2mm² and 1.71 ×1.96mm², respectively, as shown in Fig. 6.

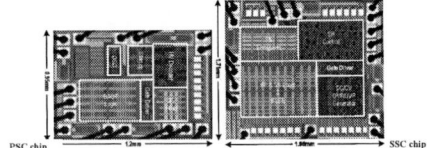

Fig. 6. Chip micrographs of PSC controller and SSC controller.

In Fig. 7, when the DATA$_{TX}$ is disturbed by the voltage change from 4.8V to 3.6V during the transmission in the main transformer, the DATA$_{TX}$ can still be detected at the PSC to turn on the primary side power MOSFET. Fig. 7(a) shows that PSC decodes the DATA$_{TX}$ to trigger the SW periods. In Fig. 7(b) during steady-state, the proposed SSEE technique adjusts T$_{on}$ from 6μs to 7.4μs and Dc from 3μs to 7μs to improve efficiency. In Fig. 7(c), T$_{on}$ does not change when the SSEE technique is not applied. In Fig. 7(d), with the spread spectrum technique, the peak amplitude of the spectrum can be reduced from 1.59V to 0.4V, which is a 73% reduction. Table I shows that the proposed flyback converter has the highest error recovery rate and the highest 100% I$_{LOAD}$ efficiency, which are due to the DSSS technique and the CCM operation, respectively.

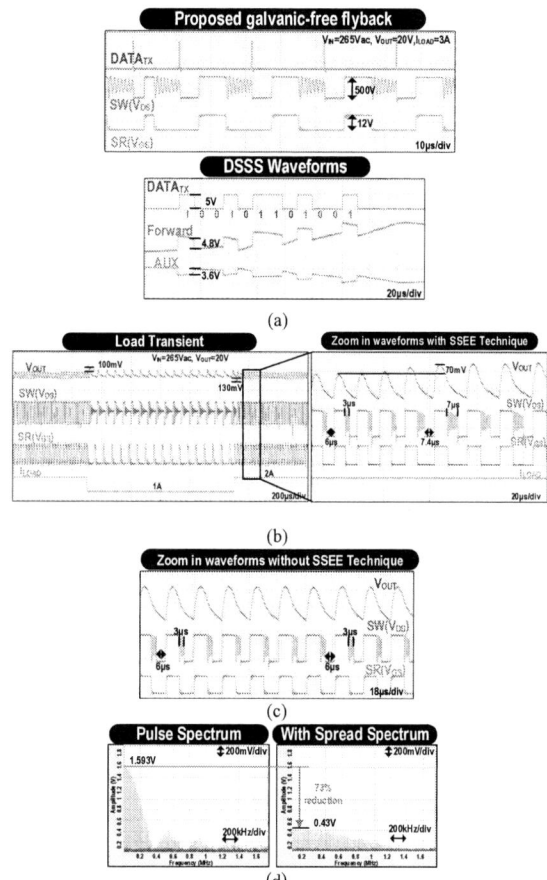

(a)

(b)

(c)

(d)

Fig. 7. (a) PSC decodes DATA$_{TX}$ to trigger SW (top left); waveforms of DSSS operation. (b) When I$_{LOAD}$ changes from 1A to 2A, T$_{on}$ is adjusted from 6μs to 7.4μs by SSEE technique. (c) T$_{on}$ does not change when SSEE technique is not applied. (d) Pulse spectrum and spectrum after spreading.

In Fig. 8, with the SSEE technique, the flyback converter has a 94% efficiency at medium load, which is higher than that of without the SSEE technique. In contrast, [2] has lower load efficiency due to the fixed frequency control. The DAOT with the SSEE technique adjusts T$_{on}$ which enables the switching frequency to have a smaller range than that of constant on-time

(COT) techniques. The error recovery rate chart shows that the recovery rate can reach 100% if only one bit is disturbed during transmission. As the disturbing bits increase, the recovery rate will decrease. However, the recovery rate can still reach 15.51% even if 30% of the transmitted bits are disturbed.

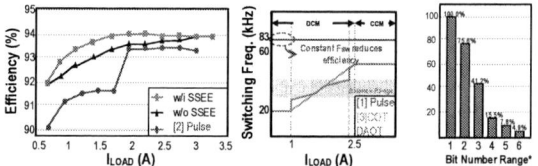

Fig. 8. Efficiency versus output load, switching frequency versus output load, and error recovery rate versus bit number range

Table I: comparison table with prior arts.

	This Work	ISSCC'20[2]	TIA'17[3]	InnoSwitch3[4]
Technology	180 nm BCD	180 nm BCD	NR	NR*
Topology	Adaptive and DSSS	ZVS by SR	ZVS by SR	QR
Power MOSFET	Si	Si	Si	GaN
Signal Isolation	Not Needed	Pulse Transformer	Not Needed	Magnetic
IC Used in Converter	PSC, SSC	PSC, SSC	FPGA, XMC45000 MCU	INN3368C-H301
Regulation Loops	Internal CC/CV	Internal CC/CV	External	External
Max. Power	65W	60W	65W	60W
Input Voltage	90-265Vac	85-265Vac	120-230Vac	85-265Vac
Max. Output V/I	20V/3.25A	20V/3A	20V/3A	20V/3A
Max. Efficiency	94%	93.5%	85%	93.8%
100% I$_{LOAD}$ Efficiency	93.8%	93.2%	89.2%	NR
15% I$_{LOAD}$ Efficiency	92%	89.1%	88.8%	NR
Error Recovery Rate30*	15.5%	0%	0%	0%
Die Size [mm2]	PSC: 1.15, SSC: 3.36	PSC: 1.44, SSC: 6.73	NR	NR

NR*: Not Reported
Error Recovery Rate30*: #(recovery)/#(30% possibility of error occur) * 100%
Bit Number Range* (k): Σ(n=1 to k) #(all possibilities of n error occur)

V. Conclusions

The proposed galvanic-free secondary-side controlled flyback converter can reduce power modules. The bulky pulse transformer and photo-coupler can be eliminated by using the main transformer to send the feedback control signal. The DSSS technique can have good encryption and improve the bit error recovery rate to 15.5% to tolerate large noise interference. Digital adaptive on-time control can improve light-load and heavy-load efficiencies to 92% and 93.8%, respectively.

Acknowledgment

The authors would like to thank Kevin Chuang and Andy Ho, Chip-GaN Power Semiconductor Corporation, Hsinchu, Taiwan, and would also like to thank Qualcomm University Research Projects for their help and support. Thanks for the help of DIODES Inc. Taiwan.

References

[1] P.H. Liu, "Design Consideration of Active Clamp Flyback Converter with Highly Nonlinear junction Capacitance," *IEEE Applied Power Electronics Conference and Exposition (APEC)*, pp. 783-790, Mar. 2018.

[2] W. Chang, K. Lin, C. Lee, L. Lo, J. Lin and T. Yang, "ZVS Flyback-Converter ICs Optimizing USB Power Delivery for Fast-Charging Mobile Devices to Achieve 93.5% Efficiency," *ISSCC*, 2020, pp. 294-296.

[3] A. Connaughton, et. al., "Investigation of a Soft-Switching Flyback Converter With Full Secondary Side-Based Control," *in IEEE Trans. Industry Applications*, vol. 53, no. 6, pp. 5587-5601, Nov.-Dec. 2017.

[4] Power Integration, "InnoSwitch™ 3-Pro Family," Datasheet, Rev. G, July 2019. Accessed on Nov. 29, 2019, <https://ac-dc.power.com/products/innoswitch-family/innoswitch3-pro/?language=en>.

[5] J. Park, Y. Roh, Y. Moon and C. Yoo, "A CCM/DCM Dual-Mode Synchronous Rectification Controller for a High-Efficiency Flyback Converter," *in IEEE TPE*, vol. 29, no. 2, pp. 768-774, Feb. 2014.

[6] H. Chen and J. Xiao, "Determination of Transformer Shielding Foil Structure for Suppressing Common-Mode Noise in Flyback Converters," *in IEEE Transactions on Magnetics*, vol. 52, no. 12, pp. 1-9, Dec. 2016.

A Half-Bridge Gate-Driver for high-efficient Boost Converter Applications with single-sided ZVS and an adaptive Ringing Suppression Technique

Michael Hanhart, Jonas Zoche, Jan Grobe, Léon Weihs, Leo Rolff, Ralf Wunderlich and Stefan Heinen

Chair of Integrated Analog Circuits and RF Systems

RWTH Aachen University

Kopernikusstrasse 16, D-52074 Aachen, Germany

Email: mailbox@ias.rwth-aachen.de

Abstract—This paper proposes an integrated bootstrapped Half-Bridge (HB) gate driver, which is used in a synchronous boost converter for photovoltaic applications. The HB is operated with single-sided zero-voltage-switching (ZVS) at the low-side (LS) or high-side (HS), dependent on the inductor current sign, to minimize switching losses. The gate driver architecture is based on segmented and distributed output stages with split outputs for turn-on and turn-off. During turn-on, the gate current is limited to 360 mA initially and gets boosted to 1.2 A after the Miller-Plateau (MP) has been actively detected. This minimizes the di/dt in the reverse recovery phase and, secondly, reduces the turn-on time by 26 %. A non-overlap time between LS and HS stage is guaranteed by a supervisor circuit. The boost converter achieves 98.6 % peak efficiency.

Index Terms—ZVS, Ringing, switching-speed, Efficiency, BbM, Half-Bridge, Boost Converter, Miller-Plateau

I. INTRODUCTION

The demand for high frequency, high power and high efficiency energy conversion is continuously increasing. This is a consequence of new GaN devices, which feature superior material characteristics with less parasitics for high voltage (HV) applications used in circuits for renewable energy and in the automotive sector. However, lower supply voltages in the advanced nodes request an ever-increasing need for high current DC-DC converters. Both trends demand for gate drivers which can control power switches securely, reliably and efficiently. That incorporates a trade-off between losses and EMI. Minimizing parasitics plays an immanent role for higher switching speed. For that purpose, a three-level driver with fully integrated capacitors has been proposed by [1]. In [2] a GaN switch is driven with a small current to reduce the di/dt, followed by a larger current, when the MP is reached. The MP sensing is based on the emulation of the threshold voltage during reverse conduction, which makes the concept not applicable to MOSFET switches. Thus, [3] uses a dynamic feedback delay compensation technique, to detect the MP. [4]

The authors acknowledge the financial support by the German Federal Ministry for Economic Affairs and Energy under 03ET1417B (Project EnEff: Stadt EnQM) as well as EUROPRACTICE MPW and design tool support. The manufacturing has been sponsored by the HiSilicon sponsorship program.

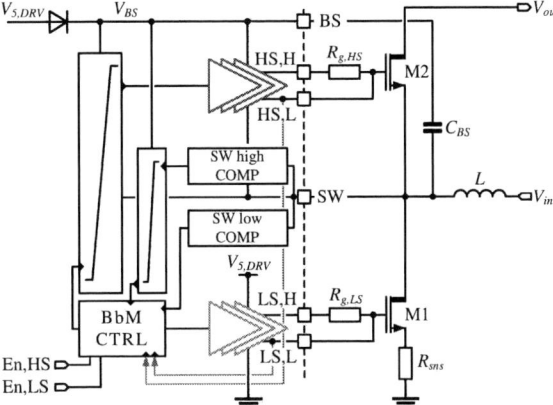

Fig. 1. Half bridge gate driver topology embedded in a boost configuration.

proposes a gate driver with a closed loop approach to regulate the V_{sw} slope by applying two different gate currents. The sensing utilizes an external capacitive divider at the switch pin. [5] employs ZVS in an isolated DC-DC converter to reduce switching losses.

II. PROPOSED ARCHITECTURE

This paper proposes a bootstrapped HB driver with single-sided ZVS for highly efficient boost converter applications. The HB topology is depicted in Fig. 1, the key building blocks are the main HV level shifter, LS and HS driver stage and a break-before-make (BbM) circuit. To achieve ZVS, the switch pin voltage is monitored by charge based HV-tolerant comparators. Positive inductor currents automatically charge the switch node to V_{out} when the LS is disabled. Thus, the HS is turned on with minimal V_{ds} and therefore, near zero losses. For near zero currents, a timer enables hard switching of the HS. The output voltage range covers up to 50 V.

The level shifter implementation, shown in Fig. 2, is based on pulsed driving signals with active feedback, to set and reset a latch in the HV-domain. Low voltage (LV) current mirrors are utilized for the core sensing cell. The level-shifted

978-1-6654-8495-4/22 $31.00 © 2022 IEEE

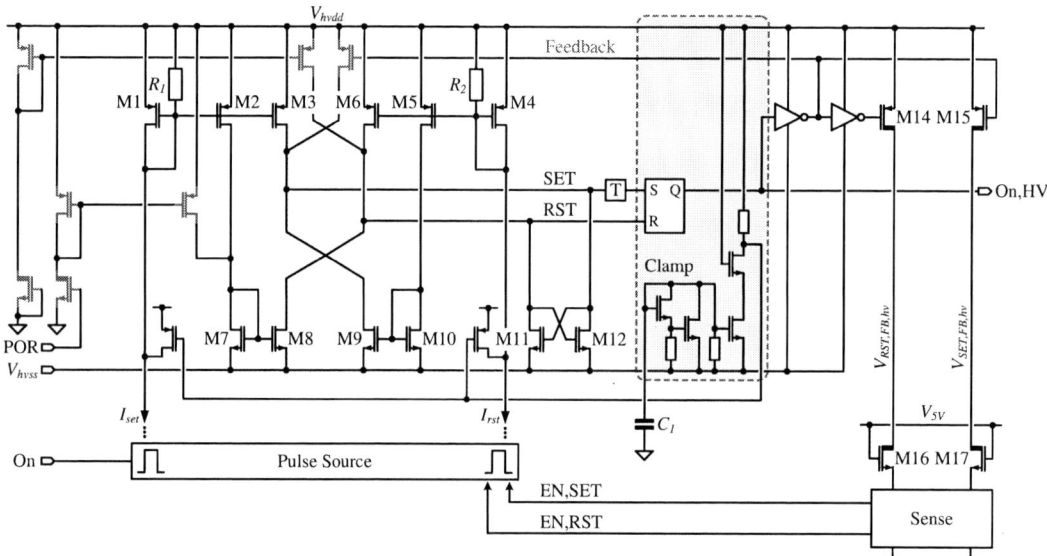

Fig. 2. Simplified level shifter implementation with enhanced CM tolerance and feedback controlled pulse length.

on-signal is latched and fed back to the LV-domain, hence, I_{set} and I_{rst} are turned off immediately. This ensures a reliable and energy efficient operation. M7 – M10 enhance the immunity against common mode noise injected by the switch pin moving. Current induced by the power-on-reset circuit is compensated by a dummy tail. Thus, a HS turn-on signal can be transferred without glitches at any point in time for positive voltage slopes up to $10\,V/ns$ across PVT variations. Negative switch pin transitions are not critical for any control signals, but may lead to overvoltage, which in turn are minimized by the pictured clamp circuit. Fig. 5 shows the worst case timing relation between rising switch pin and level shifter turn-on. For that purpose, the timing has been deliberately adjusted to show the largest effect on SET and RST potentials in the HV domain. This proves that the output stage operates reliably under all conditions, even when the HS cannot be soft-switched and the switch node has already been charged to V_{out} by the inductor. The rising edge delay measures $7.9\,ns$ due to filtering, while the turn-off is not delayed further and amounts to $1.8\,ns$.

The driver concept uses two methods to reduce the switching speed, and thus the ringing of the switch pin. During turn-on, the gate is charged with a fixed current to reduce the di/dt in the reverse recovery phase of the power stage. When a decreasing V_{gs}-slope is detected, the gate current is boosted up to $1.2\,A$ to quickly reach the full supply voltage, which reduces the losses in the switch. The sense circuit omits any further external components in contrast to [4]. Additionally, no complex mixed signal control loop like in [3] is needed. Fig. 3 depicts the block diagrams of the core driver - each circuit is segmented into two substages. By separating the outputs into two pins, the gate charge- and discharge characteristic can be influenced independently. Hence, a low

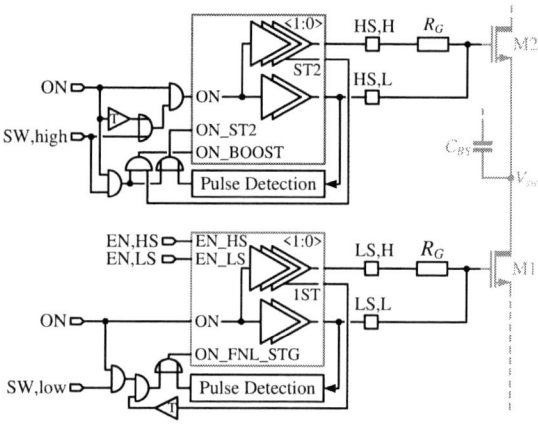

Fig. 3. LS and HS driver block level implementation.

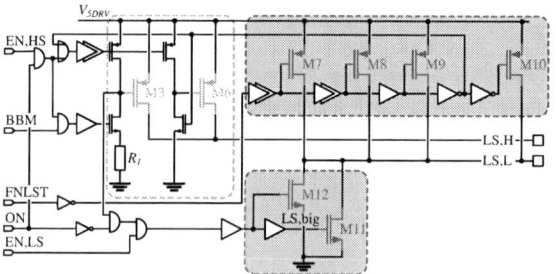

Fig. 4. Simplified multistage LS driver circuit implementation.

impedance NMOS path with $2\,A$ sinking capability protects the external FETs from parasitic turn-on during fast switch pin transients, whereas a $2.2\,\Omega$ gate resistor slows down the turn-on. Fig. 4 shows the simplified LS-driver implementation.

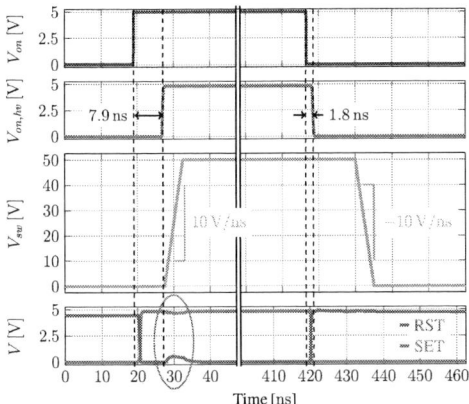

Fig. 5. Post layout level shifter simulation for worst case switch pin timing.

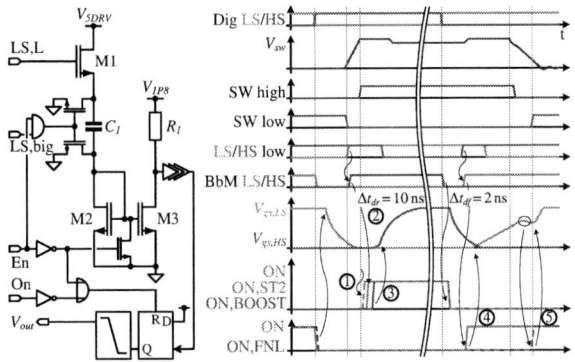

Fig. 6. Dynamic gate sensing and timing diagram of the half bridge driver.

The gate driver consists of distributed output stages. The use of multiple transistors and their delayed turn-on lead to a smeared current profile, which reduces supply current spikes. Each driver stage has an additional BbM input, which is driven by a supervisor circuit to prevent any shoot-through current in the driver itself. M3 turns on slowly, which limits the initial gate charge current. Nevertheless, the potential at *LS,H* will rise faster compared to the gate voltage of the external FET, that is directly sensed at the second output *LS,L*. When the pulse detection (PD) circuit senses a decaying or negative slope, the MP is reached, hence I_g is boosted by turning on the PMOS transistors M7-M10. The BbM concept ensures that I_L charges the switch pin capacitance to V_{out} before the HS is turned on with ZVS. Hence, the switching losses are further minimized. Fig. 6 illustrates the functional timing diagram.

III. MEASUREMENT RESULTS

Fig. 7 shows the 14 ns HS driver delay during turn-off, which includes the BbM circuit, as well as the level shifter. HS_{on} and LS_{on} are the internal digital control signals, outputted by measurement IOs. Fig. 8 depicts the difference between enabled and disabled PD circuit for the LS turn-on. For that purpose the HB is employed in a boost converter

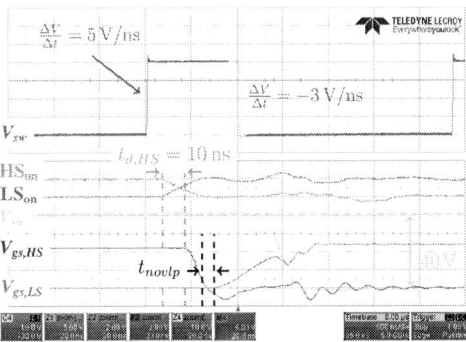

Fig. 7. V_{gs} for HS turn-off and LS turn-on and V_{sw} transient.

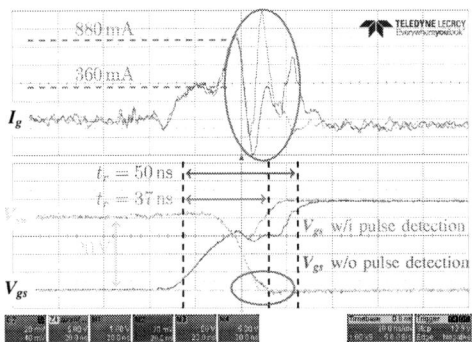

Fig. 8. V_{gs}, I_g and V_{sw} for different driver settings.

with 300 kHz switching frequency and 6.1 A inductor current. Both FETs (Infineon BSZ037N06LS5 4.2 mΩ $R_{RD,on}$) are operated with 2.2 Ω gate resistors. The LS gate is charged with 360 mA until the FET starts to conduct, which in turn discharges the switch pin. The reverse recovery current of the HS body diode, in conjunction with the parasitic inductances of the HB lead to ringing of I_D. With the activated V_{gs}-PD, the rise time of the gate-source voltage is shortened by 26 % and amounts to 37 ns. The V_{sw}-slope changes accordingly, from -0.6 V/ns to -1.5 V/ns, while remarkably, the ringing at the switch node does not increase. The gate current shows a second peak, when the PD enables the boost stage, however the measurement is disturbed by common mode noise, injected by I_s. The measured switch pin voltage slope during active turn-on of the LS ($I_L = 4.5$ A) is smaller (-3 V/ns) compared to the rising slope (5 V/ns), which is only forced by 7.7 A inductor current - the HS is still switched off due to the ZVS architecture - as pictured in Fig. 7. This shows the ability of the driver concept to limit the di/dt significantly, omitting large gate-resistors.

Fig. 9 pictures different configurations of the LS driver, including various driving strength settings as well as enabled and disabled V_{gs}-PD. It can be clearly observed that the final gate charging is decoupled from the initial gate current, which is not possible with standard drivers and gate resistors. The classical approach limits the current, especially in the last

978-1-6654-8495-4/22 $31.00 © 2022 IEEE

TABLE I
PERFORMANCE COMPARISON OF STATE-OF-THE-ART PUBLICATIONS

Param.	This work	ISSCC 2018 [1]	ISSCC 2021 [3]	VLSI 2017 [2]	ISSCC 2019 [4]
Topology	Half-Bridge Boost	Half-Bridge isolated DC-DC	Half-Bridge	Half-Bridge Buck	Driver
Switch Type	MOSFET	GaN	GaN	GaN	MOSFET
V_{LV}	7 - 24 V	48 V	-	-	-
V_{HV}	10 V - 50 V	150 V - 400 V	600 V	5 V - 40 V	(*)
Soft Switching	HS or LS ZVS	ZVS	-	-	-
Driving Method	Gate I-control	Single output	Gate I-control	Gate I-control	2 gate resistors
Driver Feedback Concept	Analog V_{GS} slope sensing	-	MP Sensing Dynamic FB delay compensation	Adaptive MP Sensing	Discrete-time analog
Peak Efficiency	98.6 % (14 V) 96.8 % (40 V)	90.2 % (400 V, 1 MHz) 88.6 % (150 V, 2 MHz)	-	81.4 %	-
Die Size	2.45 mm²	9 mm²	-	0.9 mm²	8.23 mm² (†)
Process	0.18 µm	1.45 µm 700 V BCD	0.5 µm 600 V SOI	0.35 µm BCD	0.13 µm HV-CMOS

(*) Single driver, no half-bridge   (†) Full chip, core size smaller

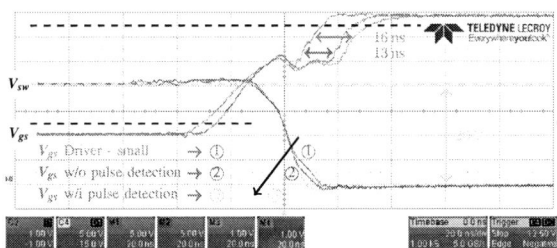

Fig. 9. V_{gs}, I_g and V_{sw} for different driver settings.

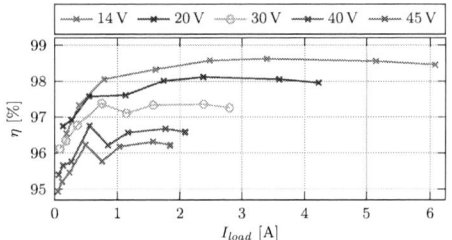

Fig. 10. Boost converter efficiency for different output voltages.

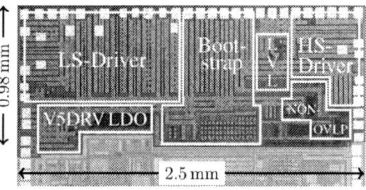

Fig. 11. Die micrograph of the fabricated Half-Bridge driver.

efficiency operation, which has been demonstrated in a boost application with 98.6 % η_{peak}. The segmented drivers feature separate outputs to support asymmetric charging and discharging of the gates. Moreover, the internal gate current limiting reduces the di/dt during the reverse recovery phase. Therefore, external gate resistors can be reduced significantly. In addition, the topology is capable of boosting the gate current after the MP to charge the gate as quickly as possible.

All in all, the presented driver architecture is suitable to lower the initial switching speed to reduce ringing. It can be extended in future designs to support various gate sizes by adapting the output current. Even advanced gate current profiles are feasible.

phase of the turn-on, when the gate is only charged to the full supply voltage. This in turn is not critical in terms of ringing, but increases the losses unnecessarily.

The peak efficiency of the boost converter lies at 98.6 % and stays well above 94.5 % for the entire operating range across V_{out}, as shown in Fig. 10. In light load conditions, the controller operates in pulse-frequency modulated DCM. Fig. 11 shows the fabricated design, embedded in a boost controller for photovoltaic applications. Table I compares the proposed HB design to various state-of-the-art driver architectures.

IV. CONCLUSION

An integrated HB gate driver with single-sided ZVS has been presented in this paper. The HV level shifter withstands 10 V/ns switch pin slope common mode noise. The integrated BbM circuit guarantees non-overlapping driver control and minimizes the switching losses. Moreover, active switch pin monitoring enables single sided ZVS, that guarantees high-

REFERENCES

[1] A. Seidel and B. Wicht, "A Fully Integrated Three-Level 11.6nC Gate Driver Supporting GaN Gate Injection Transistors," in *2018 IEEE International Solid - State Circuits Conference - (ISSCC)*, 2018, pp. 384–386. DOI: 10.1109/ISSCC.2018.8310345.

[2] Y. Chen, X. Ke, and D. B. Ma, "A 10MHz 5-to-40V EMI-Regulated GaN Power Driver with Closed-Loop Adaptive Miller Plateau Sensing," in *2017 Symposium on VLSI Technology*, 2017, pp. C120–C121. DOI: 10.23919/VLSIT.2017.7998137.

[3] J. Zhu, D. Yan, S. Yu, W. Sun, G. Shi, S. Liu, and S. Zhang, "A 600V GaN Active Gate Driver with Dynamic Feedback Delay Compensation Technique Achieving 22.5% Turn-On Energy Saving," in *2021 IEEE International Solid- State Circuits Conference (ISSCC)*, vol. 64, 2021, pp. 462–464. DOI: 10.1109/ISSCC42613.2021.9365974.

[4] S. Kawai, T. Ueno, and K. Onizuka, "A 4.5V/ns Active Slew-Rate-Controlling Gate Driver with Robust Discrete-Time Feedback Technique for 600V Superjunction MOSFETs," in *2019 IEEE International Solid-State Circuits Conference - (ISSCC)*, 2019, pp. 252–254. DOI: 10.1109/ISSCC.2019.8662534.

[5] L. Cong and H. Lee, "A 2MHz 150-to-400V Input Isolated DC-DC Bus Converter with Monolithic Slope-Sensing ZVS Detection achieving 13ns Turn-On Delay and 1.6W Power Saving," in *2018 IEEE International Solid - State Circuits Conference - (ISSCC)*, 2018, pp. 382–384. DOI: 10.1109/ISSCC.2018.8310344.

A 4 to 40V Wide Input Range and Energy Re-Cycling High Power LiDAR Driver for 5% Efficiency Enhancement and 300m Long-distance Object Detection

Si-Yi Li[1], Zheng-Lun Huang[1], Sheng Cheng Lee[1], Ke-Horng Chen[1], Kuo-Lin Zheng[2], Ying-Hsi Lin[3], Shian-Ru Lin[3], Tsung-Yen Tsai[3]

[1]National Yang Ming Chiao Tung University, [2]Chip-GaN Power Semiconductor Corporation, Hsinchu, Taiwan [3]Realtek Semiconductor Corp, Hsinchu, Taiwan

Abstract—**In automotive applications, the battery voltage may drop to a low voltage of 4V when a sudden power-on occurs, and immediately rise to a high voltage of 40V during a loading pump. Thus, the paper presents a 4 to 40V wide input range high power LiDAR driver based on an enhancement Gallium Nitride (eGaN) switch. In addition, in the case of high di/dt, the proposed protection re-cycling (PRC) technique is used to re-direct the charge at the gate of the eGaN switch to a storage capacitor, which not only protects the GaN switch but also improves the efficiency by 5%. Experimental results show that the resultant long-distance object detection can be up to 300 meters.**

Keywords—Enhancement Gallium Nitride (eGaN), high di/dt, protection re-cycling (PRC) technique

I. INTRODUCTION

The use of adaptive cruise control (ACC) has become more popular in today's road vehicles to automatically control vehicles' speed. The battery voltage in automotive applications may drop to a low voltage of 4V when a sudden power-on occurs, and will immediately rise to a high voltage of 40V during a load dump. Therefore, precise light detection and ranging (LiDAR) systems are needed to maintain a safe distance from the vehicle ahead. In Fig. 1(a), the required detection specification in LiDAR systems hovers around 300 m for a 30 cm object, with a reflectivity of 10%, to ensure the safety of autonomous vehicles in high-speed driving [1]. Consequently, the required optical power ranges from 100W to 200W, and the pulse repetition frequency (PRF) is up to 1MHz to fill the front-facing field of vision. In other words, the laser pulse width is reduced to several nanoseconds to improve the detection precision and distance resolution, since a laser pulse t_p that is too wide will cause the reflections to overlap, as shown in Fig. 1(b) highlighted as 2. In contrast, well-controlled pulses that are narrow enough can allow the LiDAR system to more precisely identify each object. That is, fast switching switches and high switching and high power gate drivers are required to ensure proper operation of the LiDAR systems. Thus, eGaN switches are widely used to provide high switching operation. However, high switching operation will cause high dv/dt effect, which may induce large voltage spike at the gate terminal of eGaN switches. , Therefore, a safe and high switching gate driver is needed in eGaN switches.

The FET-controlled driver in Fig. 2(a) uses a large discharge capacitor C_{BUS}, which, compared to a laser pulse, can store much more charge that is required in every period [1] - [2]. Due to low voltage stress components, the peak laser diode current is limited in high power applications. Another capacitive discharge driver in Fig. 2(b) uses a smaller C_{BUS} and a correspondingly larger stray inductance L_{stray} [3]. Although the pulse shape is

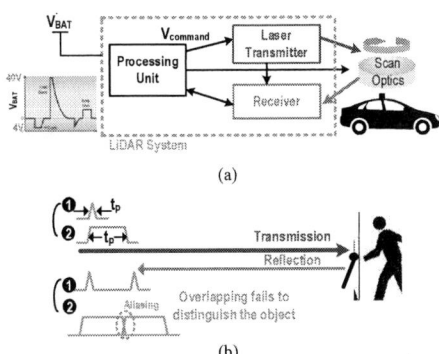

(a)

(b)

Fig. 1. (a) LiDAR system. (b) Operating principle of the LiDAR system.

uniform in every operation cycle, the capacitive discharge driver needs an extremely high V_{BUS} to generate a high driving current I_{LDPK}. High voltage swing and high power have become the challenges for fast turn-on of eGaN switches. Additionally, the gate-to-source (V_{GS}) of the eGaN switch needs to be less than 6V to avoid overstressing and damaging the eGaN switch. The coupling effect due to the high dv/dt in the case of turn-off eGaN switches, sets a trade-off between high dv/dt and high switching operation. In this paper, the proposed protection re-cycling (PRC) technique can alleviate the voltage stress at the gate by re-cycling the excess charges from the gate back to the input source for safety and efficient switching.

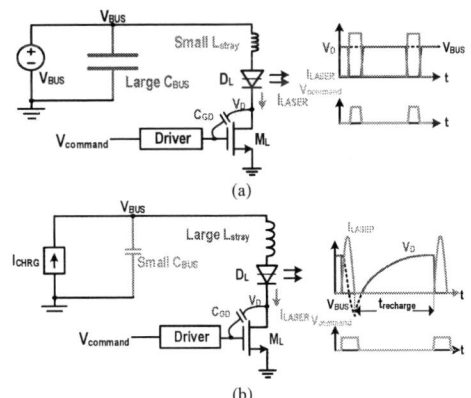

(a)

(b)

Fig. 2. Conventional laser diode drivers. (a) FET-controlled driver. (b) Capacitive discharge driver.

The organization of this paper is as follows. Section II shows the fundamental operation of the proposed LiDAR driver. To maintain correct supply for LiDAR systems, the wide input range boost (WIRB) power stage and the PRC technique are proposed. Circuit implementations are shown in Section III. Experimental results are shown in Section IV to prove this work. Finally, conclusions are made in Section V.

II. PROPOSED LiDAR DRIVER

In automotive applications, the battery voltage may drop to a low voltage of 4V when a sudden power-on occurs, and will immediately rise to a high voltage of 40V during a load dump. Therefore, the supply voltage for the LiDAR needs to tolerate a wide input range from 4 to 40V. Fig. 3 illustrates the proposed LiDAR driver with the WIRB power stage and the PRC technique for the low-side eGaN switch in high power LiDAR systems [4] - [7]. The WIRB power stage is a boost type and has more advantages compared with prior arts, as shown in Table I. In the case of wide input variation from 4 to 40V, the WIRB power stage can regulate and maintain high voltage V_{BUS} to ensure the correct operation of the LiDAR system. On the other hand, the PRC technique can protect the low side eGaN switch and re-cycle the coupling charges at the gate to improve efficiency.

A. Boost type WIRB power stage

In Fig. 4(a), during phase 1 ($\varphi1$) when S_1 and S_4 are on, the inductor current charging slope increases to $2V_{IN}/L$ to provide fast storing energy. During phase 2 ($\varphi2$) in Fig. 4(b), the inductor releases energy to the output, and the flying capacitor is recharged. The inductor current discharging slope is similar to the conventional slope, which is '($V_{IN}-V_{BUS}$)/L'. The maximum voltage across each switch is listed in Fig. 5. Compared to conventional designs, the advantage is that the maximum voltage stress is only V_{BUS}, which occurs in S_4 and S_5 during $\varphi2$ and $\varphi1$, respectively. In contrast, the maximum voltage stress of the dual-path step-up converter in [7] is raised to $2V_{OUT}-V_{IN}$. Moreover, the proposed WIRB power stage can use lower voltage components to reduce silicon area. With lower voltage components, the switching operation of lower voltage components can be faster than that of high voltage devices. In

conclusion, with the increase of the inductor current charging slope and the fast operation of lower voltage devices, the proposed WIRB power stage can rapidly react to the variation of input voltage from 4 to 40V.

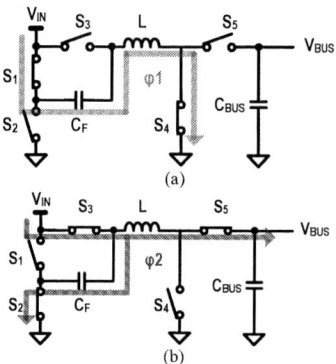

Fig. 4. Operation principle of the proposed WIRB power stage. (a) Phase $\varphi1$. (b) Phase $\varphi2$.

	S1	S2	S3	S4	S5
Φ1	0	V_{IN}	V_{IN}	0	V_{BUS}
Φ2	V_{IN}	0	0	V_{BUS}	0

Fig. 5. Max. voltage across each switch in proposed boost converter.

B. PRC technique

Since M_L turns off under high switching operation, the coupling effect through the parasitic gate-to-drain capacitance (C_{GD}) will cause large voltage spikes at the gate. Worst-case scenario is the abnormal turn on of the low side M_L. Another serious effect is the damage to the eGaN switch due to low overstress capability of the eGaN gate terminal. To effectively solve these critical problems, the proposed PRC technique is proposed to mainly redirect the injecting current from the gate to the storage capacitor C_{store} to reduce the coupling gate voltage and improve the reliability of low side eGaN switch M_L. Moreover, since high switching operation will induce large

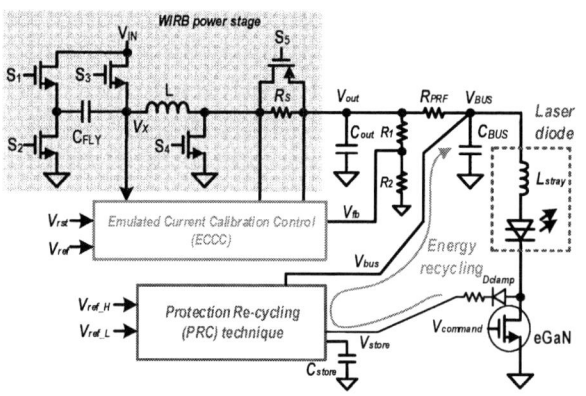

Fig. 3. Overall architecture of the proposed LiDAR driver with the PRC technique for wide input voltage range and recycling energy.

Table I: Comparison of boost type WIRB power stage and prior arts

	[5] ISSCC'20	[7] ISSCC'18	[6] ISSCC'21	[4] ISSCC'20	This Work
Operating mode	buck/boost	boost	boost	boost	WIRB
Control scheme	Current mode	Current mode	Adaptive Hysteresis Control	Voltage mode	Emulated Current Calibration Control
Input voltage(V)	2.5 - 5	2 - 4.2	6	2 - 4.4	4 - 40
Output voltage(V)	0.4 - 9	3 - 5	30	9 - 20	70
Inductor	2.2μH	4.7μH	4.7μH	2x3.3μH	4.7μH
Output Capacitor	4.7μF	10μF	4.7μF	4.7μF	4.7μF
Flying Capacitor	10μF	10μF	2.2μF	2*470nF	2.2μF
Conversion Ratio	1+D	$\frac{2-D}{2-2D}$	$\frac{2-D}{1-D}$	$\frac{2}{1-D}$	$\frac{1+D}{1-D}$
Ideal Conversion Ratio Range	1 - 2	1 - ∞	2 - ∞	4 - ∞	1 - ∞
Number of switch	5	5	3	6	5
Switching Frequency (MHz)	0.8 - 3	1	0.4 - 2.3	2	2
Maximum Voltage Stress	V_{IN}	$2V_{OUT}-V_{IN}$	$V_{OUT}-V_{IN}$	V_{OUT}	V_{OUT}
Conversion Ratio Limited	Yes	No	Yes	Yes	No

reverse diode current and noise, the emulated current calibration control (ECCC) circuit can reject interference through the emulated inductor current signal to improve the system reliability.

The control circuit implementation of the PRC technique is shown in Fig. 6, which is composed of three high-speed comparators and a voltage level shifter. The timing diagram of the PRC is shown in Fig. 7. When $V_{command}$ changes from low to high, I_{Laser} is raised up to drive the laser diode. V_{BUS} falls to about $2V_{LDF}-V_{BUS(steady)}$, where $V_{BUS(steady)}$ is the steady state of the V_{BUS} voltage, and I_{Laser} will approach zero but ringing will occur at $t=t_2$, where V_{LDF} is the forward voltage of the laser diode. Once the drain voltage V_D is larger than the threshold voltage V_{TH}, the high-speed comparator rapidly turns on S_{P1} to suppress the voltage stress on V_D, which is caused by large Ldi/dt. Block 1 uses the high-to-low $V_{command}$ signal to decide when to turn on S_{P1} to recycle the charge on the drain of the eGaN power FET to the C_{store}. The charge in the C_{store} can continue to increase after several LiDAR pulses. Block 2 uses two reference voltages, V_{Ref_H} and V_{Ref_L}, to decide whether the energy stored in the capacitor is enough to supply V_{BUS}. When V_{store} is higher than V_{Ref_H}, the comparator will trigger the SR latch to let V_{ER} go high, which means the capacitor has stored enough energy to achieve fast recovery of V_{BUS}. In addition, an AND gate is added to Block 2 to avoid S_{P1} and S_{P2} to turn on at the same time. After S_{P1} is turned off, S_{P2} is turned on to discharge C_{store} until V_{store} is lower than V_{Ref_L}.

Fig. 8 shows that the recycling energy is used to boost the V_{BUS}. During $\varphi1$ while S_{P1} is on, the excess charges are stored at the C_{store} in Fig. 8(a). On the other hand, during $\varphi2$, S_{P2} turns on to redirect the energy to the bond wire inductance L_{wire} in Fig. 8(b). When S_{P2} turns off, the stored energy is transferred back to the V_{BUS} for fast recovery.

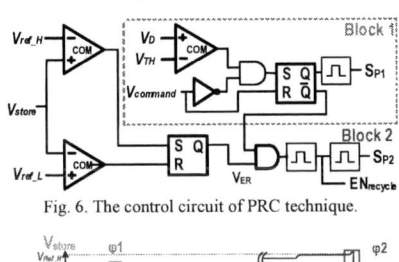

Fig. 6. The control circuit of PRC technique.

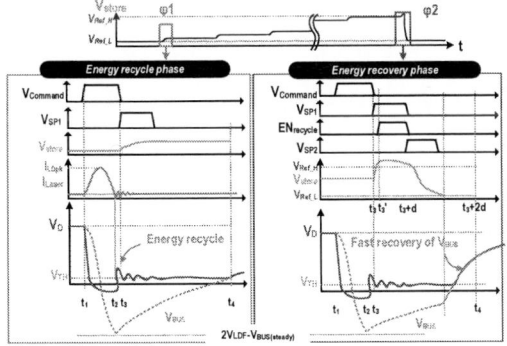

Fig. 7. Operation sequence of PRC technique. (a) Energy recycle phase. (b) Energy recovery phase.

III. CIRCUIT IMPLEMENTATIONS

In order to fast react to large input voltage variations, the ECCC circuit is proposed, which has the function of inductor current emulation for high noise immunity.

A. ECCC circuit

The ECCC circuit, shown in Fig. 9, can emulate the inductor current to reduce the noise sensitivity of the PWM circuit, allowing reliable control for the system. To emulate the inductor current, C_{Ramp} is determined by $(g_m*L)/(R_S*A)$, where g_m is the transconductance value of the current source, and A is the amplifier gain. Two sample-and-hold (S/H) circuits are used to store the current and the previous states of the inductor current through the on-resistance Rs of switch S_5. The emulated ramp signal is based on the pedestal voltage V_{SH_Pre}. The comparison result of the two S/H values (V_{SH} and V_{SH_Pre}) can adjust the calibration current I_{cal} from the current calibration circuit in Fig. 10 to charge the RAMP capacitor C_{Ramp}.

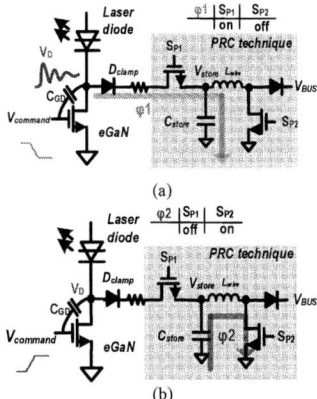

Fig. 8. (a) Operation of storage in the PRC. (b) Operation of energy transferring in the PRC.

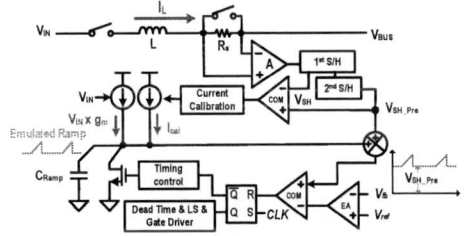

Fig. 9. Proposed ECCC circuit.

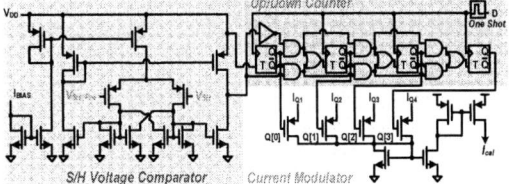

Fig. 10. Current calibration circuit.

978-1-6654-8495-4/22 $31.00 © 2022 IEEE

B. High Speed Comparator

Since the pulses of a laser diode driver always operate within a few nanoseconds, the implementing circuit must have fast transient capability to improve the efficiency and speed of the operation. Adjusting the output impedance at V_{N1}, the high-speed comparator can reduce propagation delay, as shown in Fig. 11. When the feedback voltage V_{fb} is higher than the reference voltage V_{ref}, V_{N1} drops and the current in M_9 increases. Contrarily, the voltage at V_{N2} increases, and M_{11} sources less current than that in M_{12}, which is mirrored from M_9. Therefore, the voltage of V_{GT} decreases. M_{14} works as a clamp, which can prevent V_{GT} from dropping too low, and it can source the current difference between M_9 and M_{11}. Then, V_{TB} and V_{RT} will drop if the current in M_{14} increases to supply enough current for M_9. Hence, the gate-source voltage of M_5 and M_6 is increased, which will decrease their R_{on} values and reduce the preamplifier gain. Thus, the voltage swing at N_1 decreases, which can reduce the current in M_9 and the propagation delay, due to the low voltage swing at V_{N1} and V_{N2}.

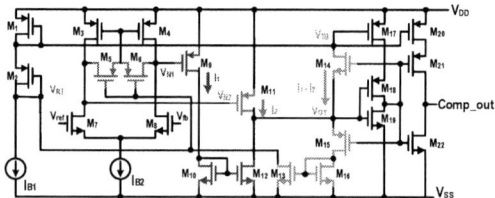

Fig. 11. Circuit implementation of the high-speed comparator.

IV. EXPERIMENATL RESULTS

The proposed boost converter with the PRC technique for high power LiDAR system is fabricated in a 0.25μm process, using the GaN device EPC 2016C, which the chip micrograph is shown in Fig. 12. The waveforms of V_{BUS} and V_D are shown in Fig. 13. When the LiDAR system operates, V_{BUS} and V_D will fall to -44V and 0V, respectively. Since the PRC technique suppresses V_D from 18.9V to 9.3V by redirecting the charge back to V_{BUS}, V_{BUS} is raised up rapidly to enable V_D to follow V_{BUS}. With the recycling energy, the peak efficiency can be enhanced by 5% from 85 % to 90% at full power condition.

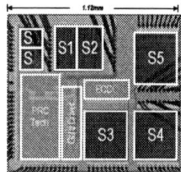

Fig. 12. Chip micrograph

Fig. 14(a) shows that V_{BUS} can be well-regulated at 80V by the ECCC technique. Fig. 14(b) shows the improved line regulation when V_{IN} changes from 4V to 40V, and vice versa, which meets the requirements of automotive applications ($\Delta V_{BUS}/V_{BUS}$<3%). Table II shows that the proposed LiDAR has the highest V_{BUS} voltage and driving current, while operating at 5MHz, which can ensure precise object detection. The peak efficiency is about 90%.

Table II: Comparison table with prior arts.

	[3] PCIM Europe'18	[1] APEC'18	[2] ISSCC'19	This Work
Process	N/A	180nm	55nm	0.25μm
Transistor Type	GaN	Customized LDMOS	GaN (EPC2040)	GaN (EPC2016C)
Laser Diode Driver Type	Capacitive Discharge	FET Controlled	FET Controlled	Capacitive Discharge
Pulse Repetition Frequency (PRF)	N/A	20MHz	200MHz	5MHz
Bus Voltage	75V	18V	10V	60V-70V
Peak of Laser Current Pulse	26A	5A	26A	40A
Peak efficiency	86%*	87%*	83%*	90%

*: SIMULATION REULTS.

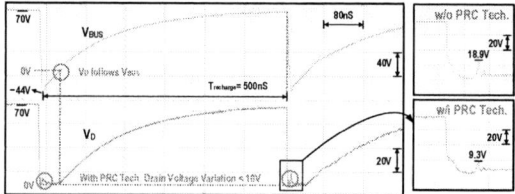

Fig. 13. Waveform of reduced V_D due to the PRC technique.

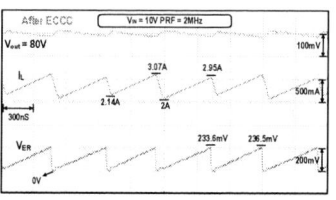

(a)

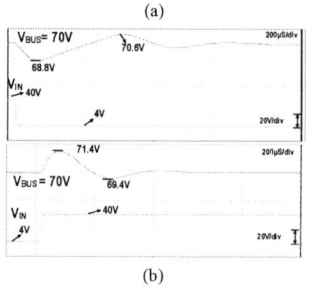

(b)

Fig. 14. (a) The waveform of the proposed WIRB power stage. (b) Regulated V_{BUS} in case of a wide input change from 4V to 40V.

V. CONCLUSIONS

This paper presents a 4 to 40V wide input range high power LiDAR driver to drive a low side eGaN switch. In the case of high di/dt, the proposed PRC technique is used to re-direct the charge from the gate of the eGaN switch to a storage capacitor, which not only protects the GaN switch but also improves the efficiency by 5%, as shown in the experimental results, and succeeds in precisely detecting objects in automotive applications.

REFERENCES

[1] E. Abramov, et al., *IEEE JESTPE*, pp. 3001-3013, Sep. 2020.

[2] Y.-S. Ma, et al., *IEEE ISSCC Dig. Tech. Papers*, pp. 466-467, Feb. 2019.

[3] J. Glaser, *PCIM Europe*, pp. 662-669, Jun. 2018.

[4] M. Huang, et al., *IEEE ISSCC Dig. Tech. Papers*, pp. 198-199, Feb. 2020.

[5] J. Baek, et al., *IEEE ISSCC Dig. Tech. Papers*, pp. 202-203, Feb. 2020.

[6] Y.-A. Lin, et al., *IEEE ISSCC Dig. Tech. Papers*, pp. 272-273, Feb. 2021.

[7] S.-U. Shin, et al., *IEEE ISSCC Dig. Tech. Papers*, pp. 430-431, Feb. 2018.

High-Voltage CMOS RF Switch with Active Biasing

Valentyn Solomko[1], Semen Syroiezhin[2,3], Danial Tayari[1], Jochen Essel[2] and Robert Weigel[3]

[1]Infineon Technologies AG, Neubiberg, Germany
[2]eesy-ic GmbH, Erlangen, Germany
[3]University of Erlangen-Nuremberg, Institute for Electronics Engineering, Erlangen, Germany
Email: valentyn.solomko@infineon.com

Abstract—**In this paper a stacked MOSFET (metal-oxide-semiconductor field effect transistor)-based RF (radio-frequency) switch comprising an active biasing circuit is demonstrated. The active biasing circuit regulates the DC voltage at the source-drain nodes of the stack, thus compensating for the self-biasing effect when the switch is exposed to a high-power RF signal. Such operating point regulation improves the robustness of the OFF-state switch against the applied RF voltage. Two high-voltage RF switches – a reference one with the passive resistive bias network and the proposed one comprising the active biasing – have been designed and implemented in a dedicated 130 nm bulk-CMOS (Complementary Metal-Oxide-Semiconductor) RF switch technology. The device with the active biasing circuit has demonstrated 6 V higher RF breakdown voltage comparing to the reference switch (62 V versus 56 V respectively). All other relevant RF parameters of the switch remain unchanged.**

Index Terms—**Antenna tuning, aperture tuning, compensation, high-voltage switch, leakage, MOSFET switch**

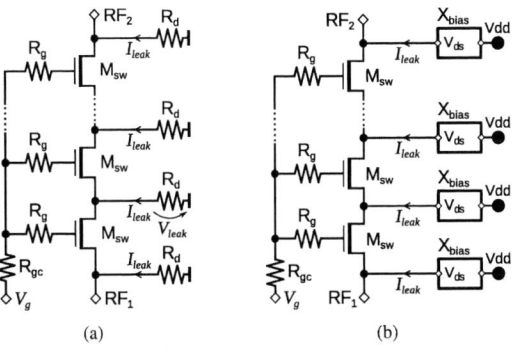

Fig. 1. Schematic diagrams of (a) – conventional stacked MOSFET-based RF switch with passive resistive bias network, (b) – proposed RF switch with active biasing.

I. INTRODUCTION

Antenna tuning circuits widely used in modern cellular handheld equipment are based on high-voltage RF switches. These devices may be exposed to RF voltages with the magnitude reaching 90 V when the maximum transmitted power is applied to built-in antennas.

Commercial switches for antenna tuning applications are typically implemented in dedicated SOI- or bulk-CMOS technologies on high-ohmic silicon substrates. They are constructed from multiple MOSFETs stacked in series to achieve high voltage handling in OFF-state. The amount of stacked FETs depends on the breakdown voltage of a single transistor and typically ranges between 15 and 35 for antenna tuners in state-of-art fabrication processes. A widespread switch topology, comprising stacked transistors M_{sw}, gate and drain-source resistive biasing networks implemented with high-ohmic resistors R_g–R_{gc} and R_d, is shown in Fig. 1(a).

Achieving the target voltage handling in a highly-stacked MOSFET switch is associated with a number of challenges and trade-offs. One important challenge is the minimization of a self-biasing effect, which adversely influences the voltage handling. This effect occurs at high RF excitation voltages and is caused by the gate-induced drain leakage current (GIDL) in MOS transistors. This work focuses on addressing this challenge.

II. PROBLEM STATEMENT AND EXISTING SOLUTIONS

It is well known that an NMOS transistor operating in a deep cut-off region (meaning when the gate-source voltage is well below the threshold level) exhibits leakage current flowing from the drain and source terminals into the body if the body is more negatively biased than the source/drain terminals [1], [2]. Such bias condition typically occurs in OFF-state MOSFET switches. The GIDL current exponentially grows with the applied negative gate-source voltage, which consists of the bias voltage and the RMS (root mean square) RF voltage across the switch. The current I_{leak} in Fig. 1(a) denotes the mentioned leakage. In general, the leakage current may differ along the stack depending on the uniformity of RF voltage division in stack.

When flowing through the high-ohmic bias resistors R_d, the current I_{leak} shifts the DC voltages at drain and source nodes to more negative values, thus increasing the gate-source bias voltage of the transistors. Such change in the operating point results in reduction of drain-source AC breakdown voltage of individual transistors in stack, which in turn reduces the overall voltage handling capability of the switch.

Significant research effort has been devoted to address the leakage current problem in analog CMOS switches. Among the solutions reported in literature are adding a negative replica current into the leaky nodes generated by matched devices

978-1-6654-8495-4/22 $31.00 © 2022 IEEE

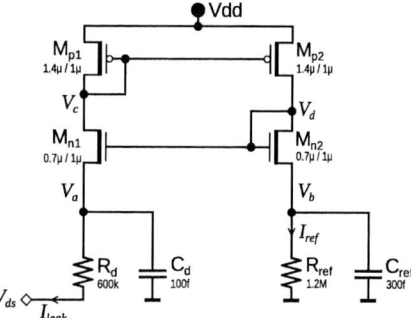

Fig. 2. Schematic of the active biasing block X_{bias} in Fig. 1(b).

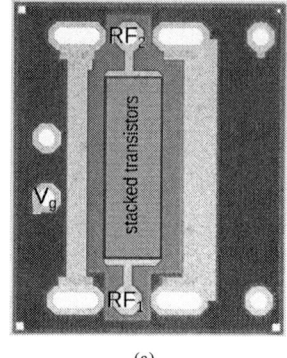 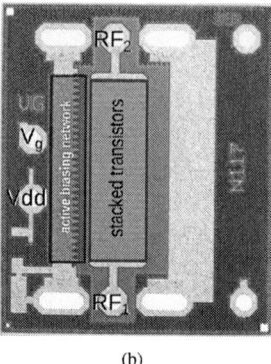

(a)   (b)

Fig. 3. Die photographs of the fabricated SPST RF switches with (a) – conventional resistive biasing and (b) – proposed active biasing.

[3], body-guarding technique [4] and others. Unfortunately all these solutions can hardly be applied to RF switches due to excessive complexity when individually applied to each of many transistors in stack, unacceptably high current consumption and linearity degradation. Bias resistance minimization or inductive biasing [5] substantially increas the power losses in the switch and reduce its operating frequency range.

To the best of authors' knowledge there are no solutions in research literature reporting circuit-based leakage compensation techniques for linear MOSFET-based RF switches. This work is aimed at filling the gap, proposing and verifying in hardware a circuit for compensation of self-biasing effect.

III. RF SWITCH WITH ACTIVE BIASING

A top level circuit diagram of the RF switch with active biasing is presented in Fig. 1(b). The bias resistors at the source and drain nodes are substituted by an active circuit element X_{bias} representing the core of proposed solution. On a functional level, the X_{bias} cell is forcing the source/drain nodes to sustain zero DC voltage while allowing the leakage current I_{leak} to flow into the MOSFET when operating at large RF excitation conditions.

The circuit implementation of X_{bias} is shown in Fig. 2. As in the conventional switch, the high-ohmic resistor R_d remains an interface element to the stack, guaranteeing the high linearity and providing the same small-signal impedance at RF frequencies to the switch transistors. The other terminal of R_d is coupled to a four-transistor voltage mirror formed by the NMOS and PMOS pairs $M_{n1} - M_{n2}$ and $M_{p1} - M_{p2}$. The four-transistor cell is known in the art and can be found in textbooks [6].

The circuit equalizes the voltages V_a and V_b by forcing matched currents through R_d and R_{ref} generated by the current mirror pair $M_{p1} - M_{p2}$. When I_{leak} and I_{ref} are scaled proportionally to the respective resistors, then $V_{ds} = 0$ V because of the same voltage drop across R_d and R_{ref}. The shunt capacitor C_d provides an AC ground at the node V_a to protect the transistors from excessive RF voltage stress and suppress any nonlinear distortion from coupling into the switch branch. Leakage current in the left branch and R_{ref} in the right

one serve as the mechanism/path for slow discharge of shunt capacitors after removing the applied RF excitation.

IV. HARDWARE PROTOTYPES

Two integrated circuits comprising high-voltage SPST (single pole single throw) switches have been designed and fabricated in a dedicated 130 nm bulk-CMOS RF switch technology from Infineon. The demonstrators are aimed at functional verification and performance evaluation of the proposed concept, particularly the impact of active biasing circuit on linearity and breakdown voltage of the switch. The photographs of fabricated ICs are shown in Fig. 3, wherein one circuit comprises the proposed active biasing network and the other one is implemented in a conventional way with all-passive biasing network serving as a reference for comparison. Both switches are designed to achieve around 60 V RF voltage handling.

The circuits were implemented according to schematic diagrams in Fig. 1. The stack size is N = 28, all switch transistors M_{sw} and gate bias resistors are configured as $W_{sw} / L_{sw} = 7$ mm / 120 nm, $R_g = 300$ kΩ. Such transistor has an OFF-state capacitance of $C_{off} = 2.45$ pF and can handle up to $V_{ds.max} = 2.8$ V RF voltage applied across the channel when biased at $V_{gs} = -2.2$ V. According to simulations, the total leakage current into the source and drain terminals of the transistor operating at such peak excitation conditions is 2.2 μA. In the conventional switch with passive biasing, Fig. 1(a), the leakage through R_d shifts the gate-source bias point and reduces the breakdown voltage per transistor down to around $V_{ds.max.leak} = 2.5$ V, which is 0.3 V lower than $V_{ds.max}$.

The active biasing circuit implemented as shown in Fig. 2 is connected to each drain/source node along the stack in Fig. 1(b), resulting in total 29 cells for the complete switch. The parameters of used devices are provided in the circuit diagram.

The maximum leakage current that can be compensated by the active biasing circuit is $I_{leak.max} = V_{a.max}/R_d$, where $V_{a.max} = V_{dd} - V_{t.Mp1} - V_{ov.Mp1} - V_{ov.Mn1}$ defines the

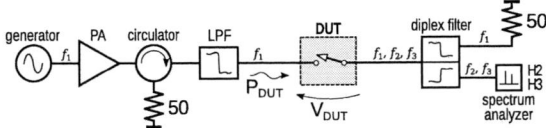

Fig. 4. Harmonic distortion measurement setup used for evaluation of the fabricated prototype ICs.

maximum voltage at the node V_a for which M_{p1} and M_{n1} are operating in saturation region at the drain current of $I_{leak.max}$. The $V_{t.Mp1}$, $V_{ov.Mp1}$ and $V_{ov.Mn1}$ are the threshold and overdrive voltages for the respective transistors. According to simulations, at least 3 μA of leakage can be compensated when the circuit is supplied from $V_{dd}=3.3$V at nominal process corner, which is enough to compensate for 2.2 μA leakage mentioned in previous paragraph.

In order to quantify the expected improvement in RF voltage handling of the switch with active biasing, a difference in breakdown voltages for switches operating in shunt mode (wherein one terminal is grounded and the other one is exposed to RF) is calculated according to analysis presented in [7]:

$$\Delta V_{rf.bd} = $$
$$= \sqrt{\frac{C_{off}}{C_{sub}}} \frac{\sinh\left(2N\alpha\right)}{\cosh\left(2N-1\right)\alpha} \left(V_{ds.max} - V_{ds.max.leak}\right), \quad (1)$$

where $\alpha = \operatorname{arsinh}\left(\frac{1}{2}\sqrt{C_{sub}/C_{off}}\right)$, C_{off} and C_{sub} are the OFF-state drain-source capacitance and substrate capacitance of a single transistor in stack.

The methodology for numerical estimation of C_{sub} has been presented in the authors' earlier paper [8]. Assuming the face-up orientation of the IC in measurements (equivalent to wire-bonded assembly configuration considered in [8]) and layout dimensions of $200 \times 22\,\mu\text{m}^2$ for a single transistor in stack, the substrate capacitance is calculated as $C_{sub}=2.6$ fF. After substituting the values into (1), the difference in breakdown voltages between the designed switches is obtained to be $\Delta V_{rf.bd}=6.7$ V.

V. MEASUREMENT RESULTS

The fabricated SPST testchips (with passive and active biasing) have been measured on wafer for RF breakdown voltage. The measurements have been conducted in open-reflection configuration, where the excitation signal has been applied at one pole of the OFF-state switch and the 2-nd and 3-rd order harmonic distortions have been measured at the other pole of the switch. The measurement setup is demonstrated in Fig. 4 (for more details the reader is referred to [9]). The active biasing circuit has been supplied from an external 3.3 V voltage source.

The measured harmonics curves for the OFF-state switches are demonstrated in Fig. 5(a). The excitation RF signal power P_{DUT} at 900 MHz is swept between 30 dBm and 40 dBm as indicated on the lower x-axis. The corresponding RF voltage magnitude V_{DUT}, considering open-reflection load conditions

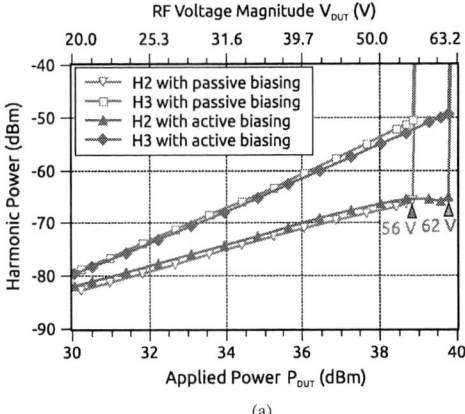

(a)

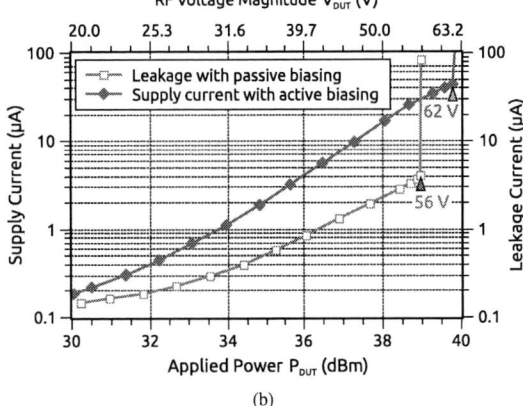

(b)

Fig. 5. Measured SPST switches (a) – harmonic distortions, (b) – leakage and current consumption.

for the applied RF signal, is listed on the upper axes of the graph and is calculated as:

$$V_{DUT} = 2 \cdot 10^{\frac{P_{DUT}-10}{20}}. \quad (2)$$

The equation (2) holds true because the RF voltage at the output of the switch IC is substantially lower (factor 30) than the voltage at the input. Under such excitation conditions the DUT operates in quasi-shunt mode assumed for voltage handling calculation in Section IV.

The measured second (H2) and third (H3) harmonics at 1800 MHz and 2700 MHz respectively are plotted in Fig. 5(a). The abrupt increase in harmonics level indicates the break-down point for the switch. The proposed switch with active biasing achieves the breakdown level of 62 V, which is 6 V higher than the reference device with passive biasing, reaching 56 V peak RF voltage handling. The difference of 6 V is slightly lower than the predicted $\Delta V_{rf.bd}=6.7$ V, which can be attributed to higher-order inaccuracies in the calculation model, particularly in estimation of C_{sub} which is influenced by the metal ground frame surrounding the switch in the layout.

978-1-6654-8495-4/22 $31.00 © 2022 IEEE

The improvement is achieved at no penalty in linearity of the switch as indicated by almost overlapping H2 and H3 harmonics curves. The total leakage current of the switch with passive biasing as well as the current consumption of the active-biased switch are plotted in Fig. 5(b). At low excitation RF power levels the leakage current as well as the supply current of active biasing are approaching sub-μA range. At peak RF drive voltage the supply current of the active biasing network reaches $40\,\mu$A, which is a price paid for the improved RF voltage handling capabilities.

As expected, both switches demonstrate the identical measured isolation in OFF-state (not shown in the paper), indicating no penalty in the small-signal response of the switch with active biasing.

VI. CONCLUSION

A high-voltage MOSFET-based RF switch with active biasing for compensating the negative influence of gate-induced drain leakage on voltage handling is presented in this paper. The measurements on hardware prototypes prove the efficiency of the proposed approach, demonstrating 10% improvement in breakdown voltage for the 60 V-class devices. Measurements also show that the improvement is achieved at no penalty in other important RF parameters, particularly in the linearity of the switch.

The proposed active biasing may complement other state-of-art techniques utilized in stacked switches (for example, RF voltage equalization [10]), thus further improving the performance of such devices. The proposed switches with active biasing can be utilized in any application requiring high-voltage handling capabilities, particularly in antenna tuning switches for cellular handheld devices.

REFERENCES

[1] A. Adan and K. Higashi, "Off-state leakage current mechanisms in bulk-Si and SOI MOSFETs and their impact on CMOS ULSIs standby current," *IEEE Transactions on Electron Devices*, vol. 48, no. 9, pp. 2050–2057, 2001.

[2] S. M. Sze, *Physics of Semiconductor Devices*. New York: Wiley, 1981.

[3] L. Wong, S. Hossain, and A. Walker, "Leakage current cancellation technique for low power switched-capacitor circuits," in *ISLPED'01: Proceedings of the 2001 International Symposium on Low Power Electronics and Design (IEEE Cat. No.01TH8581)*, 2001, pp. 310–315.

[4] J. J. Su, K. S. Demirci, and O. Brand, "A low-leakage body-guarded analog switch in 0.35-μm BiCMOS and its applications in low-speed switched-capacitor circuits," *IEEE Transactions on Circuits and Systems II: Express Briefs*, vol. 62, no. 10, pp. 947–951, 2015.

[5] P. Katzin, B. E. Bedard, M. B. Shifrin, and Y. Ayasli, "High-speed, 100+ W RF switches using GaAs MMICs," *IEEE Transactions on Microwave Theory and Techniques*, vol. 40, no. 11, pp. 1989–1996, 1992.

[6] P. R. Gray, P. J. Hurst, S. H. Lewis, and R. G. Meyer, *Analysis and Design of Analog Integrated Circuits*. New York: Wiley, 2001.

[7] Y. Zhu, O. Klimashov, and D. Bartle, "Analytical model of voltage division inside stacked-FET switch," in *2014 Asia-Pacific Microwave Conference*, 2014, pp. 750–752.

[8] V. Solomko, O. Oezdamar, R. Weigel, and A. Hagelauer, "Model of substrate capacitance of MOSFET RF switch inspired by inverted microstrip line," in *2021 European Solid-state Circuits and Devices Conference (ESSCIRC)*, 2021, pp. 207–210.

[9] O. Özdamar, R. Weigel, A. Hagelauer, and V. Solomko, "Linearity analysis of high-voltage RF switches for antenna tuning applications," *IEEE Transactions on Microwave Theory and Techniques*, pp. 14–23, 2021.

[10] O. Oezdamar, R. Weigel, A. Hagelauer, and V. Solomko, "Considerations for harmonics distribution in aperture-tuned inverted-F antenna for cellular handheld devices," *IEEE Transactions on Microwave Theory and Techniques*, vol. 68, no. 10, pp. 4122–4130, 2020.

Switching Time Acceleration for High-Voltage CMOS RF Switch

Semen Syroiezhin[1,3], Oguzhan Oezdamar[2], Robert Weigel[3] and Valentyn Solomko[2]

[1]eesy-ic GmbH, Erlangen, Germany
[2]Infineon Technologies AG, Neubiberg, Germany
[3]University of Erlangen-Nuremberg, Institute for Electronics Engineering, Erlangen, Germany
Email: semen.syroiezhin@eesy-ic.com

Abstract—**A high-voltage MOSFET-based RF switch with improved switching time is presented in this paper. The improvement is achieved by adding an auxiliary circuitry distributed along the stack which substantially speeds up the charging and discharging of gate oxide of the transistors. The auxiliary network is enabled by a delay-based control circuit defining acceleration time-window for switching transient. An RF switch comprising the proposed solution has been implemented in a dedicated 65 nm CMOS switch technology. The measured hardware demonstrates the improvement in switching time from 19.2 μs to 1.6 μs in OFF-to-ON direction and from 0.6 μs to 0.2 μs in ON-to-OFF direction compared to the state-of-art implementation. The improvement is achieved at no penalty in key RF characteristics of the device. Particularly, both conventional and proposed switches are able to withstand up to 48 dBm RF power in OFF-state and demonstrate identical small- and large-signal response.**

Index Terms—**Antenna tuning, cellular user equipment, fast switching, harmonics, high-voltage switch, switching speed, switching time acceleration.**

I. INTRODUCTION

High-voltage, high-linearity switches in dedicated CMOS technologies find a use in various RF front-end applications such as adaptive antenna tuning [1], power amplifiers load modulation [2], T/R switching, to mention only a few.

A particular case of such device is a shunt switch with the schematic diagram shown in Fig. 1. It consists of multiple MOSFETs stacked in series and a bias network. The high-ohmic resistive interconnects between gate and body (not shown) terminals as well as drains and sources of the transistors in stack provide the ability for the switch to achieve low power losses, high linearity and RF voltage handling. These resistors together with gate-channel and metal interconnect capacitances of the transistors create a time constant which delays switching events and cannot be decreased by a straightforward resizing of the components without substantial penalty in important RF parameters like linearity, insertion loss, bandwidth and others.

Several approaches for overcoming the aforementioned design tradeoff have been reported in literature. The inductor-based biasing network proposed in [3] ultimately solves the switching speed problem but suffers from severe power loss, linearity and bandwidth degradation, not mentioning the unacceptable silicon area utilization when applied to high-stacked switch devices. The solution presented in [4] is very efficient

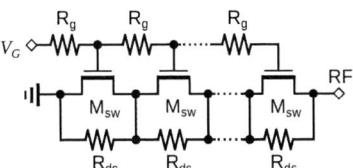

Fig. 1. Schematic diagram of the state-of-art RF switch.

in terms of chip area usage, however reduces the switching time by some 50%, which might not be enough for many demanding applications. Bypassing the bias resistors with a full-stacked switch works efficiently only for low-voltage RF switches [5]. The circuit demonstrated in [6] has limitations when changing the state into OFF direction and introduces nonlinear distortion to the switch.

In this work we propose a switching time acceleration circuit which operates with wide range of positive and negative bias voltages, occupies reasonably small silicon area and sustains low level of harmonic distortions and power losses in the high-stacked MOSFET switch.

II. PROPOSED SWITCH WITH ACCELERATION

The proposed topology comprises two additional switch branches which short high-ohmic bias resistors during switching event and substantially reduce the RC time constant of the shunt switch. These branches are coupled to control terminals V_N and V_P as shown in Fig. 2(a). The V_P stays normally at high voltage corresponding to ON-state of the transistor while V_N is biased with the negative voltage, which allows keeping all auxiliary devices in OFF-state. The V_P branch and all connected to it PMOS devices serve as the main charge delivery path for gates of the switch.

The time diagram in Fig. 2(b) demonstrates voltage transients at the nodes for corresponding switching events. All bias resistors coupled to V_P node get shorted during switching from OFF- to ON-state because of the positive V_N pulse and consequent transition of the NMOS auxiliary devices N_1 into ON-state. The negative pulse at V_P during ON $\rightarrow$ OFF switching phase again shorts all auxiliary gate resistors R_p by means of N_1.

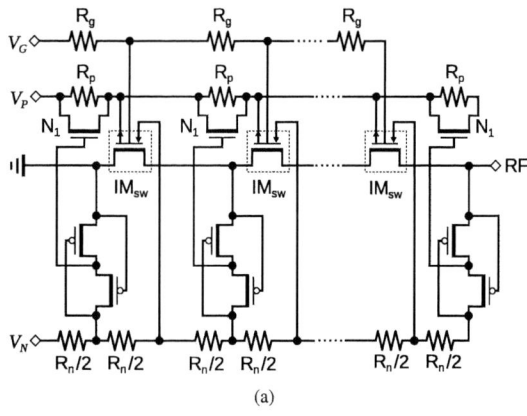

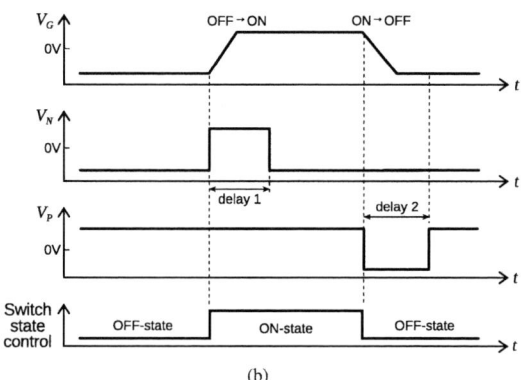

Fig. 2. (a) – Schematic diagram of the RF switch with proposed switching time acceleration, (b) – corresponding time diagram.

The detailed circuit diagram of the subcells IM_{sw} is shown in Fig. 3. Despite the complexity of the circuit, by far the largest device is the main switch transistor M_{sw} while the other transistors are sized almost at minimum and insignificantly contribute to the total switch silicon area. The gate recharging of M_{sw} is performed via N_3 and either P_2 (for switching ON) or P_3 (for switching OFF). The positive pulse at V_N increases the gate-source voltage of N_3 and source-gate voltage of P_2, which establish the recharging path. During the negative pulse at V_P the switching of the main device to OFF state occurs through N_3 and diode connected device P_3. The transmission gate made of N_2 and P_1 provides the low-ohmic path to the channel of M_{sw}. Both N_3 (controlled by the terminal V_n) and the diode-connected P_3 devices allow unidirectional charging. The rising edge of V_P and falling edge of V_N do not discharge the main device M_{sw}.

III. Hardware Demonstrator Design

In order to prove functionality and demonstrate performance benefits of the proposed solution a high-voltage RF switch incorporating the switching time acceleration circuit has been designed in a dedicated 65 nm CMOS switch technology on a high-ohmic substrate. The acceleration branch is arranged in exact accordance to the schematics described in Section II.

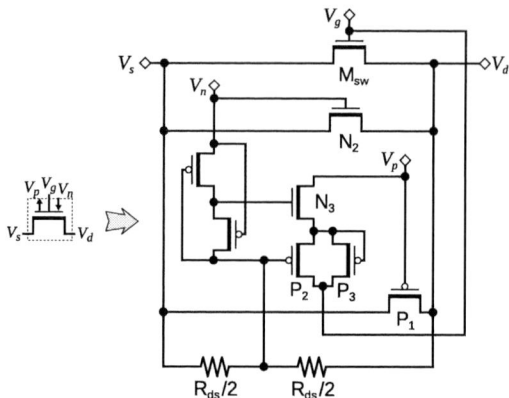

Fig. 3. Schematic diagram of the cell IM_{sw} in Fig. 2(a).

The switch consists of 26 identical stacked elements. The size of the main switch transistor M_{sw} is $W_{Msw}/L_{Msw} = 7.3\,\text{mm}/0.15\,\mu\text{m}$. It occupies around 92% of the total stack area and is by far the largest device in switch. The bias resistors are configured as follows: $R_g = 13\,\text{k}\Omega$, $R_p = R_n = R_{ds} = 20\,\text{k}\Omega$.

Among the performance-critical transistors in the switching time acceleration circuit are bypass devices N_1 sized as $W_{N1}/L_{N1} = 30\,\mu\text{m}/0.15\,\mu\text{m}$, P_1 sized as $W_{P1}/L_{P1} = 20\,\mu\text{m}/0.26\,\mu\text{m}$ and N_2 sized as $W_{N1}/L_{N1} = 10\,\mu\text{m}/0.15\,\mu\text{m}$. Their width defines the total resistance of the charge/discharge path for the gate oxide of M_{sw}. The other transistors are configured to have minimum or close to minimum channel size and are considered to be uncritical for the overall functionality and performance of the switch.

The auxiliary control voltage pulses V_P and V_N sketched in the time diagram of Fig. 3(b) are generated by a CMOS inverter-based delay cell (not discussed in detail due to its straightforward implementation) and have the pulse width of $1.5\,\mu\text{s}$ at the nominal PVT (process-voltage-temperature) corner.

For the sake of comparison a state-of-art switch according to Fig. 1 has been designed and placed on the same die as a reference. The switch transistor M_{sw} has identical size as its counterpart in switching time acceleration implementation, namely $W_{Msw}/L_{Msw} = 7.3\,\text{mm}/0.15\,\mu\text{m}$. The values of bias resistors are set as: $R_g = R_{ds} = 10\,\text{k}\Omega$. They are more low-ohmic than the respective bias resistors in the acceleration switch. The switches are designed to handle at least 80 V peak RF voltage in OFF-state.

The floorplan of hardware demonstrator is arranged as shown in the fabricated die photograph in Fig. 4. The chip comprises four switch throws coupled between the RF pins and a common ground as well as the power management and control unit designed to provide required control voltages for the switches. The device is programmed by means of a 2-wire RFFE MIPI interface. The design details of the power management and control units are outside the scope of this

978-1-6654-8495-4/22 $31.00 © 2022 IEEE

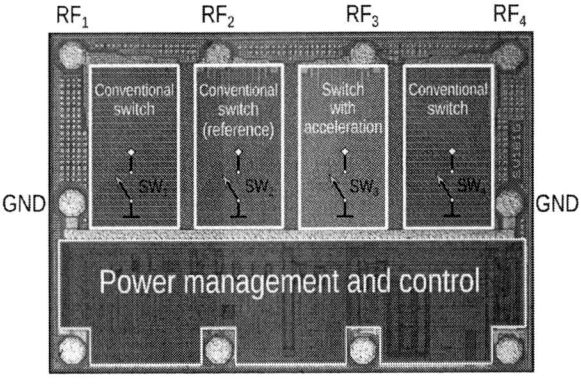

Fig. 4. Die photograph.

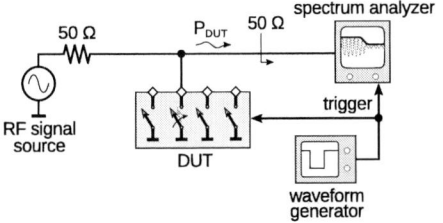

Fig. 5. Measurement setup for switching time evaluation.

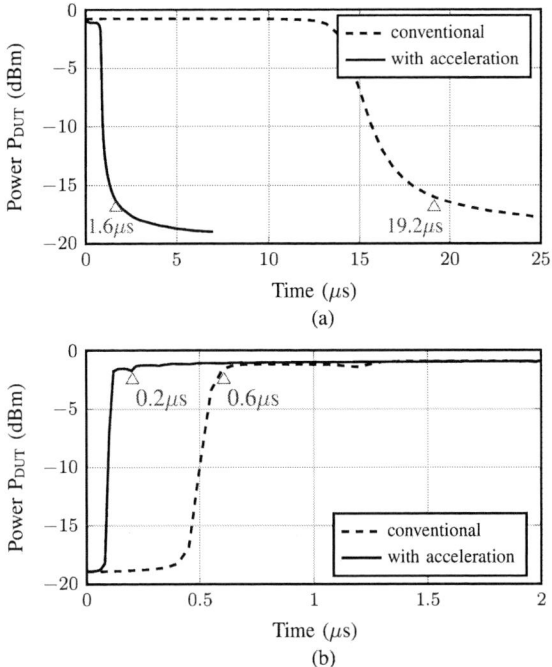

Fig. 6. Measured power envelope of RF signal P_{DUT} for (a) – OFF→ON switching transition and (b) – ON→OFF switching transition.

work and are not discussed further in the paper.

The shunt switch SW_3 attached to RF_3 comprises the acceleration circuit, while all other three switches are designed in a conventional way. The neighboring switch SW_2 on the left from SW_3 is chosen as a reference for measurements for the reason of layout symmetry, which shall minimize any deviations in small- and large-signal characteristics as a factor of switch placement in the layout.

IV. MEASUREMENT RESULTS

The fabricated integrated circuit has been flip-chip bonded into a $1.5 \times 1.1 \, mm^2$ 10-pin leadless plastic package and mounted on a printed-circuit board for measurements. For switching time evaluation the DUT (device under test) has been attached to a $50 \, \Omega$ thru line on the board. The measurement setup is portrayed in Fig. 5.

A continuous-wave RF excitation signal at 824 MHz and 0 dBm has been applied to the thru line coupled to a spectrum analyzer. The spectrum analyzer has been operating is zero-span mode, recording the power envelope of the signal at 824 MHz in time domain. The DUT and the spectrum analyzer have been simultaneously triggered by a waveform generator to synchronize the 'change-state' command event and start of the power envelope recording. An uncertainty in the range of 10 ns related to control circuit reaction time is disregarded as being well below the switching time of designed DUT.

The obtained envelope curves in both switching directions are shown in Fig. 6. The 0 second time point represents the 'change-state' command event and trigger signal for the

spectrum analyzer. In OFF-state of the switches the recorded power is –0.75 dBm. The insertion loss of 0.75 dB corresponds to power losses in the board thru line as well as power and mismatch losses caused by the DUT. In the OFF-state the settled power is –19 dBm, dropped primarily due to the strong mismatch loss in the thru line shorted by the shunt switch to ground.

In this work we consider the switch to reach OFF-state when the signal power crosses the level of 1 dB below the settled OFF-state power, meaning –1.75 dBm. In the other direction the shunt switch is considered to be ON when the measured power is 3 dB above the final value, namely when the power envelope crosses –16 dBm point. The markers placed at the respective power levels indicate the switching time for both conventional and proposed fast switches. In the OFF→ON direction the proposed solution reaches –16 dBm level in 1.6 μs, while the conventional switch requires 19.2 μs to get into the ON-state, resulting in improvement of switching time by the factor of 12. In the other direction the proposed switch with acceleration reaches the OFF-state three times faster, in just 0.2 μs versus 0.6 μs for the conventional implementation.

The nonlinear performance of the switches have been evaluated to prove that the switching time acceleration circuit added to the intrinsic stack does not degrade the linearity of the switch. The third harmonic in a one-tone test serves as a sensitive indicator of the nonlinear performance. The harmonic measurement setup on a functional level is somewhat similar to the one shown in Fig. 5, where in the spectrum analyzer

978-1-6654-8495-4/22 $31.00 © 2022 IEEE

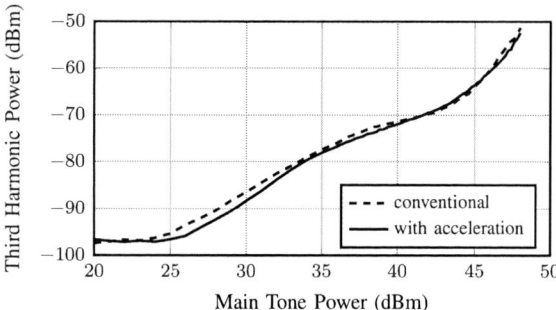

Fig. 7. Measured third harmonic of the OFF-state switch DUTs.

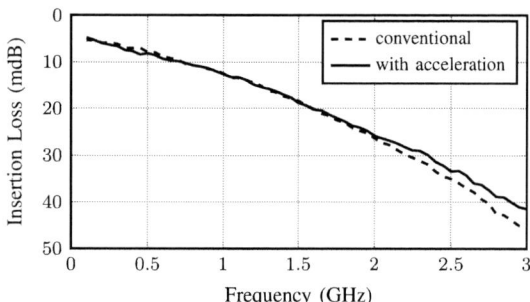

Fig. 8. Measured insertion loss of the OFF-state switches.

TABLE I
PERFORMANCE COMPARISON OF HIGH-VOLTAGE RF SHUNT SWITCHES

Parameter	[6]	[4]	[8]	this work
CMOS node	45 nm	130 nm	N.A.	65 nm
Power and peak voltage handling	30 dBm 9.5 V	40 dBm 32 V	48.4 dBm 83 V	48 dBm 80 V
IL at 900 MHz	N.A.	0.14 dB	0.012 dB	0.013 dB
H3@P_{DUT} = 26 dBm	N.A.	−85 dBm	−95 dBm	−96 dBm
Switching time	1 ns	6 μs	15 μs	1.6 μs

state-of-art design with no penalty in key RF characteristics. A basic performance comparison of the designed switch with shunt devices of the same class is presented in Table I. The 1 ns switching time reported in [6] is achieved at the expense of voltage handling and linearity (even thought it is not quantified in the paper).

The ability to settle the gate voltage to within one threshold voltage of a MOS device can be seen as a drawback of proposed solution in technologies with the ratio between the threshold voltage and the target gate bias voltage exceeding 5%. A programmable biasing circuit can be used as a workaround in this case. The target bias voltage will be increased by 10−20% in the switching phase, which can partially compensate the drawback of the circuit.

REFERENCES

[1] V. Solomko, O. Oezdamar, R. Weigel, and A. Hagelauer, "Penta-band inverted-F antenna tuned by high-voltage switchable RF capacitors," *IEEE Open Journal of Antennas and Propagation*, vol. 2, pp. 375–384, 2021.
[2] F. Carrara, C. D. Presti, F. Pappalardo, and G. Palmisano, "A 2.4-GHz 24-dBm SOI CMOS power amplifier with fully integrated reconfigurable output matching network," *IEEE Transactions on Microwave Theory and Techniques*, vol. 57, no. 9, pp. 2122–2130, 2009.
[3] P. Katzin, B. Bedard, M. Shifrin, and Y. Ayasli, "High-speed, 100+W RF switches using GaAs MMICs," *IEEE Transactions on Microwave Theory and Techniques*, vol. 40, no. 11, pp. 1989–1996, 1992.
[4] V. Solomko, O. Oezdamar, R. Weigel, and A. Hagelauer, "CMOS RF switch with fast discharge feature," *IEEE Solid-State Circuits Letters*, vol. 4, pp. 68–71, 2021.
[5] N. Ilkov, "Switch using an accelerating element," Patent US7 868 683B2.
[6] C. S. Levy, P. M. Asbeck, and J. F. Buckwalter, "A CMOS SOI stacked shunt switch with sub-500 ps time constant and 19-Vpp breakdown," in *2013 IEEE Compound Semiconductor Integrated Circuit Symposium (CSICS)*, 2013, pp. 1–4.
[7] O. Oezdamar, R. Weigel, A. Hagelauer, and V. Solomko, "Considerations for harmonics distribution in aperture-tuned inverted-F antenna for cellular handheld devices," *IEEE Transactions on Microwave Theory and Techniques*, vol. 68, no. 10, pp. 4122–4130, 2020.
[8] SKY19237-001: Tripple-throw shunt high-voltage switch datasheet. Accessed: Apr. 1, 2022. [Online]. Available: https://www.skyworksinc.com/en/Products/switches/SKY19237-001

is tuned to measure the third harmonic of the applied signal after DUT. For more details on the harmonics measurement setup the reader is referred to [7].

The DUT in OFF-state has been excited by a 900 MHz signal with the power swept between 20 dBm and 48 dBm (the 48 dBm power in 50 Ω load impedance corresponds to 80 V peak RF voltage across the DUT). The third harmonic power at 2700 MHz has been recorded and plotted in Fig. 7. As can be observed from the graphs, both curves are almost overlapping, indicating that the switch with acceleration circuit does not adversely impact the linearity of the device. Some minor difference between the curves lies within the measurement and fabrication tolerances for different instances of assembled DUTs.

The small-signal performance of the switches have also been evaluated. Particularly, the insertion loss curves in OFF-state are demonstrated in Fig. 8. The curves are running very close to each other, with the accelerated switch exhibiting slightly lower losses at high frequencies, which might partly be attributed to higher total shunt resistance of the biasing network. In ON-state both switches behave identically, demonstrating the resistance of 2.15 Ω at low frequencies.

V. CONCLUSION

This work proposes and experimentally verifies the switching time acceleration circuit for high-voltage CMOS-based RF switches. According to measurements, the proposed solution demonstrates up to 12 times faster switching speed than the

A Low-Loss 77 GHz Sub-Sampling Passive Mixer Integrated in a 28-nm CMOS Radar Receiver

Alexandre Flete[#*1], Christophe Viallon[*2], Philippe Cathelin[#3], Thierry Parra[*4]

[#] STMicroelectronics, France
[*] LAAS-CNRS, Université de Toulouse, CNRS, UT3, Toulouse, France
[1]alexflete@sfr.fr, [2]cviallon@laas.fr, [3]philippe.cathelin@st.com, [4]parra@laas.fr

Abstract — This paper demonstrates that using sub-sampling mixer topology can be the right choice to implement a millimeter-wave (mmW) receiver. In a 77 GHz radar receiver, it allows to reduce the burden of a 77 GHz LO distribution chain, in term of area and consumption, while preserving RF performances. Demonstration is done from two topologies of ×3 sub-sampling mixers implemented in 28-nm FD-SOI CMOS technology. The best proposed passive mixer exhibits a -2.1 dB conversion gain (G_{cv}), a 1dB input-referred compression power (ICP1dB) of +3.5 dBm, and a 16.8 dB Single Side Band Noise Figure (NF_{SSB}). The mixer LO signal shaping consumption is 32 mW on a 1.2 V supply while the passive mixer core does not require any DC power.

Keywords — *Millimeter-wave, Sub-harmonic, FD-SOI.*

I. INTRODUCTION

By preventing crashes and identifying potential dangers even under low visibility conditions, the 77 GHz automotive radar drastically improves driving safety. Using mm-wave frequencies enables a good circuit integration and a fine range resolution, thanks to wide modulation bandwidths. Nevertheless, using a 77 GHz frequency often requires complex and power-hungry LO distribution chains. Conventional radar architectures are mostly based on a 38.5 GHz VCO followed by a frequency doubler and high consumption 77 GHz drivers [1],[2]. At the scale of a full chip including several receivers, this approach is complex and results in extra power consumption and extra circuit area. A sub-sampling mixer driven with a LO frequency sub-multiple of the RF frequency ($f_{LO}{\approx}f_{RF}/n$ with n a natural integer) would drastically simplify the picture. Passive sub-sampling mixers as described in [3] are popular for low frequency receivers as the sub-harmonic down-conversion is performed with low conversion losses and high linearity. The aim of this paper is to demonstrate that it can be extended to mmW frequencies. Hence, a ×3 sub-sampling mixer is proposed and implemented with a LNA in a 77 GHz receiver showing that this solution is compliant with the high requirements of automotive radar application. Section II introduces the sub-sampling mixer principle. Then, two mixer topologies relying on different mixer cores and associated 26 GHz LO signal shapers are detailed in section III. Measured performances of the 28-nm FD-SOI CMOS test chips validating this topology are given in section IV. Finally, section V draws the conclusions of this work.

II. SUB-SAMPLING MIXER PRINCIPLE

In an ideal sampling operation, every RF signal close to a sampling frequency harmonic ($n.f_{LO}$) is converted at a f_{IF} of $|f_{RF}-n.f_{LO}|$ without any conversion losses. Consequently, using sub-sampling with $f_{LO}{\approx}f_{RF}/n$ appears as a good solution to perform a frequency conversion with a LO frequency sub-multiple of the RF frequency without prohibitive conversion losses.

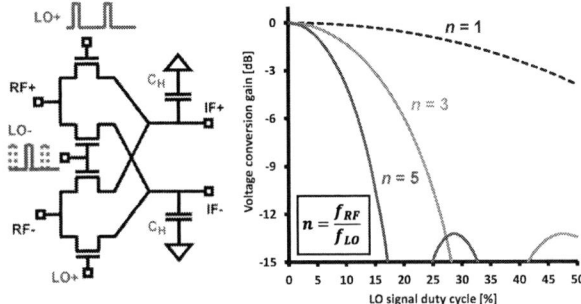

Fig. 1. sub-sampling mixer topology and conversion gain.

Fig.1 illustrates the sub-sampling mixer principle in a double-balanced configuration. When loaded by a hold capacitor (C_H) and driven by a low duty cycle LO signal, the passive mixer acts as a voltage sampler. The voltage conversion gain (G_{cv}) of the mixer of fig.1 is calculated with a similar approach as in [4] and represented versus the LO signal duty cycle (D) for different odd ratios $n = f_{RF}/f_{LO}$. Even n ratios are not considered because the double balanced topology rejects RF signals around the even LO harmonics resulting into a zero-conversion gain. Fig.1 shows that passive sub-sampling mixer allows a sub-harmonic frequency conversion with low conversion losses when D is low enough. Furthermore, passive mixers are highly linear as long as transistors switching time remains small [5]. Thus, driving mixing transistors with a LO signal having sharp transitions results in a high linearity.

Fig.1 highlights that G_{cv} drastically decreases when increasing either n or D. Because the generation of duty cycles below 15% becomes difficult when frequencies are increasing, a n ratio of 3 has been chosen to keep the conversion gain above -3 dB. Therefore, the 77 GHz sub-sampling mixer will use a 26 GHz LO frequency. The proposed sub-sampling mixer implementation in 28-nm FD-SOI CMOS will be detailed in the next section.

III. CIRCUIT DESCRIPTION AND IMPLEMENTATION

A. 26 GHz LO pulse shaper based on AND logic gates

Using a 26 GHz LO frequency in conjunction with the 28-nm FD-SOI technology, providing high speed CMOS transistors, open the way to a digital approach for the pulse shaper as presented in fig.2.a. In the differential pulse shaper of fig.2.a, number of stages between inverter chains A/C and B/D are different to create the required time delay between output square signals. Then, the AND gate turns both out of phase square signals into a pulsed signal. Inverters and AND gates are specifically designed for the application. The high body effect in FD-SOI technologies (about 85mV/V), enables to tune the threshold voltage (V_{th}) of transistors by setting the body voltage. This feature is used to implement a duty cycle tuning to the pulse shaper.

978-1-6654-8495-4/22 $31.00 © 2022 IEEE

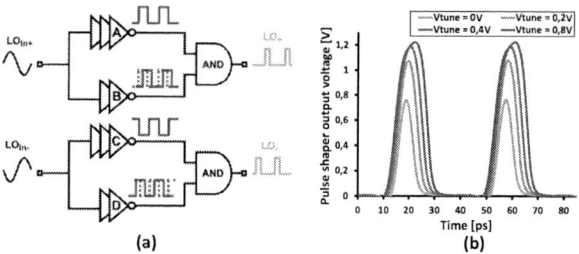

Fig. 2. (a) 26 GHz pulse shaper (b) simulated LO waveforms.

In each inverter, a tuning voltage (V_{tune}) is applied to the body of NMOS and PMOS transistors with opposite sign leading to a variation of the inverter time delay. From this variation, the duty cycle of LO pulsed signals, at the output of AND gates, is tuned as it can be depicted in Fig.2.b.

The LO pulse shaper supply voltage V_{DD} is 1.2 V. This value gives the necessary LO swing for a good mixer linearity. In logic gates, the V_{GS} stays low (no current) when V_{DS} is high, so the transistor does not enter in high Hot Carrier Injection and stays in a safe operating area even under a 1.2 V supply. Transient simulation of the pulse shaper output voltage LO+ is given in fig.2.b for different V_{tune} ranging from 0 to 0.8 V. A duty cycle range from 19% to 33% is thus available without decreasing the LO voltage swing below 1.1 V. The AND gate fall time is the limiting factor to address duty cycles lower than 19%".

B. Implementation of a 77 GHz sub-sampling mixer

The double balanced passive sub-sampling mixer described in fig.1 has been implemented with the 26 GHz LO pulse shaper of fig.2 in a 28-nm FD-SOI CMOS test chip. The test chip is depicted in fig.3.a and fig.3.b is a microphotograph of the manufactured chip. The circuit area is 0.81 mm × 0.54 mm. The chip is sticked and wire-bonded on a PCB test board supporting DC supplies and IF outputs while the RF and LO signals are provided by RF probes. As a differential 77 GHz signal synthesizer was not available, a passive balun is implemented on the RF access. This balun also provides a 50 Ω matching and exhibits a voltage gain of 8.1 dB corresponding to the voltage transformation ratio.

Two Op-amp in follower configuration (0-dB gain) are implemented as an output IF buffer, providing a high resistive impedance R_{IF} (10kΩ) in parallel with C_H to preserve the sampling operation ($1/(2.\pi.R_{IF}.C_H) \ll f_{RF}$). A mixing transistor size of W/L = 15 μm / 30 nm is set to get the best noise/linearity trade-off for the mixer. This sizing ensures a good linearity since LO signal rising and falling edges are not significantly soften by a too high mixing transistor gate capacitance.

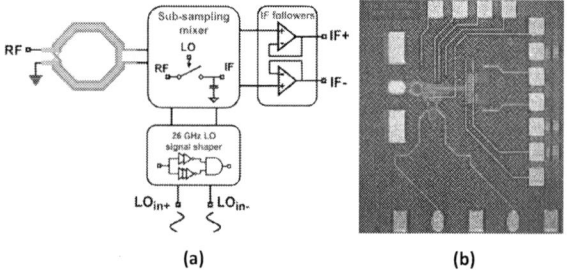

Fig. 3. (a) Block diagram of the test chip (b) manufactured chip.

At the same time, the ON-state resistor r_{on} of transistors remains low enough to limit the thermal noise contribution. Finally, the hold capacitor value C_H is set to 300 fF to properly store the signal sampled value without limiting the circuit bandwidth. Fig.4 shows the layout of the sub-sampling mixer core with the 26 GHz LO pulse shaper. The circuit area is 90 μm x 70 μm. It can be noticed that this sub-sampling mixer remains very compact as it does not require inductors on the LO chain.

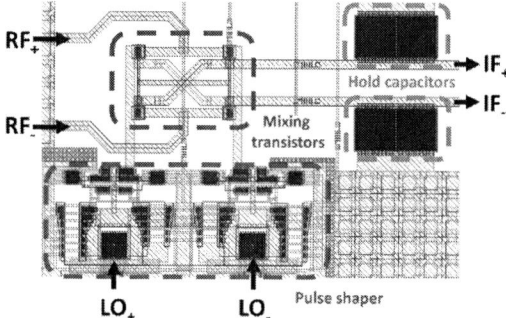

Fig. 4. Layout of the sub-sampling mixer.

Stand-alone mixer performances (excluding RF balun and output buffer) are summarized in table 1. They result from an harmonic balance and a non-linear noise simulation. f_{RF} and f_{LO} are respectively set to 78.02 GHz and 26 GHz.

TABLE I. SUB-SAMPLING MIXER SIMULATED PERFORMANCES

f_{RF} / f_{LO} [GHz]	G_{cv} [dB]	ICP1dB [dBm]	NF_{SSB} [dB]	P_{dc} [mw]
78 / 26	-3.5	+3	13.3	36

As previously stated, the sub-sampling mixer conversion gain is limited by the LO signal duty cycle lower values resulting from the use of AND gates in 26 GHz pulse shapers (fig.2). An implementation getting rid of the AND gates is proposed as an alternate solution in the next section.

C. Co-integration of LO AND gates and mixing transistors

To overcome the AND gate fall time limitation, the AND function can be merged with the mixing function. This solution is presented in fig.5. The implementation of this innovative mixer has been presented in [6] with the associated simulated performances. Present paper provides additional details, as well as measurement results validating all simulations. Its aim is to compare this approach with the more conventional solution relying of the 26 GHz pulse shaper integrating AND logic gates.

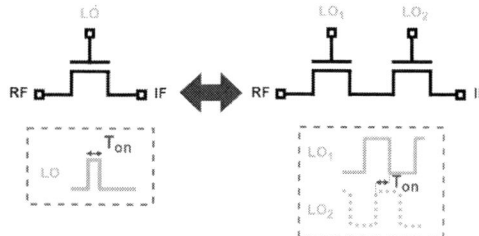

Fig. 5. Co-integration of the AND function and mixing transistors.

978-1-6654-8495-4/22 $31.00 © 2022 IEEE

When a mixing transistor driven by a low duty cycle LO signal (fig.1) is replaced by two series transistors directly driven by time delayed square waves (fig.5) the sub-sampling mixer operating principle remains the same. The conversion gain is the same with both solutions as soon as the ON-time (T_{ON} in fig.5) of mixing transistors is similar. On the other hand, the co-integration depicted in fig.5 allows to remove the AND gates. Consequently, lower duty cycles can be reached leading to a better conversion gain. The sub-sampling mixer based on this co-integration of the AND function and the mixing function has been implemented in a test chip similar to the one of fig.3. The time delay tuning between LO_1 and LO_2 generated square waves is also available in this second version. In the following sections the mixer presented in III.B is referred as mixer 1. The mixer co-integrating the AND function to the mixer core as described in III.C is called mixer 2. For both mixers, the sizing of the mixing transistors stays similar. Nevertheless, the two series mixing transistors on each mixer path are implemented following a specific layout strategy as illustrated in fig.6.

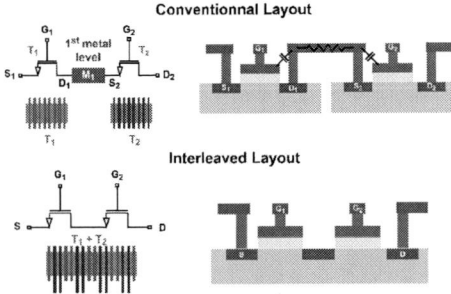

Fig. 6.   Layout of the NMOS interleaved series mixing transistors

Using an interleaved layout for series mixing SOI CMOS transistors rather than a conventional approach enables to merge their source and drain. This way, parasitic capacitors and resistors coming from the connection between both transistors via first metal level (M_1) are removed. Simulated performances of this version of the stand-alone sub-sampling mixer are reported in table 2.

TABLE II.   SUB-SAMPLING MIXER 2 SIMULATED PERFORMANCES

f_{RF} / f_{LO} [GHz]	G_{cv} [dB]	ICP1dB [dBm]	NF_{SSB} [dB]	P_{dc} [mw]
78 / 26	-2.5	+3	14.4	36

Compared to mixer with AND gates (mixer 1), the simulation results in table 2 confirm that the co-integration of AND gates and mixing transistors allows to reach a lower LO duty cycle resulting in a higher conversion gain. The next section presents measured performances of both sub-sampling mixers.

IV.   MEASUREMENT RESULTS

The measurements presented in this section are performed with a 0-dBm LO input power and include the RF input balun and the output buffer performances. The measured consumption of the LO shaping circuit on a 1.2V supply is 32 mW for both mixers. The duty cycle tuning integrated to inverter chains is used to set the best conversion gain. All measurements corresponding to the sub-sampling mixer test chip of fig.3 (mixer 1) are reported in fig.7.

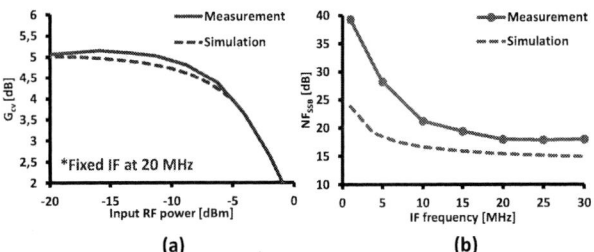

Fig. 7.   Mixer with AND gates (a) Measured G_{cv} compression curve at a f_{RF} of 78 GHz (b) Measured NF_{SSB} at a f_{RF} of 78 GHz.

The G_{cv} compression obtained versus RF input power is plotted in fig.7.a. f_{RF} and f_{LO} are respectively set to 78.02 GHz and 26 GHz. The extracted G_{cv} and ICP1dB are respectively 5.1 dB and -5.3 dBm. If the simulated voltage transformation ratio of the RF input balun contribution (8.1 dB) is removed from measurements, G_{cv} becomes equal to -3 dB with an ICP1dB of + 2.8 dBm. Fig.7.b shows the NF_{SSB} with f_{LO} of 26 GHz and the IF frequency ranging up to 30 MHz. The measured value of the NF_{SSB} is 17.9 dB at an IF offset of 30 MHz. When the noise contribution of the output stage is removed, the NF_{SSB} of the mixer, including the input balun, is 16.9 dB.

The same measurements are performed on the mixer with co-integration AND/mixer (mixer 2). Corresponding results are presented in fig.8. The input RF balun implemented with this second mixer is different than in mixer with AND gates and presents a voltage gain of 4.8 dB. With the previous approach, overall measurements of G_{cv} and ICP1dB are 2.7 dB and -1.5 dBm, respectively. It translate into a G_{cv} of -2.1 dB and an ICP1dB of +3.3 dBm for the standalone mixer, after balun de-embedding. Fig.8.b shows the NF_{SSB} in the same conditions as in fig.7.b. For this version of the mixer, measured values are close to simulated ones above 20 MHz. The measured value of NF_{SSB} is 18.4 dB at an IF offset of 30 MHz. Removing the noise contribution of the output stage results in a NF_{SSB} of 16.8 dB.

For both mixers, the measured NF_{SSB} is higher than the simulated one at low frequencies, highlighting an extra $1/f$ noise. In [7], the influence of various noise sources on a CMOS inverter is discussed showing that the noise coming from the inverter input and voltage supply is amplified, resulting in an important output jitter. In the presented LO signal shaper, at least five stages are included in each inverter chain. Since simulations do not consider noise for the LO source and voltage supplies, large discrepancies between simulated and measured NF_{SSB} values can be observed.

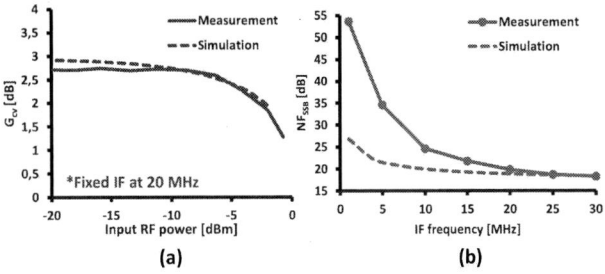

Fig. 8.   Mixer with co-integration AND/mixer (a) Measured G_{cv} compression curve (b) Measured NF_{SSB}, at a f_{RF} of 78 GHz.

TABLE III.   MEASURED PERFORMANCES FOR BOTH VERSIONS OF THE SUB-SAMPLING MIXER

Mixer	f_{RF}/f_{LO} [GHz]	G_{cv} [dB]	ICP1dB [dBm]	NF_{SSB} [dB]	P_{dc} [mW]
Mixer 1	78 / 26	-3	+2.8	16.9	32
Mixer 2	78 / 26	-2.1	+3.3	16.8	32

The table III sums up the measured performances for the two versions of the sub-sampling mixer. They both show very low conversion losses, especially when considering that they are passive and subharmonic mixers, as well as a good linearity. The second mixer, which is merging the AND function and the mixing one, presents the best performances, as expected since lower duty cycles are achievable. Simulated NF_{SSB} values in fig.7 and fig.8 exceed those reported in table I and II because they include input balun losses which are difficult to extract.

Because stand-alone sub-harmonic passive mixers are not reported in the literature related to CMOS 77 GHz radar receivers, a meaningful comparison between the proposed mixers and other existing solutions is a bit difficult. To do so, both proposed sub-sampling mixers have been implemented with a variation of the LNA presented in [8] in a 77 GHz front-end (LNA / mixer). As a result, a proper comparison with the state of the art can be provided. The measured front-end performances are summarized and compared with other radar receivers in table IV. The receiver front-ends including mixer 1 and mixer 2 will be respectively called Rx_1 and Rx_2.

TABLE IV.   COMPARISON WITH THE STATE OF THE ART.

	f_{RF}/f_{LO} [GHz]	G_{cv} [dB]	ICP1dB [dBm]	NF_{SSB} [dB]	Area [mm²]	P_{dc} [mW]
Rx_1	78 / 26	8.9	-11	11.9	0.2	42
Rx_2	78 / 26	9.5	-12	11.5	0.2	42
[1]	78 / 78	16	-7.8	8.7	NA	NA
[2]	78 / 78	18/60	-7/-30	20/11	Rx 0.3 LO 0.4	Rx 31 LO 58
[9]	78 / 39	14.5	-29	10.5	0.58	57
[10]	78 / 39	16	-20	13	2.33	28.5

The 77 GHz radar receivers in [1] and [2] are based on a conventional architecture using a 38.5 GHz VCO followed by a frequency doubler and 77 GHz LO drivers. [1] shows the best characteristics of this table but the LO power consumption and the circuit area dedicated to the LO chain are not given. However, [2] gives an idea of the power consumption of a conventional LO chain (about 60 mW) and allows to estimate the related chip area (0.4 mm2). Receivers based on sub-sampling mixers can use a 26 GHz LO VCO without requiring a frequency tripler and 77 GHz drivers. The mixer is also very compact as it does not require inductors. Consequently, proposed receivers based on sub-sampling mixers save DC power and silicon area while keeping decent performances when compared to conventional radar receivers. Finally, the comparison between this work and radar receivers based on sub-harmonic Gilbert cells in [9] and [10] highlights that using sub-sampling mixers lead to better overall performances with a higher ratio between f_{RF}/f_{LO}.

V.   CONCLUSION

A 3x sub-sampling passive mixer is proposed as a solution for a 77 GHz receiver implementation in a 28 nm FD-SOI CMOS. To overcome frequency limitations due to the complexity of AND gates required for the OL pulse shaper, an innovative solution based on the co-integration of the AND function and the mixing function is used.

Measurements demonstrate that a good conversion gain associated to a high linearity can be obtained using sub-harmonic passive mixers based on a sub-sampling operation. A higher f_{RF}/f_{LO} ratio than other sub-harmonic receivers is observed, also. Furthermore, when compared to conventional radar receivers, the LO chain is greatly simplified, saving area and reducing power consumption, while good performances are kept.

REFERENCES

[1]   T. Arai et al., « A 77-GHz 8RX3TX Transceiver for 250-m Long-Range Automotive Radar in 40-nm CMOS Technology », IEEE J. Solid-State Circuits, vol. 56, nᵒ 5, Art. nᵒ 5, 2021, doi: 10.1109/JSSC.2021.3050306.

[2]   D. Pan et al., « A 76–81-GHz Four-Channel Digitally Controlled CMOS Receiver for Automotive Radars », IEEE Trans. Circuits Syst. Regul. Pap., vol. 68, nᵒ 3, p. 1091-1101, 2021, doi: 10.1109/TCSI.2020.3042976.

[3]   H. Pekau et J. W. Haslett, « A 2.4 GHz CMOS sub-sampling mixer with integrated filtering », IEEE J. Solid-State Circuits, vol. 40, nᵒ 11, p. 2159-2166, nov. 2005, doi: 10.1109/JSSC.2005.857364.

[4]   A. R. Shahani, D. K. Shaeffer, et T. H. Lee, « A 12-mW wide dynamic range CMOS front-end for a portable GPS receiver », IEEE J. Solid-State Circuits, vol. 32, nᵒ 12, p. 2061-2070, 1997, doi: 10.1109/4.643664.

[5]   E. W. Lin et W. H. Ku, « Device considerations and modeling for the design of an InP-based MODFET millimeter-wave resistive mixer with superior conversion efficiency », IEEE Trans. Microw. Theory Tech., vol. 43, nᵒ 8, p. 1951-1959, 1995, doi: 10.1109/22.402285.

[6]   A. Flete, C. Viallon, P. Cathelin, et T. Parra, « A ×3 Sub-Sampling Mixer for a 77 GHz Automotive Radar Receiver in 28 nm FD-SOI CMOS Technology », in 2022 IEEE 22nd Topical Meeting on Silicon Monolithic Integrated Circuits in RF Systems (SiRF), 2022, p. 52-54. doi: 10.1109/SiRF53094.2022.9720061.

[7]   M. S. Illikkal, J. N. Tripathi, et H. Shrimali, « Analysing the Impact of Various Deterministic Noise Sources on Jitter in a CMOS Inverter », in 2019 6th International Conference on Signal Processing and Integrated Networks (SPIN), 2019, p. 208-211. doi: 10.1109/SPIN.2019.8711770.

[8]   R. Paulin, D. Pache, P. Cathelin, P. Garcia, S. Scaccianoce, et M. De Matos, « 4.6dB/5.5dB NF and 6.5dB/21.5dB gain 2 stages and 6 stages 76-81GHz LNAs for radar application in 28nm RF CMOS FD-SOI », in 2019 IEEE Asia-Pacific Microwave Conference (APMC), 2019, p. 479-481. doi: 10.1109/APMC46564.2019.9038496.

[9]   J. Jang, J. Oh, C.-Y. Kim, et S. Hong, « A 79-GHz Adaptive-Gain and Low-Noise UWB Radar Receiver Front-End in 65-nm CMOS », IEEE Trans. Microw. Theory Tech., vol. 64, nᵒ 3, p. 859-867, 2016, doi: 10.1109/TMTT.2016.2523511.

[10]   V. H. Le et al., « A CMOS 77-GHz Receiver Front-End for Automotive Radar », IEEE Trans. Microw. Theory Tech., vol. 61, nᵒ 10, p. 3783-3793, 2013, doi: 10.1109/TMTT.2013.2279368.

An FDD Auxiliary Receiver with a Highly Linear Low Noise Amplifier

Jin Jin[+], Simone Lecchi[*], Rinaldo Castello[*], and Danilo Manstretta[*]

[+]National ASIC System Engineering Research Center, Southeast University, Nanjing, China

[*]Department of Electrical, Computer and Biomedical Engineering, University of Pavia, Pavia, Italy

Email: danilo.manstretta@unipv.it

Abstract—An auxiliary receiver is proposed, including a high dynamic range low-noise amplifier and second-order baseband filter to improve the compression point with low power dissipation. The receiver has high input impedance and it can be placed at the transmitter output without loading effects. Implemented in a 28nm CMOS technology it occupies an active area of 0.5 mm^2. The receiver has a measured NF of 6 dB and it can withstand up to +7 dBm continuous waveform (CW) signal at 80 MHz offset with less than 1-dB gain compression. The signal path and the clock generation circuits consume 27 mW and 20 mW at 2 GHz respectively.

Index Terms—receiver front-end, LNA, SAW-less.

I. INTRODUCTION

Modern mobile communications standards must sustain an increasing demand for larger capacity, higher data rates and lower latency. Even though 5G has expanded into the millimeter-wave regime with FR2 bands, with several GHz of available bandwidth, sub-6 GHz bands are still the backbone of the network, providing ubiquitous coverage and data rates up to several Gb/s thanks to the introduction of an increasing number of frequency bands and the use of carrier aggregation, high-order modulation, and multiple-input and multiple-output [1]. This trend has made transceivers increasingly more complex and requires a growing number of off-chip SAW filters to mitigate interference and self-interference. Tunable duplexers [2], [3] could address this issue, but their transmitter-to-receiver isolation is inferior [4] when compared to duplexers working on a single band, which is typically in the order of 55 dB [5]. Considering a transmitter with an output power of 27 dBm, an out-of-band noise of -152 dBc/Hz, and duplexer isolation of 55 dB, the transmitter noise leakage is -180 dBm/Hz at the receiver, resulting in a very small sensitivity loss. Active self-interference cancellation can be used to relax duplexer isolation requirement. The final goal of this work is to design an active self-interference canceller that can be used to relax the duplexer isolation requirement to less than 45 dB.

II. FDD SYSTEM AND AUXILIARY ARCHITECTURE

A. Self-interference Cancellation Architecture

Different architectures have been proposed in the literature [4] for self-interference cancellation. In Fig.1 the architecture assumed in this work is reported. It consists of a main receiver, an auxiliary receiver, which captures the transmitted signal together with the associated noise and distortion components,

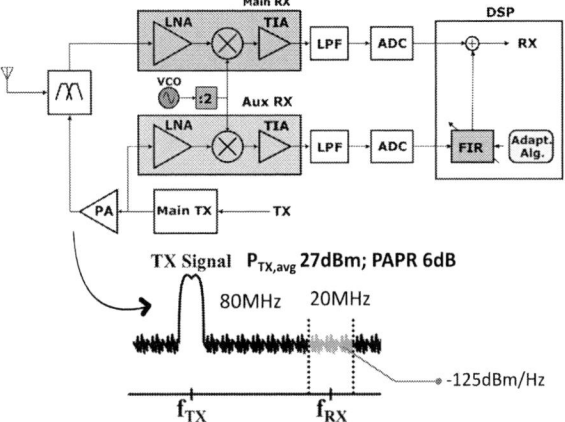

Fig. 1. FDD system with self-interference cancellation Architecture.

and an adaptive digital filter, that combines the two signal and performs noise cancellation. This allows wideband noise cancellation even in the presence of significant frequency-selectivity of the leakage component due to the duplexer characteristic as well as multiple antenna reflections. The most critical component of this architecture is the auxiliary receiver, which must have low noise and high compression. The ratio between the transmit signal and the residual noise after cancellation, sets the ratio between the auxiliary receiver compression point and its noise. With 45 dB duplexer isolation, the transmitter noise leakage at the receiver would be $-170\ dBm/Hz$, with a severe sensitivity loss. To bring this back to $-180\ dBm/Hz$ requires 10 dB of cancellation (which seems a reasonable target [6]) and an auxiliary receiver with a ratio between noise and 1dB compression point (P_{1dB}) of at least $-167\ dBc/Hz$ (considering 6 dB of margin for the peak-to-average power ratio). As an example, 1dB compression point of 3 dBm and a noise of $-164\ dBm/Hz$ (approximately 10 dB NF) would be required. A suitable figure of merit (FOM) for the auxiliary receiver was defined in [7] for RF band-pass filters, as $FOM = P_{1dB}/(PN_{1Hz}P_{dc})$, where PN_{1Hz} is the noise is 1 Hz of bandwidth and P_{dc} is the power consumption.

B. Proposed Auxiliary Receiver

Auxiliary receivers call for a high blocker compression point. Mixer-first architectures are good candidates [8], [9].

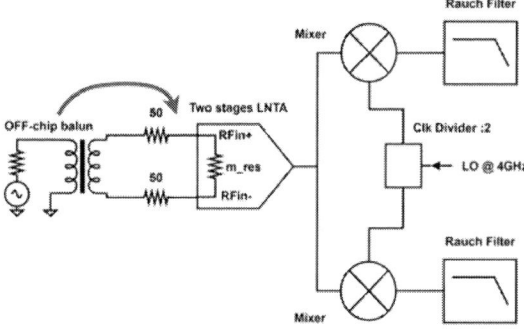

Fig. 2. Architecture of the proposed auxiliary receiver.

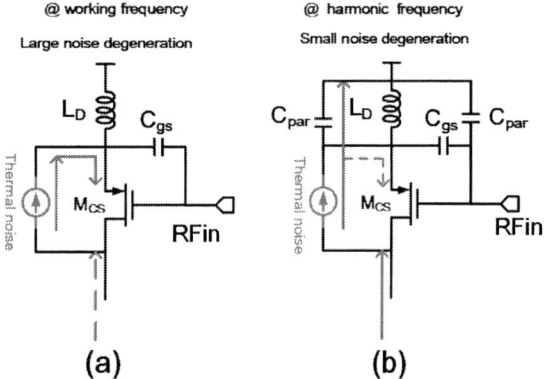

Fig. 3. Inductive source degeneration topology. (a) M_{CS} thermal noise at working frequency;(b) M_{CS} thermal noise at harmonic frequency.

However, they suffer from local oscillator (LO) leakage and low input impedance at the frequency of the TX. The auxiliary receiver is likely to be integrated with the transmitter. Hence, it should have high RF input impedance to avoid TX loading and we adopt the LNA-first receiver topology in this work, as shown in Fig.2. A high input impedance LNA provides gain and additional isolation between LO and input port. The RF current from the LNA is down-converted by a 4-phase 25% duty-cycle passive mixer. Finally, Rauch TIAs convert the signal current to voltage and perform second-order filtering.

III. CIRCUIT IMPLEMENTATION

A. Highly-linear LNA Design

Source degeneration (SD) is a common technique to improve linearity of common source (CS) amplifiers. Fig.3 shows the typical structure of inductive SD-LNAs. An added benefit of inductive degeneration is that the source voltage can exceed the supply limit, which helps in improving the compression point. The effective input voltage of CS transistors (Vgs) at resonance point can be written as:

$$V_{gs} = \frac{V_{RF_{in}}}{sL_D g_{mCS}} \quad (1)$$

where $V_{RF_{in}}$ represents the voltage of the input port. The input impedance of the SD-LNA at resonance is:

$$R_{in} = \omega_T L_D \quad (2)$$

where $\omega_T = g_{mCS}/C_{gs}$ represents the cut-off frequency of the transistor. In the main receiver, 50-Ω input matching limits the value of L_D. An additional inductor in series with the gate of M_{CS} can provide extra design freedom. In this way, the L_D can be reduced to increase the V_{gs}, hence increasing the LNA gain. On the contrary, for the auxiliary receiver, high input impedance and strong source degeneration to increase the linearity are required. Fortunately, these requirements can be simultaneously satisfied by increasing either L_D or g_{mCS}. An approximate expression of the LNA noise figure (NF) is given below:

$$NF \approx 1 + \gamma g_{mCS} R_s (\frac{\omega_o}{\omega_T})^2 = 1 + \gamma R_s \frac{\omega_o^2}{C_{gs}^2 g_{mCS}} \quad (3)$$

where the loss of L_D is neglected and ω_o is the LNA working frequency. In practical designs, the resonance frequency determines C_{gs} once L_D is selected. Hence, NF can be improved by increasing g_{mCS}.

In an LNA with resonated SD, transistor noise grows moving away from the resonance frequency. Fig.3(a) and 3(b) give an intuitive explanation of this phenomenon. At the working frequency, where L_D and C_{gs} resonate, the large impedance at the source causes the transistor thermal noise to partially recirculate within the transistor. In contrast, the noise current appears at the output node without significant attenuation at the harmonic frequencies, where the source impedance is much lower. Fig.4 reports the output noise spectrum of the proposed SD-LNA. We can find low in-band noise at 2 GHz, but much higher noise at higher frequencies due to large M_{CG} noise current.

The 4-phase 25% duty-cycle mixer exhibit an harmonic rejection ratio (HRR) of 10 dB and 14 dB at the 3rd and 5th harmonics, respectively. Considering the large increase in LNA noise at the harmonics, noise folding caused by the mixer leads to a considerable NF degradation for this LNA.

We can employ eight-phase LO to improve HRR [10], but consuming considerable power in the divider. Since the frequency distance between harmonics is large, an LC tank with a moderate quality factor can reduce folding. A transformer is used to implement this LC filter while introducing current gain. As shown in Fig.5, the LNA adopts a two-stage topology. In the main stage, $M_{CSn1,2}$ shares the bias current with $M_{CSp1,2}$, increasing the total transconductance. Thanks to the stacked topology, only one differential degeneration inductor is required. At the output of the the first stage, two transformers absorb the current at the drain terminal of p and n transistors and provide the current gain. The turn ratio of the transformers is 5:3. From simulations, 3rd harmonic rejection improves by more than 13 dB. The second stage acts as a current buffer, where common gate (CG) transistors $M_{CGn1,2}$ and $M_{CGp1,2}$ are utilized to absorb the current from main

978-1-6654-8495-4/22 $31.00 © 2022 IEEE   310

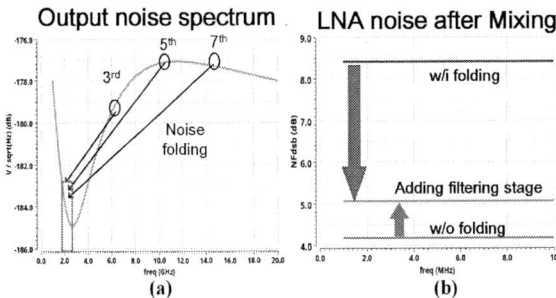

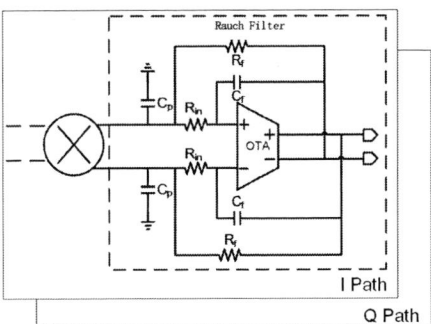

Fig. 4. (a) Output noise spectrum of the proposed SD-LNA. (b) LNA Noise figure after down-conversion.

Fig. 6. The diagram of TIA.

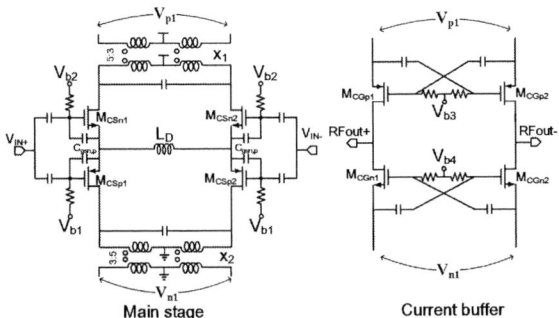

Fig. 5. proposed a highly linear LNA with improved HRR.

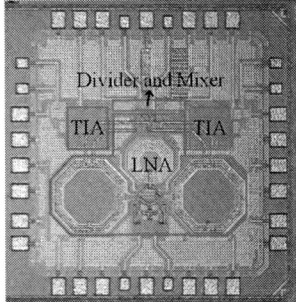

Fig. 7. Chip microphotograph.

stage. Moreover, the cross-coupling technique is implemented to increase the effective g_m of the CG stage. Note that the CG stage's noise spectrum also increases with frequency. However, in our design, the bias current for the second stage is much less than the first stage, which means much less high frequency noise. In this way, the noise-folding effect is much reduced.

Fig.4(b) compares the NF after down-conversion of the proposed topology with that of a conventional SD-LNA. The blue and red line represent the NF of a conventional SD-LNA with and without noise folding, showing a large increase due to noise-folding. The pink line shows the much lower noise of our topology thanks to noise filtering.

The total LNA bias current is 9.5 mA, of which 8 mA go to the SD-LNA stage and the rest to the second stage. The g_m of each transistor in the first and second stage is 90 mS and 14.4 mS respectively. L_D is 4.9 nH. In this configuration, we can achieve a 4.1 dB NF and +13.4 dBm 1dB blocker compression point for a low impedance load. The LNA input impedance is 200 Ω in the TX band.

B. Mixer and TIA Design

Since a large blocker current will flow into the passive mixer, the on-resistance (R_{sw}) of the mixer becomes critical. A moderate R_{sw} could cause a large voltage swing across the LNA output. In the extreme case, mixer clipping may occur. In our design, the size of a single switch transistor is 32 μm/30 nm. With 1.2 V_{pp} clock, the on-resistance can be reduced to 7 Ω according to simulation. Fig.6 shows the structure of the propose Rauch filters. In our case, blockers from TX

are 80 MHz away from the center frequency of the desired signal. Thus, we need to minimize the input impedance at this frequency, which is mainly set by the input capacitor C_p. However, a large C_p will degrade the loop bandwidth of the TIA and increase the noise due to the OTA. After optimization, C_p was set to 60 pF as a compromise between noise and linearity. Simulation shows that the TIA exhibits a 15 dBm 1 dB compression point referred to the LNA input.

IV. MEASUREMENT RESULTS

The chip was realized in a 28nm CMOS technology and has an active area of 0.5 mm^2. A chip microphotograph is shown in Fig. 7. The chip was wire-bonded on a board for testing purposes and an external balun together with on-chip 50 Ω resistors were used to provide an impedance-matched differential input signal, as shown in Fig. 2. The losses due to balun and input matching were de-embedded from the measurements since they will not be present when the circuit is integrated with the TX. The measured down-conversion gain versus output frequency for $f_{LO} = 2\ GHz$ is reported in Fig. 8(a). In-band down-conversion gain is 25.2 dB. Thanks to second order filtering in the Rauch filter, the front-end provides an attenuation of over 35 dB at the TX channel 80 MHz away from the RX band. Fig. 8(b) reports the measured noise spectrum, corresponding to an integral NF of 6 dB, which is very close to the value expected from simulations. The large signal handling capability of the receiver was tested measuring the 1dB in-band gain reduction as a function of the blocker power (B1dB) at 80 MHz frequency offset (Fig.8(c)). 1dB gain compression occurs at 7 dBm, i.e. 4 dB earlier than in

TABLE I
PEFORMANCES SUMMARY AND COMPARSION WITH STATE-OF-THE-ART RECEIVERS.

	This work	ESSCIRC'17 [6]	TMTT'16 [11]	ISSCC'17 [12]	RFIC'17 [8]	JSSC'19 [13]	ISSCC'21 [14]
Tech (nm)	28	28	65	28	45	65	40
Architecture	LNA-first RX	N-Path-LNA RX	N-Path filter	N-Path RX	Mixer-first	Quantized RX	N-Path LNA RX
Area(mm^2)	0.5	0.12	2.4	0.49	0.8	0.25	0.6
Z_{in} (Ω)	200	High	50	50	50	50	50
Freq. (GHz)	1.9-2.1	0.7-2	0.1-1.4	0.1-2	0.2-8	0.7-1.4	0.4-3.2
Gain (dB)	25.3	3	12-18	21	27-30	20.8-36.8	36
BW (MHz)	20	30	10	13	20	20	160
NF (dB)	6	6.2-9.6	3-4.2	4-10.3	2.3-5.4	1.9-14.6	2.7-3.6
B_{1dB} (dBm)	7	5	8	13	12	-8.5-10.5	-6
Δf/BW	4	1.7		6.3	4	5	1.7
NF_{1dB} deg.(dBm)	-7	5	0	1	-5	N/A	-10
$P_{diss}(mW)$	27	16.2	44	30	50	14	58.5
$P_{dig}(mW)$	20	10-48	5-27	8-66	60*	20-50	35*
FoM (dBm)	159.3	151.6*	161.1	158.1	161.5	151.9	148.1

* at working frequency 2 GHz.

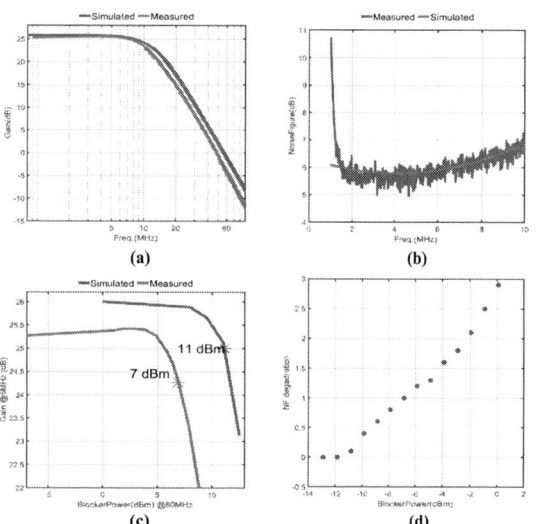

Fig. 8. Measured performance: a) gain, b) NF, c) gain compression, d) blocker NF degradation.

pre-tapeout simulations. This discrepancy was traced back to the interconnect parasitic resistance (about 25 Ω) between the mixer and the baseband filter, that was not correctly extracted. This parasitic resistance can be easily corrected using a metal interconnect with a lower resistivity. Fig. 8(d) reports the NF degradation in the presence of a CW blocker at 80 MHz offset frequency. We can observe less than 3 dB NF degradation for a 0 dBm blocker. Table I compares the proposed auxiliary receiver with other blocker-tolerant receivers, all are matched to 50 Ω with the exception of [6] that is a high input impedance auxiliary receiver. N-path based and mixer-first receivers with impedance matched input achieve the highest FoM. This work comes close to these solutions even if it has a much higher input impedance and potentially less LO leakage, while it shows 7.7 dB improved FoM compared with previous auxiliary receiver in [6]. Post-layout simulations with a lower resistance mixer interconnect give over 3 dB improvement of B1dB,

which corresponds to a FOM that is the best ever reported.

REFERENCES

[1] S. Sadjina, et al., "A Survey of Self-Interference in LTE-Advanced and 5G New Radio Wireless Transceivers," *IEEE Transactions on Microwave Theory and Techniques*, vol. 68, no. 3, pp. 1118-1131, March 2020.

[2] A. Goel, B. Analui and H. Hashemi, "Tunable Duplexer With Passive Feed-Forward Cancellation to Improve the RX-TX Isolation," *IEEE Transactions on Circuits and Systems I: Regular Papers*, vol. 62, no. 2, pp. 536-544, Feb. 2015, doi: 10.1109/TCSI.2014.2360759.

[3] D. J. Simpson, et al., "Single-/Multi-Band Bandpass Filters and Duplexers With Fully Reconfigurable Transfer-Function Characteristics," *IEEE Transactions on Microwave Theory and Techniques*, vol. 67, no. 5, pp. 1854-1869, May 2019.

[4] J. Zhou, T. Chuang, T. Dinc and H. Krishnaswamy, "Integrated Wideband Self-Interference Cancellation in the RF Domain for FDD and Full-Duplex Wireless," *IEEE Journal of Solid-State Circuits*, vol. 50, no. 12, pp. 3015-3031, Dec. 2015.

[5] R. Vazny, et al., "An interstage filter-free mobile radio receiver with integrated TX leakage filtering," *Proc. IEEE Radio Freq. Integr. Circuits Symp. (RFIC)*, May 2010, pp. 21–24..

[6] D. Montanari et al., "An FDD Wireless Diversity Receiver With Transmitter Leakage Cancellation in Transmit and Receive Bands," *IEEE Journal of Solid-State Circuits, vol. 53, no. 7, pp. 1945-1959*, July 2018.

[7] W. B. Kuhn, D. Nobbe, D. Kelly and A. W. Orsborn, "Dynamic range performance of on-chip RF bandpass filters," *IEEE Transactions on Circuits and Systems II: Analog and Digital Signal Processing*, vol. 50, no. 10, pp. 685-694, Oct. 2003, doi: 10.1109/TCSII.2003.818364.

[8] Y. Lien, E. A. M. Klumperink, B. Tenbroek, J. Strange, and B. Nauta, "Enhanced-selectivity High-linearity Low-noise Mixer-first Receiver with Complex Pole Pair due to Capacitive Positive Feedback," *IEEE Journal of Solid-State Circuits*, vol. 53, no. 5, pp. 1348–1360, 2018.

[9] T. Y. Liu and A. Liscidini, "20.9 A 1.92mW filtering transimpedance amplifier for RF current passive mixers," *2016 IEEE International Solid-State Circuits Conference (ISSCC)*, 2016, pp. 358-359.

[10] Y. Xu, J. Zhu and P. R. Kinget, "A Blocker-Tolerant RF Front End With Harmonic-Rejecting N-Path Filter," *IEEE Journal of Solid-State Circuits*, vol. 53, no. 2, pp. 327-339, Feb. 2018, doi: 10.1109/JSSC.2017.2778273.

[11] M.N. Hasan, et al., "Tunable Blocker-Tolerant On-Chip Radio-Frequency Front-End Filter With Dual Adaptive Transmission Zeros for Software-Defined Radio Applications," *IEEE Trans. on Microwave Theory and Techniques*, vol. 64, no. 12, pp. 4419-4433, Dec. 2016.

[12] Y. Lien, et al., "24.3 A high-linearity CMOS receiver achieving +44dBm IIP3 and +13dBm B1dB for SAW-less LTE radio," *2017 IEEE International Solid-State Circuits Conference (ISSCC)*, 2017, pp. 412-413.

[13] J. Musayev and A. Liscidini, "A Quantized Analog RF Front End," *IEEE Journal of Solid-State Circuits*, vol. 54, no. 7, pp. 1929-1940, July 2019.

[14] M. A. Montazerolghaem, et al., "A 3dB-NF 160MHz-RF-BW Blocker-Tolerant Receiver with Third-Order Filtering for 5G NR Applications," *2021 IEEE Int. Solid-State Circuits Conference (ISSCC)*, pp.98-100.

A 433-MHz Wireless Burst-Chirp Modulation Transmitter with Adaptive Duty-Cycle Control and Precharge Mechanism

Jing-Siang Chen, Chun-Ting Chang, and Yu-Te Liao

Department of Electronics and Electrical Engineering, National Yang Ming Chiao Tung University

Hsinchu, Taiwan

Abstract—**This paper presents an energy-efficient 433 MHz wireless transmitter with burst-chirp modulation. To compromise the design trade-offs between output power and power consumption, the transmitter (TX) adopts a duty-cycle controlled class-E power amplifier (PA), whose switching duty-cycle ratio is less than 50%. Through this method, the impedance transformation ratio at the output of the PA is reduced, thus improving efficiency. Furthermore, a digital controller directly modulates the oscillator, generating chirp pulses. Moreover, the precharge mechanism is proposed to improve settling time, allowing short-pulse data transmission time. The TX IC is fabricated using 180-nm CMOS technology and occupies an active area of 1.73 mm². The design achieves 8.71/Mbit normalized system efficiency, while the output power is -1.18 dBm at a 0.8-V supply, and the maximum system efficiency is 45%.**

Keywords: burst-chirp modulation, CMOS, duty-cycle control loop, power amplifier, power-efficient, transmitter

I. INTRODUCTION

Advanced wireless sensor networks require a wide range of sensing devices for seamless connection and data acquisition. Long-distance communication and good power efficiency are critical for implementing wireless sensing nodes. Recently, the need to design a power-efficient wireless transmitter has attracted much attention in various applications, such as environmental monitoring, indoor positioning, and biomedical wearables. To achieve good power efficiency, frequency multiplication architectures, such as capacitive coupling [1] and edge combiner [2], reduce the number of circuits operated at high frequency, thus lowering power consumption. Although power consumption can be reduced, these architectures sacrifice output power, communication distance, and system efficiency.

In addition, other high-efficiency transmitters have been developed using direct power transfer transmitters (DPTT) [3][4] and switching mode power amplifiers (PAs) [5]-[7]. DPTT usually adopts an oscillator output as an antenna directly without power-hungry PAs and a phase/frequency locked loop. However, this architecture suffers from limited output power and signal fluctuations due to environmental interference. The output power is approximately -8 to -20 dBm, which is generally insufficient for long-distance data communication. A class-E PA [5] combines zero-voltage switching (ZVS) and zero-voltage derivative switching (ZDS) conditions to achieve higher efficiency than other linear PAs. Moreover, class-D PAs [6] or E/F2 PA circuits can improve power conversion efficiency [7]. However, class-E/F2 PAs provide lower output power than conventional class-E PAs. The switching losses of class-D PAs increase at high frequencies. Switch-mode PAs have low output power and efficiency at reduced supply voltage and do not benefit from a reduced power supply for power savings. Fig. 1 compares the output power and normalized efficiency of the state-of-the-art

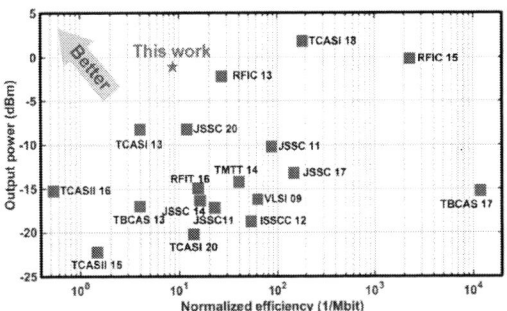

Fig. 1. Comparison of the output power and power efficiency of the state-of-the-art transmitters (Normalized Efficiency = (P_{dc} (mW) / DR (Mbps))/P_{out} (mW))

high-efficiency transmitters. The power efficiency decreases with the increase of output power, which is related to the transmission distance. Therefore, it is quite challenging to maintain large output power and good power efficiency at the same time at low supply voltage and power consumption.

In recent years, burst-chirp modulation methods have been proposed in radar transmitters to enhance communication distance [8][9]. The burst data communication reduces the average power consumption in each data transfer. The spread spectrum leads to a bit error rate lower than that of the frequency shift keying (FSK) method at the same signal-to-noise ratio. Therefore, burst-chirp modulation provides the advantages of a long communication distance and tolerance to interference compared to FSK modulation [8][9]. This work proposes a burst-chirp modulation transmitter with a duty-cycle control loop and a precharge mechanism for high power efficiency and fast settling time in burst-mode data communication.

This paper is organized as follows. Section II describes the design of the proposed burst-chirp wireless transmitter. Measurement results are shown in Section III. Finally, Section IV provides a brief conclusion.

II. PROPOSED BURST-CHIRP MODULATION TX

A. Duty-cycle control

Fig. 2 shows a conventional class-E PA. The class-E PA has optimal output power and load resistance according to ZVS and ZDS conditions [10]. The relationship between output power and optimal load resistance can be expressed as [10]:

$$P_{out} = 0.5768 \frac{V_{DD}^{\,2}}{R_L} \qquad (1)$$

978-1-6654-8495-4/22 $31.00 © 2022 IEEE

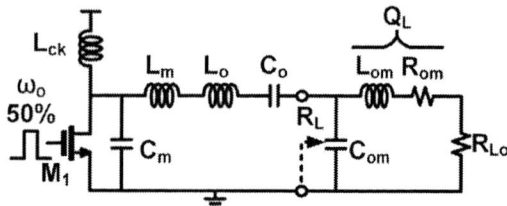

Fig. 2. Schematic of conventional class-E power amplifier with impedance-matching network

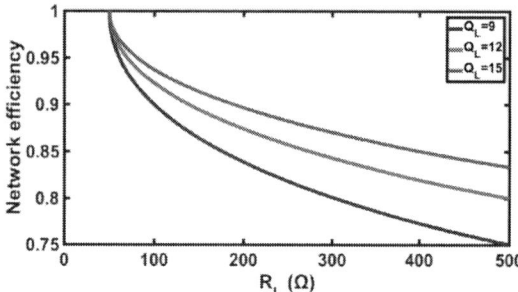

Fig. 3. Relationship between Network Efficiency and R_L

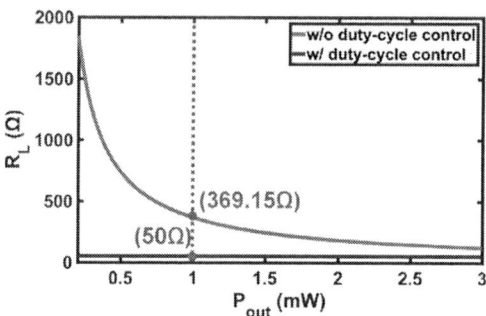

Fig. 4. Simulated result of the load resistance in the output power range

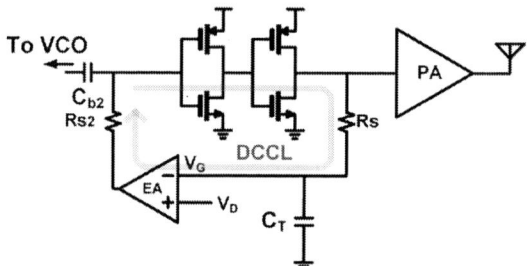

Fig. 5. Schematic of the duty-cycle control scheme

Assuming that V_{DD} is 0.8 V and the output power is 0 dBm, R_L can be obtained as 369.152Ω. Low output power requires a large load resistance when the supply voltage is defined. The transformation from the load resistance R_L to the output load resistance R_{LO} (e.g. 50-Ω), impedance transformation ratio (ITR), is

$$ITR = \frac{R_L}{R_{LO}} \qquad (2)$$

Moreover, network efficiency can be obtained as follows:

$$Network\ eff. = \frac{Q_L}{Q_L + \sqrt{\frac{R_L}{R_{LO}} - 1}} \qquad (3)$$

where Q_L is the quality factor of the inductor. Fig. 3 plots the network efficiency over R_L in different Q_L. This shows that the lower R_L, the higher the achievable network efficiency. The best network efficiency occurs at ITR = 1. To achieve good power efficiency at a fixed output power level, a duty-cycle control method for the power transistor was proposed, and the optimum duty-cycle of PA is determined by the required output power and V_{DD} [11]. The optimal output impedance is calculated as follows:

$$R_L = T \frac{V_{DD}^2}{P_{out}} \qquad (4)$$

T is derived in (5), where D is the duty cycle, θ_x, θ_F, θ_c and θ_D are the function of D, respectively. From (4) and (5), D can be selected to achieve the optimal R_L at a fixed V_{DD} and P_{out}. Fig. 4 shows the simulated results of the load resistance in the output power range with and without duty-cycle control. At an output power of 0 dBm, the load resistance is reduced by 7.38 times, and can directly match to 50 Ω, thus eliminating the impedance matching network that sacrifices power efficiency.

B. Precharge mechanism

Fig. 5 shows the duty-cycle control circuit diagram. The duty-cycle control loop (DCCL) consists of a buffer stage, a low-pass filter, and an error amplifier circuit (EA). The buffer stage, composed of inverters, is inserted between the oscillator and the power transistor to enlarge the signal swings to rail-to-rail. The EA senses the output voltage (V_G) of the inverter and adjusts the input bias voltage of the first inverter, maintaining V_G close to V_D. Thus, the duty cycle of the power transistor is adjusted by the bias voltage. The DCCL can reduce power consumption and optimize power efficiency according to output power.

To further reduce system power consumption, burst-chirp modulation is employed. Burst-chirp modulation requires fast activation and ramp frequency in cases of increase or decrease for data '0' and "1," respectively. However, the DCCL needs a large capacitor that suppresses the RF carrier at the EA in the feedback loop to eliminate self-mixing-caused DC drifts. Therefore, with DCCL, the settling speed could be slow under stability considerations. The slow settling time constrains the TX data rate (DR).

In this work, we propose a precharged architecture to overcome the long settling time due to the large capacitance in the loop. The open-loop transfer function of the DCCL can be defined as follows:

$$T_{open}(s) = T_{buffer}(s) \times \frac{1}{1 + \frac{s}{\omega_{p5}}} \times \frac{1}{1 + \frac{s}{\omega_{p6}}} \times T_{EA}(s) \qquad (6)$$

$$T = \frac{[\sin(\theta_x(D)) + \sin(\theta_D(D) - \theta_x(D))]^4}{2\{-\theta_F(D)\theta_c(D)\sin(\theta_x(D)) + 2\theta_c(D)\sin(\frac{\theta_D(D)}{2})\sin(\theta_c(D)) + [\theta_F(D)\cos(\frac{\theta_D(D)}{2}) + 2\sin(\frac{\theta_D(D)}{2})]\cos(\theta_c(D))\}^2} \qquad (5)$$

978-1-6654-8495-4/22 $31.00 © 2022 IEEE

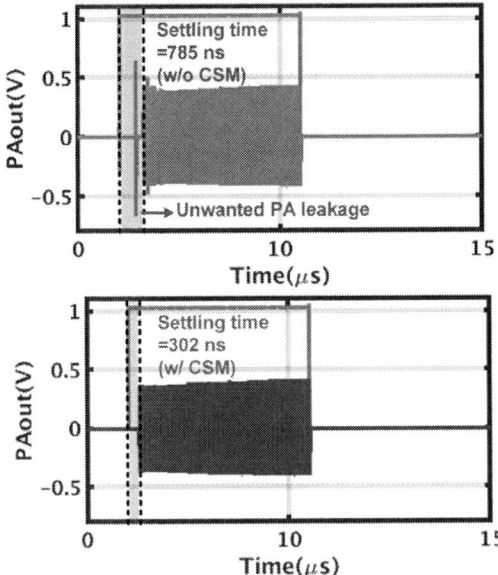

Fig. 6. Simulated result of the settling response comparison

where ω_{p5} and ω_{p6} are the poles of the R_SC_T and $R_{S2}C_{b2}$ in the filter, and T_{buffer} and T_{EA} are the transfer functions of the buffer stage and EA. The settling time is limited by the charge time of the large capacitor (C_T) for loop stability. In this work, the voltage of C_T can be charged to close to V_D before the data-ready signal. To reduce the settling time, the voltage of V_G is quickly pulled close to V_D before the loop is enabled. Fig. 6 shows the simulation results for the settling response. The settling time is reduced by 2.6 times by the precharge mechanism when compared to the original design.

C. Proposed burst-chirp modulation transmitter

Fig. 7 shows a schematic of the proposed burst-chirp modulation transmitter, consisting of duty-cycle control and precharge topology. The large DCCL capacitor (C_T) is split into C_s and C_b for improving stabilization and settling speed. First, the LC digitally controlled oscillator (LC-DCO) is locked to an initial frequency f_o by a fixed voltage (V_{DAC}) to keep the LC-DCO frequency unchanged. In Phase 1, V_{tm} and PA_{en} have no signals, and the analog MUX output voltage (V_{DG}) is pulled down, setting V_G to ground and PA off. In Phase 2, the sensor data are modulated through a chirp generator (CG), and then V_{tm} is pulled to a high-voltage level. R_S and C_S are connected to the loop to maintain loop stability. At the same time, the voltage of V_{tm} starts to charge C_b to V_D. In Phase 3, V_{tm} changes from a high-level signal to a low-level signal, and C_b is switched to the loop again to stabilize the loop. The PA_{en}, generated by the CG, becomes a high-level signal, starting to output a burst-chirp signal, where the DR is set to 100 kb/s, and the period of Phase 3 is 4 μs. After 4 μs transmission time, the PA is turned off by the signal from the CG. Furthermore, the CG directly modulates the 4-bit capacitor bank of the LC-DCO, creating 8 step-up or -down frequency changes according to data "1" and "0", respectively, in 4 μs, to achieve a wide frequency spread and an enhanced DR.

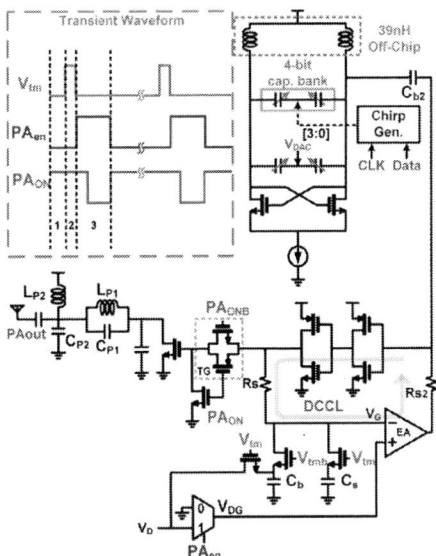

Fig. 7. Schematic of the proposed burst-chirp modulation TX

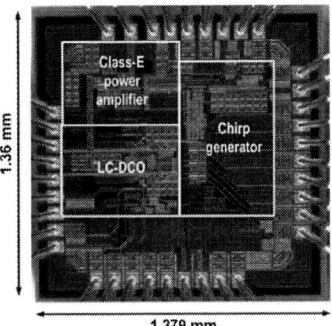

Fig. 8. Chip micrograph

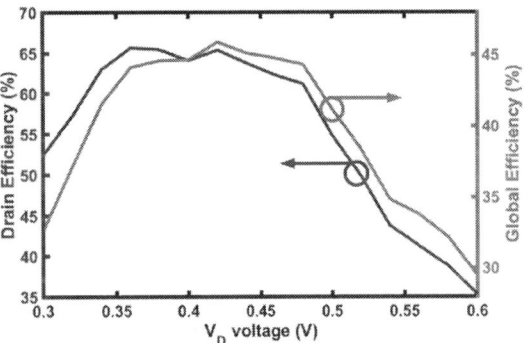

Fig. 9. Measured PA drain efficiency and global efficiency versus V_D

III. MEASUREMENT RESULTS

The proposed 433-MHz burst-chirp modulation transmitter was fabricated in a 180-nm CMOS process. Fig. 8 shows the chip micrograph, which includes the core circuit of the duty-cycle control class-E PA, LC-DCO, and the chirp generator. The total chip area is 1.73 mm². Fig. 9 depicts the measured system and PA drain efficiency with a duty cycle from 30% (0.3V) to 60% (0.6V). The peak PA drain efficiency is 65.34% when the output power is -1.18 dBm. In addition,

978-1-6654-8495-4/22 $31.00 © 2022 IEEE

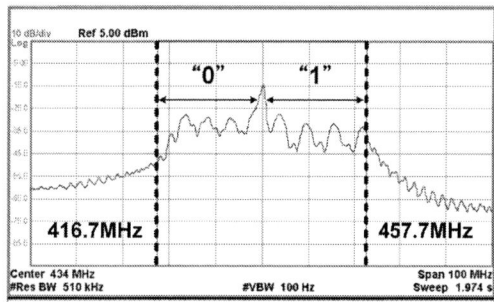

Fig. 10. Measured PA output spectrum with burst-chirp modulation

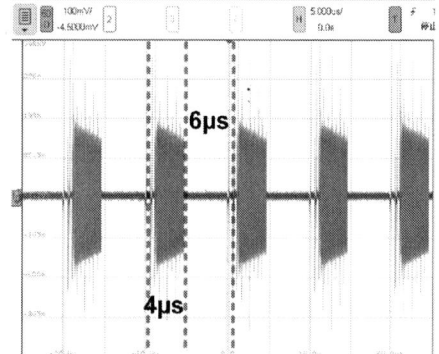

Fig. 11. Measured transient output waveform of TX with a 100 kb/s DR

the TX achieves a system efficiency greater than 45%. Fig. 10 shows the output spectrum of TX at 100 kb/s with PRBS-7 random data. The center burst-chirp frequency is 433 MHz. Data '0' linearly decreases the frequency to 416.7 MHz, and data '1' increases the frequency linearly to 457.7 MHz within 4µs, respectively. Fig. 11 shows the measured transient output waveform of TX at a DR of 100 kb/s. Table I shows a comparison of our results with those of other works. Compared to the high-efficiency DPPT-based 433 MHz TX [3], the proposed burst-chirp modulation TX achieved better normalized efficiency while maintaining a larger output power. Compared to TX with PAs [5], [12], [13], this work improves efficiency by using a duty-cycle controlled class-E power amplifier and burst-chirp modulation.

IV. CONCLUSION

This design adopts a burst-chirp modulation in the 433 MHz transmitter and uses duty-cycle control and a precharge mechanism for improving power efficiency and settling speed. Thus, the design achieves 8.71/Mbit normalized efficiency at an output power of -1.18 dBm and a supply of 0.8 V.

ACKNOWLEDGMENTS

The authors would like to thank the Taiwan Semiconductor Research Institute for chip fabrication and the Ministry of Science and Technology, Taiwan, for financial support under the grant MOST 111-2628-E-A49-013.

Table I. Performance summary and comparison

	[3]	[5]	[12]	[13]	This Work
Proc(nm)	90	130	130	130	180
Supply(V)	0.6	0.5/1.2	0.7-1.2	0.6/1	0.8/1
Mod.	FSK	BFSK	BFSK	BFSK	Burst chirp
Freq. (MHz)	429	433	433	480	433
DR (Mbps)	0.2	0.001	0.08	1	0.1
Power (mW)	0.39	2.34[*3]	1.361[*3]	0.144[*3]	1.66
Pout (dBm)	-8.2	0.04	-2	-20	-1.18
NE Eff. (1/Mbit)[*1]	12.9	2318	26.9	14.4	8.71[*2]
Area(mm²)	0.154[*4]	1.1	0.41	0.217[*4]	1.73

[*1] NE Eff. = $(P_{dc}$ (mW) / DR (Mbps))/P_{out} (mW)
[*2] Because of the burst chirp, the PA only generates a burst chirp pulse of 4µs in the pulse period of 10 µs (DR=100 kb/s), P_{dc} = (Power dissipation*0.4) [*3] Without PLL power dissipation [*4] Core area

REFERENCES

[1] N. Dau, Y. Chen and Y. Liao, "A 145µW 315MHz harmonically injection-locked RF transmitter with two-step frequency multiplication techniques," *IEEE MTT-S International Microwave Symposium*, 2017, pp. 1781-1783.

[2] J. Pandey and B. P. Otis, "A Sub-100 µW MICS/ISM Band Transmitter Based on Injection-Locking and Frequency Multiplication," in *IEEE Journal of Solid-State Circuits*, vol. 46, no. 5, pp. 1049-1058, May 2011.

[3] C. -Y. Chiu, Z. -C. Zhang and T. -H. Lin, "Design of a 0.6-V, 429-MHz FSK Transceiver Using Q-Enhanced and Direct Power Transfer Techniques in 90-nm CMOS," in *IEEE Journal of Solid-State Circuits*, vol. 55, no. 11, pp. 3024-3035, Nov. 2020.

[4] Y. Shi, X. Chen, H. -S. Kim, D. Blaauw and D. Wentzloff, "A 606µW mm-Scale Bluetooth Low-Energy Transmitter Using Co-Designed 3.5×3.5mm² Loop Antenna and Transformer-Boost Power Oscillator," *IEEE International Solid- State Circuits Conference*, 2019, pp. 442-444.

[5] N. E. Roberts, M. C. Kines and D. D. Wentzloff, "A 380µW Rx, 2.6mW Tx 433MHz FSK transceiver with a 102dB link budget and bit-level duty cycling," *IEEE Radio Frequency Integrated Circuits Symposium*, 2015, pp. 171-174.

[6] A. Ba, V. K. Chillara, Y. -H. Liu, H. Kato, K. Philips and R. B. Staszewski, "A 2.4GHz class-D power amplifier with conduction angle calibration for −50dBc harmonic emissions," *IEEE Radio Frequency Integrated Circuits Symposium*, 2014, pp. 239-242.

[7] M. Babaie, F. Kuo, H. R. Chen, L. Cho, C. Jou, F. Hsueh, M. Shahmohammadi and R. B. Staszewski, "A Fully Integrated Bluetooth Low-Energy Transmitter in 28 nm CMOS With 36% System Efficiency at 3 dBm," in *IEEE Journal of Solid-State Circuits*, vol. 51, no. 7, pp. 1547-1565, July 2016.

[8] F. Chen, Y. Li, D. Liu, W. Rhee, J. Kim, D. Kim and Z. Wang, "A 1mW 1Mb/s 7.75-to-8.25GHz chirp-UWB transceiver with low peak-power transmission and fast synchronization capability," *IEEE International Solid-State Circuits Conference*, 2014, pp. 162-163.

[9] Y. -H. Liu *et al.*, "A680 µW Burst-Chirp UWB Radar Transceiver for Vital Signs and Occupancy Sensing up to 15m Distance," *IEEE International Solid- State Circuits Conference*, 2019, pp. 166-168.

[10] N. O. Sokal and A. D. Sokal, "Class E-A new class of high-efficiency tuned single-ended switching power amplifiers," in *IEEE Journal of Solid-State Circuits*, vol. 10, no. 3, pp. 168-176.

[11] M. Silva-Pereira and J. Caldinhas Vaz, "A Single-Ended Modified Class-E PA With HD2 Rejection for Low-Power RF Applications," in *IEEE Solid-State Circuits Letters*, vol. 1, no. 1, pp. 22-25, Jan. 2018.

[12] K. Natarajan, D. Gangopadhyay and D. Allstot, "A PLL-based BFSK transmitter with reconfigurable and PVT-tolerant class-C PA for medradio & ISM (433MHz) standards," *IEEE Radio Frequency Integrated Circuits Symposium*, 2013, pp. 67-70.

[13] M. A. A. Ibrahim and M. Onabajo, "A Low-Power BFSK Transmitter Architecture for Biomedical Applications," in *IEEE Transactions on Circuits and Systems I: Regular Papers*, vol. 67, no. 5, pp. 1527-1540, May 2020.

A Pipeline ADC with Negative *C*-assisted SC Amplifier Canceling Gain Error and Nonlinearity

Woojin Jang[1,2], Gyeong-Gu Kang[1], Yong Lim[2], and Hyun-Sik Kim[1]

[1]School of Electrical Engineering, KAIST, Daejeon, Korea, [2]Samsung Electronics Co., Ltd., Hwaseong, Korea
E-mail: {wjinjang, hyunskim}@kaist.ac.kr

Abstract—A 12-b pipeline ADC based on a new negative *C*-assisted MDAC is presented. Proposed negative *C* scheme aids to cancel gain error and nonlinearity by generating negative charges at the summing node during the amplification phase. A design strategy for ensuring stability even when the negative *C* coexists is also introduced in this work. In addition, a highly G_m-linearized low-gain amplifier is proposed for negative *C* implementation. The prototype ADC fabricated in a 28nm CMOS achieves 54.7dB SNDR and 71.8dB SFDR without any calibration at 320MS/s, demonstrating that the proposed scheme improves SFDR by +32dB across the Nyquist band.

Keywords—*Analog-to-digital converter (ADC), multiplying digital-to-analog converter (MDAC), gain enhancement, negative capacitance, nonlinearity cancellation, pipeline ADC*

I. INTRODUCTION

In modern CMOS technology, it has become more challenging to implement high-gain OTAs for ADCs due to the reduction of supply voltage and the intrinsically low gain of transistors. Instead of cascode architecture, multi-stage amplifiers have been often utilized for low-voltage operation, but they require significant power overhead as well as a complicated frequency compensation. Thus far, many studies [1]–[7] were reported to compensate for non-idealities of low-gain OTA, such as gain error and nonlinearity.

Fig. 1 shows simplified diagram examples of closed-loop switched-capacitor (SC) amplifiers for pipeline ADCs with existing correction schemes. Assuming the SC amplifier has a non-ideal transfer function of $f(\cdot)$, including gain error and distortion, the on-chip calibration approaches [2], [4] are capable of correcting the errors in both analog and digital domains with an inverse transfer function, $f^{-1}(\cdot)$. However, long convergence times are hard to be unavoidable in conjunction with additional digital hardware resources. Several off-chip calibrations [1], [3] can be applied although these require higher system cost. Alternative solutions using the correlated level shifting (CLS) [5] and digital amplifier (DA) [6] were demonstrated as illustrated on the right-side of Fig. 1, to minimize the errors of $f(\cdot)$. On the other hand, the extra clock phase to realize CLS or DA could complicate the ADC design and reduce the conversion speed. The gain error shaping (GES) technique [7], inspired by the noise-shaping notion, was recently introduced. However, it necessitates the use of oversampling ratio (OSR), and its additional feedback behavior may limit the maximum available ADC speed.

In this paper, we present a new negative *C*-assisted SC amplifier technique that successfully cancels the non-idealities caused by the limited gain of OTAs in the pipeline ADC. It has not only a low cost without on-/off-chip

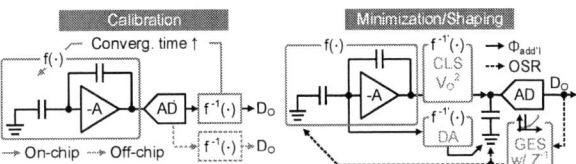

Fig. 1. Simplified diagrams for residue amplifier's error suppresion techniques [1]–[7].

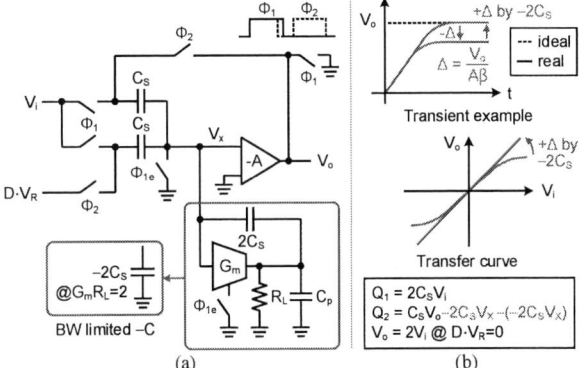

Fig. 2. (a) Proposed negative *C*-assisted MDAC and (b) its residue output.

calibration, but also the potential to maximize the sampling rate due to the elimination of extra clock phase and OSR.

II. PROPOSED NEGATIVE *C*-ASSISTED SC AMPLIFIER

A. Proposed Concept

The proposed approach is primarily based on the negative capacitance (negative *C*) technique, as shown in Fig. 2. In a conventional SC amplifier without the proposed negative *C*, non-zero V_X, which is caused by the OTA's low gain (A), creates an error charge of $-2C_S \cdot V_X$, resulting in the V_o distortion. On the other hand, by connecting the bandwidth-limited negative *C* (equivalently $-2C_S$ from $G_m R_L = 2$) at the summing node (V_X) of the main SC amplifier, compensation charges are additionally driven into V_X, which cancel the error charges of $-2C_S \cdot V_X$ during amplification phase. As a result, even with a low-gain OTA (i.e., $V_X \neq 0$), a desirable output V_o (= $2V_i$) is achievable without gain error and nonlinearity.

B. Stability

As shown in Fig. 2, non-idealities due to low gain of the OTA can be effectively corrected without any calibration if the negative *C* is equivalent to $-2C_S$. However, a concern may arise as to a stability issue in the proposed technique

978-1-6654-8495-4/22 $31.00 © 2022 IEEE

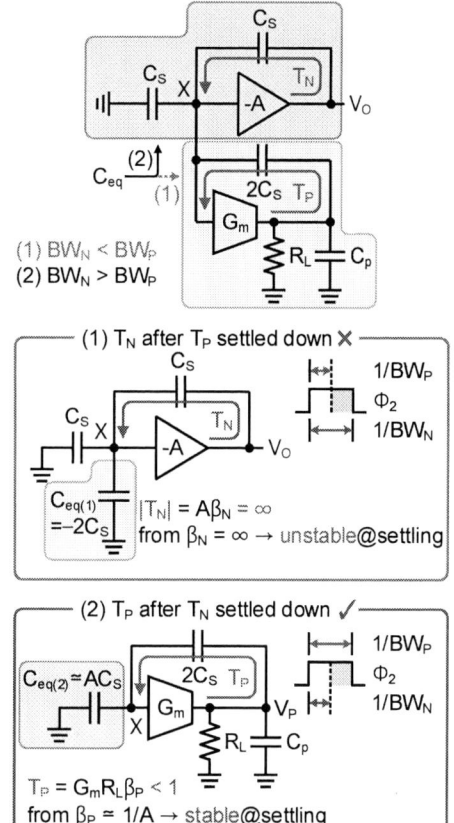

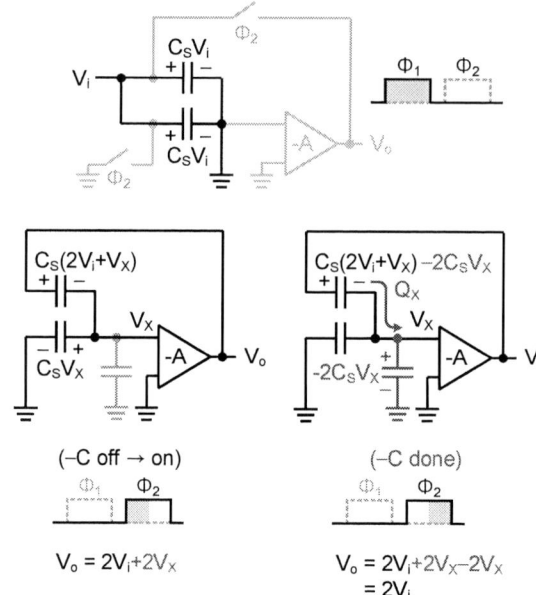

Fig. 3. Design approach for stable operation of negative C circuit.

Fig. 4. Detailed operation during sampling and amplification phase.

because the feedback factor β (from V_o to V_x) of the main SC amplifier becomes infinite owing to the negative C circuit. In this section, a design strategy to ensure stability will be presented even if the negative C is perfectly matched to $-2C_S$.

Fig. 3 shows the design approach for the stable operation of the negative C-assisted SC amplifier. To guarantee stability, the bandwidth (BW_P) of the positive feedback loop (T_P) in the negative C circuit must be lower than the BW_N of the main SC amplifier's negative feedback loop (T_N) during the amplification phase. If BW_P is wider than BW_N, the negative C circuit completes "$C_{eq(1)} = -2C_S$" too early before the main T_N has settled. In this scenario, the feedback factor β_N (from V_o to V_x) of T_N becomes to be theoretically infinite due to $C_{eq(1)}$ connected to the node X, leading to unstable V_o.

On the contrary, when BW_P is lower than BW_N, the feedback factor β_P (from V_P to V_x) for the perspective of the negative C circuit's T_P is lowered to roughly $1/A$ in virtue of "$C_{eq(2)} = A \cdot C_S$" by T_N. The open-loop gain ($G_m R_L$) of the negative C circuit is designed to 2 for the generation of $-2C_S$. Therefore, the loop gain T_P, determined as $G_m R_L \beta_P$, can be ensured to be less than unity; it implies the positive feedback T_P is stable. In this design, the OTA (A) of the main SC amplifier is implemented with a ring amplifier [8], and it enters the steady-state after settling. Accordingly, once the T_N's task is finished, even if $\beta_N \approx \infty$, the stability is unaffected, and only the error charge of $-2C_S \cdot V_x$ is canceled. In summary, the criterion of "$BW_N > BW_P$" guarantees the stability of the proposed negative C-assisted SC amplifier.

Another point of concern is the right-half-plane (RHP) pole visible from T_N if |negative C (= $C_{eq(1)}$)| is unwantedly greater than $2C_S$. However, since the negative C is effectively valid and cancels the capacitances after T_N enters DC steady state, instability caused by the RHP pole formed later is hard to occur. This is another reason why the requirement of "$BW_N > BW_P$" must be satisfied in our design.

C. Detailed Operation

As shown in Fig. 4, the proposed negative C-assisted MDAC has two behavioral phases as the conventional MDAC: sampling phase and amplification phase. The input V_i is sampled on two sampling capacitors (C_S) during the sampling phase (Φ_1). At the beginning of the amplification phase (Φ_2), the MDAC acts like a typical SC amplifier before the negative C is fully effective. The input charge sampled on the bottom C_S is transferred to the feedback C_S, but error charges due to non-zero V_X produced by the finite gain (A) remain on each C_S. The negative C is then activated slightly later due to the lower bandwidth, which does not precede the typical SC-amplifying operation, ensuring stability.

While the negative C circuit is being settled, the negative charge Q_X (= $-2C_S \cdot V_X$) is delivered from the feedback capacitor to the activated negative C because the charge of bottom C_S is held, and the Q_X has to be absorbed from somewhere. This charge absorption (or shift) by the negative C cancels the gain error and nonlinearity, yielding the ideal output ($V_o = 2V_i$), as described in Fig. 2. Note that the time to cancel the error by the negative C can be shortened further under the limitation of "$BW_P < BW_N$", although the Φ_2 phase in Fig. 3 and Fig. 4 is depicted as being separated into two equal sub-steps for easier understanding. Also, it can be designed to considerably overlap with the conventional amplification phase.

The proposed scheme benefits from greatly relaxed performance requirements of the OTA, minimized loss of sampling rate due to the absence of an extra clock phase, and low cost without on-/off-chip calibration. Furthermore, because this technique does not need to be combined with the bottom-plate sampling [4], it can be widely applied to not just ADCs, but also SC-amplifier-based standalone DACs.

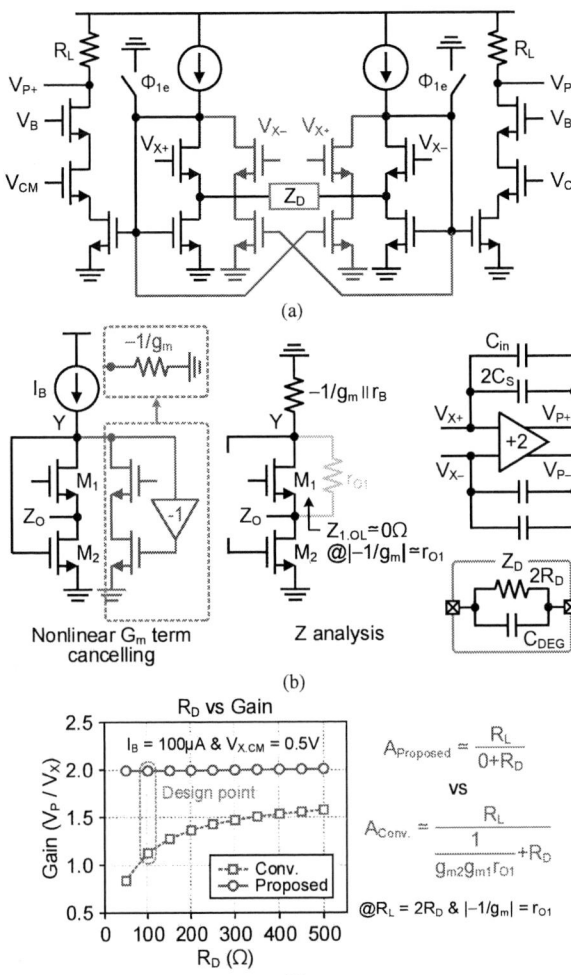

(a)

(b)

(c)

Fig. 5. (a) Low-gain amplifier with the proposed locally positive feedback loops in negative C circuit, (b) impedance analysis and fully-differential negative C implementation, and (c) simulated gain with respect to R_D.

D. Implementation of Negative C

A fully differential ring amplifier (ringamp) designed with a minimum channel length was adopted for the main OTA with passive SC CMFB [8] in this work. The open-loop DC gain (A) of the ringamp was simulated to be about 40dB, which is not high enough to obtain >7b residue accuracy [3]. To compensate for the gain error due to low A, the equivalent $C_{eq(1)}$ of the negative C needs to be equal to $-C_S \cdot A_{CL}$, where A_{CL} is the MDAC's closed-loop gain ($A_{CL} = 2$). Since $C_{eq(1)}$ is determined as $2C_S \cdot (1 - G_m R_L)$ in Fig. 2 and Fig. 3, the open-loop gain ($G_m R_L$) of the negative C circuit must be set to 2.

Fig. 5 shows the proposed design to achieve the aimed gain of 2 ($= G_m R_L = R_L/R_D$) by using a flipped voltage follower (FVF) in the negative C circuit. Because swing of V_X is somewhat suppressed even if the main OTA has gain of 40dB, a low gain amplifier for negative C can have a small output swing, which allows the exploiting of cascode and FVF. However, many FVF-based residue amplifiers have been shown in earlier studies to have a nonlinear relationship between the degenerative resistor R_D and the load resistor R_L. Calibration was thus necessary to get the appropriate residue transfer gain, or they were utilized to process only the low-resolution residue stage [9].

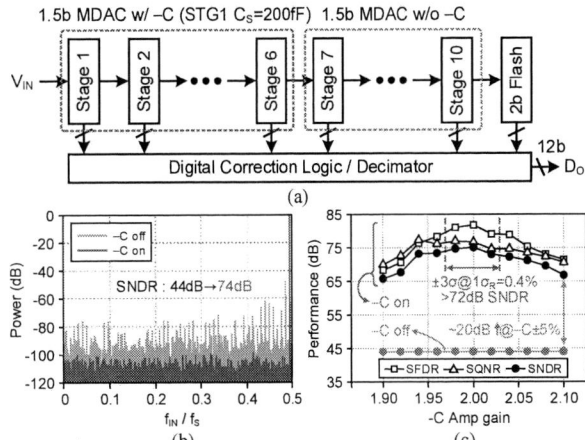

(a)

(b)

(c)

Fig. 6. (a) Top block diagram of 12-b pipeline ADC, (b) behavioral FFT results with/without negative C under $A_{OL} = 40$dB, and (c) simulated SNDRs with gain variation of negative C amplifier in 1st MDAC.

To mitigate such nonlinear dependency in the FVF structure, two locally positive feedback loops are inserted between two internal nodes in the proposed design, as depicted in Fig. 5(a). The added feedback loop additionally provides "$-1/g_m$" at node Y and can be examined with open-loop impedance (Z_{OL}), as displayed in Fig. 5(b). Without $-1/g_m$, $Z_{1.OL}$ is approximately $(r_B + r_{O1})/(g_{m1}r_{O1}) \approx r_{O1}$, where r_B ($\approx g_{m1}r_{O1}r_{O2}$) is the output resistance of the cascode current source I_B. Despite the fact that the closed-loop Z_O is reduced to $1/g_{m1}g_{m2}r_{o1}$ [9], non-zero Z_O is the dominant factor causing nonlinearity of G_m (ideally $1/R_D$) in FVF. Meanwhile, $Z_{1.OL}$ can be significantly reduced when "$-1/g_m$", closely set to r_{O1} ($|-1/g_m| \approx r_{O1}$), is coupled in parallel with r_B. It contributes to further lowering the closed-loop Z_O, and the FVF-based amplifier's voltage gain is thus more purely dictated to be R_L/R_D. Fig. 5(c) shows the simulated open-loop gain (V_P/V_X) of the negative C circuit with respect to R_D. It can be clearly observed in Fig. 5(c) that the gain loss is highly insensitive to R_D because of ultra-low Z_O reduced by the "$-1/g_m$" effect. In this work, R_L ($= 2R_D$) is designed to be 200Ω, considering BW_P of $2\pi(2C_S R_L)^{-1}$ and the mismatch between R_L and R_D.

To further negate main and auxiliary amplifiers' input parasitic capacitances, C_{in} is purposefully placed in the negative C, as shown in the right-side of Fig. 5(b). In addition, the capacitor (C_{DEG}) in Fig. 5(b) is there to compensate for the bandwidth degradation caused by the inserted positive feedback at internal nodes. Owing to limitation of matching accuracy between R_L and R_D, the negative C increases β_N (–6dB in the conventional MDAC) to about 42dB in maximum, and the loop-gain T_N ($= A\beta_N$) of 80dB is achievable, which is enough to amplify residue within >11-b accuracy.

III. IMPLEMENTATION OF 12-BIT PIPELINE ADC

A. Architecture

A 12-b pipeline ADC was implemented as proof of the concept prototype to demonstrate the effectiveness of the proposed negative C-assisted MDAC scheme. Fig. 6(a) shows the overall architecture. The ADC consists of ten 1.5-b MDAC stages and a 2-b flash back-end stage. The negative C technique is applied to the first six MDACs. The MDAC stages and C_S are scaled down by a factor of 2 up to stage 4

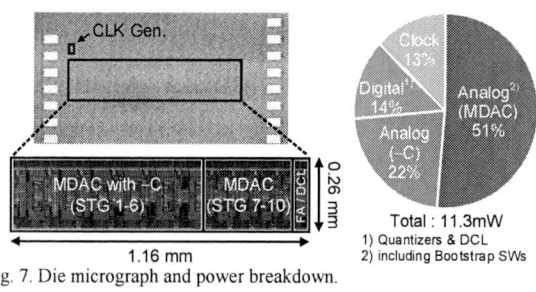

Fig. 7. Die micrograph and power breakdown.

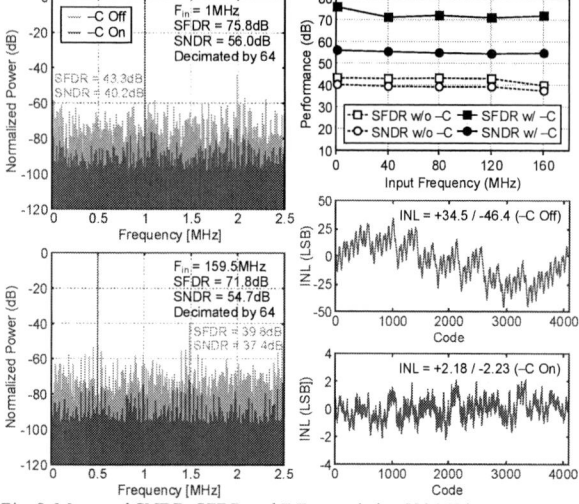

Fig. 8. Measured SNDR, SFDR and INL sampled at 320 MS/s

TABLE I. PERFORMANCE SUMMARY AND COMPARISON

	JSSC'18 [1]	VLSIC'17 [3]	JSSC'08 [5]	ISSCC'17 [6], [10]	JSSC'20 [7]	**This work**
Process	28nm	28nm	180nm	28nm	40nm	**28nm**
Architecture	Pipeline	Pipeline	Pipeline	Pipeline SAR	Pipeline SAR	**Pipeline**
Res.[bit]	12	12	12	12	N/A	**12**
Gain Error Suppression	Off-chip Cal.	Off-chip Cal.	CLS	DA	GES	**Negative C**
F_S [MS/s]	280	600	20.2	160	100	**320**
F_{in} [MHz]	137.5	293	10	80	1	**1/159.5**
SNDR[a] [dB]	64	56.3	61	61.1	75.8	**56.0/54.7**
SFDR[a] [dB]	77	69.2	68	N/A	92.9	**75.8/71.8**
SFDR↑[b] [dB]	34	26.5[c]	N/A	10.0[d]	12.4	**32.5/32.0**
Supply [V]	1.0	0.9	1.2	0.7	1.0	**1.0**
Power [mW]	13[e]	14.2[e]	7.5	1.9	1.54	**11.3**
FoM$_W$ [fJ/c.s.]	35.8	44.3	405.0	12.8	12.2	**68.5/79.6**
Area [mm²]	0.22	0.62	2.3	0.097	0.061	**0.31**

[a] At F$_{in}$, [b] Improvement by the proposed technique, [c] Estimated from FFT figure [3]
[d] From simulated result [10], [e] Excluding calibration

band. The measured INL is +2.18/-2.23LSB at 12-b level with the negative C activation.

Table I summarizes the proposed ADC performance in comparison with previous state-of-the-art ADCs focusing on accuracy enhancement. The presented ADC shows a Walden FoM of 79.6fJ/conversion-step for a Nyquist frequency input. This work, which presents a novel approach to resolving gain error and nonlinearity, obtains the fastest sampling speed (F_S) among non-calibrated gain error suppression ADCs while offering a competitive +32dB SFDR increment comparable to that of the off-chip calibrated ADCs [1], [3].

REFERENCES

[1] R. Sehgal, F. Van Der Goes and K. Bult, "A 13-mW 64-dB SNDR 280-MS/s pipelined ADC using linearized integrating amplifiers," in *IEEE J. Solid-State Circuits*, vol. 53, no. 7, pp. 1878-1888, July 2018.

[2] A. M. A. Ali *et al.*, "A 16-bit 250-MS/s IF sampling pipelined ADC with background calibration," in *IEEE J. Solid-State Circuits*, vol. 45, no. 12, pp. 2602-2612, Dec. 2010.

[3] J. Lagos, B. Hershberg, E. Martens, P. Wambacq and J. Craninckx, "A single-channel, 600Msps, 12bit, ringamp-based pipelined ADC in 28nm CMOS," in *IEEE Symp. on VLSI Circuits*, June. 2017, pp. C96-C97.

[4] Y. Miyahara, M. Sano, K. Koyama, T. Suzuki, K. Hamashita and B. Song, "A 14b 60 MS/s pipelined ADC adaptively cancelling opamp gain and nonlinearity," in *IEEE J. Solid-State Circuits*, vol. 49, no. 2, pp. 416-425, Feb. 2014.

[5] B. R. Gregoire and U.-K. Moon, "An over–60 dB true rail-to-rail performance using correlated level shifting and an opamp with only 30 dB loop gain," in *IEEE J. Solid-State Circuits*, vol. 43, no. 12, pp. 2620–2630, Dec. 2008.

[6] K. Yoshioka *et al.*, "A 0.7V 12b 160MS/s 12.8fJ/conv-step pipelined-SAR ADC in 28 nm CMOS with digital amplifier technique," in *Proc. IEEE Int. Solid-State Circuits Conf. (ISSCC)*, Feb. 2017, pp. 478–479.

[7] C. Hsu, T. R. Andeen and N. Sun, "A pipeline SAR ADC with second-order interstage gain error shaping," in *IEEE J. Solid-State Circuits*, vol. 55, no. 4, pp. 1032-1042, April 2020.

[8] B. Hershberg *et al.*, "A 4-GS/s 10-ENOB 75-mW ringamp ADC in 16-nm CMOS with background monitoring of distortion," in *IEEE J. Solid-State Circuits*, vol. 56, no. 8, pp. 2360–2374, Aug. 2021.

[9] K.-J. Moon, D.-S. Jo, W. Kim, M. Choi, H.-J. Ko and S.-T. Ryu, "A 9.1-ENOB 6-mW 10-bit 500-MS/s pipelined-SAR ADC with current-mode residue processing in 28-nm CMOS," in *IEEE J. Solid-State Circuits*, vol. 54, no. 9, pp. 2532-2542, Sept. 2019.

[10] K. Yoshioka *et al.*, "Digital amplifier: a power-efficient and process-scaling amplifier for switched capacitor circuits," in *IEEE Transactions on Very Large Scale Integration (VLSI) Systems*, vol. 27, no. 11, pp. 2575-2586, Nov. 2019.

to save power. To adjust the dead-zone of the ringamp, externally voltage-biased CMOS resistors are used [3]. All transistors designed with a minimum channel length are low-V_T devices for a supply of 1V.

B. Behavioral Verification

Fig. 6(b) and Fig. 6(c) show the behavioral simulation results with the negative C variations in 1st MDAC under OTA's gain (A_{OL}) of 40dB. An SNDR of >72dB, including only quantization noise, was achieved within resistance (R_D & R_L) variation of estimated ±3-sigma; It validates that the proposed technique satisfies the performance criteria of 11-b residue accuracy for 1st-stage MDAC. In Fig. 6(c), the stable results for gain (R_L/R_D) > 2 also confirm that the proposed technique does not cause instability even if the negative C cancels the capacitance more than intended, as discussed in Section II-B. Despite a ±5% variation in negative C value, an improvement of ≥20dB in SNDR can be expected.

IV. MEASUREMENT RESULTS

The prototype 12-b pipeline ADC was fabricated in a 28-nm CMOS process, as shown in Fig. 7. The core area of ADC is 0.31 mm², including the negative C circuits occupying less than 8% of the total area. The full-scale signal range is 1.6V$_{pp,d}$. It consumes 11.3mW with a 1.0V supply, including 8.3mW by analog blocks (2.5mW for negative C), 1.5mW by clock circuits, and 1.5mW by digital circuits at a sampling rate of 320-MS/s.

Fig. 8 shows the measured output spectrum for a 1MHz and a 159.5MHz input signal. At the Nyquist input frequency, the ADC achieves SNDR of 54.7dB and SFDR of 71.8dB, respectively. The proposed negative C technique is proven with the SFDR improvement of +32dB across the Nyquist

A 2GHz 2-bit Continuous-Time Delta Sigma ADC with 2GHz chopper achieving 12nV/sqrt(Hz) 1/f noise at 153kHz and -104.7dBc THD in 30MHz BW

P. Cenci[1], H. Brekelmans[1], S. Bajoria[1], M. Ganzerli[1], B. Burdiek[2], R. Rutten[1], Y. Gao[1], M. Bolatkale[1], P. Swinkels[3], L. Breems[1]

[1]NXP Semiconductors, Eindhoven, the Netherlands, [2]NXP Semiconductors, Hamburg, Germany, [3]NXP Semiconductors, Nijmegen, the Netherlands

Abstract—**This paper presents a 2GHz 2-bit continuous-time Delta Sigma ADC employing a 2GHz chopper to achieve low 1/f noise while maintaining high spectral purity over a 30MHz bandwidth. The proposed ADC achieves 12nV/(sqrt(Hz)) noise density at 153kHz, 78.5dB SNDR, -104.7dBc THD and better than 122dBFS SFDR (excluding HDx) in 30MHz BW. It is fabricated in a 28nm CMOS process consuming 61.4mW.**

Keywords—ADC, Delta Sigma, low flicker noise, high linearity, GHz chopper, wide bandwidth, 28nm, calibration

I. INTRODUCTION

The decreasing feature size of advanced CMOS technologies allows for higher speed data converters thanks to the reduced device parasitic and consequently a smaller latch time constant and wider bandwidth circuits. As a result, high-resolution $\Delta\Sigma$ ADCs at ever increasing speeds and wider bandwidths are emerging [1]. At the same time, a significant increase of the device 1/f noise, with order of magnitude process variation, is observed which can become a dominant noise contribution at tens up to hundred MHz bandwidth. Chopping in a $\Delta\Sigma$ modulator is a well-known and effective technique to reduce 1/f noise but has some serious challenges. For the highest 1/f reduction it is beneficial to chop the amplifier at a low frequency in order to have an accurate 50% chopping clock duty cycle for example. However, the chopper clock frequency and its harmonics will be located at positions where the $\Delta\Sigma$ modulator quantization noise is high, which can degrade the in-band noise and spurious performance due to downfolding of quantization noise. A solution to this problem is to use a FIR DAC as the main feedback DAC of the modulator that can create notches in the quantization noise spectrum at the frequency bands that can be downfolded by the chopping operation. However, this solution effectively works for narrowband applications (25-250kHz) but is not suitable for wideband $\Delta\Sigma$ ADCs with tens of MHz bandwidth.

In this paper the design of a 30MHz bandwidth, 78.5dB dynamic range highly linear ADC in 28nm CMOS is presented that achieves -105dBc THD and >122dBFS (non-harmonic) SFDR for analog and digital broadcast radio application. To mitigate 1/f noise the proposed ADC has an embedded chopper running at the 2GHz sampling rate of the modulator. At the sampling frequency (Fs), the quantization noise is highly suppressed due to the noise shaping, avoiding noise degradation due to the chopper non-idealities. Clocking the chopping circuitry at such a high rate poses some

important challenges. The chopping will increase thermal noise and reduce the amplifier loop gain, the latter degrading the linearity performance of the modulator. Moreover, it is difficult at 2GHz to guarantee an accurate 50% duty cycle to effectively remove the 1/f noise. The design in this paper has overcome the abovementioned problems and it is demonstrated that a multi-GHz chopper can be embedded inside a high-performance wideband $\Delta\Sigma$ ADC without degrading the noise, linearity and spectral purity.

II. CT $\Delta\Sigma$ ADC ARCHITECTURE WITH GHZ CHOPPER

The implemented 4th order continuous-time (CT) $\Delta\Sigma$ modulator includes four integrators, internal feedforward paths, 2-bit quantizer, excess loop delay compensation (ELD), and a feedback DAC. The flash quantizer offset voltages and DAC mismatch errors are digitally calibrated in the background [6]. Fig. 1 shows the simplified schematic of the ADC with a chopped first integrator including the blocks used for the calibration of quantizer and DAC. The loopfilter implements two complex-conjugate pole pairs with a cascade of four RC integrators employing pseudo differential 3-inverter-based amplifiers. Fig. 2 shows the amplifier schematic including chopper switches with a timing diagram.

The in-band noise performance of the CT$\Delta\Sigma$ modulator is degraded by 1/f noise contributed mainly by the first inverter I1 of the first integrator. To reduce 1/f noise one can choose large devices for I1 but this leads to a loss of gain-bandwidth due to larger parasitic capacitance and device operation at reduced current densities, in turn leading to loss of linearity of the first integrator crucial for overall ADC performance. An alternative technique consists of chopping the first inverter I1 as shown in Fig. 1. The signal is up-converted to chopping frequency (Fc) (and odd harmonics), amplified and again down-converted by the output chopper. Since flicker noise of I1 at Fc is almost absent, the chopping prevents flicker noise of first inverter from degrading the SNR. Flicker noise of I1 will still be up converted to Fc by output chopper switches but sufficiently attenuated inside A_1. Chopping of I1 avoids need for large MOS devices, favorable for the gain-bandwidth, nevertheless effectively amplifying signal at Fc will reduce the available open loop gain of A_1.

III. NON-IDEALITIES OF CHOPPER CLOCK

The duty-cycle of the chopping clock driving the chopper switches and its phase alignment with respect to DAC1 clock is of importance for maintaining low 1/f noise and high spectral purity over the wide bandwidth. Deviations from 50%

978-1-6654-8495-4/22 $31.00 © 2022 IEEE

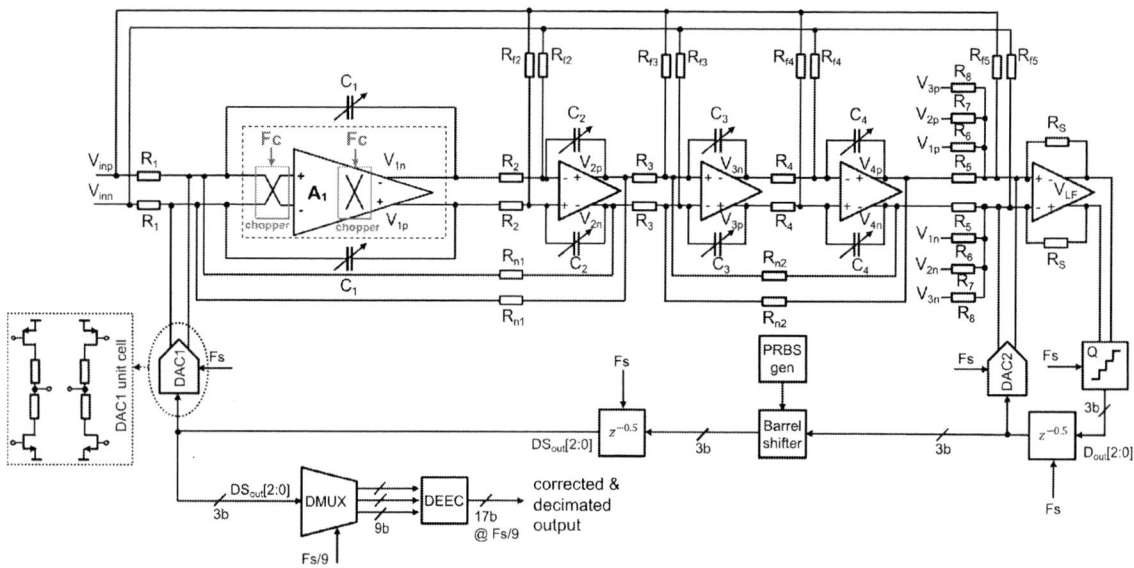

Fig. 1: Architecture of the 4th order CT $\Delta\Sigma$ modulator with embedded GHz chopper.

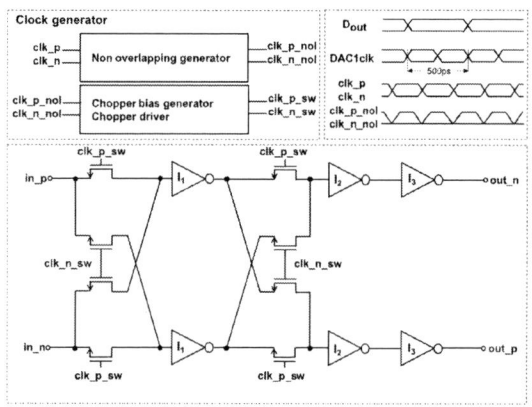

Fig. 2: Block diagram of chopped amplifier with time diagram

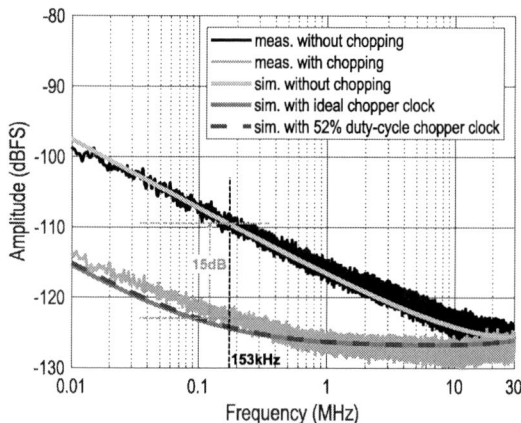

Fig. 3: Frequency spectrum of simulated and measured noise without input signal with chopper ON/OFF (RBW=153Hz).

chopper clock duty-cycle limit the 1/f noise attenuation due to non-zero direct transfer of I1 and subsequent 1/f noise leakage in signal path.

As shown in Fig. 2, a clock generator is used that converts the differential input clock (50%±2%) to non-overlapping chopper clock signals (clk_p_nol and clk_n_nol) to avoid shorting of amplifier internal nodes, maintain a high gain that ensure high linearity without compromising the noise transfer function (NTF) of the modulator. Chopper clocks are phase shifted by quarter of a period with respect to DAC1 clock such that the DAC1 output feedback current is settled and does not introduce inter-symbol interference (ISI).

Fig. 3 shows the simulated voltage noise density of the 4th order loopfilter for an unchopped and chopped amplifier A_1 with 50% and 52% duty-cycle of the chopper clock overlayed with measurement results. The 1/f noise at 153kHz is suppressed by 15dB when a 50% duty-cycle chopping clock is used. A chopper clock with a 52% duty-cycle will increase 1/f noise by approx. 0.8dB at 10kHz. The thermal noise of the amplifier at multiples of the chopping frequency is folded due to the chopping operation, slightly increasing the in band

thermal noise of the ADC. However, when amplifier A_1 is chopped, the noise density is reduced across the full ADC bandwidth since the 1/f noise contribution is still dominant up to several tens of MHz.

IV. COPPER BIAS GENERATOR AND CHOPPER DRIVER

A non-overlapping clock signal drives the chopper transistors via an AC coupling network. The supply voltage of the inverter-based amplifier (delivered by an on chip LDO not shown in Fig. 1&2) varies with process and temperature to achieve a constant gm biasing. The bias voltage of the chopper transistors also needs to track this variation. A dedicated biasing circuit, shown in Fig. 4, uses a replica of the chopper switch to establish a Vgs voltage at which the chopper transistors have a very large but well defined channel resistance R_{off}. This is realized by a regulation loop that equates the NMOS R_{off} to a value V_R/I_R. Via a voltage-to-current converter and a resistor R_2 approximately half the supply voltage is added on top of the V_{off} voltage. The result is a DC biasing voltage V_{ref} around which the bootstrapped clock signal operates. The regulation of the chopper switch

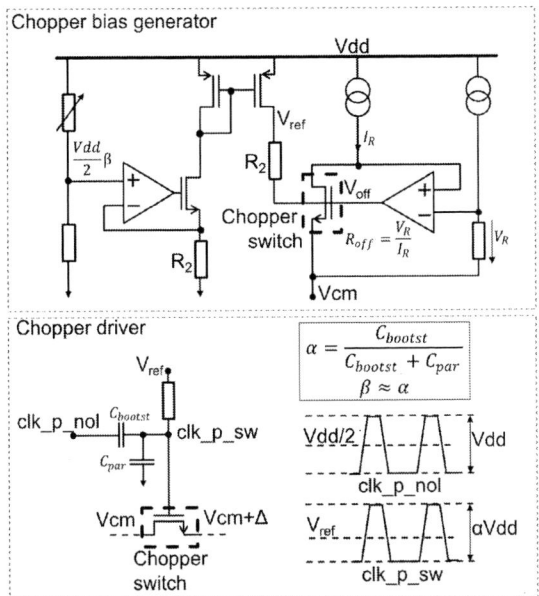

Fig. 4: Chopper bias generator and chopper driver

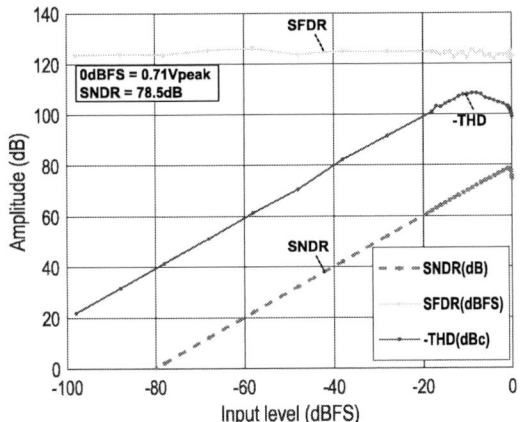

Fig. 5: Measured SNDR/THD/SFDR vs. input level with a 9.334MHz input (SFDR does not include HDx).

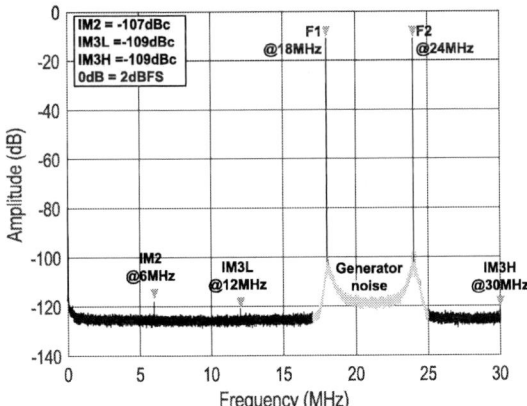

Fig. 6: Frequency spectrum of a two-tone input signal of -9dBFS at F1=18MHz and F2=24MHz (RBW=321.5Hz) (worst typical sample).

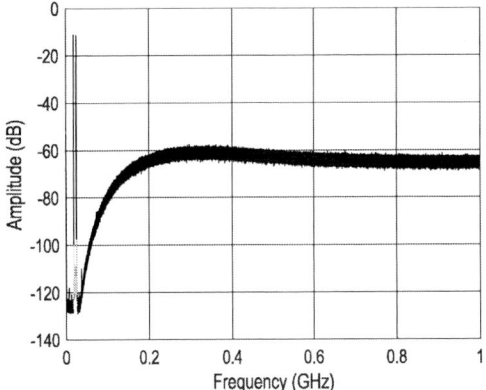

Fig. 7: Frequency spectrum of a two-tone input signal of -9dBFS at F1=18MHz and F2=24MHz (RBW=321.5Hz) (Full spectrum).

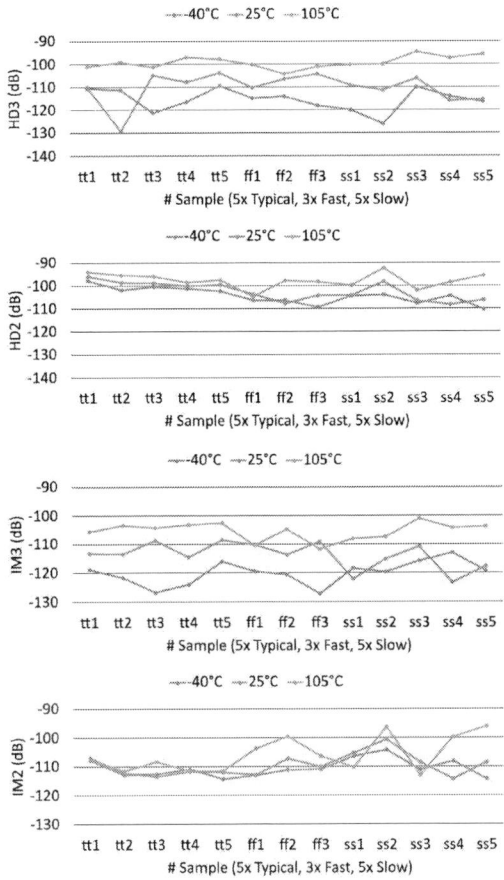

Fig. 8: HD2, HD3, IM2 and IM3 over corner sample and temperature. HD2 and HD3: one-tone of -5dBVsup at 9.334MHz. IM2 and IM3: two-tones of -11dBVsup at F1=20MHz and F2=24MHz. (RBW=0.4Hz).

"OFF" state ensures proper turn OFF of chopper switches under all PVT conditions, without unnecessarily compromising the R_{on} in conducting state. Since the part of V_{ref}, which replicates the clock swing, $\beta*Vdd/2$ is slightly smaller than actual damped bootstrapped clock swing $\alpha*Vdd/2$, where the damping is caused by parasitic capacitance. Residual conductivity of not fully turned OFF chopper switches leads to increased intermodulation distortion and increased noise. Bias voltage V_{ref} can be adjusted with a programmable resistor divider.

V. CALIBRATION

The feedback DAC consists of 3-unit elements using a time-interleaved return-to-open scheme [6], any mismatch between the elements degrades the spectral purity of the ADC. Per clock cycle, the unit elements are randomly shuffled, turning distortion into noise. This noise is reduced by a digital correlation-based DAC error estimation and correction (DEEC) block [6]. Offsets between the 3 comparators degrade the linearity of the ADC. Each comparator has an offset DAC that is compensating the offset for that comparator. Every clock cycle, the 3 reference levels are randomly shuffled. The offset code is determined by a closed loop iterative system that equalizes the 1/0 decision ratios for all comparator outputs [6].

VI. MEASUREMENT RESULTS

The modulator core occupies 0.12mm² area and is fabricated in TSMC 28nm CMOS. Fig. 5 shows the measured SNDR/-THD/SFDR versus input amplitude for 9.334MHz input signal with 2GHz chopper enabled. This modulator achieves 78.5dB SNR & SNDR and better than 122dBFS SFDR (no HDx included). When the chopper is disabled SNR & SNDR decrease to 77.4dB while the SFDR does not change. Fig. 6 shows the measured intermodulation performance (chopper enabled) of the worst typical sample for a two-tone -9dBFS input signal at 18MHz and 24MHz. The measured third order intermodulation (IM3) and second order intermodulation (IM2) distortion are -109dBc and -107dBc respectively. The increased noise floor around the input signal tones (light grey) is due to signal generator noise. Fig. 7 reports the same measurement as in Fig. 6 focusing on the high frequency NTF, showing no out of band peaking and demonstrating stable behavior. Fig. 8 show the HD3, HD2, IM2 and IM3 for three different corners (5 samples for Typical, 3 samples for Fast and 5 sample for Slow) and 3 temperatures (-40°C, 25°C and 105°C). HD3 and HD2 measurements are with one single tone of -3dBFS at 9.334MHz and IM3 and IM2 measurements are with two tones of -9dBFS at 20MHz and 24MHz. Fig. 3 shows the measured low frequency ADC output spectrum. The 1/f noise corner is 140 kHz, limited by residual 1/f noise of I2 and I3 of the first amplifier and the subsequent amplifiers. The power consumption of the modulator (including the DMUX and clocking and chopper driver) is 61.4mW. Table I summarizes the measured performance with chopper enable and disable.

VII. CONCLUSIONS

This work demonstrates the implementation of a 2GHz CT $\Delta\Sigma$ modulator with the fastest reported chopping frequency (Fc=Fs) which is 3 orders of larger than state-of-the-art ADCs. The ADC achieves -105dBc THD in 30MHz BW and >122dBFS SFDR (non-harmonic). This performance is enabled by using a 2GHz chopper frequency (Fc=Fs), a chopper clock generator and driver which tracks PVT for optimum chopper performance, and the calibration of the feedback DAC and the comparators.

ACKNOWLEDGMENT

The authors thank Yu Lin, Georg Kisters, Bert Oude Essink, Chenming Zhang, Ferdinand Manalac, John Aerts, Vinoth Ilamurugan, Hendrik van der Ploeg, Peter Prieshof.

TABLE I.   COMPARISON WITH STATE-OF-THE-ART ADCs.

	This work		[4]	[3]	[5]	[7]
Technology	28nm		180nm	180nm	65nm	65nm
Fs (MHz)	2000		32	6.144	8	2200
BW (MHz)	30		0.25	0.024	0.024	25
OSR	33.33		64	128	166.7	44
DAC architecture	2b R-DAC		8-tap FIR DAC	12-tap FIR DAC	6-tap tri-level FIR DAC	1b R-DAC
Chopping freq. (MHz)	2000 (Fs)	Chopper disabled	2 (Fs/16)	0.256 (Fs/24)	0.67 (Fs/12)	n.a.
Peak SNDR (dB)	78.5	77.4	105.3	100.9	99.4	77
Peak SNR (dB)	78.5	77.4	108.2	101.7	101	77
THD (dBc)	-104.7	-104.4	-107.8	-107.4	-111.8	-101
NSD (dBc/Hz)	-153.3	-152.2	-160.8	-144.6	-140.7	-151
Power (mW)	61.4	59.2	24	0.265	0.134	41.4
FoM$_S$ (dB)	165.4	164.4	175.5	180.5	181.9	164.8
FoM$_W$ (fJ/conv)	148.8	162.8	319.0	60.9	36.6	143.1

$FoM_S = SNDR_{max} + 10 \cdot \log 10(BW/Power)$, $FoM_W = Power/(2^{ENOB} \cdot 2 \cdot BW)$

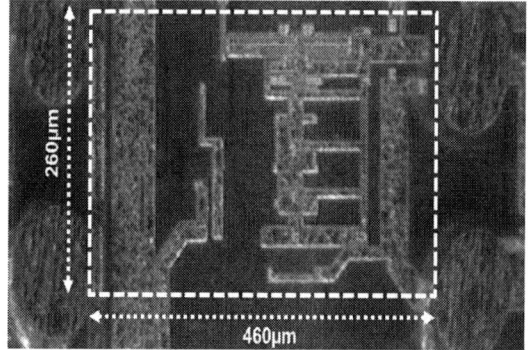

Fig. 9: Chip Micrograph.

REFERENCES

[1] M. J. Knitel, P. H. Woerlee, A. J. Scholten and A. Zegers-Van Duijnhoven, "Impact of process scaling on 1/f noise in advanced CMOS technologies," International Electron Devices Meeting 2000. Technical Digest. IEDM (Cat. No.00CH37138), 2000, pp. 463-466.

[2] S. Billa, A. Sukumaran and S. Pavan, "Analysis and Design of Continuous-Time Delta–Sigma Converters Incorporating Chopping," in IEEE Journal of Solid-State Circuits, vol. 52, no. 9, pp. 2350-2361, Sept. 2017.

[3] S. Billa, S. Dixit and S. Pavan, "Analysis and Design of an Audio Continuous-Time 1-X FIR-MASH Delta–Sigma Modulator," in IEEE Journal of Solid-State Circuits, vol. 55, no. 10, pp. 2649-2659, Oct. 2020.

[4] R. Theertham, P. Koottala, S. Billa and S. Pavan, "Design Techniques for High-Resolution Continuous-Time Delta–Sigma Converters With Low In-Band Noise Spectral Density," in IEEE Journal of Solid-State Circuits, vol. 55, no. 9, pp. 2429-2442, Sept. 2020.

[5] M. Jang, C. Lee and Y. Chae, "A 134-µW 99.4-dB SNDR Audio Continuous-Time Delta-Sigma Modulator With Chopped Negative-R and Tri-Level FIR-DAC," in IEEE Journal of Solid-State Circuits, vol. 56, no. 6, pp. 1761-1771, June 2021

[6] M. Bolatkale et al., "A 28nm 6GHz 2b Continuous-Time $\Delta\Sigma$ ADC with −101 dBc THD and 120MHz Bandwidth Using Digital DAC Error Correction," 2022 IEEE International Solid- State Circuits Conference (ISSCC), 2022, pp. 416-418.

[7] L. Breems et al., "A 2.2 GHz Continuous-Time Delta Sigma ADC With −102 dBc THD and 25 MHz Bandwidth," in IEEE Journal of Solid-State Circuits, vol. 51, no. 12, pp. 2906-2916, Dec. 2016.

6GS/s 8-channel CIC SAR TI-ADC with Neural Network Calibration

Evelyn Ware, Justin Correll, Seungjong Lee, and Michael Flynn
Electrical Engineering and Computer Science, University of Michigan
Ann Arbor, Michigan, USA
Email: evware@umich.edu, correllj@umich.edu, snjnlee@umich.edu, mpflynn@umich.edu

Abstract— **In this paper we introduce an area efficient time-interleaved charge-injection-cell SAR ADC. The prototype TI-ADC interleaves 8 CIC SAR channels for a sampling rate of 6Gs/s and a compact area of 0.00608mm². A neural network calibration algorithm corrects multiple error sources in the time-interleaved ADC. The neural calibration method effectively improves the average measured ENOB of the TI-ADC from 4.1 bits to 5.49 bits.**

Keywords—Charge injection, calibration, neural, interleaving

I. INTRODUCTION

Time-interleaving of SAR ADCs is an essential technique for high-speed analog-to-digital conversion. Traditional interleaved SAR ADCs require a large die area and have limited conversion speed due to the overhead of multiple switched-capacitor DACs. Additionally, difficulties with matching between interleaved ADC channels limit performance [1]. Common ADC non-idealities such as gain, offset, and timing errors cause signal-dependent errors in the time-interleaved output necessitating correction with sophisticated calibration techniques. We tackle these challenges with: (1) a time-interleaved charge-injected cell (CIC) SAR ADC which benefits from hardware sharing of CIC cells for a small area and (2) a hardware-friendly neural-network calibration scheme.

Machine learning provides a new approach to ADC calibration. Traditional digital and analog calibration methods for time-interleaved ADCs require an initial estimate of channel mismatches before compensation can be applied [2]-[3]. Additionally, these methods typically only target one source of error, so multiple calibration schemes are necessary to achieve a high-performance ADC. In contrast, a single neural network can correct multiple errors by directly learning a mapping to the desired output without first characterizing the error mechanisms present in the ADC. Neural-network calibration also offers flexibility as network parameters can easily be retrained to deal with changes in error conditions. Prior work on neural network calibration of ADCs has shown promising results. However, these works focus on calibrating single-channel ADCs [4]-[5] or are limited to simulation [6]-[7].

The very compact size of the CIC SAR ADC makes it an attractive candidate for time interleaving. Moreover, unlike a switched capacitor SAR, the CIC cells are reused in different bit trials enabling area efficiency. This work realizes the first time-interleaved CIC SAR ADC aided by two essential innovations: (1) sharing of CIC cells between sub-ADCs and (2) neural network correction of the ADC output due to CIC element and sub-ADC mismatch.

This work applies a neural network calibration algorithm to a 6 GS/s 8 channel 6-bit TI-CIC SAR ADC to correct for distortion, gain, and offset mismatch. Additionally, this work introduces a novel low area TI-ADC architecture using charge-injection-cell DACs. A neural network corrects gain mismatch, timing errors, offset mismatch and distortion in the CIC TIADC.

II. CHARGE-INJECTION-CELL TI-ADC ARCHITECTURE

A. Charge Injection Cell SAR ADC

In a conventional switched-capacitor DAC, the sampled charge is redistributed across a bank of capacitors to produce the DAC output. In a CIC SAR ADC, the differential input signal samples onto a pair of capacitors. The CIC cells inject quanta of charge under control of a timing signal to produce the DAC outputs. CIC cell operation allows for better control of the DAC timing, enabling interrupted settling and CIC cell reuse over successive decisions [8].

Figure 1 shows the charge-injection-cell. Charge from the sampling capacitors injects into the charge-reservoir capacitor, implemented by a long-channel NMOS, M3. Two NMOS switches (M1 and M2) control charge transfer from the sampling capacitors to the charge reservoir. A reset switch resets the charge reservoir between conversion steps, enabling hardware reuse. The sign of the sampled input determines the MSB decision. After that, the SAR conversion proceeds in a set-and-down fashion – at each step, CIC cells subtract charge from the positive or negative capacitance.

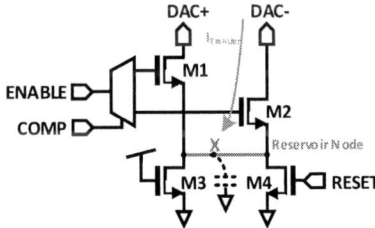

Fig. 1. A single charge injection cell (CIC) based on [8].

Critical advantages of the CIC SAR ADC are: (1) CIC cell hardware sharing and (2) interrupted settling. Interrupted settling allows faster operation than with a SC SAR ADC. In traditional SAR ADCs, the DAC settling time limits conversion speed and causes distortion if settling from previous conversion steps continues into subsequent steps. The charge injection DAC enables interrupted settling eliminating residual settling after a conversion step [8]. In addition, we can share CIC cells because we can reuse them after every bit trial. In contrast, a conventional SC SAR ADC requires dedicated capacitors for each step.

B. Time-Interleaved CIC SAR ADC

This work introduces the first time-interleaved CIC SAR ADC. We extend the hardware reuse enabled by the CIC operation to multiple interleaved sub-ADCs. We expand the CIC DAC hardware reuse concept for a single ADC [8] by sharing CIC DAC cells between multiple sub-ADCs to construct an area-efficient time-interleaved ADC. Prior work has shown the effectiveness of hardware reuse in ADCs for power and area savings [9]. This work introduces an 8 channel TI-ADC SAR with CIC DAC reuse.

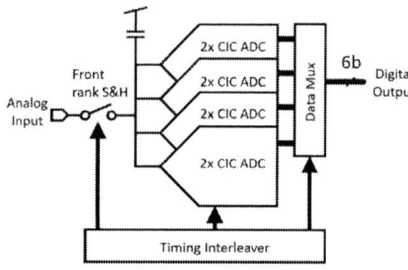

Fig. 2. Four CIC SAR ADC pairs for the 8x interleaved ADC.

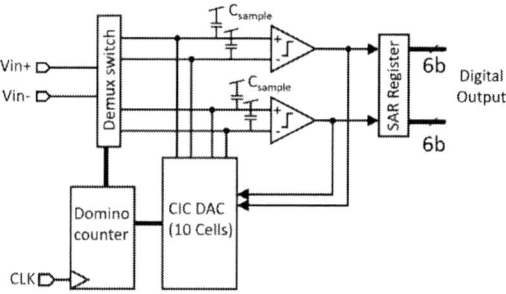

Fig. 3. Sharing of CIC DAC between two interleaved SAR sub-ADCs.

The 8x interleaved ADC is built with four units (Figure 2) – each unit functions as two sub-ADC. As shown in Figure 3, each ADC channel has its own comparator, sampling capacitors, and SAR logic but two ADC channels share a single CIC DAC. The CIC DAC is implemented with 10 unit charge injection cells, connected to the DAC lines of the two shared ADCs through NMOS switches. Each sub-ADC pair produces a complete output every four clock cycles. For example, in the overall ADC, sub-ADC 0 and sub-ADC 4 share the same DAC. One sub-ADC samples and generates the 3 MSBs while the other sub-ADC

produces the 3 LSBs. The sharing of CIC hardware provides a significant reduction in area. Furthermore, hardware sharing helps reduce the mismatch between sub-ADCs.

Figure 4 shows the timing diagram for the 10 CIC DACs shared between a pair of sub-ADCs. Sub-ADC A samples and decides the MSB in the first clock cycle. In the subsequent two cycles, 8 CIC units transfer 16 units of charge for the MSB-1 step. The 4th cycle transfers 8 units of charge for MSB-2. Sub-ADC B determines the 3 LSBs during these four clock cycles, using 2 CIC units. The first two steps transfer 4 units for LSB+2 and so on.

Each ADC channel operates at a sampling rate of 750MHz, resulting in an interleaved ADC that operates at 6GS/s. A front-rank sampler, which was successfully implemented in [10], reduces timing skew errors between ADC channels.

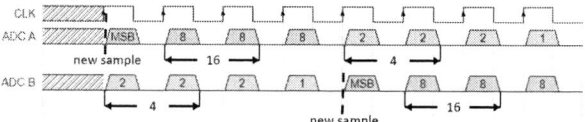

Fig. 4. Timing diagram of CIC DAC sharing between a pair of sub-ADCs. One sub-ADC uses 8 CIC cells while the other employs 2.

III. Neural Network Calibration Algorithm

A. Neural Calibration Architecture

A trained neural network calibrates the TI-ADC. The network takes as an input the converted digital value from each TI-ADC channel rescaled to be between 0 and 1. The network outputs the calibrated digital signal for each ADC channel.

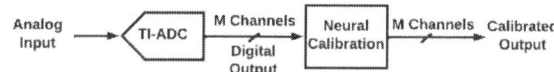

Fig. 5. TI-ADC with neural calibration.

B. Neural Network Algorithm

Neural networks can effectively learn non-linear functions between two sets of data. Their basic structure consists of an input layer, a series of hidden layers that perform data computations, and an output layer. Each hidden layer multiplies the input by a matrix of learned weights and then passes it through a non-linear activation function, such as a rectified linear unit (ReLU) or tanh function. This allows the network to learn complex non-linear function mappings.

Convolutional neural networks (CNNs) perform convolution between inputs and learned convolutional weights. 2D CNNs are commonly used in image classification to extract relevant features from image data that has spatial dependence. In this work we apply a 1D CNN, which applies convolution along the time dimension of the ADC output. This time-based CNN allows the network to learn time-dependent features of the data, which makes it well suited for correcting the time-dependent errors present in a TI-ADC.

In this work, we train a feedforward network with a 1D convolutional layer followed by two fully connected layers to map the output of the TI-ADC to an ideal quantized version of the signal. In this way, we can calibrate multiple error sources present in the ADC without having to first characterize them explicitly

The input to the network is the scaled output of each of the M channels from the TI-ADC. The hidden convolutional and fully connected layers learn the features of the ADC errors. The 1D convolutional layer performs convolution over the time dimension of each channel of the TI-ADC, which allows the network to learn time-dependent errors such as time-skew. The dense connections in the subsequent fully connected layers allow the network to correct gain and offset errors by learning relationships between channels of the TI-ADC, to normalize the gain and offset of each individual channel. The final layer produces the M-channel calibrated output. ReLU activation and batch normalization are applied between layers to improve calibration accuracy.

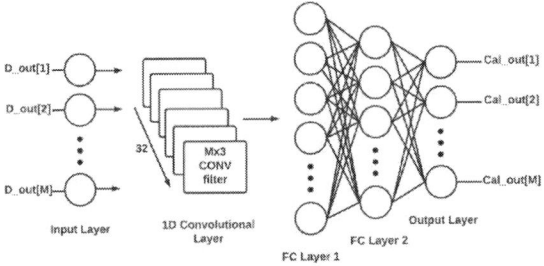

Fig. 6. Neural network architecture for ADC calibration.

C. Network Training

The training set for the network is obtained through recording the TI-ADC output for full-scale sine-wave inputs applied to the IC. 250 sine-wave frequencies are selected from a random uniform distribution over the decimated Nyquist range of the ADC. Each frequency sequence contains 2^{16} samples. An ideal quantizer produces ground truth signals.

During training, the network calculates the mean-squared error (MSE) between the non-ideal ADC output and ideal quantizer and uses a back propagation algorithm to minimize the error by updating the network weights. The ADAM (adaptive moment estimation) optimizer with a learning rate of 0.01 and minibatch size of 50 is used to train the network for 10000 epochs or until convergence is reached. To validate the network, testing is performed on signals at different frequencies from the training set. The ENOB and SNDR of the calibration network output are compared to the raw TI-ADC output.

IV. EXPERIMENTAL RESULTS

The prototype 6GS/s 8 channel 6-bit TI-CIC SAR ADC is fabricated in 28nm CMOS. The design has a total die area of

0.00608mm² and a total power consumption of 15.3mW. The TI-ADC output is decimated by a factor of 65 to simplify the digital output interface.

The pre-calibration performance of the CIC SAR TI-ADC is limited by gain and offset mismatch and non-linear distortion. Offset errors are mainly caused by comparator offset mismatch between the SAR ADC channels. Furthermore, we deliberately manually add additional mismatches to the sub-ADCs by individually adjusting the comparator reference voltages. This additional mismatch provides a further challenge for the neural-network calibration scheme.

Input signals are recorded from the Nyquist zone in the range 646 MHz to 692 MHz. The measured ENOB within this range is 4.1 bits and SNDR is 26.5 dB. A comparison of the pre and post calibration FFT is shown in Figure 8.

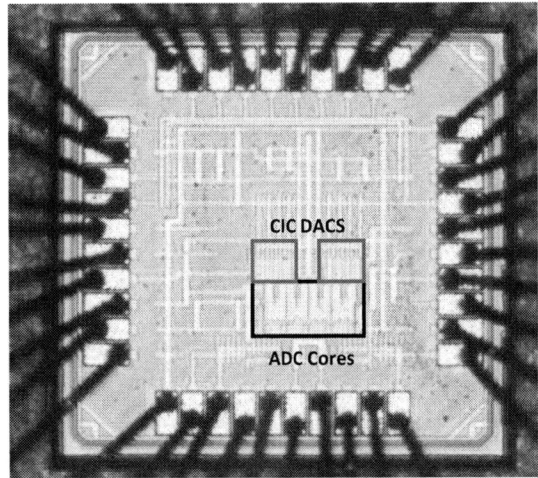

Fig.7. Die photo 28nm CMOS 8-channel CIC SAR TI-ADC.

TABLE 1. Performance summary and comparison to similar works.

	This work	[11]	[12]	[13]
Architecture	TI-ciSAR	TI-SAR	TI-SAR	TI-SAR
Technology [nm]	28	55 LP	28	28
Channels	8	8	2	8
Resolution [bits]	6	7	7	8
Supply [V]	1	1.2/1.5	1.2/0.9	1
Speed [GS/s]	6	5	2	10
Power [mW]	15.3	38	7.62	21
Area [mm²]	0.00608	0.69	0.0082	0.022

The trained network has a 1D convolutional layer with 32 length 3 filters, followed by fully connected layers containing

128 and 64 neurons. The total number of trainable parameters in the network is only 13,864, making it compact enough to be implemented in hardware.

A plot of the measured ENOB from the ADC output and after neural calibration is shown in Figure 13. The ENOB improves from 4.1 bits to >5 bits across the frequency range.

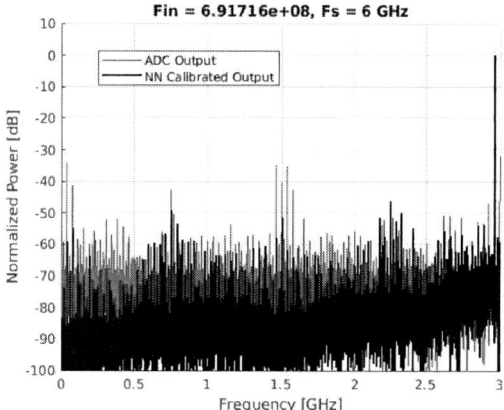

Fig. 8. FFT of raw CIC SAR TI-ADC and calibrated outputs.

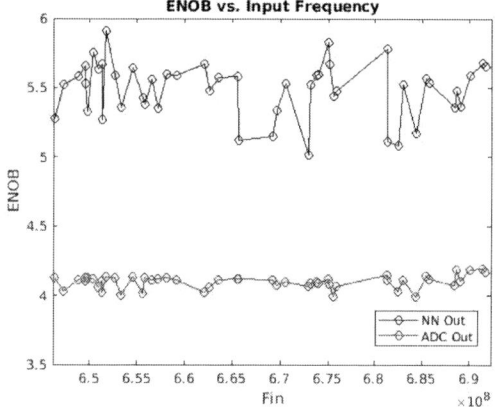

Fig. 9. Measured raw ENOB for CIC SAR TI-ADC and calibrated output.

V. CONCLUSIONS

This paper introduces a novel TI-ADC architecture implementing the charge injection cell DAC, which provides exceptional area and power efficiency. The CIC SAR TI-ADC uses 8 interleaved channels to produce an output at 6GS/s, with an area of only 0.00608 mm². A neural-network calibration system corrects multiple sources of error in the TI-ADC, including gain, offset, and distortion. The CIC SAR TI-ADC is verified experimentally in 28nm and neural calibration is applied to correct time-interleaved errors. The neural calibration algorithm provides significant performance improvements with a low complexity overhead.

REFERENCES

[1] P. Nuzzo, C. Nani, S. Saponara, L. Fanucci and G. Van der Plas, "Mixed-Signal Design Space Exploration of Time-Interleaved A/D Converters for Ultra-Wide Band Applications," 2008 Design, Automation and Test in Europe, 2008, pp. 1390-1393.

[2] S. Lee, A. P. Chandrakasan and H. Lee, "22.4 A 1GS/s 10b 18.9mW time-interleaved SAR ADC with background timing-skew calibration," 2014 IEEE International Solid-State Circuits Conference Digest of Technical Papers (ISSCC), 2014, pp. 384-385.

[3] B. Razavi, "Design Considerations for Interleaved ADCs," in IEEE Journal of Solid-State Circuits, vol. 48, no. 8, pp. 1806-1817, Aug. 2013. [4] X. Peng et al., "A Neural Network-Based Harmonic Suppression Algorithm for Medium-to-High Resolution ADCs," 2021 5th IEEE Electron Devices Technology & Manufacturing Conference (EDTM), 2021, pp. 1-3.

[5] H. Deng, Y. Hu, and L. Wang, "An efficient background calibration technique for analog-to-digital converters based on neural network," *Integration*, vol. 74, pp. 63–70, 2020.

[6] Y. Qiu, J. Zhou, Y. Liu and Y. Huangfu, "A Novel Calibration Method of Gain and Time-skew Mismatches for Time-interleaved ADCs Based on Neural Network," 2019 IEEE MTT-S International Wireless Symposium (IWS), 2019, pp. 1-3.

[7] A. Samiee, Y. Zhou, T. Zhou, and B. Jalali, "Deep analog-to-digital converter for wireless communication," AI and Optical Data Sciences II, 2021.

[8] K. D. Choo, J. Bell and M. P. Flynn, "27.3 Area-efficient 1GS/s 6b SAR ADC with charge-injection-cell-based DAC," 2016 IEEE International Solid-State Circuits Conference (ISSCC), 2016, pp. 460-461.

[9] S. Ryu, B. Song and K. Bacrania, "A 10-bit 50-MS/s Pipelined ADC With Opamp Current Reuse," in IEEE Journal of Solid-State Circuits, vol. 42, no. 3, pp. 475-485, March 2007.

[10] M. Palm, D. Mastantuono, R. Strandberg, L. Sundstrcom and S. Mattisson, "A 12b, 1 GSps TI pipelined-SAR converter with 65 dB SFDR through buffer linearization and gain mismatch correction in 28nm FDSOI," ESSCIRC 2017 - 43rd IEEE European Solid State Circuits Conference, 2017, pp. 179-180.

[11] Y. -H. Chung, C. -Y. Hu and C. -W. Chang, "A 38-mW 7-bit 5-GS/s Time-Interleaved SAR ADC with Background Skew Calibration," 2018 IEEE Asian Solid-State Circuits Conference (A-SSCC), 2018, pp. 243246.

[12] W. Jiang, Y. Zhu, C. -H. Chan, B. Murmann and R. P. Martins, "A 7-bit 2 GS/s Time-Interleaved SAR ADC With Timing Skew Calibration Based on Current Integrating Sampler," in IEEE Transactions on Circuits and Systems I: Regular Papers, vol. 68, no. 2, pp. 557-568, Feb. 2021.

[13] E. Swindlehurst et al., "An 8-bit 10-GHz 21-mW Time-Interleaved SAR ADC With Grouped DAC Capacitors and Dual-Path Bootstrapped Switch," in IEEE Solid-State Circuits Letters, vol. 2, no. 9, pp. 83-86, Sept. 2019.

A 56 GS/s 8-bit 0.011 mm^2 4x Delta-Interleaved Switched-Capacitor DAC in 16 nm FinFET CMOS

Pietro Caragiulo, Athanasios Ramkaj, Amin Arbabian and Boris Murmann

Department of Electrical Engineering, Stanford University, Stanford, CA 94305 USA

E-mail: pietroc@stanford.edu

Abstract— **This paper presents a compact 4x delta-interleaved switched-capacitor (SC) digital-to-analog converter (DAC) for digital-intensive transmitter architectures. To minimize area and leverage the strengths of FinFET technology, the implementation departs from the traditional current steering approach and consists mainly of inverters and sub-femtofarad switched capacitors. The DAC's architecture is based on parallel charge redistribution, and separates level generation, pulse timing and output power generation. The 16 nm FinFET 8-bit prototype occupies only 0.011 mm^2 while running at 56 GS/s and providing up to 0.27 V$_{pp}$ signal swing across its differential 100 Ω load. It achieves an IM3 ≤ −48 dBc and an SFDR ≥ 42 dB within its ≈10 GHz output bandwidth while consuming 280 mW from a single 0.85 V supply.**

Index Terms—**Digital-to-analog converter (DAC), interleaving, switched-capacitor circuits.**

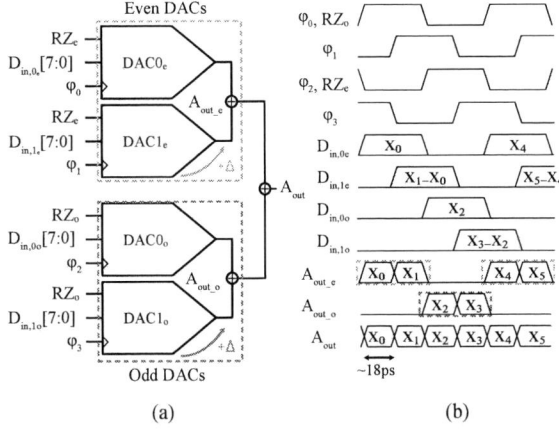

Fig. 1. (a) Block diagram and (b) timing diagram of a 4x delta-interleaved DAC.

I. INTRODUCTION

Digital-to-analog converters (DACs) are ubiquitous in modern digital communication systems [1], [2]. These applications typically leverage the latest FinFET CMOS nodes for maximum digital density and seek to integrate the data converters on the same substrate with low area overhead. Unfortunately, advanced FinFET processes exhibit increased interconnect parasitics [3], which complicates the design of traditional high-speed current-steering (CS) DACs. Specifically, FinFET technology does not shine at switching the typical milliampere-level currents at high speed due to its ultra-thin base metal layers and high contact/via resistances. Overcoming these issues through metal straps and via pillars essentially degenerates the circuit density and speed to that of a previous technology node. The end result is that the area of a high-speed DAC may even increase when ported to a FinFET technology [4], [5].

Our work has explored high-speed, moderate-resolution (8-bit) switched-capacitor (SC) DACs as a more scaling-friendly alternative to the traditional CS approach. We previously reported 16 nm FinFET CMOS prototypes operating at 14 and 28 GS/s [6], [7], demonstrating significant area reductions over state-of-the-art designs. This is achieved using SC arrays with sub-femtofarad unit capacitors, buffered into standard 50 Ω loads using inverter-based drivers. This approach leads to a FinFET-friendly, compact layout that additionally helps with mitigating timing errors that are difficult to manage in a more dispersed CS fabric.

The present work extends our SC DAC approach to 56 GS/s and achieves further area reductions, amounting to a 10x advantage over FinFET designs in the same speed category. To achieve the high output update rate, we employ four sub-DACs arranged in a "delta interleaving" scheme that operates without an output multiplexer. Section II of this paper explains the delta interleaving concept in more detail, while Section III describes the prototype IC architecture and circuit details. The measured results are summarized in Section IV and Section V concludes the paper.

II. DELTA INTERLEAVING CONCEPT

Fig. 1 illustrates our 4x delta-interleaved DAC approach. The converter is composed of four cores, each operating on one phase (ϕ_i) of a quarter-rate quadrature clock to achieve an output update interval of ≈18 ps. The cores' outputs hierarchically combine in groups of two [indicated as even/odd in Fig. 1 (a)] and within each group, the output of DAC0 is directly proportional to the input code, X_i, while the output of DAC1 adds/subtracts the difference, Δ, between two consecutive codes. Each group's output is returned to zero by asserting the signal $RZ_{e/o}$ and is finally combined to generate the DAC output at the aggregate update rate. Contrary to conventional time-interleaving (see e.g., [7]), delta interleaving does not

978-1-6654-8495-4/22 $31.00 © 2022 IEEE

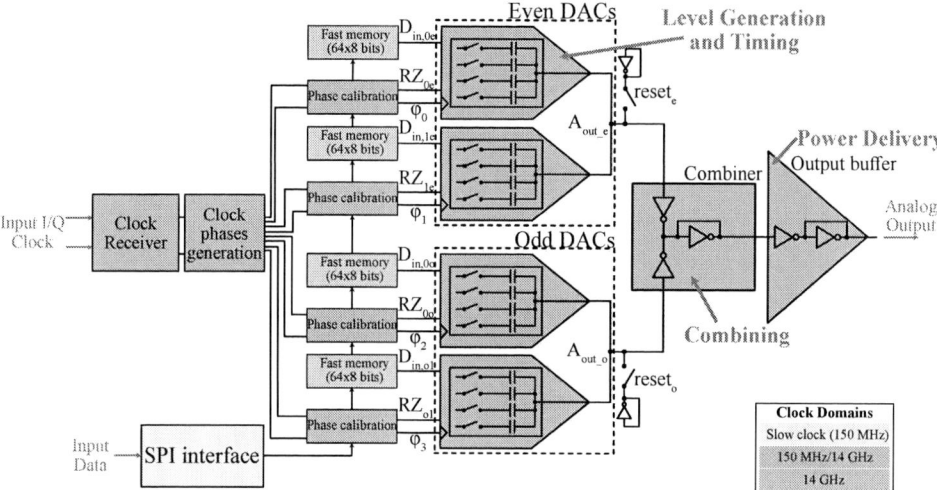

Fig. 2. 4x delta-interleaved DAC architecture (analog half-circuits shown for simplicity; actual implementation is pseudo-differential).

require an analog multiplexer and hence does not translate the sub-DAC offsets into a time varying sequence (no offset interleaving spurs). However, similar to a time-interleaved DAC, timing nonidealities, such as skew and clock duty-cycle mismatch, lead to spurs at the output. Nonetheless, for low signal frequencies, the timing spurs of a delta-interleaved DAC are smaller than in a time-interleaved DAC because two cores process the difference between consecutive samples. A downside of the delta-interleaving approach is that the difference operation doubles the quantization noise.

III. CHIP ARCHITECTURE AND CIRCUIT DETAILS

Fig. 2 details the architecture of our proof-of-concept chip. It contains four identical switched-capacitor DAC cores, which generate the analog levels by independently processing the digital data provided by four on-chip custom fast memories. These memories with a sample depth of 256 are needed for characterization purposes only and are based on a true single-phase-clock (TSPC) bit-cell [8].

Each DAC core contains a segmented array composed of 256 unit capacitor cells [Fig. 3 (a)], in which the 6 LSBs are binary and the 2 MSBs are thermometer encoded. It has a footprint of only $10\,\mu\text{m} \times 12\,\mu\text{m}$ achieving significant area reduction over our prior work [6], [7], thanks to an improved unit-cell [single-ended schematic shown in Fig. 3 (b)-(c)]. This cell contains only five transistors and includes the logic to perform the RZ operation and data re-timing, an inverter driver and a 0.5 fF multi-layer metal-oxide-metal (MOM) capacitor. Inside such a small array, the simulated maximum skew of the clocks used for re-timing is $\leq 70\,\text{fs}$, which is low enough to provide nearly glitch-free operation without re-timing the analog output. Any additional timing skew or duty-cycle imperfection in the signal provided to the unit-cell can be adjusted using the phase calibration circuit

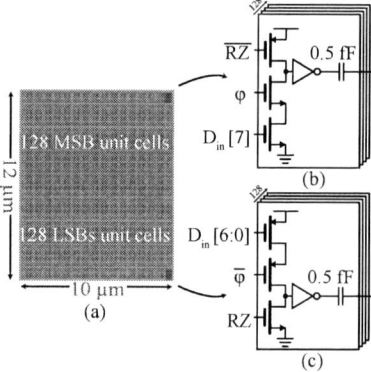

Fig. 3. SC DAC core (a) layout, and schematics of (b) the MSB and (c) LSB unit cells.

(based on programmable delay lines) with a step of 50 fs. During the RZ phase, the array's analog output is returned to zero by driving the cap array to mid-code while keeping the $A_{out,e/o}$ nodes floating.

The cores' outputs are combined and buffered by an inverter-based circuit [7] that forms the full-rate output and provides $50\,\Omega$ impedance matching. The half-circuit schematic is shown in Fig. 4 (implementation is pseudo-differential). The combiner's voltage gain is ≈ 1.5, set by the relative size of the drivers and load transistors. The $200\,\Omega$ series resistance provides active peaking which increases bandwidth at the expense of SNR degradation. The header and footer transistors are added mainly for programmability and standby control. They do not have a significant effect on distortion, which is dominated by large-signal g_{ds} nonlinearities (gradual transition to triode region), as opposed to derivatives of g_m [9].

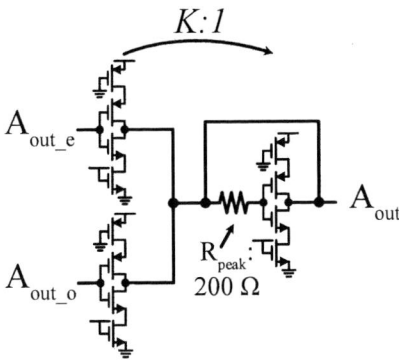

Fig. 4. Half circuit of the inverter-based combiner (implementation is pseudo-differential).

During the reset phase (reset$_{e/o}$), a diode-connected inverter sets the sub-DAC's output common-mode voltage around mid-supply. The reset circuit is a replica of the combiner's input stage to match the midpoint-voltage over process, voltage, and temperature (PVT).

IV. MEASUREMENT RESULTS

The proof-of-concept DAC was fabricated in TSMC's 16 nm FinFET CMOS process. Fig. 5 shows the die photo and floorplan. The DAC occupies only 0.011 mm^2 excluding the fast memories (needed for testing only). The 1.3 mm x 1.3 mm bare die is directly attached to a four-layer FR4 printed circuit board (PCB). Supply, ground and SPI interface control signals are wire-bonded, while the 14 GHz input quadrature clocks and the differential analog outputs are wafer probed using 40 GHz GSGSG 100 μm pitch probes. The quadrature input clocks are generated by an external balun (Marki Microwave QH-0440) and a R&S SMF100A signal generator. The DAC output is measured using a R&S FSW43 spectrum analyzer via a Marki Microwave BAL-0067 wideband balun. The total measured power consumption is 280 mW (clock RX: 15 mW, fast memories: 95 mW, SC DAC cores: 70 mW, phase generation and calibration: 90 mW, combiner and buffer: 10 mW) when operating at 56 GS/s from a 0.85 V supply.

The measured spectrum for a single-tone input at 9.4 GHz and 56 GS/s update rate is shown in Fig. 6. Here, the skew and duty-cycle distortion in the clocks (ϕ_i) and return-to-zero (RZ) signals were adjusted using the on-chip phase calibration circuits via a manual sweep. In a real use case, these knobs can be tuned via a periodic foreground calibration that is managed at the system level (e.g., via loopback in a transceiver). The overlaid output power curve [$P_{out}(f)$] shows that the DAC has a bandwidth of ≈10 GHz, limited by the capacitive load of the output pads which were sized (85 μm x 50 μm) to ease direct wafer probing. The swing at the output of the balun is about -18 dBm at low frequencies, which corresponds to 0.27 V$_{pp}$ (-10.5 dBm) across the differential 100 Ω load after de-embedding the measurement setup losses (dominated by the balun). The measured noise spectral density

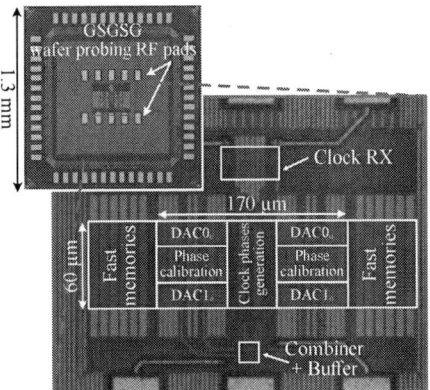

Fig. 5. Die photo.

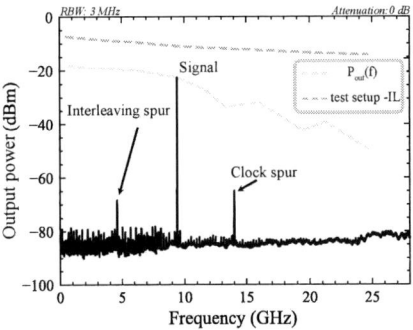

Fig. 6. Measured output spectrum at 56 GS/s for a single-tone input at 9.4 GHz. The red line is the DAC's output power vs. frequency and includes the test setup insertion loss (IL, green line in dB).

(NSD) at the balun output is -151 dBm/Hz and dominated by the combiner's peaking resistor noise. The spectrum shows a spur at the 14 GHz clock frequency, arising from decay time mismatch of the floating nodes A$_{out,e/o}$. The measured SFDR versus signal frequency is shown in Fig. 7, overlaid with the SFDR due to the interleaving spurs (SFDR$_{IS}$) and clock spur (SFDR$_{clk}$) alone. The SFDR is ≥42 dB within the DAC's output bandwidth and limited by the SFDR$_{clk}$ and output buffer nonlinearities.

The measured spectrum for two-tone test at the band-edge is shown in Fig. 8. This is a more relevant test for applications that are limited by in-band distortion. The measured IM3 vs. signal frequency is shown in Fig. 9 indicating third-order intermodulation products (IM3) at −48 dBc. Table I shows a performance summary and comparison with comparable FinFET DACs. The DAC described in this work is nearly 10x smaller than its closest competitor while achieving similar SFDR within 10 GHz output bandwidth. The power consumption is also competitive, while the output swing is smaller. The latter is partly due to our lower supply voltage and the design

978-1-6654-8495-4/22 $31.00 © 2022 IEEE

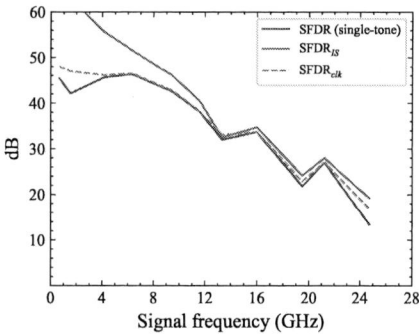

Fig. 7. Measured SFDR, SFDR$_{IS}$ and SFDR$_{clk}$ versus signal frequency at 56 GS/s.

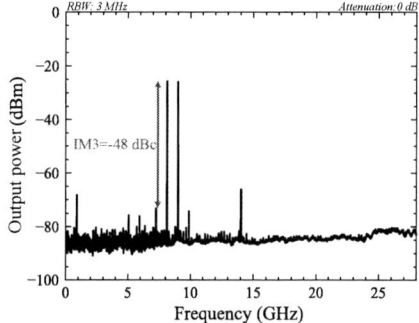

Fig. 8. Measured output spectrum at 56 GS/s for a two-tone test (includes loss from test setup).

choices made in the combiner. Depending on the application, other combiner architectures can be considered and in a system with heterogeneous integration, additional power gain may be achieved using a different transistor type (e.g., Gallium Nitride layer on top).

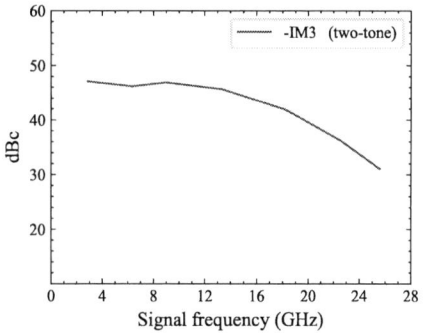

Fig. 9. Measured IM3 versus signal frequency at 56 GS/s.

TABLE I: Comparison with state-of-the-art DACs

	Greshishchev [5] BCICTS 2019	Menolfi [1] ISSCC 2018	This Work
Resolution (bits)	8	8	8
f$_{sample}$ (GS/s)	60	56	56
V$_{supply}$ (V)	0.9/1.9	0.95	0.85
Area (mm^2)	0.31	0.095	0.011
Swing (V$_{pp\text{-}diff}$)	0.8	0.92	0.27
SFDR $\leq$ 10 GHz (dB)	33	42	42
IM3 (dBc)	N/A	N/A	-48
Power (mW)	560	286	280
Technology (nm)	7 FinFET	14 FinFET	16 FinFET
Topology	CS	SST	SC
Interleaving Factor	2	4	4
T/I calibration	N/A	On-chip	On-chip

V. CONCLUSION

We presented a 56 GS/s 8-bit scaling-friendly SC DAC that employs an innovative delta-interleaving scheme. It mainly consists of inverter-based circuits and achieves a nearly 10x area reduction over prior-art high-speed FinFET DACs. The DAC's buffer circuitry limits the output swing (0.27 V$_{pp}$ into 100 Ω) as well as the intermodulation distortion (–48 dBc), and hence is a good target for future innovation.

ACKNOWLEDGMENTS

This work was supported by the Defense Advanced Research Projects Agency (DARPA) grant FA8650-18-1-7895. Silicon fabrication was provided by the TSMC university shuttle program. P.C. was supported in part by the William R. Hewlett and Sang Samuel Wang Stanford Graduate Fellowship. The authors thank the Stanford System Prototyping Facility and Sawson Taheri for PCB design support, and Diakopto Inc. for access to the ParagonX parasitic analysis and visualization tool.

REFERENCES

[1] C. Menolfi et al., "A 112Gb/s 2.6pJ/b 8-Tap FFE PAM-4 SST TX in 14nm CMOS," in *ISSCC Dig. Tech. Papers*, San Francisco, CA, Feb. 2018, pp. 104–106.

[2] T. Cameron, "Architectures and Technologies for the 5G mmWave Radio," in *ISSCC Dig. Tech. Papers*, San Francisco, CA, Feb. 2018.

[3] H. -S. P. Wong et al., "IC Technology – What Will the Next Node Offer Us?" in *2019 IEEE Hot Chips 31 Symposium (HCS)*, 2019, pp. 1–52.

[4] Y. M. Greshishchev et al., "A 56GS/S 6b DAC in 65nm CMOS with 256x6b memory," in *ISSCC Dig. Tech. Papers*, San Francisco, CA, Feb. 2011, pp. 194–196.

[5] Y. M. Greshishchev et al., "A 60 GS/s 8-b DAC with > 29.5dB SINAD up to Nyquist frequency in 7nm FinFET CMOS," in *2019 IEEE BiCMOS and Compound semiconductor Integrated Circuits and Technology Symposium (BCICTS)*, 2019, pp. 1–4.

[6] P. Caragiulo et al., "A Compact 14 GS/s 8-bit Switched-Capacitor DAC in 16 nm FinFET CMOS," in *Symposium on VLSI Circuits*, Honolulu, HI, June 2020, pp. 1–2.

[7] P. Caragiulo et al., "A 2x Time-Interleaved 28-GS/s 8-Bit 0.03mm^2 Switched-Capacitor DAC in 16-nm FinFET CMOS," *IEEE Journal of Solid-State Circuits*, vol. 56, no. 8, pp. 2335–2346, 2021.

[8] J. Yuan et al., "High-speed CMOS circuit technique," *IEEE Journal of Solid-State Circuits*, vol. 24, no. 1, pp. 62–70, 1989.

[9] K. Zheng et al., "An Inverter-Based Analog Front-End for a 56-Gb/s PAM-4 Wireline Transceiver in 16-nm CMOS," *IEEE Solid-State Circuits Letters*, vol. 1, no. 12, pp. 249–252, 2018.

978-1-6654-8495-4/22 $31.00 © 2022 IEEE

A 28GHz Area-Efficient CMOS Vector-Summing Phase Shifter Utilizing Phase-Inverting Type-I Poly-Phase Filter for 5G New Radio

Minzhe Tang, Yi Zhang, Jian Pang, Atsushi Shirane, and Kenichi Okada
Department of Electrical and Electronic Engineering, Tokyo Institute of Technology
2-12-1-S3-28, Ookayama, Meguro-ku, Tokyo, 152-8552, Japan
Tel & Fax +81-3-5734-3764
Email: tang. mz@ssc.pe.titech.ac.jp

Abstract—A 28GHz CMOS passive vector-summing phase shifter is demonstrated in this paper. The proposed design consists of a phase-inverting type-I poly-phase filter and a 45° switch-type phase shifter. By adopting the phase-inverting poly-phase filter for I/Q generation, I/Q weight assignment and summation, the proposed design can realize ± 22.5° phase tuning with fine resolution in each phase quadrant. Together with the 45° phase shifting provided by the switch-type phase shifter, the proposed design can achieve a 360° phase coverage in a compact area of 0.1mm². Transformers are utilized for the inter-stage and output matching, which shrink the area compared with transmission lines. In measurement, the fabricated passive vector-summing phase shifter realizes a root mean square (RMS) phase error smaller than 1.77° from 26.5GHz to 29.5GHz and reaches 0.06° at 28GHz with the help of the fine-tuning in phase-inverting polyphase filter. The measured RMS gain error of the design is smaller than 0.27dB in the desired frequency range.

Keywords—*Phase-inverting, Type-I poly-phase filter, Switch-type phase shifter, I/Q generation, Vector modulation, High phase resolution, Area-efficient.*

I. Introduction

To support the exponential growth of the wireless data traffic, the 28GHz band located from 26.5 to 29.5 GHz with maximum 400-MHz channel bandwidth are specialized in the 5G new radio (NR). Due to the short wavelength characteristic, millimeter-wave experiences large atmosphere absorption and high free-space path loss (FSPL) during transmission. The phased-array system has become an essential building block that can perform directional transmission using beamforming technique and improve the transceiver link budget by enhancing the equivalent-isotropic-radiated-power (EIRP) [1]. The design of phase shifter is critical in phased-array transceiver regarding the required gain/phase errors, phase resolution, power consumption, and chip area [2].

Phase shifter can be implemented through either passive or active approaches. The switch-type phase shifter (STPS) is one of the passive approaches. It can switch between high-pass (HP) and low-pass (LP) state to realize the phase shifting with a fixed step. By cascading several stages as shown in Fig. 1 (a), it can achieve 360° phase shifting with low phase/gain errors. However, such method suffers from the trade-off between chip area and resolution [3]. In 5G transceivers design, phase shifter with high phase resolution is required for beam pointing accuracy. Large chip area is needed to realize high phase resolution for phase shifter based on STPSs. For active approaches, the active vector-summing phase shifter (VSPS) can achieve phase shifting with high resolution, low gain/phase errors and a more compact area than passive STPS.

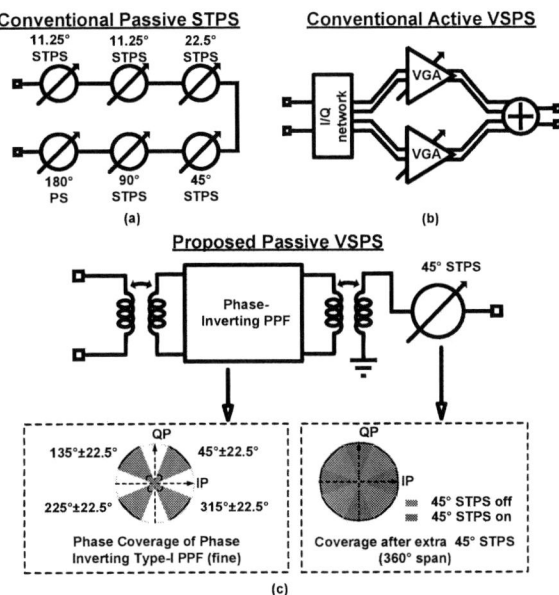

Fig. 1. (a) Conventional active VSPS, (b) passive STPS, and (c) proposed passive VSPS base on phase-inverting t PPF.

However, it induces extra nonlinearity to the RF links. Fig. 1 (b) presents the structure of the conventional active VSPS. Generally, it consists of an I/Q generator followed by I/Q variable gain amplifiers (VGAs) and an output adder. The VGAs modulate the magnitudes of the I/Q phasors. Then, the modulated I/Q phasors are combined by the adder and the desired phasor is achieved. The utilization of two VGAs is not power-efficient and occupies a large area due to their passive components. To realize high resolution phase shifting with minimized area and power, a passive VSPS using phase-inverting poly-phase filter (PPF) is proposed in this work for I/Q generation, I/Q weight assignment, and vector summation. Benefit from the proposed phase-inverting PPF, no I/Q VGAs or attenuations are adopted in the designed. Based on the proposed topology, a passive VSPS with <1° phase resolution within 0.1mm2 chip area is achieved.

II. Circuit Implementation

To maintain a power-efficient phase shifter with high resolution and small chip area, the passive phase-inverting VSPS is proposed in this paper depicted in Fig. 1 (c). The proposed design consists of one-stage phase-inverting PPF that can generate ±22.5° phase shifting at the center of each

978-1-6654-8495-4/22 $31.00 © 2022 IEEE

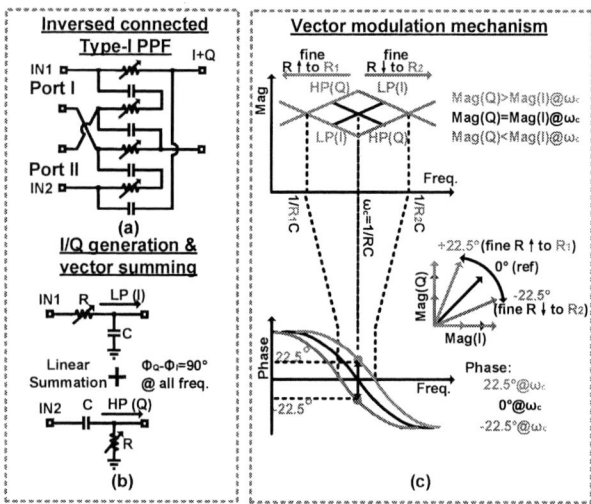

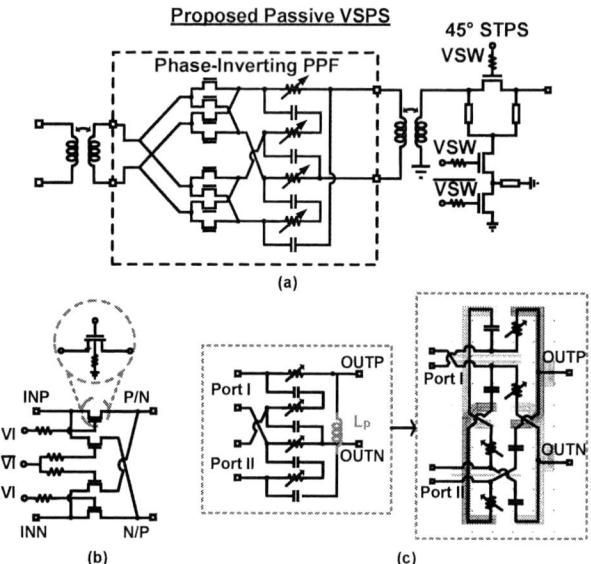

Fig. 2. (a) Structure of inversed-connected Type-I PPF. (b)Type-I PPF simplification together with principle of I/Q generation and vector summing. (c) Vector modulation mechanism that realizes ±22.5° phase shifting.

four phase quadrants. It is then followed by a STPS to generate 45° phase shift with low gain/phase errors and group delay. After cascading the 45° STPS, the proposed design can achieve a full 360° phase coverage. The conventional active VSPS divides the I/Q generation, I/Q weight assignment, and vector summation into different building blocks, which is not power-efficient and occupies large chip area. The proposed passive VSPS can realize those functions in one phase-inverting PPF, which occupies a much smaller area. It also benefits from the passive STPS and phase-inverting PPF that can realize a 360° phase coverage and consume zero DC power. In contrast to the passive STPS, the proposed passive VSPS can achieve a much smaller size with fine phase tuning.

Fig. 2 shows the operation principle of the inversed-connected type-I PPF. In Fig. 2 (a), port II can be seen as virtual ground when port I is injected with a differential input signal, and vice versa [4]. Therefore, Fig. 2 (a) can be simplified to Fig. 2 (b). IN1 and IN2 experience a low-pass network and a high-pass network, respectively, which result in the phasors I and Q at the output node. The phase between I/Q output phasors always has 90° difference over all frequency span and the magnitude of them only equal at the angular frequency of $1/RC$. Summing these two phasors linearly using the superposition theorem at the center frequency of $\omega_c = 1/RC$ will result in an output phasor locates at the center phase between them. If R is increased to R_1, the decreased pole causes the magnitude of I phasor decrease and the magnitude of Q phasor increase at ω_c. Under this circumstance, summing these two phasors can result in a positive phase shifting. If R is decreased to R_2, the increased pole causes the magnitude of I phasor increase and the magnitude of Q phasor decrease at ω_c. In this case, summing these modulated I/Q phasor can obtain a negative phase shifting. As a result, by modifying the pole around the desired center frequency ω_c, the magnitude of the 90° I/Q phasors can be modulated oppositely, which can realize the similar function as conventional I/Q VGAs. After summing the modulated I/Q phasors, the phase-shifted phasor at the desired center frequency can be obtained as shown in Fig. 2 (c). The S_{21} of the proposed design is calculated as

Fig. 3 (a) Detail of proposed passive VSPS, which can realize ±22.5° in four phase quadrants. (b) Structure of phase-inverting switches in phase-inverting PPF. (c) Details of tuning resistor and centrosymmetric layout topology for inversed-connected type-I PPF.

$$S_{21} = \frac{2(1+j\omega_c RC)Z_0 R}{3(1+j\omega_c RC)Z_0 R + R^2 + Z_0^2(1+j\omega_c RC)^2} \quad (1)$$

where ω_c is the desired center frequency and Z_o is the 50Ω characteristic impedance. The phase tuning can be achieve by altering the resistor value. A $\pm 22.5°$ phase range is selected for the proposed PPF design under the consideration of gain variation and insertion loss.

Fig. 3 (a) presents the detailed circuit of the proposed passive VSPS. The phase-inverting PPF consists of two parallel-connected phase-inverting switches followed by one stage inversed-connected type-I PPF. The details of the switches are shown in Fig. 3 (b). These switches are adopted to select the input signal for inversed connected type-I PPF and to expand the phase shifting into four phase quadrants. The transistors used as the switches are bulk-floating connected to improve the insertion loss [5]. Conventional PPF design suffers from phase/gain imbalance. In detail, the long asymmetric connection in layout can introduce parasitic inductance [6]. To minimize the phase/gain imbalance of the proposed inversed-connected type-I PPF, a highly centrosymmetric layout topology is adopted as shown in Fig. 3 (c). The novel phase-inverting PPF can realize a $\pm 22.5°$ phase shift in each phase quadrant. In addition, the realization is explained from Fig. 4 (a) to (d). The phase-inverting switches can either directly inject a phase-thru-signal (THU) or a phase-flipped-signal (FLP) to the port I and port II in the inversed-connected type-I PPF. We take Fig. 4 (a) as an example to explain the input selection method. If the THU is connected to port I of the inversed-connected PPF, a direct I-phase signal IP can be obtained at the output. Based on the symmetric property of the proposed PPF, a direct Q-phase signal QP is obtained by selecting the port II with FLP. If $R = 1/\omega_c C$, the IP and QP output phasors with the same magnitude

978-1-6654-8495-4/22 $31.00 © 2022 IEEE

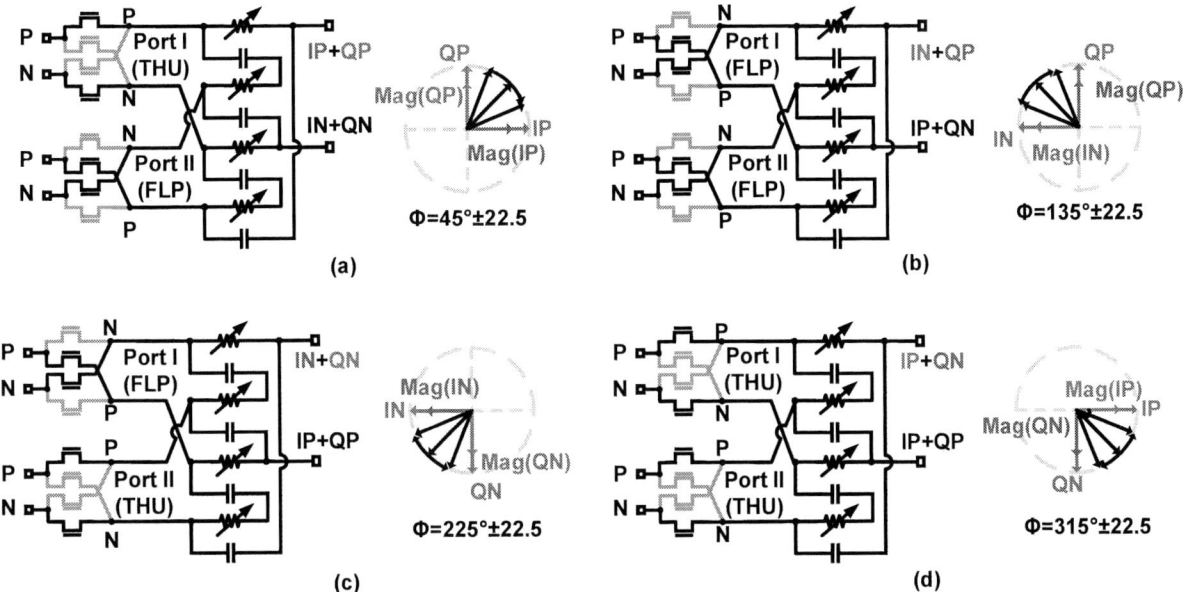

Fig. 4. Proposed input selection method generating ±22.5° phase shifting around 45°, 135°, 225°, and 315°.

are added linearly. The 45° phasor between IP and QP can be obtained. After applying the same principle, the 135°, 225°, and 315° phasors can be obtained by input the port I and II with FLP/FLP, FLP/THU, and THU/THU, respectively. According to (1), a ±22.5° phase tuning range around 45°, 135°, 225°, and 315° phasors are selected and realized at the desired center frequency by tuning the resistors in the proposed PPF.

To achieve a full 360° phase coverage, the STPS is connected after the proposed PPF to generate the 45° phase shift. Then the rest ±22.5° phase tuning range at 0°, 90°, 180°, and 270° are achieved by flipping the input and tuning the resistors of the phase-inverting PPF according to the same method. Thanks to this phase tuning method, the proposed passive VSPS can realize a fine phase tuning resolution in a small chip area compared to conventional VSPS.

III. Measurement Resutls

The proposed passive VSPS is fabricated in a 65nm CMOS technology. Fig. 5 shows the die micrograph of the proposed design, which only includes one phase-inverting PPF, one-stage 45° STPS, and two transformers. With the help of adopting the phase-inverting PPF for I/Q generation, I/Q weight assignment and vector summing, the proposed passive VSPS only occupies a chip area of 0.1mm2 and consumes 0 DC power. The chip is measured on wafer by N5247B PNA-X. Fig. 6 shows the relative phase responses with a 11.25° phase step. The proposed design realizes a 360° phase coverage over the desired frequency band. The insertion loss of the design is presented in Fig. 7 with an average value of -13dB. Fig. 7 also shows the measured RMS phase/gain errors against the frequency under the same control code. The RMS phase error is less than 1.77° within the frequency range. Thanks to the fine-tuning of the phase-inverting PPF the RMS phase error reaches a minimum value of 0.06° at 28GHz. With the help of vector modulation technique based on the phase-inverting PPF, the measured RMS gain error is less than 0.27

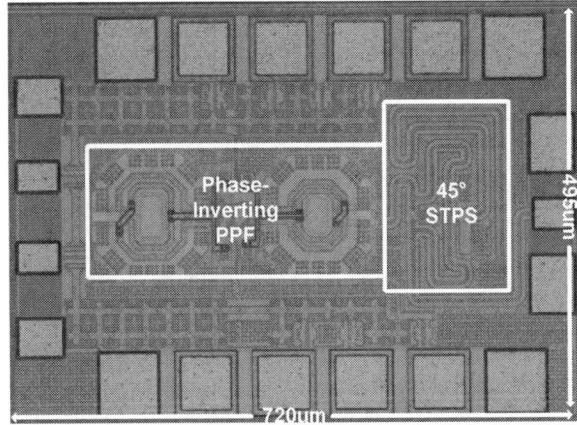

Fig. 5. Micrograph of the proposed passive VSPS.

within 26.5 to 29.5GHz. The return losses of the proposed design are smaller than 10dB as depicted in Fig. 8. Table I compares the proposed passive VSPS with state-of-the-art phase shifter. This work consumes 0 dc power, realizes a more compact chip area and high phase resolution with a comparable RMS phase/gain errors performance.

IV. Conclusion

This paper introduces a passive VSPS, which employs the phase-inverting PPF, realizing 45° phase coverage in each phase quadrant. Combined with a 45° STPS, it performs a 360° phase coverage over 26.5 to 29.5GHz. The phase-inverting PPF integrated the function of I/Q generation, magnitude modulation and vector summing in one simple block, achieving low RMS gain/phase errors and high phase resolution within a compact chip area.

978-1-6654-8495-4/22 $31.00 © 2022 IEEE

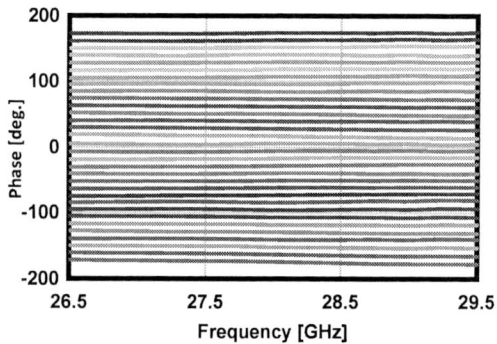

Fig. 6. Measured relative phase response with accurate 11.25° phase step at 28GHz.

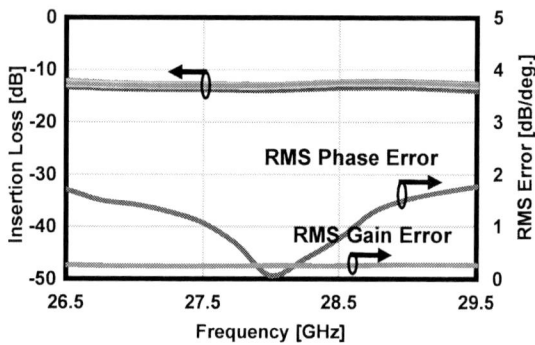

Fig. 7. Measured insertion loss and RMS gain and phase errors of the proposed passive VSPS.

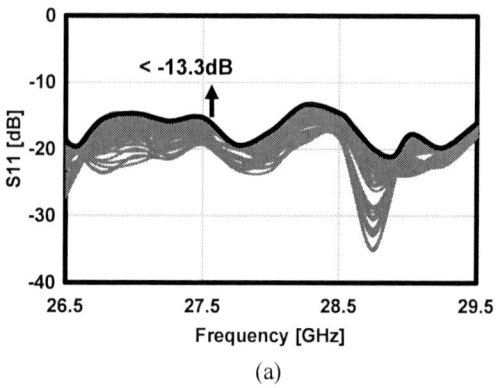

(a)

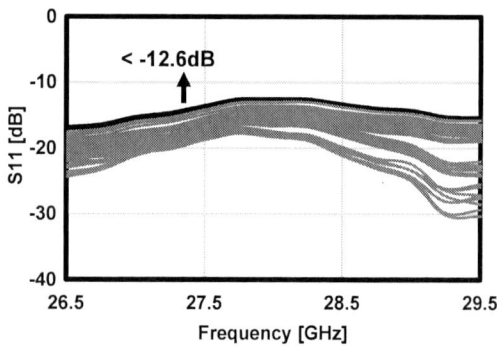

(b)

Fig. 8 Measured return loss of the proposed passive VSPS.

Table I. Performance comparison with state-of-the-art

	This Work	[7]	[8]	[9]	[10]
Process	65nm CMOS	65nm CMOS	90nm SiGe BiCMOS	65nm CMOS	0.12um SiGe BiCMOS
Freq. (GHz)	26.5~29.5	33~39	20~27	27.4~28.6	28~40
Type	Passive VSPS		Active VSPS		STPS
Core Area (mm^2)	0.1	0.14	0.126	0.32	0.11
Power (mW)	0	0	10	25.2	0
Phase Res. (deg.)	<1	2.6	5	5.625	22.5
RMS Phase Error (deg.)	<1.77	<1.1	0.9 @f_0	<0.54	<22.5*
RMS Gain Error (dB)	<0.27	<0.28	1.5dB Var. @f_0	<0.13	<4*
Insertion Loss (dB)	-13	-17.3	0*	-3.2*	-13
IP1 (dBm)	9	N/A	-16	N/A	10

* Estimate from paper

ACKNOWLEDGMENT

This work is partially supported by MIC (JPJ000254), NICT(00801, 00601), JSPS (JP20H00236), STAR, and VDEC in collaboration with Cadence Design Systems, Inc., Mentor Graphics, Inc., and Keysight Technologies Japan, Ltd.

REFERENCES

[1] [1J. Pang et al., "A 28-GHz CMOS Phased-Array Transceiver Based on LO Phase-Shifting Architecture with Gain Invariant Phase Tuning for 5G New Radio," in IEEE Journal of Solid-State Circuits, May 2019.

[2] X. Yu et al., "A 17.3-mW 0.46-mm2 26/28/39GHz Phased-Array Receiver Front-End with an I/Q-Current-Shared Active Phase Shifter for 5G User Equipment," 2021 IEEE Radio Frequency Integrated Circuits Symposium (RFIC), Atlanta, GA, USA, 2021.

[3] Yun-Chieh Chiang, Wei-Tsung Li, Jeng-Han Tsai and Tian-Wei Huang, "A 60GHz digitally controlled 4-bit phase shifter with 6-ps group delay deviation," 2012 IEEE/MTT-S International Microwave Symposium Digest, Montreal, QC, Canada, 2012.

[4] J. Kaukovuori, K. Stadius, J. Ryynanen and K. A. I. Halonen, "Analysis and Design of Passive Polyphase Filters," in IEEE Transactions on Circuits and Systems I: Regular Papers, Nov. 2008.

[5] M. Elkholy, S. Shakib, J. Dunworth, V. Aparin and K. Entesari, "Low-Loss Highly Linear Integrated Passive Phase Shifters for 5G Front Ends on Bulk CMOS," in IEEE Transactions on Microwave Theory and Techniques, Oct. 2018.

[6] F. Piri, M. Bassi, N. R. Lacaita, A. Mazzanti and F. Svelto, "A PVT-Tolerant >40-dB IRR, 44% Fractional-Bandwidth Ultra-Wideband mm-Wave Quadrature LO Generator for 5G Networks in 55-nm CMOS," in IEEE Journal of Solid-State Circuits, Dec. 2018.

[7] Y. Li et al., "A 32-40 GHz 7-bit CMOS Phase Shifter with 0.38dB/1.6° RMS Magnitude/Phase Errors for Phased Array Systems," 2020 IEEE Radio Frequency Integrated Circuits Symposium (RFIC), Los Angeles, CA, USA, 2020.

[8] J. S. Park and H. Wang, "A K-band 5-bit digital linear phase rotator with folded transformer based ultra-compact quadrature generation," 2014 IEEE Radio Frequency Integrated Circuits Symposium, Tampa, FL, USA, 2014.

[9] J. Pang, R. Kubozoe, Z. Li, M. Kawabuchi and K. Okada, "A 28GHz CMOS Phase Shifter Supporting 11.2Gb/s in 256QAM with an RMS Gain Error of 0.13dB for 5G Mobile Network," 2018 48th European Microwave Conference (EuMC), Madrid, Spain, 2018.

[10] B.W. Min and G. M. Rebeiz, "Single-Ended and Differential Ka-Band BiCMOS Phased Array Front-Ends," in IEEE Journal of Solid-State Circuits, Oct. 2008.

A 24-30 GHz Broadband Doherty PA with a maximum 15.37 dBm P_{avg} and 14.6% PAE_{avg} in 0.13 µm SiGe for 400 MHz BW 5G NR

Hao Gao[+#], Sina Mortezazadeh Mahani[+], David Seebacher[*], Matteo Bassi[*], Gernot Hueber[+]

Silicon Austria Labs, Linz, Austria[+]
Eindhoven University of Technology, Eindhoven, The Netherlands[#]
Infineon Technologies, Austria AG, Villach, Austria[*]

Abstract—This paper presents a 24-30 GHz broadband Doherty PA (DPA) in Infineon 0.13 µm SiGe BiCMOS technology. A broadband transformer-based impedance inverter matching network (IIMN) is proposed and implemented to achieve simultaneous wide bandwidth and high efficiency within a compact size. Furthermore, a stacked PA unit cell topology is applied for power added efficiency (PAE) enhancement. In the 5G NR FR2 64-QAM modulated measurement from 24-30 GHz (5G N257 N258 bands) with 5% EVM constrain, this DPA achieves a maximum 17.24 dBm P_{avg} and 16.8% PAE with 100MHz BW. With 200 MHz BW, the maximum P_{avg} is 16.94 dBm, and PAE_{avg} is 16.6%. With 400 MHz, the maximum P_{avg} is 15.37 dBm, and PAE_{avg} is 14.6%. In 24-30 GHz (N257 N258), this PA could deliver above 12 dBm average output power (P_{avg}), and its average PAE (PAE_{avg}) keeps higher than 10% in whole bands, all reported at critical 85℃ condition.

Index Terms— Doherty PA, broadband, 5G NR, SiGe BiCMOS.

I. INTRODUCTION

The demands for wireless connectivity motivate the development of the fifth-generation (5G) communication for a significant improvement in data rate and latency in wireless communications. The 3GPP 5G new radio (NR) specifies N257 band (26.5-29.5 GHz), N258 band (24.25-27.5 GHz), N260 band (27-40 GHz) for the frequency range 2 (FR2). Especially with the development of silicon-based SiGe BiCMOS technology, a low-cost, high-performance, and high-efficiency Tx front-end is potentially the most economical solution. However, it is still challenging for a broadband mm-wave Tx front-end to cover the interested bands (N257&258:24.25-29.5 GHz) with high efficiency for 5G communication requirements.

To achieve a high-efficiency broadband mm-Wave PA, the tunable matching method-based PAs, and broadband matching method-based PAs are implemented [1-2], but those are only optimized at the P_{sat} or P-1dB. However, the high peak-to-average power ratio poses a stringent efficiency requirement at power backoff (PBO), and its average PAE. The Doherty PA could enhance PBO efficiency[3-7]. However, a conventional current-combined Doherty PA has a limitation in bandwidth. A voltage-combining Doherty PA could break its bandwidth limitation. In [8], a lumped quarter-wave impedance inverter is bulky and introduces extra loss. To achieve a

compact layout, a transformer-based series output combiner is applied in [9]. In [10], a voltage-combining single frequency Doherty PA in 22 nm FDSOI technology is presented. However, its efficiency is dropped due to a phase issue. In this work, a broadband transformer-based impedance inverter matching network (IIMN) is proposed in a voltage-combined DPA to achieve a dynamic broadband load modulation. In this new broadband IIMN, the proposed series component corrects the phase issue with phase compensation. Furthermore, the stacked PA unit cell topology is applied in the SiGe for PAE enhancement.

This paper is organized as follows. Section II presents the design methodology of a broadband transformer-based impedance inverter matching network. This broadband Doherty PA with the stack PA unit cell is discussed in Section III. Section IV shows the 5G NR FR2 64-QAM measurement. Conclusions are drawn in Section V.

II. BROADBAND DOHERTY PA WITH IIMN BW ENHANCEMENT

The broadband operation of a Doherty output network is distinguished by the flatness of maximum output power and efficiency over frequency, which can be degraded by the variation of the presented load to the main/aux amplifier or the load transformation ratio. In Fig. 1, the topologies are compared among the current-combined, voltage-combined and proposed IIMN based voltage-

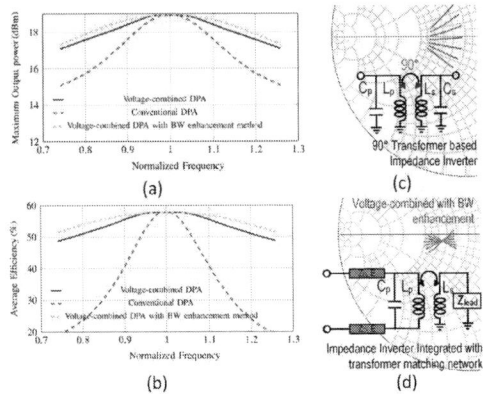

Fig. 1. DPA topologies consideration in terms of bandwidth comparison (a)-(b), and auxiliary amplifier presented impedance to the main amplifier (c)-(d), evaluated from the 24-30 GHz.

978-1-6654-8495-4/22 $31.00 © 2022 IEEE

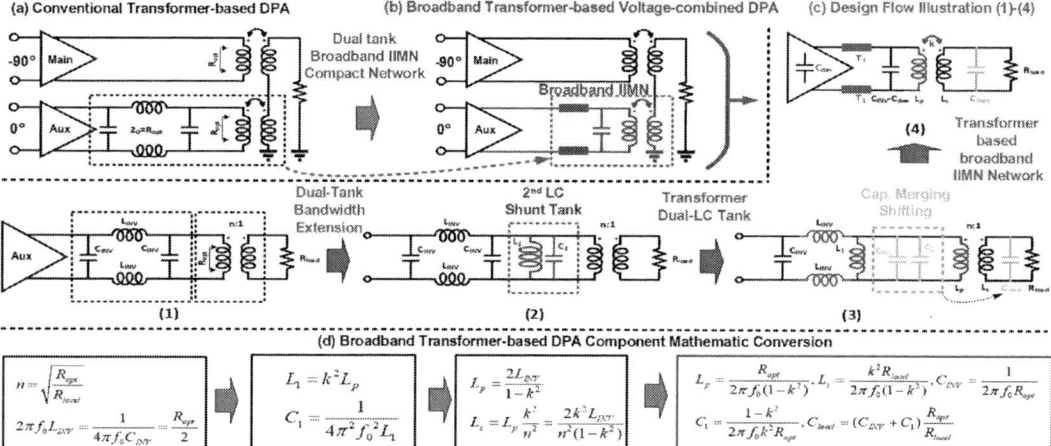

Fig. 2. Illustration of broadband IIMN based voltage combined Doherty PA with its design flow, and its mathematic conversion.

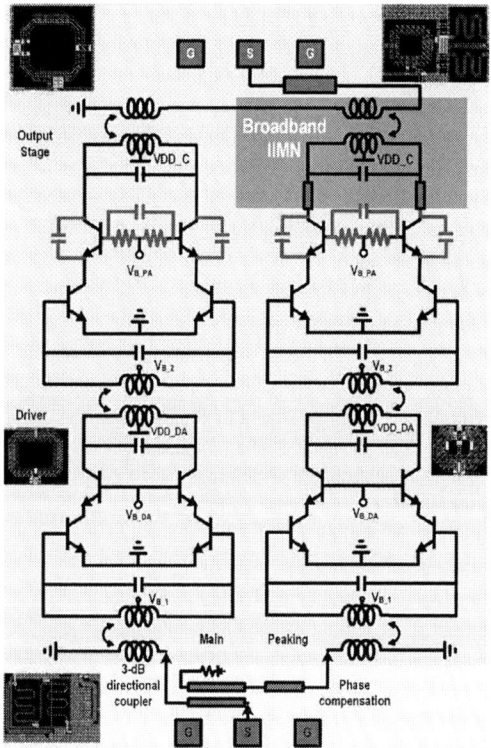

Fig. 3. Simplified block diagram of the voltage-mode broadband Doherty PA with the broadband IIMN.

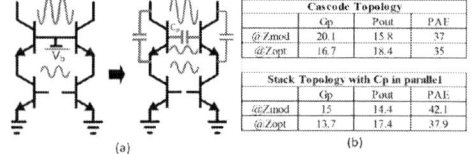

Fig. 4. Stack PA unit cell topology, and its comparison in the SiGe technology with G_p, P_{out} and PAE.

combined Doherty PA. The main bandwidth-limiting

components in a current-combining DPA architecture are two T-lines at the output and impedance transfer ratio. Voltage-combined topology performs with relatively lower load variation at power backoff, and it requires a low impedance transformation. The proposed IIMN based voltage-combined Doherty PA further reduces the Z_{opt} variation (Fig.1.d) from the transformer-based 90° inverter (Fig.1.c), and extends the bandwidth (Fig.1.a.b).

The illustration of this Doherty PA with transformer-based impedance inverter matching network (IIMN) is shown in Fig. 2. Compared to the traditional π-network-based phase inverting, a resonant shunt tank is inserted in series in the auxiliary amplifier. With this Dual-LC tank method [11], the bandwidth could be extended. The capacitors in the primary winding could be combined and shifted to the secondary winding with Norton transfer, and the inductors are also simplified to shunt and series components, as shown in Fig.2(c). With the proposed methodology, a compact transformer-based impedance inverter matching network (IIMN) could present a lower load variation to the main amplifier in all 24-30 GHz, and the series T-line (T_1) could adjust the phase.

III. BROADBAND DPA CIRCUIT IMPLEMENTATION

The schematic of the proposed 5G broadband Doherty PA is shown in Fig. 3. The circuit is implemented in Infineon 0.13μm SiGe RF BiCMOS technology with a peak f_T/f_{MAX} of 250/370 GHz. A fully differential topology in the main amplifier path and auxiliary amplifier path improves layout symmetry and common-mode signal suppression. Considering the gain and output power requirements, the main and auxiliary amplifier paths apply a driver stage and an output stage. In the driver stage, a cascode structure is employed. The stacked unit cell topology is applied in the output stage, as shown in Fig. 4. In a typical cascode topology, the base of the CB transistor is grounded. However, extra capacitors are

978-1-6654-8495-4/22 $31.00 © 2022 IEEE

inserted at the stacked transistor, which allows a voltage swing at the base. This voltage swing is in phase with the voltage swing at the collector. In this way, the voltage swing of the V_{be} and V_{ce} in the stacking stage is reduced and releases the breakdown condition. The PA unit cell's PAE is increased from 37% to 42.1%. The sizes of the PA unit cells are scaled to $0.22 \times 10 \ \mu m^2$. With this Z_{opt} for CE, its G_p is close to 11 dB, P_{out} is close to 16.7 dBm. In this combination, the partially stacked cascode with 13.7 dB G_p, 17.4 dBm P_{out}, and 37.9% PAE at Z_{opt}.

IV. MEASUREMENT

Its die microphotograph is shown in Fig. 5 with Infineon's in-house fabrication and characterization, the core area is $0.38 \times 0.8 \ mm^2$. The large-signal measurement setup is shown in Fig. 6, with 2-step power calibration. It is with the Keysight signal source and two Keysight power meters. For the modulated signal test, an R&S vector signal generator generates modulated signals for the chip, and the output is fed into an R&S FSW signal & spectrum analyzer. The measured gain and efficiency characterization (PAE) versus output power (P_{out}) at 26/27/28 GHz and a temperature of 85°C are shown in Fig. 7. The PAE in this work includes the driver stage's DC power consumption instead of the final stage only.

At 26 GHz, a P_{sat} of 23.6 dBm and an OP1dB of 23.3 dBm are achieved. Also, it shows a maximum PAE of 24.5%, a PAE of 23.85% at 1-dB compression points, and a PAE of 19% at 6-dB output power backoff (PBO). At 28 GHz, a P_{sat} of 22.55 dBm and an OP1dB of 22.4 dBm are achieved. Also, it shows a maximum PAE of 22.9%, a PAE of 21.5% at 1-dB compression points, and a PAE of 17.4% at 6-dB PBO.

The error-vector-magnitude (EVM) is measured using 5G NR FR2 64-QAM modulated signal without any digital pre-distortion. The results for 400 MHz bandwidth

as shown in Fig. 8. At 24 GHz, it shows an EVM of 4.91% with P_{avg} of 12.04 dBm, PAE_{avg} of 10.2%. At 26 GHz with 4.99% EVM, the P_{avg} is 12.6 dBm and PAE_{avg} is 11.2%. At 28 GHz with 4.93% EVM, the P_{avg} is 15.32 dBm and PAE_{avg} is 14.6%.

In the 5G NR FR2 64-QAM modulated measurement from 24-30 GHz with 5% EVM constrain, the average output power (P_{avg}) and average PAE (PAE_{avg}) are shown in Fig.9. Under 5% EVM constraints in 100/200/400 MHz signal bandwidth, this PA could deliver above 12 dBm output power from 24-30 GHz, and its efficiency keeps

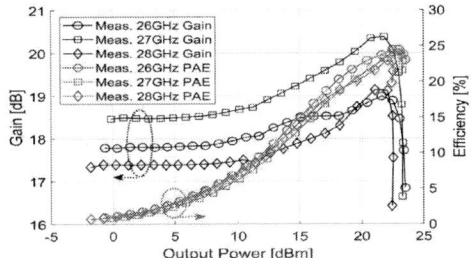

Fig. 7. Measured large-signal characterization, Gain and PAE, with output power at 85°C.

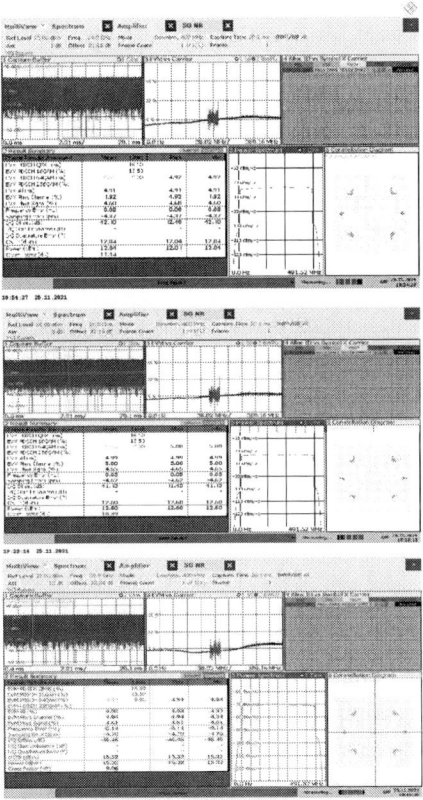

Fig. 8. 5G NR measurement at 24 GHz, 26 GHz and 28 GHz for 64-QAM, 400-MHz BW signals.

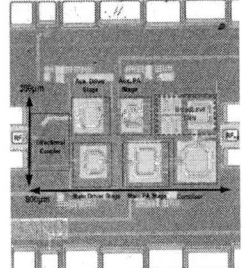

Fig. 5. Die Microphotograph in Infineon 0.13μm SiGe.

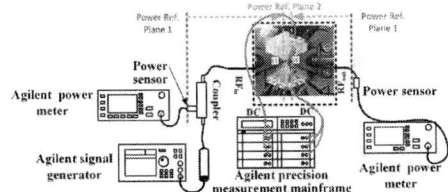

Fig. 6. Large signal measurement setup.

978-1-6654-8495-4/22 $31.00 © 2022 IEEE

TABLE I PERFORMANCE SUMMARY AND COMPARISON TO OTHER STATE-OF-ARTS

	This Work		[4] ISSCC 17		[7] ISSCC 20	[2] LSSC 21	[1] TMTT 22	[5] JSSC 13	[6] RFIC 17	[9] TMTT 21	
Technology	130nm SiGe		130nm SiGe		130nm SiGe	130nm SiGe	250nm SiGe	45nm RF-SOI	28nm Bulk CMOS	22nm FD-SOI	
Frequency (GHz)	27 (24-30 GHz)		28		28	28	28	42	32	28	
Supply (V)	2.5		1.5		2	3.3	2.5	2.5	1	2.4	
Gain (dB)	19		18.2		20.5	20.8	25.7	8	22	26.1	
P_{sat} (dBm)	23.6		16.8		28.3	16.5	18.8	18	19.8	22.5	
OP_{1dB} (dBm)	23.3		15.2		26.8	13.5	17.8	17	16	21.1	
PAE_{MAX} (%)	25.5		20.3		30.4	18.3	25	20	21	28.5	
$PAE@OP_{1dB}$ (%)	25		19.5		30.2	18	24.2	18	12.8	26.6	
PAE@6dB PBO (%)	18		13.9		21.2	N.A.	10	21	4	22.1	
Core Area (mm^2)	0.36		1.76		1.35	0.63	0.23	0.45	0.59	0.2	
QAM Modulation	64 (200M)	64 (400M)	64		64	64 (200M)	64 (400M)	64 (400M)	N.A.	64	64	256
Pout_avg@EVM (dBm)	16.94	15.37	9.2	7.2	17.5*	8.6	7.5.	N.A.	11.7	10.9	10.2
PAE_avg@EVM (%)	16.6	13.8	10.1	7.9	12.9*	N.A.	N.A.	N.A.	5.75	9.2	9
Temperature (°C)	85		25		25	25	25	25	25	25	

* 11.84 dB PAPR Condition

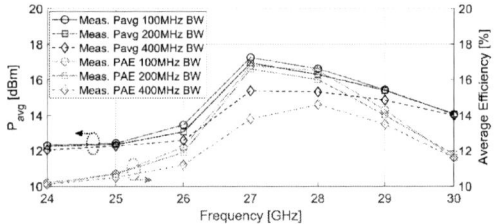

Fig. 9. Average Pout and PAE under 5% EVM in 5G NR 64-QAM modulated measurement from 24-30 GHz, all at 85°C.

higher than 10% in the full band at critical 85°C (few reported in this temperature). This DPA achieves a maximum 17.24 dBm P_{avg} and 16.8% PAE with 100MHz BW. With 200 MHz BW, the maximum P_{avg} is 16.94 dBm, and PAE_{avg} is 16.6%. With 400 MHz, the maximum P_{avg} is 15.37 dBm, and PAE_{avg} is 14.6%.

Table I summarizes the performance of the state-of-the-art PA design at similar 5G frequencies. Compared with the other PAs, this PA achieves the best average PAE in 400MHz BW with 5G NR FR2 64-QAM, and it covers the 5G N257 N258 bands.

V. CONCLUSION

A 24-30 GHz broadband Doherty PA is implemented with the Infineon 0.13 μm SiGe BiCMOS technology. A broadband transformer-based impedance inverter matching network (IIMN) and a stacked topology is applied to enhance its broadband performance. At 5G 26 GHz and 28 GHz frequency band, its OP1dB are 23.3 dBm and 22.4 dBm, PAE at 6-dB PBO are 19% and 17.4%, respectively. With 5G NR FR2 64-QAM modulated measurement with 5% EVM constrain, in 24-30 GHz frequency (5G N257 N258), this DPA achieves a maximum 17.24 dBm P_{avg} and 16.8% PAE with 100MHz BW. With 200 MHz BW, the maximum Pavg is 16.94 dBm, and PAE_{avg} is 16.6%. With 400 MHz, the maximum P_{avg} is 15.37 dBm, and PAE_{avg} is 14.6%. All reported data are measured with critical 85°C enviorment. This PA

achieves the best average PAE in 400MHz BW with 5G NR FR2 64-QAM and covers the 5G N257 N258 bands.

REFERENCES

[1] K. Ding, D. M. W. Leenaerts and H. Gao, "A 28/38 GHz Dual-Band Power Amplifier for 5G Communication," in IEEE Transactions on Microwave Theory and Techniques, 2022.

[2] P. Zhou, J. Chen, P. Yan, D. Hou, H. Gao and W. Hong, "A Broadband Power Amplifier in 130-nm SiGe BiCMOS Technology," in IEEE Solid-State Circuits Letters, vol. 4, pp. 44-47, 2021

[3] F. Wang, T.-W. Li, S. Hu, and H. Wang, "A super-resolution mixed signal Doherty power amplifier for simultaneous linearity and efficiency enhancement," IEEE J. Solid-State Circuits, vol. 54, no. 12, pp. 3421–3436, Dec. 2019.

[4] S. Hu, F. Wang, and H. Wang, "A 28 GHz/37 GHz/39 GHz multiband linear Doherty power amplifier for 5G massive MIMO applications," in IEEE Int. Solid-State Circuits Conf. (ISSCC) Dig. Tech. Papers, Feb. 2017, pp. 32–33.

[5] A. Agah, H.-T. Dabag, B. Hanafi, P. M. Asbeck, J. F. Buckwalter, and L. E. Larson, "Active millimeter-wave phase-shift Doherty power amplifier in 45-nm SOI CMOS" IEEE J. Solid-State Circuits, vol. 48, no. 10, pp. 2338–2350, Oct. 2013.

[6] P. Indirayanti and P. Reynaert, "A 32 GHz 20 dBm-PS AT transformer based Doherty power amplifier for multi-Gb/s 5G applications in 28 nm bulk CMOS" in Proc. IEEE Radio Freq. Integr. Circuits Symp. (RFIC), Jun. 2017, pp. 45–48.

[7] F. Wang and H. Wang, "A 24-to-30GHz Watt-Level Broadband Linear Doherty Power Amplifier with Multi-Primary Distributed-Active-Transformer Power-Combining Supporting 5G NR FR2 64-QAM with >19dBm Average Pout and >19% Average PAE" 2020 IEEE International Solid-State Circuits Conference - (ISSCC), 2020, pp. 362-364.

[8] P. M. Asbeck, "Will Doherty continue to rule for 5G?," 2016 IEEE MTT-S International Microwave Symposium (IMS), 2016, pp. 1-4

[9] Z. Zong et al., "A 28-GHz SOI-CMOS Doherty Power Amplifier With a Compact Transformer-Based Output Combiner," in IEEE Transactions on Microwave Theory and Techniques, vol. 69, no. 6, pp. 2795-2808, June 2021

[10] Z. Zong, X. Tang, K. Khalaf, D. Yan, G. Mangraviti, J. Nguyen, Y. Liu, and P. Wambacq, "A 28 GHz VoltageCombined Doherty Power Amplifier with a Compact Transformer-based Output Combiner in 22nm FD-SOI," IEEE Radio Frequency Integrated Circuits Symposium, pp. 299–302, 2020.

[11] Z. Chen, H. Gao, D. Leenaerts, D. Milosevic and P. Baltus, "A 29–37 GHz BiCMOS Low-Noise Amplifier with 28.5 dB Peak Gain and 3.1-4.1 dB NF," 2018 IEEE Radio Frequency Integrated Circuits Symposium (RFIC), 2018, pp. 288-291

A 24-to-44 GHz Compact Linear 5G Power Amplifier with Open-Terminated Balun in 65nm Bulk CMOS

Kyutaek Oh[*1], Hyunjin Ahn[†], Ilku Nam[*] and Ockgoo Lee[*]

[*]Department of Electrical Engineering, Pusan National University, Busan, South Korea
[†]Qualcomm Inc., San Diego, CA, USA
[1]Email : ktoh95@pusan.ac.kr

Abstract—This paper presents a broadband linear CMOS power amplifier (PA) for mm-wave 5G applications. A broadband low-loss open-terminated balun is proposed for a PA output network with a smaller size. The PA fabricated using a 65-nm bulk CMOS process shows flat saturated output power, P_{sat}, ranging from 17.5 to 19.5 dBm, within 2-dB difference, while achieving over 24.5% peak PAE for CW signals ranging from 24 to 44 GHz. It achieves a flat linear output power, P_{linear}, of 10.4/11.6/10.9/8.8 dBm and a P_{linear} PAE, PAE_{linear}, of 6.24/8.94/7.9/5.1% with -25-dB rms EVM, EVM_{rms}, and -23.3/-25.7/-25.9/-22.8-dBc ACPR at 24/28/39/44 GHz. Using the 5G NR FR2 64-QAM 8-CC×100-MHz OFDM signal (PAPR = 9.7 dB), broadband operation was demonstrated over 24–44 GHz covering n257–261 5G bands. It has a core size of 0.15 mm².

Keywords—*power amplifiers, open-terminated balun, compact, broadband, mm-wave, 5G NR FR2 bands, bulk CMOS.*

I. INTRODUCTION

With the increasing demands of data rate in wireless communication, mm-wave 5G technology is indispensable as the next-generation of long-term evolution (LTE) systems. 5G new radio (NR) FR2 signal bands are defined as n257 (26.5–29.5 GHz), n258 (24.25–27.5 GHz), n259 (39.5–43.5 GHz), n260 (37–40 GHz), and n261 (27.5–28.35 GHz) [1]. Globally, different countries and regions have adopted some of the 5G NR bands. Broadband operation is required to cover various bands in different countries. However, traditional broadband architectures of PAs have various limitations. Distributed PAs sacrifice a large size and poor passive loss owing to long transmission lines [2]. High-order filter-based PAs often suffer from a large form factor and poor passive loss in practical implementation [3], [4]. A method using variable tuning elements is limited by the loss and tuning range [5], [6]. Thus, it is desirable to achieve broadband operation in a smaller area without sacrificing the passive loss.

A balun with a short-termination is typically used as the output matching structure of the PA. A balun is generally analyzed comprising two magnetically coupled inductors and operates under a self-resonance frequency (SRF). In addition, the balun can be considered to consist of two coupled transmission lines, known as a distributed-balun, which operates in both electrically and magnetically by absorbing the parasitic capacitance. Therefore, the distributed balun operates in broadband, over the SRF. Traditionally, distributed baluns consist of two λ/4 long couplers. However, for on-chip

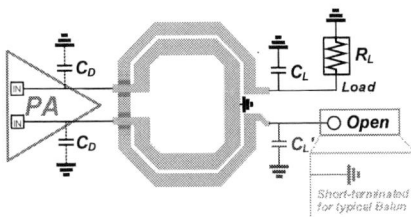

Fig. 1. Block diagram of the proposed open-terminated balun with the PA.

implementations, λ/4 transmission lines can be bulky and lossy. To reduce the electrical length, θ, of the balun, input and output capacitors can be added [4], [7]. In [4], a PA was successfully demonstrated for broadband operation by adopting distributed-balun with typical short-terminated case. The most important aspect of distributed-balun design is the selection of even- and odd-mode impedances ($Z_{0,e}$ and $Z_{0,o}$) that provide the optimum load-pull impedance (R_{opt}) at the operating frequency. $Z_{0,e}$ and $Z_{0,o}$ are determined by the input- (C_D) and output-capacitor (C_L) with θ. C_D is the incorporated device parasitic capacitor in the PA design, and is determined together with R_{opt} from the load-pull simulation. Therefore, only C_L can be considered as the design parameter in the short-terminated case. Because the implementable $Z_{0,e}$ and $Z_{0,o}$ range in the bulk CMOS process is narrow, $Z_{0,e}$ and $Z_{0,o}$ of the short-terminated balun providing R_{opt} with a smaller θ may not exist. Hence, it is desirable to add a design parameter to construct a small distributed-balun within the limited range of $Z_{0,e}$ and $Z_{0,o}$. In this work, an open-terminated balun with an additional design parameter (e.g., C_L') is proposed, as shown in Fig. 1. With more degrees of design freedom for an open-terminated asymmetric-lumped distributed balun, θ can be significantly less than λ/4 within the achievable $Z_{0,e}$ and $Z_{0,o}$.

II. PROPOSED BROADBAND POWER AMPLIFIER

The schematic and equation of the proposed broadband PA with an open-terminated balun are shown in Fig. 2. A two-coupler, a two-input capacitor (C_D), and output capacitors (C_L and C_L') are incorporated in the proposed balun. Typically, balun is short-terminated at port 4′ and, C_L' is neglected. However, the open-terminated balun has an additional output capacitor, C_L', to find the achievable $Z_{0,e}$ and $Z_{0,o}$ with a small θ. The load resistance, R_L, is 50 Ω. Using the relationship between the voltage and current in the coupler (blue dotted box in Fig. 2), the voltage and current relationship of the proposed open-

978-1-6654-8495-4/22 $31.00 © 2022 IEEE

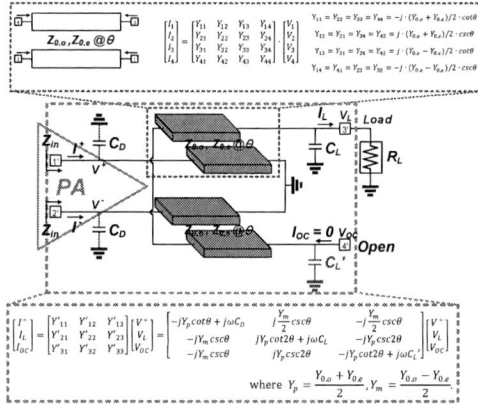

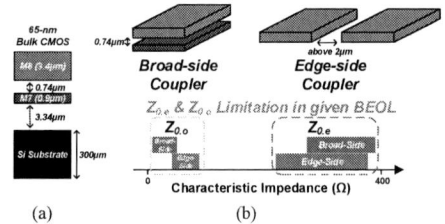

Fig. 2. Schematic and equation of the open-terminated lumped-distributed balun.

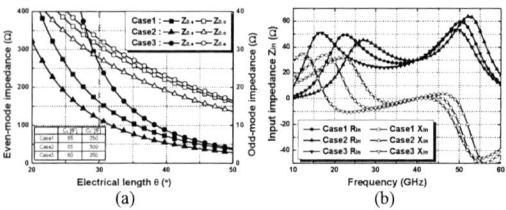

(a)   (b)

Fig. 3. (a) Metal stack-up of the 65-nm bulk CMOS and (b) limitation of $Z_{0.e}$ and $Z_{0.o}$ in broad- and edge-side coupler.

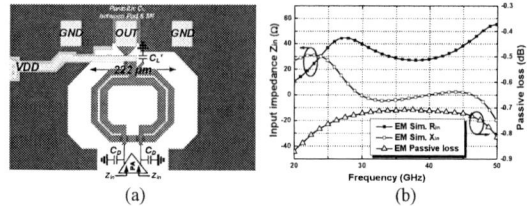

(a)   (b)

Fig. 4. (a) Calculated $Z_{0.e}$ and $Z_{0.o}$ for different C_L and C_L', and (b) simulated Z_{in} with $\theta = 30°$ for different C_L and C_L'.

terminated balun was derived including C_D, C_L, and C_L' (red dotted box in Fig. 2). This equation assumed that $V^+ = -V^-$ and $I^+ = -I^-$ at both port 1´ and 2´ in a differential PA. From the voltage and current relationship of the proposed balun, the input impedance, Z_{in}, which is the inverse of the input admittance ($Y_{in} = 1/Z_{in} = I^+/V^+$), was derived as:

$$Y_{in} = Y'_{11} - \frac{Y'_{13}Y'_{31}}{Y'_{33}} + $$
$$(Y'_{12} - \frac{Y'_{13}Y'_{32}}{Y'_{33}})\frac{R_L(-Y'_{21}Y'_{33} + Y'_{23}Y'_{31})}{R_LY'_{22}Y'_{33} - R_LY'_{23}Y'_{32} + Y'_{33}} . (1)$$

Equation (1) is a function of $Z_{0.e}$, $Z_{0.o}$, Z_{in}, C_D, C_L, C_L', and θ. If C_L' is chosen as infinity, this specific condition is the same with the case of typical short-terminated at port 4´. Considering the target output power and PAE, the PA device size and R_{opt} are determined from the load-pull simulations. Thus, C_D is determined according to the required device size. Because C_D is included in the output network, as shown in Fig. 2, Z_{in} is purely

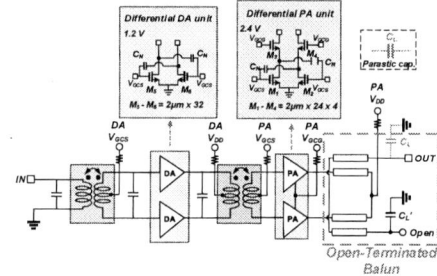

(a)   (b)

Fig. 5. (a) Layout of the proposed open-terminated balun, and (b) simulated Z_{in} and passive loss of the proposed balun.

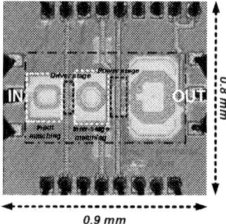

Fig. 6. Detailed schematic of the two-stage CMOS PA with the proposed balun.

Fig. 7. Die photograph of the PA with the proposed balun.

real and has to be matched with R_{opt} [4]. For the implementation of the proposed balun, providing R_{opt}, 65-nm bulk CMOS process was used in this work. In Fig. 3(a) shows the metal stack-up of the process used. As shown in Fig. 3(b), a coupler is typically designed in a broad- or edge-side manner. In a given back-end-of-line (BEOL) process, there are thick and ultra-thick metal (M7 and M8) layers. In addition, the gap of two lines in the broad-side coupler using the M7 and M8 layers is fixed at 0.74 μm, and the distance of two lines in the edge-side coupler using the M8 layer must be more than 2 μm. Because the spacing between the two lines of couplers is limited, the implementable range of $Z_{0.e}$ and $Z_{0.o}$ will be narrow. In the limited range of $Z_{0.e}$ and $Z_{0.o}$, with a smaller θ, a solution may not exist because a typical short-terminated balun at port 4´ has only one design variable, C_L, for $Z_{0.e}$ and $Z_{0.o}$. However, an additional parameter, C_L', is included in the proposed balun. Using C_L and C_L', it is possible to construct the open-terminated balun providing the target R_{opt} with a low loss within narrow range of implementable $Z_{0.e}$ and $Z_{0.o}$. In addition, by selecting the optimum values for C_L and C_L', the value of θ can be significantly less than $\lambda/4$ to minimize the required die area.

Fig. 4(a) shows the calculated $Z_{0.e}$ and $Z_{0.o}$ using (1) for exemplary design cases with $R_{opt} = 30$ Ω, $C_D = 130$ fF, and

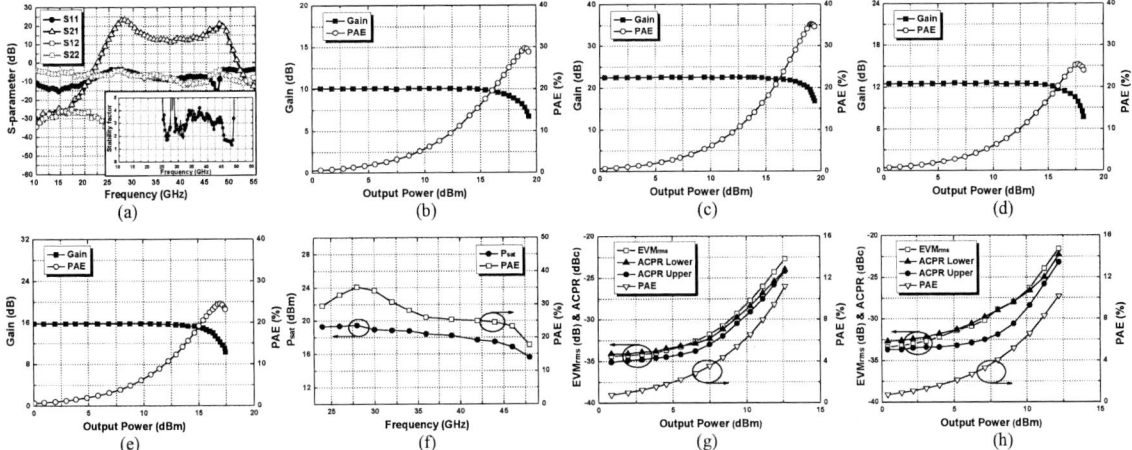

Fig. 8. Measured (a) S-parameters, large signal CW results at (b) 24 GHz, (c) 28 GHz, (d) 39 GHz, (e) 44 GHz, (f) large-signal CW performances from 24 to 48 GHz, EVM$_{rms}$ and ACPR versus P$_{linear}$ with 5G NR FR2 64-QAM 8-CC OFDM (800-MHz BW) at (g) 28 GHz and (h) 39 GHz.

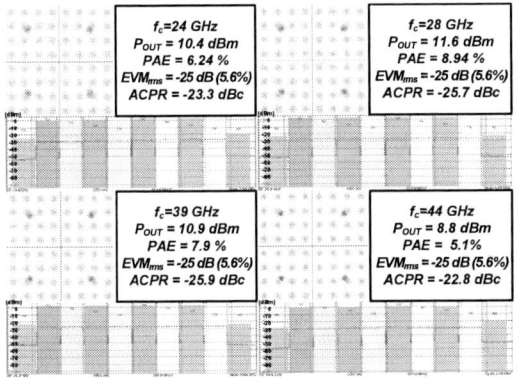

Fig. 9. Measured EVM$_{rms}$ and ACPR with 5G NR FR2 64-QAM 8-CC OFDM signals at 24/28/39/44 GHz.

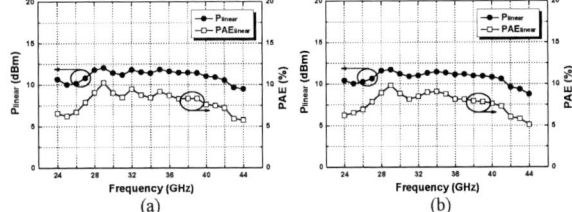

Fig. 10. Modulation results with the 5G NR performances from 24 to 44 GHz: (a) 1-CC (BW = 100 MHz) and (b) 8-CC (BW = 800 MHz).

frequency = 40 GHz. Three cases are plotted for different C$_L$ and C$_L$′ in the range of achievable Z$_{0.e}$ and Z$_{0.o}$, as shown in Fig. 4(a). Fig. 4(b) shows the simulated Z$_{in}$ based on ideal circuit elements for three cases with θ = 30°. The Z$_{in}$ in all cases are broad and the frequency range is varied according to the change of C$_L$ and C$_L$′. The layout of the proposed open-terminated balun is shown in Fig.5(a). Two C$_D$ are parasitic capacitances of a differential PA, and C$_L$ is the parasitic capacitance of the output pad and ground plate. The proposed balun is implemented in an edge-coupled configuration to achieve the required Z$_{0.e}$ and Z$_{0.o}$. By adjusting the two output capacitors C$_L$ and C$_L$′, Z$_{0.e}$ and Z$_{0.o}$ values could be obtained for small θ while adhering to the design rules for the implemented process. Fig. 5(b) shows the simulated Z$_{in}$ and passive loss of the proposed balun using the layout shown in Fig. 5(a). The simulated Z$_{in}$ is approximately 30 Ω, with a wide bandwidth. The reactance component is approximately zero. In addition, it shows that the proposed open-terminated balun achieves low passive loss of approximately 0.7 dB in broadband over 20-GHz bandwidth. Fig. 6 shows a detailed schematic of the proposed two-stage PA. The driver stage (DA) is designed using a common-source (CS) topology for a 1.2-V supply voltage. For the power stage design,

a cascode configuration is adopted for a 2.4-V supply voltage. The differential DA and power stage include two neutralized capacitors, C$_N$, for high reverse isolation. The input and inter-stage matchings are designed using a double-pole transformer topology for broadband matching [3]. The proposed broadband balun is adopted in an output matching network with a low passive loss. An anti-phasing IMD3 cancellation method is applied to improve the linearity performance.

III. MEASUREMENT RESULTS

The proposed 24 to 44 GHz broadband PA with an open-terminated balun was implemented in 65-nm bulk CMOS technology. The PA chip occupies 0.15 mm² core area, as shown in Fig. 7. To validate its performance, the implemented PA was measured using on-wafer probing with dc wire-bonding to an external PCB. Fig. 8(a) shows the measured S-parameter results including the stability factor. The stability factor is greater than one in the range of 10–55 GHz. Fig. 8(b)-(e) show the measured large signal results for continuous-wave (CW) at 24/28/39/44 GHz. The PA delivers a P$_{sat}$ of 19.3/19.5/18.2/17.5 dBm while achieving a peak PAE, PAE$_{max}$, of 29.6/35.2/25.4/24.5%. Fig. 8(f) shows the measured large-signal CW power sweep results from 24 GHz to 48 GHz. In the range of 24–44 GHz, the measurement result shows a flat P$_{sat}$ of 17.5 to 19.5 dBm, within 2-dB difference, while achieving flat PAE$_{max}$ ranging from 24.5 to 35.2%. The PA was also tested using the 5G NR FR2 64-QAM 1-/8-CC (100-/800-MHz bandwidth) OFDM signal (with a 9.7-dB PAPR) at 24/28/39/44 GHz. Fig. 8(g)-(h) show the

978-1-6654-8495-4/22 $31.00 © 2022 IEEE

TABLE I.   PERFORMANCE COMPARISON WITH STATE-OF-ART MM-WAVE TWO-STAGE PAs FOR 5G APPLICATIONS.

Ref.	This work				[4]JSSC21					[8]JSSC19			[9]JSSC21				[3]JSSC18			[6]TMTT22	
Tech.	65nm Bulk CMOS				45nm SOI CMOS					130nm SiGe			45nm SOI CMOS				28nm Bulk CMOS			28nm Bulk CMOS	
Freq.(GHz)	24	28	39	44	24	28	37	39	42	28	37	39	26	32	45	60	30	40	50	26.5	37
P_{sat} (dBm)	19.3	19.5	18.2	17.5	20	20.4	20	19.1	17.9	16.8	17.1	17	20.19	22.62	22.1	19.53	16.6	15.9	15.1	19.1	19.5
PAE_{max} (%)	29.6	35.2	25.4	24.5	38.9	45	38.7	38.6	35	20.3	22.6	21.4	19.4	41.88	29.41	22.59	24.2	18.4	14.9	31.3	30.5
Modulation scheme	5G NR FR2 64-QAM 8-CC×100-MHz OFDM (800-MHz BW)				5G NR FR2 64-QAM 2-CC×400-MHz OFDM (800-MHz BW)					64-QAM OFDM (3 Gb/s)			5G NR FR2 64-QAM OFDM (200-MHz BW)				64-QAM OFDM (675-MHz BW)			64-QAM OFDM (100-MHz BW)	
PAPR (dB)	9.7				11.78					6			9.64				8.3			7	
Freq.(GHz)	24	28	39	44	24	28	37	39	42	28	37	39	32	36	52	55	28	32	34	26.5	37
EVM_{rms} (dB)	<-25				-25.1	-25.1	-25.1	-25.1	-25.1	-27	-30.3	-28.7	-24	-24	-24	-24	<-25			-25.3	-25.1
ACPR (dBc)	-23.3	-25.7	-25.9	-22.8	-25.2	-25.6	-27.9	-26.1	-26.4	-28.4	-28.2	-29.8	-28.8	-28.1	-24.1	-24	-37.6	-34.2	-30.2	-22.2	-21.7
P_{linear} (dBm)	10.4	11.6	10.9	8.8	10.9	11.3	10.2	10.2	8.4	9.2	9.5	9.3	10.7	11.15	10.89	12.25	6.8	8.1	8.9	12.3	11.6
PAE_{linear} (%)	6.24	8.94	7.9	5.1	14.2	16.6	13.6	13.4	10.3	10.8[a]	13.5[a]	11[a]	14.5	16.54	10.2	13.5	2.9	3.9	4.4	12.5	12.7
Area (core area) (mm²)	0.72 (0.15)				1.35 (0.21)					1.76 (1.03[a])			2.46 (0.67)				0.49 (0.12[a])			0.81 (0.165)	
Topology	Open-Terminated Balun Based Broadband PA				Distributed Balun Based Broadband PA					Transformer Based Doherty Power Combining PA			Multi-Frequency Role-Exchange Coupler Doherty PA				Transformer-based high-order network			Variable tuning elements	

[a]. Graphical estimation.

EVM_{rms} and ACPR versus P_{linear} results at 28/39 GHz with the 8-CC 5G NR signal. Fig. 9 shows that the proposed PA achieves P_{linear} of 10.4/11.6/10.9/8.8 dBm with 6.24/8.94/7.9/5.1% PAE_{linear} while obtaining -25-dB EVM_{rms} and -23.3/-25.7/-25.9/-22.8-dBc ACPR at 24/28/39/44 GHz with 5G NR 8-CC signal. Fig. 10(a) and (b) show the measured modulation results with the 5G NR 1-CC×100-MHz and 8-CC×100-MHz signals. The proposed PA achieves 9.5 to 12-dBm P_{linear} while over 5.7% PAE_{linear} from 24 to 44 GHz with the 5G NR 1-CC signal under -25-dB EVM_{rms}. Using the 5G NR 8-CC signal, the proposed PA achieves 8.8 to 11.7-dBm P_{linear} while over 5.1% PAE_{linear} under -25-dB EVM_{rms}. The P_{linear} and PAE_{linear} using an 800-MHz BW signal are close to those using a 100-MHz BW signal because of wideband operation without performance degradation. The proposed PA performance is summarized and compared with state-of-the-art mm-wave silicon PAs for 5G applications in Table I. Among the bulk CMOS PAs in the comparison table, the proposed PA demonstrates the highest and broadest output power and PAE within a wide frequency range in a small core area. In addition, the PA was successfully demonstrated using the 5G NR FR2 64-QAM 8-CC OFDM signal from 24 to 44 GHz for the cover n257–261 band.

IV. CONCLUSION

This paper presents a broadband mm-wave PA with an open-terminated balun using 65-nm bulk CMOS technology. The proposed open-terminated balun incorporates asymmetric output capacitors. The proposed compact CMOS PA demonstrated a flat output power performance at 24–44 GHz. This PA achieved 17.5–19.5-dBm P_{sat}, 24.5–35.2% PAE_{max}, and 8.8–11.7-dBm P_{linear} in 59% fractional bandwidth. The PA supports the 5G NR FR2 64-QAM 8-CC OFDM signal over 24–44 GHz, covering the n257/258/ 259/260/261 5G bands. This was achieved by adopting a compact mm-wave power amplifier design for broadband 5G applications.

ACKNOWLEDGMENT

This work was supported in part by Samsung Electronics Co., Ltd., and in part by the National Research Foundation of Korea (NRF) grant funded by the Korea government (MSIT) (NRF2020R1A2B5B01001508).

REFERENCES

[1] NR; User Equipment (UE) Radio Transmission and Reception; Part 2: Range 2 Standalone, 3GPP TS 38.101-2, Version 17.3.0, Sep. 2021.

[2] O. El-Aassar and G. M. Rebeiz, "A DC-to-108-GHz CMOS SOI distributed power amplifier and modulator driver leveraging multi-drive complementary stacked cells," IEEE J. Solid-State Circuits, vol. 54, no. 12, pp. 3437–3451, Dec. 2019.

[3] M. Vigilante and P. Reynaert, "A wideband class-AB power amplifier with 29–57-GHz AM-PM compensation in 0.9-V 28-nm bulk CMOS," IEEE J. Solid-State Circuits, vol. 53, no. 5, pp. 1288–1301, May 2018.

[4] F. Wang and H. Wang, "A broadband linear ultra-compact mm-Wave power amplifier with distributed-balun output network: analysis and design," IEEE J. Solid-State Circuits, vol. 56, no. 8, pp. 2308–2323, Aug. 2021.

[5] C. R. Chappidi and K. Sengupta, "Frequency reconfigurable mm-wave power amplifier with active impedance synthesis in an asymmetrical non-isolated combiner: Analysis and design," IEEE J. Solid-State Circuits, vol. 52, no. 8, pp. 1990–2008, Aug. 2017.

[6] J. Lee, J.-S. Paek, and S. Hong, "Millimeter-wave frequency reconfigurable dual-band CMOS power amplifier for 5G communication radios," IEEE Trans. Microw. Theory Tech, vol. 70, no. 1, pp. 801–812, Jan. 2022.

[7] K. S. Ang, Y. C. Leong, and C. H. Lee, "Analysis and design of miniaturized lumped-distributed impedance-transforming balun," IEEE Trans. Microw. Theory Tech, vol. 51, no. 3, pp. 1009–1017, Mar. 2003.

[8] S. Hu, F. Wang, and H. Wang, "A 28-/37-/39-GHz linear Doherty power amplifier in silicon for 5G applications," IEEE J. Solid-State Circuits, vol. 54, no. 6, pp. 1586–1599, Jun. 2019.

[9] N. S. Mannem, T.-Y. Huang, and H. Wang, "Broadband active load modulation power amplification using coupled-line baluns: A multifrequency role-exchange coupler Doherty amplifier architecture," IEEE J. Solid-State Circuits, vol. 56, no. 10, pp. 3109–3122, Oct. 2021.

A CMOS Full-Wave Switching Rectifier with Frequency Up-Down Conversion for 5G NR Wirelessly-Powered Relay Transceivers

Sena Kato, Keito Yuasa, Michihiro Ide, Atsushi Shirane, and Kenichi Okada
Department of Electrical and Electronic Engineering, Tokyo Institute of Technology
2-12-1-S3-28, Ookayama, Meguro-ku, Tokyo 152-8552, Japan
Email: kato@ssc.pe.titech.ac.jp

Abstract—This paper presents a CMOS full-wave switching rectifier capable of mixer operation. The proposed CMOS switching rectifier allows DC power generation and frequency conversion without a power supply. The proposed CMOS switching rectifier using a center-tapped balun produces DC power with 25% efficiency, while at the same time providing frequency up-conversion with -18.9 dB and down-conversion with -11.9 dB. The size of the chip is 0.286 mm², and four chips are mounted on the phased-array antenna board. The board can be used as a wirelessly powered relay module, and the module can be driven without an external power supply. The measured EVM values are -27.4dB for Tx mode and -27.5dB for Rx mode with a 400-MHz 64QAM OFDMA-mode signal (5G NR, MCS 17).

Keywords—Wireless Power Transfer, CMOS, Rectifier, Transceiver, CMOS, Phased-Array, 5G mobile network

I. INTRODUCTION

Millimeter-wave 5G relay transceivers are promising solutions to solve the problem of the large propagation loss and the greatly limited communication range. In the millimeter-wave band communication, base stations are required to be installed more closely in order to keep the coverage area equal to or higher than that of the conventional wireless systems. However, it is difficult to install a large number of base stations due to the need for new locations for installation and incredibly increased base station costs. Thus, placing relay transceivers instead can maintain the coverage while providing higher-speed communication [1].

Battery-less relay transceivers are highly flexible for installation and can contribute to the expansion of millimeter-wave band 5G. Many battery-less devices using wireless power transfer (WPT) have been proposed along with the development of devices with a variety of functions. Battery-less devices are highly convenient and maintenance-free because they operate without an external power supply. By combining WPT with millimeter-wave band 5G relay transceivers, it is possible to create maintenance-free and can be installed anywhere at a lower cost than base stations. Since the battery-less relay transceivers operate only by supplying power from the WPT signal, the relay transceivers need to generate enough DC power from the received WPT power to operate the entire relay system.

The reported millimeter-wave band 5G battery-less relay transceiver can operate by generating 3.1 mW DC power [2][3]. This relay transceiver uses a circuit that integrated a rectifier and

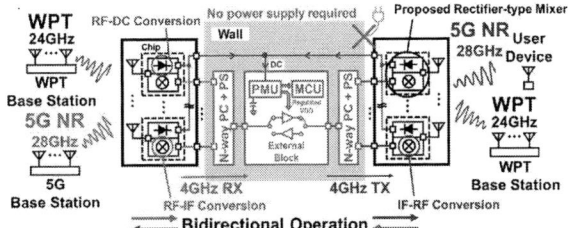

Fig.1. System block diagram of proposed 5G WPT

a mixer, called a rectifier-type mixer, which can generate DC power and perform frequency conversion at the same time. The DC conversion efficiency of the rectifier-type mixer used in the relay transceiver is 15% when the WPT signal power is 0 dBm. The up-conversion gain and down-conversion gain of the circuit are -27 dB and -17 dB, respectively. These performances are inadequate for a battery-less relay transceiver system. The low-frequency conversion gain requires a high gain amplifier. Also, the low RF-DC conversion efficiency limits the power consumption of the amplifier. In order to obtain as much signal power as possible, the rectifier-type mixer requires not only a high-frequency conversion gain but also high RF-DC conversion efficiency enough to operate an amplifier for a relay system.

This paper proposes a new rectifier-type mixer that is full-wave rectifiable and capable of single-ended input using a center-tapped balun. The rectifier is 1.5 times more efficient and has a frequency conversion gain of more than 5 dB higher than the conventional type. This paper is organized as follows. In Section II, the proposed battery-less relay transceiver is described. In Section III, the proposed circuit configuration is explained. In Section IV, the relay transceiver fabricated in this study is shown. In Section V, the measurement results are demonstrated, followed by the conclusion in Section V.

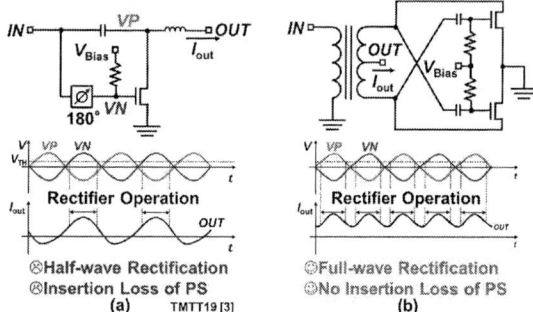

Fig.2. The circuit operation of the rectifier: (a) conventional rectifier and (b) proposed circuit

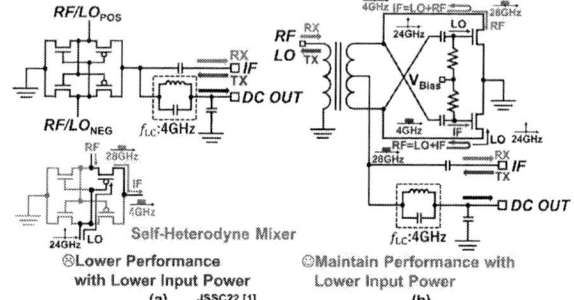

Fig.3. The circuit operation of the mixer: (a) conventional circuit (b) proposed circuit

II. WIRELESSLY POWERED RELAY TRANSCEIVER

Fig.1 shows the diagram of the proposed battery-less relay transceiver system for millimeter-wave band 5G. This relay transceiver can be mounted on a wall to send signals from a base station on one side to a device on the other side. The system contains phased-array antennas, so the system can provide stable communication to the device and the base station (BS) with beamforming. The proposed relay transceiver can operate without an external power supply by generating DC power from the signal radiated from the WPT base station. The power management unit (PMU) controls the generated DC power so that it is usable by microcontroller units (MCU) and amplifiers. Phased array antennas are used to ensure signal power. The phased-array antennas not only reduce the size but also the system power consumption.

The WPT signal is also used as an LO signal to convert frequency. The relay transceiver operates by converting the 24GHz WPT signal to DC. Moreover, the 28 GHz RF signal received from the 5G base station is converted to the 4 GHz IF signal by using the WPT signal as the LO signal. The opposite operation is performed for transmission. Inside the wall, the IF signals are phase-shifted by phase shifters (PS), combined by a power combiner (PC), and amplified and then divided to each chip for the opposite wall.

III. PROPOSED RECTIFIER AND MIXER

The rectifier-type mixer circuit needs a high-efficiency rectifier and passive mixer. The rectifier function needs to convert the WPT signal to DC with the highest possible efficiency to generate much DC power. The mixer function requires also achieving the highest possible frequency conversion gain. In this case, an active mixer is not suitable for WPT because an active mixer consumes about several tens of mW and WPT is difficult to keep the stable power supply voltage. Hence, a passive mixer that performs frequency conversion while generating DC power is required.

A. Rectifier Operation

Fig.2 (a) presents a conventional CMOS switching rectifier circuit [4]. This CMOS switching rectifier circuit is more efficient than the diode-connected rectifier [5]. The CMOS switching rectifier works by making the transconductance of the NMOS negative. First, the phases of the signals input to the gate and drain are shifted by 180°. When VN exceeds the threshold

voltage of the NMOS, the VP is a negative voltage, so the drain-source voltage becomes negative and the transconductance becomes negative. Then the drain current becomes negative and the current flows to the inductor side. By repeating this process, current flows to the OUT side regularly and DC power can be obtained. Furthermore, the rectifier can be biased to the gate to increase efficiency. The rectifier operates as a half-wave rectifier. Half-wave rectifiers are less efficient because they operate as rectifiers for a shorter time. The efficiency is also degraded by the insertion loss due to the need to use a phase shifter.

Fig.2 (b) shows the proposed rectifier circuit. The conventional rectifier can operate as a full-wave rectifier by simply placing two rectifiers and making the differential input signals. However, two conventional rectifiers require two large inductors and a 180° phase shifter with high insertion loss. To solve this problem, a center-tapped balun is used. The center-tapped balun not only acts as a balun but also as two inductors, allowing the signal to be taken out of the balun's center tap. The proposed circuit can be described as a CMOS switching rectifier capable of full-wave rectification operation without the need to use a phase shifter. The proposed circuit can generate DC power at 1.5 times higher efficiency than the conventional rectifier.

B. Mixer Operation

Fig.3 (a) shows a reported rectifier-type mixer circuit [2][6]. The circuit composed of NMOS and PMOS operates as a full-wave rectifier to obtain DC power. 24 GHz differential signals are input from RF/LO_{POS} and RF/LO_{NEG}. By simultaneously inputting 28 GHz differential signals, the PMOS operates as a self-heterodyne mixer: the 28 GHz RF signal input to the drain of the PMOS is converted to the 4 GHz IF signal at the source by the 24 GHz LO signal input to the gate. Since MOS transistors cannot bias to the gate, the gate-source voltage cannot exceed the threshold voltage. The MOS will not operate unless the threshold is exceeded, so low WPT input power will drop conversion efficiency and frequency conversion gain.

Fig.3 (b) presents the proposed rectifier-type mixer. The proposed rectifier-type mixer can also operate as a self-heterodyne mixer for RX mode. For RX operation, the RF signal differentiated by the balun is input to the drain, and a similarly differentiated LO signal is input to the gate. This causes the MOS to act as a heterodyne mixer, returning the IF signal to the

978-1-6654-8495-4/22 $31.00 © 2022 IEEE

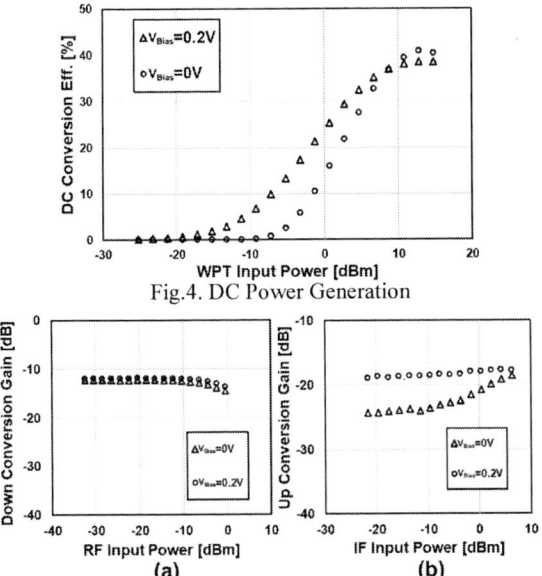

Fig.4. DC Power Generation

Fig.5. Mixer measurement results (a) down conversion and (b) up conversion

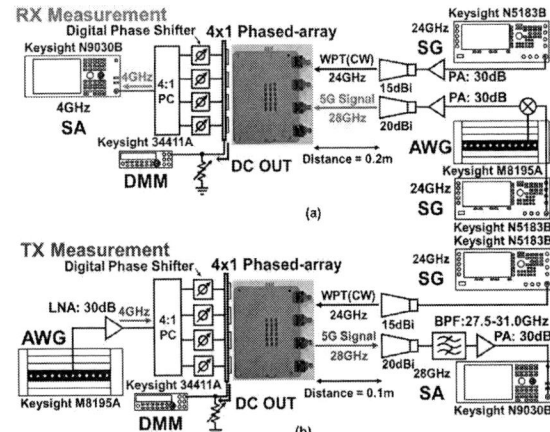

Fig.6. OTA measurement setup for (a) RX mode and (b) TX mode

drain side. This signal passes through the balun and then through the capacitor to be output as an IF signal. On the other hand, for TX operation, the input IF signal is input to the gate through capacitors and balun. At the same time, a differential LO signal is input to the drain. The MOS acts as backscatter and the RF signal is returned to the drain side. The generated differential RF signal is output as a single-ended RF signal by the balun.

In the rectifier type mixer circuit, the IF signal and DC power are mixed in the output signal because the rectifier operation and the mixer operation are performed at the same time. To separate these, a capacitor for DC cut is included on the IF input/output side, and a filter for cutting 4 GHz is placed on the DC output side.

IV. TRANSCEIVER IMPLEMENTATION

The fabricated relay transceiver board contains four antennas. The antennas are dual-band U-slot patch antennas [2][3]. The antennas enable simultaneous receiving of the 24 GHz WPT signal and the 28 GHz 5G NR modulated signal. RF connectors are connected to each IF signal input/output port.

The spacing of the phased-array antenna is d=5.3 mm, which is half of the wavelength for 28 GHz signal, thus the relay transceiver can steer the beam in a wide range. However, the proposed mixer receives the 24 GHz LO signal as well as the 28GHz modulated signal. For RX mode, the phase difference of the IF signals of adjacent chips φ_{IF} are given by

$$\varphi_{IF} = \pi \left(\sin \theta_{RF} - \frac{\lambda_{RF}}{\lambda_{LO}} \sin \theta_{LO} \right) \qquad (1)$$

where $\lambda_{RF}, \lambda_{LO}, \theta_{RF}, \theta_{LO}$ represent the wavelength and incident angle of 28GHz, 24GHz signal, respectively [2]. On the other hand, since the IF is up-converted by the LO in the

TX mode, the IF signals of adjacent chips φ_{IF} are expressed in (2).

$$\varphi_{IF} = \pi \left(\sin \theta_{RF} + \frac{\lambda_{RF}}{\lambda_{LO}} \sin \theta_{LO} \right) \qquad (2)$$

Thus, we need to be noted that the formula changes by receiving the LO signal as well.

V. MEASUREMENT RESULTS

Fig.4 shows the measured RF-DC conversion efficiency of the proposed rectifier. The measurement is conducted with on-wafer probe equipment. Without gate bias, that is, when VG is 0 V, the DC conversion efficiency is 15 % when the input power is 0dBm. Biasing gates can increase efficiency in the lower input power range. With gate bias, the DC conversion efficiency is 25 %, which is 10 % increase in efficiency when the input power is 0dBm.

Fig.5 (a) and (b) show the measured results of the down and up conversion gain of the proposed mixer in the RX and TX mode, respectively. Both results are shown when the LO input power is 0dBm. With gate bias, the down-conversion gain is -22 dB, and the up-conversion gain is -19 dB, increasing gain by 1dB and 5 dB, respectively.

Fig.6 shows the over-the-air (OTA) measurement setup in TX and RX modes with a 5G NR modulated signal for beam pattern and EVM measurement. Fig.7 shows the measured beam pattern in the azimuth plane. The beam is swept from -45° to 45° by 15°. The 4x1 array antennas allow the beam to steer from -45° to +45°. The OFDMA-mode modulated signal is generated by Keysight arbitrary waveform generator (AWG) M8195A. The received signal is analyzed by the signal analyzer (SA) N9030B with vector signal analyzer (VSA) 89600. The WPT signal is generated by analog signal generator (SG) N5183B. The generated DC power is measured by digital multimeter (DMM) and variable load resistors. Fig.8 shows the measured constellations and EVMs for TX and RX modes. The 400 MHz standard-compliant 5G NR OFDMA-mode signal is used for both TX and RX measurement. The proposed 4-array relay

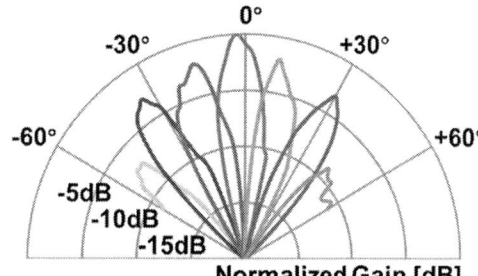

Fig.7. Measured beam pattern

Mode	TX (4-elements)			RX (4-elements)		
Modulation	QPSK	16QAM	64QAM	QPSK	16QAM	64QAM
MCS	5G NR MCS4	5G NR MCS10	5G NR MCS17	5G NR MCS4	5G NR MCS10	5G NR MCS17
Bandwidth and Spectrum	400MHz			400MHz		
Constellation						
EVM	-27.4dB	-27.4dB	-27.4dB	-28.0dB	-27.9dB	-27.5dB

Fig.8. Measured EVMs and constellations for Tx and Rx modes

system can send the modulated signal using 64QAM with -27.4 dB. Also, the module can receive the modulated signal using 64QAM with -27.5 dB.

Fig.9 shows the die and board photo. The chip was fabricated in a standard 65nm Si CMOS process. The chip area of the proposed rectifier-type mixer is 0.286 mm². The prototype phased-array relay transceiver board contains 4 patch antennas surrounded by the dummy antennas.

Table.1 summarizes a comparison with the state-of-the-art millimeter-wave rectifiers. The power conversion efficiency (PCE) of the proposed rectifier at 0dBm input power is 25%, which is higher than the conventional rectifier. Furthermore, even with input power as low as -4 dBm, DC conversion can be performed with 16% efficiency. The rectifier-type mixer has up conversion gain of 8 dB and down conversion gain of 5 dB higher than the conventional type.

VI. CONCLUSION

This paper has presented the CMOS full-wave switching rectifier type mixer. The proposed CMOS switching rectifier type mixer using a center-tapped balun can convert with 25 % efficiency when the input power is 0 dBm. The proposed rectifier-type mixer circuit allows up conversion gain of -18.9 dB and down conversion gain of -11.9 dB to be achieved. The 4x1 array dual-band U-slot patch antenna board with 4 chips containing the proposed rectifier-type mixer circuit. Based on the results of EVM measurement, this module can send and receive the standard-compliant 5G NR OFDMA-mode modulated signal with 400-MHz bandwidth with this module. The proposed battery-less relay transceiver will be widely used thanks to its high convenience and low cost.

ACKNOWLEDGEMENT

This work is partially supported by the MIC/SCOPE #192203002 and #192103003, JSPS (JP20H00236), MIC (JPJ000254), NICT(00601), STAR,

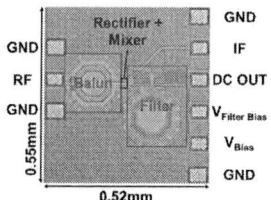

Fig.9. Die and board photograph

Table. 1 Performance comparison with rectifiers

	This work	JSSC 2021 [1]	TMTT 2019 [2]	IMS 2014 [5]
Process	65nm CMOS			
Rectifier Type	Full wave CMOS Switching	Full wave CMOS	Half wave CMOS Switching	Half wave Dual Diode
Frequency	24,28GHz	24,28GHz	35GHz	35GHz
Pout(@Max PCE)	18.8mW@80Ω	17.5mW@220Ω	11.6mW@50Ω	20mW
Max PCE	38.4%	15%	36%	34%
PCE(@Pin=0dBm)	25%	15%	10%(1)	0.5%(1)
PCE(@Pin=-4dBm)	16%	10%	3%(1)	0%(1)
Up Conv. Gain	-18.9dB	-27dB(1)	N/A	N/A
Down Conv. Gain	-11.9dB	-17dB(1)	N/A	N/A
Number of Array	4x1	4x8	N/A	N/A
Bandwidth	400MHz	400MHz	N/A	N/A

(1)Estimated based on the graphs.

and VDEC in collaboration with Cadence Design Systems, Inc., Mentor Graphics, Inc., and Keysight Technologies Japan, Ltd.

REFERENCES

[1] K. Sakaguchi, T. Yoneda, M. Iwabuchi, T. Murakami, "MMWAVE MASSIVE ANALOG RELAY MIMO," in ITU Journal on Future and Evolving Technologies, Volume 2 (2021), Issue 6.

[2] M. Ide, A. Shirane, K. Yanagisawa, D. You, J. Pang and K. Okada, "A 28-GHz Phased-Array Relay Transceiver for 5G Network Using Vector-Summing Backscatter With 24-GHz Wireless Power and LO Transfer," in IEEE Journal of Solid-State Circuits, vol. 57, no. 4, pp. 1211-1223, April 2022, doi: 10.1109/JSSC.2021.3137336.

[3] M. Ide, A. Shirane, K. Yanagisawa, D. You, J. Pang and K. Okada, "A 28-GHz Phased-Array Relay Transceiver for 5G Network Using Vector-Summing Backscatter with 24-GHz Wireless Power and LO Transfer," 2021 Symposium on VLSI Circuits, 2021, pp. 1-2, doi: 10.23919/VLSICircuits52068.2021.9492431.

[4] P. He and D. Zhao, "High-Efficiency Millimeter-Wave CMOS Switching Rectifiers: Theory and Implementation," in IEEE Transactions on Microwave Theory and Techniques, vol. 67, no. 12, pp. 5171-5180, Dec. 2019, doi: 10.1109/TMTT.2019.2936566.

[5] S. Ladan and K. Wu, "35 GHz harmonic harvesting rectifier for wireless power Transmission," 2014 IEEE MTT-S International Microwave Symposium (IMS2014), 2014, pp. 1-4, doi: 10.1109/MWSYM.2014.6848572.

[6] K. Kotani, A. Sasaki and T. Ito, "High-Efficiency Differential-Drive CMOS Rectifier for UHF RFIDs," in IEEE Journal of Solid-State Circuits, vol. 44, no. 11, pp. 3011-3018, Nov. 2009, doi: 10.1109/JSSC.2009.2028955.

INTIACC: A 32-bit Floating-Point Programmable Custom-ISA Accelerator for Solving Classes of Partial Differential Equations

Paul Xuanyuanliang Huang, Daniel Jang, Yannis Tsividis, and Mingoo Seok
Department of Electrical Engineering
Columbia University
New York, NY, USA
Email: xh2373@columbia.edu

Abstract— **We propose a numerical integration accelerator (INTIACC) that speeds up the solution of partial differential equations (PDEs) for scientific computing. In contrast to recent works, INTIACC applies to a variety of PDEs and boundary conditions, has enhanced nonlinear function capability, supports high-order integration algorithms, and uses floating-point arithmetic for orders of magnitude smaller solution error. With all the benefits, our test chip still achieves 40X speed-up over prior accelerators and orders of magnitudes over CPU and GPU based systems.**

Keywords—Digital Accelerator, Scientific Computing, Partial Differential Equations

I. Introduction

PDEs are the core mathematical tool for modeling a wide range of scientific and engineering phenomena, making it of great interest to accelerate their numerical solution. Today's primary choices for solving PDEs are multi-core CPUs [5, 6] and GPUs, but it still takes a long time and a large amount of energy to solve them. This limitation has motivated multiple accelerator architectures [1-4, 7]. However, they are still severely limited in mainly four ways. First, they can only handle ordinary differential equations (ODEs) [3] or a single type of PDE (Poisson's equation) [1, 2], leaving a large number of essential PDEs inapplicable. Second, they only support constant (i.e., Dirichlet) boundary conditions [1-4]. Third, some accelerators do not support the computation of nonlinear terms such as x^2, $\sin(x)$, and $1/x$, which frequently arise in real-world problems [1, 2]. Fourth, they have low computation precision and limited dynamic range, using analog current signals [3] or 4-16-bit fixed-point arithmetic [1, 2].

II. Proposed Accelerator

A. Architectural Features

INTIACC contains 16 processing elements (PEs), each of which is a fully programmable custom-ISA RISC processor. The programmable architecture allows INTIACC to solve a wide range of PDEs (Fig. 1) and to support Dirichlet, Neumann, and time-dependent boundary conditions. Also, each core employs the 32-bit floating-point ALU and multiplier, allowing it to compute high-precision solutions without dynamic range issues (Fig. 2). Also, the accelerator supports high-order integration algorithms such as the Runge-Kutta 4th order (RK4). Furthermore, it can compute almost any nonlinear terms by evaluating them with numerical routines (i.e., algorithms) programmed in each PE, such as the Goldschmidt fast division for $1/x$, Newton-Raphson for $\sqrt{x}$, and Taylor approximation for $\sin(x)$. As compared to using look-up tables (LUTs) [3, 4], the algorithmic method is orders-of-magnitude more efficient in memory usage (Fig. 3). We also adopt a hybrid local and global clock scheme, where each PE operates at a fast local clock (570 MHz), with only the slow global clock (10-50 MHz) distributed across PEs, significantly saving clock power dissipation. We prototyped INTIACC in 65nm (Fig. 12). It vastly outperforms the previous accelerators in applicability and precision while improving speed by 40X (Fig. 14). As compared to the CPU/GPU-based system, INTIACC achieves one to three orders of magnitude improvement both in speed and energy (Fig. 13 bottom right).

	Equation	Class	Boundary Conditions	JSSC16 [3]	JSSC20 [2]	ISSCC21 [1]	INTIACC
	Ordinary differential equations	ODE	Dirichlet	✓			✓
			Neumann	✗	✗	✗	✓
			Time-dependent	✗			✓
PDEs	Advection equation $\frac{\partial u}{\partial t} = -v\left(\frac{\partial u}{\partial x} + \frac{\partial u}{\partial y}\right)$	1st-order hyperbolic	Dirichlet				✓
			Neumann	✗	✗	✗	✓
			Time-dependent				✓
	Wave equation $\frac{\partial^2 u}{\partial t^2} = c^2\left(\frac{\partial^2 u}{\partial x^2} + \frac{\partial^2 u}{\partial y^2}\right)$	2nd-order hyperbolic	Dirichlet				✓
			Neumann	✗	✗	✗	✓
			Time-dependent				✓
	Heat equation $\frac{\partial u}{\partial t} = D\left(\frac{\partial^2 u}{\partial x^2} + \frac{\partial^2 u}{\partial y^2}\right)$	Parabolic	Dirichlet				✓
			Neumann	✗	✗	✗	✓
			Time-dependent				✓
	Conv-diff equation $\frac{\partial u}{\partial t} = D\left(\frac{\partial^2 u}{\partial x^2} + \frac{\partial^2 u}{\partial y^2}\right) - v\left(\frac{\partial u}{\partial x} + \frac{\partial u}{\partial y}\right) + R$	Hyperbolic parabolic	Dirichlet				✓
			Neumann	✗	✗	✗	✓
			Time-dependent				✓
	Poisson's equation $\frac{\partial^2 u}{\partial x^2} + \frac{\partial^2 u}{\partial y^2} = f$	Elliptic	Dirichlet	✗	✓	✗	✓
			Neumann	✗	✗	✗	✓
	Laplace's equation $\frac{\partial^2 u}{\partial x^2} + \frac{\partial^2 u}{\partial y^2} = 0$	Elliptic	Dirichlet	✗	✓	✓	✓
			Neumann	✗	✗	✗	✓

Fig. 1. Applicability compared to prior arts.

978-1-6654-8495-4/22 $31.00 © 2022 IEEE

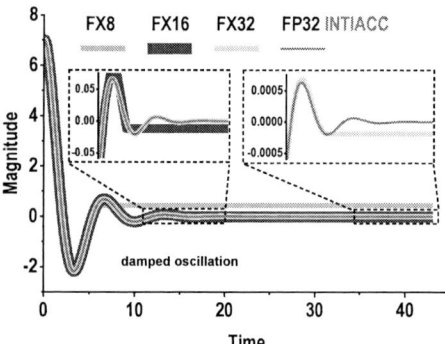

Fig. 2. FP vs. FX in dynamic range.

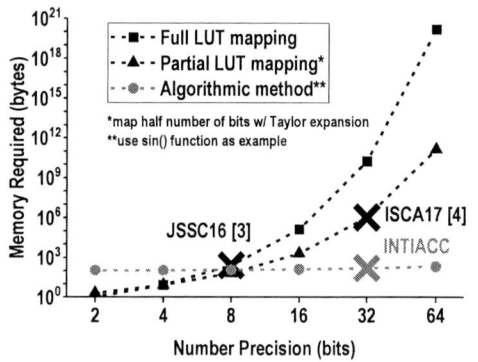

Fig. 3. Methods for mapping nonlinear functions and their memory usage.

B. Programming Model

Fig. 4 illustrates the programming model of INTIACC. First, we discretize a target PDE by applying the finite-difference method to the space dimensions and a numerical integration method (e.g., Euler, RK4) to the time dimension [8] (Fig. 4, top left). The discretization produces 16 coupled equations, each leading to a time-evolving PDE solution at one of the 16 discretized spatial grid points in the domain of interest. Then, we formulate a data flow for each discrete equation. Fig. 4 top right depicts the exemplary data flow of the first equation. It performs Euler's method (the green box), i.e., computing the solution $u_1[i+1]$ for the $(i+1)$-th time step from the solution $u_1[i]$ and a slope value $k[i]$ (the red box). We can compute $k[i]$ using $u[i]$'s from neighboring grid points or boundary conditions. Next, we write a program based on each data flow and map it on each PE (Fig. 4 bottom). The figure shows the program for PE1, which starts with getting $u[i]$ values from neighboring grid points, then evaluating $k[i]$, updating $u_1[i+1]$, and sending $u_1[i+1]$ to its neighbors. If the PE is at a boundary that has time-dependent or Neumann boundary conditions, then we add to the program the additional step of updating the boundary values (blue box in Fig. 4). Note that for PDEs that do not have the time dimension (e.g., Poisson's or Laplace's equation), we can create a similar program that performs an iterative algorithm (e.g., Jacobi's) without the time

The work is supported in part by NSF (1840763).

integration. Finally, we write the programs to the PEs through a test circuit consisting of scan chains and a global clock generator, etc.

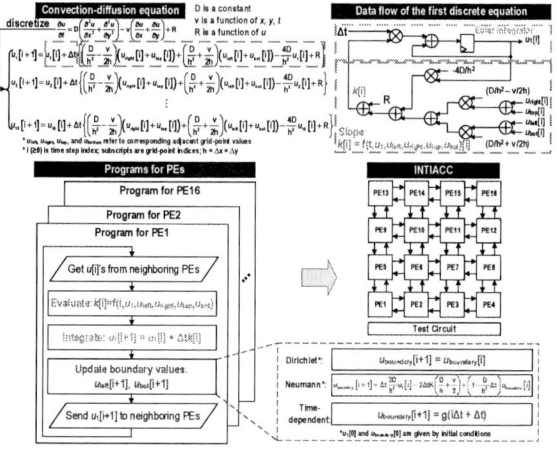

Fig. 4. INTIACC programming methodology.

C. Microarchitecture

Fig. 5 depicts the microarchitecture of the PE. The PE features the proposed hybrid local (570 MHz) and global (10-50 MHz) clock scheme for saving clock power dissipation. In this scheme, each positive edge of the global clock triggers the local clock generator to produce a local clock signal using a ring oscillator (Fig. 6-7). With this active local clock, the PE executes the instructions of the program. At the end of the program execution, as determined by the instruction, the local finish signal is asserted, which disables the local clock and resets the PC to zero, getting the PE ready for the next global clock positive edge. The D flip-flop and mux in the local clock generator ensure that the local clock always starts with the logic-0 value to avoid any setup time violation. We designed the special register file (SRF) in each PE to capture the computation results from neighboring PEs at every global clock positive edge. The captured data are transferred to the last four registers (A28~31) of the general-purpose register file (GRF), which the PE can access with the local clock (Fig. 8).

D. Custom ISA and Algorithms

We designed a custom ISA for the PE to implement numerical routines to perform high-order integration and evaluate nonlinear functions. Fig. 9 shows INTIACC's 28-bit instruction format and a summary of the custom ISA. Fig. 10 shows the pseudo-program implementation of RK4, which INTIACC can perform across four global clock cycles for each time step (it computes one slope $[k_1$ to $k_4]$ per cycle). RK4 is the de facto standard method for time integration since it is significantly more accurate than the Euler method. On the other hand, Fig. 11 left shows the pseudo-program for computing the sin(u) function using Taylor approximation. Here, the variable u is first reduced to a value between 0 and $\pi/4$ based on trigonometric identities, and Taylor expansion is applied to approximate the function. Meanwhile, an iterative algorithm gives higher accuracy for some nonlinear functions, e.g., the Goldschmidt fast division is optimal to compute the reciprocal function 1/u. For this algorithm to converge, however, it is critical to give an approximately correct initial guess. Therefore, we propose a new procedure (Fig. 11 top right), which exploits the fact that the magnitude of the mantissa part of the floating-point number (i.e., M) is between 1 and 2. Hence, the initial guess for its reciprocal (i.e., 1/M) can be any number from 0.5 to 1. While those programs look complex, the custom ISA allows us to implement them with small numbers of instructions, typically 11-33, consuming 38.5-115.5 bytes of memory (Fig. 11 bottom right).

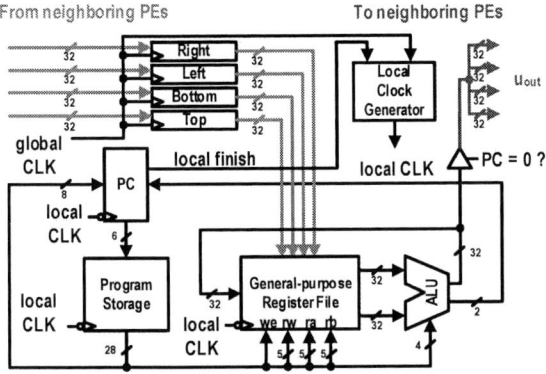

Fig. 5. Microarchitecture of INTIACC's PE.

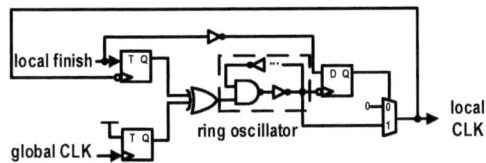

Fig. 6. Local clock generator circuit.

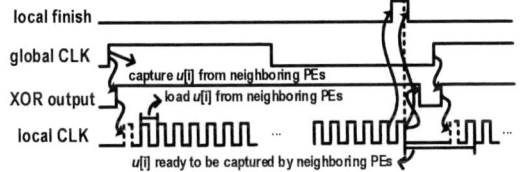

Fig. 7. Timing diagram of local clock generation.

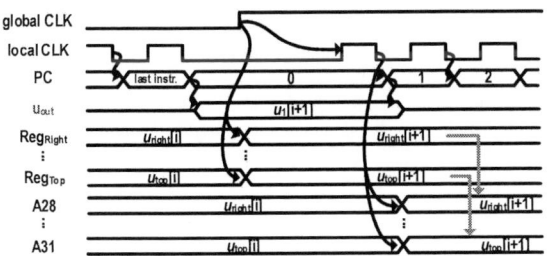

Fig. 8. Timing diagram of PE operation.

27	24	22		18 17		13 12		8 7		4 3		0
opcode		we	rw		ra		rb		pcy		pcn	
4		1	5		5		5		4		4	

Name	Mnemonic	Operation (in Verilog) (number format: +-MAN*2^EXP)	opcode (binary)
Add	add	R[rw]=R[ra]+R[rb]	0000
Subtract	sub	R[rw]=R[ra]-R[rb]	0001
Multiply	mult	R[rw]=R[ra]*R[rb]	0010
Compare and branch	cpbr	if(R[ra]>R[rb]) R[rw]=-1.0, PC=PC+pcy else R[rw]=1.0, PC+pcn	0011
Get exponential unsigned	gexpu	R[rw]=1.0*2^(EXP of R[ra])	0100
Get exponential reciprocal signed	gexprs	R[rw]=+-1.0*2^(-EXP of R[ra])	0101
Get mantissa unsigned	gmanu	R[rw]=(MAN of R[ra])*2^0	0110
Get fractional	gfrac	R[rw]=fractional part of R[ra]	0111
Get sign	gsign	R[rw]=sign of R[ra]	1000
Absolute value	abs	R[rw]=absolute value of R[ra]	1001
Pass	pass	R[rw]=R[ra]	1010
Jump	jump	PC=PC+R[ra]	1011
Square root initial guess	sqrig	If(EXP of R[ra] is even) R[rw]=(MAN of R[ra])*2^((EXP of R[ra])/2) else R[rw]=1.0*2^(((EXP of R[ra])+1)/2)	1100

Special notes:
  we: write enable (if we=1); PC = PC + pcy by default; If pcn = 1000$_{binary}$, PC = 0 and local finish = 1;

Fig. 9. instruction format and instructions of the custom instruction set architecture (ISA).

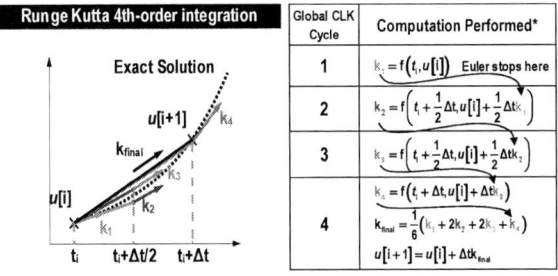

Fig. 10. Implementation of RK4 scheme on INTIACC.

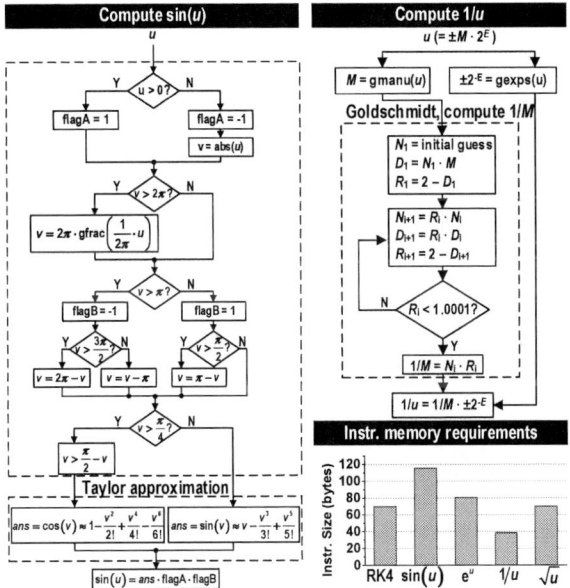

Fig. 11. Examples of numerical routines and the memory requirements.

978-1-6654-8495-4/22 $31.00 © 2022 IEEE

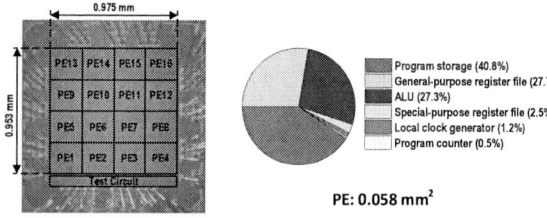

Fig. 12. Die photo and PE area breakdown.

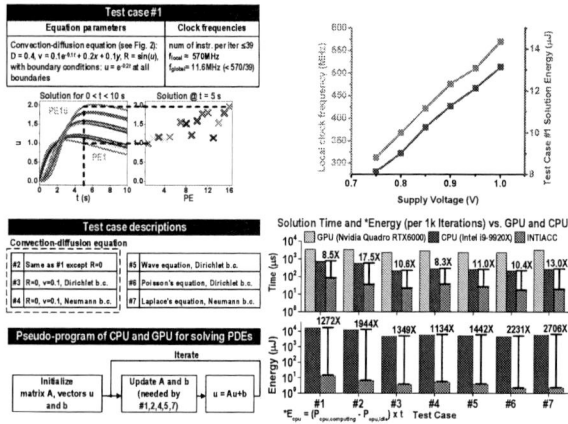

Fig. 13. Test case specifications and measurement results vs. CPU and GPU.

III. Measurement Results

We prototyped INTIACC in a 65nm CMOS with a core area of 0.975 mm² (Fig. 12). At 1 V, it operates at 570 MHz locally, offering an aggregated computation performance of 9.12 GFLOPS (FP32). Fig. 13 top left shows the equation parameters and measured solutions of our first test case of seven in total (Fig. 13 bottom left). For case #1, we chose the parameter v to depend on both space and time and the boundary conditions to change over time. We solved the equation from t = 0-10 s with a step size of 0.01 s. The global clock frequency (11.6 MHz) is set based on the maximum number of instructions required per iteration. For case#1, INTIACC consumes 13 µJ at 1 V (Fig. 13 top right).

For all test cases, we also compared INTIACC's performance with MATLAB running on an Intel i9-9920X CPU and with CUDA running on an NVIDIA Quadro RTX6000 GPU. For the test-case equations, we chose them to represent different types of PDEs with various parameter settings and boundary conditions. For instance, as compared to case #1, case #2 sets the nonlinear term R to zero, thus eliminating the computation time for evaluating sin(u). Then in cases #3 and #4, we set the parameter v to a constant and used different boundary conditions. For the programs running on the CPU and GPU, we used a conventional matrix/vector approach (Fig. 13 bottom left) [9]. The matrix A (16-by-16) contains the parameter information of the equation, while the vector b (16-by-1) includes the boundary conditions. We summarize the test results in Fig. 13 bottom right. We observed up to 17.5X speed and 2,706X energy improvements over the CPU. The speed-up stems from the efficient in-place computing of INTIACC's PEs. Finally, Fig. 14 compares INTIACC with recently published accelerators. INTIACC significantly expands the diversity of accelerable problems in equation types, boundary conditions, and applicable numerical algorithms with orders of magnitude better solution errors, dynamic range, and speed.

		JSSC 2016 [3]	JSSC 2020 [2]	ISSCC 2021 [1]	INTIACC
Technology		65nm CMOS	180nm CMOS	65nm CMOS	65nm CMOS
Analog/Digital		Mixed-signal	Mixed-signal	Digital	Digital
Number Precision		8bit Fixed-Point	5bit Fixed-Point	16bit Fixed-Point	32bit Floating-Point
Applicable to different types of PDEs		No	No	No	Yes
Problem Dimension[1]		1+1D	2D	2D	2D, 2+1D, 1D, 1+1D
Variable Boundary Conditions		No	No	No	Yes
Can Compute Nonlinear Terms		Yes	No	No	Yes
Energy Efficiency Against CPU		N/A	N/A	N/A	1,100X-2,700X
Speedup Against CPU		N/A	N/A	N/A	8-17X
Numerical Routine Support		N/A	N/A	N/A	Yes
Core Area		2.0 mm²	1.868 mm²	0.462 mm²	0.975 mm²
Supply Voltage		1.2 V	1.8 V	0.6-1.2 V	0.5-1 V
Clock Frequency		N/A	200 MHz	25.6 MHz	570 MHz Local
Energy per Clock Cycle		N/A	332 pJ	94 pJ	263 pJ
Laplace's Equation[2]	Number of Iterations[2]	Not mappable	Time per iteration not mentioned	473	390
	Solution Time[3]			311.7 us	7.8 us
	Solution EDP[3]			234 uJ·us	252 uJ·us
	Solution NRMS Error			1.00E-04	3.43E-08
Conv-diff Equation[4]	Number of Time Steps	Not mappable	Not mappable	Not mappable	1000
	Solution Time				86.2 us
	Solution EDP				1.11 mJ·us

[1]space dimensions + time dimension (e.g., 2+1D means 2 space dimensions and 1 time dimension)
[2]FX16 (range -128 to 128) vs. FP32 simulation results, solving 21x21 grid points, boundary conditions: 1.992, -2.66, 56.33, -101.5
[3]For achieving the same NRMS error (i.e., 1.00E-04), with INTIACC's performance data extrapolated to 21x21 PEs
[4]equation setup specified in Fig. 5a

Fig. 14. Comparison with prior arts.

References

[1] J. Mu et al., "A 21x21 Dynamic-Precision Bit-Serial Computing Graph Accelerator for Solving Partial Differential Equations Using Finite Difference Method," ISSCC, pp. 230-231, 2021.

[2] T. Chen et al., "A 1.87-mm2 56.9-GOPS Accelerator for Solving Partial Differential Equations," IEEE JSSC, vol. 55, no. 6, pp. 1709-1718, June 2020.

[3] N. Guo et al., "Energy-Efficient Hybrid Analog/Digital Approximate Computation in Continuous Time," IEEE JSSC, vol. 51, no. 7, pp. 1514-1524, July 2016.

[4] J. Kung et al., "A Programmable Hardware Accelerator for Simulating Dynamical Systems," IEEE/ACM ISCA, pp. 403-415, 2017.

[5] P. Vivet et al., "IntAct: A 96-Core Processor With Six Chiplets 3D-Stacked on an Active Interposer With Distributed Interconnects and Integrated Power Management," IEEE JSSC, vol. 56, no. 1, pp. 79-97, Jan. 2021.

[6] K. Bryson, "NVIDIA Announces CPU for Giant AI and High Performance Computing Workloads", NVIDIA, Apr. 12, 2021.

[7] M. Zidan et al., "A General Memristor-Based Partial Differential Equation Solver," Nature Electronics, pp. 411-420, July 2018.

[8] Schiesser, W. E. (1991), The Numerical Method of Lines Integration of Partial Differential Equations, Academic Press, San Diego.

[9] Bruaset, Tveito, A., & Bruaset, A. M. (Are M. (2006). Numerical solution of partial differential equations on parallel computers / Are Magnus Bruaset, Aslak Tveito (eds.). Springer. https://doi.org/10.1007/3-540-31619-1

A Scalable Bit-Serial Computing Hardware Accelerator for Solving 2D/3D Partial Differential Equations Using Finite Difference Method

Junjie Mu[1], Chengshuo Yu[1,2], Tony Tae-Hyoung Kim[1], Bongjin Kim[3]

[1]School of Electrical and Electronic Engineering, Nanyang Technological University, Singapore 639798
[2]Institute of Microelectronics, A*STAR, Singapore 138634
[3]Department of Electrical and Computer Engineering, University of California, Santa Barbara, California 93106, United States
Email: {junjie003, e190026}@e.ntu.edu.sg, thkim@ntu.edu.sg, bongjin@ucsb.edu

Abstract—This work presents a scalable bit-serial computing hardware accelerator for solving two/three-dimensional (2D/3D) partial differential equations (PDEs). The finite difference method (FDM) approximates the PDEs by solving algebraic equations that contain finite differences between discretized solutions in a 2D/3D grid. The proposed PDE solving hardware accelerator comprises a 16×16 regular PE array, where neighboring PEs are connected based on a 2D lattice graph configuration. The regular PEs update their solutions corresponding to a 2D/3D grid based on serial bitstreams coming from their neighboring PEs. Besides the regular PEs, 4×16 boundary PEs surrounding the regular PEs store and transmit boundary conditions to the regular PEs adjacent to them. The proposed PDE solver with distributed 92Kb embedded SRAM can be reconfigured to map and solve 2D/3D Laplace and Poisson equations using a sequential layer-by-layer update method with no external memory access. A fabricated 65nm test chip occupying the core area of 0.811mm² consumes 5.2pJ and 6.5pJ, respectively, when solving 2D and 3D PDEs at 1V and 25.6MHz.

Keywords—scalable, partial differential equations (PDEs), 2D/3D Laplace/Poisson equations, finite difference method (FDM)

I. INTRODUCTION

Numerical methods are used for approximating solutions to PDEs formulating various physical phenomena, such as the propagation of heat or sound, fluid flow, electrodynamics, and quantum mechanics. Numerical computing with high precision consumes excessive energy using conventional computers. For instance, more than 320J of energy is consumed for solving a 2D Poisson equation in a 128×128 grid using GPU [1]. Hence, there are growing demands for energy-efficient hardware accelerators for solving PDEs.

Prior works [1-3] have demonstrated hardware accelerators based on in-memory computing [1] and continuous or discrete-time analog mixed-signal computing [2], [3] for solving 1D and 2D differential equations. However, they suffered from several challenges, including limited computing precisions (e.g., 5bit [1] or 8bit [2]), analog-specific nonidealities (e.g., device variations and noise susceptibilities), and data-conversion overhead. A programmable digital accelerator [4] has been introduced, but it consumes excessive energy due to frequent external memory access. A recent development [5] proposed an area- and energy-efficient accelerator to solve 2D Laplace equations with minimal communication bandwidth and a small footprint using bit-serial computing architecture. However, it has limited scalability and applications (i.e., a fixed 2D grid used to solve only 2D Laplace equations).

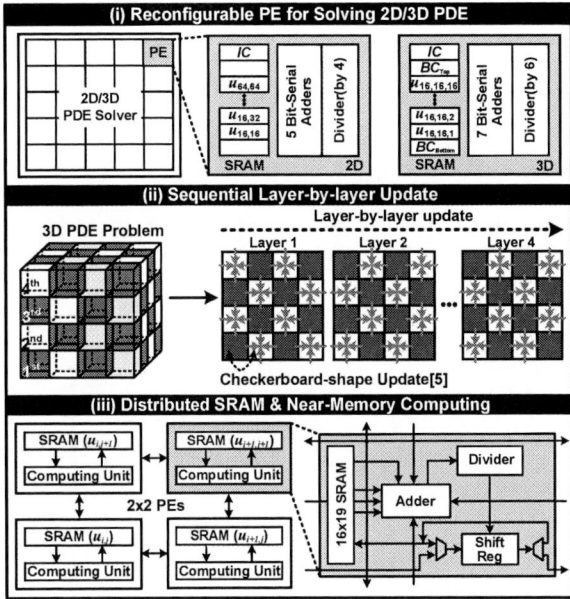

Fig. 1. Key features of the proposed PDE solving hardware accelerator.

In this work, we propose a scalable bit-serial computing accelerator for solving 2D/3D Laplace and Poisson equations using the finite difference method. Fig. 1 describes key features of the proposed work. The proposed accelerator solves 2D and 3D PDEs by reconfiguring a 2D PE array, as shown in Fig. 1(i). Embedded SRAM in each PE stores solutions corresponding to the distributed 2D grid points when solving 2D PDEs. For 3D PDE, a 3D grid is projected onto a 2D PE array of the proposed PDE solver, where solutions from a column of the 3D grid (in Z-direction) are assigned to an SRAM array embedded in each PE. A shared bit-serial computing unit in each PE sequentially computes and updates the solutions and then writes them back to the SRAM. The 3D PDE grid solutions stored in the SRAM column are updated layer by layer, as shown in Fig. 1(ii). While updating the solutions, the bit-serial computing logic is reused, maximizing the area efficiency of the proposed accelerator. The distributed SRAM embedded in each PE enables near-memory computing, minimizing energy consumption by eliminating the external memory access, as shown in Fig. 1(iii).

978-1-6654-8495-4/22 $31.00 © 2022 IEEE

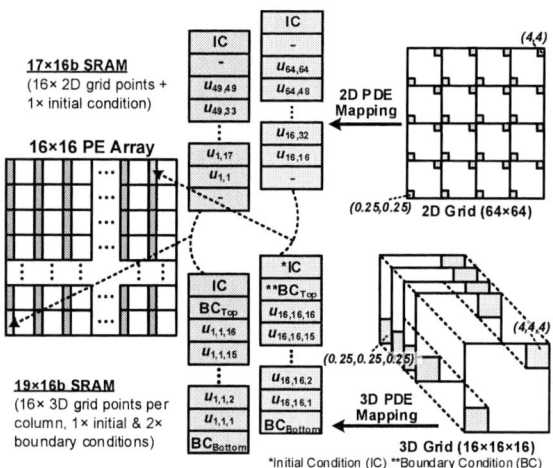

17×16b SRAM
(16× 2D grid points + 1× initial condition)

16×16 PE Array

19×16b SRAM
(16× 3D grid points per column, 1× initial & 2× boundary conditions)

*Initial Condition (IC) **Boundary Condition (BC)

Fig. 2. Mapping 2D and 3D Poisson equations into the 16×16 PE array with the step size of 0.25.

II. PROPOSED 2D/3D PDE SOLVER DESIGN

The FDM approximates the PDE solutions by iteratively calculating finite differences between solutions from nearby discrete points in a 2D or 3D grid. A 3D Poisson equation (or Laplace equation when $f(x,y,z)$ is zero),

$$\frac{\partial^2 u}{\partial x^2} + \frac{\partial^2 u}{\partial y^2} + \frac{\partial^2 u}{\partial z^2} = f(x, y, z) \quad (1)$$

can be discretized into (2) with the step size Δ ($=x_{i+1}-x_i=y_{j+1}-y_{ij}=z_{k+1}-z_k$) using FDM, where $u(x,y,z)$ is the solution.

$$\frac{u_{i+1,j,k} + u_{i-1,j,k} + u_{i,j+1,k} - 6u_{i,j,k} + u_{i,j-1,k} + u_{i,j,k+1} + u_{i,j,k-1}}{\Delta^2}$$
$$\approx f(x, y, z) \quad (2)$$

The solution is given in (3).

$$u_{i,j,k} \approx$$
$$\frac{u_{i+1,j,k} + u_{i-1,j,k} + u_{i,j+1,k} + u_{i,j-1,k} + u_{i,j,k+1} + u_{i,j,k-1} - \Delta^2 f(x,y,z)}{6} \quad (3)$$

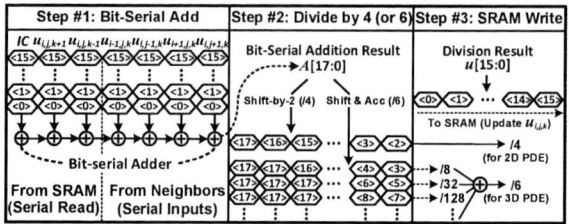

Fig. 3. Operation sequence based on a checkerboard update method.

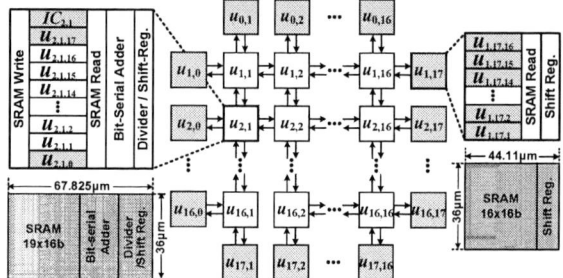

Fig. 4. Regular/boundary processing element (PE) block diagrams and layouts.

Similarly, the solution of 2D Poisson equations is approximated as follows:

$$u_{i,j} = \frac{u_{i+1,j} + u_{i-1,j} + u_{i,j+1} + u_{i,j-1} - \Delta^2 f(x, y)}{4} \quad (4)$$

The 3D/2D grid solutions are updated according to formulas (3) and (4) until the solutions converge.

Fig. 2 illustrates the mapping of 2D/3D Poisson equations into the proposed accelerator with a 16×16 PE array. A 64×64 2D grid is mapped to the proposed PE array, where solutions for a group of 16× distributed 2D grid points are stored in an SRAM array embedded in each PE (Fig. 2, top). Similarly, a 16×16×16 3D grid is mapped to the PE array, where solutions for a column of 16× 3D grid points are stored in the same SRAM column of the PE. Initial conditions of a 64×64 2D (or 16×16×16 3D) grid with the step size of 0.25 are computed and programmed into the first row of the SRAM array. Boundary conditions are assigned to the second and the last rows of the SRAM for interactions

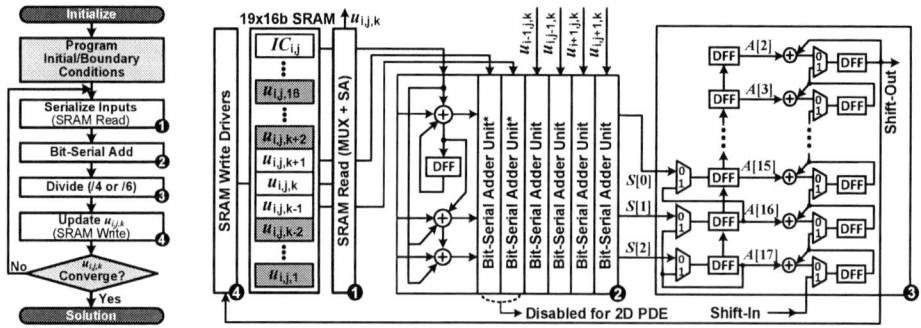

Fig. 5. Operation flow for solving 2D/3D Laplace and Poisson equations and detailed block diagrams of the proposed PDE processing element.

978-1-6654-8495-4/22 $31.00 © 2022 IEEE

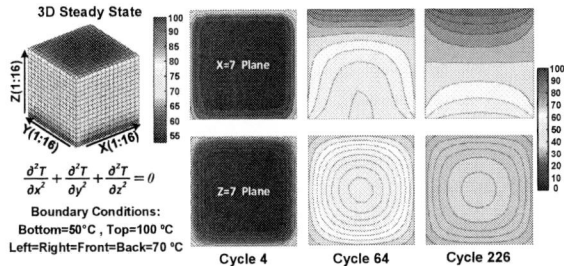

$$\frac{\partial^2 T}{\partial x^2} + \frac{\partial^2 T}{\partial y^2} + \frac{\partial^2 T}{\partial z^2} = 0$$

Boundary Conditions:
Bottom=50°C , Top=100 °C
Left=Right=Front=Back=70 °C

Fig. 6. Measured numerical solutions for a 3D Laplace equation (heat conduction).

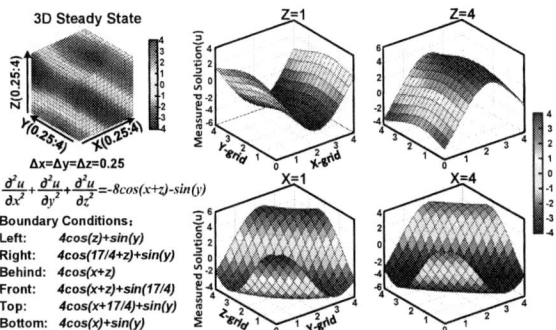

$\Delta x = \Delta y = \Delta z = 0.25$

$$\frac{\partial^2 u}{\partial x^2} + \frac{\partial^2 u}{\partial y^2} + \frac{\partial^2 u}{\partial z^2} = -8\cos(x+z) - \sin(y)$$

Boundary Conditions:
Left: $4\cos(z) + \sin(y)$
Right: $4\cos(17/4+z) + \sin(y)$
Behind: $4\cos(x+z)$
Front: $4\cos(x+z) + \sin(17/4)$
Top: $4\cos(x+17/4) + \sin(y)$
Bottom: $4\cos(x) + \sin(y)$

Fig. 7. Measured numerical solutions for 3D Poisson equation.

between the top/bottom planes to their boundary planes for 3D PDEs. Hence, these two SRAM rows are not used for 2D PDEs. After mapping the PDE solutions to the embedded SRAM, we iterate FDM operations based on the checkerboard-shape update method [5] until the solution converges.

As illustrated in Fig. 3, a 16×16 PE array is divided into two groups of PEs (even and odd PEs) based on the checkerboard shape grid configuration, and it takes two cycles to update the entire FDM grid. Assuming the solutions of odd (or even) PEs are updated in the previous cycle, the updated solutions are then serialized and transmitted to even (or odd) PEs in the next cycle. A series of operations (bit-serial addition, division, and SRAM write) is performed at the even (or odd) PEs to update their grid solutions. A divide-by-6 operation for 3D PDEs is approximated by sequentially shifting the bit-serial addition result to the right by 3, 5, 7, …, 17b and accumulating them (i.e., $1/6 \approx 1/2^3 + 1/2^5 + 1/2^7 + \dots + 1/2^{17}$). The updated solutions are written back to the SRAM and then serialized to update the solutions for their neighboring grids in the next cycle. We repeat the same process until the solution converges. Note that solving a 3D PDE involves the accumulation of seven 16bit inputs (i.e., an initial condition and solutions from six adjacent grid points) and a divide-by-6 operation. In contrast, a 2D PDE is solved by accumulating five 16bit inputs (i.e., an initial condition and four adjacent grid points) followed by a divide-by-4 operation.

The proposed solver comprises a 16×16 regular PE array and 64 boundary PEs with distributed 92Kb SRAM embedded in the PE array. The block diagrams and the layouts of the regular (left) and boundary (right) PEs are shown in Fig. 4. The boundary PEs

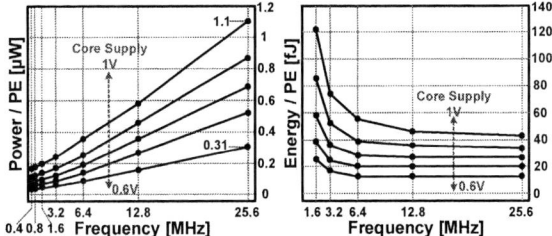

Fig. 8. Measured power and energy consumption for solving 2D PDEs.

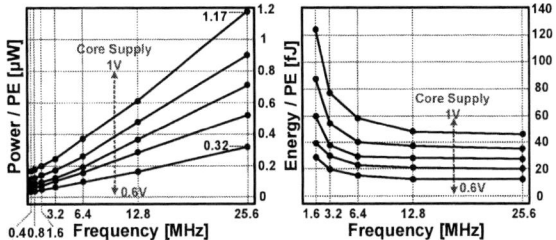

Fig. 9. Measured power and energy consumption for solving 3D PDEs.

are designed for storing boundary conditions and transmitting data to the regular PE array. A regular PE communicates with four adjacent PEs, and the shift registers in PEs are connected from left to right to transmit the data to the embedded SRAM for the initial programming and read PDE solutions to the output.

Fig. 5 shows the operation flow (left) and the detailed block diagram (right) of the proposed PE. A 2D (or 3D) PDE is first discretized into a 64×64 2D (or $16 \times 16 \times 16$ 3D) grid and mapped into the 16×16 PE array for finding numerical solutions based on FDM. After programming initial and boundary conditions, repetitive operations of the even and odd solution update cycles are processed based on the checkerboard method until the PDE solution converges. A bit-serial PE comprises a 19×16b SRAM and a bit-serial computing logic, as shown in Fig. 5 (right). The SRAM stores nineteen 16bit data, including sixteen 2D/3D grid solutions and three extra 16bit data (to store an initial condition for 2D/3D PDEs and two boundary conditions for 3D PDE). A bit-serial computing logic consists of a bit-serial adder, a shift register, and an accumulator. A bit-serial adder is comprised of seven-unit columns. The first column is dedicated to adding the initial condition, and the second and third columns are used for adding solutions from the top and bottom planes (for 3D PDEs). The next four columns on the right are used for adding solutions from the four neighboring PEs in the same plane for 2D and 3D PDEs. A shift register and an accumulator are combined and used for approximating divide-by-4/divide-by-6 operations for 2D/3D PDEs. The accumulator circuit is also used to write back the updated solutions to the SRAM and read PDE solutions by shifting them to the right.

III. Test-Chip Measurement Results

A test chip comprising 256 regular PEs and 64 boundary PEs is fabricated using 65nm. The proposed reconfigurable hardware accelerator solves 2D/3D Laplace and Poisson equations for the

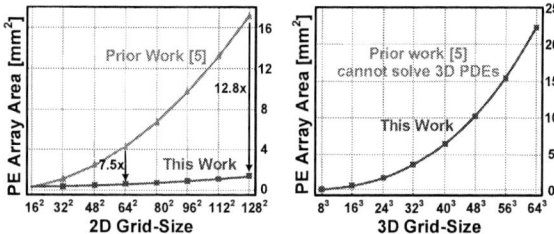

Fig. 10. Estimated PE array area versus 2D and 3D grid-size.

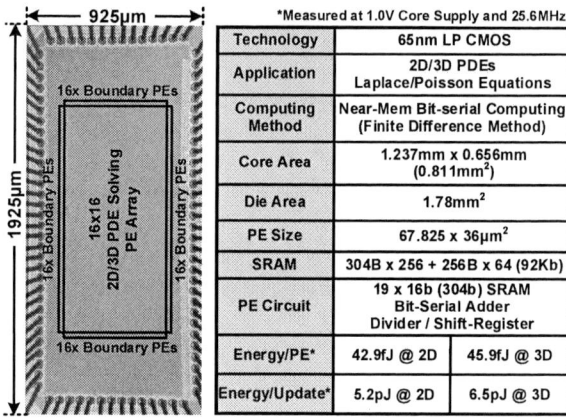

Fig. 11. Test chip die micrograph and performance summary table.

first time using a single chip. Fig. 6 shows the measured 3D heat conduction behavior using the numerical solution of the Laplace equation, where the FDM maps its discrete grid points to the test chip PE array. The temperatures at the boundaries are fixed while the heat gradually diffuses into the center. Transient temperature changes on the cross-sections of x=7 and z=7 are shown. Fig. 7 shows the measured numerical solution for the 3D Poisson equation with boundary conditions defined as a function of its 3D grid coordinates (x, y, z). The solutions with separate Z and X sections are observed when the 3D grid solutions reach a steady state.

Fig. 8 and 9 are the measured power and energy per PE with the supply voltage and the operating clock frequency swept for solving 2D and 3D PDEs. The measured power per PE is 0.32-1.17μW for solving 3D PDEs at 0.6-1V and 25.6MHz, while it consumes 0.31-1.1μW at 0.6-1V and 25.6MHz for solving 2D PDEs. The measured energy per PE ranges from 12.4fJ to 45.9fJ (or 12.1fJ to 43fJ) for 3D (or 2D) PDEs at 0.6-1V and 25.6MHz. The proposed accelerator occupies a 7.5× smaller area than the prior work [5] when mapping a 64×64 grid to the PE array for solving 2D PDEs. The area efficiency becomes more significant as the grid size increases (Fig. 10), showing the scalability of the proposed near-memory architecture for solving the PDEs with a larger 2D grid size. Table I compares the proposed accelerator with the state-of-the-arts. Fig. 11 (left) shows the die micrograph of the 65nm test chip, occupying a total chip area of 1.78mm². A summary table is shown in Fig. 11 (right).

IV. CONCLUSION

This work proposes a scalable 2D/3D PDE solver based on a bit-serial computing architecture consisting of 16×16 regular PEs and 64 boundary PEs. Each PE consists of an embedded SRAM and a shared bit-serial computing unit circuit, which minimizes the memory access energy and achieves high-density integration of the proposed PE array in a single chip. The PE is reconfigured to support 2D/3D Laplace and Poisson equations with a minimal area overhead. A test chip of the proposed PDE accelerator is implemented in 65nm CMOS technology. The bit-serial PE occupies 2441.7μm², and the total energy for updating the entire grid solutions is 5.2pJ and 6.5pJ for 2D and 3D PDEs at 1V and 25.6MHz.

ACKNOWLEDGMENT

The authors would like to thank Kevin Tshun Chuan Chai and Anh-Tuan Do (IME, A*STAR, Singapore) for their support and contributions.

REFERENCES

[1] T. Chen, J. Botimer, T. Chou, Z. Zhang, "A 1.87-mm² 56.9-GOPS Accelerator for Solving Partial Differential Equations," in IEEE Journal of solid-state circuits (JSSC), vol. 55, no. 6, pp. 1709-1718, Jan. 2020.

[2] N. Guo, Y. Huang, T. Mai, S. Patil, C. Cao, M. Seok, S. Sethumadhavan, Y. Tsividis, "Energy-efficient Hybrid Analog/Digital Approximate Computation in Continuous Time," in IEEE J. Solid-State Circuits (JSSC), vol. 51, no. 7, pp. 1514–1524, Jul. 2016.

[3] J. Liang, N. Udayanga, A. Madanayake, S. I. Hariharan, S. Mandal, "An Offset-Cancelling Discrete-Time Analog Computer for Solving 1-D Wave Equations," in IEEE Journal of solid-state circuits (JSSC), Vol. 29, No. 1, pp. 2881-2894, Jan. 2021.

[4] J. Kung, Y. Long, D. Kim, S. Mukhopadhyay, "A programmable hardware accelerator for simulating dynamical systems," in Proc. Int. Symp. Comput. Archit. (ISCA), pp. 403–415, Jun. 2017.

[5] J. Mu, B. Kim, "A 21×21 Dynamic-Precision Bit-Serial Computing Graph Accelerator for Solving Partial Differential Equations Using Finite Difference Method," in IEEE Int. Solid-State Circuits Conf. (ISSCC) Dig. Tech. Papers. pp. 406-407, Feb. 2021.

TABLE I. PERFORMANCE SUMMARY AND COMPARISON

	JSSC 2016 [2]	JSSC 2020 [1]	ISSCC 2021 [5]	This Work	
Technology (Circuit Type)	65nm (Analog)	180nm (Analog)	65nm (Digital)	65nm (Digital)	
Core Supply	1.2V	1.8V	0.6-1.2V	0.6-1V	
Computing Method	Continuous Time Analog	In-Memory Computing	Bit-Serial Computing	Bit-Serial Computing	
Core Area (mm²)	2.0	1.868	0.462	0.811	
Core Area /Grid Size (μm²)	N/A	N/A	1049.76	198.09	
Computation Precision	Fixed 8bit	Fixed 5bit	Dynamic 4/8/12/16b	Fixed 16bit	
Application	Approx. Computing	2D PDE	2D Laplace PDE Only	Laplace/Poisson PDEs	
				2D	3D
Graph Structure (Grid Size)	No	No	Yes (21×21)	Yes (64x64)	Yes (16x16x16)
Update Method (# of Cycles for Grid Update)	N/A	Hybrid Method (M Cycles)	Checker-board (2 Cycles)	Checker-board (2 cycles)	Checker-board (32 cycles)
Power Consumption	1.2mW	66.4mW	2.41mW	1.1mW	1.18mW

A 283 pJ/b 240 Mb/s Floating-Point Baseband Accelerator for Massive MU-MIMO in 22FDX

Oscar Castañeda, Luca Benini, and Christoph Studer

Department of Information Technology and Electrical Engineering, ETH Zurich, Zurich, Switzerland

Abstract—We present PULPO, a floating-point baseband-processing accelerator for massive multi-user multiple-input multiple-output (MU-MIMO) basestations (BSs). PULPO accelerates matrix-vector products, not only with a matrix but also with its Hermitian, as well as affine transforms and nonlinear projections used in iterative algorithms that outclass traditional linear methods in various applications. PULPO is integrated in a system-on-chip (SoC) with a tight integration to the system's data memory, facilitating data exchange and co-operation with 8 RISC-V cores. The fabricated accelerator achieves comparable efficiency as recently-proposed fixed-point baseband processors, while eliminating the burdens associated with fixed-point design, thus simplifying massive MU-MIMO BS development.

I. INTRODUCTION

Baseband processing for modern high-rate wireless communication systems poses significant implementation challenges in terms of power consumption and throughput. Emerging technologies, such as massive multi-user multiple-input multiple-output (MU-MIMO) [1] and millimeter-wave (mmWave) communication [2], further aggravate the situation as they require processing of high-dimensional signals acquired at hundreds of basestation (BS) antennas at rates exceeding billions of tasks per second. As a result, baseband processing at infrastructure BSs for such systems is expected to be carried out with application-specific integrated circuits (ASICs), which achieve the best energy efficiency and highest throughput. Over the last few years, a wide range of baseband-processing ASICs for massive MU-MIMO BSs have been proposed; see, e.g., [3]–[6].

A. Programmable Accelerators for Massive MU-MIMO

Baseband-processing ASICs often implement a single algorithm that is highly optimized for specific system parameters and operation conditions. Specialization limits their adaptability to time-varying system and channel conditions, which was shown to be key to low-power mmWave massive MU-MIMO baseband processing [6]. Furthermore, supporting an entirely different set of system parameters (e.g., the number of BS antennas) might even require fabrication of another ASIC. To counter these issues, flexible and programmable hardware accelerators have emerged recently [7], [8]. The work in [7] proposes a configurable systolic array comprising 64 processing elements (PEs), while [8] proposes an application-specific instruction-set

This work was supported in part by ComSenTer, one of six centers in JUMP, a SRC program sponsored by DARPA. The work of CS was also supported by an ETH Research Grant and by the US NSF under grants CNS-1717559 and ECCS-1824379. Contact author: O. Castañeda (e-mail: caoscar@ethz.ch)

The authors thank A. Di Mauro, M. Scherer, M. Eggiman, G. Rutishauser, F. Glaser, and Microelectronics Design Center for assistance with chip fabrication.

processor (ASIP) comprising 8 lanes; both of these designs use complex-valued 32-bit fixed-point datapaths. These two accelerators provide improved flexibility and adaptability over an ASIC, and improved hardware efficiency over a general processor. However, to support vastly different conditions (such as system dimensions, propagation conditions, or modulation and coding schemes) with fixed-point datapaths, extensive simulations and re-configuration/re-programming are necessary to make best use of the available dynamic range. In fact, failing to re-adjust a fixed-point datapath to a given scenario (often significantly) degrades performance; see, e.g., Sec. IV-A.

B. Contributions

We propose PULPO, a programmable floating-point baseband-processing accelerator for massive MU-MIMO BSs. PULPO supports a wide range of baseband algorithms for both the uplink and the downlink. By implementing a (transprecision) floating-point datapath, PULPO adapts to different system conditions without requiring fixed-point parameter tuning or a specialized compiler to compensate for changes in numeric representations. PULPO is tightly integrated with the data memory of a system-on-chip (SoC) containing 8 Parallel Ultra-Low Power (PULP) RISC-V cores; the SoC's architecture follows that of [9]. This tight integration enables PULPO to access the SoC's entire data memory at full bandwidth, which facilitates communication and interaction with the RISC-V cores. Hardware measurements of the 22 nm FD-SOI prototype demonstrate that PULPO offers comparable performance and efficiency to the fixed-point solutions in [7], [8], while further improving flexibility and adaptability.

II. THE PULPO BASEBAND-PROCESSING ACCELERATOR

PULPO supports several baseband algorithms for both uplink and downlink in massive MU-MIMO BSs equipped with B antennas serving U single-antenna user equipments (UEs).

A. Supported Baseband Algorithms

1) Linear Algorithms: In the uplink, the BS uses the received signals to estimate the transmitted UE symbols. To minimize complexity, one typically resorts to the linear minimum mean-squared error (LMMSE) equalizer, which simply corresponds to a matrix-vector product (MVP). In the downlink, the BS must generate a precoded vector that removes MU interference in the UE signals to be transmitted [10]. This task can be achieved with the linear zero-forcing (ZF) precoder, also applied as an MVP. Computing the LMMSE and ZF matrices requires MVPs too. Thus, MVPs are crucial tasks to be accelerated by PULPO.

978-1-6654-8495-4/22 $31.00 © 2022 IEEE

2) Nonlinear Algorithms: Linear detection and precoding algorithms are suboptimal, especially (i) in systems with a large load factor U/B [11], (ii) for BSs that utilize low-resolution digital-to-analog converters (DACs) [12], or (iii) for channels with short coherence times [13]. Thus, PULPO also supports (among many others) the following nonlinear algorithms that alleviate these issues: (i) box-constrained (BOX) equalization achieving near-maximum-likelihood performance even in systems with large load factors [14]; (ii) biconvex x-bit precoding (CxPO) for BSs with low-resolution DACs [12]; and (iii) projection-onto-convex-hull (PrOX) for joint channel estimation and data detection (JED) [13]. PULPO further supports computing finite-alphabet equalization and precoding matrices, which enable low-power linear detection and precoding [15].

B. Operating Principle

All of the above linear and nonlinear baseband algorithms involve one or more of the following three atomic operations:

$$\mathbf{z} = \mathbf{A}\mathbf{x} + \mathbf{y} \ (\mathcal{M}), \ \mathbf{z} = \mathbf{y} - \tau\mathbf{A}^H\mathbf{x} \ (\mathcal{H}), \ \mathbf{z} = \text{prox}(\rho\mathbf{x}) \ (\mathcal{P}).$$

Here, $\mathbf{A}$, $\mathbf{x}$, $\mathbf{y}$, τ, and ρ are algorithm-specific quantities. For example, LMMSE equalization only needs an MVP ($\mathcal{M}$) with $\mathbf{A}$ being the equalization matrix, $\mathbf{x}$ the signals received by the BS, and $\mathbf{y} = \mathbf{0}$. The C1PO and PrOX algorithms iterate MVPs ($\mathcal{M}$) followed by projection ($\mathcal{P}$), resulting in iterations of the form $\mathbf{x}^{(t+1)} = \text{prox}(\rho\mathbf{A}\mathbf{x}^{(t)})$; see [12], [13] for the specific choices for ρ, $\mathbf{A}$, and $\mathbf{x}^{(0)}$. The BOX and C2PO algorithms iteratively perform all three operations ($\mathcal{M}$), ($\mathcal{H}$), and ($\mathcal{P}$) as follows: $\mathbf{x}^{(t+1)} = \text{prox}(\rho(\mathbf{x}^{(t)} - \tau\mathbf{A}^H(\mathbf{A}\mathbf{x}^{(t)})))$, where ρ and τ are system-dependent tunable parameters. For all of these algorithms, the prox$(\cdot)$ operator is an element-wise clipper.

PULPO accelerates exactly the atomic operations ($\mathcal{M}$), ($\mathcal{H}$), and ($\mathcal{P}$) for any problem dimension that fits the available memory. To achieve this goal at the highest efficiency, one must support (i) MVPs with a matrix $\mathbf{A}$ and also its Hermitian $\mathbf{A}^H$ without requiring explicit transposition of $\mathbf{A}$ in memory, (ii) hardware-support for sequencing these atomic operations for iterative algorithms and for varying system dimensions, and (iii) flexibility for different prox$(\cdot)$ operators. We next detail a VLSI architecture that satisfies all of these requirements.

III. VLSI Architecture

Fig. 1(a) illustrates the SoC comprising 8 PULP RISC-V cores with $128\,\text{kB}$ of data memory that is shared with the PULPO baseband-processing accelerator. PULPO consists of 16 PEs connected in a unidirectional ring network. As detailed in Fig. 1(b), each PULPO PE is a configurable complex-valued floating-point multiply-accumulate unit equipped with a projection unit that is able to perform the most common prox$(\cdot)$ operators. The PE's complex-valued multiplier is pipelined, and can be configured to conjugate one of the operands or to compute two real-valued multiplications in parallel.

A. Cannon's Algorithm for MVPs

The PEs implement Cannon's algorithm [16] to efficiently calculate MVPs: Each PE is associated with a different row of

the matrix $\mathbf{A} \in \mathbb{C}^{N \times N}$, and all PEs operate simultaneously on a different entry of the vector $\mathbf{x} \in \mathbb{C}^N$ (and thus on a different column of their respective row). The PEs then circularly exchange their vector $\mathbf{x}$ entries (blue path in Fig. 1) to complete the MVP $\mathbf{A}\mathbf{x}$ in $\mathcal{O}(N)$ clock cycles. The advantage of Cannon's algorithm is that it can also be used to multiply a vector $\tilde{\mathbf{x}} \in \mathbb{C}^N$ by the Hermitian $\mathbf{A}^H$ with the same matrix readout pattern, thus avoiding the need of rewriting the memory to explicitly transpose $\mathbf{A}$; this is achieved with minimal overhead by the PEs circularly exchanging the partial products $a_{j,i}^*\tilde{x}_j$ (red path in Fig. 1) instead of the entries of the vector $\tilde{\mathbf{x}}$.

B. Higher-Dimensional Problems and Memory Access

Since PULPO incorporates 16 PEs, it natively executes MVPs with 16×16 matrices. To support larger square and non-square matrices, we divide them in blocks of 16×16. MVPs with wider matrices are computed by accumulation via the $\mathbf{y}$ vector in ($\mathcal{M}$) and ($\mathcal{H}$); MVPs with taller matrices are simply computed in a time-multiplexed manner. Intermediate matrix dimensions are handled through zero-padding.

To perform tasks with matrices as large as the SoC's memory can support, each PULPO PE is tightly-coupled to a memory bank (SRAM) of the shared data memory: Each PULPO PE has a read/write connection to a specific SRAM that bypasses the tightly-coupled-data-memory (TCDM) interconnect used to route communication between RISC-V cores and data memory (green and orange paths in Fig. 1). To avoid interfering with the cores' memory accesses, PULPO's requests (orange paths) have low priority and will only reach a memory bank if no core is trying to access that bank. To reduce memory requests, PULPO has an internal memory that stores four 16×16 blocks.

C. Operating PULPO

Tightly coupling PULPO with the data memory requires the operands $\mathbf{A}$, $\mathbf{x}$, and $\mathbf{y}$ to be aligned with the memory banks. Furthermore, to simplify readout logic for PULPO, we store the matrix $\mathbf{A}$ in a so-called *skewed* fashion (see Fig. 1(c)). These two data-arrangement requirements are handled by the RISC-V cores when receiving or generating the input operands. Once the operands have been arranged in memory, the cores can configure PULPO's control unit with their dimension and memory location (among other parameters; see Fig. 1(d)), and start PULPO's operation. PULPO offers hardware support to automatically and independently sequence through the ($\mathcal{M}$), ($\mathcal{H}$), and ($\mathcal{P}$) atomic operations, and iterate over such sequence. Specifically, PULPO can iterate over ($\mathcal{M}$) or ($\mathcal{H}$) only, as well as over the sequences ($\mathcal{M}$)→($\mathcal{P}$), ($\mathcal{H}$)→($\mathcal{P}$), and ($\mathcal{M}$)→($\mathcal{H}$)→($\mathcal{P}$), where the output $\mathbf{z}$ of each sequence step becomes the input $\mathbf{x}$ of the subsequent step. Other sequences can be implemented by using the RISC-V cores to execute the ($\mathcal{M}$), ($\mathcal{H}$), and ($\mathcal{P}$) operations in the desired order. Once finished with its task, PULPO raises an event and the cores can access the results directly from the shared data memory.

D. Exploiting Shared Memory with the RISC-V Cores

The shared data memory enables tight interaction between the RISC-V cores and PULPO. For example, PULPO's prox$(\cdot)$

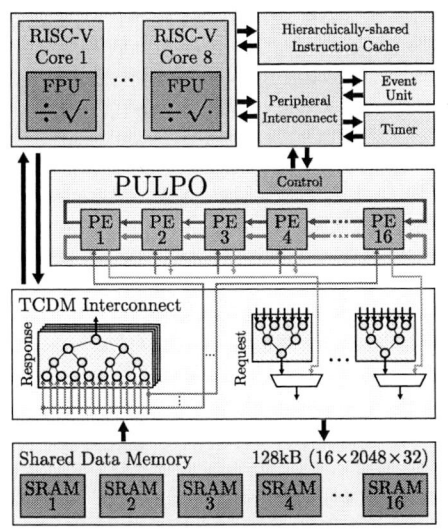

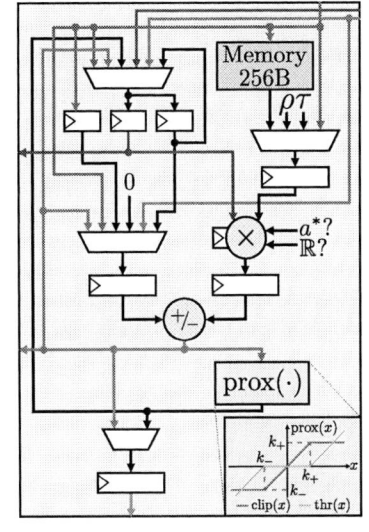

Bank 1	$(1,1)$	$(1,2)$	$(1,3)$
Bank 2	$(2,2)$	$(2,3)$	$(2,1)$
Bank 3	$(3,3)$	$(3,1)$	$(3,2)$

(c) Skewed 3×3 matrix **A**

off.	function	off.	function
00	**x** address	1C	$\rho\&\tau$ address
04	**y** address	20	k_-, k_+
08	**A** address	24	# iterations
0C	**z** address	28	dimensions
10	ρ and fix $\rho\&\tau$		and op mode
		2C	stop
14	τ	30	resume
18	τ in use	3C	start

(d) PULPO configuration words

floating-point format	number of bits			
	sgn	exp	mant	tot
binary16	1	5	10	16
bfloat16	1	8	7	16
binary8	1	5	2	8
hardware	1	8	10	19

(e) PULPO floating-point formats

(a) PULPO-SoC system overview (b) PULPO processing element (PE)

Fig. 1. Implementation details of the PULPO baseband-processing accelerator and its host SoC.

unit only supports element-wise clipping operations (required by the nonlinear algorithms from Sec. II-A2) and thresholding operations (useful, e.g., for sparse signal recovery and channel vector denoising). For other tasks, PULPO can rely on the RISC-V cores to perform any kind of projection. To this end, PULPO simply writes intermediate results to memory, raises an event, and waits to receive a resume command after the cores have calculated the projection. Another use case is early stopping: In each iteration, PULPO writes the current iterate to memory and raises an event so that the cores can keep track of the iterates, e.g., to stop PULPO upon algorithm convergence.

E. Configurable Floating-Point Precision

A key feature is PULPO's transprecision support for different floating-point formats: `binary16`, `bfloat16`, and `binary8`. To improve area efficiency, all formats share 19-bit arithmetic units supporting the maximum number of required mantissa and exponent bits (`hardware`; see Fig. 1(e)). When operating with `binary8`, PULPO only needs to access half of the memory banks. Moreover, PULPO has a real-valued mode that performs two real-valued MVPs (same matrix, different vectors) at once. The RISC-V cores' floating-point units (FPUs) additionally support `binary32` with the architecture details provided in [17]. The RISC-V cores' FPUs can also perform division and square-root operations, further extending the capabilities of PULPO.

IV. IMPLEMENTATION RESULTS

A. Bit Error-Rate (BER) Performance

Fig. 2(a) shows the uncoded BER performance of the implemented PULPO for LMMSE equalization with varying system configurations. For a fair comparison, we consider a design similar to that of [8], which uses 16-bit fixed-point (`integer16`) representation per real and imaginary parts, as well as 24-bit accumulation. To illustrate the advantages of floating-point support, we perform the following experiment: We calibrate the

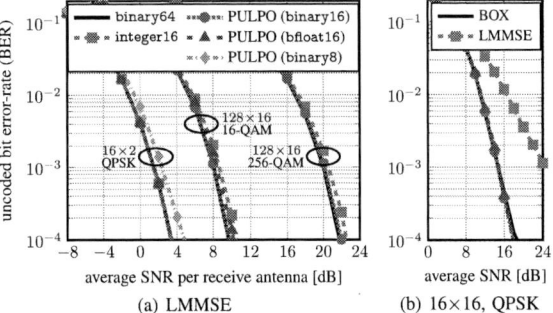

(a) LMMSE (b) 16×16, QPSK

Fig. 2. Uncoded bit-error rate (BER) vs. SNR performance of PULPO when operating with (a) LMMSE equalization and (b) BOX equalization. In this second plot, the black and red curves correspond to BOX and LMMSE with `binary64`; the other curves follow the same legend as in (a).

position of the fixed-point of `integer16` so that it achieves an acceptable BER across a wide variety of system configurations. Then, we evaluate the performance using that single fixed-point format for different system configurations. As shown in Fig. 2(a), having such a "universal" fixed-point configuration causes a performance loss of up to 0.5 dB at 0.1% BER. While a fixed-point design can recover from this performance loss, doing so would require (i) extensive simulations and (ii) to re-configure/re-program the accelerator/ASIP to shift the fixed points accordingly. In contrast, PULPO's quarter-precision (`binary16`) floating-point format achieves virtually the same performance as double-precision (`binary64`) floating-point for all system configurations, without requiring any manual re-calibration. Fig. 2(a) also shows that, as we reduce the system dimensions or modulation scheme, alternative floating-point formats with fewer mantissa bits achieve comparable performance: For a 16×2 $(B \times U)$ system, halving the floating-point bitwidth (`binary8`) results in a 1 dB loss at 0.1% BER.

Fig. 2(b) exemplifies how nonlinear algorithms can outperform linear algorithms—here, we consider a system with

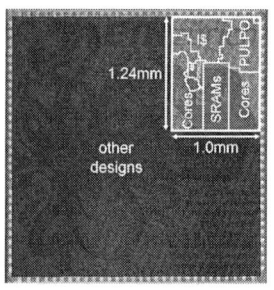
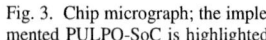
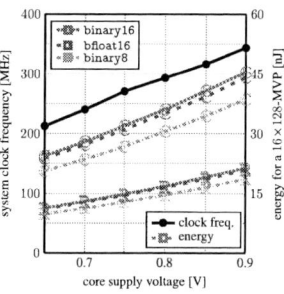

Fig. 3. Chip micrograph; the implemented PULPO-SoC is highlighted.

Fig. 4. Measured system clock frequency and energy vs. core supply.

TABLE I
COMPARISON WITH OTHER MASSIVE MU-MIMO ACCELERATORS

	PULPO	Prabhu [8]	Peng [7]	Tang [3]	Jeon [4]	Wen [5]	Castañeda [6]
Programmable B, U, alg.	**yes**	**yes**	**yes**	no	no	no	no
Floating-point support	**hw**	sw	no	no	no	no	no
Shared core memory	**yes**	no	no	no	no	no	no
BS antennas B	128	128	128	128	256	256	32
UEs U	16	8	8	32	32	32	16
Technology [nm]	22	28	28	40	28	40	65
Core supply [V]	0.8	1.0	0.9	0.9	0.9	1.1	1.0
Core area [mm²]	1.24	2.2	4.8	0.58	0.37	0.73	4.20
Max. frequency [MHz]	293	290	800	425	400	290	312
Total power [mW]	97	180	528	221	151	87	396
Throughput [Gb/s]	0.240	0.169	1.54	2.76	0.354	1.96	9.98
Accelerator area [mm²]	0.42	0.38	4.8	0.58	0.37	0.73	2.41
Accelerator power [mW]	68	91.7	528	221	151	87	290
Area eff.^{a} [Gb/s/mm²]	0.57	0.91	0.66	28.6	1.97	16.1	106
Energy^a [pJ/b]	283	214	167	19.1	208	7.10	2.13

^{a}Technology normalized to 22 nm and 0.8 V core supply.

a load factor $U/B = 1$ and obtain a 9 dB improvement at 0.1% BER. Similar examples can be found for massive MU-MIMO BS architectures with low-resolution DACs [12] or low-resolution equalizers [15]. We reiterate that PULPO meets double-precision performance and its native acceleration of $(\mathcal{M})$, $(\mathcal{H})$, and $(\mathcal{P})$ achieves an $11\times$ speed-up compared to execution on the 8 RISC-V cores with their FPUs.

B. Measurements and Comparison

Fig. 3 shows the fabricated 9 mm² chip with the presented SoC highlighted; the rest of the chip contains other designs. The PULPO-SoC occupies an area of 1.24 mm², with the PULPO accelerator, the data memory, and the 8 RISC-V cores taking 18%, 16%, and 32% of the area, respectively. At 0.8 V supply voltage and 300 K, the SoC achieves a maximum clock frequency of 293 MHz, with the critical path being between the instruction cache (I$) and a RISC-V core. When considering a 128-antenna BS communicating with 16 UEs using 256-QAM, PULPO achieves a maximum detection/precoding throughput of 240 Mb/s. For applications in which the cores are not using their FPUs, PULPO is responsible for 70% of the SoC's power.

Fig. 4 shows the frequency and energy versus the SoC's supply voltage. We consider PULPO's energy per complex-valued 16×128 MVP ($\mathbb{C}$ markers) or per real-valued 16×128 MVP ($\mathbb{R}$ markers) for different floating-point formats. We observe that binary16 dissipates the most energy, caused by the largest amount of mantissa bits. The alternative 16-bit format, bfloat16, requires 3% lower energy, while binary8

saves 16%. Two real-valued MVPs (same matrix, two vectors) are 7% more energy efficient than one complex-valued MVP.

Tbl. I compares PULPO to the other programmable massive MU-MIMO accelerators [7], [8], as well as to existing detection ASICs [3]–[6]. Since PULPO handles all critical baseband-processing operations on its own, we only take into account the area and energy used by PULPO and the shared data memory when reporting area- and energy-efficiency. After technology normalization, the three programmable accelerators are orders-of-magnitude less efficient than the ASICs [3]–[6], which is the cost of enabling programmability and adaptability. Nonetheless, PULPO's area- and energy-efficiency is only $1.6\times$ and $1.7\times$ worse, respectively, than the other two accelerators [7], [8], which is a low price to pay for the flexibility and ease-of-use provided by native floating-point support.

V. CONCLUSIONS

We have presented PULPO, the first programmable floating-point baseband accelerator for massive MU-MIMO BSs. Our accelerator is tightly integrated with the data memory of an 8-core RISC-V SoC, and supports a wide range of algorithms. Our comparison with recent fixed-point baseband accelerators [7], [8] has revealed that PULPO achieves comparable throughput and energy efficiency while avoiding fixed-point optimizations and simplifying design, thus resolving the common misconception that floating-point baseband processing is inefficient.

REFERENCES

[1] E. G. Larsson *et al.*, "Massive MIMO for next generation wireless systems," *IEEE Commun. Mag.*, vol. 52, no. 2, pp. 186–195, Feb. 2014.

[2] T. S. Rappaport *et al.*, *Millimeter Wave Wireless Communications*, 2015.

[3] W. Tang *et al.*, "A 0.58mm² 2.76 Gb/s 79.8pJ/b 256-QAM massive MIMO message-passing detector," in *IEEE Symp. VLSI C.*, Jun. 2016.

[4] C. Jeon *et al.*, "A 354 Mb/s 0.37 mm² 151 mW 32-user 256-QAM near-MAP soft-input soft-output massive MU-MIMO data detector in 28nm CMOS," *IEEE SSC Lett.*, vol. 2, no. 9, pp. 127–130, Sep. 2019.

[5] C.-C. Wen *et al.*, "A 1.96Gb/s massive MU-MIMO detector for next-generation cellular systems," in *IEEE Symp. VLSI C.*, Jun. 2020.

[6] O. Castañeda *et al.*, "A resolution-adaptive 8 mm² 9.98 Gb/s 39.7 pJ/b 32-antenna all-digital spatial equalizer for mmWave massive MU-MIMO in 65nm CMOS," in *IEEE ESSCIRC*, Sep. 2021, pp. 247–250.

[7] G. Peng *et al.*, "A 2.92-Gb/s/W and 0.43-Gb/s/MG flexible and scalable CGRA-based baseband processor for massive MIMO detection," *IEEE JSSC*, vol. 55, no. 2, pp. 505–519, Nov. 2019.

[8] H. Prabhu *et al.*, "A 1070pJ/b 169Mb/s quad-core digital baseband SoC for distributed and cooperative massive MIMO in 28 nm FD-SOI," in *IEEE Symp. VLSI C.*, Jun. 2021.

[9] D. Rossi *et al.*, "Vega: A ten-core SoC for IoT endnodes with DNN acceleration and cognitive wake-up from MRAM-based state-retentive sleep mode," *IEEE JSSC*, vol. 57, no. 1, pp. 127–139, Jan. 2022.

[10] S. Jacobsson *et al.*, "Quantized precoding for massive MU-MIMO," *IEEE Trans. Commun.*, vol. 65, no. 11, pp. 4670–4684, Nov. 2017.

[11] M. Wu *et al.*, "Large-scale MIMO detection for 3GPP LTE: Algorithms and FPGA implementations," *IEEE JSTSP*, pp. 916–929, Oct. 2014.

[12] O. Castañeda *et al.*, "1-bit massive MU-MIMO precoding in VLSI," *IEEE JETCAS*, vol. 7, no. 4, pp. 508–522, Dec. 2017.

[13] ——, "VLSI designs for joint channel estimation and data detection in large SIMO wireless systems," *IEEE TCAS-I*, pp. 1120–1132, Mar. 2018.

[14] C. Jeon *et al.*, "Mismatched data detection in massive MU-MIMO," *IEEE TSP*, vol. 69, pp. 6071–6082, Oct. 2021.

[15] O. Castañeda *et al.*, "Finite-alphabet MMSE equalization for all-digital massive MU-MIMO mmWave communication," *IEEE JSAC*, Sep. 2020.

[16] L. Cannon, "A cellular computer to implement the Kalman filter algorithm," Ph.D. dissertation, Montana State University, 1969.

[17] S. Mach *et al.*, "FPnew: An open-source multiformat floating-point unit architecture for energy-proportional transprecision computing," *IEEE TVLSI*, vol. 29, no. 4, pp. 774–787, Dec. 2020.

A Wide-input-range 918MHz RF Energy Harvesting IC with Adaptive Load and Input Power Tracking Technique

Jing-Ren Yan, Hao-Yi Kuo, Yu-Te Liao

Department of Electronics and Electrical Engineering, National Yang Ming Chiao Tung University,
Hsinchu, Taiwan

Abstract—This paper presents a high-efficiency radio frequency (RF) power harvesting IC with an extended input range. The proposed system comprises a reconfigurable rectifier, maximum power point tracking unit, load modulation circuit, and low-dropout regulator. The reconfigurable rectifier extends the high-efficiency range of the RF input power through stage number control and load modulation according to input power and load requirements. The design is fabricated using 0.18-μm CMOS technology, and the chip area is 1.15 × 0.57 mm². The proposed system can achieve an RF–DC power conversion efficiency greater than 20% in a 16.5-dB input power range with a peak value of 45%, and sensitivity of −16 dBm for a 1-V output voltage.

Keywords—RF energy harvesting, maximum power point tracking, load modulation, wide input range rectifier.

I. INTRODUCTION

The rapid development of the Internet of Things (IoT) demands the wide-range deployment of wireless sensors. These sensors must be compact and low-power to ensure seamless integration with other devices and pervasively distributed in the ambient. Radio frequency (RF) powering technology offers controllable, long-distance, and mobile power supply solutions for batteryless sensors [1].

The challenges of implementing an RF power sensor node are power fluctuations due to path loss and environmental uncertainty. Additionally, variations in the power of the RF energy harvester hinder mobile power conversion efficiency. Therefore, several RF energy harvesting rectifiers [2]-[5] have been developed to accommodate power variations, stabilize the power supply, and maintain high efficiency.

Conventional RF–DC conversion rectifiers have a small input power range due to conduction loss at low input power and reverse power loss at high input power. Thus, it is difficult to maintain a high power conversion efficiency (PCE) over a wide input power range in a standalone rectifier architecture. Fig. 1 shows the state-of-the-art rectifier with an extended input power range. Multipath rectifiers [2][3] can achieve optimal power conversion in the individual configurations according to different input power levels. However, the input impedance deviates from the number of paths used in the RF harvester, requiring an adjustable input impedance matching network. Alternatively, a reconfigurable rectifier stage number [4] [5] provides a wide input range with respect to the input power and the required output voltage. However, this architecture does not consider changes in load from system operations, such as sleep and active modes, in which power requirements can vary significantly. Furthermore, because load power impacts the conversion efficiency, switching converter-based power management units are employed to provide stable output voltage, control the rectifier load, and store the harvested energy for subsequent use [6][7]. The maximum power point tracking technique (MPPT) [6][7][8] is usually applied to determine power delivery, the number of rectifier stages, and

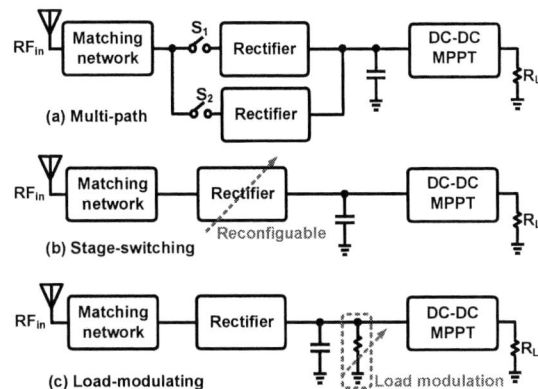

Fig. 1. Architecture of multi-path, stage-switching, and load-modulation rectifiers.

regulated output voltage. A conventional MPPT circuit detects the open-circuit voltage of a rectifier output and adjusts the load to match the impedance for maximum power transfer. However, since the rectifier is a nonlinear circuit, impedance matching is difficult to achieve at various power levels. The design in [5] adjusts the stage numbers of a rectifier to modulate the output impedance; however, the range is restricted due to limited configurations, and the output voltage is not constant. Although a current detection circuit is used to monitor the output load in analog load modulation, it usually consumes a lot of power and degrades power transfer efficiency [9]. Moreover, a regulator is usually used to stabilize the supply voltage to other functional blocks, and voltage drops between the supply and output of the regulator degrade the efficiency. Therefore, to improve end-to-end power efficiency, the output voltage of the rectifier must be close to that of the regulator to avoid power waste. Consequently, conventional rectifiers can achieve high peak efficiency but have a limited input power range (<10dB). Multipath and switch-stage-number architectures extend the high PCE range but do not provide good regulation and end-to-end efficiency.

In this paper, a wide-input-range RF-power management technique with adaptive input power and load requirement tracking is presented. The paper is organized as follows: In Section II, a reconfigurable rectifier design is described. The measurement results of the proposed rectifier are presented in Section III, and conclusions are drawn in Section IV.

II. DESIGN OF THE PROPOSED RF-POWER HARVESTING IC

Fig. 2 shows the block diagram of the proposed reconfigurable RF-power management IC. The design includes a rectifier with switchable stage numbers, MPPT unit, load adjustment circuit, and low dropout regulator (LDO). The self-biased cross-coupled rectifier can reduce the turn-on voltage of a transistor and improve sensitivity. Additionally, a four-stage rectifier is adopted to accommodate the wide input power range and required

978-1-6654-8495-4/22 $31.00 © 2022 IEEE

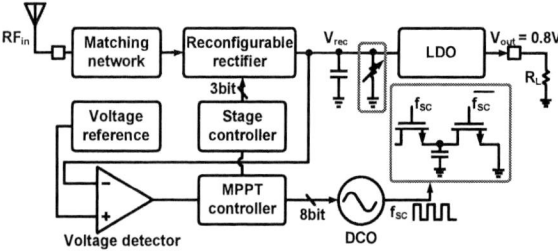

Fig. 2. Block diagram of the proposed RF–DC converter.

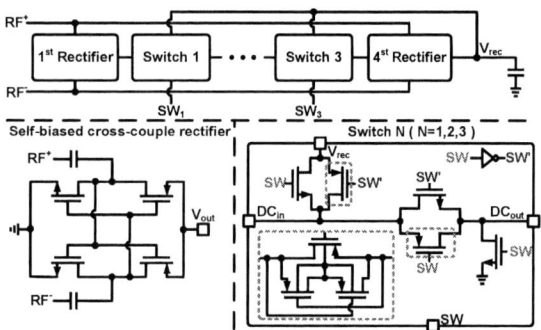

Fig. 3. Schematic diagram of a rectifier with reconfigurable stages.

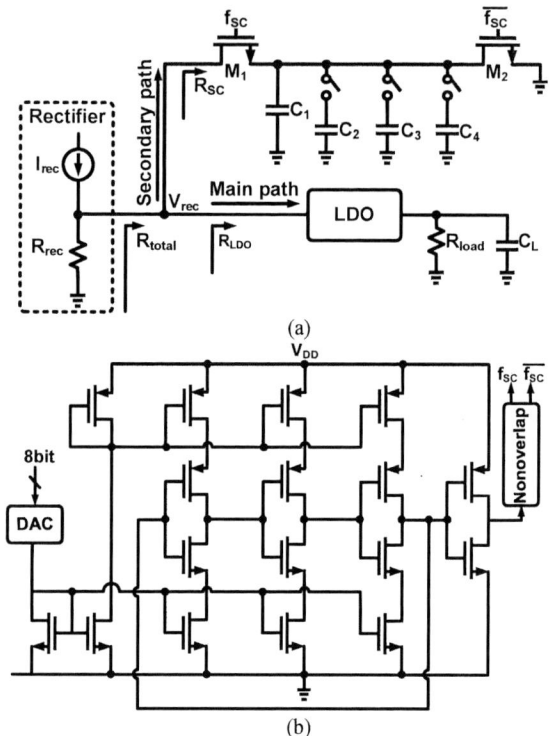

Fig. 4. Schematic of (a) the load modulation circuit (b) digitally controlled current starving oscillator for clock generation.

output voltage. Meanwhile, the power conversion efficiency and sensitivity depend on the stage numbers of the rectifier. More stages increase the output voltage but degrade the PCE. Since PCE is equal to $V_{out}^2/(P_{in}R_L)$, it varies with V_{out}, R_L, and P_{in}. Conventional open-circuit voltage detection cannot be used to effectively find the optimal efficiency of the rectifier. Therefore, in this work, a stage number controller is used to maintain the output voltage of the rectifier in a range by configuring the series or parallel numbers of the rectifier units. Load modulation can achieve output voltage regulation and efficiency optimization at different P_{in} values. The LDO regulator suppresses the voltage ripple and offers a stable supply voltage. To improve the efficiency of the LDO regulator, the rectifier voltage is kept close to the required voltage for the other circuits by load modulation and stage number adjustment.

A. Reconfigurable Rectifier Design

Fig. 3 shows the schematic of the reconfigurable rectifier, which comprises a four-stage self-biased cross-coupled rectifier and three groups of switches for stage number selection. Considering the power loss due to reverse leakage and conduction loss, the sizes of the NMOS and PMOS rectifying transistors need to be optimized. A large transistor size reduces conduction loss but increases reverse leakage and parasitic capacitance. A large parasitic capacitance increases the input voltage drop from the capacitive division to the gate of the rectifying transistor, thereby reducing the sensitivity. Note that a large capacitive load at the input degrades the quality factor of the matching network, leading to a low voltage-boosting gain. Meanwhile, a small transistor size induces a large on-resistance, resulting in a high conduction loss. Therefore, the sizes of the NMOS and PMOS rectifying transistors are 9 μm/0.2 μm and 12 μm/0.2 μm, respectively, to compromise the conduction loss and parasitic loss through simulation.

The stage selection switch is placed on the rectifying DC path instead of the RF signal path to avoid dynamic power and parasitic losses. The parallel and series configurations are controlled by transmission gates. Because a dynamic body-biasing switch reduces the leakage current, a body bias voltage is adaptively connected to the higher voltage between the drain and source voltages of the switch by adding two cross-coupled transistors to switch the body bias voltage. When the SW signal is low and SW' is high, the rectifier units are connected in series by connecting DC_{out} to DC_{in}. Otherwise, the rectifier units are connected in parallel. The rectifier is configured into 1-2-4 stages according to the input power level. A four-stage rectifier is used at low input power to increase the output voltage. At high input power, a one-stage rectifier is used to reduce the output voltage to avoid power loss.

B. Load Modulation Circuit

There is an optimal output resistance for matching the power at different loads and input power. Since the stage number connected to an RF input does not change, the input impedance is not affected by the rectifier reconfiguration. However, the output impedance varies with the stage numbers. Thus, a load modulation using a switched capacitor (SC) architecture is used to accommodate the wide-range load impedance. Fig. 4 shows the schematic of the proposed load modulation circuit. In the SC load modulator, the switching frequency (f_{SC}) and capacitance (C_{eff}) define the effective resistance, $R_{SC}=1/ f_{SC}C_{eff}$. Thus, the total load resistance is $R_{LDO}//R_{SC}$, which is set to the optimal value by adjusting the R_{SC} according to the input power and load requirement. A 2-bit capacitor array is utilized to extend the equivalent resistor range, and a current-starved ring oscillator is used to adjust the switching frequency, which is

978-1-6654-8495-4/22 $31.00 © 2022 IEEE

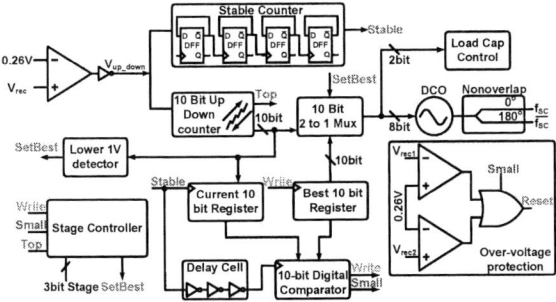

Fig. 5. Schematic diagram of the MPPT controller.

automatically controlled by the MPPT. Therefore, the rectifier power and LDO efficiency are simultaneously considered in the design.

C. Proposed MPPT Controller

Fig. 5 shows the proposed MPPT controller circuit blocks. The MPPT controller consists of a hysteretic comparator, stable counter, 10-bit up-down converter, 8-bit digital controlled oscillator (DCO), two sets of a 10-bit register, 10-bit digital comparator, stage controller, low power voltage reference, and an over-voltage protection circuit. There are four main functional blocks: stage control, load modulation, power detection, and over-voltage protection. First, the load modulator changes the load impedance by adjusting the frequency of the DCO to keep V_{rec} at approximately 1 V. Subsequently, the charge time to 1 V at the output of the rectifier is recorded and compared with different stage configurations. To ensure that the circuit can be activated, a four-stage rectifier is used and reduced in sequence until the best configuration is reached. The perturb-and-observe method is used to find the largest power transfer at the rectifier output. The rectifier output voltage (V_{rec}) is first scaled by a resistor divider and then compared to the on-chip 0.26-V voltage reference, which comprises two-stage stacked diode MOS transistors [10], by using a hysteresis comparator to generate the up or down signal, which represents a V_{rec} value higher or lower than 1 V, respectively. The up/down signal is counted by a 10-bit counter and produces the increment/decrement of the DCO frequency. The DCO signal passes through a non-overlap signal generator to control the SC load modulator. If the counter is overflowing, the output voltage cannot reach 1 V at possible switching frequencies. The capacitance is then increased to accommodate a wide range of load impedance. In the detection circuit realized using the ripple counter, a stable state is defined by eight sequential up-down pulses to avoid false detection. A stable signal rises and activates the comparison between the signal recorded in the previous state and the incoming signals. When the rise time is shorter than that of the recorded state, the digital codes replace the recorded codes in the registers and adjust the stage numbers. At the same time, the sub-1 V detector monitors whether the V_{rec} is below 1 V when switching the stage numbers. If the V_{rec} is below 1 V after switching, the system keeps the previous results and completes the search process by sending a 'small' signal to the controller. After the MPPT process, the voltage protection circuit is activated to detect sudden voltage changes across the upper and lower threshold voltages. When the V_{rec} is out of the preset range due to input

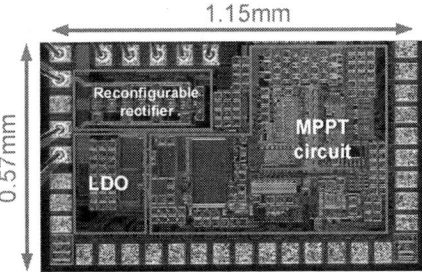

Fig. 6. Chip micrograph.

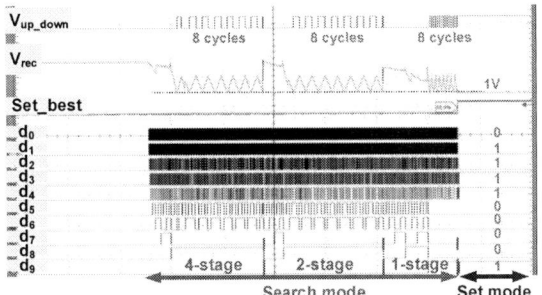

Fig. 7. Measured waveform of the MPPT controller at an input power of 1 dBm.

power or load changes, the reset signal triggers the search mode again.

III. MEASUREMENT RESULTS

The proposed wide-input-range RF energy harvesting IC is fabricated through a 0.18-μm CMOS process. Fig. 6 shows the micrograph of the design. The chip area is 1.15 × 0.57 mm². The matching network, antenna, and chip are mounted on a printed circuit board. The chip impedance is matched to 50 Ω at 918 MHz. Fig. 7 shows the measured waveforms of the MPPT controller at 0 dBm input power, where d_{0-9} are the output codes of the up-down counter. First, the stage number is set to 4, and the output ($V_{up-down}$) is compared every 8 pulses to find out the best configuration of stage numbers and load modulation. At 1-dBm input power, the best stage number is 1, which can be observed from V_{rec} regarding the switching frequency of the power stage.

Fig. 8 shows the measured PCE versus input power. The dashed lines represent the maximum efficiency as the load in the 1-2-4 stage rectifier is manually changed. The solid line is the result of automatic load modulation and adjustment of stage number using the proposed MPPT controller and load modulator, which shows a good tracking accuracy. The tracking power was limited to 4 dBm due to the load modulator range determined by the on-chip capacitor. At 918 MHz, the sensitivity was −16 dBm, the peak efficiency is 45% at −5.5 dBm input power, and the high-efficiency range (>20%) is 16.5 dB. Fig. 9 shows the measured output voltages of the designs with and without load modulation (fixed 220 kΩ). With load modulation, the output voltage is maintained at approximately 1 V, resulting in ~80% LDO efficiency when the LDO output voltage is 0.8 V and input power is increased from −16.5 to 4 dBm. Without load modulation, the input power varies from −14 to −10.5 dBm, resulting in a decrease in the LDO efficiency from 80% to 37.2%. Note that the sensitivity of the 1 V output is extended

978-1-6654-8495-4/22 $31.00 © 2022 IEEE

TABLE I.
PERFORMANCE SUMMARY AND COMPARISON OF THE PROPOSED ARCHITECTURE WITH THOSE OF PRIOR ICS

	[1]	[4]	[5]	[7]	[8]	**This work**
Freq. (MHz)	900	915	820	915	2450	**918**
Tech. (nm)	40	180	130	180	65	**180**
No. of stages	1	1-2-4-8	1-2-4-8	2-3-4-6-12	1	**1-2-4**
V_{out}	0.6-2.45	>1	0.3-0.7	0.8-2.2	0.1-0.9	**1**
Sensitivity 1 V@1 MΩ	5% ** −11 dBm	3% −14.8 dBm	7.5% −22 dBm	6% −17.8 dBm	3% * −17 dBm	**3.1% −16 dBm**
Peak PCE	47% 0 dBm	25% 0 dBm	39% −5 dBm	34.4% 1.3 dBm	48.3% −3 dBm	**45.2% −5.5 dBm**
High PCE range (>20%)	24 dB	15 dB	14 dB	13 dB	11 dB	**16.5 dB**
Area (mm²)	N/A	5.3	0.0108	0.4	0.000125	**0.6555**
Regulation	No	No	No	No	No	**Yes**
Load control	Manual	Automatic	Manual	Manual	N/A	**Automatic**

*The open circuit reached 400 mV at −17dBm, **Open circuit

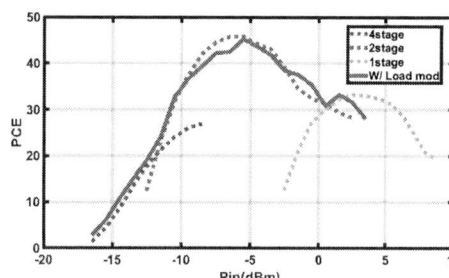

Fig. 8. Measured PCE versus P_{in}.

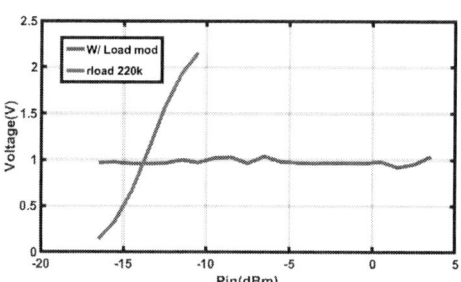

Fig. 9 Measured output voltage versus Pin of the design with load modulation and a fixed resistance of 220kΩ

by 2.5 dB while load modulation is enabled. Table I presents the performance summary and comparison of the results with those of state-of-the-art wide-range RF energy harvesting ICs. Hence, the proposed automatic power tracking RF energy harvesting IC has high power conversion efficiency in wide input power and load ranges.

IV. CONCLUSION

This paper presents a 918 MHz CMOS rectifier with automatic MPPT and load modulation schemes to achieve extension of the operating range and robustness to various power fluctuations. The design achieves −16 dBm sensitivity, a 2.5-dB improvement over the design without load modulation. Additionally, the high-efficiency range is extended to 16.5 dB, while the peak efficiency at −5.5-dBm input power is 45%. Therefore, the proposed design can enable the harvesting of RF energy for ubiquitous sensing in an extended range.

REFERENCE

[1] J. Wang, Y. Jiang, J. Dijkhuis, G. Dolmans, H. Gao and P. Baltus, "A 900 MHz RF energy harvesting system in 40 nm CMOS technology with efficiency peaking at 47% and higher than 30% over a 22dB wide input power range," IEEE European Solid State Circuits Conference, 2017, pp. 299-302.

[2] Y. Lu, H. Dai, M. Huang, M.-K. Law, S.-W. Sin, S.-P. U, M. P. Rui, "A Wide Input Range Dual-Path CMOS Rectifier for RF Energy Harvesting," IEEE Transactions on Circuits and Systems II: Express Briefs, vol. 64, no. 2, pp. 166-170, Feb. 2017.

[3] S.-H. Lin, C.-Y. Kuo, S.-Y. Lu and Y.-T. Liao, "A high-efficiency power management IC with power-aware multipath rectifier for wide-range RF energy harvesting," IEEE MTT-S International Microwave Symposium, 2017, pp. 304-306.

[4] M. A. Abouzied, K. Ravichandran and E. Sánchez-Sinencio, "A Fully Integrated Reconfigurable Self-Startup RF Energy-Harvesting System With Storage Capability," IEEE Journal of Solid-State Circuits, vol. 52, no. 3, pp. 704-719, March 2017.

[5] Z. Zeng, J. J. Estrada-López, M. A. Abouzied and E. Sánchez-Sinencio, "A Reconfigurable Rectifier With Optimal Loading Point Determination for RF Energy Harvesting From −22 dBm to −2 dBm," IEEE Transactions on Circuits and Systems II: Express Briefs, vol. 67, no. 1, pp. 87-91, Jan. 2020.

[6] G. Saini, L. Somappa and M. S. Baghini, "A 500-nW-to-1-mW Input Power Inductive Boost Converter With MPPT for RF Energy Harvesting System," IEEE Journal of Emerging and Selected Topics in Power Electronics, vol. 9, no. 5, pp. 5261-5271, Oct. 2021.

[7] Z. Zeng, S. Shen, X. Zhong, X. Li, C-Y. Tsui, A. Bermak, R. Murch, E. Sánchez-Sinencio, "Design of Sub-Gigahertz Reconfigurable RF Energy Harvester From −22 to 4 dBm With 99.8% Peak MPPT Power Efficiency," IEEE Journal of Solid-State Circuits, vol. 54, no. 9, pp. 2601-2613, Sept. 2019.

[8] P. Xu, D. Flandre and D. Bol, "Analysis, Modeling, and Design of a 2.45-GHz RF Energy Harvester for SWIPT IoT Smart Sensors," IEEE Journal of Solid-State Circuits, vol. 54, no. 10, pp. 2717-2729, Oct. 2019.

[9] C.-Y. Kuo, C.-A. Lu and Y.-T. Liao, "A 918MHz Wide-Range CMOS Rectifier with Diode-Feeding and Switch-Capacitor-Based Load Modulation Technique," IEEE Asian Solid-State Circuits Conference (A-SSCC), 2019, pp. 251-254.

[10] C.-Z. Shao, S.-C. Kuo, and Y.-T Liao, "A 1.8-nW, −73.5-dB PSRR, 0.2-ms Startup Time, CMOS Voltage References with Self-Biased Feedback and Capacitively-Coupled Schemes" IEEE Journal of Solid-State Circuits, vol. 56, no. 6, pp. 1795-1804, June 2021.

Input Nonlinear Adaptive Voltage Position Technique in the Switched-capacitor Converter with Feedforward Compensation for 87.8% Peak Efficiency under 8X Input Interference

Shu-Yung Lin[1], Sheng Cheng Lee[1], Ke-Horng Chen[1], Kuo-Lin Zheng[2], Ying-Hsi Lin[3], Shian-Ru Lin[3], Tsung-Yen Tsai[3]

[1]National Yang Ming Chiao Tung University, [2]Chip-GaN Power Semiconductor Corporation, Hsinchu, Taiwan [3]Realtek Semiconductor Corp, Hsinchu, Taiwan

Abstract—**The proposed switched capacitor converter for energy harvesting applications can have a wide input range of 0.25V to 2V. The input nonlinear adaptive voltage position (INAVP) control and feedforward compensation can decrease ΔV_{out} and the settling time in line transients. The mismatch calibration circuit can reduce the influence caused by the mismatch in different phase. The highest peak efficiency is 87.8%. The reduced efficiency variation is 11.6% in the case of 800% V_{in} perturbation.**

Keywords—*feedforward compensation, input nonlinear adaptive voltage position (INAVP) control, mismatch calibration circuit*

I. INTRODUCTION

Energy harvesting techniques have been used in many applications to sustain the long-term operation of battery-free electronics. Due to unstable energy source, the captured input voltage range is 0.25 V to 2 V, corresponding to 8X input disturbance. Switched capacitor (SC) converters can be used for energy harvesting applications to achieve compact size, but it needs to automatically operate in buck and boost modes and also achieve fast line transient response for low output voltage ripple ΔV_{out}. Fig. 1 shows that due to limited voltage conversion ratio (VCR), large input disturbance will cause voltage difference ΔV between $V_{in}*VCR$ and V_{OUT}, where V_{in} is the input voltage. The two major power losses are power charge sharing loss (P_{CS}) ($\propto \Delta V^2$) and switching power loss (P_{SW}) ($\propto \Delta V^{-2}$). A larger positive ΔV will greatly reduce the efficiency. In addition, due to limited number of VCRs, a negative ΔV will cause V_{out} to be unregulated. Although higher switching frequency and multiphase soft-charging in [1] can reduce ΔV, the switching power loss (P_{SW}) increases and ΔV_{out} increases with ΔV [1].

In [2], a large on-chip inductor is used to reduce the switching clock frequency $F_{CLK}=(LC)^{-1/2}$ to 47.5MHz. However, the on-chip inductor has higher P_{CS} due to its poor quality factor. In contrast, the multi-stage architecture in [3, 4] uses various VCRs to reduce ΔV, but at the cost of drive power. Moreover, in the case of 800% ΔV_{in} occurring in energy harvesting sources,

the efficiency may be degraded to 25%, although the multi-stage SC converter can regulate V_{out}. In order to fast react to variations of input harvesting sources, the minimization of ΔV becomes important to achieve high efficiency.

Therefore, the proposed SC converter in Fig. 2 uses the input nonlinear adaptive voltage position (INAVP) technique to position V_{out} in response to ΔV_{in}. To react to input harvesting sources when they are above, equal to, or below V_{out}, the INAVP has three voltage position slopes corresponding to buck, boost, and deep boost modes. Consequently, the INAVP regulates V_{out} within an acceptable droop voltage ($V_{out(max)}$- $V_{out(min)}$) to achieve fast line transient response. Moreover, to fit into a compact size in an energy harvesting system, the proposed SC converter in Fig. 2 utilizes the bond wire L_{bond} switching network without using any external off-chip inductors. The main feature of L_{bond} switching network is to generate the middle voltage V_{mid} to supply the SC stage and keep $V_{mid}*VCR$ constant. To rapidly regulate V_{mid} and reduce the inner loop transient response time for switching network, the inner loop feed-forward (FF) calibration control technique is utilized in the INAVP. Thus, ΔV can be kept almost constant, as shown in Fig. 3(a), at the optimum efficiency, shown in Fig. 3(b), to obtain low P_{CS} and P_{SW}.

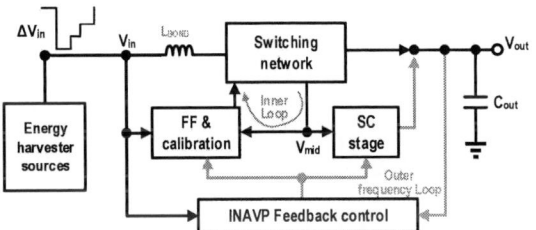

Fig. 2. Proposed INAVP SC converter and the comparison of different types of SC converters in efficiency and input transient response.

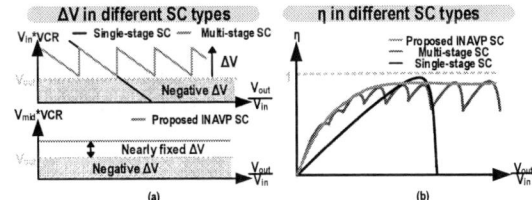

Fig. 3. (a) ΔV in different SC types. (b) η in different SC types.

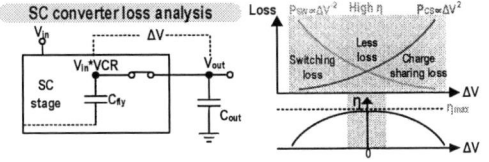

Fig. 1. The voltage difference ΔV versus switching power loss P_{SW} and power charge sharing loss P_{CS} in conventional SC converter.

The organization of this paper is as follows. Section II shows the fundamental operation of the proposed INAVP. Circuit implementations and the three operation modes of the INAVP are shown in Section III. Experimental results are shown in Section IV to prove this work. Finally, conclusions are made in Section V.

II. PROPOSED INAVP SC CONVERTER

Fig. 4(a) shows the architecture of INAVP SC converter with three interleaved power stages. There is a phase shift of 120° between each phase to reduce ΔV_{out} by a factor of three. Two-phase interleaved SC stage only needs two VCRs (1/2 and 2) for buck, boost, and deep boost operations, as shown in Fig. 4(b).

The buck mode operates in two phases $\Phi 1$ and $\Phi 2$. S_D in Fig. 4(a) blocks the high voltage V_{mid}. In Fig. 5(a), S_A will directly transfer energy from V_{in} to V_{out} in $\Phi 1A$. In $\Phi 1B$ (S_A off, S_B on), the boosted V_{mid} (>V_{in}) is supplied to 2:1 SC stage. $\Phi 1C$ and $\Phi 1D$ are used for SC stage operation. If I_{bond} is 0, the switching network shuts down and deadtime phase will occur in $\Phi 1C$. The SC stage provides energy to V_{out} in $\Phi 1D$. Four phases $\Phi 2A\sim\Phi 2D$ of the interleaving SC stage operates the same as $\Phi 1$. Consequently, ΔV is small and efficiency increases in buck mode. In boost mode, in Fig. 5(b), V_{mid} can be generated lower than V_{in} by the switching network for the interleaved 1:2 SC stages. In deep boost mode of Fig. 5(c), no current is directly flowing L_{bond} to V_{out} in ΦA, while in ΦB, L_{bond} delivers energy to V_{mid}, which is higher than V_{in} to ensure correct interleaved 1:2 SC operation in ΦC and ΦD. As illustrated in Fig. 5(d), the proposed INAVP further suppresses overshoot and undershoot voltages compared to prior arts.

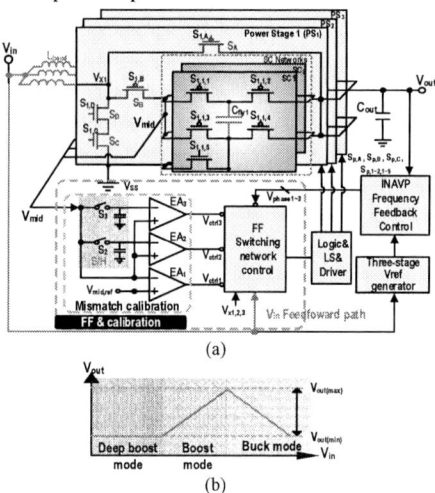

(a)

(b)

Fig. 4. (a) Architecture of INAVP SC converter with three interleaved power stages. (b) Three operation modes.

The current transfer of buck and boost modes are shown in Fig. 6(a). V_{in} directly provides energy for V_{out} in ΦA, and provides energy for V_{mid} in ΦB. If ΔV_{in} occurs, the inner loop will maintain the energy from V_{in} to C_{fly} almost constant to regulate V_{mid}. Since ΔV_{in} disturbs I_{bond} and its current slope, V_{out} will be greatly disturbed and will change the F_{sw} in the outer feedback control. Fortunately, the proposed INAVP control can rapidly regulate V_{out} to its non-linear adaptive positioning voltage level when V_{in} changes. In buck mode, the increase of V_{out} can decrease ΔV when V_{in} decreases. When the energy directly increases from V_{in} to V_{out}, the INAVP control finds the

energy balance by reducing the energy from the SC level to V_{out}. Once the proposed converter enters boost mode, INAVP control regulates V_{out} to a lower level as V_{in} decreases for fast transient response, since the increase of ΔV can directly compensate for the energy decrease from V_{in} to V_{out}. The converter enters deep boost mode when V_{in} is ultra-low. ΔV_{in} has little effect on V_{out}. F_{sw} is slightly changed since there is no current directly flowing to V_{out}. The frequency change in different modes are shown in Fig. 6(b).

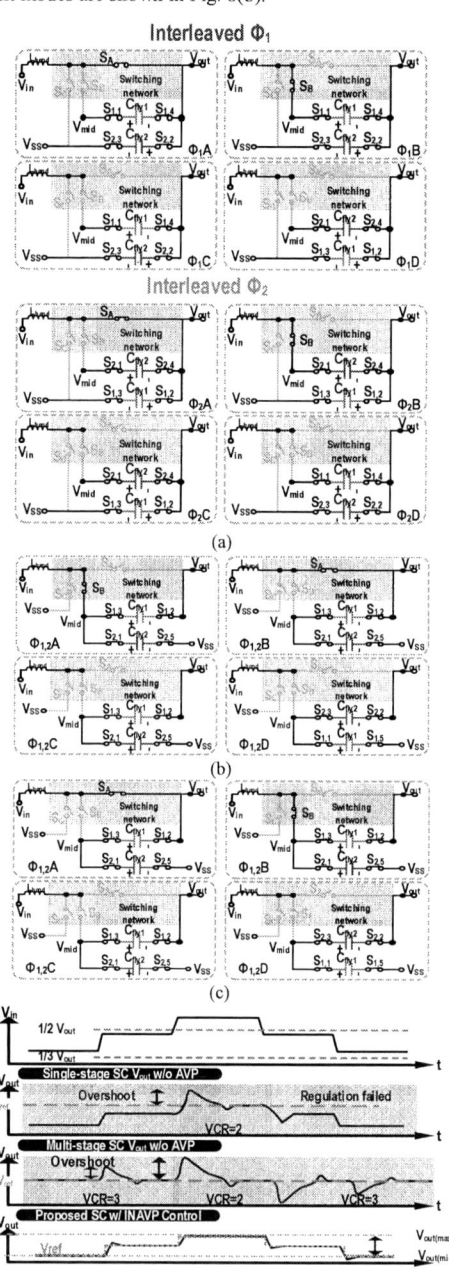

(a)

(b)

(c)

(d)

Fig. 5. (a) Buck mode in two phases $\Phi 1$ and $\Phi 2$. (b) Boost mode. (c) Deep boost mode. (d) Suppression of voltage spikes in the INAVP.

978-1-6654-8495-4/22 $31.00 © 2022 IEEE

III. CIRCUIT IMPLEMENTATIONS

A. Outer INAVP frequency feedback control

The outer INAVP frequency feedback control in Fig. 7, generates a INAVP three-stage reference voltage $V_{ref,3stage}$ to regulate V_{out} between $V_{out(max)}$ and $V_{out(min)}$, as shown in Fig. 4(b). $V_{ref,3stage}$ is compared with V_{fb} to generate a loading dependent I_{ctrl} to determine the load-dependent F_{SW} in the current controlled oscillator (CCO).

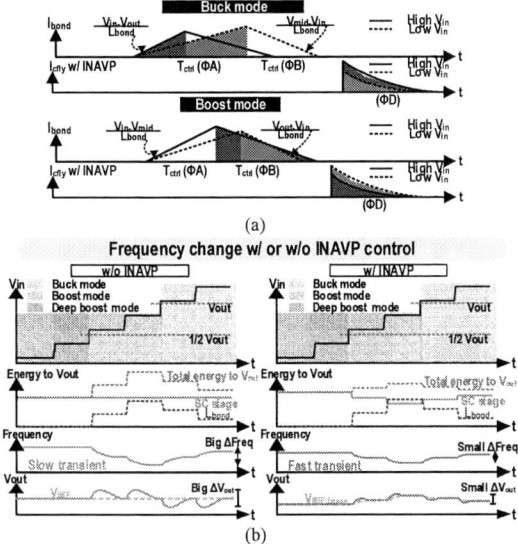

(a)

(b)

Fig. 6. (a) INAVP concept shows the current transfer in different modes. (b) Change of frequency, and V_{out} with and without the INAVP.

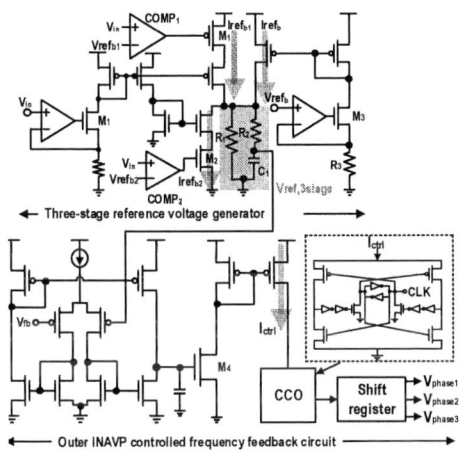

Fig. 7. Outer AVP controlled frequency feedback loop.

B. Mismatch calibration circuit

Due to mismatch, the energy delivered by each power stage will cause large ΔV_{out}, as shown in Fig. 8(a). Three interleaved phases (PSs) are used in the mismatch calibration circuit. The averaged V_{mid} will be compared with $V_{mid,ref}$ in EA_1 and will generate V_{ctrl1}. The S/H circuit stores V_{mid} in PS_2 and PS_3 and compares it with the average V_{mid} to generate V_{ctrl2} and V_{ctrl3},

respectively. V_{ctrl1}~V_{ctrl3} are used to control the voltage-controlled on-resistance MOSFET M_{res}, to adjust $I_{delay,ctrl}$ and the duration of ΦA in each phase to achieve the charge balance of different PSs. The duration of ΦA in PS will be shortened when V_{ctrl} is high, contrarily, the ΦA of PS with low V_{ctrl} will be extended. When ΦA is ended and ΦB starts, the circuit uses the zero current detector (ZCD) to shut down the switching network to avoid negative current and LC oscillation for high efficiency. PS_2 with small L_{bond} will shorten the duration of ΦA, while the duration of ΦA in PS_3 increases if L_{bond} is large. Thus, the mismatch calibration circuit in Fig. 8(c) can effectively reduce the ΔV_{OUT} by up to 83%. The rising and falling slope of I_{bond} is equal to V_{rise}/L_{bond} and V_{fall}/L_{bond}, respectively. T_{ctrl} (ΦA duration) needs to be proportional to $V_{rise}^{-1/2}$ to maintain the energy from V_{in} to V_{mid} constant in boost mode.

The FF circuit in Fig. 8(d) can reduce the V_{mid} settling time, since FF signal V_{ff} does not pass through the inner loop. Through the on/off control of M5 and M6, the I_{ff} can implement V_{rise}. As the current V_{rise}/R flows into to M7, the overdrive voltage V_{ov} of M7 is proportional to $V_{rise}^{1/2}$ by the saturation current equation.

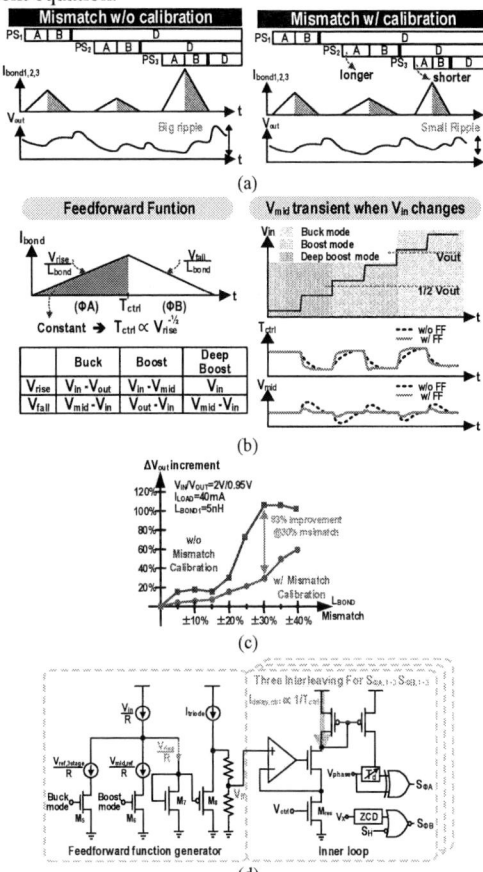

Fig. 8. (a) Large ΔV_{out} caused by bond wire mismatch can be reduced by the proposed mismatch calibration circuit. (b) Feedforward control table in different modes. Thus, V_{mid} can be kept constant in the case of ΔV_{in}. (c) Improvement of ΔV_{out}. (d) Feed forward function generator.

978-1-6654-8495-4/22 $31.00 © 2022 IEEE

Since the bias current of M_8 is small, the V_{gs} of M_8 is V_{th}. With the voltage divider, a low voltage related to T_{ctrl} is generated when the V_{ds} of M8 equals $V_{gs(M7)}$-$V_{th(M7)}$-$V_{gs(M8)}$, which is proportional to $V_{rise}^{1/2}$. Similarly, T_{ctrl} can be determined by the table in Fig. 8(b) for buck and deep boost modes. Since V_{ff} can change $I_{delay,ctrl}$ immediately without passing the inner loop feedback, the loop can achieve fast transient response as V_{in} changes.

IV. EXPERIMENATL RESULTS

The proposed INAVP SC converter with the FF technique for supplying energy harvesting systems is fabricated in a 65nm CMOS process. Fig. 9 shows the chip micrograph.

ΔV_{out} changes from 128mV to 62mV (improvement of 106%) when V_{in} changes from 0.65V to 0.95V in Fig. 10(a). The settling time reduces from 2.8µs to 1.3µs (improvement of 115%) at the load of 50mA. In Fig. 10(b), ΔV_{out} is 63mV when V_{in} changes from 1.2V to 2V in buck mode, and is 58mV when V_{in} changes from 0.25V to 0.6V in deep boost. Since V_{mid} needs to have a longer settling time in the transition of boost and buck, ΔV_{out} is 122mV and settling time is 1.8µs when V_{in} changes from 0.65V to 2V. ΔV_{out} is 68mV when V_{in} changes from 0.25V (deep boost) to 0.95V (boost). At the load of 100mA, ΔV_{out} is reduced from 35mV to 19mV (improved by 84%) by the mismatch calibration in Fig. 10(c).

In Fig. 11, when V_{in} is 1.8V, the peak efficiency is 85.2% at load of 75mA. Under 50mA, the peak efficiency is 87.8% when V_{in} is 1.2V. Table I shows the comparison table with state-of-the-arts. The proposed INAVP control can regulate V_{out}, and has the highest efficiency of 87.8% and smallest efficiency variation of 11.6% in the case of 800% ΔV_{in}. Although the efficiency in [1] is 95%, the input range is limited within 1.865 to 2.07V.

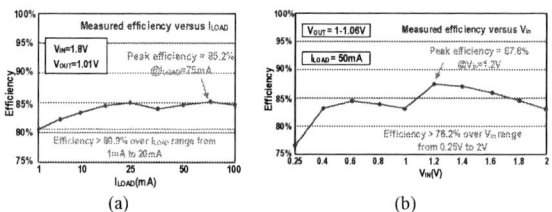

Fig. 11. (a) Efficiency versus I_{LOAD}. (b) Efficiency versus V_{in}.

Table I: Comparison with state-of-the-arts.

	[1] ISSCC'16 Butzen	[2] ISSCC'18 Jiang	[3] ISSCC'16 Lutz	[4] ISSCC'17 Butzen	[5] ISSCC'20 McLaughlin	This Work
Technology	40nm CMOS	65nm CMOS	350nm HVCMOS	28nm CMOS	180nm CMOS	65nm CMOS
Topology	Step-Down SC	Step up/ step down SC	Step up/ step down SC	Step-Down SC	Step-Down Resonant SC	Step up/ step down SC
Vin (V)	1.865-2.07	0.22-2.4	2-13	3.2	2.4-4.4	0.25-2
Vout (V)	0.9	0.85-1.2	5	0.95	1.0-2.2	1.0-1.06
C_{total}/ L_{total}	10nF N/A	14nF / N/A	3.64nF N/A	1.5nF + C_{init} N/A	10.4nF 7.7nH	0.7nF + 1.3nF 5nH (bond wire)
Δ Eff. In Vin perturbation	N/A	64%	59%	N/A	N/A	11.6%
Peak output power	3.15mW	80mW	10mW	80mW	0.87W	112mW
Peak efficiency	95%	84.1%	81.5%	82%	85.5%	87.8%
Peak Power Density	1.3uW/mm²	13.2mW/mm²	7.353mW/mm²	1.1W/mm²	97mW/mm²	38.62mW/mm²

V. CONCLUSIONS

The proposed INAVP SC converter is proposed for energy harvesting applications because the INAVP can tolerate a wide input range of 0.25V to 2V. The feedforward technique can react to the input variations to generate the reference voltage for fast line transient response. Moreover, the mismatch calibration circuit can reduce the influence caused by the mismatch in different phase. The highest peak efficiency is 87.8%. The reduced efficiency variation is 11.6% in the case of 800% V_{in} perturbation.

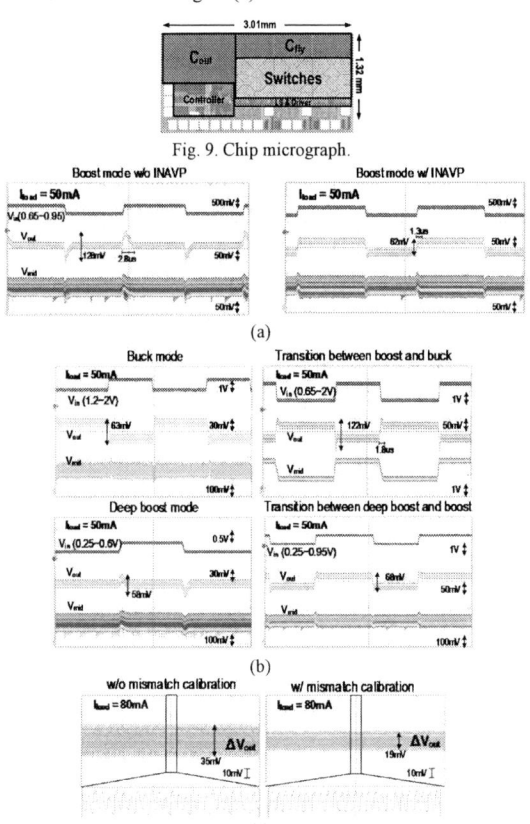

Fig. 9. Chip micrograph.

Fig. 10. (a) Comparison of line transient response w/ and w/o INAVP in boost mode (b) Buck and deep boost mode line transient (left), transition of boost to buck and deep boost to boost (right) (c) Δ V_{out} w/ and w/o mismatch calibration circuit.

REFERENCES

[1] N. Butzen and M. Steyaert, "1.1W/mm²-power-density 82%-efficiency fully integrated 3 : 1 Switched-Capacitor DC-DC converter in baseline 28nm CMOS using Stage Outphasing and Multiphase Soft-Charging," *ISSCC Dig. Tech. Papers*, 2017, vol. 60, pp. 178–179.

[2] Y. Jiang et al., "A 0.22-to-2.4V-Input Fine-Grained Fully Integrated Rational Buck-Boost SC DC-DC Converter Using Algorithmic Voltage-Feed-In (AVFI) Topology Achieving 84.1% Peak Efficiency at 13.2mW/mm2 ," *ISSCC Dig. Tech. Papers*, pp. 422- 424, Feb. 2018.

[3] D. Lutz et al., "12.4 A 10mW fully integrated 2-to-13V-input buck-boost SC converter with 81.5 % peak efficiency," *ISSCC Dig. Tech. Papers*, Jan 2016, pp. 224–225.

[4] P. H. McLaughlin, Z. Xia and J. T. Stauth, "11.2 A Fully Integrated Resonant Switched-Capacitor Converter with 85.5% Efficiency at 0.47W Using On-Chip Dual-Phase Merged-LC Resonator," *ISSCC Dig. Tech. Papers*, pp. 192-194, 2020.

[5] N. Butzen et al., "A 94.6%-Efficiency Fully Integrated Switched-Capacitor DC-DC Converter in Baseline 40nm CMOS Using Scalable Parasitic Charge Redistribution," *ISSCC Dig. Tech. Papers*, pp. 220-221, Feb. 2016.

A 385mV, 270nW, Accurate Voltage Level Detector for IoT

Omer Nechushtan*, Asaf Feldman* and Joseph Shor

Bar Ilan University, Faculty of Engineering, Ramat Gan, Israel (*equally credited authors)

emails: omernech@gmail.com, joseph.shor@biu.ac.il

Abstract—Voltage Level Detectors (VLD) are used to determine that the computing system has reached a valid operating voltage and can initiate computation. The VLD must detect its own supply voltage, which makes it a challenge for low voltage applications. In this work, a VLD is presented which can detect supplies as low as 385mV at a power of 270nW. Despite the low power, a fast detection speed of 12µs is demonstrated. The circuit occupies a small footprint of 6900µm². These characteristics make the circuit attractive for IoT applications.

Keywords—Power-on-reset, Voltage Level Detector, Reference Voltage, Power Management.

I. INTRODUCTION

In recent years, the design of ultra-low power and low voltage Integrated Circuits (ICs) has become a focus area of both industry and academia. The emergence of Internet-of-things (IoT) ICs, which require minimal energy, has accelerated this trend. Processors operating at near threshold voltages (V_{th}) have been demonstrated by industry for maximizing energy efficiency [1]. A circuit which is present in nearly every such system is the Voltage Level Detector (VLD), also known as Power-on-Reset (POR) [2-9]. The VLD senses when the supply voltage has reached its required value and then indicates an enable signal called "Power-Good" that initiates the reset flow of the system-on-chip (SoC). The accuracy of the VLD is important, since this determines the minimal voltage of the system, as is its power, since the VLD is generally an always-on circuit. One of the main challenges in designing a low voltage VLD is that it detects its own supply. At near-V_{th} voltages, there could be glitches in the "Power-Good", which could cause the system to wake up early in an undefined state, leading to logical errors and wasted power. Designing a "glitch-free" VLD has become a key feature for this supervisory circuit [5,6]. It is important that the VLD has a relative fast response since this will enhance the wakeup latency of the IC. It should also have minimal random variation and PVT (Process, Voltage, Temperature) dependence, since it is not always possible to trim the VLD, since it may wake up before the fuse unit. The lowest reported VLD trip point is 0.52V [5], to the authors best knowledge. However, near-V_{TH} processors with supply voltages as low as 280mV have been demonstrated [1]. In this paper, a glitch-free VLD is demonstrated with a trip point of 0.385V, a nominal power of 270nW. It can detect supply ramps as fast as 1µS with delay of 12µS.

II. CIRCUIT ARCHITECTURE AND DESIGN

Figure 1 shows a simplified block diagram of the Proposed VLD. A current controlled oscillator (CCO) is used to charge a switched capacitor divider. This provides a divided version of the supply voltage (V_{DIV}) to a comparator,

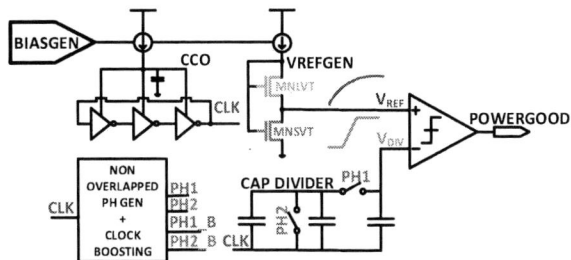

Fig. 1. Block Diagram of the proposed VLD

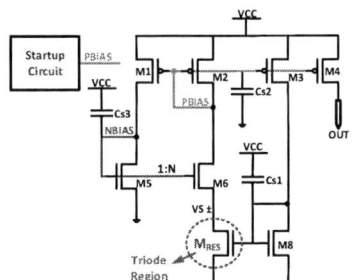

Fig. 2. Simplified schematic of the Biasgen.

which compares it to a reference voltage, V_{REF}. When the comparator output changes to logic high, it indicates that supply voltage has reached its desired level. The Proposed VLD architecture combines accurate supply voltage detection with glitch-free operation at near-V_{th} voltages. This is accomplished by charging both the voltage reference generator and the CCO with same biasing source (Biasgen). The CCO oscillations begin only when its supply reaches V_{th} levels, which delays the rise of V_{DIV}. This enables V_{REF} to be ready before the divided supply voltage V_{DIV} which prevents glitches at the output of the comparator. In addition, capacitors are placed at strategic nodes in the design to ensure that Power-Good only takes place at the desired trip-point and not beforehand. There are also small DFT (design for test) buffers (not shown), which operate off a 1.2V supply, thus enabling some visibility of the internal signals during powerup. These buffers have a sigma offset of 20mV and a bandwidth of ~ 2MHz, so they are useful for a "qualitative" probing of the analog signals at low frequency.

Figure 2 illustrates the bias generator circuit (Biasgen). The proposed Architecture is very similar to the PTAT circuit (Proportional to Absolute Temperature) in [11]. The Biasgen was designed to generate 50nA at room temperature with minimal temperature dependence. To reduce the area of the VLD, the resistor R is replaced by MOS transistor "M_{RES}" which is biased in triode regime. To reduce the sensitivity to power supply voltage as well as to achieve better current mirroring all the transistors in Fig. 2 were cascoded by self-biased cascodes [6]. When M_5 and M_6 are in subthreshold, the difference in their V_{gs} (δV_{gs}) will appear across M_{RES} to yield the following current:

$$I = \frac{\delta V_{gs}}{R_{ds.MRES}} = \frac{nKT}{q} * \left(\frac{\ln(N)}{\mu C_{ox} \frac{w}{L} (V_{gs} - V_{th})} \right) \quad (1)$$

With correct biasing of M_{RES}, the PTAT behavior of δV_{gs} can be canceled out by the temperature dependence of M_{RES} to achieve a nearly temperature independent bias current. This is useful to limit the PVT dependence of the CCO. A startup circuit [12] provides an initial current to the PMOS devices, which is then shut off automatically during regular operation. The Capacitors C_{S2}, C_{S1} and C_{S3} were placed at the Bias gates to enable a fast wakeup.

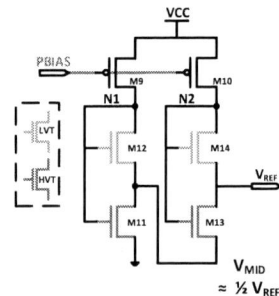

Fig. 3. Schematic of the reference voltage generator.

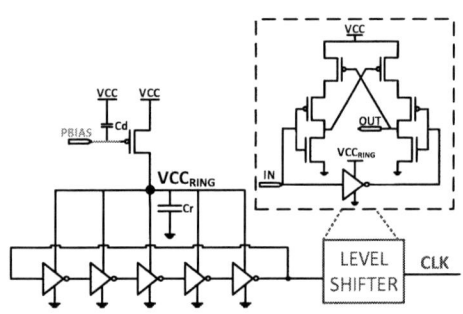

Fig. 4. Schematic of the CCO.

In several prior-art VLD's, conventional band-gap references were implemented [6-9]. However, these require high voltage (due to the diodes), large currents and significant area. In this work, a stacked 2T CMOS reference was utilized similar to [2-5, 13], as shown in Fig. 3. This reference utilizes the difference in V_{th}'s between high V_{th} (HVT) and low V_{th} (LVT) devices, available in the process, to generate a stable temperature independent reference. The stacked devices are charged by a bias from the Biasgen.

Figure 4 shows the schematic of the CCO used in this work. The frequency, F, of the oscillator will obey the following equation:

$$F = \frac{I}{C_{RO} V} \quad (2)$$

where I=bias current, C_{RO} = total capacitance in the oscillator and V=VCC$_{RING}$ which is the oscillation amplitude. VCC$_{RING}$ is a relatively low voltage, near V_{TH},

which must be level shifted up to CMOS levels by a level shifter (LS) as shown in Fig. 4. Applying the temperature independent bias described earlier to the CCO enables the CCO have low sensitivity to PVT. The current bias of the CCO is a few nA. As the CCO requires its supply level to be near V_{th}, the capacitor at VCC$_{RING}$ delays the initiation of the CLK. This enables V_{REF} to be ready before V_{DIV} which facilitates the glitch free operation.

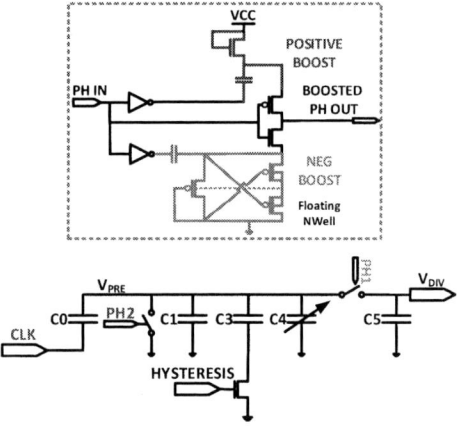

Fig. 5. Simplified schematic of the switched capacitor divider and clock boosting circuit.

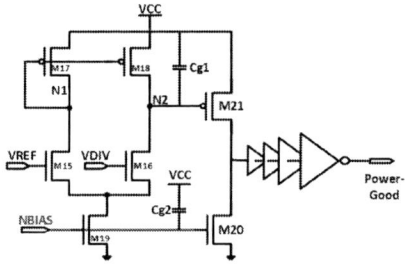

Fig. 6. Schematic of the comparator.

Figure 5 shows the divider and booster circuits. Switched capacitor dividers can be very efficient in both power and area. The impedance of the capacitor is Z = 1/sC, such that a large impedance can be realized in a very small area. A phase generator (not shown) receives the CLK signal from the CCO to generate two non-overlapping phases. A boosting circuit was utilized where necessary to boost phases both above Vcc (to 2*Vcc-V_{th}) and below Vss (~ -0.2V) (Fig. 5). The boosting circuit initiated the voltage to Vcc-V_{th} (positive) and Vss+V_{th} (negative) using a diode connected device. The signal is then boosted using a capacitor during the relevant phase edge. The switches driven by PH* are NMOS devices, and the boosting enables them to have minimal resistance when they are conducting, and low leakage when they are non-conducting. Note, that the triple well process was not used here, which limited the negative boosting to ~ -200mV. Before the CLK is present, V_{DIV} is initiated to 0V by the leakage of the switches. During PH2, V_{PRE} is initiated to 0V, and at PH1 it is boosted based on the capacitor's ratio and transferred to the comparator (Fig. 1). A digitally controlled variable capacitor, C4, enables control of the division ratio. When Power-Good is enabled, it also enables the Hysteresis, which alters the capacitor ratio. This prevents noise from toggling the Power-Good when the supply is close to the trip-point.

The Comparator shown in Fig. 6 uses a two-stage amplifier with n-type input pair. Pull up capacitors C_{g1} and C_{g2} are connected to M21 and M20 to initiate Power-Good to zero during the supply ramp. There are a series of CMOS buffers following the comparator which drive the Power-Good Signal off chip for measurement.

Figure 7 shows simulated waves (typical, 30°C, Vcc=400mV) where Vcc ramps to 20mV above the minimal trip point in 1µs. The V_{REF} signal is charged to its nominal value before the CLK begins to toggle. After the CLK appears, V_{DIV} ramps up and causes Power-Good to be enabled after 12µs.

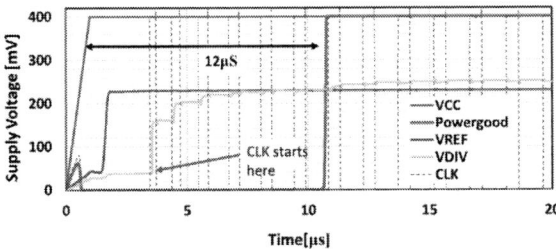

Fig. 7. Simulated VLD signals versus time at room temperature for a 1µS ramp at the minimum trim.

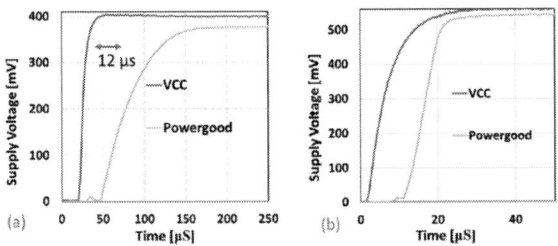

Fig. 8. Measured VLD signals versus time at room temperature for a fast ramp at ~ 20mV above the (a) minimum trim and (b) maximum trim.

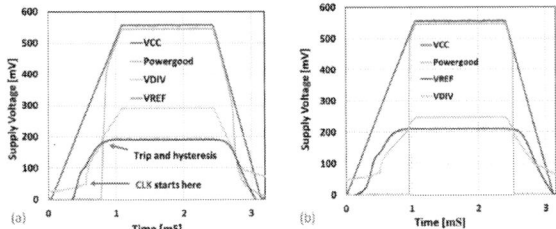

Fig. 9. Measured VLD signals versus time at room temperature for two different VLD's for a 1mS ramp at the (a) min trim and (b) max trim.

III. SI MEASURMENTS

Figure 8 shows the measured Power-Good signal for a fast 10µs supply ramp for the minimum (385mV) and maximum (508mV) trims. The Power-Good signal has a

relatively long slope (10's of µs), since the IO buffers are operating in sub-threshold and need to drive very large capacitances on the package and board so that it can be measured. In an actual system, this signal would be internal and would have a fast slope, such as that in the simulated waves (Fig. 7). The trip point was evaluated at the beginning of the Power-Good ramp to compensate for this. Figure 9 shows a measured 1ms Vcc ramp of the Power-Good, along with the internal V_{REF} and V_{DIV} signals, as measured by the DFT buffers for two different VLD's. At very low Vcc (<200mV), the signals float since there is almost no current in the chip. As the supply ramps, the V_{REF} signal is charged first, and then followed by V_{DIV}. One can observe the initiation of the CLK, as well as the hysteresis in V_{DIV} after the two signals cross. Note that the DFT buffers have limited bandwidth (~ 2MHz) and some voltage offset, such that the relative timings and voltage levels are not exact. Many such ramps were measured at different trim levels, ramp rates and temperatures, and no glitches were observed.

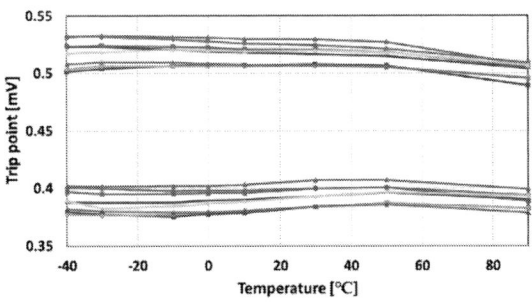

Fig. 10. Measured trip point versus temperature for 8 VLD's at 2 different voltage trims.

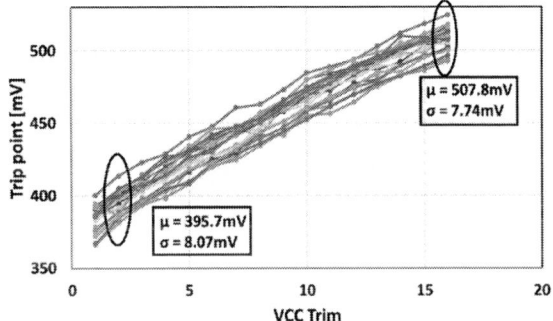

Fig. 11. Measured trip point versus trim for 24 VLD's at room temperature.

Figure 10 shows the trip vs. temperature for 8 VLD's at the two extreme trims. The nominal temperature coefficient (TC) was 114 µV/C, while the worst case was 230 µV/C. The trip point was measured for 24 VLD's at room temperature for all of the trims (Fig. 11). The trip point can be trimmed across a range between 385-508mV at ~ 9mV steps. The sigma variation is roughly 8mV. In order to evaluate the effect of noise on the trip point, the supply was set multiple times to voltages near the trip point and the number of Power-Goods was evaluated. This is in shown in Fig. 12 for 20 evaluations at 0.1mV steps near the 373mV trip point of this VLD. It is observed that there is a 0.2mV

978-1-6654-8495-4/22 $31.00 © 2022 IEEE

range for which the Power-Good did not evaluate deterministically to either 0 or 1 in all the cases. This indicates the effects of noise are relatively small. We assume that part of this is associated with the hysteresis function which effectively latches the state to Power-Good=high once V_{DIV} crosses V_{REF}. Figure 13 shows the power vs. supply voltage. At Vcc=0.41V, which is above the minimal trip point of all the measured VLDs, the power is 270nW.

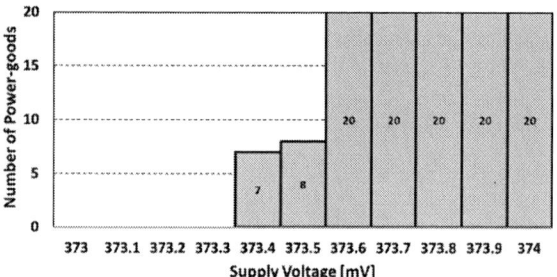

Fig. 12. Number of successful Power-Goods (out of 20 tries) versus supply voltage for the minimum trim.

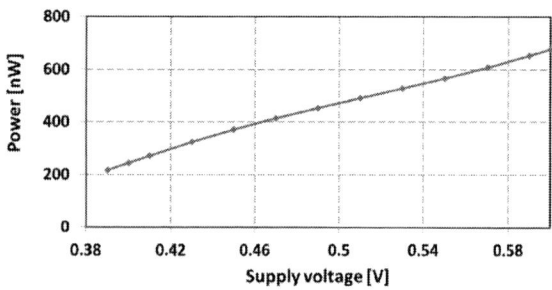

Fig. 13. Power consumption versus Supply voltage

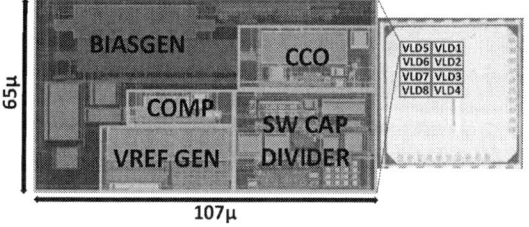

Fig. 14. Die photo and layout.

IV. CONCLUSIONS

Figure 14 shows a die photo and the layout of the VLD. A comparison to recent work is shown in Table 1. The proposed VLD achieves the lowest trip-point voltage, while exhibiting glitch-free operation at ramp rates from 1μs. It also exhibits the smallest area and is competitive in the other parameters. The delay is the fastest measured despite the low power. The trip point can be finely trimmed, and the variation is better than most of the prior-art. The low-voltage, glitch free operation was enabled by the delayed CCO wakeup which allowed the reference voltage to settle before the divided Vcc. These features make the proposed VLD very suitable for near-threshold Processors and IoT devices.

This work was sponsored by the Israel Science Foundation.

V. REFERENCES

1. S. Jain, *et. al.*, "A 280mV-to-1.2V wide-operating-range IA-32 processor in 32nm CMOS," *2012 IEEE International Solid-State Circuits Conference*, 2012, pp. 66-68.
2. C. De Roover and M. S. J. Steyaert, "Energy Supply and ULP Detection Circuits for an RFID Localization System in 130 nm CMOS," in *IEEE Journal of Solid-State Circuits*, vol. 45, no. 7, pp. 1273-1285, July 2010.
3. I. Lee, Y. Lee, D. Sylvester and D. Blaauw, "Battery Voltage Supervisors for Miniature IoT Systems," in *IEEE Journal of Solid-State Circuits*, vol. 51, no. 11, pp. 2743-2756, Nov. 2016.
4. T. Someya, et. al., "A 0.90–4.39-V Detection Voltage Range, 56-Level Programmable Voltage Detector Using Fine Voltage-Step Subtraction for Battery Management," in *IEEE Transactions on Circuits and Systems I: Regular Papers*, vol. 66, no. 3, pp. 1270-1279, March 2019.
5. T. Someya, et. al., "248pW, 0.11mV/°C glitch-free programmable voltage detector with multiple voltage duplicator for energy harvesting," *ESSCIRC Conference 2015 - 41st European Solid-State Circuits Conference (ESSCIRC)*, 2015, pp. 249-252.
6. A. Feldman and J. Shor, "An Accurate 0.55-V 2.6-μW Voltage-Level Detector," in *IEEE Solid-State Circuits Letters*, vol. 3, pp. 166-169, 2020.
7. P. Pandey. Low-voltage power-on-reset circuit with least delay and high accuracy. Electronics Letters, 51(11):856–858, 2015.
8. J. Shor and D. Zilberman, "An Accurate Bandgap-Based Power-on-Detector in 14-nm CMOS Technology," in *IEEE Transactions on Circuits and Systems II: Express Briefs*, vol. 63, no. 5, pp. 428-432, May 2016.
9. Toru Tanzawa, "A process- and temperature-tolerant power-on reset circuit with a flexible detection level higher than the bandgap voltage," *2008 IEEE International Symposium on Circuits and Systems (ISCAS)*, 2008, pp. 2302-2305.
10. B. Zhou, Y. Jin and F. Zhao, "Sub-1-V BGR and POR Hybrid Circuit With 2.25-μA Current Dissipation and Low Complexity," in *IEEE Transactions on Very Large Scale Integration (VLSI) Systems*, vol. 28, no. 10, pp. 2228-2232, Oct. 2020.
11. B. Razavi, *Design of Analog CMOS Integrated Circuits*, McGraw-Hill, Singapore, pp.377-405.
12. Q. A. Khan, S. K. Wadhwa and K. Misri, "Low power startup circuits for voltage and current reference with zero steady state current," *Proceedings of the 2003 International Symposium on Low Power Electronics and Design, 2003. ISLPED '03.*, 2003, pp. 184-188.
13. D. Zagouri and J. Shor, "A Subthreshold Voltage Reference with Coarse- Fine Voltage Trimming," *2021 IEEE International Symposium on Circuits and Systems (ISCAS)*, 2021, pp. 1-4.

Table 1 - Comparison Table

Reference	This work	2	3	4	5	6	8	9	10
Year	2022	2010	2016	2019	2015	2020	2015	2008	2020
Technology (nm)	65	130	180	250	250	65	14	70	65
Mechanism	2T	2T	2T	2T	2T	BJT	BJT	BJT	MOS
Area (mm²)	0.0069	0.085	0.01	0.086	0.0091	0.095	0.02	0.064	0.014
Power	270nW	2nW	635pW	1.2nW[2]	248pW	2.6μW	496μW[1]	1.2μW	2.25μW
Nominal Trip point (V)	0.385	0.94	3.58	NA	0.52	0.59	0.8	2	NA
Trim Range (V)	0.385-0.52	NA	NA	0.9-4.39	0.52-0.85	0.55-0.64	0.74-0.91[3]	NA	NA
Trim Step (mV)	9	NA	NA	63	49	5.9	10	NA	NA
Ramp Rate Range	1μS-DC	400ms	5μS-DC	400ms[1]	DC	10μS-DC	50μS[1]	1μS-1ms[1]	400μS-100ms
Measured Delay (Sec)	12μ	NA	1.9	7.4m	NA	200μ	NA	NA	NA
Range (°C)	-40 to 90	0 to 50	0 to 80	-20 to 60	-20 to 80	-10 to 110	-10 to 110	25 to 90	-30 to 90
Trip Point Temp Coeff (μV/°C)	114	4000	1625	280	110	292	60	292	NA
Units measured	24	NA	NA	NA	5	21	3716	NA	5
σ variation (mV)	7.8	15[1]	35[4]	NA	NA	1.68	12	25	NA

1. simulated   2. At 3.5V   3. Needs Additional 1.1V analog supply   4. Simulated Random and Process

An Extended Temperature Range ePCM Memory in 90-nm BCD for Smart Power Applications

M. Carissimi[1], C. Auricchio[1], E. Calvetti[1], L. Capecchi[1], M. Torres[1], S. Zanchi[1], P. Gupta[2],
R. Zurla[1], A. Cabrini[3], D. Gallinari[1], F. Disegni[1], M. Borghi[1], E. Palumbo[1], A. Redaelli[1], and M. Pasotti[1]
[1]STMicroelectronics, Agrate Brianza, Italy, [2]STMicroelectronics, Greater Noida, India,
[3]University of Pavia, Pavia, Italy. email: marcella.carissimi@st.com

Abstract—**This paper presents a temperature-robust embedded Phase-Change Memory (ePCM) with high cycling capability able to meet all the stringent specifications coming from the automotive environment and, more specifically, the used phase-change material (based on Ge-rich GST alloy) has been tuned to fit power ICs constraints. In order, to cope with the −40 °C to 175 °C operation requirements, a temperature-compensated write algorithm was conceived and specific circuits were added to render the statistical distribution of programming pulses equal at any temperature as it is required to obtain a uniform ageing of the cells thus ensuring an higher reliability after 100k cycling. Programming operation was optimized thanks to an improved program load that has been designed to compensate for the expected large power supply variations. Experimental characterization demonstrated a 16 ns access time over the whole temperature range.**

I. INTRODUCTION

It is demonstrated that PCMs are a promising solution for integrating embedded NVMs (eNVMs) in the back end of the line (BEOL) offering excellent performance in term of area, read and write time [1], [2]. Beside PCMs, other kind of BEOL eNVM have been proposed in the literature (e.g. magnetic [3] or resistive [4] memories), but none of these solutions seems to be adequate to guarantee operation and data retention at temperatures up to 175 °C and, for these reasons, they do not fully comply with automotive and smart-power applications needs. Indeed, when considering the design of eNVMs for such applications high cycling capability and reliability together with fast access times must be guaranteed over a wide temperature range. The key feature that ensures high-temperature operation for PCM is the availability of Ge-rich chalcogenide alloys [5], [6] that are suitable to guarantee both data retention up to $T = 175$ °C and soldering compliance [7]. Nonetheless, to fully exploit PCM advantages, custom write algorithms and read/program circuits must be developed taking into account the temperature behavior of ePCM cells during both read and program operations.

In this paper an ePCM test vehicle designed in 90 nm BCD technology including a temperature-compensated adaptive verify scheme featuring high reliability and endurance is presented. The organization of the paper is as follows. Section II presents the proposed memory architecture, while the circuits designed to address the high-temperature requirements are illustrated in Section III. Section IV finally provides measurements while conclusions are given in Section V.

II. PROPOSED MEMORY ARCHITECTURE

To cope with the stringent constraints of automotive specifications (mainly temperature and reliability), the proposed test vehicle was designed to address the problems that arise when the memory device operates at high temperature (i.e., at $T > 150$ °C). In this respect, the main issues are related to the verify operation where the temperature variation of the cell current I_{cell} requires, as shown in Fig. 1(a), a suitable reference current, I_{ref}, that must be generated to compensate the temperature effects (electrical conductivity of a PCM cell changes over temperature as in a semiconductor and, hence, its electrical resistance decreases for increasing temperature). Fig. 2 shows the temperature behavior of a 90 nm technology PCM cell resistance in both SET and RESET states (the activation energy of RESET state is about one order of magnitude higher than SET state). It is important to mention that the temperature variation of I_{cell} (in both SET and RESET states) must be taken into account to adjust the verify levels of the write algorithm so as to keep the statistical distribution of the number of applied programming pulses almost constant at any temperature. This aspect is clearly visible in Fig. 3 where the measured programming times are given as a function of temperature (programming time changes over temperature since a different number of programming pulses is required to program the cell at different temperatures). In particular, the most important aspect is to avoid over-programming conditions that adversely affect endurance which can be done by increasing the verify levels for increasing

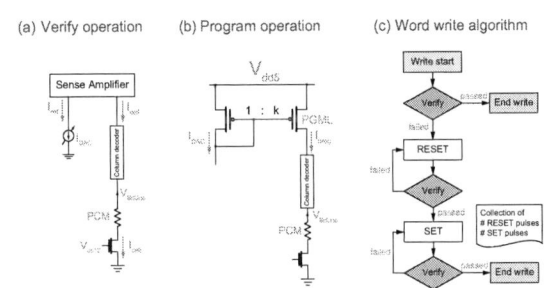

Fig. 1. Main circuits and algorithm challenges: compensating the temperature behavior of the PCM cell during verify (a), reducing the variability of the programming current in presence of V_{dd5} fluctuations (b), and uniformity of programming pulses occurrence over temperature (c).

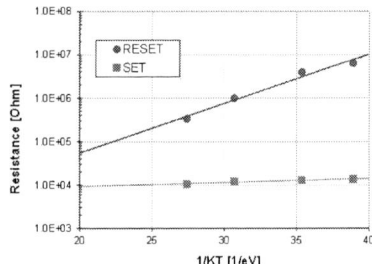

Fig. 2. Variation of the cell resistance over temperature.

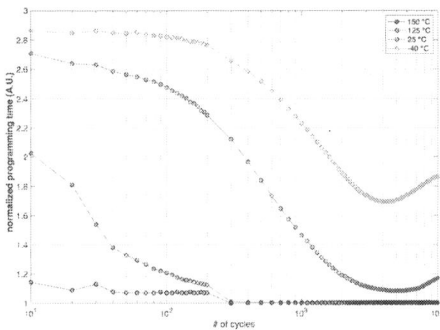

Fig. 3. Measured normalized programming time as a function of programming cycles over the temperature range $-40\,^{\circ}C$ to $150\,^{\circ}C$.

temperature. Another critical design issue (which is more significant when considering write operation) is the variation of the programming current, that is generated through a so called program load current mirror as depicted in Fig. 1(b), that suffers of accuracy limitations determined by source-to-drain voltage modulation caused by supply variations and programming conditions (e.g. cell state and temperature).

Fig. 4 illustrates the block diagram of the proposed test chip where the most significant elements, specifically designed to address the above-mentioned temperature issues, are highlighted in red. The memory array is divided in five sectors: four 64 KB sectors are dedicated to store the code (of a micro-controller), while a single 16 KB data sector is used to store the values of variables and parameters. The designed ePCM includes a finite state machine (FSM) that controls the execution of all operations required for the proper management of the memory in the different operating conditions.

III. ENHANCED VERIFY AND PROGRAM LOAD CIRCUITS

To properly control high-temperature operations, while still preserving the desired automotive-grade reliability, several circuits have been specifically designed. The most important characteristics of these circuits are reviewed in the following subsections.

A. Verify Temperature Compensation

One of the key parameters affecting ePCM cells reliability is the accuracy achieved during program. In this respect, write operations are carried out using a program-and-verify

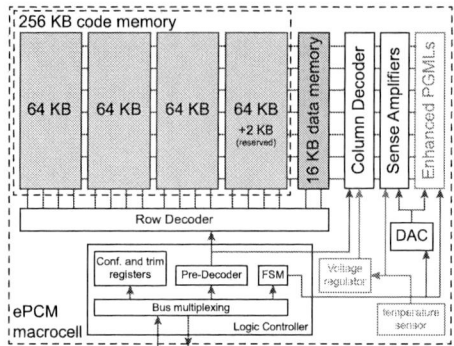

Fig. 4. ePCM macrocell blocks diagram.

technique as shown in Fig. 1(c): a programming pulse is followed by a verify operation that compares the cell current (I_{cell}) with a reference current (I_{ref}) to determine whether or not a further programming pulse is needed to reach the target cell resistance. If the cell to be programmed already contains the desired data program is skipped to minimize the average number of programming pulses as it is beneficial to optimize power consumption, cell reliability, and life span.

Verify operations are carried out on multiple cells in parallel by means of different sense amplifiers (SAs). The circuit schematic of the designed SAs is shown in Fig. 5. The programmable bias current, I_{bias}, must be set to a value $I_{bias} > I_{DAC}$ in order to guarantee the proper biasing of the SA branches in all operating conditions. The variation of the cell resistance over temperature is taken into account

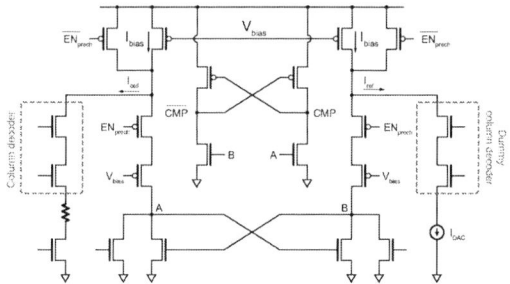

Fig. 5. Circuit schematic of the sense amplifier.

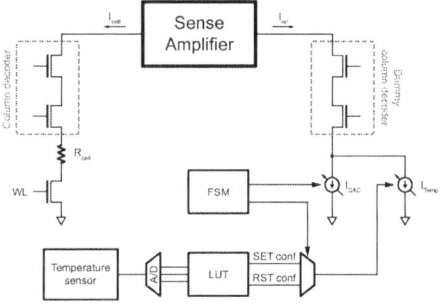

Fig. 6. Temperature compensation of verify current.

978-1-6654-8495-4/22 $31.00 © 2022 IEEE

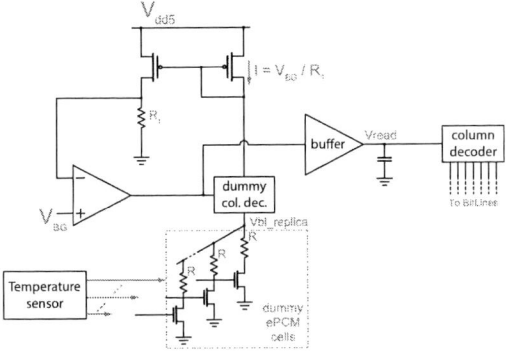

Fig. 7. Temperature compensation of PCM decoding column and PCM bit line voltage.

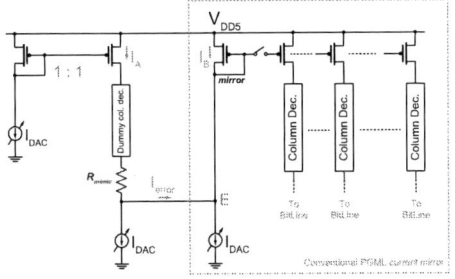

Fig. 8. Enhanced current mirror for program loads.

by adjusting I_{ref} that is generated by adding to I_{DAC} a suitable compensation current, I_{temp}, that is generated, as shown in Fig. 6, by using a look-up table (LUT) that is fed by a suitable 3-bit temperature sensor. As will be discussed in Section IV, this compensation largely improves the efficiency of the programming algorithm without compromising the read access time in all operating conditions.

Analogously, the read/verify bitline voltage (applied to the cell to generate I_{cell}) is controlled by a programmable voltage regulator, shown in Fig. 7, that generates the voltage V_{read} used to bias the gate terminal of the column decoder MOS transistor. The proposed bitline biasing scheme was conceived to guarantee the memory operation over the extended temperature range. Indeed, thanks to this approach, the bitline voltage modulation compensates for the PCM resistance variation over temperature (see Fig. 2).

B. Dynamic Program Load

As shown in Fig. 1(b), the effective programming current I_{prog} is generated through pMOS transistors PGML. This current depends on both the V_{sg} (set by the diode connected pMOS transistor) and V_{sd} voltages of the PGML pMOS. The problem is that, while V_{sg} can be considered well controlled, the source-to-drain voltage V_{sd} is instead affected by variations that occur on both the source and the drain terminals of the PGML pMOS transistors. In particular, the drain voltage (that, by neglecting the effect of the column decoder, corresponds to the bitline voltage) depends on the electrical behavior of

the cell during programming, while the voltage of the source terminal (that is connected to V_{dd5}) is affected by variations on the power supply (which, in this respect, can vary from 4.5 V up to 6.0 V). This determines a large variation of V_{sd} for the PGML pMOS transistors that implies a non-negligible channel-length modulation effect and, hence, a variation of the effective programming current (that can be significantly different with respect to the ideal target value $k \times I_{DAC}$). This phenomenon becomes particularly critical considering the large range of variation of I_{prog}. A possible solution is to increase the channel length of the PGML pMOS transistors but this worsen both the silicon area occupation and, more important, the transient response of the current mirror. To cope with above drawbacks, an enhancing circuit was designed and applied to a standard current mirror as shown in Fig. 8. The idea is to generate a current I_A (where, to mimic the load effect on the bitlines, R_{mimic} is added) that is affected by the channel-length modulation, which is compared to I_{DAC} to generate I_{error}. This compensating current is added/subtracted (at node E) to I_{DAC} to obtain the desired I_{prog}. The programming system was also designed including a circuit that minimize the delay time in the generation of the programming pulses [8].

IV. EXPERIMENTAL RESULTS

The proposed ePCM system was designed in 90-nm ST BCD technology and the micrograph of the fabricated die is shown in Fig. 9. The system, that operates with two supplies, V_{dd12} (from 1.08 V to 1.32 V) and V_{dd5} (from 2.8 V to 6.0 V), was successfully validated over the extended temperature range (from -40 °C to 175 °C). In particular, thanks to the improved circuits designed for both the SA and the bitline decoder, an access time of 16 ns was guaranteed in all corners (Fig. 10). The measured statistical distribution of write pulses occurrences for both SET and RESET pulses are similar at -40 °C, 25 °C and 175 °C when compared to traditional approach as shown in Tab. I. As evident from Tab. I, when

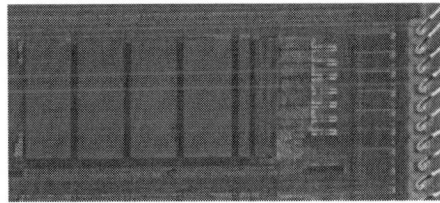

Fig. 9. ePCM macrocell micrograph.

TABLE I
STATISTICAL COMPARISON OF PROGRAMMING PULSES (PER WORD).

	proposed	conventional	both	conventional	proposed
Temperature	-40 °C	-40 °C	27 °C	175 °C	175 °C
Verify current	I_{-40}	I_{DAC}	I_{DAC}	I_{DAC}	I_{+175}
BitLine Voltage	$V_{BL}+\Delta V$	V_{BL}	V_{BL}	V_{BL}	$V_{BL}-\Delta V$
relative # of RES1	100%	100%	100%	100%	100%
relative # of RES2	20.0%	2.2%	13.7%	100%	44.7%
relative # of RES3	1.2%	0.1%	0%	99.6%	0%
relative # of SET1	100%	100%	100%	100%	100%
relative # of SET2	100%	100%	100%	65.2%	100%
relative # of SET3	58.8%	41.2%	54.1%	8.3%	55.2%

TABLE II
COMPARISON WITH THE STATE OF THE ART.

	ESSCIRC 2017 [9]	VLSI 2019 [1]	ISSCC 2019 [4]	ISSCC 2019 [3]	VLSI 2020 [10]	**This work**
Feature Size	110 nm BCD	28 nm FD-SOI	40 nm	22 nm	28 nm	**90 nm BCD**
Memory Type	PCM	PCM	ReRAM	MRAM	e-FLASH	**PCM**
Integration	BEOL	BEOL	BEOL	BEOL	FEOL	**BEOL**
Memory size	32 KB	6 MB	1.38 MB	0.88 MB	1.25 MB	**274 KB**
Oper. Temp.	−40 °C to 150 °C	−40 °C to 165 °C	−40 °C to 125 °C	−40 °C to 105 °C	−40 °C to 150 °C	**−40 °C to 175 °C**
Write Cycles	100k	100k	1k	1M	N.A.	**100k**

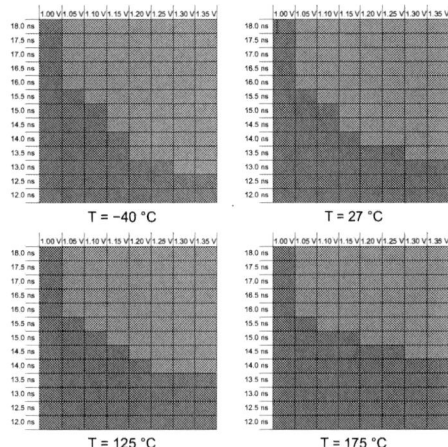

Fig. 10. Access time shmoo plot with V_{dd12} and temperature.

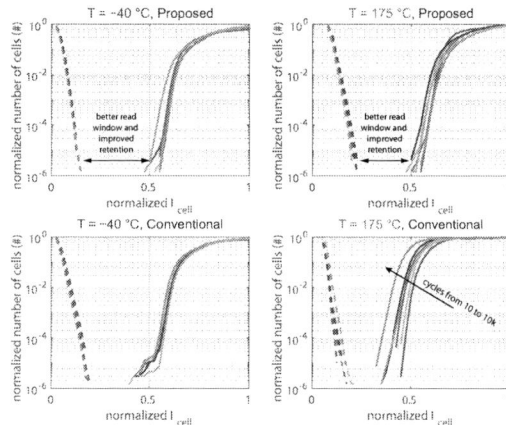

Fig. 11. Measured normalized cumulative distribution of I_{cell} obtained after cycling with conventional circuits (bottom plots) and proposed architecture (top plots). Both the programming operation and the data collection were carried out over the extended temperature range (−40 °C 175 °C).

using a conventional verify approach at 175 °C a larger number of reset pulses, together with a lower number of set pulses, are required to successfully complete the write operation. Vice-versa, the behavior at −40 °C is the opposite as the cell is less conductive in this case. Thanks to the proposed approach, it was possible to equalize the distribution of pulse occurrences over the considered temperature range thus providing a significant overall benefit in terms of endurance, data retention, and reliability as can be seen from Fig. 11 where the cell current distributions, after 10k-cycling, are provided at −40 °C and 175 °C for both improved and conventional write algorithm. Moreover, the proposed solutions also allowed improving the read window after 100k cycles. Summary and comparison with other works are given in Table II.

V. CONCLUSION

In this paper, an improved read/write architecture for an ePCM operating over an extended temperature range has been presented. Thanks to the proposed circuits, a more uniform distribution of the number of programming pulses over the considered temperature range has been obtained. This allowed limiting the stress applied to PCM material, thus ensuring a reliable cycling capability and a 16 ns access time (as guaranteed by an adequate read current window). The improved circuits allowed obtaining better control of the programming currents in the presence of power supply variations and overvoltages. The designed system demonstrated a 100k cycles operation capability which enables the 90 nm BCD technology with PCM memory for smart power applications.

REFERENCES

[1] F. Disegni et al., "Embedded PCM macro for automotive-grade microcontroller in 28nm FD-SOI," *2019 Symp. on VLSI Circuits*, pp. C204–C205.

[2] M. Carissimi et al., "2-Mb embedded Phase Change Memory with 16-ns read access time and 5-Mb/s write throughput in 90-nm BCD Technology for automotive applications," *ESSCIRC 2019 - IEEE 45th European Solid State Circuits Conference (ESSCIRC)*, 2019, pp. 135–138.

[3] L. Wei et al., "A 7Mb STT-MRAM in 22FFL finFET technology with 4ns read sensing time at 0.9V using write-verify-write scheme and offset-cancellation sensing technique," *2019 IEEE International Solid-State Circuits Conference (ISSCC) Digest of Technical Papers*, San Francisco, CA, USA, 2019, pp. 214–216.

[4] C. Chou et al., "An N40 256K 44 embedded RRAM macro with SL-precharge SA and low-voltage current limiter to improve read and write performance," *2018 IEEE Int. Solid-State Circuits Conf.*, pp. 478–480.

[5] N. Ciocchini et al., "Modeling resistance instabilities of set and reset states in phase change memory with Ge-rich GeSbTe," *IEEE Trans. on Electron Devices*, vol. 61, no. 6, pp. 2136-2144, June 2014.

[6] F. Arnaud et al., "Truly innovative 28 nm FDSOI technology for automotive micro-controller applications embedding 16 MB phase change memory," *2018 IEEE Int. Electron Devices Meeting (IEDM) Technical Digest*, San Francisco, CA, 2018, pp. 18.4.1–18.4.4.

[7] P. Zuliani et al., "High Temperature Challenge for PCM as enabler for a wide applications spectrum," *2019 IEEE 11th Int. Memory Workshop (IMW)*, 2019, pp. 1–4.

[8] R. Zurla et al., "Enhanced multiple-output programmable current pulse generator," *Proc. 2019 IEEE Int. Symp. on Circuits and Systems (ISCAS)*, Sapporo, Japan, 2019, pp. 1–5.

[9] M. Pasotti et al., "A 32KB 18ns random access time embedded PCM with enhanced program throughput for automotive and smart power applications," *2017 - 43rd IEEE European Solid State Circuits Conf.*, 2017, pp. 320–323.

[10] H. Shin et al., "A 28nm 10Mb embedded flash memory for IoT product with uultra-low power near-1V supply voltage and high temperature for grade 1 operation," *2020 IEEE Symp. on VLSI Circuits*, 2020, pp. 1-2.

On-Chip High-Resolution Timing Characterization Circuits for Memory IPs

Amit Agarwal[1], Steven Hsu[1], Mark Anders[1], Gunjan Pandya[2], Ram Krishnamurthy[1], James Tschanz[1], Vivek De[1]

[1]Circuits Research Lab, Intel Labs, [2]Corporate Memory Organization
Intel Corporation,
Hillsboro, OR 97124
amit1.agarwal@intel.com

Abstract—A fully configurable and synthesizable timing characterization test-bench for memory IPs enables high resolution clk2q, setup, hold and cycle-time delay measurements. The timing test-bench features distributed regional capture FFs, mesh based low-skew clock and setup difference measurement across regional capture FFs to minimize error, multiple data/input delay generators to handle timing permutations across memory inputs, automated relative placement/pre-routing for matched layout and XORed clock delay generators to create multiple edges for measuring read after write delay/cycle time.

I. INTRODUCTION

As process technology advances, design technology co-optimization (DTCO) is necessary to extract best power, performance, area (PPA) and V_{MIN} by exploiting new process scaling boosters (buried power rail, GAA, and complementary FET [1]) for improved silicon memory and standard cell circuit IPs. One of the key DTCO methodologies is to perform silicon IP characterization for power, performance, area (PPA) and robustness as newer process technology matures [2-3]. This also provides confidence for IP adoption and improved library models with silicon correlation to foundry customers.

Memory IPs are key building blocks in high-performance microprocessors, discrete graphics, and hardware accelerators, where the impact of timing variations and margin on clock frequency has become increasingly critical due to emerging applications such as machine learning, computer vision, and autonomous driving [4]. This paper presents a hardware test-bench to measure on-chip timing parameters with high resolution for memory IPs (such as setup, hold, clock-to-q and cycle time). Previous published works have shown techniques to measure on-chip timing parameters for sequential elements [5-6]. However, unlike sequential elements, memory IPs have additional challenges for on-chip timing measurements. These challenges include: i) multiple address, write-data, clock, and data-out inputs/outputs, ii) physical location of inputs/outputs are not in close-proximity which adds to measurement error, and (iii) complexity due to multiple input switching permutations. This paper presents a fully configurable synthesizable memory IP timing characterization test-bench fabricated in a leading-edge CMOS node, featuring distributed regional capture FFs (RCFFs) with mesh based low skew clock, main capture FF (MCFF) to measure setup difference across RCFFs, multiple data/input delay generators with high-resolution to handle timing permutations, automated relative placement/pre-routing for matched layout and XORed clock delay generators to create multiple edges for measuring read after write delay/cycle time.

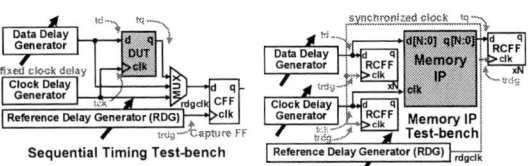

Fig. 1. Memory IP timing characterization methodology.

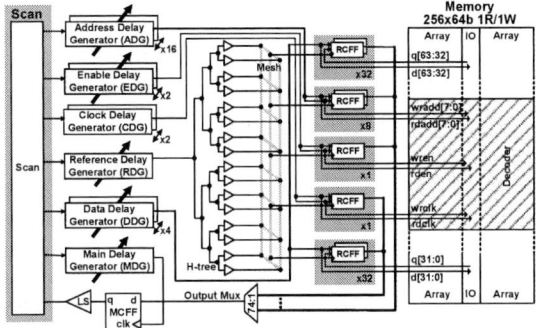

Fig. 2. Memory IP timing characterization circuits with scan, short/medium /long delay generators, clock mesh, RCFFs, MCFF and memory under test.

II. MEMORY IP TIMING CHARACTERIZATION CIRCUITS AND METHODOLOGY

Fig. 1 shows the proposed timing characterization methodology. Since memory IPs have multiple inputs/outputs whose physical locations are not in close-proximity, the capture FF in the sequential test-bench [6] is replicated and placed close to the memory I/O physical location as an RCFF. These RCFFs are driven by a synchronized variable delay clock (rdgclk). Each input to the memory is connected to a variable delay generator and is also captured by its own RCFF. The setup and clk2q of the memory IP is calculated as:

$$t_{rdg,ck} - t_{rdg,d} = (t_{ck} - t_d) + \Delta t_{rdgclk\_skew} + \Delta t_{RCFF\_setup}$$

$$t_{rdg,q} - t_{rdg,ck} = (t_q - t_{ck}) + \Delta t_{rdgclk\_skew} + \Delta t_{RCFF\_setup} \quad (1)$$

Fully synthesizable memory IP timing characterization circuits include a memory IP under test (MIPUT), variable short/medium/long delay generators, multiple RCFFs, output multiplexor, MCFF and scan chain (Fig. 2). The test-bench characterizes a 256x64b 1R/1W ported dynamic register file memory as the MIPUT. The short delay generators (ADG, EDG, CDG and DDG) feed the address, enable, clock and data inputs of the memory IP, while the medium delay generator

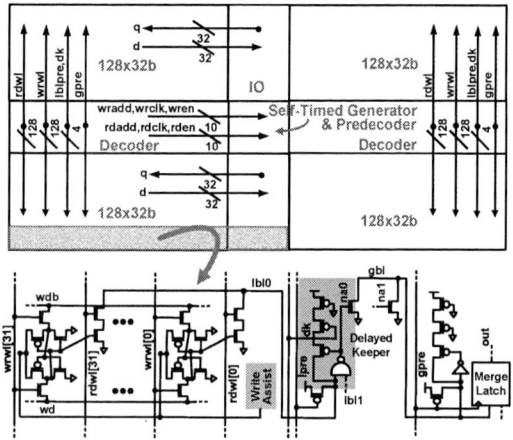

Fig. 3. 1R/1W 256x64b self-timed dynamic register file with write and delayed keeper read assist circuits.

(RDG) drives the clock input of the RCFFs. The long delay generator (MDG) is used as a variable delay clock for the final MCFF, which characterizes the setup time of RCFFs. The setup time differences (Δt_{RCFF_setup}, eq. (1)) across RCFFs due to physical locality and PVT are added/subtracted to compensate for measurement errors while capturing input/outputs of the memory IP.

Fig. 3 shows the organization of the MIPUT consisting of a 256x64b 1-read, 1-write ported register file. The register file employs a self-timed synchronous design, where all inputs are latched on the rising edge of the read/write clocks and outputs are available after the rising edge of the read clock. A fully static one-hot 8b address decoder generates the read/write select wordlines, while clocked NAND gates convert the static decoder outputs into domino-compatible select wordlines. Each dynamic local bitline (LBL) and keeper supports a single-ended read with 32 cells/bitline followed by a two-way merge via static NAND and a two-way NOR global bitline (GBL) with merge latch. The register file implements delayed keeper, which allows the LBL to evaluate before the keeper turns on, for read assist [7] and a write assist for improved V_{MIN}.

Delay generators are designed using delay cells containing 15 mux-ladder-based coarse delay elements and 15 gate-cap-based fine delay elements, selected by 4b thermometer decoders for monotonic delay (Fig. 4). Delay cell is characterized by measuring the ring oscillator frequency across all delay configurations. Saturation bits are computed to eliminate delay overlap between coarse and fine delay settings, enabling a monotonic delay. Measurement at 750mV, 25°C show ~20ps/~0.7ps resolution across coarse/fine delay settings, respectively. 12b parallel reference scan chains provide the triggering edge input to all the delay generators for 12 consecutive testing cycles (scan clock). The outputs of these delay generators provide memory inputs, resulting in 6 memory clock cycles execution capability (Fig. 4).

The clocks to the memory IP are derived using 3 variable short delay generators with a series of XOR gates (CDG). Based on programmable scan chains and variable delays (t_0-t_2)

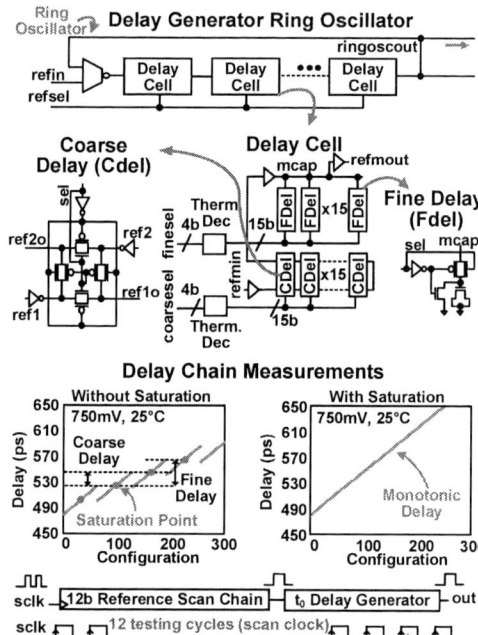

Fig. 4. Delay generator circuits with measurement and reference scan chain for 12 testing and 6 memory execution clock cycles.

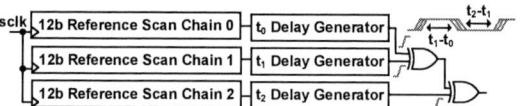

Fig. 5. 3-edge clock/en generation circuit with programmable delay pulse.

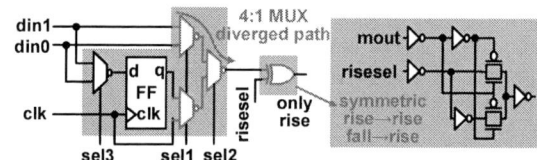

Fig. 6. RCFF with rising select XOR.

the clocks in a given testing cycle can be configured as a single edge or up to 3-edge clock with programmable delay pulse (Fig. 5). A similar scheme is implemented for write and read enable signals (EDG), providing high-resolution setup, cycle time and read-after-write delay measurements.

An RCFF can capture two inputs and contains a flip-flop, inverting symmetric mux tree and rising select XOR (Fig. 6). Mux tree has only a two-gate stage diverged path for each input, minimizing measurement error. Rising select XOR converts falling edges into rising edges with symmetrical delay switching across PVT. This enables only rising edge injection to the output MUX delay path, removing measurement error due to delay propagation and MCFF setup time mismatch between rising/falling edges. RCFF's clock input is connected to a low-skew clock tree mesh, which is driven by symmetrically placed multiple clock mesh buffers. The inputs

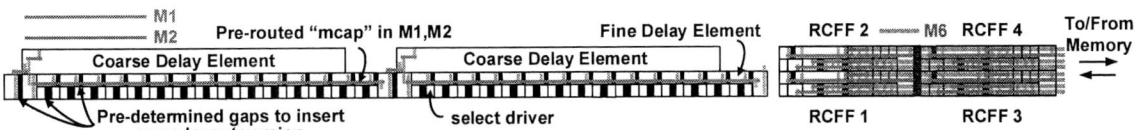

Fig. 7. Relative placement and pre-routing of delay cells and RCFFs to minimize measurement errors due to layout/routing mismatch.

TABLE I. BISECTIONAL ALGORITHM EXAMPLE
(N BITS DG REQUIRE N+2 STEPS)

Cycle	DG Bits (e.g. 5b)	DG Bits (w/ sat*)	CFF out	Next delay
1	11111	11110	1	↓
2	01111	01110	0	↑
3	10111	10110	0	↑
4	11011	11010	1	↓
5	11001	11001	1	↓
6	11000	11000	0	↑
7	11001	11001	1	--

* e.g. 2b fine-delay saturation "10"

of these clock buffers are connected through an H-tree and a fishbone routing is performed to connect the RCFF's clock pin to the clock mesh resulting in sub-ps skew. The RCFF uses this low-skew variable delay clock to capture and characterize setup/hold/clk-q timing of memory IP ($\Delta t_{rdgclk\_skew} \approx 0$, eq. (1)).

The synthesis scripts are fully reconfigurable to accept memory IPs of varying sizes and configurations. Relative placement and custom pre-routing commands are used for the critical circuits to improve layout/routing quality. The physical design scripts are implemented to enable matched layout/routing for all the delay generators and RCFFs independent of the memory IP under test (Fig. 7). RCFFs are structurally placed in columns at the edge of the memory IP based on the input/output port physical location. Pre-determined gaps are inserted between layout cells to infer same layout version across all the instantiations. These semi-automated optimizations along with low skew clock mesh reduce measurement errors due to layout/routing mismatch and physical proximity difference of memory IP I/Os.

Bisection algorithm-based testing is implemented for fast test-time and capability of testing different timing parameters by using IO scan only. This bisectional algorithm implements a binary sweep (Table 1) to calculate timing parameters with n+2 steps instead of a traditional linear sweep with 2^n steps, where n is the number of delay generator configuration bits (e.g. 12b for 5 delay cell based short delay generator).

Fig. 8 shows example steps for measuring clk2q and read address setup time. For rising setup, the measured address bit (rdadd[3]) is switched from 0 to 1, while rest of the address bits are kept constant in a given state. A 0 input data is written to wradd[3] when it is equal to 0, and 1 is written to wradd[3] when it is equal to 1. Following this data is read sequentially from both locations, creating a rising transition on the memory outputs. The address and output transitions are captured following the methodology shown in [6] to sample the setup vs. clk2q timing curve. Setup delay is extracted from the timing curve with a target delay push-out from minimum clk2q delay. This is repeated for all address input combinations for address/data rising and falling pairs.

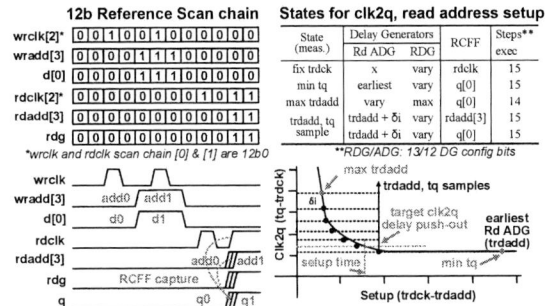

Fig. 8. Clk2q and read address setup time characterization methodology.

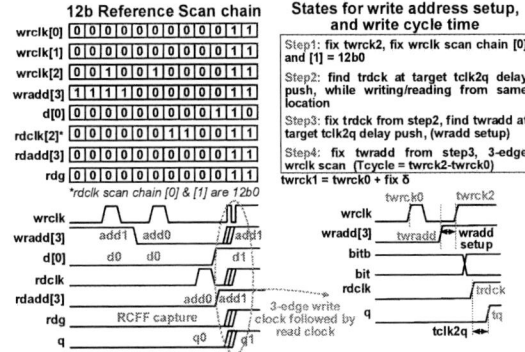

Fig. 9. Write address setup and write-cycle-time characterization methodology.

Fig. 9 shows example steps for measuring write address setup and write-cycle-time. Unlike read address setup, 0 input data is written to both address locations. The test is performed by writing 1 when wradd[3] transitions to 1, while simultaneously reading from the same location. The read clock delay is varied to find the delay setting where read delay after write has a target delay push-out from minimum read clk2q. Read clock is fixed at this delay setting. Write address delay is then increased to sample write setup vs. read clk2q curve, to extract the write setup delay. A write-cycle-time measurement is setup by providing 3-edge write clock, while fixing the read clock and write address delays as in write address setup measurement. Write-cycle-time is extracted from cycle time vs. read clk2q curve by varying the delay between two rising edges of the 3-edge write clock.

III. MEASUREMENT RESULTS

The characterization test-bench was fabricated in a leading-edge CMOS node (Fig. 10). The memory IP under test is characterized for clk2q, address/data setup/hold and cycle time at 0.75V, 25°C. Fig. 11 shows the heatmap for measured clk2q when reading a 1, where the bit read is plotted based on its physical location. Heatmap shows the delay bands of 32

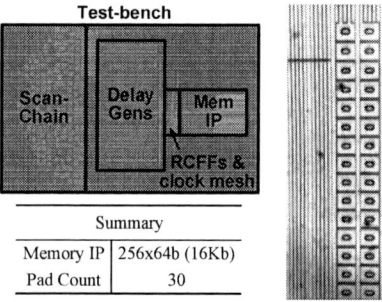

Test-bench

Summary	
Memory IP	256x64b (16Kb)
Pad Count	30

Fig. 10. Memory IP timing test-bench die micrograph and layout in a leading-edge CMOS node.

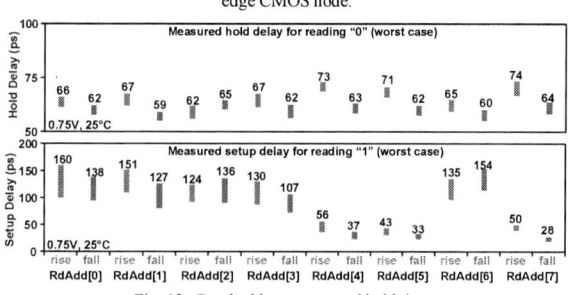

Fig. 12. Read address setup and hold time.

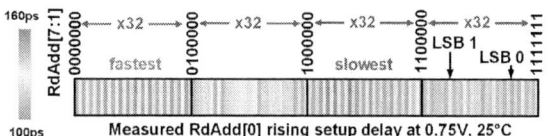

Fig. 14. RdAdd[0] setup delay dependency on other address bits.

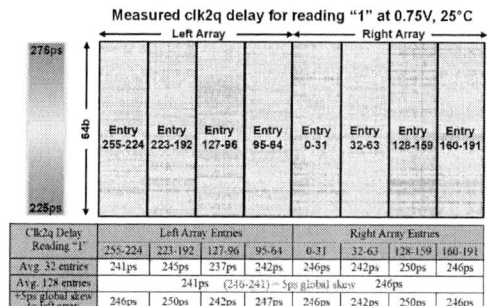

Fig. 11. Reading "1" Clk2q delay across 256x64b.

Clk2q Delay Reading "1"	Left Array Entries				Right Array Entries			
	255-224	223-192	127-96	95-64	0-31	32-63	128-159	160-191
Avg. 32 entries	241ps	245ps	237ps	242ps	246ps	242ps	250ps	246ps
Avg. 128 entries	241ps				(246-241) = 5ps global skew			246ps
+5ps global skew to left array	246ps	250ps	242ps	247ps	246ps	242ps	250ps	246ps

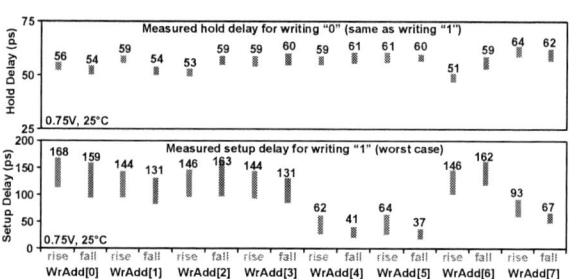

Fig. 13. Write address setup and hold time.

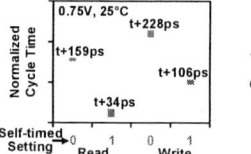

Fig. 15. Measured cycle time.

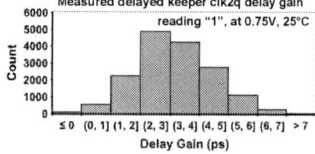

Fig. 16. Delayed keeper clk2q delay gain.

entries, which are associated with common LBL. The physical layout differences in two LBLs merged results in 4ps average delay difference across their 32 entry bundles. There is 5ps average delay global skew between left array and right array due to difference in GBL length. Read/write address rise/fall setup/hold times are measured for a given address bit (Fig. 12, Fig. 13), while keeping the rest of the address bits constant (128 permutations). Heatmap for rdadd[0] setup delay are divided in 4 delay bands, showing the dependency on two MSBs (Fig. 14). Each of those bands have delay strips, further showing dependency on LSB. These measurements can expose delay outliers in silicon and can be used to improve memory IP PPA. Cycle time for two internal memory clock self-timed settings shows 69ps-72ps delay difference between read and write cycle (Fig. 15). Measured clk2q delay with delayed keeper enabled shows ~10ps maximum gain across entire 256x64b register file, while having a ~5ps gain for worst case clk2q delay (Fig. 16), resulting in V_{MIN} improvement.

IV. CONCLUSIONS

A fully configurable and synthesizable timing characterization test-bench for memory IPs clk2q, setup, hold and cycle-time delay measurements is presented. The measurement results demonstrate the high-resolution silicon IP timing characterization capability required by DTCO to optimize PPA as process matures.

ACKNOWLEDGMENT

The authors thank Vida Ilderem for encouragement and discussions.

REFERENCES

[1] S. B. Samavedam et al., "Future Logic Scaling: Towards Atomic Channels and Deconstructed Chips," *IEEE International Electron Devices Meeting (IEDM)*, 2020, pp. 1.1.1-1.1.10.

[2] P. -L. Yang et al., "A 4-GHz universal high-frequency on-chip testing platform for IP validation," *IEEE 32nd VLSI Test Symposium*, 2014, pp. 1-6.

[3] Luo Zhihong, Zhang Yihao and H. Law, "Self-Calibrate two-step digital setup/hold time measurement," *Proceedings of International Symposium on VLSI Design, Automation and Test*, 2010, pp. 232-235.

[4] W. Gomes et al., "Ponte Vecchio: A Multi-Tile 3D Stacked Processor for Exascale Computing," *IEEE International Solid State Circuit Conference (ISSCC)*, 2022, pp. 42-44.

[5] N. Nedovic, W. W. Walker and V. G. Oklobdzija, "A test circuit for measurement of clocked storage element characteristics," *IEEE Journal of Solid-State Circuits*, vol. 39, no. 8, pp. 1294-1304, Aug. 2004.

[6] A. Agarwal et al., "25.7 Time-Borrowing Fast Mux-D Scan Flip-Flop with On-Chip Timing/Power/VMIN Characterization Circuits in 10nm CMOS," *IEEE International Solid State Circuit Conference (ISSCC)*, 2020, pp. 392-394.

[7] A. Agarwal et al., "A 32nm 8.3GHz 64-entry × 32b variation tolerant near-threshold voltage register file," *IEEE Symposium on VLSI Circuits*, 2010, pp. 105-106.

Capacitance-Based Voltage Regulation- and Reference-Free Temperature-to-Digital Converter down to 0.3 V and 2.5 nW for Direct Harvesting

Orazio Aiello[1,2], Massimo Alioto[1]

[1] National University of Singapore, Singapore   [2] Università di Genova, Italy

Abstract— **A temperature-to-digital converter for direct harvesting is proposed, where no DC-DC conversion is required between the DC harvester and the system. Temperature-induced capacitance differences are read out through ring oscillator frequency. PVT variations are suppressed by the differential nature of the temperature sensor architecture, whereas mismatch is compensated via a self-referenced calibration procedure. No reference, regulator, digital post-processing and digital direct temperature readout is needed to retain true-nW and low-V_{min} operation. A 180-nm testchip tested across corner wafers shows 7bit ENOB, 2.5-4.5nW from solar and thermal direct harvesting at 0.3-0.5 V, as representative of a very wide range of environmental conditions.**

Keywords—Temperature sensing, ultra-low-power, harvesting

I. INTRODUCTION

The fast-rising demand for ubiquitous sensor nodes has recently led to a widespread interest in directly-harvested silicon system architectures without any energy storage or intermediate DC-DC conversion between the DC harvester and the system. Many recent silicon demonstrations have indeed shown that direct harvesting drastically reduces cost through the elimination of the battery and silicon area for DC-DC conversion for harvesting. At the same time, directly-harvested systems aggressively shrink the form factor, and extend the system lifespan well beyond the shelf life of batteries (Fig. 1).

As main challenge posed by direct harvesting, the limited and highly-fluctuating harvested power and voltage mandate very low system peak power (rather than average) to avoid harvester overloading, as well as low minimum supply voltage V_{min} to reliably and uninterruptedly operate across a wide range of environmental conditions without DC-DC conversion or regulation (e.g., 0.3-0.6 V in solar cells). Being widely used, temperature-to-digital converters various in the nW power range (or below) have been recently demonstrated (e.g., [1], [2]). In this class of temperature sensors, power is reduced by orders of magnitude at moderate resolutions in the °C and sub-°C range, as opposed to designs with best-in-class performance [3].

Temperature sensors with nW power and below are based on various sensing elements such as MOS transistors [2], [4]-[7], leakage [8], and resistor [9]. Such temperature sensors generally have a rather high V_{min} in the 0.7-0.8 V range or higher, or 0.5-0.6 V in [1], [2]. In addition, their potential reductions in power and V_{min} are compromised by the related penalties associated with the need for extra peripheral circuitry such as voltage regulation [1]-[10], references (e.g., voltage, current, frequency) [1]-[10], digital post-processing for linearization and output re-scaling for direct temperature readout [2], [6], [8], [10]. In this work, a novel temperature-to-digital converter with power down to 2.5 nW and 0.3-V

supply is introduced. Its architecture is self-contained in that no extra regulation, reference or digital post-processing is needed to retain true peak nW operation (rather than average power). The proposed sensor also has direct digital temperature readout and inherent sensing element linearization. Operation under light and thermal harvesting sources is demonstrated to show its wide applicability.

II. PROPOSED NW TEMPERATURE-TO-DIGITAL CONVERTER

In this work, ultra-low voltage and power are achieved by exploiting the depletion capacitance of on-chip pn junctions as a temperature sensing element for the first time, to the best of the authors' knowledge (capacitance with humidity-sensitive dielectric was previously used for humidity sensing [10]). Capacitance readout indeed avoids any DC power and removes V_{min} limitations from the sensing element. As shown in Fig. 1, the rest of the architecture is fully digital and synthesizable. Differential capacitance digitization via ring oscillators suppresses process/voltage variations, and a self-referenced calibration inexpensively compensates mismatch.

A. Proposed Temperature-to-Digital Converter

The temperature sensor architecture is detailed in Fig. 2, where capacitors as sensing elements load two ring oscillators to determine a temperature-dependent oscillation frequency. To maximize the temperature sensitivity and hence the temperature coefficient difference, a junction capacitor C_J and a metal-oxide-metal (MOM) capacitor C_{MOM} are used for differential capacitance sensing. The two ring oscillators RO1 and RO2 respectively drive an up-counter (COUNTER1) and a down-counter (COUNTER2), where the

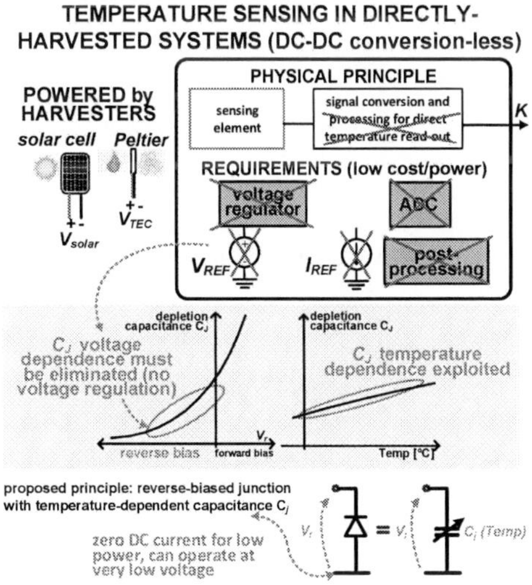

Fig. 1. Capacitor as temperature sensing element, and non-linearity of junction capacitor as challenge to be addressed under direct harvesting (i.e., highly-fluctuating supply voltage). Solutions are discussed later in the paper.

978-1-6654-8495-4/22 $31.00 © 2022 IEEE

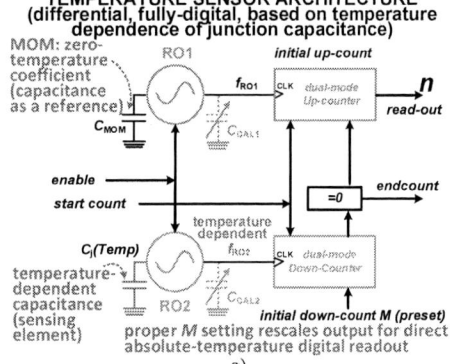

TEMPERATURE SENSOR ARCHITECTURE
(differential, fully-digital, based on temperature dependence of junction capacitance)

a)

SIGNAL CHAIN AND CONVERSIONS

b)

Fig. 2. a) Proposed temperature-to-digital converter architecture, b) derivation of its digital output through an example.

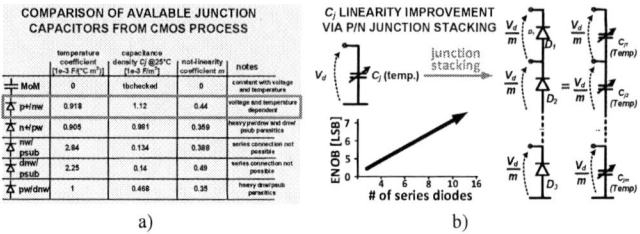

Fig. 3. a) Comparison and selection of junction capacitance as temperature sensing element, b) linearity over unregulated supply voltages improved by stacking pn junctions to reduce the voltage swing across each junction.

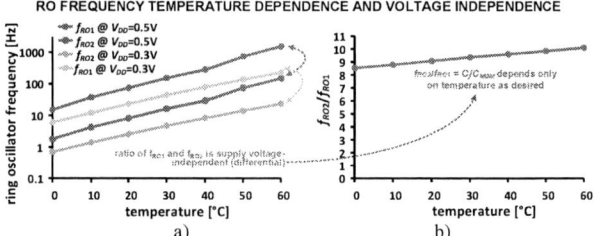

Fig. 4. Independence of f_{RO1}/f_{RO2} from a) unregulated supply fluctuations, b) temperature fluctuations

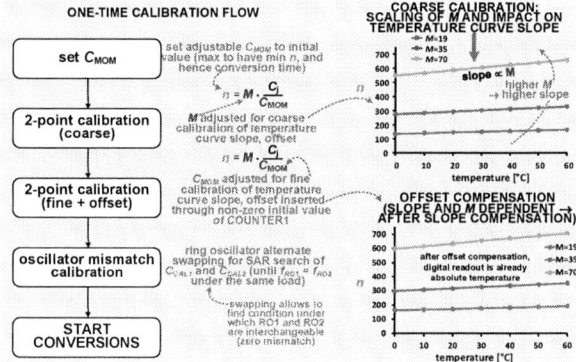

Fig. 5. 2-point calibration flow for absolute temperature direct digital readout.

latter is initialized at a pre-defined value M before starting temperature readout. RO2 drives the enable signal of the RO1 up-counter, defining the time window in which it is counted, which ends when COUNTER2 reaches 0. At the end of the window, RO2 has experienced M oscillation periods, and RO1 has oscillated a number n of times equal to the final count in COUNT1, resulting to $M/f_{RO2}=n\cdot f_{RO1}$ (f_{RO2} and f_{RO1} are the oscillation frequencies driving COUNTER2 and COUNTER1). Neglecting the effect of mismatch (see self-calibration below), the oscillation frequency f_{RO1} (f_{RO2}) is inversely proportional to its load capacitance C_{MOM} (C_j), leading to the condition $n=M\cdot C_j/C_{MOM}$. Accordingly, the final value n of COUNTER1 has the same temperature dependence as C_j, since MOM capacitors have an essentially zero temperature coefficient. As further system knob, larger values of the digital setting M improve the temperature sensitivity and reduce the time quantization error $1/n$ in the counting process, at the cost of longer conversion time.

Regarding the impact of the inherently wide supply voltage fluctuations in directly-harvested systems, the differential nature of the architecture in Fig. 2 suppresses the voltage dependence of transistors, whereas C_j voltage independence needs to be assured to suppress any voltage regulation (i.e., to make the condition $n=M\cdot C_j/C_{MOM}$ unaffected by the supply voltage V_{DD}). This can be achieved through proper choice among the available junction capacitors on chip, and proper arrangement to further improve their linearity over voltages (see Fig. 3), since C_j is alternatively charged to V_{DD} and discharged to ground. Since the non-linearity of the junction capacitance is more pronounced at heavier reverse biasing

(i.e., higher V_{DD}), it can be mitigated by reducing the voltage swing across each C_j across a full charge/discharge cycle. In turn, this can be simply achieved by stacking m reverse-biased junctions as in Fig. 3, which effectively act as a voltage divider and reduce the voltage swing to V_{DD}/m by symmetry. However, the heavy parasitic junctions associated with the deep n-well break such symmetry (Fig. 3), leaving the p+/n-well junction as best option. To keep the non-linearity of C_j within 8 bits across the range 0-60°C, $m=16$ junctions are stacked to form C_j with a resulting temperature coefficient of 0.02 K⁻¹. From the BSIM model card, C_j is set large enough (1.32 pF) to keep linearity over temperatures within the same 8-bit target. To reduce leakage in logic gates, the dual-mode logic style is employed [11]-[12].

B. Proposed Reference-Less Self-Calibration

Fig. 4 shows the measured ratio of the ring oscillator frequencies versus temperature and the harvested supply voltage. From its linear trend versus temperature, a two-point temperature calibration is sufficient to set the slope and the offset of the temperature curve. To this aim, from the flowchart in Fig. 5 the adjustable C_{MOM} is first set to the highest value of 147 fF to initially minimize the conversion time (i.e., lowest $n=M\cdot C_j/C_{MOM}$). The value M is then used as

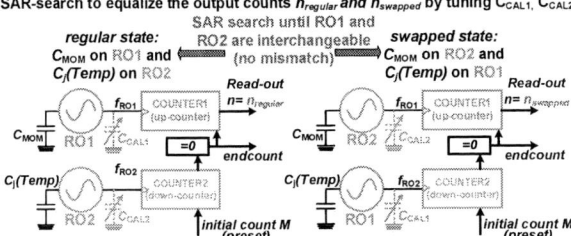

SAR-search to equalize the output counts $n_{regular}$ and $n_{swapped}$ by tuning C_{CAL1}, C_{CAL2}

mismatch suppressed when RO1 and RO2 have same frequency at same load (no need for reference): $n_{regular} = n_{swapped}$

Fig. 6. Proposed ring oscillator reference-less mismatch compensation.

a knob to correct the slope in a coarse fashion, whereas the adjustable C_{MOM} enables fine tuning. The offset is adjusted by using the initial COUNTER1 digital setting. By construction, the slope of $n=M \cdot C_j/C_{MOM}$ after calibration is clearly a power of 2 and allows direct digital readout without intermediate logic. The latter is preserved even if M is then multiplied by powers of 2 to further increase the resolution, if useful (in the following, M is set to achieve LSB=1°C).

To preserve symmetry in Fig. 2 and reject PVT variations without voltage regulation or reference, the residual mismatch between RO1 and RO2 is suppressed through their mutual calibration as in Fig. 6. The binary-scaled calibration capacitors C_{CAL} are fine-tuned, and their incremental changes affect the slope in a negligible manner to avoid further tuning iterations. To avoid any extra reference, RO1 and RO2 are made swappable in terms of both counter and load capacitor (Fig. 6). This allows to mutually calibrate the ring oscillators, so that RO1 has the same oscillation frequency as RO2 when alternatively driving the same load (i.e., swapping their role). Binary SAR search is performed at boot or run time to calibrate binary-scaled capacitors C_{CAL1} and C_{CAL2} in Fig. 6, until the difference between RO1 and RO2 under the same load converges to near-zero (sub-LSB of C_{MOM}, corresponding to 0.2LSB in temperature).

III. MEASUREMENT RESULTS

The resulting 180-nm testchip measurements in Fig. 7 show a 0.58-LSB RMS INL with M set for 8-bit nominal resolution (i.e., without accounting for noise), and an RMS noise of 0.69LSB at V_{DD}=0.5V, as reported in Fig. 7. As consequence of the above C_j linearization, the maximum INL is little sensitive to V_{DD} and is 0.9-1.4 LSB after calibration across (harvesting) voltages from 0.3 V to 0.6 V, with a ~0.3-LSB improvement over pre-calibration. The resulting post-calibration SNDR at 0.5V is 43.9 dB and is within ±3dB across voltages, leading to an ENOB of 7 bits.

From Fig. 8, the peak power consumption at room temperature is 2.5-4.5nW across voltages, thanks to the adoption of a sensing element drawing zero DC current and the simplicity of the architecture. Operating in sub-threshold, the power expectedly increases with temperature similarly to transistor leakage, and hence up to 20.9 nW in TT corner wafer at 60 °C, and 45 nW in FF. The conversion time at room temperature is 3.3 s at TT corner wafer. For completeness, the energy per conversion is 11.5nJ/conv, although power is actually the relevant metric for consumption in directly-harvested systems, since they are peak power-limited (rather than average power-limited). The die micrograph and the layout are shown in Fig. 9. Fig. 10 shows the testing setup including under solar cell and

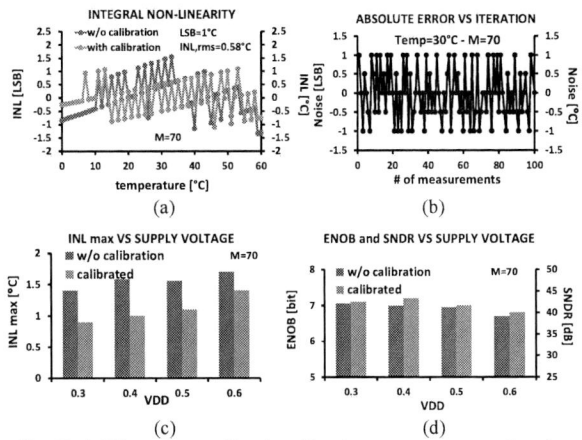

Fig. 7. a) INL pre/post-calibration, b) noise vs. measurement iteration, c) max INL error vs. V_{DD}, d) ENOB and SNDR vs. V_{DD}.

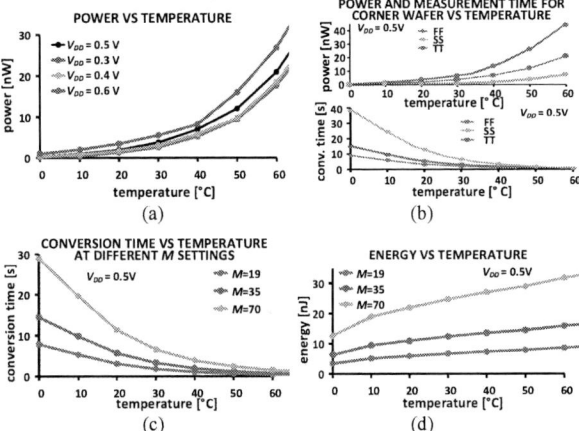

Fig. 8. a) Power vs. temperature at V_{DD}=0.3 V…0.6 V, b) impact of corner wafer on power and conversion time (FF, TT, SS wafers), c) conversion time vs temperature, d) energy/conversion vs. temperature.

thermoelectric generator (Peltier cells) harvester placed around a tea cup.

The above performance was confirmed to be achieved in full when powering the sensor with two different types of harvesters, i.e. a 1-mm² solar cell (down to very dim light of 20 lux) and a Peltier cell (>7.5°C gradient with 20 cells, and 10 cells with doubled-stacked cells). As an example of practical application, the latter was attached to a teacup to power the sensor with the temperature gradient between the tea content and the environment temperature. The circuit was experimentally validated to be able to measure the tea temperature while also being powered by it, and signal when the temperature decreases below a targeted value above the environment temperature for maximum enjoyment.

IV. COMPARISON WITH THE STATE OF THE ART

Compared to prior art (see Table I), the proposed temperature sensor uniquely enables 7-bit conversion down to 0.3 V with no voltage regulation, no recalibration at different voltages, no external reference and no post-processing for true-nW consumption or direct absolute temperature read-out.

The above properties allow direct harvesting for low-cost systems, e.g. from light and thermal sources, thanks to the low V_{min} of 0.3 V (lowest reported to date), and the fourth lowest power. From Table I, the power is higher than [1], [8],

[9] by 3.3-14.4X, whereas it is 2.6-12,200X lower than all others. It is worth noting that this work reports peak power for direct harvesting, whereas prior art reports average power, as direct harvesting was not considered.

The second-scale conversion time is higher than prior art, although still well within the range targeted in a wide range of applications where harvesting is used (e.g., building, asset and infrastructure monitoring). Two-point calibration is used like the vast majority of prior art [1], [4]-[10].

V. CONCLUSION

In this work, a temperature-to-digital converter down to 0.3 V supply and 2.5-nW peak power for directly-harvested systems has been introduced and demonstrated for resource-constrained sensor systems. Its unique ability to suppress any regulation, reference and digital post-processing ancillary circuitry allows to retain true-nW power when integrated in an actual system, while also drastically reducing the system integration effort for low cost targets. Among the other demonstrations, a self-powered tea temperature monitoring for maximum enjoyment was experimentally shown.

 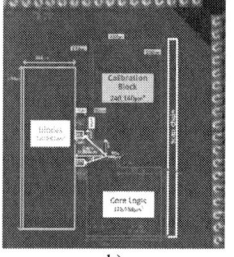

a)   b)

Fig. 9. Temperature-to-Digital Converter a) layout and b) die micrograph.

Fig. 10. Testing setup with solar and thermal harvesting demonstration.

ACKNOWLEDGMENT

The authors wish to thank the support of the Singapore Ministry of Education (MOE2019-T2-2-189 grant) and TSMC for fabrication.

REFERENCES

[1] H. Wang, P. P. Mercier, "A 763 pW 230 pJ/Conversion Fully Integrated CMOS Temperature-to-Digital Converter With +0.81°C/−0.75 °C Inaccuracy," *IEEE Journal of Solid-State Circuits*, vol. 54, no. 8, pp. 2281-2290, Aug. 2019.

[2] J. S. Y. Tan, J. H. Park, J. Li, Y. Dong, K. H. Chan, G. W. Ho, J. Yoo, "A Fully Energy-Autonomous Temperature-to-Time Converter Powered by a Triboelectric Energy Harvester for Biomedical Applications," *IEEE Journal of Solid-State Circuits*, May 2021.

[3] J. A. Angevare, Youngcheol Chae, Kofi A. A. Makinwa "A Highly Digital 2210µm2 Resistor-Based Temperature Sensor with a 1-Point Trimmed Inaccuracy of ±1.3°C (3σ) from -55°C to 125°C in 65nm CMOS," in *IEEE ISSCC Dig. Tech. Papers*, pp. 76-78, 2021.

[4] T. Someya, A. K. M. M. Islam, T. Sakurai, M. Takamiya, "An 11-nW CMOS Temperature-to-Digital Converter Utilizing Sub-Threshold Current at Sub-Thermal Drain Voltage," *IEEE Journal of Solid-State Circuits*, vol. 54, no. 3, pp. 613-622, March 2019.

[5] K. Yang, Q. Dong, W. Jung, Y. Zhang, M. Choi, D. Blaauw, D. Sylvester, "9.2 a 0.6nj -0.22/+0.19c Inaccuracy Temperature Sensor Using Exponential Subthreshold Oscillation Dependence," in *IEEE ISSCC Dig. Tech. Papers*, pp. 160–161, 2017.

[6] T. Someya, A. K. M. M. Islam, T. Sakurai and M. Takamiya, "An 11-nW CMOS Temperature-to-Digital Converter Utilizing Sub-Threshold Current at Sub-Thermal Drain Voltage," *IEEE Journal of Solid-State Circuits*, vol. 54, no. 3, pp. 613-622, March 2019.

[7] T. Someya, A. K. M. M. Islam, K. Okada, "A 6.4 nW 1.7% Relative Inaccuracy CMOS Temperature Sensor Utilizing Sub-Thermal Drain Voltage Stabilization and Frequency-Locked Loop," *IEEE Solid-State Circuits Letters*, vol. 3, pp. 458-461, 2020.

[8] D. S. Truesdell, B. H. Calhoun, "A 640 pW 22 pJ/sample Gate Leakage-Based Digital CMOS Temperature Sensor with 0.25°C Resolution," in *Proc. of CICC*, 2019, pp. 1-4.

[9] H. Xin, M. Andraud, P. Baltus, E. Cantatore, P. Harpe, "A 174 pw–488.3 nw 1 s/s–100 ks/s All-Dynamic Resistive Temperature Sensor with Speed/Resolution/Resistance Adaptability," *IEEE Solid-State Circuits Letters*, vol. 1, no. 3, pp. 70-73, March 2018.

[10] Haowei Jiang, Chih-Cheng Huang, Matthew R Chan, Drew A Hall, "A 2-in-1 Temperature and Humidity Sensor with a Single FLL Wheatstone-Bridge Front-End," *IEEE Journal of Solid-State Circuits*, vol. 55, no. 8, pp. 2174-2185, Aug 2020.

[11] O. Aiello, P. Crovetti, M. Alioto, "Capacitance-to-Digital Converter for Operation Under Uncertain Harvested Voltage down to 0.3V with No Trimming, Reference and Voltage Regulation," in *IEEE ISSCC Dig. Tech. Papers*, Feb. 2021.

[12] L. Lin, S. Jain, M. Alioto, "A 595pW 14pJ/cycle Microcontroller with Dual-Mode Standard Cells and Self-startup for Battery-Indifferent Distributed Sensing," in *IEEE ISSCC Dig. Tech. Papers*, Feb. 2018.

TABLE I. PERFORMANCE SUMMARY AND COMPARISON WITH THE STATE OF THE ART OF TEMPERATURE SENSORS

	This work	JSSC'19 [1]	SSCC'21 [2]	ISSCC'21 [3]	JSSC'19 [4]	ISSCC'17 [5]	JSSC'20[6]	SSC-L'20[7]	CICC'19[8]	SSC-L'18[9]	JSSC'20[10]
sensor type	depletion capacitance	3-transistor PTAT	MOS (Time-based)	resistor, highly-digital	MOSFET Sub Ut	MOSFET	MOSFET	Stabilized Sub Ut	Gate Leakage	Resistive	RC
technology [nm]	180	65	180	65	180	180	55	65	65	65	180
temp. range [°C]	0 - 60	0 - 100	-15 - 45	-55 - 125	-20 - 80	-20 - 100	-40 - 125	-30 - 70	-20 - 100	0 - 100	-40 – 85 °C
energy source	Light, thermal	Ext. power supply	Triboelectric Harvester	Ext. power supply	Ext. power supply	Ext. power supply	Ext. power supply	Ext. power supply	Ext. power supply	Ext. power supply	Ext. power supply
power @ room temperature [nW]	2.5 - 4.5 @0.3 - 0.6V	0.763	17	28,000	11	75	9,300	6.4	0.64	0.174	15,600
supply voltage [V]	**0.3**	0.5	0.6	0.9	0.8	1.2	0.8	0.8	0.9/1.2	1/0.65	1.5
# supply voltages	1	1	1	1	1	1	1	1	2	2	1
voltage range [V]	0.3 - 0.6	0.5 - 1	0.6 - 1	N/A	0.7 - 1.5	N/A	0.8 - 1.3	0.75 - 1.05	0.9 - 1.2	N/A	1.5 - 2
conversion time @ 25 °C [ms]	3,300	300	0.014	1	839	8	1.31	N/A	34.3	0.01	1
inaccuracy [°C]	1.75 @ 0°C	-1.53 - 1.61	-1 - +1	±1.3 (3σ)	-0.9 - 1.2	-0.22 - 0.19	±0.7 (3σ)	-1 - 0.7	-2.7 - 1.8	-1.1 - 1.5	0.55 (3σ)
resolution [°C]	1	0.3	-1 - 1	0.0128 (rms)	0.145	0.016	0.016	0.075	0.25	0.61	0.002
area [um²]	488,000	630,000	54,000	2,210	74,000	8,865	1,770	320,000	13,000	60,000	720,000
voltage regulation-less	**YES**	NO	NO	NO	NO	NO	NO	NO	NO	NO	NO
reference-less	**YES**	NO	NO	NO	NO	NO	NO	NO	NO	NO	NO
external digital-free	**YES**	**YES**	NO	**YES**	**YES**	**YES**	NO	**YES**	NO	**YES**	NO
calibration	2-point	2-point	1-point	1-point	2-point	2-point	2-point	2-point	2-point	2-point	2-point

A 24 GHz Quadrature VCO Based on Coupled PLL with -134 dBc/Hz Phase Noise at 10 MHz Offset in 28 nm CMOS

Agata Iesurum*, Davide Manente*, Fabio Padovan†, Matteo Bassi†, Andrea Bevilacqua*

*University of Padova, Italy; †Infineon Technologies, Villach, Austria

Abstract—This work breaks the trade-off between quadrature error and phase noise, typical of quadrature voltage-controlled oscillators (QVCOs), by leveraging a coupled PLL approach. Prototypes implemented in a 28 nm bulk CMOS technology show 0.9° quadrature error, and -134 dBc/Hz phase noise at 10 MHz offset from the 24 GHz carrier. The measured tuning range spans from 24 to 29.2 GHz, while the power consumption is 60 mW.

Index Terms—Quadrature voltage-controlled oscillator, CMOS, RFIC, coupled phase-locked loop (CPLL).

I. INTRODUCTION

Quadrature local oscillator signals are ubiquitous in wireless transceivers to realize carrier modulation/demodulation. Typically, they are obtained by frequency division of a local oscillator at twice the desired carrier frequency. At millimeter-wave (mmWave), such an approach leads to a high power consumption, such that alternatives are pursued. Quadrature signals may be generated using lossy passive filters, such as polyphase filters or hybrid couplers, or by forcing two voltage-controlled oscillators (VCOs) to operate in quadrature. The most popular way of attaining the latter is to appropriately couple two VCOs. Several coupling methods have been explored: parallel-coupling [1], superharmonic coupling [2], [3], in-phase injection coupling [4], and magnetic coupling [5]. All these approaches suffer from a common drawback: they show a trade-off between quadrature error and phase noise performance.

In this work, a coupled PLL architecture is leveraged to to overcome the phase noise-quadrature error trade-off. Two voltage controlled oscillators are closed in a phase-locked loop where each one is the reference for the other, making up a coupled PLL (CPLL) system (Fig. 1). A mixer is used as phase frequency detector, such that the oscillators are forced to operate in quadrature. The quadrature error is set by the parameters of the control loop, while the phase noise improves by 3 dB with respect to the noise of each VCO, similarly to a multi-core oscillator [6]–[8]. The CPLL-based QVCO architecture was first introduced in [8], while it is validated here in silicon for the first time.

II. DESIGN OF THE CPLL-BASED QVCO

Figure 1 shows a block diagram of the proposed QVCO, based on a CPLL architecture. A mixer extracts the phase difference of oscillators VCOI and VCOQ. The error signal is filtered, amplified and fed back to control the frequency

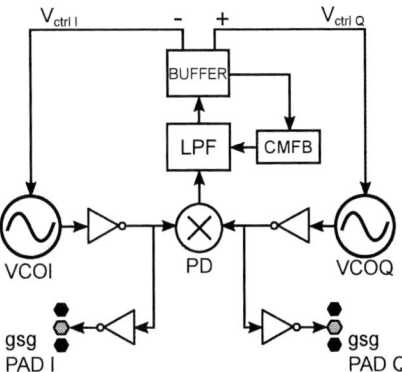

Fig. 1: Block diagram of the proposed CPLL-based QVCO.

of the two oscillators. The control is differential: when the frequency of, say, VCOI is increased, the frequency of VCOQ is correspondingly decreased. The use of a mixer as phase frequency detector implies that, at steady state, VCOI and VCOQ operate in quadrature and at the same frequency [8].

In case the resonance frequencies of the two VCO tanks differ, a departure from quadrature arises. Such an error is determined by the parameters of the control loop: it is inversely proportional to the product of the gain of the phase frequency detector, the gain of the filter, and the frequency sensitivity of the oscillator (K_{VCO}) [8]. Within the CPLL bandwidth, the phase noise of the system is reduced by 3 dB with respect to the noise of each oscillator [7], [8]. Such a behavior is similar to that of an array of coupled oscillators (i.e., a multi-core oscillator), and in stark contrast with conventional QVCOs, where quadrature coupling leads to an increase of the phase noise of each oscillator, and the coupling network further degrades the noise performance significantly [1]. Moreover, while in conventional QVCOs a stronger coupling results in a more precise quadrature sequence, but also in higher phase noise, the proposed CPLL approach allows to break the trade-off: the phase noise is dictated by the VCO performance, while the quadrature accuracy by the loop parameters. Each metric can thus be concurrently optimized.

The control loop in Fig. 1 is, as discussed, differential. Therefore, any common variation of the frequency of VCOI and VCOQ is unnoticed by the loop. This opens the door to the possibility of controlling the frequency of operation of the

978-1-6654-8495-4/22 $31.00 © 2022 IEEE

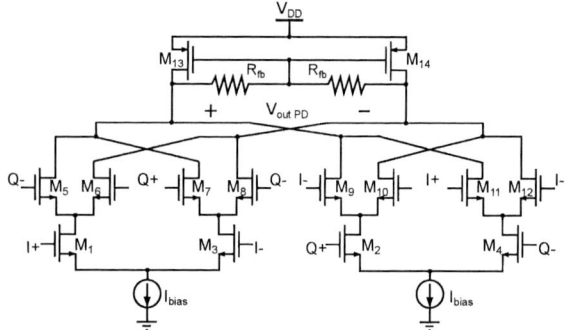

Fig. 2: Simplified schematic of the phase frequency detector.

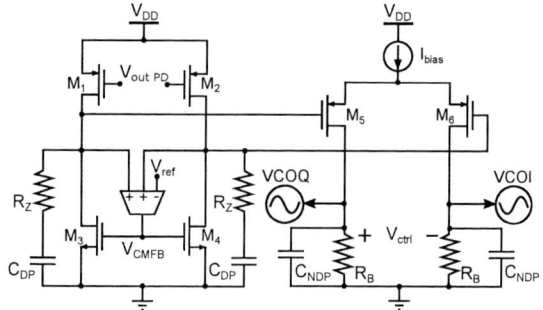

Fig. 3: Schematic of the loop filter (left) and buffer (right).

proposed QVCO by embedding it in a (common-mode) global PLL, which operates conventionally on the QVCO. At the same time, this behavior implies that, while any differential-mode noise due to the mixer and loop filter (referred to the mixer input) gets attenuated by 6 dB within the CPLL band [8], the common-mode noise generated by the control circuit (referred to the VCOs tuning knobs) is accumulated into phase noise. The design of the loop control circuits is hence a delicate task, which is discussed next.

A. Control Loop Design

The control loop of the proposed CPLL is based on a type-II PLL with a leaky integrator [8]. It is designed to correct for a mismatch of the resonance frequency of the tanks of VCOI and VCOQ up to 1%, while keeping the quadrature error below $1.5°$. The bandwidth of the loop is set to 260 MHz.

The multiplier-based phase frequency detector is implemented combining two double-balanced Gilbert cells, as illustrated in Fig. 2, to offer the same load to VCOI and VCOQ. To ensure hard switching of transistors $M_1 - M_{12}$, the devices are driven by inverter chains, which operate as LO buffers for VCOI and VCOQ, and make the phase frequency detector insensible to variations of the oscillation amplitude. Transistors $M_1 - M_{12}$ are AC-coupled to the LO buffers and biased to guarantee that the transistors of the tail generators ($I_{bias} = 750\,\mu A$ in Fig. 2) operate in saturation. Particular care was taken in the layout of the phase frequency detector to minimize any mismatch, since the offset introduced by the mixer is not corrected by the loop and translates into a systematic quadrature error.

The loop filter is made of a single-stage pseudo-differential OTA with a series R-C load (Fig. 3). The relatively large load capacitor C_{DP} = 25 pF sets the bandwidth of the control loop. The resistance $R_Z = 200\,\Omega$ introduces a zero in the loop transfer function, necessary to guarantee a $60°$ phase margin, and hence the stability of the loop. The loop filter draws $250\,\mu A$ per branch. The loop filter common-mode output voltage is set by a common-mode feedback loop (CMFB in Fig. 1), based on an auxiliary OTA, as shown in Fig. 3.

The loop filter output is buffered by a fully-differential amplifier with resistive load, as depicted in Fig. 3. The buffer

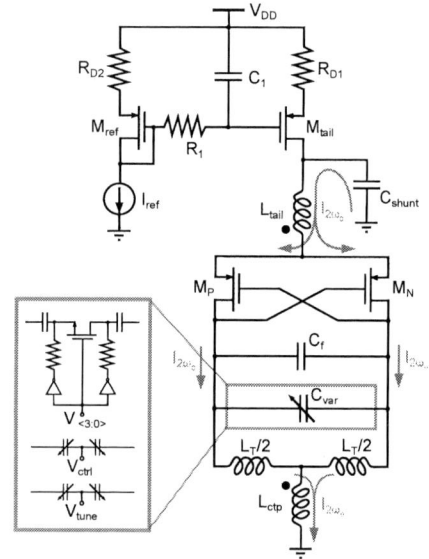

Fig. 4: Schematic of the proposed class-B VCO.

has the task of rejecting any common-mode noise arising from the mixer and the filter, which, as discussed, could impair the noise performance of the CPLL-based QVCO. The buffer draws 1.7 mA from the supply. To increase the common-mode rejection, the tail generator of the buffer is implemented with thick oxide transistors with long channel lengths, that display higher output resistance. The buffer load capacitance C_{NDP} = 540 fF filters any higher frequency noise generated in the control chain.

B. VCO Design

VCOI and VCOQ are identical. They are based on a class-B topology with tail current source, as shown in Fig. 4. The cross-coupled differential pair uses pMOS devices, to take advantage of the lower 1/f noise with respect to their nMOS counterparts.

The 70 pH tank inductor is implemented as a single-turn coil using the thickest metal of the metal stack; the trace width is 12 μm. Electro-magnetic simulations of the coil yield a quality factor of 29 at 26 GHz. The capacitance of the

978-1-6654-8495-4/22 $31.00 © 2022 IEEE   386

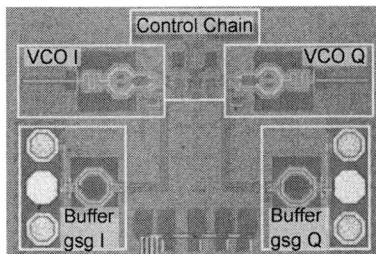

Fig. 5: Microphotograph of the realized testchip.

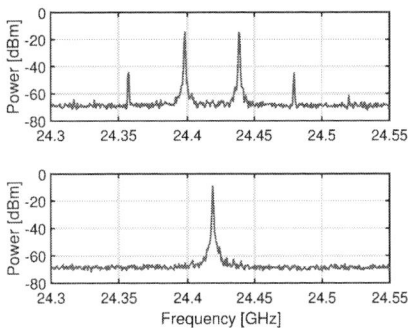

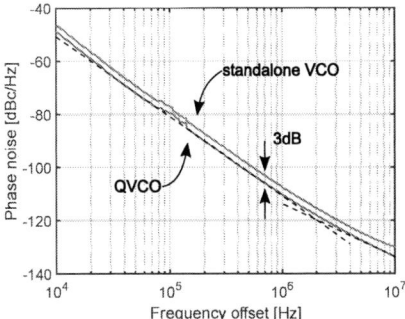

Fig. 6: Measured output spectrum of VCOI and VCOQ with the control loop off (top) and on (bottom).

tank is split into 4 parts: a 4-bit digitally-controlled capacitive bank, a fixed 86 fF capacitor, and two varactors. The losses of the tank capacitor reduce the overall resonator quality factor to approximately 12. The two varactors are identical. One varactor is driven by the CPLL control signal (V_{ctrl} in Fig. 4) to force VCOI and VCOQ operate in quadrature, as discussed. The other varactor provides continuos frequency tuning to the VCO. A global PLL would identically drive the voltage V_{tune} of both VCOI and VCOQ to set the oscillation frequency of the system.

To limit the 1/f noise upconversion, the tank common-mode resonance is set to the second harmonic [9]. To this aim, the second harmonic current loop is carefully modeled. Explicit structures L_{tail} and L_{ctp} are realized in the interconnections between the tail generator and the cross-coupled pair, and between the coil center tap and ground (Fig. 4). L_{tail} and L_{ctp} are magnetically coupled to increase the overall inductance. Capacitor C_{shunt} provides low impedance to close the second harmonic current loop. The tail generator is made of thick oxide pMOS devices with long channels to reduce the generated 1/f noise. Small degeneration resistances help to further reduce the low-frequency noise from the tail devices.

III. Measurement Results

Prototypes of the proposed CPLL-based QVCO have been fabricated in a 28 nm bulk CMOS technology. The microphotograph of the test chip is shown in Fig. 5. The QVCO core occupies approximately 0.2 mm². The QVCO operates from 24 to 29.2 GHz, showing a tuning range as wide as 20%. The QVCO draws 67 mA from the 0.9 V supply.

The VCOI and VCOQ output signals are sensed by differential buffers that drive relatively long, identical transmission lines. Amplifiers loaded by baluns perform differential to single-ended conversion and drive GSG pads (Fig. 5). This arrangement allows for direct measurement of the quadrature error by chip probing.

The spectrum of VCOI and VCOQ is first evaluated with the mixer (and hence the CPLL control loop) off. The result, illustrated in Fig. 6 (top), shows that the two oscillators have a frequency mismatch of about 40 MHz. When the control loop is enabled (Fig. 6 (bottom)), the frequency mismatch is corrected, and the oscillators are forced to operate at the same frequency, as expected.

Fig. 7: Measured phase noise sideband for QVCO (blue curve) and standalone VCO (red curve) for a 24 GHz carrier.

Figure 7 shows the measured phase noise sideband at 24 GHz. Each oscillator is biased with a 18.4 mA tail current. The measured QVCO phase noise (blue curve in Fig. 7) is -110.5 dBc/Hz at 1 MHz offset, and -134 dBc/Hz at 10 MHz offset. The $1/f^3$ corner frequency is approximately 2 MHz. The measured $1/f^3$ sideband is significantly higher than simulated, possibly due to a mistuning of the common-mode second-harmonic resonance, while the measured noise in the thermal region is as expected.

The measurement of the phase noise of a standalone VCO (with all the other blocks turned off) is also reported in Fig. 7 for comparison (red curve). It is 3 dB higher than the noise of the CPLL-based QVCO, as anticipated by the theory.

The measured phase noise of the QVCO at 1 and 10 MHz offsets is reported in Fig. 8 across the tuning range. The increase of the phase noise at 1 MHz offset with frequency is larger than simulated, corroborating the hypothesis that the second harmonic resonance frequency is lower than designed.

Continuous tuning is achieved through V_{tune}, as described in Section II-B. The oscillation frequency of the QVCO versus V_{tune} is plotted in Fig. 9. Eight sub-bands at the two extremes of the tuning range are shown. There is a considerable overlap between the sub-bands, making the presented QVCO suitable for being integrated in a frequency syntheziser.

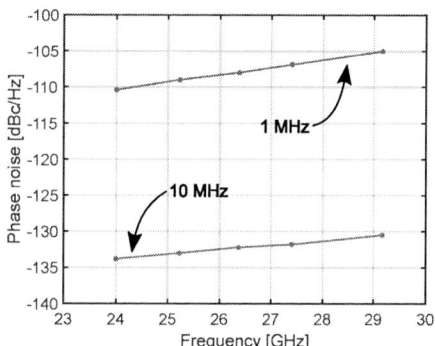

Fig. 8: Measured QVCO phase noise vs. frequency.

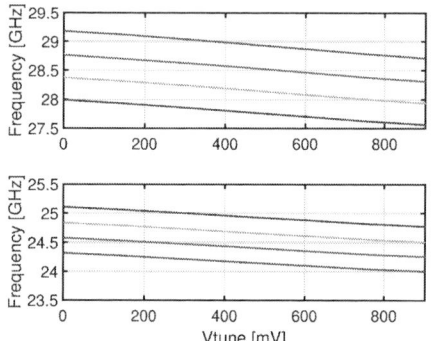

Fig. 9: Measured oscillation frequency as a function of V_{tune} for few sub-bands at the two extremes of the tuning range.

To verify the quadrature accuracy, the output signal of the two VCOs are probed and measured with a VNA. In Fig. 10, the measured quadrature error is plotted versus the oscillation frequency. The average quadrature error is $0.9°$.

The performance of the CPLL-based QVCO is summarized and compared to state-of-the-art mmWave CMOS QVCOs in Table I. The FoM of the proposed QVCO, including the

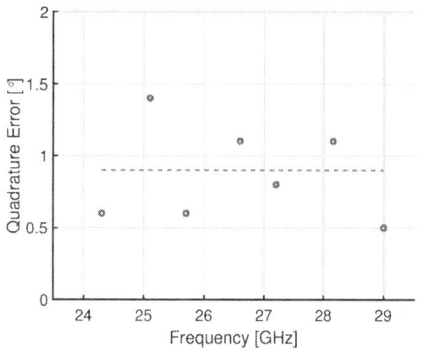

Fig. 10: Measured quadrature error vs. oscillation frequency.

TABLE I: Performance summary and comparison with the state-of-the-art.

	This work	[3]	[4]	[2]	[5]
CMOS technology	**28 nm**	28 nm	65 nm	40 nm	65 nm
V_{DD} [V]	**0.9**	0.75	1	0.9	1
Power [mW]	**60**	8.4	11.8	14	22
Frequency [GHz]	**24**	41	26	58.2	58.6
Tuning range [%]	**20**	18.4	15.4	16	4.35
PN @1MHz [dBc/Hz]	**-110.5**	-94.3	-99	-93.5 [a]	-95
PN @10MHz [dBc/Hz]	**-134**	N/A	-119 [a]	-117 [a]	-117
Eq. PN @1MHz ref. to 24 GHz [dBc/Hz]	**-110.5**	-99	-100	-101.5	-103
Eq. PN @10MHz ref. to 24 GHz [dBc/Hz]	**-134**	N/A	-120	-125	-125
FoM [dBc/Hz]	**-184**	-177	-176	-177.4	-179
Quadrature error [°]	**0.9**	0.18	0.36	5	< 1.5
Core area [mm^2]	**0.2**	0.068	0.024	N/A	0.075

[a] Graphically estimated

power consumption of the control chain, is -184 dBc/Hz. In comparison with the literature, the presented QVCO shows a much lower phase noise and a competitive FoM, while guaranteeing a small quadrature error.

IV. CONCLUSION

The use of a CPLL architecture allows to design a QVCO in which both quadrature accuracy and phase noise can be indipendently optimized. Implemented in a bulk 28 nm CMOS technology, prototypes operating from 24 to 29.2 GHz show $0.9°$ quadrature accuracy and, at the same time, a remarkable phase noise performance of -134 dBc/Hz at 10 MHz offset, while consuming 60 mW.

REFERENCES

[1] P. Andreani, A. Bonfanti, L. Romano, and C. Samori, "Analysis and design of a 1.8-GHz CMOS LC quadrature VCO," *IEEE Journal of Solid-State Circuits*, vol. 37, no. 12, pp. 1737–1747, 2002.

[2] V. Szortyka, Q. Shi, K. Raczkowski, B. Parvais, M. Kuijk, and P. Wambacq, "A 42 mw 200 fs-jitter 60 GHz sub-sampling PLL in 40 nm CMOS," *IEEE Journal of Solid-State Circuits*, vol. 50, no. 9, pp. 2025–2036, 2015.

[3] L. Zhang, N.-C. Kuo, and A. M. Niknejad, "A 37.5–45 GHz superharmonic-coupled QVCO with tunable phase accuracy in 28 nm CMOS," *IEEE Journal of Solid-State Circuits*, vol. 54, no. 10, pp. 2754–2764, 2019.

[4] N.-C. Kuo, J.-C. Chien, and A. M. Niknejad, "Design and analysis on bidirectionally and passively coupled QVCO with nonlinear coupler," *IEEE Transactions on Microwave Theory and Techniques*, vol. 63, no. 4, pp. 1130–1141, 2015.

[5] U. Decanis, A. Ghilioni, E. Monaco, A. Mazzanti, and F. Svelto, "A low-noise quadrature VCO based on magnetically coupled resonators and a wideband frequency divider at millimeter waves," *IEEE Journal of Solid-State Circuits*, vol. 46, no. 12, pp. 2943–2955, 2011.

[6] J. Buckwalter, T. Heath, and R. York, "Synchronization design of a coupled phase-locked loop," *IEEE Transactions on Microwave Theory and Techniques*, vol. 51, no. 3, pp. 952–960, 2003.

[7] S. Karman, F. Tesolin, S. Levantino, and C. Samori, "A novel topology of coupled phase-locked loops," *IEEE Transactions on Circuits and Systems I: Regular Papers*, vol. 68, no. 3, pp. 989–997, 2021.

[8] A. Bevilacqua, C. Sandner, A. Gerosa, and A. Neviani, "Quadrature VCOs based on coupled plls," in *2007 IEEE International Symposium on Circuits and Systems*, 2007, pp. 2140–2143.

[9] D. Murphy, H. Darabi, and H. Wu, "Implicit common-mode resonance in LC oscillators," *IEEE Journal of Solid-State Circuits*, vol. 52, no. 3, pp. 812–821, 2017.

A K-band Gilbert-Cell Frequency Doubler with Self-Adjusted 25% LO Duty-Cycle in SiGe BiCMOS Technology

Lorenzo Piotto [#], Guglielmo De Filippi [#] Daniele Dal Maistro [+], Simone Erba [+] and Andrea Mazzanti [#]

[#] Univeristy of Pavia, Italy

[+] Infineon Technologies AG, Villach, Austria

Abstract—**This paper presents a novel frequency doubler that further enhances the superior performance of solutions based on the Gilbert-cell mixer. A novel scheme is proposed to operate the cell with a 25 % LO duty-cycle. This technique boosts the conversion gain by generating a square-wave like output current. Moreover, the use of a quadrature generation block, commonly adopted in mixer-based frequency doublers, is not required, thus improving the operation bandwidth. The duty-cycle is automatically regulated by a low-frequency feedback loop which ensures optimal operation against input power and PVT variations. The performance of a test chip in a SiGe-BiCMOS process is presented. With a low supply voltage of 1.5 V, the chip achieves 6 dB conversion gain, 5.7 dBm peak P_{sat} and 17 % power efficiency at 20 GHz. The doubler delivers $P_{sat} > 3$ dBm over more than one octave bandwidth. Experimental results compare favorably against previously reported frequency doublers in the same frequency range.**

Index Terms—**Gilbert cell, Frequency doubler, push-push, BiCMOS, LO generation.**

I. INTRODUCTION

Frequency synthesis is a key function in RF and mm-Waves wireless transceivers. The voltage-controlled oscillator (VCO) is the core building block, but frequency multipliers are often included to extend the frequency coverage or to improve the overall synthesizer performance by designing the VCO at a lower frequency, where it may provide enhanced spectral purity and/or power efficiency.

Frequency doublers (FD) (and in general even-order multipliers realized by cascading doublers) are commonly preferred over odd-order multipliers because of their intrinsic performance advantage. Doublers can be straightforwardly implemented with the well-known push-push topology, i.e. a pair of transistors driven in class-B by anti-phase signals, which work as a full-wave rectifier. Albeit very simple, this solution has several shortcomings. First, the conversion efficiency is limited. Second, it needs a differential input while delivering a single-ended output, and it suffers from poor rejection of the fundamental frequency component, being it sensitive to any input common-mode arising from amplitude and phase unbalances on the (ideally balanced) driving signals. Different approaches are introduced to mitigate these drawbacks, as the use of injection locking to boost efficiency, at the expense of bandwidth [1]–[3], and balanced versions to increase input

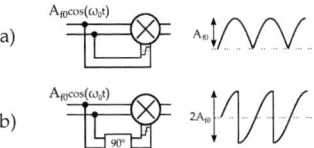

Fig. 1. Mixer based frequency doublers and the characteristic output waveforms.

common-mode rejection, in this case by either relying on complex baluns [4] or hybrid couplers [5], thus again limiting the operation bandwidth, or using balancing branches, affecting the conversion gain [6].

Frequency doublers can also be realized with analog multipliers. Shown in Fig. 1a, a Gilbert-cell mixer driven by the same sinusoidal signal at the input and LO ports provides a full-wave rectified output, with a fundamental component at twice the input frequency. Compared to the simpler push-push rectifier, a double-balanced Gilbert cell operates with fully balanced input and output signals, thus it is ideally insensitive to the input common-mode and yields a remarkable improvement on the rejection of the input-frequency and, in general, odd harmonics. On the other hand, the 2nd harmonic conversion efficiency, defined as the 2nd order Fourier coefficient normalized to the peak input amplitude A_{f0}, is still relatively low ($a_2 = 0.42$) and a strong DC offset is present on the output signal, which might also compromise performances if the load is not a short circuit at DC. If the input and LO ports of the mixer are driven by quadrature signals, as shown in Fig. 1b, the DC offset is removed, and the conversion efficiency is doubled ($a_2 = 0.85$). On the other hand, quadrature generation needs polyphase filters [7] or hybrid couplers [8] and in both cases it involves losses, area occupation, and again limits the operation bandwidth.

This paper presents a novel approach to implement a Gilbert-cell based FD. A low-frequency feedback loop senses the output DC component and introduces an offset to the LO signals that forces the mixer to operate with a ~25 % duty-cycle LO. This technique removes the output DC offset and boosts the harmonic conversion gain without the need for quadrature generation, thus avoiding a lossy, bulky and substantial bandwidth limiting block.

978-1-6654-8495-4/22 $31.00 © 2022 IEEE

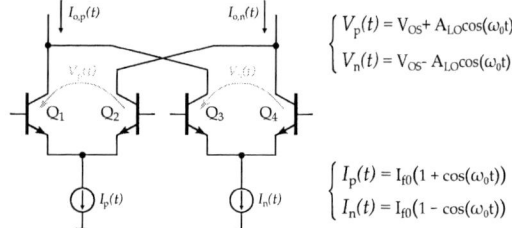

$$\begin{cases} V_p(t) = V_{OS} + A_{LO}\cos(\omega_0 t) \\ V_n(t) = V_{OS} - A_{LO}\cos(\omega_0 t) \end{cases}$$

$$\begin{cases} I_p(t) = I_{f0}(1 + \cos(\omega_0 t)) \\ I_n(t) = I_{f0}(1 - \cos(\omega_0 t)) \end{cases}$$

Fig. 2. Frequency doubler based on a Gilbert-cell mixer.

II. OPERATIONAL PRINCIPLE

The schematic of a double balanced Gilbert-cell mixer is shown in Fig. 2. The current sources $I_{p/n}(t)$ represent the tail transistors, operating in class-A with equal DC bias and signal current, I_{f0}. The switching quad transistors Q_{1-2}-Q_{3-4} are instead considered as current switches, sensitive to the sign of the LO-port voltages. The latter are assumed with a sinusoidal component of differential amplitude A_{LO} and a common-mode DC offset V_{OS} which regulates the duty-cycle, δ, of switching transistors. With the assumptions above, the differential output current $I_{o,d}(t) = \frac{1}{2}(I_{o,p}(t) - I_{o,n}(t))$ is:

$$I_{o,d} = \frac{1}{2}(I_p \cdot \text{sgn}(V_p) + I_n \cdot \text{sgn}(V_n)) \qquad (1)$$

where sgn(x) is the well-known sign function.
If $V_{OS} = 0$, the switching-quad duty-cycle is $\delta = 50\%$ and the resulting characteristic waveforms are plotted in Fig. 3a. In this situation, the Gilbert cell operates as a full-wave rectifier, with an output current $I_{o,d}(t) = \frac{1}{2}|I_p - I_n|$. The fundamental component of the output current is at twice the input frequency and with amplitude $I_2 = a_2 \cdot I_{f0}$, thus the second second-harmonic conversion gain results:

$$a_2 = \frac{4}{3\pi} \sim 0.42 \qquad (2)$$

The situation is significantly improved by regulating V_{OS} to reduce the switching-quad duty-cycle to $\delta = 25\%$, as shown by the waveforms in Fig. 3b. In this case, the output current assumes a quasi square-wave shape at twice the input frequency. During the half-period T_1, $I_{o,d} = \frac{1}{2}(+I_p - I_n) = I_{f0} \cdot \cos(\omega_0 t)$ while during T_2, $I_{o,d} = \frac{1}{2}(-I_p - I_n) = -I_{f0}$. The fundamental component of the output current is again at twice the input frequency and, approximating the waveform as a square-wave, the conversion gain is:

$$a_2 = \frac{4}{\pi} \sim 1.27 \qquad (3)$$

In this case, the second-harmonic current conversion gain is increased by 3 times (9.5 dB) over the case with 50% duty-cycle previously considered, and yet by a remarkable 1.5x (3.5 dB) over the implementation in Fig. 1b where the input and LO ports are driven by signals in quadrature phase. The offset voltage on the LO port which gives $\delta = 25\%$ switching duty-cycle can be easily calculated as:

$$V_{OS} = A_{LO}\cos(\pi\delta) = \frac{A_{LO}}{\sqrt{2}} \qquad (4)$$

showing that V_{OS} cannot be pre-set, being it dependent on the LO signal amplitude. A possible solution could be to

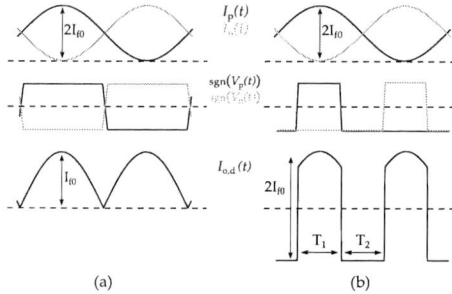

Fig. 3. Waveforms with (a) $\delta = 50\%$, (b) $\delta = 25\%$.

generate the required offset by measuring the LO swing with an envelope detector. However, this would be sub-optimal being sensitive to component spreads in the detector and not accounting for non idealities, such as the actual operation of the switching quad at limited LO amplitudes. A more robust and effective solution is suggested by noticing in Fig. 3 that, opposite to the case of $\delta = 50\%$, $\delta = 25\%$ leads to a quasi symmetric $I_{o,d}(t)$, thus with a low (almost zero) DC component. Being the mean value of the output current relatively easy to measure, a low-frequency feedback loop can be implemented to automatically adjust V_{OS} to the LO swing, in order to null the DC component of $I_{o,d}$. This technique leads to almost optimal performance, robust with respect to the actual LO amplitude. To gain further insight, Fig. 4 plots the DC and second-harmonic components of $I_{o,d}$, normalized to the input current ($a_0 = \text{mean}(I_{o,d})/I_{f0}$, $a_2 = I_{2f0}/I_{f0}$), derived from simulations of the circuit in Fig. 2 with different LO amplitude A_{LO}. With ideal switching behavior (grey curves), well approximated at large A_{LO}, a_2 is maximum at $\delta = 25\%$ but the curve is relatively flat and it remains close to the maximum at $\delta = 26\%$, when $a_0 = 0$. At lower A_{LO} (orange and red curves), $Q_{1,2}$-$Q_{3,4}$ no longer behave as ideal switches and a_2 is maximized with higher duty-cycle. Nevertheless, also in this case, a_2 is relatively flat around its maximum which remains close to the value of δ corresponding to $a_0 = 0$. Therefore, a simple loop that settles V_{OS} to force $a_0 = 0$ allows to maintain a nearly maximum second-harmonic conversion gain independently from A_{LO}. Finally, it is worth noticing that,

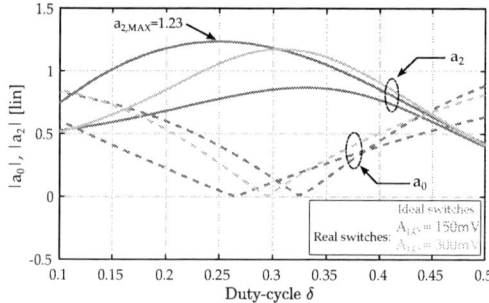

Fig. 4. Simulated DC and second-harmonic coefficients, a_0, a_2, of the differential output current $I_{o,d}$ versus δ for different A_{LO}.

regardless of the actual duty-cycle, the proposed modification of the Gilbert-cell FD preserves the fully balanced topology, thus not only the differential output is odd-harmonics free, but also the common-mode is constant and free of spurs, meaning there is no penalty over the classic solutions in Fig. 1.

III. DESIGN

The schematic of the proposed frequency doubler is shown in Fig. 5 (biasing is omitted for simplicity). The Gilbert-cell core is composed of the common-emitter transistors Q_{1-2}, operated in class A, and the switching quad Q_{3-4}, Q_{5-6}; All the transistors have the same area of $8\,\mu m \times 0.13\,\mu m$. The nominal DC current consumption is $I_B = 2 \times 7\,mA$ @ $P_{in} = 0\,dBm$ and the supply voltage is $V_{CC} = 1.5\,V$. The low-frequency feedback loop is implemented by a fully differential OTA which senses the DC output current through a pair of resistors, $R_{sense} = 10\,\Omega$, and provides a differential V_{OS} DC component to the base of Q_{3-5} and Q_{4-6}. The value of resistors R_{sense} is chosen as a compromise between low DC voltage drop and robustness against the input offset voltage of the OTA. The OTA has been designed with 1.25 V output common-mode voltage, thus allowing a maximum $V_{OS} = \pm 250\,mV$ i.e. the DC value at the base of Q_{3-5}, Q_{4-6} can range from 1 V to 1.5 V. The OTA current consumption is $200\,\mu A$ from 1.5 V supply. The loop cut-off frequency is limited to 100 MHz by miller compensation capacitors. The latter are also exploited, together with resistors in series to the OTA input, to suppress the RF signal at the R_{sense} nodes. Choke resistors $R_{big} = 1\,k\Omega$ are used to feed the base voltage of the switching quad transistors without loading the LO signal, which is coupled through capacitors $C_{big} = 400\,fF$. Resonant networks are used to provide input matching and to synthesize the optimal load impedance at the output in order to maximize the saturated power, P_{sat}.

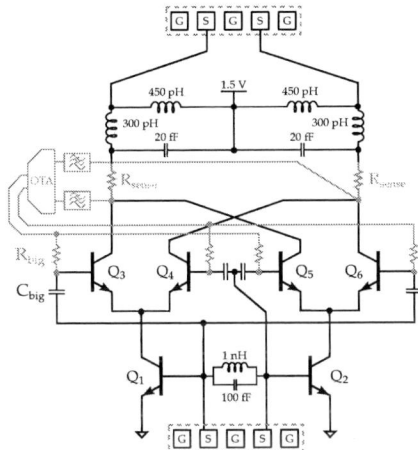

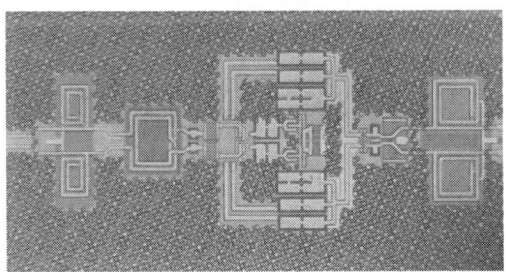

Fig. 5. Schematic of the proposed frequency doubler. Greyed components are part of the low-frequency feedback loop.

Fig. 6. Photograph of the realized frequency doubler.

IV. MEASUREMENTS

The proposed frequency doubler is realized in $0.13\,\mu m$ SiGe BiCMOS technology featuring f_t/f_{max} of 250/370 GHz. The layout is shown in Fig. 6 and the area occupation, without pads, is $180\,\mu m \times 420\,\mu m = 0.075\,mm^2$. Including pads, the overall area is $0.30\,mm^2$. For measurements, the chip is glued on a PCB which provides supply and biasing through bondwires. The input and output RF signals are interfaced with Cascade Infinity GSGSG probes. The differential input signal is provided through a pair of Agilent E8257D signal generators synched together in antiphase. On the output side, a balun is used for differential to single-ended conversion, and output power is measured with Agilent PXA N9030A spectrum analyzer. Fig. 7 shows the saturated output power P_{sat}

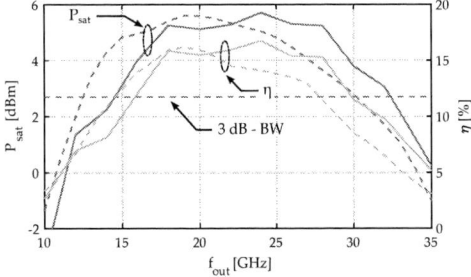

Fig. 7. Measured (solid) and simulated (dashed) output saturated power P_{sat}, and collector efficiency η vs frequency.

and the collector efficiency η versus the output frequency f_{out}. The peak output power is 5.7 dBm measured at $f_{out} = 24\,GHz$, with a corresponding peak collector efficiency of $\eta = 17\%$. The 3 dB-bandwidth on P_{sat} is more than an octave, from 14 to 32 GHz, equivalent to 85 % fractional bandwidth. Fig. 8 reports a sweep of the input power at an output frequency $f_{out} = 20\,GHz$, showing a peak conversion gain of 6 dB. The measured rejection of the driving signal (i.e. the fundamental component) is reported in Fig. 9 showing $> 40\,dB$ across the full operation bandwidth. Fig. 10 plots the output phase noise (PN) of the FD with respect to the signal generator PN showing that the degradation ΔPN corresponds to the theoretical $\Delta PN = 20\log_{10}(2) = 6\,dB$.

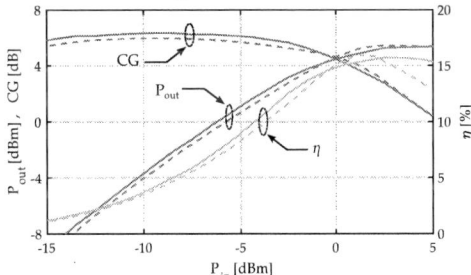

Fig. 8. Measured (solid) and simulated (dashed) output power P_{out}, Gain and collector efficiency η vs input power P_{in} @ $f_{out} = 20\,GHz$.

The measured results are finally summarized in Table-I and compared against previously reported frequency doublers operating in a similar frequency range. The table highlights a remarkable improvement of the operating (fractional) bandwidth, except for [11] which lacks in terms of conversion gain

978-1-6654-8495-4/22 $31.00 © 2022 IEEE

TABLE I
PERFORMANCE SUMMARY AND COMPARISON TABLE.

	This Work	[4]	[5]	[9]	[10]	[11]	[12]	[13]	[14]
Technology	0.13 μm SiGe	0.13 μm SiGe	65 nm CMOS	0.13 μm SiGe	0.13 μm SiGe	0.18 μm CMOS	0.11 μm CMOS	0.15 μm GaAs	0.18 μm SiGe
Topology	Self-Adjusted Gilbert Cell	Push-Push	Balanced Push-Push	Gilbert Cell	Push-Push +Amp	Push-Push	Push-Push	Gm-boosted Single-end CG	CB-CE Pair
Frequency [GHz]	14 ÷ 32	25 ÷ 40	23 ÷ 25	22 ÷ 36	27 ÷ 41	15 ÷ 36	21.8 ÷ 25.8	37 ÷ 43	26 ÷ 40
BW [%]	85	47	8.3	50	42	90	17	15	43
Conv. Gain [dB]	6	14	5	21	19.8	-10.2	-4.2	0.9	-4
P_{sat} [dBm]	5.7	13.0	5.0	11.9	8.0	-5.2	0	2.0	5.0
Peak η [%]	17.0	22.9	9.9	27.4	18.0	2.7	16.6	2.1	2.0
Fund. Rejection [dB]	>40	>35	44	32	26	33	39	6	14
P_{DC}/V_{SUP} [mW/V]	21 / 1.5	87 / 2.5	31 / -	43.1 / 2.5	32 / 1.6, 2	11 / 1	6 / 1.2	86 / 2	66 / 2
Area [mm^2]	0.30	0.50	0.47	0.48	0.34	0.32	0.45	0.72	0.56

and P_{sat}. This work also demonstrates P_{sat} and efficiency (η) comparable or better than previous works. It is worth noticing that the supply voltage of the presented doubler is 1.5 V, while the best efficiency, of [4], [9], is achieved with the benefit of a remarkably higher supply, 2.5 V, which makes more negligible the voltage headroom required to keep transistors in the active region.

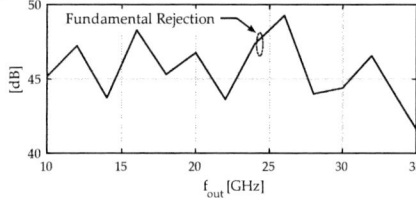

Fig. 9. Measured fundamental harmonic rejection vs frequency.

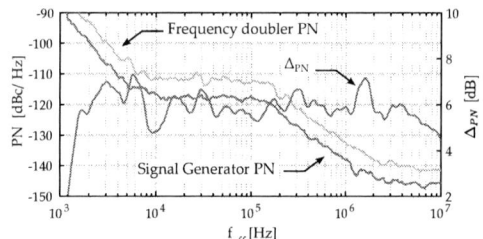

Fig. 10. Measured phase noise of the frequency doubler @ f_{out}=20 GHz vs measured phase noise of the input signal generator.

V. CONCLUSIONS

This paper presented a technique to further enhance the performance of frequency doublers realized around a Gilbert-cell mixer. A low-frequency feedback loop is implemented which regulates the bias voltage of the switching transistors to maintain a ~25 % equivalent LO duty-cycle. In this way, the output current has a quasi square-wave shape at twice the input frequency, with enhanced conversion gain. Moreover, the technique avoids the need for bandwidth-limiting quadrature generation of the driving signals, extending the operation bandwidth. Compared to previously reported frequency doublers, experimental results on a K-band test chip in SiGe BiCMOS technology demonstrated the highest operation bandwidth with a state-of-the-art power efficiency at low supply voltage.

REFERENCES

[1] E. Monaco et al., "Injection-Locked CMOS Frequency Doublers for μ-Wave and mm-Wave Applications," in *IEEE Journal of Solid-State Circuits*, vol. 45, no. 8, pp. 1565-1574, Aug. 2010.

[2] M. M. Pirbazari and A. Mazzanti, "High Gain 130-GHz Frequency Doubler With Colpitts Output Buffer Delivering Pout up to 8 dBm with 6% PAE in 55-nm SiGe BiCMOS," in *IEEE Solid-State Circuits Letters*, vol. 4, pp. 36-39, 2021.

[3] S. Hara et al., "60 GHz injection locked frequency quadrupler with quadrature outputs in 65 nm CMOS process," *2009 Asia Pacific Microwave Conference*, 2009, pp. 2268-2271.

[4] I. Ju, C. D. Cheon and J. D. Cressler, "A Compact Highly Efficient High-Power Ka-band SiGe HBT Cascode Frequency Doubler With Four-Way Input Transformer Balun," in *IEEE Transactions on Microwave Theory and Techniques*, vol. 66, no. 6, pp. 2879-2887, June 2018.

[5] S. Vehring and G. Boeck, "Truly Balanced K-Band Push-Push Frequency Doubler," *2018 IEEE Radio Frequency Integrated Circuits Symposium (RFIC)*, 2018, pp. 348-351.

[6] M. M. Pirbazari and A. Mazzanti, "E-Band Frequency Sextupler With >35 dB Harmonics Rejection Over 20 GHz Bandwidth in 55 nm BiCMOS," in *IEEE Journal of Solid-State Circuits*.

[7] J. Kaukovuori et al., "Analysis and Design of Passive Polyphase Filters," in *IEEE Transactions on Circuits and Systems I: Regular Papers*, vol. 55, no. 10, pp. 3023-3037, Nov. 2008.

[8] D. Ozis, J. Paramesh and D. J. Allstot, "Integrated Quadrature Couplers and Their Application in Image-Reject Receivers," in *IEEE Journal of Solid-State Circuits*, vol. 44, no. 5, pp. 1464-1476, May 2009.

[9] M. Frounchi, S. G. Rao and J. D. Cressler, "A *Ka*-Band SiGe Bootstrapped Gilbert Frequency Doubler With 26.2% PAE," in *IEEE Microwave and Wireless Components Letters*, vol. 28, no. 12, pp. 1122-1124, Dec. 2018.

[10] J. Li et al., "A 27–41 GHz Frequency Doubler With Conversion Gain of 12 dB and PAE of 16.9%," in *IEEE Microwave and Wireless Components Letters*, vol. 22, no. 8, pp. 427-429, Aug. 2012.

[11] P. -H. Tsai et al., "Broadband Balanced Frequency Doublers With Fundamental Rejection Enhancement Using a Novel Compensated Marchand Balun," in *IEEE Transactions on Microwave Theory and Techniques*, vol. 61, no. 5, pp. 1913-1923, May 2013.

[12] Y. Gong et al., "An Ultra-Low Power K band Balanced Frequency Doubler with a Novel Current-reused Structure," *ESSCIRC 2021 - IEEE 47th European Solid State Circuits Conference (ESSCIRC)*, 2021, pp. 443-446.

[13] K. -H. Lu et al., "A K-Band Frequency Doubler in 0.15 − μm GaAs pHEMT with an Autonomous Circuit for Stability Analysis," *2018 IEEE International Symposium on Radio-Frequency Integration Technology (RFIT)*, 2018, pp. 1-3.

[14] G. -Y. Chen et al., "A Ka-band broadband active frequency doubler using CB-CE balanced configuration in 0.18 μm SiGe BiCMOS process," *2012 IEEE/MTT-S International Microwave Symposium Digest*, 2012, pp. 1-3.

A 1.8-67GHz Divide-by-4 ILFD Using Area-Efficient Transformer-Based Injection-Enhancing Technique

Yudai Yamazaki, Jian Pang, Atsushi Shirane, Kenichi Okada
Department of Electrical and Electronic Engineering, Tokyo Institute of Technology
2-12-1-S3-28, Ookayama, Meguro-ku, Tokyo 152-8552, Japan
Email: yamazaki@ssc.pe.titech.ac.jp

Abstract— In this paper, a wide-locking-range divide-by-4 injection locked frequency divider (ILFD) using area-efficient transformer-based injection-enhancing technique is introduced. This ILFD is composed of a 4-stages differential ring oscillator and area-efficient injection-enhancing transformer. With the proposed technique, injection current is increased in wide-band frequency with low power consumption and small area. The locking range is extended in all frequency bands of 28GHz, 39GHz and 60GHz. The measured operating range is 1.8-67GHz (190%). The core area of this ILFD is 0.059mm² in a 65nm CMOS process.

Keywords—millimeter wave, 5G divide-by-4 ILFD, transformer, ring oscillator.

I. INTRODUCTION

Millimeter-wave bands of up to 60GHz are utilized in 5G high-data-rate transceivers [1, 10]. In millimeter-wave transceivers, injection locked frequency divider (ILFD) circuits are widely used in reference signal generation and millimeter-wave PLLs. Usually, only digital dividers are used for reference signal generation at few GHz frequency. However, high-division-ratio ILFD is also needed in millimeter-wave because of the limited speed of digital dividers. So, the ILFD, which is used for 5G multi-band operation with compact and low power, is required. Conventional ILFDs [4, 7, 11] are designed based on LC oscillator. They realize high-frequency division with low power consumption. However, large area is consumed by the on-chip inductors. The locking range is also narrow due to the large Q factor. [5] and [6] are designed based on multi-stages current-mode-logic (CML) divider, which have been utilized for high-division-ratio operation. However, it consumes large power in millimeter-wave band. [2-3, 9, 13-14] are ring-oscillator-based ILFDs. They can operate at high division ratio with wide locking range, low power, and small area. On the other hand, their locking ranges can degrade in millimeter-wave frequencies due to the large parasitic capacitance of transistors.

In this work, wide-locking-range divide-by-4 ILFD using area-efficient transformer-based injection-enhancing technique is introduced. The proposed circuit is designed with ring-oscillator-based ILFD, which adopts area-efficient wide-matching transformers. With the proposed technique, the injection current is increased in wide-band frequencies. The locking range is therefore extended in all frequency bands including 28GHz, 39GHz and 60GHz, with small area and low power consumption. The measured operating range is 1.8-67GHz (190%). The core area of this ILFD is 0.059mm² in 65nm CMOS process.

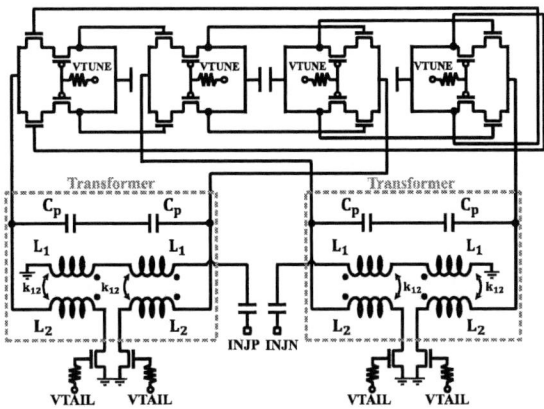

Fig. 1. Schematic of the proposed divide-by-4 ILFD using area-efficient transformer-based injection-enhancing technique.

II. CIRCUIT IMPLEMENTATION

In this work, ring-oscillator-based divide-by-4 ILFD is adopted to extend the locking range with small area and low power consumption. Fig. 1 shows the schematic of the proposed divide-by-4 ILFD. This ILFD is composed of a 4-stages differential ring oscillator. The differential input signal, whose frequency is about 4 times of the oscillation frequency, is injected from INJP and INJN. The signals are sent to each stage of ring oscillator through transformers. Then, the oscillation frequency is synchronized to divide-by-4 injection frequency. In differential stages, the 2nd-harmonic signal is generated at each common node due to the non-linearity of transistors. This harmonic signal could be used to enhance injection locking in high-division-ratio dividers [2][9]. Furthermore, to achieve multiple-bands operation, the free-run frequency of ring oscillator is changed by current tuning, depending on injection frequency.

The locking range of ILFD depends on the magnitude of injection current into the oscillator [8]. In addition, parasitic capacitance has huge influence in ring oscillator. It decreases the injection signal in high frequency due to the low-pass characteristic. In this work, with the area-efficient transformer-based injection-enhancing technique, the injection current into ring oscillator is significantly larger at all frequencies from 0 to 67GHz with small area. To achieve high-frequency division in ring-based ILFD, inductors or conventional transformers are

978-1-6654-8495-4/22 $31.00 © 2022 IEEE

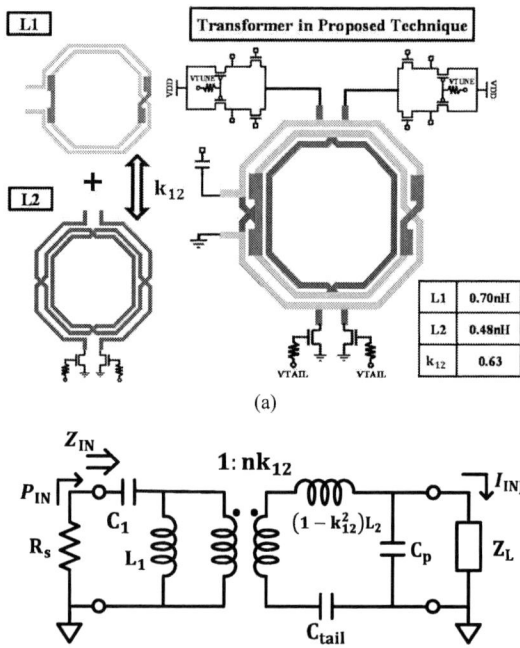

(a)

(b)

Fig. 2. (a) Layout of the transformer and (b) equivalent circuit for proposed technique.

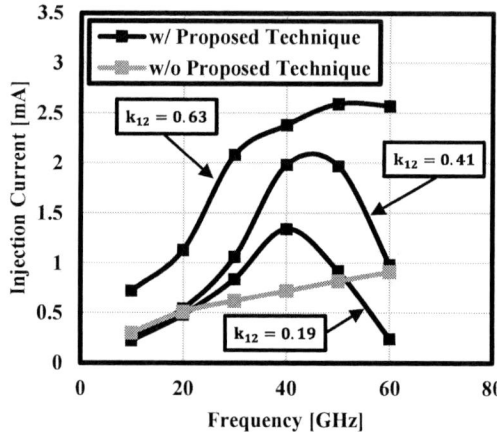

Fig. 3. Simulated injection current with proposed technique.

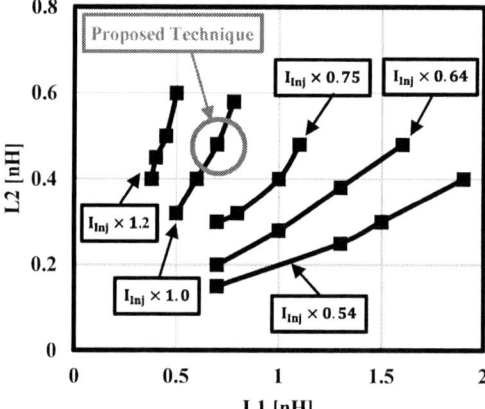

Fig. 4. Simulated results for optimization of L1 and L2.

usually needed for impedance matching with parasitic capacitance. However, they consume large chip area. To achieve high-frequency matching with small area, area-efficient transformer is adopted in this proposed technique. The layout of proposed transformer is shown in Fig. 2 (a). The value of L1, L2 and k_{12} are also shown in the figure. L1 connects the injection input port and GND. L2 is attached to the tail node of the ring oscillator. This transformer helps to enhance the injection through transformer coupling and wide-band impedance matching. Fig. 2 (b) shows the equivalent circuit from the injection node to the common node of differential stage. R_S and Z_L are the input impedance and the common-node impedance, respectively. P_{IN} is the injection power to the proposed ILFD. I_{INJ} is injection current to the differential stage in ring oscillator. The equation of input impedance Z_{IN} and injection current I_{INJ} are shown in (1) and (2) below. These equations show that the transformer operates as a fourth-order resonator with two magnetically-coupled LC tanks. This transformer-based technique achieves input matching and injection enhancing in wide-band millimeter-wave frequencies.

In this technique, injection current depends on the value of k_{12} greatly. When k_{12} is closer to 1, injection power from input

port to ring oscillator is maximized. Based on the equations of (1) and (2), the simulated injection current to differential stages with and without proposed technique is shown in Fig. 3. The ILFD without proposed technique refers to the simulated result of [9]. Injection power is constant at 0dBm. In this figure, the current values are plotted in divide-by-4 operation at 10-60GHz injection frequencies. The value of k_{12} is swept from 0.19 to 0.63 by EM simulation. At $k_{12} = 0.63$, injection current is more than twice as much as the results without the proposed technique. In addition, injection signal is increased in wide-band frequencies by matching the transformer with parasitic capacitance C_p. Fig. 4 shows the simulated injection current for optimization of L1 and L2. To optimize L1 and L2 regardless of k_{12}, the lumped transformer is used for simulation, which has a

$$Z_{IN} = \frac{\left(1 + sC_pZ_L\right)\{1 + s^2(C_1L_1 + C_{tail}L_2) + s^4C_1C_{tail}(1 - k_{12}^2)L_1L_2\} + sC_{tail}Z_L(1 + s^2C_1L_1)}{sC_1\{(1 + sC_pZ_L)(1 + s^2C_{tail}L_2) + sC_{tail}Z_L\}} \quad (1)$$

$$I_{INJ} = nk_{12} \times \frac{sC_1Z_{IN} - 1}{sC_1(Z_{IN} + R_S)^{\frac{3}{2}}} \times \frac{sC_{tail}}{(1 + sC_pZ_L)\{1 + s^2C_{tail}(1 - k_{12}^2)L_2\} + sC_{tail}Z_L} \times \sqrt{P_{IN}} \quad (2)$$

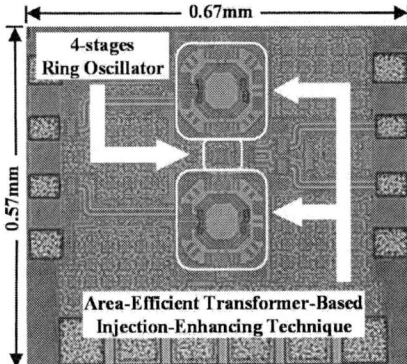

Fig. 5. Die Photo

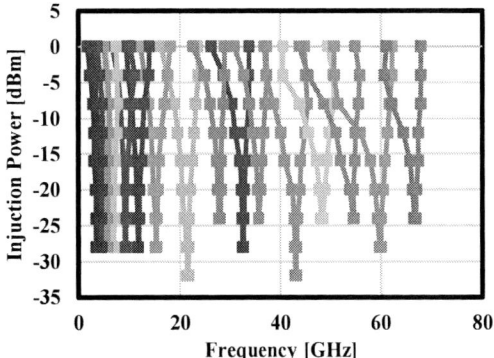

Fig. 7. Measured locking range of the proposed divide-by-4 ILFD.

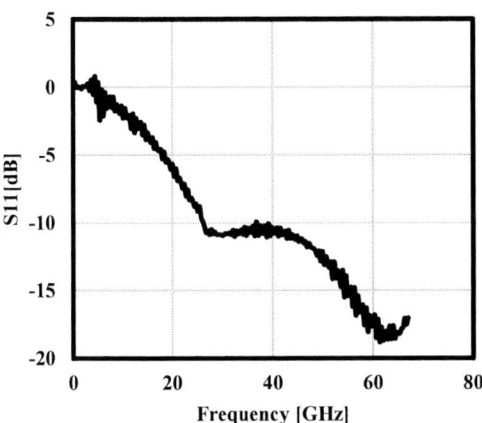

Fig. 6. Measured S11 of the proposed ILFD

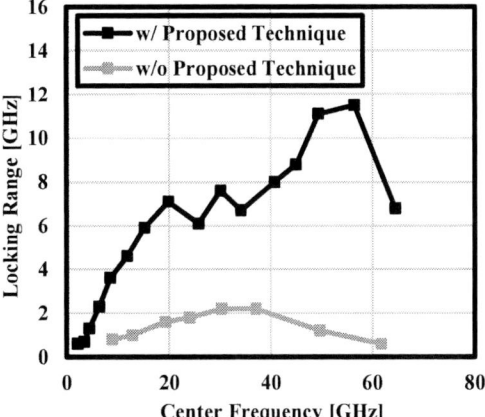

Fig. 8. Measured locking range at center frequency

constant k_{12}. They are swept from 0.38nH to 1.9nH and from 0.15nH to 0.60nH, respectively. With the proposed technique, the smaller L1 and larger L2, the more injection current increases. Ideally, it is possible to inject the largest current by minimize L1 and maximize L2 in Fig. 4. However, in the actual layout, k_{12} decreases dramatically as the L1 value decreases. Increasing L2, the area consumption of the transformer becomes large. So, L1 and L2 are optimized to 0.70nH and 0.48nH, respectively.

III. MEASUREMENT RESULTS

This proposed divide-by-4 ILFD is implemented in a 65nm CMOS process. Fig. 5 shows the die photo. The total size including pads is 0.57mm x 0.67mm, while the core area is 0.059mm^2. This ILFD operates at 1.0-V supply voltage. The measured S11 at injection ports is shown in Fig. 6. This S11 is measured in 10MHz-67GHz by a PNA-X (Keysight N5247B). The return loss is suppressed to less than -10dB in 26.5-67GHz. This result show that the proposed technique achieves the wide-range input matching with fourth-order resonation in the transformers. Thanks to this technique, the wide-band injection efficiency is greatly improved.

The measured divide-by-4 locking range is shown in Fig. 7. For the measurement of the locking range, the injection signal is provided by a signal generator (Keysight E8257D), which is connected to an external RF balun to generate differential signals. The output signal is observed by a spectrum analyzer (Keysight E4448A) in this work. The locking range is measured against different bias setting of the ring oscillator. A wide-band locking range can be achieved at 0dBm injection power with the same ILFD. The operating range is 1.8-67GHz (190%). The power consumption is 5.05mW at 20GHz divide-by-4 operation. Fig. 8 summarizes the locking range with fixed bias settings for the ring oscillator. Locking range of 26.1-33.7GHz (25.4%), 36.8-44.8GHz (19.6%), and 61.1-67GHz (10.5%) are also achieved at 0dBm injection power with fixed settings. Compared with conventional ILFD [9], this result is greatly improved with the transformer injection. Please note that, the measured locking range was limited less than 67GHz due to operating frequency range of the used signal generator. So, this ILFD could work at even higher frequency.

Table 1 compares the proposed divide-by-4 ILFD with area-efficient transformer-based injection-enhancing technique to the

TABLE 1: PERFORMANCE COMPARISON

	This Work			[2]	[9]	[4]	[5]	[6]	[7]	[11]
Process (nm)	65			65	65	180	90	130	180	180
Division Ratio	4			4	6	4	4	4	4	5
Injection Power (dBm)	0			0	0	0	0	0	0	0
Operating Range (GHz)	1.8-67			7.3-21.3	26.1-33.0	45.9-50.9	13.2-18.4	19.3-23.4	7.8-12.3	13-16.6
Operating Range (%)	190			97.9	23.3	10.33	33.0	19	44.8	24.3
Area (mm²)	0.059			0.003	0.002	0.17	0.56	0.25	0.81	0.95
Power (mW)	9.4-14.0 GHz	16.3-23.4 GHz	43.8-54.9 GHz	8.5-14.0 GHz	27.7-32.0 GHz	7.56	10.8	3.40	7.10	4.09
	3.18	5.05	11.98	3.90	3.1					
FoM$_1$ (GHz/mW)	1.45	1.41	0.93	1.41	1.39	0.66	0.48	1.21	0.63	0.88
FoM$_2$ (%/mW)	12.36	7.09	1.88	10.95	4.0	1.37	3.06	5.59	6.31	5.94

*FoM$_1$ = Locking Range (GHz)/Power (mW), FoM$_2$ = Locking Range (%)/Power (mW)

state-of-the-art ILFDs. The proposed ILFD achieves much wider locking range. Furthermore, this ILFD also performs better with fixed bias settings. The table also shows the comparison of figure of merit (FoM). FoM$_1$ and FoM$_2$ are used for the comparisons. With the proposed technique, better tradeoff is achieved from few GHz to millimeter-wave frequency range than other ILFDs.

IV. CONCLUSION

The divide-by-4 ILFD with area-efficient transformer-based injection-enhancing technique is introduced for multi-band operation. The proposed technique increases the injection current significantly and achieves wider locking range in multi-band frequencies. Measured total locking range is 1.8-67GHz (190%). The core area of this circuit is 0.059mm².

ACKNOWLEDGMENT

This work is partially supported by MIC (JPJ000254), NICT (00801, 00601), JSPS (JP20H00236), STAR, and VDEC in collaboration with Cadence Design Systems, Inc., Mentor Graphics, Inc., and Keysight Technologies Japan, Ltd.

REFERENCES

[1] J. Pang, et al., " A Power-Efficient 24-to-71 GHz CMOS Phased-Array Receiver Utilizing Harmonic-Selection Technique Supporting 36dB Inter-Band Blocker Rejection for 5G NR," IEEE International Solid-State Circuits Conference, pp. 434-436, Feb. 2022.

[2] A. Musa, et al., "A 20GHz ILFD with Locking Range of 31% for Divide-by-4 and 15% for Divide-by-8 Using Progressive Mixing," IEEE Asian Solid-State Circuits Conf. Tech. Dig., pp. 85–88, Nov. 2011.

[3] A. Garghetti, et al., "Analysis and Design of 8-to-101.6-GHz Injection-Locked Frequency Divider by Five With Concurrent Dual-Path Multi-Injection Topology," IEEE Journal of Solid-State Circuits, vol. 57, no. 5, pp.1788-1799, Jun. 2022.

[4] M. C. Chuang, et al., "A 50 GHz divide- by-4 injection lock frequency divider using matching method," IEEE Microw. Wireless Compon. Lett., vol. 18, no. 5, pp. 344–346, May 2008.

[5] Y. S. Lin, et al., "Wide-locking- range multi-phase-outputs regenerative frequency dividers using even- harmonic mixers and CML loop dividers," IEEE Trans. Microw. Theory Techn., vol. 62, no. 12, pp. 3065–3075, Dec. 2014.

[6] Z.-D. Huang, et al., "Design of 24-GHz 0.8-V 1.51-mW coupling current-mode injection-locked frequency divider with wide locking range," IEEE Trans. Microw. Theory Tech., vol. 57, no. 8, pp. 1948–1958, Aug. 2009.

[7] S.-L. Jang, et al., "High Even-Modulus Injection-Locked Frequency Dividers," IEEE Trans. Microw. Theory Techn., vol. 67, no. 12, pp. 5069–5079, Dec. 2019.

[8] B. Razavi, et al., "A Study of Injection Locking and Pulling in Oscillators," IEEE Journal of Solid-State Circuits, vol. 39, no. 9, pp. 1415-1424, Sep. 2004.

[9] T. Siribranon, et. al., "A 13.2% Locking-Range Divide-by-6, 3.1mW, ILFD using Even-Harmonic-Enhanced Direct Injection Technique for Millimeter-Wave PLLs," IEEE European Solid State Circuits Conference, pp. 404-407, Sep. 2013.

[10] J. Pang, et al., " A CMOS Dual-Polarized Phased-Array Beamformer Utilizing Cross-Polarization Leakage Cancellation for 5G MIMO Systems," IEEE Journal of Solid-State Circuits, vol. 56, no. 4, pp. 1310-1326, Apr. 2021.

[11] S.-L. Jang, et al., "Wide-Locking Range RLC-Tank Balanced-Injection Divide-by-5 Injection-Locked Frequency Dividers Based on Harmonic Mixing," IEEE Trans. Microw. Theory Techn., vol. 68, no. 3, pp. 894-903, Mar. 2020.

[12] J. Zhu, et al., "A 19–48.3-GHz 6th-Order Transformer-Based Injection-Locked Frequency Divider with 87.1% Locking Range in 40-nm CMOS," IEEE Trans. Microw. Theory Techn., vol. 68, no. 9, pp. 3053-3057, Dec. 2019.

[13] M. Baert, et al., "A PVT-Compensated 0.1-67 GHz Injection-Locked Frequency Divider with Replica-based Automatic Tuning," IEEE Radio Frequency Integrated Circuits Symposium, pp. 75-78, Jul, 2021.

[14] A. Ghilioni, et al., "Analysis and Design of mm-Wave Frequency Dividers Based on Dynamic Latches With Load Modulation," IEEE Journal of Solid-State Circuits, vol. 48, no. 8, pp. 1842-1850, Aug. 2013.

A Fully Differential 40 MHz Switched-Capacitor Crystal Oscillator with Fast Start-Up.

Wim Kruiskamp
Connectivity & Audio Business Division
Renesas Electronics Corporation
's-Hertogenbosch, The Netherlands
wim.kruiskamp.uj@renesas.com

Abstract—This paper introduces a switched capacitor crystal oscillator as a fully differential alternative to the widely used single ended Pierce oscillator. The presented 40 MHz oscillator consumes 55 µW during steady state operation at a 0.8 Vpp amplitude and a 6 pF load capacitance. Fast start-up, in typically 12 µs and consuming a start-up energy of 8 nJ, is achieved by self-timed energy injection, without requiring an additional start-up oscillator. The oscillator is realized in a 22nm FD-SOI process.

Keywords—*crystal oscillator, fully differential, switched capacitor, fast startup, phase noise, low-power circuits.*

I. INTRODUCTION

Low-power wireless systems typically use a crystal oscillator to generate an accurate low-noise reference clock for the radio. The crystal oscillator typically spends most of its time in a low-power sleep state and only briefly wakes-up during the receiving and transmitting of data by the transceiver. Traditionally, the crystal oscillator is implemented as a 3-point Pierce oscillator as depicted in Fig. 1. In this oscillator, the crystal is connected between the input and output of a self-biased inverter, with load capacitors from each crystal terminal to ground. The simplicity and good performance of this circuit has made it the de-facto standard in many RF systems for decades. This circuit however suffers from a few important drawbacks. Firstly, the oscillator of Fig. 1. is a single-ended circuit making it sensitive to common-mode crosstalk coupled to the crystal terminals which will be converted into differential noise. Secondly, the start-up of the oscillator is slow due to the high Q of the crystal. The inverter is very inefficient while the amplitude is still low, resulting in a significant start-up energy loss. Thirdly, the total capacitance required is four times higher than for a topology where the load capacitor is directly across the crystal. In this paper a fully differential crystal oscillator is proposed, which overcomes the listed drawbacks.

II. CRYSTAL OSCILLATOR ARCHITECTURE

A. Steady-state

The concept is depicted in Fig. 2. The crystal is connected to one of the three identical switched capacitors A, B, or C. During each peak of the oscillation voltage, when the current through the inductor is zero and all energy is stored in capacitors, the currently connected switched capacitor is replaced by a similar capacitor charged positively to the supply voltage. This will instantly increase the voltage across the crystal, adding energy to the system. At each valley of the oscillation voltage, the currently

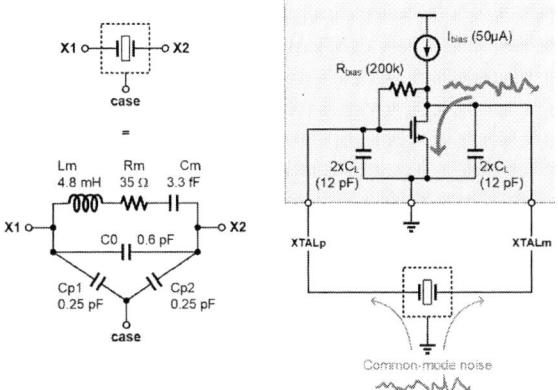

Fig. 1. Typical Pierce oscillator with 40 MHz crystal equivalent circuit.

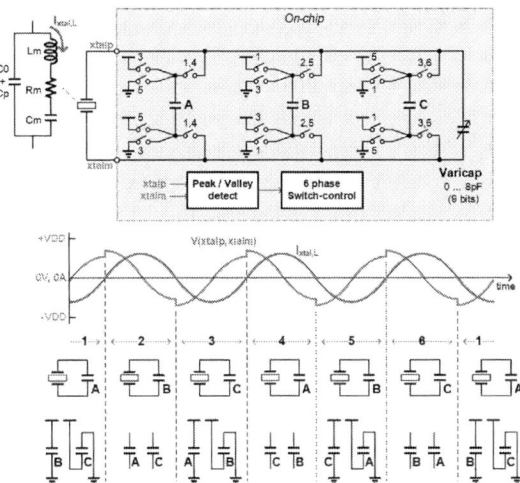

Fig 2. Switched Capacitor Crystal Oscillator Concept.

connected switched capacitor is again replaced by a similar capacitor, which this time is charged negatively to the supply voltage. This will again increase the voltage across the crystal, adding energy to the system. The steady-state amplitude settles to a value where the added power exactly compensates the power dissipation in the crystal. A 9-bit varicap with a capacitance between 0 pF and 8 pF is added to trim the frequency and to accommodate different crystals.

The switching order in Fig. 2. is optimized for low jitter during the rising edges of the crystal voltage since this crystal oscillator was designed for a radio that only looks at the rising edge of the clock signal. The two spare capacitors (B+C, A+B or C+A) are sampled to the supply voltage

978-1-6654-8495-4/22 $31.00 © 2022 IEEE

simultaneously during the odd phases. One of them is connected at the next peak. The other one, charged with opposite polarity, is connected at the next valley. Any supply voltage dependent error added at the peak is therefore largely cancelled by subtracting a similar error at the valley. This will make the rising edges, but not the falling edges, of the crystal voltage hardly sensitive to noise on the supply voltage. Fortunately, many radio systems only use the rising edge of the reference clock.

B. Start-up

Another way of looking at this concept is by decomposing the crystal voltage (blue line in Fig. 2.) into a sine wave of the harmonic oscillation and a square wave due to the switched capacitors. The square wave will appear across the inductor via Cm. The square wave toggles at the peaks and valleys of the voltage, which is at the zero-crossings of the inductor current; The square wave is always reinforcing the inductor current. In steady-state operation, this square wave only compensates for the losses in the crystal and is small in amplitude. During start-up however, the square wave should have a large amplitude. An H-bridge can generate a square wave with an amplitude equal to the supply voltage. This is larger than with the switched capacitor circuit of Fig. 2. During fast start-up, it is therefore beneficial to drive the crystal with an H-bridge. Different methods of controlling when the H-bridge toggles to achieve fast low-energy start-up have been proposed [1-4]. Each approach differs in the way they determine when to toggle the H-bridge. [2-4] make use of an accurate and stable free-running oscillator for the first tens of periods. The free-running oscillator is either periodically synchronized with the crystal oscillation [2] or replaced by a fast-locking PLL after the initial start-up [3,4]. This works very well, but the need for an accurate and stable free-running oscillator is a major drawback. [1] does not need such an oscillator, but instead determines the zero-crossing of the current by monitoring the voltage across the on-resistance of an H-bridge switch. This however brings contradicting requirements for the switch: It has to be low-impedance in order to avoid phase shift in combination with the parasitic capacitance of the crystal, yet it has to be high-impedance to get a decent amplitude across the switch. To avoid such a compromise, the proposed start-up circuit, depicted in Fig. 3, only turns on one of the high-side switches S1 and S3 for the first part of each half-period. The voltage will remain on the parasitic capacitance C0+Cp for the remainder of the half-period and the now opened high-side switch will not reduce nor influence the oscillating voltage on the crystal anymore. The inductor current will create a voltage, exactly 90 degrees out-off phase with itself, across the capacitance between the crystal pins. By differentiating and amplifying that voltage, the ideal moment to change the polarity of the H-bridge can be determined.

III. PEAK DETECTION IMPLEMENTATION

A. Start-up

The signal during start-up is a single-ended signal since one side of the crystal is connected to ground. The peak-detection is done with the circuit depicted in Fig. 4. The differentiator and the gain stages are implemented as common-source amplifiers. The current sources provide

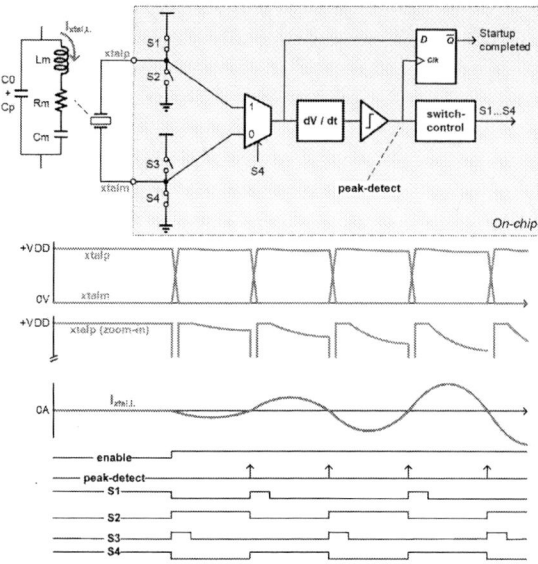

Fig. 3. Start-up circuit.

isolation to noise on the supply. CMOS inverters would be twice as efficient, but the lack of supply rejection makes them less suitable for this circuit.

The differentiator and each gain stage are auto zeroed during each drive pulse as the sine wave amplitude during the first cycle is typically less than a few mV. Startup is completed when the output of the multiplexer drops below half the supply voltage at the rising edge of the peak-detect signal, indicating that the differential peak-to-peak amplitude is equal to the supply voltage.

B. Steady-state

The peak detection during steady-state is implemented with the circuit depicted in Fig. 5. The two capacitors are connected between the crystal terminals and virtual ground

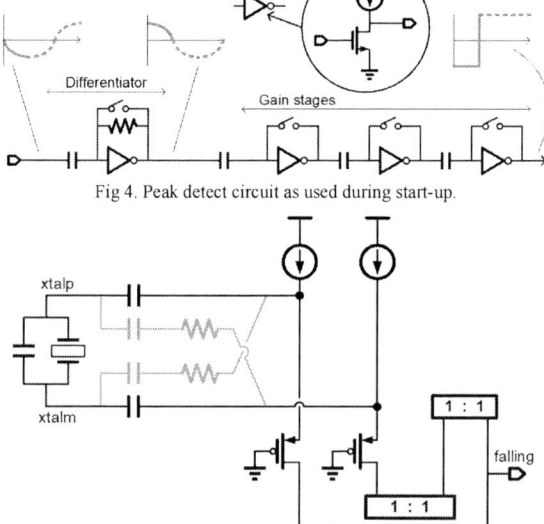

Fig 4. Peak detect circuit as used during start-up.

Fig. 5. Peak detect circuit as used during steady-state.

nodes. The current through the capacitors are compared to determine the peak and valley of the voltage across the crystal. There will be a propagation delay from peak and valley to the actual switching of the capacitors. This propagation delay is very small compared to the oscillation period but nevertheless compensated for by the two cross-coupled resistors, depicted in Fig. 5.

IV. MEASUREMENT RESULTS

The circuit is realized in a 22 nm FD-SOI process. A die-photo is depicted in Fig. 6. Even though the varicap capacitance is 4 times lower than would be required for a Pierce Oscillator, it still consumes a significant part of the total area. The circuit was evaluated with a 40 MHz crystal (Murata, RCGB40M000F1S2FR0). No calibration was required for the circuit other than trimming the varicap to a frequency of exactly 40 MHz.

The voltage across the crystal during start-up is depicted in Fig. 7. Just after start-up, the voltage is a square wave with a hardly visible sine wave on top. Nevertheless, the circuit is locking to this 40 MHz sine wave. Half-way the start-up procedure, the sine wave superimposed on the square wave is clearly visible. Toggling of the H-bridge happens exactly at the peaks of the sine wave. The startup is completed when the amplitude of the sine wave exceeds half the supply voltage. This typically takes 12 µs.

The switched capacitors are programmable between 0.25 pF and 1.75 pF. The typical value of 0.75 pF, charged with a 0.8 V supply, results in an amplitude of 0.4 V (= 0.8 Vpp). The square wave on top of the sine wave during steady state is typically invisible but can be observed just after artificially changing the switched capacitor from its minimum to its maximum value as shown in Fig. 8.

The measured phase-noise is depicted in Fig. 9. At low frequencies, the phase-noise is proportional to $(1/f)^3$. This is assumed to be 1/f noise from the peak-valley comparator in combination with the second-order behavior of the LRC branch in the crystal. The peak-valley comparator uses small transistors and the bias currents are not fully optimized for low 1/f noise.

The noise floor is due to the kT/C noise from the switched capacitors. The -153 dBc/Hz is matching the results from an ideal simulation for 0.75 pF switched-capacitors, 6 pF fixed capacitance and a 0.8 Vpp amplitude across the crystal. A higher supply voltage to charge the switched capacitors would allow for smaller capacitors to sustain a 0.8 Vpp amplitude or would yield a larger amplitude with similar capacitors. Either option would reduce the noise floor.

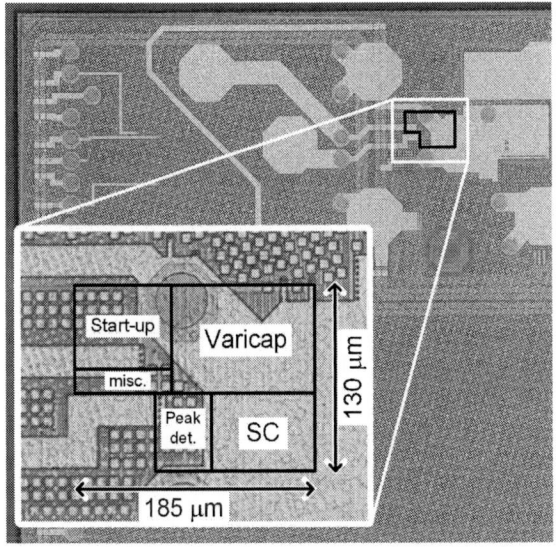

Fig. 6. Die micrograph in 22 nm FD-SOI.

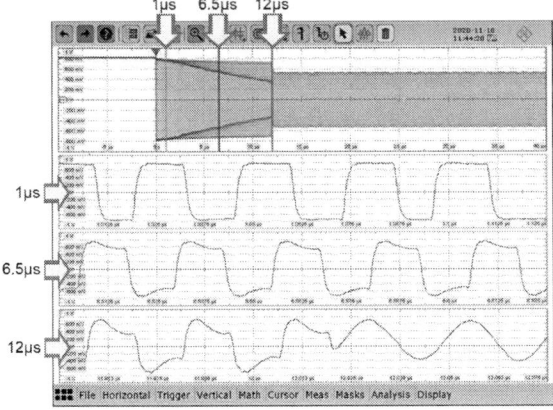

Fig. 7. Measurements of voltage across the crystal during start-up.

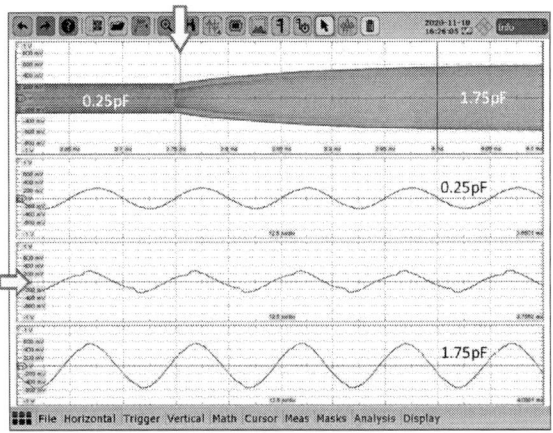

Fig. 8. Measurements of voltage across the crystal during steady-state operation.

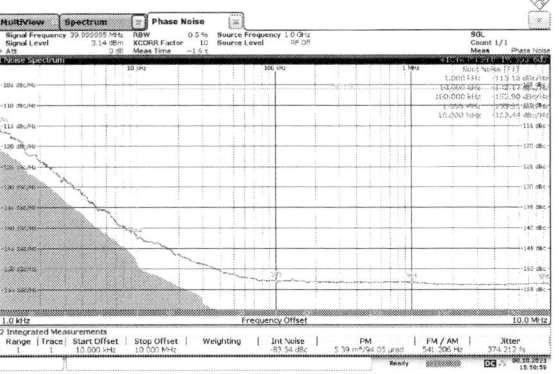

Fig. 9. Measured phase noise

TABLE 1 PERFORMANCE SUMMARY AND COMPARISON TO PRIOR WORKS

		This work	Lechevallier [1] JSSC'19		Verhoef [2] ISSCC'19	Megawer [3] ISSCC'19	Jung [4] ISSCC'22	Chang [5] JSSC'12	Rajavi [6] JSSC'17
Technology (nm)		22	22		55	65	28	65	28
Supply voltage (V)		0.8	0.8		1.2	1	N/A	1.8	1
Core area (mm^2)		0.021	0.02		0.129	0.069 **	2.51 ***	0.15	0.015
Frequency (MHz)		40	24	50	32	54	76.8	26	48
Load capacitance (pF)		6	6	7	6	6	7	8	8
Diff. amplitude (Vpp)		0.8	0.32		0.75	0.7	1.2	1	1.6
Steady-state power consumption (µW)		55	10	51	190	198	1080	2160	1500
Startup time (µs)		12	15	6	23	19	39.6	3200	N/A
Startup energy (nJ)		8	4.4	3.7	20.2	34.9	92.8	N/A	N/A
Phase-Noise (dBc/Hz)	@ 1kHz	-115.1	-123		N/A	-139.5	N/A	-136.1	-114.3
	@ 10kHz	-142.2	N/A		N/A	-149.7	N/A	-149.1	-147.7
	@ 100kHz	-152.9	N/A		N/A	-157.8	-158.2	-153.4	-155.8
	@ 1MHz	-153.2	N/A		N/A	-163.1	N/A	N/A	-158.5
RMS jitter (fs) [10kHz - 10MHz]		374	N/A		N/A	103	72	420 [10k-5MHz]	168
CMRR * (dBc/1Vpp)		-67.4	N/A		N/A	N/A	N/A	N/A	N/A
Steady-state technique		Switched capacitor	Single-ended Pierce		Single-ended Pierce	Single-ended Pierce	Single-ended Pierce	AC- and cross-coupled gm stage	Active inductor, cross-coupled gm stage
Fully differential		**Yes**	No		No	No	No	**Yes**	**Yes**
Fast start-up technique		Self-timed injection	Self-timed injection		Precisely timed injection	2-step injection	2-step injection	none	none

* CMRR = level of spurious tone with a 1 Vpp 1 MHz signal applied on the crystal case ** Off-chip load capacitance *** Complete chip

A comparison table is shown in Table 1. This work consumes an order of magnitude less power than other recently published differential crystal oscillators [5,6], and since power consumption typically scales with frequency and amplitude, the power consumption of this work is on par with the best reported single-ended oscillator [1], which lacks the common-mode rejection of a fully differential circuit.

The CMRR is evaluated by applying a 1 Vpp, 1 MHz signal at the crystal case and measuring the spurious tones at the output of an on-chip comparator, monitoring the signal across the crystal. The measured spectrum is depicted in Fig. 10. As a reference, simulation results of the single-ended Pierce oscillator of Fig. 1 show a spurious tone of -34 dBc with single-ended probing on XTALp, mainly defined by the ratio Cp1 divided by 2 CL, and an even higher spurious tone of -25 dBc with differential probing across the crystal pins, due to the gain of the inverter. The measured spurious tone of -67.4 dBc is therefore a 33 dB to 42 dB improvement in CMRR over a typical Pierce oscillator.

V. CONCLUSION

A fully symmetrical and fully differential crystal oscillator is presented as an alternative for the single ended and asymmetrical Pierce oscillator. The benefit of this fully differential oscillator is a significantly improved CMRR, which is very important for wireless systems in noisy environments. The load capacitance is connected directly across the crystal, which is an area reduction for the load capacitor of a factor four compared to the Pierce oscillator.

Also presented is a fast start-up circuit which does not require an additional start-up oscillator. The difficulty to measure a weak crystal signal across a low-Ohmic drive switch, present in other self-timed start-up circuits, is overcome by duty-cycling the drive pulse within each half period.

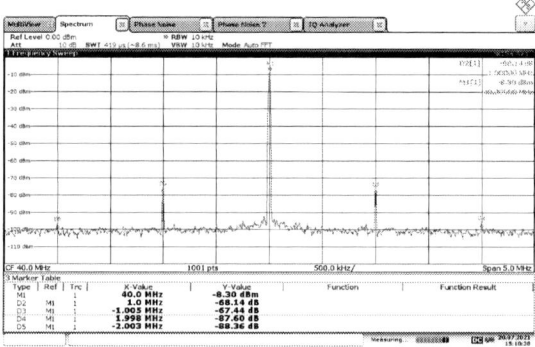

Fig. 10. Measured spectrum with a 1 MHz, 1 Vpp signal applied on the crystal case

REFERENCES

[1] J. B. Lechevallier, R. A. R. Van Der Zee, and B. Nauta, "Fast & energy efficient startup of crystal oscillators by self-timed energy injection," in IEEE JSSC, vol. 54, no. 11, pp. 3107–3117, Nov. 2019.

[2] B. Verhoef, J. Prummel, W. Kruiskamp and R. Post, "A 32MHz crystal oscillator with fast start-up using synchronized signal injection," ISSCC, pp. 304–305, Feb. 2019.

[3] K. M. Megawer, et al., "A 54MHz Crystal Oscillator With 30× Start-Up Time Reduction Using 2-Step Injection in 65nm CMOS," ISSCC, pp. 302-303, Feb. 2019.

[4] J. Jung et al., "A Single-Crystal-Oscillator-Based Clock-Management IC with 18× Start-Up Time Reduction and 0.68ppm/°C Duty-Cycled Machine-Learning-Based RCO Calibration," ISSCC, pp. 58-60, Feb. 2022.

[5] Y. Chang, et al., "A Differential Digitally Controlled Crystal Oscillator With a 14-Bit Tuning Resolution and Sine Wave Outputs for Cellular Applications," in IEEE JSSC, vol. 47, no. 2, pp. 421-434, Feb. 2012.

[6] Y. Rajavi, M. M. Ghahramani, A. Khalili, A. Kavousian, B. Kim and M. P. Flynn, "A 48-MHz Differential Crystal Oscillator With 168-fs Jitter in 28-nm CMOS," in IEEE JSSC, vol. 52, no. 10, pp. 2735-2745, Oct. 2017.

A 128ksps 120dB THD Low Noise Analog Front End

Gerard Mora-Puchalt
Analog Devices
Valencia, Spain
Gerard.Mora-Puchalt@analog.com

Gabriel Banarie
Analog Devices
Limerick, Ireland
Gabriel.Banarie@analog.com

Pawel Czapor
Analog Devices
Limerick, Ireland
Pawel.Czapor@analog.com

Adrian Sherry
Analog Devices
Limerick, Ireland
Adrian.Sherry@analog.com

Roberto Maurino
Analog Devices
Turin, Italy
Roberto.Maurino@analog.com

Jesús Bonache
Analog Devices
Valencia, Spain
Jesus.Bonache@analog.com

Italo Medina
Analog Devices
Valencia, Spain
Italo.Medina@analog.com

Abstract—**This paper describes a complete analog to digital front-end system consisting of a programmable gain two stage capacitive gain amplifier (CGA) driving a multistage delta-sigma ADC. It features a 60kHz bandwidth and 1ppm/FSR linearity at low gains (G≤8) and 3ppm/FSR at high gains (G>8). At a gain of 128, it achieves an input referred noise of 5.5nV/√Hz at a power consumption of 43mW.**

Keywords—*Capacitive Gain Amplifier (CGA), Switched capacitor circuits, chopping, delta-sigma analog-to-digital converter (ADC).*

I. Introduction

Read-out integrated circuits (ROIC) based on a CGA and followed by a ΔΣ analog-to-digital converter (ADC) are widely used in systems where a small signal, generated in a sensor, needs to be digitized with a high dynamic range [1]-[4]. Typical sensing applications are temperature or pressure sensors such as thermistors, thermocouples, resistance temperature detectors (RTD) and weight bridges. Usually, the bandwidth of these signals is well below 1kHz. In the last years however, Industry 4.0 has increased the need for industrial data acquisition systems to provide condition-based monitoring (CBM), often based on vibration analysis, and requiring higher bandwidth and excellent linearity.

This paper presents a ROIC based on a CGA and a switched capacitor ΔΣ ADC featuring a 60kHz bandwidth and 120dB THD, together with high input impedance, excellent noise/power trade-off, low offset, gain error and drift in temperature and time. Sections II and III describe the CGA and the ΔΣ ADC, respectively. Finally, section IV presents the measured results.

II. Capacitive Gain Amplifier Topology

The CGA utilizes capacitors as gain elements. Capacitors have advantages over resistors such as excellent temperature drift and stability over time and they do not contribute noise. The input signal is fully chopped, amplified by the CGA and finally dechopped (chop frequency is 512kHz). This mechanism provides signal amplification over all the frequency spectrum, including DC, and decouples the amplifier low frequency noise and offset from the input signal. Additionally, as the signal is not sampled but chopped, there is no kT/C noise penalty, hence making this architecture very suitable for ultra-low noise applications [1].

The CGA contains two gain stages. For the lower gains (0.5 to 8), the CGA only employs one gain stage and for the higher gains (16 to 128), the two stages are engaged. This

architecture allows the first stage to be optimized for ultra-low noise performance, while the second stage (where the output signal is larger due to amplification) can be optimized for high linearity.

One of the main challenges in the CGA is to define the Stage1 and Stage2 amplifier common mode input voltages. This function is performed by dividing the input capacitors C_{IP1} and C_{IN1} (Fig. 1) into two halves and performing an input signal chopping event in two steps. First, half of the input capacitors are chopped so that the CGA sees 0V differentially at the inputs and at the outputs. At this point, an autozero phase takes place to define the amplifiers common mode input voltages. And finally, the second step of the chopping event takes place. This sequence of events (half chop, autozero, half chop) needs to happen between two consecutive ADC sample events so that this function is transparent to the ADC [1].

A. Bootstrapped Switches

In the CGA architecture, the signal is amplified through the input and feedback capacitors, but it also travels through some of the CGA switches such as the input and output choppers. To achieve high linearity and excellent THD over a wider input bandwidth, a very linear switch is required, not only in terms of parasitic capacitance but also its on-resistance. To address this, the use of ultra-small, bootstrapped switches along the entire signal chain has been employed.

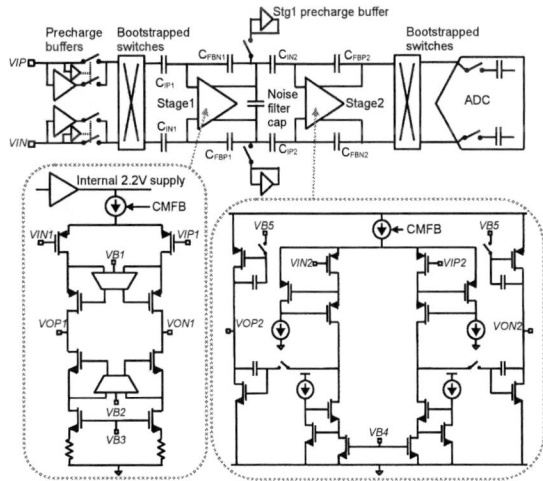

Fig. 1. ROIC schematic.

978-1-6654-8495-4/22 $31.00 © 2022 IEEE

B. First Stage Implementation

The first stage is designed for a 5.5nV/√Hz noise density target at gain of 128. It is a telescopic Operational Transconductance Amplifier (OTA) made of core 1.8V devices and powered off an internal 2.2V LDO to maximize its output swing and, therefore, its maximum stage gain achievable (gain of 128 is split into 16x in Stage1 and 8x in Stage2). This is critical to reduce the second stage noise when input referred. The choice of 1.8V devices instead of 5V devices also gives a noise advantage. The 1.8V devices have a better transconductance (g_m) versus gate capacitance ratio. Hence, g_m can be maximized by increasing the input pair width, but gate capacitance can be kept low by choosing almost minimum length input pair devices. A small gate capacitance in the input pair is important to avoid degrading the CGA noise transfer function. The telescopic amplifier also has degeneration resistors to reduce the load devices noise. The first stage common mode feedback loop (CMFB) is a switched capacitor implementation. There are two sets of CMFB capacitors that operate in different phases. During the Stage1 autozero phase, when the Stage1 amplifier input voltages are defined, the CMFB uses a set of capacitors to bias the input pair gates at a higher voltage than that of the amplification phase. This technique helps to decouple the input and the output common mode voltages and therefore maximize the telescopic amplifier output swing. To minimize noise while keeping a competitive bandwidth, the Stage1 is bandwidth-limited by filter capacitors and Stage1 output precharge buffers are used to boost its slew rate during the chopping transitions. The filtering capacitors also work as a sample-and-hold circuit to define the Stage1 precharge buffers desired output voltage during the chopping transition when slewing takes place. Stage1 precharge buffers are power cycled so that they only consume significant power during Stage1 slew condition. These buffers have been implemented as current mirror amplifiers with complementary source follower output, to offer very low output impedance and therefore a fast delivery of charge during slew.

C. Second Stage Implementation

The second stage not only amplifies the input signal but also directly drives the ADC (at a sampling frequency F_S of 8.192MHz) which makes the entire signal chain even more efficient as there is no need for additional buffers between the CGA and the ADC. The second stage amplifier is a 5V two-stage Miller compensated amplifier with a telescopic first stage and a class A output stage. The telescopic first stage employs cascode regulation to boost the loop gain. This regulation is done by forcing a gate-to-source voltage across the drain and source of the critical devices with a single device and a current source.

The output stage can switch into a class AB output stage during an initial phase of the full ADC driving cycle to improve slew and settling. During this initial phase, the now reconfigured as class AB output stage also gets disconnected from the rest of the amplifier. The structure formed by the output stage and the Miller capacitors is extremely fast to react against switched capacitor loads because any stimulus at the output translates into a stimulus at their gates immediately through the Miller capacitors. After this initial phase completes, the amplifier reconnects its stages together and the small signal settling takes place.

To drive the ADC during the chopping phase, the output stage is divided into two. One half output stage drives the ADC as described above and the other half output stage does the chopping events and the autozero. As the input signal can be as large as 5V, to speed up the chopping process, the second stage Miller capacitors are swapped, greatly speeding up the chopping slew rate phase which otherwise would be limited by charging and discharging the Miller capacitors.

D. Input Current Cancellation Scheme

The CGA employs precharge buffers during the chop event to minimize input currents. The precharge buffers are Miller compensated two-stage amplifiers. First stage is folded cascode with rail-to-rail complementary input pair. The NMOS input pair is used for higher input voltages and the PMOS input pair for lower input voltages. A current-steering circuit ensures that the amplifier exhibits a constant g_m over the entire input range. The output stage is class AB. The precharge buffers are autozeroed to minimize their offset. Low offset helps the CGA better settle after the precharge phase is complete and it also reduces the input currents. Only the first stage is autozeroed as the output stage offset is negligible when input referred by the first stage gain. The autozero switches and capacitors are connected in parallel with the folded cascode current mirror. During autozero, the output stage is disconnected from the first stage and their Miller capacitors charge is preserved. The precharge buffers are power cycled to minimize their power consumption.

The input current cancellation scheme uses a set of switches to engage the precharge buffers and other set of switches to reconnect the CGA input capacitors back to the inputs. Having a small channel area switch and good impedance balancing on both sides of the switches is also key to achieve low input currents, hence, bootstrapped switches are also employed in the precharge buffers stage together with auxiliary buffers to drive the gates and back-gates of those

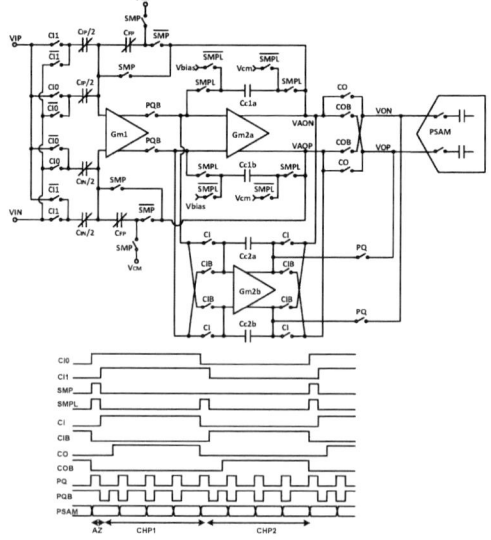

Fig. 2. CGA schematic and timing diagram.

bootstrapped switches. By having two different sets of buffers, the precharge activity is decoupled from the switches activity, hence achieving a better overall input current.

III. ADC IMPLEMENTATION

A. ADC architecture

The backend ADC is a discrete time multi stage noise shaping (MASH) 2-2-0 $\Delta\Sigma$ modulator [5], cascading two multi bit second order stages and a simple flash ADC (Fig. 3). The second order stages use a low distortion topology featuring two major benefits. First, in this topology only quantization noise is processed by the loop filter, which greatly relaxes the requirements of the switched capacitor integrators in terms of headroom and linearity. Secondly, the output of the last integrator is just a delayed version of the quantization noise. This makes the cascading of stages straightforward, as the next stage can be simply fed from the output of the last integrator. The quantization error of the first stage is measured by the second stage and subtracted in the digital reconstruction logic. Similarly, the quantization error of the second stage is converted by the final flash ADC and removed. The final recombined output only contains the quantization error of the last stage. The outputs of the 3 stages are given by:

$$V_1 = \tfrac{3}{4}\,U + (1\text{-}z^{-1})^2 Q_1 \qquad (1)$$

$$V_2 = z^{-2}\,Q_1/3 + (1\text{-}z^{-1})^2\,Q_2$$

$$V_3 = -\,z^{-2}\,Q_2 + Q_3$$

and the final recombined output is given by:

$$V = \tfrac{3}{4}\,z^{-4}\,U + 3(1\text{-}z^{-1})^4\,Q_3 \qquad (2)$$

This MASH architecture provides an aggressive 4th order shaping without stability concerns and it efficiently spreads the quantization levels across the 3 stages.

Mismatch noise from the DAC of the first loop is shaped by a 2nd order dynamic element matching (DEM) engine.

MASH modulators are intrinsically more resilient to idle tones, since any idle tone introduced in the main loop is cancelled by the following stages. Nevertheless, the cancellation is ultimately limited by the matching of the analog stages. Therefore a random dither spanning a quantizer step is introduced in the first stage. Conventionally, dither is just added in front of the quantizer, resulting in an increased headroom requirement for the loop filter. As an example, in the modulator topology of Fig 3, adding a dither with a span of one quantizer step range doubles the required headroom.

To avoid the headroom increase, a subtractive dithering scheme is used instead [6]. Dither is still added at the input of the quantizer but is then removed at its output. A dedicated dither DAC is used to remove dither at the input of the loop filter. To relax the gain matching requirements of the dither DAC, the uniform dither d spanning one quantizer step is first order shaped by a $1\text{-}z^{-1}$ transfer function.

B. First integrator of first stage

The first integrator of the first stage is implemented as a twin double sampling integrator. For sake of simplicity, a single ended schematic is shown in Fig 4, albeit the actual implementation is fully differential. p1 and p2 are non-overlapping clocks. The subtractive dither path is also shown.

This integrator topology features several advantages. Firstly, the input referred noise is low and can be shown to be well approximated by $\sqrt{\dfrac{kT}{2C_{IN}}\left(1 + \dfrac{C}{C_{IN}}\right)}$. Also, as long as the operating range of the input switches is not exceeded, the input voltage can be applied with any DC common mode. The same applies to the reference voltage.

Finally, the transfer function can be shown to be:

$$V_{OUT}(z) = \frac{1}{4}\frac{\left(1+z^{-\frac{1}{2}}\right)^2}{1-z^{-1}} \qquad (3)$$
$$\left(\frac{C_{IN}}{C}V_{IN}(z) - V_{DAC}(z)\right) + \frac{C_{DITH}}{4C}z^{-1}D(z)$$

Equation (3) features a double transmission zero at sampling frequency F_S, rejecting input signals at that frequency. The first aliasing frequency is therefore at $2\,F_S$, and the antialiasing filter can be relaxed. The output of the first integrator is resampled at F_S and the rest of the modulator operates at F_S. Notice that the dither d is applied to C_{DITH} delayed by one clock cycle and it appears directly at the output of the first integrator just scaled by $C_{DITH}/4C$. This is equivalent to applying a first order shaped dither at the input of the integrator. Thanks to the reduced headroom requirement the amplifier can be implemented as a power efficient telescopic OTA, operating from the 1.8V supply. Only the front end, switches, namely the switches directly

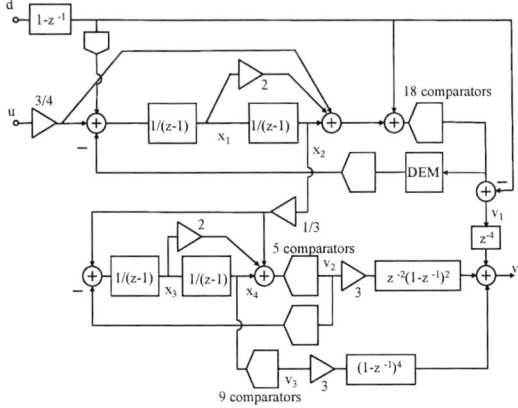

Fig. 3. ADC architecture.

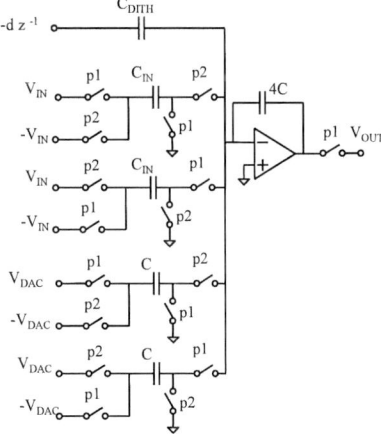

Fig. 4. Simplified first integrator schematic.

978-1-6654-8495-4/22 $31.00 © 2022 IEEE

connected to either the reference or the input signal, are operated from a 5V supply.

IV. MEASUREMENT RESULTS

The presented signal chain was manufactured in 180nm. Silicon results (Fig 5) show a performance of 120dB THD with a full-scale tone at 1kHz, while achieving 5.5nV/√Hz noise density at gain of 128 and 45nV/√Hz at gain of 1. The measured INL is 1ppm at low gains (G≤8) and 3ppm at high gains (G>8), relative to full scale range (FSR). The offset is (8μV+50μV/gain) with no need for calibration, hence keeping cost low. The gain error drift is 0.1ppm/°C typical. The CGA takes 5.4mA at low gains (from 0.5x to 8x) from 5V, and an additional 3.1mA from 2.2.V at high gains (from 16x to 128x). The ADC takes 0.9mA from a 5V supply and 2.5mA from a 1.8V supply.

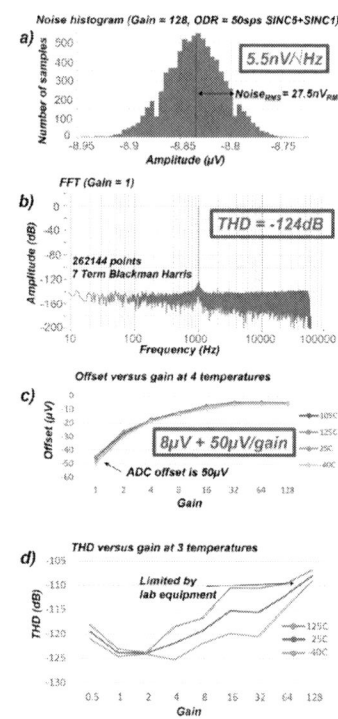

Fig. 5. Measured performance.

TABLE I. PERFORMANCE SUMMARY AND COMPARISON

	This work	[4] 2021	[2] 2019	[3] 2019	[1] 2017
Architecture	CCIA DTΔΣ	CC-CTΔΣ	CCIA DTΔΣ	CCIA CTΔΣ	CCIA DTΔΣ
Technology (nm)	180	180	130	180	180
Active Area (mm²)	2.78	1.65	0.65	0.73	0.53^b
Supply Voltage (V)	2.2/5/1.8	1.8	3	1.8	3.3
Supply Current (mA)	3.1/6.3/2.5	0.23	0.142	1.2	0.075
Bandwidth (kHz)	60	1	0.04	2	0.015
Input range (V_{pp})	8.8/gain	0.06	5.6/gain	0.02	2.5/gain
CM Input Range (V)	0.1 - 4.9	0-1.8	0-3	0-3.3	0-3
Offset (μV)	8+50/gain	1.9	1.87	7	1.8
Input Noise (nV/√Hz)	5.5	19.4	44	3.7	19
CMRR (dB)	138	140	126	134	109
FoM^a (dB)	146.6^c	155	144.6^c	151.1	141.5^bc
INL (ppm/FSR)	1-3	16	7	28	5
THD(dB)	-120	-110	-	-	-

a. FOM = $10\log(V_{FS}^2/(8\,v_n^2\,P))$, only analog power b. Without ADC c. At gain = 128

Table I compares this work to recently published ROIC based on CGA and ΔΣ converters. This work achieves the widest bandwidth, extending it by more than an order of magnitude, and the highest linearity.

V. CONCLUSIONS

A ROIC has been proposed with extended input bandwidth and excellent linearity targeted for the Industry 4.0 applications such as vibration analysis, system parametric extraction, analytics and CBM. Ultra-low noise performance is achieved by using a continuous time CGA (pushing the noise performance beyond kT/C limits) and an architecture that allows the use of core devices in the noise critical stages of the CGA and the ADC.

ACKNOWLEDGMENT

The authors would like to thank Damien McCartney, Colin Lyden, Christian Birk, Peter Hurrell, Eamonn Byrne and John Quill.

REFERENCES

[1] H. Wang, G. Mora-Puchalt, C. Lyden, R. Maurino and C. Birk, " A 19 nV/√Hz noise 2μV offset 75μA capacitive-gain amplifier with switched-capacitor ADC driving capability ", IEEE J. Solid-State Circuits, vol. 52, no. 12, pp. 3194-3203, Dec. 2017.

[2] J. Jun, S. Park, J. Kang and S. Kim, " A 22-bit read-out IC with 7-ppm INL and sub-100-μHz 1/f corner for DC measurement systems ", IEEE J. Solid-State Circuits, vol. 54, no. 11, pp. 3086-3096, Nov. 2019.

[3] H. Jiang, S. Nihtianov and K. A. A. Makinwa, " An energy-efficient 3.7 nV/√Hz bridge readout IC with a stable bridge offset compensation scheme ", IEEE J. Solid-State Circuits, vol. 54, no. 3, pp. 856-864, Mar. 2019.

[4] C. Lim et al., "A Capacitively Coupled CT Δ ΣM With Chopping Artifacts Rejection for Sensor Readout ICs," in IEEE Transactions on Circuits and Systems I: Regular Papers, vol. 68, no. 8, pp. 3242-3253, Aug. 2021.

[5] R. Brewer, J. Gorbold, P. Hurrell, C. Lyden, R. Maurino and M. Vickery, "A 100dB SNR 2.5MS/s output data ΔΣ ADC," ISSCC. 2005 IEEE International Digest of Technical Papers. Solid-State Circuits Conference, 2005.

[6] H. Zhang et al., "A 6μW 95 dB SNDR Inverter Based ΣΔ Modulator With Subtractive Dithering and SAR Quantizer," in IEEE Transactions on Circuits and Systems II: Express Briefs, vol. 66, no. 4, pp. 552-556, April 2019.

A 0.01mm^2, $0.4\text{V-}V_{\text{DD}}$, 4.5nW-Power DC-Coupled Digital Acquisition Front-End Based on Time-Multiplexed Digital Differential Amplification

Paolo Crovetti[1], Roberto Rubino[1], Pedro Toledo[1], Francesco Musolino[1], Hamilton Klimach[2], Yong Chen[3], Anna Richelli[4]

[1]*Dept. of Electronics and Telecommunications (DET), Politecnico di Torino*, Torino, Italy
[2]*Electr. Eng. Dept., Graduate Program in Microelectronics, Fed. Univ. of Rio Grande do Sul*, Porto Alegre, Brazil
[3]*State-Key Laboratory of Analog and Mixed-Signal VLSI and IME/ECE-FST, University of Macau*, Macao, China
[4]*Department of Information Engineering, University of Brescia*, Brescia, Italy
e-mail: paolo.crovetti@polito.it

Abstract—A reconfigurable, high-impedance, DC-coupled low-frequency digital acquisition front-end (DAFE) suitable to operate under a power supply voltage ranging from 0.2 to 1V down to 600 pW power is presented in this paper. Matching-indifferent DC accuracy over a rail-to-rail input range is uniquely achieved by the new time-multiplexed digital differential amplification technique at ultra-low area and without chopping and auto-zeroing. A 180nm testchip of the proposed DAFE occupies $0.00945\,\text{mm}^2$ and draws 4.5 nW at a 0.4 V supply, has a 120 Hz gain-bandwidth product, with an in-band input noise of $11.3\,\mu\text{V}_{\text{rms}}$, a $137\,\mu\text{V}$ input offset voltage standard deviation, 65.7 dB CMRR, 63.8 dB PSRR, and provides a 46.5 dB-SFDR, 6.9 bit-ENOB digitized output at -12 dBFS.

Index Terms—DC-coupled digital acquisition front-end (DAFE), time-multiplexed digital differential amplification (TMD^{2}A), CMOS, ECG, IoT.

I. INTRODUCTION

Ultra-compact, energy-quality scalable sensor interfaces operating from a low, extremely variable supply under a $< 10\,\text{nW}$ power budget are required to enable next-generation sub-millimeter remotely-powered implantable biosensors [1] and energy-autonomous IoT nodes [2]. In this scenario, AC-coupled acquisition front-ends (AFEs) with $< 10\,\text{nW}$ power [3], [4], [5] and 0.01-mm^2-scale area [6], [7] have been demonstrated for ECG, EEG and audio signal acquisition.

DC-coupled AFEs, needed e.g. in electrochemical and temperature sensors acquisition, pose further challenges, and DC accuracy (i.e. low DC offset/drifts, high DC CMRR and PSRR) is achieved by chopping [8], [6] and/or auto-zeroing [9], at the cost of much increased power, area and lower input impedance.

In this paper, a scalable, DC-coupled, biosignal digital AFE (DAFE) with inherent DC offset and drift suppression and direct digitization is demonstrated in 180nm CMOS aiming to simultaneously achieve order-of-nanowatt power, area-efficiency and decent DC accuracy over a wide supply range.

II. TIME-MULTIPLEXED DIGITAL DIFFERENTIAL AMPLIFICATION (TMD^{2}A)

The proposed DAFE entails inherently offset- and drift-free, differential-mode amplification without relying on matching,

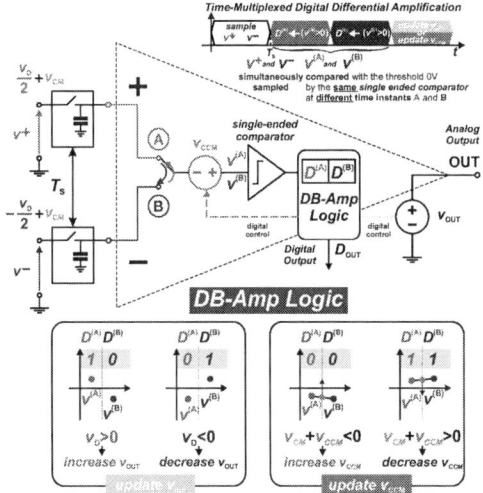

Fig. 1. Concept diagram of the Time-Multiplexed Digital Differential Amplification principle.

chopping or auto-zeroing, thanks to the new time-multiplexed digital differential amplification (TMD^{2}A) approach illustrated in Fig.1. Here, the same single-ended comparator is multiplexed in time to compare the non-inverting (v^+) and the inverting (v^-) differential analog inputs against the same threshold (assumed to be 0 V in the following), and differential-mode signal amplification is achieved as in a Digital-Based Amplifier (DB-Amp) [10].

In details, after v^+ and v^- are simultaneously sampled, the switch in Fig.1 is operated in A, so that v^+ is in series to the common-mode (CM) compensating floating voltage source (CM-FVS) v_{CCM}, and the input voltage

$$v^{(\text{A})} = v^+ + v_{\text{CCM}} = \frac{v_{\text{D}}}{2} + v_{\text{CM}} + v_{\text{CCM}}$$

is applied to the comparator. The comparator digital output is 1 if $v^{(\text{A})} > 0$ and 0 otherwise, and is stored in $D^{(\text{A})}$. Then,

---

978-1-6654-8495-4/22 $31.00 © 2022 IEEE

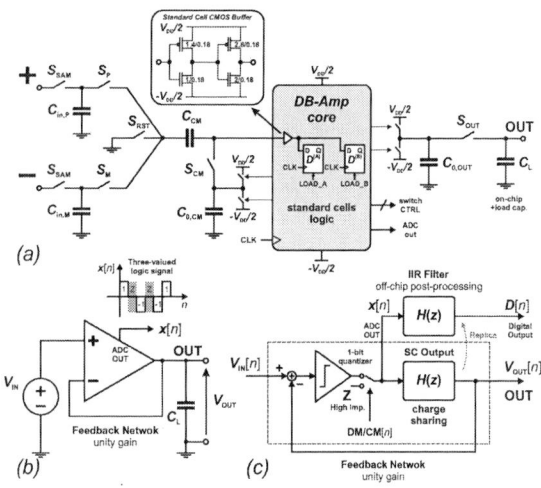

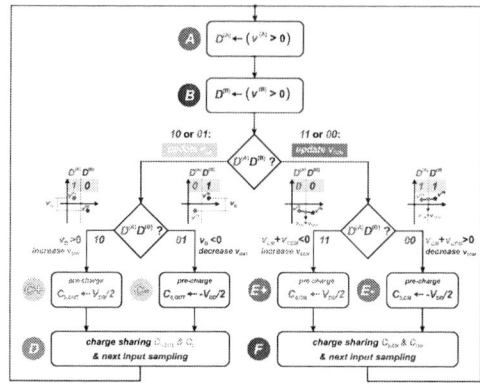

Fig. 2. Proposed SC TMD²A DAFE: (a) schematic, (b) unity-gain configuration considered for input signal digitization, (c) equivalent block diagram of the configuration in (b).

Fig. 3. State transition graph of the DB-Amp control unit.

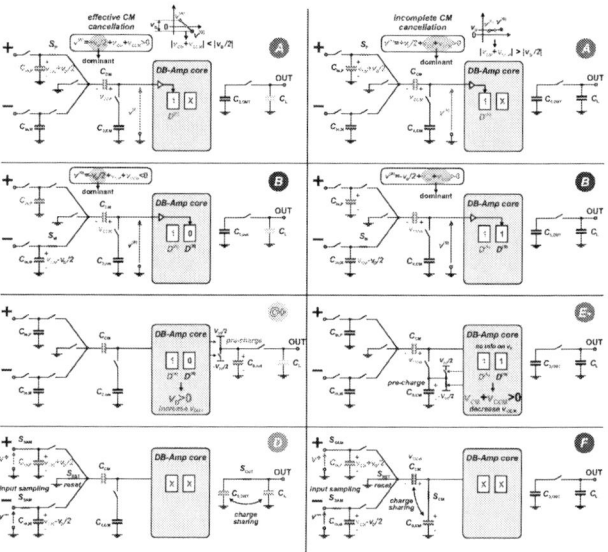

the switch is operated in B, the comparator input is

$$v^{(B)} = v^- + v_{CCM} = -\frac{v_D}{2} + v_{CM} + v_{CCM}$$

and the corresponding comparator output is stored in $D^{(B)}$.

Based on $D^{(A)}$ and $D^{(B)}$, either v_{OUT} or v_{CCM} are updated as in a DB-Amp [10], i.e. if $(D^{(A)}, D^{(B)}) = (1,0)$ $((D^{(A)}, D^{(B)}) = (0,1))$, which implies that $v_D > 0$ ($v_D < 0$), v_{OUT} is increased (decreased) by a minimum step, and v_{CCM} is kept unchanged. On the contrary, if $(D^{(A)}, D^{(B)}) = (1,1)$ $((D^{(A)}, D^{(B)}) = (0,0))$, v_{CCM} is decreased (increased) by a minimum step so that to track v_{CM}, aiming to enforce $|v_{CM} + v_{CCM}| < |v_D/2|$, as required to detect the sign of v_D in the next cycles, while v_{OUT} is hold.

A. Transistor-Level Design

In the proposed DAFE, TMD²A is realized in 180nm CMOS by switched capacitor (SC) techniques [see schematic in Fig. 2(a)]. A small standard cell CMOS digital buffer [transistor level details in Fig. 2(a)] is used as a single-ended comparator, while the input sampling capacitors, the CM-FVS and the output voltage source are implemented by pF-range Metal-insulator-Metal (MiM) capacitors, and the floating switches are designed by CMOS pass gates operated at bootstrapped voltage. The DB-Amp core is a finite state machine (FSM) automatically synthesized in standard-cell logic, whose state transition graph is shown in Fig. 3.

B. Circuit Operation

In state A (B), the input capacitor $C_{in,P}$ ($C_{in,M}$), charged at the non-inverting (inverting) input voltage v^+ (v^-) sampled at the previous cycle (see below), is connected in series to C_{CM} at the buffer input (Fig.4, states A-B in the left and right column) and its digital output is stored in the D-Flip Flop $D^{(A)}$ ($D^{(B)}$) at the next clock edge. Depending on $(D^{(A)}, D^{(B)})$, the FSM evolves to states C+/C-, to update the output voltage (left

Fig. 4. Sequence of switch configurations corresponding to the states in Fig.3: for $|v_{CM} + v_{CCM}| < |v_D/2|$, i.e. dominant differential component (left) and for $|v_{CM} + v_{CCM}| > |v_D/2|$, i.e. dominant CM component (right).

column in Fig.4), or to states E+/E-(right column in Fig.4), to update the voltage v_{CCM} across C_{CM}.

In order to increase (decrease) the output voltage, the fF-range capacitor $C_{0,OUT}$ is first pre-charged to $V_{DD}/2$ ($-V_{DD}/2$) in state C+ (C−), and then connected in parallel to C_L in state D, so that v_{OUT} is increased (decreased) by:

$$\Delta v_{OUT} = \frac{C_{0,OUT}}{C_L + C_{0,OUT}} \left(\underset{(-)}{\overset{+}{}} \frac{V_{DD}}{2} - v_{OUT} \right).$$

In the same state D, v^+ and v^- are simultaneously sampled on $C_{in,P}$ and $C_{in,M}$. In a similar fashion, in E+ (E-), the fF-range capacitor $C_{0,CM}$ is pre-charged to $V_{DD}/2$ ($-V_{DD}/2$) and then connected in parallel to C_{CM} in the next state F, to update v_{CCM} as demanded to compensate the CM input component, as well as drifts and process, voltage and temperature (PVT)-related variations in the CMOS buffer logic threshold. In the

978-1-6654-8495-4/22 $31.00 © 2022 IEEE

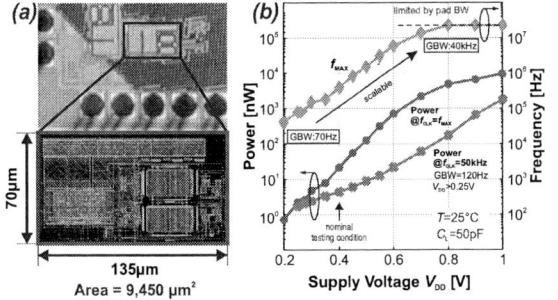

Fig. 5. Test chip micrograph (a) and (b) maximum clock frequency $f_{\max}$ and power (@$f_{\text{clk}} = f_{\max}$ and @$f_{\text{clk}} = 50\,\text{kHz}$) versus supply voltage V_{DD}.

TABLE I
MEASURED DC PERFORMANCE OF SIX DICE AT $25°$ C, $V_{\text{DD}} = 0.4$ V

	Unit	#1	#2	#3	#4	#5	#6	μ	σ
Input Offset	μV	+83	+19	+54	+230	-20	-188	+30	137
DC Gain	dB	40.8	39.9	39.6	38.0	42.7	36.3	39.6	2.2
CMRR	dB	68.0	74.0	68.5	60.4	60.0	63.4	65.7	5.4
PSRR	dB	83.9	55.3	54.0	60.6	69.0	60.1	63.8	11.1

same state F, v^+ and v^- are also sampled, as in D. Either state D or state F conclude an operating cycle, and are followed by state A in the next cycle.

As in [10], [11], the circuit operates as an analog differential amplifier with a gain-bandwidth (GBW) product proportional to the clock frequency. When connected in negative feedback under fixed capacitive load [e.g. in the unity-gain configuration as in Fig. 2(b)], the circuit operates as a delta modulator [6], [7] [see Fig. 2(c)], and a digitized version of the output can be retrieved in post-processing from the sequence of the DB-Amp FSM states by infinite impulse response (IIR) digital filtering.

III. EXPERIMENTAL RESULTS

A CMOS test-chip of the proposed DAFE has been fabricated in 180 nm and occupies merely 0.00945 mm^2 [micrograph in Fig. 5(a)]. When tested at 25°C in unity-gain feedback configuration [Fig. 2(b)] with $C_{\text{L}} = 50\,\text{pF}$, the DAFE operates under a 0.2-1 V supply voltage range with a rail-to-rail input/output swing, at a maximum clock frequency $f_{\max}$ ranging from 28 kHz to 15 MHz (upper-limited by the analog pads), corresponding to a scalable GBW from 70 Hz to 40 kHz. From Fig. 5b, the power consumption varies from 600 pW at 0.2 V to 9.58 μW at 1 V (from 2.0 nW at 0.25 V to 1.77 μW at 1 V) under $f_{\text{clk}} = f_{\max}$ ($f_{\text{clk}} = 50\,\text{kHz}$). At 0.4 V supply ($V_{\text{DD}}$) and $f_{\text{clk}} = 50\,\text{kHz}$, considered hereafter in the DAFE characterization, the power is 4.5 nW.

Based on the DC characterization at 25°C, 0.4 V V_{DD}, $C_{\text{L}} = 50\,\text{pF}$ and $f_{\text{CLK}} = 50\,\text{kHz}$ (see Tab. I), the input offset voltage of six dice ranges from -188 to $+230\,\mu$V, the open-loop DC gain from 36.3 to 42.7 dB, the CMRR from 60.0 to 74.0 dB and the PSRR from 54.0 to 83.9 dB. These results prove the reasonable DC accuracy of the proposed DAFE. Moreover, the DC input resistance of all dice is above 1 GΩ.

The time-domain analog output of sample #2 in unity-gain feedback [v_{OUT} in Fig. 2(b)] under 0.1 Hz, 360 mV$_{\text{pk-pk}}$

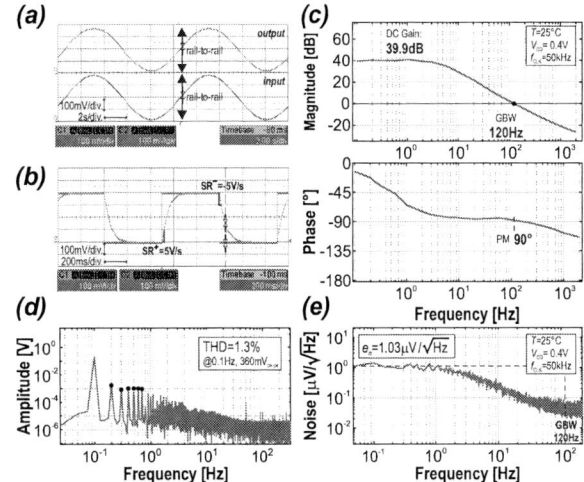

Fig. 6. Analog output testing: time-domain response in unity-gain closed-loop configuration for rail-to-rail sine (a) and square wave (b) input; (c) open-loop differential gain frequency response (magnitude and phase); (d) output voltage spectrum (test conditions as in (a)), (e) input noise spectrum (from output noise spectrum measured with $v^+ = v^- = 0$, considering 39.9 dB DC gain).

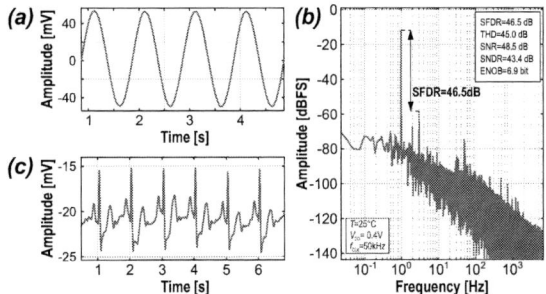

Fig. 7. ADC output testing: time-domain waveform reconstructed from the ADC output stream [Figs. 2(b)-(c)] under 1 Hz, 50 mV$_{\text{pk}}$ (-12 dBFS) sine input (b) spectrum of the reconstructed output in (a) and ADC dynamic performance; (c) reconstructed ECG waveform (ECG-20 mV electrode offset).

sine-wave input, and under a rail-to-rail square-wave input are shown in Figs. 6(a)-(b). The spectrum of the signal in Fig. 6(a) is shown in Fig. 6(c) and reveals a THD of 1.3%. Based on the open-loop frequency response reported in Fig. 6(c), the DC gain is 39.9 dB and the GBW is 120 Hz. The open-loop input-referred noise spectrum in Fig. 6(e) reveals an in-band noise power spectral density of 1.03 μV/$\sqrt{\text{Hz}}$ and an effective suppression of $1/f$ noise. The integrated input noise in the 0.05-1 Hz (0.05-120 Hz) bandwidth is 1.03 μV$_{\text{rms}}$ (11.3 μV$_{\text{rms}}$).

In Figs. 7(a)-(b), the time-domain waveform and the spectrum of a 50 mV$_{\text{pk}}$ (i.e. -12 dBFS), 1 Hz sine-wave reconstructed from the digital output stream of the DAFE post-processed off-chip by an IIR filter as in [6] is shown. The results reveal 46.5 dB SFDR, 45.0 dB THD and 43.4 dB SNDR, corresponding to 6.9 ENOB, and demonstrate the operation of the DAFE as an ADC. An ECG signal reconstructed from the digital output is shown in Fig. 7(c) and confirms the suitability of the DAFE to biosignal acquisition.

TABLE II
PERFOMANCE SUMMARY AND COMPARISON WITH LOW FREQUENCY SENSOR INTERFACES

Reference	Unit	[6]	[12]	[3]	[5]	[8]	[9]	[13]	[7]	**This Work**
Architecture		AFE	ATC	AFE	AFE	CCIA	Buffer	LNA	Δ-ADC	DAFE
Coupled		DC	DC	AC	AC	DC	DC	DC	DC	DC
Technology	nm	40	65	65	180	65	180	180	180	180
Area	mm^2	0.015	**0.006**	0.200	0.75	0.100	1.4	1	0.011	0.00945
Normalized Area	$10^6 \cdot$ F^2	9.37	1.42	47.3	23.1	23.7	43.2	30.9	0.65	**0.291**
Supply Voltage	V	0.6	0.5	0.6	1.2/0.6	1	1.8	0.2/0.8	0.6/1.2/3.3	0.4
Supply Range	V	N/A	N/A	0.5-0.7	N/A	N/A	N/A	N/A	N/A	0.2-1
Input Range	% V_{DD}	50	0.8	N/A	N/A	N/A	N/A	N/A	91	90
Power	nW	3,300	1,275	**3**	8.3$^{(e)}$	1,800	378,000	790	990	4.5
Bandwidth	Hz	150	11,000	1.5-370	470	0.5-100	**1,450,000**	670	500	120$^{(i)}$
Input Resistance	GΩ	0.05	0.031$^{(f)}$	N/A	N/A	0.8$^{(f)}$	**(g)**	0.01$^{(f)}$	2.96	> 1
DC Gain	dB	N/A	N/A	32	31-59	40	0	**57.8**	N/A	39.6$^{(i,k)}$
Input Offset	μV	N/A	N/A	N/A	N/A	1	**0.4**	N/A	N/A	137$^{(j)}$
In-band Noise (noise BW)	μV$_{rms}$ (Hz)	20$^{(c)}$ (1-150)	3.8	26 (1.5-370)	17	6.7$^{(d)}$ (0.5-100)	N/A	**0.94** 1-500	2.6$^{(c)}$	11.3$^{(c,h)}$ (0.05-120)
Noise PSD	μV$/\sqrt{\text{Hz}}$	N/A	0.036	N/A	N/A	0.021	**0.02**	0.036	N/A	1.03
NEF$^{(a)}$		147$^{(c)}$	2.2	2.1	2	3.3	7.3	**1.6**	3.5$^{(c)}$	4.1$^{(c)}$
PEF$^{(b)}$		13,012$^{(c)}$	2.4	2.6	4.8	10.9	96	**2.1**	15.2$^{(c)}$	6.7$^{(c)}$
THD	%	1	**0.2**	N/A	N/A	1.5	N/A	N/A	N/A	1.3$^{(l)}$
CMRR	dB	60	60	60	N/A	**134**	N/A	85	> 78	65.7$^{(i)}$
PSRR	dB	N/A	N/A	63	N/A	120	**125**	80/75	N/A	63.8$^{(i)}$
SFDR	dB	56	**60**	N/A	N/A	N/A	N/A	N/A	N/A	46.5
ENOB	bit	N/A	N/A	9.2	7.7	N/A	N/A	N/A	**9.7**	6.9

$^{(a)}$ $NEF = V_{\text{n,rms}}\sqrt{\frac{2I_B}{4\pi k T V_T \text{BW}}}$, where $V_{\text{n,rms}}$ is the rms input noise, I_B is the supply current and BW is the bandwidth; $^{(b)}$ $PEF = V_{DD} \cdot NEF^2$; $^{(c)}$ includes ADC quantization noise (and ADC current for NEF/PEF), $^{(d)}$ with DC servo loop, $^{(e)}$ LNA+ADC power considered, $^{(f)}$ estimated. $^{(g)}$ <1pA total input current; $^{(h)}$ Input rms noise in the 0.05-120 Hz BW estimated from the measured open-loop input-referred noise spectral density e_n in Fig.6(e) as $e_n \cdot \sqrt{BW}$; $^{(i)}$ Average of 6 samples; $^{(j)}$ Std. Dev. of 6 samples; $^{(k)}$ Open-loop gain; $^{(l)}$ Under 0.1 Hz, 90% rail-to-rail input; Best in **bold**.

IV. COMPARISON AND CONCLUSIONS

Compared with the state of the art of low-frequency AFEs in Table II, the proposed DAFE occupies the lowest normalized area (2.2X less than [7]) and works at the minimum voltage and over the widest supply range (4X wider than [3]) with rail-to-rail input/output swing, uniquely offering power-quality scalability. At 0.4 V, the power is comparable with AC-coupled AFEs [3], [5] (1.5X more than [3], in 65 nm, and 1.8X less than [5], also in 180nm), which have 3.1X-3.9X larger bandwidth and 2.3X-1.5X more in-band rms noise.

Compared to DC-coupled AFEs with similar bandwidth [6]–[8], [13], the DAFE has an inherently high input resistance (> 1 GΩ), consumes 1.2X-106X less area and 175X-733X less power, while providing a digitized output as [6], [7]. The offset, although 137X-342X larger than [8], [9], is the only reported in sub-μW AFEs. The in-band rms noise is 1.8X less than [6] and 4X more than [7], which inherently include quantization noise as the DAFE, while it is 12X more than the best [13]. The noise efficiency factor (NEF)/power efficiency factor (PEF) metrics, defined in Table II, are 1.2X worse/2.3X better than [7] and 36X/1,942X better than [6], while they are 2.6X/3.2X worse than [13], the best in Table II. The digitized output of the DAFE is slightly worse in terms of effective resolution (-0.8/-2.8 bit ENOB) compared to [3], [5], [7].

The measured results confirm that the performance of the DAFE are suitable to DC-coupled sensor interfaces for next-generation microscale biosensing and energy-autonomous IoT nodes.

REFERENCES

[1] S. Carrara, "Body Dust: Well beyond wearable and implantable sensors," *IEEE Sensors Journal*, pp. 1–1, 2020.

[2] M. Alioto, "From less batteries to battery-less alert systems with wide power adaptation down to nWs" *IEEE Design Test*, vol. 38, no. 5, pp. 90–133, 2021.

[3] P. Harpe et al., "A 0.20 mm^2, 3 nW signal acquisition IC for miniature sensor nodes in 65 nm CMOS," *IEEE Journ. of Solid-State Circ.*, vol. 51, no. 1, pp. 240–248, 2016.

[4] H. Wang et al., "A 5.5 nW battery-powered wireless ion sensing system," in *ESSCIRC 2017*, 2017, pp. 364–367.

[5] S. Jeong, et al., "Always-on 12-nW acoustic sensing and object recognition microsystem for unattended ground sensor nodes," *IEEE Journ. of Solid-State Circ.*, vol. 53, no. 1, pp. 261–274, 2018.

[6] R. Mohan, et al., "A 0.6-V, 0.015-mm^2, time-based ECG readout for ambulatory applications in 40-nm CMOS," *IEEE Journ. of Solid-State Circ.*, vol. 52, no. 1, pp. 298–308, 2017.

[7] M. R. Pazhouhandeh et al."Opamp-less sub-μW/channel Δ-modulated neural-ADC with super-GΩ input impedance," *IEEE Journ. of Solid-State Circ.*, vol. 56, no. 5, pp. 1565–1575, 2021.

[8] Q. Fan et al.,"A 1.8 μW 60 nV$/\sqrt{\text{Hz}}$ capacitively-coupled chopper instrumentation amplifier in 65 nm CMOS for wireless sensor nodes," *IEEE Journ. of Solid-State Circ.*, vol. 46, no. 7, pp. 1534–1543, 2011.

[9] T. Rooijers, J. H. Huijsing, and K. A. A. Makinwa, "An auto-zero-stabilized voltage buffer with a quiet chopping scheme and constant sub-pA input current," *IEEE Journ. of Solid-State Circ.*, pp. 1–1, 2021.

[10] P. S. Crovetti, "A digital-based analog differential circuit," *IEEE Trans. on Circ. and Syst. I: Reg. Papers*, vol. 60, no. 12, pp. 3107–3116, 2013.

[11] P. Toledo, et al. "Fully digital rail-to-rail OTA with sub-1000-μm^2 area, 250-mV minimum supply, and nW power at 150-pF load in 180 nm," *IEEE Solid-State Circ. Lett.*, vol. 3, pp. 474–477, 2020.

[12] L. B. Leene and T. G. Constandinou, "A 0.006 mm^2 1.2 μW analog-to-time converter for asynchronous bio-sensors," *IEEE Journ. of Solid-State Circ.*, vol. 53, no. 9, pp. 2604–2613, 2018.

[13] F. M. Yaul and A. P. Chandrakasan, "A sub-μW 36nV$/\sqrt{\text{Hz}}$ chopper amplifier for sensors using a noise-efficient inverter-based 0.2V-supply input stage," in *ISSCC 2016*, 2016, pp. 94–95.

A 69dBA-730μW Silicon Microphone System with Ultra & Infra-Sound Robustness

Jose Luis Ceballos
Infineon Technologies AG
Villach, Austria
Joseluis.Ceballos@infineon.com

Christopher Rogi
Infineon Technologies AG
Villach, Austria
Christopher.Rogi@infineon.com

Fulvio Ciciotti
Infineon Technologies AG
Villach, Austria
Fulvio.Ciciotti@infineon.com

Cesare Buffa
Infineon Technologies AG
Villach, Austria
Cesare.Buffa@infineon.com

Dietmar Straeussnigg
Infineon Technologies AG
Villach, Austria
Dietmar.Straeussnigg@infineon.com

Andreas Wiesbauer
Infineon Technologies AG
Villach, Austria
Andreas.Wiesbauer@infineon.com

Abstract—This paper presents a complete Silicon Microphone System with 69dB SNR (A-weighted) measured at 1Pa - 1kHz input signal (-34dBFS digital output level), with a power consumption of only 730μW for the complete product. This is greatly due to a *DAC-on-demand* approach used in the internal Delta-Sigma ADC. Ultra and Infra-sound robustness are also addressed at circuit level. Developed in a 130nm technology, the ASIC uses an area of 1.13mm² (including all auxiliary circuits).

Keywords—*Silicon Microphones, Sigma-Delta ADC, PGA*

I. INTRODUCTION

Latest consumer electronic products put stringent specs in microphones systems [1,2]. State of the art is around 69dBA SNR (A-Weighted) with power consumption below 1mW @ 1Pa-1kHz inputs. Programmability of power modes with a graceful degradation of Signal-to-Noise Ratios (SNR) is part of the design specs, together with ultra and infrasound robustness among other requirements.

This paper presents a flexible high-performance system macro specially tailored to fit this kind of specifications. The paper is organized as follows: Section II presents the basic structure. Section III presents the circuit considerations, while in Section IV measurement results are described. The paper ends with the conclusions and acknowledgments.

II. ARCHITECTURAL CONSIDERATIONS

At very high level, Figure 1 depicts the system level partitioning of the microphone: a differential MEMS is biased using the constant charge approach for linearity reasons, by means of antiparallel cascaded MOSFETs acting as biasing diodes in their cutoff (weak inversion) region. The differential input is coupled to a Programmable Gain Amplifier (PGA) in current mode with resistive feedback [3, 6], ensuring programmability of gains at constant bandwidth, related to settling constraints on the next stage capacitors. The optimized PGA differential output drives a 3rd order Switched-Capacitor 4-bits Delta-Sigma (DS) ADC with optimized zeros.

The included digital circuitry helps to cancel offsets and adds extra level of gain programmability, allowing at the same time *soft-hearable* transitions between the different modes of operation (those being clock, power and SNR

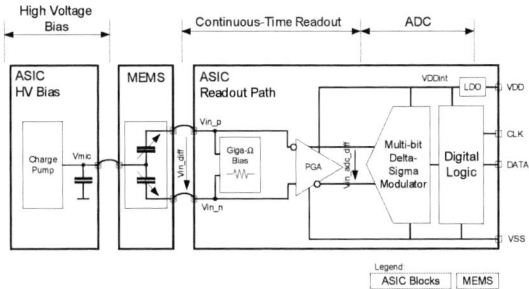

Figure 1: Simplified block diagram of the microphone system. Digital Logic also includes a 1-bit digital noise shaper and calibration.

related). The complete set of auxiliary circuits (analog and digital LDOs, charge-pump biasing, low-power bandgap reference voltage, etc.) are included in the total power consumption numbers reported.

III. CIRCUIT CONSIDERATIONS

This section will describe the followed system approach and the circuits for the different sub-blocks.

A. Programmable Gain Amplifier (PGA)

As mentioned, the current feedback approach is used to ensure enough settling in the next switched-capacitor stage (ADC input). Figure 2 shows its simplified schematic. The resistive feedback puts some constraints: for big signals, the resistor values will limit the maximum amount of current coming from the internal LDO supply (related to LDO stability, and to the maximum power consumption allowed at high level). For small signals, the resistors will also limit the thermal noise and the flicker noise SNR (this last one was maximized at expenses of extra-required area).

Class-AB driving stages are internally used to minimize static power consumption for small signals, and to have enough slewing capabilities. A big *reservoir* capacitance is placed at the output to minimize harmonic distortion due to switching spikes, and to provide a fast charge transfer to the sampling capacitors.

B. Delta-Sigma ADC

Figure 3 shows the simplified block diagram of the delta-sigma ADC used. It is a 3rd order modified feedforward [4] (2nd feedforward path is done internally with the 1st integrator

978-1-6654-8495-4/22 $31.00 © 2022 IEEE

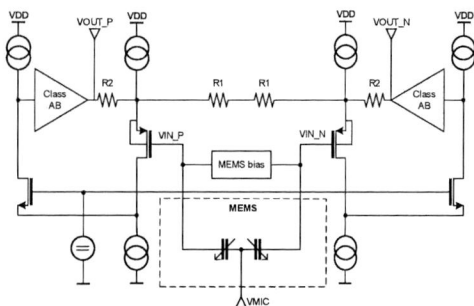

Figure 2: Programmable Gain Amplifier (PGA) current feedback. It uses resistive gain setting for the different modes of the product. Class AB is used to improve static and dynamic power consumption.

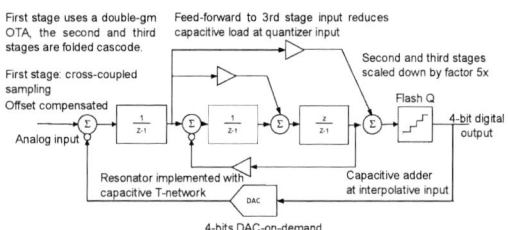

Figure 3: Delta-sigma 3rd order feedforward with FLASH quantizer.

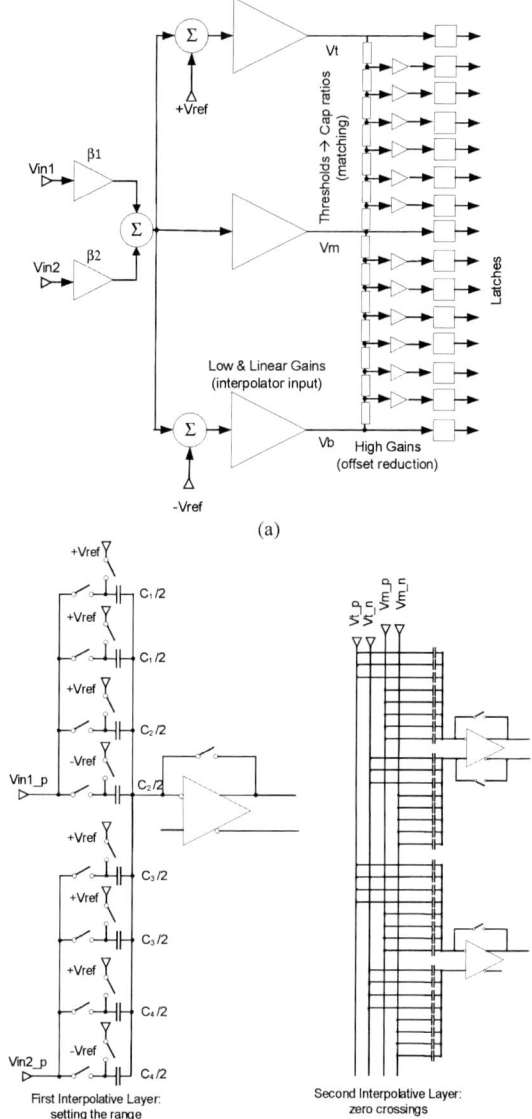

Figure 4: Flash quantizer. Two-layers interpolative switched capacitor. a) simplified block diagram, b) detail of how to generate the range setting with the top linear range amplifier, c) example of the capacitive interpolation between two of the previously described linear gains.

output to 3rd integrator input) to reduce quantizer input capacitances and wiring. A 4-bits FLASH quantizer is used to improve ultrasound responses (instead of the low-power tracking quantizer used in previous designs). The quantizer uses a 2 layers interpolative approach to reduce input wiring too. Simple internal amplifiers are used as zero-crossing detectors, helping at the same on reducing the latches' input referred offsets (Figure 4).

The first stage integrator is implemented with a double-gm OTA with switched-cap common-mode feedback [5] (Figure 5). Power optimized modes (switchable with internal commands or via automatic clock detection) are in place to have a power-on-demand system.

The feedback DAC is implemented using the concept of *DAC-on-demand* [7] (Figure 6): the idea behind is that for small signals only few DAC cells are really contributing to the feedback (all other cells' charges cancel out differentially), so there is no need to have them all ON (injecting noise and also consuming dynamic power). Noise is reduced as per equation 1, where C_{DAC} now is modulated with the input signal.

$$v_n^2 \propto \frac{KT}{C_{IN}} \left(1 + \frac{C_{DAC}}{C_{IN}} \right)$$

(1)

By means of a simple decoder out of the thermometer quantizer output, only the required cells will be turned ON. To save extra power, the reference buffer was also split in several sub-buffers (Figure 7). Segmentation was used to save even more power: the inner cells (approximately half of the DAC) are having always their respective sub-reference buffer ON (just to reduce dynamic effects when switching among those). In any case, their capacitances are switched to common mode if they are not used (to reduce noise). The other half of the DAC cells (used for higher signals) are almost OFF: they are always pre-biased with a very small current to reduce power and speed up their turn-on time.

The PGA input has a digitally assisted RBIAS [8], shown in Figure 8, to cope in a better way with big input signals in the Infrasound range. Under some circumstances, after receiving a sound shock, the extra high impedance (hundreds of Giga-Ohms) biasing takes seconds of time to recover. During that time, the asymmetrically forced output of the PGA allows the LDO noise to couple towards the ADC input. In addition, this unbalanced big signal forces the DAC on demand to contribute with more noise, which is added to the extra quantization noise due to the potential saturation of the loop integrators (on purpose intrinsic stabilization due to the feedforward structure). The shock detector uses a window comparator. Above some programmable threshold, there is a

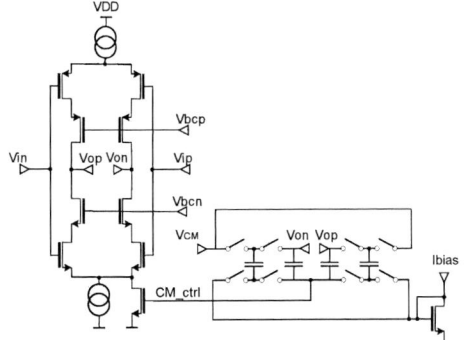

Figure 5: Double gm fully differential 1st Stage OTA with switched capacitor common mode feedback.

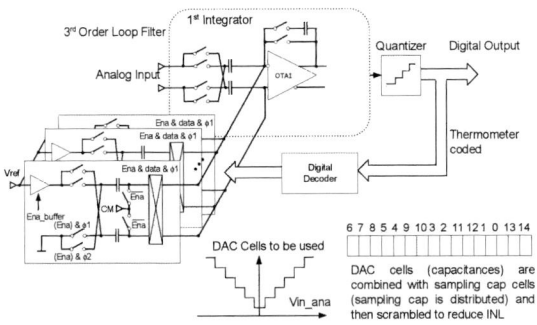

Figure 6: DAC on demand concept. Depending on the input signal, the reference buffers and the corresponding DAC capacitances are switched on/off

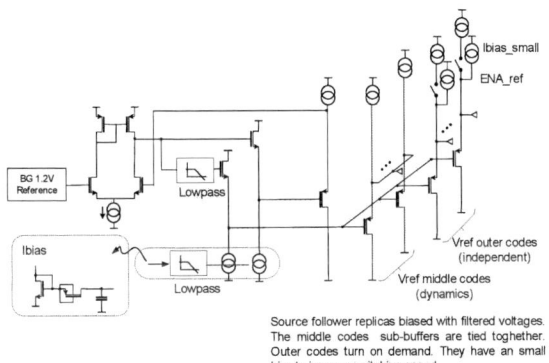

Figure 7: Reference buffer simplified schematic.

virtual short circuit of the high-impedance node to release the trapped charge. The state machine then controls a DAC, which in turn controls the gate of the bypass device over time and with optimized shape, gradually coming back to the OFF state, where it does not contribute with noise.

IV. MEASUREMENTS

Figure 9 shows the measured SNR and SNDR (ASIC + MEMS) for different operating modes (low power and high performance modes). The small inputs signal range (below 94dBSPL) is specified in terms of noise. The high level

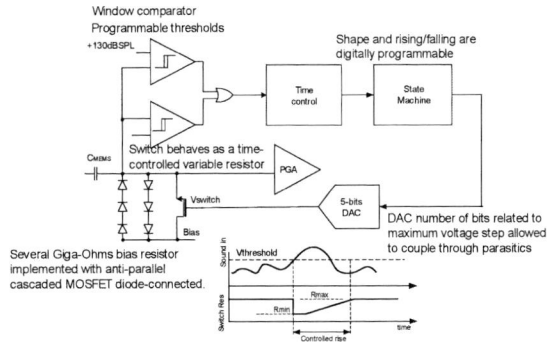

Figure 8: Digitally assisted shock detector / correction.

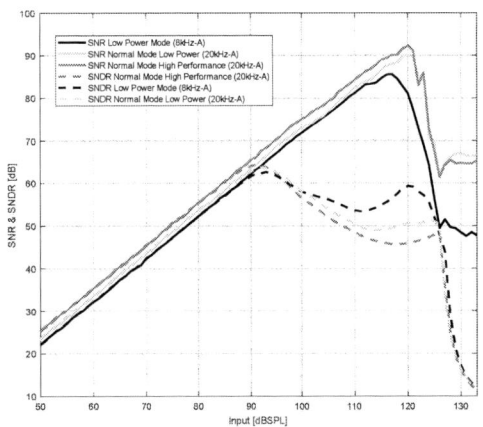

Figure 9: Measured SNR / SNDR for different performance modes (20kHz BW-A high and low performance modes clocked at 3MHz, and 8kHz BW-A low power mode clocked at 768kHz).

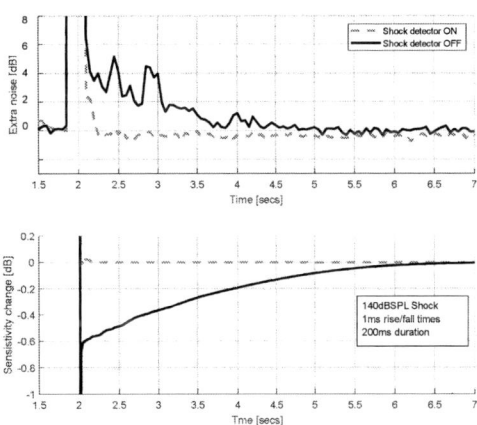

Figure 10: Measured shock recovery (Infrasound robustness) with and without the digitally assisted block enabled. Extra noise over time and sensitivity change over time shown. The shock is a 140dBSPL/200ms pulse with 1ms rise/fall times.

sound pressure input signals range is more related to harmonic distortion specs. The SNDR is worsen on purpose for big amplitudes in order to optimize the PGA and the ADC for low power in the audio band.

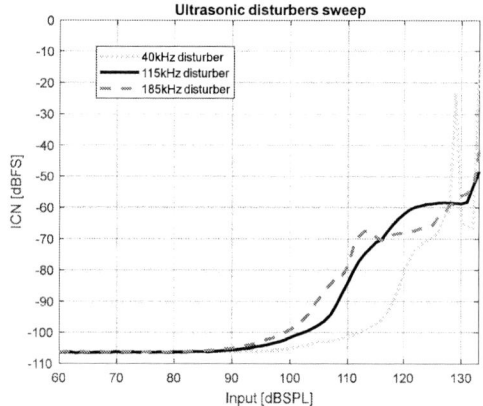

Figure 11: Measured leveled noise in presence of Ultrasound signals (amplitude swept until 140dBSPL back and forth). Several input frequencies chosen (185kHz is close to the Signal Transfer Function peaking due to the feed-forward topology).

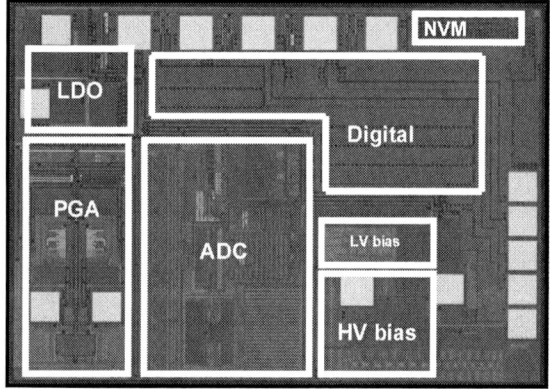

Figure 12: chip micrograph.

Figure 10 shows the shock recovery in operation (Infrasound robustness). When the digitally assisted block is enabled, the recovery time is in the hundreds of milliseconds.

Figure 11 shows the Ultrasound test and robustness, for input signals up to 140dBSPL with different frequencies, swept back and forth in the high performance mode of operation. At 185kHz the frequency signal is close to the ADC's Signal Transfer Function (STF) peaking, forcing the ASIC towards instability under worst-case conditions.

Table I shows the measured power (in 3 different operating modes) for the main blocks using external power supplies: 1.8V analog power supply and 1.2V digital supply.

At 94dBSPL input, the microphone SNRs are:

69dBA @ 3MHz clock, High Performance Mode (20kHz) [1]
67dBA @ 3MHz clock, Low Performance Mode (20kHz) [2]
66dBA @768kHz clock, Low Power Mode (8kHz) [3]

Figure 12 shows the chip micrograph of the 130nm ASIC, using 1.13 mm2 area, highlighting the main blocks.

TABLE I
CHIP POWER CONSUMPTION IN 3 DIFFERENT OPERATING MODES

Blocks	Power	Operating Modes
Analog signal path (PGA+ADC)	[1] 560μW@1.8V [2] 330μW@1.8V [3] 200μW@1.8V	[1] 3MHz clock High Performance (20kHz-Aweighted) **730μW total** -------------------- [2] 3MHz clock Low Performance (20kHz-Aweighted) **500μW total** -------------------- [3] 768kHz clock Low Power (8kHz-Aweighted) **260μW total**
Auxilay analog circuits: Bandgap, LDOs, Charge pump, etc.	[1] 60μW @1.8V [2] 60μW @1.8V [3] 30μW @1.8V	
Digital circuitry (including PAD drivers)	[1] 110μW@1.2V [2] 110μW@1.2V [3] 30μW@1.2V	

CONCLUSIONS

A low-power high-SNR microphone system is presented. Running at 3MHz it has a power consumption of only 730uW including all auxiliary circuits and presents a 69dBA SNR (A-weighted) in production (+/-1dB in rainbow lots) at -34dBFS output digital signal (1Pa$_{rms}$ sound input). The acoustic overload point (AOP - 10% THD) is 128dBSPL. Good power supply rejection is ensured using differential circuits and internal LDOs. Programmability and calibration circuits are included.

ACKNOWLEDGMENT

The authors will want to thank specially the great layout work done by Johan Kaeferboeck, Michael Druml & Rene Druml, and the excellent measurements at ASIC and system level performed by Michael Wassermann, Bernhard Toescher & Bernhard Kuttin. The fantastic concept work of Andrea Fant made possible the success of this IC and system.

REFERENCES

[1] MEMS Microphone Report 2021 Database, Omdia, London, UK, https://omdia.tech.informa.com/OM018422/MEMS-Microphone Report--2021-Database

[2] Status of the MEMS Industry 2021, Yole Développement, Lyon, France, https://www.i-micronews.com/products/status-of-the-mems-industry-2021

[3] L. Sant, E. Bach, R. Gaggl, A. Baschirotto, "A Current-Feedback Amplifier with Programmable Gain for MEMS Microphone Read-Out circuits", 2019 Austrochip Workshop on Microelectronics (Austrochip), October 2019.

[4] E. Bach et al., "A 1.8V true-differential 140dB SPL full-scale standard CMOS MEMS digital microphone exhibiting 67dB SNR", 2017 IEEE nternational Solid-State Circuits Conference (ISSCC), February 2017.

[5] S. Karmakar, B. Gönen, F. Sebastiano, R. van Veldhoven, andK. A. A. Makinwa, "A 280 μ W dynamic zoom ADC with 120 dB DR and 118 dB SNDR in 1 kHz BW," IEEE J. Solid-State Circuits,vol. 53, no. 12, pp. 3497–3507, Dec. 2018.

[6] L. Sant, M. Füldner, E. Bach, F. Conzatti, A. Caspani, R. Gaggl, A. Baschirotto, A. Wiesbauer, "A 130dB SPL 72dB SNR MEMS microphone using a sealed-dual membrane transducer and a power-scaling read-out ASIC", IEEE Sensors Journal, 2022, doi: 10.1109/JSEN.2022.3154446

[7] US Patent Application No. 17/457,577

[8] US Patent Application No. 17/451,562

A Nonuniform Sampling Lifetime Estimation Technique for Luminescent Oxygen Measurements

Ian Costanzo, Devdip Sen, John McNeill, Ulkuhan Guler

Electrical and Computer Engineering Department, Worcester Polytechnic Institute, Worcester, MA, USA
imcostanzo@wpi.edu, uguler@wpi.edu

Abstract—**This paper presents a nonuniform sampling technique for measuring the lifetime of luminescent materials for oxygen sensing. The system features a switched-capacitor circuit to implement fixed-voltage steps for quantization, enabling long integration times without saturating the front-end amplifier. A control circuit automatically tunes the light emitting diode (LED) excitation pulses to avoid overpowering or starving the front end as photodiode current varies with changes in the partial pressure of oxygen. Time gating of the front-end integrator removes the need for optical filtering. The analog front end (AFE) has a gain bandwidth product of 10 MHz and an input-referred noise of 124 μV_{rms} (measured 200 Hz - 100 kHz). The circuit was realized in 180 nm CMOS technology. The AFE and LED driver consume a maximum of 16 μJ per calculation. We have demonstrated the entire system's functionality by measuring oxygen concentrations from 0 to 240 mmHg in a controlled gas vessel. The results indicate satisfactory linearity on a Stern-Volmer plot covering the human-relevant range of 50 to 150 mmHg.**

Index Terms—**Oxygen sensor, nonuniform sampling, luminescent measurements, transcutaneous sensing, blood gases.**

I. INTRODUCTION

The ability to monitor the respiration of at-risk patients in real-time at home is critical to preventing respiratory failures. Care providers can assess the effectiveness of respiration by measuring blood gases, namely oxygen (O_2) and carbon dioxide (CO_2) [1]. Recently, luminescent gas sensing has reemerged as an attractive method to create miniaturized noninvasive transcutaneous blood gas monitors [2]–[6]. These medical devices measure the partial pressure of O_2 and CO_2 molecules diffusing through the skin, directly correlated with arterial O_2 and CO_2 [1]. The efforts in the field are new, requiring specialized analog front ends (AFE) with lightweight estimation algorithms for accurate sensor readings.

This work presents a nonuniform sampling technique and an offset immune luminescent lifetime estimation algorithm for O_2 sensing. The partial pressure of O_2 (PO$_2$) closely correlates with the lifetime of the luminescence quenched by O_2 [7]. Lifetime measurements are preferred over intensity measurements because of their robustness to factors such as variations in the optical path, reducing the vulnerability to motion artifacts and skin color differences [7].

The presented system 1) features a specialized AFE that obtains the luminescent lifetime information and 2) leverages time-gated excitation, removing the need for optical filters to extract the luminescent response from the excitation light. The lifetime (τ) is calculated with 3) the proposed offset immune algorithm using the time differences between equal

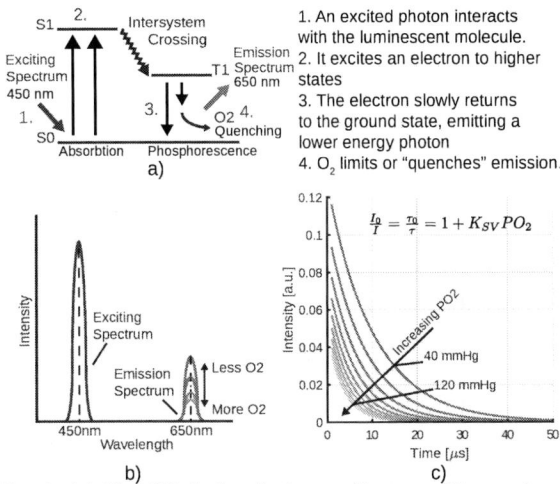

Fig. 1: (a) Simplified electrical state diagram, (b) wavelength shift, (c) τ and intensity of the emission quenched by O_2 [8].

voltage steps. 4) The measured mean error is as accurate as 1.9% without post-processing. 5) The system is showcased by measuring the PO$_2$ in a gas vessel.

The rest of the paper is organized as follows. Section II introduces the basics of luminescent O_2 sensing and explains the proposed lifetime estimation algorithm. Section III describes the hardware implementation. Section IV discusses the measurement results. Section V concludes the paper.

II. LUMINESCENT OXYGEN SENSING AND LIFETIME ESTIMATION ALGORITHM

Fig. 1 illustrates the basics of the luminescent O_2 sensing mechanism. The O_2-sensitive dye consists of functional groups known as luminophores. When the luminophores are exposed to relatively high-energy photons (450 nm), electrons in the luminophores jump to higher energy levels. After some time, the electrons try to return to the ground state. For luminescent materials, the excited electrons can enter a triplet state, preventing the electron from returning directly to the ground state. When the electron returns to the ground state from this triplet state, a photon of a longer wavelength is emitted (650 nm). This emission is "quenched" or suppressed by O_2 [9]. The dynamics of this interaction are described by the Stern-Volmer equation, shown inside of Fig. 1c. The intensity (I) and τ are related to PO$_2$ through a rate constant, K_{SV} and intensity/lifetime with no oxygen (I_0/τ_0).

978-1-6654-8495-4/22 $31.00 © 2022 IEEE

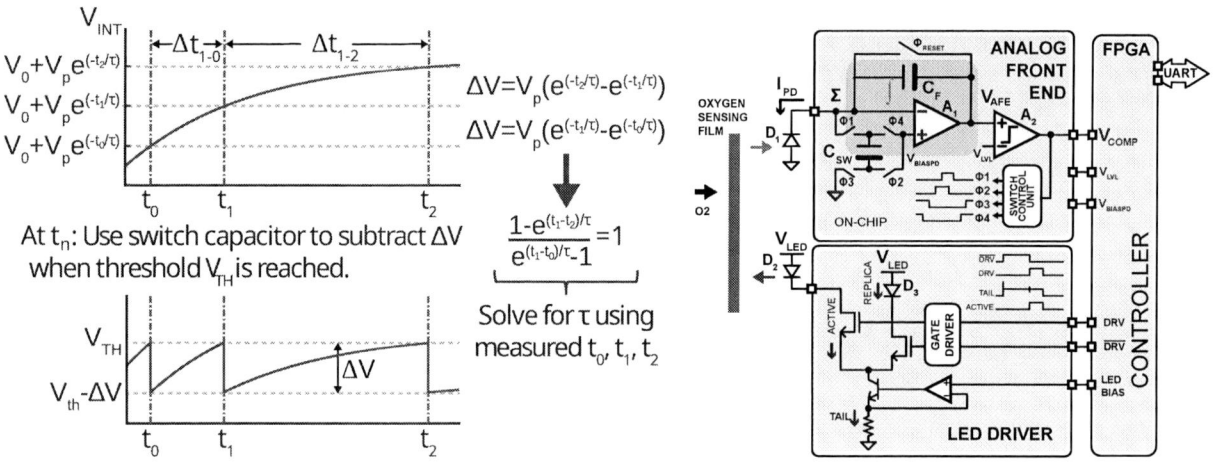

Fig. 2: Lifetime estimation algorithm.

Fig. 3: Diagram of the O$_2$ measurement system.

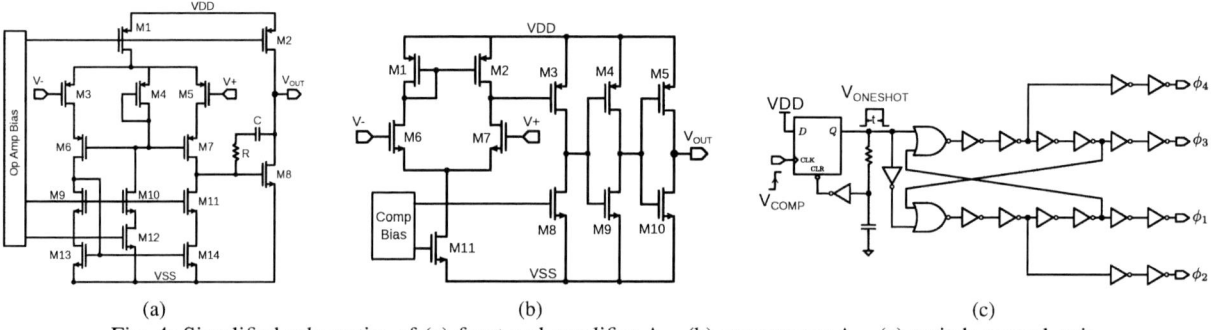

Fig. 4: Simplified schematics of (a) front-end amplifier A$_1$, (b) comparator A$_2$, (c) switch control unit.

Fig. 2 visualizes the proposed lifetime estimation algorithm. We use the time differences (Δt) between fixed-voltage steps (ΔV) to extract the time constant (τ) of the decaying exponential. This algorithm, easily realizable in hardware, is a modification of rapid lifetime determination [10]. The integration of an exponential does not change the time-constant (τ), the key parameter we want to measure. Integrating the current generated by the photodiode captures the exponential signal. When the integrated signal voltage reaches a pre-set threshold, the system records the time of the level crossing (e.g., t_0, t_1), and the integration proceeds without interruption.

Taking the difference between two level crossings eliminates the integration constant, as seen in Fig. 2. This operation removes any offset errors in the system. The quotient of two differences removes the amplitude component of the exponential. The result is a transcendental equation that is only dependent on the level crossings' sample times, t_0, t_1, and t_2, and the unknown luminescent time constant, τ. A basic root-finding algorithm calculates τ within a few iterations. This lightweight, offset immune algorithm requires only three samples, relaxing the data rate requirements.

III. SYSTEM ARCHITECTURE

Fig. 3 shows a diagram of the nonuniform sampling system. The system's operation is divided into an excitation phase and a readout phase. During the excitation phase, the light-emitting diode (LED) driver creates a pulse at a set intensity and duration, exciting the O$_2$-sensitive film with blue light ($\lambda = 450nm$). The integrator, A_1 and C_F, is set to unity-gain with a switch (ϕ_{RESET}) to avoid saturating the AFE with the high-intensity LED burst. C_F is only reset during the excitation phase, minimizing the impact of reset noise. In addition, the time-domain signal gating via ϕ_{RESET} removes the need for optical filtering to extract the relatively weak luminescent signal from the powerful excitation signal.

In the readout phase, a photodiode, D_1, captures the emitted red photons ($\lambda = 650nm$), generating a current (I_{PD}) that matches the decay of the photons, depicted in Fig. 7c. A_1 and C_F integrate this current to generate a voltage. The integrator is assisted with a nonuniform switched-capacitor technique to improve the dynamic range of its output, V_{AFE}. The comparator, A_2, detects the level crossings and sends a signal (V_{COMP}) to the switch control unit (SCU). The SCU in turn sends four non-overlapping signals ($\phi_1 - \phi_4$) to drive the switched capacitor, C_{SW}. This action injects a fixed amount of charge into the summing node of the integrator, Σ, as seen in Fig. 3. The charge injection causes a fixed voltage subtraction, ΔV, at V_{AFE}, allowing A_1 to continue integrating the photocurrent, as illustrated in Fig. 2. The circuit details of A_1, A_2, and SCU are given in Fig. 4.

The digital logic was implemented on a Xilinx XC7A35T

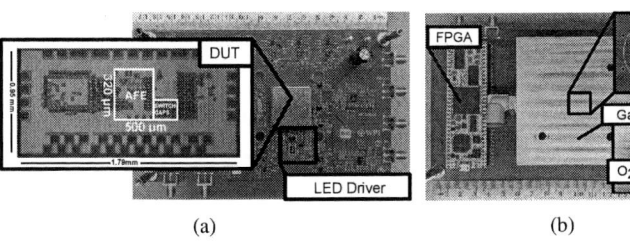
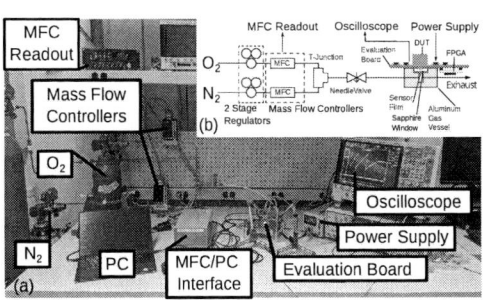

Fig. 5: (a) Micrograph and top side of eval board. (b) Bottom side of eval board with FPGA and gas test vessel.

Fig. 6: a) Gas experiment physical testbench and (b) diagram.

FPGA, which controls power-up sequence, excitation and readout phase timing, front-end bias, and LED driver power. A significant role of the FPGA was the automatic "gain control" for the LED driver. The FPGA counts the number of comparator (A_2) pulses and adjusts the LED drive to ensure the number of pulses is the same for each measurement. For low PO_2, the luminescent sensor emits many red photons for a given amount of exciting blue photons, so the LED drive current is set weaker. Conversely, the LED drive increases to excite more luminophores for high PO_2. If the number of pulses from the AFE is less than a set threshold, the LED bias increases by one bit or vice versa. If the pulses match the threshold, there is no change. The automatic LED drive tuning expands the O_2 measurement dynamic range.

Fig. 4 contains simplified schematics of the AFE sub-blocks. The front-end amplifier, A_1, is a telescopic cascode with PMOS inputs with a common source output stage [11]. The time constant of the front end should be at least one order of magnitude less than the luminescent decay time to reduce the error induced from the front end to less than 1% [12]. The comparator, A_2, performs the level crossing detection function of the sampler. The circuit detects when the output of the integrator reaches the threshold and triggers the switched-capacitor circuit to inject a fixed amount of charge into the summing node of the op-amp and generate the voltage step (ΔV), depicted in Fig. 2.

To achieve narrow LED pulses, we employed a current steering technique inspired by emitter-coupled logic [13]. A regulated current source controls the tail current that sets the LED drive based on a bias voltage provided by the controller. The controller turns on the replica current path when the main control loop sends the command to excite the O_2 sensing film. This action wakes up the tail current source and establishes the LED drive current. Once the replica path establishes the LED driver current, the controller quickly switches to the active channel to excite the luminophores. The driver can set an LED pulse intensity up to 1 A with 10-bit resolution and pulse width as narrow as a few hundred nanoseconds.

IV. MEASUREMENT RESULTS

We implemented the AFE in TSMC's 180 nm 1P6M CMOS process. The inset in Fig. 5a is the die micrograph. The AFE active area is 0.16 mm^2. Fig. 5 displays the evaluation board

with the FPGA daughter card installed. We have conducted four majors tests with the prototype: 1-2) the bandwidth and input-referred noise to evaluate key front end performance; 3) a transient test to assess the hardware implementation of the τ-estimation algorithm; and 4) a gas test to demonstrate the system's O_2 measurement performance.

During a 40 μs excitation phase (30 μs replica, 10 μs active), the auxiliary LED driver uses a maximum of 15 μJ to excite the sensor at high PO_2. During the excitation and readout phase (100 μs), the AFE uses 130 nJ to collect three samples for a τ calculation. The input-referred noise and the bandwidth of A_1 are given in Figs. 7a and 7b. The integrated noise is 124 uV_{rms}, measured over a bandwidth from 200 Hz to 100 kHz, and the 1/f corner is $\sim$ 10 kHz. The measured gain-bandwidth product of the front-end amplifier is 10 MHz.

We calculated the τ employing the proposed lifetime estimation algorithm using the time differences between comparator pulses, V_{COMP}. To assess the accuracy and precision of the technique, we fed the AFE with an exponentially decaying current with a controlled time constant, i_{test} in Fig. 7c. We measured each time constant ten times. We also calculated the lifetime by fitting an exponential model to the same stimulus current waveform for comparison. Fig. 8 shows the error between the proposed algorithm and the exponential model. For time constants ranging from 1 to 20 μs, the relevant range for human transcutaneous O_2 estimation, the mean error is $\sim$ -1.9%, with the error bounds $+1\%/-4\%$. Increased error for $t < 1\mu s$ is due to the comparator response delay. With a ΔV of 0.7 V and a C_F of 2 pF, the sampler requires a minimum of 3 pC to perform a calculation.

The in vitro gas test is a ratiometric mass flow rate experiment, explained in detail in [7]. We used the testbench, shown in Fig. 6, to test the O_2 sensor over a range of PO_2. To control the PO_2, we mixed O_2 and nitrogen (N_2) in controlled ratios at 760 mmHg (1 atm). MKS 1179A mass flow controllers (MFC) set the gas mixture ratios. For fixed volume, temperature, and pressure, the partial pressure of a gas is directly related to the mass of the gas in the mixture. The PO_2 was swept from 0 to 228 mmHg in 14.25 mmHg steps. A PC software controls the mass flow set points and the data readout.

The calculated time constants are plotted against PO_2 in Fig. 9a, and Fig. 9b is the Stern-Volmer plot of the same data. The deviation from the linear relationship at relatively high

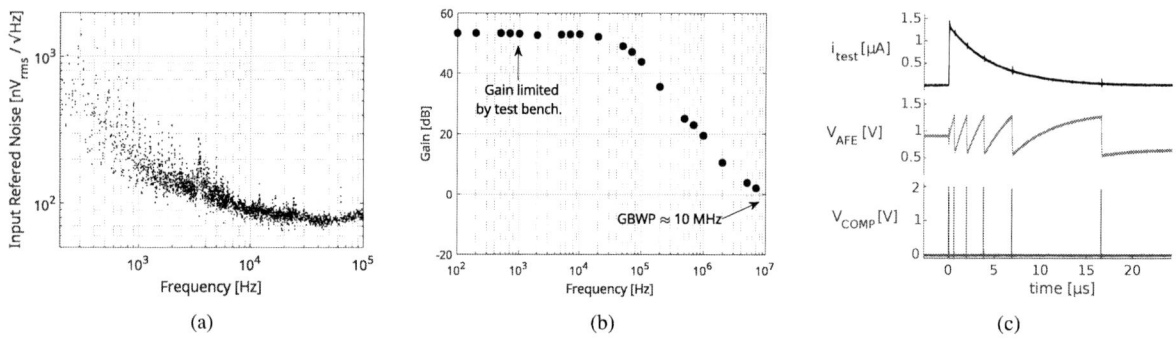

Fig. 7: Measured (a) input refered noise, (b) AC response, and (c) transient response.

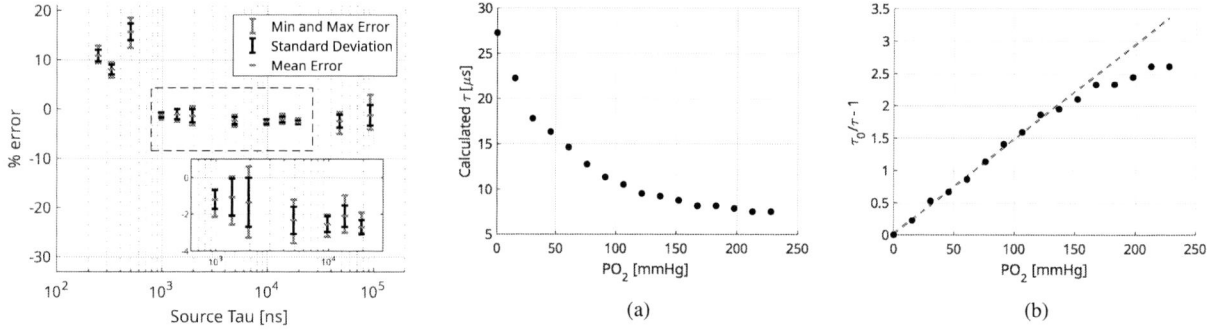

Fig. 8: Lifetime measurement error.

Fig. 9: (a) Calculated τ from gas test. (b) Stern-Volmer plot.

PO_2 is due to some luminophores being inaccessible to be quenched by O_2 [14]. The nonuniform sampler can measure PO_2 from 0 to 150 mmHg, covering the human-relevant range, with a 2.2 mmHg mean error, lower than the FDA standard of 5 mmHg [15]. Table I compares this work to prior art. A device for invasive deep tissue oxygenation measurement is presented in [16], with a slower sampling rate and smaller technology node, leading to lower power consumption.

V. CONCLUSION

This work demonstrates for the first time, a hardware implementation of a nonuniform sampling technique and algorithm to extract the τ of luminescent events, for the emerging application of noninvasive transcutaneous O_2 measurements.

ACKNOWLEDGEMENTS

This work was supported in part by NSF ECCS-2143898.

TABLE I: Comparison of luminescent oxygen sensors.

	[4] CICC '20	[2] ACS '18	[16] ISSCC '20	This work ESSCIRC '22
Tech (nm)	180	PCB	65	180
Area (mm²)	1.04	-	3.842	0.16 (AFE)
Power (μW)	632	-	140	835
V_{DD}/V_{LED} (V)	1.8/5	-/-	1.2/-	1.8/5
Technique	Time	Intensity	Phase	Time
TransZ Gain	59 - 943 kΩ	-	-	2 pF
Bandwidth	80 kHz	-	-	10 MHz*
Sample Rate	-	-	360 SPS	5 MSPS†
O_2 Error (mmHg)	-	-	-	2.2
Noise (μV) ‡	-	-	-	124

* Gain bandwidth product, † Maximum, ‡ Input referred

REFERENCES

[1] I. Costanzo *et al.*, "Respiratory monitoring: Current state of the art and future roads," *IEEE Rev. Biomed. Eng.*, vol. 15, pp. 103–121, 2022.

[2] C. J. Lim *et al.*, "Wearable, luminescent oxygen sensor for transcutaneous oxygen monitoring," *ACS Applied Materials & Interfaces*, vol. 10, no. 48, pp. 41 026–41 034, Jul 2018.

[3] W. Weteringen *et al.*, "Novel transcutaneous sensor combining optical tcPO2 and electrochemical tcPCO2 monitoring with reflectance pulse oximetry," *Med. Biol. Eng. Comput.*, vol. 58, no. 2, pp. 239–247, 2020.

[4] I. Costanzo *et al.*, "An integrated readout circuit for a transcutaneous oxygen sensing wearable device," in *CICC*, 2020, pp. 1–4.

[5] T. B. Tufan *et al.*, "A fluorescent thin film-based miniaturized transcutaneous carbon dioxide monitor." in *BioCAS*, 2021, pp. 1–5.

[6] U. Guler *et al.*, "Emerging blood gas monitors: How they can help with COVID-19," *IEEE Solid-State Circuits Mag.*, vol. 12, pp. 33–47, 2020.

[7] I. Costanzo *et al.*, "A noninvasive miniaturized transcutaneous oxygen monitor," *IEEE TBioCAS*, vol. 15, no. 3, pp. 474–485, 2021.

[8] M. Y. Berezin *et al.*, "Fluorescence Lifetime Measurements and Biological Imaging," *Chem. Rev.*, vol. 110, no. 5, pp. 2641–2684, May 2010.

[9] M. Quaranta *et al.*, "Indicators for optical oxygen sensors," *Bioanalytical Reviews*, vol. 4, no. 2, pp. 115–157, 2012.

[10] R. J. Woods *et al.*, "Transient digitizer for the determination of microsecond luminescence lifetimes," *Anal. Chem.*, vol. 56, no. 8, pp. 1395–1400, 1984, pMID: 6465515.

[11] J. R. Baker, *CMOS Circuit Design, Layout, and Simulation*, 3rd ed. Piscataway, NJ: IEEE Press, 2010.

[12] H. Johnson *et al.*, *Calculation of Rise Time.* Prentice-Hall, Inc., 1993, pp. 399–407.

[13] M. Thompson *et al.*, "High power laser diode driver based on power converter technology," *IEEE Trans. Pow. Elec.*, vol. 12, pp. 46–52, 1997.

[14] J. R. Lakowicz, *Principles of fluorescence spectroscopy*, 3rd ed. New York: Springer, 2006.

[15] "Cutaneous Carbon Dioxide (PcCO2) and Oxygen (PcO2) Monitors - Guidance," U.S. Food and Drug Administration, Dec 2018.

[16] S. Sonmezoglu *et al.*, "34.4 a 4.5mm3 deep-tissue ultrasonic implantable luminescence oxygen sensor," in *ISSCC*, 2020, pp. 454–456.

A 2.4GHz Full-Duplex Transceiver with Broadband (+120MHz), Linearity-Calibrated and Long-Delayed Self-Interference Cancellation

Xichen Li, Yi-Hsiang Huang, Fucheng Yin, Jacques C. Rudell
{xcli, yihsianh, yin4444, jcrudell}@uw.edu
Department of Electrical and Computer Engineering, University of Washington, Seattle, WA, USA

Abstract—A 2.4GHz Full-Duplex (FD) transceiver which employs multiple self-interference (SI) cancellation techniques featuring wideband SI suppression, highly-linear cancelers and the ability to cancel a measured long-delay spread SI up to 171ns is presented. A prototype chip was implemented in TSMC 40nm process and works with an Altera FPGA to complete the radio backend digital calibration and cancellation. The FD chip integrates an electrical-balanced duplexer (EBD) with tuning impedance, two broadband 5-complex tap true-time delay RF cancelers, and a baseband mixed-signal cancellation path operating with an FPGA to complete a long-delay spread SI canceler. 62dB of SI suppression was measured across a cancellation bandwidth (BW) of +120MHz for delay spreads up to 0.3ns. 18-23dB of SI suppression with an 80MHz cancellation BW was achieved for delay spreads between 0.4ns and 171ns. The entire analog system allows for digital calibration of the canceler linearity and frequency response. The RF canceler has a post-linearity calibration IIP_3 of +42dBm, while the RX has a measured maximum gain of 40dB, 6.8dB NF and -21dBm IIP_3.

Index Terms—Radio frequency integrated circuits, interference suppression, power amplifiers, phase-locked loops.

I. INTRODUCTION

The increased usage of the S and C-bands for 5G, IoT, and WiFi applications continues to motivate research on more efficient utilization of this highly congested spectrum. In-band Full Duplex (FD) communication allows transceivers to simultaneously transmit and receive using the same carrier frequency, thus providing one method to increase spectral efficiency. Key challenges arise when the transceiver attempts to receive a signal at the sensitivity limit in the presence of its own transmitter, which is delivering a high-output power signal to the antenna, appearing as self-interference (SI). Numerous research efforts have made significant progress with respect to mitigating the undesired SI presented to the receiver through the use of cancellation filters [1]–[3], integrated circulators [4], and code-domain modulation techniques [5]. While there have been significant contributions in the design of integrated cancellation and suppression circuits/systems for FD radios in the analog front-end (AFE), there remain significant challenges to address before practical transceivers can be produced for commercial applications. This paper address three of the remaining challenges surrounding the implementation of an AFE for FD transceivers:

- Broadband SI cancellation (+120MHz) at RF.
- Realizing a long-delay spread (171ns) SI canceler.
- Enhanced linearity for the RF cancelers.

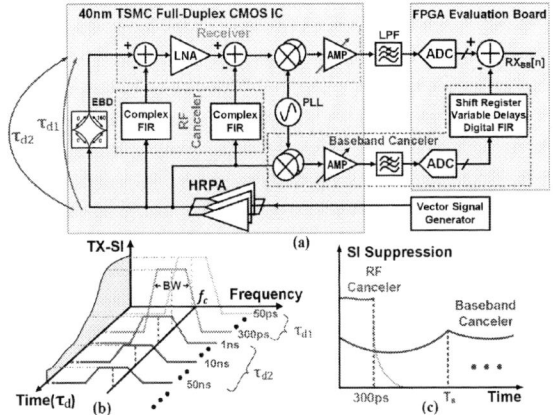

Fig. 1. (a) Full duplex transceiver architectural diagram of multi-path cancellation; both AFE chip and FPGA back-end, (b) Illustration of TX-SI with different delay spreads, (c) Desired SI suppression versus delay spread.

Section II describes the proposed radio AFE architecture which includes the mixed-signal long-delay spread canceler and two broadband, linearity calibrated RF cancelers. Section III presents the implementation details of the radio air interface and the complex tap RF cancelers. Measurement results are given in Section IV, with concluding comments in Section V.

II. FULL DUPLEX AFE RADIO ARCHITECTURE

A. Multi-path Hybrid Self-Interference Cancellation

The key challenge surrounding the implementation of a high-performance SI canceler is the presence of multiple time-delayed SI signal paths over the channel bandwidth of interest including a direct-coupling path and a combination of other environmental reflections. Fig.1(b) is an illustration of the SI as the delay/time-of-flight(τ_d) increases. Typically the strongest SI peak arrives within several hundred pico-seconds from direct coupling at the antenna interface (shown in red in Fig.1), while the peaks with longer delay are associated with environmental reflections which produce an SI power greater than the RX sensitivity up to 100ns (shown in blue). To address a wide range of SI characteristics in amplitude, phase, and delay, multiple cancellation paths are implemented on the FD AFE chip, each of which serves a different role with respect to canceling SI with varying delay and amplitude. The SI cancellation implemented on this radio starts at the receiver

978-1-6654-8495-4/22 $31.00 © 2022 IEEE

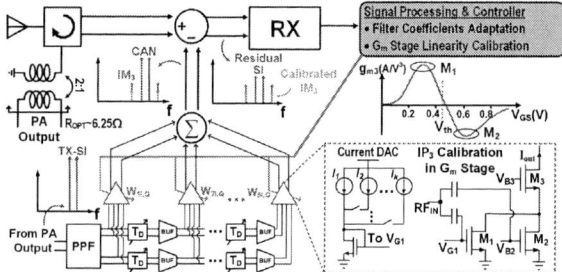

Fig. 2. One of two analog complex tap FIR filters with linearity calibrated G_m stages. The bias circuitry for $V_{B2,3}$ and the complementary PMOS pair are not shown in the figure.

input, shown in Fig.1(a), with the EBD, followed by pre- and post-LNA 5-complex tap continuous-time feed-forward cancelers (FFC) which exclusively suppress the short-delay spread SI $\tau_{d1}<300ps$. The SI with a longer delay-spread $400ps<\tau_{d2}$ is suppressed using a baseband (BB) mixed-signal cancellation path that samples the power amplifier (PA) output, then down-converts the signal to baseband, where it is taken off-chip and passed to the digital domain using an FPGA board which includes 12-bit 150MS/sec ADCs. The cancellation path is completed in the digital domain with variable delay FIR filter before subtracted from the receive path. The SI with delay between $300ps<\tau_d<400ps$ is suppressed by a combination of the RF and BB cancellation paths.

In the digital portion of the BB cancellation path, a shift register, implemented as a cascade of flip-flops, generates a variable delay up to 200ns without gain variations as compared to purely analog BB cancellation path [1], [2]. The delay spread of the mixed signal canceler is frequency-independent and is determined by the ADC sampling rate and the number of flip-flops. Moreover, the adaptation of the mixed-signal canceler's digital FIR filter uses a least mean square (LMS) algorithm converging in less than $100\mu s$ which represents a factor of $100\times$ faster than the adaptation algorithm in [2].

B. RF Feedforward Canceler with Linearity Calibration

The two RF FFCs for short-delay and high amplitude SI are implemented using an FIR filter complex tap coefficients to achieve deeper and broader bandwidth (120MHz) SI cancellation as compared to FIRs with real filter coefficients [6]. The linearity of the FFC is limited by the G_m-stage in the canceler which also sets the value of the filter tap coefficients. To minimize the IM_3 distortion generated by the G_m-stage, a digitally controlled linearity calibration technique is used in the complex FIR filter, see Fig.2. The G_m-stage uses an auxiliary transistor (M_1) biased in the subthreshold region producing a positive 3rd order nonlinear coefficient which cancels with the negative 3rd order nonlinearities associated with the main transistor (M_2) biased in the saturation region [7], ultimately enhancing the IIP_3 of the overall FFC by 6dB post-calibration. The linearity calibration is performed by digitally tuning the biasing current from sources $I_1,I_2\ldots I_k$ using an I_{DAC} for different PVT corners.

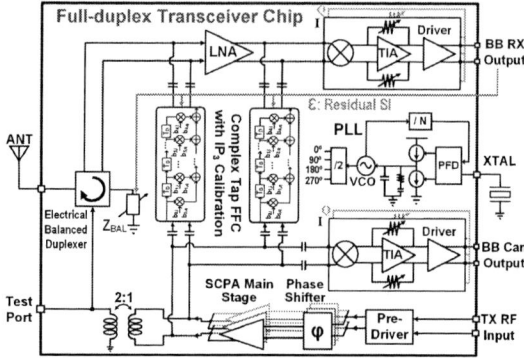

Fig. 3. 40nm FD transceiver chip block diagram with EBD, balanced impedance, receiver, pre- and post-LNA complex-FIR-based FFCs, phase-locked loop (PLL), and mixed-signal BB cancellation path.

III. Implementation Details

A. Transceiver RF Front-end Interface

Fig.3 is the block diagram of the 40nm TSMC FD chip. Three RF SI suppression paths and the analog portion of the BB cancellation path were integrated on chip. Fig.4 shows circuit implementation of the AFE where all cancellation paths are supplied with the PA output to capture the TX noise and nonlinearities. A harmonic-rejection(HR) PA with a pre-driver, phase shifter, and switched-capacitor stage combines the signal of three parallel drivers [8] with phase difference of $0°,45°,90°$, and gain ratios of $1,\sqrt{2},1$. This configuration effectively suppresses the possibility of producing intermodulation distortion from the PA 3rd and 5th order harmonics intermodulating as they pass through the FFC, resulting in IM_3 in the signal BW (Fig.4). The phase shifter including 7 coarse and 1 fine tuning stage based on inverter-delay covers a phase-tuning range between $0°$ and $90°$ at 2.4GHz, with a resolution smaller than $1.5°$. The balancing impedance, Z_{bal}, is integrated with the EBD and implemented with a 4th order RLC filter containing three tunable capacitors to match the antenna impedance over wide range(VSWR 1.5:1) and reduce the TX-RX group delay variation over broad bandwidth ($\pm 5\%$ over 200MHz). A capacitor shorts the center tap of the EBD secondary to ground, minimizing the TX-RX common-mode leakage, further suppressing the differential distortion products [9]. The LNA is implemented with a current reuse G_m stage, driving the passive down-conversion mixer to improve the noise figure(NF) and linearity of the receiver. The cross-coupled LNA input devices allow noise cancelling of $MN_{1,2}$ and $MP_{3,4}$ [10] and further rejection of EBD common-mode leakage.

B. Complex-Tap FIR Filter based RF Canceler

The inputs to the complex 5-tap FIR filters are connected to the PA output which has an R_{opt} of 6.25Ω transformed by a matching network (Fig.2). A RC-passive poly-phase filter (PPF) generates differential signals with quadrature phases that are delayed and buffered. The maximum delay of each tap implemented by an RC-CR all-pass filter is 68ps. The signals

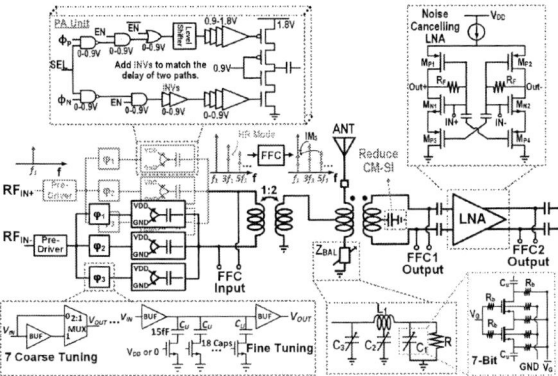

Fig. 4. FD radio air interface: HRPA, EBD with Z_{bal} network and LNA.

with different delay are complex vector weighted by an 8-bit G_m-stage which uses a regulated-cascode to bias M_3 (Fig.2), improving the linearity and output impedance. The canceler linearity is further enhanced by tuning the gate voltage of the auxiliary transistor M_1 biased in subthreshold region through a 5-bit I_{DAC}. The added bias circuit and auxiliary transistor have negligible impact on FFC power and noise, since M_1 operates in subthreshold. Each canceler presents a high-input ($2.2k\Omega$) and output impedance($1k\Omega < R_{Out,FFC}$) to prevent loading the PA output and LNA input.

IV. MEASURED PERFORMANCE

The transceiver chip was fabricated in TSMC's 40nm CMOS process and occupies 3.2mm^2(Fig.5) while operating off a 0.9V and 1.8V supply. An Altera Cyclone III FPGA with a Terasic data conversion card emulates the digital backend to control the EBD Z_{bal} and both RF FFCs, in addition to completing the mixed-signal long-delay spread cancellation path (Fig.6). The complex tap RF filter coefficients were calibrated using a pseudo-blind search algorithm, the adaptation time of which was estimated to be 10^{10} times faster than a conventional blind search algorithm for an 8-bit 5-complex tap filter. The 8-tap digital FIR filter coefficients of the mixed-signal BB canceler were adapted using a LMS algorithm. Transmission lines of varying length were attached between the power splitter output and RX input to emulate long-delay spread SI.

The integrated SI cancelers were tested with a 120MHz QPSK modulated signal (Fig.7a). The measured overall on-chip SI suppression is 62dB with 120MHz cancellation bandwidth (BW), where the EBD and two cancelers provide 34dB, 17dB, 11dB cancellation, respectively. Fig.7b shows the SI suppression of each path referred to the input carrier frequency. The maximum on-chip cancellation of a single-tone input is 78dB. The measured long-delay spread SI cancellation versus time resulting from the mixed-signal BB canceler is shown in Fig.8. 18-23dB SIC is achieved across 80MHz BW between 0.4-171ns (Fig.8a). A typical measured BB SIC spectrum resulting from a 13.9ns delay spread, before and after

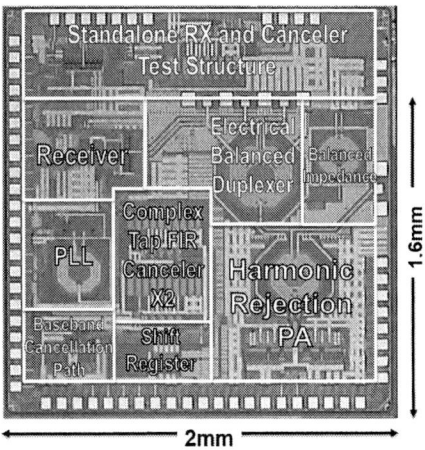

Fig. 5. 40nm TSMC FD die photo. 2mm × 1.6mm FD transceiver with standalone test structure at the top.

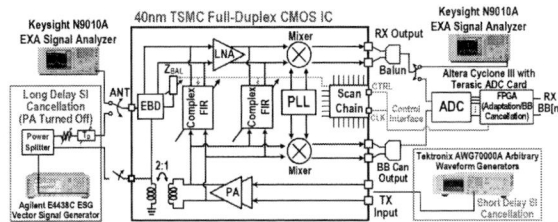

Fig. 6. Measurement set-up which interfaces the AFE chip with test equipments and an FPGA.

cancellation, is shown in Fig.8b. The required adaptation time of the BB canceler is approximately 60μs.

The measured canceler IIP$_3$ (Fig.9a) referred to the PA output is 42dBm after linearity calibration which represents a 6dB enhancement for the uncalibrated FFC. The receiver operates from 2.1-2.5GHz with a measured gain of 40dB and power consumption of 25mW (excluding the output 50Ω driver). The measured RX double-side band NF is 6.8dB (3.7dB from the EBD loss), with a NF degradation of 1.7dB (Fig.9b) in FD mode. The measured EBD TX-ANT insertion loss is 4.4dB and the RX S$_{11}$ is better than -15dB between 2.0-2.7GHz. The PA has a measured maximum output power of 19.1dBm. The harmonic suppression of the 3rd and 5th

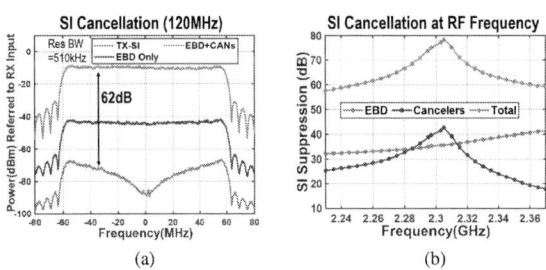

Fig. 7. (a) Measured integrated short-delay SI suppression using a multi-carrier QPSK signal (P_{ANT}=14dBm). ANT port is terminated into a load(VSWR: 1.15). (b) Measured SI suppression at the carrier frequency.

978-1-6654-8495-4/22 $31.00 © 2022 IEEE

TABLE I
PERFORMANCE SUMMARY AND COMPARISON TABLE WITH PRIOR FD PUBLICATIONS.

	This Work	ISSCC [2]	JSSC [1]	ISSCC [3]	JSSC [5]	RFIC [11]
Architecture	Integrated EBD + Mixes Signal BB + Linearity Calibrated RF	Stacked-capacitor Switched-cap FIR	RF+Analog BB Adaptive Filter	Double-RF Adaptive Filter	Code-domain IAD N-path	Noise Cancelling I/Q Path
Technology	CMOS 40nm	CMOS 65nm	CMOS 40nm	CMOS 40nm	GF 45nm SOI	CMOS 180nm
Frequency(GHz)	2.1-2.5	0.1-1	1.7-2.2	1.6-1.9	0.4-1.1	0.37-0.42
Integrated SIC(dB)/BW(MHz)	62 / 120, 68 / 40	42 / 40	50 / 42	65 / 80	49.5 / 13	20 / 50
Baseband SIC(dB)/BW(MHz)	23 / 80	12 / 40	18 / 40	NA	NA	NA
Canceler Delay Spread(ns)	RF:0.3ns, BB:200ns	RF:7.75ns, BB:85ns	RF:0.26ns, BB:130ns	RF:0.2ns	>5	NA
RF Canceler IIP$_3$ (dBm)	42	NA	36	40	NA	NA
Canceler Power(mW)	33(RF) + 3.2^a(BB)	118.4(RF)+14.4(BB)	3.5(RF) + 8(BB)	14.3	<40	64-154
IB RX IIP$_3$(dBm)/P$_{1dB}$(dBm)	-21 / -31(Gain of 40dB)	-14 / -25(Gain of 24dB)	-5 / -15	NA	+23.1 / +12.1	NA
RX NF (dB)	6.8 (3.7 from EBD)	3.7	4	8.1	2.6-5.6	3.5
SIC NF Degradation(dB)	1.7 (RF*2)	0.8(LP) / 2.8(HP)	1.55	1.6	NA	1.5/3.8
Active Area (mm^2)	3.2	7.2	3.5	4	1.12	6.5

^{a}-Power consumption of the on-chip analog portion of the BB canceler. ^{b}-Calculated by assuming 20dB initial TX-RX isolation.

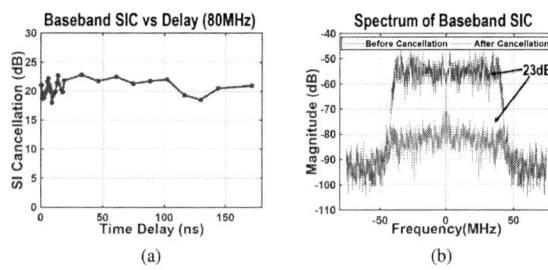

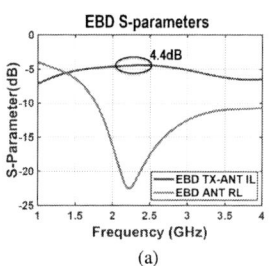

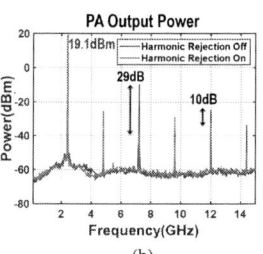

Fig. 8. (a) Measured BB SI cancellation versus delay using a 80MHz QPSK modulated signal. (b) Measured spectrum of BB SIC with a 13.9ns delay.

Fig. 10. (a) Measured EBD antenna port return loss and TX-ANT loss. (b) Measured PA output spectrum.

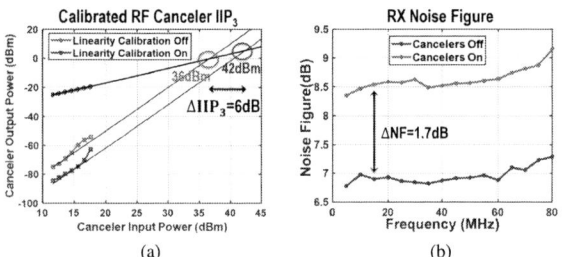

Fig. 9. (a) Measured standalone canceler linearity pre- and post-calibration. (b) Measured receiver NF with cancelers turned off/on.

order harmonics is 29dB and 10dB, respectively. The PLL consumes 10mW with a phase noise of -116dBc/Hz at 1MHz offset. Table.I provides a comparison with a prior art on FD radios. The FD-AFE in this work integrates the antenna interface and achieves the longest measured delay spread (171ns) SI cancellation across broad BW using a mixed-signal BB canceler, than any other integrated AFEs.

V. CONCLUSION

A multi-path SI cancellation network demonstrates the ability to provide a broadband cancellation of both short- and long-delay spread SI. The 2.4GHz FD transceiver achieves 23dB long-delay spread SI cancellation across 80MHz, 62dB integrated SI suppression across 120MHz, and enhances the FFC's IIP$_3$ of +6dB through the use of a linearity calibration.

ACKNOWLEDGMENT

The authors would like to thank SRC-Intel, CDADIC and Qualcomm for their contributions as well as Christopher Hull and Farhana Sheikh for their advice.

REFERENCES

[1] T. Zhang et al., "Wideband dual-injection path self-interference cancellation architecture for full-duplex transceivers," IEEE J. Solid-State Circuits, vol. 53, no. 6, pp. 1563–1576, 2018.

[2] A. Nagulu et al., "Full-duplex receiver with wideband multi-domain FIR cancellation based on stacked-capacitor, N-path switched-capacitor delay lines achieving >54dB SIC across 80Mhz BW and >15dBm TX power-handling," in IEEE ISSCC, Feb. 2021, pp. 100–102.

[3] K. D. Chu et al., "A broadband and deep-TX self-interference cancellation technique for full-duplex and frequency-domain-duplex transceiver applications," in IEEE ISSCC, Feb. 2018, pp. 170–172.

[4] A. Nagulu et al., "Multi-watt, 1-Ghz CMOS circulator based on switched-capacitor clock boosting," IEEE J. Solid-State Circuits., vol. 55, no. 12, pp. 3308–3321, 2020.

[5] H. Alshammary et al., "A code-domain RF signal processing front end with high self-interference rejection and power handling for simultaneous transmit and receive," IEEE J. Solid-State Circuits., vol. 55, no. 5, pp. 1199–1211, 2020.

[6] A. E. Sayed et al., "A hilbert transform equalizer enabling 80 Mhz RF self-interference cancellation for full-duplex receivers," IEEE Trans Circuits and Systems I, vol. 66, no. 3, pp. 1153–1165, 2019.

[7] V. Aparin et al., "Modified derivative superposition method for linearizing FET low-noise amplifiers," IEEE Trans. Microw. Theory Techn., vol. 53, no. 2, pp. 571–581, 2005.

[8] T. Zhang et al., "A low-noise reconfigurable full-duplex front-end with self-interference cancellation and harmonic-rejection power amplifier for low power radio applications," in IEEE ESSCIRC, Sep. 2017, pp. 336–339.

[9] X. Li et al., "Effects of receiver input impedance on nonlinear distortion in full-duplex radios," in IEEE ISCAS, May 2021, pp. 1–5.

[10] M. El-Nozahi et al., "An inductor-less noise-cancelling broadband low noise amplifier with composite transistor pair in 90nm CMOS technology," IEEE J. Solid-State Circuits., vol. 46, pp. 1111–1122, 2011.

[11] M. Essawy et al., "A noise-cancelling self-interference canceller with +7 dBm self-interference power handling in 0.18μm CMOS," in IEEE RFIC, Aug. 2021, pp. 95–98.

A 3-10GHz 21.5mW/Channel RX and 8.9mW TX IR-UWB 802.15.4a/z 1T3R Transceiver

Elbert Bechthum, Minyoung Song, Gaurav Singh, Erwin Allebes, Charis Basetas, Pepijn Boer, Arjan Breeschoten, Stefan Cloudt, Johan Dijkhuis, Ming Ding, Sherwin Gatchalian, Yuming He, Johan van den Heuvel, Martijn Hijdra, Paul Mateman, Bernard Meyer, Gert-Jan van Schaik, Mohieddine El Soussi, Bart Thijssen, Stefano Traferro, Evgenii Turin, Peter Vis, Nick Winkel, Peng Zhang, Yao-Hong Liu and Christian Bachmann

IMEC NL, Eindhoven, The Netherlands
Email: elbert.bechthum@imec.nl

Abstract—**Using IR-UWB for accurate battery-powered localization requires low energy consumption and high interference resilience. The presented IR-UWB 802.15.4z transceiver features low power consumption thanks to its inverter-based RX architecture and polar TX. The two-stage distributed PLL enables simultaneous multi-channel reception, reducing the energy consumption and measurement time of localization. It consumes 8.9mW in TX mode and 21.5mW/ch. in RX mode while achieving -33dBm OOB blocker tolerance.**

I. INTRODUCTION

Impulse-Radio Ultra-wide band (IR-UWB) systems offer distinctive advantage compared to narrowband radios in terms of ranging accuracy [1]. This is aided by the new IEEE 802.15.4z standard which offers several improvements over IEEE 802.15.4a by introducing PHY-level security, mandating coherent mode operation and enabling shorter packet lengths. However, these improvements come at the cost of increased circuit and system complexity, resulting in a higher power dissipation. The high-power dissipation in existing IEEE 802.15.4a/z IR-UWB radios (>100mW) [2], [3] poses limitations for battery-powered applications. Additionally, the frequency band is shared with cellular and WiFi radios, requiring high tolerance for out-of-band (OOB) blockers.

For accurate ranging with conventional UWB radios, many tags and anchors share the same radio frequency channel while using localization techniques such as Time-Difference of Arrival (TDoA). These systems typically use time-division multiplexing which requires an extensive measurement time, (leading to high energy consumption), or limits the number of available tags and anchors.

This work proposes a 1x3 Single-Input Multiple-Output (SIMO) 802.15.4a/z compatible IR-UWB transceiver with multi-channel reception capability for enhanced ranging/communication and enabling Angle-of-Arrival (AoA) detection while consuming 2 to 10-times less power than prior-arts and supporting a 3-10GHz frequency band with >-33dBm OOB tolerance.

II. SYSTEM ARCHITECTURE

The system architecture of the proposed IR-UWB TRX is shown in Fig. 1. It contains a system clock generator,

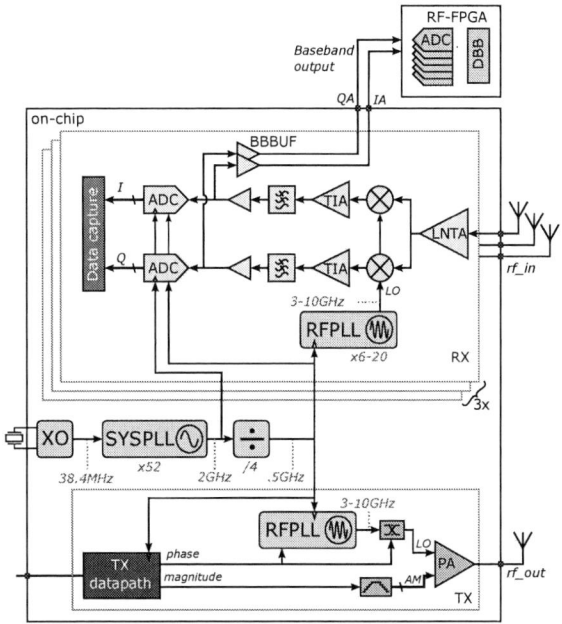

Fig. 1: IR-UWB transceiver system overview

along with an energy-efficient polar TX and 3 RXs with self-contained PLLs. To meet the IEEE 802.15.4 TX spectral mask, the transmitter is implemented using an asynchronous digital pulse-shaper and a digital PA [4]. The RX consists of a two-stage LNTA (Low-noise transconductance amplifier), passive mixer, TIA (Transimpedance amplifier), LPF and ADC (Fig. 1). All amplifiers in the RX consist of unit inverter-based gm cells regulated by a self-biased current regulator. The ADC is a 2GSps 6bit 2x time-interleaved (TI) ADC, which is a trade-off between TI complexity and slice sample rate. The two MSB bits are resolved using a loop-unrolled SAR architecture. The remaining 4 bits are generated by an asynchronous digital slope ADC [5]. The offset of the ADC comparators is corrected using foreground calibration to compensate for mismatch and PVT corners.

The output of each RX chain is captured on an on-chip memory for offline evaluation of the whole receiver chain.

978-1-6654-8495-4/22 $31.00 © 2022 IEEE

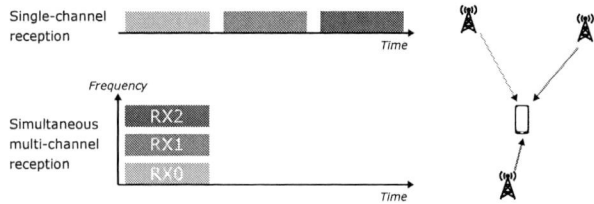

Fig. 2: Using simultaneous multi-channel reception, the ranging time is reduced significantly

Additionally, the output of the LPF is also buffered out of the chip to interface with measurement equipment and with the RX digital baseband implemented on an RF-FPGA via the ADCs on-board the FPGA.

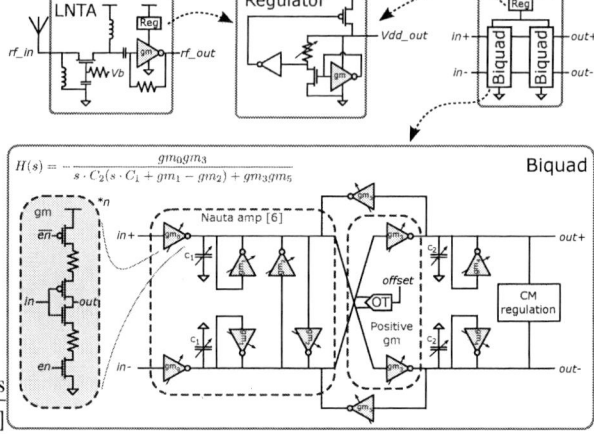

Fig. 3: Schematic details of the RX, where a low power consumption is achieved by inverter-based amplifiers, e.g. the LPF Biquad

III. CLOCK GENERATION

In the 802.15.4a/z standard, the center frequencies of all the channels are integer multiples of 499.2MHz, enabling a clock-generation approach using two cascaded integer PLLs when using specific XO frequencies, e.g. 38.4MHz. The distributed cascaded PLL architecture, consisting of a system clock PLL (SYSPLL) and multiple TX/RX local PLLs (RFPLLs), is proposed to enable low-power clock distribution. Moreover, this architecture enables simultaneous reception from three transmitters operating on different frequency channels. Thanks to the multi-channel reception, the ranging throughput is increased 3x, which in turn allows for a threefold larger wireless-node network, or results in a threefold reduction in the measurement time or results in a threefold increase in the update rate, leading to a significant improvement in precision. See Fig. 2 for an illustration of the advantages of simultaneous multi-channel reception with respect to single-channel reception. Furthermore, since all RFPLLs are phase-synchronous with the SYSPLL, accurate AoA measurements can be performed. With the reference clock of 38.4MHz from the crystal oscillator (XO), the SYSPLL provides a 1996.8 MHz clock for the RX ADCs and a divided-by-4 499.2MHz clock as the system clock and the reference of the TX and RX RFPLLs.

The SYSPLL employs a LC oscillator to meet the stringent jitter requirement for the ADC ($<2ps_{rms}$), and to supply a clean reference for the RFPLL. To minimize the area- and power-overhead of the distributed cascaded-PLL architecture, the RFPLL is implemented as an injection-locked ring oscillator (ILRO) [4], using an injection frequency of 499.2MHz. The ILRO loopfilter bandwidth is dimensioned such that the overall PLL jitter is dominated by the reference clock. Since the reference clock comes from the low-noise LC-based SYSPLL, the cascaded-PLL can achieve the required low jitter ($<2ps_{rms}$) for high ranging accuracy. The RFPLL covers an output frequency range of 3GHz to 10GHz with 499.2MHz steps.

IV. RECEIVER

Fig. 3 shows key design details of the RX chain. To achieve both broadband (3-10G) and high gain (>30 dB), a common-gate (CG) LNA and an inverter cell are employed at the first and second stage of the LNTA, respectively. The analog baseband (ABB) including the TIA and the LPF has a 5th order Gaussian-up-to-6dB filter characteristic for group-delay flatness, anti-aliasing of the ADC, and blocker rejection. The up to 30dB of adjacent channel filtering provides the blocker tolerance necessary to facilitate multi-channel reception.

The TIA converts the down-converted current to a voltage signal, applying a first-order filtering. The TIA uses a three-stage inverter-based amplifier with two compensation paths for stability and common-mode feedback loop. The LPF is based on two cascaded biquads, each consisting of two gain stages. An inverter-based gm-C filter topology is chosen for wide-band and low noise operation while maintaining low-power consumption. The two poles are implemented by the capacitance at the outputs of the two stages. The gm elements consist of multiple parallel inverter stages, which can be individually enabled and disabled. The configurability in the gm-elements and capacitors enable tuning of the filter characteristic and gain. Each biquad also contains an offset-trimming DAC (OTDAC). The output of the LPF is buffered to drive the ADC and provide a constant common-mode voltage to the ADC.

The inverter cell is sensitive to the PVT variation and supply noise/distortion, and its gm is not linear. A self-biased supply regulator-based gm biasing is proposed for sufficient PVT tolerance and PSRR. The unit cell used in each amplifier is employed to generate a voltage reference dependent on the regulator output. The sensed voltage is then fed back to the power transistor, regulating the regulator output.

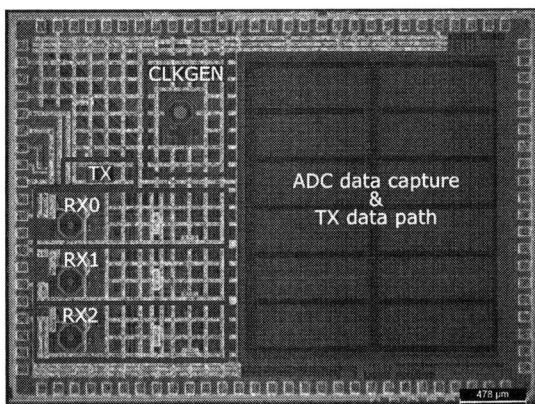

Fig. 4: Die photo

V. MEASUREMENTS

The proposed architecture is implemented in 28nm CMOS, see Fig. 4 for the die photo. The fully integrated transceiver including system clocking, transmitter and 3 RXs occupies 1.06mm2. All circuits operate from a 0.9V supply.

Fig. 5 shows the RFPLL performance, achieving <2ps jitter.

The TX output spectrum is shown in Fig. 6, demonstrating compliance to the IEEE 802.15.4 spectral mask. The maximum peak TX output power at 6.5GHz is -1dBm, and -6.5dBm to 0dBm over the whole 3-10GHz band (including PCB losses). When transmitting an 802.15.4z packet at 6.5GHz, the power consumption is 8.9mW.

RX measurement results are shown in Fig. 7. The blocker tolerance (at 3dB noise figure degradation) is -33dBm, -26dBm and -21dBm at the adjacent channel, the alternate channel and far out (1.5GHz offset) respectively. The in-band blocker tolerance is -45dBm. The receiver BER curve for an IEEE 802.14.4z HPRF packet at 27.2Mb/s at channel 5 (6.5GHz) shows a sensitivity of -83dBm (BER=10^{-3}).

After initial calibration, the 2GS/s ADC ENOB is 5.7bit up to 200MHz input frequency, and 5.4 bit at Nyquist. The ADC power consumption is 3.3mW, resulting in FOMW$_{LF}$=32fJ/c.s. and FOMW$_{HF}$=42fJ/c.s., which is similar as state-of-the-art low-resolution high sample-rate ADCs [7].

A breakdown of the power consumption is shown in Fig. 8. The system clocking, containing the crystal oscillator, SYSPLL and clock distribution, consumes 1.6mW, 1.8mW and 2.2mW for the TX, 1RX and 3RX mode respectively. The RFPLL, which is present in each RX and in the TX consumes 2.4mW. The power consumption of each receiver including ADC and RFPLL is 20.8mW. Hence, the power consumption of the complete transceiver including system clocking in the 1RX and 3RX mode is 22.6mW and 64.6mW (21.5mW/channel), which is 4-5x better than the lowest-power state-of-the-art 802.15.4z radios listed in the comparison table of Table I. Since only the low-power system clock generation is shared between RXs, the power consumption per RX in the 3RX mode is almost identical to the 1RX mode. The sensitivity FOM of this work is >2dB better than state-of-the-art.

Fig. 9 highlights the simultaneous reception of three different channels, where crosstalk and adjacent-channel leakage (ACL) are the main challenges. These are measured by applying an 802.15.4z packet at various channels at the receiver inputs. Both the crosstalk and ACL are >20dB below the desired signal, providing the required SNR for accurate

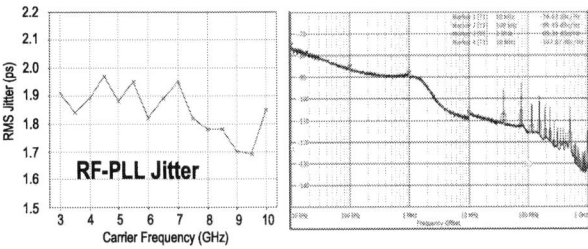

Fig. 5: RFPLL jitter is <2ps over the complete RX band, and the RFPLL phase noise at 7GHz

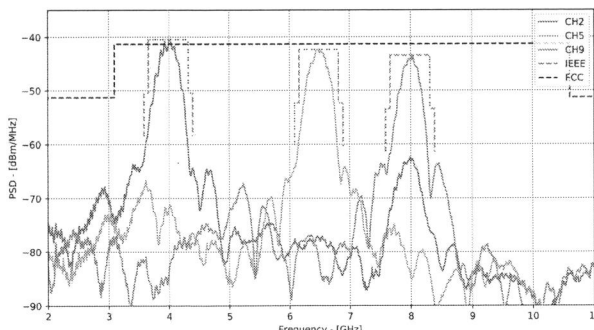

Fig. 6: TX output spectrum for CH2, CH5 and CH9

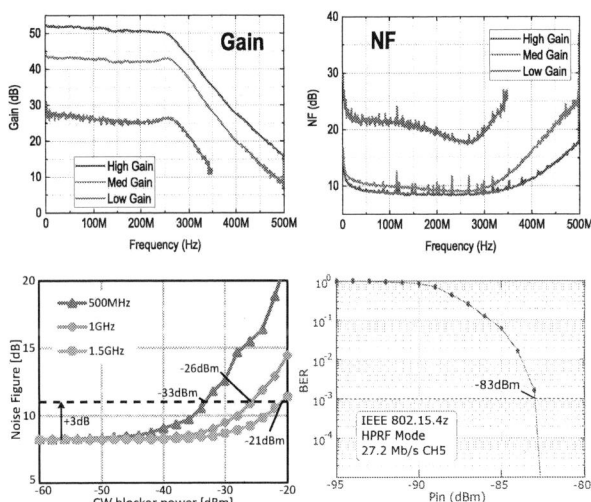

Fig. 7: RX measurements: gain, noise figure (NF), blocker power vs. NF at various offset values (all at 8GHz), and BER curve at at 27.2Mb/s (at 6.5GHz)

978-1-6654-8495-4/22 $31.00 © 2022 IEEE

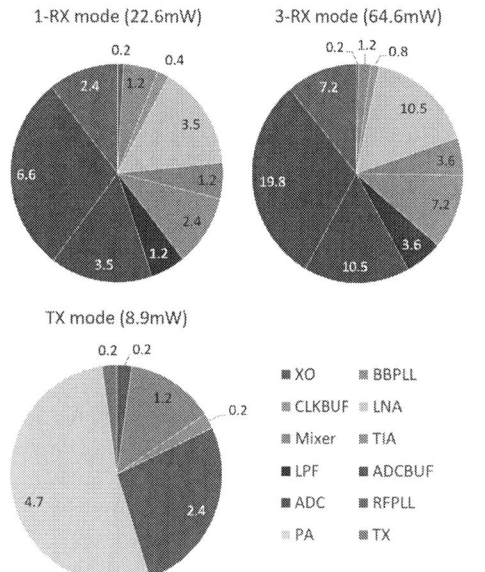

Fig. 8: Power-consumption break-down for various modes

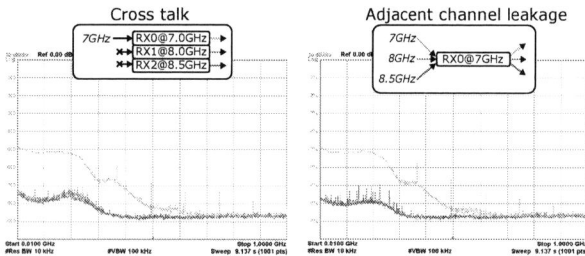

Fig. 9: Simultaneous multi-channel reception requires low cross-talk and low ACL (measured with 802.15.4z packet)

simultaneous ranging measurements. This work is the only state-of-the-art 802.15.4z RX of Table I that supports fast ranging through simultaneous multichannel reception.

Table I summarizes the performance of this work and compares it with state-of-the-art coherent UWB transceivers.

VI. CONCLUSION

This paper presented a transceiver for the IEEE 802.11.4z standard, featuring lowest power consumption, excellent interference rejection and the capability of simultaneous multi-channel reception for fast, accurate and low-energy localization.

REFERENCES

[1] S. Gezici and H. V. Poor, "Position Estimation via Ultra-Wide-Band Signals," *Proc. IEEE*, vol. 97, no. 2, 2009.
[2] *DW3000 product family user manual*, Qorvo, March 2021, v1.1.
[3] R. Chen *et al.*, "A 6.5-to-10GHz IEEE 802.15.4/4z-Compliant 1T3R UWB Transceiver," in *IEEE Int. Solid- State Circuits Conference (ISSCC)*, vol. 65, 2022.
[4] E. Allebes *et al.*, "21.2 A 3-to-10GHz 180pJ/b IEEE802.15.4z/4a IR-UWB Coherent Polar Transmitter in 28nm CMOS with Asynchronous Amplitude Pulse-Shaping and Injection-Locked Phase Modulation," in *IEEE Int. Solid- State Circuits Conference (ISSCC)*, vol. 64, 2021.
[5] P. Harpe *et al.*, "A 0.8mW 5bit 250MS/s time-interleaved asynchronous digital slope ADC," in *IEEE Asian Solid-State Circuits Conference*, 2010.
[6] B. Nauta, "A CMOS transconductance-C filter technique for very high frequencies," *IEEE J. Solid-State Circuits*, vol. 27, no. 2, pp. 142–153, 1992. [Online]. Available: https://bit.ly/3lN3l4u
[7] B. Murmann. ADC Performance Survey 1997-2021. [Online]. Available: http://web.stanford.edu/~murmann/adcsurvey.html
[8] J. Ko and R. Gharpurey, "A Pulsed UWB Transceiver in 65 nm CMOS With Four-Element Beamforming for 1 Gbps Meter-Range WPAN Applications," *IEEE J. Solid-State Circuits*, vol. 51, no. 5, 2016.
[9] D. Morche *et al.*, "Double-Quadrature UWB Receiver for Wide-Range Localization Applications With Sub-cm Ranging Precision," *IEEE J. Solid-State Circuits*, vol. 48, no. 10, 2013.

TABLE I: Performance summary and comparison with coherent UWB transceivers

		This work	[3] R. Chen ISSCC'22	[2] DW3000 (2021)	[8] J. Ko, JSSC16	[9] D. Morche, JSSC13
Technology		28nm CMOS	130nm SOI & 40nm CMOS	-	65nm CMOS	130nm CMOS
802.15.4 support		**a/z**	**a/z**	**a/z**	-	-
Architecture		**3RX+TX**	**3RX+TX**	RX+TX	RX+TX	RX
Supply voltage	[V]	**0.9**	1.25/3.3	3.0	-	1.2
Area (AFE only)	[mm²]	**1.06**	1.68+2.9*	-	1.7(TRX)	5.8▽
Frequency band	[GHz]	**3.0-10**	6.5-9.5	6.5, 8.0	6.25-8.5	3.2-4.7
RX						
Simultaneous multi-channel RX		**Y**	N	N	N	N
Sensitivity	[dBm]	**-83 @27.2Mb/s CH5**	-102 @0.85Mb/s CH5	-94.3 @6.8Mb/s CH5	-69 @1GS/s	-95 @1Mb/s
Sensitivity FOM†	[dB]	**-173.8**	-171.9	-168.9▽	-165.3	-168.0
Blocker tolerance at 0.5/1.0GHz	[dBm]	**-33/-26**	-46* @0.69GHz offset	**-30/-30***	N/A	N/A
AFE power consumption	[mW]	**22.6 (1RX), 64.6 (3RX)=21.5mW/ch**	114 (1RX), 263 (3RX)=87.5mW/ch	<234▽	211+32	50
Noise Figure	[dB]	**8.3 @8.5GHz**	**4.2–6.2**	-	5 - 8	7.5
TX						
Peak output power	[dBm]	**-6.5 - 0**	**10**	-	-	N/A
AFE average power consumption	[mW]	**8.9**	54.8	144▽	221	N/A

*Estimated from figure; ▽Full chip including DBB; †Sensitivity+10·log10(power consumption/data rate)

Gap in pagination due to withheld paper.

Pages 425-428

A 58 GHz Bandwidth, and less than 1.8% THD, Mach-Zehnder Driver, in 28 nm CMOS Technology

Nicola Cordioli, Danilo Manstretta, Rinaldo Castello

Department of Electrical, Computer and Biomedical Engineering, University of Pavia
Pavia, Italy, nicola.cordioli01@universitadipavia.it, danilo.manstretta@unipv.it, rinaldo.castello@unipv.it

Abstract—This work presents a linear Mach-Zenhder modulator driver. The key features of this design are a distortion which goes with the inverse of the input amplitude and a wide bandwidth. The two things make the reported driver suitable for Coherent Optical Applications. The mainstream technology for these applications is BiCMOS, however, in this case, a 28 nm CMOS technology has been used, aiming for higher integration level and lower cost. The resulting Total Harmonic Distortion (THD) for 1.5 V peak-to-peak differential output swing is always below 1.8%, with a -3 dB bandwidth of 58 GHz. The complete transmitter provides gain ranging from 10 dB to 20 dB in a continuous way. The overall power consumption at 20 dB is 297 mW, with a single supply voltage of 2.4 V.

Index Terms—CMOS integrated circuits, high-speed integrated circuits, Mach-Zehnder Modulator Driver, linear circuits.

I. INTRODUCTION

Optical links are the best candidate to satisfy the ever-growing demand for higher bit-rate networks. Coherent optical modulations use multi-level formats, that allow higher bit rate for a given baud-rate, leading to better spectral efficiency. In the transmitter, digital data is converted to analog signals, which modulate the IQ components of the Optical Carrier. This is made possible by an Electro-Optical Modulator (EOM), that combines laser beams with different phases and magnitudes, depending on the electrical signals. For our design, the EOM is a Mach-Zehnder Modulator (MZM), with a half-wave voltage V_π of 1.5 V. V_π defines the amplitude of the signal to be provided by the driver, with good linearity and wide bandwidth [1]. Since the digital-to-analog converter (DAC) output amplitude is not known a priori, the driver must adapt its gain. CMOS is not widely used in coherent optical systems, mainly because of the higher break-down voltage, g_m/I_D and $f_t g_m$ product of BiCMOS. Nonetheless, CMOS has some features, such as scalability and integration of analog and digital parts, that might lower the costs of highly integrated systems. Furthermore, in scaled technologies, p- and n-type MOS transistors have similar performance. This allows to use complementary structures, to compensate the poor MOS efficiency. The circuit, shown in Fig. 1, uses four stages, with the first one providing the whole gain programmability. The third one is used to buffer the main DRV, to preserve the overall bandwidth. The main DRV is the most critical block for linearity, having to provide a large signal to the MZM. In the following, the proposed idea (Section II), circuit design (Section III), simulation and measurement results (Section IV) are reported.

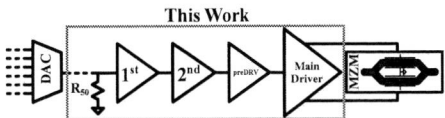

Fig. 1: Block diagram of the proposed driver.

II. HARMONIC DISTORTION AND VARIABLE GAIN

For multi-level modulation schemes, such as 16QAM, a THD lower than 2% is typically specified [2]. Concerning the linearity, the most challenging block is the output stage, where the use of CMOS is most critical. To get a high transistor f_t, a large overdrive is used, resulting in low intrinsic gain. This makes the non-linear output conductance to dominate the distortion when the swing is large [3]. Furthermore, ensuring gain variability with low distortion, makes the design of the variable gain amplifier (VGA) challenging. Our work intends to overcome these intrinsic limitations of CMOS. The proposed solution is reported in Fig. 2a. It consists of a cascode (to avoid transistor break-down, given the high supply) amplifiers where the voltage at the gate of the common-gate (CG) device is lowered until the common-source (CS) is pushed into triode. In this way the transconductance is reduced, but the voltage-to-current characteristic is more linear. G_m reduction can be mitigated thanks to the pn topology. Fig. 2b shows how the main DRV distortion is reduced working in triode, and the drain-to-source (V_{ds}) voltage used. Furthermore, a VGA can be implemented varying the voltage at the drain of the CS transistor. The CS gate-to-drain gain is given by (1), where $r_{o,CS}$ is the CS output resistance, and $g_{m,CG}$ is the CG transconductance. Since the CS is in triode, both $1/r_{o,CS}$ and $g_{m,CG}$ are higher than $g_{m,CS}$ (if CG and CS have the same size and bias current).

$$A_{V,gd} = -\frac{g_{m,CS}}{g_{m,CG} + \frac{1}{r_{o,CS}}} \qquad (1)$$

The resulting attenuation between gate and drain reduces the impact of non-linear terms in the voltage-to-current characteristic, making the gain more linear. The stage transconductance is $G_m \cong g_{m,CG} \times A_{V,gd}$, hence lowering the CS V_{ds} for the same V_{gs}, reduces $A_{V,gd}$, and so the G_m. This behaviour is used to implement the VGA. The circuit of Fig. 2a is used in the first, second and last stages with the first acting as a VGA. Gain control (through biasing) allows to fix linearity

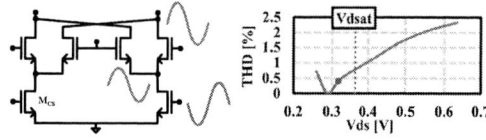

(a) Cascoded transconductor. (b) Main driver THD versus V_{ds}. The blue spot is the working point in this case.

Fig. 2: THD considerations about the proposed technique.

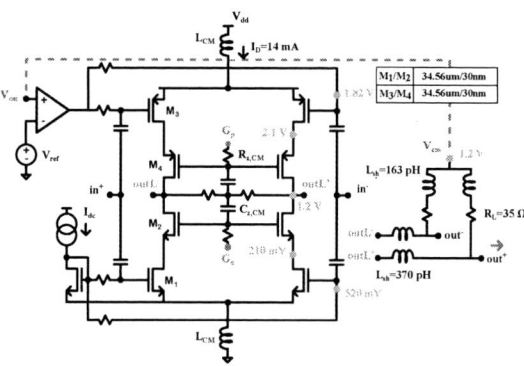

Fig. 3: 1st stage VGA.

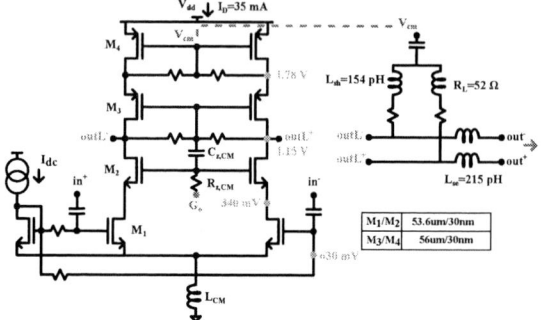

Fig. 4: 2nd stage of amplification.

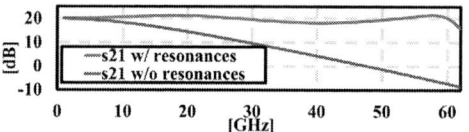

(a) Simulated differential mode transmitter s_{21} with and without RTRN and TRN.

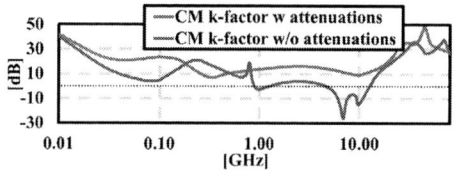

(b) An always above zero CM k-factor (unconditioned stability) is obtained attenuating the CM transconductance and impedance

Fig. 5: Simulated differential mode S-parameters and CM k-factor.

degradations introduced by PVT variations. Hand tuning will be replaced by a feedback loop made by a photodiode and a peak detector to perform automatic corrections. Large bandwidth is achieved using resonant loads in the first two stages. This, together with the use of a pn follower as pre-DRV, helps to compensate the gain drop at high frequency in the last big stage.

III. CIRCUIT DESIGN

This section gives a transistor level description of each stage. Design choices, with benefits and shortcomings, are reported.

A. 1st Stage Variable Gain Amplifier

The 1st stage, shown in Fig. 3, uses a pn inverter-like structure with a resonant load, that implements the ideas proposed in Section II. The DC current is set biasing M1 through a current mirror. The CG gate voltages are controlled by off-chip signals G_p and G_n to give the full 10 dB gain variation. This keeps the signal levels in the following stages independent from the gain setting. Attenuation is realized pushing the CSs progressively more into triode. Hence, as described in Section II, linearity improves while the input signal level is increased. The resonant load, realized with inductors and resistors, has three resonances and it is called Reverse Triple Resonant Network (RTRN). Thanks to the explicit reactive components, the starting bandwidth can be widely extended. [4] gives an exhaustive description of this technique. Since the loading circuit is purely differential, the common-mode (CM) gain is larger than the differential one,

making CM stability more critical. Due to this, the CM output impedance and transconductance are reduced. Feedings the CM out voltage to the gate of M2 via capacitor $C_{z\text{-}CM}$, transistor M2 goes from CG to diode connection for CM signals at $\omega_{z,CM} = 1/R_{z,CM}C_{z,CM}$, lowering the output impedance. Inductor L_{CM} degenerates the CM transconductance. (Fig. 5b).

B. 2nd stage amplifier

For the second stage, a n-only pseudo-differential structure (Fig. 4) has been preferred for its lower CM gain. As for the VGA, the off-chip voltage at the CG gate controls the CS bias and, with it, its distortion. The load is made by explicit resistors and inductors, which resonate with the parasitic capacitances, boosting the bandwidth. The circuit, called Triple Resonant Network (TRN), is different from the previous one and it is analysed in [5]. In summary, the TRN shows a higher peak gain at relatively low frequency while the RTRN peaks at high frequency. The simulations reported in Fig. 5a show the bandwidth enhancement given by these networks. Even if the n-only circuit has a small CM gain, a degeneration inductor L_{CM} and a zero are used to further

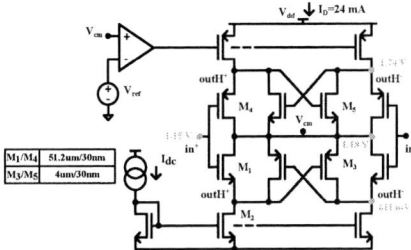

Fig. 6: Pre-driver source follower.

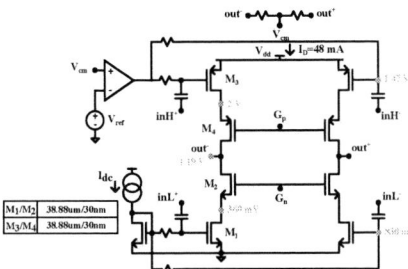

Fig. 7: Main Driver.

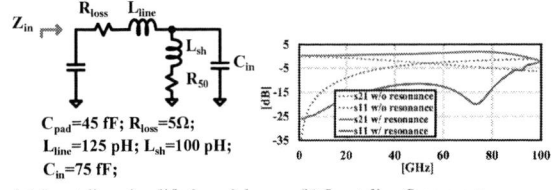

(a) Input line simplified model. (b) Input line S-parameters.

Fig. 8: Input line simplified model and simulated S-parameters.

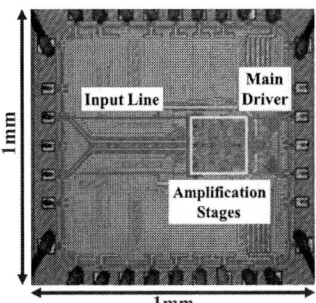

Fig. 9: Fabricated die photo.

reduce it. At frequencies above $\omega_{z,CM} = 1/R_{z,CM}C_{z,CM}$ $r_{out,CM}$ is decreased (Fig. 5b).

C. Pre-driver stage

A source follower drives the main DRV large input capacitance. The pn structure shown in Fig. 6 save current. To reach high f_t the transistors are biased with high overdrive and current. This lowers the current source output impedances giving significant attenuation in the follower. To reduce losses, a negative resistance, realized with two transistors in positive loop, is added as loading, whose effect is shown in (2).

$$A_{V,n} = \frac{g_{m,M1}}{(g_{ds,M2} - g_{m,M3}) + g_{m,M1}} \quad (2)$$

The resulting smaller input signal for the same output amplitude, improves the stage linearity.

D. Main Driver

This stage must drive the load impedance of the MZM with a 1.5 V_{ppdiff} highly linear signal. The circuit, shown in Fig. 7, is a pseudo-differential pn cascoded amplifier, AC coupled with the rest of the chain. AC coupling allows to optimize the DC gate-to-source voltage of the CS for best linearity. The CSs work in triode, with the drain terminals squeezed by the CGs. Most of the power is burned by this stage, so a careful layout is required to minimize parasitics. The biasing current I_{dc} can be changed to optimize linearity.

E. Input Line

Since we cannot yet merge analog and digital parts on one chip, input matching is required to avoid back-reflection

between the DAC and the transmitter (also for testing purposes). s_{11} should be lower than -10 dB over the entire bandwidth. The matching network has been realized with passive components. A simplified model is reported in Fig. 8a. L_{line} is the parasitic inductance of the input line (proportional to its length), and L_{sh} is an explicit inductor in series with the 50 Ω matching resistor. C_{pad} models the pad and the ESD protections, while C_{in} is the input capacitance of the first stage. At $\omega_{se} \cong 1/\sqrt{L_{line}((C_{in}C_{pad})/(C_{in} + C_{pad}))}$, the T-line inductance resonates with the series of C_{pad} and C_{in}, making Z_{in} an high impedance. The quality factor of this resonance is reduced thanks to the line series resistance R_{loss}, which is inversely proportional to its width. For a given C_{pad} and C_{in}, an optimum line width can results in a broadband 50 Ω matching. While improving matching, the input line lowers the s_{21} at high frequency, affecting the bandwidth. For this reason, an inductor is placed in series with the 50 Ω matching resistor. The resulting RTRN network, improves both s_{21} and s_{11} at high frequency, as shown in Fig. 8b.

IV. SIMULATIONS AND MEASUREMENTS

The circuit was fabricated in TSMC 28-nm CMOS hpc+ technology (Fig. 9), exploiting low-voltage devices. Fig. 10 shows the measured s_{21} and s_{11}, changing the gain from 20 to 10 dB. A return loss below -10 dB over the entire bandwidth is obtained. No shape variations is detected among the curves while changing the gain. s_{21} variations are bounded to 3 dB up to 58 GHz. An electro-opto (EO) transfer function (normalized at the 1 GHz gain) is reported, simulated loading the measured device with a MZM model. Parasitic (capacitive and inductive contributors) coming from the modulator itself and interconnections are embedded into the used model. Group

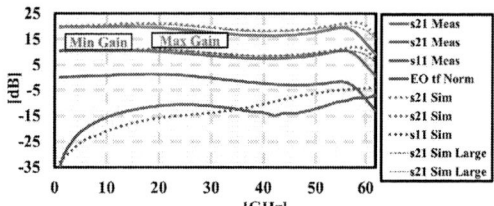

Fig. 10: Measured and simulated driver S-parameters.

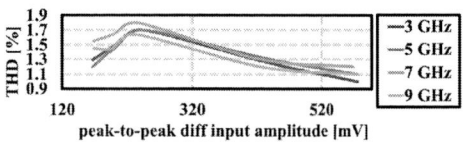

(a) THD with a fixed output amplitude of 1,5 V_{ppdiff}.

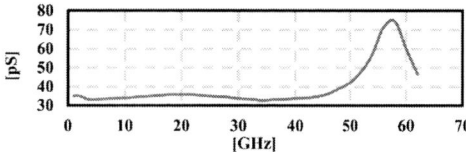

(b) Measured group delay over frequency.

Fig. 11: Measured THD and group delay.

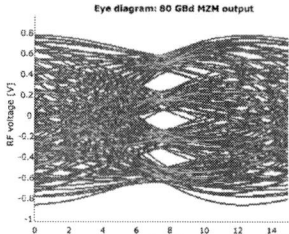

(a) 80 GBaud, PAM-4 eye-diagram without digital pre-equalization.

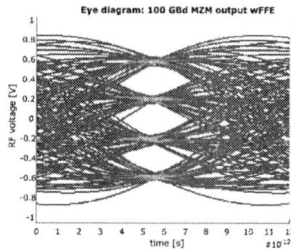

(b) 100 GBaud, PAM-4 eye-diagram with digital pre-equalization.

Fig. 12: Simulated PAM-4 eye-diagrams at the MZM model output.

TABLE I: Comparison with State of the Art.

	This Work	ISSCC 2020 [5]	ISSCC 2019 [6]	IMS 2017 [7]	IMS 2016 [8]
Technology	28 nm CMOS	65 nm CMOS	130 nm SiGe BiCMOS	55 nm SiGe BiCMOS	0.25 μm InP DHBT
Electrical BW [GHz]	58	48	40	57.5	67
Gain(1GHz) [dB]	10-20 (continuous)	13-22.5	20-30	18.8	0.9-10.7
Supply [V]	2.4	2.6	3.6	2.5/3.3/6	5
THD [%]	1.8	2.2	3.6	6	
P_{dc} [mW/Ch]	297	225	1000	820	840
Diff. Out Swing [V]	1.5	1.5	6	6	1.8
Integrated Noise (80GHz) [mV$_{rms}$]	3.28 (20 dB) 2.89 (10 dB)	-	-	-	-

Delay (GD) versus frequency is shown in Fig. 11b. The GD stays almost constant over frequency increasing smoothly by about 40 pS at the edge. Furthermore, as shown in Fig. 10, the simulated large signal (1.5 V$_{ppdiff}$ swing) and small signal s$_{21}$ are in good agreement and very close to the measured one. Fig. 11a gives the measured THD at 3, 5, 7, 9 GHz, varying the gain from 20 to 10 dB. THD is always below 1.8%, for a constant output swing of 1.5 V$_{ppdiff}$ and the trend towards better distortion at lower gain is visible. Power consumption at 20 dB is 297 mW. At high and low gain the output noise integrated over 80 GHz is 3.28 and 2.89 mV$_{rms}$ (a SNR above 37 dB). Fig. 12 shows simulated eye-diagrams for 80 GBaud (Fig. 12a) and 100 GBaud (Fig. 12b) PAM-4 signals. The amplitude is lower than the 1.5 V$_{ppdiff}$ at the MZM input, because of the model used. CMOS is seldom used at such wide bandwidths, so mostly SiGe-BiCMOS and InP-DHBT works are reported in Table I. The only CMOS implementation to our knowledge is in 65 nm [2], and swing efficiency, distortion and bandwidth are in our favour.

V. CONCLUSIONS

This work proves CMOS to be a viable technology for coherent optical links. The tested driver results to be suitable for a 16 QAM modulation. To our knowledge, we achieved the largest bandwidth and the lowest harmonic distortion for the same output swing, among all CMOS solutions. For some features, SiGe-BiCMOS and InP-DHBT drivers are outperformed. A gain control technique, which results in better linearity at larger input voltage swings has been demonstrated.

VI. ACKNOWLEDGMENTS

The authors acknowledge the technical contribution of F. Vecchi, L. Romanò, D. Montanari, D. Caccioli and M. Dalla Santa from the Milano Huawei R&D center.

REFERENCES

[1] B. Razavi, "Design of Integrated Circuits for Optical Communications", Hoboken, NJ, USA: Wiley, 2012.

[2] T. Jyo, M. Nagatani, J. Ozaki, M. Ishikawa, H. Nosaka, "A 48 GHz BW 225 mW/ch Linear Driver IC with Stacked Current-Reuse Architecture in 65nm CMOS for Beyond-400 Gb/s Coherent Optical Transmitters," *2020 IEEE International Solid-State Circuits Conference* February 2020.

[3] P. Wambacq, G. G. E. Gielen, P. R. Kinget, and W. Sansen, "High-Frequency distortion analysis of analog integrated circuits," *IEEE Trans. Circuit Syst. II, Analog Digit. Signal Process.*, vol. 46, no. 3, pp. 335-345, Mar. 1999.

[4] C. Liao, and S. Liu, "40 Gb/s Transimpedance-AGC Amplifier and CDR Circuit for Broadband Data Receivers in 90 nm CMOS," *IEEE Journal of Solid-State Circuits*, vol. 43, no. 3, Mar. 2008.

[5] S. Galal, and B. Razavi, "40-Gb/s Amplifier and ESD Protection Circuit in 0.18-um CMOS Technology," *IEEE Journal of Solid-State Circuits*, vol. 39, no. 12, Dec. 2004.

[6] Abdelrahman H. Ahmed, D. Lim, A. Elmoznine, Y. Ma"A 6V Swing 3.6% THD >40 GHz Driver with 4.5xBandwidth Extension for a 272 Gb/s Dual-Polarization 16-QAM Silicon Photonic Transmitter," *2019 IEEE International Solid-State Circuits Conference*, February 2020.

[7] A. Zandieh, P. Schvan, S. P. Voinigescu,"57.5 GHz Bandwidth 4.8 V$_{pp}$ Swing Linear Modulator Driver for 64 GBaud m-PAM Systems," *in 2017 IEEE MTT-S International Microwave Symposium (IMS)*, June 2017.

[8] H. Wakita, M. Nagatani, K. Kurishima, M. Ida,"An Over-67-GHz-Bandwidth 2 V$_{ppd}$ Linear Differential Amplifier with Gain Control in 0.25-μm InP DHBT Technology," *n 2016 IEEE MTT-S International Microwave Symposium (IMS)*, May 2016.

A 136GΩ-Input-Impedance Active Electrode for Non-Contact ECG Using Auto-Calibrated Positive Feedback and Capacitance Scaling in Femtofarad Resolution

Peizhuo Wang, Tianxiang Qu, Liangbo Lei, Zhiliang Hong and Jiawei Xu

State Key Laboratory of ASIC and System, Fudan University, Shanghai, China.

Email: jwxu@fudan.edu.cn

Abstract—**This paper presents a 10.46µW, 136GΩ ultra-high-input-impedance, 700mV$_{pp}$ linear-input-range active electrode (AE) for non-contact ECG sensing. The femtofarad-resolution calibration against internal parasitic capacitance is realized via an excitation-signal-free auto-calibrated positive feedback loop, as well as the capacitance fine scaling and dummy input structure. The external parasitic capacitance on PCB is bootstrapped by active shielding. As a result, the AE exhibits an equivalent input capacitance of 19.5fF (average of 10 samples), a 2.7x-34x improvement on the state of the art.**

I. INTRODUCTION

Non-contact ECG measurement through capacitive coupling eliminates the direct skin contact for maximum user comfort and safety. However, low coupling capacitance, in the order of 1pF to 1nF, exacerbates the design challenges of the readout circuits. To reduce the noise and interference coupling to the high-impedance input terminals, the electrode is often co-integrated with a voltage buffer in the same module (Fig. 1), known as an active electrode (AE). However, the AE must exhibit ultra-high input impedance (Z_{IN}) greater than a few GΩ in the entire ECG signal bandwidth (0.5-100Hz) to prevent signal attenuation and a significant drop of the common-mode rejection ratio (CMRR) between a pair of AEs. This can be a challenging target given the fact that multiple sources contribute to parasitic capacitance (Fig.1), including external stray capacitance of the PCB, on-chip parasitic capacitance of the bond pad, ESD and the buffer itself.

Prior art designs have employed multiple technique to realize an ultra-high-input-impedance buffer or amplifier (Fig.2). Active shielding can increase the Z_{IN} of AEs [1], but it is only effective in cancelling the external board-level parasitic capacitance, while the on-chip parasitic capacitance (C_{INT}) remains a Z_{IN}-limiting factor. Bootstrapping the gate capacitance of an input pair [2][3] or the capacitance of an ESD diode [4] by an additional buffer can also achieve a high Z_{IN}. However, the cost is a complicated shielding scheme along all critical input signal traces, leading to significant layout efforts. Alternatively, a positive feedback loop (PFL) [5][6] can be adjusted to provide a majority of the input current, thereby reducing both C_{EXT} and C_{INT}. Nevertheless, these capacitance are ill-defined, and prior art PFLs [3][4] still rely on the manual trial-and-error calibration of the feedback capacitor array (C_{PF}) to null C_p. Frequent manual calibration is unpractical during the real time ECG recording. A self-calibrated PFL [5] was proposed to explore the optimal control words (CW) of C_{PF}, but this design involves a

square-wave excitation signal for calibration, as well as the manual trimming to achieve a 10-100fF resolution for the least significant bits (LSBs).

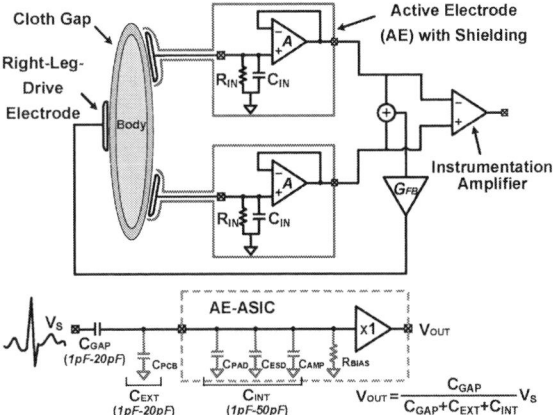

Fig. 1. Architecture of the AE-based non-contact ECG recording (top) and the parasitic capacitance contributors (bottom).

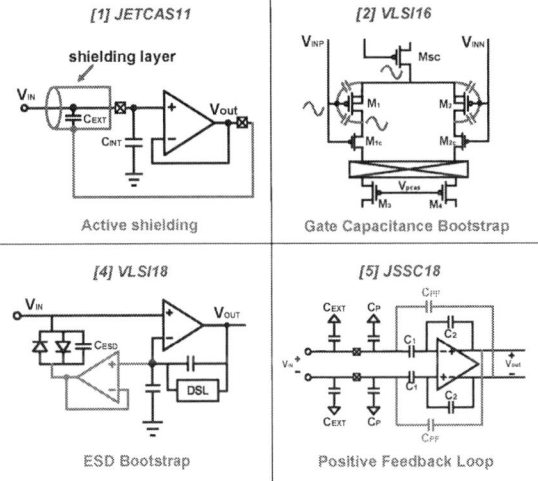

Fig. 2. Prior art impedance boosting techniques overview

978-1-6654-8495-4/22 $31.00 © 2022 IEEE

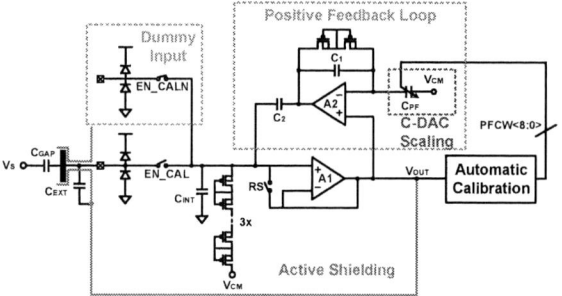

Fig. 3. Ultra-high-input-impedance AE with an auto-calibrated PFL, capacitance fine scaling, input dummy and active shielding.

II. SYSTEM ARCHITECTURE

As shown in Fig.3, the AE core includes a self-biased voltage buffer and an auto-calibrated PFL. The internal on-chip parasitic capacitance (C_{INT}) is compensated by the PFL and dummy input. The PFL contains a C-DAC array to reduce the impact of its switch parasitic and so ensures a femtofarad-level fine compensation. The external parasitic capacitance (C_{EXT}) is nulled by active shielding.

A. Auto-Calibrated Positive Feedback Loop

The AE has two operation modes: calibration and acquisition. In the calibration mode, the AE is isolated from the input pad and internally biased to V_{CM} through three pseudo resistors R_{IN} (Fig.4). The total input parasitic capacitance C_{INT} is compensated by the PFL, which controls the process of sourcing an input bias current to C_{INT} such that impedance boosting is achieved. However, the parasitic capacitance of input pad and ESD cannot be calibrated. To solve this issue, a dummy input pad with ESD [5] is tied to the buffer during the calibration. Hence, the I/O parasitic capacitance is also compensated in the first order. This technique offers high precision while being more power efficient compared to the existing solutions, i.e. by driving an estimated I/O capacitor with an amplifier [3][6], or by bootstrapping the ESD capacitance with a buffer [4].

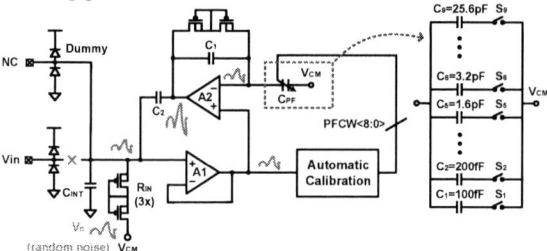

Fig. 4. Oscillation-detection based PFL calibration

Although I/O parasitic capacitance can also be calibrated by the PFL without using the dummy input structure, the drawbacks are twofold. First, the input signal source must be removed from input pad whenever calibration is performed. Second, the PFL needs to compensate more parasitic capacitance while maintaining a high resolution, resulting in a substantially larger area of the feedback capacitor array C_{PF}.

Prior art self-calibrated PFL [5] requires a calibration voltage source to generate the square test signal. Instead, this work uses an auto-calibrated PFL without any excitation signal by making use of the AE's intrinsic stability, where an overcompensated PFL results in oscillation. This is illustrated in Fig.4, the random noise V_n propagates in the feedback system, the loop gain equation can be expressed as:

$$\frac{V_{OUT}}{V_n} = \frac{\frac{1}{R_{IN}} + sC_{PF}}{s(C_{INT} - \frac{C_{PF}}{C_1}C_2) + \frac{1}{R_{IN}}} \quad (1)$$

R_{IN} is the high impedance pseudo resistors to bias the AE. C_{INT} represents the total on-chip parasitic capacitance. C_{PF} and C_1 form the capacitance fine scaling network, and C_2 is the actual feedback capacitor determining the compensation current. A dominant pole can be found at

$$\omega_p = \frac{1}{R_{IN}(\frac{C_{PF}}{C_1}C_2 - C_{INT})} \quad (2)$$

Based on the feedback theory, the system remains stable only if this pole is located in the left half plane, indicating a negative feedback. However, when the PFL is dominant, oscillation occurs and the pole moves to the right half plane. Since R_{IN}, C_1, C_2 and C_{INT} are rather static, adjusting C_{PF} can prevent oscillation while simultaneously achieving high input impedance.

B. SAR Based Oscillation Detection Circuit

To determine the optimum compensation current of the PFL, an oscillation detector is placed at the AE's output (Fig.5). If the buffer oscillates, the toggling of AE output triggers the window comparator. Otherwise, the AE maintains a steady DC within the thresholds of $\pm V_{TH}$, indicating the amount of PFL current can be further increased to boost the input impedance. It is worth noting that V_{TH} is reconfigurable from 50mV to 200mV with respect to V_{CM} to ensure reliable oscillation detection. A reference generator in Fig.5 produces V_{TH} that is independent of the absolute accuracy of replica resistors and bias currents.

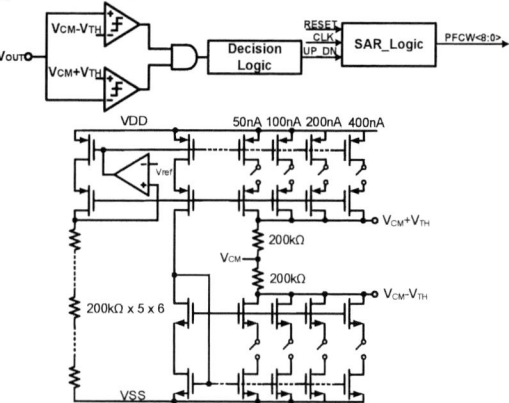

Fig. 5. Automatic calibration logic and reference ($V_{CM} \pm V_{TH}$) generator.

The auto-calibration is based on a successive approximation (SAR) logic that is faster compared to the linear search logic in [5][6]. Fig.6 shows the AE output with respect to the C_{PF} control words during the calibration. When the SAR completes its 9 clock cycles, the calibration is done and the optimum C_{PF} control words corresponding to the highest input impedance are frozen.

In the subsequent acquisition mode, the AE is switched back to the input pad for ECG measurement. The total internal parasitic

978-1-6654-8495-4/22 $31.00 © 2022 IEEE

capacitance (C_{INT}), including the pad and ESD capacitance, has been already calibrated, and so the off-chip parasitic capacitance (C_{EXT}) at board level is cancelled by active shielding. This leads to minimized residual input capacitance. In case it is sensitive to PVT variation, multiple calibration can be easily made without disconnecting the AE module from the human body.

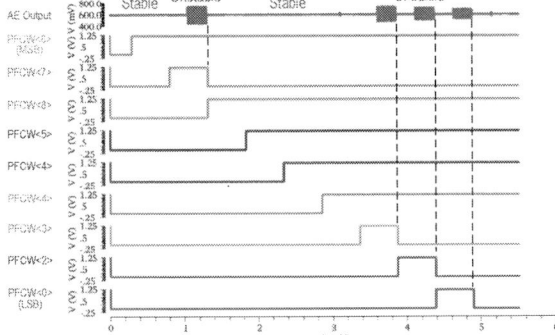

Fig. 6. Waveform illustration of AE output and the C_{PF} control words.

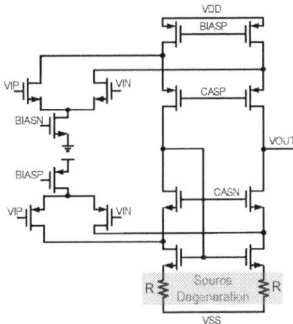

Fig. 7. Complementary input buffer with source degeneration.

C. Capacitance Fine Scaling

The residual input capacitance after calibration is limited by the mismatch between the input and dummy IOs, as well as the LSB of C_{PF}. In prior art, it is very difficult to implement a 9-bit C-DAC with a sub-10fF LSB [1][3]-[6], considering parasitic capacitance from the C-DAC switches, amplifiers and layout traces.

This work use a capacitance scaling technique to implement a parasitic-insensitive PFL (Fig.4). Instead of adapting C_2 directly, the buffer output is amplified by a noninverting amplifier with a gain of $1+C_{PF}/C_1$. C_{PF} is used to determine the amount of feedback current, and so the PFL's equivalent feedback capacitance C_{eq}. In this way, it is possible to utilize C_{PF} with a 100fF LSB to realize a sub-10fF equivalent of C_{eq}. This is a 5x-30x improvement on the state-of-the-art PFL solutions [1][3]-[6]. The auto-calibrated PFL technique is broadly applicable to various amplifiers.

D. Low Noise Rail-to-Rail Input Buffer

For non-contact ECG sensing, to ensure the AE not saturated by motion artifact or 50/60Hz interference, a large input common-mode (CM) range of at least 600mV$_{pp}$ is necessary according to the IEC standard. In Fig.7, the buffer employs a complementary input pair to provide a rail-to-rail input CM range of 700mV$_{pp}$.

Although chopping can eliminate $1/f$ noise, considering that the switched-capacitor resistance can result in low input impedance penalty, we deliberately use large-area input pairs (W/L=320/6µ) to reduce $1/f$ noise. Although the complementary input transistors with large size contribute significant parasitic capacitance, it can be fully compensated by the auto-calibrated PFL. In spite of the capacitance fine scaling, the 9-bit C_{PF} of 51.1pF can effectively cover the maximum on-chip parasitic capacitance of 5.11pF while still retaining a sub-10fF equivalent resolution. Therefore, the AE can achieve both low noise and ultra-high input impedance at the same time.

III. MEASUREMENT RESULTS

Manufactured in a 0.18µm CMOS process, the AE prototype occupies an area of 0.186 mm^2 (Fig.8). The power consumption is 10.46µW at a 1.2V supply. Fig.9 illustrates the measurement setup for input impedance and CMRR. By applying a 100mV$_{pp}$, 60Hz sinusoidal to the AE through a 100fF coupling capacitor and then measuring the AE's output signal, the equivalent input capacitance $C_{in,eq}$ and the input impedance at 60Hz can be derived.

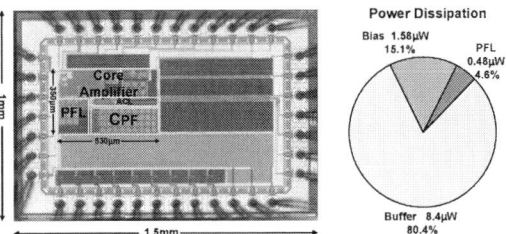

Fig. 8. Chip photograph and measured AE power breakdown.

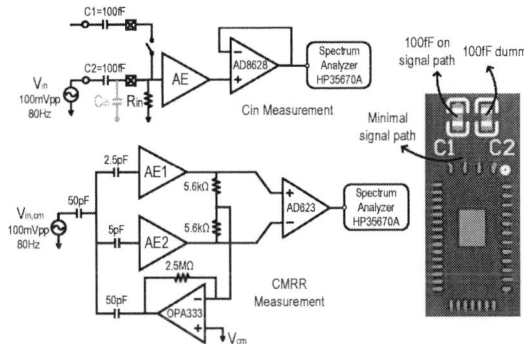

Fig. 9. Measurement circuits of C_{in} and CMRR (left), where external coupling capacitors (100fF, 2.5pF and 5pF) are placed very close to the AE with minimal signal path to reduce noise coupling. Fully symmetric layout of the 100fF coupling capacitors is required to reduce capacitance mismatch between input I/O and dummy I/O (right).

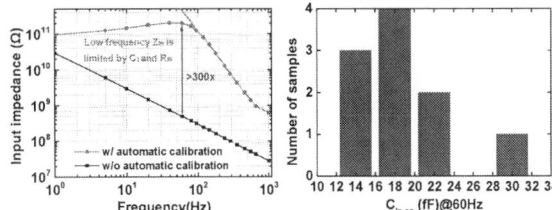

Fig. 10. R_{IN} versus frequency (left) and histograms of C_{IN} (right).

978-1-6654-8495-4/22 $31.00 © 2022 IEEE

In Fig.10, the left shows the measured input impedance versus frequency. After enabling the PFL, the AE exhibits the maximum input impedance of $200G\Omega$ at 60Hz (i.e. $C_{in,eq}=13.3fF$), a 300x improvement on the result without calibration. Moreover, the AE maintains its high input impedance above $100G\Omega$ throughout the ECG bandwidth. The right plot shows the histogram of the input capacitance $C_{in,eq}$ after calibration, where 9 out of 10 samples are below 23fF. Fig.11 shows the input and the output voltages of the AE, before and after calibration, respectively. The auto-calibrated PFL can significantly reduce the AE's gain loss.

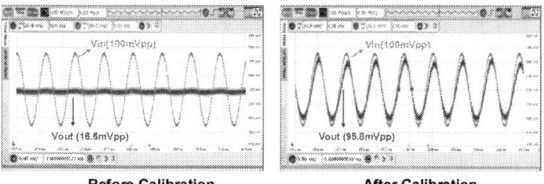

Before Calibration   **After Calibration**

Fig.11. Buffer's input (pink) and output (blue) voltages, before (left) and after (right) auto-calibration.

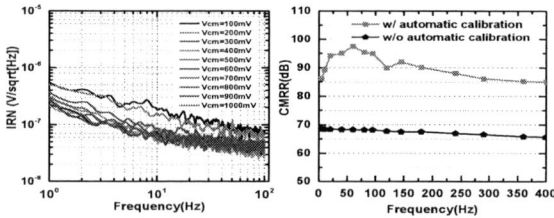

Fig. 12. Measured input referred noise versus V_{CM} (left), and measured CMRR (with RDL) under imbalanced coupling capacitors (right).

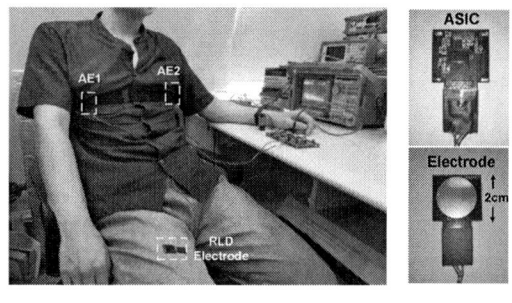

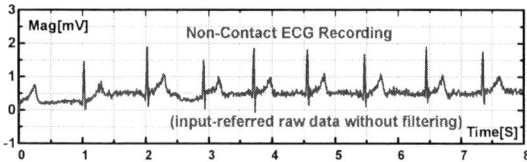

Fig. 13. Non-contact ECG recording with miniaturized AEs and RLD.

Fig.12 shows the input referred noise (IRN) of the AE across different input CM voltages. The integrated noise at the default $V_{CM}=600mV$ is $0.72\mu V_{rms}$ (0.5-100Hz). Within a wide input CM range of 200-900mV, the AE exhibits a similar IRN and it only increases slightly at $V_{CM}=100mV$ and 1000mV. The bottom plot shows the measured CMRR of a pair of AEs with the Right Leg Drive (the RLD in Fig.1). After enabling the auto-calibrated PFL, the CMRR was improved from 68dB to 95dB. Please note that two

coupling capacitors (2.5pF and 5pF) were deliberately selected in the CMRR test to mimic the capacitance mismatch between skin and electrodes.

A non-contact ECG measurement on the subject was validated through a miniaturized AE system (Fig.13). The IC is bonded on the bottom of a 4-layer PCB, the copper plate on top implements the electrode of a 2cm diameter. Two intermediate layers for active shielding and grounding ensure good immunity to interference. By attaching two AE boards inside a chest strap and a RLD board on the leg, the measured raw ECG signal in Fig.13 shows good signal quality and demonstrates the necessity of impedance calibration.

Table I compares this AE with recently developed ultra-high input impedance amplifiers and buffers. Thanks to the excitation-signal-free auto-calibrated PFL with capacitance fine scaling. The AE achieves the lowest input capacitance of 19.5fF (average), i.e. an ultra-high input impedance of $136G\Omega$ at 60Hz, this is a 2.7x-34x improvement on the state of the art [1]-[6]. Moreover, the AE also exhibits $0.72\mu V_{rms}$ input referred noise over a $700mV_{pp}$ input CM range, and a high CMRR of 95dB with the largest impedance mismatch.

Table I Comparison with ultra-high input impedance amplifiers.

Parameters	[1] JETCAS11	[5] JSSC18	[4] VLSI18	[2] VLSI16	[3] ISSCC20	[6] VLSI21	This work
Process	0.35μm	0.18μm	0.18μm	0.18μm	0.18μm	0.11μm	0.18μm
V_{dd}	3.3V	0.8V	5V	1.8V	1.8V	1/1.5V	1.2V
Power	15.84μW	0.26μW	90μW	108μW	3.35μW	3.83μW	10.46μW
R_{IN}@50Hz	50GΩ	50GΩ	4GΩ	6.7GΩ	50GΩ	2GΩ	136GΩ@60Hz(avg) 200GΩ@60Hz(max)
C_{IN}	60fF	63fF	800fF	475fF	63fF	1.6pF	19.5fF(avg) 13.3fF(min)
Noise (IRN)	N/A	8.26μV$_{rms}$ (0.5-400Hz)	3.7μV$_{rms}$ (0.5-100Hz)	0.67μV$_{rms}$ (0.5-100Hz)	3.15μV$_{rms}$ (0.5-5kHz)	0.36μV$_{rms}$ (0.5-300Hz)	0.72μV$_{rms}$ (0.5-100Hz)
Excitation Signal Free	No	No	No	No	No	No	Yes
Fully Auto Calibration	No	No	No	No	No	Yes	Yes
CMRR with Impedance Mismatch	N/A	N/A	N/A	78dB (ΔR=1MΩ)	102dB (ΔR=1MΩ)	N/A	95dB ΔC=2.5pF (ΔR=1.2GΩ)

IV. CONCLUSIONS

The AE proposed in this paper combines diverse techniques, including auto-calibrated PFL, capacitance scaling, dummy input and active shielding, to null multiple parasitic capacitance in high precision. As a result, the AE achieves ultra-high input impedance of $136G\Omega$ at 60Hz. Furthermore, the oscillation-detection based calibration obviates the need of an excitation source and manual adjustment of the PFL. A capacitively-coupled non-contact ECG measurement with miniaturized AE boards has been successfully demonstrated.

REFERENCES

[1] Y. M. Chi et al, "Ultra-High Input Impedance, Low Noise Integrated Amplifier for Noncontact Biopotential Sensing," vol. 1, no.4, pp. 526-535, *IEEE JETCAS*, Dec. 2011.

[2] X. Zhou et al., "A Wearable Ear-EEG Recording System Based on Dry-Contact Active Electrodes," *Symp. VLSI Circuits*, pp. 260–261, June. 2016.

[3] S. Zhang et al, "A 130dB CMRR Instrumentation Amplifier with Common-Mode Replication," *Digest of ISSCC*, pp. 356-358, Feb. 2020

[4] M. Chen et al., "A 400GΩ Input-Impedance, 220mV$_{pp}$ Linear-Input-Range, 2.8V$_{pp}$ CM Interference-Tolerant Active Electrode for Non-Contact Capacitively Coupled ECG Acquisition," *Symp. VLSI Circuits*, pp. 129-130, June. 2018.

[5] J. Lee et al., "An Ultra-High Input Impedance Analog Front End Using Self Calibrated Positive Feedback," *IEEE JSSC*, vol. 53, no. 8, pp. 2252–2262, Aug.2018.

[6] Y. Park et al., "A 3.8-μW/Ch, 15-GΩ Total Input Impedance Chopper Stabilized Amplifier with Dual Positive Feedback Loops and Auto-calibration Scheme," *Symp VLSI Circuits*, pp.13-19, June.2021.

A Chopper Biopotential Instrumentation Amplifier With DSL-Embedded Input Stage Achieving 109 dB CMRR and 400 mV DC Offset Tolerance

Surachoke Thanapitak[1] and Chutham Sawigun[2]

Email: surachoke.tha@mahidol.ac.th, chutham.sawigun@imec.be

[1]Department of Electrical Engineering, Mahidol University, Nakhon Pathom, Thailand

[2]Circuits for Neural Interfaces Group, imec, Heverlee, Belgium

Abstract— **This paper presents a bio-potential chopper current-balanced instrumentation amplifier (CBIA) that uses a dc-isolated, source-coupled-pair input stage with embedded dc-servo loops. Despite chopping a large electrode offset, gate-source voltage of each input transistor does not vary excessively. As a result, for typical bio-potential input voltages, less than 1 percent THD can be achieved. Hence, excessive power consumption dedicated for linearization of the input stage is no longer needed. This results in 90 % power reduction compared with the most recent CBIA. Fabricated in a 0.35-micron CMOS, the proposed CBIA maintained CMRR over 80 dB up to 0.3 V input offset (109 dB for zero offset). The chip occupies 0.41 sq. mm and consumes 1.19 microwatt from a 3 V supply. The input noise density, input impedance and noise efficiency factor of 91 nV per square-root-hertz, 469 Megaohms at 5 Hz and 2.22 NEF are obtained, respectively. Compared with other recent state-of-the-art CBIAs, this proposed IA attains the lowest NEF and PEF.**

Keywords—amplifier, bio-potential, chopping, CMOS, CMRR, electrode offset, instrumentation amplifier, low-noise.

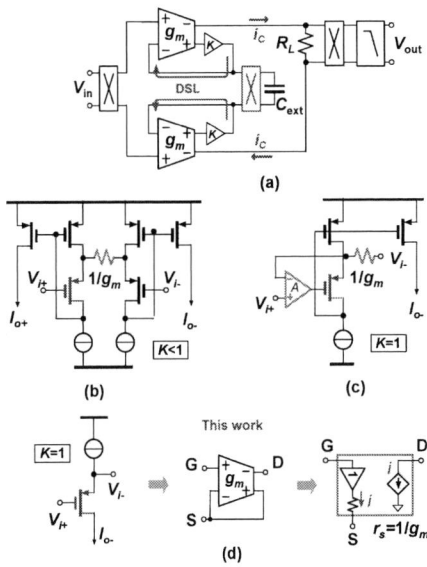

Fig. 1 General model of chopper CBIA and its g_m variants.

I. INTRODUCTION

Bio-potential signals, for example electroencephalogram (EEG) and electrocardiogram (ECG), can be employed as an indicator for cardiac and brain conditions. These signals play an important role in wearable devices [1] which have the capability to continuously monitor and provide information for personalized preventive healthcare [2] . ECG and EEG can be measured by using electrodes attached to the skin [3] with amplitudes approximately one hundred times smaller than the dc electrode offset (EO) and common-mode interference (CMI). Both EO and CMI can be as large as 300 mV [3]. An instrumentation amplifier (IA) for measuring such bio-potential thus needs to attain sufficiently high EO and common-mode rejections and low input-referred noise (IRN).

A chopping technique is usually employed to remove the $1/f$ noise and enhance the CMRR of the IA. However, typical chopper IAs turn the large EO and the required biopotential voltage into a larger modulated signal. Linearization is thus needed for the input stage of the IA to minimize distortion. The undesired EO will be down-converted back at the output. A dc-servo loop (DSL) is then used to remove the EO. Unfortunately, these require additional power and area [4, 5]. The current-balanced IA (CBIA) of [6] could dispense with the bulky DSL by inserting an off-chip capacitor in series with the degenerated resistor of the linearized input stage. However, the excessive power consumption remains.

This work proposes a chopper CBIA using an inherent current-reuse DSL to save power and area. Although the degenerated resistor is dispensed, reasonable EO rejection can still be achieved with considerable power reduction (>90% reduced from [6]). In the next section, the power-efficient CBIA will be described. Design consideration targeting for ECG signal acquisition is presented in Sec. III. Performance of this proposed CBIA is demonstrated and benchmarked with other recent state-of-the-art IAs in Sec. IV. Finally, the conclusion is summarized in Sec. V.

II. PROPOSED CHOPPER CBIA

A. Traditional CBIAs

Fig. 1 (a) shows the generalized macro-model of the chopping CBIA. It comprises input transconductance (TC) and output transimpedance (TI) stages, three sets of chopping switches and DSLs. The TI stage is realized by two transconductors (g_ms) each with feedback factor K. External capacitor (C_{ext}) and chopping switches are inserted between the transconductors. Together they form two DSLs to remove dc input components. After chopping at the input, the up-modulated V_{in} (small bio-potential (V_{bio}) and large EO) will be demodulated by the blue switches to appear across C_{ext}.

After blocking by C_{ext}, only ac current (i_c) can be conveyed to the TI stage (made by R_L). It will be further demodulated and filtered to obtain an offset-free V_{out}. For $K = 1$, the highpass cutoff equals $0.5g_m/C_{ext}$. The voltage

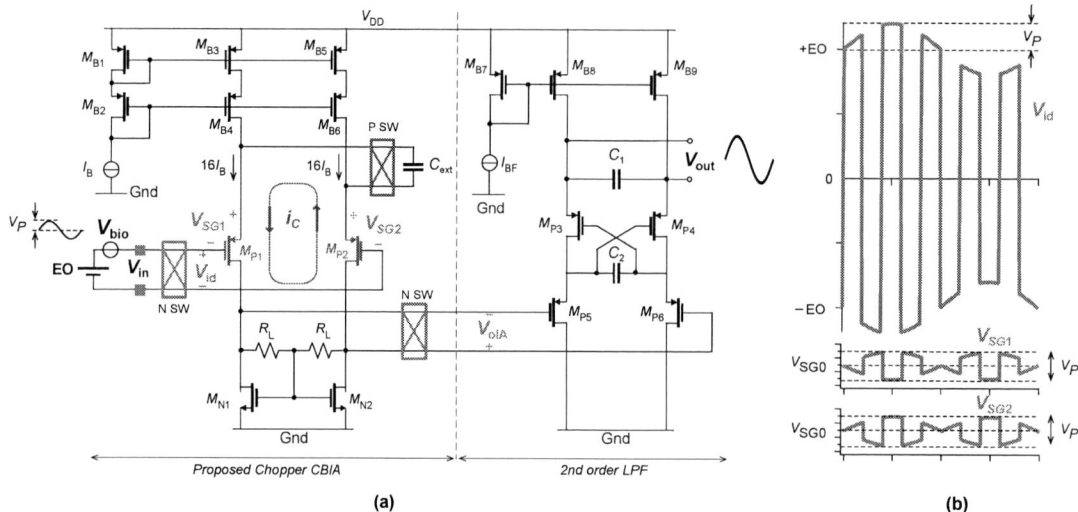

Fig. 2 The proposed CBIA (a) transistor circuit level and (b) voltage signal (from simulation) at source-gate of input transistors.

gain is $0.5g_m R_L$ and the lowpass cutoff is defined by the lowpass filter (LPF).

Recent developments of g_m circuits for CBIA with their simplified versions are shown in Fig. 1(b) [7] and 1(c) [6]. The flipped voltage followers (FVFs) are used as input elements that provide high input impedance and sufficient low output impedance to drive a degenerative resistor ($1/g_m$). It is common for the circuits in Fig. 1(b) and 1(c) to have additional branches for current mirroring to transfer the resistor's current to TI stage. To achieve the required IRN and LR, considerable power is allocated here. As the FVFs need to drive a resistive load with sufficient LR, minimum bias current of $LR \times g_m$ is required. To enhance this drivability, [6] even added an extra amplifier in front of the FVF. As discussed earlier, the large LR is required to accommodate the undesired EO whereas the required biopotential voltages have only a few millivolts of amplitude.

B. Proposed CBIA

We propose to use a single transistor and one bias current only to replace the g_m block as depicted in Fig. 1(d). For small-signal operation, the single transistor is equivalent to a feedback g_m block with $K = 1$. It is also equivalent to a 2nd-generation current conveyor with r_s of $1/g_m$. This inherent source resistance r_S is sufficient to form a DSL, if V_{i+} does not drive V_{SG} to fluctuate beyond $5mV_p$. A degenerated resistor is no longer required here. The proposed IA circuit is shown in Fig. 2(a). R_L is split into two components to serve as the TI stage and common-mode control. As C_{ext} blocks dc, neglecting all second-order effects, constant I_B is always flowing through source of M_1 and M_2 making the static source-gate voltage for both transistors ($V_{SG0} = n_p U_T \ln(I_B/I_0)$), where n_p is the weak inversion slope factor, U_T is thermal voltage and I_0 is the zero-V_{GS} drain current) always constant regardless of the EO. The input and the blue switches respectively steer V_G and V_S of each transistor in the same direction. Although V_G swings as high as 2EO (see the simulated v_{id} waveform in Fig. 2(b)), V_S follows V_G with the same magnitude superimposed on V_{SG0}. Only a small ac component ($V_P = V_{bio}$) will be added to

V_{SG0} as shown in the simulated $V_{SG1,2}$ in Fig. 2(b). This allows the modulated i_c to flow to the TI stage. The waveforms of $V_{SG1,2}$ also imply that only small modulated V_{bio} (5 mV in the case of ECG) is appeared on the source-gate of input transistors without large components from EO. Acceptable distortions can then be achieved. Therefore, the degenerative resistor (that was traditional required) is no longer needed here. The excessive power spent to drive the resistor can thus be saved.

Cascode current mirrors ($M_{B1\text{-}B6}$) are used to provide current sources $16I_B$ with very high output resistance. The voltage transfer characteristic of this proposed IA is

$$\frac{v_{oIA}}{v_{in}} = \frac{sR_L C_{ext}}{1 + s(g_{mP12}/C_{ext})}, \qquad (1)$$

where $g_{mP12} = 8I_B/n_p U_T$ is the transconductance of the TC stage. The midband gain and highpass cutoff frequency are described by $K_{mid} = 0.5g_{mP12}R_L$ and $f_H = g_{mP12}/2\pi C_{ext}$, respectively.

To achieve a low-power bandwidth control of the IA, we employed a 2nd-order LPF of [8] which provides a cutoff frequency $f_L = g_m/(2\pi\sqrt{C_1 C_2})$, where $g_m = I_{BF}/n_p U_T$ is the transconductance of $M_{P3\text{-}6}$.

III. DESIGN CONSIDERATION

The TI stage (M_{N1} and M_{N2}) is designed to operate in strong inversion while the TC stage is biased in weak inversion. Noise contribution from the cascode current mirrors (M_{B1}-M_{B6}) can be neglected for frequencies above f_L since the noise from the two branches will be combined after C_{ext} is shorted. As a result, the IRN of this CBIA is dominated by the shot noise from M_{P1} and M_{P2}. The IRN can be estimated by $\overline{v_n^2} = (n_p kT)^2/2qI_B$ where q, k and T are the electric charge, the Boltzmann constant and the absolute temperature, respectively. This leads to noise efficiency factor NEF = $\sqrt{\overline{v_n^2}\, I_{Tot}/(\pi U_T \times kT \times f_L)} \cong 3$, where I_{Tot} is the total IA current consumption and $n_p \cong 1.3$.

For practical measurements that deal with a maximum ECG amplitude of 4 mV [9], only pure M_1 and M_2 without a degeneration resistor are sufficient to keep THD below 1%. We designed this amplifier targeting for ECG detection via dry electrodes. We set $I_B = 0.19$ µA and $R_L = 10$ MΩ. M_1 and M_2 are sized large with high aspect ratios in weak inversion. M_3 and M_4 are also large but with small aspect ratios in strong inversion. These are to achieve low IRN. The LPF maintains 100 Hz bandwidth by setting $I_{BF} = 1$ nA. The other design parameters are specified in TABLE I.

TABLE I. DESIGN PARAMETERS

Transistor	W [µm]	L [µm]
M_{B1}, M_{B2}	5	4
M_{B3}, M_{B4}, M_{B5}, M_{B6}	80	4
M_{P1}, M_{P2}	120	1
M_{N1}, M_{N2}	4	100
M_{P3}, M_{P4}, M_{P5}, M_{P6}	12.5	5
M_{B7}, M_{B8}, M_{B9}	24	6
Capacitor	Value	
C_{ext}, C_1, C_2	3.3 µF, 6.5 pF, 13.5 pF	

IV. MEASURED RESULTS

The fabricated IA occupies 0.41 mm^2 area in a 0.35µm CMOS process as shown in Fig. 3. Approximately three-quarters of the chip area is dominated by a poly-resistor. With chopping frequency of 1 kHz, Fig. 4 (a) shows magnitude responses of differential gain (A_{dm}), common-mode gain (A_{cm}) and power supply gain (A_{VDD}) of the proposed IA when EO = 0. This reveals CMRR and PSRR of 109.2 dB and 83.4 dB, respectively. When EO is non-zero, the CMRR is degraded. Fig. 4(b) shows the CMRR measured for 9 samples and for the applied EO of 0 and 0.3 V showing that the minimum CMRR of 80 dB can be maintained over the passband.

The CMRR spotted at 50 Hz for various values of EO can be seen from Fig. 5(a) for cases of with and without chopping. The histogram for the case of chopping is also illustrated in Fig. 5(b) (102.1±6.6 dB and 85.9±1.9 dB for EO of 0 and 0.3 V, respectively). THD versus EO is also shown in Fig. 6(a) for the input frequency of 10 Hz where the worst case of 0.95% is found at EO = −0.4V with the input amplitude of 4 mV$_p$. The midband gain obtained from different EOs is reported in Fig. 6(b), where the gain variation below 0.5 % can be maintained within EO = ±0.4 V.

The 1/f noise is clearly reduced by chopping as can be seen from Fig. 7(a). The IRN integrated over 100 Hz equals 0.91 µV$_{rms}$. The input impedance (Z_{in}) was measured for two cases: with and without chopping. These are shown in Fig. 7(b). Z_{in} becomes highly resistive when chopping is applied. The value of 469 MΩ remains flat from 1 Hz up to 100 Hz for the chopping case. Beyond these frequencies, Z_{in} tends to be capacitive for both cases. Considering the requirement of an IA for dry-electrode ECG recording [9] (CMRR > 77 dB, Z_{in} > 80 MΩ, IRN < 5 µV$_{rms}$, EO = ±0.3 V and V_{inp} = 4 mV), all characteristics of the proposed IA meet well with the requirements, in spite of

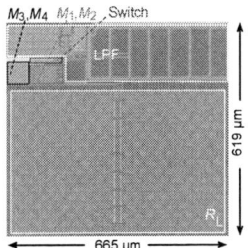

Fig. 3 Fabrication of the proposed chopper CBIA.

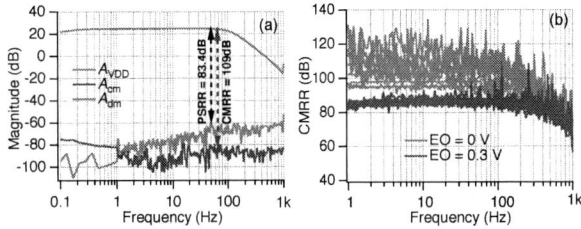

Fig. 4 Magnitude responses of (a) A_{dm}, A_{cm}, A_{VDD} and (b) multi-chip CMRR for different EOs.

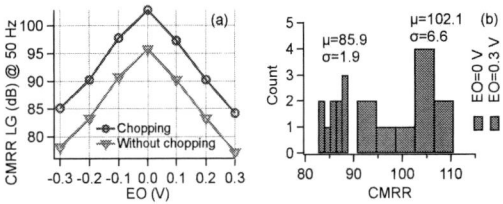

Fig. 5 CMRR@50Hz: (a) against different EOs and (b) its histogram for chopping case.

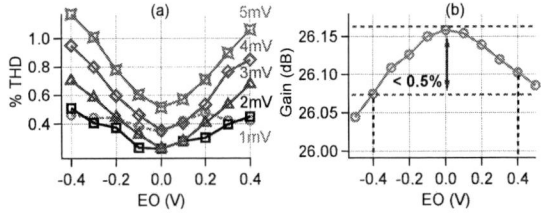

Fig. 6 For different EO: (a) THD @ 10 Hz and (b) the midband gain.

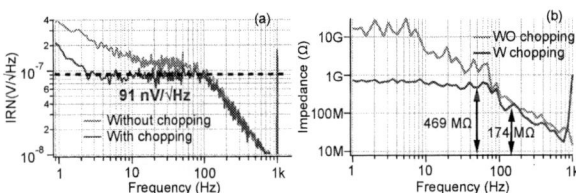

Fig. 7 For with/without chopping cases: (a) IRN and (b) input impedance.

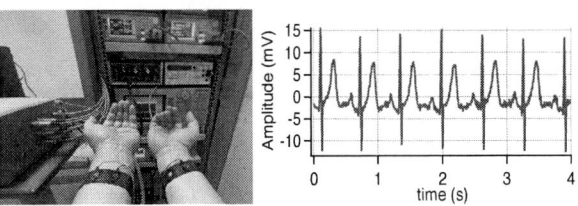

Fig. 8 Dry electrodes ECG measurement setup and ECG signal recording.

978-1-6654-8495-4/22 $31.00 © 2022 IEEE

TABLE II. PERFORMANCE COMARISON WITH STATE-OF-THE-ART IAS.

	This work	ASSCC'20 [6]	CICC'21 [10]	VLSI'21 [11]	VLSI'21 [12]	TBCAS'18 [13]	JSSCC'15 [7]
Process(μm)	0.35	0.18	0.065	0.11	0.055	0.18	0.18
V_{DD}(V)/I_{DD}(μA)	3/0.396	1.2/11.45*	1.2/1.9	1/3.83	1.2/8.8	1.2/35.8	1.2/13.3
P_{Tot}(μW)	**1.19**	13.74*	2.3	3.83	44.2	43	15.96
Gain (dB)/IRN(μV)	26/0.91	34/0.59	27.5/5.75	61-74/0.36	NA/2.88	32-60/1.2	28-36/0.61
BW(Hz)	0.3-100	<1-100	0.5-250	0.5-300	DC-10k	0.5-100	0.5-150
NEF/PEF	**2.22/14.8**	7.83*/73.55*	19.3/463	1.54/2.54	9.8§/115§	28.1/951.2	7.15/61.2
Z_{in} (Ω)@f_{in}(Hz)	469M@50	380M@50	NA	2G@50	283M@10	720M@50	>500M@50
EOmax(V)	±0.4	±0.5	±0.315	NA	±0.07	±0.3	±0.4
CMRR(dB)@EO(V)	**109@0, 85.9@0.3**	105@NA	NA	92@NA	NA	100@NA	>110@NA
%THD@V_{in}(mV$_p$)/EO(V)	0.9@5/0.3	0.13@9/NA	NA	NA	0.44@10/0▫	NA	NA
Area(mm²) IA/C_{ext} (F)	0.41/3.3μ	0.2/10μ	0.31†/0	0.74/82p	0.0077/0	0.97†/1μ	0.6†/1μ
Structure	CBIA	CBIA	CCIA	CCIA+PGA	CBIA	CBIA	CBIA

*Excluding the active lead bias circuit †Estimated from chip photograph §Entire channel ▫1kHz input frequency.

consuming 396 nA total current only (> 3 μA for all traditional CBIAs). The ECG measurement setup (bipolar readout configuration) and the recorded ECG signal are demonstrated in Fig. 8. where PQRST wave can be visible from this measurement.

For performance benchmarking from Table II, this work exhibits the best NEF and PEF among other CBIAs. Dispensing active elements required for DSL and input stage linearization, this work becomes the most power efficient CBIA that is able to achieve sufficient EO tolerance.

V. CONCLUSION

A chopper CBIA capable of withstanding ±0.4 V electrode offset and maintaining > 80 dB without traditionally required input stage linearization has been demonstrated in this work. The proposed design achieves significant power reduction compared with the most recent state-of-the-art CBIA. The given experimental results performed on a human subject well confirm the applicability of the proposed IA.

ACKNOWLEDGMENT

This project is funded by the National Research Council of Thailand (NRCT, N41A640163) and Faculty of Engineering, Mahidol University. The human subject experiment in this work was approved by Mahidol University Central Institutional Review Board (MUCIRB 2021/120.1703).

REFERENCES

[1] X. Teng, Y. Zhang, C. C. Y. Poon, and P. Bonato, "Wearable Medical Systems for p-Health," *IEEE Reviews in Biomedical Engineering*, vol. 1, pp. 62-74, 2008.

[2] R. Matthews, N. J. McDonald, P. Hervieux, P. J. Turner, and M. A. Steindorf, "A Wearable Physiological Sensor Suite for Unobtrusive Monitoring of Physiological and Cognitive State," in *2007 29th Annual International Conference of the IEEE Engineering in Medicine and Biology Society*, 22-26 Aug. 2007 2007, pp. 5276-5281.

[3] L. Yan and J. Bae, "Challenges of Physiological Signal Measurements Using Electrodes: Fudamentals to Understand the Instrumentation," *IEEE Solid-State Circuits Magazine*, vol. 9, no. 4, pp. 90-97, 2017.

[4] Q. Fan, F. Sebastiano, J. H. Huijsing, and K. A. A. Makinwa, "A 1.8 μW 60 nV/√ Hz Capacitively-Coupled Chopper Instrumentation Amplifier in 65 nm CMOS for Wireless Sensor Nodes," *IEEE Journal of Solid-State Circuits*, vol. 46, no. 7, pp. 1534-1543, 2011.

[5] N. V. Helleputte et al., "18.3 A multi-parameter signal-acquisition SoC for connected personal health applications," in *2014 IEEE International Solid-State Circuits Conference Digest of Technical Papers (ISSCC)*, 9-13 Feb. 2014 2014, pp. 314-315, doi: 10.1109/ISSCC.2014.6757449.

[6] J. Xu and Z. Hong, "A 2-Electrode ECG Amplifier with 0.5% Nominal Gain Shift and 0.13% THD in a 530mVpp Input Common-Mode Range," in *2020 IEEE Asian Solid-State Circuits Conference (A-SSCC)*, 9-11 Nov. 2020 2020, pp. 1-4.

[7] N. V. Helleputte et al., "A 345 μW Multi-Sensor Biomedical SoC With Bio-Impedance, 3-Channel ECG, Motion Artifact Reduction, and Integrated DSP," *IEEE Journal of Solid-State Circuits*, vol. 50, no. 1, pp. 230-244, 2015.

[8] T. Zhang et al., "15-nW Biopotential LPFs in 0.35- μm CMOS Using Subthreshold-Source-Follower Biquads With and Without Gain Compensation," *IEEE Transactions on Biomedical Circuits and Systems*, vol. 7, no. 5, pp. 690-702, 2013.

[9] M. J. Burke and D. T. Gleeson, "A micropower dry-electrode ECG preamplifier," *IEEE Transactions on Biomedical Engineering*, vol. 47, no. 2, pp. 155-162, 2000.

[10] G. Mu, D. Ye, L. Lyu, X. Zhao, and C. J. R. Shi, "An 8-Channel Analog Front-End with a PVT-Insensitive Switched-Capacitor and Analog Combo DC Servo Loop Achieving 300mV Tolerance and 0.64s Recovery Time to Electrode-DC Offset for Physiological Signal Recording," in *2021 IEEE Custom Integrated Circuits Conference (CICC)*, 25-30 April 2021 2021, pp. 1-2.

[11] Y. Park, J. H. Cha, S. H. Han, J. H. Park, and S. J. Kim, "A 3.8-μW/Ch, 15-GΩ Total Input Impedance Chopper Stabilized Amplifier with Dual Positive Feedback Loops and Auto-calibration Scheme," in *2021 Symposium on VLSI Circuits*, 13-19 June 2021 2021, pp. 1-2.

[12] S. Wang et al., "A 77-dB DR 16-Ch 2nd-order Δ-$\Delta\Sigma$ Neural Recording Chip with 0.0077mm²/Ch," in *2021 Symposium on VLSI Circuits*, 13-19 June 2021 2021, pp. 1-2.

[13] J. Xu et al., "A 665 μW Silicon Photomultiplier-Based NIRS/EEG/EIT Monitoring ASIC for Wearable Functional Brain Imaging," *IEEE Transactions on Biomedical Circuits and Systems*, vol. 12, no. 6, pp. 1267-1277, 2018.

A Direct Sensor Readout Circuit Using VCO-Driven Chopping with 42dB SNR at 800μVpp Input

Yingjie Chen[1], Marino D Guzman[1], Beomsoo Park[1,2], Nima Maghari[1]
{chenyingjie, mguzman, beomsoo0927}@ufl.edu, maghari@ece.ufl.edu
[1]Department of Electrical and Computer Engineering, University of Florida, Gainesville, FL, U.S.
[2]Qualcomm Inc., San Diego, CA, USA

Abstract— This work presents a new voltage-controlled-oscillator (VCO)-based sensor readout chip for sub-millivolt signal sensing applications. The proposed sensor front-end utilizes a new multi-path chopping topology driven by VCO output phases that reduces the modulation noise magnitude by spreading the noise power over a relatively large frequency range. It removes the need for extra filters required by the conventional chopping technique, reducing the front-end area and complexity. A linearization technique for the VCO is developed to improve the linear operating range. The prototype was implemented in a 65nm process. It processes an 800μVpp input signal and achieves 42dB SNR with 1KHz bandwidth while consuming 1.84μW.

I. INTRODUCTION

The demand for highly sensitive and area-efficient sensors is driven by the tendency of using Internet-of-things (IoT) devices in different areas such as biopotential signal detection, remote sensors, and portable monitoring devices. Among many challenges in designing such systems, area and power reduction are some of the critical ones. Low-frequency sensor front-ends often use chopping to modulate the 1/f noise and offset and suppress them by using a low pass filter. The filter must attenuate the modulated noise and output ripple over the band of interest before quantization, therefore can lead to larger passive component sizes or an increase in silicon area. Alternatively, the chopping frequency can be increased to relax the filter design at the cost of power consumption. However, filtering is not the only solution. Work presented in [1] suppresses the output ripple by using internal feedback. Unfortunately, introducing the feedback loop requires the addition of extra components whose non-idealities need to be handled as well, which may increase the design complexity of the system as well as the power and area. Another approach pursued in [2] reduced the magnitude of the modulated tones and ripple by spreading the power over a range of frequencies. Nonetheless, the design required a random number generator incorporated in the clocking scheme which may increase the overall design overhead.

The aspect of having a highly digital implementation and taking advantage of Moore's law in terms of density make VCO-based structures popular, such as using VCO-based quantizers to achieve analog-to-digital (A/D) conversion [3]. In most of the VCO-based quantizers, the nonlinearity of the VCO in the voltage-to-current conversion is the bottleneck of the performance. Various digital background calibration schemes are pursued to address this issue [4][5]. While many of these techniques are viable, they tend to increase silicon area and power consumption [6][7]. As an alternative, reducing the input

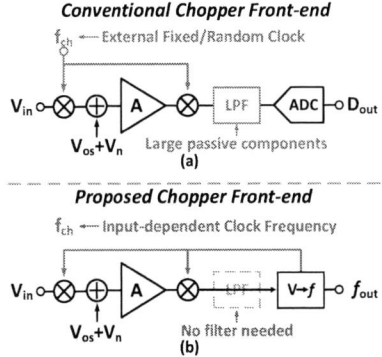

Figure 1. System-level diagram of (a) conventional chopper front-end, (b) proposed structure.

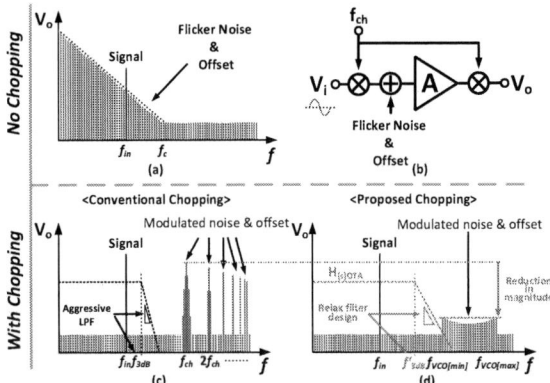

Figure 2. Spectrum comparison between the conventional and proposed chopping schemes.

range of the VCO can help improve its linearity. This method has been widely applied in the VCO-based delta-sigma modulators [8], at the cost of possible dynamic range reduction of the overall system.

This work proposes a chopping scheme that uses the back-end VCO quantizer to generate the chopping clocks. The chopping frequency is varied as the input changes, creating a spread spectrum chopping without using a frequency randomizer. A new switched-capacitor feedback linearization technique is proposed to enhance the linearity over the dynamic range of the VCO. Section II discusses the principle and implementation of the proposed VCO-driven chopping and the VCO linearization technique. Section III summarizes the

978-1-6654-8495-4/22 $31.00 © 2022 IEEE

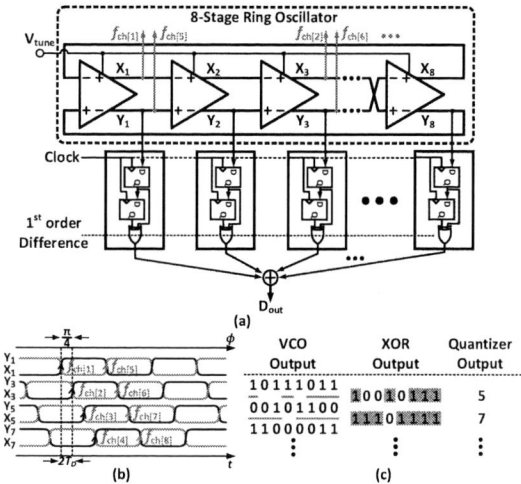

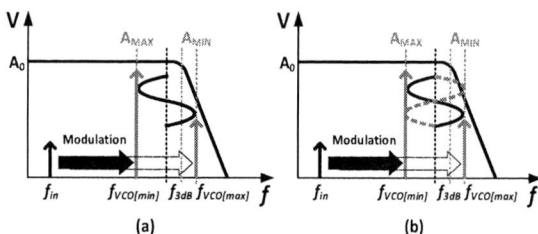

Figure 5. (a) Gain variation caused by unfixed signal modulation & (b) making the system pseudo-differential to mitigate the nonlinearity.

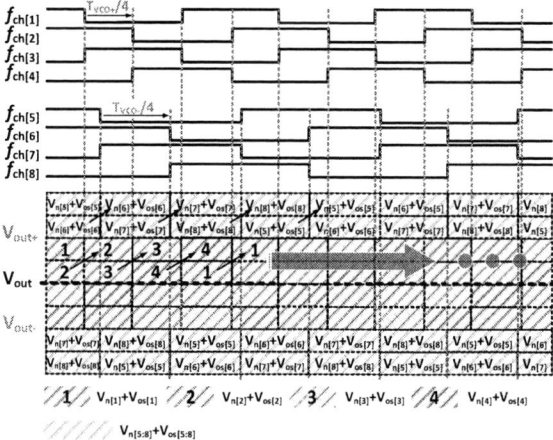

Figure 3. (a) 8-stage VCO-based quantizer and its (b) phase outputs and quantization (c).

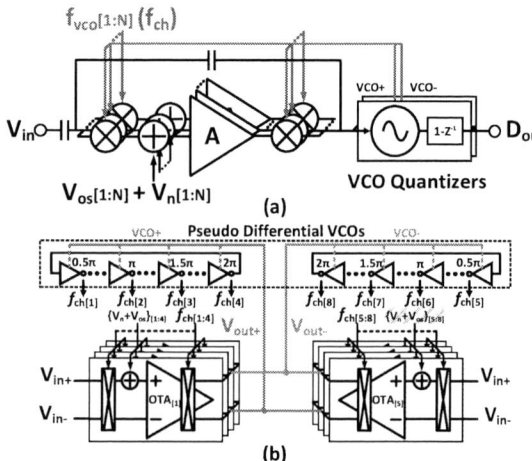

Figure 4. Implementation of the proposed VCO-driven chopping in (a) system-level & (b) its connection details.

measurement results and compares the proposed work with the state-of-the-art sensor readout circuit with similar input levels. Finally, Section IV concludes this work.

II. ARCHITECTURE

A. System-level Overview

Unlike conventional chopping (**Fig.1a**) and spread spectrum chopping, where an external signal is needed to drive the chopper, this work shown in **Fig.1(b)** generates the chopping signal internally from the voltage-to-frequency (V-F) converter. The VCOs are used here to provide chopping signals and as well as analog-to-digital conversion of the amplified signal.

Fig.2 displays the spectrum for the conventional and proposed chopping schemes. In conventional chopping, the flicker noise (V_n) and offset (V_{os}) will be modulated to the chopping frequency f_{ch} and its harmonics. Chopping only modulates V_n and V_{os} while the power remains the same. Thus,

This work was funded by NSF under EAGER-1723366.

Figure 6. Timing diagram and operation principle of the proposed VCO-driven chopping scheme under DC input.

without suppression, the in-band modulated tones will degrade the system performance. The proposed chopping scheme uses a VCO to generate input and quantization noise dependent chopping frequency. Thus, the modulated noise power is spread over a frequency range reducing the magnitude of modulated tones (**Fig.2d**). Unlike [2], the proposed technique doesn't require an additional frequency randomizer. The reduction of the modulated tone magnitude eliminates the need for a low-cutoff high-order suppression filter, which improves the area efficiency and the bandwidth of the system. The chopping frequency range in the proposed scheme depends on the probability density function of the amplifier's output, which will show itself as raised noise spectra as illustrated in **Fig.2d**. The chopping frequency range is further randomized by circuit noise and the quantization noise. Therefore, even for a small magnitude input, due to other noise sources mentioned above, the VCO will generate a wide range of chopping frequencies.

B. VCO Driven Chopping

Fig.3 describes the operation of a VCO-based quantizer in a 3-bit example. The transition of delay cells within a clock cycle can be observed by comparing samples of their current and previous states via XOR operation. The total number of transitions within one clock period yields the quantization of the V_{tune} voltage [3]. In this example, as shown in **Fig.3b**, 8 outputs with $\pi/4$ phase shift can be extracted from this 8-stage pseudo-differential VCO directly, which can be used as chopping signals in a multi-path fashion, as described in [9].

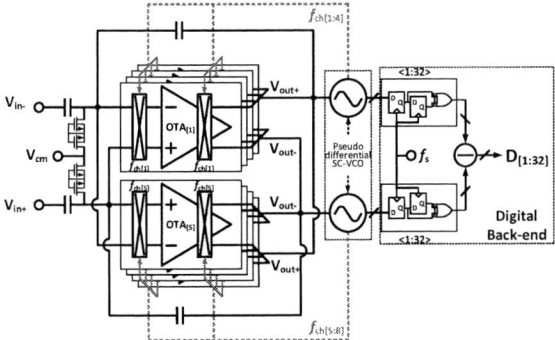

Figure 7. Schematic level of the proposed readout circuit.

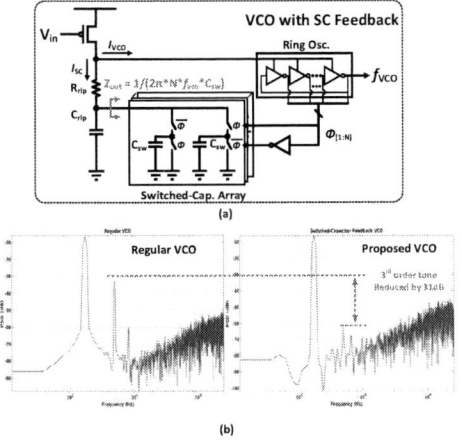

Figure 8. (a) Proposed switched-capacitor feedback VCO and (b) the simulation results of the VCO quantizers

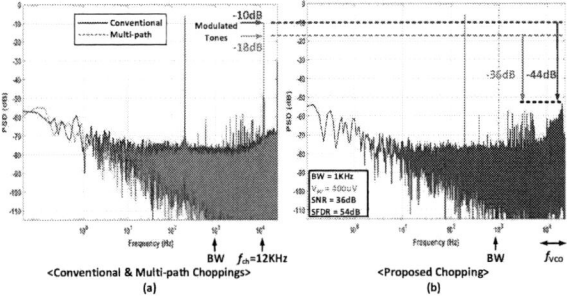

Figure 9. Measured spectrum comparison between (a) the conventional & multi-path chopping and (b) the proposed chopping.

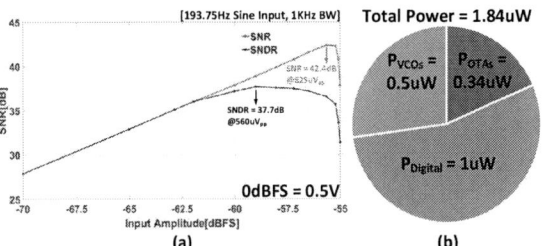

Figure 10. (a) System dynamic range, (b) Measured power breakdown.

Figure 11. Chip micrograph.

The implementation of the proposed chopping scheme is illustrated in **Fig.4a**. Two VCOs are used in pseudo-differential configuration to form the quantizer and directly provide the chopping signals to the OTA slices. This multi-path chopping scheme ensures that the input signal will get processed by the amplifier all the time while the conventional chopping scheme suffers from the signal runaway issue[2]. At the same time, this multi-path chopping scheme will reduce the chopping sensitivity to the uneven clock duty cycle. Different from [3], the proposed VCO-driven chopping scheme can generate the multiple chopping signals with π/N phase shift inherently from the VCO delay cells without having the extra clock generation module, which makes this approach more power and area efficient.

Depending on the amplifier bandwidth and the VCO frequency range, the system may encounter a 2nd order nonlinearity due to the amplifier gain variation caused by varying chopping frequencies, as demonstrated in **Fig.5**. To overcome this issue, a dual chopping structure is implemented where each side of the pseudo-differential VCOs will generate the chopping signals for half of the OTA slices. Therefore, a system-level pseudo-differential can be achieved which eliminates the 2nd order nonlinearity.

The implementation details are illustrated in **Fig.4b**, where

8 OTA slices are used for signal amplification along with two 16-delay-cell VCOs to provide the chopping signal and compose the 4-bit quantizer. The timing and operation of the proposed chopping scheme with a DC input are shown in **Fig.6** for simplicity. Here, each VCO delay cell with $\pi/4$ phase shift directly controls the chopping of one OTA slice in an asynchronous fashion. Due to the pseudo-differential nature of the VCO, each VCO directly controls half of the OTA slices. When DC input is applied to the system, the pseudo-differential VCOs will operate at two different frequencies thus each half of the OTA slices will be chopped at a different speed. As shown in **Fig.6**, the chopping frequency for each half is different and the 1/f noise and offset from each OTA slice is chopped at different phases and rotates between V_{out+} and V_{out-}. In this scheme, the average noise at the output over time remains at zero and the basic chopping function is accomplished.

Due to the multi-path nature of the chopping, each side of the output will always see two noise and offsets chopped by VCO_+ and two chopped by VCO_-. Among those chopped noise and offsets, the possibility of occurrence of each OTA slice's noise and offset is the same. As illustrated in **Fig.6**, initially, for the half of OTAs that are chopped by VCO_+, V_{out+} sees the combination of the noise and offset from $OTA_{[1]}$ and $OTA_{[2]}$. After 4 chopping events, the sequence of the combination of

978-1-6654-8495-4/22 $31.00 © 2022 IEEE

Table I. Comparison with state-of-the-art sensors

Specifications	This Work	JSSC 2017[8]	JSSC 2015[10]	CICC 2014[11]	JSSC 2018[12]	CICC 2015[13]
Technology	65nm	40nm	65nm	65nm	65nm	65nm
Architecture	Open-loop +VCO	VCO-based CTDSM	VCO-based	Beat Freq.	Analog-Time Converter	Beat Freq.
Fs(KHz)	48	4200	1	300	>2000**	50
BW(Hz)	1K	5K	500	10K	11K	1.2K
Supply(V)	0.5	1.2	0.5	1.2	0.5	1.2
Power(μW)	0.34(A)/1.5(D)	17	2.3	36	1.28	38
Input V_{pp}(mV)	0.8	8	1	2	4	10
RMS Noise(μV_{rms})	6.12*	6.35*	3.67*	14.18*	3.8	59.5*
IRN(nV/$\sqrt{Hz}$)	193*	32	58	100	36	1734*
Area(mm²)	0.0567	0.0145	0.025	0.258	0.006	0.096

* Estimated based on SNR/SNDR
** Estimated from measured spectrum

noise and offset from $OTA_{[1-4]}$ starts to repeat. The noise and offset from the other half of the OTAs have the same feature but rotate at a different speed. This 1-bit rotation of the noise and offset forms the weighted averaging pattern, similar to other VCO-based quantizers [3], which gives a 1st-order shaping to the difference between each OTA's noise and offset. The schematic level diagram of the proposed readout circuit is shown in **Fig.7**.

C. Switched-Capacitor Based VCO Linearization Technique

The VCO quantizer improves the in-band SNR due to its inherent 1st-order noise shaping. However, as the back-end of an open-loop system, the linearity of the VCO quantizer will directly determine the dynamic range of the system. Conventional VCOs suffer from the non-linearity issue caused by the inherently non-linear transconductance (g_m) of the control transistor and its varying drain-source voltage across the input. Here, the non-linearity of the VCO is suppressed using a new switched-capacitor (SC) feedback technique, as shown in **Fig.8a**. The control node of the VCO is connected to a switched-capacitor array whose switching signal comes from each output of the delay cells. When a surge of input current causes the VCO frequency to be increased, the equivalent impedance of the SC circuit will reduce, pulling more current through the SC path from the ring oscillators. Thus, negative feedback is formed and enhances the linearity. The more elements of the switch-capacitor array are used, the larger the loop gain becomes. Therefore, all outputs of the delay cells are used to control the array. A pair of passive components R_{rip} and C_{rip} are used to form a lowpass filter aiming to mitigate the glitches caused by frequent switching events. Simulation results (**Fig.8b**) show this SC feedback topology improves the SFDR of the VCO quantizer by more than 30dB.

III. MEASUREMENT RESULTS

The proposed design was fabricated in a GP 65nm CMOS process. Measurement results indicate that the prototype chip achieves 54dB SFDR and 36.3dB SNDR for a 400μVpp input signal, shown in **Fig.9b**. An attenuator was used to reduce the input signal magnitude during the measurement due to the lack of an audio signal generator. Compared to the conventional and multi-path chopping schemes also implemented on this chip, the proposed chopping structure reduces the modulation tone by 44dB and 36dB respectively, as shown in **Fig.9a**. The dynamic range plot in **Fig.10a** shows the peak SNR and SNDR is 42.4dB and 37.7dB with 825μVpp and 560μVpp input, respectively. The achieved RMS noise is 6.12μVrms and the total power consumption is 1.84μW with a 0.5V supply (**Fig.10b**), and the chip dimensions are shown in **Fig.11**. A large portion of the power in digital could have been reduced if the fabrication was

implemented in the LP process. Table I. summarizes the performance and compares it with the state-of-the-art sensor readout circuits with similar input levels. This prototype chip consumes one of the least power as a whole readout system and shows its capabilities in sub-microvolt signal sensing.

IV. CONCLUSION

This work has provided a new chopping scheme that uses the quantizer of the system to realize noise and offset modulation for sensor applications. This chopping scheme inherits the merit of the spread spectrum chopping that eliminates aggressive filter design meanwhile doesn't rely on the extra digital module to realize the function, which makes it more power and area-efficient solution. A new VCO linearization technique is also presented which can be extended to all VCO-based structures at any frequency band.

REFERENCES

[1] Rong Wu, K. A. A. Makinwa and J. H. Huijsing, "A chopper current-feedback instrumentation amplifier with a 1mHz $1/f$ noise corner and an AC-coupled ripple-reduction loop," 2009 IEEE International Solid-State Circuits Conference - Digest of Technical Papers, 2009, pp. 322-323,323a.

[2] J. Brown, U.S. Patent 5115202, 1991.

[3] M. Z. Straayer and M. H. Perrott, "A 12-Bit, 10-MHz Bandwidth, Continuous-Time $\Sigma\Delta$ ADC With a 5-Bit, 950-MS/s VCO-Based Quantizer," in IEEE Journal of Solid-State Circuits, vol. 43, no. 4, pp. 805-814, April 2008.

[4] S. Rao et al., "A Deterministic Digital Background Calibration Technique for VCO-Based ADCs," in IEEE Journal of Solid-State Circuits, vol. 49, no. 4, pp. 950-960, April 2014.

[5] K. Ragab and N. Sun, "A 12-b ENOB 2.5-MHz BW VCO-Based 0-1 MASH ADC With Direct Digital Background Calibration," in IEEE Journal of Solid-State Circuits, vol. 52, no. 2, pp. 433-447, Feb. 2017.

[6] W. Jiang et al., "A ±50-mV Linear-Input-Range VCO-Based Neural-Recording Front-End With Digital Nonlinearity Correction," in IEEE Journal of Solid-State Circuits, vol. 52, no. 1, pp. 173-184, Jan. 2017.

[7] J. Huang and P. P. Mercier, "A 112-dB SFDR 89-dB SNDR VCO-Based Sensor Front-End Enabled by Background-Calibrated Differential Pulse Code Modulation," in IEEE Journal of Solid-State Circuits, vol. 56, no. 4, pp. 1046-1057, April 2021.

[8] C. Tu, Y. Wang and T. Lin, "A Low-Noise Area-Efficient Chopped VCO-Based CTDSM for Sensor Applications in 40-nm CMOS," in IEEE Journal of Solid-State Circuits, vol. 52, no. 10, pp. 2523-2532, Oct. 2017.

[9] S. Pavan, "Improved Chopping in Continuous-Time Delta–Sigma Converters Using FIR Feedback and N-Path Techniques," in IEEE Transactions on Circuits and Systems II: Express Briefs, vol. 65, no. 5, pp. 552-556, May 2018.

[10] R. Muller et al., "A Minimally Invasive 64-Channel Wireless μECoG Implant," in IEEE Journal of Solid-State Circuits, vol. 50, no. 1, pp. 344-359, Jan. 2015.

[11] B. Kim, S. Kundu, S. Ko and C. H. Kim, "A VCO-based ADC employing a multi-phase noise-shaping beat frequency quantizer for direct sampling of Sub-1mV input signals," Proceedings of the IEEE 2014 Custom Integrated Circuits Conference, 2014, pp. 1-4.

[12] L. B. Leene and T. G. Constandinou, "A 0.006 mm2 1.2 μ W Analog-to-Time Converter for Asynchronous Bio-Sensors," in IEEE Journal of Solid-State Circuits, vol. 53, no. 9, pp. 2604-2613, Sept. 2018.

[13] S. Kundu, B. Kim and C. H. Kim, "Two-step beat frequency quantizer based ADC with adaptive reference control for low swing bio-potential signals," 2015 IEEE Custom Integrated Circuits Conference (CICC), 2015, pp. 1-4.

Mixed-Signal Compensation of Tripolar Cuff Electrode Imbalance in a Low-Noise ENG Analog Front-End

Rémi Dekimpe and David Bol

ICTEAM institute, Université catholique de Louvain (UCLouvain), Louvain-la-Neuve, Belgium
Email: {remi.dekimpe, david.bol}@uclouvain.be

Abstract—Due to their low amplitude, electroneurogram (ENG) signals are particularly subject to external muscle artefacts and intrinsic electronic noise. However, achieving a high signal-to-noise ratio is a challenge for implanted systems that have a limited power budget. This work presents a low-power analog front-end which features low intrinsic noise and high interference rejection. The proposed mixed-signal feedback loop for tripolar cuff electrode imbalance compensation provides an interference rejection of 56 dB with a negligible power overhead. The instrumentation amplifier achieves a gain of 91.5 dB, an input-referred noise of 1.35 μV, an input offset voltage below 1 μV, and digitally-tunable imbalance compensation with 7 bits of resolution over a ±20% range. The results are validated on the ICare microcontroller system-on-chip, a 22-nm fully-depleted silicon-on-insulator prototype.

Index Terms—Biomedical electronics, CMOS integrated circuits, analog front-end, electroneurogram, low noise, imbalance.

I. INTRODUCTION

The vagus nerve is a cranial nerve which plays a key role in the autonomic nervous system. It carries bidirectional information between the medulla oblongata in the brainstem and multiple organs such as the heart, the lungs, the liver, and many others. Vagus nerve stimulation has already proven useful for the treatment of epilepsy and depression, and has demonstrated promising results for other disorders [1]. The interest around vagus nerve sensing, i.e., vagus nerve electroneurogram (VENG), has also been growing for diagnostic purposes. It has been used previously for epileptic seizures detection [2] or the extraction of respiratory markers [3].

However, the design of VENG acquisition systems, as the one represented in Fig. 1, encounters several challenges. Firstly, compared to other electrical biosignals, neural signals have a very low amplitude below 20 μV and a high bandwidth up to 20 kHz [2]. The input instrumentation amplifier (IA) thus requires a high gain above 90 dB and a very low input-referred noise below 2 μV RMS. A low DC offset voltage is also necessary to avoid saturation with the high gain. Secondly, the VENG signal is subject to interferences from surrounding electrical biosignals from the heart and muscles. Due to their high amplitude, these artefacts can saturate the amplifier if they are not properly rejected. Tripolar cuff electrodes can

This work was supported by the Fonds de la Recherche Scientifique - FNRS under Grant n° CDR J.0014.20 and Research Fellowship of R. Dekimpe.
ARM Academic Access provided the ARM Cortex-M4 IP.

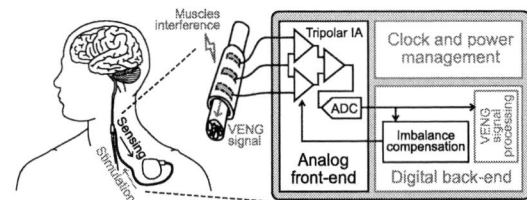

Fig. 1. VENG acquisition system with tripolar cuff electrode input. The analog front-end interfaces the cuff electrode and compensates the imbalance. The digital back-end (DBE) estimates the imbalance to control the compensation feedback loop and can perform additional VENG signal processing to infer clinical information. The clock and power management unit generates all clocks and voltage references used in the system.

be used to linearize the external artefact and reject it using a true tripole or quasi-tripole amplifier configuration. Unfortunately, this linearization may not be ideal due to electrode imbalance, which requires additional compensation techniques [4]. Finally, in the case of long-term implantable devices, the limited battery capacity, usually in the order of 1 Ah, imposes a power consumption around 100 μW to last several years.

In this work, we propose an analog front-end (AFE) for VENG signal acquisition with a new mixed-signal imbalance compensation method. The AFE features a low input-referred noise of 1.35 μV RMS and a power consumption of 104 μW. Section II presents the mixed-signal imbalance compensation scheme which provides high signal-to-interference ratio and allows duty-cycled operation for reduced power overhead. In Section III, the IA with gain balancing is described and optimized for low-noise, high-gain and low-offset performance. Section IV finally details the AFE prototype embedded in a 22-nm fully-depleted silicon-on-insulator (FD-SOI) microcontroller SoC codenamed ICare, with its measurement results.

II. IMBALANCE ESTIMATION AND COMPENSATION

The linearization effect inside the tripolar cuff electrode enables the rejection of external artefacts. Indeed, as the resistivity inside the cuff is assumed constant and as A, B, and C electrodes are equally spaced, V_{AB} and V_{CB} voltages have opposite values as shown in Fig. 2(a). By adding those voltages, the artefact current I_{art} is cancelled and only the ENG signal (V_{ENG}) remains. The most common amplifier configurations for instrumentation amplifiers to perform this cancellation are the true tripole and quasi-tripole architectures

978-1-6654-8495-4/22 $31.00 © 2022 IEEE

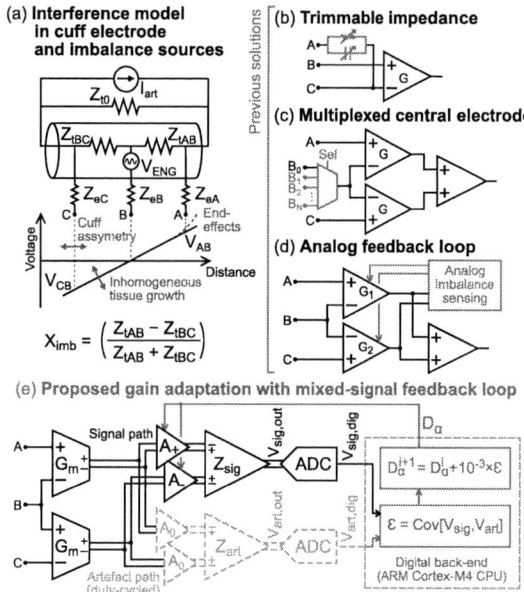

Fig. 2. Tripolar cuff electrode imbalance model and compensation methods. In the lumped-element model (a), V_{ENG} and I_{art} are the sources of ENG signal and external muscle artefacts. The impedances model the tissue resistance outside (Z_{t0}) and inside (Z_{tAB}, Z_{tBC}) the cuff electrode. Trimmable impedance (b), multiplexed central electrodes (c) and analog feedback loop (c) techniques have been proposed in the literature. This work proposes a mixed-signal compensation scheme (e) with duty-cycled artefact sensing path.

[4]. Unfortunately, the linearization assumption is not always verified due to cuff non-idealities such as inhomogeneous scar tissue growth, cuff asymmetry, and non-linearity of the electric field close to the cuff extremities. These effects lead to an imbalance X_{imb} between the Z_{tAB} and Z_{tBC} impedances which induces an imperfect cancellation of the artefact.

Modifications to the tripole amplifier configurations have been proposed to enable the compensation of electrode imbalance. Demosthenous et al. [5] added a trimming impedance network in series with one of the electrodes in the quasi-tripole configuration (Fig. 2(b)). This solution however requires large off-chip capacitors up to 500 nF. ElAnsary et al. [6] used a multiplexer with multiple selectable electrodes to modify the sensing location inside the cuff (Fig. 2(c)). The resolution of this solution is limited by the number of electrodes and the minimal manufacturing dimensions. The previous solutions do not discuss how the estimation of the imbalance is performed prior to setting the actual compensation. Demosthenous and Triantis [7] proposed an analog adaptive tripole (AAT) configuration in which the individual amplifier gains can be tuned (Fig. 2(d)). The gain control is performed by a bias voltage generated from an analog imbalance sensing module. This solution however requires the imbalance sensing to remain active, which consumes 1.4 mW (20% of the total power).

In this work, we propose a new imbalance compensation scheme based on a mixed-signal artefact estimation and compensation loop (Fig. 2(e)). In the signal path, two transconductance stages (G_m) followed by current amplifiers ($A_\pm$) and

a transimpedance stage (Z_{sig}) amplify the signal. All blocks are fully-differential to improve the common-mode rejection. Similarly to the AAT [7], the imbalance compensation is performed by symmetrically tunable gains. However, digital control signals D_α are used here to enable robust and duty-cycled artefact estimation. The gain adaptation is performed at the current amplification stage by two complementary gains $A_\pm = A_0(1 \pm \alpha)$. When the imbalance remains small, the resulting signal obtained along the ENG signal sensing path before digitization is approximately equal to

$$V_{sig,out} \approx 2G_m A_0 Z_{sig} \left[V_{ENG} - V_{art,res} \right],$$
$$V_{art,res} = \frac{I_{art} Z_{t0}}{2}(X_{imb} + \alpha), \tag{1}$$

where it can be seen that the ENG signal V_{ENG} can be properly recovered, i.e. the residual artefact interference $V_{art,res}$ is zero, if the gain adaptation α matches the imbalance X_{imb}.

In order to properly set the gain control D_α, the artefact is measured through a separate artefact sensing path using fixed-gain current amplifiers (A_0) and a second transimpedance (Z_{art}). Reusing the same G_m stages allows to save power and additionally compensate mismatch between the transconductances. Note that as V_{AB} is subtracted from V_{CB} instead of added, by inverting the wires compared to the signal sensing path, the output signal of the artefact path is equal to

$$V_{art,out} = G_m A_0 Z_{art} I_{art} Z_{t0}. \tag{2}$$

The covariance between the outputs of the signal and artefact paths is computed in the digital back-end (DBE) by an ARM Cortex-M4 CPU over a 200-sample window and is used as an error signal ε, as shown in Fig. 2(e). When the $A_\pm$ gains are correctly adapted and the artefact is effectively removed from the signal, the covariance is close to zero. When the gain adaptation does not match the imbalance, the artefact is mixed with the signal and the covariance increases in absolute value. The sign of the covariance provides information on the direction of the imbalance. The gain control is performed iteratively by a first-order infinite-impulse-response (IIR) filter integrating the error until it reaches zero.

Thanks to the separate sensing paths, the gain in the signal path can be set higher as the feedback loop is robust to artefact-related saturation. Assuming maximum amplitudes of 20 µV and 500 µV for the ENG signal and the artefact [7], the gains are set as follows: $G_m = 4.2$ mS, $A = 0.25$, $Z_{sig} = 18$ MΩ, and $Z_{art} = 590$ kΩ. The analog-to-digital conversion is implemented with time-based ADCs using voltage-controlled oscillators. They have a resolution of 10 bits with a power consumption of 0.8 µW per ADC at a 32-kHz sampling rate.

The main limitation of the mixed-signal imbalance compensation is the resolution of the gain control, which limits the worst-case artefact attenuation, i.e., when the imbalance X_{imb} sits between two successive codes. From Eq. (1), it can be seen that the worst-case signal-to-interference ratio (SIR) is equal to

$$SIR_{out} = \frac{V_{ENG}}{V_{art,res}} = \frac{4V_{ENG}}{I_{art} Z_0 \alpha_{LSB}}, \tag{3}$$

978-1-6654-8495-4/22 $31.00 © 2022 IEEE

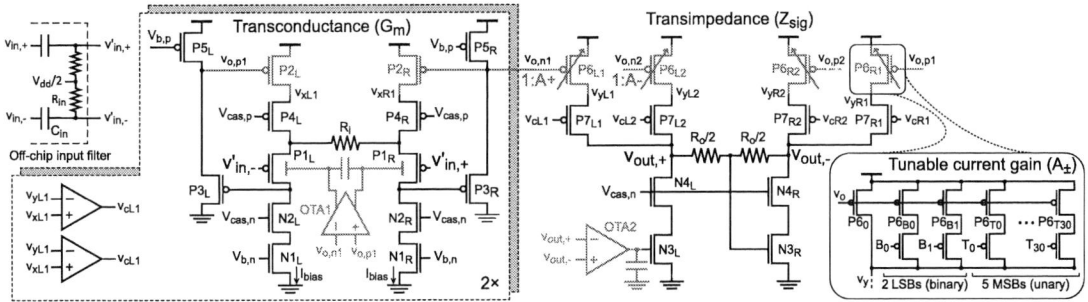

Fig. 3. Instrumentation amplifier circuit. The blue color highlights the current mirrors with tunable gain $A_{\pm}$ for the imbalance compensation. The green color shows the offset correction feedback loops with OTA_1 and OTA_2.

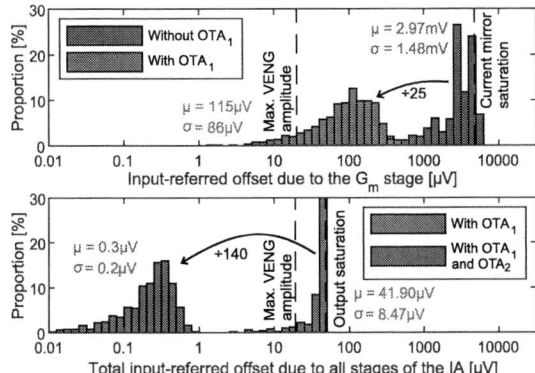

Fig. 4. Monte-carlo offset simulation results before and after correction. OTA_1 and OTA_2 avoid saturation in the G_m and Z_{sig} respectively. The resulting offset is lower than 1 µV, which is well below the signal level.

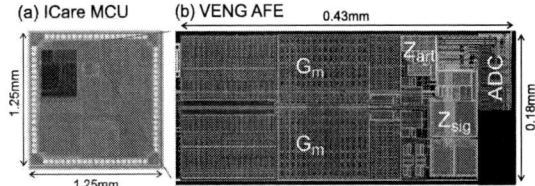

Fig. 5. ICare chip (a) and AFE (b) layouts. The G_m stages take up 65% of the space due to the noise requirement on the input stage of the IA. Time-based ADCs occupy only 3500 µm² thanks to their digital-only implementation.

with α_{LSB} the gain control step size. In order to reach a minimum signal-to-interference ratio (SIR) of 20 V/V, a gain resolution α_{LSB} of 0.35% is required. This corresponds to an artefact rejection of 56 dB. With a nominal gain of 91.5 dB (37.5 kV/V), this corresponds to a resolution of 130 V/V, with 128 steps (7 bits) over a ±20% imbalance range.

III. LOW-NOISE INSTRUMENTATION AMPLIFIER

A current-balancing topology is used for the IA as shown in Fig. 3 [8]. The input differential voltage v'_{in} after the external high-pass filter is copied to the terminals of resistor R_i by transistors P_1 in flipped voltage-follower structures, which generates a current variation between the branches. The G_m transconductance is equal to $(R_i + g_{d,P1}/(g_{m,P1}g_{m,P2}))^{-1}$. The current variation is copied from transistors P_2 to P_6 with a current gain $A = (W/L)_{P6}/(W/L)_{P2}$. The tunable current gain is implemented by digitally controlling PMOS switches (B_{0-1}, T_{0-30}) to modify the width of transistors P_6. The current from the two G_m blocks is summed at the nodes v_{out} and the impedance R_o generates the output differential voltage.

Given the low amplitude of the VENG signal, the instrumentation amplifier in the AFE must achieve high noise efficiency in order to reach a sufficient signal-to-noise ratio. A genetic optimization algorithm similar to [8] performs the sizing of the devices in the signal sensing path to reach the optimal noise-power trade-off. The resulting design achieves an input-referred noise (IRN) of 1.3 µV RMS and consumes an average power of 102 µW when the artefact path is disabled.

Due to the large gain of 91.5 dB, the input-referred offset voltage of the amplifier needs to remain in the µV range. In the considered topology, the offset arises mainly from the mismatch between the input pair P_1, the current sources N_1 and the current mirrors P_2-P_6. Without an appropriate solution, this mismatch leads to a large DC current being amplified and results in the saturation of the amplifier. As the VENG signal bandwidth starts at 500 Hz, the offset can be filtered out by introducing two DC feedback loops in the G_m and Z_{sig} stages. Those feedback loops work by matching the DC levels of the output signals ($v_{o,p}$ and $v_{o,n}$ in G_m, and $v_{out,+}$ and $v_{out,-}$ in Z_{sig}). The first feedback loop (OTA_1) acts on the back-gate of the input transistor pair P_1, which modulates their respective threshold voltage V_{th}. The second feedback loop (OTA_2) directly controls the front-gate of the bottom left NMOS N_3 to regulate the bias current. The low-frequency pole at 500 Hz in the feedback loop is set by the amplifiers with low conductance and on-chip capacitors. Simulation results show that the offset is effectively reduced below 1 µV as shown in Fig. 4.

IV. MEASUREMENT RESULTS

The AFE along with the proposed compensation scheme have been prototyped in a 1.6 mm² chip in 22-nm FD-SOI technology from GlobalFoundries. As shown in Fig. 5, the VENG AFE only takes up 0.077 mm² while the rest of the chip is occupied by the DBE and the clock and power management unit which are not discussed in this paper. Measurements of the AFE show a good matching with expected performance, with a gain of 91.5 dB and a bandwidth up to 16 kHz. The

978-1-6654-8495-4/22 $31.00 © 2022 IEEE

TABLE I
ENG AFE COMPARISON

	This work	ElAnsary [6] JSSC 2021	Demosthenous [5] Sensors 2013	Demosthenous [7] JSSC 2005	Ng [9] JSSC 2019	Schönle [10] JSSC 2018	Lee [11] TCAS-II 2018
Technology	22nm FD-SOI	0.13µm CMOS	0.35µm CMOS	0.8µm BiCMOS	0.18µm CMOS	0.13µm CMOS	0.13µm CMOS
Electrode configuration	Tripole	Quasi-tripole	Quasi-tripole	Tripole	Bipole	Bipole	Quasi-tripole
Compensation resolution	7 bits	3 bits	7 bits	Continuous	-	-	-
Compensation range	±20%	±45%	±80%	±40%	-	-	-
Artefact rejection [dB]	56	25	44	60	-	-	-
Power (/channel) [µW]	104	0.14	27 / 930	7200	34	285	11 / 176
Gain [dB]	91.5	N/A	6 - 47	87	55 / 75	18 / 66	44
IRN [µV RMS]	1.4	24.7	2.95 / 0.7	0.29	1.9	1.2	3 / 1
Power efficiency factor	27	13	24 / 43	60	33	4100	9 / 15
ADC	10b time-based	8b $\Sigma\Delta$	-	-	10b SAR	14b SAR	-

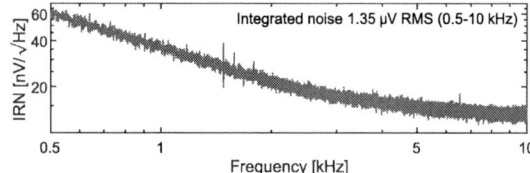

Fig. 6. Measured noise power spectrum density for the signal sensing path.

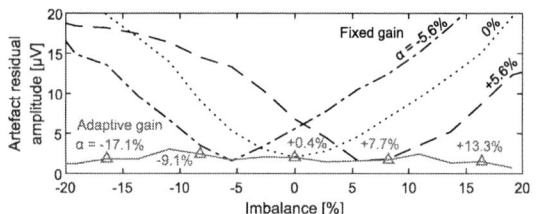

Fig. 7. Artefact interference and imbalance compensation. The gain compensation α in blue shows the inferred imbalance estimated and corrected by the feedback loop.

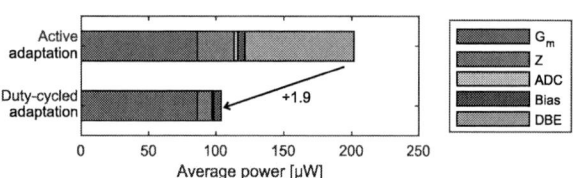

Fig. 8. Power breakdown of the VENG sensing SoC with active and duty-cycled compensation. The G_m stages dominate the power consumption for noise requirements.

measured noise distribution shown in Fig. 6 corresponds to an IRN of 1.35 µV. Compared to the state of the are (Table I), the proposed AFE provides excellent resolution for imbalance compensation, while having a low power consumption, a high gain, and a competitive noise efficiency.

With fixed gains at different D_α configurations, Fig. 7 shows that the artefact only cancels at specific imbalance values. When enabling the gain adaptation with the proposed mixed-signal feedback loop, the system is able to accurately estimate the imbalance and compensate it with the appropriate setting to keep the residual amplitude close to the measurement noise level. The response to a change in imbalance takes less than 250 ms to settle, thus allowing very low duty-cycle (< 1%). As shown in Fig. 8, while the feedback loop increases the power consumption by 93%, it can be disabled after estimation, thus making its impact on average power consumption negligible. The average power consumption with duty-cycled gain adaptation is 104 µW.

V. CONCLUSION

The mixed-signal scheme proposed in this work provides a solution to tripolar cuff electrode imbalance with low power and area overhead, which achieves high compensation resolution and robust imbalance estimation. A high artefact rejection of 56 dB reduces the residual artefact interference below noise level. The prototyped AFE achieves high gain and noise efficiency which is of the utmost importance for sensing low-amplitude ENG signals. The proposed AFE was successfully validated on a 22-nm FD-SOI prototype chip.

REFERENCES

[1] R. L. Johnson and C. G. Wilson, "A review of vagus nerve stimulation as a therapeutic intervention," *J. Inflamm. Res.*, vol. 11, p. 203, 2018.

[2] L. Stumpp et al., "Recording of spontaneous vagus nerve activity during Pentylenetetrazol-induced seizures in rats," *J. Neurosci. Methods*, vol. 343, p. 108832, 2020.

[3] C. Sevcencu et al., "A respiratory marker derived from left vagus nerve signals recorded with implantable cuff electrodes," *Neuromodulation*, vol. 21, no. 3, pp. 269–275, 2018.

[4] I. F. Triantis et al., "On cuff imbalance and tripolar ENG amplifier configurations," *IEEE TBE*, vol. 52, no. 2, pp. 314–320, 2005.

[5] A. Demosthenous et al., "An integrated amplifier with passive neutral-ization of myoelectric interference from neural recording tripoles," *IEEE Sensors*, vol. 13, no. 9, pp. 3236–3248, 2013.

[6] M. ElAnsary et al., "Bidirectional Peripheral Nerve Interface With 64 Second-Order Opamp-Less $\Delta\Sigma$ ADCs and Fully Integrated Wireless Power/Data Transmission," *IEEE JSSC*, vol. 56, no. 11, pp. 3247–3262, 2021.

[7] A. Demosthenous and I. F. Triantis, "An adaptive ENG amplifier for tripolar cuff electrodes," *IEEE JSSC*, vol. 40, no. 2, pp. 412–421, 2005.

[8] R. Dekimpe and D. Bol, "A Configurable ULP Instrumentation Ampli-fier With Pareto-Optimal Power-Noise Trade-Off Achieving 1.93 NEF in 65nm CMOS," *IEEE TCAS-II*, vol. 68, no. 7, pp. 2272–2276, 2021.

[9] K. A. Ng et al., "A wireless multi-channel peripheral nerve signal acquisition system-on-chip," *IEEE JSSC*, vol. 54, no. 8, pp. 2266–2280, 2019.

[10] P. Schönle, F. Glaser, T. Burger, G. Rovere, L. Benini, and Q. Huang, "A multi-sensor and parallel processing SoC for miniaturized medical instrumentation," *IEEE JSSC*, vol. 53, no. 7, pp. 2076–2087, 2018.

[11] B. Lee and M. Ghovanloo, "An adaptive averaging low noise front-end for central and peripheral nerve recording," *IEEE TCAS-II*, vol. 65, no. 7, pp. 839–843, 2018.

A Full D-band Multi-Gbit RF-DAC in 90 nm SiGe BiCMOS based on Passive Vector Aggregation

Tim Maiwald[#], Akshay Visweswaran[†], Klaus Aufinger[$], and Robert Weigel[#]

[#]Institute for Electronics Engineering, Friedrich-Alexander-University Erlangen-Nürnberg, Cauerstr. 9,
91058 Erlangen, Germany
[†]Imec, Kapeldreef 75, B-3001, Leuven, Belgium
[$]Infineon Technologies AG, Am Campeon 1-12, Neubiberg, Germany
[1]tim.maiwald@fau.de

Abstract—This paper presents the first fully integrated RF-DAC for realizing and passively scaling QPSK-signals for generating higher-order modulations over the entire D-band (110-170 GHz). Vector-symbol generation in the RF-DAC is accomplished via compact distributed passives on chip. In this modulator architecture, a sinusoidal local-oscillator (LO) signal is split and weighted by a backward-wave directional coupler to meet length requirements for a complex-valued RF-signal scaling vector. QPSK vector-modulation is then accomplished with a Lange Coupler, two Marchand baluns and a switch octet operating on four LO phases (0°, 90°, 180° and 270°), which is driven directly with the data streams. Higher-order modulation schemes are subsequently created by summing RF outputs of individual QPSK unit cells. A 16-QAM modulator, based on two-way summing, prototyped in Infineon's advanced 90 nm SiGe BiCMOS technology demonstrating a datarate of 6.4 Gbps with a measured EVM of <8% over the entire D-band. Simulations show a maximum datarate of 60 Gbps, however, the measured datarate is currently limited by the test setup comprising an in-house low-cost FPGA clocked at 1.6 GHz. The chip consumes 120 mW from a 2.1 V supply and occupies 0.36 mm² of core area.

Keywords—RF-DAC, D-band, direct modulation, QAM, 5G, 6G.

I. INTRODUCTION

It is clear from the evolution of the 5th generation (5G) mobile-communication standard that mm-wave and sub-THz wireless links will eventually enter the realm of commercial wireless communication. Frequency bands above 100 GHz, such as the D-band ranging from 110-170 GHz, are currently experiencing a growth in R&D interests, as they promote novel and compact system demonstrators in addition to the efficient use of large aggregate bandwidths of several tens of GHz. The two most popular approaches to quadrature amplitude modulation (QAM) at these high frequencies are [1]–[3]:

(1) Baseband signal generation with fast digital-to-analog converters or M-PAM transmitters followed by quadrature (I/Q) upconversion, and

(2) direct-modulators or RF-DACs using binary input(s), which are then directly multiplied with a carrier signal or used to control amplitude and phase of a carrier signal.

At D-band frequencies, the availability of large bandwidths facilitates the use of low-complexity direct modulation without the processing overhead of spectral shaping, and thereby, maximizing throughput for a given clock rate. By combining power-efficient BPSK/QPSK RF-DAC unit cells, the datarate

(per channel) can be increased via higher-order constellation schemes, as shown in [4]. In this work, we present the first demonstration of a fully integrated, coupler-based method of realizing and scaling QPSK-signals to create higher order modulations at D-band frequencies. As opposed to traditional current-steering DACs that sum binary-weighted signal currents, in the modulator architecture proposed in this work, a sinusoidal local-oscillator (LO) signal is split and weighted by a backward-wave directional coupler to meet length requirements for a complex-valued RF-signal vector. In the 16-QAM modulator demonstrated in this work (see Fig.1a), the direct and coupled outputs of the coupler are fed to QPSK-modulator unit-cells. The unit cell comprises a Lange coupler for quadrature-phased outputs, each feeding a Marchand balun for anti-phased drive of a transistor pair in the switch octet operating on the four resulting signal phases (0°, 90°, 180° and 270°). The switch octet is driven directly with the binary data streams, thereby achieving phase control required for QPSK modulation. Higher order modulation schemes are created by vector aggregation, i.e., summing the RF outputs of individual QPSK unit cells (we use a Wilkinson combiner).

A 16-QAM modulator operating over the entire D-band is prototyped in Infineon's advanced 90 nm SiGe BiCMOS technology and provides proof of concept. The 90 nm SiGe process offers high-speed npn-transistors with an f_T of 300 GHz and f_{MAX} of 505 GHz, respectively. The BEOL offers four thin copper layers for digital routing and an aluminum pad

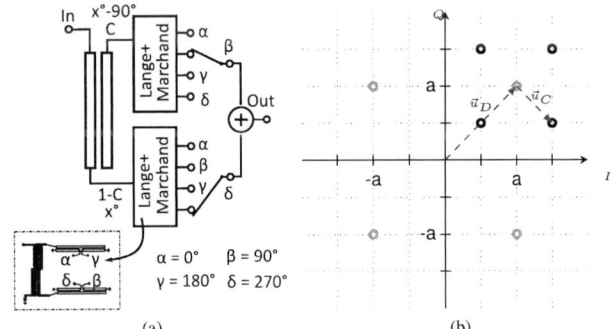

Fig. 1. RF-DAC concept based on passive vector-generation and aggregation (a) schematic of 16-QAM modulator and (b) as vector symbol representation.

978-1-6654-8495-4/22 $31.00 © 2022 IEEE

layer. Additionally, there are two (M$_7$ & M$_6$) 2 μm thick, and one (M$_5$) 0.8 μm thick copper layers for RF signal routing.

II. Passive Phase and Symbol Generation Concept

As shown in Fig. 1(b), the symbols of a QAM-constellation diagram can be assembled by a linear combination of signal vectors with fixed and weighted magnitudes as well as one of four, QPSK phases (0°, 90°, 180°, 270°). In this figure, the vector addition of two weighted sub-QPSKs is used to create a 16-QAM constellation.

As indicated in Fig. 1(a), the scaling or weighting of the vectors $|\vec{u}_C|$ and $|\vec{u}_D|$ is accomplished by a backward-wave directional coupler by adjusting its power coupling value C. This component is called *scaling coupler* in the following. The constant 90° phase difference between the coupled and the direct port of ideal backward-wave couplers needs to be considered for symbol-mapping, but unlike improper coupler parameters (e.g. C, phase- and amplitude imbalances), it does not disturb the shape of the constellation diagram itself. In case of lossless and ideal couplers, C is the part of the input signal power, which is transferred to the coupled port, consequently the other part $(1 - C)$ arrives at the direct port of the coupler. For the QPSK scheme shown in Fig. 1 with a symbol spacing of $2a$, the vector length to one symbol point is $|\vec{u}_D| = \sqrt{a^2 + a^2} = a\sqrt{2}$. To realize a uniform QAM constellation, the symbol spacing gives a accordingly. This leads to a voltage vector length of $|\vec{u}_C| = \sqrt{(0.5a)^2 + (0.5a)^2} = a/\sqrt{2}$ and, thus, to the following magnitude and power relation of the vectors:

$$\frac{|\vec{u}_D|}{|\vec{u}_C|} = \frac{\sqrt{2}\not{a}\sqrt{2}}{\not{a}} = 2 \;\Rightarrow\; \frac{|\vec{u}_D|^2}{|\vec{u}_C|^2} = \frac{1-C}{C} = 4(\approx 6\,dB)$$

The required power coupling factor C for the coupler results in

$$C = \frac{1}{5} \Rightarrow C_{dB} = 10\log_{10}(\frac{1}{5}) \approx -7dB.$$

The block diagram of the chip architecture in Fig. 2 shows how four quadrature phases of a QPSK unit-cell are created using a Lange coupler and two Marchand baluns. The Lange coupler generates quadrature signals, which drive Marchand baluns for anti-phased signal outputs. Thus, four signals with the same

amplitude and a relative phase spacing of 90° are interfaced to a switch octet. The switch octet is driven directly with the data streams, thereby achieving phase control required for QPSK modulation. A 16-QAM signal is then generated by combining the output signals from two QPSK-modulator unit-cells using a differential Wilkinson combiner.

III. Broadband On-Chip Coupler Designs

Metal density rules and top-level layout constraints limit the symmetry of on-chip couplers. All couplers that are part of this design are subject to the theory of asymmetric uniform coupled lines in inhomogeneous media, and their design is, therefore, a trade-off between bandwidth, matching, transmission imbalances in phase and amplitude and directivity.

Fig. 3 shows the layout (Fig. 3(a)), break-out test structures (Fig. 3(b)) and measurement results for the designed scaling coupler (Figs. 3(c) & (d)). The coupled lines are made in the thick upper metals M$_6$ and M$_7$. As depicted in Fig. 3(a), C is mainly controlled by the offset d between the lines. The measured and simulated relative magnitudes in Fig. 3(c) shows an amplitude relation of the coupled and direct port of 5.2±0.3 dB over the entire D-band. Simulations and measurements are in good agreement, and the ≈1 dB offset to the ideal curve in the amplitude is attributed to a miscalculation during the concept phase and can be corrected in a redesign. The phase imbalance in Fig. 3(d) of up to 10° (5° in simulation) is mainly caused by unequal phase velocities for the common-(c) and differential-(π) modes on the coupled lines and affect the EVM marginally. Standalone measurements of the Lange Coupler also show phase and amplitude imbalances of <1 dB and <8°, respectively.

The required interconnection of Lange couplers and Marchand baluns causes relative imbalances in phase and

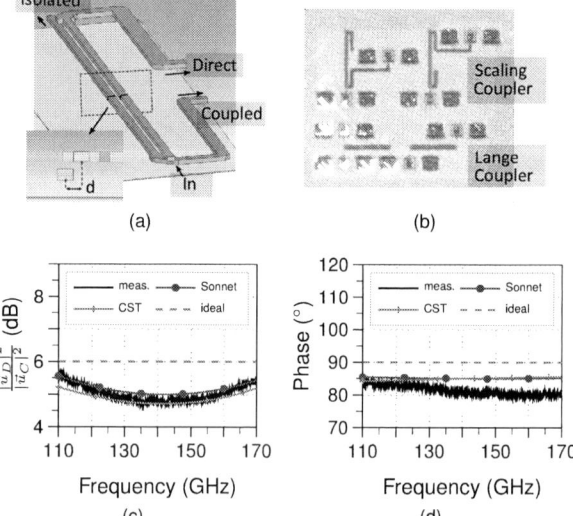

(a)

(b)

(c)

(d)

Fig. 3. (a) Scaling coupler layout (3D-view), (b) micrograph of standalone coupler test structures, and simulation and measurement results for the coupler (c) relative magnitudes and (d) relative phases of the scaling coupler.

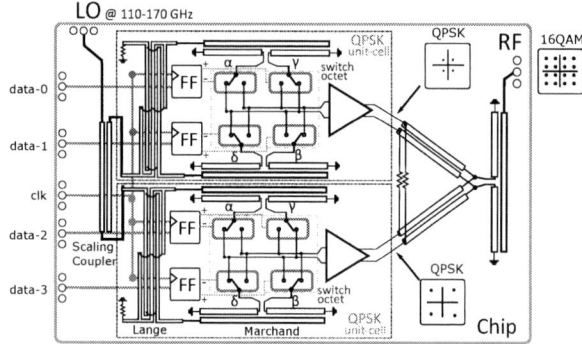

Fig. 2. Block diagram of the 16-QAM modulator prototype IC.

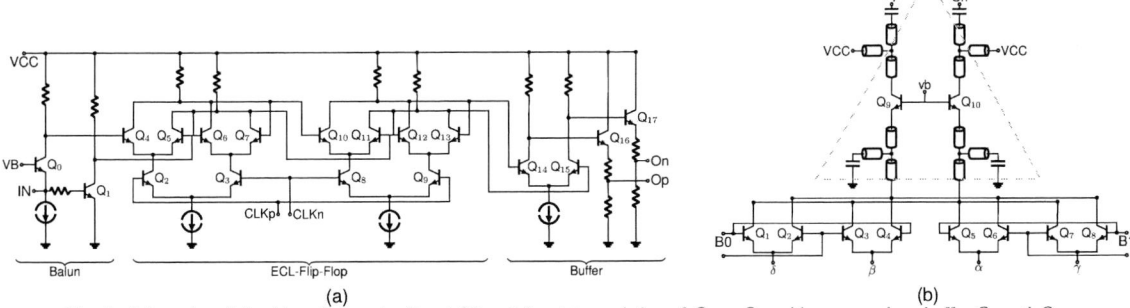

Fig. 4. Schematics of the (a) re-timing-circuit and (b) switch octet, consisting of $Q_1 - Q_8$, with common base buffer Q_9 and Q_{10}.

amplitude, which are evaluated by simulation and shown in Fig. 5. Over the entire D-band imbalances of <3 % and <0.75 dB are achieved for phase and amplitude, respectively.

IV. CIRCUIT DESIGN

As shown in Fig. 2, besides the passive components, a QPSK unit-cell consists of two re-timing cells for each of the two binary inputs and one switch octet for discrete vector-phase control. The schematics of these building blocks are shown in Fig. 4. The outputs of the re-timing blocks are directly connected and, hence, DC-coupled to the transistor bases of the switch octet.

Re-timing circuit: An active balun converts the single-ended data signal for the differential high-speed Flip-Flop, which is composed of two ECL-latches. A nearly identical balun feeds the differential clock signal to the Flip Flop. The proceeding buffer works as limiter and level shifter for the switch octet and minimizes the transition time of the switches.

Switch octet: The transistors $Q_1 - Q_8$ create a QPSK signal by switching the four input phases (α, β, γ, δ) that are provided directly by the Marchand baluns. By carefully choosing the transistor sizes and bias point, the phase terminals present a broadband 50 Ω input impedance and very fast switching behaviour. Transistors Q_9 and Q_{10} form the core of a buffer, which is impedance matched using T-type transmission-line input- and output matching networks, which create a flat frequency response over 110-170 GHz.
The performance bottleneck of this prototype is the amplitude imbalance of the active baluns beyond a baudrate of 15 Gsps, which corresponds to a maximum datarate of 60 Gbps for 16-QAM modulation. All parasitic impedances associated with the core RF transistors are EM-simulated using Keysights Momentum™. The matching networks are EM-simulated with Sonnets™ EM-simulator. Calibre™-PEX is used for the digital/mixed signal re-timing circuitry.

V. MEASUREMENTS AND RESULTS

A chip-foto and the test setup are depicted in Fig. 6. The circuit consumes 120 mW from a 2.1 V supply, which is confirmed by measurements. The chip is glued and wire-bonded on a PCB with FR-4 substrate for concept demonstration. The D-band signals are probed using Infinity™ GSG probes of 100 μm pitch. Calibrated measurements of the standalone modulator are performed up to a maximum LO input power of -10 dBm to avoid compression of the spectrum analyser frequency extender module (SAX). The observed LO-referred conversion loss of 7 dB is in line with simulation results. Data and clock signals are fed via SMA connectors and length-compensated microstrip lines on PCB. To create well

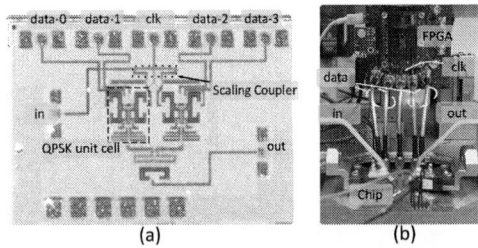

Fig. 6. (a) Chip micrograph of the 16-QAM prototype IC, occupying 0.36 mm^2, and (b) photograph of the probed test setup.

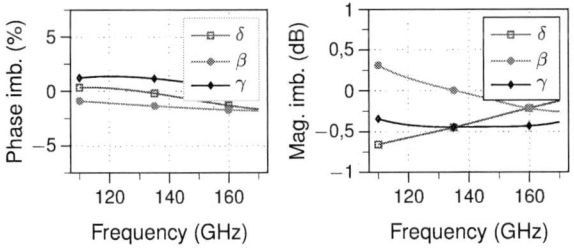

Fig. 5. Simulated phase imbalances and magnitude imbalances for the phases β, γ and δ with respect to phase α.

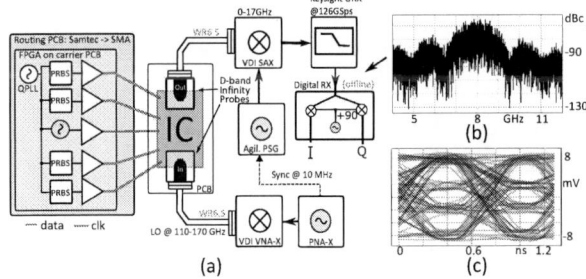

Fig. 7. (a) Block diagram of the measurement setup, (b) spectrum of the down-converted signal at 8 GHz IF and (c) eye diagram of the demodulated 16-QAM signal.

978-1-6654-8495-4/22 $31.00 © 2022 IEEE

Table 1. State-of-the-art comparison with other RF-DACs.

Ref.	f_c (GHz)	R (Gbps)	Tech. (nm)	Mod	Pow. (mW)	Area (mm²)	EVM/ BER
[3]	135	12	45*	64-Q.	1255^p	>1$^\vartheta$	25.3$^{\%e}$
[4]	110	20	180‡	16-Q.	520$^{p,+}$	3.17	-15^{dB}
[5]	94	20	32*	QPSK	220^p	0.25	24$^{\%e}$
[6]	35-130	30	130‡	4-ASK	90	0.45$^\vartheta$	1e-8^B
[7]	220	20	55‡	OOK	63$^+$	0.32$^\vartheta$	n.a
This	110-170	6.4/60*	90‡	16-Q.	120	0.36	<8$^{\%e}$

*simulated; †CMOS; *SOI CMOS ; ‡SiGe BiCMOS HBT; $^+$internal LO;
$^\vartheta$estimated with chip foto; ^{p}with rf (power) amplification;
$^{\%e}$ EVM in %; ^{dB}EVM in dB; ^{B}BER.

synchronized high-speed data- and clock signals, we used the GTX transmitters of a Kintex-7 FPGA with internal PRBS pattern generators. This FPGA was shipped and pre-mounted on a board with Samtec high-speed connectors. Another PCB routes the high-speed signals to SMA connectors and provides power supply circuitry and programming interfaces. The maximum rate supported by this assembly with multiple lossy signal transitions is 1.6 GHz, which limits the demonstrated datarate to 6.4 Gbps.

As shown on Fig. 7(a), the modulated D-band signal is down-converted with a VDI SAX harmonic mixer to an IF of 8 GHz and sampled by a 12 bit, 126 Gsps Keysight oscilloscope. Offline post-processing consists of phase adjustment, demodulation, matched filtering and sampling. A sinc-shaped spectrum of the down-converted signal is shown in Fig. 7(b). The eye-diagram of the demodulated signal is shown in Fig. 7(c). Measured constellation plots at different carrier frequencies are shown in Fig. 8. Phase- and amplitude imbalances of the couplers are observable by slight but constant rotations and shifts within the constellation points. Further minor deviations to the ideal constellation points are attributed to noise, non-ideal sampling time and minimal

frequency offsets in the measurement setup. The measured EVM stays below <8% over the entire D-band.

The work is compared with state-of-the-art in Table 1. It is the first demonstration of an RF-DAC for realizing and passively scaling QPSK-signals for generating higher-order modulations over the entire D-band. It has the potential for high datarates, and achieves broadband performance with high signal integrity at a low power consumption.

VI. CONCLUSION

This paper presents a scalable RF-DAC concept, which utilizes the broadband properties of on-chip distributed passives (Lange couplers, backward-wave couplers and Marchand baluns) to create QAM signals over the entire D-band. This low-complexity architecture for high frequencies supports high datarates and vector aggregation for creating higher-order modulations, it is amenable with modern integrated technologies and benefits from the inherent device matching of QPSK unit cells. The concept is verified via a 16-QAM modulator prototyped in 90 nm SiGe BiCMOS. The prototype demonstrates a data rate of 6.4 Gbps with a measured EVM of <8% over the entire D-band. Simulations show a maximum datarate of 60 Gbps, however, the measured datarate is currently limited by the test setup comprising an in-house low-cost FPGA clocked at 1.6 GHz. The chip consumes 120 mW from a 2.1 V supply and occupies 0.36 mm² of core area. Full D-band standalone characterization of the backward-wave and Lange couplers show amplitude imbalances of <1 dB and quadrature phase errors of <8° over 110-170 GHz.

ACKNOWLEDGMENT

This work was partly supported by the European Union (ESCS16104) and the German Federal Ministry of Education and Research (16ESE0214) as part of the H2020-ECSEL-2016-1 project TARANTO.

REFERENCES

[1] S. Callender et al., "A fully integrated 160Gb/s D-band transmitter with 1.1 pj/b efficiency in 22nm FinFET technology," in *2022 IEEE International Solid- State Circuits Conference (ISSCC)*, vol. 65, pp. 78–80.

[2] N. Dolatsha et al., "17.8 a compact 130GHz fully packaged point-to-point wireless system with 3D-printed 26dBi lens antenna achieving 12.5Gb/s at 1.55pj/b/m," in *2017 IEEE International Solid-State Circuits Conference (ISSCC)*, pp. 306–307.

[3] S. Shopov et al., "A D -band digital transmitter with 64-QAM and OFDM free-space constellation formation," vol. 53, no. 7, pp. 2012–2022.

[4] H. Wang, H. Mohammadnezhad, and P. Heydari, "Analysis and design of high-order QAM direct-modulation transmitter for high-speed point-to-point mm-wave wireless links," vol. 54, no. 11, pp. 3161–3179.

[5] H. Al-Rubaye and G. M. Rebeiz, "A 20 Gbit/s RFDAC-based direct-modulation W-band transmitter in 32nm SOI CMOS," in *2016 IEEE Compound Semiconductor Integrated Circuit Symposium (CSICS)*, pp. 1–4.

[6] F. Strömbeck, Z. S. He, and H. Zirath, "Multi-gigabit RF-DAC based duobinary/PAM-3 modulator in 130 nm SiGe HBT," in *2020 15th European Microwave Integrated Circuits Conference (EuMIC)*, pp. 257–260.

[7] B. Hadidian et al., "An energy efficient fully integrated 20Gbps OOK wireless transmitter at 220GHz," in *2021 IEEE Custom Integrated Circuits Conference (CICC)*, pp. 1–2.

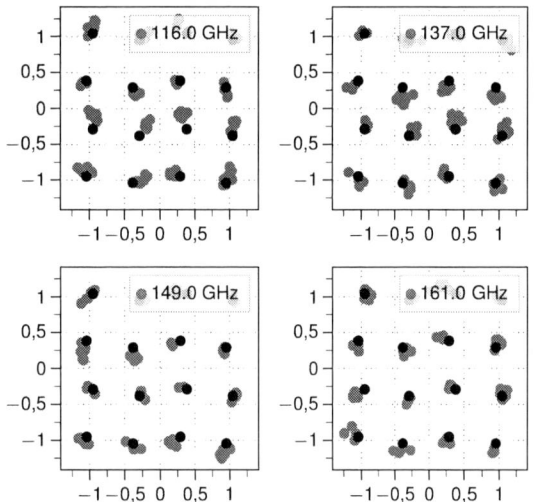

Fig. 8. Measured 16-QAM constellations at a data rate of 6.4 Gbps at 116, 137, 149 and 161 GHz. EVM degradation due to the 5° imbalance of the input coupler of is taken into account for ideal constellation points.

A 111-149-GHz, Compact Power-combined Amplifier With 17.5-dBm P_{sat}, 16.5% PAE in 22-nm CMOS FD-SOI

Jeff Shih-Chieh Chien
Department of Electrical and Computer Engineer ing
University of California, Santa Barbara
Santa Barbara, USA
schien@ucsb.edu

James F . Buckwalter
Department of Electrical and Computer Engineering
University of California, Santa Barbara
Santa Barbara, USA
buckwalter@ucsb.edu

Abstract— **This paper presents a 4-way, 3-stage power-combined CMOS SOI power amplifier (PA) operating from 111-149 GHz. Pseudo-differential stages and 4-way parallel power combining increases the output power from power cells matched to 50 Ohms. Using sub-quarter wavelength (SQWL)-type coupled-line balun (CLB), low-loss output matching networks achieve higher power-added efficiency (PAE). Fabricated in 22-nm FD-SOI CMOS technology, the PA occupies 0.113mm² core area generates 17.5-dBm saturated output power (P_{sat}) and 16.5% peak PAE. To the author's knowledge, this work demonstrates the highest PAE in a CMOS SOI process at D-band.**

Keywords—millimeter-wave, D-band, power amplifier, FD-SOI.

I. INTRODUCTION

D-band (110-170 GHz) communication systems have gained more attention for a potential candidates for automotive radar and beyond-5G communication systems in recent years due to the wide available bandwidth and the reduced wavelength for compact phased array systems [1]. Operating in upper millimeter-wave bands suggests improved range resolution for radar but the circuit implementation in CMOS is challenged by limited f_{max}/f_T of the RF-FETs. With low gain at D-band, multiple stages and multi-way power combining in PA designs is required to reach 20 dBm output power. Since antenna spacing of a phased-array transmitter ranges from 1.4 mm to 0.9 mm across D-band, the total chip area of a single element should be smaller than 1.96mm² to 0.81mm². A compact, high-efficiency PA design is desirable to deliver power for compensating high path loss at D-band while constrained by the available chip area.

Several D-band PA in silicon have been published in the past few years based on compact designs but with P_{sat} of only 15dBm [2] and 12.5dBm [3], respectively. In [4], the efficiency is limited. However, [5]-[7] demonstrate PAs with more than 50-mW P_{sat}, but the area also scales with power since multi-way power combiners and multiple transmission lines are used.

In this work, a compact 4-way-combined PA is introduced based on a multi-stage, pseudo-differential power cell with neutralization capacitance. The power cell layout is carefully re-designed to optimize power and efficiency performance. A coupled-line balun (CLB) provides low-loss impedance transformation as well as differential to single-ended

transformation. A power splitter and combiner are in parallel topology and realized by co-designing with RF pad impedance. Implemented in a 22-nm CMOS FD-SOI process, the proposed PA operates over the bandwidth of 111-149 GHz with the peak P_{sat} of 17.5 dBm and peak PAE of 16.5%.

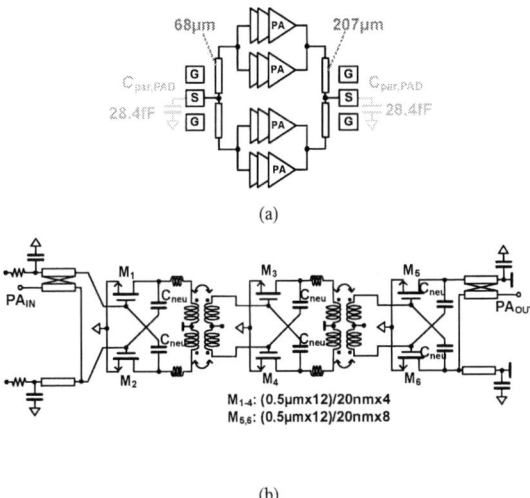

(a)

(b)

Fig. 1. (a) Four-way combined PA block diagram, (b) schematic of three-stage PA.

II. DESIGN OF FOUR-WAY COMBINED PA

The block diagram of the proposed 4-way combined PA is shown in Fig. 1(a). To reduce the area, microstrip line is used in the splitter and combiner design and the GSG RF pad capacitance is incorporated into the splitter and combiner. Fig. 1(b) shows the detailed schematic of PA in each path. The three power stages use a power cell ratio of 1:1:2 to ensure the last stage could be driven to saturation. The input and output balun are formed with SQWL-type CLB while interstage matching are based on transformer for compact area considerations.

A. Power cell design in 22-nm FD-SOI CMOS

Transistor layout plays an important role in mm-wave PA design since the parasitic resistance from the routing and capacitance between metals determine maximum available gain (G_{max}). The expressions for f_T and f_{max} are

$$f_T = \frac{g_m}{2\pi(C_{gs}+C_{gd})}, f_{max} = \sqrt{\frac{f_T}{8\pi r_g C_{gd}}}, \quad (1)$$

and the gate-drain capacitance should be minimized for higher maximum available gain. Although minimizing r_g and C_{gd} presents a trade-off, reducing C_{gd} is a priority since the neutralization capacitor (C_{neu}) will be added to both the input and output node in the differential power cell design. In most cases, C_{neu} is equal to C_{gd} for achieving best stability condition and, therefore, the total capacitance on the drain node will become

$$C_{d,total} = C_{db} + C_{ds} + 2C_{gd}. \quad (2)$$

The increase of the capacitance pushes the power and PAE contours of the transistor to a high voltage-standing-wave ratio on the Smith chart. This will also further deteriorate the output matching loss and bandwidth. With the above mentioned approach, a single transistor layout is shown in Fig. 2(a). Using the first metal as base, the source is connected to global ground using the second metal layer. This could not only reduce the resistance to ground but also prevent the fringe capacitance with gate and drain. The gate is connected to the upper thin metal layer from C3 to C5 and only drain node is connected to the thick metal layer for reducing fringe capacitance between gate and drain. The simulated G_{max} comparison between different models are shown in Fig. 3. The optimized layout with M1 model provides higher f_{max} than the C3 model in PDK.

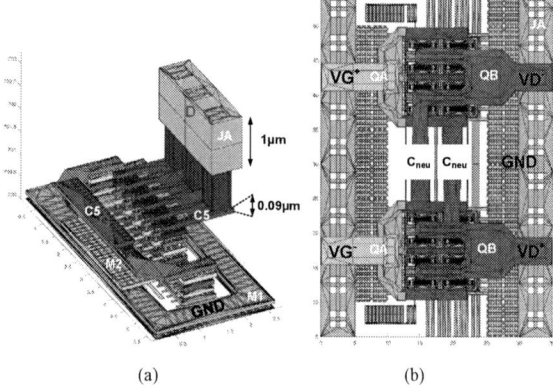

Fig. 2. (a) Single transistor layout, (b) differential power cell layout with 8 multipliers on each side.

To further increase output power of a single PA path, pseudo-differential with 8-fold gate connections on each side is designed as shown in Fig. 2(b). To improve the small signal stability, two C_{neu} are added between two cells. The load-pull simulation of the proposed power cell layout with 6 dBm input power is shown in Fig. 4 with 33.6% peak PAE and 12.2 dBm output power at $Z_{opt} = 18+j*32\Omega$. The simulated P_{sat} of this power cell is around 13.4 dBm under 1.1-V supply.

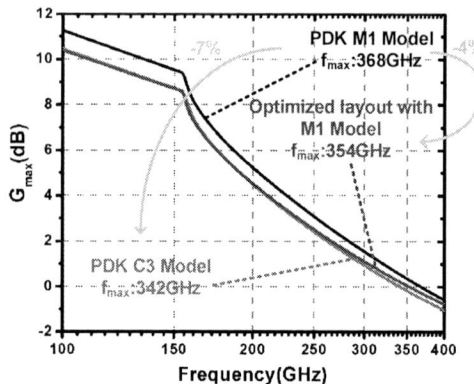

Fig. 3. G_{max} comparison between different transistor layout.

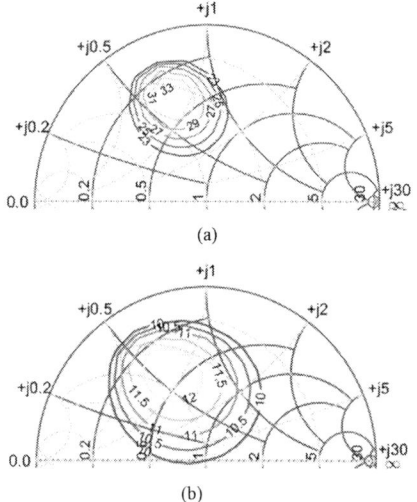

Fig. 4. (a) PAE contour and (b) Power contour of power cell in Fig. 2(b).

B. CLB and parellel power combiner design

The output balun plays an important role in mm-wave differential PA design. One popular implementation is based on transformer balun [2][3][6][7]. However, the loss of a transformer balun increases dramatically due to lower in-band quality factor as well as the drop of coupling factor since the shunt inductance required to resonate out the power cell capacitance reduces at high frequency and targeted output power. With the fixed metal stack, [2] proposed a high-k sandwiched transformer. However, the coupling factor is still limited to 0.73 and this constrains the minimum feasible loss at the output matching network.

In [8], the authors proposed a SQWL balun that achieves excellent insertion loss (IL) with impedance transformation. However, the bottom transmission line of the SQWL balun needed to be far away from ground so that the desired impedance could be fulfilled. This makes the topology less compact and hard to incorporate with multi-way power combiner. Also, at high frequency, the finite qualify factor of the bypass capacitor at two ends of bottom plates increases the loss of the whole balun.

Here, we propose the CLB as shown in Fig. 5. With the bended structure, the AC ground could be leveraged and supply voltage could be fed on left or right direction which makes multi-way power combining simpler. The simulated G_{max} of the proposed CLB is -0.51 dB as shown in Fig. 5, the equivalent inductance of two coils are 87.4pH and 82.4pH and the coupling factor is 0.82 at 140 GHz. To achieve higher output power, 4-way parallel power combiner is shown in Fig. 6. The design methodology follows

Step 1) Optimize the single path balun with the targeted P_{sat} and highest achievable PAE.

Step 2) Place two balun in parallel closely and figure out Z_{opt} that provides similar P_{sat} and PAE as Step1. The result in our implementation is Z_{opt} = 17.1+j*7.8.

Step 3) Figure out the RF pad impedance Z_{pad} with pre-fill metal dummies.

Step 4) Figure out the transmission line impedance and length to match Z_{opt} and $2Z_{pad}$.

With the above mentioned steps, 4-way parallel power combiner is designed with simulated 1.2dB insertion loss including RF pad loss at 140GHz.

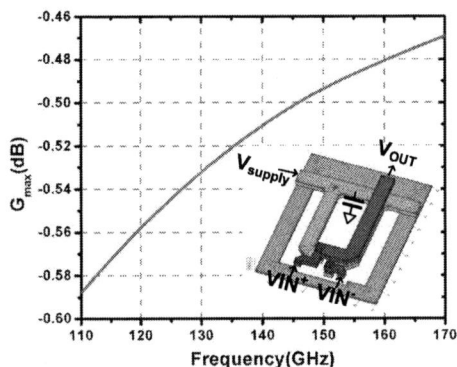

Fig. 5. G_{max} of the proposed CLB.

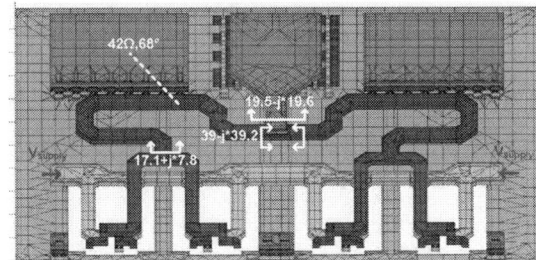

Fig. 6.   Layout and impedances of four-way output combiner.

III. MEASUREMENT RESULTS

The proposed PA is implemented in a GlobalFoundries 22-nm CMOS FD-SOI technology and illustrated in the microphotograph in Fig. 7. The core area of the PA is 450 µm by 250 µm while the overall chip size limited by probe spacing is 620 µm by 600 µm. The PA gate is biased with 0.5 V and the collector supply is 1.1V.

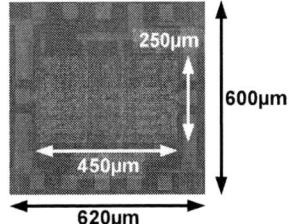

Fig. 7.   Die photograph of the fabricated PA.

The S-parameters and power measurement are performed with D-band (110-170 GHz) VDI frequency extenders and D-band 100µm pitch waveguide probes. Calibration is done with CS-15 calibration substrate to the probe tip. The measured and simulated S-parameters are plotted in Fig. 8. The measured peak gain is 13.5dB with 111-149 GHz 3-dB bandwidth. Power measurements are plotted at 140 GHz as shown in Fig. 9. The 16% peak PAE is measured at 15.7 dBm output power while still having 10-dB power gain at 140 GHz. The output power and PAE are also measured in Fig. 10 across the D-band. The measured peak PAE is 16.5% at 114 GHz and 17.5 dBm peak P_{sat} at 124 GHz. Table 1 summarizes the state-of-the-art D-band PAs in silicon for the comparison of the presented work. The PA demonstrates the highest PAE in CMOS SOI while also demonstrating compact area and wideband operation.

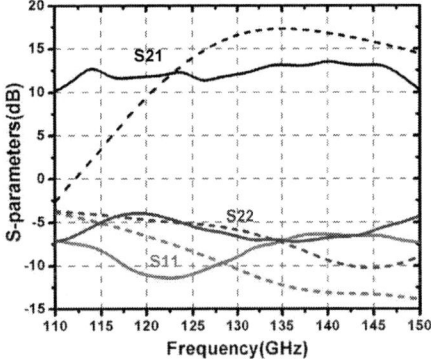

Fig. 8.   Measured (solid) and simulated (dashed) S-parameters.

IV. CONCLUSION

We demonstrated a D-band high efficiency PA with compact area and wideband operation. The design includes optimized pseudo differential power cell design, CLB and 4-way power combiner for achieving compact layout. With 3-dB bandwidth covering 111-149 GHz, peak PAE is 16.5% at 114 GHz and 17.5 dBm peak P_{sat} at 124 GHz is measured.

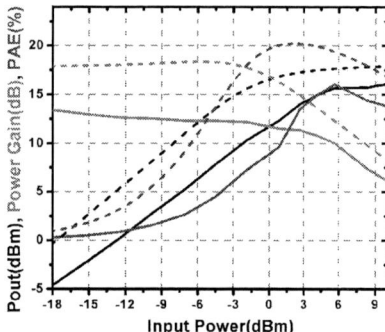

Fig. 9. Measured (solid) and simulated (dashed) ouptut power, power gain and PAE versus input power at 140 GHz.

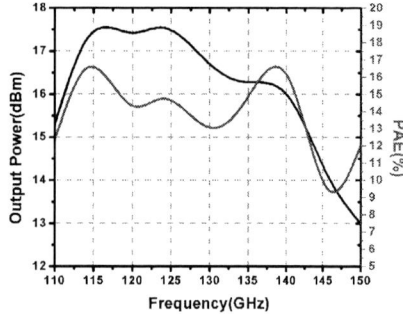

Fig. 10. Measured output power and PAE over 110-150 GHz.

ACKNOWLEDGMENT

This work was supported by the Semiconductor Research Corporation (SRC) and DARPA under the JUMP program (ComSenTer). The authors would also like to thank Kenneth Barnett and Ned Cahoon at GlobalFoundries for access to 22-nm FD-SOI process through the University Partnership Program.

REFERENCES

[1] M. Elkhouly et al., "Fully Integrated 2D Scalable TX/RX Chipset for D-Band Phased-Array-on-Glass Modules," *2022 IEEE International Solid-State Circuits Conference (ISSCC)*, 2022, pp. 76-78.

[2] B. Philippe and P. Reynaert, "24.7 A 15dBm 12.8%-PAE Compact D-Band Power Amplifier with Two-Way Power Combining in 16nm FinFET CMOS," *2020 IEEE International Solid- State Circuits Conference - (ISSCC)*, 2020, pp. 374-376.

[3] X. Tang, J. Nguyen, G. Mangraviti, Z. Zong and P. Wambacq, "A 140 GHz T/R Front-End Module in 22 nm FD-SOI CMOS," *2021 IEEE Radio Frequency Integrated Circuits Symposium (RFIC)*, 2021, pp. 35-38.

[4] S. Daneshgar and J. F. Buckwalter, "Compact Series Power Combining Using Subquarter-Wavelength Baluns in Silicon Germanium at 120 GHz," in *IEEE Transactions on Microwave Theory and Techniques*, vol. 66, no. 11, pp. 4844-4859, Nov. 2018.

[5] H. Lin and G. M. Rebeiz, "A 110–134-GHz SiGe Amplifier With Peak Output Power of 100–120 mW," in *IEEE Transactions on Microwave Theory and Techniques*, vol. 62, no. 12, pp. 2990-3000, Dec. 2014.

[6] A. Visweswaran, B. Vignon, X. Tang, S. Brebels, B. Debaillie and P. Wambacq, "A 112-142GHz Power Amplifier with Regenerative Reactive Feedback achieving 17dBm peak Psat at 13% PAE," *ESSCIRC 2019 - IEEE 45th European Solid State Circuits Conference (ESSCIRC)*, 2019, pp. 337-340.

[7] S. Li and G. M. Rebeiz, "High Efficiency D-Band Multiway Power Combined Amplifiers With 17.5-19-dBm Psat and 14.2-12.1% Peak PAE in 45-nm CMOS RFSOI," in *IEEE Journal of Solid-State Circuits*.

[8] H. Park, S. Daneshgar, Z. Griffith, M. Urteaga, B. Kim and M. Rodwell, "Millimeter-Wave Series Power Combining Using Sub-Quarter-Wavelength Baluns," in *IEEE Journal of Solid-State Circuits*, vol. 49, no. 10, pp. 2089-2102, Oct. 2014.

[9] Y. Hsiao, Z. Tsai, H. Liao, J. Kao and H. Wang, "Millimeter-Wave CMOS Power Amplifiers With High Output Power and Wideband Performances," in IEEE Transactions on Microwave Theory and Techniques, vol. 61, no. 12, pp. 4520-4533, Dec. 2013.

[10] I. Petricli, D. Riccardi and A. Mazzanti, "D-Band SiGe BiCMOS Power Amplifier With 16.8dBm P₁dB and 17.1% PAE Enhanced by Current-Clamping in Multiple Common-Base Stages," in *IEEE Microwave and Wireless Components Letters*, vol. 31, no. 3, pp. 288-291, March 2021.

[11] X. Li et al., "A 110-to-130GHz SiGe BiCMOS Doherty Power Amplifier With Slotline-Based Power-Combining Technique Achieving >22dBm Saturated Output Power and >10% Power Back-off Efficiency," *2022 IEEE International Solid- State Circuits Conference (ISSCC)*, 2022

Table 1. Performance Comparison of CMOS FinFET, SOI and SiGe D-band PAs

Ref.	Technology	Frequency (GHz)	Gain(dB)	P_{sat}(dBm)	Peak PAE (%)	Core Area (mm²)	$\frac{P_{sat}}{Area}$(W/mm²)	$\frac{Gain}{Area}$(dB/mm²)
This work	**22-nm FD-SOI**	**111-149**	**13.5**	**17.5**	**16.5**	**0.113**	**0.5**	**119**
[2]	16-nm FinFET	110-128	20.5	15	12.8	0.041	0.77	500
[3]	22-nm FDSOI	124-154	33.6	12.5	10.8	0.024	0.74	1400
[4]	90-nm SiGe	110-140	7.7	22	3.6	0.348*	0.46	22
[5]	90-nm SiGe	114-134	20	17	12.5	0.858*	0.058	23
[6]	0.13μm SiGe	112-142	34	17	13	0.35	0.14	97
[7]	45-nm SOI	130-151	24	17.5	13.4	0.43	0.13	56
[9]	65-nm CMOS	125-155	15	13.2	14.6	0.26*	0.08	46
[10]	55-nm SiGe	125-159	24	17.6	17.5	0.18	0.32	133
[11]	130-nm SiGe	107-135	21.8	22.7	18.7	0.58	0.32	37.6
*Graphically estimated								

978-1-6654-8495-4/22 $31.00 © 2022 IEEE

A D-Band mm-wave spectroscopy TX and RX in 28 nm CMOS with 15.6 dBm EIRP and 17.1 dB NF with integrated antennas

Gabriel Guimaraes[#1], Patrick Reynaert[#2],

[#]KU Leuven, Departement Elektrotechniek ESAT/MICAS, Leuven, Belgium
[1]gabriel.guimaraes@esat.kuleuven.be, [2]patrick.reynaert@kuleuven.be

Abstract — **This paper presents a pair of CMOS D-band transmitter and receiver ICs to be used in mm-Wave spectroscopy systems. The TX consists of an on-chip VCO and divide chain for external frequency locking, a tripler, a power amplification chain and a power-combining on-chip antenna. It operates from 123 GHz to 142 GHz and has a peak EIRP of 15.6 dBm and 0.9 dBm of measured total radiated power (TRP). The heterodyne RX has an on-chip VCO and divide chain as well as an LNA, a passive mixer and a programmable IF amplifier. It operates at the same range as the TX. Its measured peak isotropic gain is 62.4 dB and isotropic noise figure is 3.1 dB at 134.7 GHz, which translate to a gain of 48.4 dB and NF of 17.1 dB when the antenna directivity is discounted. A spectroscopy system demonstration is built with the developed ICs and it is used to detect the 136.906 GHz absorption line of evaporated isopropyl alcohol.**

Keywords — **mm-wave CMOS, THz spectroscopy, On-chip antenna, mm-Wave sensing.**

I. INTRODUCTION

Advancements in circuit technology and design have allowed the development of mm-wave and THz circuits in CMOS technologies. Nevertheless the frequency region above 100GHz, although promising, is still relatively unexplored. To close this gap a number of applications have been proposed that benefits from this high carrier frequency. Communication and sensing applications, namely spectroscopy, imaging and radar have become more and more common in the scientific literature as they make their way into market. In the spectroscopy bracket, gas rotational spectroscopy is the main explored field since many polar gasses have the peak of their absorption spectra in the middle of the terahertz range.

Instead of the pulse-based time-domain spectroscopy done in optical that require fine true-time delays, circuits can implement many different architectures to probe the gas absorption lines. One of these techniques is cavity-coupled spectroscopy as presented in [1], where a pulse of mm-wave signal is inserted into a cavity and the decaying characteristics of the pulse are analyzed in the frequency-domain to identify gasses inside of the cavity. In a more common technique, the High-Q gas absorption spectrum can be detected with frequency-modulated continuous wave signals, as illustrated in Fig. 1. This architecture lends itself well to radio transmitters and is called wavelength modulation spectroscopy. It was implemented in several works on the mm-wave or Terahertz range [2], [3], [4].

The operating frequency of spectroscopy systems range from bellow 100GHz to THz frequencies. The rotational spectrum of different gases have different regions where its

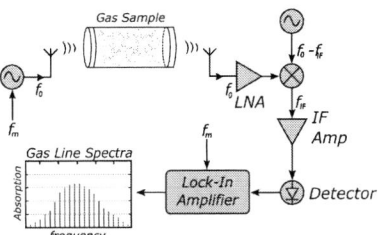

Fig. 1. Diagram of a heterodyne wavelength modulation spectroscopy system. A FM-modulated continuous wave probing signal is converted to an AM-modulated signal through the absorption line of the gas. This signal envelope is extracted and correlated with the modulating reference for high SNR detection.

absorption lines are most absorbent, but as a rule of thumb most gases present their optimum at the middle of the terahertz range (> 500GHz and < 1.5THz). These frequencies are way above the f_{max} of CMOS and even most SiGe technologies. A terahertz transceiver can be designed above f_{max} but without amplification at these frequencies. On the other hand, a lower frequency system can compensate the lower absorption of the gas with higher output power due to a power amplifier last architecture at TX and lower NF due to a low-noise amplifier at the RX in order to achieve similar or even better sensitivity depending on the technology parameters and specific gas.

This paper presents a CMOS transmitter and receiver systems at D-band for terahertz gas spectroscopy. The chips include a VCO and divider chain for external frequency stabilization and modulation, as well as RF front-ends, antennas and controllable-gain base-band. Sections II and III explains in detail TX and RX architecture and design respectively. Section IV depicts the chip measurements and results and a spectroscopy demonstration and section V concludes the paper.

II. TX DESIGN

A. Architecture

The architecture of the mm-Wave transmitter is displayed in Fig. 2. An on-chip VCO generates a signal at around 45 GHz that is tripled to 135 GHz, this signal is then amplified by an power amplifier chain that splits and re-combines at the on-chip multi-port slot antenna, the signal is radiated through the back-side of the substrate by means of a silicon lens in order to avoid substrate modes.

To stabilize the VCO and produce the required frequency-modulation, an external PLL board is used. The

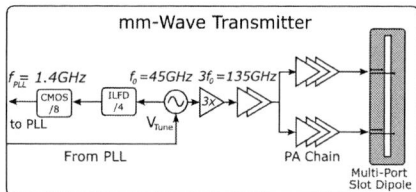

Fig. 2. Architecture diagram of the 135GHz transmitter.

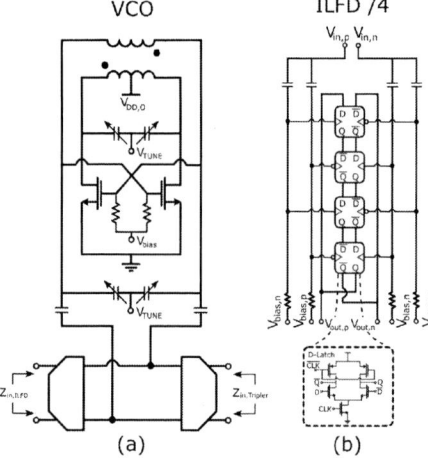

Fig. 3. Schematic of select circuits of the frequency generation chain (a) Voltage Controlled Oscillator (VCO) (b) Injection-Locked Frequency Divider by 4 (ILFD/4).

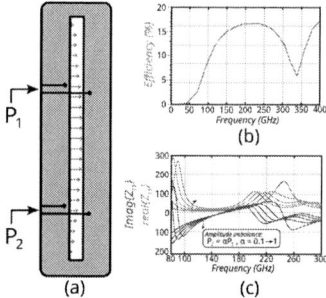

Fig. 4. (a) Multi-port slot antenna (b) Simulated antenna efficiency (c) Simulated input impedance under input amplitude mismatch between ports.

parasitics and allows a high-frequency input. The absence of a regenerative pair limits lower frequency operation range but it is not relevant since that limit is significantly lower than the lower frequency expected from the VCO. The divider differential output is fed to two CMOS flip-flop based asynchronous divide-by-8 for a balanced load, that are then buffered to the output.

C. Tripler, PA and Antenna

The tripler stage is a neutralized differential pair driven to saturation and matched by transformers where the input is matched at 45 GHz and output at 135 GHz. Extra care is taken to place a decoupling capacitor close to transformer center-tap since the second harmonic common-mode impedance can influence the tripler output power performance.

The PA stage is a cascade of transformer-matched neutralized-pairs. The first three stages are the same size. The signal is splitted after the second stage, the last and second to last stages are scaled to increase the maximum saturated power of the chain. The simulated output power of each branch at 135 GHz is 7 dBm after the matching network, delivered to a 32 Ohm antenna input impedance.

The multi-port antenna combines the output of the two amplifier branches and radiates this signal. It is a slot-dipole antenna with two feed-ports. This topology was proposed and modelled in [7]. The simulated radiation efficiency of the antenna is presented in Fig. 4 and is 13% or $-8.8dB$. The antenna is sized off-resonance for wider bandwidth input impedance around 32 Ohm. The input impedance variation due to amplitude input mismatch is also reduced drastically by sizing the antenna off-resonance as seen in Fig. 4, thus reducing the system sensitivity to gain mismatches between the two amplifier branches.

III. RX Design

A. Architecture

The RX architecture is presented in Fig. 5. The frequency synthesis is exactly the same as the TX. On the receiving path, the signal is received by a single-port slot dipole and amplified by a three-stage LNA before being downconverted by a passive mixer to an IF (10-200MHz). On the IF path the signal is amplified by a two-stage digitally programmable VGA with

VCO signal is pre-scaled by an injection-locked frequency divide-by-4 and is further divided by a CMOS divide-by-8 for a low-GHz range off-chip signal for frequency-phase comparison.

B. Frequency Generation

A transformer-based tank VCO is used as shown in Fig. 3. This tuned-input tuned-output topology can provide better phase-noise than a single tuned tank, but it is used due to the additional degree of freedom of having different DC-voltages at drain and gate of the transistor and is thus able to optimize the VCO output power in order to saturate the tripler stage following it. A single large continuous range is used with varactors at gate and drain in order to simplify the external PLL capture and locking. This comes at a cost of poor phase-noise performance, but as shown in [5], the phase-noise is not the limiting factor in sensitivity of a spectroscopy system, since its impact on the overall system SNR is dependant on the gas absorption and for small absorption (the case with low gas concentrations) it is much smaller than the receiver noise. The VCO drain signal is fed to two differential-pair buffers through AC-coupling capacitors, one of these buffers is matched to the input impedance of the tripler stage, and the other to the impedance of the injection locked divider.

The topology of the injection locked divider is the one presented in [6]. The compact latch has very small layout

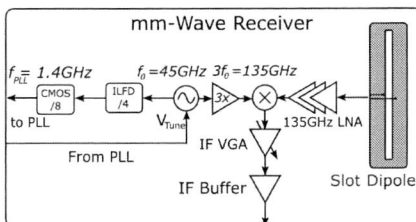

Fig. 5. Architecture diagram of the 135 GHz receiver.

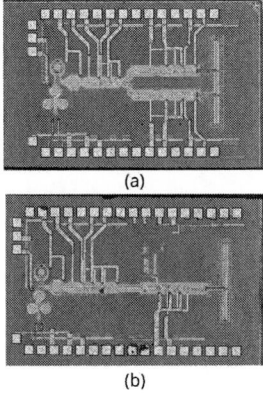

Fig. 6. Chip micrographs of (a) Transmitter (b) Receiver.

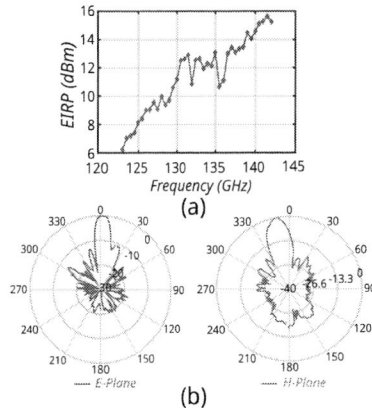

Fig. 7. TX Measurements (a) EIRP (b) Radiation Pattern.

10 gain modes before being buffered to drive a 100 ohm differential load. The VGA is made to ensure optimal SNR independent of signal input power variation. This variation can be due to different gas chamber lengths, different TX powers and losses at the signal coupling from the chamber to chips mainly due to lens misalignment.

B. Circuit Design

A passive double-balanced mixer is used for signal downconversion after the LNA. The IF amplifier The circuit consists of a differential common-source amplifier with a switchable binary resistor load. The amplifier has an AC-coupled input with a lower corner frequency of 10MHz, which simplifies the input common mode definition for cascading stages as it is defined by an external voltage. It is followed by a source follower buffer to drive the external 100 Ohm differential load.

IV. MEASUREMENTS

The chips were manufactured in a 28 nm bulk CMOS technology. The die micrographs are presented in Fig. 6. The chips were assembled on top of an undoped silicon wafer piece for bonding of the pads to PCB. An undoped hemispherical silicon lens was attached to the backside of the wafer piece for damping substrate modes and better efficiency of the mm-wave radiated or received signal. The TX die is attached to a lens with a diameter of 15 mm while the RX die lens has a diameter of 12 mm. There is no particular reason for these sizes besides the lens availability at the time for these particular

samples, and the circuits performance is relatively unaffected by variations of hemispherical lenses.

A. TX and RX performance

The TX was characterized by down-converting its radiated signal by a radiometer FS-Z170 subharmonic mixer, captured by a standard gain horn antenna at a far-field distance. The mixer conversion loss was characterized by using a characterized signal source (a R&S ZVA extender whose power was measured with a Erickson PM5 calibrated power meter). The measured TX EIRP is presented in Fig. 7, the peak EIRP at 141.5 GHz is 15.6 dBm. The measured radiation pattern is also presented in 7. The directivity is calculated to be 14.7 dB, which implies a total radiated power (TRP) of 0.9 dBm.

The receiver was similarly characterized at far-field. A R&S ZVA140 extender is used as a signal source with a standard gain antenna and the receiver isotropic gain is characterized. The notches in the isotropic gain that limit bandwidth are due to lens half-bowtie resonances as described in [8]. This is due to the use of a lens without anti-reflection coating. The receiver isotropic noise figure is characterized as in [9], where the noise floor at the receiver output is measured and referred to the input of the chain by using the previously measured isotropic gain and then compared to a 50 Ohm ideal noise source. These results are presented in Fig. 8. The measured Isotropic gain ranges from 27 dB to 62.4 dB while the measured minimum isotropic NF ranges from 3.1 dB to 6.3 dB at 134.7 GHz. The NF can be calculated by degrading the isotropic NF by the measured antenna directivity of 14 dB which results in an minimum NF of 17.1 dB. This noise figure is impacted by the low efficiency of the on-chip antenna at this frequency (simulated as -9.2dB).

In table 1 we compare this system with previous works from the literature.

B. Spectroscopy Demonstration

A wavelength spectroscopy setup was assembled with the TX and RX transmitting through colimating lenses into a

978-1-6654-8495-4/22 $31.00 © 2022 IEEE

Table 1. Comparison of the Silicon-based mm-Wave and THz Spectroscopy Systems

	This work	T-TST 2017 [1]	T-TST 2021 [2]	JSSC 2018 [3]	IRMMW-THZ 2015 [4]
Technology	28nm CMOS	65nm CMOS	28nm CMOS	65nm CMOS	0.13m SiGe BiCMOS
Frequency (GHz)	123 - 142	95 - 105	180 - 190	220 - 320 †	480-500
Total Radiated Power (dBm)	0.9*	6	-2.22	-0.55*	-7
SSB Noise Figure (dB)	17.1*	9.2	/×	14.6*	/
Total Power Consumption (mW)	TX: 288; RX: 213	TX: 226 RX: 126	TX: 350	TX: 1700 RX: 1700	TX: 1100 RX: /

† Chip has 10 independent sources that double as transmitters and receivers, best power and NF were used for comparison.
* Antenna loss is included, directivity is de-embedded.
× TX Only.

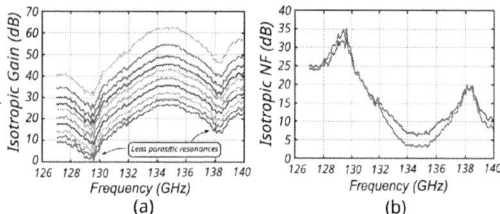

Fig. 8. RX Measurements: (a) Isotropic gain over gain settings and (b) maximum and minimum isotropic noise figure.

45 cm gas chamber that is pumped to low pressure with a turbo-molecular vacuum pump. The on-chip VCOs are stabilized through ADI41513 external PLL boards and the TX signal is frequency modulated at $f_m = 50$kHz and with a frequency deviation of $f_{dev} = 615$ kHz by a lock-in amplifier, which then measures the received envelope signal. A small amount of isopropyl alcohol is evaporated into the chamber using the vacuum pump for measurement. The setup is presented in Fig. 9 (a). The measurement was performed at 10 Pa and is presented in 9 (b) where the line at 136.906 GHz is shown. The presented line is the average of 3 measurements each with 0.1 s of integration time for each frequency point.

V. CONCLUSION

This works presented a transmitter and receiver pair for D band mm-wave spectroscopy. The transmitter achieves a maximum of 15.6 dBm EIRP and 0.9 dBm total radiated power while consuming 288 mW. The receiver achieves a minimum isotropic NF of 3.1 dB or NF of 17.1 dB while consuming 213 mW. A gas spectroscopy demonstration was performed detecting the 139.906 GHz line of evaporated isopropanol.

ACKNOWLEDGMENT

The authors wish to thank Analog Devices Inc., Limerick, Ireland for supporting this research.

REFERENCES

[1] A. Tang, B. Drouin, Y. Kim, G. Virbila, and M.-C. F. Chang, "95–105 ghz 352 mw all-silicon cavity-coupled pulsed echo rotational spectroscopy system in 65 nm cmos," *IEEE Transactions on Terahertz Science and Technology*, vol. 7, no. 3, pp. 244–249, 2017.

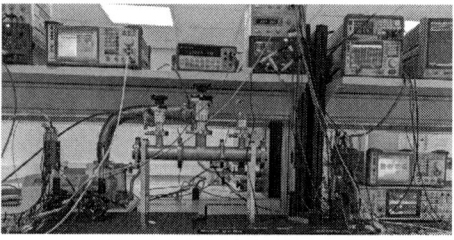

(a)

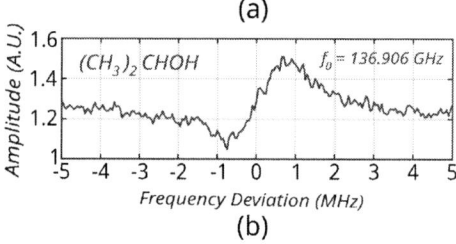

(b)

Fig. 9. Measurement of Isopropanol spectral absorption: (a) Measurement setup (b) 136.906 GHz absorption line.

[2] D. J. Nemchick, H. Hakopian, B. J. Drouin, A. J. Tang, M. Alonso-delPino, G. Chattopadhyay, and M.-C. F. Chang, "180-ghz pulsed cmos transmitter for molecular sensing," *IEEE Transactions on Terahertz Science and Technology*, vol. 11, no. 5, pp. 469–476, 2021.

[3] C. Wang and R. Han, "Dual-terahertz-comb spectrometer on cmos for rapid, wide-range gas detection with absolute specificity," *IEEE Journal of Solid-State Circuits*, vol. 52, no. 12, pp. 3361–3372, 2017.

[4] K. Schmalz, P. Neumaier, R. Wang, J. Borngräber, W. Debski, M. Kaynak, D. Kissinger, and H.-W. Hübers, "500 ghz sensor system in sige for gas spectroscopy," in *2015 40th International Conference on Infrared, Millimeter, and Terahertz waves (IRMMW-THz)*, 2015, pp. 1–2.

[5] C. Wang, B. Perkins, Z. Wang, and R. Han, "Molecular detection for unconcentrated gas with ppm sensitivity using 220-to-320-ghz dual-frequency-comb spectrometer in cmos," *IEEE Transactions on Biomedical Circuits and Systems*, vol. 12, no. 3, pp. 709–721, 2018.

[6] M. Vigilante and P. Reynaert, "A 25–102ghz 2.81–5.64mw tunable divide-by-4 in 28nm cmos," in *2015 IEEE Asian Solid-State Circuits Conference (A-SSCC)*, 2015, pp. 1–4.

[7] T. Chi, S. Li, J. S. Park, and H. Wang, "A multifeed antenna for high-efficiency on-antenna power combining," *IEEE Transactions on Antennas and Propagation*, vol. 65, no. 12, pp. 6937–6951, 2017.

[8] A. Boriskin and R. Sauleau, "Drastic influence of the half-bowtie resonances on the focusing and collimating capabilities of 2-d extended hemielliptical and hemispherical dielectric lenses," *Journal of the Optical Society of America. A, Optics, image science, and vision*, vol. 27, pp. 2442–9, 11 2010.

[9] D. Simic, K. Guo, and P. Reynaert, "A 420-ghz sub-5-m range resolution tx–rx phase imaging system in 40-nm cmos technology," *IEEE Journal of Solid-State Circuits*, vol. 56, no. 12, pp. 3827–3839, 2021.

978-1-6654-8495-4/22 $31.00 © 2022 IEEE

A Joint Space/Time Modulation Lens-Coupled 230-GHz Terahertz Source with 40°/43°2-D Beam Steering for Fast High-Resolution Imaging/Sensing Applications

Hossein Jalili[#1], Yuqi Liu[#], Tzu-Yuan Huang[#], Hua Wang[$2]

[#]Georgia Institute of Technology, Atlanta, GA 30308, USA
[$]ETH Zurich, Zurich, Switzerland

[1]hjalili3@gatech.edu, [2]huawang@ethz.ch

Abstract—This paper presents an integrated reconfigurable terahertz source array which employs a new method of joint space/time modulation to perform 2-D scalable and wide-range beam steering without phase shifters in a lens-coupled radiation set up. The reconfigurable array circuit can be activated in a variety of ways and is capable of simultaneous multi-beam/-frequency radiation. The chip is implemented in a 22nm CMOS-SOI process with a total area of $2.3 \times 2.3 mm^2$, consumes 21-62.5 mW power per cell from a 0.8-V supply voltage, and provides a total of 40°/43° 2-D scanning range at 230-GHz frequency.

Keywords—beam steering, terahertz source, reconfigurable array, time modulation, harmonic oscillator.

I. INTRODUCTION

Implementation of Terahertz (THz) integrated circuits in low-cost silicon technologies presents opportunities in a variety of applications including but not limited to imaging and sensing. It is therefore desirable to implement THz high-performance sources in terms of imaging speed and quality. The image resolution can be enhanced by improving the directivity and signal-to-noise-ratio (SNR) of radiation as well as the scanning resolution. Fast imaging can also be realized through electronic beam steering of the radiated THz power. However, as we move toward the THz spectrum, the physical limitations of fabrication processes limit the circuit performance. Achievable radiation power levels are restricted by the maximum oscillation frequency (f_{max}) of the process, pushing implementations toward array structures for improved power and directivity. Such coherent arrays of THz sources, however, are power hungry and would lead to slow imaging speed since mechanical scanning of the object would be necessary [1]. Phased arrays can be employed to improve the imaging speed by providing electronic scanning. However, large power-hungry array sizes would be necessary for high radiation directivity and EIRP [2]. Furthermore, implementation of wide-range phase shifters is quite challenging at THz spectrum, mainly due to highly lossy varactors which significantly compromise the already limited output power level [3]. Silicon lenses are often used with THz integrated sources to boost the EIRP, but would limit the scanning range of a conventional phased array [4]. Arrays of non-coherent sources have also been demonstrated in conjunction with silicon lens, where the independent sources in the array provide the same number of beam directions.

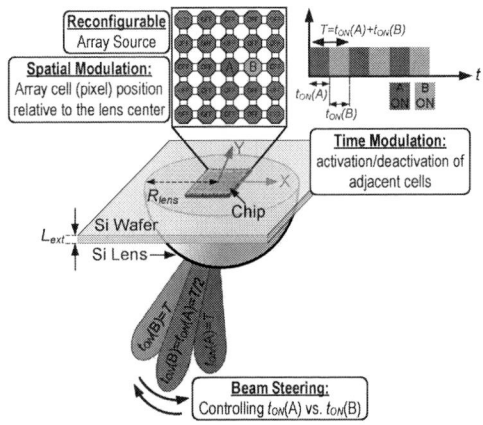

Fig. 1. Using spatial and time modulation for scalable FOV and high-resolution 2-D beam steering.

This method, however, leaves uncovered "dead zones" between individual firing angles limiting the resolution and therefore image quality [5].

In this work, we present a reconfigurable 230-GHz terahertz source array fabricated in GlobalFoundries 22nm CMOS-SOI process. As shown in Fig. 1, the implemented circuit combines spatial and time modulation schemes to perform 2-D fine-resolution beam-steering with high directivity through a lens-coupled radiation structure without employing phase shifters. The implemented source provides a field-of-view (FOV) of 40°/43°which is scalable with the size of the array. Section II of this paper describes the circuit structure and its main principles of operation and section III presents the experimental results from characterizing the implemented prototype chip.

II. CIRCUIT STRUCTURE AND OPERATION

Fig. 1 shows the main operation principle of this work. It consists of a reconfigurable array circuit that acts as the source for generation and radiation of THz power through the chip substrate and a silicon lens. The methods that have been employed for beam steering and implementation of the reconfigurable array circuit are described in the following subsections.

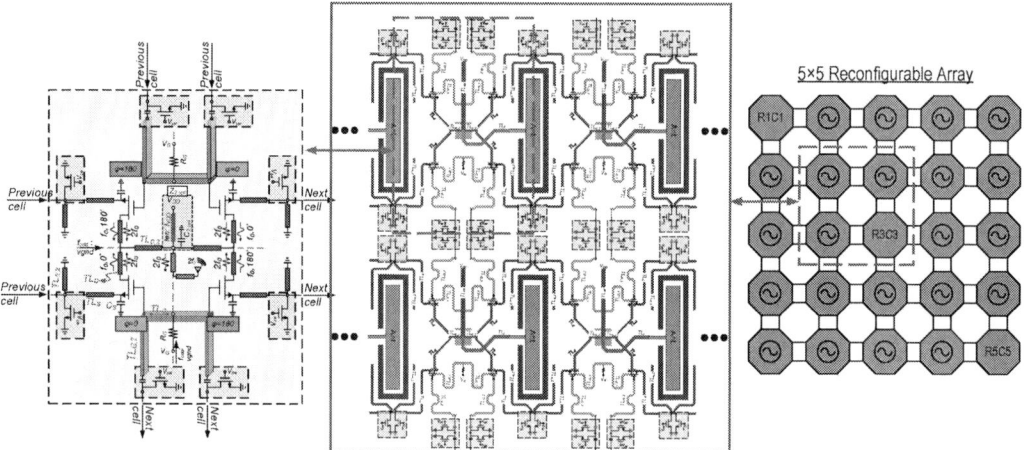

Fig. 2. Structure of the implemented 5×5 array circuit (right), schematic of the harmonic oscillator unit cell (left), and layout and coupling scheme of unit cells inside the array (middle).

A. Beam Steering Method

The firing angle of each pixel is determined by its location relative to the silicon lens, thus providing a set of discrete steering angles through this "spatial modulation", as shown for pixels A and B in Fig. 1. To cover the angles between the two individual directions of these adjacent cells, their activation can be modulated in time with $t_{ON}(A)$ and $t_{ON}(B)$ denoting their respective activation slots within the modulation period. The direction of the peak of the the time average radiated power can then be steered toward either side by controlling the activation duty cycle. A similar approach can be taken by modulating the activation of two adjacent cells in the vertical direction for steering the beam in the y direction. Therefore, the combination of the spatial modulation by the lens and time modulation of pixel activation enables continuous 2-D beam steering covering the dead zones in an imaging system set up.

B. Circuit Structure

Fig. 2 shows the structure of the implemented 5×5 array as well as the schematic and layout of individual source circuits. The array is made of 25 individual unit cells that are coupled together in a reconfigurable structure. Each individual unit cell consists of a harmonic oscillator in which second harmonic power at 230-GHz is extracted, combined and fed to an on-chip folded slot antenna for radiation through the chip substrate and silicon lens. Inside the oscillator, each transistor is oscillating at fundamental frequency in an out-of-phase relation with its adjacent transistors, therefor creating virtual grounds in both horizontal and vertical planes of symmetry. The fundamental power and its odd harmonics are therefore canceled before the antenna while the desired second harmonic power that is generated through the transistors' nonlinearities combine in-phase and are radiated by the antenna.

The unit cells of the array are coupled in the horizontal (x) direction through source transmission lines TL_S and in the vertical (y) direction through gate transmission lines

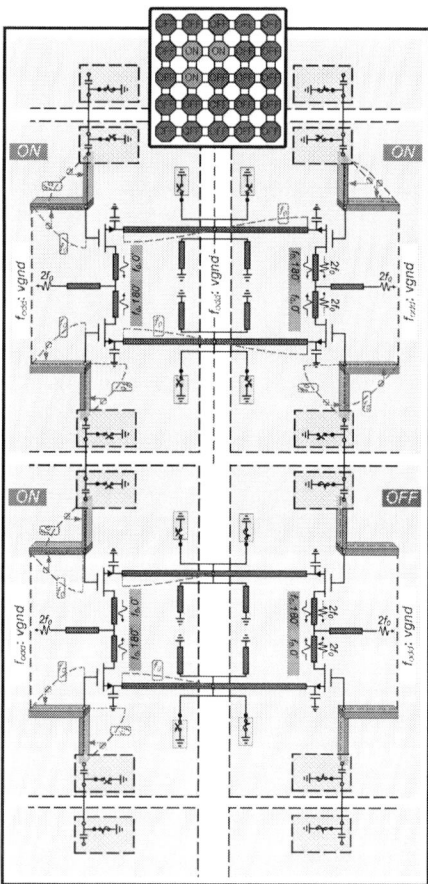

Fig. 3. Reconfigurable coupling between ON and OFF unit cells and automatic adjustment of switches for coupling and termination in ON and OFF cells respectively.

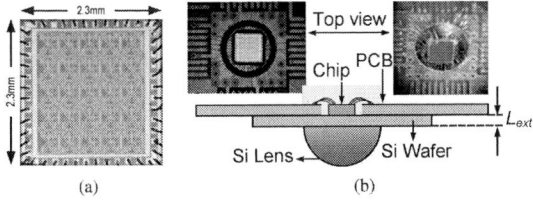

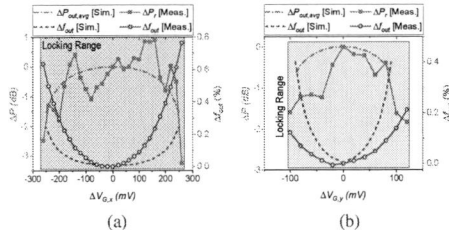

Fig. 4. (a) Chip photograph; (b) prototype assembly.

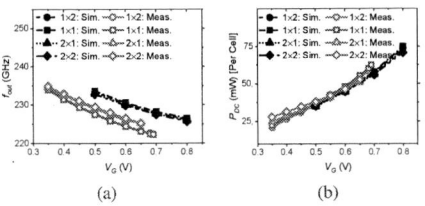

Fig. 6. Locking ranges for (a) 1×2 activation (x-direction); (b) 2×1 activation (y-direction).

Fig. 5. (a) Measured and simulated frequency; (b) Measured and simulated power consumption.

$TL_{G,2}$. Switches (highlighted in yellow boxes in Fig. 2) are incorporated at the connection points between adjacent cells. These switches are controlled by V_{sw} voltages that are generated on chip by inverting the gate bias voltages (V_G) of transistors. Therefore, when a cell is activated by application of an ON gate bias voltage, these switches are automatically opened, enabling coupling and injection locking to the adjacent cell. On the other hand, if the cell is deactivated by application of 0 gate bias voltage, these switches are closed, providing terminations to the neighboring cell and preventing the OFF cell from loading the activated section. This is demonstrated in Fig. 3, which shows how the switches are configured between On and OFF cells and how they facilitate injection coupling between ON cells as well as termination and load suppression between ON/OFF cells in both x and y directions.

It is important to note that when two adjacent cells are activated, the switches will be placed at the nodes of fundamental frequency standing waves, as shown in Fig. 3. Therefore, the loss and parasitics that are associated with these switch transistors do not disturb the circuit operation and its performance.

Based on the described reconfigurable coupling scheme, in the implemented array circuit, individual, multiple adjacent, or a subsection of the array can be activated. The activated pixels can also phase and frequency couple together if activated at the same time. Furthermore, the loading effects of the deactivated cells are suppressed, thus maintaining the frequency of operation and performance in different modes of activation.

III. EXPERIMENTAL RESULTS

The chip photograph is shown in Fig. 4a. It was fabricated in a GlobalFoundries 22nm CMOS-SOI process and occupies an area of 2.3×2.3 mm^2. As shown in Fig. 4b, the chip is mounted inside a PCB opening to a silicon wafer that is attached to the back of the board. A high-resistivity

hemispherical silicon lens with 10-mm diameter is used in this measurement. The thickness of the wafer between the chip and silicon lens (L_{ext}) plays an important role in the radiation characteristic of the circuit, as will be further discussed in this section. This value is controlled by addition of extra wafers between the chip and the lens. A VDI WR4.3 horn antenna is used on the receiver side to capture the radiated power and a VDI WR4.3 MixAMC module is employed to downcovert and characterize the received power by mixing it with the 24th harmonic of an applied LO signal. Fig. 5a and 5b show the measured and simulated frequency and power consumption (per cell) of the circuit respectively in 4 different modes of activation (single cell, 1×2, 2×1, and 2×2). The frequency of the radiated power can be tuned from 222 to 235 GHz using the gate bias voltage (V_G) resulting in a 5.7% tuning range. The chip consumes 21 to 62.5 mW per activated cell from a 0.8-V supply voltage. The results in Fig. 5a and 5b verify the robustness of the reconfigurable coupling structure by showing that the circuit performance does not change noticeably in different activation modes.

Fig. 6a and 6b show the measured and simulated locking ranges of the circuit in the 1×2 (x direction) and 2×1 (y direction) activation modes respectively with application of a differential mismatch voltage (ΔV_G). It can be seen that the adjacent activated cells remain locked with $<-260mV<\Delta V_G<260mV$ and $<-120mV<\Delta V_G<120mV$ in the x and y directions respectively. They also present the relative variations of frequency (Δf_{out}) and received power (ΔP_r) compared to $\Delta V_G=0$, showing small performance variation and reasonable tolerance of the reconfigurable coupling scheme. Fig. 7b shows the measued illumination pattern for four independent activation modes (with distinct

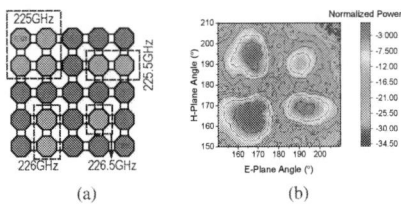

Fig. 7. (a) Simultaneous activation of 4 different independent configurations and (b) the measured illumination pattern.

978-1-6654-8495-4/22 $31.00 © 2022 IEEE

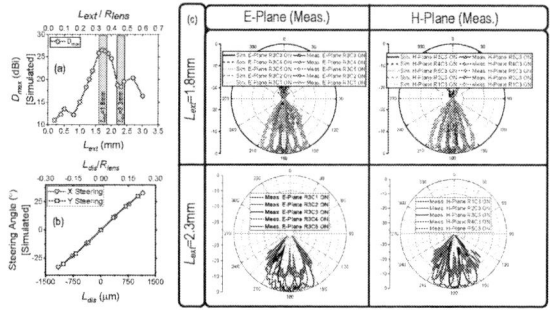

Fig. 8. Simulated (a) maximum directivity versus wafer thickness (L_{ext}), and (b) steering angle versus antenna displacement (L_{dis}) relative to the lens center. (c) Measured radiation patterns of single pixels in the E and H planes with $L_{ext} = 1.8mm$ and $L_{ext} = 2.3mm$.

Table I Comparison with the State of the Art

	Array Size	Radiation	Beam Steer. Method	Tech.	f_c (GHz)	Scan Range (°)	T.R. (%)	P_{DC} (W)	Area (mm²)
This work	5×5	Folded Slot Ant. + Si Lens (R_{lens}=5mm)	Space/ Time Mod.	22nm CMOS-SOI	230	40/43 (2D)	5.7	0.021-0.062[1]	5.29
ISSCC 2020 [2]	4×4	Patch Ant.	Phase Shifting	65nm CMOS	416	60/60 (2D)	1.7	1.45	4.1
ISSCC 2020 [4]	6×6	Folded Monopole + Si Lens (R_{lens}=5mm)	Phase Shifting	40nm CMOS	566.7	30 (1D)	0.7	1.278	0.68
ISSCC 2020 [5]	8×8	Circular Slot Ant. + Si Lens (R_{lens}=7.5mm)	Lens Coup.	130nm SiGe	420	68²(2D) (discrete)	0.7	6.9[4]	12.6
JSSC 2020 [1]	25	Folded Slot Ant. + Si Lens	R_{lens}= 12.5mm	N/A	65nm CMOS	459	0	8.9	1.18	3.94
			R_{lens}= 5mm	N/A						

[1] Per Cell   [2] 7.5° beam spacing   [3] All pixels ON

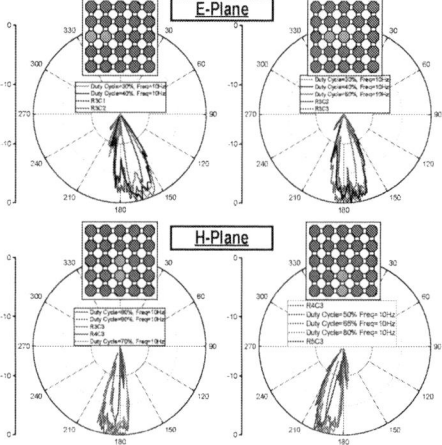

Fig. 9. Measured radiation patterns with beam steering by time modulated activation of adjacent cells in the horizontal (x) or vertical (y) directions.

frequencies) (single cell, 1×2, 2×1, and 2×2) shown in Fig. 7a. This verifies the expected reconfigurable coupling and activation within the circuit as well as the capability of the circuit for simultaneous multi-beam/-frequency radiation from the chip. It also shows how the direction of the radiated power can be controlled and the beam can be steered in 2-D by the choice of the position of the activated sub-section inside the array.

Fig. 8a shows the simulated maximum directivity (D_{max}) of single antenna versus the thickness of the wafer between chip and silicon lens (L_{ext}). It can be seen that directivity is maximized at $L_{ext} \approx 1.8mm$ and can be reduced by increasing L_{ext} to other values such as 2.3mm. This plays an important role for the beam steering behavior of the circuit. Fig. 8b shows the simulated steering angle of each antenna in the array versus its displacement from the lens center (L_{dis}) in both x and y direction, showing the spatial modulation of the radiation direction by the lens. Fig. 8c shows the measured radiation patterns of single pixels in both the E-plane and H-plane with $L_{ext} = 1.8mm$ and

2.3mm where $RmCn$ denotes the pixel in the mth row and n^{th} column. The measured results verify the expected spatial modulation characteristic of radiation in Fig. 8b and the change in directivity in Fig. 8a.

To characterize the beam steering behavior of the circuit, $L_{ext} = 2.3mm$ is chosen in order to provide sufficient overlap between radiation beams of adjacent cells so that beamforming is possible by superposition of these two beams. To perform time modulation, a set of inverted square waves with 10-Hz frequency (V_G and $\bar{V}_G$) is applied to the gate biases of the adjacent cells, therefore activating one and deactivating the other at any given time, similar to the scheme depicted in Fig. 1. By changing the duty cycle of V_G, we can control and change the portion of modulation period in which cells are ON/OFF. Examples of the recorded results are shown in Fig. 9 for both E and H planes showing that the peak of the net average power steers toward the firing angles of the individual pixels. Using this method, a total 2-D scanning range of 40°/43°is achieved. Furthermore, maximum FOV of the array can be scaled using the same method by increasing the size of the array and expanding the lens-related spatial modulation. A comparison between this work and the state of the art is presented in Table I, showing extended, scalable and continuous scanning range compared to other similar lens-coupled implementations.

REFERENCES

[1] Hossein Jalili, Omeed Momeni, "A 0.46-THz 25-Element Scalable and Wideband Radiator Array With Optimized Lens Integration in 65-nm CMOS," *IEEE Journal of Solid-State Circuits*, vol. 55, no. 9, pp. 2387 - 2400, Sept. 2020.

[2] H. Saeidi, et. al., "29.9 A 4×4 Distributed Multi-Layer Oscillator Network for Harmonic Injection and THz Beamforming with 14dBm EIRP at 416GHz in a Lensless 65nm CMOS IC," *IEEE ISSCC, 2020*.

[3] Hossein Jalili, Omeed Momeni, "23.2 A 436-to-467GHz Lens-Integrated Reconfigurable Radiating Source with Continuous 2D Steering and Multi-Beam Operations in 65nm CMOS," *IEEE ISSCC, 2021*.

[4] Kaizhe Guo; Patrick Reynaert, "29.2 A 0.59THz Beam-Steerable Coherent Radiator Array with 1mW Radiated Power and 24.1dBm EIRP in 40nm CMOS," *IEEE ISSCC, 2020*.

[5] R. Jain, P. Hillger, J. Grzyb and U. R. Pfeiffer, "29.1 A 0.42THz 9.2dBm 64-Pixel Source-Array SoC with Spatial Modulation Diversity for Computational Terahertz Imaging," *IEEE ISSCC, 2020*.

Gap in pagination due to withheld paper.

Pages 465-468

A 0.9-nA Temperature-Independent 565-ppm/°C Self-Biased Current Reference in 22-nm FDSOI

Martin Lefebvre, Denis Flandre and David Bol
ICTEAM Institute, Université catholique de Louvain, Louvain-la-Neuve, Belgium
Email: {martin.lefebvre, denis.flandre, david.bol}@uclouvain.be

Abstract—**Temperature-independent current references operating in the nA range are rarely area-efficient due to the use of large resistors occupying a significant silicon area at this current level. In this paper, we introduce a nA-range constant-with-temperature (CWT) current reference relying on a self-cascode MOSFET (SCM), biased by a proportional-to-absolute-temperature voltage with a CWT offset voltage. The proposed reference has been fabricated in a 22-nm fully-depleted silicon-on-insulator (FDSOI) technology and, as a result of using an SCM, occupies a silicon area of 0.0132 mm^2 at least 4× smaller than state-of-the-art CWT references operating in the same current range. It consumes 5.8 nW at 0.9 V and achieves a 0.9-nA current with a line sensitivity of 0.39 %/V and a temperature coefficient of 565 ppm/°C.**

I. INTRODUCTION

Designing current references that are both area-efficient and robust to process, voltage and temperature (PVT) variations is a challenging task, which is not equally difficult depending on the target current. For example, Fig. 1(a) highlights the absence of area-efficient solutions for the generation of a constant-with-temperature (CWT) nA-range current. This gap originates from the fact that conventional CWT current references [Fig. 1(b)] consist in applying a reference voltage to a voltage-to-current converter, which is usually implemented by a gate-leakage transistor [1] or a resistor [2], [3]. These converters are respectively well suited to the generation of a pA- or μA-range current, in the sense that they occupy a reasonable silicon area at this current level. Unfortunately, the generation of a nA-range current would lead to a much larger area, as the resistance of the gate-leakage transistor (resp. resistor) needs to be decreased (resp. increased) by placing devices in parallel (resp. series). Besides, proportional-to-absolute-temperature (PTAT) current references operating in the nA range, such as [4], are area-efficient because they rely on a self-cascode MOSFET (SCM) biased with a PTAT voltage [Fig. 2(a)], hence generating a current proportional to a so-called specific sheet current and thus, to temperature.

In this work, we propose a nA-range CWT current reference based on an SCM, which differs from [4] by the addition of a CWT offset to the PTAT voltage biasing the SCM [Fig. 2(b)]. This offset voltage is obtained as the threshold voltage difference ΔV_T between two transistors of the same type, one of them being reverse body-biased to reduce its threshold voltage. The current reference is fabricated in GlobalFoundries (GF) 22-nm fully-depleted silicon-on-insulator (FDSOI) CMOS

This work was supported by the Fonds de la Recherche Scientifique (FRS-FNRS) of Belgium under grant n°CDR J.0014.20.

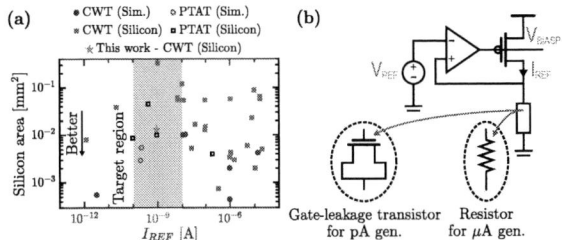

Fig. 1. (a) Trade-off between area and current, based on prior art, highlighting the absence of area-efficient solutions for the generation of a CWT nA-range current. (b) Conventional CWT current references are based on a reference voltage applied to a gate-leakage transistor or a resistor, which are respectively well suited to the generation of a pA- or μA-range current.

technology. Measurements over 20 dies demonstrate a 0.9-nA current with a 0.39-%/V line sensitivity (LS) from 0.9 to 1.8 V, a 565-ppm/°C temperature coefficient (TC) from -40 to 85°C, and a 9.20-% variability (σ/μ) due to combined mismatch and process variations at ambient temperature. The reference consumes 5.8 nW at $V_{DD,\min} = 0.9$ V and 25°C, while occupying a silicon area of 0.0132 mm^2. The paper is structured as follows. Section II explains the operation principle of the proposed current reference based on its governing equations. Sections III and IV respectively detail the methodology used to size the reference, and discuss the measurement results. Section V finally offers some concluding remarks.

II. OPERATION PRINCIPLE AND GOVERNING EQUATIONS

To generate a nA-range current, the proposed reference [Fig. 2(b)] relies on a moderate-inversion SCM formed by M_{1-2}, requiring the advanced compact MOSFET (ACM) model [5] to describe the transistor I-V relationship. In this model, the drain current is given by

$$I_D = I_{SQ} S \left(i_f - i_r \right), \tag{1}$$

where $I_{SQ} = \frac{1}{2}\mu C'_{ox} n U_T^2$ is the specific sheet current, μ the carrier mobility, C'_{ox} the normalized gate oxide capacitance, n the subthreshold slope factor, U_T the thermal voltage, $S = W/L$ the transistor aspect ratio, and i_f, i_r the forward and reverse inversion levels. In addition, the relationship between inversion level, gate and source/drain voltages is captured by

$$V_P - V_{S/D} = U_T \left[\sqrt{1 + i_{f/r}} - 2 + \log\left(\sqrt{1 + i_{f/r}} - 1\right) \right], \tag{2}$$

where $V_P = \frac{V_G - V_{T0}}{n}$ is the pinch-off voltage, V_{T0} is the threshold voltage at zero V_{BS}, and voltages are referred to the body of the transistor, which is actually the transistor's

978-1-6654-8495-4/22 $31.00 © 2022 IEEE

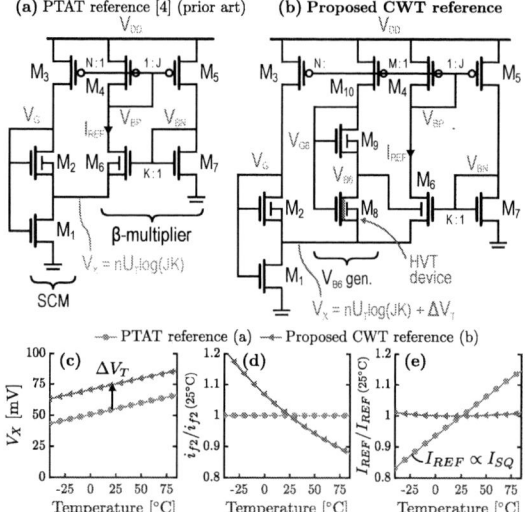

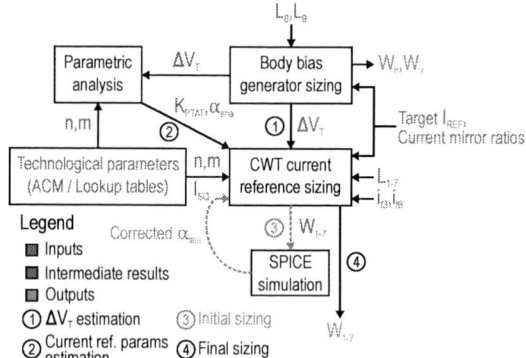

Fig. 3. Four-step flowchart of the design and sizing methodology.

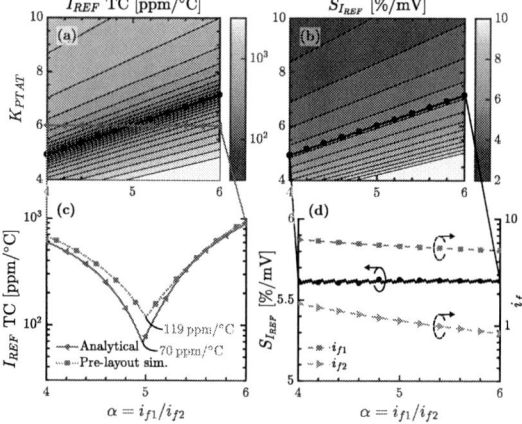

Fig. 2. The proposed circuit generates a CWT current by biasing an SCM using a PTAT voltage with a CWT offset. (a) PTAT reference generating a current proportional to I_{SQ} [4]. (b) Proposed CWT reference, with M_6 biased with $V_{BS6} > 0$, leading to $\Delta V_T = V_{T7} - V_{T6} = \gamma_b V_{BS6} > 0$. Analytical equations for (c) the voltage applied to the SCM, (d) the inversion level of M_2, and (e) the reference current as a function of temperature, with (d) and (e) normalized by their value at 25°C.

Fig. 4. Analytical sizing of the current reference for $\Delta V_T = 17.6$ mV, $m = 1$ and $n = 1.25$, computed by steps 1 and 2 of the sizing methodology. (a) I_{REF} TC and (b) $S_{I_{REF}}$ as a function of K_{PTAT} and α. (c) I_{REF} TC vs. α for $K_{PTAT} = 6$, based on analytical and pre-layout simulation results. (d) $S_{I_{REF}}$ and inversion level of M_{1-2} along the I_{REF} TC valley.

back gate in FDSOI. Applying these equations to the SCM and defining $\alpha \triangleq i_{f1}/i_{f2}$, voltage V_X is given by

$$V_X = nU_T \left[\left(\sqrt{1 + \alpha i_{f2}} - \sqrt{1 + i_{f2}} \right) + \log \left(\frac{\sqrt{1 + \alpha i_{f2}} - 1}{\sqrt{1 + i_{f2}} - 1} \right) \right]. \quad (3)$$

Note that the current flowing from M_6 (and M_8) only impacts the required ratio S_1/S_2. Next, applying the ACM equations to the β-multiplier formed by M_{6-7} and assuming these transistors operate in weak inversion, i.e., $i_f \ll 1$, leads to

$$V_X = \Delta V_T + nU_T \log (K_{PTAT}), \quad (4)$$

where $K_{PTAT} = JK$, J and K being current mirror ratios. With the effect of reverse body bias on the threshold voltage modeled as $V_T = V_{T0} - \gamma_b V_{BS}$, where γ_b is the body factor, we obtain $\Delta V_T = \gamma_b V_{BS6}$.

In Fig. 2(b), our work differs from [4] by adding a threshold voltage difference $\Delta V_T = V_{T7} - V_{T6}$ to V_X, obtained by biasing the body of M_6 with a non-zero V_{BS6}. Voltage V_{BS6} is generated by a voltage reference formed by M_{8-10} relying on the V_T difference between high- and regular-V_T (RVT) devices, and can be made temperature-independent by tuning the ratio S_9/S_8. Given that γ_b does not depend on temperature at first order in FDSOI, ΔV_T is a CWT offset voltage. Then, the reference current is found by applying (1) to M_2, giving

$$I_{REF}(T) = I_{SQ2}(T) \, i_{f2}(T) \, (S_2/N), \quad (5)$$

in which $I_{SQ2}(T) \propto T^{2-m}$, assuming $\mu = \mu(T_0)(T/T_0)^{-m}$ with m the temperature exponent of carrier mobility. The temperature dependence of i_{f2} is determined by solving the non-linear relation found by equating (3) and (4). Using a purely PTAT voltage as in [4], i_{f2} is temperature-independent and leads to $I_{REF} \propto I_{SQ}$ [Figs. 2(c) to 2(e)]. In the

proposed reference, we add a CWT offset to the PTAT voltage [Fig. 2(c)], causing i_{f2} to vary with temperature. With a proper choice of K_{PTAT} and α, the temperature dependence of i_{f2} [Fig. 2(d)] can compensate that of I_{SQ2}, making I_{REF} temperature-independent [Fig. 2(e)]. Next, we define two important characteristics of the SCM, namely the sensitivity of I_{REF} to V_X computed as $S_{I_{REF}} = (1/I_{REF})(dI_{REF}/dV_X)$, which depends on i_{f2} through

$$S_{I_{REF}} = \frac{2}{i_{f2} n U_T} \left[\frac{\alpha}{\sqrt{1 + \alpha i_{f2}} - 1} - \frac{1}{\sqrt{1 + i_{f2}} - 1} \right]^{-1}, \quad (6)$$

and the required ratio of M_{1-2} aspect ratios, i.e.,

$$\frac{S_1}{S_2} = \frac{I_{SQ2}}{I_{SQ1}} \frac{1 + M + N}{N} \frac{1}{\alpha - \beta}, \quad (7)$$

where $\beta \triangleq i_{r1}/i_{f2}$. Finally, in what follows, the LS and TC are computed using the box method, i.e.,

$$\text{LS} = \frac{(I_{REF,\max} - I_{REF,\min})}{I_{REF,avg} (V_{DD,\max} - V_{DD,\min})} \times 100 \, \%/\text{V}, \quad (8)$$

$$\text{TC} = \frac{(I_{REF,\max} - I_{REF,\min})}{I_{REF,avg} (T_{\max} - T_{\min})} \times 10^6 \, \text{ppm}/°\text{C}. \quad (9)$$

978-1-6654-8495-4/22 $31.00 © 2022 IEEE

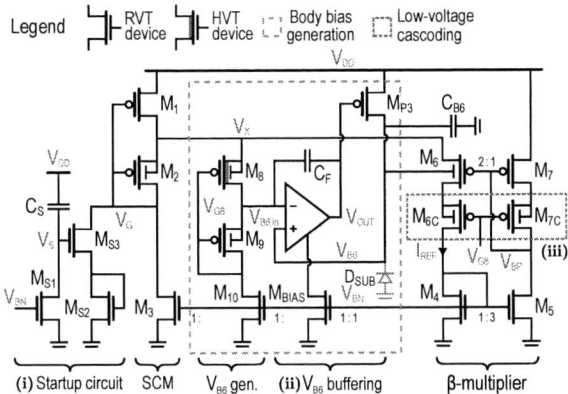

Legend: RVT device | HVT device | Body bias generation | Low-voltage cascoding

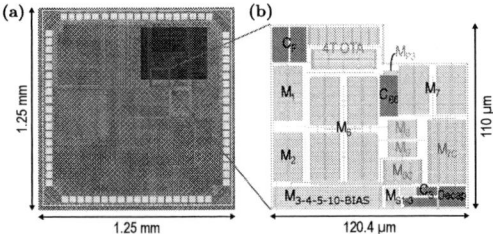

(i) Startup circuit | SCM | V_{B6} gen. | (ii) V_{B6} buffering | β-multiplier

Fig. 5. Schematic of the proposed circuit with a pMOS SCM implemented in GF 22-nm FDSOI, which includes a startup circuit, a low-voltage cascode, and the generation and buffering of the reverse body-bias voltage of M_6.

Fig. 6. (a) Chip microphotograph with overlaid layout and (b) layout of the proposed current reference in GF 22-nm FDSOI.

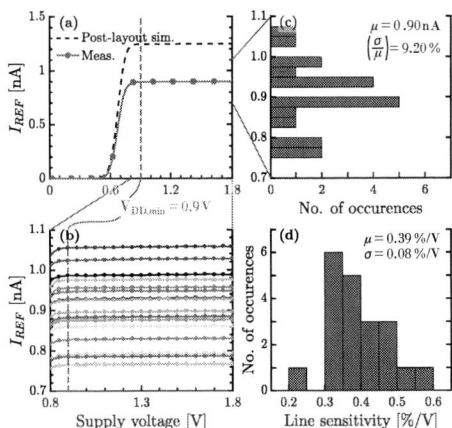

Fig. 7. (a) Post-layout-simulated and measured average I_{REF} at 25°C, with (b) details of the 20 dies. Measured histograms of (c) I_{REF} and (d) LS.

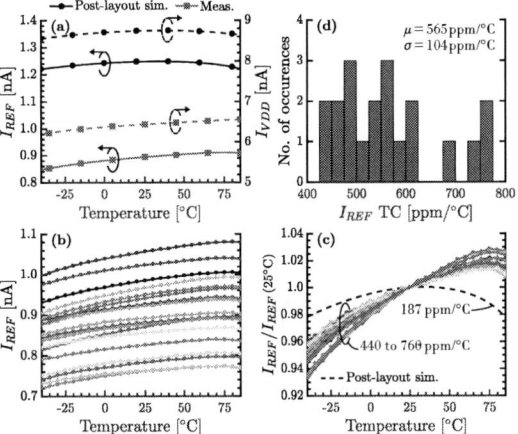

Fig. 8. (a) Post-layout-simulated and measured temperature dependence of the average I_{REF} and I_{VDD}. Measured temperature dependence of I_{REF} (b) without and (c) with normalization by the value at 25°C, for the 20 dies. (d) Measured histogram of I_{REF} TC.

III. DESIGN AND SIZING METHODOLOGY

The design and sizing methodology is summarized by the flowchart in Fig. 3 and can be divided into four main steps: (i) the V_{B6} generator sizing, determining the CWT offset ΔV_T, (ii) the parametric analysis providing the values of (K_{PTAT}, α) that minimize the TC of I_{REF}, (iii) an initial sizing of the current reference for different values of α, based on the equations in Section II, and (iv) a final sizing based on the value of α minimizing the TC of I_{REF} in pre-layout simulations. Fig. 4(a) demonstrates the existence of an I_{REF} TC valley, corresponding to a linear relationship between K_{PTAT} and α, but also to an iso-sensitivity curve in Fig. 4(b), in which $S_{I_{REF}}$ is computed from (6). Note that only a change in ΔV_T can change the location of this valley. For the choice of $K_{PTAT} = 6$ made for our design [Fig. 4(c)], a purely analytical approach yields a 70-ppm/°C TC for $\alpha = 4.98$, which is in line with the pre-layout simulation results based on the initial sizing, giving a 121-ppm/°C TC for $\alpha = 5$. Nevertheless, the initial sizing and subsequent SPICE simulations are performed for a range of α values, to ensure that the minimum TC location has been correctly identified. Moreover, Fig. 4(d) proves that along the I_{REF} TC valley, $S_{I_{REF}}$ remains constant around 5.6 %/mV while i_{f1} and i_{f2} at 25°C tend to decrease for larger values of (K_{PTAT}, α) but remain in moderate inversion, i.e., $i_f \in [1; 100]$.

Fig. 5 depicts the complete schematic of the proposed reference, implemented with RVT pMOS transistors to limit the area overhead entailed by the different body voltages. Several elements have been added to this schematic: (i) a dynamic startup circuit, (ii) a simple 6T buffering stage for V_{B6}, to cope with the leakage of the parasitic nwell/psub diode D_{SUB} at high temperature, and (iii) a low-voltage cascode to improve the LS by reducing the impact of M_6 V_{DS} variations.

IV. MEASUREMENT RESULTS

The proposed reference was manufactured in GF 22-nm FDSOI as part of the ICare biomedical microcontroller unit. The chip microphotograph is shown in Fig. 6(a), with the reference layout in Fig. 6(b). Currents were measured with a Keithley K2636A SMU, and an Espec SH-261 climatic chamber was used to sweep temperature from -40 to 85°C.

Figs. 7 and 8 respectively depict the measured dependence of I_{REF} to supply voltage and temperature variations, while Table I compares our simulation and silicon results to the state of the art. First, Fig. 7(a) represents I_{REF} as a function

TABLE I
COMPARISON TABLE OF TEMPERATURE-INDEPENDENT NA-RANGE CURRENT REFERENCES.

	Huang [6]	Far [7]	Kim [8]	Cordova [9]	Aminzadeh [10]	Kayahan [11]	Dong [12]	Ji [13]	Wang [14]	Wang [2]	Huang [3]	Lee [15]	Lefebvre **This work**
Publication Year	ISCAS 2010	ROPEC 2015	ISCAS 2016	ISCAS 2017	AEU 2022	TCAS-I 2013	ESSCIRC 2017	ISSCC 2017	VLSI-DAT 2019	TCAS-I 2019	TCAS-II 2020	JSSC 2020	ESSCIRC **2022**
Type of work	Sim.	Sim.	Sim.	Sim.	Sim.	Silicon	Silicon	Silicon	Silicon	Silicon	Silicon	Silicon	Sim. / Silicon
Number of samples	N/A	N/A	N/A	N/A	N/A	90	32*	10	10	16	10	10	N/A / 20
Technology	$0.18\mu m$	$0.18\mu m$	$0.13\mu m$	$0.18\mu m$	$0.18\mu m$	$0.35\mu m$	$0.18\mu m$	$0.18\mu m$	$0.18\mu m$	$0.18\mu m$	$0.18\mu m$	$0.18\mu m$	22nm FDSOI
I_{REF} [nA]	2.05	14	27	10.86	6.7	25	35.02	6.64	6.46	9.77	11.6	1	1.25 / 0.9
Power at $V_{DD,min}$ [nW]	5.1	150	N/A	30.5	51	28500	1.02	9.3	15.8	28	48.64	4.5/14^b	7.84 / 5.81
Area [mm²]	N/A	0.0102	N/A	0.01	0.46	0.0053	0.0169	0.055	0.062	0.055	0.054	0.332	0.0132
Supply range [V]	0.85 – 2.2	1 – 3.3	1.2	0.9 – 1.8	1.1 – 1.8	5	1.5 – 2.5	1.3 – 1.8	0.85 – 2	0.7 – 1.2	0.8 – 2	1.5 – 2	0.9 – 1.8
LS [%/V]	1.35	0.1	N/A	0.54	0.03	150	3	1.16	4.15	0.6	1.08	1.4	0.26 / 0.39^d
Temperature range [°C]	0 – 150	0 – 70	-30 – 150	-20 – 120	-40 – 120	0 – 80	-40 – 120	0 – 110	-10 – 100	-40 – 125	-40 – 120	-20 – 80	-40 – 85
TC [ppm/°C]	91	20	327	108	40.33	128/250*	282	680/283†	138‡	149.8	169	289/265^b	187 / 565^d
I_{REF} var. (process) [%]	7.5	N/A	3.7	15.8/11.6†	N/A	8/1.22*	4.7	N/A	N/A	+11.7/-8.7^d / +17.6/-10.3^d	N/A	+9.9/-9.5	9.20
I_{REF} var. (mismatch) [%]	N/A	5.8	N/A	N/A	0.70	1.4 (sim.)	1.6	4.07/1.19*	3.33	1.6	4.3	1.26/0.25†	6.39
Trimming	No	No	No	Yes	No	No	No	No	Yes	No	Yes	Yes	No / No

* 16 dies in TT corner, 4 dies for each FF, SS, SF and FS corner. * Simulated and measured values. † Before and after trimming. ‡ Best TC, the average/median one is not reported. $^\circ$ Estimated from figures. ^{d} Mean measured value across the 20 dies. ^{b} For 25 and 2.5 minutes between two calibrations.

---

of the supply voltage, and exhibits a 0.9-V $V_{DD,min}$ and 0.9-nA I_{REF}, compared to 1.25 nA in simulation, likely due to global process variations. Details of the 20 dies are shown in Fig. 7(b), with I_{REF} variability characterized in Fig. 7(c). The measured variability of 9.20 % is larger than the 6.39-% simulated one due to global die-to-die (D2D) process variations, whereas only local mismatch is accounted for in simulation. Lastly, Fig. 7(d) illustrates the histogram of LS and presents an average of 0.39 %/V, comparable with the 0.26 %/V obtained in simulation. Second, Fig. 8(a) illustrates the evolution of I_{REF} and the supply current I_{VDD} with temperature. I_{REF} presents a 2nd order behavior, with a slightly PTAT trend in measurement, and the current reference consumes 8.8 nA (resp. 6.4 nA) at 1.8 V in simulation (resp. in measurement). The temperature dependence of I_{REF} across the 20 dies is depicted in absolute value in Fig. 8(b) and normalized by the value of I_{REF} at 25°C in Fig. 8(c). In addition, Fig. 8(c) highlights the TC increase from 187 ppm/°C in simulation to 565 ppm/°C in measurement due to global D2D process variations, an issue which could be solved by adding a calibration scheme to the proposed reference. Finally, Fig. 8(d) shows that the TC has a standard deviation of 104 ppm/°C and reaches a minimum value of 440 ppm/°C.

V. CONCLUSION

We demonstrated a 0.9-nA current reference in 22-nm FDSOI, which relies on an SCM biased by a PTAT voltage with a CWT offset, realized as the threshold voltage difference between two transistors of the same type. This structure consumes 5.8 nW at 0.9 V and leads to a 0.0132-mm² area, a 0.39-%/V LS thanks to cascoding, and a 565-ppm/°C TC. Its main drawback is the relatively large 9.20-% variability due to its sensitivity to mismatch and process variations, which calls for a TC calibration scheme to be added in further work.

ACKNOWLEDGMENTS

The authors would like to thank Pierre Gérard for the measurement testbench, Eléonore Masarweh for the micropho-tograph, and ECS group members for their proofreading.

REFERENCES

[1] H. Wang and P. P. Mercier, "A 3.4-pW 0.4-V 469.3ppm/°C Five-Transistor Current Reference Generator," *IEEE Solid-State Circuits Lett.*, vol. 1, no. 5, pp. 122–125, May 2018.

[2] L. Wang and C. Zhan, "A 0.7-V 28-nW CMOS Subthreshold Voltage and Current Reference in One Simple Circuit," *IEEE Trans. Circuits Syst. I, Reg. Papers*, vol. 66, no. 9, pp. 3457–3466, Aug. 2019.

[3] Q. Huang, C. Zhan, L. Wang et al., "A -40 °C to 120 °C, 169 ppm/°C Nano-Ampere CMOS Current Reference," *IEEE Trans. Circuits Syst. II, Exp. Briefs*, vol. 67, no. 9, pp. 1494–1498, Sept. 2020.

[4] E. M. Camacho-Galeano, C. Galup-Montoro, and M. C. Schneider, "A 2-nW 1.1-V Self-Biased Current Reference in CMOS Technology," *IEEE Trans. Circuits Syst. II, Exp. Briefs*, vol. 52, no. 2, pp. 61–65, Feb. 2005.

[5] A. I. A. Cunha, M. C. Schneider, and C. Galup-Montoro, "An MOS Transistor Model for Analog Circuit Design," *IEEE J. Solid-State Circuits*, vol. 33, no. 10, pp. 1510–1519, Oct. 1998.

[6] Z. Huang, Q. Luo, and Y. Inoue, "A CMOS Sub-1V Nanopower Current and Voltage Reference with Leakage Compensation," in *Proc. IEEE Int. Symp. Circuits Syst.*, 2010, pp. 4069–4072.

[7] A. Far, "Subthreshold Current Reference Suitable for Energy Harvesting: 20ppm/°C and 0.1%/V at 140nW," in *2015 IEEE Int. Autumn Meet. Power Electron. Comput.*, 2015, pp. 1–4.

[8] T. Kim, T. Briant, C. Han et al., "A Nano-Ampere 2nd Order Temperature-Compensated CMOS Current Reference Using Only Single Resistor for Wide-Temperature Range Applications," in *Proc. IEEE Int. Symp. Circuits Syst.*, 2016, pp. 510–513.

[9] D. Cordova, A. C. de Oliveira, P. Toledo et al., "A Sub-1 V, Nanopower, ZTC Based Zero-V_T Temperature-Compensated Current Reference," in *Proc. IEEE Int. Symp. Circuits Syst.*, 2017, pp. 1–4.

[10] H. Aminzadeh and M. M. Valinezhad, "A Nano-Power Sub-Bandgap Voltage and Current Reference Topology with No Amplifier," *AEU - Int. J. Electron. Commun.*, vol. 148, p. 154174, Mar. 2022.

[11] H. Kayahan, Ö. Ceylan, M. Yazici et al., "Wide Range, Process and Temperature Compensated Voltage Controlled Current Source," *IEEE Trans. Circuits Syst. I, Reg. Papers*, vol. 60, no. 5, pp. 1345–1353, Mar. 2013.

[12] Q. Dong, I. Lee, K. Yang et al., "A 1.02 nW PMOS-Only, Trim-Free Current Reference with 282ppm/°C from -40°C to 120°C and 1.6% within-wafer inaccuracy," in *IEEE 43rd Eur. Solid-State Circuits Conf.*, 2017, pp. 19–22.

[13] Y. Ji, C. Jeon, H. Son et al., "5.8 A 9.3 nW All-in-One Bandgap Voltage and Current Reference Circuit," in *Proc. IEEE Int. Solid-State Circuits Conf.*, 2017, pp. 100–101.

[14] J. Wang and H. Shinohara, "A CMOS 0.85-V 15.8-nW Current and Voltage Reference without Resistors," in *2019 Int. Symp. VLSI Design Autom. Test*, 2019, pp. 1–4.

[15] S. Lee, S. Heinrich-Barna, K. Noh et al., "A 1-nA 4.5-nW 289-ppm/°C Current Reference Using Automatic Calibration," *IEEE J. Solid-State Circuits*, vol. 55, no. 9, pp. 2498–2512, Sept. 2020.

A Compact 2.5-nJ Energy/Conversion NPN-Based Temperature-to-Digital Converter with a Fully Current-Mode Processing Architecture

Antonio Aprile*, Michele Folz†, Daniele Gardino†, Piero Malcovati*, Edoardo Bonizzoni*

*Department of Electrical, Computer and Biomedical Engineering, University of Pavia, Italy
†TDK InvenSense, Assago, Italy

Abstract—This paper presents a NPN-based Temperature-to-Digital Converter (TDC) designed for thermal drift compensation of MEMS devices. The circuit, featuring a fully current-mode processing architecture, has been fabricated in a standard 180-nm CMOS technology, with a silicon area of 0.08 mm², and consumes 2.5 nJ of energy per conversion, with a nominal supply voltage of 1.8 V. The measured resolution of the TDC is equal to 158 mK and it exhibits a worst case inaccuracy of 1.39 % in the −20 °C to 80 °C temperature range.

I. INTRODUCTION

The Micro-Electro-Mechanical-Systems (MEMS) field has experienced a remarkable growth in the last years, due to the increasing exploitation of MEMS devices in everyday life sensors, supported by a concurrent technological development, thus driving the research target toward systems with higher robustness with respect to environmental effects. One of the main challenges in this framework is, indeed, to overcome the impact of the ambient temperature on the performance of MEMS devices, considering that the micro-structures used as sensing elements are characterized by particularly fast temperature variations, that result in a significant degradation of the reliability of the sensed quantity. Therefore, in order to cancel this effect in high precision applications, MEMS devices require a responsive and energy-efficient monitoring of the temperature, to compensate for the drift of the parameters of interest [1]–[3].

This paper presents a NPN-based Temperature-to-Digital Converter (TDC) designed for this application and fabricated in a standard 0.18-μm CMOS process. Since it needs to operate within a compact battery-supplied system, the main targets of this work have been the minimization of the energy per conversion and of the occupied silicon area. The resulting circuit features a resolution of 158 mK and a worst case inaccuracy of 1.39 % in the −20 °C to 80 °C temperature range. With a nominal supply voltage of 1.8 V, the measured energy per conversion is as low as 2.5 nJ and the whole sensor occupies 0.08 mm² of silicon area.

II. ARCHITECTURE AND CIRCUIT DESCRIPTION

The proposed TDC is based on a fully current-mode processing approach: except for the digital output stage, every signal is in the form of an electrical current. Considering the extreme simplicity in summing and scaling signals in the

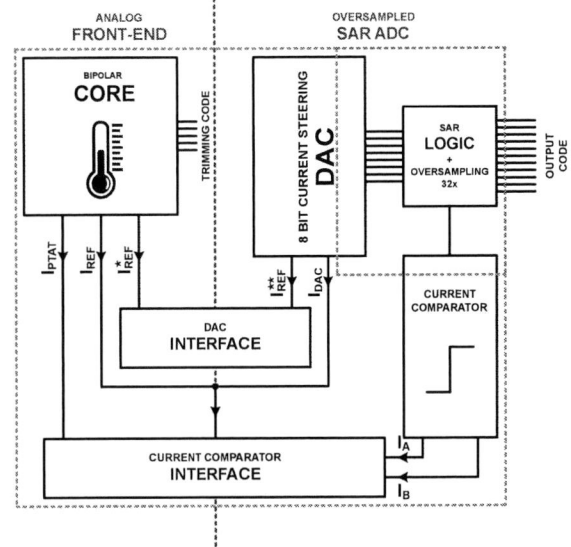

Figure 1. Block diagram of the proposed TDC

current domain, this approach allows sparing several adding and amplifying circuits, inherently required in its voltage-domain counterpart. Moreover, it opens the possibility to operate with reduced supply voltage constraints, making it suitable for more scaled fabrication technologies.

Fig. 1 shows the block diagram of the TDC. It consists of an Analog Front-End (AFE) circuit followed by an oversampled SAR ADC. Two dedicated interface circuits connect the AFE circuit to the ADC. The main block of the AFE circuit is the bipolar core, which is responsible for the generation of all the signals needed for the temperature-to-digital conversion: a proportional to absolute temperature (PTAT) signal, which contains the information to be converted, and two differently weighted replicas of a reference signal, with respect to which the A/D conversion is carried out. The ADC consists of an 8-bit current steering DAC, a current comparator and a SAR logic circuit, whose output code is decimated to achieve the desired resolution. The majority of the blocks process signals in the current domain (green frame), while just the logic part (blue frame) operates in the voltage domain.

978-1-6654-8495-4/22 $31.00 © 2022 IEEE

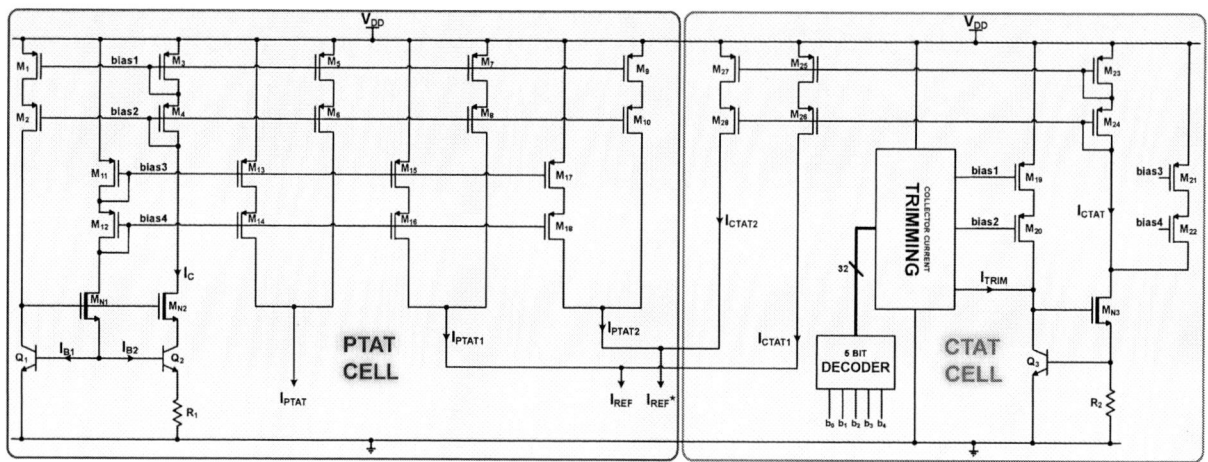

Figure 2. Schematic diagram of the bipolar core circuit

A. Analog Front-End Circuit

Fig. 2 shows the schematic diagram of the bipolar core; it consists of two NPN-based sections called *PTAT Cell* and *CTAT Cell*, respectively. The first provides three replicas (I_{PTAT}, I_{PTAT1}, I_{PTAT2}) of a PTAT current, while the second provides two replicas (I_{CTAT1}, I_{CTAT2}) of a complementary to absolute temperature (CTAT) current. Transistors Q_1 and Q_2 have been designed with a 1 : 8 emitter area ratio, implying that the current flowing into R_1 is PTAT. The value of the β factor of the bipolar transistors available within the adopted technology lies between 6 and 7, resulting in a quite poor performance of these devices. This makes a base current recovery mechanism necessary, considering that the linearity over temperature of the signal $\Delta V_{BE}/R_1$ degrades significantly at the collector, due to the thermal behavior of the base current and to its relative weight. Therefore, a reconstruction scheme has been implemented to make a replica of the emitter current of Q_2 available to be processed. The native transistors M_{N1} and M_{N2} have been used as current buffers and to bias the collectors of both Q_1 and Q_2 at the same voltage. Transistor M_{N1} senses the sum of I_{B1} and I_{B2}, while M_{N2} senses I_C. In both cases, a cascode structure mirrors these contributions, which are then recombined at the output with a proper scaling factor (α), according to

$$I_{PTAT} = \alpha I_C + \frac{\alpha}{2}\left(I_{B1} + I_{B2}\right), \qquad (1)$$

where the additional scaling factor $1/2$ for the second term has been selected to implement the average of I_{B1} and I_{B2}. This approach has been adopted also for the generation of I_{PTAT1} and I_{PTAT2}, which are replicas of I_{PTAT}, needed for the generation of the reference currents for the ADC.

Current I_C is also mirrored (M_3-M_4/M_{19}-M_{20}) to bias Q_3, the NPN transistor on which the *CTAT Cell* is based. In addition to this, a 5-bit trimming structure has been implemented to make this biasing current programmable, according to the idea presented in [4]. Moreover, the same base current recovery

contribution discussed above is mirrored (M_{11}-M_{12}/M_{21}-M_{22}) and injected into the drain of M_{N3}, to make current I_{CTAT} equal to V_{BE}/R_2. Two properly-weighted replicas of I_{CTAT} (I_{CTAT1} and I_{CTAT2}) are obtained by means of two additional cascode current mirrors and summed to the corresponding PTAT signals, to obtain the two reference signals (I_{REF} and $I_{REF}^\star$) required for the temperature-to-digital conversion.

B. Temperature-to-Digital Conversion

Referring to Fig. 1, the signals generated by the bipolar core circuit are collected by two interface circuits, which are basically nMOS current mirrors. Consequently, $I_{REF}^{\star\star}$ is a scaled version of $I_{REF}^\star$, while I_A and I_B are replicas of I_{PTAT} and I_{REF} + I_{DAC}, respectively, I_{DAC} being the DAC output current. The impedances of the output branches of the current comparator interface circuit have been designed to be identical, to avoid a detrimental systematic offset in the comparator [5].

Current $I_{REF}^{\star\star}$ is provided as reference to a pMOS 8-bit current steering DAC, whose schematic diagram is shown in Fig. 3, in which every branch is marked with its mirroring factor with respect to the reference. This circuit is controlled by the binary code provided by the SAR logic: depending on the configuration of ($b_0, b_1, \cdots, b_7$), a larger or smaller replica of $I_{REF}^{\star\star}$ is provided as output current (I_{DAC}), while a smaller or larger replica is absorbed by a diode connected transistor. To optimize the area of the circuit and the matching among the branches corresponding to the different binary weights, this mechanism has been implemented with the three sections highlighted in Fig. 3. The *Split Stage* block has been introduced to obtain a size restoration passing from the *MSBs* to the *LSBs* section. The maximum required mirroring factor with this solution (16) is indeed 16 times smaller with respect to a standard 8-bit binary approach (256). Moreover, with this solution, the matching requirements of the DAC have been achieved with more ease.

The operation of the ADC is based on a successive approximation algorithm: the subsequent decisions of the current

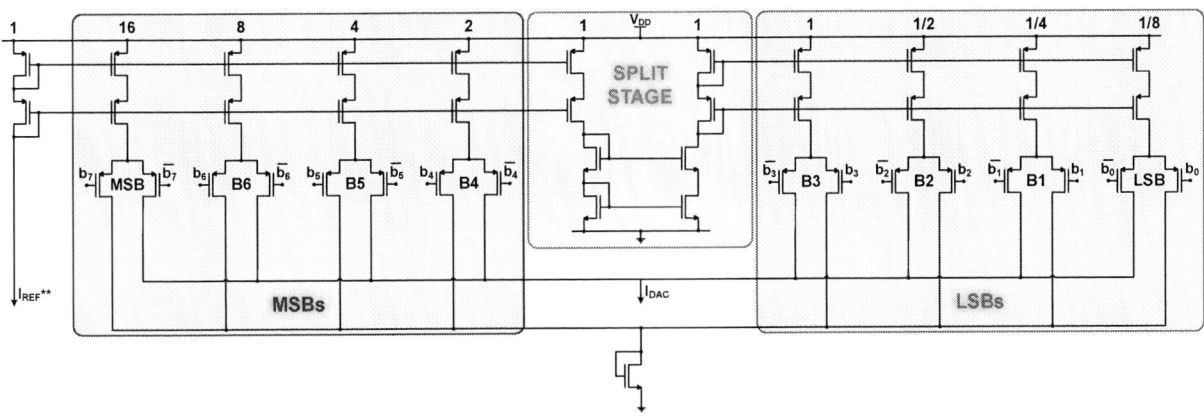

Figure 3. pMOS cascode current-steering DAC with split stage

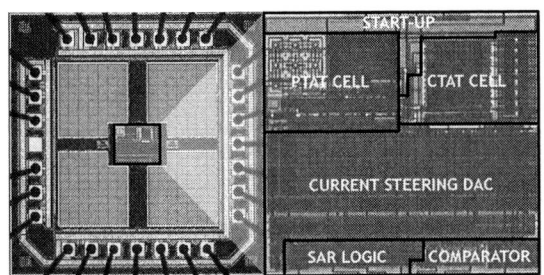

Figure 4. Chip micrograph with magnified view of the TDC active area

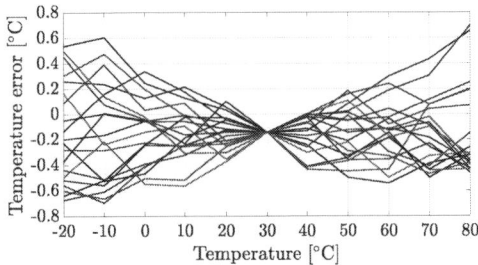

Figure 5. Inaccuracy of the TDC (20 samples) in the temperature range −20 °C to 80 °C

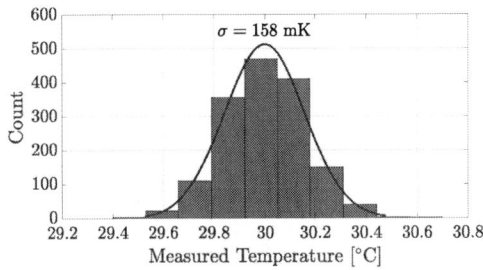

Figure 6. Temperature resolution evaluation considering and oversampling factor of 32

comparator about the sign of the difference $I_A - I_B$ drive the SAR logic block, which controls the 8-bit current steering DAC, with the purpose of tracking the input PTAT signal downshifted by I_{REF}. After 8 iterations of this procedure, the output code is obtained, according to

$$D_{OUT} \propto \frac{I_{PTAT} - I_{REF}}{I_{REF}} = \left(\frac{I_{PTAT}}{I_{REF}} - 1 \right). \qquad (2)$$

To achieve higher resolution, exploiting oversampling, the intrinsic 8-bit digital codes are decimated and filtered. Even considering a fixed temperature, indeed, D_{OUT} is still affected by an uncertainty due to the presence of electronic noise. This uncertainty allows achieving a more refined code on the basis of the results of multiple conversions on which decimation and filtering are carried out. In the proposed TDC, the 8-bit output codes are oversampled by a factor 32.

III. MEASUREMENT RESULTS

The proposed TDC has been fabricated in a 180-nm CMOS process. The chip area, including pads, is 1700 μm × 1700 μm, but the active area of the TDC is just 268 μm × 298 μm. Fig. 4 shows the chip micrograph and a magnified view of the active area in which the main circuit blocks have been highlighted.

We characterized 20 different samples in a thermal chamber, in which a temperature sweep from −20 °C to 80 °C with steps of 10 °C has been carried out. The collected temperature characteristics have been analyzed and the measured samples have been trimmed, exploiting the 5-bit programmability of the collector current of Q_3. Besides the measurement at room temperature (30 °C), used to achieve offset compensation, the TDCs have been trimmed on the basis of the measurement at the upper bound of the temperature range (80 °C). The resulting temperature inaccuracy is reported in Fig. 5. The worst case error across the whole temperature range, considering the complete set of analyzed samples, turns out to be equal to ±0.697 K, which corresponds to a relative inaccuracy of 1.39 %, while the integral non-linearity is equal to ±0.88 K.

For improving the resolution, we exploited oversampling,

Table I
Performance summary and comparison with state-of-the-art BJT-based temperature-to-digital converters

Parameter	This work	[7]	[8]	[9]	[10]	[11]	[12]
Technology Feature Size (F) [nm]	**180**	130	55	130	28	180	65
Area (A) [mm^2]	**0.08**	0.09	0.01	0.29	0.01	0.2	0.003
Temperature Range (T_R) [°C]	**−20 to 80**	−40 to 125	−40 to 125	−40 to 125	−25 to 125	25 to 45	−10 to 110
Supply Voltage [V]	**1.8**	1.2	1.0	3.3	1.8	1.0	1.3
Measured Samples	**20**	16	12	16	76	20	25
Relative Inaccuracy (i_A) [%]	**1.39**	0.65	2.06	0.57	2.47	2.0	2.25
Number of Trimming Points (n_T)	**1**	1	0	1	0	1	2
Power Consumption [μW]	**39.6**	39.7	37	313.5	18.8	1.1	111.8
Conversion Time [ms]	**0.064**	1.3	32.8	5.12	8.2	500	4.10
Energy per Conversion (E_C) [nJ]	**2.5**	52	1200	1600	150	550	460
Resolution (R) [mK]	**158**	60	20	16	150	10	130
Resolution FoM = $E_C R^2$ [nJ K^2]	**0.06**	0.19	0.49	0.41	3.5	0.06	7.7
Global FoM = $(1 + n_T) \frac{\sqrt{A}}{F} E_C i_A \frac{R}{T_R}$ [nJ] [6]	**0.18**	0.55	6.66	7.35	13.2	27.2	28.2

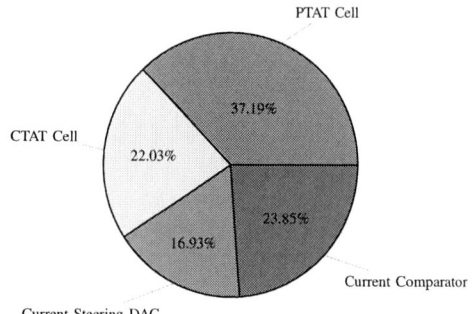

Figure 7. Power consumption breakdown of the TDC

by filtering 32 8-bit output codes to obtain a more refined digital word. To evaluate the resulting resolution, we acquired 50 000 8-bit readings, corresponding to more than 1500 refined samples, at a temperature of 30 °C, and computed the standard deviation, which turned out to be 158 mK, as shown in. Fig. 6.

Considering that the TDC is clocked at 4 MHz, the resulting conversion time, filtering 32 samples, is equal to 64 μs. The whole system drains 22 μA from a 1.8-V power supply. This means that the energy amount required to achieve 158-mK resolution is as low as 2.5 nJ. Fig. 7 shows the power breakdown among the main blocks of the TDC. The power consumption of the interface circuits has been added to the neighboring blocks, while the SAR logic power consumption has been omitted, considering that its contribution is negligible.

Tab. I summarizes the performances of the proposed TDC and compares them with recently published relevant BJT-based works in this field. The presented system turns out to be excellent in terms of energy per conversion, while it is comparable with existing solutions for the other parameters.

IV. Conclusions

In this paper, a fully current-mode based TDC for MEMS applications has been presented. The device, fabricated in a standard 180-nm CMOS technology, occupies an area of less than 0.08 mm^2. Consuming 39.6 μW from a 1.8-V power supply, it provides temperature-dependent digital samples with 158 mK of resolution, requiring an energy per conversion of 2.5 nJ. The relative inaccuracy over the temperature range −20 °C to 80 °C is 1.39 %.

References

[1] S. A. Zotov et al., "High Quality Factor Resonant MEMS Accelerometer With Continuous Thermal Compensation" in *IEEE Sensors J.*, vol. 15, no. 9, pp. 5045-5052, 2015.

[2] C. Chiang et al., "A Novel, High-Resolution Resonant Thermometer Used for Temperature Compensation of a Cofabricated Pressure Sensor" in *2012 S-S Sensors, Actuators, and Microsystems Workshop*, pp. 54-57.

[3] S. Zaliasl et al., "A 3 ppm 1.5 × 0.8 mm 2 1.0 μA 32.768 kHz MEMS-Based Oscillator" in *IEEE J. of Solid-State Circuits*, vol. 50, no. 1, pp. 291-302, 2015.

[4] A. Aprile et al., "A One-Point Exponential Trimming Technique for an Effective Suppression of Process Spread in BJT-based Temperature Processing Circuits" accepted in *ISCAS 2022*, pp. 1-4.

[5] G. M. Yin et al., "A high-speed CMOS comparator with 8-b resolution" in *IEEE J. of Solid-State Circuits*, vol. 27, no. 2, pp. 208-211, 1992.

[6] A. Aprile et al., "An Extensive Investigation and Analysis of Temperature-to-Digital Converter FoMs" in *ICECS 2021*, pp. 1-4.

[7] Z. Huang et al., "A BJT-Based CMOS Temperature Sensor With Duty-Cycle-Modulated Output and ±0.54°C (3σ) Inaccuracy From -40 °C to 125 °C" in *IEEE Trans. on Circuits Syst. II, Exp. Briefs*, vol. 68, no. 8, pp. 2780-2784, 2021.

[8] Z. Tang et al., "An Untrimmed BJT-Based Temperature Sensor With Dynamic Current-Gain Compensation in 55-nm CMOS Process" in *IEEE Trans. on Circuits Syst. II, Exp. Briefs*, vol. 66, no. 10, pp. 1613-1617, 2019.

[9] Z. Tang et al., "A CMOS Temperature Sensor With Versatile Readout Scheme and High Accuracy for Multi-Sensor Systems" in *IEEE Trans. on Circuits and Syst. I, Reg. Papers*, vol. 65, no. 11, pp. 3821-3829, 2018.

[10] Y. Hsu et al., "5.9 An 18.75μW dynamic-distributing-bias temperature sensor with 0.87°C (3σ) untrimmed inaccuracy and 0.00946 mm^2 area" in *ISSCC 2017*, pp. 102-103.

[11] M. Law et al., "A 1.1 μW CMOS Smart Temperature Sensor With an Inaccuracy of ±0.2 °C (3σ) for Clinical Temperature Monitoring" in *IEEE Sensors J.*, vol. 16, no. 8, pp. 2272-2281, 2016.

[12] O. Bass and J. Shor, "A Miniaturized 0.003 mm^2 PNP-Based Thermal Sensor for Dense CPU Thermal Monitoring" in *IEEE Trans. on Circuits and Syst. I, Reg. Papers*, vol. 67, no. 9, pp. 2984-2992, 2020.

A 5.4-mW 50-MHz 29.3-dBm-IIP3 Fourth-Order Low-Pass Filter

Liangbo Lei, Cong Tao, Zhipeng Chen, Zhiliang Hong, Yumei Huang
State Key Laboratory of ASIC and System, Fudan University, Shanghai, China
Email: yumeihuang@fudan.edu.cn

Abstract—This paper presents a continuous-time 4th-order low-pass filter with 50MHz cut-off frequency. The proposed two-stage opamps in the filter use push-pull output stage to enhance the in-band linearity. High linearity of +29.3dBm IIP3 is achieved at 30MHz. Simple R-C compensation is adopted for high unity-gain-bandwidth of the opamp. The R-C compensation elements in the output stage is also served as the second integrator of the biquad for power reduction. The low noise characteristics of the passive integrator improve the noise performance of the filter. The in-band output integrated noise is -61.7dBm. The SNR for a -41dB-THD is 72.1dB. This filter draws only 3mA from 1.8V supply voltage. The filter prototype has been fabricated in 180nm CMOS process. The achieved high figure-of-merit of 164.8dB·J⁻¹ validates the good power efficiency of the filter.

Index Terms—Acitve-RC filters, high linearity, power efficiency, operational amplifier, push-pull output

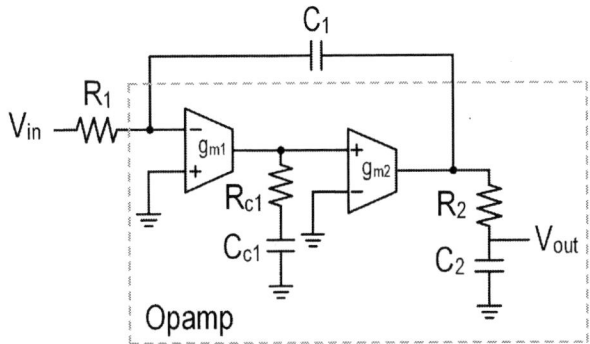

Fig. 1. Implementation of integrators in the biquad.

I. INTRODUCTION

Active-RC filters are widely used in communication systems due to the great dynamic range. For 5G receivers in the sub-6GHz frequency band, the required filter cut-off frequency can reach 50MHz. The opamps in the filter usually needs a unity-gain-bandwidth (UGB) above 500MHz and the filter typically consumes tens of milliwatt, which cannot be negligible [1]. In [2], an active-g_m-RC topology is adopted to save power consumption. Only one opamp is needed to implement the conjugate poles in this architecture. And the opamp do not need high UGB. However, the dynamic range of the filter will deteriorate as the linearity decrease near the cut-off frequency. Source-follower-based filter in [3] also have the same problems, although the power consumption is low. Good linearity performance over the entire passband is critical for 5G receivers. It is challenging to design highly linear low-pass filters with low power consumption.

In this design, the first active integrator uses opamp with high UGB to guarantee the filter performance. And a passive R-C integrator is used to synthesize the the second pole in Tow-Thomas biquad for power reduction. The two-stage opamps in this design use push-pull output stage to achieve good power efficiency and enhanse the filter's linearity. Left-half-plane (LHP) zeros are generated by a simple R-C series to guarantee the loop stability and improve the UGB of the opamp. The R-C series at the output stage of the opamp also serves as the passive integrator in the biquad.

This paper is organized as follows. Section II presents the biquad design considerations and design details of the opamp.

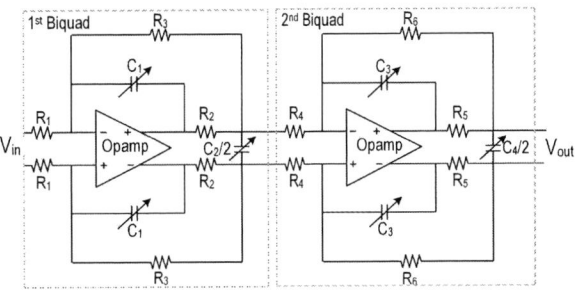

Fig. 2. Diagram of the 4th-order filter.

The measurement results is given in Section III. And Section IV gives the conclusion about this paper.

II. FILTER DESIGN CONSIDERATION

In conventional Tow-Thomas biquad, two opamps with high UGB are used to implement the two integrators [1]. In order to reduce the power consumption of the biquad, the second opamp-based integrator is replaced by the passive R-C integrator (R_2-C_2) in this work as shown in Fig. 1. R_2-C_2 is also used to generate a LHP zero. The two LHP zeros generated by R_{c1}-C_{c1} and R_2-C_2 respectively are used to ensure the loop stability of the biquad. This compensation scheme enables the opamp to have a larger UGB compared with Miller compensation [4].

978-1-6654-8495-4/22 $31.00 © 2022 IEEE

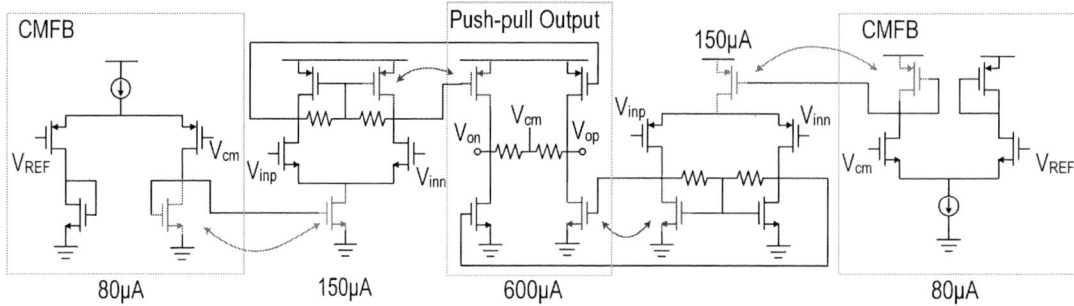

Fig. 3. Schematic of the opamp transistor-level implementation, not including compensation elements.

Fig. 2 shows the diagram of the proposed filter. Two biquads are cascaded to synthesize the 4th-order Butterworth low-pass transfer function. The transfer function of the 1st biquad is:

$$H(s) = \frac{R_3}{R_1} \cdot \frac{1}{\frac{s^2}{\omega_0^2} + \frac{1}{Q\omega_0}s + 1}$$

$$\omega_0 = \frac{1}{\sqrt{R_2 R_3 C_1 C_2}}, Q = \frac{\sqrt{R_2 R_3}}{R_2 + R_3 + \frac{R_2 R_3}{R_4}} \sqrt{\frac{C_2}{C_1}} \quad (1)$$

where ω_0 is the cut-off frequency of the 1st biquad and Q is the quality factor. The load resistance and capacitance of the overall filter can affect the cut-off frequency and quality factor of the 2nd biquad. In a conventional receiver link, the low-pass filter is usually followed by a variable gain amplifier. The virtual ground caused by the closed-loop mechanism loads the filter with passive elements, just as the input resistance of the 2nd biquad R_4. Therefore, the transfer function can be determined by the ratio of passive elements. In this design, the filter is followed by a test buffer with an input capacitance much smaller than C_4 (about 26pF). And the effect on the filter transfer function is negligible.

A. Linearity

The opamp in the active integrator directly determines the filter's performance because it is the only active device in the biquad. Two-stage opmap with push-pull output stage is adopted to enhance the linearity performance, as shown in Fig. 3. According to the analysis in [5], the nonlinear currents generated by g_{m2} in Fig. 1 dominate the distortion in the 1st biquad. And the distortion of the biquad is proportional to $g''_{m2}/g_{m2}^4 R_{out}^3$, where R_{out} represents the output resistance of the opamp. The push-pull output stage can increase the value of $g_{m2}^4 R_{out}^3$, thus improving the linearity performance. Using transistors with long channel length can further improve the linearity of the filter. However, it will increase the capacitive load of the first stage of the opamp. In this design, the channel lengths of NMOS and PMOS are 240nm and 180 nm, respectively.

B. Noise

In traditional Tow-Thomas biquad, the second active integrator increases the output noise power spectral density (PSD)

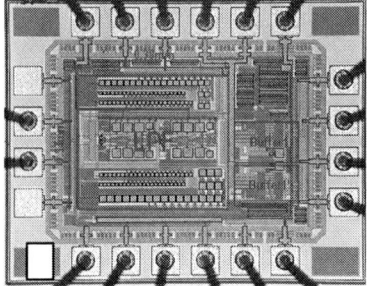

Fig. 4. Die photo of the proposed filter.

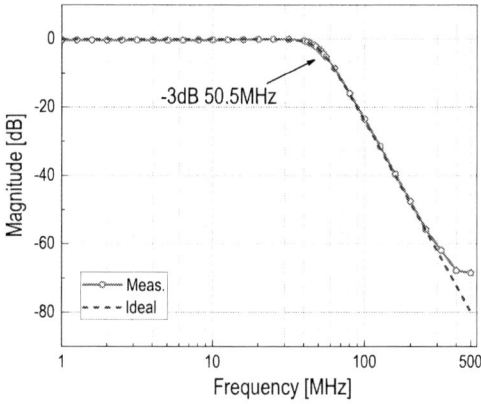

Fig. 5. Measured filter magnitude response.

of the biquad near the cut-off frequency [2]. As the order of the filter increases, its output noise PSD will have a peak near the cut-off frequency deteriorating dynamic range. The noise sources in the passive integrators are only R_2 and R_5 shown in Fig. 2. The R_2 and R_5 are set as 330Ω and 500 Ω, respectively, for an acceptable output signal swing while not excessively increasing the load effect on the opamp. Therefore, the output noise PSD peak can be reduced, and dynamic range of the filter is improved.

978-1-6654-8495-4/22 $31.00 © 2022 IEEE   478

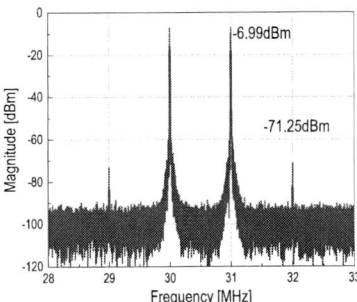

Fig. 6. Measured output spectrum of LPF+Buffer with -7dBm 30MHz & 31MHz two-tone input.

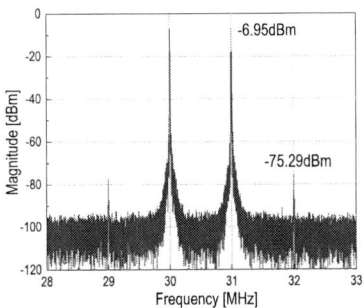

Fig. 7. Measured output spectrum of Buffer with -7dBm 30MHz & 31MHz two-tone input.

C. Opamp biasing

To stabilize the quiescent current of the push-pull output stage, two global common-mode feedback circuits and local common-mode feedback at the first stage of the opamp are employed, as shown in Fig. 3. Transistors with the same color have the same channel length. If the common-mode gain of the two-stage opamp is larger than 0dB, the common-mode loop of the biquad will oscillate [1]. The two local common-mode feedback can further suppress the opamp's common-mode gain below -10dB, which greatly guarantees the common-mode loop stability. The global common-mode feedback circuits draws only 160μA.

III. MEASUREMENT RESULTS

The core area of the proposed filter based on the 180nm CMOS is 0.13mm², as shown in Fig. 4. This filter draws only 3mA from 1.8V supply voltage. Two integrated test buffers are used to evaluate the filter gain, linearity and noise performance.

Fig. 5 shows the measured filter gain, which matches well with the ideal 50MHz Butterworth filter frequency response. The cut-off frequency of the filter can be tuned by using variable capacitors. And the ripple of less than 0.25dB at the cut-off frequency is achieved.

Fig. 6 shows the measured output spectrum of LPF + Buffer with -7dBm 30MHz & 31MHz two-tone input. Fig. 7 shows the output spectrum of Buffer with the same input signal. The

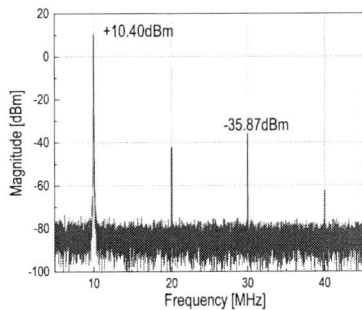

Fig. 8. Output spectrum for input signal of +10.4dBm at 10MHz.

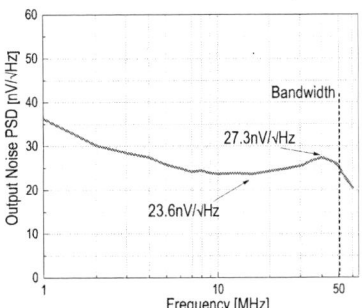

Fig. 9. Measured filter output noise PSD.

input signal with 282mV$_{pp}$ is used to accurately demonstrate the linearity performance of the proposed filter. According to the cascade IIP3 formula [1], the IIP3 at 30MHz of the filter is about +29.3dBm. This shows the filter still has good linearity at the passband edge. Fig. 8 shows the output spectrum for input signal of +10.4dBm at 10MHz, resulting in -41dB Total-Harmonic-Distortion (THD). Both single tone and double tone test validates the filter's linearity performance.

The output noise PSD is illustrated in Fig. 9. And the output noise PSD is between 23.6nV/$\sqrt{\text{Hz}}$ and 27.3nV/$\sqrt{\text{Hz}}$ in the range of 10MHz-50MHz. Thanks to the low-noise passive integrator, no large peaking of output noise PSD occurs near the cut-off frequency. The noise performance is improved to some extent and the in-band (1MHz-50MHz) output integrated noise power is -61.7dBm, achieving 72.1dB dynamic range.

Table I gives the filter performance summary and comparison with other state-of-the-art. The Figure-of-Merit (FoM) used in this paper is defined as [5]:

$$\text{FoM} = 10 \cdot \log_{10} \frac{\text{SFDR}^2 \cdot f_0 \cdot N}{\text{Power}} \cdot \frac{f_{\text{IM3,LOW}}}{f_{\text{POLES}}} \quad (2)$$

$$\text{SFDR} = \frac{2}{3}(\text{IIP3} - P_{\text{noise}})$$

where N is the order, $f_{\text{IM3,LOW}}$ is the lower third-order inter-modulation tone, and f_{POLES} is the poles frequency, and P_{noise} is the integrated input-referred noise power. This design has achieved FoM of 164.8dBJ^{-1}, which is even better to the Source-follower-based filter [3].

978-1-6654-8495-4/22 $31.00 © 2022 IEEE   479

TABLE I

PERFORMANCE SUMMARY AND COMPARISON TO PRIOR WORKS

Parameters	This Work	[1]	[3]	[5]	[6]	[8]	[9]	[10]
Order&TF*	4-B	6-B	5-B	4-B	2-B	5-C	5-C	4-B
Topology	Opamp-RC	SF-based	Opamp-RC	Opamp-RC	Opamp-RC	Opamp-RC	Opamp-RC	Opamp-RC
CMOS Tech. [nm]	180	180	180	180	28	130	180	65
Supply Voltage [V]	1.8	1.8	1.3	1.8	1.1	1	1.8	1.2
Power [mW]	5.4	0.4	0.65	12.6	0.3	7.5	5.6	9.6
Gain [dB]	0	0	0	0	20	0	0	0
-3dB Bandwidth [MHz]	50	10	20	22.5	40	20	20	5
IRN [μV_{RMS}]	182.8	200	68.4	87	5.05	232	980	80
SNR [dB]	72.1	-	76	63	69.5	63.7	60.3	-
f_{IM3} [MHz]	30	4	10	4	10	0.27	6	1
IIP3 [dBm]	29.3	14.4	11.5	17.6	-6	31	43.3	19
SFDR [dB]	60.7	52.2	54.5	59.8	-	-	60.3	-
Area [mm^2]	0.13	0.21	0.12	0.35	0.024	1.53	0.2	0.8
Power/Pole [mW]	1.35	0.067	0.13	3.15	0.15	1.5	1.12	2.4
FOM [dBJ^{-1}]	164.8	150.4	157.8	150.8	158.4	143.2	152	144.8

*C=Chebyshev, B=Butterworth.

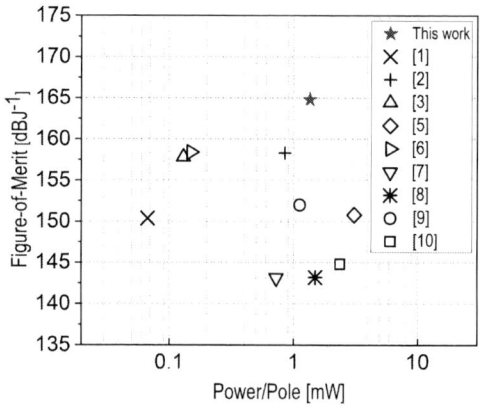

Fig. 10. Figure-of-Merit vs. Power/Pole.

Fig. 10 shows the FoM versus (Power/pole), implying that the filter compares well than other works. Although the proposed filter still has higher power consumption compared to source-follower-based filters, it achieves good linearity performance over the entire passband, which is critical for 5G receivers.

IV. CONCLUSION

In this paper, a 4th-order 50MHz low-pass filter with 72.1dB dynamic range has been presented. Two-stage opamp with push-pull output stage is adopted to guarantee the linearity performance. And passive R-C integrator is used to reduce power consumption and improve the noise performance of the filter. The achieved high FoM of $164.8dBJ^{-1}$ proves the good power efficiency. And it can be well applied to 5G receivers with high linearity requirement.

ACKNOWLEDGMENT

The author would like to thank Professor Jiawei Xu for his help in chip fabrication. This project was sup-ported by National Key Research and Development Project (2018YFB2201003).

REFERENCES

[1] L. Ye, C. Shi, H. Liao, R. Huang and Y. Wang, "Highly Power-Efficient Active-RC Filters With Wide Bandwidth-Range Using Low-Gain Push-Pull Opamps," in IEEE Transactions on Circuits and Systems I: Regular Papers, vol. 60, no. 1, pp. 95-107, Jan. 2013.

[2] S. D'Amico, V. Giannini and A. Baschirotto, "A 4th-order active-G/sub m/-RC reconfigurable (UMTS/WLAN) filter," in IEEE Journal of Solid-State Circuits, vol. 41, no. 7, pp. 1630-1637, July 2006.

[3] Y. Xu, H. Hu, J. Muhlestein and U. -K. Moon, "A 77-dB-DR 0.65-mW 20-MHz 5th-Order Coupled Source Followers Based Low-Pass Filter," in IEEE Journal of Solid-State Circuits, vol. 55, no. 10, pp. 2810-2818, Oct. 2020.

[4] G. Pini, D. Manstretta and R. Castello, "Analysis and Design of a 20-MHz Bandwidth, 50.5-dBm OOB-IIP3, and 5.4-mW TIA for SAW-Less Receivers," in IEEE Journal of Solid-State Circuits, vol. 53, no. 5, pp. 1468-1480, May 2018.

[5] M. De Matteis, A. Pipino, F. Resta, A. Pezzotta, S. D'Amico and A. Baschirotto, "A 63-dB DR 22.5-MHz 21.5-dBm IIP3 Fourth-Order FLFB Analog Filter," in IEEE Journal of Solid-State Circuits, vol. 52, no. 7, pp. 1977-1986, July 2017.

[6] M. De Matteis, A. Baschirotto and E. A. Vallicelli, "309-µW 40-MHz 20-dB-Gain Analog Filter in 28nm-CMOS," ESSCIRC 2021 - IEEE 47th European Solid State Circuits Conference (ESSCIRC), 2021, pp. 339-342.

[7] M. S. Savadi Oskooei, N. Masoumi, M. Kamarei and H. Sjoland, "A CMOS 4.35-mW +22-dBm IIP3 Continuously Tunable Channel Select Filter for WLAN/WiMAX Receivers," in IEEE Journal of Solid-State Circuits, vol. 46, no. 6, pp. 1382-1391, June 2011.

[8] H. Amir-Aslanzadeh, E. J. Pankratz and E. Sanchez-Sinencio, "A 1-V +31 dBm IIP3, Reconfigurable, Continuously Tunable, Power-Adjustable Active-RC LPF," in IEEE Journal of Solid-State Circuits, vol. 44, no. 2, pp. 495-508, Feb. 2009.

[9] S. V. Thyagarajan, S. Pavan and P. Sankar, "Active-RC Filters Using the Gm-Assisted OTA-RC Technique," in IEEE Journal of Solid-State Circuits, vol. 46, no. 7, pp. 1522-1533, July 2011.

[10] Y. Wang, L. Ye, H. Liao, R. Huang and Y. Wang, "Highly Reconfigurable Analog Baseband for Multistandard Wireless Receivers in 65-nm CMOS," in IEEE Transactions on Circuits and Systems II: Express Briefs, vol. 62, no. 3, pp. 296-300, March 2015.

A 135 GHz 32 Gb/s Direct-Digital Modulation 16-QAM Transmitter in 28 nm CMOS

Carl D'heer, Patrick Reynaert

KU Leuven ESAT-MICAS, Leuven, Belgium

{carl.dheer, patrick.reynaert}@esat.kuleuven.be

Abstract — This paper presents a 135 GHz direct-digital modulation 16-QAM transmitter employing a Cartesian architecture with separate phase and amplitude modulation. The transmitter, implemented in 28 nm CMOS, demonstrates a peak output power of 0 dBm. A maximum data rate of 32 Gb/s is achieved using 16-QAM without any equalization or pulse shaping. Star-QAM and QPSK modulations lead to data rates of 27 Gb/s and 24 Gb/s, respectively. The transmitter consumes 130 mW of DC power.

Keywords — 16-QAM, 6G, CMOS, direct-digital modulation, wireless communication, mm-wave, QPSK, sub-THz, transmitter.

I. Introduction

The ever-increasing demand for higher data rate communication links has pushed the carrier frequency into the mm-wave and THz spectrum due to its large available bandwidth. Furthermore, implementing the RFICs in CMOS is attractive to realize cheap and power-efficient integrated systems. However, as the operation frequency approaches the technology's f_{max}, the performance of the active devices drops steeply as does the efficiency of the implemented circuits. In 28 nm CMOS ($f_{max} \approx 280$ GHz), operating in the D-band (110-170 GHz) seems to be suitable for high-speed wireless transceivers, offering a large bandwidth while still supporting adequate gain and noise performance.

As the desired data rate approaches 50 Gb/s and above, simple modulations no longer suffice and more spectrally efficient modulation schemes, like QPSK, star-QAM and 16-QAM, have to be considered. Traditional transmitter architectures [1] typically employ a linear upconverter and rely on an external multi-bit (> 6b) digital-to-analog converter (DAC) with high sample rate (> 10 GBaud) to generate high-order constellations. However, at such sample rates these DACs consume significant power (~100 mWs) [2] limiting the power efficiency of the overall system. Moreover, the limited linearity of these modulators further deteriorates the transmitter efficiency by requiring predistortion or equalization blocks. Finally, the design challenges and area overhead associated with these circuits yield a very costly system.

Direct-digital modulation (DDM) transmitter architectures have been proposed [2]–[6] to construct high-order modulations with 1-bit digital data interfaces, avoiding power-hungry multi-bit DACs. However, existing digital modulators may suffer from data-dependent distortion [4]–[6], IQ distortion due to load modulation [2], [6], or LO imbalance-induced distortion [3], requiring equalization

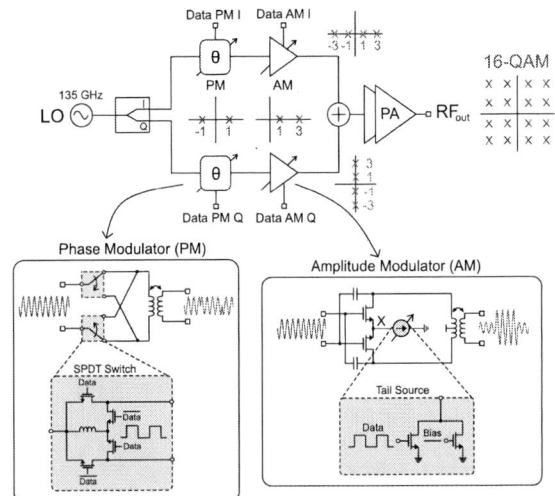

Fig. 1. Block diagram of the proposed direct-digital modulation 16-QAM transmitter with implemented phase and amplitude modulator.

or digital pre-distortion to avoid EVM degradation at high speeds. In addition, extensive and power consuming I/Q LO generation circuitry is needed to generate linear high-order constellations [2], [3], leading to a large chip area and low power efficiency.

In this paper, a 135 GHz power-efficient direct-digital modulation architecture is proposed capable of generating various high-order constellations at high speeds with a low EVM, even without offline equalization. Section II illustrates the modulation principle and transmitter architecture. The circuit implementations are described in section III and section IV discusses the measurements results of the fabricated transmitter.

II. Transmitter Architecture

Figure 1 presents the block diagram and modulation principle of the transmitter. Two 135 GHz quadrature LO signals are generated on-chip after which modulation is performed in two separate steps in both the I and Q branches. First, the phase of each LO signal is modulated by a switch-based phase modulator and subsequently its amplitude is modulated by a two-level amplitude modulator. Both modulated signals are combined and amplified to form a 16-QAM constellation. Additionally, star-QAM modulation can be obtained by using identical data streams for both

978-1-6654-8495-4/22 $31.00 © 2022 IEEE

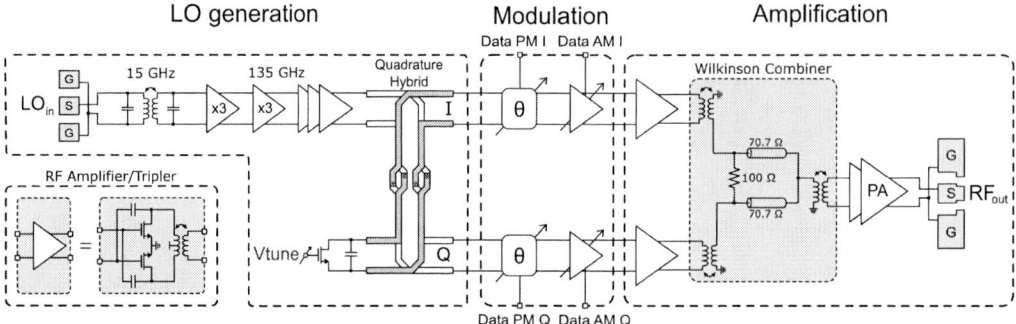

Fig. 2. Circuit diagram of the 135 GHz transmitter.

amplitude modulators in the I and Q branch. Finally, QPSK is achieved by keeping both amplitudes constant.

Special care is taken to ensure a low EVM 16-QAM constellation in case of process variation or modeling inaccuracies. IQ imbalances can be corrected thanks to a tunable IQ hybrid output phase and variable gain in I and Q stages. The symmetrical phase modulator ensures a balanced output with high switching speed. The proposed amplitude modulator with adjustable gain levels and built-in kickback cancellation enables high modulation speeds while minimizing data-dependent distortion. Finally, the well isolated power combiner with broadband amplifying network ensures minimal load-modulation effects and ripples in the frequency response. Combined with a compact LO generation network, the proposed Cartesian architecture supports power- and area-efficient high-order digital modulation with robust constellation linearity

III. TRANSMITTER CIRCUIT DESIGN

A. LO Generation

The complete transmitter schematic is shown in Fig. 2. All tripler and amplifier blocks are implemented by means of a capacitively neutralized pseudo-differential pair. The 135 GHz quadrature LO signals are generated from a 15 GHz external LO reference by means of two triplers, three buffers and a coupled-line quadrature hybrid. Any possible IQ phase and magnitude imbalance can be compensated with a variable impedance at the isolation port of the hybrid and tunable buffers in the I and Q branches after the modulators.

B. Modulators

The quadrature LO signals are modulated in phase and amplitude by two separate modulators. Figure 1 also shows the circuit-level implementation of both modulators. The phase modulator is realized by means of two cross-connected single-pole double-throw (SPDT) switches, each implemented as a switched-inductor SPDT (SI-SPDT) switch to minimize the insertion loss and footprint of the switch [7]. This topology ensures a high switching speed with minimal phase error or magnitude imbalance. The amplitude modulator consists of a neutralized differential pair and two tail transistors. One is driven by digital data to switch between the high and

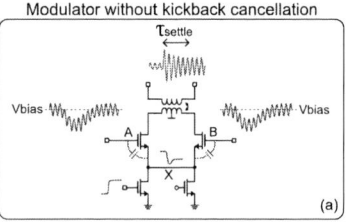

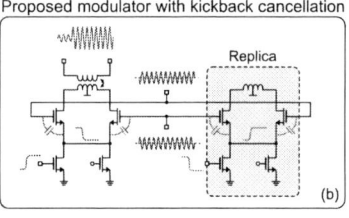

Fig. 3. Schematic and principle of the amplitude modulator without (a) and with (b) kickback cancellation (neutralization capacitors omitted for clarity).

low gain state producing amplitude modulation. The other one is controlled by a bias voltage which determines the low gain level and consequently also the amplitude ratio of the modulated signal. By choosing the appropriate bias voltage, optimal spacing between the constellation points can be guaranteed resulting in a robust constellation with low EVM. Note that the amplitude ratio is controlled by a DC voltage and does not switch at the baud rate.

The switching nature of the amplitude modulator results in a high modulated output power which reduces the number of power amplifying stages required to reach the desired transmitted output power. This not only enhances power efficiency, but also improves modulation bandwidth by minimizing the number of bandwidth-limiting interstage matching networks. In addition, since the constant tail source ensures the modulator is never fully turned off, long settling times of the tail (X) node voltage are avoided which improves the obtainable switching speed. However, as Fig. 3a illustrates, due to the steep edges of the digital switching signal, serious kickback charge is injected into the tail (X) and gate (A,B) nodes of the modulator. This would lead to significant unwanted settling of the common mode input voltage whenever the modulator switches, severely impacting

978-1-6654-8495-4/22 $31.00 © 2022 IEEE   482

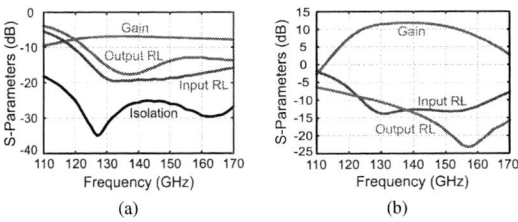

Fig. 4. Simulated S-parameters of the Wilkinson combiner (a) and amplification chain after the modulators (b).

the speed and dynamic range of the modulation. To improve the dynamic behavior of the modulator, a replica modulator is added which is driven by an inverted version of the data (Fig. 3b). Consequently, data coupling to the gate nodes is cancelled, drastically improving the switching speed. Furthermore, adding the replica reduces the variation of the modulator's input impedance while switching and leads to a more constant supply current, both resulting in less data-dependent distortion.

C. Amplification

Both modulated signals are amplified by separate buffers and added using a balun-based differential Wilkinson power combiner [2]. The combiner must have a flat frequency response and good isolation between the input ports to minimize the interaction between the I and Q branches, and thus avoid load modulation effects which degrade the EVM. Fig. 4(a) shows 7 dB of insertion loss and provides > 25 dB of isolation over more than 40 GHz between the input ports.

Finally, the output signal is amplified by a two-stage power amplifier and matched to the GSG output pads. Each amplifier stage is implemented as a pseudo-differential pair with cross-coupled MOM capacitors. The GSG pads are custom designed to be compatible with both a flip-chip process and 50 μm pitch GSG probes for evaluation purposes. By stagger tuning and compensating the frequency-dependent gain roll-off of the active devices, a broadband frequency response is obtained which is essential to support high symbol rates. A simulated 3-dB bandwidth of 35 GHz is achieved from the output of the modulators to the transmitter output with no in-band ripple, as shown in Fig. 4(b).

IV. MEASUREMENT RESULTS

The presented 135 GHz transmitter is implemented in a 28 nm CMOS technology and the die micrograph is shown in Fig. 5. The chip occupies an area of 2.29 mm². The setup to measure the transmitter is depicted in Fig. 6. For CW measurements, the transmitter output signal is brought off chip via a 50 μm pitch GSG probe and downconverted using an external D-band mixer. The IF mixer output is evaluated with a spectrum analyzer after calibration by a PM5 power meter. For modulated measurements, a custom IQ receiver chip is used to downconvert the modulated signal to baseband. The baseband receiver outputs are analyzed with a high-speed oscilloscope. The data inputs for the transmitter are rail-to-rail

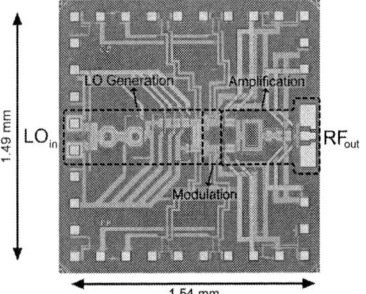

Fig. 5. Die micrograph of the proposed transmitter.

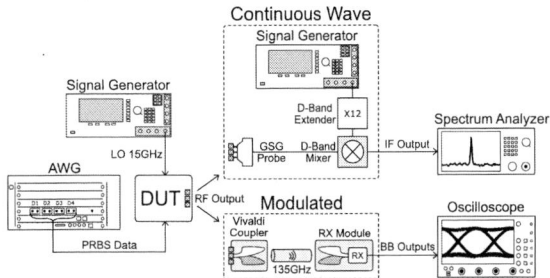

Fig. 6. Measurement setup.

binary PRBS7 sequences generated with an arbitrary waveform generator and the receiver LO signal is derived from a low frequency reference locked to that of the transmitter. The transmitter and receiver chips are flip-chip integrated with wideband Vivaldi couplers on PCB and connected via 1 m of dielectric waveguide [8].

The measured output power versus LO input power is shown in Fig. 7(a), demonstrating a saturated output power at 135 GHz of 0 dBm with less than 3 dB variation over 10 GHz of bandwidth. To characterize the upconversion response of the fabricated transmitter, one of the phase modulators is driven by a 50 % duty cycle square wave while the other modulators are kept constant. The power of the fundamental sidebands around the 135 GHz carrier were then observed while sweeping the frequency of the square wave input. Figure 7(b) shows this upconversion response normalized to the value at the center frequency. Driving one of the amplitude modulators instead or all modulators simultaneously yields a similar response.

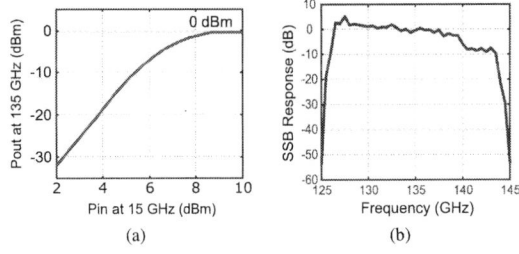

Fig. 7. Measured TX output power versus LO input power (a) and normalized upconversion response (b).

Table 1. Comparison of high-speed, sub-THz silicon-based transmitters.

Reference	IMS 2020 [1]	JSSC 2019 [2]	JSSC 2020 [3]	JSSC 2018 [4]	CICC 2020 [5]	ISSCC 2022 [6]	**This Work**
Technology	130nm SiGe	180nm SiGe	65nm CMOS	45nm SOI	28nm CMOS	22nm FinFET	**28nm CMOS**
Carrier Frequency (GHz)	240	115	170	138	113	140	**135**
Modulation Scheme	16-QAM	16-QAM	8-PSK	16-QAM	16-QAM	16-QAM	**16-QAM**
Modulator D/A Interface	BB I/Q DAC	Direct-digital	Direct-digital	Direct-digital	Direct-digital	Direct-digital	**Direct-digital**
Peak Output Power (dBm)	12[3]	3	-3.3	13.2 (EIRP)	0 (EIRP)	0.8	**0**
Area (mm²)	7	3.17	7.8	4.2	4.72[2]	4	**2.29**
Power Consumption (mW)	1237	520	560	1255	470	173	**130**
Unequalized Data Rate (Gb/s)[1]	-	-	-	5	20 (QPSK)	-	**32**
Equalized Data Rate (Gb/s)[1]	60	20	15	7[4]	72	160	**-**
Unequalized TX Efficiency (pJ/bit)	-	-	-	251	23.5	-	**4.1**
Equalized TX Efficiency (pJ/bit)[5]	20.6	26	37	179	6.5	1.1	**-**

[1] Corresponding to a BER < 2e-3   [2] Area of full transceiver   [3] Simulated value   [4] Using digital pre-distortion   [5] Excluding DSP power

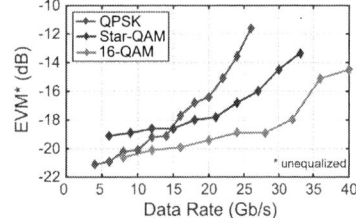

Fig. 8. Unequalized EVM versus symbol rate.

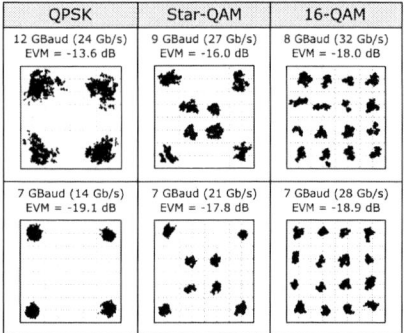

Fig. 9. Measured unequalized constellations for various modulation schemes.

Figure 8 shows the measured EVM versus data rate for all three supported modulations and Fig. 9 presents the measured constellations for these modulation schemes. A maximum data rate of 32 Gb/s is achieved for 16-QAM modulation. Using star-QAM and QPSK leads to maximum data rates of 27 Gb/s and 24 Gb/s, respectively. No pulse shaping, equalization or calibration of the equipment was performed to obtain these data rates. The results thus include all impairments (e.g., limited bandwidth and dispersion) caused by the receiver chip, PCB couplers, and dielectric waveguide. Applying offline equalization could significantly boost the achievable data rate, but would severely degrade the system efficiency. All constellations have an EVM corresponding to a BER < 1e-3. The transmitter consumes 130 mW from a 0.9 V supply while transmitting modulated data.

Table 1 compares this work to other state-of-the-art high-speed sub-THz silicon-based transmitters capable of generating higher-order modulations. The proposed transmitter shows competitive data rates for multiple modulation schemes (16-QAM, star-QAM, QPSK) while demonstrating an excellent energy efficiency, even without offline equalization.

V. Conclusion

This paper presented a power-efficient 135 GHz direct-digital modulation transmitter employing a Cartesian architecture with separate phase and amplitude modulation. The transmitter, implemented in 28 nm CMOS, demonstrated high-speed linear 16-QAM modulation without the need for high-speed multi-bit DACs paving the way for more energy efficient communication systems.

References

[1] M. H. Eissa, N. Maletic, E. Grass, R. Kraemer, D. Kissinger, and A. Malignaggi, "100 Gbps 0.8-m Wireless Link based on Fully Integrated 240 GHz IQ Transmitter and Receiver," in *2020 IEEE/MTT-S International Microwave Symposium (IMS)*, 2020, pp. 627–630.

[2] H. Wang, H. Mohammadnezhad, and P. Heydari, "Analysis and Design of High-Order QAM Direct-Modulation Transmitter for High-Speed Point-to-Point mm-Wave Wireless Links," *IEEE Journal of Solid-State Circuits*, vol. 54, no. 11, pp. 3161–3179, 2019.

[3] P. Nazari, S. Jafarlou, and P. Heydari, "A CMOS Two-Element 170-GHz Fundamental-Frequency Transmitter With Direct RF-8PSK Modulation," *IEEE Journal of Solid-State Circuits*, vol. 55, no. 2, pp. 282–297, 2020.

[4] S. Shopov, O. D. Gurbuz, G. M. Rebeiz, and S. P. Voinigescu, "A *D* -Band Digital Transmitter with 64-QAM and OFDM Free-Space Constellation Formation," *IEEE Journal of Solid-State Circuits*, vol. 53, no. 7, pp. 2012–2022, 2018.

[5] A. Townley, N. Baniasadi, S. Krishnamurthy, C. Sideris, A. Hajimiri, E. Alon, and A. Niknejad, "A Fully Integrated, Dual Channel, Flip Chip Packaged 113 GHz Transceiver in 28nm CMOS supporting an 80 Gb/s Wireless Link," in *2020 IEEE Custom Integrated Circuits Conference (CICC)*, 2020, pp. 1–4.

[6] S. Callender, A. Whitcombe, A. Agrawal, R. Bhat, M. Rahman, C. C. Lee, P. Sagazio, G. Dogiamis, B. Carlton, M. Chakravorti, S. Pellerano, and C. Hull, "A Fully Integrated 160Gb/s D-Band Transmitter with 1.1 pJ/b Efficiency in 22nm FinFET Technology," in *2022 IEEE International Solid- State Circuits Conference (ISSCC)*, vol. 65, 2022, pp. 78–80.

[7] C. D'heer and P. Reynaert, "A High-Speed 390GHz BPOOK Transmitter in 28nm CMOS," in *2020 IEEE Radio Frequency Integrated Circuits Symposium (RFIC)*, 2020, pp. 223–226.

[8] K. Dens, J. Vaes, S. Ooms, M. Wagner, and P. Reynaert, "A PAM4 Dielectric Waveguide Link in 28 nm CMOS," in *ESSCIRC 2021 - IEEE 47th European Solid State Circuits Conference (ESSCIRC)*, 2021, pp. 479–482.

A 135 GHz 24 Gb/s Direct-Digital Demodulation 16-QAM Receiver in 28 nm CMOS

Carl D'heer, Patrick Reynaert

KU Leuven ESAT-MICAS, Leuven, Belgium

{carl.dheer, patrick.reynaert}@esat.kuleuven.be

Abstract — This paper presents a 135 GHz direct-digital demodulation 16-QAM receiver. A direct-conversion RF front-end is combined with an analog PAM-4 decoder to achieve efficient 16-QAM demodulation without needing a high-speed multi-bit ADC. The receiver, implemented in 28 nm CMOS, demonstrates a conversion gain of 25 dB and integrated DSB noise figure better than 10 dB. A 24 Gb/s 16-QAM signal was detected and directly demodulated on-chip with a power consumption of 175 mW.

Keywords — 16-QAM, 6G, CMOS, direct-digital demodulation, wireless communication, mm-wave, PAM-4, sub-THz, receiver.

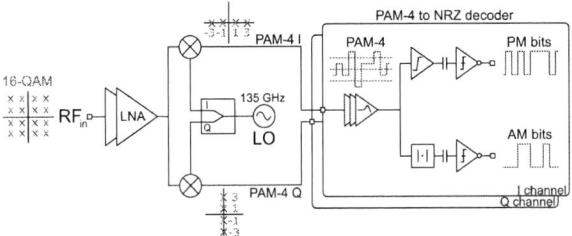

Fig. 1. Block diagram of the proposed 135 GHz receiver architecture with PAM-4 decoder.

I. INTRODUCTION

Recent research towards the next generation of wireless communication (6G) has sparked interest in the sub-THz frequency band (100-300 GHz) as a means to achieve energy-efficient high-speed communication rates [1]–[7]. Despite the sub-optimal high-frequency performance, a CMOS IC implementation is preferred for maximum integration potential and low-cost high-volume production. In 28 nm CMOS ($f_{max} \approx 280$ GHz), the D-band (110-170 GHz) strikes a good balance between available bandwidth and active device performance.

To support vastly increasing data rates with realistic signal bandwidths, research has been trending towards higher-order modulation schemes, such as 16-QAM and above. Traditional receiver architectures convert the modulated RF signal down to multi-level baseband [6] or IF data outputs [7]. These outputs are subsequently demodulated in the digital domain after being sampled by a high-resolution analog-to-digital converter (ADC) sampling at multiple times the symbol rate. As symbol rates start exceeding 5 GBaud, such high-speed ADCs become challenging to design with sufficient resolution, while their large power consumption ($\sim$100 mWs) degrades the energy efficiency of the entire system [1]. In most works, these ADCs are replaced by an expensive, watt-level oscilloscope to demodulate their data with an acceptable bit-error rate (BER). Furthermore, any high-speed digital signal processing (DSP) such as equalization is implemented offline and not included in the power budget of the receiver. To facilitate future energy-efficient high-speed systems, the necessary resolution and sample rate of the ADC can be significantly reduced by performing the demodulation in the analog domain and by substantially simplifying the DSP.

Direct-digital demodulation receivers aim to perform efficient demodulation on-chip, thereby alleviating the aforementioned challenges associated with high-speed multi-bit ADCs. Recently, a demodulation architecture was proposed capable of directly converting a modulated 8-PSK signal into its output bits using an RF-correlation method with little digital logic [1]. Direct-digital demodulation receivers for modulations with even higher spectral efficiency have not yet been reported at multi-GBaud speeds. This paper proposes a 135 GHz receiver architecture supporting direct-digital 16-QAM demodulation completely in the analog domain without requiring a multi-bit ADC or complex DSP. Section II introduces the receiver architecture with the proposed demodulation scheme. The receiver circuit design is described in section III and section IV discusses the measurement results of the fabricated receiver.

II. RECEIVER ARCHITECTURE

Figure 1 shows the high-level block diagram and basic principle of the proposed 135 GHz receiver which consists of an RF front-end and two PAM-4 decoders. The RF front-end implements a classic direct-conversion IQ receiver comprised of a low-noise amplifier (LNA) with subsequent IQ downconversion mixers. The mixers are driven by a 135 GHz LO signal generated from a low-frequency reference by an on-chip multiplier chain. Demodulation of the downconverted I and Q baseband signals is performed on-chip by PAM-4 decoders.

PAM-4 decoders are commonly used in non-ADC-based PAM-4 wireline receivers to recover the LSB and MSB from the received PAM-4 signal [8], [9]. In this paper, an analog PAM-4 decoder is implemented to efficiently decode the PAM-4 data from the I and Q channels of the 16-QAM signal. After downconversion, each PAM-4 signal is first conditioned by four linear amplification stages, including one continuous-time equalizer. Then the decoder separately

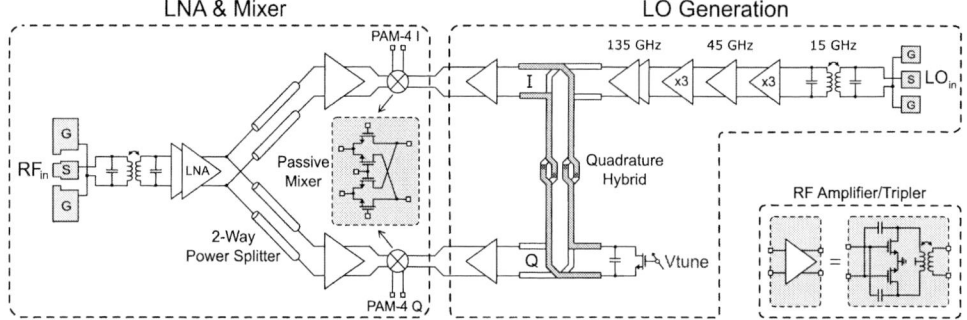

Fig. 2. Circuit diagram of the 135 GHz RF front-end.

extracts the Gray-coded amplitude-modulated (AM) bits and phase-modulated (PM) bits from its 4-level input[1].

The PM bits can readily be decoded by a limiting amplifier and a sign-check comparator. Demodulation of the AM bits is less straightforward, similar to the challenging LSB demodulation in wireline receivers [8]. Traditional decoders employ two comparators with two distinct nonzero thresholds which need to be generated adaptively to ensure optimal LSB decoding performance. The design of such comparators, combined with necessary adaptive threshold generation, is challenging at high speeds and power consuming. Inspired by [8], this paper proposes a rectifier-based demodulation method. After rectification and AC coupling of the PAM-4 signal, the AM bits can be extracted using a simple sign-check comparator, obviating the need for multiple nonzero-threshold comparators or any additional logic. Furthermore, all blocks have an unclocked implementation, thus working in continuous time. Removing all clocking circuits saves additional power and area while avoiding timing issues at high clocking speeds [9]. Finally, all unclocked sign-check comparators can be implemented efficiently using a CMOS inverter, showing a clear power advantage compared to CML comparators while supporting multi-GBaud operation.

III. RECEIVER CIRCUIT DESIGN

A. RF Front-End

The schematic of the RF front-end is shown in Fig. 2 and includes a wideband three-stage LNA with two-way power splitting and two downconversion mixers driven by on-chip generated quadrature LO signals. The GSG input pads are custom designed to be compatible with both a flip-chip process and 50 μm pitch GSG probes for characterization purposes. The wideband LNA is implemented as a cascade of three neutralized pseudo-differential pair stages matched by means of stagger-tuned transformer networks. The transistor sizes are optimized to provide adequate gain with simultaneous noise and gain matching to attain a low receiver noise figure. The signal is split into separate I and Q paths by a parallel power

splitter after the second stage. The LNA achieves 10 dB of simulated gain, including the splitting loss, with a 3-dB BW of 37 GHz and a minimum NF around 8 dB. Both signals are then downconverted by passive double-balanced mixers driven by 135 GHz quadrature LO signals. The mixer switch size is optimized for the expected LO power of 0 dBm, resulting in a simulated mixer voltage conversion gain of -1 dB. The high-frequency quadrature LO signals are generated from a 15 GHz external LO reference by means of two triplers, buffer amplifiers and a coupled-line quadrature hybrid. Any possible IQ phase and magnitude imbalance can be calibrated out with a variable impedance at the isolation port of the hybrid and tunable buffers in the I and Q branches following the hybrid.

B. Baseband and Demodulation

Figure 3 shows the block diagram and implemented circuits of the baseband and demodulation block. Each downconverted PAM-4 signal is first amplified by a cascade of baseband amplifiers and a continuous-time linear equalizer (CTLE). The baseband amplifiers are implemented as a resistively-loaded pseudo-differential pair with inductive peaking to extend the

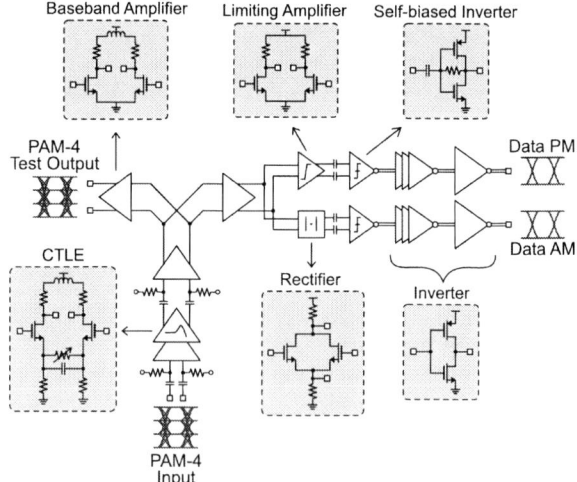

Fig. 3. Circuit diagram of the PAM-4 decoder in the I and Q channels, including the baseband amplifiers and demodulation circuitry.

---

[1]The PAM-4 levels {+3, +1, -1, -3} are represented by the following 2-bit coding scheme (AM bit , PM bit): {(1,1), (0,1), (0,0), (1,0)}. For both (I/Q) 16-QAM components, the AM bit denotes a high/low amplitude level while the PM bit indicates a positive/negative phase.

978-1-6654-8495-4/22 $31.00 © 2022 IEEE

bandwidth of the baseband chain. A compact, staggered layout configuration for all inductors is adopted to minimize the bandwidth degradation caused by long inter-stage routing, as can be seen in Fig. 4. The CTLE functions as a variable-gain amplifier and provides programmable frequency peaking to further compensate for bandwidth-limiting circuits in the receiver chain. The amplified PAM-4 signals are sent to the decoder for further demodulation and are also brought off-chip for testing purposes. The amplifier chain provides 23 dB of simulated gain with 22 GHz of baseband BW to the input of the decoder.

To extract the phase bits, the PAM-4 signals are fed into a limiting amplifier after which a sign-check comparator outputs a binary signal determined by the polarity of the PAM-4 input. To perform the demodulation of the amplitude bits, the PAM-4 input is first processed by a rectifier which maps the four levels onto two amplitude levels. After AC coupling, the amplitude bits can easily be recovered by another self-biased sign-check comparator. Finally, all demodulated signals are passed through a differential inverter chain to provide clean, binary signals and to drive the instrument load. The inverter chain is scaled up to drive the load and weak cross-coupled inverters are added along the chain to maintain signal balance.

IV. MEASUREMENT RESULTS

The presented receiver was fabricated in a 28 nm CMOS technology. The die micrograph is shown in Fig. 4 and occupies an area of $3.61\,\text{mm}^2$. The measurement set-up is depicted in Fig. 5. For CW measurements, a D-band signal is generated by means of a D-band extender and brought on-chip via a $50\,\mu\text{m}$ pitch GSG probe. The input was calibrated using a PM5 power meter. For modulated measurements, a custom transmitter chip is used which can generate a 16-QAM modulated signal at 135 GHz. PRBS7 data sequences are generated with an arbitrary waveform generator and the carrier is derived from a low frequency reference locked to that of the receiver. The transmitter and receiver chips are flip-chip integrated with wideband Vivaldi couplers on

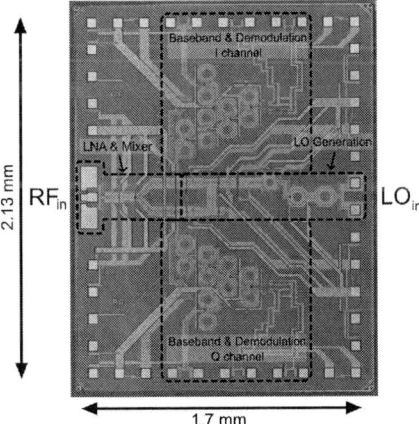

Fig. 4. Die micrograph of the proposed receiver.

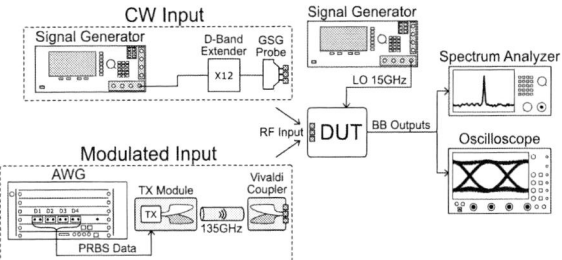

Fig. 5. Measurement setup.

PCB and linked via 1 m of dielectric waveguide [10]. The baseband outputs of the receiver are connected to a spectrum analyzer or oscilloscope for spectral and time-domain analysis, respectively.

The measured receiver conversion gain (to the PAM-4 test output) and input return loss are shown in Fig. 6(a). The receiver is well matched to $50\,\Omega$ across the desired frequency band and demonstrates a maximum gain of about 25 dB with a 3-dB BW of 16 GHz. The DSB NF is given in Fig. 6(b) and shows a minimum value of 6 dB. The integrated NF over a 15 GHz band is 9.8 dB.

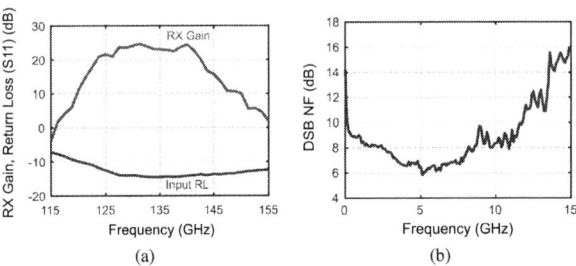

Fig. 6. Measured conversion gain and input return loss (a) and DSB noise figure (b) of the receiver.

To demonstrate the performance of the RF front-end and baseband circuitry, a 16-QAM modulated signal at 135 GHz was used as the RX input and the PAM-4 test outputs were observed. Figure 7 shows the eye diagram and reconstructed 16-QAM constellation measured at a symbol rate of 6 GBaud. The functionality of the PAM-4 decoder was verified by observing the demodulated AM and PM outputs bits and comparing them to the input data supplied to the transmitter chip to perform BER measurements. A 5 GBaud (20 Gb/s) 16-QAM signal was successfully demodulated with a BER below 1e-6 (limited by the memory of the oscilloscope). Figure 8 shows the eye diagrams of the demodulated amplitude and phase bits at this symbol rate. At 6 GBaud (24 Gb/s) a BER below 1e-3 was achieved.

The modulated measurements were performed without any equalization, pulse shaping or predistortion, and thus include all impairments (e.g., limited bandwidth and dispersion) caused by the transmitter chip, PCB couplers, and dielectric waveguide. The receiver consumes 175 mW from a 0.9 V supply while demodulating, including all $50\,\Omega$ buffers.

978-1-6654-8495-4/22 $31.00 © 2022 IEEE

Table 1. Comparison of high-speed, sub-THz direct-digital demodulation receivers.

Reference	ISSCC 2017 [2]	JSSC 2020 [3]	RFIC 2014 [4]	ESSCIRC 2021 [5]	JSSC 2019 [1]	This Work	
Technology	55nm SiGe	65nm CMOS	45nm SOI	40nm CMOS	55nm SiGe	**28nm CMOS**	
Carrier Frequency (GHz)	130	180	155	272	125	**135**	
Modulation Scheme	OOK	MSK	QPSK	QPSK	8-PSK	**16-QAM**	
Demodulator	Envelope Detector	PLL	IQ Zero-IF	Costas-Loop	Multi-Phase RF-Correlator	**PAM-4 Decoder**	
Data Rate (Gb/s)	12.5	12.5	20	8	36	**20**	**24**
BER	1e-6	4e-5	1e-6	2e-5	1e-6	**1e-6**	**1e-3**
Conversion Gain (dB)	/	8	23	/	32	**25**	
Noise Figure (dB)	/	18.6 (SSB)	/	/	10.3 (DSB)	**6 (DSB)**	
Area (mm^2)	3.2*	0.9	3.92*	10	8.75	**3.61**	
Power Consumption (mW)	38	160	345*	1240	200	**175**	
RX Efficiency (pJ/bit)	3.04	12.8	17.25*	155	5.56	**8.75**	**7.29**

*includes the full transceiver

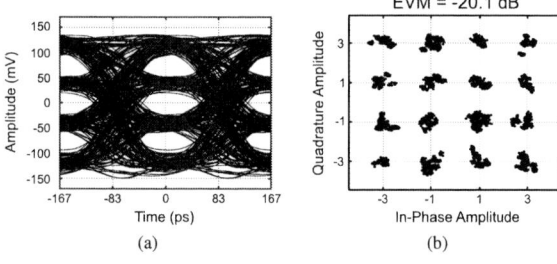

Fig. 7. Measured eye diagram (a) and reconstructed constellation (b) of a 6 GBd 16-QAM signal at the PAM-4 test output.

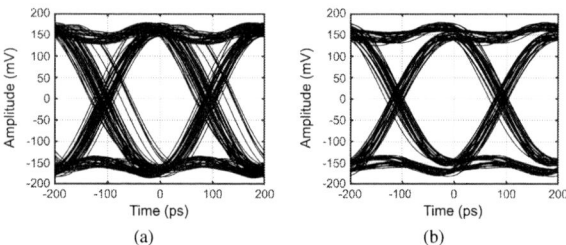

Fig. 8. Measured eye diagrams of the AM (a) and PM (b) demodulated NRZ outputs at 5 GBd 16-QAM input.

Table 1 compares the presented receiver to other state-of-the-art high-speed sub-THz receivers with direct-digital demodulation schemes. This work shows the highest-modulation order with direct-digital demodulation and achieves the highest data rate in a CMOS technology with excellent energy efficiency.

V. CONCLUSION

This paper proposed a 135 GHz receiver architecture capable of directly demodulating a 16-QAM signal at high speeds in the analog domain. The receiver was implemented in 28 nm CMOS and its performance was verified with measurements. The efficient, high-speed operation without the need of a multi-bit ADC or any extensive DSP confirms that the proposed 16-QAM receiver is a promising solution for future high-speed energy efficient communication systems.

REFERENCES

[1] H. Mohammadnezhad, H. Wang, A. Cathelin, and P. Heydari, "A 115–135-GHz 8PSK Receiver Using Multi-Phase RF-Correlation-Based Direct-Demodulation Method," *IEEE Journal of Solid-State Circuits*, vol. 54, no. 9, pp. 2435–2448, 2019.

[2] N. Dolatsha, B. Grave, M. Sawaby, C. Chen, A. Babveyh, S. Kananian, A. Bisognin, C. Luxey, F. Gianesello, J. Costa, C. Fernandes, and A. Arbabian, "17.8 A Compact 130GHz Fully Packaged Point-to-Point Wireless System with 3D-Printed 26dBi Lens Antenna Achieving 12.5Gb/s at 1.55pJ/b/m," in *2017 IEEE International Solid-State Circuits Conference (ISSCC)*, 2017, pp. 306–307.

[3] S. Dong, I. Momson, S. Kshattry, P. Yelleswarapu, W. Choi, and K. K. O, "A 10-Gb/s 180-GHz Phase-Locked-Loop Minimum Shift Keying Receiver," *IEEE Journal of Solid-State Circuits*, vol. 56, no. 3, pp. 681–693, 2021.

[4] Y. Yang, S. Zihir, H. Lin, O. Inac, W. Shin, and G. M. Rebeiz, "A 155 GHz 20 Gbit/s QPSK transceiver in 45nm CMOS," in *2014 IEEE Radio Frequency Integrated Circuits Symposium*, 2014, pp. 365–368.

[5] S. Lee, S. Hara, R. Dong, K. Takano, S. Amakawa, T. Yoshida, and M. Fujishima, "A 272-GHz CMOS Analog BPSK/QPSK Demodulator for IEEE 802.15.3d," in *ESSCIRC 2021 - IEEE 47th European Solid State Circuits Conference (ESSCIRC)*, 2021, pp. 415–418.

[6] M. H. Eissa, N. Maletic, E. Grass, R. Kraemer, D. Kissinger, and A. Malignaggi, "100 Gbps 0.8-m Wireless Link based on Fully Integrated 240 GHz IQ Transmitter and Receiver," in *2020 IEEE/MTT-S International Microwave Symposium (IMS)*, 2020, pp. 627–630.

[7] K. K. Tokgoz, S. Maki, J. Pang, N. Nagashima, I. Abdo, S. Kawai, T. Fujimura, Y. Kawano, T. Suzuki, T. Iwai, K. Okada, and A. Matsuzawa, "A 120Gb/s 16QAM CMOS millimeter-wave wireless transceiver," in *2018 IEEE International Solid - State Circuits Conference - (ISSCC)*, 2018, pp. 168–170.

[8] Q. Pan, L. Wang, X. Luo, and C. P. Yue, "A Low-Power PAM4 Receiver With an Adaptive Variable-Gain Rectifier-Based Decoder," *IEEE Transactions on Very Large Scale Integration (VLSI) Systems*, vol. 28, no. 10, pp. 2099–2108, 2020.

[9] I. Petricli, H. Zhang, E. Monaco, G. Albasini, and A. Mazzanti, "A 112 Gb/s PAM-4 RX Front-End With Unclocked Decision Feedback Equalizer," *IEEE Transactions on Circuits and Systems II: Express Briefs*, vol. 68, no. 1, pp. 256–260, 2021.

[10] K. Dens, J. Vaes, S. Ooms, M. Wagner, and P. Reynaert, "A PAM4 Dielectric Waveguide Link in 28 nm CMOS," in *ESSCIRC 2021 - IEEE 47th European Solid State Circuits Conference (ESSCIRC)*, 2021, pp. 479–482.

978-1-6654-8495-4/22 $31.00 © 2022 IEEE

A Low-Power and Energy-Efficient D-Band CMOS Four-Channel Receiver with Integrated LO Generation for Digital Beamforming Arrays

Ethan Chou, Nima Baniasadi, Hesham Beshary, Meng Wei, Emily Naviasky, Lorenzo Iotti, Ali Niknejad
Berkeley Wireless Research Center (BWRC), University of California, Berkeley, CA

Abstract—This work presents a D-band receiver characterized inside a four-channel transceiver with integrated local oscillator generation and distribution, intended for use in digital beamforming arrays. The receiver leverages a low-noise amplifier with an active balun to minimize its noise figure, while the local oscillator is optimized for low power consumption to enable scaling to larger arrays. A prototype implemented in 28-nm CMOS is flip-chip packaged onto an organic interposer with patch antenna arrays. A wireless downlink with the proposed receiver channel capable of supporting a data rate in excess of 12 Gb/s with QPSK and 16-QAM modulation, while achieving one of the lowest DC power consumption levels per element and energy-per-bit at 98 mW/element and 8.1 pJ/bit, respectively, demonstrates competitive performance and the highest level of integration compared to other D-band CMOS receiver wireless links.

Index Terms—CMOS transceiver, D-band transceiver, millimeter-wave, PLL/LO generation and distribution.

I. INTRODUCTION

As demand for higher wireless data capacity continues to grow, carrier frequencies have moved towards the mm-wave band, where a modest fractional bandwidth yields a high absolute bandwidth. Moreover, spatial multiplexing and directivity at mm-wave frequencies incorporating narrow-beam antennas allow more efficient data transmission and higher spectrum utilization. The severe propagation path loss at mm-wave frequencies necessitates using large arrays that provide directivity gains, higher transmitted EIRP, and higher effective signal-to-noise ratio (SNR) at the receiver. Such receive systems operating at D-band have been reported in previous works [1–3]. However, DC power consumption per array element is high for scaling to larger arrays, where the power consumption per element must be minimized.

In this work, the focus is low-power and energy-efficient design of sub-array elements to enable many receiver elements to be utilized and co-packaged together, with minimum concern for thermal effects. Circuit implementations such as an active balun and subharmonic local oscillator (LO) distribution have been carefully optimized to improve noise performance and power consumption, with large-array system tradeoffs in mind. We demonstrate integration of a complete four-channel transceiver and the LO generation and distribution intended for digital-beamforming applications, illustrated in Fig. 1. The focus of this paper is the receiver (RX) and LO generation and distribution. The RX and LO are fully characterized in over-the-air link measurements and summarized in this work.

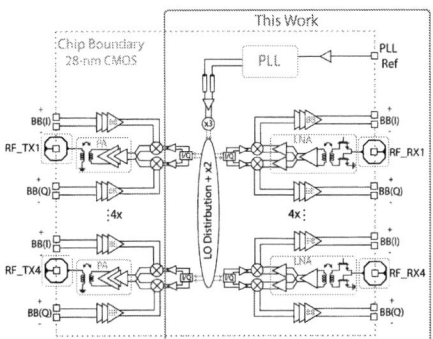

Fig. 1: Block diagram of the D-band 4-channel transceiver.

II. ARCHITECTURE AND CIRCUIT IMPLEMENTATION

A. Receiver Front-end and Baseband Chain

The RX implementation is shown in Fig. 2 and consists of a wideband low-noise amplifier (LNA) with an active balun input stage and I/Q splitter, double-balanced passive mixers, and broadband variable-gain baseband amplifiers.

The active balun replaces the conventional passive balun, which converts the single-ended input to differential signals before the LNA. Passive baluns are lossy and contribute significantly to the overall LNA noise figure (NF). An active balun, consisting of a common-source stage in parallel with a common-gate stage, avoids the passives losses by providing power gain and thus improving the NF. Noise from the active devices is suppressed by maximizing the common-mode output impedance of the subsequent transformer-based matching network. The remaining LNA gain stages are implemented as common-source differential pairs, unilateralized with dummy MOS capacitors to obtain high reverse isolation at mm-wave. The transistors are biased for optimal trade-off between minimum noise and power gain at $200\mu A/\mu m$. The amplification stages utilize low-k, broadside-coupled transformers for inter-stage matching so that only the top-most redistribution metal layer is used for both primary and secondary coils. These transformers achieve an inductor quality factor of $Q>25$ and low insertion loss at 140 GHz. The I/Q splitter leverages inductive terminations with the effective shunt capacitance of short ($l<\lambda/4$) transmission lines to synthesize a low-loss, 1:2 impedance transformation network with a wideband equiripple

978-1-6654-8495-4/22 $31.00 © 2022 IEEE

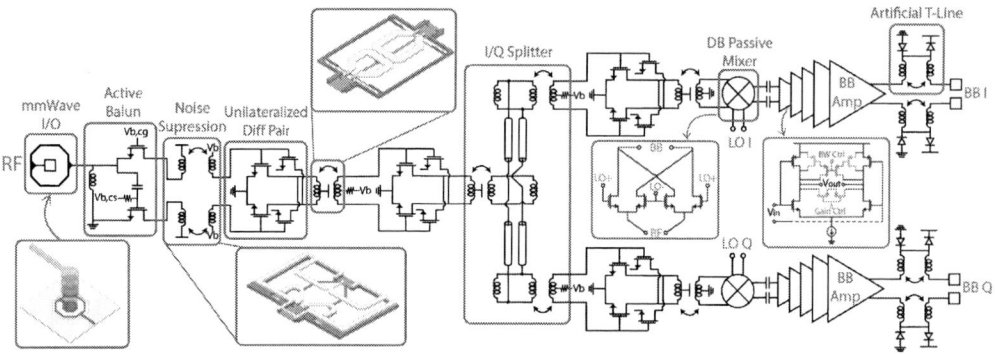

Fig. 2: RX implementation with 3-D EM view of the low-loss mm-wave package-to-chip transition and inter-stage matching networks.

Chebyshev filter response, using cascaded high-k transformers to maintain wide bandwidth. The overall simulated LNA, including the mm-wave interposer-to-chip I/O transition to mixer inputs, achieves a gain of 21 dB, 3-dB BW of 13 GHz, and NF of 6.4 dB at 140 GHz. The broadband mm-wave transition contributes a full-wave 3-D EM-simulated loss of 1 dB. Passive mixers are then used for direct down-conversion to minimize power consumption and reduce flicker noise.

The cascaded unit cell of the tapered baseband amplification chain is implemented with complementary differential pairs and folded active inductors to increase g_m efficiency and bandwidth, respectively. The gate voltage of a triode transistor load is used to tune the gain. Finally, series inductors are used to form an artificial transmission line with the pad and ESD capacitances to obtain high bandwidth. The overall baseband amplification has a simulated gain tuning range of $2-32$ dB and 3-dB BW > 10 GHz across all gain settings.

B. LO Generation and Distribution

The LO implementation is shown in Fig. 3. The LO generation consists of a central 23.3 GHz integer-N, Type-II analog phase-locked loop (PLL). Because phase noise in wideband mm-wave systems can be dominated by the far-out phase noise within the signal bandwidth and close-in phase noise can be filtered with a carrier recovery and phase noise tracking loop for large beamforming arrays [4], this moderate-performance but low-power PLL architecture is used. Moreover, by running the PLL in this lower frequency range, the oscillator benefits from higher quality factor tank components to achieve lower phase noise performance and higher tuning range with lower power consumption. The voltage-controlled oscillator (VCO) is implemented as a resistively-coupled dual-core cross-coupled topology with inductive degeneration to further reduce PLL phase noise. In simulation, this VCO achieves $21-27$ GHz frequency range and phase noise of < -110 dBc/Hz at 1 MHz offset. Following the PLL, a tail-switching injection-locked frequency tripler [5] multiplies the LO frequency to 70 GHz.

The LO is then distributed to eight elements. Considering the frequency-dependent performance of the amplifiers' G_{max} and P_{sat}, the losses of the transmission line routing, and power consumption of the push-push frequency doublers, an optimized portion of the power splitting stages is performed at 70 GHz to maximize the DC power consumption efficiency. Qualitatively, implementing more splitting stages at 70 GHz increases the number of required doublers, but this penalty is outweighed by the benefits of more power-efficient amplifiers and lower-loss splitters at lower subharmonic frequency. Similar to the RX I/Q splitter, the non-isolating power splitters employ inductive terminations for impedance matching to minimize loss. After frequency-doubling at each pair of elements and splitting at 140 GHz, quadrature generation using a branchline hybrid with 100 Ω characteristic impedance is performed at each element. The hybrid adopts an input and output impedance matching technique [6] to reduce the driver current consumption for a given voltage swing with minimal impact on quadrature accuracy, but with transformer-based instead of π matching networks for bandwidth enhancement. Finally, transformer-coupled LO buffers drive the mixer inputs.

III. MEASUREMENTS

The four-channel transceiver test chip, shown in Fig. 4, was fabricated in a standard 28-nm CMOS process and flip-chip packaged onto an 8-layer organic build-up substrate interposer with integrated wideband patch antennas for over-the-air measurements. The antenna has a simulated peak gain of ~6 dBi. The measured insertion loss of the antenna-to-chip routing is 1.4 dB. The interposer is mounted onto a PCB as a ball grid array module. A single RX element occupies an area of 1.2 mm^2, including signal and multiple supply and ground pads. While the test chip integrates four receive channels, only a single RX channel was enabled for the following measurements. Performance across the RX elements is consistent. The measured power consumption for the PLL amortized over the number of elements is 3 mW/element, while the remaining LO distribution and frequency multiplication circuits consume a measured 40 mW/element. The RX consumes a measured 55 mW/element.

The LO generation is measured by using a 2.82 GHz PLL reference and analyzing a divide-by-8 output signal of the PLL, as shown in Fig. 5a). The measured PLL locking range is $19.8-26.6$ GHz, or ~29%. Phase noise of the free-running

978-1-6654-8495-4/22 $31.00 © 2022 IEEE

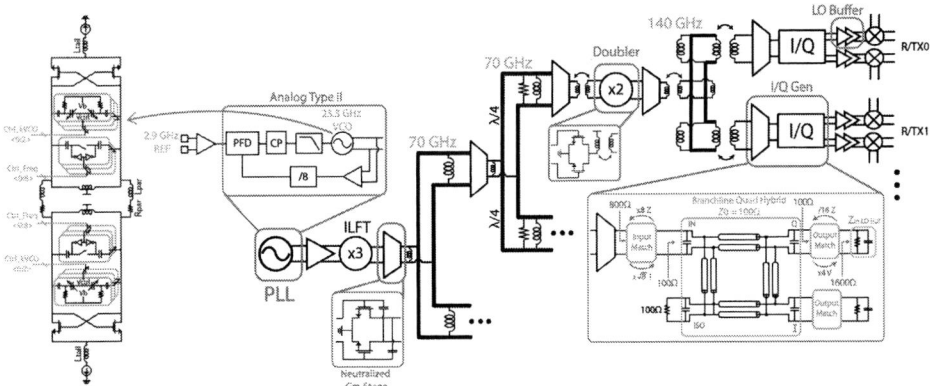

Fig. 3: Schematic of the LO chain implementation, highlighting 23.3 GHz VCO and 140 GHz quadrature generation network.

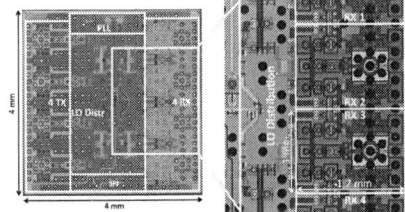

Fig. 4: Chip micrograph (left) and detailed view of the RX and LO distribution portion (right).

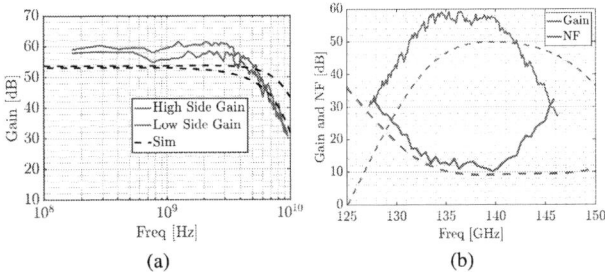

Fig. 7: (a) Measured (solid) and simulated (dashed) RX IF conversion gain for low and high-side injection and (b) SSB NF.

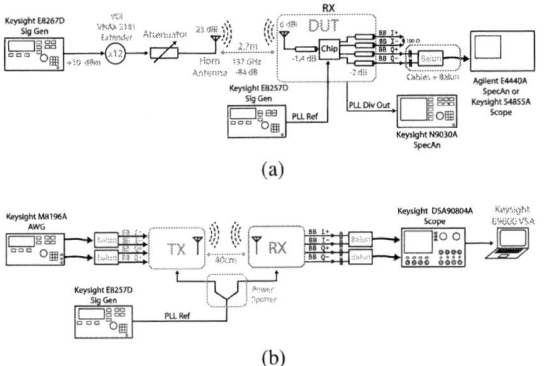

(a)

(b)

Fig. 5: (a) Setup for continuous-wave measurements and (b) wireless downlink measurements.

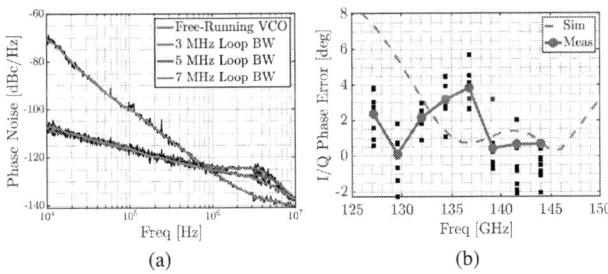

(a)

(b)

Fig. 6: (a) Measured phase noise of PLL divide-by-8 output at 2.82 GHz under various loop bandwidths and (b) average quadrature phase error over carrier frequency without calibration. The black squares represent multiple measured iterations that were averaged.

VCO and locked PLL measured using this divided output signal are shown in Fig. 6a) for various PLL loop bandwidth settings. Quadrature phase accuracy is measured by comparing the quadrature outputs of a received 48 MHz IF tone captured on a scope using the setup in Fig. 5a). The measured quadrature phase accuracy without calibration versus carrier frequency is shown in Fig. 6b). The measured image-rejection ratio without calibration ranges between 28 to > 40 dBc across the same 127−145 GHz frequency band.

The RX is characterized over-the-air at a 2.7m transmission distance using a WR-6 ×12 frequency extender and horn antenna as a D-band continuous-wave (CW) source, as shown in Fig. 5a). The frequency of the CW source is swept with a fixed LO frequency to measure the IF conversion gain and bandwidth. To decouple the performance of the integrated RX from the measurement setup, the measured extender output power, path loss, cable loss, and on-package antenna gain and routing losses are used for back-calculation from the measured response. Single-sideband (SSB) NF is measured by observing the output noise floor at room temperature on the spectrum analyzer (SA), with resolution bandwidth BW_{res}, of only one of the RX's I/Q channels over various LO frequencies. After ensuring that the noise contribution of the SA itself is significantly lower than the noise introduced by the RX, the SSB NF can be calculated from the output noise floor F_{out} as

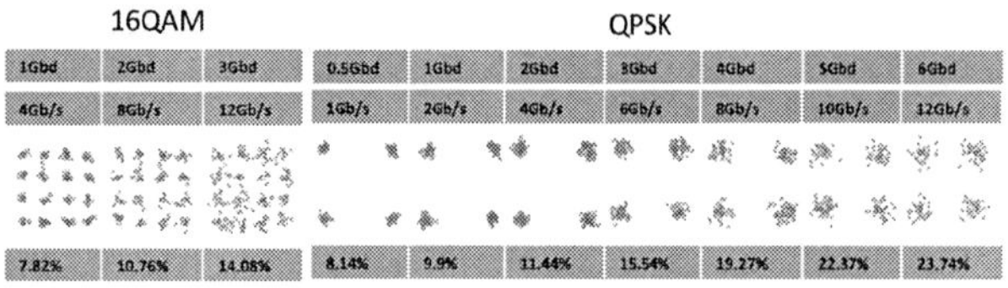

Fig. 8: Measured QPSK and 16-QAM constellations and EVM for various data rates and BER $< 10^{-3}$.

follows:

$$NF_{SSB} = (F_{out} - G_{RX}) + 174\text{dBm/Hz} - 10\log(BW_{res})$$

where G_{RX} is the measured conversion gain of the RX referred to its chip input after accounting for the on-package RF routing and antenna efficiency losses. The RX CW characterization is summarized in Fig. 7, exhibiting 3-dB IF bandwidth of up to 5 GHz and SSB NF of approximately 10.8 dB at 136 GHz. The DSB NF should ideally be 3 dB lower.

To characterize the wireless downlink performance of a single receiver channel, the setup in Fig. 5b) is used with a 135 GHz carrier frequency. A modulated baseband signal generated by an arbitrary waveform generator is directly up-converted and transmitted by the D-band integrated transmitter on a separate but identical test chip. The transmitter has a measured EIRP of ∼5 dBm. The link distance is 40 cm to maximize the RX's SNR while maintaining adequate linearity. The down-converted baseband signals are then demodulated with Keysight VSA9600 software with internal equalization. Measured over-the-air constellations and error-vector magnitude (EVM) for QPSK and 16-QAM modulation and various data rates for a bit-error rate (BER) $< 10^{-3}$ are shown in Fig. 8. The wireless link supports data rates in excess of 12 Gb/s for both QPSK and 16-QAM.

Table I reports comparison to other CMOS D-band receiver links. The proposed RX array achieves the lowest DC power per element, which includes LO power. The highest level of integration is achieved with a packaged wireless link and on-chip PLL and LO generation/distribution. The relatively high EVM at low symbol rates can partly be attributed to the higher phase noise of an on-chip PLL, compared to using a frequency multiplier chain or injection-locked oscillator (ILO) driven by an off-chip instrument-class signal source.

IV. Conclusion

The proposed receiver channel has been demonstrated to be capable of supporting a high-capacity wireless downlink with one of the lowest power consumption levels and competitive energy efficiencies. An on-chip PLL for LO generation, LO distribution for multiple elements, and packaging with patch antenna arrays provide a high level of integration not found in similar works. The low power consumption and high level of integration enable realizing larger arrays by tiling to further increase capacity and enable multi-user communications.

Acknowledgment

This work was supported in part by the ComSenTer research center, under the JUMP program, a Semiconductor Research

TABLE I: Comparison with CMOS D-band RX wireless links.

	This work	[1]	[2]	[7]	[3]
CMOS Node	**28nm bulk**	65nm bulk	45nm SOI	28nm bulk	28nm bulk
Freq [GHz]	**135**	140	147	150	113
No. of Elements	**4**	4	8	1	2
IF 3-dB BW [GHz]	**4-5**	N/A	2-3	10§	10*
Peak Gain [dB]	**58**	43	27.5	70	44*
NF [dB]	**7.8♯**	11.0*	6.4^	N/A	8.0*♯
Modulation	**16-QAM**	BPSK	16-QAM	QPSK	16-QAM
Data Rate [Gb/s]	**12**	10	10	12	80
Energy Eff. [pJ/bit]	**8.1**	23.0†	14.5	49.2	5.9
LO Generation	**PLL**	Mult. Chain	Mult. Chain	Mult. Chain	ILO
Antenna Integration	**Organic Substrate**	PCB	Quartz Superstrate	Bond-wire	PCB
DC Power per Element [mW]	**98**	230†	145	591	470

* Simulated   † Including TX power   ♯ Calculated w/ NF$_{DSB}$=NF$_{SSB}$-3dB
∧ Measured from standalone probed LNA   § Estimated from plot
Energy Efficiency = (DC Power per Ele. [mW])/(Data Rate [Gb/s]) [pJ/bit]

Corporation program sponsored by DARPA. The authors thank the TSMC University Shuttle program for chip fabrication, Integrand EMX for EM simulation, and Ken Nishimura and Keysight Technologies for equipment and support.

References

[1] S. Ma, J. Lin, C. J. Ma, and H. Yu, "A 140 GHz Transceiver for 4×4 Beamforming Short-Range Communication in 65nm CMOS," in *IEEE Asia-Pacific Microwave Conference*, Singapore, Dec. 2019.

[2] S. Li, Z. Zhang, B. Rupakula, and G. Rebeiz, "An Eight-Element 140-GHz Wafer-Scale IF Beamforming Phased-Array Receiver With 64-QAM Operation in CMOS RFSOI," *IEEE J. Solid-State Circuits*, Aug. 2021.

[3] A. Townley *et al.*, "A Fully Integrated, Dual Channel, Flip Chip Packaged 113 GHz Transceiver in 28nm CMOS supporting an 80 Gb/s Wireless Link," in *IEEE Custom Integrated Circuits Conference*, Boston, MA, Mar. 2020.

[4] G. LaCaille, A. Puglielli, E. Alon, B. Nikolic, and A. Niknejad, "Optimizing the LO distribution architecture of mm- wave massive MIMO receivers," 2019. [Online]. Available: http://arxiv.org/abs/1911.01339.

[5] L. Iotti, G. LaCaille, and A. Niknejad, "A 57–74-GHz Tail-Switching Injection-Locked Frequency Tripler in 28-nm CMOS," in *IEEE Eur. Solid State Circuits Conf. (ESSCIRC)*, Cracow, Poland, Sep. 2019.

[6] E. Naviasky, L. Iotti, G. LaCaille, E. Alon, and A. Niknejad, "A 71-to-86-GHz 16-Element by 16-Beam Multi-User Beamforming Integrated Receiver Sub-Array for Massive MIMO," *IEEE J. Solid-State Circuits*, vol. 56, pp. 3811–3826, Dec. 2021.

[7] Z. Chen *et al.*, "A 122-168GHz Radar/Communication Fusion-Mode Transceiver with 30GHz Chirp Bandwidth, 13dBm Psat, and 8.3dBm OP1dB in 28nm CMOS," in *IEEE Symposium on VLSI Circuits*, Kyoto, Japan, Jul. 2021.

A Reconfigurable Digital Beamforming V-Band Phased-Array Receiver

Deniz Dosluoglu*, Kun-Da Chu[†], Diego Pena-Colaiocco*, Ivan Zhao[‡], Visvesh Sathe* and Jacques C. Rudell*

*Department of Electrical and Computer Engineering, University of Washington, Seattle, WA, USA

[†]Apple, San Diego, CA, USA [‡]Boeing, Seattle, WA, USA

Abstract—**A scalable 4-element V-band digital beamforming phased-array receiver (RX) implemented in TSMC 28-nm CMOS integrates both an analog front-end and beamforming functions in the digital back-end (DBE). Each element includes a mixer-first front-end (MFFE), reconfigurable-resolution, oversampled in-phase and quadrature analog-to-digital converters (ADCs), adjustable finite impulse response (FIR) filters and beamforming weights applied at the DBE. The DBE can store pre- and post-beamforming data in a 0.5 Mb SRAM. The MFFE has an S_{11} lower than -10 dB across an 8-GHz bandwidth centered around 52.5 GHz, a maximum gain of 27.1 dB and a minimum noise figure of 7.8 dB. The ADC has a maximum SNDR of 21 dB over a 100-MHz bandwidth when configured as a third-order continuous-time delta-sigma modulator. The chip was configured for experiments with low resolution (1-bit) ADCs which use a digitally-controlled dithering method to enhance the array gain. Although the RX only has 4 elements, the intent is to study the ability to recover information using low-resolution ADCs in a massively-arrayed front-end (FE) [1], [2]. Beamsteering at 0° and 45° is demonstrated, and a maximum QPSK data rate of 400 MS/s with a minimum error vector magnitude (EVM) of -16.2 dB is achieved. The power and area consumption for one element including the FE, ADCs and custom digital interface is 96 mW and 0.18 mm², respectively. The area for the entire system is 3.3 mm².**

Index Terms—**Digital beamforming, millimeter-wave (mm-wave), phased-array, receiver**

I. INTRODUCTION

There is a growing body of research investigating millimeter-wave (mm-Wave) receivers (RXs) to support data rates on the order of gigabits/s (Gbps) [2]–[5]. Phased-array beamforming radios provide directionality and array gain to compensate the high path loss at the mm-Wave bands. Digital beamforming (DB) provides several advantages over conventional analog beamforming, including steering of multiple beams, improved adaptivity, and tunability, features that continue to be the focus of implementing next-generation mm-Wave systems [6]. However, DB requires an entire RX chain for every antenna element. This requirement poses a large obstacle for scalability due to challenges of maintaining performance and power-efficiency across a multi-GHz signal bandwidth using conventional RX and analog-to-digital (ADC) architectures.

The RX described in this paper includes an ADC with programmable resolution down to 1-bit when configured as an open-loop comparator. In this mode, the power consumption of N 1-bit ADCs is lower than N multi-bit ADCs at the same sampling rate. The low effective number of bits associated with a single comparator is compensated by utilizing massive

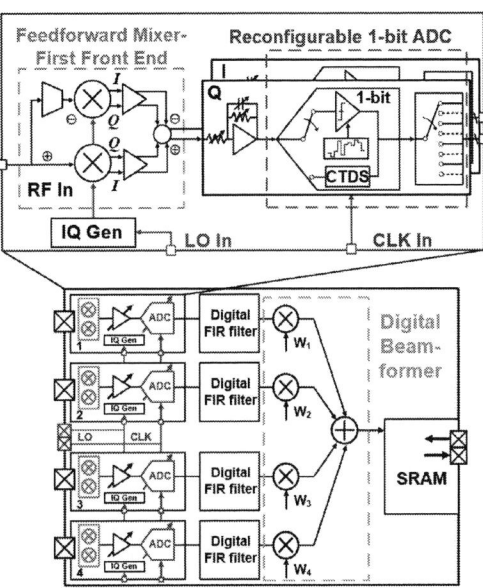

Fig. 1. Block-level diagram of the DB mixer-first RX.

arrays (not implemented on this chip) and dithering methods. A component of this paper examines the differences between data conversion using a Delta-Sigma ADC and a single-bit high-speed comparator.

The following sections describe the implementation of a mixer-first V-band DB CMOS phased-array RX utilizing low-resolution reconfigurable ADCs. Section II provides a top-level description of the system and highlights key circuits in the RX chain. Section III details the measurements of the front-end (FE) and ADC standalone test structures as well as system-level data. The benefits of dithering a 1-bit ADC and the ability to perform on-chip DB are demonstrated. Section IV summarizes the paper with concluding comments.

II. SYSTEM ARCHITECTURE

The proposed 4-element direct-conversion (DC) RX is shown in Fig. 1. This design employs a mixer-first front-end (MFFE), baseband amplification with low-pass filtering, a reconfigurable 8x oversampled (OS) 1-bit ADC, and custom digital timing blocks to interface with the synthesized on-chip digital beamformer (BX) and data demodulator. Both the local oscillator (LO) and high-speed ADC clock (CLK) are supplied from off-chip, then routed to each of the four RX elements using an H-tree distribution strategy. CLK is frequency-divided down to Nyquist rate and provided to the digital back end

978-1-6654-8495-4/22 $31.00 © 2022 IEEE

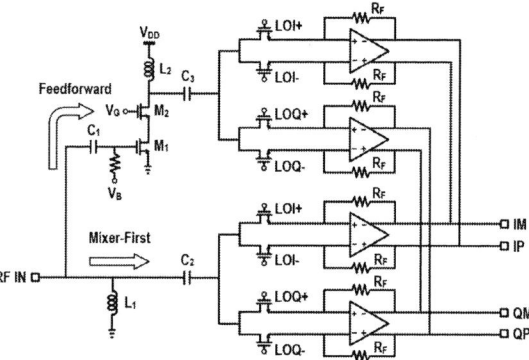

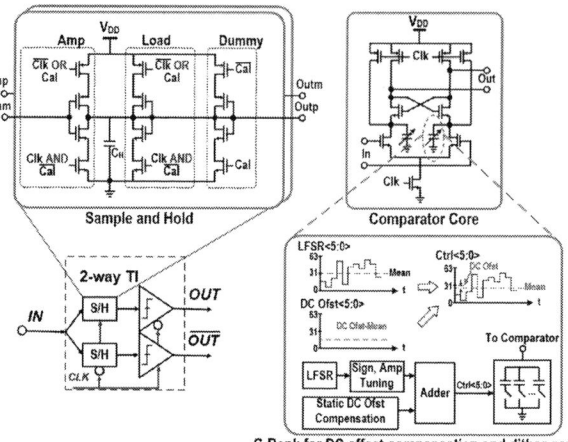

Fig. 3. High-speed comparator implementation with offset cancellation and dither injection circuitry used for the 1-bit open-loop ADC mode.

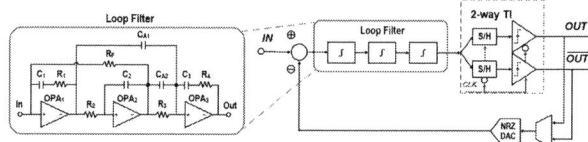

Fig. 4. ADC in CTDSM Mode.

Fig. 2. Mixer-first Direct-Conversion Front-End with an Auxilliary Gm path. (DBE). Data from the output of the decimator, finite-impulse response (FIR) filter and BX can be stored on a 0.5 Mb SRAM. This work uses the BX described in [7].

A. Mixer-First Front-End

The MFFE is employed to improve both the power and area efficiency while maintaining an acceptable linearity across a wide bandwidth compared to the conventional DC RX architecture [8]. All of these attributes are desired to implement an analog FE in a massive DB array that both achieves high data rates and performs spatial filtering in the digital domain. A feedforward path is utilized to improve the FE gain and compensate for the relatively poor noise performance of MFFEs. This RX FE has two paths to amplify the signal: the main mixer-first (MF) path and the auxiliary feedforward Gm path. The mixer-first path is designed to perform the input match while the Gm path provides additional gain. The signal in the main MF path is summed constructively with its amplified replica at the BB output (Fig. 2). In contrast, the noise contributed by the equivalent 50-ohm impedance from the MF path presented at the input has the opposite phase at the Gm input, and is thus suppressed at the transimpedance amplifier output after subtraction. This method is similar to the RF noise-canceling approach described in [9].

B. Reconfigurable ADC

The reconfigurable ADC can operate in two modes (see Fig. 3 and Fig. 4): (1) as an open-loop oversampled 1-bit digitizer in scenarios where low resolution is permissible (e.g. when the input signal is QPSK-modulated [1]) and (2) as a 3rd-order continuous-time delta-sigma modulator (CTDSM) which can achieve a higher resolution as needed.

The samplers in both ADC modes are 2-way time-interleaved (Fig. 3) and clocked at 4x the Nyquist rate to achieve an oversampling rate (OSR) of 8. The sample and hold circuit (S/H) is implemented with an inverter-based input stage with a replica self-biased load to maintain a large output swing at a high sampling rate (F_s) [10]. The binary-weighted 6-bit capacitive bank (C-bank) in the comparator core performs two functions: 1) DC offset calibration and 2) dither injection for the 1-bit mode ADC. When running DC offset calibration, the added dummy branch in the S/H is enabled to provide a stable impedance for the comparator looking back into the S/H. After calibration, the capacitive loading of internal nets in the comparator is continuously altered at the OSR by a uniformly-distributed random bit stream generated by a linear-feedback shift register (LFSR). Since the quantization noise is greater than the RX thermal noise in 1-bit mode, this dither generation helps de-correlate the quantization noise between ADCs to recover the array gain. A 6-flip-flop (LFSR6) or 18-flip-flop LFSR (LFSR18) can be enabled to change the sequence period. The dither amplitude based on simulations can be adjusted from 0.41LSB to 0.019LSB via gating in series with the LFSR output taps, and a static offset can be applied through a pipelined 6-bit adder. Increasing the dither amplitude involves increasing the capacitive load of the comparator pre-amplfier which, in turn, reduces the maximum F_s and increases the minimum comparator offset.

The 3rd-order CTDSM improves the resolution of a 1-bit quantizer utilizing feedback to shape the in-band quantization noise. A capacitive feedforward architecture was implemented for the loop filter. The excess loop delay is compensated by tuning resistors R_1, R_4 and R_f (Fig. 4).

III. MEASUREMENTS

The chip was fabricated in TSMC 28HPC+RF CMOS process and is shown in Fig. 5. Each sub-RX consumes an area of 0.18 mm^2, including I/O pads. In addition to the four RXs, DBE and LO and clock routing, the chip includes pads and test structures to characterize the MFFE and ADC.

A. Front-End Test Structure

The RX FE operates from 48 GHz to 60 GHz achieving a maximum gain of 27.1 dB and a minimum noise figure (NF) of 7.8dB (Fig. 6a). An $S_{11} < -10dB$ was measured over an

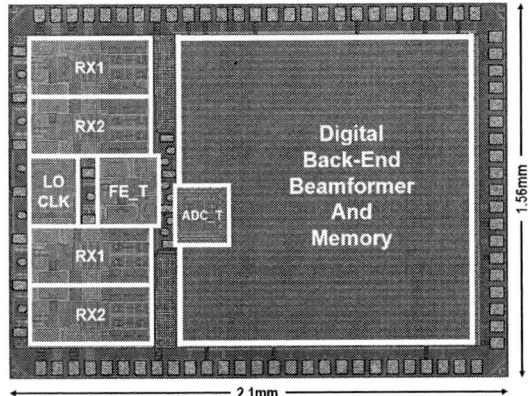

Fig. 5. Die Photo of the 28nm CMOS 4-element DB, total area 3.3 mm^2.

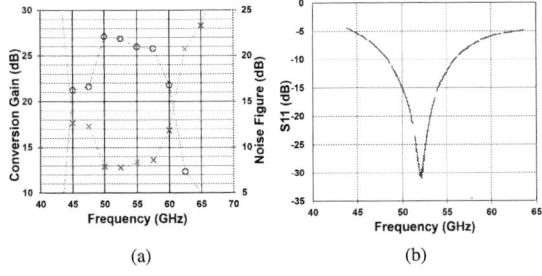

Fig. 6. (a) Measured NF and gain from a single-element MFFE (b) Single-element measured S_{11}

8-GHz bandwidth (Fig. 6b). An input $IP3 > -18dBm$ was measured over the bands of interest.

B. ADC Test Structure

Fig. 7a illustrates the effect dithering has on a 1-bit quantizer in a four-element array. A full-scale input 10-MHz tone was sampled at 2.4 GHz. Without dithering, the distortion noise dominates. The signal distortion is reduced after applying the maximum amplitude 6-bit LFSR6-generated dither sequence at the cost of raising the noise floor (see Table I). However, since the period of this dither sequence is less than the pre-decimated sample size dithering spurs can be observed in the spectrum. To ensure the injected noise spectrum is flat, LFSR18 is utilized to generate a sequence with a period greater than the maximum sample that can be stored on the chip. Only six of the 18 bits are tapped and applied to the C-bank.

The measured magnitude of the frequency response for a 100-MHz full-scale input tone is shown in Fig. 7b for the 3rd order CTDSM. A maximum signal-to-noise and distortion ratio (SNDR) of 21 dB was achieved.

TABLE I
SNDR AND POWER BREAKDOWN OF FIG. 7A

Dither Mode	SNDR (dB)	Signal Bin Power (dBm)	Total Distortion Power (dBm)	Total Noise Power (dBm)
No Dither	9.1	-24.9	-34.2	-50
LFSR6	9.5	-25.7	-39	-37.5
LFSR18	10.2	-25.7	-38.8	-39.2

C. Phased Array System Measurements

The setup for the phased-array measurement is shown in Fig. 8. An AWG was used to generate modulated signals

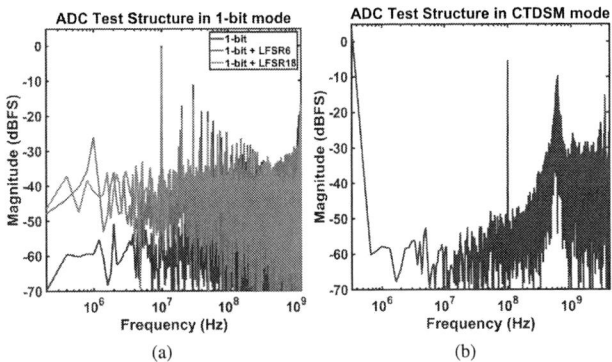

Fig. 7. (a) Measured Output Spectrum of ADC test structure in 1-bit mode with a Full-Scale 10-MHz tone, F_s = 2.4 GHz, (b) Measured Output Spectrum of ADC Test Structure in CTDSM Mode with Full-Scale 100-MHz tone, F_s = 8 GHz

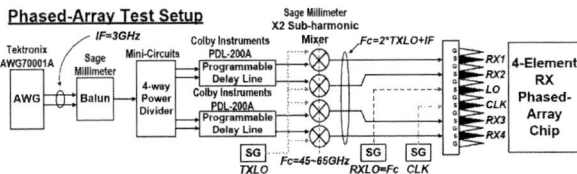

Fig. 8. Test Setup for System Level Phased-Array Measurements

with an intermediate frequency (IF) of 3 GHz. The phase differences between the RX inputs were generated using programmable delay lines to emulate the different delays associated with the incident angles of incoming signals, with respect to the 3-GHz IF. The IF signals were then up-converted to the carrier frequency (52.5 GHz) and supplied to the left side of the chip (Fig. 5) along with the RX LO and CLK using a custom 13-contact probe.

The impact of dithering the ADCs in 1-bit mode is illustrated in Fig. 9 using data collected from the on-chip memory post-beamforming. Without dithering, the measured array gain for the open loop ADCs is less than 6 dB for I and Q. Applying the maximum amplitude LFSR18 dither sequence restores the array gain, as illustrated by the +6-dB SNDR boost post-beamforming at 0°, Table II. Each LFSR is seeded differently so the dither sequences applied to each ADC are uncorrelated.

Two beam patterns steered at 0° and 45° (Fig. 10) were measured using a 200-MHz input tone in 5° steps and using

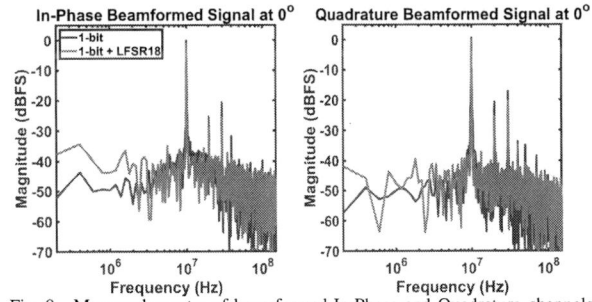

Fig. 9. Measured spectra of beamformed In-Phase and Quadrature channels applying a 52.51 GHz tone at RF IN and enabling the 1-bit ADC mode

TABLE II
SNDR Before and After Beamforming

	1-bit (mean)	1-bit BX	1-bit + LFSR18 (mean)	1-bit + LFSR18 BX
$SNDR_I$ (dB)	9.5	14	7.5	15
$SNDR_Q$ (dB)	10.8	12.8	9.3	16

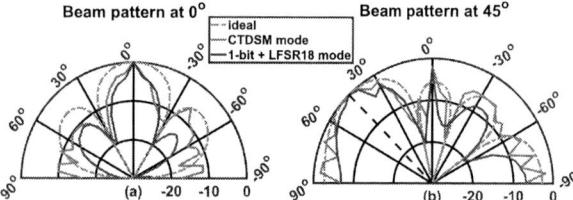

Fig. 10. (a) Beam pattern at $0°$ (b) Beam pattern at $45°$

post-beamformed chip data in both the 1-bit + LFSR18 and CTDSM ADC modes. A 400-MS/s QPSK signal applied at $0°$ was recovered at the output of the ADC decimation filters. After an unsigned to signed conversion off-chip, the data was fed back and beamformed on-chip. The beamformed output was then demodulated in MATLAB. The constellations using ADCs in 1-bit + LFSR18 and CTDSM modes are shown in Fig. 11, with calculated EVMs of -13.4 and -16.2 dB, respectively. Per-element power consumption is illustrated in Fig. 12. The measured power consumption of the ADC in 1-bit + LFSR18 (Dither) mode is eclipsed by that of the custom digital timing blocks required to downsample the ADC bits prior to being processed by the DBE. However, the ADC itself is not a power bottleneck in scaling this architecture. The DBE power consumption during beamforming is 373.2 mW with a VDD = 1.2 V and clock rate of 1.2 GHz (divided down from the 4.8 GHz CLK supplied at the FE). Table III compares this work with prior mm-Wave phased-array RXs.

IV. Conclusion

This work demonstrates the implementation and system-level measurements of a four-element DB RX and motivates utilization of MFFE and 1-bit ADC architectures in massive DB arrays. Demodulation of a 400 MS/s QPSK signal, on-chip beamforming and the impact of dithering to recover the system array gain were demonstrated. We achieve a signal bandwidth 4x greater than measured in [2] while reducing per-element power and area consumption.

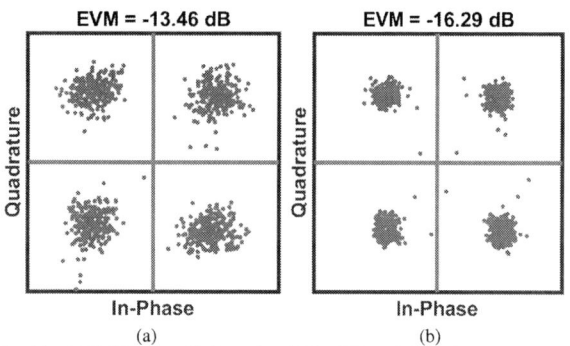

Fig. 11. (a) QPSK constellation using 1-bit + LFSR18 (b) QPSK constellation using CTDSM

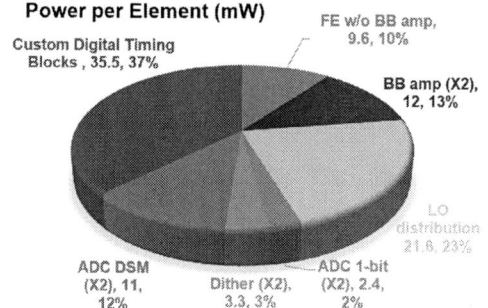

Fig. 12. Phased-Array RX Power Breakdown Per Element

TABLE III
Comparison Table

		This Work	Lu et al. JSSC'21 [2]	Pellerano et al. ISSCC'19 [3]	Park et al. ISSCC'20 [4]	Naviasky et al. ISSCC'21 [5]
	Technology	28nm	40nm	22nm FinFET	28nm	28nm
	Frequency (GHz)	47-57	28	71-76	39	71-86
	Elements per IC	4	16	4	16	16
Beam Former	Type	Digital	Digital	Analog	Analog	Analog
	Phase Res (bits)	10	10	8	4	5
RX FE	Gain (dB)	27.1	21	36.7		35
	NF (dB)	7.8-12	7	6	4.2-4.6	9-11
	BB BW (MHz)	400	100	2000	800 / 100	2000
ADC	Number of ADCs	8	64	-	-	-
	SNDR	21	32	-	-	-
Data Mod.	Constellation	QPSK	4QAM	16QAM	64QAM	16QAM
	Data Rate (MS/s)	400	5	1000	-	250
	EVM (dB)	-16.2	-18	-24	-34.8 / -37.9	-19.2
	Power per Element (mW)	96*	177**	42	39	107
	Area (mm²)	0.18 (1 RX)	0.48 (1 RX)	1.26 (1 RX)	1.875 (1 RX)	16

*MFFE+BB Amps+ADC+Digital Timing   **Front End+ADC+PLL+BSP

Acknowledgment

This work has been supported by the NSF (1408575), Boeing, Qualcomm and CDADIC. We also thank Professor Subhanshu Gupta for his advice.

References

[1] J. Mo *et al.*, "Capacity Analysis of One-Bit Quantized MIMO Systems With Transmitter Channel State Information," *IEEE Transactions On Signal Processing*, vol. 63, no. 20, pp. 5496–5512, 2015.

[2] R. Lu *et al.*, "A 16-Element Fully Integrated 28-GHz Digital RX Beamforming Receiver," *IEEE Journal of Solid-State Circuits*, vol. 56, no. 5, pp. 1374–1386, 2021.

[3] S. Pellerano *et al.*, "A Scalable 71-to-76GHz 64-Element Phased-Array Transceiver Module with 2x2 Direct-Conversion IC in 22nm FinFET CMOS Technology," in *IEEE ISSCC*, Feb. 2019, pp. 176–178.

[4] H. C. Park *et al.*, "A 39GHz-Band CMOS 16-Channel Phased-Array Transceiver IC with a Companion Dual-Stream IF Transceiver IC for 5G NR Base-Station Applications," in *IEEE ISSCC*, Feb. 2020, pp. 76–78.

[5] E. Naviasky *et al.*, "A 71-to-86-GHz 16-Element by 16-Beam Multi-User Beamforming Integrated Receiver Sub-Array for Massive MIMO," *IEEE Journal of Solid-State Circuits*, vol. 56, no. 12, pp. 3811–3826, 2021.

[6] S. H. Talisa *et al.*, "Benefits of Digital Phased Array Radars," *Proc. IEEE*, vol. 104, no. 3, pp. 530–543, 2016.

[7] D. Pena-Colaiocco *et al.*, "An Optimal Digital Beamformer for mm-Wave Phased Arrays with 660MHz Instantaneous Bandwidth in 28nm CMOS," in *IEEE ISSCC*, Feb. 2022, pp. 1–3.

[8] P. Song *et al.*, "mm-Wave Mixer-First Receiver With Selective Passive Wideband Low-Pass Filtering," *IEEE Journal of Solid-State Circuits*, vol. 56, no. 5, pp. 1454–1463, 2021.

[9] D. Murphy *et al.*, "A Blocker-Tolerant, Noise-Cancelling Receiver Suitable for Wideband Wireless Applications," *IEEE Journal of Solid-State Circuits*, vol. 47, no. 12, pp. 2943–2963, 2012.

[10] O. Mattia *et al.*, "An 80 GS/s 5.5 ENOB Time-Interleaved Inverter-Based CMOS Track-and-Hold," *Electronics Letters*, vol. 56, no. 7, pp. 328–329, 2020.

A 112-Gb/s Single-Ended PAM-4 Transceiver Front-End for Reach Extension in Long-Reach Link

Xiongshi Luo [1], Xuewei You [1], Jiahan Fu [1], Zhenghao Li [1], Liping Zhong [1], Taiyang Fan [1], Zhang Qiu [1], Wenbo Xiao [1], Yong Chen [2], and Quan Pan [1]

[1] School of Microelectronics Engineering, Research Center of Integrated Circuits for Next-Generation Communications
Southern University of Science and Technology, Shenzhen, 518055, China

[2] State-Key Laboratory of Analog and Mixed-Signal VLSI and IME/ECE-FST, University of Macau, Macao, China
Email: panq@sustech.edu.cn

Abstract—**This paper presents a 112-Gb/s single-ended (SE) PAM-4 transceiver front-end for the reach-extension module in a 130 nm SiGe BiCMOS technology. The transmitter front-end is based on a differential-to-SE driver where the negative capacitance scheme is introduced to extend its bandwidth. The receiver front-end features a low-mismatch SE-to-differential (S2D) amplifier and an inductor-reuse continuous-time linear equalizer (CTLE). In the S2D, both the asymmetric reused inductor and capacitance compensation techniques are implemented to eliminate the mismatch at the pseudo-differential outputs. In the CTLE, both the inductor reuse and boosted current reuse techniques are adopted to save area, and further boost the maximum peaking frequency and equalization range. Our SE link demonstrates the highest 112-Gb/s PAM-4 data rate at a 20-dB channel loss with an energy efficiency of 1.81 pJ/bit.**

Keywords—*SiGe BiCMOS, long-reach link, PAM-4, reach-extension module, single-ended (SE), transceiver (TRX).*

I. INTRODUCTION

With the sharp increase of global internet traffic, wireline transceivers (TRXs) demand a higher data rate per lane. Recently, IEEE P802.3 Beyond 400G Ethernet Study Group has been discussing the signaling schemes of 200 Gb/s per lane. Limited by the passive link, especially the bandwidth (BW) limitation of the connectors and packages, the conventional ADC-DSP-based PAM-4 scheme [Fig. 1(a)] might not be able to support 224 Gb/s per lane for the long-reach link. Currently, there are three potential schemes: 1) a more advanced PAM scheme [1], which requires better linearity and SNR; 2) a single-ended (SE) MIMO scheme, which requires crosstalk cancellation between the differential line; 3) XSR/VSR SerDes + reach-extension TRX module [Fig. 1(b)]. In the last case, the TRX module is implemented for differential-to-SE and SE-to-differential conversion, where the SE signal is transmitting in a SE passive channel composed of the coaxial cables and connectors. The SE signaling could significantly increase the BW density of I/Os, which can efficiently reduce the complexities of chip package and communication equipment panels. Yet, scaling the performance to realize the SE 112-Gb/s PAM-4 signaling – required by the beyond 400G reach-extension module – is not straightforward. Prior SE links were presented in [2-5], and the recorded highest data rate is limited to 40 Gb/s. This work proposes a low-power SE 112-Gb/s PAM-4 TRX front-end, which is suitable for beyond 400G reach-extension modules. The TRX provides 20-dB equalization capability and extends the reach by four meters with cables.

II. TOP ARCHITECTURE OF PROPOSED SE TRX FRONT-END

Fig. 2 depicts the top architecture of the proposed SE TRX front-end link supporting the PAM-4 format. The

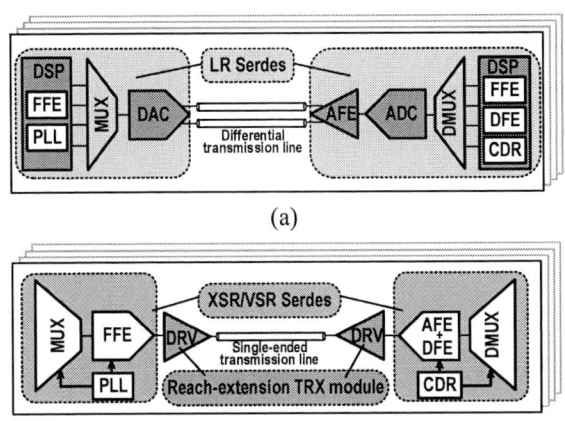

Fig. 1. (a) Conventional ADC-DSP-based long-reach link, and (b) XSR/VSR SerDes + reach-extension TRX module for the long-reach link.

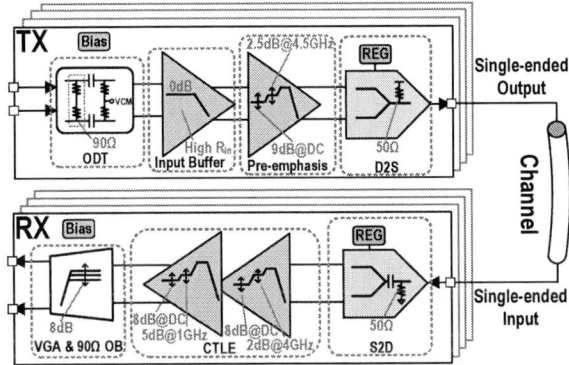

Fig. 2. Top architecture of our SE link based on the proposed TRX front-end.

transmitter (TX) consists of a 90-Ω termination block, a differential input buffer, a pre-emphasis equalizer, and a differential-to-SE (D2S) output driver. The receiver (RX) incorporates a SE-to-differential (S2D) amplifier with a 50 Ω termination, a continuous-time linear equalizer (CTLE), and a 90-Ω variable gain output buffer. The influence of the power distribution network (PDN) is vital to the overall performance. To alleviate the power supply noise and power bounce, two fast-response on-chip low dropout regulators with low output impedance are realized to supply the D2S driver and the S2D amplifier, respectively. The TX and RX are connected via the SE coaxial cables and short traces on the printed circuit board (PCB).

The most crucial sub-block in the TX is the wide-BW D2S [Fig. 3(a)]. To keep an adequate gain in the D2S driver, the transistor size needs to be large, resulting in significant

(a)

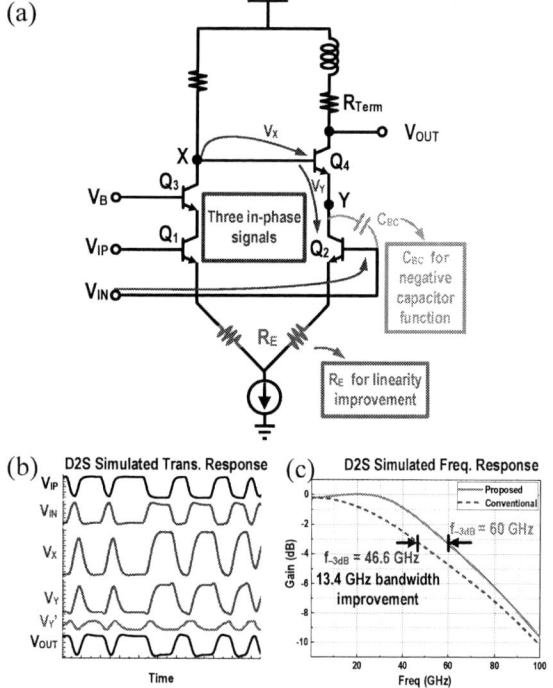

Fig. 3. (a) Proposed D2S driver at the TX side, simulated (b) transient and (c) frequency response.

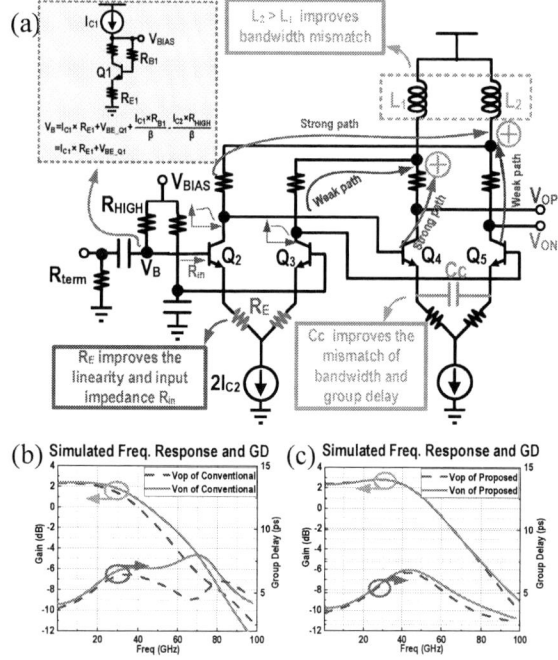

Fig. 4. (a) Proposed S2D at the RX side, simulated (b)frequency response and (c) group delay of the S2D.

parasitics. To address this issue, we develop the D2S driver with one novel negative capacitor technique. The negative capacitor is generated by the in-phase amplification. The input signal V_{IP} is amplified by the cascode amplifier formed by Q_1 and Q_3, and then fed into the base of Q_4. Compared to the input signal V_{IN}, signal V_X has a larger amplitude.

(a)

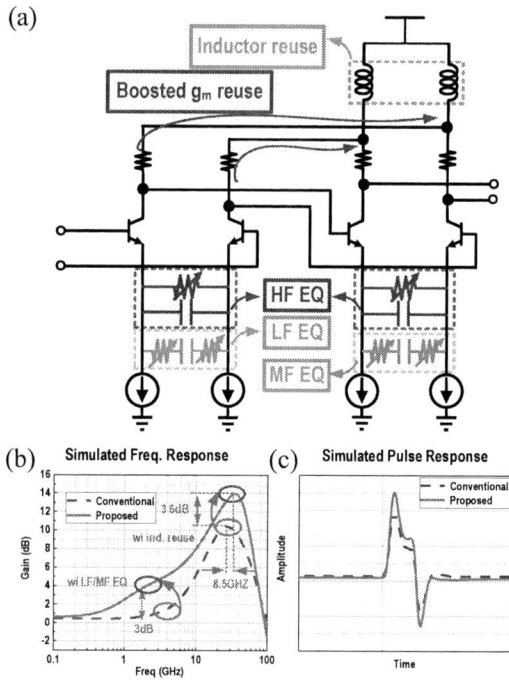

Fig. 5. (a) Proposed CTLE at the RX side, simulated (b) frequency response and (c) pulse response of the CTLE.

Therefore, the signal at node Y is determined by V_X and Q_4, forcing V_Y to follow V_X, which means that V_X, V_Y, and V_{IN} are in phase [Fig. 3(b)]. As V_Y and V_{IN} are in phase, the capacitor (C_{BC}) operates as a negative capacitor, achieving significant BW boosting. R_E is inserted to improve the linearity of the D2S. A conventional D2S driver with Q_4 working as a cascode transistor whose base connects to V_B is designed for comparison purpose. As shown in Fig. 3(b), the signal ($V_{Y'}$) at node Y' (e.g., the same position as node Y) is reverse-phase with V_{IN} and has less swing. The post-layout simulations [Fig. 3(c)] show that the conventional D2S driver provides a 0-dB DC gain and a 46.6-GHz BW, while the proposed D2S has the same DC gain with a 60-GHz BW, indicating a BW-enhanced ratio of 1.3.

The two key sub-blocks in the RX front-end are the S2D amplifier and the CTLE shown in Figs. 4 and 5, respectively. After the 50-Ω input termination, both high-pass and low-pass filters generate the required RX input reference voltages and reject low-frequency common-mode reference voltage noise. To overcome the process, voltage, and temperature (PVT) variations and supply a stable reference voltage, a β-independent bias generator is designed in Fig. 4(a). By ensuring $I_{C1} \times R_{B1}$ equals $I_{C2} \times R_{HIGH}$, the generated bias voltage is independent of β. R_E is inserted to improve the linearity and increase the equivalent input impedance (R_{in}), which also avoids affecting the corner frequency of the high-pass filter. The conventional CML-based S2D suffers from BW mismatch and group delay mismatch due to its built-in pseudo-differential characteristic, leading to eye skew between their two SE outputs [2]. To solve this problem, three design schemes are proposed in Fig. 4(a). The first is the inductor-reuse technique, which realizes the addition of a strong current path and weak current path, achieving a better bandwidth balance. Moreover, these two inductors (L_1 and L_2) are designed to be different values, which can further

978-1-6654-8495-4/22 $31.00 © 2022 IEEE   498

improve the bandwidth mismatch of the outputs. However, due to asymmetric inductors, the group delay deviation of differential outputs becomes more severe. Therefore, the compensation capacitor C_C is employed to ameliorate the group delay mismatch, which also improves the bandwidth mismatch. Our S2D greatly eliminates the BW mismatch up to 70 GHz, and it also achieves a much smaller group delay mismatch as shown in Figs. 4(b) and (c), respectively.

Shown in Fig. 5(a) is the presented inductor-reuse CTLE. It adopts a two-stage topology and achieves area-saving and power-efficient equalization by reusing on-chip inductors and the boosted current from the first stage. Differing from inductor reuse [6], our inductor-reuse CTLE increases the maximum boosting without the extra power consumption and additional input parasitics. Compared to the conventional two-stage source-degenerated CTLE [5] along with separate inductors on both stages, our structure reduces the inductance by 75%. To realize a flat frequency response with different channels, RC networks with different values at the source-degenerated nodes compensate for the channel losses at the high/middle/low-frequency range, respectively. Our high/middle/low network gives a 5-dB equalization tuning range at the low frequency (i.e., ~1 GHz), a 2-dB equalization tuning range at the middle frequency (i.e., ~4 GHz), and an 8-dB equalization tuning range at the high frequency in each stage by gain tuning, respectively (Fig. 2). Compared with the conventional counterpart [5] with separate inductors, our CTLE can successfully not only further extend the maximum peaking frequency by 8.5 GHz and the maximum equalization range by 3.6 dB, but also enhance the equalization at the middle frequency by 3 dB [Fig. 5(b)]. The pulse response also demonstrates the preferable equalization capability of the proposed CTLE [Fig. 5(c)].

The common noise sources in the SE systems mainly come from the power supply noise, power and ground bounce, simultaneous switching output noise, crosstalk, inter-symbol interference (ISI), etc. For beyond 100-Gb/s data rate applications, the influence of the PDN is vital to the overall performance. In this work, for the power supply noise and power bounce, two fast-response regulators with low output impedance are added to accommodate the rapid pseudo-random binary sequence (PRBS) data and isolate the supply network from core circuits. The regulators can provide suppression of the crosstalk between lanes introduced by the shared power supply network. For the ground bounce, the through-silicon via (TSV) technique with the backside metallization is introduced to alleviate the parasitic ground inductance. The crosstalk among different SE lanes in this work is mainly handled by delicating layout skills and signal isolations.

III. Measurement Results

The proposed TRX front-end was fabricated in a 130-nm SiGe BiCMOS technology and tested with the chip on board (CoB). Fig. 6 shows the chip microphotographs and power breakdowns. The core circuit area of the TX and RX are 0.25 × 0.28 mm² and 0.27 × 0.2 mm², respectively.

The measured S-parameters and bathtub curves of both TX and RX are shown in Fig. 7. With a −1.5-dB@28 GHz input cable, PCB traces, Rosengerber RF connectors, a −5.5-dB@28 GHz output cable, a DC block, and a female-to-male coaxial connector, the total channel loss during the TX standalone measurement is −8-dB@28 GHz. The TX chip

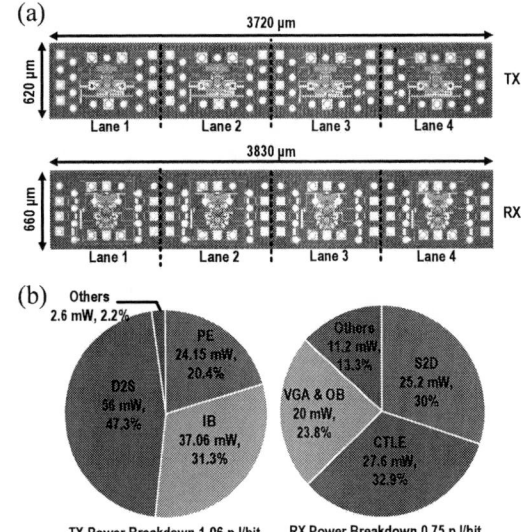

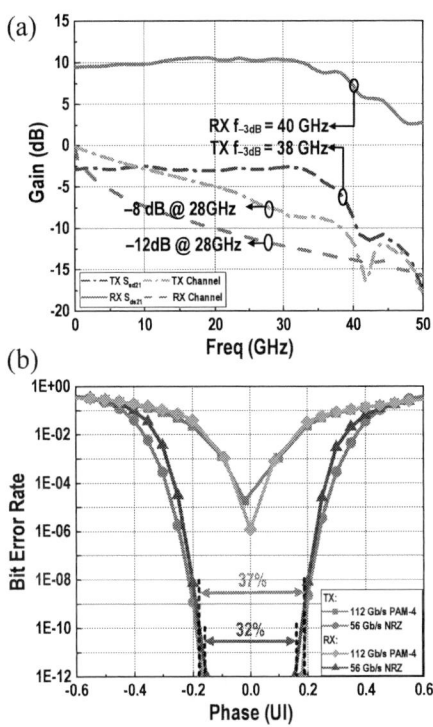

Fig. 6. (a) Chip micrographs and (b) power breakdowns.

Fig. 7. (a) Measured channel loss applied in TX and RX, and measured S parameter of both the TX and RX with channels, (b) measured TX and RX bathtub curves with channels.

along with the channel can achieve a −3-dB BW of 38 GHz. For a 56-Gbaud eye, the TX achieves a bit error rate (BER) of less than 10^{-12} for NRZ and 10^{-4} for PAM-4, respectively. With Rosengerber RF connectors and PCB traces, and a −11-dB@ 28 GHz input cable, the total channel loss during the RX standalone measurement is −12-dB@28 GHz. The RX chip along with the channel can realize a −3-dB BW of 40 GHz. For a 56-Gbaud eye, the RX scores a BER of less than 10^{-12} for NRZ and 10^{-5} for PAM-4, respectively. Figs. 8(a) and 8(b) show the measured eye diagrams at 56 Gbaud of the standalone TX and RX with the respective channel. Fig. 9(a)

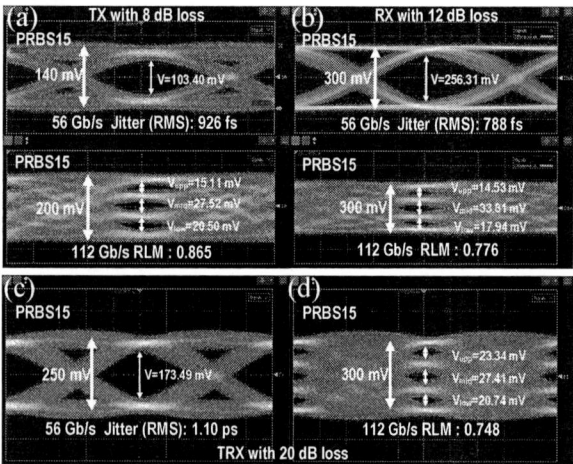

Fig. 8. Measured eye diagrams of (a) TX with 8-dB loss, (b) RX with 12-dB loss, and (c), (d) TRX with 20-dB loss.

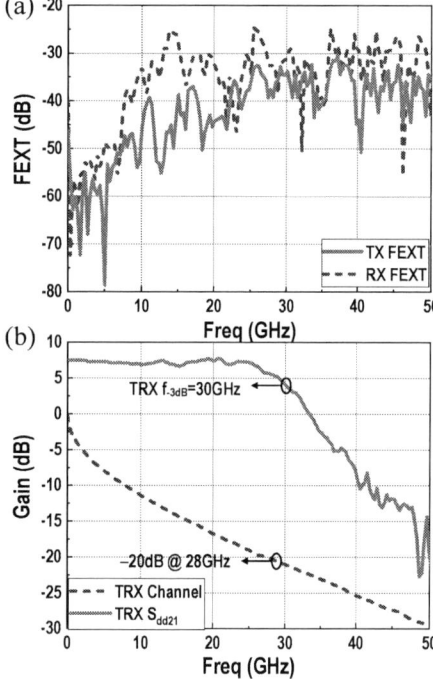

Fig. 9. (a) Measured FEXT of both the TX and RX and (b) measured total channel loss and S parameter of our TRX front-end link.

shows the measured crosstalk performance. The measured far-end crosstalk (FEXT) of adjacent lanes is less than −30 dB in the TX and less than −25 dB in the RX, respectively.

The SE TX outputs from the board are looped back to the SE RX inputs via lossy channels, each of which is composed of two Rosenberg RF connectors, two RF cables, and one SE PCB trace. In the 112-Gb/s TRX full-link measurement setup, the overall SE insertion loss from TX to RX is −20-dB@28 GHz, and the measured −3-dB BW of the TRX is 30 GHz as shown in Fig. 9(b). Figs. 8(c) and 8(d) show the measured TRX data eyes at 56-Gb/s NRZ and 112-Gb/s PAM-4, respectively. Table I summarizes the performance and its comparison with other state-of-the-art SE links. The TX and

RX have an energy efficiency of 1.06 pJ/bit and 0.75 pJ/bit at 112 Gb/s, respectively. To our knowledge, this link achieves the highest data rate of the SE PAM-4 signal in both CMOS and SiGe BiCMOS technologies.

CONCLUSION

This paper reported an area- and power-efficient SE TRX front-end supporting the PAM-4 data in a 130-nm SiGe BiCMOS and the measured data rate is up to 112 Gb/s. The negative capacitor in the TX front-end, the asymmetrical inductor, group delay mismatch calibration, and inductor reuse in the RX front-end are combined to achieve a high equalization capability of 20 dB, better power efficiency of 1.81 pJ/bit, and high-performance D2S-to-S2D conversion. Our work verified the viability of designing a SE beyond 100-Gb/s/lane TRX front-end for the high-speed link in the data center.

TABLE I. PERFORMANCE SUMMARY AND COMPARISION WITH PRIOR SE LINKS

	ISSCC 2019[2]	ISSCC 2020[3]	ISSCC 2021[4]	ESSCIRC 2018[5]	This work
Date Rate (Gb/s)	27	32	40	24 / 0.3125	112
Architecture	Digital& Analog	Analog	Analog	Analog	Analog
Signaling	PAM-3	PAM-4	NRZ	NRZ	PAM-4
Application	MCM	MCM	MCM	Coaxial Link	Coaxial Link
Packaging	Bonding	N/A	Flip-chip	Flip-chip	Bonding
Equalization	1-tap DFE	3-tap FFE&2-tap DFE	CTLE	CTLE	CTLE
Technology	CMOS 28 nm	CMOS 65 nm	CMOS 7 nm	SiGe 130 nm	SiGe 130 nm
Loss (dB) @ Nyquist	7.1	11.6	8	18.3	20
Active Area per Lane (mm²)	0.026	0.0087	0.66	0.23	0.148
Power Efficiency (pJ/bit)	1.03	0.97	1.7	5.81	1.81

REFERENCES

[1] E. Chong, et al., "A 112Gb/s PAM-4, 168Gb/s PAM-8 7bit DAC-based transmitter in 7nm FinFET," *IEEE 47th European Solid-State Circuits Conference (ESSCIRC)*, pp. 523−526, Sep. 2021.

[2] H. Park, et al., "A 3-bit/2UI 27Gb/s PAM-3 single-ended transceiver using one-tap DFE for next-generation memory interface," *IEEE International Solid-State Circuits Conference (ISSCC)*, pp. 382−384, Feb. 2019.

[3] P. -W. Chiu and C. Kim, "A 32Gb/s digital-intensive single-ended PAM-4 transceiver for high-speed memory interfaces featuring a 2-tap time-based decision feedback equalizer and an in-situ channel-loss monitor," *IEEE International Solid-State Circuits Conference (ISSCC)*, pp. 336−338, Feb. 2020.

[4] K. McCollough, et al., "A 480Gb/s/mm 1.7pJ/b short-reach wireline transceiver using single-ended NRZ for die-to-die applications," *IEEE International Solid-State Circuits Conference (ISSCC)*, pp. 1−3, Feb. 2021.

[5] A. Manian, et al., "A simultaneous bidirectional single-ended coaxial link with 24-Gb/s forward and 312.5-Mb/s back channels," *IEEE 44th European Solid-State Circuits Conference (ESSCIRC)*, pp. 178−181, Sep. 2018.

[6] A. Atharav and B. Razavi, "A 56Gb/s 50mW NRZ receiver in 28nm CMOS," *IEEE International Solid-State Circuits Conference (ISSCC)*, pp. 192−194, Feb. 2021.

A 2×50 Gb/s Single-Ended MIMO PAM-4 Crosstalk Cancellation and Signal Reutilization Receiver in 28 nm CMOS

Liping Zhong, Hongzhi Wu, Weitao Wu, Wenbo Xiao, Xiongshi Luo, Dongfan Xu, Xuxu Cheng, Zhenghao Li, Taiyang Fan, and Quan Pan

School of Microelectronics Engineering, Research Center of Integrated Circuits for Next-Generation Communications

Southern University of Science and Technology, Shenzhen, 518055, China

Email: panq@sustech.edu.cn

Abstract—**This paper presents a dual-coupled single-ended (SE) receiver with crosstalk cancellation and signal reutilization (XTCR) technique in 28 nm CMOS process. The asymmetric-inductor SE to differential converter (S2D) is proposed to eliminate the mismatch at its pseudo-differential outputs. The conventional and the proposed active crosstalk extraction (AXTE) techniques are adopted to one of the two SE lanes, respectively. Compared to the conventional one, the proposed AXTE technique can compensate for higher insertion loss without extra power consumption, and improve the measured horizontal and vertical eye openings by 17% and 18% at 28 Gb/s NRZ, respectively. Moreover, the PAM-4 XTCR scheme is realized for the first time, and the measurement results show that it achieves an efficiency of 0.85 pJ/bit/lane for 2×50 Gb/s with a pair of 10-inch PCB traces.**

Keywords—crosstalk cancellation, PAM-4, single-ended, signal reutilization, SE MIMO, wireline receiver.

I. INTRODUCTION

With the recent surge in the demand for high data rates, wireline communication transceiver (TRX) faces new challenges. Considering the various limitations from circuits, connectors, packages, etc., it is extremely difficult to achieve 200 Gb/s per differential link. Therefore, it is worthwhile for researchers to explore interesting ideas that break through these limitations. Nowadays, mainstream links are with differential input-outputs, requiring both terminals P and N, and their corresponding return references to transmit one-way signals. However, these existing differential links can be easily used as single-ended (SE) multiple-input multiple-output (MIMO) transmission to improve I/O efficiency and data throughput.

The idea of the SE MIMO four pulse-amplitude modulation (PAM-4) scheme has been discussed in [1], where delivering two SE signals over one differential or differential-like link is highly promising for high-speed data transmission. The SE MIMO PAM-4 scheme has three main advantages. First, it relaxes the link bandwidth (BW) requirements by half, especially the rigorous limitations from connectors and packages. Second, the link data rate can be improved up to 2X for the same BW. Third, the pin efficiency can be doubled. However, the SE MIMO TRX must possess the crosstalk cancellation (XTC) capability to improve the signal integrity within the SE MIMO link. Although various SE XTC designs based on NRZ format have been reported [2-8], their performance is unsuitable to meet current standards due to low data rates and high power consumption. The recorded highest data rate for SE NRZ XTC receiver is 12 Gb/s [3]. If SE MIMO PAM-4 TRXs with XTC capability can operate at PAM-4 2×50 Gb/s via a pair of differential channels, this would push communication

data rates to upgrade from 200G to 400G. Moreover, this performance promotion can exactly reuse the existing infrastructure with little additional cost.

This paper proposes a SE MIMO PAM-4 crosstalk cancellation and signal reutilization (XTCR) receiver (RX), realizing 2×50 Gb/s data transmission. The active crosstalk extraction (AXTE) and the asymmetric inductor peaking technique are proposed to optimize the circuit, making it suitable for supporting signaling requirements of 400G Electrical Interfaces. Moreover, this work also provides a promising method to realize 200 Gb/s per differential or differential-like link in more advanced technologies.

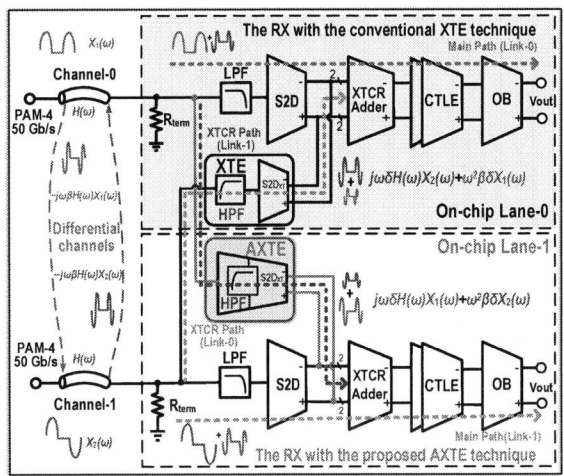

Fig. 1. Top architecture of the proposed SE MIMO XTCR link.

II. ARCHITECTURE

Fig.1 shows the top architecture of the proposed XTCR link. The link consists of a pair of differential channels and two SE lanes. One main path and one XTCR path are parallel to form one SE lane. Each lane consists of a crosstalk extraction (XTE), a low pass filter (LPF), a SE to differential converter (S2D), a XTCR adder, a two-stage continuous-time linear equalizer (CTLE), and an output buffer (OB). The difference between Lane-0 and Lane-1 is the XTE sub-block. The proposed AXTE sub-block is adopted to extract the crosstalk energy from Lane-0 to Lane-1. To make a fair comparison, the conventional crosstalk extraction (XTE) sub-block in [3] is also used to extract the crosstalk from Lane-1 to Lane-0. Note that in order to reduce crosstalk trails, the delay of the XTE should be optimized to be consistent with the S2D. A resistor is added in the main path to form the LPF with the input capacitance of S2D. To suppress the simultaneous switching noise (SSN), the RX employs the

978-1-6654-8495-4/22 $31.00 © 2022 IEEE

S2D as the first stage, and 4-stage differential circuits as the succeeding stages. Different from [4], this work implements XTC in the XTCR adder to reduce pseudo-differential signal mismatch and crosstalk trails. The XTCR adder adjusts the ratio of the XTCR path to the main path by controlling the tail currents. A two-stage CTLE is adopted to provide sufficient compensation at middle- and high-frequencies. The OB circuit is used to drive the following circuits or testing equipment.

Fig. 1 also shows the principles of far-end crosstalk (FEXT) generation and XTCR. Due to inductive and capacitive couplings from adjacent channels, the FEXT will be introduced into the victim channels when the aggressor channels transmit data to the far end. The aggressor channels $(X_1(\omega)H(\omega))$ generate the pulse signal $(-j\omega\beta H(\omega)X_1(\omega))$ associated with its data. The pulse signal impedes the data motion of the victim channel $(H(\omega)X_2(\omega))$. Note that the victim channels are mainly affected by the high-frequency component of the aggressor's signal, and the effect of the low-frequency component is not critical. To eliminate the pulse signal, the high pass filter (HPF) extracts the aggressor's signal to generate the forward pulse signal $(j\omega\delta H(\omega)X_1(\omega)+\omega^2\beta\delta X_2(\omega))$ for XTCR. This forward pulse signal forms a 90° phase shift when the victim's signal passes through the LPF, and adds together with the victim's signal. In addition, $\omega^2\beta\delta X_2(\omega)$ is the boosting signal associated with the victim signal, which is used to eliminate the inter-symbol interference noise.

crosstalk signals from SE to differential. However, the HPF resistor (R_{HPF}) used in the conventional XTE topology will leak part of the crosstalk energy and introduce additional parasitic capacitance. Therefore, the AXTE technique is used in this design to avoid energy leakage. As shown in Fig. 2(b), C_{HPF} is inserted between the pseudo-differential source nodes of the S2D$_{XT}$ to form an active HPF. From the previous analysis, a crosstalk pulse signal is generated when the signal jumps occur (from 0 to 1 level or vice versa). It is clear that most of the useful energy originates from the high-frequency portion of the crosstalk signals. In each period, one pulse of the main path data will generate two pulses of the crosstalk signal. Therefore, the effective frequency range of the crosstalk signal is from $1/(2T_b)$ to $1/T_b$. For the targeted data rate of 28 GBaud, the majority of the crosstalk signal interested is in the frequency range from 14 to 28 GHz.

Fig. 2(c) shows that, compared to the XTE between 14 GHz and 28 GHz, the simulated high-frequency response of the proposed AXTE is improved by 2-3 dB. The AXTE improves the efficiency of crosstalk energy reutilization without increasing circuit complexity and extra power consumption.

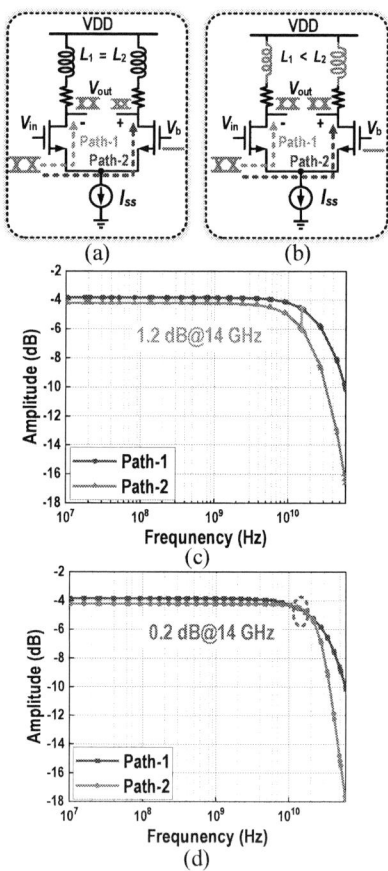

Fig. 3. Schematics of (a) the conventional S2D, (b) the proposed asymmetric-inductor S2D, simulated frequency responses of (c) the conventional S2D, and (d) the asymmetric-inductor S2D.

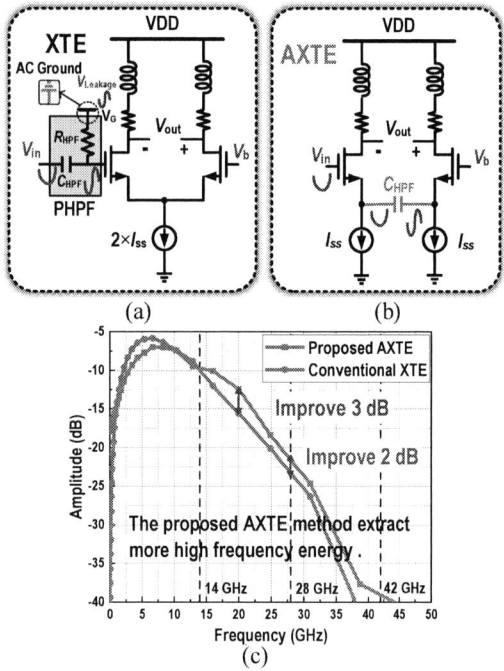

Fig. 2. Schematics of (a) the conventional XTE circuit, (b) the proposed AXTE circuit, and (c) simulated frequency responses of AXTE and XTE.

III. CIRCUIT IMPLEMENTATION

A. Crosstalk Extraction (XTE) Circuit

Fig. 2(a) and (b) depict schematics of the conventional XTE and the proposed AXTE circuit, respectively. The XTE sub-block consists of a HPF and a S2D$_{XT}$. The HPF collects the crosstalk energy and the S2D$_{XT}$ converts the collected

B. Asymmetric-Inductor S2D

As known, the SE topology is vulnerable to the power supply noise and the SSN. To mitigate these noises, the current mode logic (CML) S2D architecture is adopted. Nevertheless, the CML S2D suffers from pseudo-differential

978-1-6654-8495-4/22 $31.00 © 2022 IEEE

signal mismatch. Compared to path-1 in Fig. 3(a), the parasitics are heavier in path-2, which dominates the pseudo-differential signal mismatch.

To alleviate the stated problem, in Fig. 3(b), the asymmetric-inductor ($L_1<L_2$) S2D is proposed to both improve the BW and mitigate the mismatch. Fig. 3(c) and Fig. 3(d) show the simulated frequency responses of the conventional S2D and the proposed asymmetric-inductor S2D, respectively. The conventional S2D has a 1.2 dB offset in frequency response between path-1 and path-2 at 14 GHz, while the proposed S2D significantly reduces the undesired mismatch by 83.3%.

C. XTCR Adder and 2-Stage CTLE

Fig. 4(a) depicts the schematic of the XTCR adder. The main path signal and the XTCR path signal are combined in the XTCR adder by adjusting the ratio of tail currents. The total current is fixed to maintain the common-mode output point. The 7-bit tuning ratio is adjusted by 3 single-pole double-throw switches. This allows a precise adjustment while reducing the number of control bits.

Fig. 4(b) shows the schematic of the CTLE core. To achieve a large equalization range for better fitting the inverse profile of the lossy channel, this 2-stage CTLE provides a 5 dB DC gain tuning range, a 2.5 dB equalization tuning range at the middle frequency (i.e., ~1 GHz), and a 14 dB equalization tuning range at the high frequency (i.e., ~14 GHz), respectively.

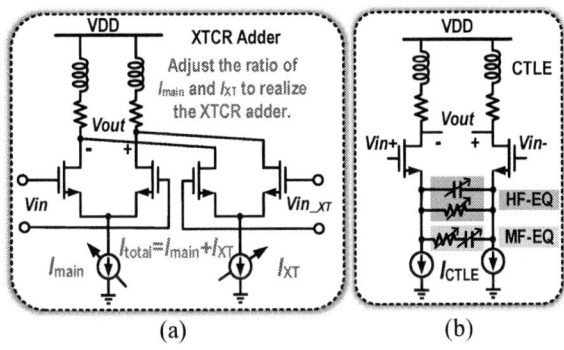

Fig. 4. Schematics of (a) the 7-bit XTCR adder and (b) the CTLE with middle and high-frequencies compensation.

IV. MEASUREMENT RESULTS

Fig. 5(a) and Fig. 5(b) depict the chip microphotograph and power breakdown. It is fabricated in 28 nm CMOS process, occupying a core area of 0.07 mm². The power consumption is 85 mW at 1 V power supply.

Fig. 6(a) shows the insertion loss (IL) and FEXT measurement results of the 10-inch PCB traces. The differential IL and the FEXT of the 10-inch PCB traces are −12 dB and −10.5 dB at 14 GHz, respectively. Due to the presence of the FEXT, the SE IL is −22.5 dB at 14 GHz. Note that the PCB traces are commonly designed as a pair of differential PCB traces in order to imitate the scenario of backplane communications, which are not custom-designed. The spacing between the differential PCB traces is 1-to-2 W (W is the traces width). Other differential PCB traces with

different IL and FEXT values are also successfully verified during the chip measurements.

Fig. 6(b) depicts the measured frequency responses in three different cases with the 10-inch PCB traces. When XTCR sub-blocks are disabled, the measured −3 dB RX BW is only 8.5 GHz. With the conventional XTE technique, the measured −3 dB BW is 11.8 GHz, while the proposed AXTE technique provides 13 GHz BW, which demonstrates a larger equalization capability.

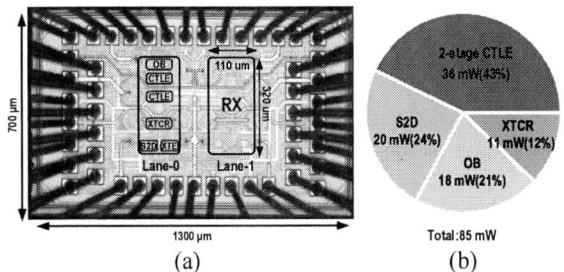

(a)   (b)

Fig. 5. (a) Chip microphotograph, and (b) power breakdown.

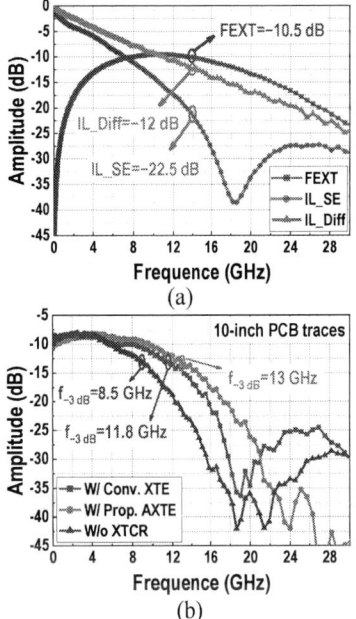

Fig. 6. (a) Measured IL and FEXT of the 10-inch PCB traces, and (b) measured frequency responses of the XTCR RX in three different cases with the 10-inch PCB traces.

Fig. 7 shows measured RX eye diagrams (XTCR enabled and disabled). Two independent PRBS-9 input data are imported to the differential 10-inch traces. When the XTCR is disabled, both the 28 Gb/s NRZ and 50 Gb/s PAM-4 eyes are totally closed. When the conventional XTE is enabled, the measured horizontal and vertical eye openings are 9% and 40% at 28 Gb/s NRZ, and 21% and 16% at 50 Gb/s PAM-4 respectively. When the proposed AXTE XTCR is enabled, the measured horizontal and vertical eye openings are 26% and 58% at 28 Gb/s NRZ, and 24% and 17% at 50 Gb/s PAM-4, respectively. Note that all settings, including the equalization coefficients, are all the same to ensure a fair comparison.

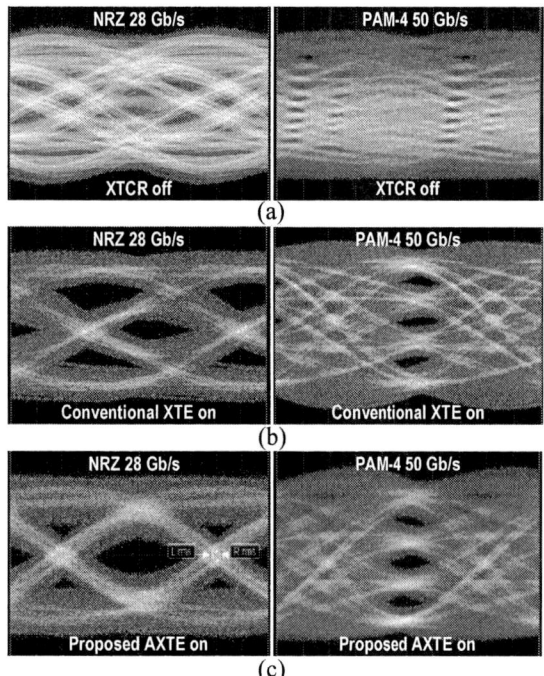

Fig. 7. Measured data eyes with the 10-in PCB traces under NRZ 28 Gb/s and PAM-4 50 Gb/s: (a) with XTCR off, (b) with conventional XTE on, and (c) with the proposed AXTE on.

Fig. 8 shows the measured RX bath curves with the 10-inch traces. The bath curves show high BER ($>4\times10^{-1}$) when the XTCR is disabled, regardless of 28 Gb/s NRZ or 50 Gb/s PAM-4. 8(a) shows that the proposed AXTE RX achieves 28 Gb/s NRZ data with a 30% UI margin at BER<10^{-12}, while the conventional XTE RX has only a 20% UI margin. Fig. 8(b) shows that the proposed AXTE RX achieves 50 Gb/s PAM-4 data with BER<10^{-5}, while the conventional XTE RX can only achieve BER<10^{-4}.

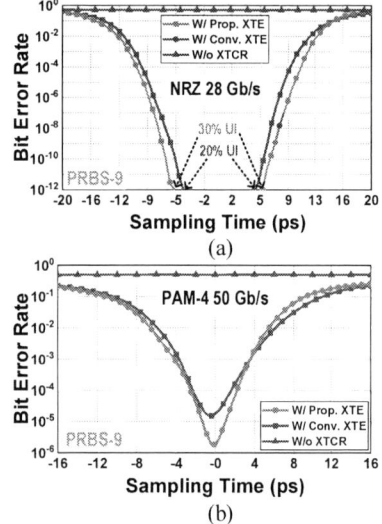

Fig. 8. Measured BER bathtub curves with the 10-inch PCB traces: (a) 28 Gb/s NRZ, and (b) 50 Gb/s PAM-4.

Table I summarizes the performance and its comparison with other state-of-the-art work. This SE MIMO XTCR RX achieves an energy efficiency of 0.85 pJ/bit/lane. Advanced modulation (PAM-4) XTCR RX is realized for the first time, and to the best of our knowledge, this work achieves the highest data rate and the best efficiency.

TABLE I.   PERFORMANCE SUMMARY AND COMPARISON

Reference	[2]	[3]	[4]	This work	
Technology	65 nm	65 nm	32 nm (SOI)	28 nm	
XTC Type	TX FIR	Analog-IIR (RX)	Analog-IIR (RX)	Analog-IIR (RX)	
	\	XTE	XTE	Proposed AXTE	
SE Channel Attenuation (dB)	10.2 @2 GHz	11 @6 GHz	30 @3.5 GHz	22.5 @14 GHz	
Data Rates (Gb/s/lane)	NRZ	NRZ	NRZ	NRZ	PAM-4
	4	12	7	28	50
Horizontal Eye Opening (%UI↑)	46	41.4	/	58	24
Vertical Eye Opening (%↑)	42	34.3	/	26	17
Bit Error Rate	<10^{-12}	<10^{-12}	<10^{-12}	<10^{-12}	<10^{-5}
Energy Efficiency (pJ/bit/lane)	1.4 (TX)	0.96*	8 (RX)	0.85 (RX)	
Area (mm²/lane)	0.008	0.036	0.012	0.035	

* The energy efficiency of only the XTCR adder is presented.

V. CONCLUSION

This paper reported a 2×50 Gb/s SE MIMO PAM-4 XTCR RX in 28 nm CMOS. The RX with the AXTE and the asymmetric inductor peaking technique achieves a high compensation capability of 22.5 dB, better power efficiency of 0.85 pJ/bit/lane, and high-performance cancellation and signal reutilization. The SE MIMO RX provides a promising method for beyond 400 Gb/s Ethernet communications.

REFERENCES

[1] IEEE 802.3 Beyond 400 Gb/s Ethernet Study Group Meeting, https://grouper.ieee.org/groups/802/3/B400G/public/

[2] K. Han-Gon, et, al., "An 8Gb/s/µm FFE-Combined Crosstalk-Cancellation Scheme for HBM on Silicon Interposer with 3D-Staggered Channels," in IEEE Int.Solid-State Circuits Conf (ISSCC). Dig. Tech. Papers, pp. 128−130, Feb. 2020.

[3] O. Taehyoun, H. Ramesh, "A 12-Gb/s Multichannel I/O Using MIMO Crosstalk Cancellation and Signal Reutilization in 65-nm CMOS," IEEE Journal of Solid-State Circuits, vol. 48, no. 6, pp. 1383−1397, Jun. 2013.

[4] A. Cosimo, et, al., "An Eight-Lane 7-Gb/s/pin Source Synchronous Single-Ended RX With Equalization and Far-End Crosstalk Cancellation for Backplane Channels," IEEE Journal of Solid-State Circuits, vol. 53, no. 3, pp. 861−872, Mar. 2018.

[5] A. Tajalli, et, al., "A 1.02-pJ/b 20.83-Gb/s/Wire USR Transceiver Using CNRZ-5 in 16-nm FinFET," IEEE Journal of Solid-State Circuits, vol. 55, no. 4, pp. 1108−1123, Apr. 2020.

[6] K. Shih-Yuan, L. Shen-Iuan, "A 7.5-Gb/s One-Tap-FFE Transmitter With Adaptive Far-End Crosstalk Cancellation Using Duty Cycle Detection," IEEE Journal of Solid-State Circuits, vol. 48, no. 2, pp. 391−404, Feb. 2013.

[7] K. Seon-Kyoo, et, al., "A 5 Gb/s Single-Ended Parallel Receiver With Adaptive Crosstalk-Induced Jitter Cancellation," IEEE Journal of Solid-State Circuits, vol. 48, no. 9, pp. 2118−2127, Sep. 2013.

[8] H. Kyu-Dong et, al., "A 6.5-Gb/s 1-mW/Gb/s/CH Simple Capacitive Crosstalk Compensator in a 130-nm Process," IEEE Transactions on Circuits and System-II: Express Briefs, vol. 60, no.6, pp. 302−306, Jun. 2013.

A 56-Gb/s PAM4 Transceiver with False-Lock-Aware Locking Scheme for Mueller-Müller CDR

Fumihiko Tachibana, Huy Cu Ngo, Go Urakawa, Takashi Toi, Mitsuyuki Ashida, Yuta Tsubouchi, Mai Nozawa, Junji Wadatsumi,
Hiroyuki Kobayashi, and Jun Deguchi
Kioxia Corporation, Kawasaki, Japan
E-mail: fumihiko.tachibana@kioxia.com

Abstract—**This paper presents a 56-Gb/s PAM4 transceiver using an ADC-based RX with a false-lock-aware locking scheme for Mueller-Müller (MM) CDR. After the false-lock-aware locking scheme, a clock phase is adjusted to achieve maximum eye height by using a post-1-tap parameter for an FFE in the CDR loop. A PLL uses an area-efficient "glasses-shaped" inductor. The RX comprises an AFE, a 28-GS/s 7-bit time-interleaved SAR ADC, and a DSP with a 31-tap FFE and a 1-tap DFE. A TX is based on a 7-bit DAC with a 4-tap FFE. The transceiver is fabricated in 16-nm CMOS FinFET technology, and achieves a BER of less than 1e-7 with a 30-dB loss channel. The measurement results show that the MM CDR escapes from false-lock points, and converges to near the optimum point for large eye height.**

Keywords—*baud-rate CDR, PAM4, false-lock-aware, phase adjustment*

I. INTRODUCTION

ADC-based PAM4 receivers (RXs) have been used for high-speed wireline applications beyond 50 Gb/s [1]-[5]. Most ADC-based RXs use baud-rate clock and data recovery (CDR) such as Mueller-Müller (MM) CDR, which does not need edge data. However, the absence of edge data makes it difficult for the CDR to converge to the optimal lock point. In particular, a NRZ-like transition within the PAM4 data pattern introduces false-lock points near its edges, leading to a higher BER. However, to the best of our knowledge, there is no proposal to escape from the false-lock points generated by the PAM4 data pattern.

In this paper, we propose a technique for escaping from the false-lock points. This paper is organized as follows. Section II proposes the techniques for the MM CDR. Section III describes a transceiver architecture. Section IV shows the experimental results. Finally, conclusion is presented in Section V.

II. PROPOSED TECHNIQUES FOR THE MM CDR

Proposed techniques for the MM CDR consist of two parts. First: a false-lock-aware locking scheme for escaping from the false-lock points. Second: a clock phase adjustment using a post-1-tap parameter for a feed forward equalizer (FFE) in the CDR loop (CDR FFE). Two techniques are described in the following subsection.

A. False-lock-aware Locking Scheme

Fig. 1 shows simulation results of MM phase detector (PD) output with PAM4 signaling, the eye diagram after the CDR FFE, and the histograms at correct/false lock points, where PD_OUT is the output of the MM PD. Fig. 2 shows the simplified block diagram of the CDR loop, where cdr_tap is the tap parameter for the CDR FFE, REFC is a

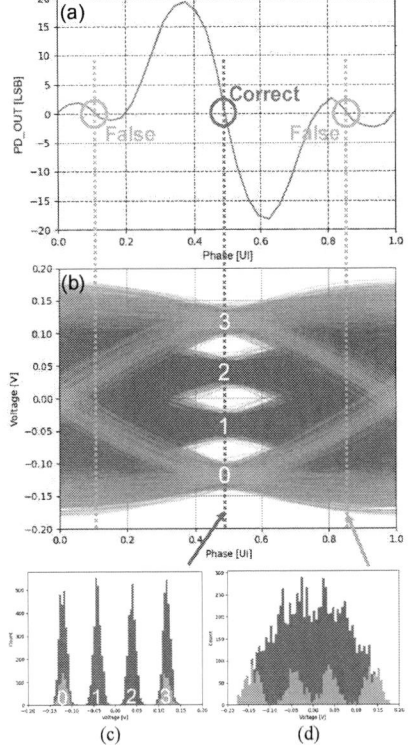

Fig. 1. Simulation results of (a) MM PD output, (b) eye diagram, (c) histogram at the correct lock point, and (d) histogram at the false-lock point.

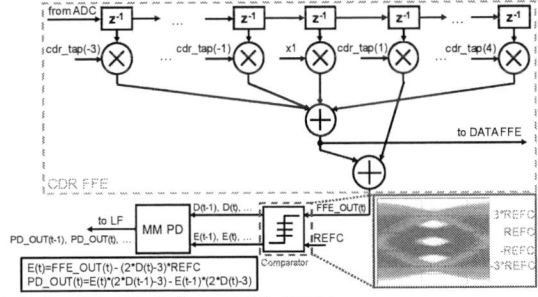

Fig. 2. Simplified block diagram of the CDR loop.

reference value (REF) for a comparator in the CDR loop, FFE_OUT is the output of the CDR FFE, D is the converted data value from FFE_OUT, and E is the difference between FFE_OUT and expected value obtained from D and REFC, respectively. As shown in Figs. 1(a) and (b), there is a correct lock point at around the center of the eye, at which the acquired data (blue plots) have the expected quad-modal distribution as shown in Fig. 1(c). On the other hand, there are two other stable points near the data edges, although the acquired data at these points do not have a quad-modal

978-1-6654-8495-4/22 $31.00 © 2022 IEEE

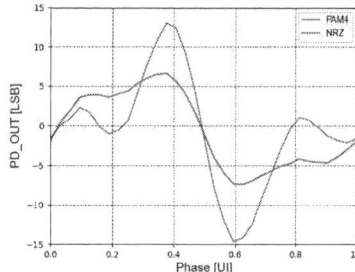

Fig. 3. Simulation results of PD_OUT with the comparator in NRZ mode or PAM4 mode (before the adaptation of the FFE parameters).

distribution as shown in Fig. 1(d). To investigate why these false-lock points exist, only the transitions between 0 and 3 (orange lines) are extracted from the PAM4 data pattern, as shown in Fig. 1(b). When only the extracted transitions are considered, there are two phase points near the data edges where the acquired data have a quad-modal distribution as shown in Fig. 1(d), so the comparator misjudges these points as PAM4 data. This incurs incorrect stable points near the data edges, and the CDR might converge to a false-lock point.

Because the false-lock points are generated by misjudgments at the additional stable points, these cases can be avoided by setting the comparator to the NRZ mode. Fig. 3 shows simulation results of PD_OUT with PAM4 signaling using the comparator with NRZ/PAM4 modes. As shown in Fig. 3, the false-lock points disappear and only the correct lock point exists with the NRZ mode. Although the NRZ mode is the effective way to avoid false-lock points, the gain of the MM PD is lower than that with the PAM4 mode, resulting in lower tracking bandwidth (BW) of the jitter tolerance (JTOL).

To achieve higher gain with the PAM4 mode while still avoiding the false-lock points, we propose a sequence that uses both comparator modes as follows. Step 1: the CDR locks to the correct lock point using the comparator with the NRZ mode. Step 2: the comparator is switched to the PAM4 mode to have higher gain than that with the NRZ mode. Because the CDR locks to the correct lock point before switching to the PAM4 mode, the clock phase gets difficult to fall into false-lock points. Since the proposed sequence just switches the comparator modes according to the steps, it does not need to evaluate the BER for detecting false-lock points, reducing the time for adaptation.

Note that the least mean square (LMS) algorithm using the output of the comparator with the NRZ mode fails in adapting the FFE parameters because the comparator converts data level of 1 and 2 to 0 and 3, respectively. However, appropriate adaptation of the FFE parameters during the NRZ mode makes the correct lock point more stable, further reducing the risk of falling into false-lock points in step 2. To activate the LMS algorithm while in step 1, the comparators connected to the MM PD are switched between NRZ and PAM4 modes according to the steps, and those connected to the adaptation circuit with the LMS algorithm remain unchanged.

B. Clock Phase Adjustment Using the Post-1-tap Parameter of the CDR FFE

After the false-lock-aware locking scheme, the clock phase is adjusted to the point where it achieves maximum eye height. We propose using the post-1-tap parameter of the

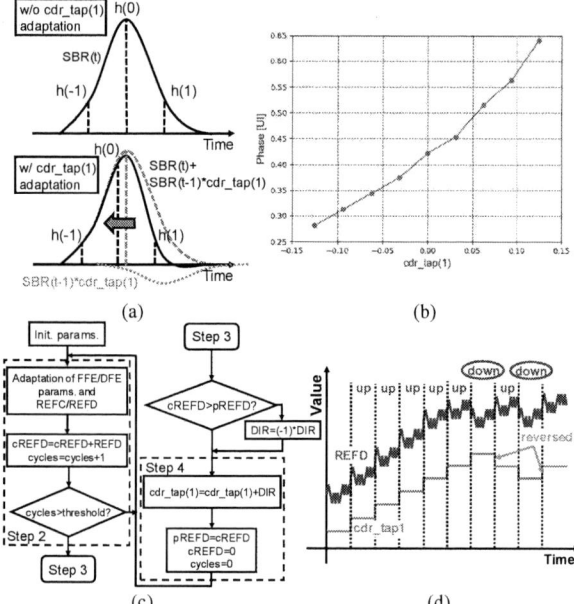

Fig. 4. Clock phase adjustment via the cdr_tap(1) adaptation.

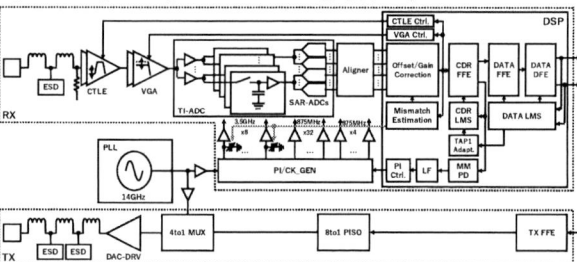

Fig. 5. Transceiver block diagram.

CDR FFE (cdr_tap(1)) as the fitting parameter instead of adding the weights of "Early" and "Late" [6] or the offset value [7] to the MM PD. Theoretically, the MM CDR locks to the point where the pre-1-cursor (h(-1)) is equal to the post-1-cursor (h(1)). On the other hand, the LMS algorithm adapts the CDR FFE parameters in order to make the pre/post-cursors zero. As a result, the lock point is not always the optimal point for low BER if the pre-1-tap parameter of the CDR FFE (cdr_tap(-1)) and cdr_tap(1) are adapted by the LMS algorithm. If cdr_tap(1) is adapted independently, the waveform of single bit response (SBR) is controlled, and locked phase value is also controlled (red arrow) as shown in Fig. 4(a). Fig. 4(b) shows the simulation results of the relation between cdr_tap(1) and the locked phase value. Whereas h(-1) is not equalized with [6], [7], both h(-1) and h(1) can be equalized with proposed technique, resulting in larger eye height for the CDR.

cdr_tap(1) is adapted independently to achieve maximum REF for the comparator in a decision feedback equalizer (DFE), that is equivalent to maximum eye height. The cdr_tap(1) adaptation is executed as follows. Step 1: initialize the direction of the cdr_tap(1) adaptation, the previous REF for the comparator in the DFE (pREFD), and the current REFD (cREFD). Step 2: the LMS algorithm adapts the FFE parameters except for cdr_tap(1), the DFE parameters, REFC, and REFD for a certain period of time. At the same time, REFD is integrated to cREFD at every cycle. Step 3: cREFD is compared with pREFD, meaning that the averaged REFD in this period is compared with that

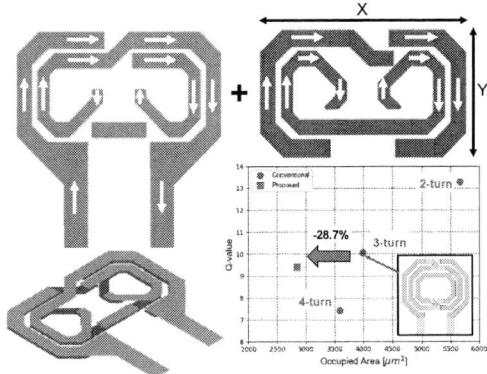

Fig. 6. Glasses-shaped inductor and simulation results of Q-value at 14 GHz.

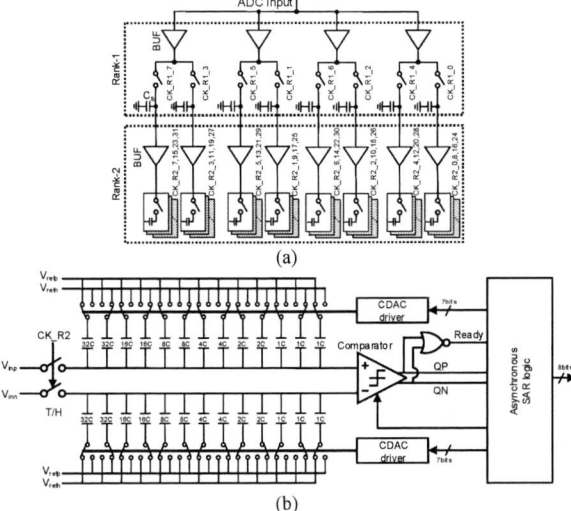

(a)

(b)

Fig. 7. TI-ADC architecture.

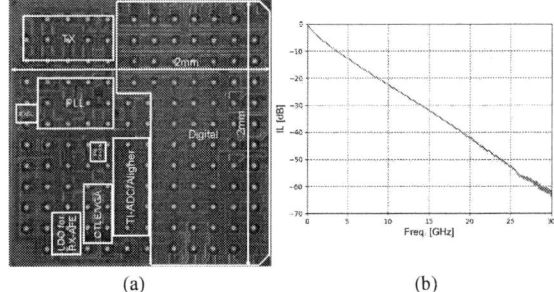

(a)   (b)

Fig. 8. (a) Die micrograph and (b) insertion loss used for external loopback tests.

high-Q spiral inductors need large area for the low-jitter performance. To reduce the occupied area of the inductor, a custom "glasses-shaped" inductor has been designed as shown in Fig. 6. The current direction of the most part is the same as the adjacent wire, resulting in larger inductance by mutual inductance. However, the current direction near the center is opposite, that reduces total inductance. To relax the impact of the center part, the wire near the center is designed to be short and widely spaced as possible. Using the proposed inductor, the occupied area of the inductor can be reduced while maintaining the required inductance and Q-value. Fig. 6 also shows the relation between occupied area and simulated Q-values at 14 GHz. Compared with conventional spiral inductors with the same inductance, the occupied area of the proposed inductor (Area=X*Y) is reduced by 28.7% at the moderate Q-value of 9.4, which is enough for the target clock jitter. In addition, a noise-cancelling charge pump [8] is also implemented in the PLL.

As shown in Fig. 5, the DSP consists of an offset/gain mismatch correction, the 3-pre/4-post-tap CDR FFE, the MM PD, a loop filter (LF), a PI controller, a 4-pre/26-post-tap FFE for data decision (DATA FFE), and a 1-tap DFE for data decision (DATA DFE). It also includes a mismatch estimation circuit, an adaptation circuits with the LMS algorithm for the FFE/DFE parameters and REFC/REFD, an adaptation circuit for cdr_tap(1), and a CTLE/VGA controller. The output of the CDR FFE without cdr_tap(1) equalization is sent to the DATA FFE/DFE to reduce the bit width of the DATA FFE parameters. The offset/gain mismatches are corrected in the digital domain whereas the skew mismatches are corrected in the analog domain.

Fig. 7(a) shows the block diagram of the TI-ADC. A hierarchical architecture with four rank-1 buffers is used in the 7-bit 32-way TI-ADC. Each rank-1 buffer is followed by two track-and-hold circuits (THs) running at 3.5 GS/s, and each TH is connected to a rank-2 buffer followed by four sub-ADCs. Each sub-ADC uses a dynamic comparator in an asynchronous loop with 8 successive approximation cycles (1-bit redundancy for testing 8-bit operation) as shown in Fig. 7(b).

in the previous period. If cREFD is larger than pREFD, then the direction is unchanged; otherwise, the direction is reversed (from +1 to -1 or from -1 to +1). Step 4: according to the direction, cdr_tap(1) is either incremented or decremented. At the same time, cREFD is saved to pREFD, and cREFD is reset to zero. Then return to step 2. Figs. 4(c) and (d) show a flowchart of the steps and the example of the cdr_tap(1) adaptation, respectively. Because the FFE is already implemented in the CDR loop, no additional computational cost except for the cdr_tap(1) adaptation is required for adjusting the clock phase.

III. TRANSCEIVER ARCHITECTURE

The transceiver architecture is shown in Fig. 5. To improve the signal quality, an LC ladder filter assisted ESD protection is used for the transmitter (TX). A 7-bit DAC with a function of the 1-pre/2-post-tap FFE is used for a TX. An RX analog front end (RX-AFE) consists of a T-Coil assisted on-die termination, a 2-stage continuous-time linear equalizer (CTLE) and a 2-stage variable-gain amplifier (VGA). The RX-AFE is followed by a time-interleaved ADC (TI-ADC).

An LC-type VCO is applied to a PLL for the low-jitter clock distributed to the TX and a phase interpolator (PI) followed by a clock generator for the ADC/DSP. In general, it is important for VCOs to achieve the low-jitter performance with small occupied area. However, conventional

IV. MEASUREMENT RESULTS

A test chip is fabricated in 16-nm CMOS FinFET technology, and a die micrograph of the test chip is shown in Fig. 8(a). The chip area of the transceiver is 4.0 mm² including test circuits. Fig. 8(b) shows the plots of channel insertion loss used for external loopback tests. The total loss from ball to ball is 30 dB at 14 GHz. Figs. 9(a) and (b) show the TX output and the RX recovered eye diagram at 56 Gb/s. Fig. 10(a) shows the bathtub curve obtained from the

978-1-6654-8495-4/22 $31.00 © 2022 IEEE

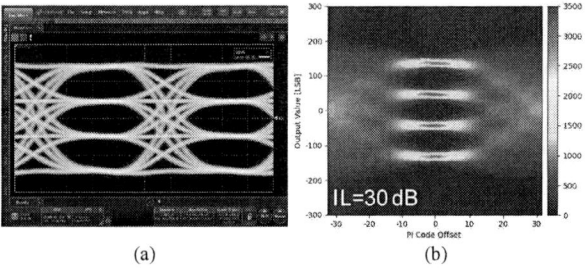

(a)   (b)

Fig. 9. (a) TX output and (b) recovered eye diagram at 56 Gb/s PRQS7.

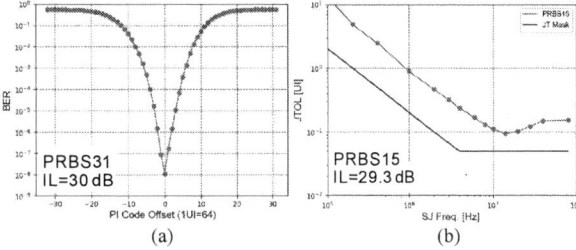

(a)   (b)

Fig. 10. (a) Bathtub curve and (b) JTOL at 56 Gb/s.

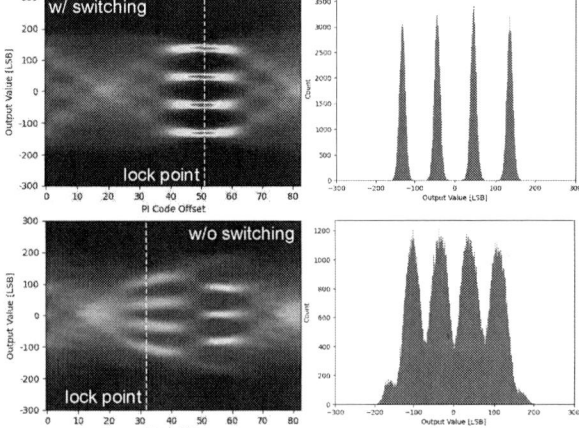

Fig. 11. Eye diagram and histogram at lock point with/without the false-lock-aware locking scheme at 56 Gb/s PRQS7 from the external loopback tests (IL=30 dB).

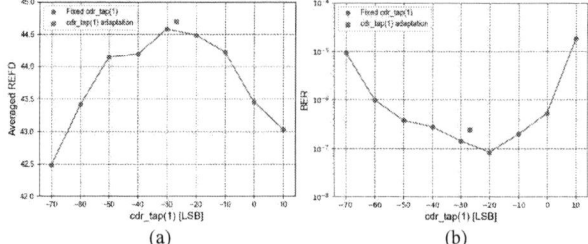

(a)   (b)

Fig. 12. Plots of (a) averaged REFD and (b) BER against cdr_tap(1) at 56 Gb/s PRBS31 from the external loopback tests (IL=30 dB).

TABLE I. PERFORMANCE SUMMARY AND COMPARISON

	This work	[1]	[2]	[3]
Technology	16nm	16nm	16nm	7nm
Power Supply [V]	0.8/0.9/1.0/1.8	0.9/1.2/1.8	0.85/0.9/1.2/1.8	0.9/0.75
Data Rate [Gb/s]	56	56	56	56
IL [dB]	30	31	32	42.5
Area/lane [mm²]	4.0 (incl. test circuit)	1.4	2.2	0.468
Power/lane (excl. DSP) [mW]	365	550	325	146
Power Efficiency [pJ/bit]	6.5	9.8	5.8	2.6
BER	1.0E-07	1.0E-15	1.0E-12	1.0E-07
False-Lock-Aware	Yes	-	-	-
Phase Adjustment	Yes	-	Yes	-

adaptation is at around the value where the highest REFD is obtained. As a result, the obtained BER is close to the lowest value. Table I shows a comparison with previous work at 56 Gb/s. Compared with previous work, our work includes both false-lock-aware locking scheme and phase adjustment scheme.

V. CONCLUSION

In this paper, a technique is proposed to allow the MM CDR to escape from false-lock points by switching the comparator modes for the MM PD. A test chip is fabricated in 16-nm CMOS FinFET technology, and it is confirmed that the MM CDR escapes from false-lock points successfully, and converges to near the optimum point for large eye height.

ACKNOWLEDGMENTS

This paper is based on results obtained from a project, JPNP20017, commissioned by the New Energy and Industrial Technology Development Organization (NEDO).

REFERENCES

[1] Y. Frans et al., "A 56-Gb/s PAM4 Wireline Transceiver Using a 32-way Time-Interleaved SAR ADC in 16-nm FinFET," IEEE JSSC, pp. 1101-1110, April 2017.

[2] P. Upadhyaya et al., "A Fully Adaptive 19-58-Gb/s PAM-4 and 9.5-29-Gb/s NRZ Wireline Transceiver With Configurable ADC in 16-nm FinFET," IEEE, JSSC, pp. 18-28, Jan. 2019.

[3] M. Pisati et al., "A 243-mW 1.25-56-Gb/s Continuous Range PAM-4 42.5-dB IL ADC/DAC-Based Transceiver in 7-nm FinFET," IEEE JSSC, pp. 6-18, Jan. 2020.

[4] Y. Krupnik et al., "112-Gb/s PAM4 ADC-based SERDES Receiver With Resonant AFE for Long-Reach Channels," IEEE JSSC, pp. 1077-1085, April 2020.

[5] J. Im et al., "A 112-Gb/s PAM4 Long-Reach Wireline Transceiver Using a 36-Way Time-Interleaved SAR ADC and Inverter-Based RX Analog Front-End in 7-nm FinFET," IEEE JSSC, pp. 7-18, Jan. 2021.

[6] M.-C. Choi et al, "A 0.1-pJ/b/dB 28-Gb/s Maximum-Eye Tracking, Weight-Adjusting MM CDR and Adaptive DFE with Single Shared Error Sampler," IEEE Symp. VLSI Circuits, 2020.

[7] R. Dokania et al., "A 5.9pJ/b 10Gb/s Serial Link with Unequalized MM-CDR in 14nm Tri-Gate CMOS," IEEE ISSCC, 2015.

[8] G. Urakawa et al., "A Noise-Canceling Charge Pump for Area Efficient PLL Design," IEEE Symp. RFIT, 2020.

external loopback tests. The BER of less than 1e-7 is achieved with PRBS31. Fig. 10(b) shows the JTOL with PRBS15 at IL=29.3 dB. In the JTOL measurements, an arbitrary waveform generator outputs the PAM4 signal to the RX. As shown in Fig. 10(b), 10-MHz tracking BW is achieved, and it meets target JTOL.

Fig. 11 shows the equalized eye diagram and the data histogram at the lock point with and without the proposed false-lock-aware locking scheme when the clock phase is set to the false-lock point before the adaptation sequence. With proposed scheme, the MM CDR can successfully escape from the false-lock point, and the distribution of each data level is separated. On the other hand, the distribution of each data level is overlapped without this scheme. From this result, it is confirmed that the false-lock-aware locking scheme is the effective way to find the correct lock point. Fig. 12 shows the acquired BER and the averaged REFD against cdr_tap(1) with 10 different adaptation sequences, where blue plots show the results with fixed cdr_tap(1), and a red plot shows the result with cdr_tap(1) obtained by the cdr_tap(1) adaptation. Note that the averaged REFD indicates the average of REFD near the end of adaptation sequence. As shown in Fig. 12, converged cdr_tap(1) with the cdr_tap(1)

A 64Gb/s Downlink and 32Gb/s Uplink NRZ Wireline Transceiver with Supply Regulation, Background Clock Correction and EOM-based Channel Adaptation for Mid-Reach Cellular Mobile Interface in 8nm FinFET

Soo-Min Lee[1], Jihoon Lim[1], Jaehyuk Jang[1], Hyoungjoong Kim[1], Kyunghwan Min[1], Woongki Min[1], Hyeonji Han[1], Gyusik Kim[1], Jaeyoung Kim[1], Chulho Kim[1], Sejun Jeon[1], Jinhoon Park[1], Hyunsu Chae[1], Sangwook Han[1], Hiep Pham[2], Xingliang Zhao[2], Qilin Gu[2], Chih-Wei Yao[2], Sangho Kim[1], Jongwoo Lee[1]

[1]Samsung Electronics, Hwaseong, Korea. [2]Samsung Electronics, San Jose, USA.
Email: soomin28.lee@samsung.com

Abstract — **A 32Gb/s/lane NRZ wireline transceiver for the cellular mobile interface has been implemented. The maximum data bandwidth is 64Gb/s downlink and 32Gb/s uplink which meets the bandwidth requirement of the next FR2 standard. The master-slave LDO regulation is used to supply the stable power at each transmitter and receiver that operates independently. The dedicated clock correction and the channel adaptation have been performed to satisfy the time margin of 0.28UI at bit error rate (BER) of 10^{-12}.**

Keywords— NRZ, transceiver, RFIC, Modem, cellular mobile interface, supply regulation, clock phase correction, eye-opening monitor, CTLE

I. INTRODUCTION

The recent 5G Frequency Range (FR) 2 has extended the frequency band to 28~39GHz and the total bandwidth has been around 1 GHz with 10 carrier components. It results in increasing the bandwidth requirement of the baseband transmission between a RFIC and a modem, called the cellular mobile interface. The total bandwidth of a downlink (DL) and an uplink (UL) for FR2 should be higher than 62Gb/s and 24Gb/s, respectively. Using a single-ended signaling with the larger number of lines [1, 2] is no longer attractive solution since the possible routing width for the off-chip signals on PCB is getting narrow. In addition, the problem by the coupled noise to the adjacent sensitive signals in RFIC is getting worse. To overcome those issues, this work implemented the 32Gb/s/lane differential signaling transceiver with supporting the maximum 64Gb/s DL and 32Gb/s UL. Because the channel loss at Nyquist frequency is still low, around 12dB at 16GHz, a multilevel signaling such as PAM-4 [3-5] is not used in this work.

To reduce the power consumption of the mobile device, the power hungry transmitter (TX) and receiver (RX) in the cellular mobile interface are turned off when no data is transferred. Since the supply is shared among all transmitters and receivers, the TX and RX have a different supply voltage whenever one of them changes the operation mode from active to stand-by or vice versa. To supply the stable power at each TX and RX without increasing package cost, the dedicated supply regulation is used in this work. The clock correction including the multiphase error correction and the clock data recovery (CDR) is preceded within a training period to compensate for the phase shift by voltage and temperature variation for a long-time stand-by. According to

the mobile devices such as a cell phone and a tablet, the channel length of the cellular mobile interface is different. An eye opening monitor (EOM) function is initially performed to optimize the use of the RX equalizer.

II. PROPOSED ARCHITECTURE

Figure 1 shows the block diagram of the implemented 32Gb/s transceiver: two lanes for DL and one lane for UL. The unidirectional interface is adopted to support not only a time-division duplexing (TDD) for the recent FR2 but also a frequency-division duplexing (FDD) for the low-band application. A differential 8GHz clock from a common phase locked loop (PLL) is converted to the 4-phase quarter-rate clocks by a poly phase filter (PPF) in each TX and RX. Since the phase error among the quarter-rate clocks affects the overall performance such as the TX output jitter and the RX time margin, the clock correction is necessary. This work uses the dedicated phase error detectors; the IQ detector (IQD) for the PPF, the duty cycle detector (DCD) and the quadrature error detector (QED) for the TX, QED and the edge-clock error detector (XED) for the RX. The finite state machine (FSM) generates the correction codes for the clock related blocks. The master-slave low-drop out (LDO) regulators are used for both TX and RX.

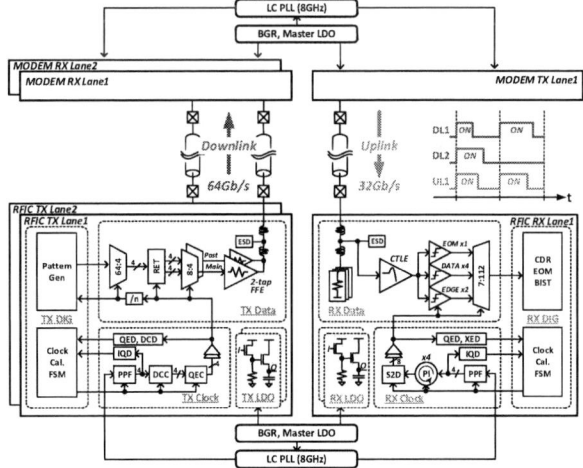

Fig. 1. Implemented 64Gb/s downlink, 32Gb/s uplink wireline transceivers for cellular mobile interface.

To compensate for inter-symbol interference, the TX uses a 2-tap feed-forward equalizer (FFE) and the RX uses a continuous-time linear equalizer (CTLE). A T-coils are used for both TX and RX pads to compensate for the parasitic effect of the ESD diode. All of the clock correction and the signal loss compensation have been performed to satisfy the time margin specification at the RX samplers larger than 0.1UI at bit error rate (BER) of 10^{-12}.

III. CIRCUIT IMPLEMENTATION

This section will describe the implemented circuits including the low-drop out (LDO) regulator, the multiphase clock generator, the CTLE with a negative capacitance, and the sampling phase locked loop (SA-PLL).

A. Master-slave LDO regulator

This work uses the mater-slave LDO regulation (Fig. 2). A closed-loop regulator in the master LDO generates the common gate voltage (V_{G1}) and it is transferred to all of the open-loop slave LDOs. The two-stage source followers reduce the kick-back noise from the LDO output (V_{OUT}) to V_{G1}. The size of the segmented 2nd-stage source follower in each slave LDO is trimmed independently to generate the appropriate V_{OUT} to each TX and RX. To achieve the better power-supply rejection ratio (PSRR) at the mid-frequency range ~100MHz, the large gate capacitance (C_{G2}) is required. However, it takes a long time with the large C_{G2} for V_{OUT} to be steady-state when the output current (I_{OUT}) starts to flow. To eliminate the trade-off between PSRR and a settling time, the switched capacitor which consists of S1, S2 and C_{G2} is used. First, the replica 1st-stage charges the C_{G2} through S1 while V_{OUT} settles to the desired value quickly. After that, the C_{G2} is connected to the gate of the 2nd-stage source follower through S2. Both switches (S1 and S2) are intentionally turned off during the short switching time interval to remove the charge sharing effect which reduces the gate voltage value (V_{G2}) thus increases the settling time of V_{OUT}. This work achieved 20dB PSRR at 100MHz with C_{G2} of 20pF.

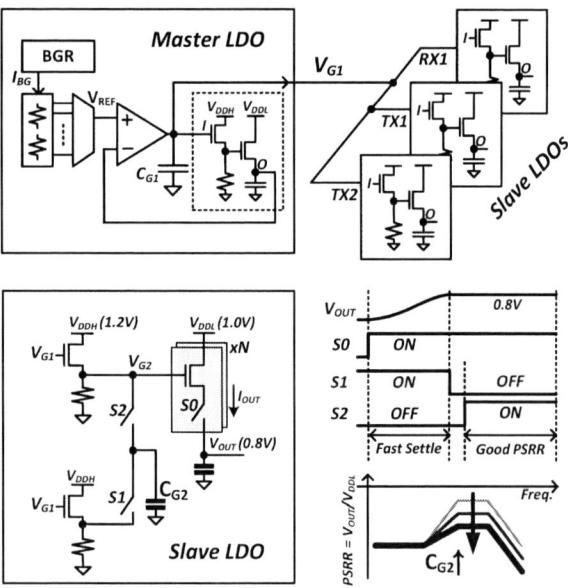

Fig. 2. Circuit diagram of master-slave LDO regulator.

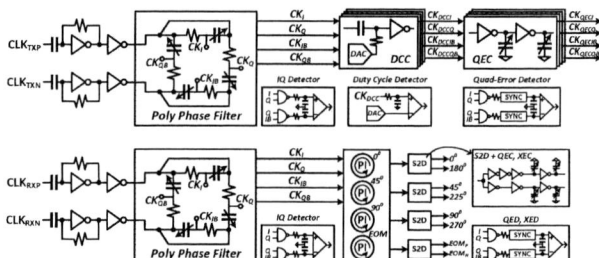

Fig. 3. Multiphase clock generation at transmitter and receiver.

B. Multiphase clock generator

A circuit diagram of the multiphase clock generation is shown in Fig. 3. The PPF adjusts the capacitance to correct the 90°-degree output clocks (CK_Q, CK_{QB}). The duty cycle is adjusted by the ac-coupled inverter with the trimmed input common-mode. Since the IQD and DCD operate during a long-time initial training period, the resistance and capacitance of a low-pass filter (LPF) is large enough so that the output voltage ripple is quite small. In RX, the phase interpolator (PI) and the single-to-differential converter (S2D) generate the multiphase clocks for the CDR and EOM operations. The quadrature error corrector (QEC) and the edge-clock error corrector (XEC) use the delay lines with the shunt-capacitors [5].

Figure 4 describes the background QEC and XEC with the proposed time-to-voltage converter (T2V). Initially, the offset correction for the comparator is required. Two T2V inputs the same clock signals (CLK0, CLK90) to generate the same output voltage. The optimal offset code is found by observing the output transition from low to high or vice versa. At the 2nd step, the phase difference between CLK270 and CLK0 is compared with that between CLK0 and CLK90. The comparator output determines the update direction of the CLK90 phase. This operation results in CLK0 in the exact center between CLK90 and CLK270. Similarly, the CLK180 phase is updated by comparing two phase differences; CLK90 – CLK180 and CLK180 – CLK270. After the 3rd step is completed, CLK0 and CLK 180 are out-of-phase each other.

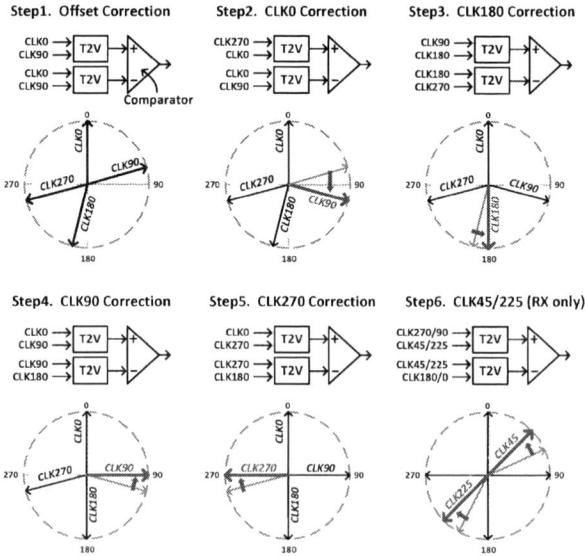

Fig. 4. Background multiphase error correction.

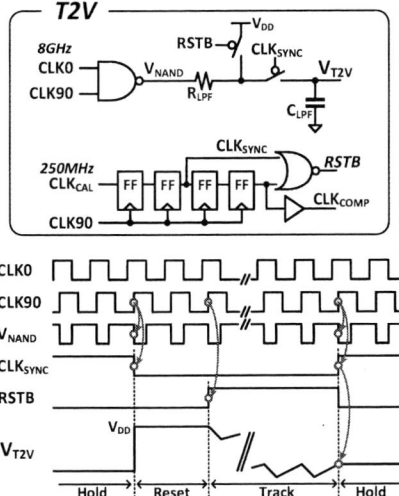

Fig. 5. Synchronized time-to-voltage (T2V) converter.

CLK90 and CLK270 are out-of-phase after the 4th and 5th steps are finished. For the RX, two edge clocks (CLK45, CLK225) are further corrected.

To do background error correction with the fast update period of 4ns, the small resistance (R_{LPF}) and capacitance (C_{LPF}) are required. To avoid the detection error caused by the large voltage ripple on the LPF output (V_{T2V}), the synchronized sampling is proposed (Fig. 5). Both the NAND output (V_{NAND}) and the synchronized clock (CLK_{SYNC}) are generated by the same clock such as CLK90. Thus, it guarantees that CLK_{SYNC} samples the V_{T2V} voltage at the same point every cycle.

C. CTLE with correction procedure

Figure 6 shows the initial RX correction procedure. After correcting the phase error of the above mentioned multiphase clock signals, the offset correction for all of samplers in the RX (Fig. 1) is performed. For that, the differential inputs of CTLE are set to be the same common-mode voltage. After finishing the offset correction, the random data is received through the lossy channel. With trimming the parameters of CTLE, the clock-data recovery (CDR) and quadrature/edge clock error correction are operated. After confirming that the CDR is locked and the recovered data has the promised pattern, the EOM function measures and stores the eye-opening area of the CTLE output with the used settings which includes the bias current (I_{MAIN}, I_{NC}), the degeneration resistance (R_S), the degeneration capacitance (C_S) and the load resistance (R_L) as shown in Fig. 7.

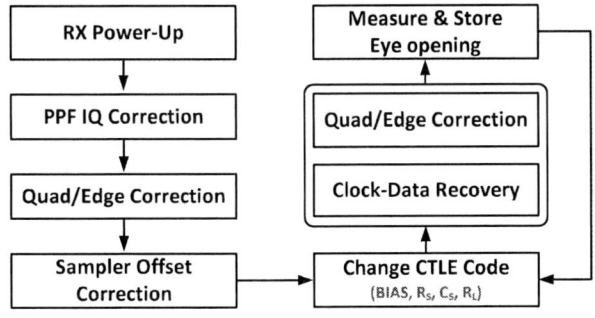

Fig. 6. RX correction procedure.

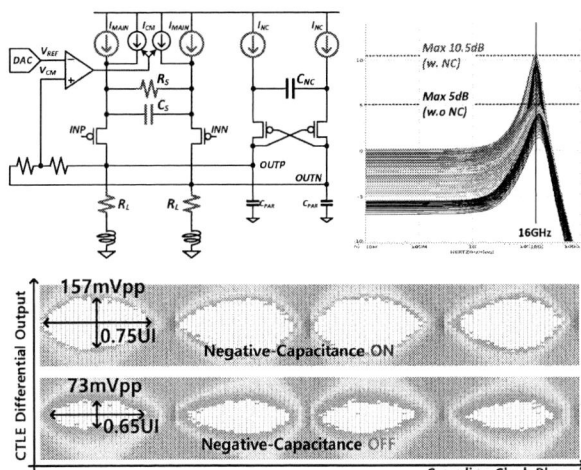

Fig. 7. CTLE with negative capacitance and EOM results.

The negative capacitance (C_{NC}) helps to increase the high-frequency gain of the CTLE by 5.5dB at 16GHz. The EOM results show that the NC improves the horizontal and vertical opening by 0.1UI and 84mVpp. The common-mode feedback circuits set the output common-mode of the CTLE to be ~200mV which is the appropriate value for the following PMOS-input samplers.

D. SA-PLL

The sampling phase locked loop (SA-PLL) which generates the low-jitter clock signal of 8GHz (Fig. 8). Compared with a charge pump PLL and an injection-locked PLL, the SA-PLL has the lower in-band noise performance because of the high detection gain of the sampling phase detector (PD). Since the bandwidth of SA-PLL is large, the power consumption of LC-VCO can be reduced to ~1mW. The dual loop system enables using a small integration capacitance so that the power consumption of the Gm cell is also decreased.

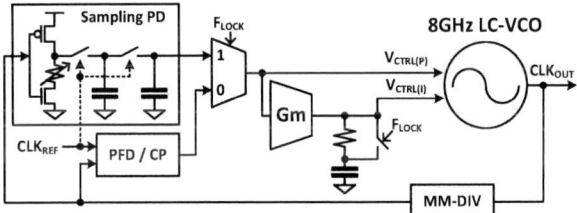

Fig. 8. Sampling phase locked loop (SA-PLL).

IV. MEASUREMENT RESULTS

The proposed transceiver chip was fabricated in a 8nm CMOS process (Fig. 9). Since this chip is for the RFIC (Fig. 1), it consists of 2 TX and 1 RX. The total transceiver with including BIAS and PLL occupies 0.85mm².

Figure 10 shows the power-on/off time of the TX outputs. With the proposed switched operation of the LDO, the on/off time was reduced to 8ns/6ns. At the measurement, the rising and falling time of the switch signals (S0, S2) is increased since those signals are output through the large resistance paths on the test board.

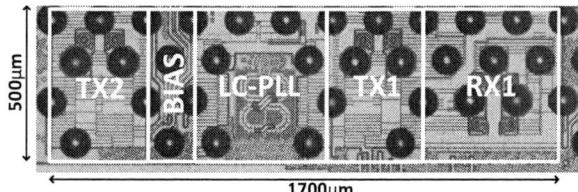

Fig. 9. Chip micorgraph.

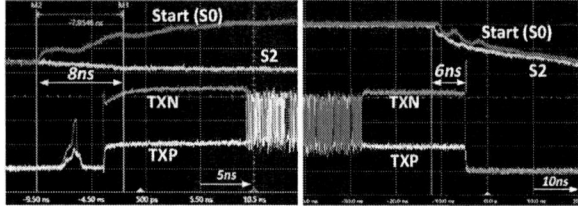

Fig. 10. Measured power-on/off transient results.

Figure 11 shows the phase noise spectrum of the SA-PLL output. The output clock was divided by 2. The integrated RMS jitter is 370fs, when integrated from 1kHz to 100MHz. The reference spur of -56dBc at 52MHz contributes only 10fs.

Figure 12 shows the measured TX eye-diagram with the lossy channel whose insertion loss is 6dB at 16GHz. The total jitter of 9.31ps was achieved with the 2-tap FFE. To verify the function of the overall transceiver, the TX output was connected to the RX through the longer lossy channel whose insertion loss is 12dB at 16GHz. After correcting the multiphase error of the clock signals and optimizing the parameters of CTLE, the time margin of 0.28UI at BER of 10^{-12} was achieved (Fig. 12). It shows that the developed transceiver meets the design specification of 0.1UI without any forward error correction.

Figure 13 shows the 32Gb/s energy breakdown. The total energy efficiency is 3.02pJ/b. The performance of the proposed work was compared with other 32Gb/s designs (Table I). This work shows the lowest energy efficiency of 2.07pJ/b with including only TX and RX. In addition, the better time margin is achieved with the supply regulation.

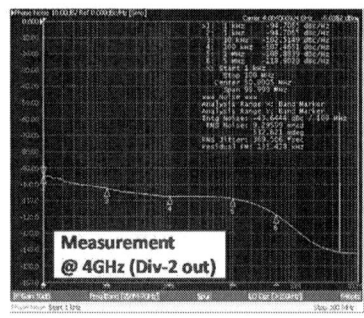

Fig. 11. Measured phase noise of SA-PLL.

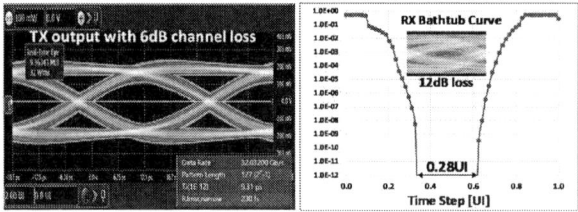

Fig. 12. Measured TX eye-diagram and RX bathtub curve at 32Gb/s.

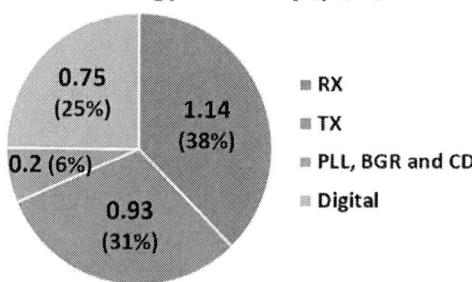

Energy Efficiency [pJ/b]

- RX
- TX
- PLL, BGR and CD
- Digital

Fig. 13. Energy efficiency at 32Gb/s.

TABLE I.   PERFORMANCE SUMMARY AND COMPARISON

	This Work	ISSCC 2019[3]	ISSCC 2018[4]	JSSC 2017[5]
Process	8nm FinFET	28nm FDSOI	65nm	65nm
Data Rate [Gb/s]	32	32	32	32
Signaling	NRZ	PAM4	PAM4	PAM4
VDD [V]	1.2 / 1.0 / 0.8	0.9 / 0.75	1.2	1.2 / 1.0
TX EQ	2-tap FFE	3-tap FFE	2-tap FFE	2-tap FFE
RX EQ	CTLE	CTLE	CTLE, 1-tap DFE	1-tap FFE, 2-tap IIR DFE
Supply Regulation	YES	NO	NO	NO
Energy efficiency [pJ/b]	2.07* 3.02**	2.9*	4.16*	5.5*
Channel loss	12dB @ 16GHz	30dB @ 8GHz	23dB @ 8GHz	13.5dB @ 8GHz
Time margin [UI] at BER=10^{-12}	0.28	< 0.1***	< 0.15****	0.06
Area [mm²]*	0.39	0.4	0.188	0.074

* TX, RX
** BGR, PLL, clock distribution and digital calibration included
*** Measured with 8dB,18dB, and 30dB lossy channel
**** Measured with 18dB, 23dB lossy channel

V.   CONCLUSION

The dedicated supply regulation, phase error correction and the EOM-based equalization procedure were used to implement the 32Gb/s/lane transceiver for the next FR2 cellular mobile interface. The master-slave LDO regulation reduces kick-back noise between the TX and RX which operate independently. To achieve both the better PSRR and faster settling time, the switched capacitor was proposed in the slave LDO. The background multiphase error correction with the synchronized time-to-voltage converter was proposed to compensate for the phase error with the fast update period of 4ns. The RX used the eye-opening monitor function to optimize the parameter values of the CTLE. The proposed transceiver chip in a 8nm CMOS process worked at 32Gb/s with a 12dB lossy channel. The time margin at BER of 10^{-12} was 0.28UI.

REFERENCES

[1] Jongwoo Lee, et al., "A Sub-6GHz 5G New Radio RF Transceiver Supporting EN-DC with 3.15Gb/s DL and 1.27Gb/s UL in 14nm FinFET CMOS," ISSCC, pp.354-355, 2019.

[2] Sangwook Han, et al., "An RF Transceiver with Full Digital Interface Supporting 5G New Radio FR1 with 3.84Gbps DL/1.92Gbps UL and Dual-Band SNGSS in 14nm FinFET CMOS," SOVC, 2020.

[3] Aurangozeb, et al., "A 32Gb/s 2.9pJ/b Transceiver for Sequence-Coded PAM-4 Signalling with 4-to-6dB SNR Gain in 28nm FDSOI CMOS," ISSCC, pp. 480-481, 2019.

[4] L. Tang, et al., "A 32Gb/s 133mW PAM-4 transceiver with DFE based on adaptive clock phase and threshold voltage in 65nm CMOS," ISSCC, pp. 114-116, 2018.

[5] A. Roshan-Zamir, et al., "A Reconfigurable 16/32 Gb/s Dual-Mode NRZ/PAM-4 SerDes in 65-nm CMOS," JSSC, vol. 52, no. 9, pp. 2430-2447, Sep., 2017.

A 2.17-pJ/b 5b-Response Attack-Resistant Strong PUF with Enhanced Statistical Performance

Kunyang Liu, Gen Li, Zihan Fu, Xuanzhen Wang, and Hirofumi Shinohara

Waseda University, Kitakyushu, Japan   Email: kylau@aoni.waseda.jp; shinohara.hiro@waseda.jp

Abstract—**This article presents a Strong physically unclonable function (PUF) with a 5-bit response output, which alleviates the critical issue of huge challenge-response pair (CRP) consumption for one authentication. Attack resistance is enhanced not only by improving the substitution-permutation network (SPN) using variable secret look-up tables (LUTs) and a complex permutation XOR box but also by mitigating LUT data collision and coupling Strong PUF functions between adjacent operation rounds to protect an individual round from attack. As a result, response bitstreams pass all NIST SP800-22 randomness tests even using highly correlated challenge inputs, and mainstream modeling attacks with up to 200-Mb training samples cannot achieve higher accuracy than random guess. An automatic data write-back circuit allows the PUF to be fully stabilized by hot carrier injection (HCI) burn-in without exposing sensitive LUT data. The architecture realizes a 1-cycle/round operation and results in 2.17 pJ/b energy and 0.625 bit/cycle throughput.**

Keywords—authentication, hardware security, IoT, modeling attack, physically unclonable function (PUF)

I. INTRODUCTION

Strong physically unclonable function (PUF) [1,2] is a promising solution for lightweight Internet-of-Things (IoT) device authentication. Compared with a Weak PUF, a Strong PUF has a massive challenge-response pair (CRP) space and can perform authentication without the aid of a separated cryptographic engine. Modeling attack vulnerability has long been a worrisome issue for Strong PUFs. Two recent works [3,4] made a breakthrough in modeling attack resistance by using architectures based on substitution-permutation network (SPN). However, several critical challenges still need to be solved before Strong PUFs can be widely used.

First of all, the latest research [3,4] has only 1-bit response output by XORing all the data at the final stage for high attack resistance. To generate enough response bits for one authentication, e.g., 100 bits, 100 CRPs need to be processed. This results in unignorable latency and energy consumption for receiving and processing these 100 CRPs and 10K challenge bits (assuming a challenge has 100 bits). The actual CRP space is also shrunk by 100 times. For practical use, a Strong PUF with a multi-bit response is necessary.

Attack resistance is another challenge. This should be considered in two aspects. First, simply expanding the response bit number would lead to security degradation. Therefore, innovations are necessary to realize an attack-resistant multi-bit response Strong PUF. Second, in addition to empirical modeling attack results, statistical and structural flaws of Strong PUF architecture must be taken care of, because they often lead to vulnerability to attacks based on cryptanalysis or heuristic approaches [5]. Therefore, improvement in statistical

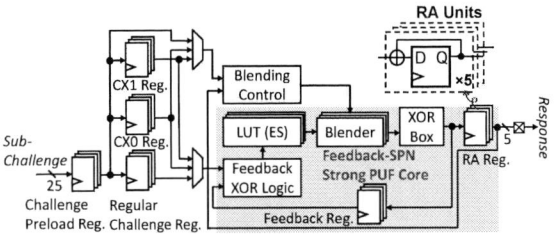

Fig. 1. Overall architecture and dataflows.

performance is a crucial measure of a PUF's attack-resistant quality.

Energy and throughput are also critical to resource-constraint IoT devices. Previous attack-resistant Strong PUFs tend to use very complicated architectures with multiple entropy source (ES) layers. Multiple accesses to these ESs in a one-round operation are energy and throughput consuming. An efficient but secure architecture is desirable.

Last but not least, stability is a fundamental requirement. Bit errors would lead to authentication failure no matter how other parts are well designed. Though recent designs succeed to realize a sub-1% bit error rate (BER), these methods either need to read out sensitive ES data in the secret look-up table (LUT) [3] or require a very long burn-in time [4] and de facto "zero" BER is still not achieved. It is challenging to realize a fully-stabilized Strong PUF without accessing its secret data.

In this work, innovations are implemented to tackle these challenges. Architecture and techniques will be introduced in Sections II and III, respectively. Experimental results collected from test chips in a 130-nm CMOS process will be shown in Section IV. Conclusions will be drawn in Section V.

II. ARCHITECTURE

The proposed Strong PUF has a 100-bit challenge input and a 5-bit response output. Correspondingly, the CRP space is 2^{100} (1.27×10^{30}). The 100 challenge bits are not processed simultaneously. Instead, the Strong PUF performs multi-round operations with feedback, and in each round, a 25-bit sub-challenge is processed. In this way, required hardware resources are reduced, and the CRP space is adjustable by changing the round number to meet different demands.

The overall architecture is shown in Fig. 1. The sub-challenge first goes through two layers of registers before entering the Strong PUF core. These registers are used for several techniques that will be explained in Section III. The Strong PUF core is a feedback-SPN architecture, whose details are shown in Fig. 2.

This research is supported by ROHM Co., Ltd. and VDEC, The University of Tokyo, in collaboration with Cadence and Mentor Graphics.

978-1-6654-8495-4/22 $31.00 © 2022 IEEE

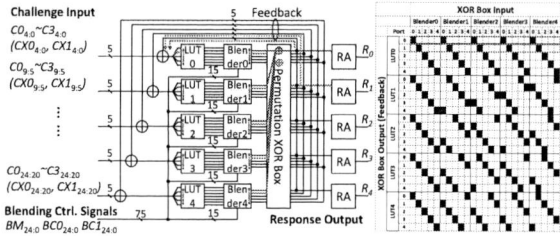

Fig. 2. Feedback-SPN core architecture and the XOR box function.

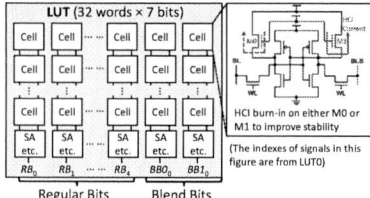

Fig. 3. LUT structure and the circuit of LUT bitcell.

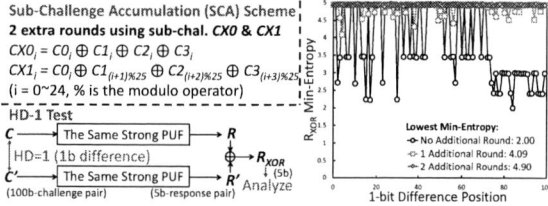

Fig. 4. Sub-challenge accumulation and HD-1 test.

In the first round, the sub-challenge $C0_i$ (i=0–24) serves as the address to the five parallel LUTs. Each LUT is arranged in a 32×7b organization (Fig. 3). These LUTs work as the secret nonlinear substitution layer and are implemented using hybrid SRAM Weak PUF [6] as the ES. Nonlinear substitution is of central significance to security, and the implementation using Weak PUF-based LUTs make it unpredictable and unique to each Strong PUF. Afterwards, the 35-bit LUT output goes through five parallel 7-to-5 blender circuits. The next stage is an XOR box, which works as the permutation layer. The function of permutation is to amplify the influence of substitution, making any 1-bit change affect multiple bits in the following layers and greatly alter the response output. The XOR box is composed of 25 5XOR gates. Every XOR gate has inputs from all the 5 LUTs. Subsequently, the 25-bit XOR box output is fed back and bitwise-XORed with the next-round 25-bit sub-challenge $C1_i/C2_i/C3_i$ (3 successive regular rounds). This makes every LUT have influence on every sub-challenge bit in the next round. Parallel to the feedback, the response accumulation (RA) units accumulate different combinations of LUT/blender outputs across all rounds in GF(2). Every unit contains information from all the 5 LUTs. The 5-bit RA data after multi-round operation are the final response output.

The Strong PUF performs a 2-phase operation. The registers work at the first phase, and LUTs are accessed at the second phase. Only 1 cycle is needed for 1-round operation. In total, the Strong PUF performs an 8-round operation, including 4 regular rounds, 2 additional rounds for the sub-challenge

accumulation technique, which will be explained in detail in Section III, and two rounds for sub-challenge preload and final response generation, respectively.

III. TECHNIQUES

A. XOR Box and Response Accumulation

The XOR box strengthens the permutation function. Each bit from the blenders connects to 5 XOR gates and influences 5 bits through feedback. In previous designs, permutation is implemented with hardwire connection, and each bit has only a 1-bit influence. In this case, a second stage of multi-bit output LUT is needed to diffuse the influence. This leads to additional clock cycles and energy consumption for accessing LUTs. In addition, statistical non-ideality of LUTs that are based on Weak PUF ES might weaken the permutation function. By using the XOR box, an enhanced permutation is realized with only combinational logic gates, which at the same time improve throughput and energy efficiency.

RA stores the history of substitution and permutation across all the rounds and gives them direct influence on the final response, reducing information waste.

B. Sub-Challenge Accumulation

The sub-challenge accumulation (SCA) improves statistical performance by dealing with two randomness problems. The first one is the LUT collision. Considering two challenges with a 1-bit difference, the corresponding 5-bit output LUT has a $(1/2)^5$=1/32 probability to get identical outputs. This leads to degradation of uniformity in the final response and provides a hint for an attack. The second one is that if the 1-bit challenge difference is in the last round, it only influences one LUT throughout the whole operation because there is no further feedback to diffuse it.

As a countermeasure, SCA adds two additional rounds X0 and X1 that follow the four regular rounds (Round#0–3). The sub-challenges of these two rounds (i.e., $CX0$ and $CX1$) are generated by reusing the 100 challenge bits in different combinations, as shown in Fig. 4. In hardware, $CX0$ and $CX1$ are realized using simple shift registers ($CX0$ and $CX1$ register in Fig. 1). By this technique, any 1-bit challenge difference not only influences the corresponding regular round but also the two additional rounds. This reduces the overall LUT collision probability from $(1/32)^1$ to $(1/32)^3$. The additional rounds also diffuse the sub-challenge bits in the last regular round, improving randomness. The number of additional rounds can be adjusted to meet actual requirements. A hamming distance (HD)-1 test shown in Fig. 4 implies the effectiveness of SCA, which improves the lowest min-entropy of the 5-bit output from 2.00 to 4.90 (0.98/bit).

C. LUT Blending

LUT blending realizes two features. First, it improves the complexity of the system by turning the fixed LUT substitution function into a variable function, and the function is different round by round depending on the control signals, which contain the information of sub-challenges. Second, it couples adjacent rounds to protect an individual round from attacks.

978-1-6654-8495-4/22 $31.00 © 2022 IEEE

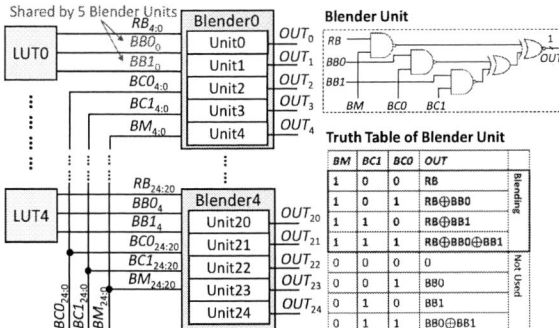

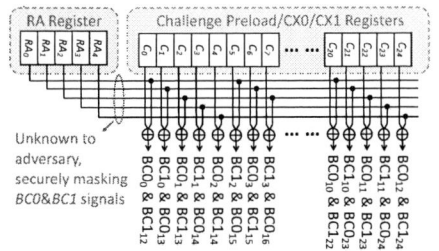

Fig. 5. Block diagram of the blenders, the circuit of a blender unit, and the truth table.

Fig. 6. Generation of blending control signals. *BC0* and *BC1* are protected by secure masking using internal RA register.

LUT blending is implemented using the blender circuits. Fig. 5 shows the block diagram of the blenders, the circuit of a blender unit, and the truth table. Blenders are set after LUTs, and each blender blends the 7-bit LUT data into 5 bits. Among the 7-bit LUT data, five of them are regular bits (RB), and the other two are blend bits (BB). One blender contains five blender units. Each unit corresponds to one RB, but the two BBs are shared by the five units. The basic concept of blending is to perform XOR on each of the five RBs using the two shared BBs. The combinations of XOR are different depending on the control signals (*BM*, *BC1*, and *BC0*), as the truth table shows. By including the information of sub-challenges in the control signals, the LUT substitution function becomes different in every round due to sub-challenge difference, greatly obfuscating the Strong PUF. At the maximum, each LUT has $(2^2)^5=2^{10}$ combinations of substitution function.

Furthermore, to set a high barrier to prevent an adversary from analyzing each round independently, the sub-challenge of the next round instead of the current round is used to generate the blending control signals. Without this scheme, a sub-challenge only influences the current round, and the current round has influence on the next round through feedback. However, it cannot influence the previous round. With this scheme, a sub-challenge now also influences the previous round by controlling its blending, thereby tightly coupling adjacent rounds. In hardware, it is realized by adding the challenge preload register to form two serial register stages as shown in Fig. 1. In addition, to protect the blending control signals, *BC0* and *BC1* are masked using the internal RA register data, which are unknown to attackers (see Fig. 6) and changing in every round.

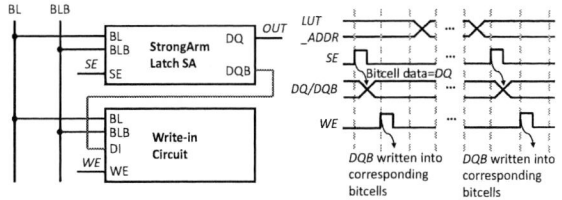

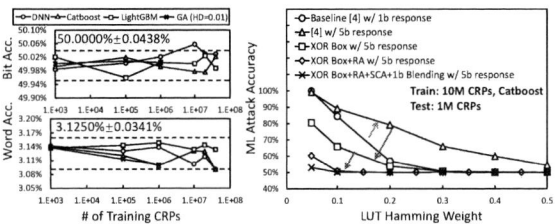

Fig. 7. A write-back circuit that can write in the inverse golden data automatically for HCI burn-in without leaking the secret LUT information.

Fig. 8. Bit-/word-wise modeling attack accuracy and the modeling accuracy in the stress test with low LUT haming weights.

D. Automatic Write-Back Circuit

Successful Strong PUF operation relies on error-free LUTs. The LUTs based on hybrid SRAM Weak PUF are compatible with reverse-direction hot carrier injection (HCI) burn-in to realize "zero" bit error (e.g., $<10^{-6}$ BER) within minutes [6]. However, it requires reading out the LUT secret data and writing in the inverse data during testing. This could be risky because a Strong PUF will be broken if LUT data are leaked.

In this work, an automatic write-back circuit is implemented to solve this problem. The block diagram and the operation waveform are shown in Fig. 7. The idea is to connect *DQB*, which is inverse to the sense amplifier (SA) read-out data *DQ*, to the input of the write-in circuit. Before burn-in, the LUT is started up once. The generated LUT data are regarded as the golden data. For each LUT address, the SAs first read out the bitcell data. Afterward, without changing the address, the write-in circuits directly write *DQB* into the bitcells. By traversing LUT addresses, the inverse data are written without leaking LUT information. Although the start-up data might include noise-induced random bit flips, these flipping bitcells are with very small native mismatch so that they can be stabilized to both "0" and "1" easily.

IV. EXPERIMENTAL RESULTS

The modeling attack results with mainstream algorithms in Fig. 8 show that even with up to 40M training CRPs (200Mb responses, 10× larger than [4]), the bit accuracy remains at 50.03%, and the 5-bit word accuracy stays at 3.13%, which are both within the 95% confidence interval (CI) boundary of random guess. Furthermore, a modeling attack stress test using low hamming weight (HW) LUTs that are not realistic in real life was performed. A model of [4] was built as the baseline for comparison. The result implies that if the response of the baseline is simply extended to 5 bits, the attack resistance will degrade. However, in this work, innovations such as XOR box, RA, SCA, and blending successfully achieve better attack

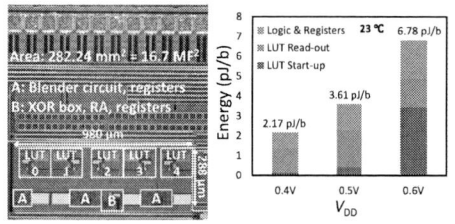

Fig. 9. Chip micrograph and measured energy.

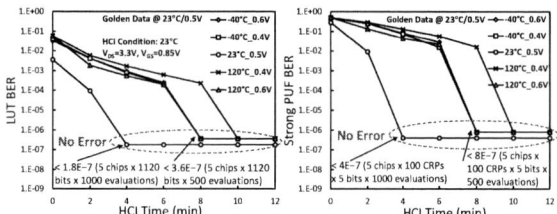

Fig. 10. LUT BER and Strong PUF BER vs. HCI burn-in time.

Test Name	Stream Length	Runs No.	Pass Rate	Avrg. P-Value	Pass?
Frequency	5M	5	5/5	0.610	Pass
Block Frequency	5M	5	5/5	0.530	Pass
Runs	5M	5	5/5	0.353	Pass
Longest Run of Ones	5M	5	5/5	0.305	Pass
Rank	5M	5	5/5	0.427	Pass
FFT	5M	5	5/5	0.364	Pass
Non-Overlapping Template Matching	5M	5	5/5	0.512	Pass
Overlapping Template Matching	5M	5	5/5	0.372	Pass
Universal	5M	5	5/5	0.504	Pass
Linear Complexity	5M	5	5/5	0.517	Pass
Serial	5M	5	5/5	0.635	Pass
Approximate Entropy	5M	5	5/5	0.604	Pass
Cumulative Sums	5M	5	5/5	0.538	Pass
Random Excursions	5M	5	5/5	0.489	Pass
Random Excursions Variant	5M	5	5/5	0.519	Pass

The Consecutive Incremental Challenges for Statistical Tests

(In Decimal)	C3	C2	C1	C0
1st 0.25M	0	0	0	0~249999
2nd 0.25M	0	0	0~249999	250000
3rd 0.25M	0	0~249999	250000	250000
4th 0.25M	0~249999	250000	250000	250000

Fig. 11. NIST test results, autocorrelation, HW, and HDs.

resistance even with 5-bit response. In the extreme case of HW=0.05 (only 5% of LUT data are "1"), modeling attack on the proposed Strong PUF only reaches 53.0% accuracy, in contrast with the 99.9% accuracy on the baseline architecture.

Five chips fabricated in a 130-nm CMOS process were tested. Fig. 9 shows the chip micrograph and measured energy results. The energy of the Strong PUF is 2.17 pJ/b (3.61 pJ/b) at 0.4 V (0.5 V) with a 0.32 MHz (1.57 MHz) frequency. Fig. 10 shows that HCI burn-in using the write-back circuit successfully achieves "zero" bit error with 3.6E−7 worst LUT BER and 8E−7 worst Strong PUF BER across VT corners, respectively. They are calculated based on the pessimistic assumption that bit error would occur at the next measurement.

Fig. 11 shows the statistical performance. Consecutively incremental challenges are used to generate response bitstreams for randomness tests. Even with highly correlated challenges, this work passes all the NIST SP800-22 randomness tests, showing excellent statistical performance. Autocorrelation of these bitstreams are close to the ideal value of zero. The distribution of inter-PUF HD, intra-PUF HD, and HW are all close to ideal.

	This Work	ISSCC'21 [4]	VLSI'20 [3]	VLSI'17 [2]	VLSI'17 [1]
Technology	130nm	130nm	14nm	28nm	130nm
Type	Feedback SPN	Feedback Cascaded SPN	Cascaded SPN	SRAM Access Sequence	Sub-Threshold Current
CRP Consumption for 1 Authentication [a]	20	100	100	100/5m*	100
NIST Tests Passed with Correlated Challenges?	YES	NO	NO	—	—
Modeling Attack — Accuracy	50.03%	50.26%	50.8%/~50%[c]	89%	60%
Modeling Attack — Training CRPs (Response bits)	40M (200M)	20M (20M)	6M (6M)	10K	10K (10K)
Stabilization Technique — Method	Reverse HCI Burn-in	In-Cell HCI Burn-in	Challenge Pruning	—	Dynamic Thresholding
Stabilization Technique — Without Secret Data Readout?	YES	YES	NO	—	NO
Stability — Worst BER	"0"[b]	0.73%	0.26%	3.2%	0.4%
Stability — Test V_{DD} (V)	0.4–0.6	1.4–1.8	0.65–0.85	0.7	1.05–1.32
Stability — Test Temp. (°C)	−40—120	0~120	0~100	27	−20~80
Throughput — (Bit/Cycle)	0.625	0.02	0.07[d]	—	—
Throughput — (Mbps)	0.2	0.02	106	1100	0.006
Energy (pJ/bit)	2.17	54.9	97	0.088	11

a. Assuming one authentication requires 100-bit response and only one Strong PUF macro is used.
b. No error in 5 chips x 500 evaluations x 500b response (8E−7 BER).
c. With 10% adversarial CRPs.   d. Calculated based on the frequency and throughput in [3].
e. Seq-5 mode. m is the number of column in SRAM array.

Fig. 12. Comparison table.

In comparison to attack-resistant Strong PUFs [1,3,4] (Fig. 12), this work is the only one with a multi-bit response while achieving the best attack resistance and statistical performance. The energy consumption is the lowest, and it has a 0.625 bit/cycle throughput that is 31× better than [4]. Furthermore, it achieves "0" BER across VT corners while not required to read out secret LUT data for stabilization.

V. Conclusion

A Strong PUF for lightweight IoT authentication has been presented. It has a 5-bit response architecture to reduce the required resource consumption for one authentication. Meanwhile, it realizes improved attack resistance proved by modeling attacks trained with 40M CRPs, modeling attack stress tests, and close-to-ideal statistical performance. By optimizing architecture, this work achieves low energy of 2.17 pJ/b and high throughput of 0.625 bit/cycle. The write-back circuit allows the PUF to be fully stabilized by HCI burn-in without needing to read out sensitive LUT data.

References

[1] X. Xi et al., "Strong subthreshold current array PUF with 2^65 challenge-response pairs resilient to machine learning attacks in 130nm CMOS," VLSI Circuits, 2017, pp. C268-C269.

[2] S. Jeloka et al., "A sequence dependent challenge-response PUF using 28nm SRAM 6T cell," VLSI Circuits, 2017, pp. C270-C271.

[3] V. Suresh et al., "A 0.26% BER, 10^28 challenge-response machine-learning resistant Strong-PUF in 14nm CMOS featuring stability-aware adversarial challenge selection," VLSI Circuits, 2020, pp. 1-2.

[4] K. Liu et al., "A modeling attack resilient Strong PUF with feedback-SPN structure having <0.73% bit error rate through in-cell hot-carrier injection burn-in," ISSCC, 2021, pp. 502-503.

[5] J. Delvaux, "Machine-learning attacks on PolyPUFs, OB-PUFs, RPUFs, LHS-PUFs, and PUF_FSMs," IEEE Trans. Inf. Forensics Secur., vol. 14, no. 8, pp. 2043-2058, Aug. 2019.

[6] K. Liu, X. Chen, H. Pu, and H. Shinohara, "A 0.5-V hybrid SRAM physically unclonable function using hot carrier injection burn-in for stability reinforcement," JSSC, vol. 56, no. 7, pp. 2193-2204, Jul. 2021.

A $183F^2$ Gate Leakage-Based Physically Unclonable Function With Area Efficient Current Tilting-Based Masking Scheme

Beomsoo Park[1,2] and Nima Maghari[1]

Email : beomsoo0927@ufl.edu, maghari@ece.ufl.edu

[1]Department of Electrical and Computer Engineering, Univesity of Florida, Gainesville, FL, USA

[2]Qualcomm Inc., San Diego, CA, USA

Abstract— This paper presents a gate leakage-based physically unclonable function (PUF) with an area efficient current tilting-based masking scheme to enhance the overall bit error rate (BER). The implemented masking scheme only requires checking the validity of the PUF at the nominal condition and is embedded within the readout circuit enabling to minimize the overall area. Designed with a 65nm process, the unit PUF size and core energy are $183F^2$ and 0.014fJ/bit as it leverages gate leakage current. The total area per a unit PUF cell considering all the peripheral circuits is $927F^2$. Experimental results show that the prototype PUF achieves a native instability and BER of 6.89% and 0.77%, respectively, at nominal condition of 1.2V and 27ºC. The BER is improved to 0% at nominal condition with even a small tilt that masks 17.3% of the PUF bits. A masking ratio of 41.8% allows to maintain the BER of 0% across supply voltage and temperature range of 0.9V~1.5V and 0~100ºC, respectively.

Keywords—PUF, gate leakage, masking scheme, BER.

I. INTRODUCTION

With the increasing demand for Internet of Things devices, a compact security solution has become essential. Physically unclonable functions (PUFs) leveraging process and fabrication variations to provided stable and unique keys/IDs have emerged as an attractive approach as it consumes considerably less amount of power and area than conventional approaches. The main challenge in using PUFs for security is maintaining a consistent PUF output despite PVT variations and device aging to achieve high reliability. While error-correcting code (ECC) helps obtain 100% reliability, the complexity of the ECC design significantly increases to reduce the bit error rate (BER) to 0% [1].

In order to minimize the design effort for ECC, many lossless stabilization techniques such as temporal/spatial majority voting (TMV, SMV), remapping, reconfiguration, among others have been proposed to improve the BER [2-4]. However, the efficiency of these techniques is limited or requires exhaustive testing across different supply voltages or temperatures. Instead, recent designs propose masking schemes that require testing at only one condition (typically nominal condition) to detect and discard the bits that possibly become dark (unstable) bits [2,5,6]. However, these schemes typically remove a large portion of PUF cells and some even require a significant amount of additional area to employ the masking scheme. Consequently, area efficiency is one of the main concerns regarding masking schemes.

This work presents a PUF that utilizes gate leakage current as an entropy source where the PUF unit size and

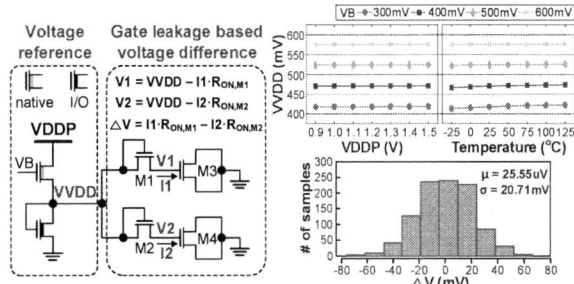

Fig. 1. Concept of the gate leakage-based PUF with 2-transistor voltage reference.

core energy is only $183F^2$ and 0.014fJ/bit, respectively. An area efficient masking method based on current tilting mechanism is embedded within the readout circuit which allows for the total area per a unit PUF cell to be smaller than $950F^2$. With a masking ratio of 41.8%, the BER is maintained at 0% across all valid supply voltage and temperature variations and device aging. Section II discusses the concept of the proposed gate leakage-based PUF and the overall architecture. Section III describes the operation of the proposed PUF with the current tilting-based masking scheme. Followingly, measurement results are shown in section IV and conclusion is made in section V.

II. PUF UNIT CELL DESIGN AND ARCHITECTURE

The concept of utilizing gate leakage current as an entropy source for the proposed PUF is shown in Fig. 1. Leakage/sub-threshold currents are attractive characteristics to utilize for PUF designs due to the large variations depending on the threshold voltage and the device size of the transistors [2, 4, 7]. Moreover, the low current consumption and small area characteristics make them highly desirable to use as the entropy source. Since the current consumed through the gates of M3 and M4 (I1 and I2 of Fig. 1) is static and extremely low, a simple 2-transistor voltage reference is implemented using native and I/O transistors to minimize the variation of the virtual voltage (VVDD) [8]. Simulation results reflect that VVDD varies less than 10mV across voltage and temperature of 0.9~1.5V and -25~125ºC, respectively. The gate leakage currents, I1 and I2, along with the diode connected transistors of M1 and M2 generate the necessary voltage difference.

The overall PUF architecture with 18 × 64 PUF array is shown in Fig. 2. Each PUF unit cell is consisted with the diode connected transistors of M1 and M2 and row selection switches (in total four transistors per unit cell) that enable the whole PUF array to be evaluated with a single readout

978-1-6654-8495-4/22 $31.00 © 2022 IEEE

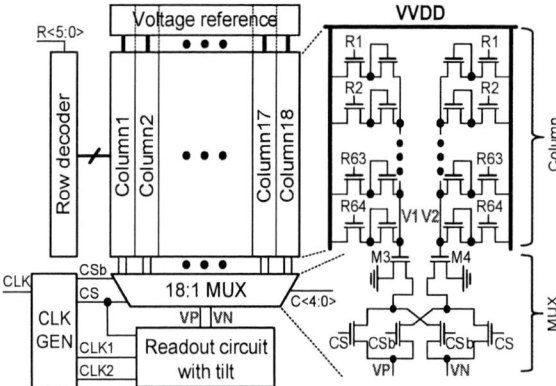

Fig. 2. Overall architecture of the proposed gate leakage-based PUF.

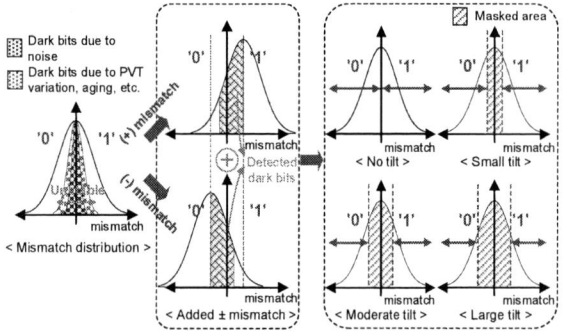

Fig. 3. Concept of the implemented masking scheme.

circuit. The 2-transistor voltage reference is properly sized to support a single column array which is consisted with 64 PUF unit cells. The M3 and M4 transistors that provides the gate leakage current shown in Fig. 1 is shared among each column and is embedded in the MUX. Therefore, each column overall consists of a 2-transistor voltage reference, 64 PUF unit cells, and a MUX unit cell. A single readout circuit that includes the proposed current tilting-based masking scheme (discussed in the next section) evaluates the entire PUF array sequentially.

III. OPERATION OF THE PROPOSED PUF

A. Concept of the Implemented Masking Scheme

The concept of the implemented masking scheme is illustrated in Fig. 3. To detect and remove the dark bits due to environmental shifts, a differential offset (mismatch) using a current tilt mechanism is added both in the negative and positive directions consecutively. If the generated voltage difference is smaller than the amount of tilt, the PUF output is dominated by the added mismatch. However, when the generated voltage difference is large, the PUF output is constantly a 0 or 1 regardless of the added mismatch. Thus, the proposed masking scheme allows to determine whether a PUF cell is stable or not without having to observe the PUF output or comparing it to the evaluated output without any tilt. Depending on the amount of mismatch added, a larger number of potential dark bits is masked which significantly improves the worst case BER.

B. Operation of the PUF with Current Tilt Mechanism

The proposed gate leakage-based PUF with the current tilt mechanism embedded within the readout circuit is shown

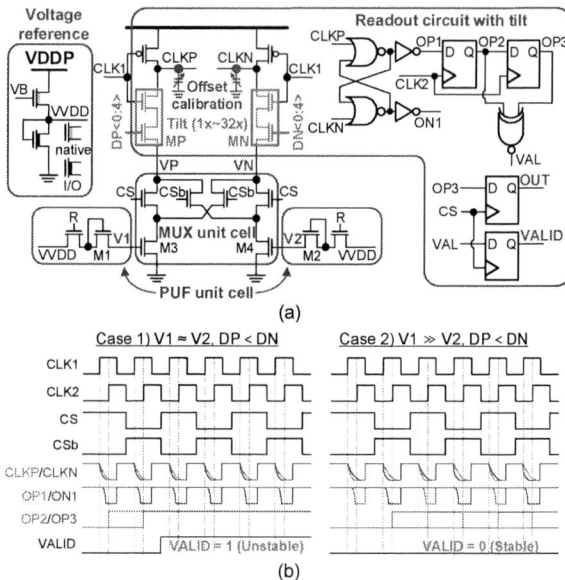

Fig. 4. Overall PUF circuit and timing diagram. (a) PUF with current tilt mechanism embedded readout circuit and (b) detailed examples for PUF validity check.

in Fig. 4(a). The additional row selection switches (gate connected to R) along with the MUX allows for one of the PUF cells to be evaluated at a time. Transistors, M3 and M4, of Fig. 1 that provides the necessary gate leakage current are employed as the tail current source of two different inverter branches. The voltage of V1 and V2 are converted into current through M3 and M4 and depending on the difference, one branch is pulled to ground faster than the other. The current tilt mechanism is employed on the pull-down NMOS switches, MP and MN (marked in red), with control signals DP<0:4> and DN<0:4> which changes the size of MP and MN from 1× to 32×. Depending on the signal, CS, the MUX unit cell enables V1 and V2 to sequentially be evaluated by both branches. This allows to determine whether the voltage difference is sufficient or not to be considered as a stable or unstable bit.

The detailed operation for validity check (determining dark bits) is shown in Fig. 4(b). The validity check is performed with an equal amount of mismatch added and subtracted from the equilibrium value of DP and DN (DP/DN=20 at equilibrium). Two main cycles are used for validity check, first with CS high and second with CS low. When CS is high, the original branches are selected for evaluation and when CS is low, the branches are swapped. The outputs from the two cycles, OP2 and OP3, are XNORed to create a VALID flag deciding the validity of the PUF bit. If the difference between V1 and V2 is small (case 1), even a small tilt between the two branches shows the result of VALID=1 indicating an unstable bit. For a large voltage difference (case 2) however, the larger voltage between V1 and V2 always sinks more current despite the tilted switches of MP and MN and consequently VALID equals 0 indicating a stable bit.

C. Proposed Masking Scheme Flow Diagram

The overall flow diagram of the masking scheme is summarized in Fig. 5(a). First, an offset calibration is performed to cancel the offset between the two inverter branches with the variable capacitors (marked in blue) at

978-1-6654-8495-4/22 $31.00 © 2022 IEEE

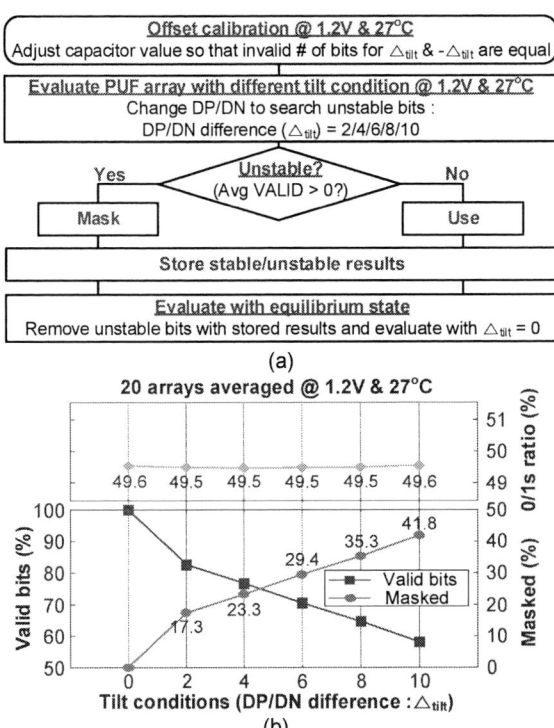

(a)

(b)

Fig. 5. Proposed masking scheme flow diagram. (a) Flow diagram and (b) number of valid bits and 0/1s ratio with different tilt conditions.

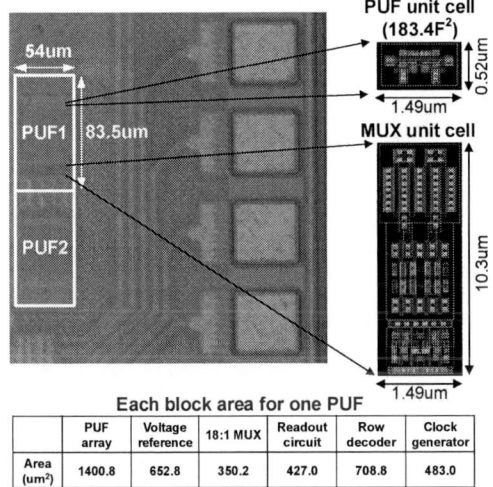

Fig. 6. Die photo and layout of the PUF and MUX unit cell.

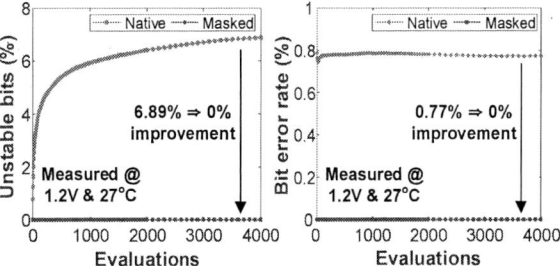

Fig. 7. Instability and BER evaluated at the nominal condition without (native) and with (masked) masking schemes.

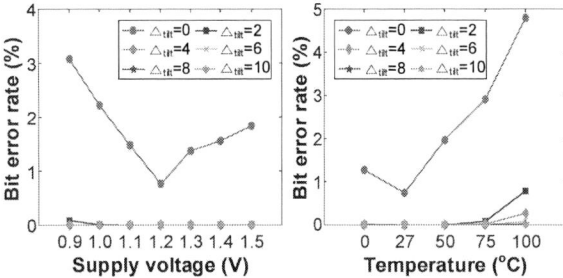

Fig. 8. BER results with supply voltage and temperature variation using different tilt conditions.

nodes CLKP/CLKN of Fig. 4(a). The implemented current tilting mechanism not only helps to detect the hidden unstable bits but allows to compensate the offset of the readout circuit as well. Ideally without any offsets, the detected number of invalid bits for a whole PUF array should equal when applying either $+\Delta$tilt or $-\Delta$tilt. For example, the two cases of DP/DN=18/22 (Δtilt = 4) and DP/DN=22/18 (Δtilt = -4) ideally should detect the same number of unstable bits. Using this characteristic, the capacitors are properly adjusted to equal the number of unstable bits for $\pm\Delta$tilt (with same value but opposite direction) to remove the offset for proper evaluation.

Followingly, five different Δtilt values from 2 to 10 are applied to check the validity of the PUF bits at the nominal condition of 1.2V and 27°C. The PUF array at each Δtilt is evaluated for 4000 cycles and the bits that have an average VALID flag value higher than 0 are masked. The results from each Δtilt case are stored to remove the unstable bits and a final evaluation at the equilibrium state of Δtilt=0 is performed. Using 20 different arrays (1152 PUF bits per array), the average masked ratio for each Δtilt is shown in Fig. 5(b) where at the largest tilt value of Δtilt=10, 41.8% bits are masked. As the same amount of mismatch is added in both the negative and positive direction from the equilibrium value, the 0s and 1s ratio is near the ideal value of 50% regardless of the tilt condition.

IV. MEASUREMENT RESULTS

The prototype chip is fabricated in a 65nm CMOS process. The die photo and layout of the PUF and the MUX unit cell are shown in Fig. 6. 20 PUF arrays from 10 dies (2 arrays per die, each with 1152 PUF bits) are evaluated with a 4000× repetition. The unit PUF cell size is 1.49um × 0.52um which is equivalent to 183.4F². The total area per a unit PUF cell including the readout circuit and all other peripheral blocks is 927F². The nominal condition stated in this section is supply voltage of 1.2V and temperature of 27°C.

Fig. 7 shows the evaluated stability and BER at nominal condition. The native instability and BER are 6.89% and 0.77%, respectively. With even a small tilt of Δtilt=2, resulting with a masked ratio of 17.3%, both the instability and BER are reduced to 0% at nominal condition. Fig. 8 shows the evaluated BER against supply voltage and temperature variations. Evaluating the PUF over different supply voltage and temperature is essential as it likely changes the measured output of the PUF. In this design, supply voltage of 0.9~1.5V and temperature range 0~100°C is tested for evaluation. The native worst case BER is 4.79% which occurs when testing the PUF at 100°C. In order to improve the BER, different tilt values from 2 to 10 are used for evaluation. With a small tilt of Δtilt=2, the worst case

TABLE I. COMPARISON WITH STATE-OF-THE-ART PUFS USING MASKING SCHEMES.

	This work	ISSCC'21 Y. He	ISSCC'21 J. Park	ISSCC'20 Y. Choi	JSSC'20 K. Liu	TCAS'20 Y. Shifman	JSSC'17 S. Satpathy	JSSC'16 J. Li
Technology (nm)	65	65	40	28	130	65	14	65
PUF cell area/bit (F²)	183.4	594.2	21675.0	3699.0	372.8	3001.0	9405.0	726.6
Total area/bit (F²)	926.4	--	--	5057.2	2969.0	--	--	1756.2
Native unstable bits (# of evaluations)	6.89% (4k)	2.83% (2k)	0.4% 1) (10k)	--	2.14% (2k)	19.6% 2) (120k)	26.37% 2) (5k)	6.54% (500)
Native BER	0.77%	0.29%	0.027%	--	0.21%	2.24%	--	--
Stabilization technique	Current tilt based masking	ASCH	TMV, masking	Validity check, masking	VSS bias, masking	Capacitive preselection	TMV, burn-in, masking	TMV, masking
Masking ratio	41.8% (worst : 47%)	27%	3.64%	25%	67.4%	59%	20%	25% (worst : 30%)
Tested Condition — Temp (°C)	0 ~ 100	-40 ~ 125	-40 ~ 125	-40 ~ 150	-40 ~ 120	-10 ~ 85	25 ~ 110	0 ~ 80
Tested Condition — Supply (V)	0.9 ~ 1.5	0.7 ~ 1.4	0.7 ~ 1.4	0.81 ~ 0.99	0.8 ~ 1.4	0.8 ~ 1.2	0.55 ~ 0.75	0.6 ~ 1.2
Worst Case — BER before stabilization	4.79%	4.20%	--	10.50%	5.80%	4.00%	5.76%	--
Worst Case — BER after stabilization	0%	0%	0.0019%	0.97%	0%	0%	1.42%	--
Core energy/bit (fJ/bit)	0.014	0.057	39	--	128	16	4	

1) Estimated from figures   2) Under voltage/temperature variation

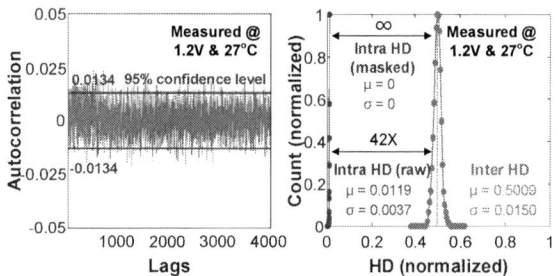

Fig. 9. Uniqueness and randomness measurement results using autocorrelation and intra/inter die hamming distance.

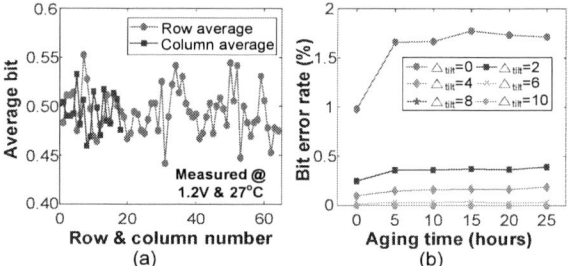

Fig. 10. (a) Digital average bits across rows and columns. (b) Accelerated aging test result.

BER is improved by 6× to 0.78% and using a large tilt of Δtilt=10, the worst case BER is reduced to 0% with a masking ratio of 41.8%.

The uniqueness and randomness of the PUF is shown in Fig. 9. The autocorrelation function shows a 95% confidence level within ±0.0134, showing good uniformity. The separation provided by the raw intra and inter hamming distance (HD) is 42× and with masking applied, the separation is significantly improved as the mean value of intra HD becomes 0. The digital bits averaged across rows and columns are shown in Fig. 10(a) where the results are around the ideal value of 0.5. This shows that despite the column shared circuit components, no systematic bias exists revealing the independency from layout pattern. In addition, the proposed PUF passes all applicable randomness tests of the NIST SP 800-22. Aging is another factor to consider regarding environmental shifts. An accelerated aging test is performed with 4 PUF arrays at 1.38V (15% higher than nominal condition) and 125°C for 25 hours as shown in Fig. 10(b). Without using any masking, the BER at worst case is 1.78% and with Δtilt=10, the BER is reduced to 0%. Comparison with the state-of-the-art designs using masking schemes is shown in Table I.

V. CONCLUSION

An area efficient PUF based on gate leakage current is presented in this paper. A current tilting mechanism is embedded in the readout circuit to mask the potential unstable bits at the nominal condition. With a masking ratio of 41.8%, the BER of 0% is achieved across a wide range of supply voltage and temperature variation. As many parts of the circuits are shared, the core PUF unit area is only 183F². The total area per a unit PUF cell is smaller than 950F² even including all other peripheral blocks.

REFERENCES

[1] S. Taneja et al., "Fully synthesizable PUF featuring hysteresis and temperature compensation for 3.2% native BER and 1.02 fJ/b in 40 nm," in *IEEE Journal of Solid-State Circuits*, vol. 53, no. 10, pp. 2828–2839, Oct. 2018.

[2] D. Li and K. Yang, "A 562F² physically unclonable function with a zero-overhead stabilization scheme," *IEEE ISSCC Dig. Tech. Papers*, pp. 400-402, Feb. 2019.

[3] Y. Choi et al., "Physically unclonable function in 28nm fdsoi technology achieving high reliability for aec-q 100 grade 1 and iso 26262 asil-b," *IEEE ISSCC Dig. Tech. Papers*, pp. 426-428, Feb. 2020.

[4] J. Lee et al., "A 445F² leakage-based physically unclonable function with lossless stabilization through remapping for IoT security," *IEEE ISSCC Dig. Tech. Papers*, pp. 132–134, Feb. 2018.

[5] K. Liu et al., "A 373-F² 0.21%-native-BER EE SRAM physically unclonable function with 2-D power-gated bit cells and VSS bias-based dark-bit detection," in *IEEE Journal of Solid-State Circuits*, vol. 55, no. 6, pp. 1719-1732, June 2020.

[6] Y. Shifman et al., "An SRAM-Based PUF with a capacitive digital preselection for a 1E-9 key error probability," in *IEEE Transactions on Circuits and Systems I: Regular Papers*, vol. 67, no. 12, pp. 4855-4868, Dec. 2020.

[7] Y. He et al., "An automatic self-checking and healing physically unclonable function (PUF) with <3×10-8 bit error rate," *IEEE ISSCC Dig. Tech. Papers*, pp. 506-508, Feb. 2021.

[8] M. Seok et al., "A portable 2-transistor picowatt temperature-compensated voltage reference operating at 0.5 V," in *IEEE Journal of Solid-State Circuits*, vol. 47, no. 10, pp. 2534-2545, Oct. 2012.

Logic-embedded Physically Unclonable Functions for Synthesizable and Periphery-free Implementation for Low Area and Design Cost IoT Security

Seonho Kim[1]*, Changyoun Im[2]*, Jongmin Lee[2], Soyoun Jeong[1], Jaerok Kim[2], and Yoonmyung Lee[2]

[1]Semiconductor and Display Engineering, Sungkyunkwan University, Suwon, Korea
[2]Electrical and Computer Engineering, Sungkyunkwan University, Suwon, Korea

*Equal Contribution
kimm315@g.skku.edu, icy0928@g.skku.edu

Abstract— **Novel logic-embedded physically unclonable functions (Logic-ePUF) are proposed to significantly reduce design/area cost with minimal modification on standard cells used for semi-custom design flow. By replacing scan flip-flops and buffers in scan chain with proposed flip-flop-ePUF (FF-ePUF) and buffer-ePUF (BUF-ePUF), efficient PUF implementation is enabled without additional readout periphery. The proposed Logic-ePUF not only operates as a standard cell but also as a PUF. To generate secure key, proposed Logic-ePUF operates under PUF mode by comparing the difference of switching voltages (V_M) of 1st and 2nd stage inverters. The proposed FF-ePUF and BUF-ePUF are fabricated in 28nm FDSOI process to evaluate effectiveness and performance. The proposed FF-ePUF achieved 178ppm BER through reconfiguration and tilting with 332-447F^2/bit area overhead. And the proposed BUF-ePUF reduced 41% of area overhead compared to conventional method.**

Keywords—Physically unclonable function (PUF), hardware security, periphery-free, synthesized implementation, cost-efficient, Internet of Things (IoT)

I. INTRODUCTION

Demand for IoT security is rapidly increasing with the ever-growing usage of IoT devices. Physically Unclonable Functions (PUFs) have been popular hardware security primitive for authentication and key generation in IoT applications. Keeping pace with IoT security needs, the performance of PUFs have gradually improved, and numerous methods have been introduced to reduce costs. Recently, many PUFs have been proposed to achieve smaller area per bit for cost-effectiveness. For example, a NAND-structured PUF [1] achieved excellent cost-efficiency with the 20F^2 area per bit. However, the additional area required for a precise readout peripheral circuit was approximately 20 times larger than PUF cell array itself, which is a critical disadvantage in IoT applications. While a PUF utilizing RRAM [2] also achieved high cost-efficiency with compact RRAM device size, the required large peripheral circuits for readout, which are 6.4 times larger than the PUF cell array itself, imposed a significant area cost. An SRAM-based PUF was proposed that reuses the SRAM as a TRNG and PUF [3], but it requires complex column peripheral circuits for TRNG/PUF operation. To improve the stability of such SRAM-based PUFs, additional capacitors [4] or transistors [5] are required. The inverter chain-based PUF [6] has low power and good stability, but the overhead for the peripheral circuit is still significant. These earlier PUF designs [1-6] commonly adopt a stand-alone PUF module structure, where peripheral circuits are added to the key-generating PUF cell array. Such designs typically incur high area costs with additional peripheral circuits and a high design cost for full custom implementation of the PUF module. They are also

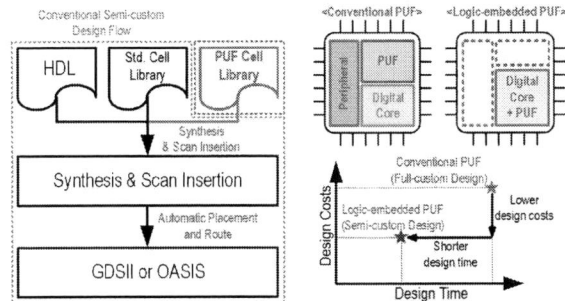

Fig. 1. Proposed logic-embedded PUF(Logic-ePUF) scheme

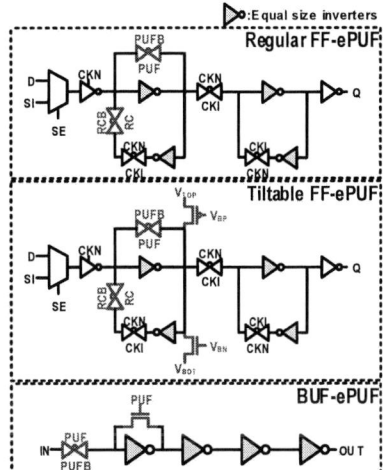

Fig. 2. Proposed Logic-ePUF schematic
(top) flip-flop-embedded PUF (bottom) buffer-embedded PUF

less immune to physical attacks, such as reverse engineering, since the location of the PUF cells can be easily specified. While works such as [7] introduced the possibility of a synthesizable PUF with digital circuitry, it was demonstrated as an independent array separated from the main circuitry since the PUF cell had no logic function. Therefore, it is still vulnerable to physical attacks.

In this paper, a logic-embedded PUF (Logic-ePUF) is proposed to generate random keys from plain digital circuits with minimal area and design overhead while maintaining their original functionality. Three types of Logic-ePUFs are implemented by modifying two types of standard cells (Fig.2): a scan flip-flop-embedded PUF (FF-ePUF) and a buffer-embedded PUF (BUF-ePUF). These are essential components for constructing a scan chain, which is widely used for design-for-testability (DFT) in general digital circuits. The FF-ePUF and BUF-ePUF utilize series-connected identically sized inverters for switching threshold

978-1-6654-8495-4/22 $31.00 © 2022 IEEE

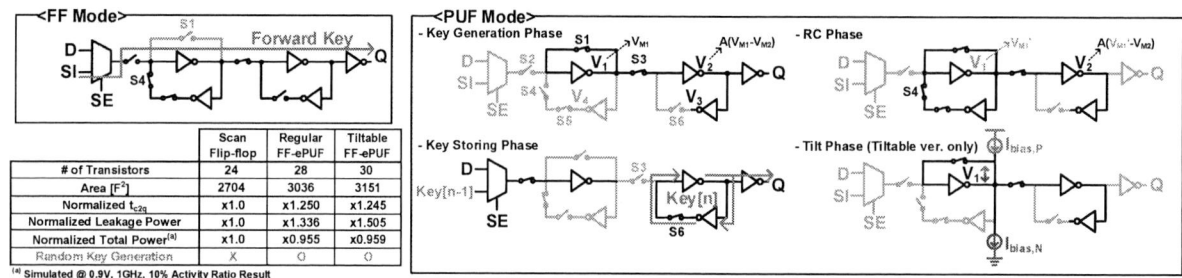

Fig. 3. Operation detail of regular/tiltable FF-ePUF

	Scan Flip-flop	Regular FF-ePUF	Tiltable FF-ePUF
# of Transistors	24	28	30
Area [F²]	2704	3036	3151
Normalized t_{c2q}	x1.0	x1.250	x1.245
Normalized Leakage Power	x1.0	x1.336	x1.505
Normalized Total Power[a]	x1.0	x0.955	x0.959
Random Key Generation	X	O	O

[a] Simulated @ 0.9V, 1GHz, 10% Activity Ratio Result

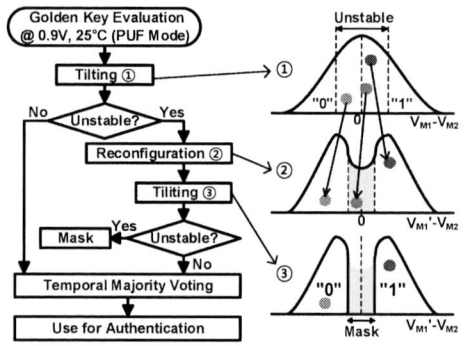

Fig. 4. Stabilization algorithm of tiltable FF-ePUF

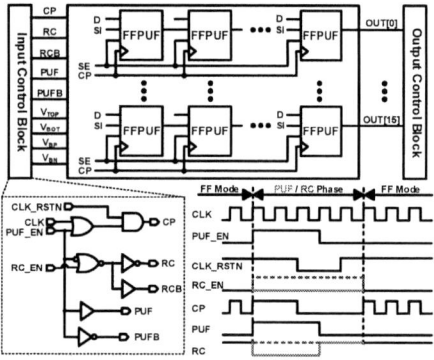

Fig. 5. Control scheme of proposed FF-ePUF

voltage comparison, that can be formed by adding a few extra switches to original standard cells, as an entropy source. By maximizing reuse of the existing circuitry in standard cells and scan chains, the overhead for key generation and readout could be minimized. The proposed PUF cells were characterized using commercial EDA tools so that they can be used in standard digital design flows. A PUF module can be easily implemented by replacing some of the flip-flops or buffer cells in the scan chains with Logic-ePUF cells when the digital block is synthesized, significantly reducing the design time and cost. Thanks to the use of existing scan chains for PUF-generated key readout, no additional peripheral readout circuits are required for the PUF module implementation, making it periphery-free.

II. LOGIC-EMBEDDED PUF

A. Flip-flop-embedded PUF

The circuit-level details of the FF-ePUF are shown in Fig.3. The regular version of the FF-ePUF (Regular FF-ePUF) requires two extra transmission gates that adds 332F²

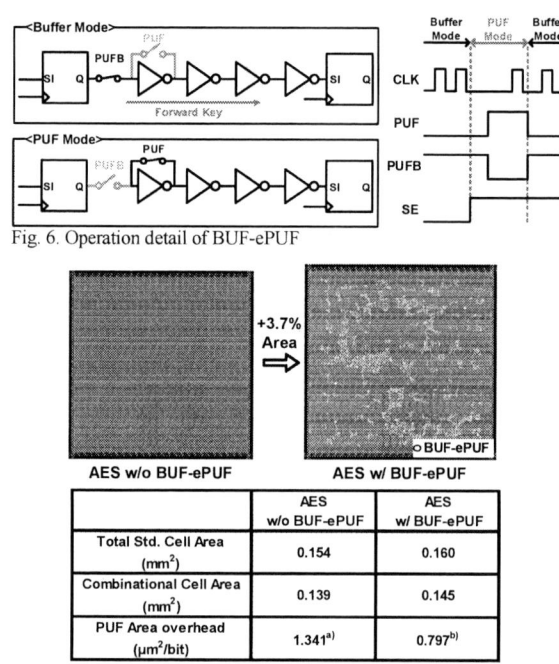

Fig. 6. Operation detail of BUF-ePUF

	AES w/o BUF-ePUF	AES w/ BUF-ePUF
Total Std. Cell Area (mm²)	0.154	0.160
Combinational Cell Area (mm²)	0.139	0.145
PUF Area overhead (µm²/bit)	1.341[a]	0.797[b]

[a] (Peripheral Area/bit)[4]+(Area of BUF-ePUF bit cell)
[b] Increased Area/bit

Fig. 7. AES area comparison with/without BUF-ePUF

area overhead compared to the original FF. There are two modes of operations. In PUF mode, S_1 and S_3 are closed (by PUF_EN) to generate the switching threshold voltage of the 1st stage inverter (V_{M1}) at V_1, and the mismatch of V_{M1} and V_{M2} is amplified at V_2, and V_3 for key generation, in similar manner as in [6] (Key Generation Phase). After the evaluation, S_3 is opened and S_6 is closed (by CLK_RSTN) so that V_2 and V_3 can keep the key value with the slave latch of the flip-flop (Key Storing Phase). By switching to FF mode, the generated keys can be serially extracted through nominal scan chain operation. However, key values can be unstable when the mismatch between V_{M1} and V_{M2} is small. In this case, the proposed PUF can be reconfigured by closing S_4 to back-to-back connect two inverters (RC Phase). The new voltage formed at V_1 (V_{M1}') is now determined by V_{M1} and V_{M4}, creating deviation of the threshold voltage difference (V_{M1}'-V_{M2}). The cells that have altered responses in the RC Phase can be masked since their threshold voltage differences (V_{M1}-V_{M2}) are determined to be small. To stabilize regular FF-ePUF, the cells whose response in Key Generation phase and RC phase are different are identified as unstable cells and discarded. To further improve stability, a tilting scheme for dark-bit detection is proposed that uses two additional transistors and incurs additional area overhead of 115F² (Tiltable FF-ePUF in Fig.3). By sweeping

978-1-6654-8495-4/22 $31.00 © 2022 IEEE

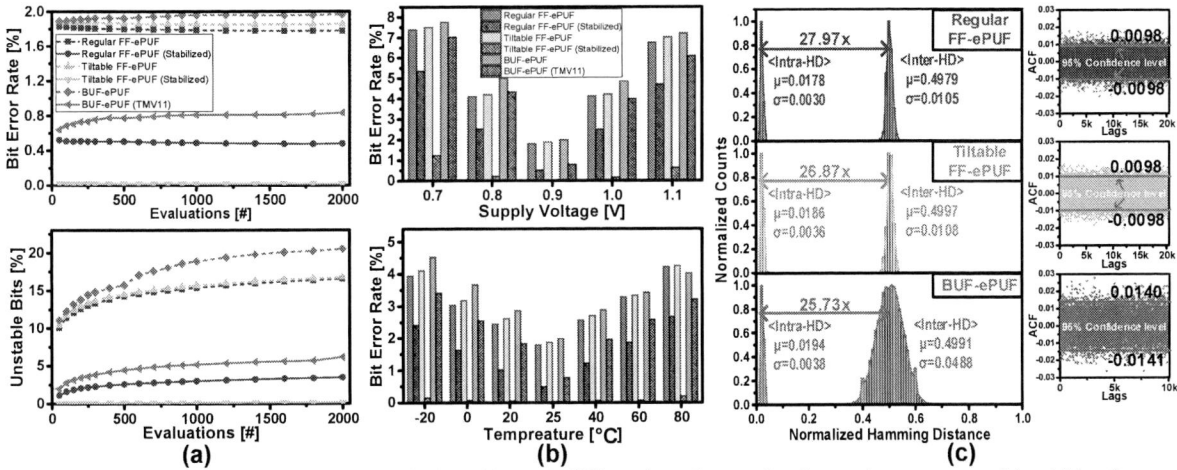

Fig. 8. Measured (a) reproducibility with repeated evaluations, (b) reproducibility under various supply voltage and temperature condition, (c) inter/intra-hamming distance and autocorrelation function of proposed ePUFs

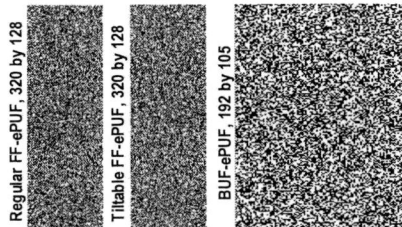

Fig. 9. Speckle pattern of FF-ePUF and BUF-ePUF

the amount of current sourced/sunk ($I_{bias,P}/I_{bias,N}$) at node V_1 until the key value is flipped, the relative threshold voltage difference ($V_{M1}-V_{M2}$) and stability of the generated key can be gauged (Tilt Phase). Applying the reconfiguration(RC)/tilting algorithm shown in Fig.4 allows significant improvement with respect to stability while minimizing the number of masked cells. A 128×16 test array of each FF-ePUF version is implemented for evaluation (Fig.5).

B. Buffer-embedded PUF

The second type of Logic-ePUF, a BUF-ePUF, is shown in Fig.6. A BUF-ePUF cell is created out of a buffer cell with 4 inverters in series by adding two switches (a pass gate and a transmission gate) with $337F^2$ area overhead. In a typical digital circuit with scan chains, buffers are required between scan FFs to prevent hold time violations. Therefore, these buffers can be replaced with BUF-ePUFs. The BUF-ePUF cell also operates in two modes. During buffer mode (PUF=0), the PUF cell operates as a regular buffer. During PUF mode (PUF=1), the PUF cell generates a 1-bit key by comparing the switching threshold voltages of the first two inverters. This key is then stored in the following scan flip-flop, and the keys can be serially extracted through nominal scan chain operation while setting BUF-ePUFs in buffer mode (PUF=0). Two types of 128-bit AES cores [8] (with/without BUF-ePUF) are implemented with identical performance targets in a test chip for evaluation (Fig.7). The total area of the AES Core with the PUF is 3.74% larger than that of the AES Core without a PUF. To compare the area efficiency of the conventional and proposed methods, the PUF area overhead is defined as the additional area per bit required to generate a random key. The PUF area overhead in the AES Core with the BUF-ePUF is measured to be 0.797μm²/bit. For a conventional approach where an independent PUF module is built, the PUF area overhead is estimated to be 1.341μm²/bit, considering the PUF cell and peripheral area ratio in [6], which is more than 68% larger than that of the proposed scheme.

III. MEASUREMENT RESULT

Two test chips for the proposed Logic-ePUF are fabricated in 28nm FDSOI process (Fig.11), and the bit error rate (BER) and unstable bits with 2000 repeated evaluations are shown in Fig.8(a). The regular FF-ePUF and tiltable FF-ePUF have native BER/unstable bits of 1.78%/16.56% and 1.86%/16.77%, respectively, at the nominal condition (0.9V, 25°C). After applying the proposed RC/tilting algorithm, the BER/unstable bits are reduced to 0.0178%/0.1780%. For the BUF-ePUF, the native BER/unstable bits are 1.97%/20.49% at the nominal condition. The BER and unstable bits of the BUF-ePUF are further reduced to 0.83% and 6.15%, respectively, with TMV11. The stability of the FF/BUF-ePUF cells is evaluated using the BER under voltage and temperature variation (Fig.8(b)). The maximum BER of the regular/tiltable FF-ePUF is 7.378%/7.486% with supply voltage variation and 4.219%/4.245% with temperature variation. By applying the RC/tilting algorithm, the worst case BER for voltage and temperature variation can be improved to 1.210% (6.2×) and 0.1809% (23.5×), respectively. With voltage and temperature variation, the maximum BERs of the BUF-ePUF are 7.739% and 4.517%, respectively. The uniqueness and reproducibility of the regular/tiltable FF-ePUF are verified as shown in Fig.8(c) by inter-hamming distances (inter-HDs) of 49.79%/49.97% and intra-HDs of 1.78%/1.86%. The identifiability is also confirmed by 27.97/26.87× separation between inter- and intra-HD for each version. The inter-HD and intra-HD values of the BUF-ePUF are 49.91% and 1.94%, respectively, resulting in 25.73× separation. The respective autocorrelation function (ACF) bounds with 95% confidence are -0.00984/-0.00976 and 0.00979/0.00971 for regular/tiltable FF-ePUF and -0.01412/0.01402 for BUF-ePUF. Speckle patterns in Fig.9 confirm the random distribution of the PUF responses. Effectiveness of the stabilization technique is also verified: by applying RC/tilting algorithm, 21%/33% less cells had to be masked to achieve comparable BER/unstable bits compared to the case where only masking is applied (Fig.10).

978-1-6654-8495-4/22 $31.00 © 2022 IEEE

TABLE I. SUMMARY OF MEASUREMENT RESULTS AND COMPARISON WITH RECENTLY PROPOSED PUFs

	This work (0.9V, 25°C)			ESSCIRC' 19 [1]	ISSCC' 19 [2]	ISSCC' 21 [3]	TCAS' 20 [4]	CICC' 20 [5]	ISSCC' 21 [6]	JSSC' 20 [7]
	Regular FF-ePUF	Tiltable FF-ePUF	BUF-ePUF							
Unified Functions	PUF+Flip-flop		PUF+Buffer	PUF	PUF	PUF+TRNG+SRAM	PUF	PUF+SRAM	PUF	PUF
Technology	28nm FDSOI			180nm	130nm	28nm	65nm	130nm	65nm	40nm
Net Area/bit (Cell Area/bit) [F^2]	332 (3036)	447 (3151)	337 (1124)	20 (20)	169 (169)	-	3001 (3001)	- (497)	594 (594)	3644 (3644)
Peripheral Area/bit[a] [F^2]	0	0	0	2330	773	-	-	1811	586	-
Total Area/bit[a] [F^2]	332	447	337	2350	942	1125[b]	> 3001	> 1811	1180	> 3644
Synthesizable	Yes			No	No	No	No	No	No	Yes
Logic Embeddable	Yes			No	No	No	No	No	No	No
Physical Location Attack Resilience	High			Low	Low	Low	Low	Low	Low	Low
Evaluations	2k			2k	-	2k	120k	1k	2k	5k
Native BER/Unstable Bits [%]	1.78 / 16.56	1.86 / 16.77	1.97 / 20.49	0.13 / 1.47	< 0.00061	1.78(LSB), 3.84(MSB) / 11.9(LSB), 36.5(MSB)	2.24 / 19.6[c]	0.29 / 2.71	0.29 / 2.83	0.81[c] / 3.48
Stabilization Technique	TMV, RC, Masking	TMV, RC, Tilt, Masking	TMV	TMV	-	-	Tilt, Masking	HCI Burn-in	RC, Skew, Masking	-
Masking Ratio	10.6%	32.0%	-	-	-	-	59%	-	27%	-
Stabilized BER/Unstable Bits	0.48% / 3.47%	0.0178% / 0.1780%	0.83% / 6.15%	0.06% / 0.53%	-	-	1.4E-9[d] / -	~0%[e] / ~0%[e]	0 (<0.34E-8) / ~0%[e]	-
Throughput [Mb/s]	18300	17500	5681	0.01	-	10160	-	56	22750	24000
Tested Condition	Voltage [V]	0.7-1.1			1.6-2.0	1.4-2.2	0.8-1.0	0.8-1.2	0.5-0.7	0.7-1.4	0.8-1.0
	Temperature [°C]	-20-80			-40-100	25-150	-10-75	-10-85	-40-120	-40-125	-40-125
Core Energy [fJ/bit]	305[f]	308[f]	1265.6[g]	3.00E+7	3028	78[h]	16	15.4	0.057	56.5

[a]: Total Area/bit = Net Area/bit + Peri. Area/bit   [b]: Including SRAM macro area   [c]: Worst case   [d]: Averaged across multiple V/T conditions
[e]: 3 minutes burn-in   [f]: Calculated with simulated Power   [g]: Including power consumption of another cells in AES Core   [h]: @ 0.8V (Nominal Voltage = 0.9V)

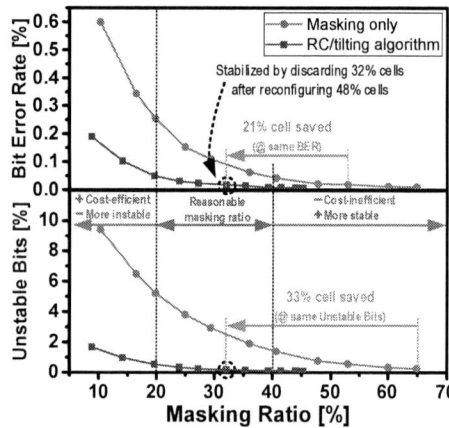

Fig. 10. Effectiveness of RC/tilting algorithm

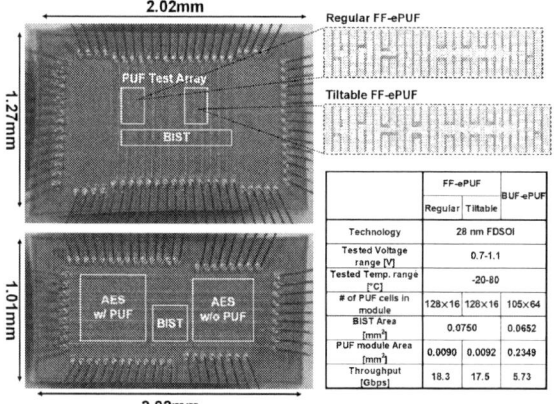

Fig. 11. Chip micrograph of Logic-ePUF

The evaluation results are compared with prior state-of-the-arts in Table I. The proposed Logic-ePUF unified standard cell and PUF with a small area increment of 332-447F^2/bit compared to conventional standard cells. Thanks to its periphery-free structure, there is no additional peripheral area overhead, resulting in 2.1-11.0× smaller total area required per bit compared to other works. Thanks to its synthesizability, the Logic-ePUF can be implemented on chip quickly at a lower cost than full custom PUF macro designs. Also, the randomly distributed Logic-ePUF cell location provides better physical attack resilience compared with prior arts.

ACKNOWLEDGMENT

This research was supported in part by NRF Korea (2022R1A2B5B02002350) and Samsung Electronics. The EDA tool was supported by IDEC, Korea.

REFERENCES

[1] J. Lee, et al., "A 20F² Area-Efficient Differential NAND-Structured Physically Unclonable Function for Low-Cost IoT Security," ESSCIRC, 2019

[2] Y. Pang, et al., "A Reconfigurable RRAM Physically Unclonable Function Utilizing Post-Process Randomness Source With <6×10⁻⁶ Native Bit Error Rate," ISSCC, 2019

[3] S. Taneja, et al., "Unified In-Memory Dynamic TRNG and Multi-Bit Static PUF Entropy Generation for Ubiquitous Hardware Security," ISSCC, 2021

[4] Y. Shifman, et al., "An SRAM-Based PUF With a Capacitive Digital Preselection for a 1E-9 Key Error Probability," IEEE TCAS-I, Dec. 2020

[5] K. Liu, et al., "A 0.5-V 2.07-fJ/b 497-F² EE/CMOS Hybrid SRAM Physically Unclonable Function with < 1E-7 Bit Error Rate Achieved through Hot Carrier Injection Burn-in," CICC, 2020

[6] Y. He, et al., "An Automatic Self-Checking and Healing Physically Unclonable Function (PUF) with <3×10⁻⁸ Bit Error Rate," ISSCC, 2021

[7] S. Taneja, et al., "Fully Synthesizable PUF Featuring Hysteresis and Temperature Compensation for 3.2% Native BER and 1.02 fJ/b in 40 nm," IEEE JSSC, Oct. 2018

[8] AES.[Online] https://opencores.org/projects/tiny_aes

Configurable Energy-Efficient Lattice-Based Post-Quantum Cryptography Processor for IoT Devices

ByungJun Kim, Jaehan Park, Seunghyun Moon, Kiseo Kang, Jae-Yoon Sim
Postech Integrated Circuits and Systems Laboratory
Department of Electronic and Electrical Engineering, POSTECH
Pohang 37673, Korea
kbj1213@postech.ac.kr

Abstract— This work presents a configurable lattice-based post-quantum cryptography processor suitable for lightweight edge devices. To reduce hardware cost and energy consumption, it employs a look-up-table-based modular multiplication for the number-theoretic transform and a real-time processing for polynomial sampling. Implementation in 28nm CMOS shows 15.4x and 14.5x reductions of gate count and on-chip memory size, respectively, compared to the previous state-of-the-art implementation at the cost of only 54% in energy.

Keywords— *post-quantum cryptography, crypto processor, lattice problem, number-theoretic transform, quantum computer*

I. INTRODUCTION

As quantum technology advances to the realization of a large-scale quantum computer [1], the vulnerability of today's public-key protocols of Rivest-Shamir-Adleman (RSA) and elliptic curve cryptography (ECC) has led extensive research on the development of quantum-secure algorithms [2][3]. The standardization process of post-quantum cryptographic protocols initiated by the National Institute of Standards and Technology (NIST) has brought to an announcement that the first standard is about to be released. The NIST selected seven finalists and eight alternates as candidates in the 3rd Round of the post-quantum cryptography standardization process [4]. The lattice-based protocols, using learning with errors (LWE), account for more than half of them. The LWE problems have been theoretically proved to hold the hardness of finding a secret vector s satisfying the equation b=As+e, where matrix A and vector b are open to the public with noise vector e sampled from discrete distributions (Fig. 1). The Ring-LWE, which is an implementation-friendly variant of LWE, uses polynomials for the vectors while still keeping the hardness against the quantum attacks [5].

There have been software implementations on public-key encryption (PKE), key encapsulation method (KEM), and digital signature operations with Ring-LWE. However, the Ring-LWE causes excessively time-consuming computations compared with the present RSA and ECC protocols. There are two major factors of the heavy computation load. First, the arithmetic multiplications require modular polynomial operations in butterfly-computing number-theoretic transform (NTT) based on Fast Fourier Transform (FFT). Second, the polynomial sampling for LWE requires complicated calculations for secure hash algorithm-3 (SHA-3) processing.

ASIC implementations on the cryptography processor for Ring-LWE problems [6][7][8] have achieved orders of magnitude speedups and energy savings. Considering the post-quantum cryptography is under the first standardization

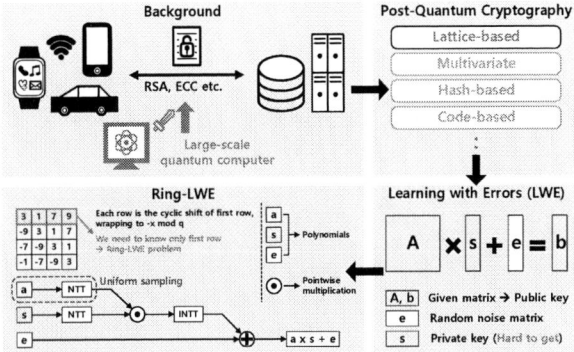

Fig. 1. Lattice-based cryptography based on Ring-LWE problem.

process, new Ring-LWE algorithms other than the current finalists could come with better performance in the future. Thus, having maximum configurability for processing general Ring-LWE problems is also important in the design of ASIC. Though [6][7] implemented configurability for already proposed post-quantum cryptographic algorithms, their architecture would eventually fail to support possible variants. Recently, an ASIC implementation supporting all the post-quantum cryptographic protocols with full configurability was reported, but at the cost of large hardware resources [8]. Considering that edge devices are more vulnerable to attack and the security threat would be more serious in personal users, the implementation of a flexible post-quantum crypto core in compact hardware is also important for widespread use in IoT applications.

This work presents a compact post-quantum crypto-processor implemented in the smallest die area. The proposed architecture provides flexibility of processing general Ring-LWE protocols including any variants of lattice-based post-quantum cryptography. Implementation in 28nm CMOS achieves 15.4x and 14.5x reductions of gate count and on-chip memory size, respectively, compared to the previous state-of-the-art implementation [8], with an increase of only 54% in energy.

II. SYSTEM ARCHITECTURE

Fig. 2 shows the overall block diagram of the proposed crypto-processor. There are three SRAM banks which store the polynomials for A, s, and e in the lattice problems. Each SRAM has 6kB so that the processor handles polynomials of length up to 2048. The processor core calculates two major performance-limiting tasks of sampling and NTT with a polynomial sampler and a modular ALU, respectively. The constant factors for negacyclic convolution in NTT are stored

978-1-6654-8495-4/22 $31.00 © 2022 IEEE

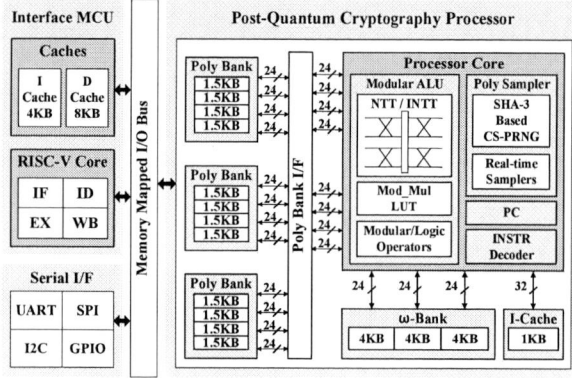

Fig. 2. Overall chip architecture.

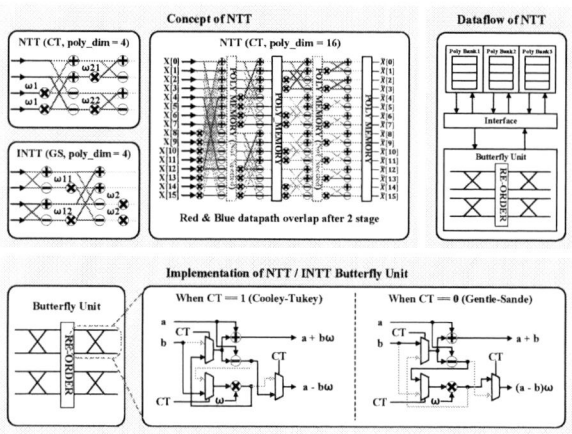

Fig. 3. Proposed architecture and dataflow of NTT.

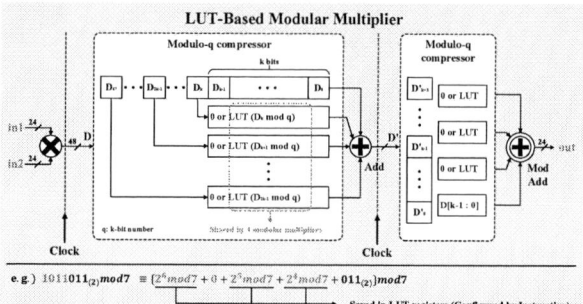

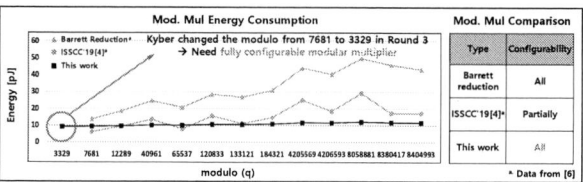

Fig. 4. Architecture of the proposed LUT-based modular multiplier and performance comparison with the previous works.

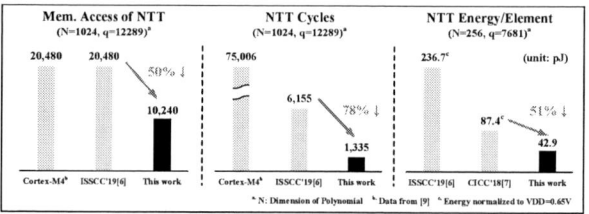

Fig. 5. Performance of the proposed NTT/INTT operations.

in ω-Bank. The negative wrapped convolution eliminates the overhead of zero-padding and bit-reversal in NTT. By merging the NTT and multiplication by powers of ψ ($\omega= \psi^2$), the ω-Bank size is reduced to only 12kB. The processor is programmed with a 32-bit instruction set to support any lattice-based algorithms in a generalized control of modular arithmetic functions with registers and polynomials, sampling, NTT/INTT and SHA-3 processing. The instructions are configured by an on-chip RISC-V microcontroller unit (MCU) through a memory-mapped I/O interface.

III. NTT

A. Butterfly Operations

The NTT and INTT are processed by Cooley-Tukey (CT) and Gentle-Sande (GS) butterfly configurations, respectively. Each addition or subtraction operation in the butterfly necessitates memory access for temporary storing and fetching. It results in a significant latency in the processing of NTT/INTT. We propose an NTT/INTT processing scheme which reduces the memory access by half. As shown in Fig. 3, we can observe a unique connection of a bijection between a set of 4 inputs and a set of 4 outputs over two consecutive butterfly stages for given CT or GS configuration. The 4-to-4 bijection can be regarded as a unit and processed in one cycle. Since the 4-to-4 bijections are computationally separated, there is no need of memory access for temporary storing and fetching. Hardware implementation for the processing unit is

realized by insertion of a reordering MUX stage between two butterflies because the reordering can define any bijection mapping.

B. Configurable Modular Multiplier

The heaviest load in NTT processing is the modular multiplication which becomes the major bottleneck of latency. We propose a LUT-based modular multiplication which is completed in two additional clock cycles compared with the normal multiplication (Fig. 4). The proposed multiplier is equipped with two stages of the modulo-q compressor. D represents the given initial result of normal multiplication (in1 x in2), which has the maximum bit-width of 24 bits. Assuming the number of bits of given modulo q is k, the k-bit LSBs of D is a modulo-q number. If the modulo-q output for each MSB is initially stored in a LUT, we can simply fetch the contents of the LUT and add them according to the bit states of MSBs. This processing effectively reduces the number of bits while preserving the output in modulo-q. Cascading another stage of the modulo-q compressor further reduces the number of bits. The final output is then calculated using a modular adder. As a result, the modular multiplication can be performed in a minimum latency. In addition, the proposed modular multiplier can be applied for any integer q ($< 2^{24}$) once the LUT contents are set for the given q by instructions.

C. Evaluation

Fig. 5 shows evaluated performance on a benchmark in the NIST Round 1 protocols. The proposed processing scheme for NTT/INTT shows reductions of 50% in memory access, 78% in the number of clock cycles and 51% in energy consumption compared to the previous best in hardware compactness [6].

978-1-6654-8495-4/22 $31.00 © 2022 IEEE

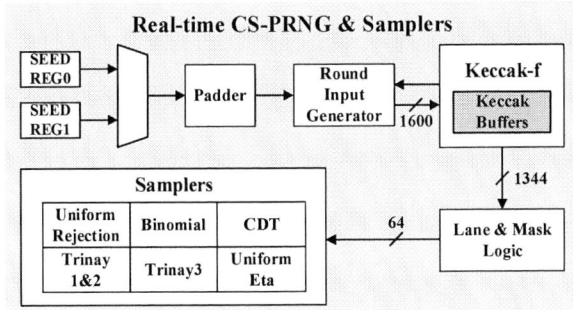

Fig. 6. Architecture of the SHA3-based sampler.

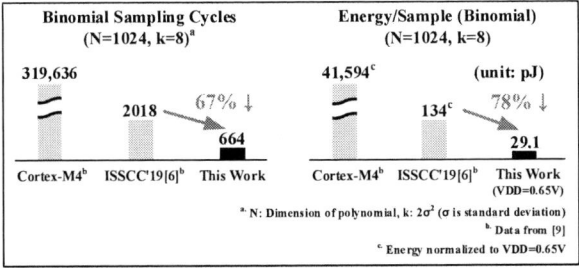

Fig. 7. Sampling performance comparison with previous works.

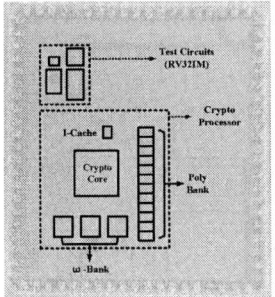

Technology	28 nm LP
Die size	1.8mm x 1.96mm
Supply voltage	0.65-1.35V
Frequency	35-190MHz
SRAM usage	31kB
Core area	0.18mm²
Logic gates	123.8k
Hash function	SHA3-224/256/384/512
CS-PRNG	SHAKE-128/256
Average Power	1.21mW at 0.65V & 35MHz (for Kyber1024)

Fig. 8. Die photo and chip summary.

IV. SAMPLING

The algorithms based on the LWE problem rely on statistical properties of polynomial sampling which takes approximately 70% of computations [6]. Therefore, the hardware design for the sampling also significantly affects speed and energy performance.

A. Real-time Random Number Generator

The sampling is performed by a cryptographically secure pseudorandom number generator (CS-PRNG) and a post-processing distribution sampler. The CS-PRNG computes SHA-3 functions using a 1600b Keccak-f core which supports various SHA-3 modes (Fig. 6). The processing for CS-PRNG takes 24 cycles and generates a 1344b / 1088b output (SHAKE-128 / SHAKE-256). We employ a Keccak buffer to store the previous 1344b / 1088b so that the Keccak core can work for the next output. It makes continuous utilization of Keccak-f core. The buffer contents are flushed to the sampler through a 64b channel (4x16b or 2x32b).

B. Sampling Modules

There are a number of algorithms for LWE problems which work with different samplers. The hardness of attacking the lattice-based cryptography is affected by the statistical distribution of the error vectors. The proposed sampler

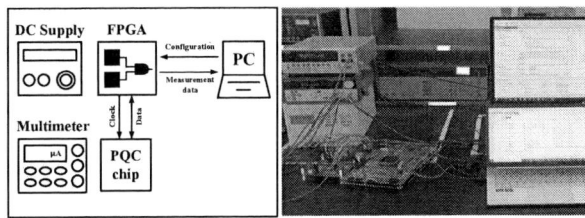

Fig. 9. Test platform.

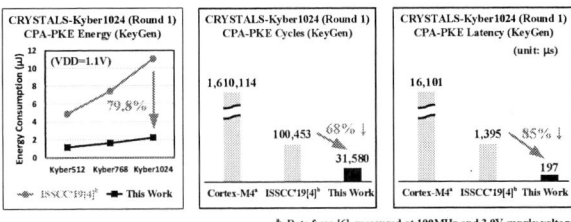

a. Data from [6], measured at 100MHz and 3.0V supply voltage
b. Data from [6], measured at 72 MHz and 1.1V supply voltage

Fig. 10. Measurement result benchmarking CRYSTALS-Kyber protocol.

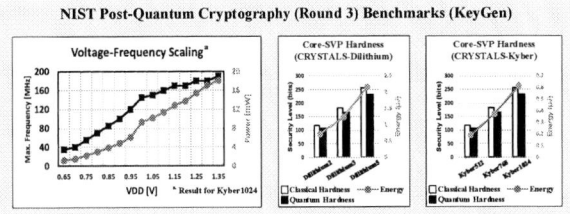

Fig. 11. Measurement result benchmarking NIST Round 3 protocols.

supports various distributions which should be performed through a number of modular subtractions. The subtractors in the modular ALU for NTT are reused since the sampling does not need to work in parallel with NTT.

C. Evaluation

Fig. 7 shows evaluated performance on a benchmark in the NIST Round 1 protocols. The proposed sampler shows reductions of 67% in the number of clock cycles and 78% in energy consumption per sampling compared to the previous best in hardware compactness [6].

V. MEASUREMENT RESULTS

The proposed crypto-processor is implemented using 28nm LP CMOS technology in a die area of 1.8mm x 1.96mm. Fig. 8 shows the chip micrograph. Fig. 9 shows the test environment. An external FPGA is used for configuring the instructions. The results are then transmitted to the FPGA to check the functionalities. The chip consumes 1.21mW from a supply voltage of 0.65V while processing the Kyber1024 protocol at 35MHz. The measurement results are compared to the previous state-of-the-art compact implementation [6]. Fig. 10 shows measurements with a benchmark of the CRYSTALS-Kyber. The proposed chip shows reductions of 79% in energy consumption, 68% in the number of clock cycles and 85% in latency for the CRYSTALS-Kyber1024. Fig. 11 shows voltage-frequency scaling, security level, and energy performance for NIST Round 3 protocol benchmarks. The chip shows full functionalities from 35 to 190MHz (0.65-1.35V). As the security level of the post-quantum cryptographic protocols increases, the processor's energy consumption also increases.

TABLE I. PERFORMANCE COMPARISON

	Cortex-M4	CICC'18 [7]	ISSCC'19 [6]ᵉ	TCAS-I'20[10]	ISSCC'22 [8]	This work
Overall Chip Specification						
Technology	-	40nm GP	40nm LP	28nm	28nm	28nm LP
Implementation	-	ASIC	ASIC	-	ASIC	ASIC
Supply Voltage [V]	3.0	0.9	0.68-1.1	0.9	0.9	0.65-1.35
Core Area [mm^2]	-	2.05	0.28	-	3.6	0.18
Frequency [MHz]	100	300	12-72	300	500	35-190
Logic gates (k)	-	-	106	979	1900	123
On-chip Memory [kB]	-	340	40.25	12	448	31
Poly. Dimension	All*	64 to 2048	64 to 2048	64 to 2048	-	64 to 2048
q Configuration	All*	All ($< 2^{24}$)	A Few of q ($< 2^{24}$)	All ($< 2^{16}$)	All ($< 2^{24}$)	All ($< 2^{24}$)
Binomial Sampling Performance (N = 256)						
# Cycles	52603	-	505	411	-	191
Energy [nJ]	32350ᵃ	-	58.2ᵇ	26.8	-	14.22ᵈ
NTT Performance (N = 256, q = 7681)						
# Cycles	22031	160	1289	45	32	305
Energy [nJ]	13550ᵃ	31	165.98ᵇ	24.6	-	20.75ᵈ
Kyber-512-CCA-KEM (KeyGen & Encaps & Decaps)						
# Cycles	2733731	-	348526	144431	10432	83148
Energy [μJ]	1681	-	26.59ᵇ	12.8	3.4	5.22ᵈ

ᵃ· Measured at 100MHz and 3.0V supply voltage.

ᵇ· Measured at 72MHz and 1.1V supply voltage.

ᶜ· Including bit-reversal process

ᵈ Measured at 160MHz and 1.1V supply voltage

ᵉ Some data from [9]

VI. CONCLUSION

This work proposes a compact hardware architecture for post-quantum crypto-processor with a flexibility of supporting general lattice-based protocols. The proposed architecture employs a look-up-table-based modular multiplication for the number-theoretic transform and a real-time processing for polynomial sampling. It enables processing any lattice-based protocols with a q of less than 2^{24} while requiring an on-chip memory of only 31kB. Implementation in 28nm CMOS achieves 4.19x, 5.09x, and 9.31x improvements in the number of cycles, energy consumption and latency, respectively, compared to the previous best in hardware compactness. It also shows 15.4x and 14.5x reductions of gate count and on-chip memory size when compared to the state-of-the-art implementation.

ACKNOWLEDGMENT

This research was supported by the National Research Foundation funded by the Korea Ministry of Science and ICT under Grant 2019R1A5A1027055 and IDEC.

REFERENCES

[1] J. Park, et al., "A Fully Integrated Cryo-CMOS SoC for Qubit Control in Quantum Computers Capable of State Manipulation, Readout and High-Speed Gate Pulsing of Spin Qubits in Intel 22nm FFL FinFET Technology," ISSCC, pp. 208-209, Feb. 2021.

[2] I. Verbauwhede, J. Balasch, S. S. Roy, and A. V. Herrewege, "Circuit Challenges from Cryptography," ISSCC, pp. 428-429, Feb. 2015.

[3] Z. Liu, K. K. R. Choo, and J. Grossschadl, "Securing Edge Devices in the Post-Quantum Internet of Things Using Lattice-Based Cryptography," in IEEE Communications Magazine, vol. 56, no.2, pp. 158-162, Feb. 2018.

[4] NIST, NIST PQC Round 3, Nov. 2020.

[5] V. Lyubashevsky, C. Peikert, and O. Regev, "On ideal lattices and learning with errors over rings," in Proc. EUROCRYPT, May 2010, pp. 1–23.

[6] U. Banerjee, A. Pathak, and A. P. Chandrakasan, "An Energy-Efficient Configurable Lattice Cryptography Processor for the Quantum-Secure Internet of Things," ISSCC, pp. 46-47, Feb. 2019.

[7] S. Song, W. Tang, T. Chen, and Z. Zhang, "LEIA: A 2.05mm^2 140mW Lattice Encryption Instruction Accelerator in 40nm CMOS," CICC, pp. 46-47, Apr. 2018.

[8] Y. Zhu, et al., "A 28nm 48KOPS 3.4μJ/Op Agile Crypto-Processor for Post-Quantum Cryptography on Multi-Mathematical Problems," ISSCC, pp. 514-515, Feb. 2022.

[9] U. Banerjee, T. S. Ukyab, and A. P. Chandrakasan, "Sapphire: A Configurable Crypto-Processor for Post-Quantum Lattice-based Protocols," IACR Trans. On CHES, pp. 17-61, Oct. 2019.

[10] G. Xin, et al., "VPQC: A Domain-Specific Vector Processor for Post-Quantum Cryptography Based on RISC-V Architecture," TCAS-I, vol. 67, no. 8, pp. 2672-2684, Aug. 2020.

Fine-Grained Electromagnetic Side-Channel Analysis Resilient Secure AES Core with Stacked Voltage Domains and Spatio-temporally Randomized Circuit Blocks

Meizhi Wang[*1], Sirish Oruganti[*1], Shanshan Xie[1], Raghavan Kumar[2], Sanu Mathew[2], Jaydeep P Kulkarni[1]

[1]University of Texas, Austin, TX   [2]Intel Circuit Research Lab, Hillsboro, OR   [*]Equally Contributing Author

{wang.mz, sirishoruganti, sxie}@utexas.edu, {raghavan.kumar, sanu.k.mathew}@intel.com, jaydeep@austin.utexas.edu

Abstract— **In this work, we demonstrate a novel unique design approach specifically targeting improved resilience against the fine-grained electromagnetic (EM) side-channel analysis (SCA) attacks. Fine-grained EM SCA captures high SNR EM signature with tiny probes (1mm diameter) compared to coarse-grained EM (~10mm diameter), making it more potent and leading to a higher threat. The EM-SCA critical circuit blocks are voltage-stacked to share a current loop, and a push-pull voltage regulator (VR) balances the mismatch current between the two stacked blocks. Dataflow and current loops are spatially and temporally randomized to obscure the EM side-channel signatures. Measurement results from a 128-bit Parallel Advanced Encryption Standard (AES) core fabricated in 65nm CMOS shows MTD improvement of 1507X for fine-grained EM SCA. Coarse-grained EM and Power SCA MTDs also show an improvement of 122X and 657X respectively, which may be improved further by combining prior reported SCA resilient techniques.**

I. MOTIVATION

Crypto-circuits are vulnerable to Side-Channel Analysis (SCA) attacks due to ease of physical access to die and terminals and unintentional data-dependent leakage paths. Figure 1 illustrates the three most common SCA measurements; the fine-grained EM SCA measurement has the highest attack efficacy among the three methods. Power SCA techniques are less effective in obtaining a high SNR side-channel data compared to EM emission-based SCA techniques; as for power SCA, the attacker has access to a few pins drawing current, whereas, in an EM-SCA attack, the entire die/package area is available for scanning with H-field probes. Among EM-SCA approaches, coarse-grained EM measurements with a 10mm diameter probe collecting comprehensive signatures from the entire die are incapable of denoising and filtering out critical information from other non-critical or countermeasure modules. However, in fine-grained EM-SCA [1], minuscule probes of ~1mm diameter can scan at ~100um steps across the die area to gather EM emanations in multiple orientations and precisely identify high SNR EM signatures generated by the cryptographic core's SCA critical vulnerable modules.

Recent research proposes countermeasures against power and coarse-grained EM SCA attacks [1-5], but their effectiveness against fine-grained EM SCA (1mm probe) is rather limited due to the presence of spatio-temporally static current loops having a good 1:1 correlation with the underlying compute activity. A fine-grained EM SCA attack is highly potent on such blocks having access to static current-loop activity as it can scan the entire die area to avoid protection modules (if present) and precisely target only the EM SCA weak spots. Hence, fine-grain EM SCA attack countermeasures could benefit from scrambling the localized current loops. This work proposes data-dependent spatial and temporal randomization of current loops by enabling multiple computations within a single current-loop through charge-recycling using a voltage stacking approach to attenuate the correlation between crypto-sensitive data and the EM emission signatures (Fig. 2).

II. ARCHITECTURE DESCRIPTION

Advanced Encryption Standard (AES) is a widely adopted cryptographic algorithm for data encryption. Fig. 2.a presents the encryption flow: the plaintext is first XORed with the initial key and then proceeds with ten rounds of substitute and permutation transformations. The SubByte, MixColumn, and AddRoundKey operations (hereafter called 'SMA') compute directly on the plaintext and intermediate ciphertext results, are more susceptible to SCA attacks. ShiftRow operation does not involve arithmetic computation or register update, making it considerably less leaky. In the baseline design (Fig.2.b), the entire 128-bit Parallel AES core operates on a single supply voltage (VDD). The proposed approach (Fig. 2.c) splits the SMA operations, which are the most vulnerable, into four identical smaller SMA blocks. These blocks operate with voltage stacking and data shuffling mechanisms to mitigate EM and power side-channel information leakage.

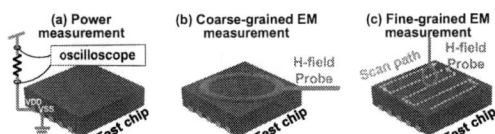

Fig. 1. (a) Power measurement monitors the entire chip's current signature, unable to target specific modules (b) Coarse-grained EM probe measures the total EM emissions from the entire chip, cannot isolate information at sub-block level (c) Fine-grained EM probe scans the entire chip to locate high SNR EM signatures having higher attack success rates

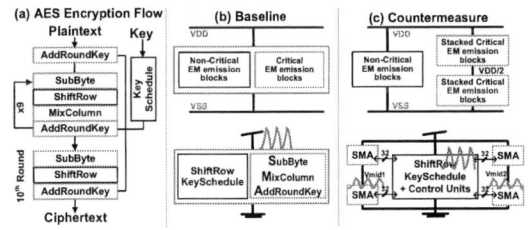

Fig. 2. (a) AES encryption flow with critical EM emission operations highlighted (b) All Modules powered by VDD/VSS in the baseline (c) Critical EM blocks are voltage stacked with reduced voltage swing, sharing current loops

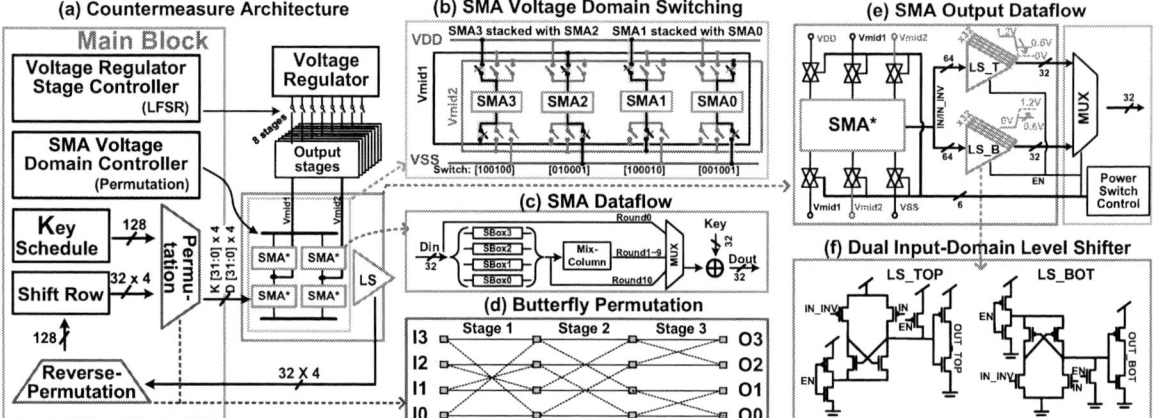

Fig. 3. (a) The proposed architecture overview (b) SMA power switches' configuration for one-stacking scenario (c) Each SMA block computes 32-bit data independently (d) Four-input permutation network [7] (e) SMA outputs level shifted to VDD-VSS swing and sent to the main block (f) Level shifter schematic for the stacked top and bottom domain blocks

The proposed countermeasure architecture, named as Voltage-Stacked AES (VS-AES), consists of the Main block, distributed SMA blocks, level shifters, and a voltage regulator (VR) (Fig. 3.a). The Main block controls the VR operation, SMA voltage-domain randomization, and performs KeySchedule and ShiftRow operations. The SMA blocks operate on two stacked voltage domains (VDD-Vmid, and Vmid-VSS) and are spatially distributed in physical placement around the main block as separate circuit blocks. This modularization enables multiple computations within a single current loop and permutable spatial distribution of the data flow. Having comparable average current signatures in all SMA blocks simplifies voltage stacking since the balanced currents help in the regulation of the intermediate node by not causing major perturbations during transient events. An internal VR based on [6] regulates the middle voltage rail (Vmid) and balances the current mismatch, avoiding an additional external supply and the associated vulnerability arising from exposing the mid-rail to the attacker. Since the VS-AES design works with a single VDD supply, it allows seamless integration with other circuit blocks in any ASIC/SOC design. The VR also facilitates spatial and temporal randomization of current loops, providing greater resilience against fine-grained EM SCA.

A. Spatio-Temporal Dataflow Randomization

The proposed approach implements a high degree of randomness in temporal and spatial domains to make the current loops dynamic. Temporal randomization comes from the stacking configuration (Fig.3.b) changed before every encryption. Spatial randomness is achieved by shuffling SMA blocks dataflow at every clock cycle. The VR, while regulating the Vmid rails, also intentionally scrambles the current loops in spatial distribution (Section II-B). Therefore, there is a high level of entropy in EM emissions, even for measurements taken at the same location with the same input data.

In this design, a total of 576 permutations in spatial and voltage-stacked domains are possible, thereby effectively mitigating EM correlation weak-spots detection and SCA attacks. The 128-bit plaintext is split into four 32-bit (4-byte)

sections, which are passed to the four SMA blocks for independent computation (Fig. 3.c). The input is permuted every clock cycle by a three-stage butterfly network [7] (Fig. 3.d) using the least significant 6 bits of data from the intermediate cyphertext as the permutation key. Since AES operations are performed byte-by-byte, splitting the SMA operation into four such sections does not affect data integrity and can be performed without the need for any special decomposition and reconstruction of the complete ciphertext. The four SMA blocks are placed on the four extreme corners of the main block at the maximum possible separation to attenuate strong local EM signatures. The voltage stacking configuration (Fig.3.b) is permuted before every encryption using a key from an off-chip True Random Number Generator (TRNG). Level shifters are used to interface the data from stacked modules to the main block (Fig. 3.e,f). The proposed spatially and temporally randomized dataflow prevents precise data tracking and revelation of key bytes by the fine-grained EM SCA attacks.

B. Voltage Stacking and Voltage Regulator Design

For stacked voltage domains, maintaining a stable middle rail is critical for the precise operation of circuit blocks. Using an external Vmid supply essentially negates the SCA mitigation

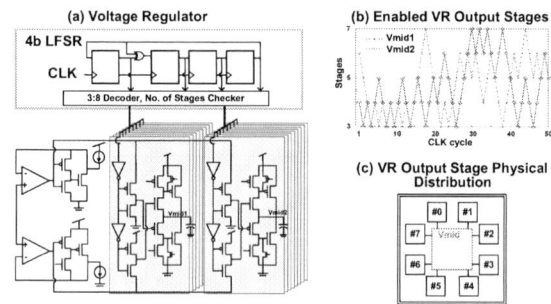

Fig. 4. (a) Voltage regulator (VR) [6] schematic and a 4-bit LFSR to randomize VR output stage configuration (b) VR output stage permutation vs CLK cycle (c) Spatially distributed VR output stages across the design

techniques proposed, as it exposes the different current signature between the stacked domains to the external world. Hence, an internal voltage regulator (VR) is employed to regulate the two middle voltage rails, one between a pair of stacked SMA blocks.

The secure VS-AES core has only one pad for a global VDD and one for VSS; all internal rails are regulated deriving from the single external VDD. The VR (Fig. 4) maintains Vmid1 and Vmid2 nodes close to VDD/2 to support any unbalanced current between the two stacked SMA blocks. The VR is comprised of a bias generation stage and a set of push-pull output stages. The bias generation stage generates the bias voltages for the output stages based on the reference voltage. The output stages set the Vmid voltage from these bias voltages, and any sink(source) load current is counteracted by local negative feedback loops within the push(pull) portions of the output stages, thereby maintaining a steady Vmid rail. Multiple output stages can be connected in parallel for improved regulation. VS-AES leverages this parallel connectivity to scramble the current flow at every clock cycle. Decoupling capacitors are also added on Vmid to provide current during sharp transient events.

In VS-AES, each Vmid node is driven by eight independently controllable, spatially distributed VR output stages. A 4-bit LFSR serves as the randomization key for this permutation, enabling or disabling exactly one VR stage on each Vmid per clock cycle while maintaining a minimum of three ON stages per Vmid rail. Randomizing the VR's spatial configuration every clock cycle modulates the physical current flow path and location of the current loops. This generates a complex and dynamically changing EM signature on Vmid rails. Thus, stacking multiple compute activities in spatially and temporally randomized current loops (Fig. 5) significantly uncorrelates the fine-grained EM signature, thereby mitigating SCA attacks.

Internal VDD and VSS terminals of all four SMA blocks are probed for testing purposes. For the baseline mode (Fig.5.a), the SMA blocks operate on the full-swing VDD-VSS domain. Fig. 5.b,c,d show representative voltage stacking configurations that are permuted before every encryption and

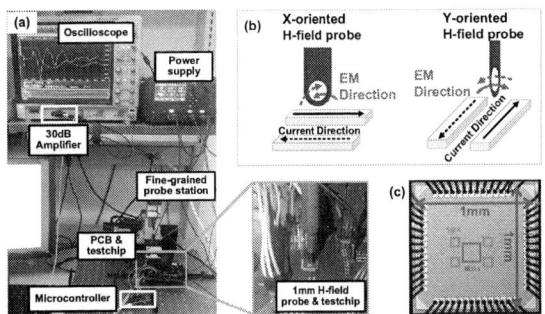

Fig. 6. (a) Fine-grained EM measurement setup (b) Illustration of x-oriented probe and y-oriented probe on top of the traces (c) 65nm CMOS die micrograph with the main and SMA block placements highlighted

VR stage randomization during encryption operations. It was observed that the voltage domain transition is clean and does not impede the performance of the AES core.

III. MEASUREMENT RESULTS

The proposed secure AES core was fabricated in 65nm CMOS. The baseline AES is tested by disabling the VR, all randomization features, and switching all SMA blocks to VDD-VSS rails, so that all AES modules are powered by a single voltage domain. Through this method, we quantify the minimum Measurement to Disclosure (MTD) with and without the proposed VS-AES countermeasure design, keeping all other measurement conditions/setup the same. The locations of the main block and spatially distributed SMA blocks are marked in the die photograph (Fig.6.c). Fine-grained EM SCA attacks were performed on the test-chip to prove and quantify the efficacy of the proposed voltage stacking and spatio-temporally randomization approach. Coarse-grained EM and power SCA attacks were also performed to check unintentional data leakage through other side channels while implementing the proposed countermeasures. In the fine-grain Correlation EM side-channel Analysis (CEMA) measurement setup (Fig. 6), a 1mm diameter H-field probe scans the entire package with a gradually shrinking probe step size to initially identify vulnerable locations (weak spots) where the EM signature is highly correlated with the AES computation, generating a fine-

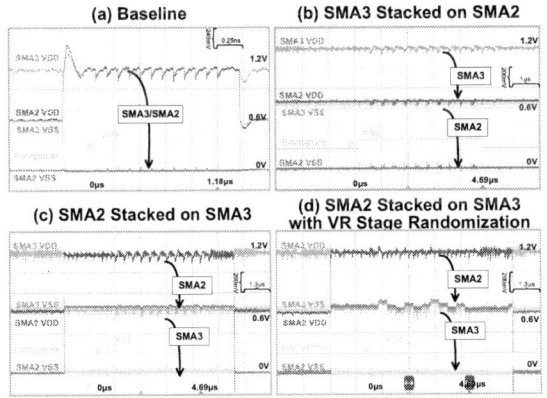

Fig. 5. Oscilloscope measurement results of SMA3's and SMA2's power rails in the: (a) baseline mode (b) default stacking mode (c) voltage domain randomization mode (d) VR output stage randomization mode

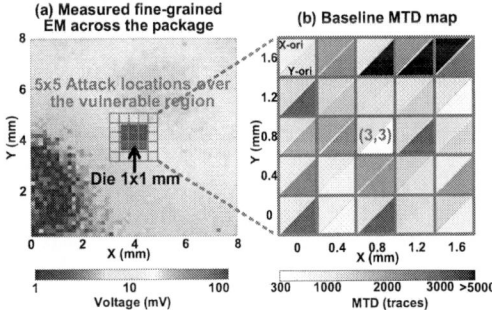

Fig. 7. (a) Measured x-oriented fine-grained EM map across the package at one time instance (b) Fine-grained Correlation EM Analysis (CEMA) MTD map at 25 attack locations for the baseline design; location (3,3) is the most vulnerable attack point

978-1-6654-8495-4/22 $31.00 © 2022 IEEE

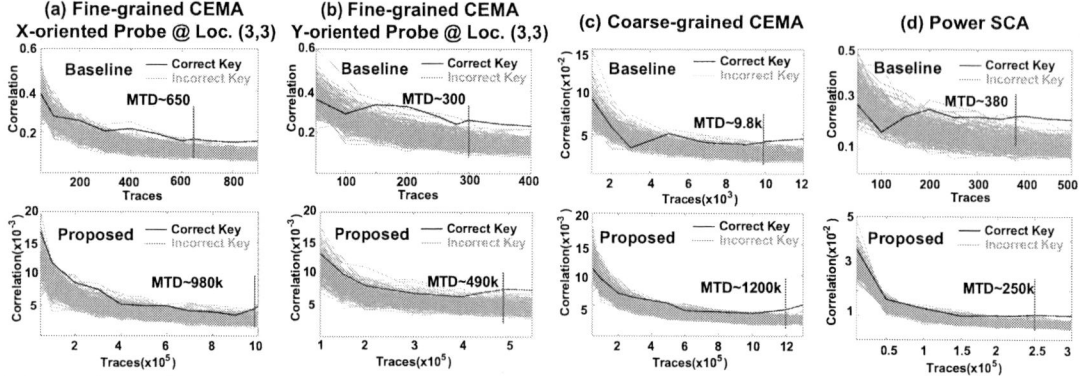

Fig. 8. (a) MTD increases by 1507X using x-oriented probe at location (3,3) (b) MTD increases by 1633X using y-oriented probe at (3,3) (c) Coarse-grained CEMA MTD increases by 122X (d) Correlation Power side-channel Attack (CPA) MTD increases by 657X

grained EM heatmap (Fig. 7.a). Data from both x and y-oriented probes capture EM signatures from all metal patterns on the die (Fig. 6.b), on which a subsequent iterative search determines the most vulnerable attack region and probe orientation [1]. CEMA spanning a 5x5 grid across the vulnerable region on the baseline design was performed, which shows an MTD of only 650/300 along the x/y orientation at the grid location (3,3) (Fig. 7.b). The low MTD numbers for fine-grained CEMA confirm that fine-grained EM-SCA attacks are highly potent compared to other attack modes, which is what VS-AES aims to mitigate. With the same CEMA protocol, voltage stacking the SMA blocks without spatio-temporal randomization improves the MTD by 4.6X/13X along the x/y-orientation; the proposed approach with randomization increases the MTD to 980K (1507X better) in x-orientation (Fig.8.a) and to 490K (1633X better) in y-orientation (Fig.8.b) at (3,3) weak-spot location.

To check whether the proposed approach causes any unintentional degradation of MTD in other prevailing SCA attack methods, a coarse-grained CEMA was performed with a larger (10mm diameter) H-field probe placed 1mm above the package, for which the MTD improved by 122X with all proposed countermeasures. Correlation Power side-channel Attack (CPA) was also performed by measuring the voltage across a 1Ω resistor in series with the VDD pin, for which the MTD improved by 657X, owing to the charge-recycling through voltage stacking which reduces the correlation between current and computation activity (Fig. 8.c,d).

The proposed VS-AES, when all countermeasures are enabled, consumes similar power compared to the baseline design at iso-VDD conditions. The power saving due to stacked SMA blocks is diminished by the VR and the switch matrix overhead circuits. However, voltage stacked SMA blocks incur frequency degradation due to smaller voltage swing across them, resulting in ~3X frequency reduction compared to the baseline design. These measurements (summary in Table 1). demonstrate the effectiveness of the proposed approach against fine-grain EM SCA attacks by performing multiple parallel computations within a single current loop through voltage stacking (charge recycling) and by spatio-temporal randomization of current loops. The contributions of this work specifically encompass techniques to mitigate fine-grained EM

SCA vulnerabilities, and VS-AES can be seamlessly integrated with prior techniques [8,9] to achieve further MTD improvements for coarse-grained CEMA and CPA attacks.

TABLE I: Measurement Summary

Technology	65nm CMOS	Fine-grained CEMA	
AES Datapath	128bit-parallel	Probe size	1mm dia.
Power (mW)		Resolution	400 µm
Baseline (1.1V)	13.64	Baseline x-ori MTD	650
VS AES (1.2V)	15.24	VS AES x-ori MTD	980k
Area (mm²)		Improvement	1507X
Baseline*	0.0421	Baseline y-ori MTD	300
VS AES	0.06254	VS AES y-ori MTD	490k
VR	0.016	Improvement	1633X
Level Shifter	0.0032	Coarse-grained CEMA	
Power switches	0.00124	Probe size	10mm dia.
CPA		Baseline MTD	9.8k
Baseline MTD	380	VS AES MTD	1.2M
VS AES MTD	250k	Improvement	122X
Improvement	657X	*Baseline contains Main block & SMA blocks	

ACKNOWLEDGMENTS

This research is supported by Intel Security Research Center (RARE), Intel Rising Star Faculty Award, and Silicon Labs. Authors thank TSMC for chip fabrication, Ge Li and Vishnuvardhan Iyer for the lab support.

REFERENCES

[1] V. Iyer et al. "A Systematic Evaluation of EM and Power Side-Channel Analysis Attacks on AES Implementations," ISI 2021
[2] A. Singh et al., "Improved Power/EM Side-Channel Attack Resistance of 128-Bit AES Engines With Random Fast Voltage Dithering," JSSC 2019
[3] A. Singh et al., "25.3 A 128b AES Engine with Higher Resistance to Power and Electromagnetic Side-Channel Attacks Enabled by a Security-Aware Integrated All-Digital Low-Dropout Regulator," ISSCC 2019
[4] D. Das et al., "27.3 EM and Power SCA-Resilient AES-256 in 65nm CMOS Through >350× Current-Domain Signature Attenuation," JSSC 2020
[5] R. Kumar et al., "A SCA-Resistant AES Engine in 14nm CMOS with Time/Frequency-Domain Leakage Suppression using Non-Linear Digital LDO Cascaded with Arithmetic Countermeasures," VLSI Symposium 2020
[6] S. Rajapandian, K. L. Shepard, P. Hazucha and T. Karnik, "High-voltage power delivery through charge recycling," JSSC 2006
[7] X. Yang et al. "Fast subword permutation instructions based on butterfly network," Proc. SPIE 3970, Media Processors 2000
[8] M. Wang et al., "Physical Design Strategies for Mitigating Fine-Grained Electromagnetic Side-Channel Attacks," CICC 2021
[9] M. Wang et al., "Galvanically Isolated, Power and Electromagnetic Side-Channel Attack Resilient Secure AES Core with Integrated Charge Pump based Power Management," CICC 2021

Gap in pagination due to withheld paper.

Pages 533-537

Integrated Optical Phased Arrays on Silicon

Farshid Ashtiani
Department of Electrical and
Systems Enigineering
University of Pennsylvania
Philadelphia, USA
farshid@seas.upenn.edu

Firooz Aflatouni
Department of Electrical and
Systems Enigineering
University of Pennsylvania
Philadelphia, USA
firooz@seas.upenn.edu

Abstract—**Integrated optical phased arrays (OPA) have many applications from LiDAR and imaging to sensing and free-space optical communication. Despite recent advances, due to the large size of photonic devices compared to the wavelength of operation, realizing 2-D OPAs with sub-wavelength element pitch is challenging and continues to be an open area of research. In this paper, a review of various OPA implementations using silicon platforms is presented.**

Keywords—optical phased array, beam-forming, photonic integrated circuits, silicon photonics

I. INTRODUCTION

Integrated silicon photonic platforms offer high optical confinement and low-loss interconnects, as well as compatibility with standard CMOS fabrication processes. These features have enabled systems for a variety of application from communication and signal processing, to sensing, imaging, and ranging, to name a few. In some applications such as LiDAR [1,2], imaging systems [3,4], projection systems [5], and free-space communication [6,7], one- or two-dimensional optical beam-forming and -steering is required. This can be realized using integrated optical phased arrays (OPA), which enable small-size, low-power, and fast electronic-assisted beam-steering, eliminating the need for any mechanical movements.

In an antenna array, beam-forming is achieved by setting the relative phase between the antennas. Fig. 1a shows the conceptual schematic of an N×N antenna array, where the input signal is distributed among the elements (antennas) and the phase of each signal is controlled using a per-element phase shifter (resulting in N^2 phase shifters within the array). By properly setting the relative phase between the array elements, a beam can be formed in the far-field that can be steered in two dimensions. Spatial resolution (set by the beam width and steering resolution) and beam-steering range are important performance metrics of an OPA. The former is a function of the aperture size and the relative phase control resolution between the elements and the latter depends on the element pitch and antenna size. Therefore, it is desired to implement an OPA with a large number of closely-spaced (compared to the wavelength of operation) elements to achieve a large steering range and a high spatial resolution. Unlike microwave phased arrays, in OPAs, the size of the antennas and optical phase shifters are often larger than the operation wavelength and while a large aperture size could be realized, reducing the antenna pitch is challenging. Fig. 1b shows the effect of the number of elements on the beam width (for a given element pitch of $\lambda/2$), which can be reduced by increasing the number of elements resulting in a higher spatial resolution. Moreover, smaller element pitch results in a larger steering range by suppressing the undesired lobes (Fig. 1c). In the next section, we review some of the previously utilized methods for implementation of OPAs for 2-D beam-steering.

II. 2-D BEAM-STEERING METHODS

A. 1-D array of grating couplers with wavelength tuning

The radiation angle of a beam emitted form a grating structure is wavelength dependent [8]. As a result, 2-D beam-steering can be achieved by using a 1-D array of grating

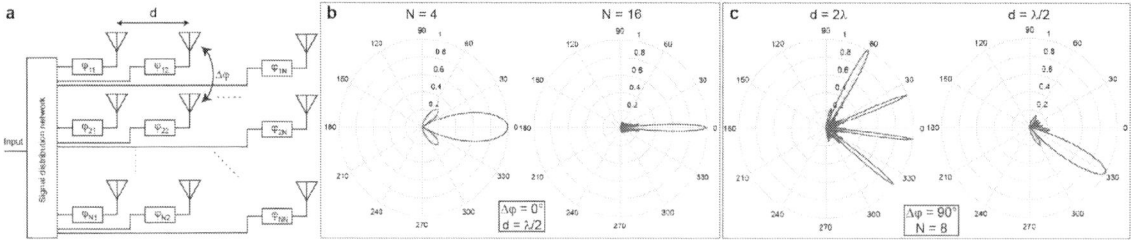

Fig. 1 – (a) Conceptual block diagram of a 2-D antenna array. (b) Simulated effect of the number of elements on the beam width. (c) Simulated effect of reducing the element pitch on undesired lobes and the beam-steering range.

US Defense Advanced Research Projects Agency MOABB Program (FA8650-18-1-7828), US National Aeronautics and Space Administration ESI Program (80NSSC19K0214).

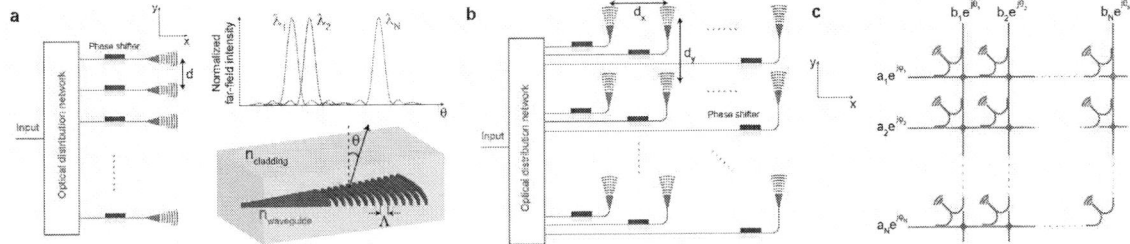

Fig. 2 – (a) Block diagram of a 1-D OPA used for 2-D beam-steering. Each element is implemented using a grating coupler. Beam-steering in X and Y directions can be achieved using wavelength tuning and relative phase adjustment, respectively. (b) An NxN 2-D OPA with per-element phase control. (c) An NxN OPA with 2N off-aperture phase control. Beam-steering is performed by individually setting the amplitude and phase of waves in rows and columns.

structures, serving as the photonic antennas, together with wavelength tuning [9-12]. Fig. 2a shows the conceptual diagram of an N-element 1-D OPA, where each photonic antenna is implemented using a grating coupler. In this figure, the input optical signal is split into N branches and the phase of each signal can be set using an optical phase shifter. Each of the phase adjusted signals is coupled to a grating coupler. By properly setting the relative phase between the adjacent elements of the array, the far-field beam can be steered along the Y axis, where the steering range is inversely proportional to the element spacing (d). To maximize the steering range, d should be equal to half of the wavelength of operation.

Using the Bragg equation [8], the polar angle of radiation (θ) of a grating coupler can be written as

$$\theta = \sin^{-1}\left(\frac{n_{waveguide}}{n_{cladding}} - \frac{\lambda}{n_{cladding}\Lambda}\right), \quad (1)$$

where $n_{waveguide}$, $n_{cladding}$, λ, and Λ are the refractive index of the waveguide medium, the refractive index of the cladding, wavelength of the input light in free-space, and the grating period, respectively. In this case, for given waveguide and cladding materials (*e.g.*, silicon and silicon dioxide) as well as the grating period, the radiation angle can be tuned by changing the wavelength of light. In Fig. 2a, the beam can be steered along the X axis by changing λ as conceptually shown by the graph. This approach enables OPAs with small element pitch (approaching sub-wavelength) and lower power consumption. However, for 2-D beam-steering using such OPA, a single wavelength operation is not possible and a widely tunable laser is required. Also, utilizing such OPA for certain applications such as free-space communication is challenging. Furthermore, the wavelength steering range is often much smaller than the phase controlled steering range.

OPAs with different element pitch and steering ranges have been demonstrated using this approach. Typically, a wavelength tuning range of tens of nanometers is used for less than 20° of steering range. In [9], a 512-element OPA with an element pitch of 1.3 μm is implemented and using thermal phase shifters, a steering range of about 70° in one direction is achieved. A multi-pass structure for phase shifters reduces the power consumption per element (3.7 mW for 2π phase shift). Moreover, tuning the optical wavelength by 23 nm results in a beam-steering range of 6° in the other direction. A larger scale wavelength-tuned OPA

is demonstrated in [12], where 8192 elements are placed with a pitch of 1 μm. This smaller element pitch results in a steering range of 100° in one direction. In the other direction, by tuning the wavelength over a 120 nm range, the beam can be steered by 17°.

B. NxN OPAs with N^2 phase control

In some applications such as free-space communication, where single wavelength operation is required, wavelength tuning is not desired. In such applications, a 2-D OPA, where the phase of individual elements is adjusted, can be used [5,13-14]. Per-element phase adjustment also allows for more complex beam formation and control. In Fig. 2b, the optical input is split into N^2 branches and each signal is coupled to a grating coupler (serving as an optical antenna) after phase adjustment. By properly adjusting the phase shifters the phase difference in X and Y directions can be set independently such that beam-steering in both directions is performed. Here, the beam-steering range in X and Y directions depends on the element pitch along the X and Y directions (*i.e.* d_x and d_y, respectively). While such OPA architecture allows for single wavelength operation, the element pitch is limited by the optical (and electrical) routing as well as the size of the phase shifters (Fig. 2b). Various implementations of this architecture have been demonstrated.

In [5], a 4x4 OPA with 16 p-doped-intrinsic-n-doped (PIN) off-aperture phase shifters is demonstrated. The large element pitch of about 50 μm results in a steering range of 1.8° using a single wavelength 1550 nm laser source and PIN modulators enable steering time of under 5 ns.

In [14], an 8x8 OPA with per-element thermo-optic phase shifters within the aperture and an array of directional couplers with varying length for uniform distribution of light between the elements is implemented. An element pitch of 9 μm is achieved that enables a steering range of about 10°. In this work, the minimum pitch is limited by the per-element thermal phase shifters and optical and electrical routing within the aperture. Furthermore, thermal cross-talk and relatively high power consumption of thermal phase shifters could limit the scalability of such systems.

In [13] a 128-element 2-D OPA with a non-uniform sparse aperture is implemented, enabling a large grating lobe suppression and a relaxed optical routing. Using this

978-1-6654-8495-4/22 $31.00 © 2022 IEEE

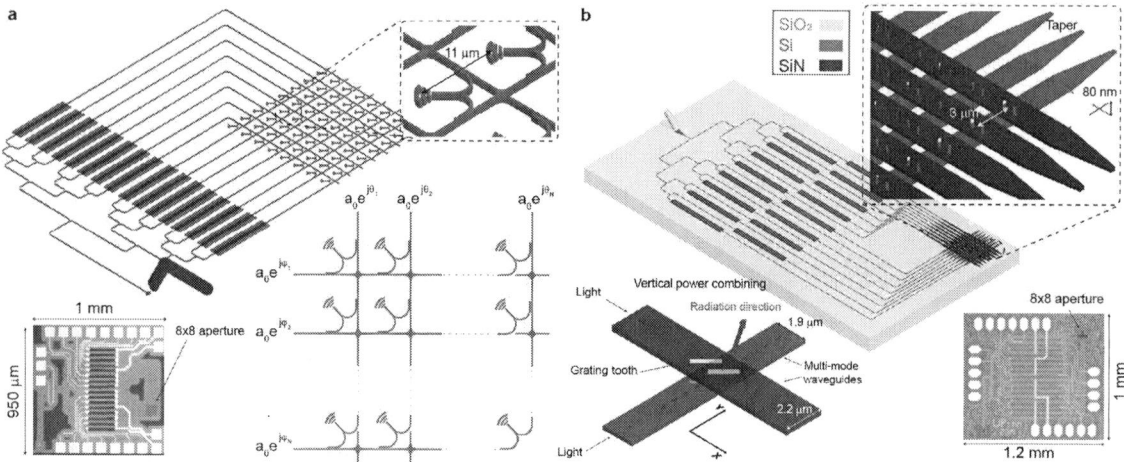

Fig. 3 – (a) An **8x8** OPA architecture with 16 off-aperture phase shifters for 2-D beam-steering. By controlling the phase of waves in rows and columns, 2-D beam-steering is performed. The chip is fabricated using AMF 180 nm silicon photonics process. (b) An **8x8** OPA with vertically-aligned tilted grating structures on silicon and silicon nitride layers as the photonic antennas. Element pitch of 3 μm is achieved.

architecture, a minimum element pitch of about 5.5 μm is achieved and a beam-steering range of 16° is demonstrated.

C. NxN OPAs with 2N off-aperture phase shifters

To achieve the largest beam-steering range, the element pitch should be half of the wavelength of operation (*e.g.*, 775 nm for optical wavelength of 1550 nm). However, the size of photonic devices, such as waveguides, directional couplers, and phase shifters is comparable with the wavelength, which results in a large element spacing. Reducing the number of required phase shifters and placing them outside the OPA aperture can significantly reduce the element pitch. Furthermore, such approach lowers the number of waveguides required to route the optical signals to the photonic antennas and eliminates electrical routing within the aperture.

In [15], a 2-D NxN OPA architecture is proposed, where only 2N phase shifters are used to set the dynamic relative phase between the elements. As illustrated in Fig. 2c, in this architecture, the phase of the optical waves in row and column optical waveguides (*i.e.*, φ_i and θ_j) are set using 2N off-aperture phased shifters. In this case, at each emitting element, a superposition of the waves in the row and column waveguides is generated, whose phase is set by the amplitude of the electric field of waves in rows and columns, the off-aperture dynamically controllable phase of the waves in the row and column waveguides, and the static relative phase between elements. The static phase of elements, set by optical devices between adjacent element on each row and column, can be designed to adjust the side-lobes (that create the main beam in the far-field). By properly adjusting these parameters, one can set the relative phase between the elements to perform 2-D beam-steering. In addition, to reduce the routing complexities, lower number of phase shifters significantly reduces the overall power consumption of the OPA. Compared to the conventional NxN OPAs with per-element phase shifters, this architecture reduces the power consumption by a factor of N/2. Note that in

this architecture only a single beam can be generated and steered. In [15], an 8x8 OPA with 16 off-aperture phase shifters for 2-D beam-steering is demonstrated. As shown in Fig. 3a, the input light is split into 16 signals that are individually phase shifted using thermal phase modulators, making 8 row and 8 column signals. Assuming uniform optical power distribution (*i.e.*, $a_i = b_i = a_0$), the phase of the emitting elements along the Y direction, φ'_i, depends on the electronically tuned phase of rows (φ_i) and the unequal static phase difference between elements on columns and, similarly, the phase of the emitting elements along the X direction, θ'_j, depends on the electronically tuned phase of columns (θ_j) and the unequal static phase difference between elements on rows. Here, the static phases are set by the devices (*e.g.* waveguide crossings, variable length couplers, and waveguides) between elements on rows and columns. The static phase of elements were individually optimized to uniformly distribute the optical power between the main and side lobes, increasing the power of the formed beam. The formed beam can be steered by adjusting the relative phase between elements in rows and columns (φ_i and θ_j) using the off-aperture optical phase shifters, where, with an element spacing of 11 μm, a 2-D beam-steering range of about 7° is achieved.

To further reduce the element pitch, a more compact photonic antenna structure and a dense optical routing are required, which could be achieved by using fabrication processes that offer more than one photonic device layers. In [16], a multilayer NxN OPA for 2-D beam-steering is presented. As shown in Fig. 3b, each photonic antenna is implemented by vertical alignment of grating structures on silicon and silicon nitride layers. The grating structures are tilted to align the beams radiated by the grating structure on each layer. This approach eliminates the need for some photonic devices within the aperture (such as directional couplers and waveguide crossings) and results in a smaller element pitch. In [16], an element pitch

TABLE I.   COMPARISON BETWEEN 2-D BEAM-STEERING SYSTEMS IMPLEMENTED USING VARIOUS METHODS

	[16]	[15]	[6]	[14]	[13]	[5]	[12]	[11]	[10]	[9]
Implementation method	2-D reduced phase control	2-D reduced phase control	2-D	2-D	2-D sparse array	2-D	1-D + wavelength tuning	1-D + wavelength tuning	1-D + wavelength tuning	1-D + wavelength tuning
Array size	64	64	64	64	128	16	8192	256	16	512
Element pitch[a]	2 λ	7.3 λ	21.3 λ	5.8 λ	3.6 λ	32 λ	0.64 λ	0.9 λ	2.9 λ	0.84 λ
Steering range	23°	7°	1.6°	10°	16°	1.8°	100°	67°	20°	70°
Number of phase shifters	16	16	64	64	128	16	8192	256	16	512 (multi-pass)
Utilized wavelength tuning range	N/A	N/A	N/A	N/A	N/A	N/A	120nm (17°)	100nm (16°)	100nm (14°)	23nm (6°)
Power consumption per element[b]	2.42 mW	1.75 mW	21 mW	9 mW	10.6 mW	N/A	0.3 mW	5.2 mW	430 mW	3.7 mW
Technology	180nm Si/SiN Ph	180nm SiPh	180nm RF SOI	SiPh	SiPh + CMOS	SiPh	SiPh + CMOS	SiPh + CMOS	SiPh	SiPh

[a] The operation wavelength of all implementations is 1550 nm.   [b] Power required for 2π phase shift.

of 3 µm is achieved and a beam-steering range of about 23° is demonstrated.

III. SUMMARY AND PERFORMANCE COMPARISON

Table I compares the performance of various OPA implementations. As shown in this table, using a 1-D array together with wavelength tuning generally results in smaller element pitch since it significantly reduces the photonic routing complexities. Therefore, relatively large steering ranges in one direction can be achieved. However, the steering range in the other direction is limited. Also, due to the need for wavelength tuning to perform 2-D optical beam-steering, this method is not suited for single-wavelength operation. Moreover, a widely tunable laser is also required that makes the system implementation and integration more challenging.

Using a 2-D array of grating couplers where the individual phases are controlled enables single-wavelength operation, but can result in larger element spacing while arbitrary beam formation and symmetric beam spot can be achieved.

Finally, other 2-D array architectures such as non-uniform sparse arrays and off-aperture phase shifting with reduced number of phase shifters can result in more compact arrays by relaxing the photonic and electrical routing complexities. Furthermore, implementing multi-layer photonic antennas in fabrication processes that offer multiple layers of photonic devices enables very compact elements, small element pitch, and therefore, a large steering range.

ACKNOWLEDGMENT

Works reported in [15] and [16] were partially funded under Defense Advanced Research Projects Agency MOABB Program (FA8650-18-1-7828) and partially under National Aeronautics and Space Administration ESI Program (80NSSC19K0214).

REFERENCES

[1] C. V. Poulton, et al., "High-performance integrated optical phased arrays for chip-scale beam steering and LIDAR," in Conference on Lasers and Electro-Optics, OSA Technical Digest (online) (Optica Publishing Group, 2018), paper ATu3R.2.

[2] C. V. Poulton, et al., "Coherent solid-state LIDAR with silicon photonic optical phased arrays," Opt. Lett. 42, 4091-4094 (2017).

[3] F. Aflatouni, B. Abiri, A. Rekhi, and A. Hajimiri, "Nanophotonic coherent imager," Opt. Express 23, 5117-5125 (2015).

[4] T. Fukui, Y. Kohno, R. Tang, Y. Nakano, and T. Tanemura, "Single-pixel imaging using multimode fiber and silicon photonic phased array," J. Lightwave Technol. 39, 839-844 (2021).

[5] F. Aflatouni, B. Abiri, A. Rekhi, and A. Hajimiri, "Nanophotonic projection system," Opt. Express 23, 21012-21022 (2015).

[6] H. Abediasl and H. Hashemi, "Monolithic optical phased-array transceiver in a standard SOI CMOS process," Opt. Express 23, 6509-6519 (2015).

[7] S. J. Spector, et al., "Broadband imaging and wireless communication with an optical phased array," in Conference on Lasers and Electro-Optics, OSA Technical Digest (online) (Optica Publishing Group, 2018), paper SM3I.7.

[8] C. R. Doerr, L. Chen, Y. Chen, and L. L. Buhl, "Wide Bandwidth Silicon Nitride Grating Coupler," IEEE Photon. Technol. Lett. 22, pp. 1461-1463 (2010).

[9] S. A. Miller, et al., "Large-scale optical phased array using a low-power multi-pass silicon photonic platform," Optica 7, 3-6 (2020).

[10] J. K. Doylend, et al., "Two-dimensional free-space beam steering with an optical phased array on silicon-on-insulator," Opt. Express 19, 21595-21604 (2011).

[11] S. Chung, M. Nakai, S. Idres, Y. Ni and H. Hashemi, "Optical phased-array FMCW LiDAR with on-chip calibration," 2021 IEEE International Solid-State Circuits Conference (ISSCC), 2021, pp. 286-288.

[12] C. V. Poulton, et al., "8192-element optical phased array with 100° steering range and flip-chip CMOS," in Conference on Lasers and Electro-Optics, OSA Technical Digest (Optica Publishing Group, 2020), paper JTh4A.3.

[13] R. Fatemi, A. Khachaturian and A. Hajimiri, "A nonuniform sparse 2-D large-FOV optical phased array with a low-power PWM drive," in IEEE Journal of Solid-State Circuits, vol. 54, no. 5, pp. 1200-1215 (2019).

[14] J. Sun, E. Timurdogan, A. Yaacobi, E. S. Hosseini, and M. R. Watts, "Large-scale nanophotonic phased array," Nature, vol. 493, no. 7431, pp. 195–199, 2013.

[15] F. Ashtiani and F. Aflatouni, "N × N optical phased array with 2N phase shifters," Opt. Express 27, 27183-27190 (2019).

[16] F. Ashtiani and F. Aflatouni, "Coherent multilayer photonic nanoantenna array with off-aperture phase adjustment," ACS Photonics 8 (12), 3433-3439 (2021).

Full-Duplex Wireless for (Joint-) Communication and Sensing

Hany Abolmagd[*‡], Raghav Subbaraman[†‡], Dinesh Bharadia[†], and Sudip Shekhar[*],
[*] University of British Columbia, Vancouver, BC. hanyaa@ece.ubc.ca sudip@ece.ubc.ca
[†] University of California, San Diego, CA. rsubbaraman@eng.ucsd.edu dineshb@eng.ucsd.edu
[‡] Equally credited authors

Abstract—**In-band simultaneous full-duplex (FD) has seen rapid research in the last decade. We review the state-of-the-art in FD radios for wireless communication (including WiFi and cellular), low-power wide area networks, relays, military and other applications. We also review the use of FD in sensing, including consumer and automotive radars. We then describe how FD can usher the next generation of wireless systems employing joint communication and sensing, enabling high-directional beam steering/control and ubiquitous IoT.**

I. INTRODUCTION

6G is evolving towards high-bandwidth (BW) highly-directional communication on one end and sensing of the environment on the other end. The highly-directional communication requires beam-training/beam-steering, which a sensing via a radar would enable passively; such sensing can be achieved jointly with the communication signals with FD radios.

Another objective of 6G is to enable low-power communications for IoT applications, more popularly to support thousands of nodes with several kilometers of range, but at low-data rates. A key challenge for such systems lies in distributing the link capacity among the thousands of nodes, which requires synchronization of those nodes for efficiency. Synchronization often requires extensive energy at the nodes, going against the low-power requirements for IoTs. FD-driven synchronization can mitigate this challenge.

We review the circuits for FD wireless for communication and sensing applications in Section II and III, respectively. We then describe our vision for joint communication and sensing in 6G leveraging FD, and circuit design challenges that we must overcome to realize that.

II. FD FOR WIRELESS COMMUNICATION

Cellular and WiFi: The main challenge in FD systems is the large self-interference (SI) signal that originates from the transmitter (TX) to the receiver (RX) which can easily be 100 dB larger than the RX sensitivity [1]. A large SIC requires a combination of antenna interface isolation, and cancellation at RF, digital baseband (BB), and sometimes even analog BB. A large isolation is extremely important, since it eases up the cancellation requirements for the RF front-end circuits. Antenna interface isolation must be achieved using a circulator or electrical balanced duplexer (EBD) - both require impedance tuners to maintain their isolation as the

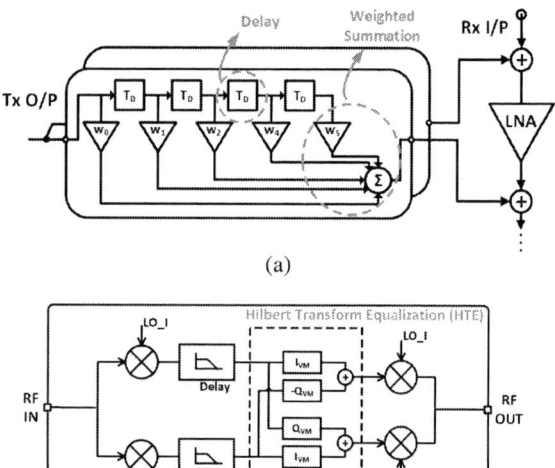

(a)

(b)

Fig. 1. RF SI cancellers. (a) A real-tap FIR equalizer with delay elements implemented with RF all-pass filter [3] or n-path switched-cap elements [4]. (b) A complex-tap equalizer with frequency translation circuits for BB VM and delay implementation [5].

antenna impdeance varies. EBDs suffer from a minimum of 3 dB insertion loss at both the TX and RX, which render their adoption difficult for cellular/WiFi where link budgets are stringent. Therefore, a lot of work has been done on on-chip circulators, since their off-chip counterparts are bulky. Power-handling, quality-factor of the passives, and jitter due to the use of LO in the circuit implementation [2] are some crucial challenges that remain to be solved.

For cancellation, amplitude, phase, and delay of the SI channel must be replicated in the canceller path for the broadband cancellation needed for cellular/WiFi. The cancellation BW, on-chip delay and the amount of cancellation are usually traded-off in IC prototypes. Cancellation can be done using multi-tap FIR equalization with RF delay elements before and/or after the LNA, as shown in Fig. 1(a). Smaller RF delay using all-pass filters [3], [6] and relatively larger delays using n-path time-interleaved switched-capacitor elements [4] have been demonstrated. RF delay range/resolution is often limited due to its tradeoff with noise, power, area and loss. Real-valued weights limit

978-1-6654-8495-4/22 $31.00 © 2022 IEEE

cancellation too. FIR equalization and delay can be further supplemented at analog BB [4], [7]. Since downconversion to BB is done using quadrature LO, complex weights are easier to implement as well [4]. A canceller leveraging frequency-translation circuits (RF to BB to RF), shown in Fig. 1(b) [5], [8], does not require changes to the RF front-end of the radio. Amplitude and phase (vector modulation (VM)) and delay - all can be implemented at BB with lower power and area, but the tradeoff between cancellation and BW remains. A review for various canceller prototypes can be found in [9]–[11].

For most prototypes, the cancellation degrades when the setup is changed from a controlled environment to a wireless channel, since a practical wireless channel changes with time [1]. Thus, the next generation of cancellers must implement a higher resolution/range for the amplitude, phase and delay, together with closed-loop adaption. A recent work in [12] shows a fully-integrated self-adaptive circuity that enables real-time RF cancellation for a 2×2 MIMO RX. The cancelers must maintain a low insertion-loss and large power handling as well, with minimal added noise and distortion.

Relays: A relay can serve as a repeater that regenerates the transmitted signal and sends it to the RX to extend the network coverage and capacity [13]. The symmetric nature of relay communication suggests maximum benefit of the FD throughput gain [13] while minimizing network relaying overhead. FD relays can also add more functionality to the system via modifying the signal before forwarding it to the RX. The throughput gain can be increased by modifying the repeater signal to add constructively with the direct signal from the source when it reaches the destination [14]. Although attractive in various network architectures, the prototypes implemented so far have focused on WiFi/cellular applications [14], [15]. Thus, the challenges described earlier remain pertinent.

LPWAN: Low-power wide area network (LPWAN) promises a wide-scale IoT deployment [17]. Gateways in LPWAN must support thousands of end-nodes while allowing low-power operation at these end-nodes. However, gateways

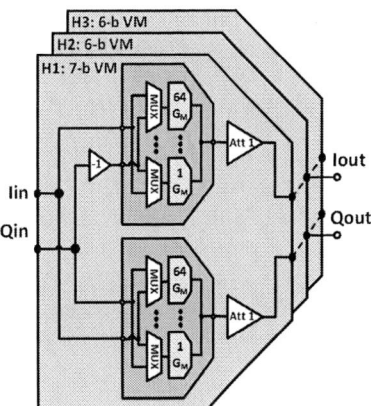

Fig. 3. Resolution of a VM can be boosted through a hierarchical implementation. A 16-bit-effective nested-VM is shown, with 3 stages of 7-b, 6-b and 6-b, respectively.

today suffer from packet collisions due to the hidden nodes problem: two transmitting end-nodes cannot hear each other and transmit at the same time, which limit the network throughput and increase the power consumed by the end-node devices (Fig. 2 (a)). Many medium access control (MAC) protocols have been investigated to eliminate the hidden nodes problem, but proven practically inefficient due to their overhead in such a low-throughput network [18]. By leveraging FD in a central gateway (FD-gateway), we can simultaneously receive an uplink (UL) transmission from an end-node and communicate a downlink (DL) acknowledgment (ACK) packet which would be heard by all end-nodes and keep them in a no-transmission state, thus improving network throughput by reducing collisions and saving significant power in many end-node IoT devices (Fig. 2 (b)).

Such FD-gateways for LPWAN require a total SIC of 140 dB for channels that have delay spread of several 10s of nanoseconds. However, the channel BW is less than 1 MHz. Since gateways permit a larger form factor, an antenna pair can be used to achieve a reliable isolation of 40 dB. A 60 dB frequency-translating RF canceller is implemented in [16]. The constraint of impedance matching and delay realization of [5] is relaxed using a delay circuit that provides a high impedance to the input port. A hierarchical cancellation circuit is used to implement a 16-b of amplitude/phase control in a nested VM with power consumption and area comparable to that of an 8-b conventional VM (Fig. 3). The nested VM is a promising hierarchical technique for FD cancellers in other applications as well to relax the range/resolution tradeoff.

Other wireless communication applications A device-to-device (D2D) link, in which mobile devices communicate directly without going through the basestation, can utilize FD operation to increase the data rate and reduce the latency, enabling ultra-reliable low-latency communication services [19]. The next-generation digital TV (DTV) broadcast can employ FD operation to transmit control and backhaul signals together with the TV service,

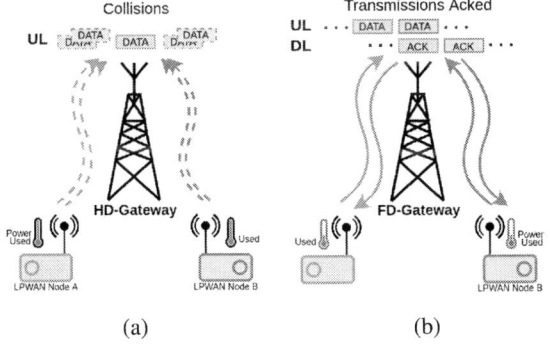

Fig. 2. (a) A half-duplex (HD) gateway in conventional LPWAN without ACK signal transmitted: collisions occur and resources are wasted. (b) A FD-gateway can transmit a downlink ACK while receiving data: collisions avoided improving network performance and energy efficiency [16]

opening the door for enhanced data services including IoT, emergency warning, and connected car [20]. Occupying a BW of 6 MHz at the DTV frequency (e.g. 500 MHz), the typical SIC requirements are moderate [20]. Underwater Acoustic Network (UAN), with applications in marine exploration and rescue missions, can utilize FD operation to increase the throughput, reduce the latency and power consumption, and avoid node collision [21]. UAN requires a BW of few kHz at a carrier frequency of 10-20 kHz with multi-path reflections in the range of milliseconds.

Military applications FD promises to improve the spectral efficiency of tactical communication networks, a significant advantage in the congested military spectrum. It also permits combining tactical communication with electronic warfare. New operating scenarios include simultaneous communication and jamming, simultaneous interception and jamming, and simultaneous interception and communication. In the first two scenarios, an FD-equipped device can prevent an adversary from controlling an unmanned air vehicle (UAV), activating improvised explosive devices (IEDs), or receiving tactical information while the FD device can be either receiving an important tactical information, e.g. its position, or even detecting the opponent's control signals. A prototype in [22] created a radio shield around the FD device, i.e. allowing secure communication while sending a jamming signal to prevent unauthorized interception. The same radio shield approach can be used in civilian security applications like secure energy transfer, implanted medical devices communication and unauthorized drone restriction. The prototype in [23] demonstrated the second scenario to intercept the radio control signal for an IED or UAV and blocked its operation, while the prototype in [24] further used the intercepted signal to adapt the jamming waveform and improve the jamming power efficiency.

Military communications networks have a lower operating frequency (typically in HF, VHF and lower UHF) and tactical communication BW [25]. FD requirements include high power handling, higher cancellation, and reliability and stability generally required for military applications.

III. FD FOR WIRELESS SENSING

Perhaps the most familiar application of FD in sensing is its use in continuous-wave (CW) radars. Radars probe the environment with a waveform and record its reflections to gather information such as distances to objects and their velocity. Since CW radars analyze environmental reflections while simultaneously transmitting the probing signal, SIC is necessary to ensure that the RX is not overwhelmed. The SI is the undesired reflection originating from the antenna interface and chip/board leakage, sometimes called 'spillover'. The SI and noise from the radar's TX affects the linearity, link budget and signal sensitivity of the radar system, ultimately limiting the range and noise performance. Most radar prototypes typically leverage digital BB cancellation and antenna interface isolation.

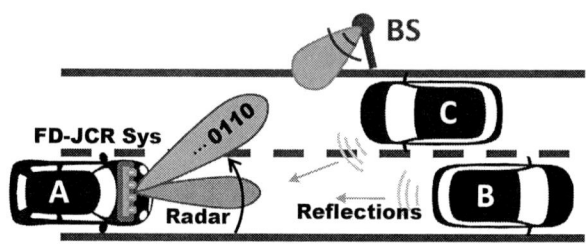

Fig. 4. Joint Communication and Radar (JCR) operation enabled by FD operation. The car 'A' uses a mmWave phased array to communicate with the base station BS while using a separate receive beam to listen for reflections from other objects in the environment. Since the transmit communication signal is cancelled, 'A' is able to ascertain angular location and speeds of cars 'B' and 'C'.

Consumer Electronics Radar: Small form-factor radars are increasingly being used in smartphones, enabling automation and augmenting IoT devices with environment-sensing capabilities and short-range touch-free gesture recognition. The Soli chip from Google is based on a mmWave FMCW radar [26]. To realize a better range for the same TX power, UWB radars operate in the microwave range (<10 GHz), with GHz BW [27], [28].

Automotive radar: Automotive radars typically support larger range and finer resolution, leading to higher transmit powers and number of TX/RX antennas [29], [30]. They operate at mmWave frequencies, employing an FMCW waveform in the 76-81 GHz range. SIC is implemented at BB analog in [30] or BB digital in [29], realizing 15 dB and 30 dB of cancellation, respectively.

Augmenting existing radars with RF/analog SIC can significantly improve their transmit power and RX sensitivity and therefore range of operation. Since the radars must integrate the entire RX/TX chains with multiple RX/TX antennas in a low-power fashion, one crucial challenge lies in handling multi-antenna SIC [31], [32]. High power-handling for automotive radars and large GHz BW for consumer UWB radars pose further SIC challenges.

IV. FD FOR JOINT COMMUNICATION AND SENSING

FD operation enables the convergence of both communication and sensing at no additional spectrum cost, making deployment attractive.

Backscatter communication and sensing: The next-generation IoT devices depend on ubiquitous low-power, wireless network coverage. Conventional radios with TX/RX front-ends consume significant power, leading to frequent battery changes or forced intermittent operation. Backscatter communication techniques avoid the conventional TX/RX front-ends. Instead, they modulate data on top of ambient signals. Backscatter is useful for environment sensing (force, temperature, human occupancy, vital sign monitoring, etc.) and conventional communication (cameras, distributed IoT sensors and displays, etc.) [33]–[35].

978-1-6654-8495-4/22 $31.00 © 2022 IEEE

Typical backscatter deployments involve three major components: 1) the IoT device/backscatter tag that modulates data on ambient signals, 2) the radio TX that generates the ambient excitation signal, and 3) the radio RX that receives the modulated backscatter. Often, the TX/RX are on the same unit - the backscatter reader, typically the size of a hand-held device. An SIC is required at the backscatter reader to prevent the excitation signal from interfering with the feeble reflections from the backscatter tag.

Commercial implementations of backscatter like RFID use techniques such as polarization isolation or EBDs to limit the SI. However, extending the range of these devices requires much higher SIC, so that the low-power backscattered signals can be decoded even when it is close to the noise floor. For example, an analog SIC budget of 78 dB is estimated for a narrowband backscatter operation even when the reflections are in an adjacent channel 3 MHz away [36]. Since the backscatter data-rates are low, it takes significant amount of time to transfer information from the tag to the reader. In addition, some readers are mobile, deployed in vehicles or UAVs, requiring stable SIC.

A recent work [36] has considered coarse/fine impedance tuner as a proposed solution for RF cancellation of a CW tone and its phase noise, enabling >60 dB SIC for LoRa backscatter. However, extending FD to multiple channels, or wideband backscatter applications like WiFi [35] remains a challenge, requiring more SIC over wider BW than demonstrated.

Joint Communication and Radar (JCR) Sensing: JCR allows for rich environment sensing and monitoring simultaneously with communication while reusing hardware and spectrum. These systems have an FD communication TX, that is also able to simultaneously receive and estimate the reflections of the transmitted signal from the environment. The RX takes advantage of the wideband communication signal waveform, along with the high data-rate ADC and BB processing to deliver better radar performance than low-BW FMCW radar [37]. Fig. 4 describes the operation of a possible JCR configuration.

FD design for JCR has to address the requirements of both communication and sensing. First, the design has to support high BW (1-2 GHz) required for communication [37]. Second, JCR systems work best at mmWave frequencies due to the higher resolution available for radar sensing. Such an operation requires development of mmWave and multi-antenna FD systems [38].

An interesting concern is the SIC budget for JCR systems. Unlike communication systems where few dB of degradation to the noise floor significantly degrades throughput, radar systems can provide at least short-range performance even without perfect SIC. Signal processing techniques like cross-correlation (with the known transmit waveform) can improve the resolution of reflections even in the presence of non-trivial SI [37]. To add to that, a slowly varying SIC is permissible in these systems, since the goal of radar is to track changes in the SI channel! That is, beyond cancelling the direct leakage from the TX to RX through the chip, board and antennas, we do not need to cancel any of the reflected paths. In fact, it is these very reflections that radar systems track. We also envision a JCR system that characterizes strong reflections, then cancels them using FD to expose weaker reflectors hidden within, making it a powerful tool for dynamic environment sensing.

V. Conclusion

FD cancellers have now demonstrated very high resolution in amplitude and phase matching of SI signal, large on-chip delays and several taps of FIR equalization. Research opportunities remain in improving the antenna interface isolation (in certain applications), multi-antenna cancellation techniques, and demonstrating reliable adaptive operation for dynamic wireless channels. Many exciting applications exist in communication, sensing and joint communication-and-sensing that can leverage FD to solve the 6G directivity and ubiquity requirements.

References

[1] Bharadia, Dinesh *et al.*, *SIGCOMM*, 2013.
[2] Nagulu, Aravind *et al.*, *IEEE J. Microwaves*, 2021.
[3] Chu, Kun-Da *et al.*, in *ISSCC*, 2018.
[4] Nagulu, Aravind *et al.*, in *ISSCC*, 2021.
[5] El Sayed, Ahmed *et al.*, *TCAS I*, 2019.
[6] Zhang, Tong *et al.*, *JSSC*, 2018.
[7] Nagulu, Aravind *et al.*, *JSSC*, 2021.
[8] El Sayed, Ahmed *et al.*, in *RFIC*, 2017.
[9] Katanbaf, Mohamad *et al.*, *IEEE Microwave Magazine*, 2019.
[10] Kolodziej, Kenneth E. *et al.*, *TMTT*, 2019.
[11] Chen, Tingjun *et al.*, *IEEE J. Communications*, 2021.
[12] Cao, Yuhe *et al.*, *JSSC*, 2020.
[13] Liu, Gang *et al.*, *IEEE Communications Surveys Tutorials*, 2015.
[14] Bharadia, Dinesh *et al.*, *SIGCOMM*, 2014.
[15] Korpi, Dani *et al.*, *IEEE Trans. Antennas and Propagation*, 2017.
[16] Abolmagd, Hany *et al.*, *JSSC*, 2022.
[17] Centenaro, Marco *et al.*, *IEEE Wireless Communications*, 2016.
[18] Subbaraman, Raghav *et al.*, *ACM Mobicom 2022*.
[19] Chai, Xiaomeng *et al.*, *IEEE Access*, 2016.
[20] Zhang, Liang *et al.*, *IEEE Trans. Broadcasting*, 2021.
[21] Qu, Fengzhong *et al.*, *IEEE Network*, 2017.
[22] Riihonen, Taneli *et al.*, in *ICMCIS*, 2019.
[23] ——, in *ICMCIS*, 2018.
[24] Pärlin, Karel *et al.*, in *ICMCIS*, 2018.
[25] ——, in *IEEE Communications Magazine*, 2021.
[26] Nasr, Ismail *et al.*, *JSSC*, 2019.
[27] Novelda, "Novelda XeThru X4 Ultra Wideband Impulse Radar Transceiver SoC," https://novelda.com/technology/datasheets/.
[28] Le, Calvin *et al.*, *IEEE TGRS*, 2009.
[29] Giannini, Vito *et al.*, *ISSCC*, 2019.
[30] Guermandi, Davide *et al.*, *ASSCC*, 2016.
[31] Dastjerdi, Mahmood Baraani *et al.*, in *ISSCC*, 2019.
[32] Mondal, Susnata *et al.*, in *ISSCC*, 2020.
[33] Gupta, Agrim *et al.*, *USENIX NSDI*, 2021.
[34] Josephson, Colleen *et al.*, *ACM SIGCAS*, 2021.
[35] Dunna, Manideep *et al.*, *USENIX NSDI*, 2021.
[36] Katanbaf, Mohamad *et al.*, *USENIX NSDI*, 2021.
[37] Kumari, Preeti *et al.*, *IEEE Trans. Signal Processing*, 2020.
[38] Barneto, Carlos Baquero *et al.*, *IEEE Wireless Communications*, 2021.

Reconfigurable Intelligent Surfaces Enabled by Silicon Chips for Secure and Robust mmWave and THz Wireless Communication

Kaushik Sengupta[†, 1] Suresh Venkatesh[†], Hooman Saeidi[†], Xuyang Lu

[†]Electrical and Computer Engineering, Princeton University, Princeton, New Jersey, USA - 08540.
[*]Electrical Engineering, University of Michigan-Shanghai Jiao Tong Joint Institute, Shanghai, China.

Abstract— **The initial deployment of 5G millimeter-Wave (mmWave) networks have shown that while they can sustain Gb/s wireless links with spatial multiplexing at low latencies, they are also highly susceptible to blockages, channel disruptions, and fading due to the nature of their directive beams. Robust mmWave coverage requires high densification of base stations that is prohibitively expensive and complex. Reconfigurable intelligent surfaces (RISs) have emerged as a technology to allow smart reconfiguration of the radio propagation environment by creating more favorable transmission characteristics. Constituted as an array of reconfigurable scattering or antenna elements, such passive surface (with near zero DC power consumption) are scalable and widely deployable. Traditionally referred to as reflect/transmit arrays in the microwave community, these arrays, when combined with silicon ICs in their new avatar, can now allow frequency scaling into the mmWave/THz, rapid programmability, scalability, amplification on-demand (for active surfaces), and sensing. Here, we present the case for such surfaces, design challenges, recent state-of-the-art work, and their impact for future wireless networks.**

Keywords— **RIS, IRS, reconfigurable intelligent surface, millimeter-wave, terahertz, reflectarray, transmitarray, 5G, 6G, beamforming, phased array, blockage.**

I. INTRODUCTION

High throughput millimeter-wave (mm-Wave) networks have shown to be susceptible to physical channel disruptions including blockages and channel propagation variations [1]. In case of line-of-sight (LOS) blockage, current 5G standards rely on the presence of a secondary strong non-line-of-sight (nLOS) path to the user— the absence of which can lead to outage. The concept of Reconfigurable intelligent surface (RIS) in a network is illustrated in Fig. 1. Such surfaces can direct incident beams to users in a programmable fashion, and overcome blockage issues through rapid and smart reconfiguration of the radio environment [2]–[4], both in indoor environment or outdoor setting. The principle of operation relies on engineering the reflected/transmitted characteristics of the surface with periodically placed resonant scatterers or antennas and tunable loads—enabling far field control with elements spaced at $\lambda/2$ (λ being the wavelength), and near/Fresnel/far field control with sub-wavelength spacings, (such as with metasurfaces). Unlike repeaters, passive versions of such surfaces are not affected by receiver noise, quantization noise of digitizers, nonlinearities, and are by principle, full-duplex—they are scalable and consume near-zero DC power. With simultaneously spatial and temporal control, such surfaces can exhibit non-reciprocal behavior and impress physical layer security on these channels [5], [6].

We would like to thank ONR, ARO, and AFSOR for funding support.

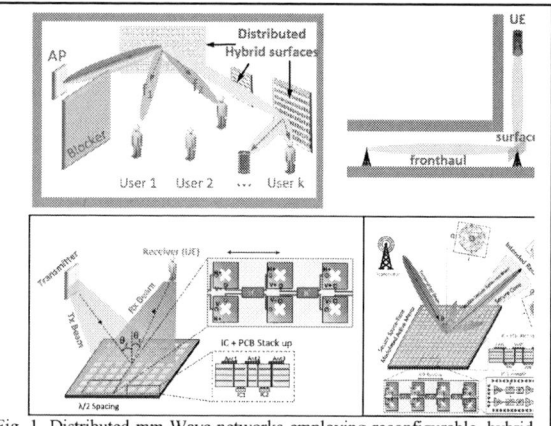

Fig. 1. Distributed mm-Wave networks employing reconfigurable, hybrid surfaces—these surfaces allow smart programming of the radio environment, overcome blockages and channel disruptions enabling robust and scalable networks for 5G and beyond.

Taxonomy matters. These electromagnetic (EM) surfaces, if viewed through the lens of technology, is not new. Reflectarrays were introduced in 1960 [8], and gained popularity later with planar MMICs [9], [10]. However, their resurgence in the form of distributed intelligent surfaces to address the critical issues in mmWave/THz networks has opened new opportunities that are both exciting and challenging. Current state of the art reflect/transmit arrays realized typically with discrete elements cannot scale to mmWave and THz networks. New architectures are needed to enable frequency scaling to mmWave/THz including enhancements in bandwidth, efficiency, scalability, beam synthesis resolution, and allow embedded sensing and control algorithms—hybrid surfaces with embedded silicon ICs can be a potential solution in this direction. Fig. 2 shows the landscape of state-of-the-art surfaces across the electromagnetic spectrum from RF-to-photonics. While their nomenclatures are different across the spectrum, there is a scope for cross-fertilization of design ideas with this generalization, including AI-based methods for synthesis of these scatterers, and there are opportunities to leverage them at mmWave and THz as well [7], [11], [12]. In this paper, we discuss the application of such surfaces in a wireless communication context, design challenges and considerations, discuss state of the art passive and active surfaces across mmWave and THz, and future opportunities of such reconfigurable hybrid surfaces enabled with silicon ICs.

978-1-6654-8495-4/22 $31.00 © 2022 IEEE

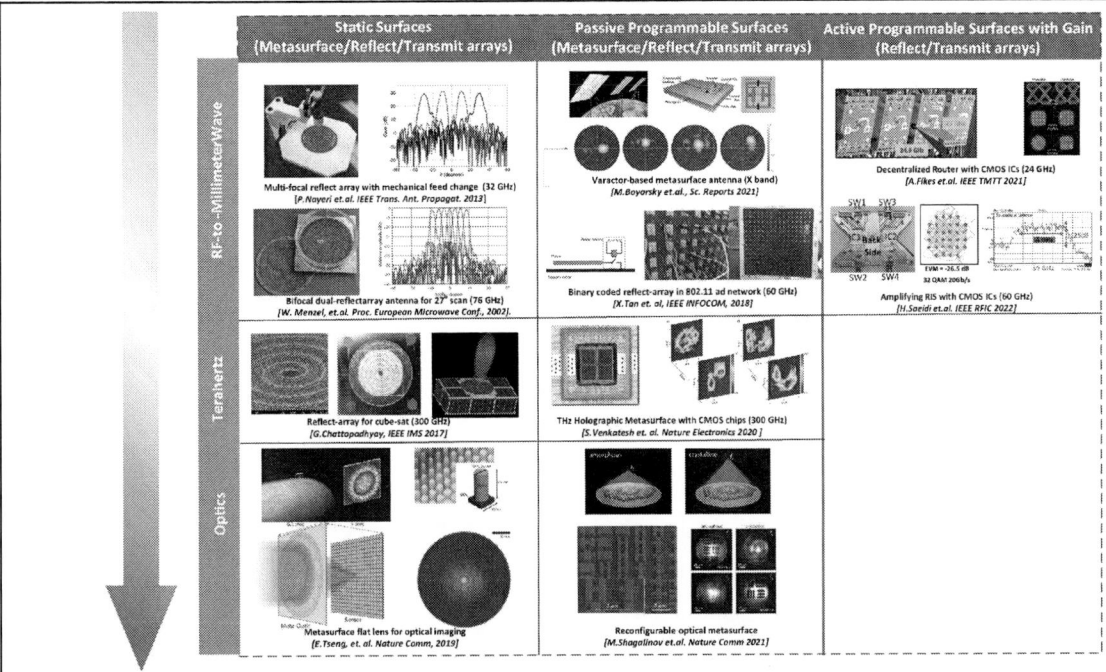

Fig. 2. Technology landscape of general class of planar electromagnetic surfaces achieving field transformation at different length scales (λ/2 or deep sub-wavelength) across the electromagnetic spectrum from RF-to-Terahertz-to-Photonics. While the underlying physical principle might be slightly different across the spectra, their operation principle cam be abstracted out creating a unified design methodology, either analytically or with AI-based models [7].

II. WIRELESS LINKS WITH RIS

The ability to incorporate smart surfaces that can direct an incoming beam precisely in the intended direction(s) changes the nature of the power budget for the wireless links [1]. While the details are beyond the scope of this paper, it is easy to demonstrate their usefulness in a single multi-path event (eg. from a metal reflector), where the receiver power (P_r) can be expressed as

$$P_r \propto P_t \left(\frac{\lambda}{4\pi}\right)^2 \left(\frac{1}{R}\right)^4 \qquad (1)$$

where P_t is the transmitted power and R is the user LOS distance. If the reflector is replaced by a metasurface with N elements that allows the signals to combine coherently in space, the power at the UE can be enhanced significantly

$$P_r \propto \alpha P_t \left(\frac{\lambda}{4\pi}\right)^2 \left| \frac{1}{R} + \sum_{i=1}^{N} \frac{1}{r_i+d} \right|^2 = \alpha P_t (N+1)^2 \left(\frac{\lambda}{4\pi R}\right)^2 \quad (2)$$

where α is the effective loss of the surface, d is the distance of the transmitter to the surface, and r_i is the effective distance from each element on the surface to the UE [4]. Expectedly, such surfaces can allow reversion back to the square law model and also increase the power at the user by a factor of $(N+1)^2$, due to the coherent combination of the reflected waves. This is a significantly large improvement, since a surface with N=100 (i.e. $100\lambda \times 100\lambda$), can lead to a 40 dB enhancement in user power. Unlike a phased array, such surfaces do not need full mm-Wave transceivers, frequency synthesis, signal distribution, synchronization, and therefore, far more easily

scalable to such large apertures. Such surfaces in an indoor environment has been shown to render orders of magnitude improvement in SNR at the user in nLOS locations [15].

III. MMWAVE AND THZ SURFACES

A. Current State of the Art

Reflectarrays in the RF and microwave frequency range have been realized with antennas connected with tunable loads utilizing varactor diodes, PIN diode switches, ferro-electric devices, and MEMS switches [10], [18]. Binary-coded surfaces are popular to simplify control of the surface for optimal beamcodes, but it comes at the cost of SNR. Examples of such works with large scalability has been demonstrated in reflectarrays with p-i-n didoes in the X and Ku band reaching up to 1600 elements [19] element loss in the 1-1.5 dB range . In [20], a 224-element reflect array with patch antennas and relay switches was implemented to demonstrate the effectiveness of such smart surfaces in a 60 GHz 802.11ad mmWave network. With a three-party beam search control algorithm, the work showed optimal reflect array positioning in an indoor environment for achieving close to 50% reduction.

B. Challenges in RIS for 5G and beyond

Notwithstanding these advancements, there is an unmet need for smart surfaces for mmWave/THz networks. Design challenges include

- mmWave/THz operation and scalability: Current

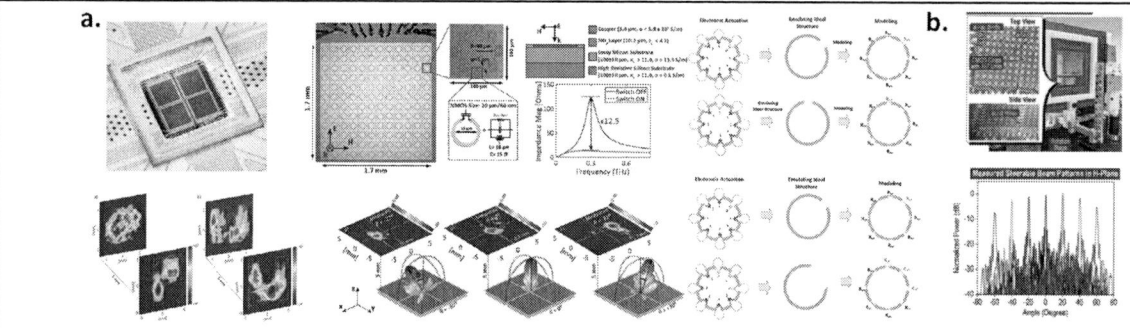

Fig. 3. Reconfigurable THz RIS. a. 576-element programmable THz metasurface (300 GHz) with CMOS tiles with 84 unique amplitude/phase states that is programmable at GHz speed [13]. b. : A 98×98-Unit 265GHz CMOS reflectarray with in-Unit digital beam Shaping and squint correction [14].

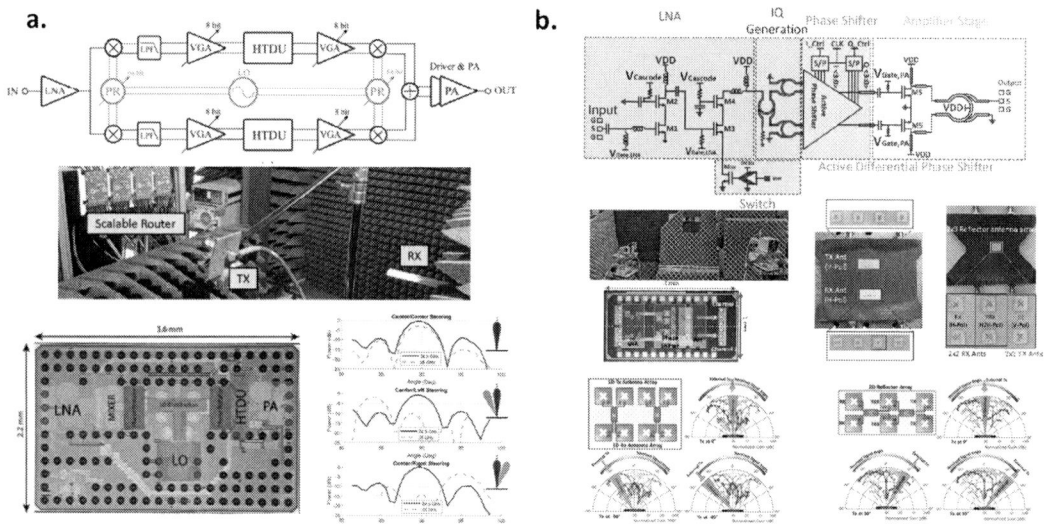

Fig. 4. Hybrid active mmWave Surfaces: a. A 4-element relay array with integrated frequency translational circuits. [16]. b. Active synthesizer-free mmWave surface with packaged dual-feed antennas, integrated amplification, and phase shift and no frequency translation for scalability [17].

surfaces realized commercially available discrete components (diodes, varactors etc) cannot address future mmWave/THz bands. Hybrid surfaces enabled with silicon ICs can play an important role [13]-[17].

- Efficiency and packaging: Losses induced from scattering from the surface directly affects the power available at the user—minimization is critical with low-loss antennas, and high quality factor circuits.
- Independent amplitude/phase control: Typically the resonant structures exploit a Lorentzian profile that couples amplitude and phase control. To separate them for beam optimization, novel higher-order EM structures are needed [13].
- Stability, noise and linearity: For Actives surfaces, improved isolation between Tx to Rx is needed to avoid stability concerns. Noise (including quantization noise) and linearity needs to be taken into account
- Self interference : For active surfaces, multi-path echo and intrinsic delay can be important with cyclic prefix

~ 500 ns for 120 KHZ sub-carrier spacing.

- Beam synthesis, channel estimation, latency and control: The design of such surfaces needs to take into consideration the network architecture, latency requirements, and the control plane algorithms. As an example, binary surfaces trade-off efficiency and latency, but this may not be optimal for SNR or throughput limited regime.

C. Passive THz RIS

At At THz frequencies, typical switching ratios scales as $|Z_{off}/Z_{on} \sim f_t/f|$ making it challenging to realize programmable surfaces. Prior works have demonstrated reconfigurability through the use of heterogeneous novel materials, phase change materials, 2D materials, or photoconductive semiconductors, none of which are scalable [13]. However at THz, CMOS chip dimensions becomes multiple wavelengths long, and therefore, each chip can integrated an end-to-end sub-array. In [13], a new CMOS integrated active meta-element operating beyond f_t at 300 GHz (Fig. 3a). The unique meta-element structure resonated out distributed parasitics of the coupled FETs, while the latter reconfigured the meta-element itself. With such a coupled co-design methodology, a 576 element metasurface was realized with tiled 2x2 CMOS chips demonstrating nearly 25 dB of amplitude control-phase coverage of 270° and transmission loss of 3.5 dB (substrate loss being 2.0 dB) (Fig. 3a). Collectively, the surface demonstrated THz beamforming and holographic projections (with sub-wavelength near-field control), that are reconfigurable at GHz speed. Such a co-design approach was demonstrated in a more recent work with 98×98 elements operating at 256 GHz with digital beam shaping and squint correction [14] (Fig. 3b). While these early works have demonstrated the potential to think about new forms of chip-enabled passive beamforming architectures at THz, it shows the path towards THz surfaces that can even embed sensing, and eventually decision making capabilities.

D. Active mmWave RIS

To mitigate the multiplicative path losses between Tx-to-Surface and Surface-to-Rx, surfaces that allow on-surface amplification on-demand, has been shown to theoretically enable an order of magnitude gain in channel capacity [21]. Such active hybrid surfaces can be enabled with silicon ICs as demonstrated in the 4-element relay with integrated down-conversion and up-1conversion chains, phase shifters and true time delay elements operating at 25 GHz [16] (Fig. 4a). Exploiting polarization orthogonality, the relay demonstrated 26 dB gain with dual-beam routing capabilities. Incorporating separate transceivers for each chain that requires frequency translation, LO generation, synchronization, however, comes at the cost of scalability. Another approach in [17] shows an active 1D and 2D array with packaged dual feed probe-fed patch antennas where each silicon IC integrates two chains (Fig. 4a). Each Rx to Tx path has no frequency translation and comprises of a LNA with 5-bit controlled phase shifter, PA with Psat of 4.2dBm at 60GHz and a total noise figure of 5-6 dB in the band. The 2D array exploited the same antenna for transmission and reception utilizing polarization orthogonality and supports up to 20 Gbps with 32-QAM constellation (Fig. 4a). In addition, each element is incorporated with controllable temporal modulation that can impress physical layer security in the re-transmitted signals, overcoming eavesdropper attacks [5]. While these results demonstrate an approach towards hybrid active surfaces as relays, there are still several challenges in scaling up such surfaces to hundreds and thousands of elements, the scale at which they are useful in deployment. Improved isolation between Tx/Rx elements is necessary to enhance gain with new antenna designs. Energy efficient architectures for high frequency signal processing are critical to balance scalability, deploybility with amplification and link coverage. The design is inextricably linked with the network architecture and control plane, which necessitates a close co-design of the end-to-end system.

IV. CONCLUSION

Reconfigurable intelligent surfaces, can address some of the pressing challenges in mmWave 5G, and allow new capabilities in future mmWave/THz networks. To enable scalable passive and active surfaces at such high frequencies, it is important to leverage the capability of silicon ICs to process these signals. In doing so, it opens up a new class of smart surfaces that can embed more functionalities such as beam-optimization, localization, distributed signal processing, and embed physical layer security, while enabling new applications in computational mmWave imaging, and sensing.

REFERENCES

[1] Q. Wu et al., "Intelligent reflecting surface aided wireless communications: A tutorial," IEEE Trans. on Comm., 2021.

[2] K.Sengupta, T. Nagatsuma, and D.M. Mittleman, "Terahertz integrated electronic and hybrid electronic–photonic systems," Nat Electron 1, 622-635 (2018).

[3] M. H. Dahri, et. al "A review of wideband reflectarray antennas for 5G communication systems," IEEE Access, vol. 5, 2017.

[4] E. Basar, et. al., "Wireless communications through reconfigurable intelligent surfaces," IEEE access, vol. 7, pp. 116 753–116 773, 2019.

[5] S. Venkatesh et al., "Secure space–time-modulated millimetre-wave wireless links that are resilient to distributed eavesdropper attacks," Nat. Elec., vol. 4, no. 11, pp. 827–836, 2021.

[6] X. Lu et al., "4.6 Space-Time Modulated 71-to-76GHz mm-Wave Transmitter Array for Physically Secure Directional Wireless Links," in IEEE ISSCC, 2020, pp. 86–88.

[7] W. Ma, et. al "Deep learning for the design of photonic structures," Nature Photonics, vol. 15, no. 2, pp. 77–90, 2021.

[8] D. Berry, R. Malech, and W. Kennedy, "The reflectarray antenna," IEEE Trans. Ant. and Propagat., vol. 11, no. 6, pp. 645–651, 1963.

[9] M. E. Bialkowski, A. W. Robinson, and H. J. Song, "Design, development, and testing of x-band amplifying reflectarrays," IEEE Trans. Ant. and Propagat., vol. 50, no. 8, pp. 1065–1076, 2002.

[10] S. V. Hum and J. Perruisseau-Carrier, "Reconfigurable reflectarrays and array lenses for dynamic antenna beam control: A review," IEEE Trans.on Ant. and Propagat., vol. 62, no. 1, pp. 183–198, 2013.

[11] E. Karahan, A. Gupta, U. Khankhoje, and K. Sengupta, "Deep learning based modeling and inverse design for arbitrary planar antenna structures at rf and millimeter-wave," in 2022 IEEE APS-URSI. IEEE, 2022.

[12] Z. Liu, E. Karahan, and K. Sengupta, "Deep learning enabled inverse design of 30-94 GHz Psat,3dB SiGe PA supporting concurrent multi-band operation at multi-gbps," IEEE MWCL., vol. 32, no. 6, 2022.

[13] S. Venkatesh et al., "A high-speed programmable and scalable terahertz holographic metasurface based on tiled CMOS chips," Nature Electronics, vol. 3, no. 12, pp. 785–793, 2020.

[14] N. M. Monroe et. al, "Electronic thz pencil beam forming and 2d steering for high angular-resolution operation: A 98 times 98-unit 265 GHz CMOS reflectarray with in-unit digital beam shaping and squint correction," in IEEE ISSCC, vol. 65, 2022, pp. 1–3.

[15] I. F. Akyildiz, C. Han, and S. Nie, "Combating the distance problem in the millimeter wave and terahertz frequency bands," IEEE Communications Magazine, vol. 56, no. 6, pp. 102–108, 2018.

[16] A. Fikes et al., "Programmable active mirror: A scalable decentralized router," IEEE Trans. Microw. Theory Techn., vol. 69, no. 3, 2020.

[17] S. H. L. X. Venkatesh, Venkatesh and K. Sengupta, "Active tunable millimeter-wave reflective surface across 57-64 GHz for blockage mitigation and physical layer security," IEEE RFIC Symp., 2021.

[18] P. Nayeri, F. Yang, and A. Z. Elsherbeni, "Beam-scanning reflectarray antennas: A technical overview and state of the art." IEEE Antennas and Propagation Magazine, vol. 57, no. 4, pp. 32–47, 2015.

[19] H. Yang, F. Yang, X. Cao, S. Xu, J. Gao, X. Chen, M. Li, and T. Li, "A 1600-element dual-frequency electronically reconfigurable reflectarray at x/ku-band," IEEE transactions on antennas and propagation, vol. 65, no. 6, pp. 3024–3032, 2017.

[20] X. Tan, Z. Sun, D. Koutsonikolas, and J. M. Jornet, "Enabling indoor mobile millimeter-wave networks based on smart reflect-arrays," in IEEE INFOCOM 2018-IEEE Conference on Computer Communications. IEEE, 2018, pp. 270–278.

[21] Z. Zhang, L. Dai, X. Chen, C. Liu, F. Yang, R. Schober, and H. V. Poor, "Active RIS vs. passive RIS: Which will prevail in 6G?" arXiv preprint arXiv:2103.15154, 2021.

Gap in pagination due to withheld paper.

Pages 550-553

A 130μW Three-Step DT Incremental ΔΣ ADC Achieving 107.6dB DR and 99.3dB SNDR with Zoom and Extended-Range Counting

Lairong Fang, Yijie Li, Yao Zhang, Shuwen Zhang, Xiaoyang Zeng, Zhiliang Hong and Jiawei Xu
State Key Laboratory of ASIC and System, Fudan University, Shanghai, China.
Email: jwxu@fudan.edu.cn

Abstract—**This paper presents an incremental ΔΣ ADC (IADC) utilizing a three-step conversion. By combining the zoom-in and extended-range (ER) counting techniques, the IADC significantly reduces the oversampling ratio (OSR) and power consumption of a standard incremental ΔΣ counterpart. Clocked at 512kHz, the prototype IADC achieves a 107.6dB DR, a 104.9dB peak SNR, and a 99.3dB peak SNDR for a conversion rate of 2kS/s while consuming 130μW from a 1.8V supply. This corresponds to a state-of-the-art Schreier FoM$_{DR}$ of 176.5dB.**

I. INTRODUCTION

High resolution ADCs are required in many applications, such as smart sensor interfaces, precision instrumentation and biomedical signals acquisition. Delta-sigma (ΔΣ) ADCs can achieve very high resolution by means of oversampling and noise-shaping, but they are less energy efficient. Furthermore, Conventional free-running ΔΣ ADCs cannot process the time-multiplexed signals due to memory in the loop. Incremental ΔΣ ADCs (IADCs) solve this issue through the regular resets of the loop filters, thereby enabling both high resolution and Nyquist-rate conversion while being less sensitive to idle tones [1]. As a result, power efficient IADCs are well suited for multi-channel or multimodal sensor interfaces.

To date, hybrid ADC architectures, combining the merits of SAR and ΔΣ modulator (ΔΣM), can considerably improve the power efficiency by means of noise-shaping SAR [2] or zoom-in ΔΣ ADCs [3][4]. Similar multi-step quantization strategies are also applicable to the IADCs, either by recycling the analog residue of the ΔΣM loop filters [5][6], or by performing a coarse SAR conversion prior to the incremental ΔΣM to complement its resolution requirement [7]. For example, a two-step IADC utilized extended-range (ER) counting by a linear-exponential conversion to realize a 100.8dB SNDR [6], but the relatively high OSR of 246 cycles in the first step requires power-hungry loop filters. A zoom-in IADC [7] achieved an 119.8dB SNR by employing a coarse 6-bit SAR conversion followed by a fine 15-bit ΔΣ conversion. However, this IADC suffers from a low bandwidth (BW) of 12.5Hz due to a large OSR of 2000.

This paper presents a three-step IADC, combining the zoom and ER counting, to improve its resolution and BW in a power efficient manner. As illustrated in Fig.1, the IADC has three conversion steps: 1) Zoom-SAR, 2) Incremental-ΔΣM, and 3) ER-SAR. The input signal V$_{IN}$ is sequentially digitized by these sub-ADCs whose outputs are weighted and summed to form the overall output DOUT. The ΔΣM has a reconfigurable OSR, and the ER-SAR is duty cycled to provide flexible bandwidth and power options for different sensor arrays.

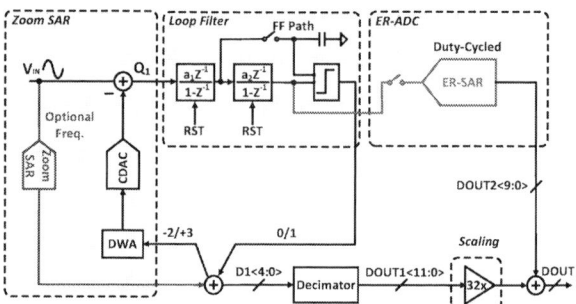

Fig. 1. Concept of the three-step IADC with Zoom SAR, incremental ΔΣ, and duty-cycled ER-SAR.

II. SYSTEM ARCHITECTURE

The IADC architecture is illustrated in Fig.1. It operates in discrete time (DT) for the precise alignment of three sub-ADCs, ensuring robustness against PVT variations in particular during the residue sampling and digital outputs reconstruction.

In the first step, an asynchronous 5-bit Zoom-SAR digitizes the input signal V$_{IN}$ in a coarse manner referring to a reference V$_{REF}$. Digital outputs of the Zoom-SAR are used to adjust the reference voltage of the succeeding ΔΣM, ensuring V$_{in}$ always fall into the reference bounds. The use of Zoom-SAR reduces input swing of the ΔΣM and alleviates its slewing requirements.

In the second step, the 2nd-order single-bit incremental ΔΣM (Fig.1) is engaged to digitize the input signal. Output bit stream of the ΔΣM is combined with the Zoom-SAR outputs to yield D1<4:0> (Fig.1), which is passed through a decimation filter to generate the filtered output DOUT1<11:0>. The incremental-ΔΣM has a finite impulse response, and so it is possible to use simple cascade of integrators (CoI) for decimation. The Zoom-SAR is only used to determine the ΔΣM's reference, thus the accuracy of the overall Zoom-in ΔΣM is mainly determined by the ΔΣM, in particular the nonlinearity of the C-DAC [3].

In the third step, at the end of ΔΣM conversion, the 10-bit duty-cycled ER-SAR takes its input from the output of the 2nd integrator of ΔΣM to complete the residue conversion and yield DOUT2<9:0> (Fig.1). To generate the overall output DOUT, DOUT1<11:0> is first multiplied by 32 and then summed with DOUT2<9:0> due to different data weights.

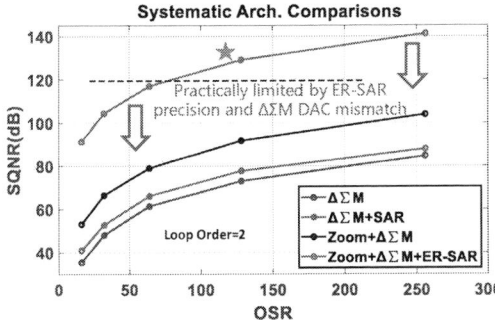

Fig. 2. Theoretical comparison among diverse ΔΣM architectures.

To illustrate the benefits of a three-step IADC, diverse ΔΣM ADCs based on ideal models are simulated and compared in Fig. 2. These ADCs are built on a 2nd-order single-bit ΔΣM with an optional 5-bit Zoom SAR and a 10-bit ER-SAR for comparison. Fig. 2 shows that the proposed three-step IADC architecture can theoretically achieve a ~120dB SQNR with a moderate OSR of 64, while the ΔΣM and two-step IADCs cannot meet this target in spite of a higher OSR of 256. This indicates that an ER-SAR can be approached as an attempt to realize a high dynamic range IADC with a relatively low OSR and power overhead. However, in practice, the actual SQNR would be considerably limited by the DAC mismatch of Zoom ΔΣM, as well as the circuit noise and nonlinearity of the ER-SAR.

Since the Zoom-SAR is only used to adjust the reference of ΔΣM and hence has almost no impact on the residue calculation. According to [10], the quantization error of the IADC can be denoted by

$$Q_{IADC} = \hat{V}_{RES} - V_{RES} = \frac{2}{a_1 \cdot a_2 \cdot OSR \cdot (OSR-1)} \cdot Q_{ER\text{-}SAR} \quad (1)$$

where $\hat{V}_{RES}$ and V_{RES} are the digitized estimate and the actual value of the residue, respectively; a_1, a_2 are the coefficients of the loop filters; $Q_{ER\text{-}SAR}$ is the quantization error of the ER-SAR. This equation implies that the IADC's overall resolution can be improved greatly even by using a moderate OSR. For example, an OSR=64 can reduce the quantization error ideally by a factor of ~2000.

Combining the Zoom-SAR and the ER-SAR considerably reduces the power overhead of the ΔΣM core by leveraging a low OSR of 16-128. Compared to conventional IADCs using a large OSR of 2000 [9] or using 256 OSR cycles [6], the three-step IADC operates at a 2x-16x lower OSR, allowing a relaxed GBW requirements of loop integrators. However, reducing the OSR deteriorates DAC mismatch and quantization noise. In this work (Fig.1), the former is tackled by data weighted averaging (DWA) and the latter is reduced by adding the duty-cycled ER-SAR.

Fig.3 shows the circuit implementation and timing diagrams of the IADC. The Zoom-SAR operates at the same speed as the ΔΣM. Their outputs are combined during each ΔΣ clock (Φ1) (after Zoom-SAR conversion ends). The ER-SAR only samples the ΔΣM's residue once per IADC conversion (Fig.3). Hence, the outputs of three sub-ADCs are also merged at a decimation speed of 1k-8kS/s. After the residue sampling, the IADC resets the charges on the integration capacitors.

In the zoom-in ΔΣM, sampling clocks of Zoom-SAR (Φ2) and ΔΣM (Φ1) are deliberately separated (Fig.3), leaving half clock cycle to prevent crosstalk [4]. The C-DAC in the ER-SAR employs custom-designed finger capacitor arrays with an LSB of 2.5fF to mitigate the tradeoff between KT/C noise and the C-DAC capacitance, which are 80μV and 2.56pF, respectively.

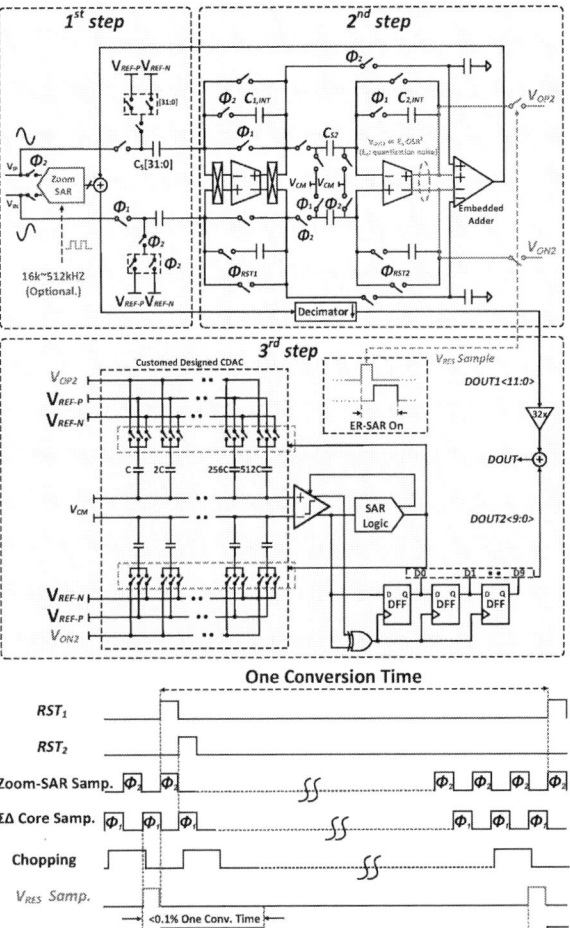

Fig. 3. Block and timing diagrams of the three-step IADC.

The ER-SAR is duty cycled and runs at the Nyquist rate of 1k-8kHz, thereby consuming insignificant power. Referring to Fig.3, in phase Φ2 (one cycle before integrator reset), the ER-SAR samples the residue. The total residue processing time is less than 0.1% of one IADC conversion time.

III. IADC FUNCTIONAL BLOCKS

A. Zoom-SAR ADC

Fig.4 shows the schematic of Zoom-SAR, it provides a 5-bit coarse digital output to determine the ΔΣM's reference, so the accuracy requirement is not stringent. However, redundant bits are required to tolerate possible over-ranging errors [3]. In this work, the positive and negative references of the ΔΣM are set to (M+3)×LSB and (M-2)×LSB, respectively, to provide redundancy of 4 LSBs, where LSB is the quantization step of the Zoom-SAR, i.e. 112.5mV for a full-scale input of 3.6V, and M is used to adjust the zoom-in range. As a result, utilizing four redundancy bits results in a higher input swing of 450mV for the succeeding $\Delta\Sigma M$.

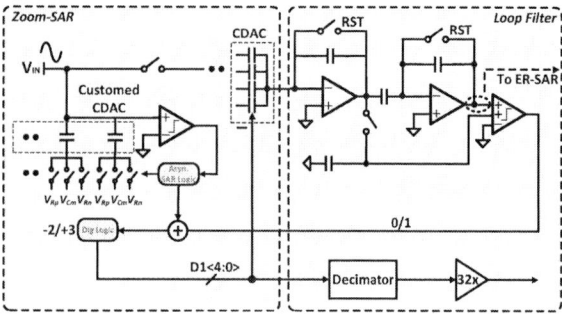

Fig. 4. Zoom-SAR and the $\Delta\Sigma M$ core implementations.

B. $\Delta\Sigma M$ Core

The $\Delta\Sigma M$ core is built on a 2nd-order, single-bit quantizer, feedforward (FF) architecture. The FF structure benefits from a low output swing of the OTA because the loop integrators only contain the quantization noise without signal components [9]. Discrete-time operation provides PVT robustness because the integrator coefficients mainly depend on the capacitors ratio. To accurately catch the $\Delta\Sigma M$ residue, discrete time (DT) is a better choice, but the OTA inside loop filters usually needs a higher bandwidth to complete the full settling in the integration phase [10]. Hence, relaxing the GBW requirement of the OTA and the associated power by using a low OSR is motivated.

Fig.5 shows the circuit details. The 1st-integrator utilizes a telescopic OTA with a PMOS input biased at 40μA, not only ensuring a high DC gain of 80dB, but also offering high current efficiency [9]. Thanks to the relaxed OSR, the OTA can settle to the desired accuracy within half clock cycle. To mitigate $1/f$ noise and offset, the OTA is chopped at $f_s/2$=256kHz. The 2nd-integrator employs a current-reused OTA to increase its power efficiency (Fig.5).

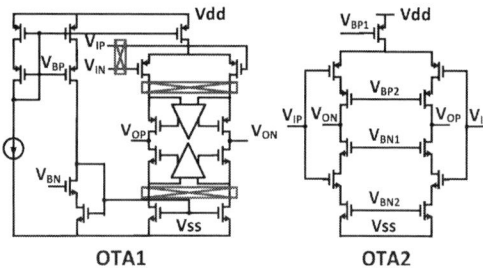

Fig. 5. Circuit details of the loop integrators.

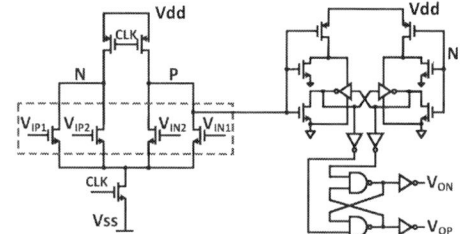

Fig. 6. The 4-input single-bit quantizer.

To implement the FF scheme in the $\Delta\Sigma M$ without using an active adder, the outputs of integrators are directly tied to a 4-input quantizer (Fig.6). It includes a preamplifier preceding the latch to reduce kickback noise.

C. ER-SAR ADC

SAR ADCs often exhibits medium resolution of 10-12 bits due to the C-DAC nonlinearity, unless mismatch error shaping (MES) [2] or calibration is used. To make the best use of the ER-SAR ADC's power efficiency, we set its resolution to 10 bits. A higher resolution ER-SAR is less efficient in obtaining more ENOB [10] but would result in a larger chip area. In this work, the ER-SAR is duty-cycled, limiting its average power to 2.4μW.

The output swing of the 2nd-integrator is about 20% of V_{REF}, so we reduce the ER-SAR's reference voltage by 4 to fit for the input signal range. As a result, a gain compensation of 1/4 is required after the ER-SAR. This was merged with the reference scaling of 32x in digital domain.

IV. MEASUREMENT RESULTS

The IADC prototype, including bandgap reference circuits and a decimation filter, was manufactured in a standard 0.18μm CMOS process (Fig.7). Clocked at 512kHz (OSR=128), the IADC consumes 130μW at 1.8V analog and 1.2V digital supply. Most of the power are occupied by the loop filters of the $\Delta\Sigma M$ (Fig.7), rather than the Zoom-SAR and the ER-SAR.

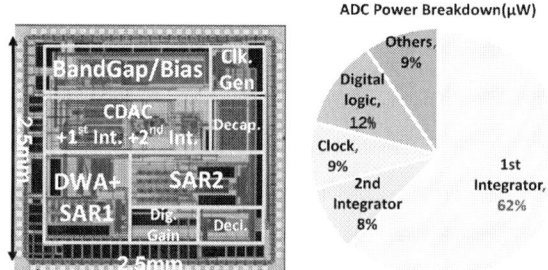

Fig. 7. Chip photograph and power breakdown.

Fig.8 (top) shows the measured DR and SNDR versus input signal amplitudes. The IADC achieves 107.6dB DR, 104.9dB peak SNR, and 99.3dB peak SNDR in 1kHz BW. Fig.8 (bottom) shows the ENOB at multiple OSRs with/wo the ER-SAR. It can either improve ENOB at the same OSR, or significantly reduce the required OSR to reach the same resolution. In these tests, both f_s and OSR are proportionally scaled to ensure the same BW. The maximum ENOB improvement is 25.6dB at OSR=16. In Fig.9 (top), the measured PSD with/wo the ER-SAR indicate that SNDR are 99.3dB and 91.9dB, respectively. Fig.9 (bottom) shows that when the DWA is off, the SNDR drops to 78.8dB due to C-DAC mismatch. Fig.10 shows the measured DNL and INL errors of +0.69/-0.60 and +0.95/-0.87 LSB, respectively, indicting no missing code at 17-bit resolution.

Table I compares this IADC with the state-of-the-art single- and two-step DT-IADCs. This work achieves the highest DR of 107.6dB and SNR of 104.9dB with a lower OSR while retaining a highly competitive FoM_{DR} of 176.5dB.

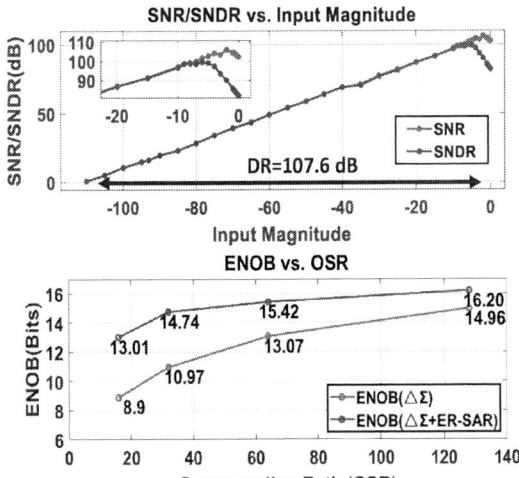

Fig. 8. Measured DR/SNDR versus input magnitude (top); ENOB vs. OSR for 1kHz BW (bottom).

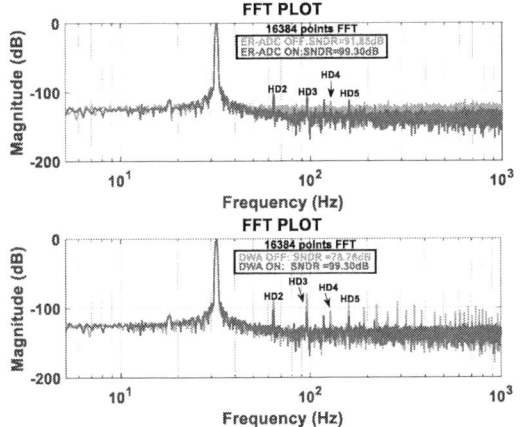

Fig. 9. Measured spectrum with -5dBFS input signal, with/wo ER-ADC (top); with/wo DWA (bottom).

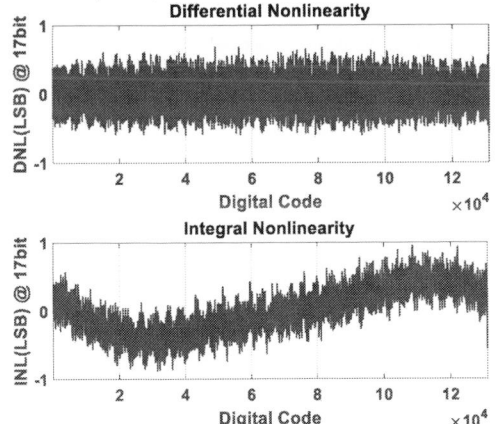

Fig.10. Measured DNL (top) and INL (bottom) in 17-bit code.

Table I. Comparison with state-of-the-art DT incremental ADCs

	This Work	JSSC'20 [5]	JSSC'19 [6]	ISSCC'18 [7]	TCASI'21 [8]
Supply(V)	1.8/1.2	1.8	1.2	3	1.5
Tech(nm)	180	180	65	180	180
Arch	Zoom IDSM +ER-SAR	IDSM+SAR	Linear-Exp. IDSM	IDSM	IDSM+Binary Counting
Power(μW)	130	25.4	550	1098	33.2
OSR	16-128	100	256	150	266
BW(kHz)	0.5-4	2.04	20	100	1.2
DR(dB)	107.6 (1k BW)	102.2	101.8	91.5	100.2
SNR$_{max}$(dB)	104.9	96	NaN	88.2	97.1
SNDR$_{max}$(dB)	99.3	95.5	100.8	86.6	96.6
FOM$_S^1$(dB)	176.5	181.2	NaN	171.1	175.8
FOM$_S^2$(dB)	173.8	175	177.4	167.8	172.7

FOMs^1=DR+10log(BW/Power); FOMs^2=SNR+10log(BW/Power)

V. CONCLUSIONS

This paper presents a power efficient DT incremental ADC for multi-channel sensors. By using a three-step conversion of Zoom-SAR, incremental $\Delta\Sigma M$ and a duty-cycled ER-SAR, the IADC largely reduces the required OSR and power dissipation to reach high resolution. Moreover, utilizing the DT operation and digital gain compensation ensures a reliable alignment of the three sub-ADCs. Among the recent state-of-the-art hybrid DT IADCs, this work achieves the highest DR of 107.6dB, as well as a competitive FoM$_{DR}$ of 176.5dB.

ACKNOWLEDGEMENT

This work was financially supported by the Joint R&D Fund of Beijing Smartchip Microelectronics Technology Co., Ltd

REFERENCES

[1] Z. Tan, C. -H. Chen, Y. Chae and G. C. Temes, "Incremental Delta-Sigma ADCs: A Tutorial Review," *IEEE Trans. Circuits Syst. I*, vol. 67, no. 12, pp. 4161-4173, Dec. 2020.

[2] Y. Shu et al., "An Oversampling SAR ADC with DAC Mismatch Error Shaping Achieving 105 dB SFDR and 101 dB SNDR Over 1 kHz BW in 55 nm CMOS," *IEEE J. Solid-State Circuits*, pp.2928-2940, Dec. 2016.

[3] S. Karmakar et al, "A 280 μW Dynamic Zoom ADC with 120 dB DR and 118 dB SNDR in 1 kHz BW," *IEEE J. Solid-State Circuits*, vol. 53, no. 12, pp. 3497-3507, Dec. 2018

[4] E. Eland et al., "A 440-μW, 109.8-dB DR, 106.5-dB SNDR Discrete-Time Zoom ADC with a 20-kHz BW." *IEEE J. Solid-State Circuits,* vol. 56, no. 4, pp.1207-1215, Jan. 2021.

[5] S. Mohamad et al, "A 102.2-dB, 181.1-dB FoM Extended Counting Analog-to-Digital Converter with Capacitor Scaling," *IEEE J. Solid-State Circuits*, vol.55, no. 5, pp.1351-1360, May 2019.

[6] B. Wang et al., "A 550-μW 20-kHz BW 100.8-dB SNDR Linear-Exponential Multi-Bit Incremental ΣΔ ADC with 256 Clock Cycles in 65-nm CMOS," *IEEE J. Solid-State Circuits*, pp.1161-1172, April. 2019.

[7] P. Vogelmann et al., "A 1.1mW 200kS/s Incremental ΔΣ ADC with a DR of 91.5dB Using Integrator Slicing for Dynamic Power Reduction," *ISSCC Dig. Tech Papers*, pp.236-238, Feb, 2018,

[8] S. Kuo et al., "A Multi-Step Incremental Analog-to-Digital Converter with a Single Opamp and Two-Capacitor SAR Extended Counting." *IEEE Trans. Circuits Syst. I*, vol. 68, no.7, pp.2890-2899, July. 2021.

[9] Y. Chae, K. Souri, and K. A. A. Makinwa, "A 6.3μW 20-Bit Incremental Zoom-ADC with 6ppm INL and 1μV Offset," *IEEE J. Solid-State Circuits*, vol. 48, no. 12, pp. 3019–3027, Dec. 2013.

[10] A. Agah et al., "A High-Resolution Low-Power Incremental ADC with Extended Range for Biosensor Arrays," *IEEE J. Solid-State Circuits*, vol.45, no. 6, pp. 1099–1110, Jun. 2010.

Gap in pagination due to withheld papers.

Pages 558-569

ESSCIRC 2022 AUTHOR INDEX

A

Abolmagd, Hany 542
Adamopoulos, Christos 77
Adragna, Claudio 17
Aflatouni, Firooz 538
Afshari, Ehsan 533
Agarwal, Amit 377
Agnew, Megan 193
Ahn, Hong Keun 97
Ahn, Hyunjin 341
Ahsan, Ishtiaq 105
Aiello, Orazio 381
Albuquerque, Edgar 177
Alioto, Massimo 33, 381
Allebes, Erwin 421
Anders, Jens 185, 225
Anders, Mark 377
Anderson, Martin 237
Andrieu, François 125, 129
Angizi, Shaahin 153
Antolini, Alessio 109
Aprile, Antonio 473
Arbabian, Amin 329
Arnal, Victor M. 225
Arrigo, Domenico 17
Ashida, Mitsuyuki 505
Ashtiani, Farshid 538
Aufinger, Klaus 449
Auricchio, Chantal 373
Aussenac, François 125

B

Babaie, Masoud 53
Bachmann, Christian 421
Bae, Jin-Hee 121
Baek, Kwang-Hyun 241
Bajoria, Shagun 321
Banarie, Gabriel 401
Baniasadi, Nima 489
Barajas, Carlos Chavez 197

Barbot, Justine 125
Barik, Gourab 209
Basetas, Charis 421
Bassi, Matteo 337, 385
Bechthum, Elbert 421
Becker, Joachim 201
Bédécarrats, Thomas 45
Bedjaoui, Messaoud 125
Benini, Luca 105, 273, 357
Benndorf, Jens 265
Bersano, Fabio 49
Bertrand, Benoit 45
Beshary, Hesham 489
Bevilacqua, Andrea 385
Bharadia, Dinesh 542
Blawat, Meinolf 265
Blume, Holger 265
Boer, Pepijn 421
Bol, David 157, 445, 469
Bolatkale, Muhammed 253, 321
Boletti, Alessio 177
Bonache, Jesús 401
Bonizzoni, Edoardo 221, 473
Borghi, Massimo 373
Bose, Pradip 269
Bourgeois, Guillaume 129
Braakman, Floris 49
Brändli, Matthias 57
Brebels, Steven 61
Breems, Lucien 321
Breeschoten, Arjan 421
Brekelmans, Hans 321
Brenneis, Andreas 73
Brew, Kevin 105
Brivio, Stefano 117
Brooks, David 269
Brugger, Jürgen 161
Bruschi, Paolo 257
Buca, Dan 165
Buchbinder, Sidney 77
Buckwalter, James F. 453
Buffa, Cesare 409

Bugalho, Ricardo 177
Buj, Christel 193
Burdiek, Bernard 321
Burghartz, Joachim 185

C

Cabrini, Alessandro 373
Caglar, Alican 61
Cagli, Carlo 129
Calhoun, Benton H. 465
Calmon, Francis 193
Calvetti, Emanuela 373
Campagna, Riccardo 12
Canegallo, Roberto Antonio 109
Cantatore, Eugenio 181
Cao, Peng 277
Capecchi, Laura 373
Capellini, Giovanni 165
Carabasse, Catherine 125
Caragiulo, Pietro 329
Cardoso Paz, Bruna 45
Carissimi, Marcella 109, 373
Carloni, Luca 269
Casse, Gianluigi 197
Castañeda, Oscar 357
Castello, Rinaldo 25, 309, 429
Catania, Alessandro 257
Cathelin, Philippe 305
Ceballos, Jose Luis 409
Cenci, Pierluigi 321
Chae, Hyunsu 509
Chan, Chi-Hang 566
Chan, Victor 105
Chandramoorthy, Nandhini 269
Chang, Chun-Ting 313
Chanrion, Emmanuel 45
Chatterjee, Baibhab 209
Chen, Chao 558
Chen, Cheng 229
Chen, Jing-Siang 313
Chen, Ke-Horng 285, 293, 365
Chen, Yiming 562
Chen, Yingjie 441
Chen, Yong 405, 425, 497
Chen, Yong-Tai 81
Chen, Zhipeng 477
Cheng, Xuxu 501

Cherubini, Giovanni 105
Cherupally, Sai Kiran 89, 153
Chi, Baoyong 233
Chien, Jeff Shih-Chieh 453
Choi, Sam 105
Choo, Min-Seong 249
Chou, Ethan 489
Chu, Kun-Da 493
Ciciotti, Fulvio 409
Cloudt, Stefan 421
Cochet, Martin 269
Coignus, Jean 125
Conti, Francesco 273
Cordioli, Nicola 429
Corinto, Fernando 117
Correll, Justin 325
Costanzo, Ian 413
Coutier, Caroline 193
Covi, Erika 137
Craninckx, Jan 61
Crovetti, Paolo 405
Cueto, Olga 129
Cyrille, Marie-Claire 129
Czapor, Pawel 401

D

D'Andragora, Alessio 197
D'Heer, Carl 481, 485
Dafallah, Hind 237
Daghbashyan, Mesrop 237
Dal Maistro, Daniele 389
De Filippi, Guglielmo 389
De Franceschi, Silvano 45
De Palma, Franco 49
De, Vivek 41, 377
Deguchi, Jun 505
Dehaene, Wim 69
Dei, Michele 257
Dekimpe, Rémi 445
Deng, Wei 233
Di Biase, Ignacio 197
Di Marco, Mauro 117
Dijkhuis, Johan 421
Ding, Ming 421
Ding, Yu-Chun 81
Disegni, Fabio 373
Dos Santos, Maico Cassel 269

Dosluoglu, Deniz 493
Drost, Andreas 133
Droste, Dirk 12
Du, Xirui 562
Duong, Quang T. 137

E

Eckmann, Bruno 189
Egger, Urs 105
Ek, Staffan 237
El Soussi, Mohieddine 421
Eland, Efraïm 253
El-Homsy, Victor 45
Erba, Simone 389
Escudero, Manuel 117
Essel, Jochen 297

F

Fan, Deliang 153
Fan, Taiyang 497, 501
Fang, Lairong 554
Fargas Cabanillas, J.M. 77
Feldman, Asaf 369
Flandre, Denis 469
Flete, Alexandre 305
Flynn, Michael 325
Folz, Michele 473
Forti, Mauro 117
Francese, Pier Andrea 57
Franchi Scarselli, Eleonora 109
Frank, Alexander 185
Freye, Florian 149
Fu, Jiahan 497
Fu, Zihan 513

G

Galdi, Ivano 12
Gallinari, Daniele 373
Ganzerli, Marcello 321
Gao, Hao 337
Gao, Yihan 321
Gardino, Daniele 473
Garofalo, Angelo 273
Gasparini, Leonardo 189
Gastaldi, Carlotta 161

Gatchalian, Sherwin 421
Gaurav Kumar, K. 209
Gemmeke, Tobias 149
Gerlach, Lukas 265
Ghanavati Nejad, Tayebeh 237
Ghisu, Davide Ugo 221
Gielen, Georges 261
Gießelmann, Timo 12
Gilhotra, Yatin 205
Giri, Davide 269
Glorieux, Olivier 125
Gnudi, Antonio 109
Goh, Youngin 113
Golanski, Dominique 193
Gönen, Burak 253
Grenmyr, Andreas 121
Grenouillet, Laurent 125
Grobe, Jan 289
Gruijć, Miloš 73
Grützmacher, Detlev 121
Gu, Jie 93
Gu, Qilin 509
Guimaraes, Gabriel 457
Guler, Ulkuhan 413
Gunaputi Sreenivasulu, Aishwarya 53
Guo, Shiyu 93
Gupta, Priyanka 373
Gutheit, Tim 9
Guzman, Marino D. 441
Gys, Benjamin 61

H

Ha, Kyoungjoon 249
Hägglund, Robert 237
Ham, Bumsub 97
Ham, Donhee 225
Han, Hyeonji 509
Han, Sangwook 509
Han, Yi 121
Hanhart, Michael 289
Harpe, Pieter 181
He, Yuming 421
Heim, David 57
Heinen, Stefan 289
Helleboid, Rémi 193
Herrmann, Ingo 73
Hersche, Michael 105

Hijdra, Martijn 421
Hinton, Henry 225
Hirtzlin, Tifenn 129
Hong, Zhiliang 277, 433, 477, 554
Hsu, Steven 377
Huang, Chao-Tsung 81
Huang, Dan 558
Huang, Paul Xuanyuanliang 349
Huang, Tzu-Yuan 461
Huang, Wei-Cheng 285
Huang, Yi-Hsiang 417
Huang, Yumei 477
Huang, Zheng-Lun 293
Hueber, Gernot 337
Hurtado, Luis 65
Hwang, Junghyeon 113

I

Ide, Michihiro 345
Ielmini, Daniele 133
Iesurum, Agata 385
Iizuka, Tetsuya 245
Ilderem, Vida 41
Im, Changyoun 521
Innocenti, Giacomo 117
Ionescu, Adrian M. 49, 161
Iotti, Lorenzo 489
Ivanisevic, Nikola 237

J

Jadot, Baptiste 45
Jalili, Hossein 461
Jang, Daniel 349
Jang, Iksu 213
Jang, Jaehyuk 509
Jang, Woojin 317
Jayant, Krishna 209
Jeon, Dongsuk 145
Jeon, Sanghun 113
Jeon, Sejun 509
Jeong, Deog-Kyoon 249
Jeong, Sangsu 145
Jeong, Soyoun 521
Jeong, Yeongseok 113
Jia, Haikun 233
Jia, Tianyu 93, 269

Jiang, Hongwu 101
Jin, Jin 309
Jin, Yuemin 558
Joo, Sunghwan 97
Ju, Yuhao 93
Jung, Seong-Ook 97, 113
Jung, Sung Jun 85
Jutte, Peter 185

K

Kaiser, Md. Abdullah-Al 173
Kamaei, Sadegh 161
Kämpe, Andreas 237
Kang, Gyeong-Gu 317
Kang, Kiseo 525
Kao, Yi-Hsiang 285
Karlsson, Patrik 237
Karmakar, Shoubhik 253
Karnik, Tanay 41
Karrenbauer, Jens 265
Karunaratne, Geethan 105
Kato, Sena 345
Kiene, Gerd 53
Kim, Bongjin 353
Kim, Byungjun 525
Kim, Byungsub 213
Kim, Chanho 213
Kim, Chulho 509
Kim, Dohyung 97
Kim, Gyusik 509
Kim, Hyoungjoong 509
Kim, Hyun-Sik 317
Kim, Jaerok 521
Kim, Jaeyoung 509
Kim, Minki 113
Kim, Sangho 509
Kim, Seonho 521
Kim, Soosung 85
Kim, Tony Tae-Hyoung 241, 353
Klein, Simon 265
Klemt, Bernhard 45
Klimach, Hamilton 405
Kneip, Adrian 157
Ko, Dong Han 113
Ko, Han-Gon 249
Ko, Jaehyun 213
Kobayashi, Hiroyuki 505

Kootte, Bart 185
Kossel, Marcel 57
Kramnik, Danielius 77
Krishnamurthy, Ram 377
Krüger, Daniel 225
Kruiskamp, Wim 397
Kulkarni, Jaydeep P. 529
Kumar, Prem 77
Kumar, Raghavan 529
Kuo, Hao-Yi 361
Kwak, Jin Woong 281
Kwon, Sangwoo 85

L

Lancaster, Suzanne 137
Langenegger, Jovin 105
Lanius, Christian 149
Le Gallo, Manuel 105
Lecchi, Simone 309
Lee, Hyunseung 85
Lee, Jae W. 85
Lee, Jongmin 521
Lee, Jongwoo 509
Lee, Junghyup 97
Lee, Ki-Beom 97
Lee, Mingyen 562
Lee, Ockgoo 341
Lee, Seungjong 325
Lee, Sheng Cheng 293, 365
Lee, Sooeun 213
Lee, Soo-Min 509
Lee, Sumin 97
Lee, Yoonmyung 521
Lee, Young Kyu 113
Lefebvre, Martin 157, 469
Lei, Ka-Meng 225
Lei, Liangbo 433, 477
Li, Dengquan 550
Li, Gen 513
Li, Hanyue 181
Li, Ning 105
Li, Shuo 465
Li, Si-Yi 293
Li, Wantong 101
Li, Xichen 417
Li, Xueqing 562
Li, Yijie 554

Li, Yongfu 229
Li, Yuan-Jin 285
Li, Zhenghao 497, 501
Liao, Yu-Te 313, 361
Lico, Andrea 109
Lim, Jihoon 509
Lim, Sehee 113
Lim, Yong 317
Lin, Chun-Yeh 81
Lin, Kai-Pin 81
Lin, Shian-Ru 285, 293, 365
Lin, Shu-Yung 365
Lin, Ying-Hsi 285, 293, 365
Ling, Bill 173
Liu, Angqi 253
Liu, Hongzhuo 233
Liu, Jiaxin 562
Liu, Kunyang 513
Liu, Shubin 550
Liu, Xia 161
Liu, Xinjian 465
Liu, Yao-Hong 421
Liu, Yaxin 229
Liu, Yongpan 562
Liu, Yuqi 461
Liu, Zhenhong 229
Liu, Zixuan 93
Loscalzo, Erik Jens 269
Lou, Jie 149
Lu, Danzhu 277
Lu, Xuyang 546
Luo, Xiongshi 497, 501

M

Ma, D. Brian 281
Ma, Shunli 425
Ma, Zhouchen 229
Maghari, Nima 441, 517
Maiwald, Tim 449
Makinwa, Kofi A. A. 253
Malcovati, Piero 221, 473
Mamdy, Bastien 193
Manente, Davide 385
Manfredini, Giuseppe 257
Manstretta, Danilo 309, 429
Mantovani, Paolo 269
Marano, Vincenzo 17

Martins, R.P. 566
Massari, Nicola 73, 197
Masud, Mohammad Ayaz 65
Mateman, Paul 421
Mathew, Sanu 529
Maurino, Roberto 401
Mazzanti, Andrea 221, 389
McNeill, John 413
Medina, Italo 401
Mehrotra, Shubham 253
Meli, Valentina 129
Meng, Jian 89
Menolfi, Christian 57
Meunier, Tristan 45
Meyer, Bernard 421
Mikolajick, Thomas 137
Milozzi, Alessandro 133
Min, Kyunghwan 509
Min, Woongki 509
Moldovan, Clara 161
Moon, Changjae 213
Moon, Seunghyun 525
Mora-Puchalt, Gerard 401
Moreno García, Manuel 189
Morf, Thomas 57
Mortemousque, Pierre-André 45
Mortezazadeh Mahani, Sina 337
Mu, Junjie 353
Mueller, Peter 57
Murmann, Boris 329
Musolino, Francesco 405
Myny, Kris 69

N

Nadalini, Alessandro 273
Naghavi, S.M. Hossein 533
Nam, Ilku 341
Narayanan, Vijay 105
Naviasky, Emily 489
Nechushtan, Omer 369
Neuner, Thomas 133
Ngo, Huy Cu 505
Nicholson, Isobel 193
Niebojewski, Heimanu 45
Niegemann, David 45
Niknejad, Ali 489
Niknejad, Tahereh 177

Nilsson, Robert 237
Novaresi, Lara 221
Nozawa, Mai 505
Nygren, Peter 237

O

Oehme, Michael 165, 169
Oezdamar, Oguzhan 301
Oh, Kyutaek 341
Ok, Injo 105
Okada, Kenichi 333, 345, 393
Oliveira, Luis 177
Oppliger, Fabian 49
Oropallo, Maria Vittoria 57
Ortmanns, Maurits 201
Oruganti, Sirish 529
Osada, Masaru 245
Oshinubi, Dayo 73
Ottaviani, Stefano 221
Overwater, Ramon W.J. 53

P

Padovan, Fabio 385
Palm, Mattias 237
Palumbo, Elisabetta 373
Pan, Quan 497, 501
Pandya, Gunjan 377
Pang, Jian 333, 393
Park, Beomsoo 441, 517
Park, Jaehan 525
Park, Jeongwoo 145
Park, Jihoon 213
Park, Jinhoon 509
Parmesan, Luca 73
Parra, Thierry 305
Parvais, Bertrand 61
Pasotti, Marco 109, 373
Pelgrims, Jonas 69
Pellegrini, Sara 193
Pellerano, Stefano 41
Pelzers, Kevin 181
Pena-Colaiocco, Diego 493
Perenzoni, Matteo 73, 189, 197
Perilli, Luca 109
Perotti, Matteo 273
Perruchot, François 45

Pettingell, John 197
Pham, Hiep 509
Piazza, Gianluca 65
Piotto, Lorenzo 389
Piotto, Massimo 257
Poggio, Martino 49
Pollmann, Eric H. 205
Popović, M.A. 77
Potocnik, Anton 61
Pozzi, Rachela 17
Prathapan, Mridula 57
Pulicelli, Fulvio 17
Pulvirenti, Francesco 17

Q

Qi, Liang 229
Qiu, Zhang 497
Qu, Tianxiang 433
Quaglia, Fabio 221

R

Radu, Ionut 49
Rahimi, Abbas 105
Ramesh, Anirudh 77
Ramkaj, Athanasios 329
Read, James 101
Redaelli, Andrea 373
Reich, Stefan 201
Reiser, Daniel 133
Ren, Junyan 425
Reynaert, Patrick 457, 481, 485
Ria, Andrea 257
Richelli, Anna 405
Rideau, Denis 193
Rieger, Stefan B. 201
Rochus, Veronique 217
Rogi, Christopher 409
Rolff, Leo 289
Rossi, Davide 273
Roy, David 193
Rubino, Roberto 405
Rudell, Jacques C. 417, 493
Ruffino, Andrea 57
Rutten, Robert 321

S

Saeidi, Ali 161
Saeidi, Hooman 546
Saikia, Jyotishman 89
Säll, Erik 237
Sathe, Visvesh 493
Saulnier, Nicole 105
Savoia, Alessandro Stuart 221
Sawan, Mohamad 229
Sawigun, Chutham 437
Scarlino, Pasquale 49
Schäfer, Sören 165
Schleipen, Jean 185
Schönewald, Sven 265
Schulze, Jörg 165, 169
Schüttler, Martin 201
Schwarz, Daniel 165, 169
Sebastian, Abu 105
Sebastiano, Fabio 53
Seebacher, David 337
Seidel, Lukas 165, 169
Selijak, Andrej 197
Sen, Devdip 413
Sen, Shreyas 209
Sengupta, Kaushik 546
Seo, Jae-Sun 89, 153
Seo, Seong Hoon 85
Seok, Mingoo 349
Shekhar, Sudip 542
Shen, Yi 550
Shen, Yuting 181
Shepard, Kenneth 205, 269
Sherry, Adrian 401
Shin, Hunbeom 113
Shinohara, Hirofumi 513
Shirane, Atsushi 333, 345, 393
Shor, Joseph 369
Sicre, Mathieu 193
Sideris, Constantine 173
Silva, Jose 177
Silvestre, Claire 105
Sim, Jae-Yoon 525
Singh, Gaurav 421
Slesazeck, Stefan 137
Solomko, Valentyn 297, 301
Song, Minyoung 421
Song, Yeonggeun 249

Song, Yi-Qiao 225
Sousa, Veronique 129
Spieth, Christian 169
Spiga, Sabina 117
Sporer, Markus 201
Sridharan, Amitesh 153
Stadtmann, Tim 149
Stefanov, André 189
Stojanović, Vladimir 77
Straeussnigg, Dietmar 409
Strandberg, Roland 237
Strohm, Thomas 73
Studer, Christoph 357
Subbaraman, Raghav 542
Suh, Han-Sok 89
Sun, Jui-Hung 173
Sun, Nan 141
Sun, Shiyan 233
Sundström, Lars 237
Swaminathan, Karthik 269
Swinkels, Paul 321
Symanzik, Horst 12
Syroiezhin, Semen 297, 301

T

Tabruyn, Dries 217
Tachibana, Fumihiko 505
Tang, Jiajie 233
Tang, Minzhe 333
Tang, Wenjun 562
Tang, Yiqiao 225
Tao, Cong 477
Tao, Sha 237
Tavakoli Taba, Morteza 533
Tayari, Danial 297
Taylor, Alan 197
Taylor, Jon 197
Terenzi, Marco 221
Tesi, Alberto 117
Thanapitak, Surachoke 437
Tharayil Narayanan, Aravind 237
Thijssen, Bart 421
Thiney, Vivien 45
Tien, Kevin 269
Toi, Takashi 505
Toledo, Pedro 405
Tombesi, Gabriele 269

Tonietto, Davide 1
Tontini, Alessandro 73
Tornow, Marc 133
Torres, Mattia 109, 373
Tortorella, Yvan 273
Trabelsi, Ahmed 129
Traferro, Stefano 421
Triozon, Francois 125
Tsai, Tsung-Yen 285, 293, 365
Tschanz, James 41, 377
Tsividis, Yannis 349
Tsubouchi, Yuta 505
Turin, Evgenii 421

U

Ulm, Markus 12
Urakawa, Go 505
Urdampilleta, Matias 45

V

Valente, Luca 273
Van Assche, Jonah 261
van Den Heuvel, Johan 421
van der Struijk, Mariska 181
Van Helleputte, Nick 217
van Schaik, Gert-Jan 421
Van Winckel, Steven 61
Varela, João 177
Vaxelaire, Nicolas 125
Veldhoven, Robert van 253
Venkataramanaiah, Shreyas Kolala 89
Venkatesh, Suresh 546
Verbauwhede, Ingrid 73
Verecken, Julien 157
Viallon, Christophe 305
Vianello, Elisa 129
Vinet, Maud 45
Vis, Peter 421
Visweswaran, Akshay 449
Wadatsumi, Junji 505

W

Wambacq, Piet 61
Wang, Guoxing 229
Wang, Hua 461

Wang, Huan-Ching 81
Wang, Imbert 77
Wang, Li-Wei 81
Wang, Meizhi 529
Wang, Peizhuo 433
Wang, Xuanzhen 513
Wang, Zhihua 233
Wanitzek, Maurice 169
Ware, Evelyn 325
Wei, Dong 425
Wei, Gu-Yeon 269
Wei, Meng 489
Weigel, Robert 297, 301, 449
Weihs, Léon 289
Wellman, John-David 269
Weng, Chi-Wen 81
Wiesbauer, Andreas 409
Winkel, Nick 421
Wong, Oi-Ying 217
Wu, Hongzhi 501
Wu, Qixiu 233
Wu, Tianxiang 425
Wu, Weitao 501
Wu, Yan 229
Wunderlich, Ralf 289

X

Xi, Fengben 121
Xiao, Shulan 209
Xiao, Wenbo 497, 501
Xie, Shanshan 529
Xu, Dongfan 501
Xu, Jiawei 277, 433, 554
Xu, Zule 245

Y

Yamazaki, Yudai 393
Yan, Jing-Ren 361
Yang, Huazhong 562
Yang, Jun 558
Yang, Xiangxing 141
Yang, Xiaofeng 566
Yang, Zekun 562
Yao, Chih-Wei 509
Yee, My Chien 237
Yeo, Injune 89

Yin, Fucheng 417
Yin, Guodong 562
Yin, Heyu 205
Yoon, Dong-Hyun 241
You, Xuewei 497
Yu, Chengshuo 353
Yu, Shimeng 101
Yuasa, Keito 345
Yue, Jinshan 562

Z

Zanchi, Stefano 373
Zarghami, Majid 189
Zarkos, Panagiotis 77
Zeng, Xiaoyang 277, 554
Zhang, Aoyang 225
Zhang, Fan 153
Zhang, Jeff Jun 269
Zhang, Jiayuan 121
Zhang, Jincheng 425
Zhang, Peng 421
Zhang, Shuwen 554
Zhang, Yao 554
Zhang, Yi 333
Zhang, Yichi 89
Zhang, Zhiru 89
Zhao, Ivan 493
Zhao, Jian 229
Zhao, Qing-Tai 121
Zhao, Xin 550
Zhao, Xingliang 509
Zhao, Yan 558
Zheng, Kuo-Lin 285, 293, 365
Zhong, Junlin 566
Zhong, Liping 497, 501
Zhou, Linfeng 229
Zhou, Mufeng 562
Zhu, Yan 566
Zhu, Zhangming 550
Zoche, Jonas 289
Zota, Cezar 57
Zuckerman, Joseph 269
Zurla, Riccardo 373

IEEE
445 Hoes Lane
Piscataway, NJ 08854-4141

ISBN 978-1-6654-8495-4